PCI DESIGN HANDBOOK

Precast and Prestressed Concrete

6th Edition

MNL 120-04

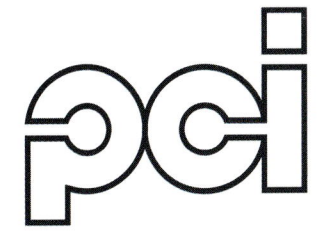

209 W Jackson Blvd, Suite 500
Chicago, Illinois 60606 - 6938
312-786-0300   Fax 312-786-0353

Copyright © 2004
By Precast/Prestressed Concrete Institute
First Edition, first printing, 1971
First Edition, second printing, 1972
Second Edition, first printing, 1978
Second Edition, second printing, 1980
Third Edition, first printing, 1985
Fourth Edition, first printing, 1992
Fifth Edition, first printing, 1999
Sixth Edition, first printing, 2004

All rights reserved. This book or any part thereof may not be reproduced in any form without the written permission of the Precast/Prestressed Concrete Institute.

ISBN 0-937040-71-1

Printed in U.S.A.

PCI DESIGN HANDBOOK

Precast and Prestressed Concrete

6th Edition

PCI Industry Handbook Committee

Kim Seeber, P.E., Chairman

Neal S. Anderson, S.E., P.E.	Phillip J. Iverson, P.E.	Steven H. Peterson, P.E.
Craig T. Barrett, P.E.	James E.A. King, P.E.	Courtney B. Phillips, S.E., P.E.
E. Fred Brecher, P.E.	Jason Krohn, P.E.	Joseph Retzner, P.E.
Ned M. Cleland, P.E., Ph.D.	Jason P. Lien, P.E.	Kurt L. Salm, S.E., R.A.
Thomas J. D'Arcy, P.E.	Rafael A. Magana, P.E.	A. Fattah Shaikh, P.E., Ph.D.
Greg Force, P.E.	Michael J. Malsom, S.E., P.E.	Irwin J. Speyer, P.E.
Harry A. Gleich, P.E.	Leslie D. Martin, P.E.	Helmuth Wilden, P.E.
Timothy Holien, P.E.	Robert F. Mast, P.E.	Charles E. Wynings, P.E.
Gary Householder, P.E.	Joseph A. Miller, P.E.	
Pat Hynes, P.E.	Stephen Pessiki, Ph.D.	

Consulting Members

Alex Aswad, P.E., Ph.D. Paul D. Mack, P.E. Jagdish C. Nijhawan, P.E.

Editor

Leslie D. Martin, P.E.
Christopher J. Perry, S.E., P.E.

DEDICATION

The PCI Industry Handbook, Sixth Edition, is dedicated to Les Martin by the members of the Industry Handbook Committee. Les was the technical editor of the First, Second and Third Editions and is the technical editor of this Sixth Edition. Les has also been a member of every Industry Handbook Committee. Without his tireless and unrelenting efforts and his dedication to excellence, this document would not have been possible. Les, you are admired and respected by this committee and the industry as a whole. Thank you.

FOREWORD

The Precast/Prestressed Concrete Institute, a non-profit corporation, was founded in 1954 for the purpose of advancing the design, manufacture, and use of structural precast/prestressed concrete and architectural precast concrete in the United States and Canada. To meet this purpose, PCI continually disseminates information on the latest concepts, techniques and design data to the architectural and engineering professions through regional and national programs and technical publications.

The First Edition of the *PCI Design Handbook* was published in 1971 with its primary focus on structural products and buildings. To fill a void in the design of architectural precast concrete, the *PCI Manual for Structural Design of Architectural Precast Concrete* was published in 1977. In 1978, the Second Edition of the *PCI Design Handbook* was published. In keeping with the tradition of continually updating information, an Industry Handbook Committee was formed in 1979 to develop the Third Edition, which was published in 1985. That edition provided, in a single source, information on the design of both architectural precast concrete and structural precast/prestressed concrete. Both the Fourth Edition, published in 1992, and the Fifth Edition, published in 1999, continued to present both architectural and structural products and systems. This emphasis is maintained in the Sixth Edition.

Since 1999, the committee has continued to monitor technical progress within the industry, with particular assistance from the many committees of PCI responsible for a variety of specific topics. This Sixth Edition is the culmination of those efforts, and presents current industry practice.

The members of the committee listed on the title page have made significant contributions of their time and expertise. In addition, PCI committees have provided a review of specific areas. Many individuals within the industry have also provided advice and comment. The Institute offers all involved in this process a special note of recognition and appreciation.

The final review phase consisted of a Blue Ribbon Review Committee, made up of Plant Engineers, Consulting Engineers and Professors. Each member of the Blue Ribbon Review Committee is a recognized leader in the analysis and design of precast/prestressed concrete products.

In addition, a comprehensive editorial and technical review of the Handbook was carried out by George Nasser and Michael Bielema of PCI staff. PCI and the Industry Handbook Committee extends its appreciation to all individuals for their invaluable review and input.

Changes have been made throughout the document. Some of particular importance include:

- Updated to the ACI 318-02 Building Code, other current PCI publications, and publications of other technical associations.
- Chapter 1 includes updated photographs that illustrate recent projects.
- Chapter 2 includes new load tables for 15 ft wide double tees.
- Chapter 3 has major changes in seismic analysis and design to reflect significant state-of-the-art revisions contained in IBC 2003.
- Chapter 4 includes the latest torsion design procedures based on the Zia and McGee approach as updated by Zia and Hsu.
- Chapter 5 has expanded and detailed stripping, handling and bracing analysis procedures.
- Chapter 6 has been updated to reflect the latest headed stud design criteria, advanced steel component design procedures and the Instantaneous Center Method of weld analysis.
- Chapter 7 is updated to include the state of the art on architectural precast concrete.
- Chapter 8 has been revised to include updated information on tolerances.
- Chapter 9 has major revisions that reflect current practice on such items as fire resistive design and vibration in concrete structures.
- Chapter 10 includes updated guide specifications on the CD-ROM.
- Chapter 11 has many new design aids.
- A searchable CD is included with this book. This CD has the full text of the Handbook and many of the cited PCI references.

Substantial effort has been made to ensure that this Handbook is accurate. However, PCI cannot accept responsibility for any errors or oversights in the use of material or in the preparation of engineering plans. The designer must recognize that no handbook or code can substitute for experience and engineering judgment. This publication is intended for use by professional personnel competent to evaluate the significance and limitations of its contents and able to accept responsibility for the application of the material it contains.

The Institute considers each new edition of the Handbook to be a living document. The user is encouraged to offer comments to PCI and suggestions for improvements to be incorporated in the next edition. Questions concerning the source or derivation of any material in the Handbook should be directed to the PCI Technical Director.

Members of the Blue Ribbon Committee

Jeff Butler, P.E.
Mohammed Habib, P.E.
Dan Kuchma, Ph.D.
Karen Laptas, P.E.
William Mancini, S.E., P.E.
Bradley E. Means, P.E.
J. Robert Norris, P.E.
Mike Owings, S.E., P.E.

Robert J. Peterman, P.E., Ph.D.
Joey Rowland, P.E.
Arturo Schultz, Ph.D.
Jim Sirko, P.E.
John Stanton, P.E., Ph.D.
Peter Troiani, S.E., P.E.
Georgina Wolfstahl, P.E.
Heidi Ziemann, P.E.

GUIDE TO CHAPTERS

Chapter 1—Precast and Prestressed Concrete: Applications and Materials
1–1 to 1–32

Chapter 2—Preliminary Design of Precast, Prestressed Concrete Structures
2–1 to 2–58

Chapter 3—Analysis and Design of Precast, Prestressed Concrete Structures
3–1 to 3–124

Chapter 4—Design of Precast and Prestressed Concrete Components
4–1 to 4–132

Chapter 5—Product Handling and Erection Bracing
5–1 to 5–36

Chapter 6—Design of Connections
6–1 to 6–102

Chapter 7—Selected Topics for Architectural Precast Concrete
7–1 to 7–22

Chapter 8—Tolerances for Precast and Prestressed Concrete
8–1 to 8–34

Chapter 9—Thermal, Acoustical, Fire and Other Considerations
9–1 to 9–86

Chapter 10—Specifications and Standard Practices
10–1 to 10–36

Chapter 11—General Design Information
11–1 to 11–58

Index

CHAPTER 1
PRECAST AND PRESTRESSED CONCRETE: APPLICATIONS AND MATERIALS

1.1 General 1–2
    1.1.1 Background 1–2
    1.1.2 Features and General Principles 1–2
    1.1.3 Common Products 1–3

1.2 Applications 1–5
    1.2.1 Building Structures 1–5
        1.2.1.1 Residential Buildings 1–5
        1.2.1.2 Office Buildings 1–6
        1.2.1.3 Warehouses and Industrial Buildings 1–6
        1.2.1.4 Other Building Structures 1–9
    1.2.2 Parking Structures 1–9
    1.2.3 Justice Facilities 1–10
    1.2.4 Precast Concrete Cladding 1–12
    1.2.5 Stadiums/Arenas 1–14
    1.2.6 Bridges 1–16
    1.2.7 Other Structures 1–17

1.3 Materials 1–19
    1.3.1 Concrete 1–19
        1.3.1.1 Compressive Strength 1–19
        1.3.1.2 Tensile Strength 1–20
        1.3.1.3 Shear Strength 1–20
        1.3.1.4 Modulus of Elasticity 1–20
        1.3.1.5 Poisson's Ratio 1–20
        1.3.1.6 Volume Changes 1–20
    1.3.2 Grout, Mortar and Drypack 1–21
        1.3.2.1 Sand-Cement Mixtures 1–21
        1.3.2.2 Non-Shrink Grouts 1–22
        1.3.2.3 Epoxy Grouts 1–22
    1.3.3 Reinforcement 1–22
        1.3.3.1 Prestressing Tendons 1–22
        1.3.3.2 Deformed Reinforcing Bars 1–23
        1.3.3.3 Structural Welded Wire Reinforcement 1–23
    1.3.4 Durability 1–23
        1.3.4.1 Freeze-Thaw and Chemical Resistance 1–24
        1.3.4.2 Protection of Reinforcement 1–25
        1.3.4.3 Protection of Connections 1–27
        1.3.4.4 Sulfate Attack 1–28

1.4 Production Process 1–29
    1.4.1 Structural Components vs. Architectural Components 1–29
    1.4.2 Long Line Forms vs. Single Piece Forms 1–29
    1.4.3 Prestressed Members vs. Non-Prestressed Members 1–29
    1.4.4 Pretensioned Members vs. Post-Tensioned Members 1–30
    1.4.5 Production Limitations 1–30

1.5 References 1–31

PRECAST AND PRESTRESSED CONCRETE: APPLICATIONS AND MATERIALS

1.1 General

1.1.1 Background

The growth of precast and prestressed concrete is a story of the vision and daring of a few notable people. These people took a new idea and maximized its potential by modifying and improving existing methods, conceiving new methods, and inventing new devices, all with a focus on mass production techniques. An excellent portrayal of the beginnings and the growth of precast and prestressed concrete in North America and the early pioneers is given in a series of papers [1] developed to commemorate the 25-Year Silver Jubilee of the founding of the Prestressed Concrete Institute which is now known as the Precast/Prestressed Concrete Institute (PCI).

The single most important event leading to the launching of the precast/prestressed concrete industry in North America was the construction in 1950 of the famed Walnut Lane Memorial Bridge (Figure 1.1.1) in Philadelphia, Pennsylvania. From a technical perspective it is surprising, and from a historical perspective it is fascinating, that the Walnut Lane Memorial Bridge was constructed in prestressed concrete. There was very little published information on the subject and there was a total lack of experience with linear prestressing in this country. Furthermore, the length of bridge spans (the main span of the structure was 160 ft long) involved would have been a daring venture in the late 1940s anywhere in the world. The bridge became a reality through a fortunate sequence of events, and the vision, courage and persistence of a few extraordinary individuals. [1]

Following completion of the Walnut Lane Memorial Bridge, American engineers and the construction industry enthusiastically embraced prestressed concrete. Many of the early applications remained in bridge construction, such as the lower Tampa Bay crossing now known as the Sunshine Skyway. Simultaneously, American engineers and constructors were conceiving new devices, improving techniques and developing new materials.

The 1950s were the years that brought into focus the 7-wire strand, precasting, long-line beds (Figure 1.1.2), admixtures, high strength concrete, vacuum concrete, steam curing and many other innovations. With these developments, coupled with the technical and logistic support provided by the Prestressed Concrete Institute, which was chartered in 1954, the industry grew and the applications of precast and prestressed concrete began to appear in an impressive variety of structures.

Development of standard products was one of the major activities through the 1950s and the 1960s. Early in the 1960s, the government sponsored "Operation Breakthrough" led to the introduction of different high rise precast building types for housing. As part of this program, a significant testing program was conducted by the Portland Cement Association to establish design principles to prevent progressive failure. These rules were adopted in the ACI Building Code.

In the late 1970s, low relaxation strand was introduced which reduced the loss of prestress force due to creep in the strand, thus allowing more efficient use of prestressing, resulting in longer spans and smaller sections. Larger strand sizes have been made available as well, such as 0.6 in. diameter strand.

In the field of bridges, there was the development of spliced girders, segmental bridges, cable stayed bridges and cantilevered girder bridges.

During the 1980s, it was recognized that durability is an important aspect of a structure. The precast and prestressed concrete industry responded by taking advantage of one of its natural strengths. Plant-cast concrete is more durable than site-cast concrete because it can be cast with lower water-cementitious materials ratios under controlled conditions. This natural durability was enhanced with the development of admixtures that make the concrete matrix more impermeable and that inhibit steel corrosion. Pretopped tees were developed for parking structures to maximize the benefits of the durability of precast, prestressed concrete at the wearing surface.

The past decade has seen the development of more efficient structural sections and more complex architectural shapes and surface treatments. The high demands of owners and architects for quality finishes has led to the development of new surface textures and surface treatments. Thin-set brick and stone-faced panels and textures and colors of infinite variety have been developed.

1.1.2 Features and General Principles

Precasting concrete in PCI certified plants ensures the manufacture of high quality architectural and structural products. Precasting also facilitates production of a wide variety of shapes and sizes, and the use of prestressing substantially extends the span capability of the products. Similarly, prestressing, defined by ACI as "...internal stresses (that) have been introduced to reduce potential tensile stresses in concrete resulting from loads," can be used to enhance the structural capabilities of a concrete member. These capabilities enable architects and engineers to achieve highly innovative and economically competitive buildings and other structures.

This Handbook serves as the primary reference for the design and use of precast and prestressed concrete structures. This section enumerates some of the important and unique features of precast and prestressed concrete. These include the following:

1. Construction speed.
2. Plant-fabrication quality control.
3. Fire resistance and durability.
4. With prestressing: greater span-to-depth ratios, more controllable performance, less material usage.
5. With architectural precast concrete: wide variety of highly attractive surfaces, shapes, finishes and colors.
6. Thermal and acoustical control.
7. All weather construction.

To fully realize these benefits and thereby gain the most economical and effective use of the material, the following general principles are offered:

1. Precast concrete is basically a "simple-span" material. However, continuity can be, and often is, effectively achieved with properly conceived connection details.
2. Sizes and shapes of members are often limited by production, hauling and erection considerations.
3. Concrete is a massive material. This is an advantage for such matters as stability under wind loads, thermal changes, acoustical vibration and fire resistance. Also, the high dead-to-live load ratio will provide a greater safety factor against gravity overloads.
4. Maximum economy is achieved with maximum repetition. Standard or repetitive sections should be used whenever possible.
5. Successful use is largely dependent on an effective structural layout and carefully conceived connection details.
6. The effects of restraint of volume changes caused by creep, shrinkage and temperature change must be considered in every structure.
7. Architectural precast panels can be used as cladding as well as loadbearing members. They can be used to support loads in both the vertical and lateral directions.
8. Prestressing improves the economy and performance of precast members, but is usually only feasible with standard shapes which are capable of being cast in long-line beds.

1.1.3 Common Products

While precast and prestressed concrete can be manufactured in a variety of customized sizes and shapes, maximum economy is achieved by using the common products that have evolved in the industry. Among the more prevalent of these products are double tees (Figure 1.1.2) and hollow-core slabs (Figure 1.1.3). Double tees are efficient for spans in the range of 40 to 90 ft, although longer spans are possible with deeper sections. Hollow-core slabs are available in a variety of widths ranging from 16 in. to 12 ft and are used for spans up to about 40 ft. Figure 1.1.4 shows cross sections of these and other commonly used products. The I-beam, box beam and bulb tee are used in bridge construction. The inverted tee, ledger beam and rectangular beam are used for structural framing to support deck members. Square or rectangular columns, with or without corbels, are an integral part of the column-beam-deck framing that makes rapid, all-weather erection possible.

Piles are manufactured in a variety of shapes, including round, square, hexagonal and octagonal, as well as rectangular sheet piles. Channel slabs are used to support heavy floor or roof loads in short and medium span ranges. Stadium riser units (Figures 1.2.35 to 1.2.40) have gained prominence with the recent boom in stadium/arena construction. These components are typically cast as single, double, or triple step units and offer spans up to 60 ft, depending on member depth. Precast concrete modular products have also established a strong presence in the construction industry, most noticeably as cell units for correctional facilities (Figures 1.2.26 to 1.2.29). Single, double and quad-cells can be manufactured, finished and furnished at a precaster's facility, significantly reducing on-site labor and schedule demands.

Precast and prestressed concrete products are designed in accordance with the latest engineering standards and produced in plants where PCI's Plant Certification Program is an integral part of plant production. PCI Plant Certification (Figure 1.1.5) assures specifiers of a manufacturing plant's audited capability to produce quality products. A minimum of two unannounced plant inspections are performed each year by specially trained engineers to evaluate compliance with current performance standards. These performance standards can be found in the PCI quality control manuals, MNL-116 [2] and MNL-117. [3]

Figure 1.1.1
Walnut Lane Memorial Bridge: Recipient of the 1978 ASCE's *Outstanding Civil Engineering Achievement Award*.

Figure 1.1.2
Long-line prestressed double tee casting bed.

Figure 1.1.3
Erection of hollow-core deck members.

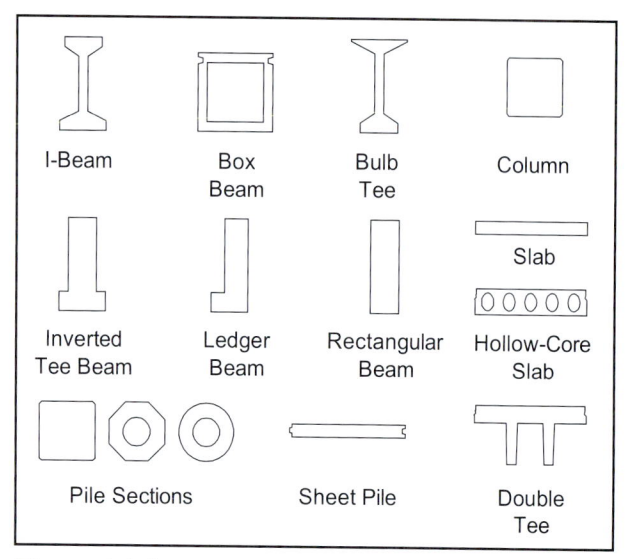

Figure 1.1.4
Common precast and prestressed concrete products.

Figure 1.1.5
Assurance of quality through PCI Plant Certification.

These manuals are considered industry standards for quality assurance and control. A plant that does not achieve a minimum required score loses certification. Plant certification is a prerequisite for PCI membership. Plants in Canada subscribe to a similar plant certification program based on Canadian Standards Association (CSA) A251 Standard. For more information on the PCI Plant Certification Program, see Section 9.5.

The high quality standard products noted above form the basis for configuring a wide variety of framing systems for buildings, bridges and other structures. The following pages provide an overview of some of the applications. For other applications and also for products and practices suitable in different geographical zones, contact with the local producers is highly recommended. A list of PCI Producer Members as well as other information pertinent to the industry can be obtained by contacting the Precast/Prestressed Concrete Institute or accessing the PCI web site at www.pci.org.

Design of precast and prestressed concrete products and structures is usually based on the ACI Building Code (ACI 318). There are, however, cases where substantial experience and/or research results suggest alternative design approaches for improved structural performance and economy. Therefore, PCI publishes the "PCI Standard Design Practice." This publication, which appears in Section 10.5 of this Handbook, is regularly updated to reflect code changes, research results and new experience. The long-term objective is to augment the ACI Building Code with common industry practices to develop the best designs taking advantage of the many special features of precast and prestressed concrete.

1.2 Applications

The developments in products, materials and techniques noted in the previous section have made precast/prestressed concrete competitive in a variety of residential, commercial, industrial, transportation and many other types of structures. A few examples of applications to different types of structures are given in this section.

1.2.1 Building Structures

Owners, developers, and designers quickly recognize the many inherent qualities of precast and prestressed concrete which make it suitable for many types of building structures. Precast and prestressed concrete structures, assembled from high quality plant produced products, provide superior flexibility for achieving the required degrees of fire resistance, sound control, energy efficiency and durability. The availability of various materials and finishes makes it possible to render virtually any desired aesthetic character. The construction speed possible with precast and prestressed concrete minimizes on-site labor costs, reduces the cost of interim financing, and thus provides important overall economy to the owner or developer.

1.2.1.1 Residential Buildings

Precast and prestressed concrete enjoys broad acceptance in low-rise and mid-rise apartment buildings, hotels, motels, and nursing homes. The superior fire resistance and sound control features are specifically recognized by owners and developers.

Two-hour fire containment within each living unit provides safety for adjacent units. With this type of high quality precast concrete housing, fire insurance rates are reduced and often higher incomes can be generated because of the safe, soundproof high quality environment and lifestyle offered (Figures 1.2.1, 1.2.4, and 1.2.30).

Figure 1.2.1
Precast concrete walls and floors ensure fire containment and lower insurance rates.

Figure 1.2.2
High quality concrete in precast concrete walls and prestressed hollow-core floors provide sound resistance, fire safety and reduced maintenance cost in multi-family housing.

The hollow-core slab is a standard product in this type of construction. Figure 1.2.2 shows a typical apartment building with hollow-core floors, loadbearing precast concrete walls and a durable maintenance free exterior spandrel panel.

1.2.1.2 Office Buildings

Significant time savings usually result from the choice of a totally precast concrete structure. The superstructure is prefabricated while the on-site foundations are being built. Details of this type of construction are shown in Figure 1.2.3. Potential delays are reduced with the complete building system being supplied under one contract. Erection of large precast concrete components can proceed even during adverse weather conditions to quickly enclose the structure. Loadbearing architectural precast panels provide a finished exterior at the same time that the superstructure is erected. The prestressed floors provide an immediate working platform to allow the interior tradesmen an early start on the mechanical, electrical and interior finishing work. The quality finishes and fast schedules result in early occupancy, tenant satisfaction and reduced financing costs making precast and prestressed concrete buildings very suitable for office buildings (Figures 1.2.5, 1.2.9, 1.2.10, 1.2.31 and 1.2.33).

There are many uses of precast and prestressed concrete in office building construction, from total building systems to single products like precast concrete stairs. Precast and prestressed concrete beams, columns and floors are used in frame systems; shear walls can be used alone or in conjunction with beams and columns to resist lateral loads. Precast concrete stairs, along with being economical, provide immediate safe use of stairwells. Architectural precast concrete is used with all types of framing systems. It provides an economical, fire resistant, soundproof, durable, maintenance-free cladding that allows the architect much freedom of expression and results in beautiful facades. Architectural precast concrete is discussed more fully in Section 1.2.4 and in Chapter 7.

1.2.1.3 Warehouses and Industrial Buildings

The ability of prestressed concrete to span long distances with shallow depths and carry heavy loads is particularly suitable for warehouses and industrial buildings (Figures 1.2.11 to 1.2.15). Standard prestressed concrete walls, insulated or non-insulated, are very economical for warehouse and light manufacturing applications. Total precast systems with prestressed roof diaphragms and precast shear walls can provide owners with a complete structural package.

Figure 1.2.3
Details of a loadbearing column-beam-double tee floor building. Spandrels can be loadbearing if required.

Figure 1.2.4
Precast concrete provides a quiet, safe, comfortable and high quality place to live.

Figure 1.2.5
This total precast concrete office building is an example of the beauty and economy that can be achieved using precast concrete. Real savings can be achieved by allowing the architectural facade to also perform structurally.

Figure 1.2.7
Using precast concrete beams and columns with precast floor slabs provides a quick, economical structure. The addition of an architectural precast concrete facade provides the owner with the security of a single source of responsibility for the structure.

Figure 1.2.6
Architectural precast concrete panels were used as loadbearing elements for this city municipal center.

Figure 1.2.8
By varying minor details like rustication, architectural precast concrete exteriors can be made expressive and economical.

Figure 1.2.9
Office buildings quite often utilize precast concrete spandrels.

Figure 1.2.10
The light industrial office structure is erected and put into service quickly using insulated precast concrete sandwich panels.

Figure 1.2.11
Precast and prestressed concrete flat panels are used for warehouse walls. Hollow-core slabs may also be used in this application.

In heavy industrial projects, prestressed floor units capable of carrying the typical heavy floor loads can be combined with other precast components to construct versatile, corrosion-resistant structural systems. The precast and prestressed concrete framing can be designed to accommodate a variety of mechanical systems and to support bridge cranes for industrial uses. High quality precast concrete provides protection against fire, dampness and a variety of chemical substances.

Figure 1.2.12
Erection of precast concrete wall panels is fast and can be performed year round even in cold climates. Roof diaphragms and supporting beams and columns are often also precast concrete.

Figure 1.2.13
Precast concrete insulated sandwich wall panels provide excellent insulation for cold storage facilities. Smooth interior finishes make these panels an excellent choice for food processing plants.

The smooth surfaces achievable in precast concrete make it ideal for food processing, wet operations, computer component manufacturing, as well as many other types of manufacturing and storage operations where cleanliness is of concern. Clear spans of 40 and 90 ft are possible using hollow-core

Figure 1.2.14
Total precast concrete structures provide many answers for heavy industrial applications. High quality plant produced concrete provides excellent corrosion resistance and durability. Heavy loads can be easily handled by prestressed concrete beams and slabs.

Figure 1.2.15
These identical water tanks benefit from the durability, economy and aesthetic capability of precast concrete.

Figure 1.2.16
This university research facility illustrates the different capabilities of precast and prestressed concrete.

Figure 1.2.17
A precast concrete facade provides aesthetic features as well as resistance to blast loads for this United States courthouse.

Figure 1.2.18
Precast concrete offers an alternative for residential housing. Thin brick are inset in panels during production.

1.2.1.4 Other Building Structures

The many benefits of precast and prestressed concrete make it suitable for many other types of buildings in addition to residential, office, and industrial buildings. Applications abound in educational institutions, commercial buildings such as shopping malls, and public buildings including hospitals, libraries, and airport terminals (see Figures 1.2.16 and 1.2.17). Precast and prestressed concrete have also been effectively used in numerous retrofit projects (see Figures 1.2.16 to 1.2.18).

1.2.2 Parking Structures

Architects, engineers, developers and owners have made precast and prestressed concrete the material of choice for their commercial, municipal and institutional parking needs. Though classified and constructed as buildings, parking structures are unique; in some ways, they may be compared to bridges with multiple decks. They are subjected to moving loads from automobile traffic and the roof

slabs and double tees, respectively. Even longer spans to about 150 ft can be obtained with bridge-type girders or special double tees.

level of a parking structure is exposed to weather in much the same way as a bridge deck. Furthermore, they are usually not enclosed and, thus, the entire structure is subjected to ambient weather conditions. Also, exposure to deicing salts in northern climates or to salt-laden atmospheres in coastal locations requires consideration of durability to ensure long-time performance.

The controlled conditions of a precast concrete plant assures the parking structure owner of the quality concrete and workmanship that provides long-term durability. The low water-cementitious materials ratio concrete that precast concrete manufacturers use has been proven to increase resistance to deterioration. Studies [4] have also shown that accelerated curing makes precast concrete more resistant to chloride penetration than field cured concrete.

These inherent durability characteristics along with low cost, rapid erection in all weather conditions, unlimited architectural expression and long clear spans make precast concrete the natural choice for parking structures.

Through surveys of existing structures and other experiences, and through research and development, significant improvements have been achieved in the engineering state-of-the-art of parking structures. [5] This accumulated experience and knowledge has been assembled in a comprehensive publication by PCI [6] that includes recommendations on planning, design, construction and maintenance.

Figures 1.2.19 and 1.2.20 show typical precast concrete parking structures with architectural loadbearing spandrels and stair tower walls. Long-span double tees are shown in Figure 1.2.21 bearing on an innovative "light" wall (litewall) that adds openness and a feeling of security. Figure 1.2.22 shows a precast, prestressed double tee being erected into a loadbearing spandrel which is pocketed to reduce load-induced torsion. In Figure 1.2.23 the double tee is being set down on an interior inverted tee beam spanning from column to column.

Vertical expansion of precast concrete parking structures can be economically accomplished with special cranes that carry new members over erected decks as shown in Figure 1.2.25. This erection method also allows precast concrete to be considered for expansion of non-precast concrete structures.

1.2.3 Justice Facilities

Justice facilities encompass many types of building occupancies. These include jails, prisons, police stations, courthouses, juvenile halls, and special mental health or drug abuse centers (Figures 1.2.26 to 1.2.29).

Precast concrete has proven to be the favored material for justice facilities because it has many inherent benefits that are important to these building types. In addition to the benefits noted previously, namely, fire resistance, durability, speed of erection and flexibility for aesthetics, precast and prestressed concrete is ideal for building in the desired level of physical security coupled with accommodation of

Figure 1.2.19
Long clear spans with architectural loadbearing spandrels make precast concrete an economical and aesthetically preferable parking structure solution.

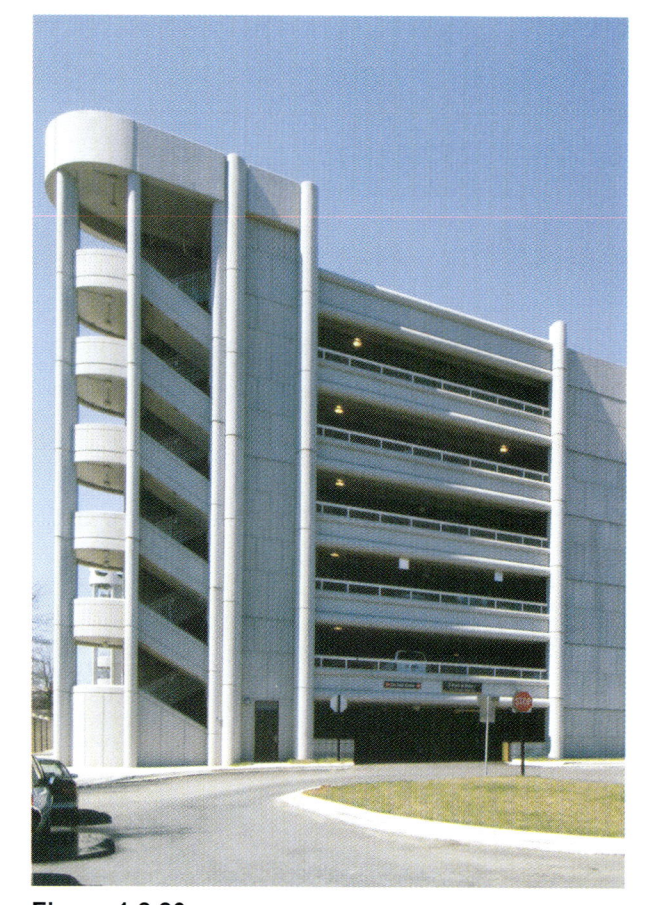

Figure 1.2.20
Stair towers acting as shear walls resist lateral loads and provide architectural features.

Figure 1.2.21
Opened loadbearing walls provide the security of visibility and openness as well as carrying vertical loads and lateral loads.

Figure 1.2.22
Erection of precast concrete components is fast and independent of climate. Pocketed loadbearing spandrels accept the double tee, which then bears on the interior inverted tee beam.

Figure 1.2.23
An interior L-beam is deep enough to also act as the car stop.

Figure 1.2.24
A lateral system without walls is provided by moment frames.

security hardware and communication systems. The thickness and reinforcement of precast wall and slab systems, designed for gravity and lateral loads and volume changes, are sufficient even for maximum security requirements. Typical precast products such as loadbearing insulated wall panels, cell walls and floor slabs are employed frequently in justice facilities. Also, the modular nature of precast and prestressed concrete products facilitates pre-installation of necessary security and communication hardware in the plant, greatly simplifying field installation work as well as saving valuable time along the project's critical path.

The use of precast concrete box modules in a one-, two-, or four-cell format has greatly reduced field labor, erection time, overall construction time, punch list problems, multi-trade confusion and, ultimately, risk. These units can be quickly stacked and can be substantially complete requiring little finishing work (see Figure 1.2.28).

Given the serious shortage of justice facilities, savings in total project time is often a critical consideration in selection of a structural material and system for these projects. Case histories show that total precast concrete projects have resulted in saving one to two years of construction time over the estimated schedule for competing systems. These experiences and the other considerations noted above have led to rapid growth in the use of precast and prestressed concrete in justice facilities projects.

Figure 1.2.25
Special erection equipment is available for adding floors to existing parking structures.

Figure 1.2.26
The repetitive nature of cell layout and the typical quick occupancy requirements make precast concrete an ideal solution to prison overcrowding.

1.2.4 Precast Concrete Cladding

Architectural precast concrete cladding (Figures 1.2.5 to 1.2.18 and 1.2.30 to 1.2.34) provides many degrees of freedom for architectural expression with the economy of mass production of precast elements. The cladding may serve only as an enclosure for the structure, or may be designed to support gravity loads as well. Cladding may also contribute to resistance to lateral loads. [7]

Architectural precast concrete can be cast in almost any color, form or texture to meet aesthetic and practical requirements. [8] Special sculptured effects can provide such visual expression as strength and massiveness, or grace and openness. Design flexibility is possible in both color and texture by varying aggregate and matrix color, size of aggregates, finishing processes and depth of exposure.

Figure 1.2.27
Precast double-cell module being lifted into place.

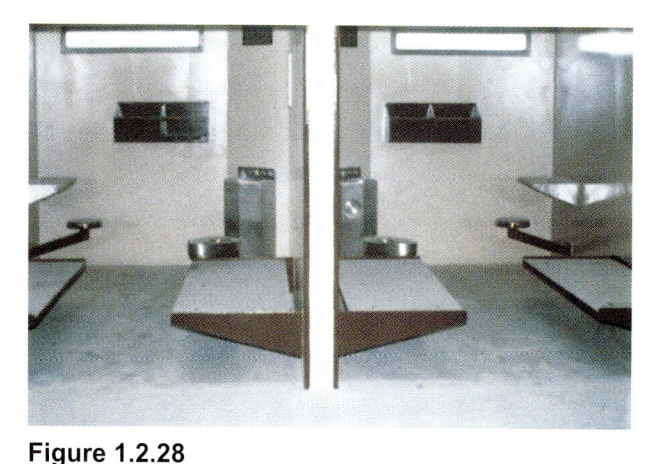

Figure 1.2.28
Box module cells are often supplied with electric conduit and boxes, furniture, window and door embedments, and mechanical and plumbing chases cast in.

Figure 1.2.29
Interior finishes of prisons are durable and maintenance free. Soundproof and fire resistant, precast concrete adds to security and limits vandalism.

Figure 1.2.30
The Paramount building in San Francisco is 39 stories tall. It is the tallest precast, prestressed concrete framed building in a high seismic region.

Figure 1.2.31
Architectural precast concrete provides a striking facade for the landmark Hearst Tower in Charlotte, North Carolina.

Figure 1.2.32
The moldability of architectural precast concrete allows for an infinite number of sizes and shapes.

PCI has developed a guide to assist designers in selecting colors and textures for architectural precast concrete. [25] Additional flexibility of aesthetic expression is achieved by casting various other materials, such as veneers on the face of precast concrete panels. Natural stone, such as polished and thermal-finished granite, limestone, marble, and clay products such as brick, tile and terra-cotta have been frequently used as veneer materials. [8]

In addition to the freedom of aesthetic expression achievable with loadbearing or non-loadbearing architectural precast concrete, there are a number of other important functional and construction advantages. Insulated wall panels consist of two concrete wythes with a continuous layer of rigid insulation sandwiched between them. These types of panels contribute substantially to the overall thermal efficiency of a building. In cast-in-place concrete construction, precast concrete cladding panels are sometimes used as permanent concrete formwork, thus becoming an integral part of the structure. Off-site pre-assembly of all components comprising a total wall system, including window sash and glazing, can also be very cost effective. For more comprehensive information on product and design, see Ref. 8.

Figure 1.2.33
Repetition of size and shape, which allows multiple form use, is a key to economy.

Figure 1.2.34
Spandrel panels spanning from column to column can be load-bearing or solely architectural.

Glass fiber reinforced concrete (GFRC) has been adopted for use in producing strong, thin, lightweight architectural cladding panels. [9] GFRC is a portland cement-based composite reinforced with randomly dispersed, alkali-resistant glass fibers. The fibers serve as reinforcement to enhance flexural, tensile, and impact strength of concrete. A major

Figure 1.2.35
This major sports stadium was built using total precast concrete after evaluating other systems.

benefit of GFRC is its light weight which provides for substantial economy resulting from reduced costs of product handling, transportation, and erection, and which also results in lower seismic loads.

1.2.5 Stadiums/Arenas

Large stadiums and arenas (Figures 1.2.35 to 1.2.40) are impressive structures. Often these projects are built on tight schedules to accommodate some important sporting event. Precast and prestressed concrete has been the overwhelming choice for many of these projects. The technique of post-tensioning precast segments together has allowed this versatile material to form complex cantilever arm and ring beam construction which supports the roofs of these structures. Post-tensioning is also commonly employed to minimize the depth of precast concrete cantilevered raker beams which carry the seating and provide unhindered viewing of the playing surface.

Mass produced seating units have been manufactured in a variety of configurations and spans to provide for quick installation and long lasting service. Consult local producers for available riser sections.

Long-spans and the ability to eliminate costly field formwork makes precast and prestressed concrete the best choice for many components of stadium construction, especially seating which can be standardized to take advantage of repeated form utilization. Components that would require tall scaffolding towers to field-form, such as raker beams and ring beams, can be simplified by precasting these units in a plant, delivering them to the site and lifting them into place. Pedestrian ramps, mezzanine floors, concession, toilet, and dressing room areas can all be framed and constructed using precast and prestressed concrete elements.

Figure 1.2.36
Precast concrete risers, raker beams, columns and mezzanine floors play a prominent role in stadium construction.

Figure 1.2.38
Special framing is handled easily with precast concrete, eliminating expensive field formwork.

Figure 1.2.37
Fast track construction allows compressed schedules. Site preparation and component manufacturing can take place simultaneously. Multiple erection crews speed erection.

Figure 1.2.39
Repetition, simplicity and multiple form use allow the columns, raker breams and riser units to be framed economically with total precast concrete.

Figure 1.2.40
Smaller grandstands and huge stadiums all benefit from the use of precast and prestressed concrete.

1.2.6 Bridges

Bridge construction gave the prestressing industry its start in North America. Precast and prestressed concrete is now the dominant structural material for short- to medium-span bridges. With its inherent durability, low maintenance and assured quality, precast and prestressed concrete is a natural product for bridge construction. The ability to quickly erect precast concrete components in all types of weather with little disruption of traffic adds to the economy of the job. For short spans (spans to 100 ft), use of box sections and double tee sections have proven economical. However, the most common product for short- to medium-spans is the I-girder. Spans to 150 or 160 ft are not uncommon with I-girders and bulb tees. Spliced girders allow spans as much as 300 ft. [10] Even longer spans (300 ft and up) can be achieved using precast box girder segments which are then post-tensioned together in the field. Using cable stays, the spanning capability of precast and prestressed concrete has been increased to over 1000 ft.

An important innovation in bridge construction has been the use of precast concrete in horizontally curved bridges. A study commissioned by PCI [11] documents the technical feasibility and the economic viability of this application.

Another application of precast and prestressed concrete in bridge construction includes the use of precast deck panels. [12] Used as stay-in-place forms, the panels reduce field placement of reinforcing steel and concrete resulting in considerable savings. The panels become composite with the field-placed concrete for live loads.

Figures 1.2.41 to 1.2.45 show some of the applications mentioned above. [29]

Figure 1.2.41
Quality, plant-produced precast and prestressed concrete results in durability. Low maintenance, economy and the ability to span long distances make precast and prestressed concrete the preferred system for bridges of all spans. Here, prestressed concrete piling easily resists the marine environment.

Figure 1.2.42
The length of this railroad bridge with extensive repetition, form usage and large quantity of product made it worthwhile to create a special section to span from pile cap to pile cap. Precast concrete pile caps eliminate expensive over water formwork. Precast and prestressed concrete piling adds speed to the project as well as durability in a harsh marine environment.

Figure 1.2.43
Standard AASHTO shapes offer immediate availability, economy, durability and low maintenance.

Figure 1.2.44
Standardization allows optimized bridge designs.

Figure 1.2.45
The use of large, hollow edge beams with solid floor slabs and field post-tensioning provided a 110-ft span with only a 14-in. structure depth in this through-girder solution.

1.2.7 Other Structures

The inherent qualities of precast and prestressed concrete noted in previous sections and the high degree of design flexibility also make it ideal for a wide variety of special applications. Properties, such as corrosion resistance, fire resistance, durability and fast installation, have been used to good advantage in the construction of poles, piles, railroad ties, storage tanks, retaining walls and sound barriers. [31–38] Where repetition and standardization exist, precasting components can economically provide the quality of plant manufactured products while eliminating expensive and risky field procedures. These applications are too numerous to categorize here separately; only a few examples are given in Figures 1.2.46 to 1.2.53.

Figure 1.2.46
Precasting this arched culvert and its footings and wing walls allowed quick erection with little site disruption and no field formwork.

Figure 1.2.47
Precast concrete soundwall provides aesthetics and protects residential neighborhood from highway noise.

Figure 1.2.48
A short span bridge over an electrified railroad blends well with a historic residential area. The bridge sits on original stone abutments nearly 200 years old.

Figure 1.2.49
Precast and prestressed utility poles provide low maintenance, durable, economical poles capable of carrying heavy line loads.

Figure 1.2.50
This breakwater/pier structure for the United States Navy incorporates piles, deck panels, utilidor lids, and vertical panel elements. Economic analysis shows precast and prestressed concrete to be the best solution.

Figure 1.2.51
This large circular storage tank is under construction. Vertically prestressed precast concrete wall segments are braced temporarily. Cast-in-place concrete joints and circumferential post-tensioning complete the structure.

Figure 1.2.52
Precast and prestressed concrete has proven to be a viable alternative to timber for railroad ties.

Figure 1.2.53
Famous amusement park building in background has precast and prestressed roof and walls. Monorail guideway is also precast and prestressed concrete.

1.3 Materials

This section provides a brief review of properties of the major materials used in precast and prestressed concrete. Also included is a discussion of durability of concrete. Refs. 13–24 and 29 provide more complete information on these subjects.

1.3.1 Concrete

Concrete properties such as stress-strain relationship, tensile strength, shear strength and bond strength are frequently expressed in terms of the compressive strength of concrete. Generally, these expressions have been empirically established based on experimental data of concretes with compressive strengths less than 6000 psi. These expressions are given in this section and are applicable to most precast and prestressed concrete structures because it is usually specified in the 5000 to 6000 psi compressive strength range.

Use of the equations given in this section are reasonable for compressive strengths up to 10,000 psi. For strengths in excess of 10,000 psi, the recommendations given in Refs. 13 and 14 should be followed.

Concrete with compressive strengths higher than 8000 psi has been used in columns of high-rise buildings, and in prestressed concrete piling and bridge girders. Often, higher strength is a result of using high performance concrete to enhance durability. Interest in a more widespread use of such concrete is growing. The strict use of ACI standard mix design concepts will not always result in the achievement of high strength concrete.

Another type of concrete gaining widespread acceptability is self-consolidating concrete (SCC). Self-consolidating concrete is an advanced approach to the production of highly flowable, self-leveling concrete that can be placed without vibration equipment and without segregation. SCC requires a high performance superplasticizer to achieve and maintain desired workability. SCC can be made with standard available raw materials. However, in order to achieve the unique rheological properties of SCC, special attention must be paid to the mix design process. [30]

1.3.1.1 Compressive Strength

The compressive strength of concrete, made with aggregates of adequate strength, is governed by either the strength of the cement paste or the bond between the paste and the aggregate particles. At early ages, the bond strength is lower than the paste strength; at later ages, the reverse may be the case. For given cementitious materials and acceptable aggregates, the strength that may be developed by a workable, properly placed mixture of cementitious materials, aggregates, and water (under the same mixing, curing, and testing conditions) is influenced by (a) the ratio of water to cementitious materials, (b) the ratio of cementitious materials to aggregate, (c) grading, surface texture, shape and strength of aggregate particles, and (d) maximum size of the aggregate. Mix factors, partially or totally independent of water-cementitious materials ratio, which also affect the strength are (a) type and brand of cement, (b) amount and type of admixture or pozzolan, and (c) mineral composition.

Compressive strength is measured by testing 6 x 12 in. cylinders in accordance with ASTM C39. The precast concrete industry also uses 4 x 8 in. cylinders and 4 in. cube specimens. Adjustment factors may need to be applied to these non-standard specimens to correlate with the standard 6 x 12 in. cylinders.

Because of the need for early strength gain, Type III cement is often used by precasters so that forms may be reused daily. Structural precast concrete and often architectural precast concrete is made with gray cement that meets ASTM C150. Type III and Type I white and buff portland cements are frequently used in architectural products. These are usually assumed to have the same characteristics (other than color) as gray cement. Pigments are also available to achieve colored concrete, and have little effect on strength at the recommended dosages. Cement types, use of other cementitious materials, and experience with color should be coordinated with the local producers.

Higher strength concrete mixes (over 6000 psi) are available in many areas. Local suppliers should be contacted to furnish mix and design information.

Initial curing of precast concrete takes place in the form, usually by covering to prevent loss of moisture and sometimes, especially in structural prestressed products, by the application of radiant heat or live steam. Curing, in addition to the initial curing cycle of approximately 12 hours, has been shown to rarely be necessary to attain the specified strength. Control techniques for the most effective and economical accelerated curing are reported in PCI Technical Report TR 1-82. [15]

When concrete is subjected to freezing and thawing and other aggressive environments, air entrainment is often specified (see Section 1.3.4.1). In some concrete mixes, a slight reduction of strength may occur with air entrainment.

1.3.1.2 Tensile Strength

A critical measure of performance of architectural precast concrete is its resistance to cracking, which is dependent on the tensile strength of concrete. Non-prestressed reinforcement does not prevent cracking, but controls crack widths after cracking occurs. Tensile stresses in prestressed concrete, which could result in cracking, are permitted by ACI 318-02. [16]

Flexural tensile strength is measured by the modulus of rupture. It can be determined by test, but testing is not required or recommended by codes. ACI 318-02 prescribes the tensile strength (modulus of rupture):

$$f_r = 7.5\lambda\sqrt{f'_c} \qquad \text{(Eq. 1.3.1)}$$

For lightweight concrete, if the splitting tensile strength, f_{ct}, is specified, the value of $f_{ct}/6.7$ may be substituted for $\lambda\sqrt{f'_c}$. Otherwise:

Normal weight concrete: $\lambda = 1.0$
Sand-lightweight concrete: $\lambda = 0.85$
All lightweight concrete: $\lambda = 0.75$

1.3.1.3 Shear Strength

Similar to tensile strength, the shear strength of concrete can also be related to its compressive strength. The equations for shear strength specified in ACI 318-02 are given in Chapter 4 of this Handbook. The shear strength of lightweight concrete is determined by modifying the equations in the same manner as for tensile strength as described above.

1.3.1.4 Modulus of Elasticity

Modulus of elasticity, E_c, is the ratio of normal stress to corresponding strain in tension or compression. It is the material property which determines the deformability of a concrete member under load. Thus, it is used to calculate deflections, axial shortening and elongation, buckling and relative distribution of applied forces in composite and non-homogeneous structural members.

The modulus of elasticity of concrete and other masonry materials is not as well defined as, for example, steel. It is, therefore, defined by some approximation, such as the "secant modulus." Thus, calculations which involve its use have inherent imprecision, but this is seldom critical enough to affect performance. While it may be desirable in some rare instances to determine modulus of elasticity by test, especially with some lightweight concretes, the equation given in ACI 318-02 is usually adequate for design:

$$E_c = w^{1.5}33\sqrt{f'_c} \qquad \text{(Eq. 1.3.2)}$$

where:
E_c = modulus of elasticity of concrete, psi
w = unit weight of concrete, pcf
f'_c = compressive strength, psi

For concrete compressive strengths greater than 6000 psi, Eq. 1.3.2 may predict a higher modulus of elasticity than is actually achieved. Alternative equations are given in Refs. 13 and 14.

1.3.1.5 Poisson's Ratio

Poisson's ratio is the ratio of transverse strain to axial strain resulting from uniformly distributed axial load. It generally ranges between 0.11 and 0.27, and, for design, is usually assumed to be 0.20 for both normal weight and lightweight concrete.

1.3.1.6 Volume Changes

Volume changes of precast and prestressed concrete are caused by variations in temperature, shrinkage due to air-drying, and by creep caused by sustained stress. If precast concrete is free to deform, volume changes are of little consequence. If these members are restrained by foundations, connections, steel reinforcement, or connecting members, significant forces may develop over time.

The volume changes due to temperature variations can be positive (expansion) or negative (contraction), while volume changes from shrinkage and creep are only negative.

Precast concrete members are generally kept in yard storage for a period of time. Thus, much of the shrinkage and creep will have taken place by the time of erection and completion of connections. However, connection details and joints must be designed to accommodate the changes which will occur after the precast member is erected and connected to the structure. In most cases, the shortening that takes place prior to making the final connections will reduce the shrinkage and creep strains to manageable proportions.

Temperature Effects. The coefficient of thermal expansion of concrete varies with the aggregate used as shown in Table 1.3.1. Range for normal weight concrete is 5 to 7×10^{-6} in./in./°F when made with siliceous aggregates and 3.5 to 5×10^{-6} in./in./°F when made with calcareous aggregates. The range for structural lightweight concretes is 3.6 to 6×10^{-6} in./in./°F depending on the type of aggregate and amount of natural sand. For design, coefficients of 6×10^{-6} in./in./°F for normal weight concrete and 5×10^{-6} in./in./°F for sand-lightweight concrete are frequently used. If greater

Table 1.3.1 Coefficients of linear thermal expansion of rock (aggregate) and concrete [17]

Type of rock (aggregate)	Average coefficient of thermal expansion × 10⁻⁶ in./in./°F	
	Aggregate	Concrete
Quartzite, Cherts	6.1 – 7.0	6.6 – 7.1
Sandstones	5.6 – 6.7	5.6 – 6.5
Quartz Sands & Gravels	5.5 – 7.1	6.0 – 8.7
Granites & Gneisses	3.2 – 5.3	3.8 – 5.3
Syenites, Diorites, Andesite, Gabbros, Diabas, Basalt	3.0 – 4.5	4.4 – 5.3
Limestones	2.0 – 3.6	3.4 – 5.1
Marbles	2.2 – 3.9	2.3
Dolomites	3.9 – 5.5	
Expanded Shale, Clay & Slate		3.6 – 4.3
Expanded Slag		3.9 – 6.2
Blast Furnace Slag		5.1 – 5.9
Pumice		5.2 – 6.0
Perlite		4.2 – 6.5
Vermiculite		4.6 – 7.9
Barite		10.0
Limonite, Magnetite		4.6 – 6.0
None (Neat Cement)		10.3
Cellular Concrete		5.0 – 7.0
1:1 (Cement:Sand)[b]		7.5
1:3 (Cement:Sand)[b]		6.2
1:6 (Cement:Sand)[b]		5.6

a. Coefficients for concretes made with aggregates from different sources vary from these values, especially those for gravels, granites, and limestones. Fine aggregates generally are the same material as coarse aggregates.
b. Tests made on 2-year old samples.

accuracy is needed, tests should be made on the specific concrete.

Since the thermal coefficient for steel is also about 6×10^{-6} in./in./°F, the thermal effects on precast and prestressed concrete may be evaluated by treating it as plain concrete.

Shrinkage and Creep. Precast concrete members are subject to air-drying as soon as they are removed from molds or forms. During exposure to the atmosphere, the concrete slowly loses some of its original water causing a shrinkage volume change to occur.

When concrete is subjected to a sustained load, the deformation may be divided into two parts: (1) an elastic deformation which occurs immediately, and (2) a time-dependent deformation which begins immediately and continues at a decreasing rate over time. This time-dependent deformation is called creep.

Creep and shrinkage strains vary with relative humidity, volume-surface ratio (or ratio of area to perimeter), level of sustained load including prestress, concrete strength at time of load application, amount and location of steel reinforcement, and other characteristics of the material and design. When high strength concretes are used, different values of shrinkage and creep may be needed. The joints between precast members typically are detailed to relieve such strains.

1.3.2 Grout, Mortar and Drypack

When water, sand and a cementitious material are mixed together without coarse aggregate, the result is called grout, mortar or drypack, depending on consistency. These materials have numerous applications: sometimes only for fire or corrosion protection, or for cosmetic treatment; other times to transfer loads through horizontal and vertical joints.

Different cementitious materials are used:

1. Portland cement.
2. Shrinkage-compensating portland cement.
3. Expansive portland cement made with special additives.
4. Gypsum or gypsum/portland cements.
5. Epoxy-cement resins.

In masonry mortar about one-half of the portland cement is replaced with lime. This improves its bonding characteristics but reduces its strength. Such mortar should not be used indiscriminately as a substitute for grout. Care should be taken to be sure that chlorides are not present in grout material. Quality control of grout is as important as that of concrete. Site-mixed grout should be made and tested at regular intervals according to ASTM C109 which parallels ASTM C39 for concrete. For more general information on grout, see Ref. 18.

1.3.2.1 Sand-Cement Mixtures

Most grout is a simple mixture of portland cement, sand and water. Proportions are usually one part portland cement to two to three parts sand. The amount of water depends on the method of placement of the grout.

Flowable grouts are used to fill voids that are either formed in the field or cast into the precast member. They are used at joints that are heavily congested but not confined, thus requiring some formwork. These grouts usually have a high water-cementitious materials ratio (typically about 0.5), resulting in low strength and high shrinkage. There is also a tendency for the solids to settle, leaving a layer of water on the top. Special ingredients or treatments can improve these characteristics.

For very small spaces in confined areas, grout may be pumped or pressure injected. The confinement must be of sufficient strength to resist the pressure. Less water may be used than for flowable grouts, hence less shrinkage and higher strength.

A stiffer grout, or mortar, is used when the joint is not totally confined, for example in vertical joints between wall panels. This material will usually develop strengths of 3000 to 6000 psi, and have much less shrinkage than flowable grouts.

Drypack is the common name used for very stiff sand-cement mixes. They are used if a relatively high strength is desired, for example, under column base plates. Compaction is attained by hand tamping.

When freeze-thaw durability is a factor, the grout should be air-entrained. Air content of plastic grout or mortar of 9% to 10% is required for adequate protection.

Typical portland cement mortars have very slow early strength gain when placed in cold weather. Heating the material is usually not effective, because the heat is rapidly dissipated to surrounding concrete. Thus, unless a heated enclosure can be provided, special proprietary mixes, usually containing gypsum, may be indicated. Mixes containing a high percentage of gypsum are known to deteriorate under prolonged exposure to water.

1.3.2.2   Non-Shrink Grouts

Shrinkage can be reduced, or more appropriately, compensated for by the use of commercially available non-shrink, high-strength pre-mixed grouts. These mixes expand during the initial hardening to offset the subsequent shrinkage of the grout. Since the non-shrink grouts are primarily proprietary, their chemical composition is usually not available to study their potential effects on the interfacing materials, such as reinforcement and inserts in the connection. Thus, it is advisable that manufacturers' recommendations should be carefully followed. For a general reference on the characteristics and methods of testing of non-shrink grouts, see Ref. 19.

1.3.2.3   Epoxy Grouts

Epoxy grouts are mixtures of epoxy resins and a filler material, usually oven dried sand. These are used when high strength is desired, or when improved bonding to concrete is necessary. Ref. 20 is a comprehensive report on the subject by Committee 503 of the American Concrete Institute.

The physical properties of epoxy compounds vary widely. Also, the epoxy grouts behave very differently than the sand-cement grouts. For example, the thermal expansion of an epoxy grout can be as much as seven times the thermal expansion of sand-cement grout. Epoxy grouts may not be applicable where fire rating is required. It is, therefore, important that use of these grouts be based on experience and/or appropriate tests. Recommended tests and methods are given in Ref. 20. See Chapter 9 for additional information.

1.3.3   Reinforcement

Reinforcement used in structural and architectural precast concrete includes prestressing tendons, deformed steel bars, and welded wire reinforcement.

Fibers, which are sometimes used to control shrinkage cracks, do not have well defined structural properties and, therefore, cannot be used to replace structural reinforcement such as welded wire reinforcement. This is particularly important for structural toppings over precast concrete decks. The reinforcement in these toppings cannot be replaced with fibers.

1.3.3.1   Prestressing Tendons

Tendons for prestressing concrete may be wires, strands or high strength bars. In precast and prestressed structural concrete, nearly all tendons are 7-wire strands conforming to ASTM A416. There is limited use of 2-wire and 3-wire strands conforming to ASTM A910. The strands are pretensioned, that is, they are tensioned prior to placement of the concrete. After the concrete has reached a predetermined strength, the strands are cut and the prestress force is transferred to the concrete through bond.

Two types of strand are covered in ASTM A416 and A910: "low-relaxation" strand and "stress-relieved" strand. Low-relaxation strand is the standard in North America and is used in the load tables in Chapter 2 and in the various examples throughout this Handbook.

Sometimes, architectural precast concrete contains tendons. Depending on the facilities available at the plant, these tendons may be either pretensioned or post-tensioned. In post-tensioning, the tendons are either placed in a conduit or are coated so they will not bond to the concrete. The tendons are then tensioned after the concrete has reached the predetermined strength. The compressive forces are transferred to the concrete from the strand by end fittings on the strand, which bear directly against the end surfaces of the concrete member. When the tendons are placed in a conduit, they are usually grouted after tensioning (bonded post-tensioning). When they are greased and wrapped, or coated, they usually are not grouted (unbonded post-tensioning). For more information on post-tensioning in general, see Ref. 21.

Precast products are typically prestressed with 7-wire strand. Prestressing bars meeting ASTM A722 have been used in connections between members, and may be applicable for prestressing short members where the seating losses associated with strand anchors are not acceptable. The properties of prestressing strand, wire and bars are given in Chapter 11.

The ability of strands to properly bond should be certified by the strand supplier.

1.3.3.2   Deformed Reinforcing Bars

Reinforcing bars are hot-rolled from steel with varying carbon content. They are usually required to meet ASTM A615, A616 or A617. These specifications are of the performance type, and do not closely control the chemistry or manufacture of the bars. Bars are usually specified to have a minimum yield of 40,000 psi (Grade 40) or 60,000 psi (Grade 60). Grade 40 bars will usually have lower carbon content than Grade 60. It is sometimes possible to weld these grades of reinforcing bars after appropriate preheating depending on carbon equivalency. [27]

ASTM A706 specifies a bar with controlled chemistry that is weldable. See Chapter 6 for further discussion of welding.

In order for the reinforcing bar to develop its full strength in the concrete, a minimum length of embedment is required or the bars may be hooked. Information on bar sizes, bend and hook dimensions and development length is given in Chapter 11 and Ref. 22.

1.3.3.3   Structural Welded Wire Reinforcement

Structural welded wire reinforcement is a prefabricated reinforcement consisting of parallel, cold-drawn wires welded together in square or rectangular grids. Each wire intersection is electrically resistance-welded by a continuous automatic welder. Pressure and heat fuse the intersecting wires into a homogeneous section and fix all wires in their proper position.

Plain wires (ASTM A185), deformed wires (ASTM A497) or a combination of both may be used in welded wire reinforcement. Plain wire sizes are specified by the letter W followed by a number indicating the cross-sectional area of the wire in hundredths of a square inch. For example, W16 denotes a plain wire with a cross-sectional area of 0.16 sq in. Similarly, deformed wire sizes are specified by the letter D followed by a number which indicates area in hundredths of a square inch.

Plain welded wire reinforcement bonds to concrete by the mechanical anchorage at each welded wire intersection. Deformed welded wire reinforcement utilizes wire deformations plus welded intersections for bond and anchorage. Welded wire reinforcement for architectural precast concrete is normally supplied in flat sheets. Use of welded wire reinforcement from rolls, particularly in the thin precast sections, is not recommended because the rolled welded wire reinforcement cannot be easily and reliably flattened to the required placement tolerance. In addition to using welded wire reinforcement in flat sheets, many plants have equipment for bending sheets into various shapes, such as U-shaped stirrups, four-sided cages, etc.

Available wire sizes, common stock sizes and other information on welded wire reinforcement are given in Chapter 11 and Ref. 23.

1.3.4   Durability

Concrete durability is a concern when the structure is exposed to an aggressive environment. The designer, contractor, and owner must recognize the deleterious effects of (a) freezing and thawing cycles in a cold, wet environment, (b) chemical attack, including carbonation, (c) corrosion of embedded metals, and (d) aggregate reactivity. The ideal approach is to make the concrete impermeable, which means making the concrete as uniformly dense as possible and designing to control cracking. In this respect, precast and prestressed concrete has inherent advantages because it is produced in a controlled environment that lends itself to high quality concrete, and prestressing leads to effective crack control.

Durability is enhanced by proper mix design with low water-cementitious materials ratio, adequate cover over reinforcement, chloride exclusion, non-reactive aggregates, air entrainment, proper finishing, proper curing, and in parking structures and bridge decks, design for adequate drainage. Other measures that may incrementally improve durability include coating of non-prestressed reinforcement, low-alkali cement, surface sealers and surface membranes.

Penetrating surface sealers can improve the durability of concrete by reducing moisture and chloride penetration. Sealers have, however, little ability to bridge cracks and should not be expected to provide protection from moisture absorption or chloride penetration at cracks. Some sealers have proven to be more effective than others. Their evaluation should be based on criteria established in Ref. 26. Silane based sealers are hydrophobic and have been demonstrated to reduce chloride penetration into concrete as much as 95%.

1.3.4.1 Freeze-Thaw and Chemical Resistance

The typical dense mixes used for precast products have high resistance to freezing and thawing. Entrained air is essential to further improve freeze-thaw resistance in particularly severe environments. Freeze-thaw damage, which manifests itself by scaling of the surface, is magnified when chloride based deicing chemicals are used. Deicers can be applied indirectly in various ways, such as by drippings from the underside of vehicles.

In some food processing facilities, particularly meat packing plants, concrete surfaces may be exposed to extremely aggressive, acidic environments from brine and biological by-products and may require special surface treatments. Such facilities also frequently have "flash-freeze" rooms which are subject to frequent severe freeze-thaw cycles.

Table 1.3.2 (Table 4.2.1, ACI 318-02) provides the required air content for both severe and moderate exposure conditions, for various maximum aggregate sizes. Severe exposure is defined as a climate where the concrete may be in almost continuous contact with moisture prior to freezing, or where deicing salts come in contact with the concrete. Salt-laden air, as found in coastal areas, is also considered a severe exposure. A moderate exposure is one where deicing salts are not used, or where the concrete will only occasionally be exposed to moisture prior to freezing.

Admixtures are added to the concrete during the mixing cycle to entrain air. Tolerance on air content is ±1.5%. ACI 318-02 permits an air content one percentage point lower than the values for in Table 1.3.2 for concrete strengths higher than 5000 psi. Other studies have shown that even this amount may not be necessary for the very low water-cementitious materials ratio used in most precast products. [28]

For some concrete mixtures, such as low-slump mixtures in extruded products or gap-graded mixtures in architectural precast concrete, it is difficult to measure air content. Thus, it is recommended that a "normal dosage" of the air-entraining agent be used instead of specifying a particular range of air content percentage. For precast concrete elements constructed above grade and oriented in a vertical position, air-entraining agents should be used, but air contents as low as 3% to 4% will usually provide the required durability. The precast and prestressed concrete industry does not use air-entraining portland cements.

In addition to entrained air, other positive measures, such as adequate concrete cover over steel and rapid drainage of surface water may be essential for durability in structures exposed to weather.

Table 1.3.3 (Table 4.2.2, ACI 318-02) provides the maximum permitted water-cementitious materials ratio (or, for lightweight concrete, minimum strength) for concrete which is exposed to an aggressive environment. The water-cementitious materials ratios specified in Table 1.3.3 are calculated using the weight of cement meeting ASTM C150, C595, or C845 plus the weight of fly ash and other pozzolans meeting ASTM C618, slag meeting ASTM C989, and silica fume meeting ASTM C1240, if any, except when the concrete is exposed to deicing chemicals. For concrete exposed to deicing chemicals, the maximum weight of fly ash, other pozzolans, silica fume, or slag that is included in the concrete must not exceed the percentages of the total weight of cementitious materials given in Table 1.3.4 (Table 4.2.3, ACI 318-02). In many plants, a water-cementitious materials ratio of 0.35 is frequently used for added durability of precast, prestressed concrete products.

Table 1.3.2   Total air content for frost-resistant concrete

Normal maximum aggregate size,[a] in.	Air content, percent	
	Severe exposure	Moderate exposure
⅜	7½	6
½	7	5½
¾	6	5
1	6	4½
1½	5½	4½
2[b]	5	4
3[b]	4½	3½

a. See ASTM C33 for tolerances on oversize for various nominal maximum size designations.
b. These air contents apply to total mix, as for the preceding aggregate sizes. When testing these concretes, however, aggregate larger than ½ in. is removed by handpicking or sieving and air content is determined on the minus 1½ in. fraction of mix. (Tolerance on air content as delivered applies to this value.) Air content of total mix is computed from value determined on the minus ½ in. fraction.

Table 1.3.3   Requirements for special exposure conditions

Exposure condition	Maximum water-cementitious materials ratio, by weight, normal weight aggregate concrete	Minimum f'_c, psi, normal weight and lightweight aggregate concrete
Concrete intended to have low permeability when exposed to water	0.50	4000
Concrete exposed to freezing and thawing in a moist condition or deicing chemicals	0.45	4500
For corrosion protection of reinforcement in concrete exposed to chlorides from deicing chemicals, salt, salt water, brackish water, sea water, or spray from these sources	0.40	5000

Table 1.3.4   Requirements for concrete exposed to deicing chemicals

Cementitious materials	Maximum percent of total cementitious materials by weight[a]
Fly ash or other pozzolans conforming to ASTM C618	25
Slag conforming to ASTM C989	50
Total of fly ash or other pozzolans, slag	50[b]
Total of fly ash or other pozzolans	35[b]

a. The total cementitious material also includes ASTM C150, C595 and C845 cement. The maximum percentages also include fly ash or other pozzolans present in Type IP or I(PM) blended cement, ASTM C595; slag used in the manufacture of an IS or I(SM) blended cement, ASTM C595; present in a blended cement.
b. Fly ash or other pozzolans shall constitute no more than 25 and 10 percent, respectively, of the total weight of the cementitious materials.

1.3.4.2   Protection of Reinforcement

Reinforcing steel is protected from corrosion by embedment in concrete. A protective iron oxide film forms on the surface of the bar, wire or strand as a result of the high alkalinity of the cement paste. As long as the high alkalinity is maintained, the film is effective in preventing corrosion.

The protective high alkalinity of the cement paste may be lost in the presence of oxygen, moisture, and chlorides. Chlorides may be found in concrete aggregates, water, cementitious materials or admixtures, and hence may be present in small quanitites in the concrete when cast. Concrete of low permeability and of sufficient cover over the steel will usually provide the necessary protection against chloride penetration as a result of the presence of deicer salts.

Moisture and oxygen alone can cause corrosion. Cracking may allow oxygen and moisture to reach the embedded steel, resulting in conditions where rusting of the steel and staining of the surface may rapidly occur. Prestressing can control cracking. Non-prestressed members should be designed for low flexural stress to minimize the potential for cracking (see Section 4.2.2).

Chloride compounds as an admixture should not be used in prestressed or reinforced concrete. Consideration should be given to the chloride ion content in hardened concrete, contributed by the ingredients of the mix. Table 1.3.5 (Table 4.4.1, ACI 318-02) indicates the maximum chloride ion content. These limits are intended for use with uncoated reinforcement. Chloride ion content values can be determined in accordance with ASTM C1218.

In order to provide corrosion protection to reinforcement, concrete cover should conform to ACI 318-02, Section 7.7.3. Concrete cover is the minimum clear distance from the reinforcement to the face of the concrete. For exposed aggregate surfaces, the concrete cover is not measured from the original surface; instead, the depth of the mortar removed between the pieces of coarse aggregate (depth of reveal) should be subtracted. Attention must also be given to scoring, reveals, false joints, and drips, as these reduce the cover.

High quality concrete provides adequate corrosion protection for reinforcement for most conditions. Even in moderate to severe aggressive environments, concrete will provide adequate protection with proper attention to mix design, steel stress level and the extent of cracking under service loads, and the depth of concrete cover. Only when these protection measures are not feasible, it may be necessary to consider other ways of protecting reinforcement, such as galvanizing, corrosion inhibitors or epoxy coating. These are described as follows:

Galvanized Reinforcement. Except for exposed connections, there is rarely any technical need for galvanized reinforcement. With proper detailing and specifications, galvanizing may be superfluous.

Galvanizing is not recommended for members subjected to chlorides, because a deleterious chemical reaction can take place between the chloride and the galvanizing when the concrete is damp and

Table 1.3.5 Maximum chloride ion content for corrosion protection of reinforcement

Type of member	Maximum water soluble chloride ion (Cl⁻) in concrete, percent by weight of cement
Prestressed concrete	0.06
Reinforced concrete exposed to chloride in service	0.15
Reinforced concrete that will be dry or protected from moisture in service	1.00
Other reinforced concrete construction	0.30

chlorides are present. Therefore, the benefit obtained by galvanizing is questionable for members subjected to deicing salts and similar exposures.

Galvanized welded wire reinforcement is usually available as a stock item in selected sizes. Individual wires are galvanized before they are welded together to form the fabric; zinc at each wire intersection is burned off during welding, but the resulting ungalvanized spots have not caused noticeable problems. After welding, the fabric is shipped without further treatment. There is no current ASTM specification for galvanized welded wire reinforcement.

When galvanized reinforcement is used in concrete, it should not be coupled directly to ungalvanized steel reinforcement, copper, or other dissimilar materials. Care should be taken to provide insulating material between galvanized reinforcing and other dissimilar metals at points of contact.

The use of galvanized reinforcement close to steel forms or non-galvanized reinforcement in fresh concrete may cause "shadowing" or reflection of the reinforcement onto the final concrete surface. When galvanized and non-galvanized reinforcement or steel forms are in close proximity in fresh concrete, "galvanic" cell problems may occur during the initial processes of hydration. The reactions of zinc in concentrated alkaline material such as concrete (pH of 12.5 to 13.5) liberates hydrogen gas. This release of gas bubbles results in the shadowing or reflection of reinforcement. Ways to avoid this occurrence, by passivating either the galvanized steel or the concrete mix, are discussed in the PCI Manual on Architectural Precast Concrete. [8]

Epoxy Coated Reinforcement. Epoxy coated reinforcing bars and welded wire reinforcement are also available for use in members which require special corrosion protection. Epoxy coated reinforcing bars should conform to ASTM A775, and epoxy coated welded wire reinforcement should conform to ASTM A884. The epoxy provides enhanced protection from corrosion if the bar is uniformly coated. Bars generally are coated when straight; subsequent bending should have no adverse effect on the integrity of the coating. If the coating is removed or damaged, the reinforcement should be touched up with commercially available epoxy compounds. Epoxy coating reduces bond strength; reference should be made to Section 12.2.4 of ACI 318 for the required increase in development length. Similarly, the requirements for crack control may need to be modfied.

Supplemental items must also be protected to retain the full advantage of protecting the main reinforcement. For example, bar supports should be solid plastic and bar ties should be nylon-, epoxy-, or plastic-coated wire, rather than black wire. Epoxy coated reinforcing bars should be handled with nylon slings.

Epoxy Coated Strand. Epoxy coated strand is sometimes proposed for use in unusually aggressive environments. It is covered by ASTM A882.

In order for it to be used as bonded strand, the epoxy coating is impregnated with a grit to ensure development of bond; without grit, the epoxy coated strand has virtually no bond strength. With adequate density and a proper distribution of the grit, the bond strength of the epoxy coated strand is comparable to that of uncoated strand. Note that tests [24] have shown the transfer and the development lengths of the epoxy coated (with grit) strand to be somewhat shorter than the corresponding lengths for the uncoated strand. Also, there are differences in some other properties of the two types of strand. For example, the relaxation loss of epoxy coated strand is higher (about twice) than that of uncoated, low-relaxation strand.

The behavior of epoxy coated strand at elevated temperatures is of concern due to softening of the epoxy. Pull-out tests [24] show that there is a progressive reduction in bond strength initiating at about 125°F with a virtually complete loss of bond occurring at about 200°F. These temperatures often occur during production. This behavior necessitates a careful monitoring of concrete temperature at transfer of prestress.

Because of the uncertainties in properties noted above, particularly the behavior under elevated temperatures, it is recommended that epoxy coated strand not be used for pretensioned, prestressed concrete products.

Corrosion Inhibitors. Corrosion inhibitors are used to extend the service life of concrete structures by offering protection to embedded reinforcement against chloride induced corrosion. Corrosion inhibitors are chemical admixtures that delay the initiation of corrosion and decrease the corrosion rate of steel in concrete, without changing the concentration of the corrosive agent.

Chemical compounds classified as corrosion inhibitors by ACI 212 and ACI 222 are borates, chromates, molybdates, nitrites, and phosphates. Of these, calcium nitrite has been used successfully with precast and prestressed concrete since 1980.

1.3.4.3 Protection of Connections

Painted Steel. In most building environments, painting of exposed steel in connections is sufficient to prevent corrosion damage. There is a broad spectrum of choices of paint systems from one coat of primer to multi-coat systems using zinc rich paint or epoxy systems. Long oil alkyds have the advantage of low cost surface preparation and the ease of application and touch up. Their disadvantage is their relatively short life span in corrosive conditions. Epoxy polyamidoamines have an extended life span and are good in corrosive environments. However, they have a higher material cost and surface preparation cost. Epoxy polyamidoamines are more difficult to field touch up since they are a two-part mixture requiring controlled temperatures to apply. Any epoxy based topcoat has the disadvantage of chalking due to weathering and environmental effects, especially direct or indirect UV exposure.

For both long oil alkyd and epoxy polyamidoamine systems, the protection is lost once the surface is broken (scratch, incomplete coverage) since corrosion can start undercutting adjacent areas. Zinc-rich urethanes minimize this problem by providing galvanic protection. Zinc-rich urethane has the best corrosion resistance and life expectancy and is relatively easy to apply. The disadvantage of the zinc-rich urethane is that it comes in one color, brown. If other colors are required, epoxy or urethane paints may be used as a top coat. Consult the local precast producers in the area for paint systems commonly used.

Galvanized Steel. In corrosive environments, hot dip galvanizing of connection hardware is sometimes used. Connections should be designed to minimize or eliminate field welding if galvanized connections are used. The fumes from welded galvanized material are very toxic and present a serious threat to the welder even with the use of protective equipment. The process of welding destroys the protective coating, requiring touch up with a cold applied zinc-rich paint, so the building owner has paid for hot dip galvanizing only to end up with a painted surface.

In order to ensure that the strength of the various elements of a connection are not reduced by embrittlement during the hot dip galvanizing process, several precautions are recomended.

When items of a connection assembly require welding, such as anchor bars to plates, the following recommendations have been found to produce satisfactory results and are recommended by the American Hot Dip Galvanizers Association:

1. An uncoated electrode should be used whenever possible to prevent flux deposits.
2. If a coated electrode is used, all welding flux residues must be removed by wire brushing, flame cleaning, chipping, grinding, needle gun or abrasive blast cleaning. This is necessary because welding flux residues are chemically inert in the normal pickling solutions used by galvanizers; their existence will produce rough and incomplete zinc coverage.
3. A welding process such as metal-inert gas (MIG), tungsten-inert gas (TIG), or CO_2 shielded arc is recommended when possible since they produce essentially no slag.
4. If special process welding is not available, select a coated rod specifically designed for self-slagging, as recommended by welding equipment suppliers and refer to Item 2.

It should also be recognized that many parts of connection components are fabricated using cold rolled steel or cold working techniques, such as bending of anchor bars. In some instances, cold working may cause the steel to become strain-age embrittled. The embrittlement may not be evident until after the work has been galvanized. This occurs because aging is relatively slow at ambient temperatures but is more rapid at the elevated temperature of the galvanizing bath.

It is recognized that any form of cold working reduces the ductility of steel. Operations such as punching holes, notching, producing fillets of small radii, shearing and sharp bending may lead to strainage embrittlement of susceptible steels.

The following precautions are recommended by the American Hot Dip Galvanizers Association if cold-worked steel is to be galvanized:

1. Select steel with a carbon content below 0.25%.
2. Choose steel with low transition temperatures since cold working raises the ductile-brittle transition temperature and galvanizing (heating) may raise it even further.
3. For steels having a carbon content between 0.10% and 0.25%, a bending radius of at least three times the section thickness (3t) should be maintained. In some cases, 6t yields even better results. If less than 3t bending is unavoidable, the material should be stress-relieved at 1100°F for one hr per in. of section thickness.

4. Drill, rather than punch, holes in material thicker than ¾ in. If holes are punched, they should be punched undersize, then reamed an additional ⅛ in. overall or drilled to size.
5. Edges of steel sections greater than ⅝ in. thick subject to tensile loads should be machined or machine cut.
6. In critical applications, the steel should be hot worked above 1200°F in accordance with steelmakers recommendation. Where cold working cannot be avoided, stress-relieve as recommended in Item 3 above.

ASTM A143 "Recommended Practice for Safeguarding against Embrittlement of Hot Dip Galvanized Structural Steel Products and Procedure for Detecting Embrittlement" and CSA Specification G164 "Galvanizing of Irregularly Shaped Articles," provide guidance on cold working and stress relieving procedures. However, severe cold working of susceptible steels is better avoided, if at all possible.

Another area of concern is hydrogen embrittlement. Hydrogen embrittlement is a ductile-to-brittle change which occurs in certain high strength steels. Hydrogen released during the pickling operation, prior to hot dipping, can cause this embrittlement. This hydrogen can be absorbed into the steel during the acid pickling, but at galvanizing temperatures it is generally expelled from the steel.

Hydrogen embrittlement is not common, but precautions should be taken if the steel involved has an ultimate tensile strength exceeding approximately 150,000 psi, or if the pickling process is poorly controlled, resulting in long exposure to HCl. In those cases, grit blasting is recommended instead of acid pickling.

These precautions are also outlined in Ref. 6. An alternative to hot dip galvanizing is cold galvanizing using zinc rich paint.

Stainless Steel. In highly corrosive environments, stainless steel may be used for connections and embedments. The AISI (American Iron and Steel Institute) 200 and 300 series stainless steels contain nickel and chromium and are essentially nonmagnetic, and are referred to as austenitic stainless steels. The AISI 200 series contains manganese in addition to the nickel and chromium. The AISI 400 series contain straight chromium and is referred to as ferritic stainless steel. They are magnetic and are less ductile than the austenitic grades. Austenitic grades are weldable, thus more applicable for connections.

AISI Types 304 and 316 are the most commonly used in structural applications. These types are a low carbon modification of Type 302 for limiting of carbide precipitation during welding. Type 316 has a higher corrosion resistance than Type 304 and is usually used for chemical handling equipment. There are Types 304L and 316L which are extra low carbon modifications of Types 304 and 316, respectively, and can be used where carbide precipitation is a problem. There are a limited number of structural shapes and sizes available in stainless steel. Consult with the local steel suppliers for availability of different shapes and sizes.

Austenitic stainless steel can be welded by all common methods, and the equipment used is basically the same as that used for carbon steel.

The choice of electrodes should be made for compatibility with the base metal. E308 electrode is used with 304, E308L with 304L, E316 with 316, and E316L with 316L. When stainless steel and carbon materials are to be joined, an E309 electrode should be used.

All stainless steel SMAW electrode coverings are of the low-hydrogen type and must be protected from moisture. Electrodes should be purchased in hermetically sealed containers, which can be stored for several months without deterioration. After opening, the electrodes should be reconditioned after 4 hours of exposure.

Fillet welds should be designed using the same guidelines as for carbon steel. Joints must be clean and dry. Moisture should be removed by heating or by blowing with dry air. Slag needs to be thoroughly removed and the weld completely cleaned before starting a new pass. Only stainless steel wire brushes should be used for cleaning.

Inspection of welds should include verification of the proper electrode, proper storage of the electrode, and operator certification, in addition to the non-destructive testing required. The method and frequency of testing should be as directed by the design engineer. Typically, the testing will be visual only.

The welding of stainless steel produces more heat than conventional welding. That, and the fact that stainless steel has a coefficient of thermal expansion greater than structural steel can create adverse expansion of embedments during welding, thus special detailing is required to avoid cracking the adjacent concrete. Stainless steel embedment edges should be kept free from adjacent concrete to allow expansion during welding without spalling the concrete.

1.3.4.4 Sulfate Attack

Chemical attack of concrete materials may occur from sulfates found in ground water, soils, processing liquids, or sewage. The proper choice of cement is particularly effective in resisting sulfate attack. ACI 318-02, Section 4.3.1, and the references contained therein, provide detailed guidance.

1.4 Production Process

The methods used to produce precast concrete components vary with the type of member being manufactured. Among the factors influencing the production process are:

- Structural vs. architectural
- Long-line forms vs. tub forms
- Prestressed vs. non-prestressed
- Pretensioned vs. post-tensioned
- Production facility limitations

Discussions with local precasters during project development will allow these factors to be considered in the design of the precast components.

1.4.1 Structural Components vs. Architectural Components

Structural precast concrete members are typically produced in a process aimed toward high volume and high efficiency. Emphasis is placed on using standard shapes so reusable forms can be used. These forms are usually long steel forms designed to produce a number of members with each casting. Products commonly considered to be structural components include double tees, interior beams, hollow-core plank and interior walls. Precast concrete producers certified by PCI as structural or "commercial" fabricators must comply with the requirements of PCI Manual 116-99, Manual for Quality Control for Plants and Production of Structural Precast Concrete Products, Fourth Edition.

The flexibility of architectural precast comcete allows designers to incorporate a wide array of shapes, colors and textures into a project's appearance. This demands that the precast producer be capable of constructing complex forms, using a variety of mixes and offering an assortment of surface treatments such as sandblasting, acid etching or veneers of brick, stone or other decorative material.

Architectural precast concrete members are fabricated under the guidelines of PCI Manual 117-96, Manual for Quality Control for Plants and Production of Architectural Precast Concrete Products, Third Edition. Members manufactured in accordance with these guidelines must satisfy stringent dimensional tolerances and offer consistently superior appearance. These requirements commonly demand that the product be poured on custom-made forms with special attention given to true, flat surfaces, clean edges and consistent concrete color and texture. Fiberglass coated forms are often used to achieve the desired surface quality and all form joints are sealed to ensure clean edges.

Reinforcement is hung from prestressing strands within the member or jigs spanning across the form. This eliminates the need for chairs which might become visible on the exposed face of the member. Special care must be taken to maintain adequate cover over the reinforcement, not only to maintain durability, but also to prevent reinforcement patterns from "shadowing" through to the face of the member.

In instances where exacting tolerances and ornate shapes are not needed but special surface treatment or colors are desired, consideration may be given to using a PCI certified commercial producer capable of providing such services. These producers are designated with an "A" suffix on their classification (e.g., C3A), and have demonstrated their ability to apply special finishes to their structural products. Additional information regarding plant certification categories can be found in Section 9.5 of this Handbook.

1.4.2 Long Line Forms vs. Single Piece Forms

When producing large quantities of a standard shaped member, precasters will commonly use a "long line form" approach. These long line forms may be several hundred feet long and are designed to produce a number of pieces with each casting. In most instances, the members produced in this fashion are prestressed. Products such as beams and double tees use bulkheads at the end of each member in the form while extruded product such as hollow-core plank is placed as one continuous piece of concrete and then cut to the prescribed member lengths after curing.

Single piece forms are used with the intent of making one piece with each casting. This type of forming system readily accommodates non-prestressed or post-tensioned products, and is commonly used for architectural precast concrete members.

1.4.3 Prestressed Members vs. Non-Prestressed Members

Where member dimensions and applied loads will allow, a precaster may elect to use reinforcing bars or welded wire to reinforce the product. In these cases, a reinforcement "cage" or "mat" is usually assembled and placed in the form. Concrete is then placed and vibrated around the bar assembly. Care must be taken by the designer to check crack control and deflections when using this type of design.

Prestressed concrete members are implemented to achieve higher span-to-depth ratios and provide resistance to cracking. Prestressing is incorporated into a member in the form of pre-tensioned or post-tensioned strands (see Section 1.4.4). Prestressing is generally used as the primary reinforcement with reinforcing bars or welded wire reinforcement serving as secondary reinforcement.

1.4.4 Pretensioned Members vs. Post-Tensioned Members

In pretensioned members, strands are tensioned. Supplemental reinforcing bars and other items to be embedded in the member are secured and concrete is placed into the form. After the concrete is allowed to cure to form a bond with the pretensioned strands, the strands are cut and the prestressing force is transferred from the strands to the concrete through bond and mechanical interlock. At this point, the members are removed from the form and placed in storage.

When post-tensioning is used, the prestressing strands are encased in a conduit or greased sleeve. This "tendon"* is typically tied to a "cage" or "mat" of reinforcement, which is then placed into the form. Mechanical anchorages are placed at each end of the tendon and secured to the form or the reinforcement. After concrete has been placed and cured within the form, the prestressing force is applied to the post-tensioning strands and held at the mechanical anchorages. At this point, the prestress force has been transferred to the member, which can then be stripped out of the mold and placed in storage.

Under some circumstances a combination of pretensioning and post-tensioning may be used. This approach may be used where the self-stressing forms or tensioning abutments cannot sustain the full prestress force. In this case, the majority of the force is applied through pretensioned strands and post-tensioned tendons are used to introduce the remainder of design prestress force after the member is stripped. This approach may also be used in prestressed panels that are poured in a flat position but stand in an upright position in service. Occasionally, the full prestress force will cause cracking in the panel while it is flat in the form. By applying a portion of the prestress by post-tensioning after the panel is stripped and upright, this cracking can be avoided. Thought should be given to space restrictions at the anchorage area and the additional cost of a two-step process.

1.4.5 Production Limitations

A wide variety of products, in a variety of shapes and sizes are shown in this Handbook, and are generally available. Not all products are available in all parts of the United States, and, therefore, designers should be aware of these limitations during the project development.

The size of the piece may be limited by the handling equipment available in the manufacturing plant, or by the transporting and erection equipment available. In most cases, manufacturers are geared to produce standard products that can be transported over the road with a minimum of special permits. Unusually large pieces can be made, but the special equipment required to handle such pieces can add significantly to the cost.

The tensioning capacity of the prestressing beds may be another limitation. Some products, such as deep long-span bridge girders require massive stressing abutments and beds that can handle large tie-down forces for depressed strands. These capabilities are costly and must usually be amortized over a number of projects. For large projects with many pieces, it is sometimes economical to install special facilities which can be written off on the single project. Self-stressing forms, which transfer the large forces along the bed with heavy compression members (usually steel) built into the forms may be another solution.

Environmental concerns may also limit the scope of a precaster's operation. For example, in most cases such operations as sandblasting and acid etching must be contained. The location of the facility related to residences, businesses, streams, etc., may require special noise, dust or runoff containment. These are problems confronted by virtually all manufacturing industries, and are generally not restrictions on the design and use of precast concrete products.

---

* ACI 318-02 defines "tendon" as follows: "In pretensioned applications, the tendon is also the prestressing steel [or strand]. In post-tensioned applications, the tendon is a complete assembly consisting of anchorages, prestressing steel [or strand], and sheathing with coating for unbonded applications or ducts with grout for bonded application."

1.5 References

1. *Reflections on the Beginnings of Prestressed Concrete in America*, JR-H-81, Precast/Prestressed Concrete Institute, Chicago, IL, 1981.

2. *Manual for Quality Control for Plants and Production of Structural Precast Concrete Products*, Fourth Edition, MNL-116-99, Precast/Prestressed Concrete Institute, Chicago, IL, 1999.

3. *Manual for Quality Control for Plants and Production of Architectural Precast Concrete Products*, Third Edition, MNL-117-96, Precast/Prestressed Concrete Institute, Chicago, IL, 1996.

4. Pfeifer, Donald W., Landgren, J. R., and Perenchio, William, "Concrete, Chlorides, Cover and Corrosion," PCI JOURNAL, V. 31, No. 4, July-August 1986.

5. Chrest, A. P., Smith, M. S., Bhuyan, S., Labal, M., and Monahan, D. R., *Parking Structures — Planning, Design, Construction, Maintenance and Repair,* Third Edition, Kluwer Academic Publishers, Boston, MA, 2001.

6. *Precast, Prestressed Parking Structures: Recommended Practice for Design and Construction*, MNL-129-98, Precast/Prestressed Concrete Institute, Chicago, IL, 1998.

7. "Architectural Precast Concrete Cladding — Its Contribution to Lateral Resistance of Buildings," *Proceedings*, SP-CP, Precast/Prestressed Concrete Institute, Chicago, IL, 1990.

8. *Architectural Precast Concrete*, Second Edition, MNL-122-89, Precast/Prestressed Concrete Institute, Chicago, IL, 1989.

9. *Recommended Practice for Glass Fiber Reinforced Concrete Panels*, Fourth Edition, MNL-128-01, Precast/Prestressed Concrete Institute, Chicago, IL, 2001.

10. Abdel-Karim, A. M., and Tadros, Maher K., "Stretched-Out Precast Concrete I-Girder Bridge Spans," *Concrete International*, V. 13, No. 9, September 1991.

11. ABAM Engineers, Inc., "Precast Prestressed Concrete Horizontally Curved Bridge Beams," PCI JOURNAL, V. 33, No. 5, September-October 1988.

12. Ross Bryan Associates, Inc., "Recommended Practice for Precast/Prestressed Concrete Composite Bridge Deck Panels," PCI JOURNAL, V. 33, No. 2, March-April 1988.

13. ACI Committee 363, "State-of-the-Art Report on High Strength Concrete," ACI 363R-92, *ACI Manual of Concrete Practice, Part 1,* American Concrete Institute, Farmington Hills, MI, 1992.

14. Ahmad, Shuaib H., and Shah, S. P., "Structural Properties of High Strength Concrete and Its Implications for Precast Prestressed Concrete," PCI JOURNAL, V. 30, No. 6, November-December 1985.

15. Pfeifer, D. W., Marusin, Stella, and Landgren, J. R., "Energy-Efficient Accelerated Curing of Concrete," *Technical Report*, TR 1-82, Precast/Prestressed Concrete Institute, Chicago, IL, 1982.

16. ACI Committee 318, "Building Code Requirements for Structural Concrete (ACI 318-02)," and "Commentary," ACI 318R-02, American Concrete Institute, Farmington Hills, MI, 2002.

17. *Concrete Manual — A Manual for the Control of Concrete Construction*, Eighth Edition, Revised, U.S. Department of Interior, Bureau of Reclamation, Denver, CO, 1981.

18. "Cementitious Grouts and Grouting," EB 111T, Portland Cement Association, Skokie, IL, 1990.

19. "Corps of Engineers Specification for Non-Shrink Grout," CRD-C588-78A, U.S. Army Corps of Engineers, Vicksburg, MS, 1978.

20. "Use of Epoxy Compounds with Concrete," ACI 503R-93, *ACI Manual of Concrete Practice, Part 5*, American Concrete Institute, Farmington Hills, MI, 1993.

21. *Post-Tensioning Manual*, Fifth Edition, Post-Tensioning Institute, Phoenix, AZ, 1990.

22. *Manual of Standard Practice*, 26th Edition, Concrete Reinforcing Steel Institute, Schaumburg, IL, 1996.

23. *Manual of Standard Practice — Structural Welded Wire Reinforcement*, Sixth Edition, Wire Reinforcement Institute, McLean, VA, 2001.

24. LeClaire, Philip J., and Shaikh, A. Fattah, "Effect of Elevated Temperature on the Bond Strength of Epoxy-Coated Prestressing Strand," PCI JOURNAL, V. 41, No. 4, July-August 1996.

25. *Architectural Precast Concrete — Color and Texture Selection Guide*, Second Edition, CTG-03, Precast/Prestressed Concrete Institute, Chicago, IL, 2003.

26. "Concrete Sealers for Protection of Bridge Structures," Report No. 244, National Cooperative Highway Research Program, Washington, DC, 1981.

27. *Structural Welding Code — Reinforcing Steel*, Fifth Edition, AWS D1.4-92, American Welding Society, Miami, FL, 1998.

28. "Get the Air Out," *The Concrete Producer*, October, 2000.

29. *Bridge Design Manual*, First Edition, MNL-133-97, Precast/Prestressed Concrete Institute, Chicago, IL, 1997.

30. "Interim Guidelines for the Use of Self-Consolidating Concrete in PCI Member Plants," TR-6-03, Precast/Prestressed Concrete Institute, Chicago, IL, 2003.

31. PCI Committee on Prestressed Concrete Poles, "Guide for the Design of Prestressed Concrete Poles," PCI JOURNAL, V. 42, No. 6, November-December 1997.

32. Einea, Amin, Yamane, Takashi, Tadros, Maher K., "Full Depth Precast and Prestressed Concrete Deck Panels," PCI JOURNAL, V. 40, No. 1, January-February 1995.

33. "Recommended Practice for Design, Manufacture and Installation of Prestressed Concrete Piling," JR-382, Precast Prestressed Concrete Institute, Chicago, IL, January 1993.

34. PCI Committee on Glass Fiber Reinforced Concrete Panels, "Recommended Practice for Glass Fiber Reinforced Concrete," PCI JOURNAL, V. 26, No. 1, January-February 1981.

35. PCI Committee on Precast Prestressed Concrete Tank Construction, "State of the Art of Precast, Prestressed Concrete Tank Construction," PCI JOURNAL, V. 28, No. 4, July-August 1983.

36. PCI Committee on Precast Prestressed Concrete Storage Tanks, "Recommended Practice for Precast Prestressed Concrete Circular Storage Tanks," PCI JOURNAL, V. 32, No. 4, July-August 1987.

37. PCI Bridge Producers Committee, "Recommended Practice for Precast Prestressed Concrete Composite Bridge Deck Panels," PCI JOURNAL, V. 33, No. 2, March-April 1988.

38. PCI Committee on Prestressed Concrete Columns, "Recommended Practice for the Design of Prestressed Concrete Columns and Walls," PCI JOURNAL, V. 33, No. 4, July-August 1988.

CHAPTER 2
PRELIMINARY DESIGN OF PRECAST, PRESTRESSED CONCRETE STRUCTURES

- 2.1 General 2–2
  - 2.1.1 Notation 2–2
  - 2.1.2 Introduction 2–3

- 2.2 Preliminary Analysis 2–3
  - 2.2.1 Framing Dimensions... 2–3
  - 2.2.2 Span-to-Depth Ratios 2–4
  - 2.2.3 Connection Concepts 2–4
  - 2.2.4 Gravity and Lateral Load Resisting Systems 2–4
  - 2.2.5 Mechanisms for the Control of Volume Changes 2–7

- 2.3 Explanation of Load Tables 2–7
  - 2.3.1 Safe Superimposed Load 2–8
  - 2.3.2 Limiting Criteria 2–8
  - 2.3.3 Estimated Cambers 2–9
  - 2.3.4 Design Parameters 2–9
  - 2.3.5 Concrete Strength and Unit Weights 2–9
  - 2.3.6 Prestressing Strands 2–10
  - 2.3.7 Prestress Losses 2–10
  - 2.3.8 Strand Placement 2–10
  - 2.3.9 Columns and Loadbearing Wall Panels 2–10
  - 2.3.10 Piles 2–10
  - 2.3.11 References 2–11

- 2.4 Stemmed Deck Members
  - Double Tee Load Tables 2–12
  - Pretopped Double Tee Load Tables 2–24

- 2.5 Flat Deck Members
  - Hollow-Core Slab Load Tables... 2–31
  - Hollow-Core Slab Section Properties 2–35
  - Solid Flat Slab Load Tables.. 2–39

- 2.6 Beams
  - Rectangular Beams 2–42
  - L-Beams 2–43
  - Inverted Tee Beams 2–45

- 2.7 Columns and Loadbearing Wall Panels
  - Precast, Prestressed Columns 2–48
  - Precast, Reinforced Columns 2–50
  - Double Tee Wall Panels 2–52
  - Hollow-Core and Sandwich Wall Panels 2–53
  - Precast, Prestressed Solid and Sandwich Wall Panels 2–54
  - Precast, Reinforced Solid and Sandwich Wall Panels 2–55

- 2.8 Piles
  - Piles 2–56
  - Sheet Piles.. 2–57

PRELIMINARY DESIGN OF PRECAST, PRESTRESSED CONCRETE STRUCTURES

2.1 General

2.1.1 Notation

A = cross-sectional area

A_g = gross cross-sectional area

b = width of compression or tension face of member

b_w = web width

D = unfactored dead loads

d_b = nominal diameter of reinforcing bar or prestressing strand

e_c = eccentricity of prestress force from the centroid of the section at the center of the span

e_e = eccentricity of prestress force from the centroid of the section at the end of the span

f'_c = specified compressive strength of concrete

f'_{ci} = specified compressive strength of concrete at time of initial prestress

f_{ps} = stress in prestressed reinforcement at nominal strength of member

f_{pu} = ultimate strength of prestressing steel

f_{se} = effective stress in prestressing steel after losses

f_y = specified yield strength of non-prestressed reinforcement

h = overall depth of member

I = moment of inertia

ℓ = span

ℓ_p = development length

L = unfactored live loads

M_n = nominal moment strength of a member

M_{nb} = nominal moment strength of a compression member under balanced conditions

M_o = nominal moment strength of a compression member with zero axial load

M_u = applied factored moment at section

N = unfactored axial load

P_n = nominal axial load strength of a compression member at given eccentricity

P_{nb} = nominal axial load strength of a compression member under balanced conditions

P_o = nominal axial load strength of a compression member with zero eccentricity

P_u = factored axial load

S = section modulus

S_b = section modulus with respect to the bottom fiber of a cross section

S_t = section modulus with respect to the top fiber of a cross section

t = thickness

V_{ci} = nominal shear strength provided by concrete when diagonal cracking results from combined shear and moment

V_{cw} = nominal shear strength provided by concrete when diagonal cracking results from excessive principal tensile stress in the web

V_u = factored shear force

V/S = volume-to-surface ratio

y_b = distance from bottom fiber to center of gravity of section

y_t = distance from top fiber to center of gravity of section

y_s = distance from centroid of prestressed reinforcement to bottom fiber

δ = moment magnification factor

ϕ = strength reduction factor

ρ = A_s/bd = ratio of non-prestressed tension reinforcement

2.1.2 Introduction

This chapter presents data on the shapes that are standard in the precast and prestressed concrete industry. Other shapes and modified standard shapes with depth and width variations are also available in many areas of the country. Designers should contact the local manufacturers in the geographic area of the proposed project to determine the properties and dimensions of products available. This section, plus the design methods and aids provided in other chapters of this Handbook, should enable the designer to prepare safe and economical designs.

The text, figures and photographs of Chapter 1 show that the benefits and advantages of using precast concrete building products are limited only by the designer's imagination and knowledge of how to use the material. This chapter provides guidelines for making the most efficient use of precast products and systems.

Engineering precast concrete requires unique design considerations not required for most other building materials. The designer needs to consider manufacturing, handling, transportation and erection of the product in addition to analysis and design for in-place loads. It is not uncommon for these items to be the controlling factors in the design.

Building designs can be optimized by following these general principles:

1. Maximize repetitive and modular dimensions for plan layout and member dimensions.

2. Use simple spans whenever possible.

3. Standardize size and locations of openings in products.

4. Use standard, locally available member sizes.

5. Minimize the number of different member types and sizes.

6. Minimize the number of different reinforcing patterns in a particular member type.

7. Minimize the number of different types of connections.

8. Specify connection types that are preferred by local producers.

9. Consider the size and weight of products to avoid premium costs associated with producing, shipping and erecting oversize and overweight pieces.

10. Utilize prestressing in precast members when spans are long, when the member depth must be minimized, or when the greatest degree of crack control is desired.

11. Avoid member designs and connection designs that require skill levels of workmanship and close tolerances that are not attainable under a producer's, or erector's normal operation.

12. Avoid specifying requirements in excess of what is needed for concrete mix designs, allowable stresses, allowable cambers, deflections, and coatings on reinforcing steel, embedded hardware, and connection hardware.

13. Make use of exterior wall panels as loadbearing members and/or shear walls whenever possible.

14. Contact a local producer as early as possible during the design development stages of a project for assistance in answering the above questions.

15. Maximize form use on architectural products.

The load tables presented in this chapter are meant to be used for preliminary design. The limiting criteria for these tables are based according to the ACI 318-02 Building Code. [1] In general, the PCI Standard Design Practice, Section 10.5, has been followed.

2.2 Preliminary Analysis

Maximum economy occurs when the building is laid out to take advantage of the principles discussed above. The primary considerations in developing a preliminary layout are:

1. Framing dimensions
2. Span-to-depth ratios
3. Connection concepts
4. Gravity and lateral load resisting system
5. Mechanisms for the control of volume changes

2.2.1 Framing Dimensions

When possible, establish building dimensions by combining the modular dimensions of standard component sizes and by establishing bay sizes based on optimum component spans.

Generally, optimum framing dimensions will result when the total number of component pieces is minimized. For example, it is often more economical to cast wall panels and columns in multistory units because of the reduced number of pieces required to produce, ship and erect. When establishing maximum component sizes, also consider the maximum shipping size and weight and the producer's and erector's crane capacity.

2.2.2 Span-to-Depth Ratios

Typical span-to-depth ratios can be used to determine the approximate required depth of a flexural precast, prestressed concrete member. During preliminary analysis, it is helpful to determine beam or slab depths and the additional space required for other construction elements, including mechanical duct work or plenum, in order to establish the floor-to-floor (or roof) dimensions of a building.

Typical span-to-depth ratios of flexural precast, prestressed concrete members are:

Hollow-core floor slabs	30 to 40
Hollow-core roof slabs	40 to 50
Stemmed floor slabs	25 to 35
Stemmed roof slabs	35 to 40
Beams	10 to 20

These values are intended as guidelines, not limits. The required depth of beam or slab is influenced by the magnitude and ratio of live load to total load. Where this ratio is high, deeper sections may be needed.

For non-prestressed flexural members, span-to-depth ratios given in ACI 318-02, Section 9.5.2.1 may to be used as a preliminary guide.

2.2.3 Connection Concepts

The types of connections to be used should be determined during the preliminary analysis, as this may have an effect on the component dimensions, the overall structural behavior as discussed above, and on the erection procedure. Also, accommodations for the connection of any planned future expansions should be incorporated. Chapter 6 and other PCI publications [2] are devoted entirely to connections.

2.2.4 Gravity and Lateral Load Resisting Systems

A building system should be selected during the preliminary analysis. The gravity and lateral load resisting systems may function separately or they may be combined.

Bearing wall construction (Figures 2.2.1 and 2.2.2) and beam-column framing (Figures 2.2.3 and 2.2.4) have been successfully used for various height buildings. Resistance to lateral forces can be provided by interior shear walls (Figure 2.2.5), exterior shear walls (Figure 2.2.6), moment frames (Figure 2.2.7), or some combination of these. Limitations of the diaphragm may dictate placement of lateral force resisting elements.

Refer to Chapter 3 for lateral force resisting system analysis and design.

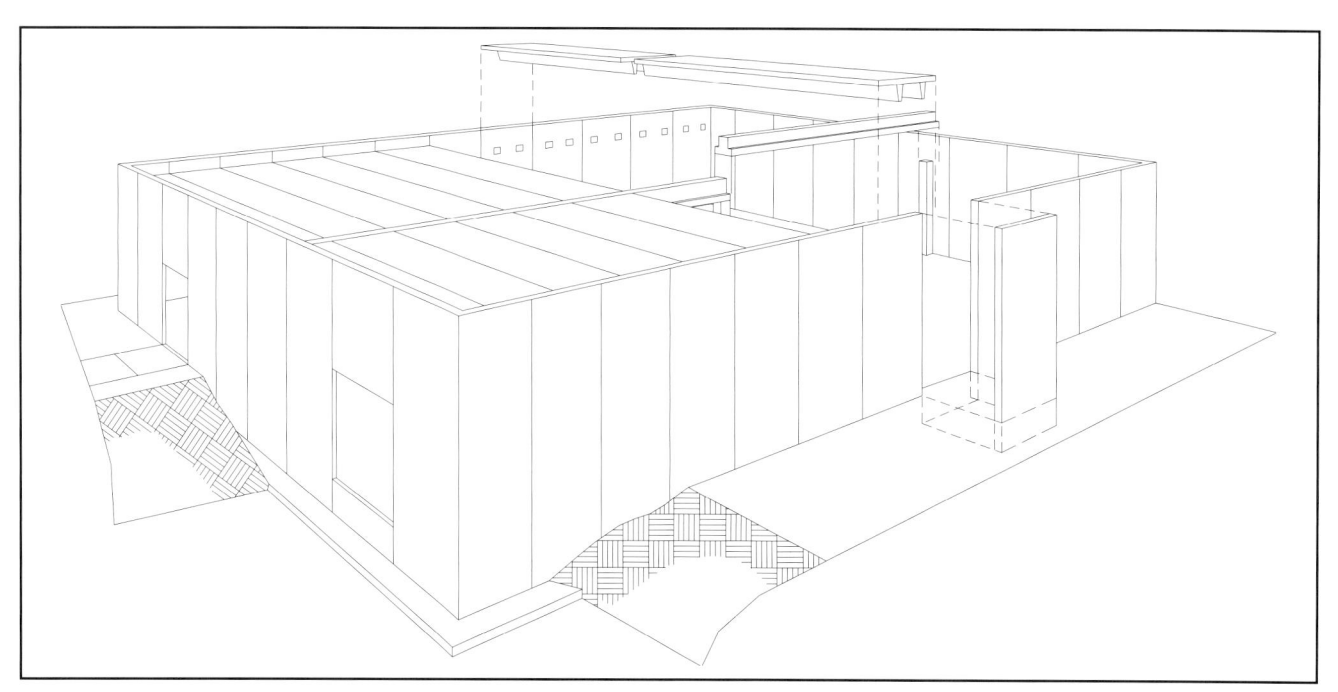

Figure 2.2.1 Single-story bearing wall construction
Provides economy by eliminating the need for a structural frame at the perimeter. The wall panels can be selected from a variety of standard sections of flat panels, and specially formed architectural precast shapes. Any of the standard precast deck units can be used for roofs.

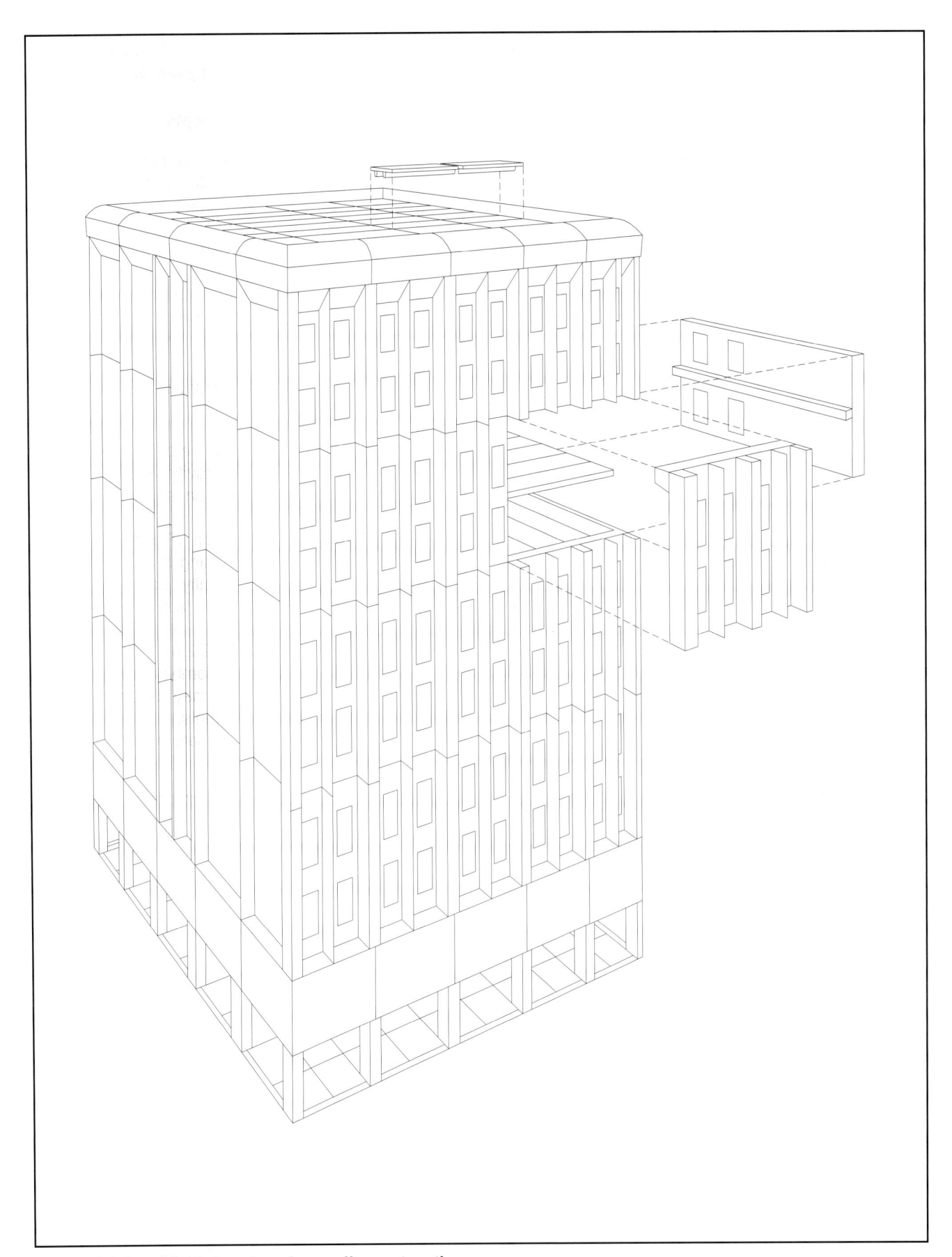

Figure 2.2.2 Multistory bearing wall construction
Precast bearing wall units can be cast in one-story or multistory units. The units may be started at the second floor level with the first floor framing consisting of beams and columns to obtain more open space on the first level.

Figure 2.2.3 Single-story beam-column construction
Any of the standard precast beam and column sections shown in this chapter can be used for single-story structures. Selection of the type of beam to be used depends on considerations such as span length, level of superimposed loads, and also on depth of ceiling construction and desired architectural expression.

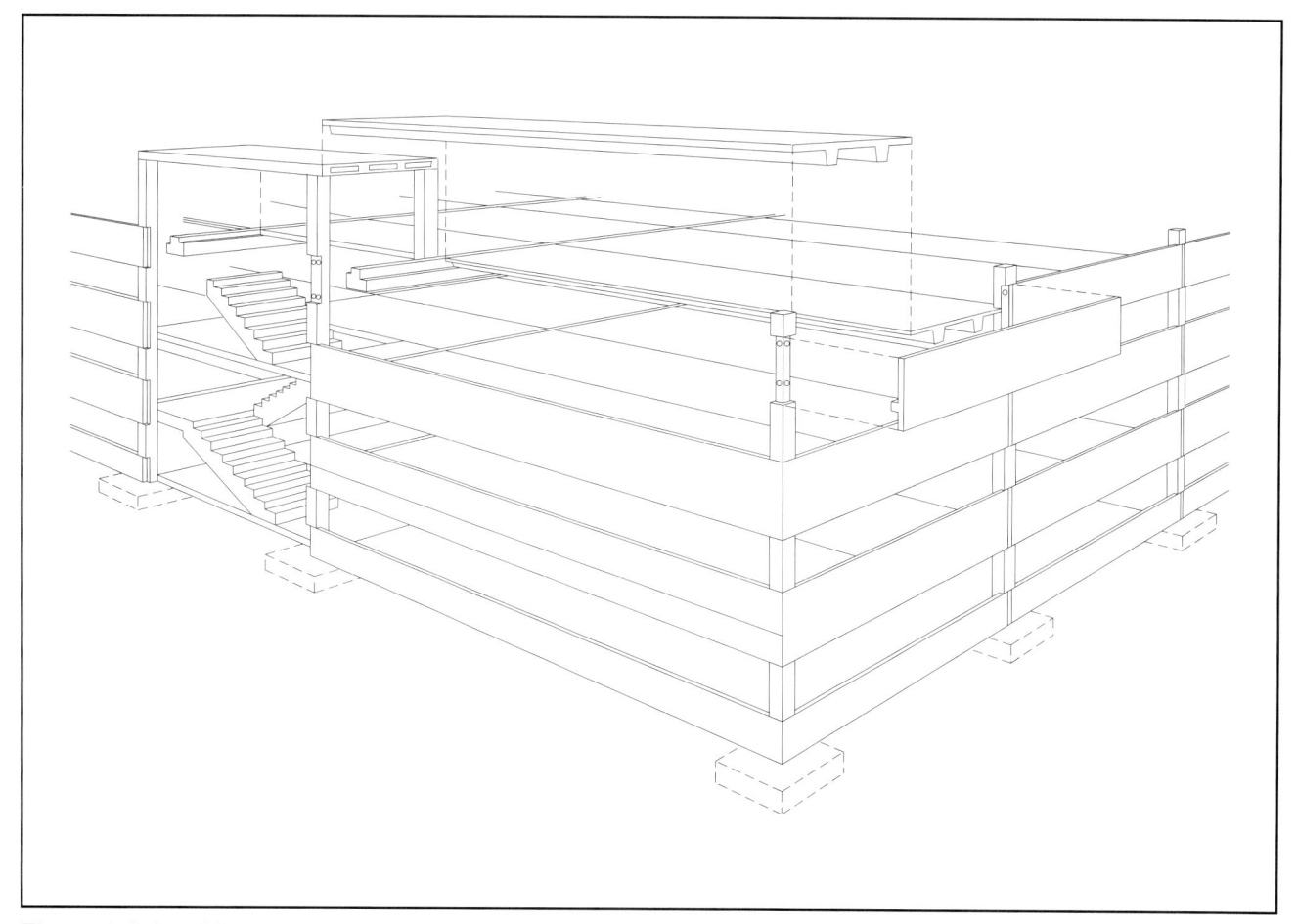

Figure 2.2.4 Multistory beam-column construction
Beam-column framing is suitable for both low- and high-rise buildings. Multistory columns with simple-span beams is the preferred method.

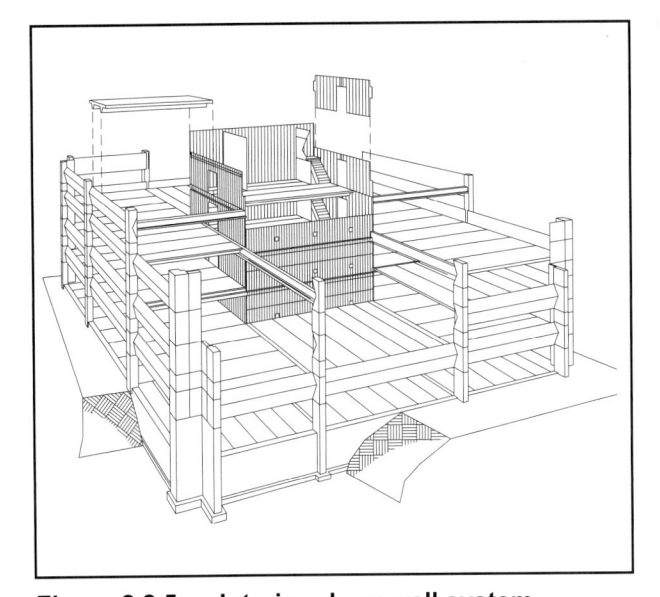

Figure 2.2.5 Interior shear wall system
Lateral loads are transmitted by floor diaphragms to a structural core of precast shear walls.

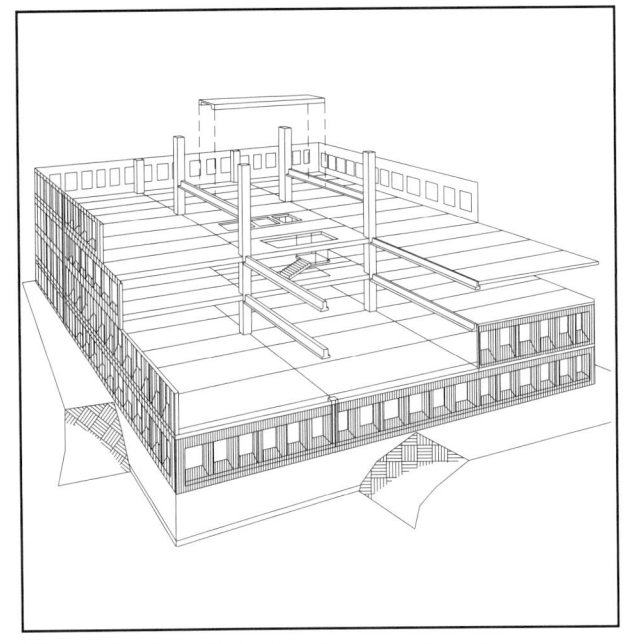

Figure 2.2.6 Exterior shear wall system
In general, the exterior shear wall system permits greater design flexibility than the interior shear wall system because it eliminates the need for a structural core. By combining gravity loadbearing function with lateral force resistance, the exterior shear wall system is, in general, more economical.

2.2.5 Mechanisms for the Control of Volume Changes

It is important to consider the effects of volume changes on a structure during the selection of the building system. Use and location of expansion joints, as well as the location of "stiff" lateral force resisting elements, can have a significant effect on how the structure responds to volume change strains.

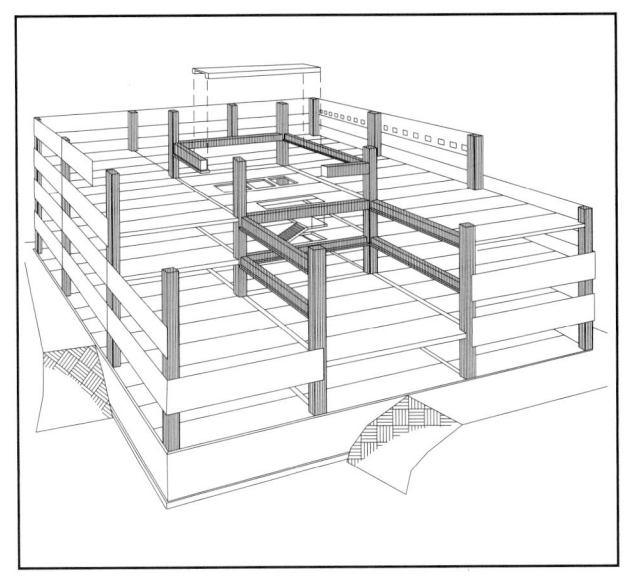

Figure 2.2.7 Moment-resisting frame system
All lateral forces are transferred to a moment-resisting frame that ties beams and columns together with rigid connections. The need for shear walls is eliminated.

Refer to Chapter 3 for recommendations on locating expansion joints and lateral force resisting elements.

2.3 Explanation of Load Tables

The load tables on the following pages show dimensions, section properties and load carrying capabilities of the shapes most commonly used throughout the industry. These shapes include double tees, hollow-core and solid flat slabs, beams, girders, columns, piles and wall panels. The dimensions of the shapes shown in the tables may vary among manufacturers. The variations are usually small and the tables given here can still be used, particularly for preliminary design.

Manufacturers will in most cases have their own catalogs with load tables and geometric properties for the members they produce. Hollow-core slabs of different thicknesses, core sizes and shapes are available in the market under various trade names. Cross sections and section properties of proprietary hollow-core slabs are shown on pages 2–35 through 2–38. Load tables on pages 2–30 through 2–34 are developed for non-proprietary hollow-core sections of thicknesses most commonly used in the industry.

Load tables for stemmed deck members, flat deck members and beams show the allowable uniform superimposed service load, estimated camber at the time of erection, and the estimated long-term camber after the time-dependent deformations have stabilized, but before the application of superimposed live load. For deck members, except pretopped double tees, the table at the top of the page gives the information for the member with no topping, and the table at the bottom of the page is for the same member with 2 in. of normal weight concrete topping acting compositely with the precast section. Values in the tables assume a uniform 2 in. topping over the full span length, and assume the member to be unshored at the time the topping is placed. Safe loads and cambers shown in the tables are based on the dimensions and section properties shown on the page, and will vary for members with different dimensions.

For beams, a single load table is used for several sizes of members. The values shown are based on sections containing the indicated number of prestressing strands. In some cases, more strands could be used. Capacity can be significantly increased with the use of a composite deck slab.

2.3.1 Safe Superimposed Load

The values for safe superimposed uniform service load are based on the capacity of the member as governed by the ACI Building Code limitations on flexural strength, service load flexural stresses, or in the case of flat deck members without shear reinforcement, shear strength. A portion of the safe load shown is assumed to be dead load for the purpose of applying load factors and determining cambers and deflections. For untopped deck members, 10 psf of the capacity shown is assumed as superimposed dead load, typical for roof members. For deck members with topping, 15 psf of the capacity shown is assumed as superimposed dead load, typical of floor members. The capacity shown is in addition to the weight of the topping. For beams, 50% of the capacity shown in the load table is assumed as superimposed dead load. Pretopped double tees are typically used for parking structures where superimposed dead loads are negligible. Thus, all of the load shown in the tables on pages 2–24 through 2–30 is live load.

Example: For an 8DT24/88-S untopped (page 2–12) with a 52 ft span, the capacity shown is 70 psf. The member can safely carry superimposed service loads of 10 psf dead and 60 psf live.

2.3.2 Limiting Criteria

The criteria used to determine the safe superimposed load and strand placement are based on the Building Code Requirements for Structural Concrete (ACI 318-02). (This is referred to as "the Code," "the ACI Building Code" or "ACI 318" in this Handbook.) For design procedures, see Chapter 4 of this Handbook. A summary of the Code provisions used in the development of these load tables is as follows:

1. Capacity governed by flexural strength:

    Load factors: 1.2D + 1.6L

    Strength reduction factor, $\phi = 0.90$

    Calculation of moments assumes simple spans with roller supports. If the strands are fully developed (see Section 4.2.3), the critical moment is assumed to be at midspan in members with straight strands, and at 0.4ℓ (ℓ = span) in products with strands depressed at midspan. (Note: The actual critical point can be determined by analysis, but will seldom vary significantly from 0.4ℓ.) Flexural strength is calculated using strain compatibility as discussed in Chapter 4.

2. Capacity governed by serviceability requirements:

    Flexural stresses immediately after transfer of prestress:

    a) Compression: $0.7f'_{ci}$ (see Section 10.5)

    b) Tension: $6\sqrt{f'_{ci}}$ (see Section 10.5)

    It is common practice to debond some straight strands a short distance to reduce transfer stresses. This causes the critical section for prestress transfer to occur further from the end of the member. The load tables for double tees and beams with straight strands assume the critical section for tension stresses is at $0.05\ell + 2\ell_d$ from the end.

    The load tables limit the double tees and beams to Class U or Class T flexural members in accordance with Section 18.3.3 of ACI 318-02 (see Table 4.2.2.1). Flat deck members are limited to Class U. Thus, the limiting stresses under service loads are:

    c) Compression:

    Extreme fiber stress in compression due to prestress plus sustained loads: $0.45 f'_c$

    Extreme fiber stress in compression due to prestress plus total load: $0.6 f'_c$

    d) Tension:

    Double tees and beams: $12\sqrt{f'_c}$

    (Deflections must be determined based on bilinear moment-deflection relationships.) (see Section 4.8.3)

Flat deck members: $7.5\sqrt{f'_c}$

Critical point for service load moment is assumed at midspan for members with straight strands and at 0.4ℓ for members with strands depressed at midspan, as described above.

3. Capacity governed by design shear strength:
Load factors: 1.2D + 1.6L
Strength reduction factor, $\phi = 0.75$

In flat deck members, use of shear reinforcement is generally not feasible, thus the capacity may be limited by the concrete shear strength. In this case, the safe superimposed load is obtained by equating the corresponding factored shear force, V_u, to the lesser of ϕV_{ci} and ϕV_{cw} (see Section 11.4.2 of ACI 318-02 and Section 4.3 of this Handbook).

For stemmed deck members and beams, the design concrete shear strength is not used as a limiting criterion since shear reinforcement can be readily provided in such members. The design of shear reinforcement is illustrated in Section 4.3. In stemmed deck members, usually minimum shear reinforcement is required (see Section 11.5.5 of ACI 318-02).

Note 1: Except as noted for debonded strands as noted above, end transfer stresses are calculated at the theoretical point of full transfer, $(f_{se}/3)d_p$.

Note 2: Release tension is not used as a limiting criteria for beams. Supplemental top reinforcement may be required as described in Section 4.2.2.2.

Note 3: For *Stemmed Deck Members*, the transverse (flange) flexural and shear strengths and the service load stresses are not considered as limiting criteria. For heavy uniform loads and/or concentrated loads, these considerations may limit the capacity or require transverse reinforcement.

Note 4: *Flat Deck Members* show no values beyond a span-to-depth ratio of 50 for untopped members and 40 for topped members. These are suggested maximums for roof and floor members, respectively, unless a detailed analysis is made.

2.3.3 Estimated Cambers

The estimated cambers shown are calculated using the multipliers given in Section 4.8.5 of this Handbook. These values are estimates, and should not be used as absolute values. Attachment of non-structural elements, such as partitions, folding doors and architectural decoration to members subject to camber variations should be designed with adequate allowance for these variations. Calculation of topping quantities should also recognize that the values can vary with camber. When choosing members, negative camber under dead load should not be allowed.

2.3.4 Design Parameters

The design of prestressed concrete is dependent on many variables. These include concrete strength at release and at 28 days; unit weight of concrete; strength grade, profile and placement of strand; jacking tension and other items. The load tables given here are based on commonly used values; somewhat higher allowable loads or longer spans may be achieved by selecting other appropriate sets of conditions. However, in these cases, consultation with local producers is recommended.

2.3.5 Concrete Strength and Unit Weights

Twenty-eight day cylinder strength for precast concrete is assumed to be 5000 psi unless otherwise indicated. Tables for units with composite topping are based on the topping concrete being normal weight concrete with a cylinder strength of 3000 psi.

Concrete strength at time of strand tension release is 3500 psi unless the value falls below the heavy line shown (Note: this area is also shown as shaded) in the load table, indicating that a cylinder strength greater than 3500 psi is required. No values are shown when the required release strength exceeds 4500 psi. Higher release strengths may be used, but check with local precasters.

The concrete strengths used in the load tables are not intended to be limitations or recommendations for actual use. Some precast concrete manufacturers may choose to use higher or lower concrete strengths, resulting in slightly different load table values. For low levels of prestress, the concrete release strength is usually governed by handling stresses.

Unit weight of concrete is assumed to be 150 pcf for normal weight and 115 pcf for lightweight.

2.3.6 Prestressing Strands

Prestressing strands are available in diameters from ¼ in. to 0.6 in., Grades 250, 270 or 300 ksi. The load tables are calculated assuming low-relaxation strand. The relaxation loss of low-relaxation strand remains linear up to an initial stress of $0.75f_{pu}$, [3], so the tables have been developed assuming this value of initial stress. Stress at transfer of prestress has been assumed at 90% of the initial stress.

In developing the load tables, the stress-strain behavior of the strand was modeled with the curves given in Design Aid 11.2.5. Other stress-strain models are also available and could be used in place of the curves used here, but would be expected to produce allowable loads which differ only slightly from those shown.

2.3.7 Prestress Losses

Prestress losses were calculated in accordance with the recommendations given in Ref. 4. This procedure includes consideration of initial stress level ($0.7f_{pu}$ or higher), type of strand, exposure conditions and type of construction. A lower limit of 30,000 psi for low-relaxation strands was used. This lower limit is reasonable; other designers may choose not to impose this limit. Additional information on prestress losses is given in Section 4.7.

2.3.8 Strand Placement

Quantity, size and profile of strands are shown in the load tables under the column headed "Strand Pattern," for example, 88-S. The first digit indicates the total number of strands in the unit, the second digit is the diameter of the strand in 16ths of an inch and the "S" indicates that the strands are straight. A "D1" indicates that the strands are depressed at midspan. Some precast producers choose to depress the strand at two points, which provides a somewhat higher capacity.

For double tees and beams, the distance from the centroid of the strand group to the bottom of the member (y_s) at the ends and midspan are shown in the load tables. Strands have been placed so that the stresses at the theoretical transfer point will not exceed those specified above.

For flat deck members, the load table values are based on prestressing strand centered 1½ in. from the bottom of the slab. Strand placement can vary from as low as ⅞ in. to as high as 2⅛ in. from the bottom, which will change the capacity and camber values shown. The higher strand placements give improved fire resistance ratings (see Section 9.3 of this Handbook for more information on fire resistance). The lower strand placement may require higher release strengths, or top tension reinforcement at the ends. The designer should contact the local supplier of flat prestressed concrete deck members for available and recommended strand placement locations.

2.3.9 Columns and Loadbearing Wall Panels

Interaction curves for selected precast, prestressed columns, precast reinforced columns and various types of commonly used wall panels are provided on 2–48 to 2–55.

These interaction curves are based on strength design for short columns. Appropriate load factors must be applied to the service loads and moments before entering the charts. The effect of slenderness must be considered as described in Section 4.9.

For prestressed columns and wall panels, curves are shown for both partially developed strands, usually appropriate for end moment capacity, and fully developed strands, usually appropriate for all conditions away from the ends. This is discussed more completely in Section 4.9. The columns and wall panels which use reinforcing bars assume full development of the reinforcing bars.

The curves are terminated at a value of $P_u = 0.80P_o$, the maximum allowable load for tied columns under ACI 318-02. Most of the wall panel curves show the lower portion of the curve only (flexure controlling). Actual design loads will rarely exceed the values shown.

The curves for double tee wall panels are for bending in the direction that causes tension in the stem. The curves for hollow-core wall panels are based on a generic section as shown. They can be used with small difference for all sections commonly marketed for wall panel use.

2.3.10 Piles

Allowable concentric service loads on prestressed concrete piles, based on the structural capacity of the pile alone, are shown in Table 2.8.1. The ability of the soil to carry these loads must be evaluated separately. Values for concrete strengths up to 10,000 psi are shown. The availability of these concrete strengths should be checked with local manufacturers. The design of prestressed concrete piles is discussed in Section 4.9.6.

Section properties and allowable service load bending moments for prestressed concrete sheet pile units are shown in Table 2.8.2. These units are available in some areas for use in earth retaining structures.

2.3.11 References

1. ACI Committee 318, "Building Code Requirements for Structural Concrete (ACI 318-02)," and "Commentary (ACI 318R-02)," American Concrete Institute, Farmington Hills, MI, 2002.

2. *Design and Typical Details of Connections for Precast and Prestressed Concrete*, Second Edition, MNL-123-88, Precast/Prestressed Concrete Institute, Chicago, IL, 1988.

3. Martin, L. D., and Pellow, D. L., "Low-Relaxation Strand–Practical Applications in Precast/Prestressed Concrete," PCI JOURNAL, V. 28, No. 4, July-August 1983.

4. Zia, Paul, Preston, H. K., Scott, N. L., and Workman, E. B., "Estimating Prestress Losses," *Concrete International*, V. 1, No. 6, June 1979.

5. PCI Committee on Prestressed Concrete Piling, "Recommended Practice for Design, Manufacture and Installation of Prestressed Concrete Piling," PCI JOURNAL, V. 38, No. 2, March-April 1993.

Strand Pattern Designation

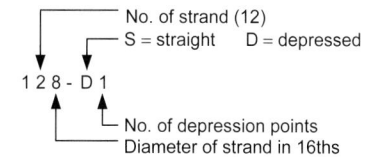

```
      ┌─ No. of strand (12)
      │  ┌─ S = straight   D = depressed
    1 2 8 - D 1
            │   └─ No. of depression points
            └──── Diameter of strand in 16ths
```

Safe loads shown include dead load of 10 psf for untopped members and 15 psf for topped members. Remainder is live load. Long-time cambers include superimposed dead load but do not include live load.

Key
186 – Safe superimposed service load, psf
0.7 – Estimated camber at erection, in.
0.9 – Estimated long-time camber, in.

DOUBLE TEE
8'-0" x 24"
Normal Weight Concrete

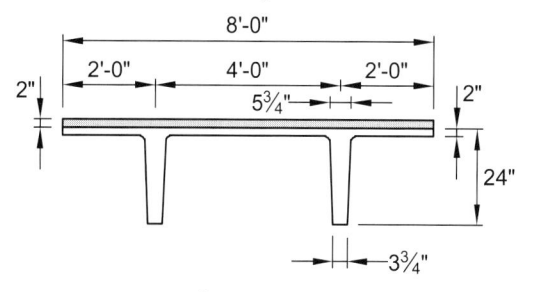

8'-0"
2'-0" | 4'-0" | 2'-0"
2"
5¾"
2"
24"
3¾"

$f'_c = 5,000$ psi
$f_{pu} = 270,000$ psi

Section Properties

	Untopped	Topped
A =	401 in.²	—
I =	20,985 in.⁴	27,720 in.⁴
y_b =	17.15 in.	19.27 in.
y_t =	6.85 in.	6.73 in.
S_b =	1,224 in.³	1,439 in.³
S_t =	3,064 in.³	4,119 in.³
wt =	418 plf	618 plf
DL =	50 psf	77 psf
V/S =	1.41 in.	

8DT24

Table of safe superimposed service load (psf) and cambers (in.) — No Topping

Strand Pattern	y_s(end) in. / y_s(center) in.	32	34	36	38	40	42	44	46	48	50	52	54	56	58	60	62	64	66	68	70	72	74	76
68-S	4.00 / 4.00	186	161	140	122	106	93	81	71	62	55	48	42	36	31	27								
		0.7	0.8	0.8	0.9	0.9	1.0	1.0	1.0	1.0	1.0	0.9	0.8	0.7	0.6	0.5								
		0.9	1.0	1.1	1.1	1.1	1.1	1.1	1.1	1.0	0.9	0.8	0.6	0.4	0.1	−0.2								
88-S	5.00 / 5.00			185	162	143	126	112	99	88	78	70	62	55	49	43	38	33	29					
				1.1	1.2	1.3	1.3	1.4	1.4	1.5	1.5	1.5	1.5	1.4	1.4	1.3	1.1	1.0	0.8					
				1.4	1.5	1.6	1.6	1.7	1.7	1.7	1.7	1.6	1.5	1.4	1.2	0.9	0.6	0.3	−0.2					
108-S	6.00 / 6.00					197	174	155	138	123	110	98	88	79	71	64	57	51	45	39	34	29		
						1.4	1.5	1.6	1.7	1.7	1.8	1.9	1.9	1.9	1.9	1.9	1.9	1.8	1.7	1.5	1.3	1.1		
						1.8	1.9	2.0	2.1	2.1	2.1	2.1	2.1	2.0	1.9	1.8	1.6	1.3	0.9	0.5	0.0			
128-S	7.00 / 7.00							159	142	128	114	101	90	81	73	66	59	53	47	42	37	32	27	
								1.8	1.9	2.0	2.1	2.1	2.2	2.2	2.2	2.2	2.2	2.1	2.0	1.9	1.7	1.5	1.2	
								2.3	2.4	2.4	2.5	2.5	2.4	2.4	2.3	2.2	2.1	1.8	1.6	1.2	0.8	0.3	−0.3	
128-D1	11.67 / 3.25												114	103	92	83	74	66	59	53	48	43	39	35
													2.5	2.6	2.6	2.7	2.7	2.6	2.6	2.5	2.3	2.1	1.9	1.6
													2.9	2.9	2.9	2.8	2.6	2.4	2.2	1.8	1.5	1.1	0.6	0.0
148-D1	12.86 / 3.50																	71	64	58	52	46	42	
																		3.1	3.1	3.0	2.9	2.7	2.5	
																		3.0	2.7	2.4	2.0	1.6	1.1	

8DT24 + 2

Table of safe superimposed service load (psf) and cambers (in.) — 2 in. Normal Weight Topping

Strand Pattern	y_s(end) in. / y_s(center) in.	28	30	32	34	36	38	40	42	44	46	48	50	52	54	56	58	60	62	64	66	68	70
48-S	3.00 / 3.00	169	141	117	97	81	67	55	45	36	29												
		0.4	0.4	0.5	0.5	0.5	0.5	0.5	0.5	0.5	0.4												
		0.4	0.4	0.4	0.4	0.3	0.3	0.2	0.0	−0.1	−0.3												
68-S	4.00 / 4.00			189	161	138	118	101	87	74	64	54	45	38	30								
				0.7	0.8	0.8	0.9	0.9	1.0	1.0	1.0	1.0	1.0	0.9	0.8								
				0.7	0.8	0.8	0.7	0.6	0.6	0.4	0.2	0.0	−0.3	−0.6									
88-S	5.00 / 5.00					188	163	142	124	108	94	82	71	62	52	42	33						
						1.1	1.1	1.3	1.3	1.4	1.4	1.5	1.5	1.5	1.5	1.4	1.4						
						1.1	1.2	1.1	1.1	1.1	1.0	0.9	0.7	0.5	0.3	−0.1	−0.5						
108-S	6.00 / 6.00							176	155	136	120	104	90	77	66	56	47	39	31				
								1.5	1.6	1.7	1.7	1.8	1.9	1.9	1.9	1.9	1.9	1.9	1.8				
								1.4	1.4	1.4	1.4	1.3	1.2	1.0	0.8	0.6	0.3	−0.1	−0.6				
128-S	7.00 / 7.00								160	140	121	104	90	77	67	57	48	41	33	27			
									1.8	1.9	2.0	2.1	2.1	2.2	2.2	2.2	2.2	2.2	2.1	2.0			
									1.6	1.6	1.6	1.5	1.3	1.1	0.9	0.6	0.3	−0.1	−0.6	−1.1			
128-D1	11.67 / 3.25												105	91	79	68	58	49	40	34	28		
													2.5	2.6	2.6	2.7	2.7	2.6	2.6	2.5	2.3		
													1.6	1.4	1.1	0.8	0.4	0.0	−0.6	−1.2	−1.8		

Strength is based on strain compatibility; bottom tension is limited to $12\sqrt{f'_c}$; see pages 2–7 through 2–10 for explanation.
Shaded values require release strengths higher than 3500 psi.

Strand Pattern Designation

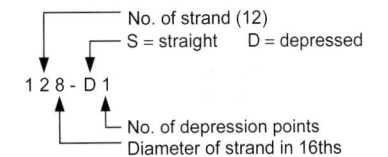

```
        ┌─ No. of strand (12)
        │ ┌─ S = straight   D = depressed
        │ │
      1 2 8 - D 1
      ↑       ↑
      │       └─ No. of depression points
      └─ Diameter of strand in 16ths
```

Safe loads shown include dead load of 10 psf for untopped members and 15 psf for topped members. Remainder is live load. Long-time cambers include superimposed dead load but do not include live load.

Key
196 – Safe superimposed service load, psf
1.2 – Estimated camber at erection, in.
1.5 – Estimated long-time camber, in.

DOUBLE TEE
8'-0" x 24"
Lightweight Concrete

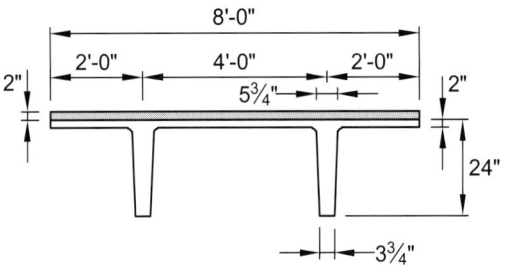

8'-0"
2'-0" | 4'-0" | 2'-0"
2"
5¾"
2"
24"
3¾"

$f'_c = 5{,}000$ psi
$f_{pu} = 270{,}000$ psi

Section Properties

	Untopped		Topped	
A =	401	in.²	—	
I =	20,985	in.⁴	29,857	in.⁴
y_b =	17.15	in.	19.94	in.
y_t =	6.85	in.	6.06	in.
S_b =	1,224	in.³	1,497	in.³
S_t =	3,064	in.³	4,927	in.³
wt =	418	plf	520	plf
DL =	40	psf	65	psf
V/S =	1.41	in.		

8LDT24

No Topping

Table of safe superimposed service load (psf) and cambers (in.)

Strand Pattern	y_s(end) in. y_s(center) in.	32	34	36	38	40	42	44	46	48	50	52	54	56	58	60	62	64	66	68	70	72	74	76	78	80
68-S	4.00 / 4.00	196 1.2 1.5	170 1.3 1.6	149 1.4 1.7	131 1.5 1.8	115 1.6 1.9	102 1.6 1.9	90 1.7 2.0	80 1.8 2.0	72 1.8 2.0	64 1.9 2.0	57 1.9 1.9	51 1.9 1.8	45 1.9 1.7	40 1.8 1.5	36 1.7 1.2	32 1.6 0.8	28 1.4 0.4								
88-S	5.00 / 5.00			194 1.8 2.3	171 1.9 2.4	152 2.1 2.6	135 2.2 2.7	121 2.3 2.8	108 2.5 2.9	97 2.6 3.0	87 2.7 3.0	79 2.7 2.9	71 2.8 2.9	64 2.8 2.8	58 2.9 2.6	52 2.9 2.5	47 2.8 2.2	43 2.8 1.9	38 2.6 1.5	35 2.5 0.9	31 2.3 0.3	28 2.0				
108-S	6.00 / 6.00					183 2.4 3.0	164 2.5 3.2	147 2.7 3.4	132 2.9 3.5	119 3.0 3.6	107 3.2 3.7	97 3.3 3.8	87 3.4 3.8	78 3.5 3.8	70 3.6 3.7	64 3.6 3.6	58 3.7 3.5	53 3.7 3.3	48 3.6 3.1	44 3.6 2.8	40 3.5 2.5	36 3.3 2.0	33 3.1 1.5	29 2.9 0.8	26 2.5 0.0	
128-S	7.00 / 7.00									110 3.7 4.3	99 3.8 4.4	89 3.9 4.4	80 4.0 4.4	72 4.1 4.3	65 4.2 4.2	59 4.2 4.0	53 4.2 3.8	49 4.2 3.6	44 4.1 3.3	40 4.0 2.9	37 3.9 2.4	34 3.7 1.9	31 3.5 1.3	28 3.2 0.6		
128-D1	11.67 / 3.25															83 4.8 5.1	76 4.9 5.0	69 5.0 4.9	62 5.1 4.7	57 5.1 4.4	51 5.1 4.0	46 5.0 3.6	42 4.9 3.0	38 4.6 2.3	34 4.3 1.4	
148-D1	12.86 / 3.50																						51 5.9 4.6	46 5.8 4.1	42 5.7 3.4	

8LDT24 + 2

2 in. Normal Weight Topping

Table of safe superimposed service load (psf) and cambers (in.)

Strand Pattern	y_s(end) in. y_s(center) in.	28	30	32	34	36	38	40	42	44	46	48	50	52	54	56	58	60	62	64	66	68	70	72	74
48-S	3.00 / 3.00	178 0.6 0.6	150 0.7 0.7	126 0.8 0.7	107 0.8 0.7	90 0.9 0.6	76 0.9 0.5	64 1.0 0.4	54 1.0 0.2	45 1.0 0.0	38 1.0 −0.2	31 1.0 −0.6	25 0.9												
68-S	4.00 / 4.00			198 1.2 1.2	170 1.3 1.2	147 1.4 1.2	127 1.5 1.3	111 1.6 1.2	96 1.7 1.2	84 1.7 1.1	73 1.8 1.0	63 1.9 0.9	55 1.9 0.7	47 1.9 0.4	40 1.9 0.1	34 1.9 −0.3	29 1.8 −0.8								
88-S	5.00 / 5.00					197 1.8 1.8	172 1.9 1.8	151 2.1 1.9	133 2.2 1.9	117 2.3 1.9	103 2.5 1.8	91 2.6 1.7	80 2.7 1.6	71 2.7 1.4	61 2.8 1.1	52 2.8 0.8	45 2.9 0.5	37 2.9 0.0	31 2.8 −0.5	25 2.8 −1.1					
108-S	6.00 / 6.00							186 2.4 2.3	164 2.5 2.3	146 2.7 2.4	129 2.9 2.4	115 3.0 2.3	102 3.2 2.2	89 3.3 2.1	76 3.4 1.9	65 3.5 1.6	56 3.6 1.3	48 3.6 0.9	41 3.7 0.5	34 3.7 −0.1	29 3.7 −0.7				
128-S	7.00 / 7.00											104 3.7 2.6	90 3.8 2.4	78 3.9 2.2	68 4.0 1.9	58 4.1 1.5	49 4.2 1.1	42 4.2 0.5	35 4.2 0.0	30 4.2 −0.7					
128-D1	11.67 / 3.25																			71 4.8 1.9	62 4.9 1.5	53 5.0 0.9	46 5.1 0.3	39 5.1 −0.5	32 5.1 −1.3	26 5.0 −2.3

Strength is based on strain compatibility; bottom tension is limited to $12\sqrt{f'_c}$; see pages 2–7 through 2–10 for explanation.
Shaded values require release strengths higher than 3500 psi.

PCI Design Handbook/Sixth Edition

Strand Pattern Designation

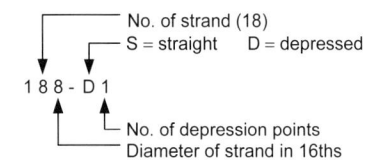

```
         No. of strand (18)
         S = straight    D = depressed
   ↓ ↓
 1 8 8 - D 1
 ↑     ↑
 │     └─ No. of depression points
 └─ Diameter of strand in 16ths
```

Safe loads shown include dead load of 10 psf for untopped members and 15 psf for topped members. Remainder is live load. Long-time cambers include superimposed dead load but do not include live load.

Key
192 – Safe superimposed service load, psf
1.1 – Estimated camber at erection, in.
1.4 – Estimated long-time camber, in.

DOUBLE TEE
8'-0" x 32"
Normal Weight Concrete

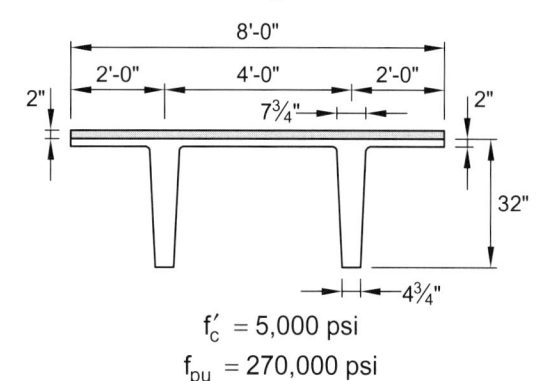

$f'_c = 5{,}000$ psi
$f_{pu} = 270{,}000$ psi

Section Properties

		Untopped		Topped	
A	=	567	in.^{2}	–	
I	=	55,464	in.^{4}	71,886	in.^{4}
y_b	=	21.21	in.	23.66	in.
y_t	=	10.79	in.	10.34	in.
S_b	=	2,615	in.^{3}	3,038	in.^{3}
S_t	=	5,140	in.^{3}	6,952	in.^{3}
wt	=	591	plf	791	plf
DL	=	74	psf	99	psf
V/S	=	1.79	in.		

8DT32

Table of safe superimposed service load (psf) and cambers (in.) — No Topping

Strand Pattern	y_s(end) in. y_s(center) in.	Span, ft
		48 50 52 54 56 58 60 62 64 66 68 70 72 74 76 78 80 82 84 86 88 90 92 94 96 98 100 102 104
128-S	7.00 7.00	192 173 156 141 127 115 104 94 85 77 69 62 56 50 44 38 33 28 1.1 1.2 1.2 1.2 1.2 1.2 1.2 1.2 1.1 1.1 1.0 0.9 0.7 0.6 0.4 0.2 0.0 –0.3 1.4 1.5 1.5 1.5 1.5 1.4 1.4 1.3 1.2 1.0 0.8 0.6 0.4 0.1 –0.3 –0.7 –1.1 –1.6
148-S	8.00 8.00	199 180 163 148 134 122 111 101 92 83 75 68 60 53 47 41 35 30 25 1.3 1.4 1.4 1.4 1.4 1.4 1.4 1.4 1.4 1.3 1.2 1.1 1.0 0.8 0.6 0.4 0.1 –0.1 –0.5 1.7 1.7 1.7 1.7 1.7 1.7 1.6 1.5 1.4 1.3 1.1 0.8 0.6 0.2 –0.1 –0.5 –1.0 –1.5 –2.1
168-S	9.00 9.00	182 165 151 137 124 113 103 93 85 76 68 61 54 48 42 36 31 26 1.5 1.6 1.6 1.6 1.6 1.6 1.6 1.5 1.4 1.3 1.2 1.1 0.9 0.7 0.5 0.2 –0.1 –0.4 1.9 1.9 1.9 1.9 1.8 1.8 1.7 1.6 1.4 1.2 0.9 0.6 0.3 –0.1 –0.5 –1.0 –1.6 –2.2
188-S	10.00 10.00	192 175 159 144 131 120 109 99 90 82 74 67 60 53 47 41 36 31 26 1.6 1.6 1.7 1.7 1.7 1.7 1.7 1.6 1.6 1.5 1.4 1.2 1.1 0.9 0.7 0.4 0.1 –0.2 –0.6 2.0 2.0 2.0 2.0 2.0 1.9 1.8 1.7 1.6 1.4 1.2 0.9 0.6 0.2 –0.3 –0.7 –1.3 –1.9 –2.6
188-D1	14.39 4.00	193 176 160 146 134 122 111 101 92 84 77 71 65 59 54 50 45 40 35 31 27 2.3 2.4 2.4 2.5 2.5 2.5 2.5 2.4 2.4 2.3 2.1 2.0 1.8 1.6 1.4 1.1 0.8 0.4 0.0 –0.4 –0.9 2.9 2.9 2.9 2.9 2.8 2.8 2.7 2.5 2.3 2.0 1.8 1.5 1.1 0.7 0.3 –0.2 –0.8 –1.5 –2.2 –3.1 –4.0
208-D1	15.50 4.25	162 148 136 124 114 104 95 87 79 72 66 61 56 51 47 43 38 34 30 26 2.7 2.7 2.8 2.8 2.8 2.7 2.7 2.6 2.5 2.3 2.1 1.9 1.7 1.4 1.1 0.7 0.3 –0.2 –0.7 –1.3 3.2 3.2 3.2 3.1 3.0 2.8 2.6 2.4 2.1 1.7 1.3 0.9 0.5 –0.1 –0.6 –1.3 –2.0 –2.9 –3.9 –4.9

8DT32 + 2

Table of safe superimposed service load (psf) and cambers (in.) — 2 in. Normal Weight Topping

Strand Pattern	y_s(end) in. y_s(center) in.	Span, ft
		42 44 46 48 50 52 54 56 58 60 62 64 66 68 70 72 74 76 78 80 82 84 86 88 90 92 94
128-S	7.00 7.00	270 240 214 190 170 152 136 121 108 97 86 76 67 56 47 38 30 1.0 1.0 1.1 1.1 1.2 1.2 1.2 1.2 1.2 1.2 1.2 1.1 1.1 1.0 0.9 0.7 0.6 1.0 1.1 1.1 1.1 1.0 1.0 0.9 0.9 0.8 0.6 0.4 0.2 0.0 –0.3 –0.7 –1.1 –1.5
148-S	7.00 7.00	289 259 232 208 187 168 152 137 123 111 100 88 77 66 57 47 39 31 1.3 1.4 1.4 1.5 1.5 1.6 1.6 1.7 1.7 1.7 1.7 1.6 1.6 1.5 1.4 1.3 1.2 1.0 1.4 1.4 1.4 1.4 1.4 1.4 1.3 1.3 1.2 1.0 0.9 0.7 0.5 0.2 –0.2 –0.6 –1.0 –1.5
168-S	8.00 8.00	288 259 233 210 190 171 155 138 124 110 98 87 76 67 57 49 40 33 25 1.4 1.5 1.6 1.7 1.7 1.8 1.8 1.9 1.9 1.9 1.8 1.8 1.7 1.6 1.5 1.3 1.2 1.0 1.5 1.6 1.6 1.6 1.6 1.5 1.5 1.4 1.3 1.1 1.0 0.7 0.5 0.2 –0.2 –0.6 –1.1 –1.6 –2.2
188-S	9.00 9.00	282 254 230 208 186 166 148 133 119 106 94 83 74 65 56 48 40 32 25 1.6 1.6 1.7 1.8 1.8 1.9 1.9 2.0 2.0 2.0 2.0 1.9 1.9 1.8 1.7 1.6 1.4 1.2 1.0 1.6 1.7 1.7 1.7 1.6 1.6 1.5 1.4 1.3 1.1 0.9 0.7 0.4 0.1 –0.3 –0.8 –1.3 –1.9 –2.5
188-D1	14.39 4.00	233 213 194 175 157 141 126 113 100 89 78 69 61 54 47 41 35 30 2.2 2.2 2.3 2.4 2.4 2.5 2.5 2.5 2.5 2.4 2.4 2.3 2.2 2.0 1.8 1.6 1.4 1.1 2.1 2.0 2.0 1.9 1.8 1.7 1.5 1.3 1.1 0.8 0.4 0.0 –0.5 –1.0 –1.5 –2.1 –2.8 –3.6
208-D1	15.50 4.25	159 143 129 115 103 92 81 72 63 55 48 42 36 31 26 2.7 2.7 2.8 2.8 2.8 2.7 2.7 2.6 2.5 2.3 2.1 1.9 1.7 1.4 1.1 2.0 1.9 1.7 1.5 1.2 0.9 0.5 0.1 –0.4 –1.0 –1.6 –2.2 –3.0 –3.8 –4.6

Strength is based on strain compatibility; bottom tension is limited to $12\sqrt{f'_c}$; see pages 2–7 through 2–10 for explanation.
Shaded values require release strengths higher than 3500 psi.

2–14   PCI Design Handbook/Sixth Edition

Strand Pattern Designation

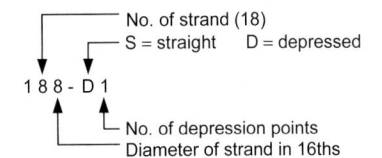

```
          No. of strand (18)
          S = straight    D = depressed
   1 8 8 - D 1
          No. of depression points
          Diameter of strand in 16ths
```

Safe loads shown include dead load of 10 psf for untopped members and 15 psf for topped members. Remainder is live load. Long-time cambers include superimposed dead load but do not include live load.

Key
186 – Safe superimposed service load, psf
2.0 – Estimated camber at erection, in.
2.4 – Estimated long-time camber, in.

DOUBLE TEE
8'-0" x 32"
Lightweight Concrete

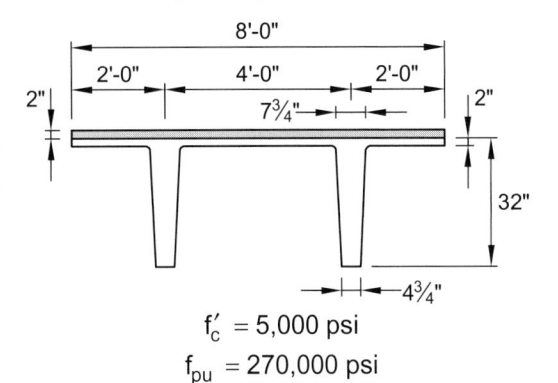

$f'_c = 5{,}000$ psi
$f_{pu} = 270{,}000$ psi

Section Properties

	Untopped	Topped
A =	567 in.^{2}	—
I =	55,464 in.^{4}	77,675 in.^{4}
y_b =	21.21 in.	24.52 in.
y_t =	10.79 in.	9.48 in.
S_b =	2,615 in.^{3}	3,167 in.^{3}
S_t =	5,140 in.^{3}	8,195 in.^{3}
wt =	453 plf	653 plf
DL =	57 psf	82 psf
V/S =	1.79 in.	

8LDT32

Table of safe superimposed service load (psf) and cambers (in.)   — No Topping

Strand Pattern	y_s(end) in. / y_s(center) in.	50	52	54	56	58	60	62	64	66	68	70	72	74	76	78	80	82	84	86	88	90	92	94	96	98	100	102	104
128-S	7.00 / 7.00	186	169	154	140	128	117	107	98	90	82	75	69	63	58	53	48	44	40	36	31	28							
		2.0	2.1	2.2	2.3	2.3	2.3	2.4	2.4	2.3	2.3	2.2	2.0	1.9	1.7	1.5	1.2	0.9	0.6	0.2									
		2.4	2.5	2.6	2.6	2.6	2.6	2.6	2.6	2.5	2.4	2.3	2.1	1.9	1.6	1.3	0.9	0.4	-0.2	-0.8	-1.4	-2.2							
148-S	8.00 / 8.00		193	176	161	147	134	123	113	104	95	87	80	74	68	62	57	52	47	43	38	34	30	26					
			2.3	2.4	2.5	2.6	2.6	2.7	2.7	2.8	2.8	2.7	2.7	2.7	2.6	2.5	2.3	2.1	1.9	1.7	1.4	1.0	0.6	0.2					
			2.9	2.9	3.0	3.0	3.0	3.1	3.0	3.0	2.9	2.8	2.7	2.5	2.3	2.0	1.7	1.3	0.8	0.3	-0.4	-1.1	-1.8	-2.7					
168-S	9.00 / 9.00				192	175	159	145	132	120	111	102	94	86	79	73	67	62	57	52	48	44	39	35	31	28			
					2.6	2.7	2.8	2.9	3.0	3.0	3.0	3.1	3.1	3.0	3.0	2.9	2.8	2.7	2.6	2.4	2.2	1.9	1.6	1.2	0.9	0.4			
					3.2	3.3	3.3	3.4	3.4	3.4	3.3	3.3	3.3	3.2	3.1	2.9	2.7	2.5	2.2	1.8	1.4	1.0	0.4	-0.2	-1.0	-1.8	-2.7		
188-S	10.00 / 10.00					185	168	154	140	128	117	107	99	91	84	77	71	65	60	55	51	47	43	39	35	31	28		
						2.8	2.9	3.0	3.1	3.1	3.2	3.2	3.2	3.2	3.2	3.2	3.1	3.0	2.8	2.7	2.5	2.2	1.9	1.6	1.2	0.8	0.3		
						3.4	3.5	3.5	3.6	3.6	3.5	3.5	3.4	3.3	3.1	2.9	2.7	2.4	2.1	1.7	1.2	0.7	0.2	-0.5	-1.3	-2.1	-3.1		
188-D1	14.39 / 4.00							172	158	146	134	124	114	105	97	90	83	76	70	64	59	54	49	46	43	40	37	34	
								4.1	4.3	4.4	4.5	4.6	4.7	4.8	4.8	4.8	4.8	4.7	4.6	4.4	4.2	4.0	3.7	3.4	3.0	2.7	2.2	1.7	
								5.0	5.1	5.2	5.2	5.2	5.2	5.1	5.0	4.8	4.5	4.2	3.8	3.4	2.8	2.2	1.5	0.9	0.2	-0.6	-1.4	-2.3	
208-D1	15.50 / 4.25												117	108	100	92	85	79	73	67	62	57	52	48	43	40	37		
													5.2	5.2	5.3	5.3	5.3	5.3	5.2	5.1	4.9	4.7	4.4	4.0	3.7	3.3	2.8		
													5.7	5.7	5.5	5.3	5.1	4.8	4.5	4.0	3.5	2.9	2.2	1.4	0.5	-0.3	-1.2		

8LDT32 + 2

Table of safe superimposed service load (psf) and cambers (in.)   — 2 in. Normal Weight Topping

Strand Pattern	y_s(end) in. / y_s(center) in.	50	52	54	56	58	60	62	64	66	68	70	72	74	76	78	80	82	84	86	88	90	92	94	96	98
128-S	7.00 / 7.00	183	165	149	134	121	109	99	89	81	73	64	56	49	42	35	29									
		2.0	2.1	2.1	2.2	2.3	2.3	2.3	2.4	2.4	2.3	2.3	2.2	2.2	2.0	1.9	1.7									
		1.8	1.8	1.7	1.7	1.6	1.5	1.3	1.1	0.9	0.6	0.3	-0.1	-0.6	-1.1	-1.7	-2.4									
148-S	8.00 / 8.00		190	172	156	141	128	115	103	92	82	73	65	57	50	44	38	32	26							
			2.3	2.4	2.5	2.6	2.6	2.7	2.7	2.8	2.8	2.8	2.7	2.7	2.6	2.5	2.3	2.1	1.9							
			2.1	2.1	2.0	2.0	1.9	1.7	1.6	1.4	1.1	0.8	0.4	0.0	-0.5	-1.0	-1.6	-2.3	-3.1							
168-S	9.00 / 9.00				192	174	156	139	124	112	100	90	80	72	64	56	50	43	38	32	27					
					2.6	2.7	2.8	2.9	3.0	3.0	3.0	3.1	3.1	3.0	2.9	2.9	2.7	2.6	2.4	2.2							
					2.3	2.3	2.3	2.2	2.0	1.9	1.7	1.4	1.2	0.8	0.4	0.0	-0.6	-1.1	-1.8	-2.5	-3.4					
188-S	10.00 / 10.00					186	167	149	134	120	107	96	86	77	69	61	54	48	42	36	31	26				
						2.8	2.9	3.0	3.1	3.1	3.2	3.2	3.2	3.2	3.2	3.2	3.1	3.0	2.8	2.7	2.5	2.2				
						2.4	2.4	2.3	2.2	2.0	1.8	1.6	1.2	1.0	0.6	0.2	-0.3	-0.9	-1.5	-2.3	-3.1	-4.0				
188-D1	14.39 / 4.00							171	155	140	127	115	104	93	84	75	67	59	52	46	39	34	29	25		
								4.1	4.3	4.4	4.5	4.6	4.7	4.8	4.8	4.8	4.8	4.7	4.6	4.4	4.2	4.0	3.7	3.4		
								3.4	3.3	3.2	3.0	2.8	2.5	2.2	1.8	1.3	0.8	0.1	-0.6	-1.5	-2.4	-3.5	-4.5	-5.6		
208-D1	15.50 / 4.25															107	96	87	78	70	62	55	49	43	37	31	26
																5.2	5.2	5.3	5.3	5.3	5.3	5.2	5.1	5.9	4.7	4.4	4.0
																2.8	2.4	2.0	1.5	0.9	0.3	-0.4	-1.3	-2.2	-3.3	-4.5	-5.8

Strength is based on strain compatibility; bottom tension is limited to $12\sqrt{f'_c}$; see pages 2–7 through 2–10 for explanation.
Shaded values require release strengths higher than 3500 psi.

PCI Design Handbook/Sixth Edition

Strand Pattern Designation

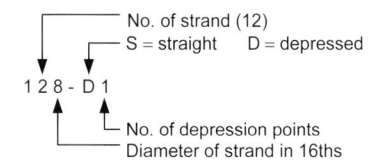

```
        ┌── No. of strand (12)
        │ ┌── S = straight   D = depressed
        │ │
      1 2 8 - D 1
          ↑   ↑
          │   └── No. of depression points
          └────── Diameter of strand in 16ths
```

Safe loads shown include dead load of 10 psf for untopped members and 15 psf for topped members. Remainder is live load. Long-time cambers include superimposed dead load but do not include live load.

Key
171 – Safe superimposed service load, psf
0.6 – Estimated camber at erection, in.
0.8 – Estimated long-time camber, in.

DOUBLE TEE
10'-0" x 24"
Normal Weight Concrete

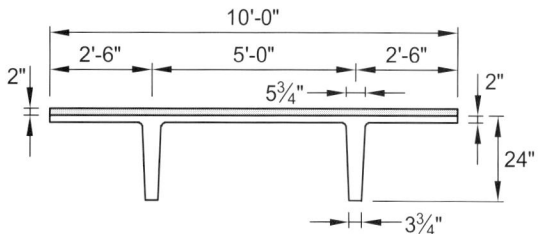

10'-0" (2'-6" | 5'-0" | 2'-6"), flange 2", stem 5¾" at top, 3¾" at bottom, depth 24"

$f'_c = 5{,}000$ psi
$f_{pu} = 270{,}000$ psi

Section Properties

	Untopped	Topped
A =	449 in.²	—
I =	22,469 in.⁴	29,396 in.⁴
y_b =	17.77 in.	19.89 in.
y_t =	6.23 in.	6.11 in.
S_b =	1,264 in.³	1,478 in.³
S_t =	3,607 in.³	4,812 in.³
wt =	468 plf	718 plf
DL =	74 psf	72 psf
V/S =	1.35 in.	

10DT24

Table of safe superimposed service load (psf) and cambers (in.) — No Topping

Strand Pattern	y_s(end) in. / y_s(center) in.	30	32	34	36	38	40	42	44	46	48	50	52	54	56	58	60	62	64	66	68	70	72	74	76	78
68-S	4.00 / 4.00	171 / 0.6 / 0.8	146 / 0.7 / 0.9	126 / 0.7 / 1.0	109 / 0.8 / 1.0	94 / 0.8 / 1.0	82 / 0.9 / 1.0	71 / 0.9 / 1.0	62 / 0.9 / 1.0	54 / 0.9 / 1.0	47 / 0.9 / 0.9	41 / 0.8 / 0.8	35 / 0.8 / 0.7	30 / 0.7 / 0.5	26 / 0.6 / 0.3											
88-S	5.00 / 5.00		193 / 0.9 / 1.2	167 / 1.0 / 1.3	146 / 1.1 / 1.4	127 / 1.1 / 1.4	112 / 1.2 / 1.5	98 / 1.3 / 1.5	87 / 1.3 / 1.5	77 / 1.4 / 1.5	68 / 1.4 / 1.5	60 / 1.4 / 1.5	53 / 1.4 / 1.4	47 / 1.3 / 1.2	41 / 1.3 / 1.0	36 / 1.2 / 0.8	32 / 1.0 / 0.5	27 / 0.9 / 0.1								
108-S	6.00 / 6.00				177 / 1.2 / 1.6	156 / 1.3 / 1.7	137 / 1.4 / 1.8	121 / 1.5 / 1.9	108 / 1.6 / 1.9	96 / 1.7 / 2.0	85 / 1.7 / 2.0	76 / 1.8 / 2.0	68 / 1.8 / 1.9	61 / 1.8 / 1.9	54 / 1.8 / 1.8	48 / 1.8 / 1.6	43 / 1.7 / 1.4	38 / 1.6 / 1.1	33 / 1.4 / 0.7	29 / 1.2 / 0.3						
128-S	7.00 / 7.00					159 / 1.6 / 2.0	141 / 1.7 / 2.1	125 / 1.8 / 2.2	112 / 1.9 / 2.3	100 / 1.9 / 2.3	90 / 2.0 / 2.3	80 / 2.0 / 2.3	72 / 2.1 / 2.2	64 / 2.1 / 2.2	58 / 2.1 / 2.1	52 / 2.1 / 1.9	46 / 2.0 / 1.7	41 / 1.9 / 1.4	36 / 1.8 / 1.1	31 / 1.6 / 0.6	26 / 1.4 / 0.1					
128-D1	11.67 / 3.25												100 / 2.3 / 2.7	90 / 2.4 / 2.7	80 / 2.5 / 2.6	72 / 2.5 / 2.5	64 / 2.5 / 2.4	57 / 2.5 / 2.2	51 / 2.4 / 1.9	46 / 2.3 / 1.6	41 / 2.2 / 1.3	37 / 2.0 / 0.9	33 / 1.8 / 0.4	30 / 1.5 / −0.2	26 / 1.2 / −0.9	
148-D1	12.86 / 3.50																	68 / 2.9 / 2.9	61 / 2.9 / 2.7	55 / 2.9 / 2.5	49 / 2.8 / 2.2	43 / 2.7 / 1.8	39 / 2.6 / 1.4	36 / 2.4 / 0.9	32 / 2.1 / 0.3	29 / 1.8 / −0.3

10DT24 + 2

Table of safe superimposed service load (psf) and cambers (in.) — 2 in. Normal Weight Topping

Strand Pattern	y_s(end) in. / y_s(center) in.	30	32	34	36	38	40	42	44	46	48	50	52	54	56	58	60	62	64	66
68-S	4.00 / 4.00	172 / 0.6 / 0.6	145 / 0.7 / 0.7	123 / 0.7 / 0.7	104 / 0.8 / 0.7	89 / 0.8 / 0.6	75 / 0.9 / 0.6	63 / 0.9 / 0.5	56 / 0.9 / 0.4	45 / 0.9 / 0.2	37 / 0.9 / 0.0	30 / 0.8 / −0.3								
88-S	5.00 / 5.00		197 / 0.9 / 0.9	169 / 1.0 / 1.0	145 / 1.1 / 1.0	125 / 1.1 / 1.0	108 / 1.2 / 1.0	93 / 1.3 / 0.9	83 / 1.3 / 0.9	69 / 1.4 / 0.7	60 / 1.4 / 0.6	51 / 1.4 / 0.4	43 / 1.4 / 0.2	34 / 1.3 / −0.2						
108-S	6.00 / 6.00				180 / 1.2 / 1.2	156 / 1.3 / 1.3	136 / 1.4 / 1.3	119 / 1.5 / 1.3	104 / 1.6 / 1.2	91 / 1.7 / 1.1	79 / 1.7 / 1.0	68 / 1.8 / 0.9	57 / 1.8 / 0.7	48 / 1.8 / 0.5	40 / 1.8 / 0.2	31 / 1.8 / −0.2				
128-S	7.00 / 7.00						160 / 1.6 / 1.5	141 / 1.7 / 1.5	124 / 1.8 / 1.5	109 / 1.9 / 1.4	93 / 1.9 / 1.3	79 / 2.0 / 1.2	68 / 2.1 / 1.0	58 / 2.1 / 0.8	48 / 2.1 / 0.5	40 / 2.1 / 0.2	33 / 2.1 / −0.2			
128-D1	11.67 / 3.25												92 / 2.3 / 1.4	79 / 2.4 / 1.2	67 / 2.5 / 0.9	57 / 2.5 / 0.6	47 / 2.5 / 0.3	39 / 2.5 / −0.2	32 / 2.4 / −1.7	
148-D1	12.86 / 3.50																	53 / 2.9 / 0.5	44 / 2.9 / 0.0	36 / 2.9 / −0.6

Strength is based on strain compatibility; bottom tension is limited to $12\sqrt{f'_c}$; see pages 2–7 through 2–10 for explanation.
Shaded values require release strengths higher than 3500 psi.

Strand Pattern Designation

```
    ┌── No. of strand (12)
    │ ┌── S = straight   D = depressed
  1 2 8 - D 1
          │ │
          │ └── No. of depression points
          └──── Diameter of strand in 16ths
```

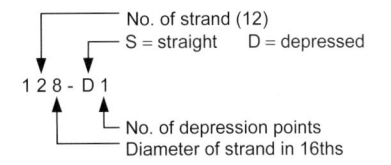

Safe loads shown include dead load of 10 psf for untopped members and 15 psf for topped members. Remainder is live load. Long-time cambers include superimposed dead load but do not include live load.

Key
179 – Safe superimposed service load, psf
1.0 – Estimated camber at erection, in.
1.3 – Estimated long-time camber, in.

DOUBLE TEE
10'-0" x 24"
Lightweight Concrete

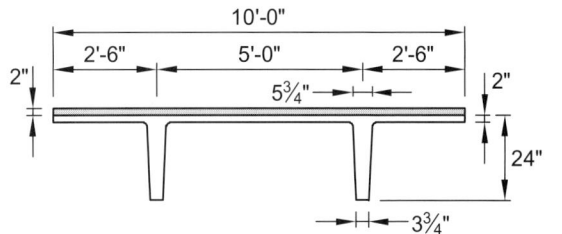

Dimensions: 10'-0" width (2'-6" + 5'-0" + 2'-6"), 2" flange, 24" depth, 5¾" web top, 3¾" web bottom.

$f'_c = 5{,}000$ psi
$f_{pu} = 270{,}000$ psi

Section Properties

	Untopped	Topped
A	449 in.²	—
I	22,469 in.⁴	31,515 in.⁴
y_b	17.77 in.	20.53 in.
y_t	6.23 in.	5.47 in.
S_b	1,264 in.³	1,535 in.³
S_t	3,607 in.³	5,761 in.³
wt	359 plf	609 plf
DL	36 psf	61 psf
V/S	1.35 in.	

10LDT24

Table of safe superimposed service load (psf) and cambers (in.) — No Topping

Strand Pattern	y_s(end) in. / y_s(center) in.	30	32	34	36	38	40	42	44	46	48	50	52	54	56	58	60	62	64	66	68	70	72	74	76	78	80
68-S	4.00 / 4.00	179 / 1.0 / 1.3	155 / 1.1 / 1.4	134 / 1.2 / 1.5	117 / 1.3 / 1.6	103 / 1.4 / 1.7	90 / 1.5 / 1.7	80 / 1.6 / 1.8	70 / 1.6 / 1.8	62 / 1.7 / 1.8	55 / 1.7 / 1.8	49 / 1.7 / 1.7	43 / 1.7 / 1.6	38 / 1.7 / 1.5	34 / 1.7 / 1.3	30 / 1.6 / 1.0	27 / 1.5 / 0.6										
88-S	5.00 / 5.00			175 / 1.6 / 2.0	154 / 1.7 / 2.2	136 / 1.9 / 2.3	120 / 2.0 / 2.4	107 / 2.1 / 2.5	95 / 2.2 / 2.6	85 / 2.3 / 2.7	76 / 2.4 / 2.7	68 / 2.5 / 2.7	61 / 2.6 / 2.7	55 / 2.6 / 2.6	49 / 2.6 / 2.5	44 / 2.6 / 2.4	40 / 2.6 / 2.2	36 / 2.5 / 1.9	32 / 2.4 / 1.6	29 / 2.3 / 1.1	26 / 2.1 / 0.6						
108-S	6.00 / 6.00				185 / 2.0 / 2.5	164 / 2.1 / 2.7	145 / 2.3 / 2.9	130 / 2.5 / 3.0	116 / 2.6 / 3.2	104 / 2.8 / 3.3	94 / 2.9 / 3.4	84 / 3.1 / 3.4	76 / 3.2 / 3.5	69 / 3.3 / 3.4	62 / 3.3 / 3.4	56 / 3.4 / 3.3	51 / 3.4 / 3.2	46 / 3.4 / 3.0	42 / 3.4 / 2.8	38 / 3.4 / 2.5	34 / 3.3 / 2.2	31 / 3.1 / 1.7	27 / 2.9 / 1.1				
128-S	7.00 / 7.00									108 / 3.2 / 3.9	98 / 3.4 / 4.0	87 / 3.5 / 4.0	78 / 3.7 / 4.0	70 / 3.8 / 4.0	63 / 3.9 / 4.0	57 / 3.9 / 3.8	51 / 4.0 / 3.8	47 / 4.0 / 3.7	42 / 4.0 / 3.5	39 / 3.9 / 3.3	35 / 3.9 / 3.0	32 / 3.7 / 2.6	29 / 3.6 / 2.2	26 / 3.3 / 1.6			
128-D1	11.67 / 3.25																72 / 4.5 / 4.7	65 / 4.6 / 4.6	59 / 4.7 / 4.4	53 / 4.8 / 4.2	48 / 4.8 / 3.9	43 / 4.8 / 3.5	39 / 4.7 / 3.0	35 / 4.5 / 2.4	32 / 4.3 / 1.7	29 / 4.0 / 1.0	27 / 3.7 / 0.2
148-D1	12.86 / 3.50																						47 / 5.6 / 4.5	43 / 5.6 / 4.0	39 / 5.5 / 3.4	35 / 5.4 / 2.8	31 / 5.2 / 2.0

10LDT24 + 2

Table of safe superimposed service load (psf) and cambers (in.) — 2 in. Normal Weight Topping

Strand Pattern	y_s(end) in. / y_s(center) in.	30	32	34	36	38	40	42	44	46	48	50	52	54	56	58	60	62	64	66
68-S	4.00 / 4.00	181 / 1.0 / 1.0	154 / 1.1 / 1.1	131 / 1.2 / 1.1	113 / 1.3 / 1.1	97 / 1.4 / 1.1	83 / 1.5 / 1.0	72 / 1.6 / 1.0	62 / 1.6 / 0.8	53 / 1.7 / 0.7	45 / 1.7 / 0.5	38 / 1.7 / 0.2	32 / 1.7 / −0.1							
88-S	5.00 / 5.00			177 / 1.6 / 1.6	153 / 1.7 / 1.6	133 / 1.9 / 1.6	116 / 2.0 / 1.6	101 / 2.1 / 1.6	89 / 2.2 / 1.5	78 / 2.3 / 1.3	68 / 2.4 / 1.1	59 / 2.5 / 0.8	51 / 2.6 / 0.5	43 / 2.6 / 0.1	35 / 2.7 /					
108-S	6.00 / 6.00				188 / 2.0 / 1.9	165 / 2.1 / 2.0	144 / 2.3 / 2.1	127 / 2.5 / 2.1	112 / 2.6 / 2.1	99 / 2.8 / 2.0	87 / 2.9 / 1.9	75 / 3.1 / 1.8	63 / 3.2 / 1.6	53 / 3.3 / 1.3	45 / 3.3 / 0.9	38 / 3.4 / 0.6	32 / 3.4 / 0.1			
128-S	7.00 / 7.00										102 / 3.2 / 2.4	88 / 3.4 / 2.2	76 / 3.5 / 2.1	65 / 3.7 / 1.8	55 / 3.8 / 1.5	46 / 3.9 / 1.1	39 / 3.9 / 0.6	33 / 4.0 / 0.1		
128-D1	11.67 / 3.25																57 / 4.5 / 1.4	49 / 4.6 / 0.9	41 / 4.7 / 0.3	34 / 4.8 / −0.4

Strength is based on strain compatibility; bottom tension is limited to $12\sqrt{f'_c}$; see pages 2–7 through 2–10 for explanation.
Shaded values require release strengths higher than 3500 psi.

PCI Design Handbook/Sixth Edition

Strand Pattern Designation

```
         No. of strand (18)
         S = straight   D = depressed
    1 8 8 - D 1
         No. of depression points
         Diameter of strand in 16ths
```

Safe loads shown include dead load of 10 psf for untopped members and 15 psf for topped members. Remainder is live load. Long-time cambers include superimposed dead load but do not include live load.

Key
189 – Safe superimposed service load, psf
1.2 – Estimated camber at erection, in.
1.5 – Estimated long-time camber, in.

DOUBLE TEE
10'-0" x 32"
Normal Weight Concrete

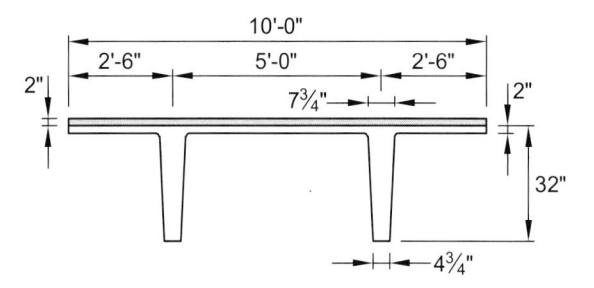

$f'_c = 5,000$ psi
$f_{pu} = 270,000$ psi

Section Properties

	Untopped	Topped
A =	615 in.²	—
I =	59,720 in.⁴	77,118 in.⁴
y_b =	21.98 in.	24.54 in.
y_t =	10.02 in.	9.46 in.
S_b =	2,717 in.³	3,143 in.³
S_t =	5,960 in.³	8,152 in.³
wt =	641 plf	891 plf
DL =	64 psf	89 psf
V/S =	1.69 in.	

10DT32 — No Topping

Table of safe superimposed service load (psf) and cambers (in.)

Strand Pattern	y_s(end) in. / y_s(center) in.	44	46	48	50	52	54	56	58	60	62	64	66	68	70	72	74	76	78	80	82	84	86	88	90	92	94	96	98	100	102
128-S	7.00 / 7.00	189 1.2 1.5	169 1.2 1.6	152 1.3 1.6	136 1.3 1.7	123 1.4 1.7	110 1.4 1.7	99 1.4 1.7	90 1.5 1.7	81 1.5 1.7	73 1.4 1.6	65 1.4 1.5	59 1.4 1.4	53 1.3 1.2	47 1.2 1.0	42 1.1 0.8	37 1.0 0.5	33 0.8 0.2	29 0.6 −0.2	25 0.4 −0.6											
148-S	8.00 / 8.00		194 1.4 1.8	174 1.4 1.9	157 1.5 1.9	142 1.6 2.0	128 1.6 2.0	116 1.7 2.0	105 1.7 2.0	95 1.7 2.0	86 1.7 1.9	78 1.7 1.8	71 1.6 1.7	64 1.6 1.6	58 1.5 1.4	52 1.4 1.1	47 1.2 0.8	42 1.1 0.5	38 0.9 0.1	34 0.7 −0.4	30 0.4 −0.9	26									
168-S	9.00 / 9.00			194 1.5 2.0	175 1.6 2.1	158 1.7 2.1	144 1.7 2.2	130 1.8 2.2	118 1.8 2.2	108 1.9 2.2	98 1.9 2.2	89 1.9 2.1	81 1.9 2.0	74 1.8 1.9	67 1.7 1.7	61 1.6 1.5	55 1.6 1.3	50 1.4 1.0	45 1.2 0.6	41 1.0 0.2	36 0.8 −0.3	33 0.5 −0.8	29 0.2 −1.4	26							
188-S	10.00 / 10.00				190 1.7 2.2	173 1.7 2.2	157 1.8 2.3	143 1.9 2.4	130 2.0 2.4	118 2.0 2.4	108 2.0 2.3	99 2.0 2.3	90 2.0 2.2	82 2.0 2.1	75 1.9 1.9	68 1.8 1.8	62 1.8 1.5	57 1.6 1.3	51 1.5 0.9	47 1.3 0.9	42 1.1 0.1	38 0.8 −0.4	34 0.5 −1.0	31 0.2 −1.7	27						
188-D1	14.39 / 4.00									132 2.7 3.4	122 2.8 3.4	112 2.9 3.4	103 2.9 3.3	95 2.9 3.3	88 2.9 3.2	81 2.9 3.0	74 2.8 2.8	68 2.8 2.6	63 2.6 2.2	58 2.5 1.9	53 2.3 1.5	49 2.1 1.1	44 1.9 0.7	41 1.6 0.2	37 1.3 −0.4	34 0.9 −1.1	30 0.5 −2.0	27 0.0 −2.9			
208-D1	15.50 / 4.25														99 3.2 3.7	92 3.3 3.6	85 3.2 3.4	78 3.2 3.2	72 3.1 3.0	67 3.1 2.7	62 2.9 2.3	57 2.7 1.9	52 2.5 1.4	48 2.3 0.9	44 2.0 0.4	40 1.7 −0.2	37 1.3 −0.9	34 0.9 −1.6	31 0.4 −2.5

10DT32 + 2 — 2 in. Normal Weight Topping

Table of safe superimposed service load (psf) and cambers (in.)

Strand Pattern	y_s(end) in. / y_s(center) in.	44	46	48	50	52	54	56	58	60	62	64	66	68	70	72	74	76	78	80	82	84	86	88
128-S	7.00 / 7.00	187 1.0 1.0	166 1.1 1.0	147 1.1 1.0	131 1.1 0.9	116 1.2 0.9	103 1.2 0.8	92 1.2 0.7	81 1.2 0.5	72 1.1 0.4	63 1.1 0.2	55 1.1 −0.1	47 1.0 −0.4	38 0.9 −0.7	30 0.8 −1.1									
148-S	8.00 / 8.00		193 1.2 1.1	172 1.2 1.1	153 1.3 1.1	137 1.3 1.1	122 1.4 1.0	109 1.4 0.9	98 1.4 0.8	87 1.4 0.7	78 1.4 0.5	68 1.3 0.3	58 1.3 0.0	49 1.2 −0.3	40 1.1 −0.7	32 1.0 −1.1	25 0.9 −1.6							
168-S	9.00 / 9.00			193 1.3 1.3	173 1.4 1.3	155 1.5 1.2	139 1.5 1.2	125 1.5 1.1	112 1.6 1.0	100 1.6 0.9	88 1.6 0.7	78 1.5 0.5	68 1.5 0.3	58 1.4 0.0	49 1.4 −0.3	41 1.3 −0.7	33 1.1 −1.2	26 1.0 −1.7						
188-S	10.00 / 10.00				189 1.5 1.3	170 1.5 1.3	153 1.6 1.3	137 1.6 1.2	122 1.6 1.1	108 1.7 1.0	96 1.7 0.9	84 1.7 0.7	74 1.6 0.5	64 1.6 0.2	56 1.5 −0.1	47 1.4 −0.5	39 1.3 −0.9	32 1.2 −1.4						
188-D1	14.39 / 4.00							183 2.1 1.9	166 2.2 1.8	151 2.2 1.8	135 2.3 1.7	121 2.4 1.5	107 2.4 1.4	95 2.4 1.2	84 2.4 0.9	74 2.4 0.6	64 2.3 0.2	57 2.2 −0.2	50 2.1 −0.6	43 2.0 −1.1	37 1.8 −1.7	32 1.7 −2.3	27 1.5 −3.0	
208-D1	15.50 / 4.25									136 2.6 1.8	122 2.6 1.7	109 2.7 1.5	97 2.7 1.3	86 2.7 1.0	76 2.7 0.7	67 2.6 0.3	58 2.6 −0.1	50 2.5 −0.6	44 2.3 −1.1	38 2.2 −1.7	33 2.0 −2.4	28 1.8 −3.1		

Strength is based on strain compatibility; bottom tension is limited to $12\sqrt{f'_c}$; see pages 2–7 through 2–10 for explanation. Shaded values require release strengths higher than 3500 psi.

Strand Pattern Designation

```
        No. of strand (18)
        S = straight   D = depressed
 1 8 8 - D 1
        No. of depression points
        Diameter of strand in 16ths
```

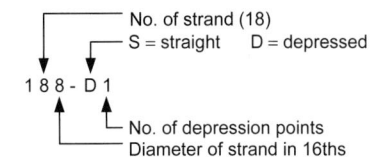

Safe loads shown include dead load of 10 psf for untopped members and 15 psf for topped members. Remainder is live load. Long-time cambers include superimposed dead load but do not include live load.

Key
180 — Safe superimposed service load, psf
1.7 — Estimated camber at erection, in.
2.2 — Estimated long-time camber, in.

DOUBLE TEE
10'-0" x 32"
Lightweight Concrete

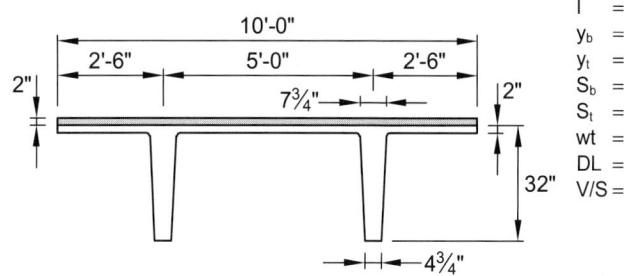

10'-0"
2'-6" | 5'-0" | 2'-6"
7¾"
2" | 2"
32"
4¾"

$f'_c = 5{,}000$ psi
$f_{pu} = 270{,}000$ psi

Section Properties

	Untopped	Topped
A =	615 in.²	—
I =	59,720 in.⁴	83,019 in.⁴
y_b =	21.98 in.	25.40 in.
y_t =	10.02 in.	8.60 in.
S_b =	2,717 in.³	3,269 in.³
S_t =	5,960 in.³	9,653 in.³
wt =	491 plf	741 plf
DL =	49 psf	74 psf
V/S =	1.69 in.	

10LDT32

Table of safe superimposed service load (psf) and cambers (in.) — No Topping

Strand Pattern	y_s(end) in. / y_s(center) in.	Span, ft
		46 48 50 52 54 56 58 60 62 64 66 68 70 72 74 76 78 80 82 84 86 88 90 92 94 96 98 100 102
128-S	7.00 / 7.00	180 163 147 134 122 111 101 92 84 77 70 64 58 53 49 44 40 37 33 30 27 1.7 1.8 1.9 2.0 2.1 2.1 2.2 2.2 2.2 2.3 2.2 2.2 2.2 2.1 2.0 1.9 1.7 1.5 1.3 1.0 0.7 2.2 2.2 2.3 2.4 2.4 2.5 2.5 2.5 2.4 2.4 2.3 2.1 2.0 1.8 1.5 1.2 0.8 0.4 −0.2 −0.7 −1.4
148-S	8.00 / 8.00	186 168 153 139 127 116 106 97 89 82 75 69 63 58 53 49 45 41 37 33 29 26 2.1 2.2 2.3 2.3 2.4 2.5 2.6 2.6 2.6 2.7 2.7 2.6 2.6 2.5 2.4 2.3 2.2 2.0 1.7 1.4 1.1 0.8 2.6 2.6 2.7 2.8 2.8 2.9 2.9 2.8 2.8 2.7 2.6 2.4 2.2 1.9 1.6 1.3 0.8 0.3 −0.3 −1.0 −1.8
168-S	9.00 / 9.00	186 170 155 142 130 118 107 99 90 83 76 70 65 59 54 50 46 42 38 34 30 27 2.3 2.4 2.5 2.6 2.7 2.8 2.9 2.9 3.0 3.0 3.0 2.9 2.9 2.8 2.7 2.6 2.4 2.2 2.0 1.7 1.4 1.0 2.9 3.0 3.1 3.1 3.2 3.2 3.2 3.2 3.1 3.0 2.8 2.6 2.4 2.1 1.8 1.4 1.0 0.5 −0.2 −0.9 −1.7
188-S	10.00 / 10.00	183 166 151 138 125 114 104 95 88 81 74 68 63 58 53 49 45 41 38 34 30 27 2.5 2.6 2.7 2.8 2.9 3.0 3.1 3.1 3.2 3.2 3.2 3.1 3.1 3.0 2.9 2.7 2.6 2.4 2.1 1.8 1.4 1.0 3.1 3.2 3.3 3.4 3.4 3.4 3.4 3.4 3.3 3.2 3.1 2.9 2.7 2.5 2.1 1.8 1.4 0.9 0.4 −0.3 −1.1 −1.9
188-D1	14.39 / 4.00	152 139 128 118 108 100 92 85 78 72 66 60 55 51 46 43 40 37 34 32 30 3.9 4.0 4.2 4.3 4.4 4.5 4.6 4.69 4.6 4.6 4.5 4.4 4.3 4.1 3.9 3.7 3.4 3.1 2.7 2.3 1.9 4.7 4.8 4.8 4.9 4.9 4.8 4.8 4.7 4.5 4.3 3.9 3.6 3.1 2.6 2.0 1.5 0.9 0.2 −0.5 −1.3 −2.2
208-D1	15.50 / 4.25	110 102 94 87 80 74 68 63 58 53 48 44 41 38 35 32 4.9 5.0 5.0 5.1 5.1 5.1 5.1 5.1 5.0 4.8 4.6 4.3 4.0 3.7 3.3 2.9 5.4 5.4 5.3 5.2 5.0 0.8 4.6 4.2 3.8 3.3 2.7 2.0 1.2 0.5 −0.3 −1.2

10LDT32 + 2

Table of safe superimposed service load (psf) and cambers (in.) — 2 in. Normal Weight Topping

Strand Pattern	y_s(end) in. / y_s(center) in.	Span, ft
		44 46 48 50 52 54 56 58 60 62 64 66 68 70 72 74 76 78 80 82 84 86 88 90 92
128-S	7.00 / 7.00	199 177 159 142 128 114 103 92 83 74 67 60 53 47 41 35 28 1.6 1.7 1.8 1.9 2.0 2.1 2.1 2.2 2.2 2.2 2.3 2.2 2.2 2.2 2.1 2.0 1.9 1.6 1.6 1.6 1.6 1.6 1.5 1.4 1.3 1.2 1.0 0.8 0.5 0.2 −0.2 −0.6 −1.1 −1.7
148-S	8.00 / 8.00	183 165 148 134 121 109 98 89 80 72 64 56 49 42 36 31 2.1 2.2 2.3 2.3 2.4 2.5 2.6 2.6 2.6 2.7 2.7 2.6 2.6 2.5 2.4 2.3 1.9 1.9 1.9 1.9 1.8 1.7 1.6 1.4 1.2 1.0 0.7 0.3 −0.1 −0.5 −1.0 −1.6
168-S	9.00 / 9.00	184 166 150 136 123 112 100 89 79 71 62 55 48 42 36 31 26 2.3 2.4 2.5 2.6 2.7 2.8 2.9 2.9 3.0 3.0 3.0 3.0 2.9 2.8 2.7 2.6 2.5 2.1 2.1 2.1 2.1 2.0 1.9 1.7 1.6 1.3 1.1 0.7 0.4 −0.1 −0.6 −1.1 −1.8 −2.5
188-S	10.00 / 10.00	182 165 149 135 121 107 95 85 76 68 60 53 46 40 34 29 2.5 2.6 2.7 2.8 2.9 3.0 3.1 3.1 3.2 3.2 3.2 3.2 3.1 3.0 2.9 2.8 2.3 2.3 2.2 2.2 2.1 1.9 1.7 1.5 1.3 1.0 0.6 0.2 −0.3 −0.8 −1.5 −2.1
188-D1	14.39 / 4.00	149 136 124 112 100 90 80 72 64 56 49 42 36 31 26 3.9 4.0 4.2 4.3 4.4 4.5 4.6 4.6 4.6 4.6 4.5 4.4 4.3 4.1 3.9 3.1 3.0 2.9 2.7 2.5 2.2 1.9 1.5 1.1 0.5 −0.1 −0.9 −1.7 −2.6 −3.6
208-D1	15.50 / 4.25	103 92 83 74 66 59 52 45 39 34 28 4.9 5.0 5.0 5.1 5.1 5.1 5.1 5.1 5.0 4.8 4.6 2.8 2.5 2.1 1.7 1.2 0.7 0.0 −0.7 −1.5 −2.5 −3.5

Strength is based on strain compatibility; bottom tension is limited to $12\sqrt{f'_c}$; see pages 2–7 through 2–10 for explanation.
Shaded values require release strengths higher than 3500 psi.

Strand Pattern Designation

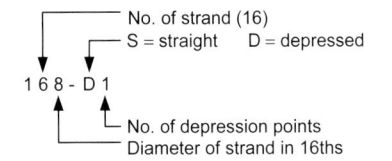

```
        No. of strand (16)
       S = straight    D = depressed
1 6 8 - D 1
         No. of depression points
         Diameter of strand in 16ths
```

Safe loads shown include dead load of 10 psf for untopped members and 15 psf for topped members. Remainder is live load. Long-time cambers include superimposed dead load but do not include live load.

Key
133 – Safe superimposed service load, psf
0.8 – Estimated camber at erection, in.
1.1 – Estimated long-time camber, in.

DOUBLE TEE
12'-0" x 28"
Normal Weight Concrete

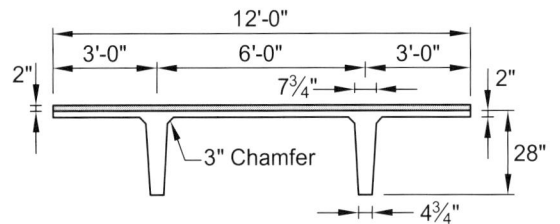

12'-0" (3'-0" | 6'-0" | 3'-0"), flange thickness 7¾", stem 4¾", depth 28", 3" Chamfer, 2" edge

$f'_c = 5{,}000$ psi
$f_{pu} = 270{,}000$ psi

Section Properties

	Untopped		Topped	
A =	640	in.²	—	
I =	44,563	in.⁴	57,323	in.⁴
y_b =	20.01	in.	22.47	in.
y_t =	7.99	in.	7.53	in.
S_b =	2,227	in.³	2,551	in.³
S_t =	5,577	in.³	7,613	in.³
wt =	677	plf	977	plf
DL =	56	psf	81	psf
V/S =	1.62	in.		

12DT28

Table of safe superimposed service load (psf) and cambers (in.) — No Topping

Strand Pattern	y_s(end) in. / y_s(center) in.	40	42	44	46	48	50	52	54	56	58	60	62	64	66	68	70	72	74	76	78	80	82	84
108-S	6.00 / 6.00	133 / 0.8 / 1.1	117 / 0.9 / 1.2	102 / 0.9 / 1.2	90 / 0.9 / 1.3	79 / 0.9 / 1.3	70 / 0.9 / 1.3	62 / 0.9 / 1.2	54 / 0.9 / 1.2	47 / 0.8 / 1.1	41 / 0.7 / 1.0	36 / 0.6 / 0.8	31 / 0.5 / 0.7	26 / 0.4 / 0.4										
128-S	7.00 / 7.00	157 / 1.0 / 1.3	139 / 1.0 / 1.4	123 / 1.1 / 1.5	109 / 1.1 / 1.5	96 / 1.1 / 1.5	86 / 1.1 / 1.6	76 / 1.1 / 1.6	67 / 1.1 / 1.6	60 / 1.1 / 1.5	53 / 1.0 / 1.4	47 / 1.0 / 1.3	41 / 0.9 / 1.2	36 / 0.8 / 1.0	31 / 0.6 / 0.8	27 / 0.4 / 0.5								
148-S	8.00 / 8.00	178 / 1.1 / 1.5	158 / 1.1 / 1.5	140 / 1.2 / 1.6	125 / 1.2 / 1.7	11 / 1.3 / 1.7	99 / 1.3 / 1.8	88 / 1.3 / 1.8	79 / 1.3 / 1.8	71 / 1.3 / 1.8	63 / 1.3 / 1.7	56 / 1.2 / 1.7	50 / 1.1 / 1.5	44 / 1.0 / 1.4	39 / 0.9 / 1.2	34 / 0.7 / 0.9	30 / 0.6 / 0.7	26 / 0.3 / 0.3						
168-S	9.00 / 9.00	196 / 1.1 / 1.6	174 / 1.2 / 1.6	155 / 1.3 / 1.7	138 / 1.3 / 1.8	123 / 1.4 / 1.9	110 / 1.4 / 1.9	99 / 1.4 / 1.9	89 / 1.4 / 2.0	79 / 1.4 / 2.0	71 / 1.4 / 1.9	64 / 1.4 / 1.9	57 / 1.3 / 1.8	51 / 1.2 / 1.6	46 / 1.1 / 1.5	40 / 1.0 / 1.2	35 / 0.8 / 1.0	30 / 0.6 / 0.7	26 / 0.3 / 0.3					
168-D1	13.00 / 3.75			180 / 1.6 / 2.3	162 / 1.7 / 2.4	146 / 1.8 / 2.5	132 / 1.9 / 2.6	119 / 2.0 / 2.6	108 / 2.0 / 2.7	98 / 2.0 / 2.7	89 / 2.0 / 2.7	80 / 2.0 / 2.6	72 / 2.0 / 2.5	64 / 1.9 / 2.4	58 / 1.8 / 2.3	52 / 1.7 / 2.1	47 / 1.5 / 1.9	42 / 1.4 / 1.7	38 / 1.2 / 1.4	34 / 0.9 / 1.1	30 / 0.6 / 0.7	26 / 0.3 / 0.2		
188-D1	14.39 / 4.00							135 / 2.1 / 2.9	122 / 2.2 / 3.0	111 / 2.2 / 3.0	100 / 2.3 / 3.0	90 / 2.3 / 3.0	81 / 2.3 / 3.0	73 / 2.3 / 2.9	66 / 2.2 / 2.8	59 / 2.1 / 2.6	53 / 2.0 / 2.4	47 / 1.8 / 2.2	43 / 1.6 / 1.9	39 / 1.4 / 1.6	35 / 1.1 / 1.3	31 / 0.8 / 0.9	28 / 0.5 / 0.4

12DT28 + 2

Table of safe superimposed service load (psf) and cambers (in.) — 2 in. Normal Weight Topping

Strand Pattern	y_s(end) in. / y_s(center) in.	40	42	44	46	48	50	52	54	56	58	60	62	64	66	68	70	72
108-S	6.00 / 6.00	127 / 0.8 / 0.8	110 / 0.9 / 0.8	95 / 0.9 / 0.8	82 / 0.9 / 0.8	70 / 0.9 / 0.7	60 / 0.9 / 0.6	51 / 0.9 / 0.5	42 / 0.8 / 0.3	35 / 0.8 / 0.1	29 / 0.7 / −0.1							
128-S	7.00 / 7.00	154 / 1.0 / 1.0	134 / 1.0 / 1.0	117 / 1.1 / 1.0	102 / 1.1 / 1.0	88 / 1.1 / 1.0	77 / 1.1 / 0.9	66 / 1.1 / 0.8	57 / 1.1 / 0.7	49 / 1.1 / 0.5	41 / 1.0 / 0.3	32 / 0.9 / 0.0						
148-S	8.00 / 8.00	177 / 1.1 / 1.1	155 / 1.1 / 1.2	136 / 1.2 / 1.2	119 / 1.2 / 1.2	105 / 1.3 / 1.1	92 / 1.3 / 1.1	80 / 1.3 / 1.0	70 / 1.3 / 0.9	60 / 1.3 / 0.8	50 / 1.2 / 0.6	41 / 1.2 / 0.3	32 / 1.1 / 0.1					
168-S	9.00 / 9.00	197 / 1.1 / 1.2	173 / 1.2 / 1.3	152 / 1.3 / 1.3	134 / 1.3 / 1.3	118 / 1.4 / 1.3	104 / 1.4 / 1.2	90 / 1.4 / 1.2	78 / 1.4 / 1.1	66 / 1.4 / 0.9	56 / 1.4 / 0.8	47 / 1.3 / 0.5	39 / 1.3 / 0.3	31 / 1.2 / 0.0				
168-D1	13.00 / 3.75			199 / 1.5 / 1.7	177 / 1.6 / 1.7	157 / 1.7 / 1.7	140 / 1.8 / 1.8	125 / 1.9 / 1.7	111 / 1.9 / 1.7	97 / 2.0 / 1.6	84 / 2.0 / 1.5	72 / 2.0 / 1.3	62 / 2.0 / 1.1	52 / 1.9 / 0.8	43 / 1.8 / 0.5	36 / 1.8 / 0.2	30 / 1.6 / −0.2	
188-D1	14.39 / 4.00						143 / 2.0 / 1.9	126 / 2.1 / 1.9	111 / 2.1 / 1.9	97 / 2.2 / 1.8	85 / 2.2 / 1.7	73 / 2.3 / 1.5	63 / 2.3 / 1.3	54 / 2.2 / 1.0	45 / 2.1 / 0.7	37 / 2.0 / 0.3	31 / 1.9 / −0.2

Strength is based on strain compatibility; bottom tension is limited to $12\sqrt{f'_c}$; see pages 2–7 through 2–10 for explanation. Shaded values require release strengths higher than 3500 psi.

Strand Pattern Designation

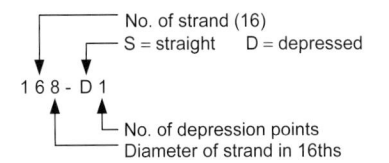

```
          No. of strand (16)
        S = straight   D = depressed
     1 6 8 - D 1
                   No. of depression points
                   Diameter of strand in 16ths
```

Safe loads shown include dead load of 10 psf for untopped members and 15 psf for topped members. Remainder is live load. Long-time cambers include superimposed dead load but do not include live load.

Key
142 – Safe superimposed service load, psf
1.4 – Estimated camber at erection, in.
1.8 – Estimated long-time camber, in.

DOUBLE TEE
12'-0" x 28"
Lightweight Concrete

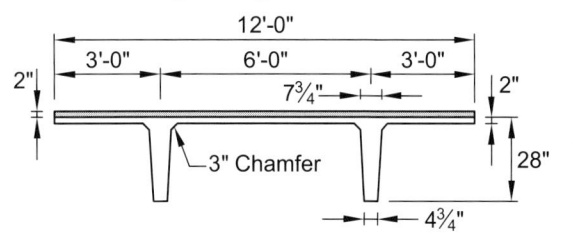

f'_c = 5,000 psi
f_{pu} = 270,000 psi

Section Properties

	Untopped	Topped
A =	640 in.²	—
I =	44,563 in.⁴	61,410 in.⁴
y_b =	20.21 in.	23.19 in.
y_t =	7.79 in.	6.81 in.
S_b =	2,227 in.³	2,648 in.³
S_t =	5,577 in.³	9,018 in.³
wt =	511 plf	822 plf
DL =	43 psf	68 psf
V/S =	1.62 in.	

12LDT28

Table of safe superimposed service load (psf) and cambers (in.) — No Topping

Strand Pattern	y_s(end) in. / y_s(center) in.	40	42	44	46	48	50	52	54	56	58	60	62	64	66	68	70	72	74	76	78	80	82	84	86	88	90
108-S	6.00 / 6.00	142 1.4 1.8	126 1.4 1.9	112 1.5 2.0	100 1.6 2.1	89 1.7 2.2	80 1.7 2.3	71 1.7 2.4	64 1.8 2.4	57 1.8 2.4	51 1.7 2.4	46 1.7 2.3	41 1.7 2.2	36 1.7 2.1	32 1.6 1.9	28 1.4 1.7	25 1.3 1.4		1.1								
128-S	7.00 / 7.00	167 1.6 2.1	148 1.7 2.3	132 1.8 2.4	118 1.9 2.5	106 1.9 2.6	95 2.0 2.7	86 2.1 2.8	77 2.1 2.8	70 2.2 2.9	63 2.2 2.9	56 2.2 2.9	51 2.1 2.9	46 2.1 2.8	41 2.0 2.7	37 1.9 2.5	33 1.8 2.4	29 1.6 2.1	26 1.4 1.7								
148-S	8.00 / 8.00	188 1.7 2.3	167 1.9 2.5	150 2.0 2.7	134 2.1 2.8	121 2.2 2.9	109 2.3 3.0	98 2.4 3.1	89 2.4 3.2	80 2.5 3.2	73 2.5 3.3	66 2.5 3.3	60 2.5 3.3	54 2.4 3.3	49 2.3 3.2	44 2.2 3.1	40 2.1 2.9	36 1.9 2.7	32 1.7 2.5	29 1.4 2.2	26 1.8						
168-S	9.00 / 9.00		183 1.9 2.6	164 2.1 2.8	148 2.2 3.0	133 2.3 3.1	120 2.4 3.2	109 2.5 3.3	98 2.6 3.4	89 2.7 3.5	81 2.7 3.6	74 2.7 3.6	67 2.8 3.6	61 2.7 3.6	55 2.7 3.5	50 2.6 3.4	45 2.5 3.3	41 2.4 3.1	37 2.2 2.9	33 2.0 2.6	30 1.8 2.3	27 1.5 1.9					
168-D1	13.00 / 3.75							156 3.0 4.0	141 3.1 4.2	129 3.3 4.4	118 3.4 4.6	107 3.5 4.7	98 3.7 4.9	90 3.8 5.0	82 3.8 5.0	74 3.9 5.1	67 3.9 5.0	61 3.9 4.9	56 3.8 4.8	50 3.7 4.6	45 3.6 4.4	41 3.4 4.1	37 3.2 3.7	33 2.9 3.3	30 2.6 2.9	28 2.3 2.5	25 1.9 2.0
188-D1	14.39 / 4.00											100 4.1 5.4	91 4.2 5.5	83 4.3 5.6	76 4.3 5.7	69 4.4 5.7	63 4.4 5.6	57 4.4 5.5	52 4.3 5.4	47 4.2 5.1	43 4.0 4.8	39 3.7 4.4	35 3.5 4.0	31 3.1 3.4	28 2.8 2.9	26 2.4 2.3	

12LDT28 + 2

Table of safe superimposed service load (psf) and cambers (in.) — 2 in. Normal Weight Topping

Strand Pattern	y_s(end) in. / y_s(center) in.	40	42	44	46	48	50	52	54	56	58	60	62	64	66	68	70	72	74	76	78
108-S	6.00 / 6.00	137 1.4 1.4	120 1.5 1.4	104 1.6 1.4	91 1.6 1.4	80 1.7 1.4	69 1.7 1.3	60 1.8 1.2	52 1.8 1.1	45 1.8 0.9	38 1.8 0.7	32 1.8 0.5	27 1.7 0.2								
128-S	7.00 / 7.00	164 1.6 1.7	144 1.7 1.7	127 1.8 1.8	111 1.9 1.8	98 2.0 1.8	86 2.1 1.7	76 2.1 1.7	67 2.2 1.6	59 2.2 1.4	51 2.2 1.3	44 2.2 1.0	38 2.2 0.8	31 2.2 0.4	25 2.1 0.1						
148-S	8.00 / 8.00	187 1.8 1.9	165 1.9 2.0	146 2.0 2.0	129 2.1 2.1	114 2.2 2.1	101 2.3 2.1	90 2.4 2.0	80 2.5 1.9	70 2.5 1.8	61 2.6 1.7	52 2.6 1.5	45 2.6 1.2	38 2.6 0.9	31 2.5 0.6	25 2.4 0.1					
168-S	9.00 / 9.00		183 2.0 2.1	162 2.1 2.2	144 2.2 2.2	128 2.4 2.2	114 2.5 2.2	101 2.6 2.2	90 2.7 2.2	78 2.7 2.1	67 2.8 1.9	58 2.8 1.7	50 2.8 1.5	42 2.8 1.2	36 2.8 0.9	29 2.8 0.5					
168-D1	13.00 / 3.75							135 3.2 3.0	121 3.3 3.0	109 3.5 3.0	98 3.6 3.0	88 3.7 2.9	77 3.8 2.8	68 3.9 2.6	59 4.0 2.3	51 4.0 2.0	43 4.0 1.6	36 3.9 1.1	30 3.9 0.6		
188-D1	14.39 / 4.00											89 4.1 3.2	79 4.3 3.0	69 4.3 2.8	60 4.4 2.6	52 4.5 2.2	45 4.5 1.9	38 4.5 1.4	32 4.5 0.8	26 4.3 0.1	

Strength is based on strain compatibility; bottom tension is limited to $12\sqrt{f'_c}$; see pages 2–7 through 2–10 for explanation.
Shaded values require release strengths higher than 3500 psi.

Strand Pattern Designation

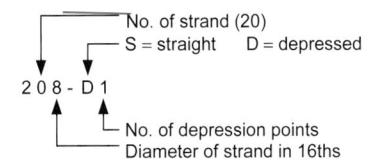

```
        ┌─ No. of strand (20)
        │  ┌─ S = straight    D = depressed
        ▼  ▼
      2 0 8 - D 1
      ▲     ▲
      │     └─ No. of depression points
      └─ Diameter of strand in 16ths
```

Safe loads shown include dead load of 10 psf for untopped members and 15 psf for topped members. Remainder is live load. Long-time cambers include superimposed dead load but do not include live load.

Key
193 – Safe superimposed service load, psf
0.8 – Estimated camber at erection, in.
1.2 – Estimated long-time camber, in.

DOUBLE TEE
12'-0" x 32"
Normal Weight Concrete

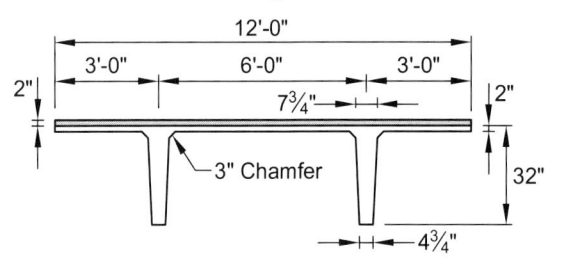

f'_c = 5,000 psi
f_{pu} = 270,000 psi

Section Properties

	Untopped	Topped
A =	690 in.²	—
I =	64,620 in.⁴	82,413 in.⁴
y_b =	22.75 in.	25.50 in.
y_t =	9.25 in.	8.75 in.
S_b =	2,840 in.³	3,232 in.³
S_t =	6,986 in.³	9,696 in.³
wt =	719 plf	1,019 plf
DL =	60 psf	85 psf
V/S =	1.70 in.	

12DT32

Table of safe superimposed service load (psf) and cambers (in.) — No Topping

Strand Pattern	y_s(end) in. / y_s(center) in.	40	42	44	46	48	50	52	54	56	58	60	62	64	66	68	70	72	74	76	78	80	82	84	86	88	90	92	94	96
128-S	7.00 / 7.00	193 0.8 1.2	171 0.9 1.2	151 0.9 1.3	135 1.0 1.4	120 1.0 1.4	107 1.0 1.4	96 1.1 1.5	85 1.1 1.5	76 1.1 1.5	68 1.0 1.4	61 1.0 1.4	54 1.0 1.3	48 0.9 1.2	42 0.8 1.1	37 0.7 0.9	33 0.6 0.7	28 0.4 0.5												
148-S	8.00 / 8.00		196 1.0 1.4	174 1.1 1.5	156 1.1 1.5	139 1.2 1.6	125 1.2 1.7	112 1.3 1.7	101 1.3 1.7	90 1.3 1.7	81 1.3 1.7	73 1.3 1.7	65 1.2 1.7	59 1.2 1.6	52 1.1 1.5	47 1.0 1.4	42 0.9 1.2	37 0.8 1.0	33 0.6 0.7	29 0.4 0.5										
168-S	9.00 / 9.00			194 1.2 1.6	174 1.2 1.6	156 1.3 1.8	140 1.3 1.8	126 1.4 1.9	114 1.4 1.9	103 1.4 1.9	93 1.4 2.0	84 1.4 2.0	75 1.4 1.9	68 1.4 1.9	61 1.3 1.8	55 1.3 1.7	49 1.2 1.6	44 1.0 1.4	40 0.9 1.1	35 0.7 0.9	31 0.5 0.6	27 0.3 0.2								
188-S	10.00 / 10.00					189 1.3 1.8	170 1.4 1.8	153 1.4 1.9	138 1.5 2.0	125 1.5 2.0	113 1.5 2.1	102 1.6 2.1	93 1.6 2.1	84 1.6 2.1	76 1.5 2.1	69 1.5 2.0	62 1.4 1.9	56 1.3 1.8	51 1.2 1.6	46 1.1 1.4	40 0.9 1.2	35 0.7 0.9	31 0.5 0.6	26 0.3 0.2						
208-D1	15.50 / 4.25											140 2.3 3.2	128 2.4 3.2	117 2.4 3.3	107 2.5 3.3	97 2.5 3.3	88 2.5 3.3	80 2.5 3.2	73 2.4 3.1	66 2.3 3.0	59 2.2 2.8	53 2.1 2.5	49 1.9 2.3	44 1.7 2.0	40 1.5 1.7	37 1.2 1.4	33 1.0 1.0	30 0.6 0.6	26 0.2 0.0	
228-D1	16.41 / 4.50														107 2.7 3.7	98 2.7 3.7	89 2.8 3.7	81 2.7 3.6	74 2.7 3.5	67 2.6 3.4	61 2.6 3.2	55 2.4 3.0	50 2.2 2.7	45 2.0 2.3	41 1.8 2.0	37 1.5 1.6	33 1.2 1.2	30 0.9 0.8	28 0.5 0.3

12DT32 + 2

Table of safe superimposed service load (psf) and cambers (in.) — 2 in. Normal Weight Topping

Strand Pattern	y_s(end) in. / y_s(center) in.	40	42	44	46	48	50	52	54	56	58	60	62	64	66	68	70	72	74	76	78	80	82	84
128-S	7.00 / 7.00	190 0.8 0.9	167 0.9 1.0	146 0.9 1.0	128 1.0 1.0	113 1.0 1.0	99 1.0 1.0	87 1.1 0.9	76 1.1 0.9	66 1.1 0.8	57 1.0 0.7	49 1.0 0.5	42 1.0 0.3	36 0.9 0.1	30 0.8 −0.2									
148-S	8.00 / 8.00		194 1.0 1.1	171 1.1 1.1	151 1.1 1.2	133 1.2 1.2	118 1.2 1.2	104 1.3 1.2	92 1.3 1.1	81 1.3 1.0	71 1.3 0.9	62 1.3 0.8	54 1.2 0.7	47 1.2 0.5	40 1.1 0.2	32 1.0 −0.1								
168-S	9.00 / 9.00			192 1.2 1.3	170 1.2 1.3	151 1.3 1.3	135 1.3 1.3	120 1.4 1.3	106 1.4 1.3	94 1.4 1.2	84 1.4 1.1	74 1.4 1.0	65 1.4 0.9	56 1.3 0.7	48 1.3 0.5	39 1.2 0.3	32 1.2 0.0							
188-S	10.00 / 10.00				187 1.3 1.4	167 1.4 1.4	149 1.4 1.4	133 1.5 1.4	119 1.5 1.4	106 1.5 1.4	94 1.6 1.3	82 1.6 1.2	71 1.6 1.1	62 1.5 0.9	53 1.4 0.7	45 1.4 0.5	38 1.3 0.2	30 1.2 −0.1						
208-D1	15.50 / 4.25											132 2.3 2.2	118 2.4 2.1	105 2.4 2.0	93 2.5 1.9	82 2.5 1.8	72 2.5 1.6	63 2.5 1.4	54 2.4 1.1	46 2.3 0.7	39 2.2 0.3	32 2.1 −0.1	27 1.9 −0.6	
228-D1	16.41 / 4.50														93 2.7 2.1	83 2.7 2.0	73 2.8 1.8	64 2.7 1.5	55 2.7 1.2	48 2.6 0.9	41 2.6 0.5	34 2.4 0.0	28 2.2 −0.5

Strength is based on strain compatibility; bottom tension is limited to $12\sqrt{f'_c}$; see pages 2–7 through 2–10 for explanation. Shaded values require release strengths higher than 3500 psi.

Strand Pattern Designation

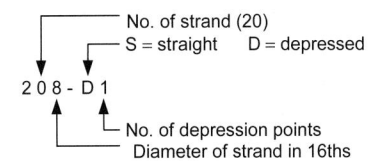

```
         No. of strand (20)
         S = straight   D = depressed
208-D1
         No. of depression points
         Diameter of strand in 16ths
```

Safe loads shown include dead load of 10 psf for untopped members and 15 psf for topped members. Remainder is live load. Long-time cambers include superimposed dead load but do not include live load.

Key
181 – Safe superimposed service load, psf
1.5 – Estimated camber at erection, in.
2.0 – Estimated long-time camber, in.

DOUBLE TEE
12'-0" x 32"
Lightweight Concrete

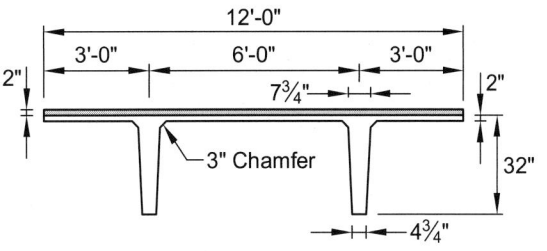

12'-0"
3'-0" | 6'-0" | 3'-0"
2"
7¾"
3" Chamfer
2"
32"
4¾"

$f'_c = 5{,}000$ psi
$f_{pu} = 270{,}000$ psi

Section Properties

	Untopped	Topped
A =	690 in.²	—
I =	64,620 in.⁴	88,305 in.⁴
y_b =	22.75 in.	26.08 in.
y_t =	9.25 in.	7.92 in.
S_b =	2,840.4 in.³	3,385.9 in.³
S_t =	6,985.9 in.³	11,149.6 in.³
wt =	551 plf	851 plf
DL =	46 psf	71 psf
V/S =	1.70 in.	

12LDT32

Table of safe superimposed service load (psf) and cambers (in.) — No Topping

Strand Pattern	y_s(end) in. / y_s(center) in.	Span, ft
		42 44 46 48 50 52 54 56 58 60 62 64 66 68 70 72 74 76 78 80 82 84 86 88 90 92 94 96 98 100 102
128-S	7.00 / 7.00	181 162 145 131 118 106 96 87 79 71 64 58 53 48 43 39 35 31 28 1.5 1.6 1.6 1.7 1.8 1.9 1.9 2.0 2.0 2.0 2.0 2.0 2.0 2.0 1.9 1.8 1.7 1.5 1.3 2.0 2.1 2.2 2.3 2.4 2.5 2.6 2.6 2.7 2.7 2.7 2.7 2.7 2.6 2.5 2.4 2.2 2.0 1.7
148-S	8.00 / 8.00	185 166 150 135 122 111 101 92 83 76 69 63 57 52 47 43 39 35 32 29 26 1.7 1.9 2.0 2.0 2.1 2.2 2.3 2.3 2.4 2.4 2.4 2.4 2.4 2.4 2.3 2.2 2.1 1.9 1.7 1.5 1.2 2.4 2.5 2.6 2.7 2.8 2.9 3.0 3.1 3.1 3.2 3.2 3.2 3.2 3.1 3.0 2.9 2.7 2.6 2.3 1.9 1.5
168-S	9.00 / 9.00	184 166 151 137 124 113 103 94 86 78 72 66 60 55 50 46 42 38 34 31 28 25 2.0 2.1 2.2 2.3 2.4 2.5 2.6 2.7 2.7 2.7 2.7 2.7 2.7 2.7 2.6 2.5 2.4 2.2 2.0 1.8 1.5 1.2 2.7 2.9 3.0 3.1 3.3 3.3 3.4 3.5 3.5 3.6 3.6 3.6 3.6 3.5 3.4 3.3 3.1 2.9 2.6 2.3 1.9 1.4
188-S	10.00 / 10.00	200 181 164 149 135 123 113 103 94 86 79 72 66 60 55 51 46 42 38 35 32 29 26 2.1 2.2 2.3 2.4 2.6 2.7 2.7 2.8 2.9 2.9 3.0 3.0 2.9 2.9 2.8 2.7 2.5 2.4 2.2 1.9 1.6 1.3 2.8 3.0 3.1 3.3 3.4 3.5 3.6 3.7 3.8 3.8 3.9 3.9 3.9 3.8 3.7 3.6 3.5 3.3 3.0 2.7 2.4 2.0 1.5
208-D1	15.50 / 4.25	107 98 91 83 77 70 65 59 54 50 45 44 38 34 31 28 26 4.4 4.5 4.6 4.7 4.7 4.8 4.8 4.8 4.7 4.5 4.3 4.1 3.9 3.5 3.2 2.8 2.4 5.9 6.0 6.1 6.2 6.2 6.2 6.1 6.0 5.8 5.6 5.3 4.9 4.4 3.9 3.4 2.8 2.2
228-D1	16.41 / 4.50	84 78 72 66 61 56 51 47 43 39 36 32 29 26 5.2 5.2 5.3 5.3 5.2 5.2 5.1 5.0 4.8 4.5 4.2 3.8 3.4 2.9 6.8 6.9 6.8 6.8 6.7 6.5 6.3 6.1 5.7 5.3 4.8 4.2 3.5 2.7

12LDT32 + 2

Table of safe superimposed service load (psf) and cambers (in.) — 2 in. Normal Weight Topping

Strand Pattern	y_s(end) in. / y_s(center) in.	Span, ft
		42 44 46 48 50 52 54 56 58 60 62 64 66 68 70 72 74 76 78 80 82 84 86 88 90
128-S	7.00 / 7.00	177 157 139 123 109 97 86 76 68 60 53 46 40 35 30 25 1.5 1.6 1.6 1.7 1.8 1.9 1.9 2.0 2.0 2.0 2.0 2.0 2.0 1.9 1.8 1.6 1.6 1.7 1.7 1.7 1.7 1.6 1.6 1.5 1.4 1.2 1.0 0.8 0.5 0.2 −0.2
148-S	8.00 / 8.00	181 161 144 128 115 102 92 82 73 65 58 51 45 39 33 27 1.7 1.9 2.0 2.0 2.1 2.2 2.3 2.3 2.4 2.4 2.4 2.4 2.4 2.4 2.3 2.2 1.9 1.9 2.0 2.0 2.0 2.0 1.9 1.9 1.8 1.7 1.5 1.3 1.1 0.8 0.4 0.0
168-S	9.00 / 9.00	181 162 145 130 117 105 94 84 76 67 59 51 44 38 32 27 2.0 2.1 2.2 2.3 2.4 2.5 2.6 2.7 2.7 2.7 2.7 2.7 2.7 2.7 2.6 2.5 2.1 2.2 2.2 2.3 2.3 2.2 2.2 2.1 2.0 1.9 1.7 1.5 1.2 0.9 0.5 0.1
188-S	10.00 / 10.00	198 177 159 143 129 116 105 94 83 73 64 56 49 42 36 30 25 2.1 2.2 2.3 2.4 2.6 2.7 2.7 2.8 2.9 2.9 3.0 3.0 3.0 2.9 2.9 2.8 2.7 2.2 2.3 2.4 2.4 2.4 2.4 2.4 2.3 2.2 2.1 1.9 1.7 1.5 1.2 0.8 0.4 −0.1
208-D1	15.50 / 4.25	98 88 78 69 61 54 47 41 35 29 4.4 4.5 4.6 4.7 4.7 4.8 4.8 4.8 4.7 4.5 3.6 3.5 3.3 3.1 2.8 2.5 2.1 1.6 1.1 0.4
228-D1	16.41 / 4.50	71 63 56 49 43 37 31 26 5.2 5.2 5.3 5.3 5.2 5.2 5.1 5.0 3.4 3.1 2.8 2.3 1.9 1.3 0.7 −0.1

Strength is based on strain compatibility; bottom tension is limited to $12\sqrt{f'_c}$; see pages 2–7 through 2–10 for explanation.
Shaded values require release strengths higher than 3500 psi.

Strand Pattern Designation

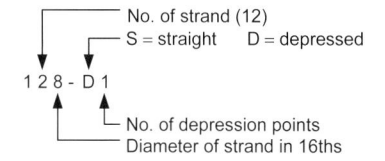

```
        No. of strand (12)
      S = straight   D = depressed
1 2 8 - D 1
          No. of depression points
     Diameter of strand in 16ths
```

Because these units are pretopped and are typically used in parking structures, safe loads shown do not include any superimposed dead loads. Loads shown are live load. Long-time cambers do not include live load.

Key
171 – Safe superimposed service load, psf
0.5 – Estimated camber at erection, in.
0.7 – Estimated long-time camber, in.

PRETOPPED DOUBLE TEE
10'-0" x 26"

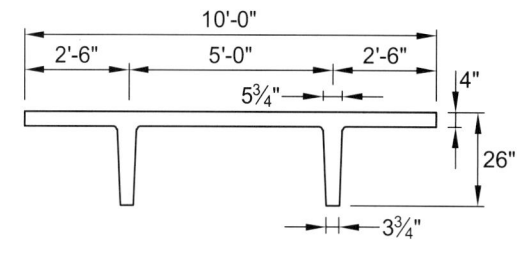

10'-0"
2'-6" 5'-0" 2'-6"
5¾"
4"
26"
3¾"

$f'_c = 5,000$ psi
$f_{pu} = 270,000$ psi

Section Properties

	Normal Weight	Lightweight
A =	689 in.²	689 in.²
I =	30,716 in.⁴	30,716 in.⁴
y_b =	20.29 in.	20.29 in.
y_t =	5.71 in.	5.71 in.
S_b =	1,514 in.³	1,514 in.³
S_t =	5,379 in.³	5,379 in.³
wt =	718 plf	550 plf
DL =	72 psf	55 psf
V/S =	2.05 in.	2.05 in.

10DT26

Table of safe superimposed service load (psf) and cambers (in.) — Normal Weight — No Topping

Strand Pattern	y_s(end) in. / y_s(center) in.	30	32	34	36	38	40	42	44	46	48	50	52	54	56	58	60	62	64	66	68	70	72
68-S	4.00 / 4.00	171 / 0.5 / 0.7	143 / 0.5 / 0.7	121 / 0.6 / 0.8	102 / 0.6 / 0.8	86 / 0.6 / 0.8	72 / 0.6 / 0.8	61 / 0.6 / 0.8	50 / 0.6 / 0.8	42 / 0.5 / 0.7	34 / 0.4 / 0.6	27 / 0.3 / 0.4											
88-S	5.00 / 5.00		196 / 0.7 / 1.0	167 / 0.8 / 1.1	143 / 0.8 / 1.2	123 / 0.9 / 1.2	106 / 0.9 / 1.3	91 / 0.9 / 1.3	78 / 0.9 / 1.3	67 / 0.9 / 1.3	57 / 0.9 / 1.2	48 / 0.8 / 1.2	41 / 0.8 / 1.0	34 / 0.7 / 0.9	28 / 0.5 / 0.7								
108-S	6.00 / 6.00				180 / 1.0 / 1.4	156 / 1.1 / 1.5	135 / 1.2 / 1.6	118 / 1.2 / 1.7	103 / 1.2 / 1.7	89 / 1.3 / 1.7	78 / 1.3 / 1.7	67 / 1.3 / 1.8	58 / 1.2 / 1.7	50 / 1.2 / 1.7	43 / 1.1 / 1.6	36 / 0.9 / 1.4	30 / 0.8 / 1.2						
128-S	7.00 / 7.00					184 / 1.3 / 1.7	161 / 1.3 / 1.8	141 / 1.4 / 1.9	123 / 1.5 / 2.0	108 / 1.5 / 2.1	95 / 1.5 / 2.1	83 / 1.6 / 2.1	73 / 1.6 / 2.1	64 / 1.5 / 2.1	56 / 1.5 / 2.0	48 / 1.4 / 1.9	41 / 1.3 / 1.7	35 / 1.1 / 1.4	29 / 0.9 / 1.1				
128-D1	11.67 / 3.25						194 / 1.4 / 2.0	171 / 1.5 / 2.1	151 / 1.6 / 2.2	134 / 1.7 / 2.3	119 / 1.8 / 2.4	105 / 1.8 / 2.4	93 / 1.8 / 2.4	82 / 1.8 / 2.3	73 / 1.7 / 2.2	64 / 1.7 / 21	56 / 1.6 / 2.0	50 / 1.4 / 1.8	43 / 1.3 / 1.6	37 / 1.0 / 1.3	31 / 0.8 / 0.9	25 / 0.5 / 0.4	
148-D1	12.86 / 3.50											127 / 2.1 / 2.8	113 / 2.1 / 2.9	101 / 2.2 / 2.9	89 / 2.2 / 2.9	78 / 2.2 / 2.9	68 / 2.1 / 2.7	59 / 2.0 / 2.6	872 / 1.9 / 2.4	46 / 1.7 / 2.2	40 / 1.5 / 1.9	36 / 1.3 / 1.6	30 / 1.0 / 1.1

10LDT26

Table of safe superimposed service load (psf) and cambers (in.) — Lightweight — No Topping

Strand Pattern	y_s(end) in. / y_s(center) in.	30	32	34	36	38	40	42	44	46	48	50	52	54	56	58	60	62	64	66	68	70	72	74	76	78
68-S	4.00 / 4.00	183 / 0.8 / 1.1	156 / 0.9 / 1.2	133 / 1.0 / 1.3	115 / 1.0 / 1.4	99 / 1.1 / 1.4	85 / 1.1 / 1.5	73 / 1.1 / 1.6	63 / 1.2 / 1.6	54 / 1.2 / 1.6	46 / 1.1 / 1.6	40 / 1.1 / 1.5	33 / 1.0 / 1.4	28 / 0.9 / 1.2												
88-S	5.00 / 5.00				180 / 1.3 / 1.7	156 / 1.4 / 1.9	136 / 1.5 / 2.0	119 / 1.6 / 2.1	104 / 1.6 / 2.2	91 / 1.7 / 2.3	80 / 1.8 / 2.3	70 / 1.8 / 2.4	61 / 1.8 / 2.4	53 / 1.8 / 2.4	46 / 1.8 / 2.4	40 / 1.7 / 2.3	35 / 1.6 / 2.2	30 / 1.4 / 1.9	25 / 1.2 / 1.6							
108-S	6.00 / 6.00				192 / 1.7 / 2.3	168 / 1.8 / 2.5	148 / 1.9 / 2.6	130 / 2.1 / 2.8	115 / 2.2 / 2.9	102 / 2.3 / 3.0	90 / 2.3 / 3.1	80 / 2.4 / 3.1	71 / 2.4 / 3.2	63 / 2.4 / 3.2	55 / 2.4 / 3.2	49 / 2.4 / 3.2	43 / 2.3 / 3.1	37 / 2.2 / 3.0	33 / 2.0 / 2.8	28 / 1.8 / 2.4						
128-S	7.00 / 7.00					196 / 2.0 / 2.8	173 / 2.2 / 3.0	153 / 2.3 / 3.2	136 / 2.5 / 3.3	121 / 2.6 / 3.5	108 / 2.7 / 3.6	96 / 2.8 / 3.7	86 / 2.9 / 3.8	76 / 2.9 / 3.8	68 / 3.0 / 3.9	61 / 3.0 / 3.9	54 / 2.9 / 3.9	48 / 2.9 / 3.8	42 / 2.8 / 3.7	37 / 2.7 / 3.6	33 / 2.5 / 3.3	29 / 2.2 / 3.0				
128-D1	11.67 / 3.25									146 / 2.8 / 3.8	131 / 3.0 / 4.0	118 / 3.1 / 4.1	106 / 3.2 / 4.3	95 / 3.3 / 4.4	85 / 3.4 / 4.5	76 / 3.5 / 4.5	68 / 3.4 / 4.4	60 / 3.4 / 4.3	53 / 3.3 / 4.1	46 / 3.2 / 3.9	41 / 3.0 / 3.6	36 / 2.8 / 3.3	33 / 2.5 / 3.0	29 / 2.2 / 2.6	26 / 1.8 / 2.2	
148-D1	12.86 / 3.50													91 / 4.0 / 5.4	81 / 4.1 / 5.4	72 / 4.2 / 5.4	65 / 4.2 / 5.4	58 / 4.2 / 5.3	51 / 4.1 / 5.1	45 / 3.9 / 4.8	40 / 3.7 / 4.4	34 / 3.4 / 4.0	31 / 3.1 / 3.5	27 / 2.8 / 3.1		

Strength is based on strain compatibility; bottom tension is limited to $12\sqrt{f'_c}$; see pages 2–7 through 2–10 for explanation.
Shaded values require release strengths higher than 3500 psi.

Strand Pattern Designation

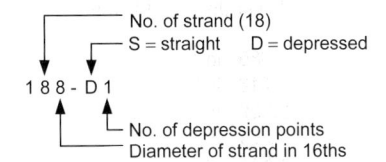

```
        No. of strand (18)
        S = straight   D = depressed
1 8 8 - D 1
        No. of depression points
        Diameter of strand in 16ths
```

Because these units are pretopped and are typically used in parking structures, safe loads shown do not include any superimposed dead loads. Loads shown are live load. Long-time cambers do not include live load.

Key
187 – Safe superimposed service load, psf
0.8 – Estimated camber at erection, in.
1.2 – Estimated long-time camber, in.

PRETOPPED DOUBLE TEE
10'-0" x 34"

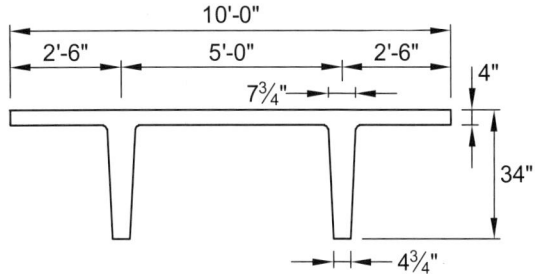

10'-0" (2'-6" + 5'-0" + 2'-6"), 4" flange, 34" overall, 7¾" top of stem, 4¾" bottom of stem.

$f'_c = 5{,}000$ psi
$f_{pu} = 270{,}000$ psi

Section Properties

	Normal Weight	Lightweight
A =	855 in.²	855 in.²
I =	80,780 in.⁴	80,780 in.⁴
y_b =	25.07 in.	25.07 in.
y_t =	8.93 in.	8.93 in.
S_b =	3,222 in.³	3,222 in.³
S_t =	9,046 in.³	9,046 in.³
wt =	891 plf	683 plf
DL =	89 psf	68 psf
V/S =	2.32 in.	2.32 in.

10DT34

Table of safe superimposed service load (psf) and cambers (in.) — Normal Weight — No Topping

Strand Pattern	y_s(end) in. / y_s(center) in.	44	46	48	50	52	54	56	58	60	62	64	66	68	70	72	74	76	78	80	82	84	86	88	90	92	94
128-S	7.00 / 7.00	187 0.8 1.2	166 0.9 1.2	147 0.9 1.2	130 0.9 1.3	115 0.9 1.3	102 09 1.3	90 0.9 1.2	79 0.9 1.2	70 0.8 1.1	61 0.7 1.0	53 0.7 0.9	46 0.6 0.7	40 0.4 0.5	34 0.3 0.3	28 0.1 0.1											
148-S	8.00 / 8.00		193 1.0 1.4	172 1.0 1.4	153 1.1 1.5	137 1.1 1.5	122 1.1 1.5	109 1.1 1.5	97 1.1 1.5	86 1.0 1.4	76 0.9 1.3	67 0.8 1.2	59 0.8 1.0	52 0.6 0.8	45 0.5 0.6	39 0.3 0.3	34 0.1 0.0	28									
168-S	9.00 / 9.00			194 1.2 1.6	174 1.2 1.7	156 1.2 1.7	139 1.3 1.7	125 1.3 1.7	112 1.3 1.7	100 1.3 1.7	90 1.2 1.6	80 1.2 1.5	71 1.1 1.4	63 1.0 1.2	56 0.9 1.0	49 0.8 0.8	43 0.6 0.5	37 0.4 0.2	32 0.2 −0.2	27 0.0							
188-S	10.00 / 10.00				192 1.3 1.8	172 1.4 1.8	155 1.4 1.9	139 1.4 1.9	125 1.4 1.9	113 1.4 1.9	101 1.4 1.9	91 1.4 1.9	82 1.3 1.8	73 1.2 1.7	65 1.1 1.5	58 1.0 1.3	51 0.9 1.1	45 0.7 0.8	39 0.5 0.5	33 0.3 0.2	27 0.0 −0.2						
188-D1	14.39 / 4.00							184 1.8 2.4	167 1.9 2.5	152 1.9 2.5	138 1.9 2.5	125 1.9 2.5	114 1.9 2.5	103 1.8 2.4	94 1.8 2.3	85 1.7 2.2	77 1.6 2.1	69 1.5 1.9	63 1.3 1.7	56 1.2 1.5	50 1.0 1.2	44 0.7 0.8	38 0.4 0.4	32 0.1 −0.1	27 −0.3 −0.6		
208-D1	15.50 / 4.25								188 2.0 2.8	172 2.1 2.8	156 2.1 2.9	143 2.2 2.9	130 2.2 2.9	119 2.2 2.9	107 2.1 2.8	96 2.1 2.7	86 2.0 2.5	78 1.9 2.4	71 1.8 2.2	65 1.6 2.0	59 1.4 1.8	53 1.2 1.5	47 1.0 1.2	41 0.7 0.7	35 0.4 0.2	30 0.0 −0.3	25 −0.4 −0.9

10LDT34

Table of safe superimposed service load (psf) and cambers (in.) — Lightweight — No Topping

Strand Pattern	y_s(end) in. / y_s(center) in.	46	48	50	52	54	56	58	60	62	64	66	68	70	72	74	76	78	80	82	84	86	88	90	92	94	96	98	100	102
128-S	7.00 / 7.00	181 1.5 2.0	162 1.5 2.1	145 1.6 2.2	131 1.6 2.2	117 1.7 2.3	106 1.7 2.3	95 1.7 2.4	85 1.7 2.3	77 1.7 2.3	69 1.7 2.3	62 1.7 2.2	55 1.6 2.0	49 1.5 1.8	44 1.4 1.6	39 1.2 1.3	34 1.0 1.0	30 0.8 0.6	26 0.6											
148-S	8.00 / 8.00	188 1.8 2.4	169 1.8 2.5	152 1.9 2.5	137 2.0 2.6	124 2.0 2.7	112 2.1 2.8	102 2.1 2.8	92 2.1 2.8	83 2.1 2.8	75 2.1 2.8	68 2.0 2.7	61 2.0 2.7	55 1.9 2.6	49 1.8 2.4	44 1.6 2.2	39 1.4 1.9	35 1.2 1.5	31 1.0 1.2	27 0.7 0.7										
168-S	9.00 / 9.00		189 2.0 2.7	171 2.1 2.8	155 2.2 2.9	141 2.3 3.0	128 2.3 3.1	116 2.4 3.2	105 2.4 3.2	96 2.4 3.2	87 2.4 3.2	79 2.4 3.2	72 2.3 3.2	65 2.2 3.1	59 2.1 3.0	53 2.0 2.8	48 1.8 2.6	43 1.5 2.3	38 1.3 2.0	34 1.0 1.6	30 0.6 1.1	26 0.6								
188-S	10.00 / 10.00			188 2.3 3.1	170 2.4 3.2	155 2.5 3.3	141 2.5 3.4	128 2.6 3.4	117 2.7 3.5	107 2.7 3.5	97 2.7 3.6	89 2.7 3.6	81 2.7 3.5	73 2.6 3.5	67 2.6 3.4	61 2.5 3.2	55 2.3 3.1	50 2.2 2.9	45 2.0 2.6	40 1.7 2.2	36 1.5 1.8	32 1.1 1.3	28 0.8 0.8							
188-D1	14.39 / 4.00						200 3.0 4.1	183 3.1 4.2	167 3.3 4.4	154 3.4 4.5	141 3.5 4.6	129 3.6 4.7	119 3.6 4.7	109 3.6 4.7	101 3.6 4.7	92 3.6 4.6	83 3.5 4.5	76 3.4 4.4	69 3.2 4.2	62 3.1 3.9	56 2.9 3.6	52 2.6 3.4	48 2.4 3.1	44 2.1 2.8	40 1.7 2.4	37 1.3 2.0	33 0.9 1.5	30 0.3 0.8	26 0.0	
208-D1	15.50 / 4.25								187 3.5 4.8	172 3.7 4.9	158 3.8 5.1	146 3.9 5.2	134 4.0 5.3	123 4.1 5.4	113 4.2 5.4	103 4.2 5.4	94 4.2 5.4	86 4.2 5.3	78 4.1 5.1	71 4.0 4.9	65 3.8 4.7	58 3.7 4.4	53 3.4 4.0	48 3.2 3.7	44 2.9 3.3	41 2.6 2.9	37 2.3 2.5	34 1.9 2.0	31 1.4 1.4	29 0.9 0.8

Strength is based on strain compatibility; bottom tension is limited to $12\sqrt{f'_c}$; see pages 2–7 through 2–10 for explanation.
Shaded values require release strengths higher than 3500 psi.

Strand Pattern Designation

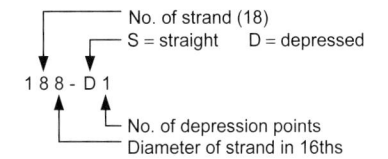

```
         No. of strand (18)
         S = straight   D = depressed
  1 8 8 - D 1
         No. of depression points
         Diameter of strand in 16ths
```

Because these units are pretopped and are typically used in parking structures, safe loads shown do not include any superimposed dead loads. Loads shown are live load. Long-time cambers do not include live load.

Key
158 – Safe superimposed service load, psf
0.8 – Estimated camber at erection, in.
1.1 – Estimated long-time camber, in.

PRETOPPED DOUBLE TEE
12'-0" x 30"

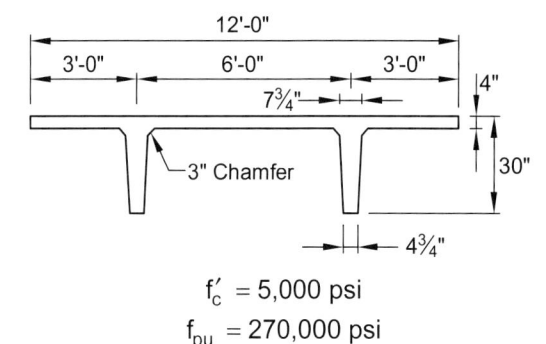

$f'_c = 5,000$ psi
$f_{pu} = 270,000$ psi

Section Properties

	Normal Weight	Lightweight
A =	928 in.^{2}	928 in.^{2}
I =	59,997 in.^{4}	59,997 in.^{4}
y_b =	22.94 in.	22.94 in.
y_t =	7.06 in.	7.06 in.
S_b =	2,615 in.^{3}	2,615 in.^{3}
S_t =	8,498 in.^{3}	8,498 in.^{3}
wt =	967 plf	741 plf
DL =	81 psf	62 psf
V/S =	2.30 in.	2.30 in.

12DT30

Table of safe superimposed service load (psf) and cambers (in.) — Normal Weight — No Topping

Strand Pattern	y_s(end) in. / y_s(center) in.	40	42	44	46	48	50	52	54	56	58	60	62	64	66	68	70	72	74	76	78	80	82
128-S	7.00 / 7.00	158 / 0.8 / 1.1	138 / 0.8 / 1.2	120 / 0.9 / 1.2	105 / 0.9 / 1.2	91 / 0.9 / 1.2	79 / 0.9 / 1.2	69 / 0.8 / 1.1	59 / 0.8 / 1.1	51 / 0.7 / 1.0	43 / 0.6 / 0.8	37 / 0.5 / 0.6	30 / 0.4 / 0.4										
148-S	8.00 / 8.00	182 / 0.9 / 1.3	160 / 1.0 / 1.3	140 / 1.0 / 1.4	123 / 1.0 / 1.4	108 / 1.0 / 1.4	95 / 1.1 / 1.5	83 / 1.0 / 1.4	73 / 1.0 / 1.4	63 / 1.0 / 1.3	55 / 0.9 / 1.2	47 / 0.8 / 1.0	41 / 0.7 / 0.9	34 / 0.5 / 0.6	29 / 0.3 / 0.4								
168-S	9.00 / 9.00		178 / 1.1 / 1.5	157 / 1.1 / 1.5	139 / 1.2 / 1.6	122 / 1.2 / 1.6	108 / 1.2 / 1.6	95 / 1.2 / 1.6	84 / 1.2 / 1.6	74 / 1.1 / 1.6	65 / 1.1 / 1.5	57 / 1.0 / 1.3	49 / 0.9 / 1.2	42 / 0.8 / 1.0	36 / 0.6 / 0.7	31 / 0.4 / 0.4							
188-S	10.00 / 10.00		194 / 1.1 / 1.6	171 / 1.2 / 1.6	152 / 1.2 / 1.7	134 / 1.3 / 1.7	119 / 1.3 / 1.8	106 / 1.3 / 1.8	93 / 1.3 / 1.8	83 / 1.3 / 1.7	73 / 1.2 / 1.7	64 / 1.1 / 1.6	56 / 1.0 / 1.4	49 / 0.9 / 1.2	43 / 0.8 / 1.0	36 / 0.6 / 0.7	29 / 0.4 / 0.4						
188-D1	14.39 / 4.00					184 / 1.6 / 2.2	165 / 1.7 / 2.3	148 / 1.7 / 2.4	132 / 1.8 / 2.4	119 / 1.8 / 2.4	107 / 1.8 / 2.4	96 / 1.8 / 2.3	86 / 1.7 / 2.2	77 / 1.7 / 2.1	68 / 1.6 / 2.0	61 / 1.4 / 1.9	54 / 1.3 / 1.7	48 / 1.1 / 1.4	42 / 0.9 / 1.1	35 / 0.6 / 0.7	29 / 0.3 / 0.2		
208-D1	15.50 / 4.25							166 / 1.9 / 2.6	149 / 2.0 / 2.7	135 / 2.0 / 2.7	121 / 2.0 / 2.7	109 / 2.1 / 2.7	97 / 2.1 / 2.7	86 / 2.0 / 2.6	76 / 1.9 / 2.4	68 / 1.8 / 2.3	61 / 1.7 / 2.1	54 / 1.5 / 1.9	48 / 1.3 / 1.7	43 / 1.1 / 1.4	37 / 0.8 / 1.0	31 / 0.5 / 0.5	26 / 0.1 / -0.1

12LDT30

Table of safe superimposed service load (psf) and cambers (in.) — Lightweight — No Topping

Strand Pattern	y_s(end) in. / y_s(center) in.	40	42	44	46	48	50	52	54	56	58	60	62	64	66	68	70	72	74	76	78	80	82	84	86	88	90
128-S	7.00 / 7.00	172 / 1.4 / 1.8	152 / 1.4 / 1.9	134 / 1.5 / 2.0	119 / 1.6 / 2.1	105 / 1.6 / 2.2	93 / 1.7 / 2.2	83 / 1.7 / 2.3	73 / 1.7 / 2.3	65 / 1.7 / 2.3	57 / 1.6 / 2.3	51 / 1.6 / 2.2	45 / 1.5 / 2.2	39 / 1.4 / 2.0	34 / 1.2 / 1.9	29 / 1.0 / 1.6	— / 1.3										
148-S	8.00 / 8.00	196 / 1.5 / 2.1	174 / 1.6 / 2.2	154 / 1.7 / 2.3	137 / 1.8 / 2.4	122 / 1.9 / 2.5	109 / 1.9 / 2.6	97 / 2.0 / 2.6	87 / 2.0 / 2.7	77 / 2.0 / 2.7	69 / 2.0 / 2.7	61 / 2.0 / 2.7	55 / 1.9 / 2.6	48 / 1.8 / 2.5	43 / 1.7 / 2.3	38 / 1.6 / 2.1	33 / 1.4 / 1.8	29 / 1.1 / 1.4									
168-S	9.00 / 9.00	192 / 1.8 / 2.4	171 / 1.9 / 2.5	153 / 2.0 / 2.6	136 / 2.1 / 2.7	122 / 2.1 / 2.8	109 / 2.2 / 2.9	98 / 2.2 / 3.0	88 / 2.3 / 3.0	79 / 2.3 / 3.0	71 / 2.3 / 3.0	63 / 2.2 / 3.0	57 / 2.2 / 2.9	50 / 2.1 / 2.8	45 / 2.0 / 2.6	40 / 1.8 / 2.4	35 / 1.6 / 2.1	31 / 1.3 / 1.7	27 / 1.0 / 1.3								
188-S	10.00 / 10.00			185 / 2.0 / 2.6	166 / 2.1 / 2.8	148 / 2.2 / 2.9	133 / 2.3 / 3.0	120 / 2.3 / 3.1	108 / 2.4 / 3.2	97 / 2.5 / 3.3	87 / 2.5 / 3.3	78 / 2.5 / 3.3	70 / 2.4 / 3.3	63 / 2.3 / 3.2	57 / 2.3 / 3.1	51 / 2.2 / 3.0	45 / 2.1 / 2.8	40 / 1.9 / 2.5	36 / 1.7 / 2.2	31 / 1.4 / 1.8	27 / 1.1 / 1.3						
188-D1	14.39 / 4.00						179 / 2.8 / 3.8	162 / 2.9 / 4.0	147 / 3.1 / 4.1	133 / 3.2 / 4.3	121 / 3.3 / 4.4	110 / 3.4 / 4.4	100 / 3.5 / 4.5	90 / 3.5 / 4.5	81 / 3.5 / 4.5	73 / 3.4 / 4.4	65 / 3.4 / 4.2	58 / 3.2 / 4.0	52 / 3.1 / 3.7	46 / 2.9 / 3.4	42 / 2.7 / 3.2	38 / 2.4 / 2.8	35 / 2.1 / 2.5	31 / 1.8 / 2.0	28 / 1.4 / 1.5		
208-D1	15.50 / 4.25											123 / 3.7 / 4.9	111 / 3.8 / 5.0	101 / 3.8 / 5.1	91 / 3.9 / 5.1	82 / 3.9 / 5.1	74 / 3.9 / 5.0	67 / 3.9 / 4.9	60 / 3.8 / 4.7	53 / 3.6 / 4.4	48 / 3.4 / 4.0	42 / 3.1 / 3.6	38 / 2.9 / 3.3	35 / 2.6 / 2.8	31 / 2.2 / 2.4	28 / 1.8 / 1.9	26 / 1.3 / 1.3

Strength is based on strain compatibility; bottom tension is limited to $12\sqrt{f'_c}$; see pages 2–7 through 2–10 for explanation.
Shaded values require release strengths higher than 3500 psi.

Strand Pattern Designation

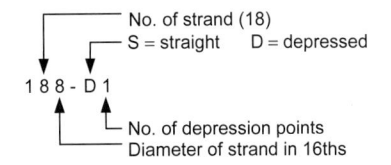

```
        No. of strand (18)
         S = straight   D = depressed
   1 8 8 - D 1
              No. of depression points
              Diameter of strand in 16ths
```

Because these units are pretopped and are typically used in parking structures, safe loads shown do not include any superimposed dead loads. Loads shown are live load. Long-time cambers do not include live load.

Key
193 – Safe superimposed service load, psf
0.7 – Estimated camber at erection, in.
1.0 – Estimated long-time camber, in.

PRETOPPED DOUBLE TEE
12'-0" x 34"

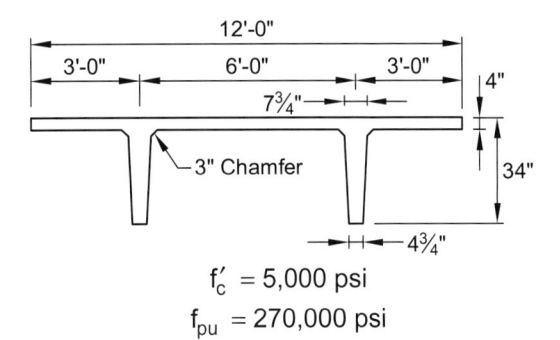

12'-0"
3'-0" | 6'-0" | 3'-0"
7¾"
4"
3" Chamfer
34"
4¾"

$f'_c = 5{,}000$ psi
$f_{pu} = 270{,}000$ psi

Section Properties

	Normal Weight	Lightweight
A =	978 in.²	978 in.²
I =	86,072 in.⁴	86,072 in.⁴
y_b =	25.77 in.	25.77 in.
y_t =	8.23 in.	8.23 in.
S_b =	3,340 in.³	3,340 in.³
S_t =	10,458 in.³	10,458 in.³
wt =	1,019 plf	781 plf
DL =	85 psf	65 psf
V/S =	2.39 in.	2.39 in.

12DT34

Table of safe superimposed service load (psf) and cambers (in.)
Normal Weight — No Topping

Strand Pattern	y_s(end) in. / y_s(center) in.	40	42	44	46	48	50	52	54	56	58	60	62	64	66	68	70	72	74	76	78	80	82	84	86	88
128-S	7.00 / 7.00	193 0.7 1.0	169 0.7 1.0	149 0.8 1.1	131 0.8 1.1	115 0.8 1.1	101 0.8 1.1	88 0.8 1.1	77 0.8 1.1	67 0.8 1.1	59 0.7 1.0	51 0.7 09	43 0.6 0.8	37 0.5 0.7	31 0.4 0.5	25 0.3 0.4	0.2 0.1									
148-S	8.00 / 8.00		197 0.9 1.2	174 0.9 1.2	154 0.9 1.3	136 1.0 1.3	120 1.0 1.4	107 1.0 1.4	94 1.0 1.4	83 1.0 1.3	73 0.9 1.3	64 0.9 1.2	56 0.8 1.1	49 0.7 1.0	42 0.6 0.8	36 0.5 0.6	30 0.3 0.4	25 0.1 0.1								
168-S	9.00 / 9.00			197 1.0 1.4	174 1.0 1.4	155 1.1 1.5	138 1.1 1.5	123 1.1 1.6	109 1.1 1.6	97 1.1 1.6	86 1.1 1.5	76 1.1 1.5	67 1.0 1.4	59 0.9 1.3	52 0.9 1.1	45 0.7 1.0	39 0.6 0.8	34 0.4 0.5	28 0.2 0.2							
188-S	10.00 / 10.00				192 1.1 1.6	171 1.2 1.6	153 1.2 1.7	137 1.3 1.7	122 1.3 1.7	109 1.3 1.8	97 1.3 1.7	87 1.2 1.7	77 1.2 1.7	69 1.1 1.6	61 1.1 1.4	53 1.0 1.3	47 0.8 1.1	41 0.7 0.8	35 0.5 0.6	30 0.3 0.3						
188-D1	14.39 / 4.00							180 1.6 2.1	162 1.6 2.2	147 1.7 2.2	132 1.7 2.2	119 1.7 2.2	108 1.7 2.2	97 1.6 2.1	88 1.6 2.1	79 1.5 2.0	71 1.5 1.9	63 1.4 1.8	57 1.2 1.6	50 1.1 1.4	45 0.9 1.1	39 0.6 0.7	34 0.4 0.3	28 0.1 –0.1		
208-D1	15.50 / 4.25							183 1.8 2.4	166 1.8 2.5	150 1.9 2.6	136 1.9 2.6	123 2.0 2.6	112 2.0 2.6	101 1.9 2.5	92 1.9 2.4	82 1.8 2.3	74 1.7 2.2	67 1.6 2.1	61 1.5 1.9	55 1.3 1.7	49 1.1 1.4	43 0.9 1.1	37 0.6 0.7	31 0.3 0.2	26 –0.1 –0.3	

12LDT34

Table of safe superimposed service load (psf) and cambers (in.)
Lightweight — No Topping

Strand Pattern	y_s(end) in. / y_s(center) in.	42	44	46	48	50	52	54	56	58	60	62	64	66	68	70	72	74	76	78	80	82	84	86	88	90	92	94	96
128-S	7.00 / 7.00	184 1.2 1.6	164 1.3 1.7	146 1.4 1.8	130 1.4 1.9	116 1.5 2.0	103 1.5 2.0	92 1.5 2.1	82 1.5 2.1	73 1.5 2.1	65 1.5 2.1	58 1.5 2.0	52 1.4 2.0	46 1.4 1.8	40 1.2 1.7	35 1.1 1.5	31 0.9 1.2	26 0.7 0.9											
148-S	8.00 / 8.00		189 1.5 2.0	169 1.6 2.1	151 1.6 2.2	135 1.7 2.3	121 1.8 2.4	109 1.8 2.4	98 1.9 2.5	88 1.9 2.5	79 1.9 2.5	71 1.9 2.5	64 1.8 2.5	57 1.8 2.4	51 1.7 2.3	45 1.6 2.2	40 1.5 2.0	35 1.3 1.7	31 1.1 1.4	27 0.9 1.1									
168-S	9.00 / 9.00			189 1.8 2.3	170 1.8 2.5	153 1.9 2.6	137 2.0 2.6	124 2.1 2.8	112 2.1 2.8	101 2.1 2.9	91 2.2 2.9	82 2.2 2.9	74 2.1 2.9	67 2.1 2.8	60 2.0 2.7	54 1.9 2.6	48 1.8 2.4	43 1.6 2.1	38 1.4 1.8	34 1.2 1.5	30 0.9 1.0	26							
188-S	10.00 / 10.00				186 2.0 2.7	168 2.1 2.8	152 2.2 2.9	137 2.2 3.0	124 2.3 3.1	112 2.4 3.1	102 2.4 3.2	92 2.4 3.2	83 2.4 3.2	76 2.4 3.2	68 2.3 3.2	62 2.3 3.1	56 2.1 3.0	50 2.0 2.9	45 1.8 2.7	40 1.6 2.4	36 1.3 2.1	32 1.0 1.7	28 1.2						
188-D1	14.39 / 4.00						195 2.6 3.5	177 2.7 3.7	161 2.8 3.8	147 3.0 3.9	134 3.1 4.1	123 3.2 4.2	112 3.2 4.2	103 3.2 4.2	94 3.2 4.2	86 3.2 4.1	78 3.1 4.0	71 3.0 3.9	64 2.9 3.7	57 2.7 3.5	53 2.6 3.3	48 2.4 3.1	44 2.1 2.9	40 1.8 2.6	36 1.5 2.2	33 1.1 1.8	29 0.6 1.2	25 0.5	
208-D1	15.50 / 4.25						198 2.9 4.0	180 3.1 4.2	165 3.2 4.3	151 3.3 4.5	138 3.4 4.6	127 3.6 4.7	116 3.6 4.8	107 3.7 4.8	98 3.8 4.9	89 3.8 4.8	80 3.7 4.7	73 3.7 4.6	66 3.6 4.4	59 3.4 4.2	53 3.3 3.9	48 3.1 3.6	44 2.9 3.4	41 2.6 3.0	37 2.3 2.7	34 2.0 2.3	31 1.6 1.8	28 1.2 1.2	

Strength is based on strain compatibility; bottom tension is limited to $12\sqrt{f'_c}$; see pages 2–7 through 2–10 for explanation.
Shaded values require release strengths higher than 3500 psi.

Strand Pattern Designation

```
         ┌─ No. of strand (20)
         │    ┌─ S = straight   D = depressed
         ▼    ▼
       2 0 8 - D 1
       ▲        ▲
       │        └─ No. of depression points
       └─ Diameter of strand in 16ths
```

Because these units are pretopped and are typically used in parking structures, safe loads shown do not include any superimposed dead loads. Loads shown are live load. Long-time cambers do not include live load.

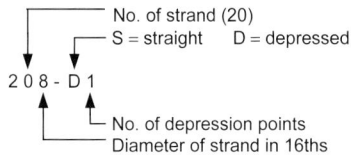

Key
194 – Safe superimposed service load, psf
0.2 – Estimated camber at erection, in.
0.4 – Estimated long-time camber, in.

PRETOPPED DOUBLE TEE
15'-0" x 26"

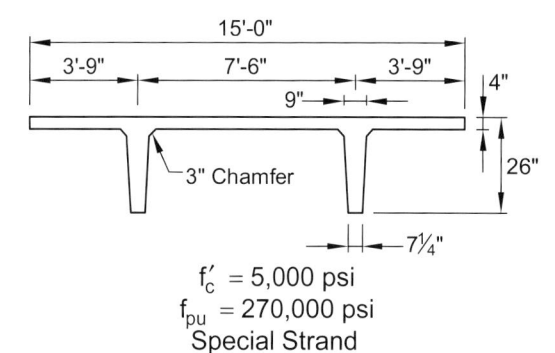

15'-0" (3'-9" | 7'-6" | 3'-9"), 9", 4", 26", 3" Chamfer, 7¼"

$f'_c = 5{,}000$ psi
$f_{pu} = 270{,}000$ psi
Special Strand

Section Properties

	Normal Weight	Lightweight
A =	1078 in.^{2}	1078 in.^{2}
I =	53280 in.^{4}	53280 in.^{4}
y_b =	19.82 in.	19.82 in.
y_t =	6.18 in.	6.18 in.
S_b =	2688 in.^{3}	2688 in.^{3}
S_t =	8618 in.^{3}	8618 in.^{3}
wt =	1122 plf	861 plf
DL =	75 psf	57 psf
V/S =	2.38 in.	2.38 in.

15DT26

Table of safe superimposed service load (psf) and cambers (in.) — Normal Weight – No Topping

Strand Pattern	y_s(end) in. / y_s(center) in.	26	28	30	32	34	36	38	40	42	44	46	48	50	52	54	56	58	60	62	64	66	68	70	72	74	76
88-S	7.00 / 7.00	194 0.2 0.4	159 0.3 0.4	132 0.3 0.4	109 0.3 0.4	90 0.3 0.4	74 0.3 0.4	61 0.3 0.4	49 0.2 0.3	40 0.2 0.2	31 0.1 0.1																
128-S	5.83 / 5.83				175 0.7 0.9	150 0.7 1.0	129 0.8 1.0	111 0.8 1.1	95 0.8 1.1	82 0.8 1.1	70 0.8 1.1	60 0.7 1.0	51 0.7 0.9	43 0.6 0.8	36 0.5 0.6	29 0.3 0.4											
168-S	5.69 / 5.69							191 1.2 1.6	167 1.2 1.7	146 1.3 1.7	128 1.3 1.8	112 1.4 1.9	99 1.4 1.9	87 1.4 1.9	76 1.4 1.9	66 1.4 1.9	58 1.3 1.8	50 1.2 1.6	43 1.1 1.4	37 0.9 1.2	31 0.7 0.9	26 0.5 0.6					
208-S	5.95 / 5.95								192 1.6 2.3	170 1.7 2.4	150 1.8 2.5	134 1.9 2.6	119 1.9 2.6	105 2.0 2.7	94 2.0 2.7	83 2.0 2.7	74 2.0 2.7	65 2.0 2.7	58 1.9 2.6	51 1.8 2.4	44 1.6 2.2	38 1.5 1.9	31 1.2 1.6	26 1.0 1.2			
248-S	6.42 / 6.42																							45 2.2 2.9	39 2.0 2.7	33 1.8 2.3	27 1.5 1.9

15LDT26

Table of safe superimposed service load (psf) and cambers (in.) — Lightweight – No Topping

Strand Pattern	y_s(end) in. / y_s(center) in.	28	30	32	34	36	38	40	42	44	46	48	50	52	54	56	58	60	62	64	66	68	70	72	74	76	78
88-S	7.00 / 7.00	172 0.5 0.6	145 0.5 0.7	122 0.5 0.7	103 0.5 0.8	87 0.6 0.8	74 0.6 0.8	63 0.6 0.8	53 0.5 0.7	44 0.5 0.7	37 0.4 0.6	30 0.4 0.5															
128-S	5.83 / 5.83				188 1.1 1.5	163 1.2 1.6	142 1.3 1.7	124 1.3 1.8	109 1.4 1.9	95 1.5 2.0	83 1.5 2.0	73 1.5 2.1	64 1.5 2.1	56 1.5 2.0	49 1.4 2.0	42 1.4 1.8	36 1.2 1.7	31 1.1 1.4	27 0.9 1.2								
168-S	5.69 / 5.69							180 2.0 2.7	159 2.1 2.9	141 2.3 3.0	126 2.4 3.2	112 2.5 3.3	100 2.6 3.4	89 2.6 3.4	79 2.7 3.5	71 2.7 3.5	63 2.7 3.5	56 2.6 3.5	50 2.6 3.4	44 2.5 3.3	39 2.4 3.2	34 2.2 2.9	30 1.9 2.5	26 1.6 2.1			
208-S	5.95 / 5.95												119 3.4 4.6	107 3.6 4.8	96 3.7 4.9	86 3.7 5.0	77 3.8 5.0	69 3.8 5.0	61 3.8 4.9	55 3.8 4.9	49 3.7 4.8	44 3.6 4.7	40 3.5 4.5	36 3.3 4.3	32 3.1 4.0	28 2.8 3.6
248-S	6.42 / 6.42																										

Strength is based on strain compatibility; bottom tension is limited to $12\sqrt{f'_c}$; see pages 2–7 through 2–10 for explanation.
Shaded values require release strengths higher than 3500 psi.

Strand Pattern Designation

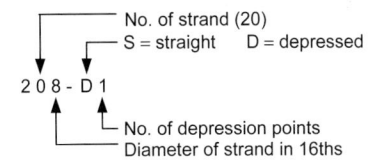

```
        ┌── No. of strand (20)
        │  ┌── S = straight   D = depressed
       208-D1
        │  └── No. of depression points
        └── Diameter of strand in 16ths
```

Because these units are pretopped and are typically used in parking structures, safe loads shown do not include any superimposed dead loads. Loads shown are live load. Long-time cambers do not include live load.

Key
169 — Safe superimposed service load, psf
0.3 — Estimated camber at erection, in.
0.4 — Estimated long-time camber, in.

PRETOPPED DOUBLE TEE
15'-0" x 30"

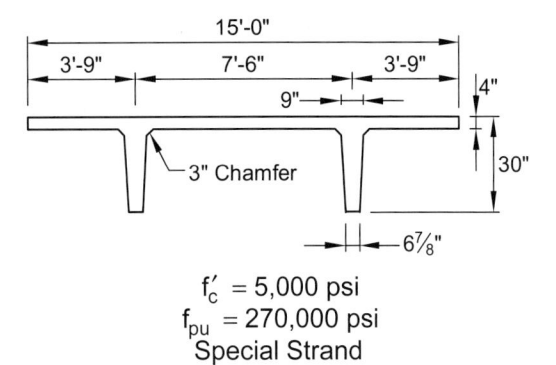

15'-0" (3'-9" | 7'-6" | 3'-9")
9", 4"
3" Chamfer
30"
6 7/8"

$f'_c = 5,000$ psi
$f_{pu} = 270,000$ psi
Special Strand

Section Properties

	Normal Weight	Lightweight
A =	1133 in.²	1133 in.²
I =	78625 in.⁴	78625 in.⁴
y_b =	22.75 in.	22.75 in.
y_t =	7.25 in.	7.25 in.
S_b =	3457 in.³	3457 in.³
S_t =	10838 in.³	10838 in.³
wt =	1180 plf	905 plf
DL =	79 psf	60 psf
V/S =	2.42 in.	2.42 in.

15DT30

Table of safe superimposed service load (psf) and cambers (in.) — Normal Weight – No Topping

Strand Pattern	y_s(end) in. / y_s(center) in.	Span, ft
		30 32 34 36 38 40 42 44 46 48 50 52 54 56 58 60 62 64 66 68 70 72 74 76 78 80 82
88-S	7.00 / 7.00	169 141 118 99 83 69 57 47 38 30 0.3 0.3 0.3 0.3 0.3 0.3 0.2 0.2 0.2 0.1 0.4 0.4 0.4 0.4 0.4 0.4 0.3 0.3 0.2 0.1

128-S	5.83 / 5.83	189 164 142 123 107 93 81 70 60 51 44 37 30
		            0.6 0.7 0.7 0.7 0.7 0.7 0.7 0.7 0.7 0.6 0.5 0.4 0.3
		            0.9 0.9 1.0 1.0 1.0 1.0 1.0 1.0 0.9 0.8 0.7 0.6 0.4

168-S	5.69 / 5.69	184 163 144 127 113 100 88 78 68 60 53 46 39 34 29
		                  1.1 1.2 1.2 1.3 1.3 1.3 1.3 1.3 1.3 1.2 1.1 1.0 0.9 0.8 0.6
		                  1.5 1.6 1.7 1.7 1.8 1.8 1.8 1.8 1.8 1.7 1.5 1.4 1.2 1.0 0.7

208-S	5.95 / 5.95	190 170 152 136 122 109 98 87 78 70 62 55 49 43 37 32 26
		                          1.6 1.7 1.8 1.8 1.9 1.9 1.9 1.9 1.9 1.9 1.8 1.8 1.6 1.5 1.3 1.1 0.8
		                          2.2 2.3 2.4 2.5 2.5 2.6 2.6 2.6 2.6 2.6 2.5 2.4 2.2 2.0 1.7 1.3 0.9

248-S	6.42 / 6.42	65 58 53 47 41 35 29
		                                                           2.4 2.3 2.2 2.1 1.9 1.7 1.4
		                                                            3.2 3.1 3.0 2.8 2.6 2.2 1.8

15LDT30

Table of safe superimposed service load (psf) and cambers (in.) — Lightweight – No Topping

Strand Pattern	y_s(end) in. / y_s(center) in.	Span, ft
		30 32 34 36 38 40 42 44 46 48 50 52 54 56 58 60 62 64 66 68 70 72 74 76 78 80 82 84 86 88 90
88-S	7.00 / 7.00	183 155 132 113 97 83 71 61 52 44 37 31 25 0.4 0.5 0.5 0.5 0.5 0.6 0.6 0.5 0.5 0.5 0.4 0.3 0.2 0.6 0.7 0.7 0.7 0.8 0.8 0.8 0.7 0.7 0.6 0.6 0.4 0.3

128-S	5.83 / 5.83	177 156 137 121 107 94 83 74 65 57 50 44 38 33 29
		            1.1 1.2 1.2 1.3 1.3 1.4 1.4 1.4 1.4 1.4 1.4 1.3 1.2 1.1 0.9
		            1.5 1.6 1.7 1.8 1.8 1.9 1.9 2.0 2.0 1.9 1.9 1.7 1.6 1.4 1.2

168-S	5.69 / 5.69	198 176 157 141 126 113 102 92 82 74 66 59 53 48 42 37 33 29 25
		            1.9 2.0 2.1 2.2 2.3 2.4 2.4 2.5 2.5 2.5 2.5 2.5 2.4 2.3 2.1 1.9 1.7 1.4
		            2.5 2.7 2.8 2.9 3.0 3.1 3.2 3.2 3.3 3.3 3.3 3.3 3.2 3.1 2.9 2.6 2.2 1.8

208-S	5.95 / 5.95	184 166 150 136 123 111 101 92 83 76 68 62 56 51 46 41 37 33 29 26
		                  2.8 2.9 3.1 3.2 3.3 3.4 3.5 3.6 3.6 3.6 3.6 3.6 3.5 3.4 3.3 3.1 2.9 2.6 2.2
		                  3.8 4.0 4.1 4.3 4.5 4.6 4.7 4.7 4.8 4.8 4.7 4.7 4.6 4.6 4.5 4.3 4.1 3.8 3.4 2.9

248-S	6.42 / 6.42	42 38 35 32 29
		                                                                                  4.4 4.2 4.0 3.8 3.5
		                                                                                  5.6 5.4 5.2 4.9 4.5

Strength is based on strain compatibility; bottom tension is limited to $12\sqrt{f'_c}$; see pages 2–7 through 2–10 for explanation.
Shaded values require release strengths higher than 3500 psi.

PCI Design Handbook/Sixth Edition

Strand Pattern Designation

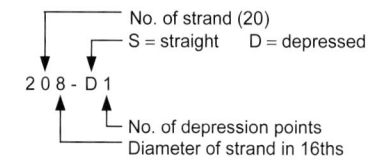

```
       ┌── No. of strand (20)
       │ ┌── S = straight   D = depressed
       ▼ ▼
    2 0 8 - D 1
    ▲     ▲
    │     └── No. of depression points
    └── Diameter of strand in 16ths
```

Because these units are pretopped and are typically used in parking structures, safe loads shown do not include any superimposed dead loads. Loads shown are live load. Long-time cambers do not include live load.

Key
174 – Safe superimposed service load, psf
0.3 – Estimated camber at erection, in.
0.4 – Estimated long-time camber, in.

PRETOPPED DOUBLE TEE
15'-0" x 34"

15'-0" (3'-9" + 7'-6" + 3'-9"), 9", 4", 34", 3" Chamfer, 6½"

$f'_c = 5{,}000$ psi
$f_{pu} = 270{,}000$ psi
Special Strand

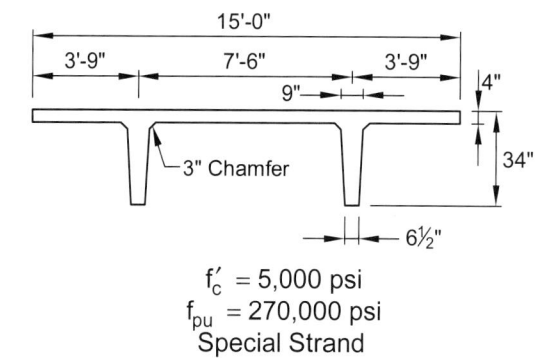

Section Properties

	Normal Weight	Lightweight
A =	1185 in.²	1185 in.²
I =	109621 in.⁴	109621 in.⁴
y_b =	25.65 in.	25.65 in.
y_t =	8.35 in.	8.35 in.
S_b =	4274 in.³	4274 in.³
S_t =	13121 in.³	13121 in.³
wt =	1234 plf	946 plf
DL =	82 psf	63 psf
V/S =	2.45 in.	2.45 in.

15DT34

Table of safe superimposed service load (psf) and cambers (in.) — Normal Weight – No Topping

Strand Pat-tern	y_s(end) in. / y_s(center) in.	Span, ft
		32 34 36 38 40 42 44 46 48 50 52 54 56 58 60 62 64 66 68 70 72 74 76 78 80 82 84 86 88
88-S	7.00 / 7.00	174 147 124 105 89 75 63 52 43 35 27 0.3 0.3 0.3 0.3 0.3 0.3 0.2 0.2 0.2 0.1 0.1 0.4 0.4 0.4 0.4 0.4 0.4 0.3 0.3 0.2 0.2 0.1

128-S	5.83 / 5.83	198 173 151 132 116 101 89 77 67 58 50 43 36 30
		         0.6 0.6 0.7 0.7 0.7 0.7 0.7 0.7 0.6 0.6 0.6 0.5 0.4 0.3
		         0.8 0.9 0.9 0.9 1.0 1.0 1.0 0.9 0.9 0.8 0.7 0.6 0.5 0.3

168-S	5.69 / 5.69	197 175 156 139 124 110 98 87 77 69 61 53 47 41 35 30 25
		               1.1 1.1 1.1 1.2 1.2 1.2 1.2 1.2 1.2 1.2 1.1 1.1 1.0 0.8 0.7 0.5 0.3
		               1.4 1.5 1.6 1.6 1.7 1.7 1.7 1.7 1.7 1.6 1.5 1.4 1.3 1.1 0.9 0.6 0.3

208-S	5.95 / 5.95	185 167 150 135 122 110 99 89 80 72 64 57 51 45 40 35 30 25
		                     1.6 1.6 1.7 1.8 1.8 1.8 1.8 1.8 1.8 1.8 1.7 1.7 1.6 1.4 1.3 1.1 0.8 0.6
		                     2.2 2.3 2.3 2.4 2.4 2.4 2.5 2.5 2.5 2.4 2.4 2.3 2.1 1.9 1.6 1.3 1.0 0.6

248-S	6.42 / 6.42	105 95 86 78 71 64 57 52 46 40 34 29
		                                  2.4 2.4 2.4 2.4 2.3 2.3 2.2 2.1 1.9 1.7 1.5 1.2
		                                  3.2 3.2 3.2 3.2 3.1 3.0 2.9 2.8 2.6 2.3 1.9 1.5

15LDT34

Table of safe superimposed service load (psf) and cambers (in.) — Lightweight – No Topping

Strand Pat-tern	y_s(end) in. / y_s(cen-ter) in.	Span, ft
		32 34 36 38 40 42 44 46 48 50 52 54 56 58 60 62 64 66 68 70 72 74 76 78 80 82 84 86 88 90 92 94
88-S	7.00 / 7.00	188 161 139 120 103 89 77 67 57 49 42 35 30 0.4 0.5 0.5 0.5 0.5 0.5 0.5 0.5 0.5 0.5 0.4 0.4 0.3 0.6 0.6 0.7 0.7 0.7 0.7 0.7 0.7 0.7 0.7 0.6 0.5 0.4

128-S	5.83 / 5.83	187 166 147 130 116 103 92 81 72 64 57 50 44 39 34 29 25
		         1.0 1.1 1.2 1.2 1.3 1.3 1.3 1.3 1.3 1.3 1.3 1.3 1.2 1.1 1.0 0.8 0.7
		         1.4 1.5 1.6 1.7 1.8 1.8 1.9 1.8 1.8 1.7 1.6 1.5 1.3 1.1 0.8

168-S	5.69 / 5.69	189 170 153 138 125 112 102 92 83 75 68 61 55 49 44 39 35 31 27
		               1.8 1.9 2.0 2.1 2.2 2.2 2.3 2.3 2.4 2.4 2.4 2.3 2.3 2.2 2.1 2.0 1.8 1.6 1.4
		               2.5 2.6 2.7 2.8 2.9 3.0 3.0 3.1 3.1 3.2 3.2 3.1 3.1 3.0 2.9 2.7 2.4 2.1 1.8

208-S	5.95 / 5.95	200 181 164 150 136 124 113 103 94 86 79 72 65 60 54 49 45 40 36 32 29 26
		                   2.6 2.7 2.8 3.0 3.1 3.2 3.3 3.4 3.4 3.5 3.5 3.5 3.4 3.4 3.3 3.2 3.0 2.9 2.6 2.3 1.9
		                   3.5 3.7 3.9 4.0 4.2 4.3 4.4 4.5 4.5 4.5 4.5 4.5 4.4 4.3 4.2 4.0 3.8 3.4 3.0 2.4

248-S	6.42 / 6.42	62 56 51 47 43 40 36 33
		                                        4.5 4.5 4.4 4.3 4.2 4.0 3.8 3.6
		                                        5.8 5.7 5.6 5.5 5.3 5.2 4.9 4.6

Strength is based on strain compatibility; bottom tension is limited to $12\sqrt{f'_c}$; see pages 2–7 through 2–10 for explanation.
Shaded values require release strengths higher than 3500 psi.

Strand Pattern Designation
76-S

▲▲▲▲
└ S = straight
└── Diameter of strand in 16ths
└─── No. of Strand (7)

Safe loads shown include dead load of 10 psf for untopped members and 15 psf for topped members. Remainder is live load. Long-time cambers include superimposed dead load but do not include live load.

Capacity of sections of other configurations are similar. For precise values, see local hollow-core manufacturer.

Key
444 – Safe superimposed service load, psf
0.1 – Estimated camber at erection, in.
0.2 – Estimated long-time camber, in.

HOLLOW-CORE
4'-0" x 6"
Normal Weight Concrete

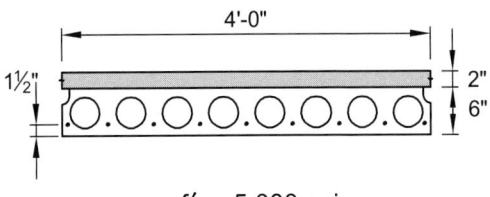

4'-0", 1½", 2", 6"

$f'_c = 5,000$ psi
$f_{pu} = 270,000$ psi

Section Properties

	Untopped	Topped
A =	187 in.²	283 in.²
I =	763 in.⁴	1,640 in.⁴
y_b =	3.00 in.	4.14 in.
y_t =	3.00 in.	3.86 in.
S_b =	254 in.³	396 in.³
S_t =	254 in.³	425 in.³
wt =	195 plf	295 plf
DL =	49 psf	74 psf
V/S =	1.73 in.	

4HC6

No Topping

Table of safe superimposed service load (psf) and cambers (in.)

Strand Designation Code	Span, ft
	10	11	12	13	14	15	16	17	18	19	20	21	22	23	24	25	26	27	28	29	30
66-S	444	382	333	282	238	203	175	151	131	114	100	88	77	68	59	52	46	40	33	28	
	0.1	0.2	0.2	0.2	0.2	0.2	0.2	0.2	0.2	0.2	0.2	0.2	0.1	0.0	−0.1	−0.2	−0.4	−0.5	−0.7		
	0.2	0.2	0.2	0.2	0.3	0.3	0.2	0.2	0.2	0.2	0.1	0.1	0.0	−0.1	−0.3	−0.5	−0.7	−0.9	−1.2	−1.5	−1.9
76-S		445	388	328	278	238	205	178	155	136	120	105	93	82	73	65	57	49	42	36	31
		0.2	0.2	0.2	0.3	0.3	0.3	0.3	0.3	0.3	0.3	0.3	0.3	0.2	0.1	0.1	0.0	−0.1	−0.3	−0.4	−0.6
		0.3	0.3	0.3	0.3	0.3	0.3	0.3	0.3	0.3	0.2	0.1	0.0	−0.1	−0.2	−0.4	−0.7	−0.9	−1.2	−1.6	−2.0
96-S		466	421	386	338	292	263	229	201	177	157	139	124	110	99	88	78	68	60	53	46
		0.3	0.3	0.3	0.4	0.4	0.4	0.5	0.5	0.5	0.5	0.5	0.5	0.5	0.5	0.4	0.3	0.3	0.1	0.0	−0.1
		0.3	0.4	0.4	0.5	0.5	0.5	0.6	0.6	0.6	0.5	0.5	0.4	0.3	0.2	0.1	−0.1	−0.3	−0.6	−0.9	−1.3
87-S		478	433	398	362	322	290	264	240	212	188	167	149	134	119	107	95	85	76	68	60
		0.3	0.4	0.4	0.5	0.5	0.6	0.6	0.7	0.7	0.7	0.7	0.8	0.8	0.7	0.7	0.7	0.6	0.5	0.4	0.3
		0.4	0.5	0.5	0.6	0.7	0.7	0.7	0.8	0.8	0.8	0.8	0.7	0.7	0.6	0.5	0.3	0.2	0.0	−0.3	−0.6
97-S		490	445	407	374	346	311	276	242	220	203	186	166	148	133	119	107	96	86	78	70
		0.4	0.4	0.5	0.5	0.6	0.7	0.7	0.8	0.8	0.9	0.9	0.9	0.9	1.0	0.9	0.9	0.9	0.8	0.7	0.6
		0.5	0.6	0.6	0.7	0.8	0.8	0.9	0.9	1.0	1.0	1.0	0.9	0.9	0.8	0.7	0.5	0.3	0.1	−0.2	

4HC6 + 2

2 in. Normal Weight Topping

Table of safe superimposed service load (psf) and cambers (in.)

Strand Designation Code	Span, ft
	12	13	14	15	16	17	18	19	20	21	22	23	24	25	26	27	28	29	30
66-S	470	396	335	285	244	210	182	158	136	113	93	75	59	46	34				
	0.2	0.2	0.2	0.2	0.2	02	0.2	0.2	0.2	0.2	0.1	0.1	0.0	−0.1	−0.2				
	0.2	0.2	0.2	0.2	0.2	0.1	0.1	0.0	−0.1	−0.2	−0.3	−0.5	−0.7	−0.9	−1.2				
76-S		461	391	334	287	248	216	188	163	137	115	95	78	63	50	38	27		
		0.2	0.3	0.3	0.3	0.3	0.3	0.3	0.3	0.3	0.3	0.2	0.1	0.1	−0.0	−0.1	−0.3		
		0.2	0.2	0.2	0.2	0.2	0.2	0.1	0.1	0.0	−0.2	−0.3	−0.5	−0.7	−0.9	−1.2	−1.5		
96-S		473	424	367	319	279	245	216	186	160	137	116	98	82	68	55	43	33
		0.4	0.4	0.4	0.5	0.5	0.5	0.5	0.5	0.5	0.5	0.5	0.4	0.3	0.3	0.1	0.0	−0.1
		0.4	0.4	0.4	0.4	0.4	0.4	0.3	0.3	0.2	0.1	−0.1	−0.3	−0.5	−0.7	−1.0	−1.4	−1.7
87-S		485	446	415	377	331	292	258	224	195	169	147	127	109	94	80	67	55
		0.5	0.5	0.6	0.6	0.7	0.7	0.7	0.7	0.8	0.8	0.7	0.7	0.7	0.6	0.5	0.4	0.3
		0.5	0.5	0.5	0.6	0.6	0.6	0.5	0.5	0.4	0.4	0.2	0.1	−0.1	−0.3	−0.5	−0.8	−1.2
97-S		494	455	421	394	357	327	288	251	219	192	168	146	127	110	95	82	70
		0.5	0.6	0.7	0.7	0.8	0.8	0.9	0.9	0.9	0.9	1.0	0.9	0.9	0.9	0.8	0.7	0.6
		0.6	0.6	0.7	0.7	0.7	0.7	0.7	0.7	0.6	0.6	0.5	0.4	0.2	0.0	−0.2	−0.5	−0.8

Strength is based on strain compatibility; bottom tension is limited to $7.5\sqrt{f'_c}$; see pages 2–7 through 2–10 for explanation.

Strand Pattern Designation
76-S

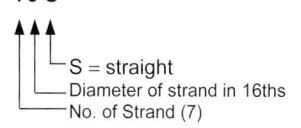

▲ ▲ ▲
   └ S = straight
  └── Diameter of strand in 16ths
 └──── No. of Strand (7)

Safe loads shown include dead load of 10 psf for untopped members and 15 psf for topped members. Remainder is live load. Long-time cambers include superimposed dead load but do not include live load.

Capacity of sections of other configurations are similar. For precise values, see local hollow-core manufacturer.

Key
458 – Safe superimposed service load, psf
0.1 – Estimated camber at erection, in.
0.2 – Estimated long-time camber, in.

HOLLOW-CORE
4'-0" x 8"
Normal Weight Concrete

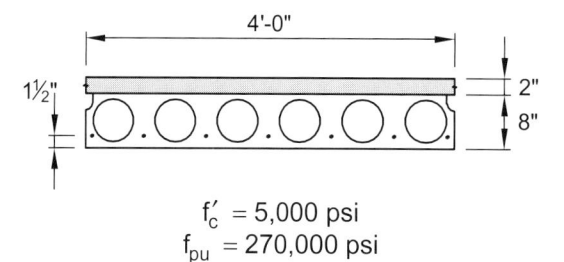

4'-0", 1½", 2", 8"

f'_c = 5,000 psi
f_{pu} = 270,000 psi

Section Properties

	Untopped	Topped
A =	215 in.²	311 in.²
I =	1,666 in.⁴	3,071 in.⁴
y_b =	4.00 in.	5.29 in.
y_t =	4.00 in.	4.71 in.
S_b =	417 in.³	581 in.³
S_t =	417 in.³	652 in.³
wt =	224 plf	324 plf
DL =	56 psf	81 psf
V/S =	1.92 in.	

4HC8

Table of safe superimposed service load (psf) and cambers (in.)
No Topping

Strand Designation Code	Span, ft
	11 12 13 14 15 16 17 18 19 20 21 22 23 24 25 26 27 28 29 30 31 32 33 34 35 36 37 38 39 40
66-S	458 415 378 346 311 269 234 204 179 158 140 124 110 98 87 77 69 61 54 48 43 38 33 29
	0.1 0.2 0.2 0.2 0.2 0.2 0.2 0.2 0.3 0.3 0.3 0.3 0.3 0.2 0.2 0.2 0.2 0.1 0.0 0.0 −0.1 −0.2 −0.3 −0.5 −0.6
	0.2 0.2 0.2 0.3 0.3 0.3 0.3 0.3 0.3 0.3 0.3 0.3 0.3 0.2 0.2 0.1 0.0 −0.1 −0.2 −0.3 −0.5 −0.7 −0.9 −1.2 −1.4
76-S	470 424 387 355 326 303 276 242 213 188 167 149 133 119 106 95 86 77 69 62 55 50 44 39 35 31 26
	0.2 0.2 0.2 0.2 0.3 0.3 0.3 0.3 0.3 0.3 0.3 0.4 0.4 0.4 0.3 0.3 0.3 0.3 0.2 0.2 0.1 0.0 −0.1 −0.2 −0.4 −0.5 −0.7 −0.9
	0.2 0.2 0.3 0.3 0.3 0.4 0.4 0.4 0.4 0.4 0.4 0.4 0.4 0.3 0.3 0.2 0.1 0.0 −0.1 −0.2 −0.4 −0.6 −0.8 −1.1 −1.4 −1.7 −2.0
58-S	464 421 384 352 323 300 280 260 244 229 211 194 177 160 144 130 118 107 97 88 80 72 66 60 54 48 42 37 32 28
	0.2 0.2 0.3 0.3 0.3 0.4 0.4 0.5 0.5 0.5 0.5 0.6 0.6 0.6 0.6 0.6 0.6 0.5 0.5 0.5 0.4 0.3 0.2 0.1 0.0 −0.4 −0.3 −0.5 −0.7 −0.9
	0.3 0.3 0.4 0.4 0.5 0.5 0.6 0.6 0.6 0.7 0.7 0.7 0.7 0.7 0.7 0.6 0.6 0.5 0.4 0.3 0.2 0.0 −0.2 −0.4 −0.6 −0.9 −1.2 −1.6 −2.0 −2.4
68-S	476 430 393 361 332 309 286 269 253 235 223 209 200 180 165 153 142 132 121 110 101 92 84 77 70 63 56 51 45 40
	0.3 0.3 0.3 0.4 0.4 0.5 0.5 0.6 0.6 0.7 0.7 0.7 0.8 0.8 0.8 0.8 0.8 0.8 0.8 0.8 0.8 0.7 0.7 0.6 0.5 0.4 0.2 0.1 −0.1 −0.3
	0.3 0.4 0.5 0.5 0.6 0.6 0.7 0.7 0.8 0.8 0.9 0.9 1.0 1.0 1.0 0.9 0.9 0.9 0.8 0.7 0.6 0.4 0.2 0.0 −0.2 −0.5 −0.8 −1.1 −1.5
78-S	488 442 402 370 341 318 295 275 259 241 229 215 203 195 180 168 157 144 135 126 118 110 101 92 84 77 70 64 58 52
	0.3 0.3 0.4 0.4 0.5 0.5 0.6 0.6 0.7 0.7 0.8 0.9 0.9 1.0 1.0 1.0 1.1 1.1 1.1 1.1 1.1 1.1 1.1 1.0 0.9 0.8 0.7 0.6 0.5 0.3
	0.4 0.5 0.5 0.6 0.7 0.8 0.8 0.9 1.0 1.0 1.1 1.2 1.2 1.2 1.3 1.3 1.3 1.3 1.2 1.2 1.1 1.0 0.8 0.7 0.5 0.3 0.0 −0.3 −0.7

4HC8 + 2

Table of safe superimposed service load (psf) and cambers (in.)
2 in. Normal Weight Topping

Strand Designation Code	Span, ft
	13 14 15 16 17 18 19 20 21 22 23 24 25 26 27 28 29 30 31 32 33 34 35 36 37 38 39 40
66-S	489 445 394 340 294 256 224 197 173 153 135 119 105 93 82 68 56 45 36 26
	0.2 0.2 0.2 0.2 0.2 0.2 0.2 0.3 0.3 0.3 0.3 0.2 0.2 0.2 0.2 0.1 0.0 −0.0 −0.1 −0.2 −0.3
	0.2 0.2 0.2 0.2 0.2 0.2 0.2 0.2 0.2 0.1 0.1 0.0 −0.1 −0.2 −0.3 −0.4 −0.6 −0.7 −0.9 −1.2 −1.4
76-S	498 457 420 387 347 304 267 235 208 184 164 146 130 116 103 88 74 62 51 41 31
	0.2 0.2 0.3 0.3 0.3 0.3 0.3 0.3 0.4 0.4 0.4 0.3 0.3 0.3 0.3 0.2 0.2 0.1 −0.0 −0.1 −0.2
	0.2 0.2 0.3 0.3 0.3 0.3 0.3 0.3 0.2 0.2 0.2 0.1 0.0 −0.1 −0.2 −0.4 −0.5 −0.7 −0.9 −1.2 −1.4
58-S	492 451 414 384 357 333 310 293 274 245 219 196 177 159 143 126 110 95 82 70 59 49 40 32
	0.3 0.3 0.3 0.4 0.4 0.5 0.5 0.5 0.5 0.6 0.6 0.6 0.6 0.6 0.5 0.5 0.5 0.1 0.3 0.2 0.1 0.0 −0.1
	0.3 0.3 0.4 0.4 0.4 0.4 0.5 0.5 0.5 0.5 0.4 0.3 0.3 0.2 0.1 −0.1 −0.2 −0.4 −0.6 −0.9 −1.2 −1.5 −1.8
68-S	463 426 393 366 342 319 299 282 267 251 239 216 195 177 158 140 124 110 97 84 73 62 53 44 36 28
	0.4 0.4 0.5 0.5 0.6 0.6 0.7 0.7 0.8 0.8 0.8 0.8 0.8 0.8 0.8 0.8 0.8 0.7 0.7 0.6 0.5 0.4 0.2 0.1 −0.1
	0.4 0.5 0.5 0.6 0.6 0.6 0.6 0.7 0.7 0.7 0.6 0.6 0.5 0.4 0.3 0.2 0.0 −0.2 −0.4 −0.6 −0.9 −1.2 −1.6 −2.0 −2.4
78-S	472 435 402 375 348 325 305 288 273 257 245 232 220 207 186 167 149 133 119 106 94 83 73 64 55 46 38
	0.5 0.5 0.6 0.6 0.7 0.7 0.8 0.9 0.9 1.0 1.0 1.0 1.1 1.1 1.1 1.1 1.1 1.1 1.0 0.9 0.9 0.7 0.6 0.5 0.3
	0.5 0.6 0.6 0.7 0.7 0.8 0.8 0.8 0.9 0.9 0.9 0.8 0.8 0.7 0.7 0.6 0.4 0.3 0.1 −0.1 −0.3 −0.6 −0.9 −1.3 −1.7 −2.2

Strength is based on strain compatibility; bottom tension is limited to $7.5\sqrt{f'_c}$; see pages 2–7 through 2–10 for explanation.

Strand Pattern Designation
48-S

- S = straight
- Diameter of strand in 16ths
- No. of Strand (4)

Safe loads shown include dead load of 10 psf for untopped members and 15 psf for topped members. Remainder is live load. Long-time cambers include superimposed dead load but do not include live load.

Capacity of sections of other configurations are similar. For precise values, see local hollow-core manufacturer.

Key
258 – Safe superimposed service load, psf
0.3 – Estimated camber at erection, in.
0.4 – Estimated long-time camber, in.

HOLLOW-CORE
4'-0" x 10"
Normal Weight Concrete

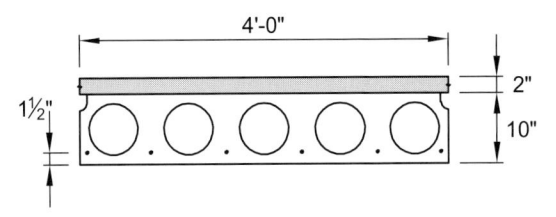

$f'_c = 5,000$ psi
$f_{pu} = 270,000$ psi

Section Properties

	Untopped	Topped
A =	259 in.²	355 in.²
I =	3,223 in.⁴	5,328 in.⁴
y_b =	5.00 in.	6.34 in.
y_t =	5.00 in.	5.66 in.
S_b =	645 in.³	840 in.³
S_t =	645 in.³	941 in.³
wt =	270 plf	370 plf
DL =	68 psf	93 psf
V/S =	2.23 in.	

4HC10

Table of safe superimposed service load (psf) and cambers (in.) — No Topping

Strand Designation Code	Span, ft																								
	20	21	22	23	24	25	26	27	28	29	30	31	32	33	34	35	36	37	38	39	40	41	42	43	44	45	46
48-S	258	234	209	187	168	151	136	123	111	100	90	82	74	66	60	54	48	43	38	34	30	26					
	0.3	0.3	0.3	0.3	0.3	0.3	0.3	0.3	0.3	0.3	0.2	0.2	0.2	0.1	0.1	0.0	-0.1	-0.2	-0.3	-0.4	-0.6	-0.7	-0.9				
	0.4	0.4	0.4	0.4	0.4	0.4	0.4	0.3	0.3	0.2	0.2	0.1	0.0	-0.1	-0.2	-0.3	-0.5	-0.7	-0.8	-1.1	-1.3	-1.3	-1.9				
58-S	267	249	237	223	211	197	179	162	148	134	122	112	102	93	85	77	70	64	58	53	48	43	39	35	30	26	
	0.4	0.4	0.4	0.5	0.5	0.5	0.5	0.5	0.5	0.5	0.5	0.4	0.4	0.4	0.3	0.2	0.2	0.1	0.0	-0.1	-0.3	-0.4	-0.6	-0.7	-0.9	-1.2	
	0.5	0.6	0.6	0.6	0.6	0.6	0.6	0.6	0.6	0.5	0.5	0.4	0.3	0.2	0.1	0.0	-0.1	-0.3	-0.5	-0.7	-1.0	-1.2	-1.5	-1.8	-2.2	-2.6	
68-S	273	255	243	229	217	206	196	187	176	162	153	141	129	118	109	100	92	84	78	71	65	60	54	49	44	39	34
	0.5	0.5	0.6	0.6	0.6	0.7	0.7	0.7	0.7	0.7	0.7	0.7	0.7	0.7	0.7	0.6	0.6	0.5	0.5	0.4	0.3	0.2	0.1	-0.1	-0.2	-0.4	-0.6	-0.8
	0.7	0.7	0.7	0.8	0.8	0.8	0.8	0.9	0.9	0.8	0.8	0.8	0.7	0.7	0.6	0.5	0.4	0.2	0.1	-0.1	-0.3	-0.6	-0.8	-1.1	-1.4	-1.8	-2.2
78-S	282	264	249	235	223	212	202	193	185	174	165	153	144	136	129	119	113	104	96	89	82	76	69	63	57	52	47
	0.6	0.7	0.7	0.8	0.8	0.9	0.9	0.9	0.9	0.9	0.9	1.0	1.0	1.0	0.9	0.9	0.9	0.8	0.8	0.7	0.6	0.5	0.4	0.3	0.1	0.0	-0.2
	0.8	0.9	0.9	1.0	1.0	1.1	1.1	1.1	1.1	1.1	1.1	1.1	1.1	1.1	1.0	1.0	0.9	0.8	0.6	0.5	0.3	0.1	-0.1	-0.4	-0.7	-1.0	-1.3
88-S	288	270	255	241	229	218	208	199	188	180	174	165	153	145	135	128	122	115	106	101	96	91	84	77	71	65	59
	0.7	0.8	0.8	0.9	0.9	1.0	1.0	1.1	1.1	1.2	1.2	1.2	1.2	1.2	1.2	1.2	1.2	1.2	1.1	1.1	1.0	0.9	0.8	0.7	0.5	0.3
	1.0	1.0	1.1	1.2	1.2	1.3	1.3	1.4	1.4	1.4	1.4	1.5	1.5	1.4	1.4	1.4	1.3	1.2	1.2	1.0	0.9	0.7	0.6	0.3	0.1	-0.2	-0.5

4HC10 + 2

Table of safe superimposed service load (psf) and camber (in.) — 2 in. Normal Weight Topping

Strand Designation Code	Span, ft																								
	20	21	22	23	24	25	26	27	28	29	30	31	32	33	34	35	36	37	38	39	40	41	42	43	44	45	46
48-S	308	287	256	228	204	183	165	148	133	119	107	96	86	74	63	52	43	34	26								
	0.3	0.3	0.3	0.3	0.3	0.3	0.3	0.3	0.3	0.3	0.2	0.2	0.2	0.1	0.1	0.0	-0.1	-0.2	-0.3	-0.4							
	0.3	0.3	0.3	0.2	0.2	0.2	0.1	0.1	0.0	-0.1	-0.2	-0.3	-0.4	-0.6	-0.8	-1.0	-1.2	-1.4	-1.7								
58-S	317	298	282	267	252	237	219	198	180	163	148	134	120	105	92	80	69	59	50	41	33	26					
	0.4	0.4	0.4	0.5	0.5	0.5	0.5	0.5	0.5	0.5	0.5	0.4	0.4	0.4	0.3	0.2	0.2	0.1	0.0	-0.1	-0.3	-0.4					
	0.4	0.4	0.4	0.4	0.4	0.4	0.3	0.3	0.2	0.1	0.0	-0.1	-0.2	-0.4	-0.5	-0.7	-0.9	-1.2	-1.5	-1.8	-2.1						
68-S	326	307	291	273	258	246	234	222	212	202	188	171	153	137	122	108	96	84	74	64	55	46	38	31			
	0.5	0.5	0.6	0.6	0.6	0.7	0.7	0.7	0.7	0.7	0.7	0.7	0.7	0.7	0.7	0.6	0.6	0.5	0.5	0.4	0.3	0.2	0.1	-0.1	-0.2		
	0.5	0.6	0.6	0.6	0.6	0.6	0.6	0.6	0.5	0.5	0.4	0.4	0.3	0.2	0.0	-0.1	-0.3	-0.5	-0.7	-0.9	-1.2	-1.5	-1.8	-2.2			
78-S	335	313	297	279	267	252	240	228	218	208	196	189	181	165	150	135	122	109	97	86	76	67	58	50	42	35	28
	0.6	0.7	0.7	0.7	0.8	0.8	0.9	0.9	0.9	0.9	0.9	1.0	1.0	1.0	0.9	0.9	0.9	0.8	0.8	0.7	0.6	0.5	0.4	0.3	0.1	0.0	-0.2
	0.7	0.7	0.7	0.8	0.8	0.8	0.8	0.8	0.8	0.8	0.7	0.7	0.6	0.5	0.4	0.3	0.2	0.0	-0.2	-0.4	-0.6	-0.9	-1.2	-1.6	-1.9	-2.3	-2.8
88-S	344	322	306	288	273	258	246	234	221	211	202	195	184	178	172	158	144	130	118	107	96	87	77	68	60	52	44
	0.7	0.8	0.8	0.9	0.9	1.0	1.0	1.1	1.1	1.2	1.2	1.2	1.2	1.2	1.2	1.2	1.2	1.2	1.1	1.1	1.0	0.9	0.8	0.7	0.5	0.3
	0.8	0.8	0.9	0.9	1.0	1.0	1.0	1.0	1.0	1.0	1.0	0.9	0.9	0.8	0.7	0.6	0.4	0.3	0.1	-0.1	-0.3	-0.6	-0.9	-1.3	-1.6	-2.0

Strength is based on strain compatibility; bottom tension is limited to $7.5\sqrt{f'_c}$; see pages 2–7 through 2–10 for explanation.

Strand Pattern Designation
76-S

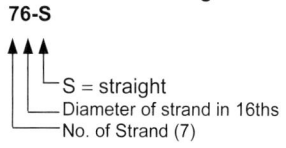

- S = straight
- Diameter of strand in 16ths
- No. of Strand (7)

Safe loads shown include dead load of 10 psf for untopped members and 15 psf for topped members. Remainder is live load. Long-time cambers include superimposed dead load but do not include live load.

Capacity of sections of other configurations are similar. For precise values, see local hollow-core manufacturer.

Key
258 – Safe superimposed service load, psf
0.3 – Estimated camber at erection, in.
0.4 – Estimated long-time camber, in.

HOLLOW-CORE
4'-0" x 12"
Normal Weight Concrete

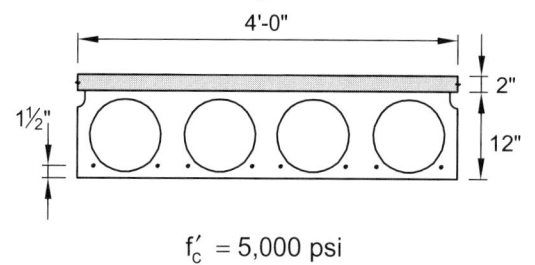

f'_c = 5,000 psi
f_{pu} = 270,000 psi

Section Properties

	Untopped	Topped
A =	262 in.²	358 in.²
I =	4,949 in.⁴	7,811 in.⁴
y_b =	6.00 in.	7.55 in.
y_t =	6.00 in.	6.45 in.
S_b =	825 in.³	1,035 in.³
S_t =	825 in.³	1,211 in.³
wt =	273 plf	373 plf
DL =	68 psf	93 psf
V/S =	2.18 in.	

4HC12

Table of safe superimposed service load (psf) and cambers (in.) — No Topping

Strand Designation Code	Span, ft
	20 21 22 23 24 25 26 27 28 29 30 31 32 33 34 35 36 37 38 39 40 41 42 43 44 45 46
76-S	258 242 228 215 204 194 175 159 144 131 120 109 99 90 82 75 68 62 56 51 46 41 37 33 30 26
	0.3 0.3 0.3 0.3 0.3 0.3 0.3 0.3 0.3 0.3 0.3 0.3 0.3 0.2 0.2 0.1 0.1 0.0 0.0 −0.1 −0.2 −0.3 −0.4 −0.5 −0.6 −0.8
	0.3 0.4 0.4 0.4 0.4 0.4 0.4 0.4 0.4 0.3 0.3 0.3 0.2 0.1 0.1 0.0 −0.1 −0.2 −0.3 −0.5 −0.6 −0.8 −1.0 −1.2 −1.5 −1.7
58-S	258 242 228 215 204 194 182 174 167 157 151 143 138 128 118 108 100 92 84 78 71 66 60 55 51 46 42
	0.4 0.4 0.4 0.4 0.5 0.5 0.5 0.5 0.5 0.5 0.5 0.5 0.5 0.5 0.5 0.5 05 0.4 0.4 0.3 0.3 0.2 0.1 0.1 0.0 −0.1 −0.3 −0.4
	0.5 0.5 0.5 0.6 0.6 0.6 0.6 0.6 0.6 0.6 0.6 0.6 0.6 0.5 0.5 0.4 0.4 0.3 0.2 0.1 −0.1 −0.2 −0.4 −0.6 −0.8 −1.0 −1.2
68-S	264 248 234 221 210 200 191 180 173 163 157 149 144 137 133 127 121 112 106 98 93 87 81 75 69 64 59
	0.4 0.5 0.5 0.5 0.6 0.6 0.6 0.7 0.7 0.7 0.7 0.7 0.7 0.8 0.8 0.7 0.7 0.7 0.7 0.6 0.6 0.5 0.5 0.4 0.3 0.2 0.1
	0.6 0.6 0.7 0.7 0.8 0.8 0.8 0.9 0.9 0.9 0.9 0.9 0.9 0.9 0.9 0.8 0.8 0.7 0.7 0.6 0.5 0.3 0.2 0.1 −0.1 −0.3 −0.5
78-S	273 257 243 230 216 206 197 186 179 169 163 155 150 143 136 133 127 121 115 110 105 98 93 86 82 78 72
	0.5 0.6 0.6 0.7 0.7 0.8 0.8 0.8 0.9 0.9 0.9 0.9 1.0 1.0 1.0 1.0 1.0 1.0 1.0 1.0 1.0 0.9 0.9 0.8 0.7 0.7 0.6
	0.7 0.8 0.8 0.9 0.9 1.0 1.0 1.1 1.1 1.1 1.2 1.2 1.2 1.2 1.2 1.2 1.2 1.1 1.1 1.0 1.0 0.9 0.8 0.7 0.5 0.3 0.2
88-S	282 266 252 236 225 212 203 192 185 175 169 161 153 149 142 136 130 127 121 116 114 107 102 95 91 84 81
	0.6 0.7 0.7 0.8 0.8 0.9 0.9 1.0 1.0 1.1 1.1 1.2 1.2 1.2 1.3 1.3 1.3 1.3 1.3 1.3 1.3 1.3 1.3 1.2 1.2 1.1 1.1
	0.8 0.9 1.0 1.0 1.1 1.2 1.2 1.3 1.3 1.4 1.4 1.5 1.5 1.5 1.5 1.6 1.6 1.5 1.5 1.4 1.4 1.3 1.2 1.1 1.0 0.8

4HC12 + 2

Table of safe superimposed service load (psf) and cambers (in.) — 2 in. Normal Weight Topping

Strand Designation Code	Span, ft
	20 21 22 23 24 25 26 27 28 29 30 31 32 33 34 35 36 37 38 39 40 41 42 43 44 45 46
76-S	295 276 258 246 232 217 196 177 160 144 131 118 107 96 87 78 71 63 56 50 43 35 28
	0.3 0.3 0.3 0.3 0.3 0.3 0.3 0.3 0.3 0.3 0.3 0.3 0.3 0.2 0.2 0.1 0.1 0.0 0.0 −0.1 −0.2 −0.3 −0.4
	0.3 0.3 0.3 0.3 0.3 0.3 0.2 0.2 0.2 0.1 0.1 0.0 −0.1 −0.2 −0.3 −0.4 −0.5 −0.7 −0.8 −1.0 −1.2 −1.4 −1.7
58-S	292 273 258 243 229 216 205 195 186 175 167 160 151 139 127 116 106 97 89 81 72 63 55 47 39 32 26
	0.4 0.4 0.4 0.4 0.5 0.5 0.5 0.5 0.5 0.5 0.5 0.5 0.5 0.5 0.5 0.5 0.5 0.4 0.4 0.3 0.3 0.2 0.1 0.1 0.0 −0.1 −0.3 −0.4
	0.4 0.4 0.4 0.4 0.4 0.5 0.5 0.4 0.4 0.4 0.4 0.4 0.3 0.3 0.2 0.1 0.0 −0.1 −0.2 −0.3 −0.5 −0.7 −0.9 −1.1 −1.3 −1.6 −1.9 −2.2
68-S	301 282 264 249 235 222 211 201 192 181 173 166 157 152 144 139 132 127 117 107 98 88 78 69 61 53 46
	0.4 0.5 0.5 0.5 0.6 0.6 0.6 0.7 0.7 0.7 0.7 0.7 0.8 0.8 0.7 0.7 0.7 0.6 0.6 0.5 0.5 0.4 0.3 0.2 0.1
	0.5 0.5 0.6 0.6 0.6 0.6 0.6 0.7 0.6 0.6 0.6 0.6 0.6 0.5 0.5 0.4 0.3 0.1 −0.1 −0.2 −0.4 −0.6 −0.8 −1.0 −1.3 −1.6
78-S	310 291 273 258 244 231 217 207 198 187 179 172 163 158 150 142 138 131 128 122 116 113 102 92 83 74 66
	0.5 0.6 0.6 0.7 0.7 0.8 0.8 0.8 0.9 0.9 0.9 1.0 1.0 1.0 1.0 1.0 1.0 1.0 1.0 1.0 0.9 0.9 0.8 0.7 0.7 0.6
	0.6 0.6 0.7 0.7 0.8 0.8 0.8 0.8 0.9 0.9 0.8 0.8 0.8 0.7 0.6 0.6 0.5 0.4 0.3 0.1 −0.1 −0.2 −0.5 −0.7 −0.9
88-S	319 300 282 267 250 237 226 213 204 193 185 175 169 161 156 148 141 137 131 125 122 117 111 106 103 94 86
	0.6 0.7 0.7 0.8 0.8 0.9 0.9 1.0 1.0 1.1 1.2 1.2 1.2 1.3 1.3 1.3 1.3 1.3 1.3 1.3 1.3 1.2 1.2 1.1 1.1
	0.7 0.8 0.8 0.9 0.9 1.0 1.0 1.1 1.1 1.1 1.1 1.1 1.0 1.0 0.9 0.9 0.8 0.7 0.6 0.4 0.3 0.1 −0.1 −0.3

Strength is based on strain compatibility; bottom tension is limited to $7.5\sqrt{f'_c}$; see pages 2–7 through 2–10 for explanation.

HOLLOW-CORE SLABS

Figure 2.5.1 Section Properties – Normal Weight Concrete Dy-Core

Trade Name: Dy-Core ®
Equipment Manufacturer: Elematic Inc., Waukesha, Wisconsin

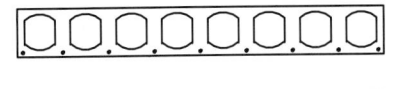

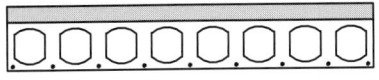

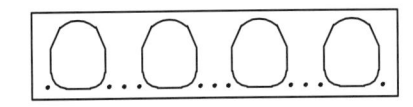

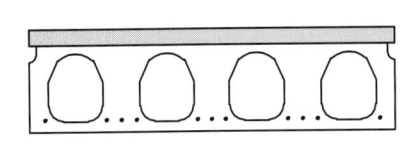

Section width x depth	Untopped A in.²	Untopped y_b in.	Untopped I in.⁴	Untopped wt psf	With 2 in. topping y_b in.	With 2 in. topping I in.⁴	With 2 in. topping wt psf
4'-0" x 6"	142	3.05	661	37	4.45	1,475	62
4'-0" x 8"	193	3.97	1,581	50	5.43	3,017	75
4'-0" x 10"	215	5.40	2,783	56	6.89	4,614	81
4'-0" x 12"	264	6.37	4,773	69	7.89	7,313	94
4'-0" x 15"	289	7.37	8,604	76	9.21	13,225	101

Note: All sections are not available from all producers. Check availability with local manufacturers.

Figure 2.5.2 Section Properties – Normal Weight Concrete Dynaspan

Trade Name: Dynaspan ®
Equipment Manufacturer: Dynamold Corporation, Salina, Kansas

Section width x depth	Untopped A in.²	Untopped y_b in.	Untopped I in.⁴	Untopped wt psf	With 2 in. topping y_b in.	With 2 in. topping I in.⁴	With 2 in. topping wt psf
4'-0" x 4"	133	2.00	235	35	3.08	689	60
4'-0" x 6"	165	3.02	706	43	4.25	1,543	68
4'-0" x 8"	233	3.93	1,731	61	5.16	3,205	86
4'-0" x 10"	260	4.91	3,145	68	6.26	5,314	93
8'-0" x 6"	338	3.05	1,445	44	4.26	3,106	69
8'-0" x 8"	470	3.96	3,525	61	5.17	6,444	86
8'-0" x 10"	532	4.96	6,422	69	6.28	10,712	94
8'-0" x 12"	615	5.95	10,505	80	7.32	16,507	105

Note: All sections are not available from all producers. Check availability with local manufacturers.

HOLLOW-CORE SLABS

Figure 2.5.3   Section Properties – Normal Weight Concrete   Flexicore

Trade Name: Flexicore ®
Licensing Organization: The Flexicore Co. Inc., Dayton, Ohio

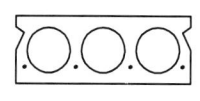

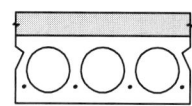

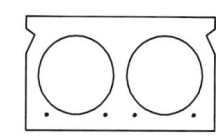

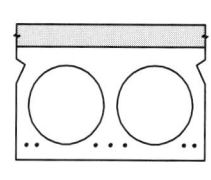

Section width x depth	Untopped				With 2 in. topping		
	A in.^{2}	y_b in.	I in.^{4}	wt psf	y_b in.	I in.^{4}	wt psf
1'-4" x 6"	55	3.00	243	43	4.23	523	68
2'-0" x 6"	86	3.00	366	45	4.20	793	70
1'-4" x 8"	73	4.00	560	57	5.26	1,028	82
2'-0" x 8"	110	4.00	843	57	5.26	1,547	82
1'-8" x 10"	98	5.00	1,254	61	6.43	2,109	86
2'-0" x 10"	138	5.00	1,587	72	6.27	2,651	97
2'-0" x 12"	141	6.00	2,595	73	7.46	4,049	98

Note: All sections are not available from all producers. Check availability with local manufacturers.

Figure 2.5.4   Section Properties – Normal Weight Concrete   Spancrete

Trade Name: Spancrete ®
Licensing Organization: Spancrete Machinery Corp., Milwaukee, Wisconsin

Standard Spancrete ®

Section width x depth	Untopped				With 2 in. topping		
	A in.^{2}	y_b in.	I in.^{4}	wt psf	y_b in.	I in.^{4}	wt psf
4'-0" x 4"	138	2.00	238	34	3.14	739	59
4'-0" x 6"	189	2.93	762	46	4.19	1,760	71
4'-0" x 8"	258	3.98	1,806	63	5.22	3,443	88
4'-0" x 10"	312	5.16	3,484	76	6.41	5,787	101
4'-0" x 12"	355	6.28	5,784	86	7.58	8,904	111
4'-0" x 15"	370	7.87	9,765	90	9.39	14,351	115

Ultralight Spancrete ®

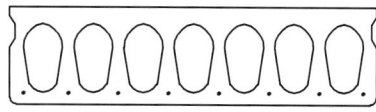

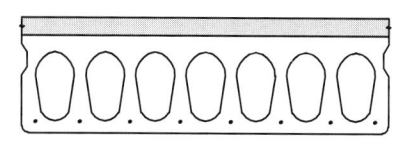

4'-0" x 8"	246	4.17	1,730	60	5.41	3,230	85
4'-0" x 10"	277	5.22	3,178	67	6.58	5,376	92
4'-0" x 12"	316	6.22	5,311	77	7.66	8,410	102

Note: Spancrete is also available in 40 in. and 96 in. widths. All sections are not available from all producers. Check availability with local manufacturers.

HOLLOW-CORE SLABS

Figure 2.5.5 Section Properties – Normal Weight Concrete Span Deck

Trade Name: Span Deck ®
Licensing Organization: Fabcon, Incorporated, Savage, Minnesota

Section width x depth	Untopped				With 2 in. topping		
	A in.²	y_b in.	I in.⁴	wt psf	y_b in.	I in.⁴	wt psf
4'-0" x 8"	246	3.75	1,615	62	5.55	2,791	87
4'-0" x 12"	298	5.87	5,452	75	8.01	7,856	100
8'-0" x 8"	477	3.73	3,236	60	5.53	5,643	85
8'-0" x 12"	578	5.86	10,909	72	7.98	15,709	97

Note: All sections are not available from all producers. Section properties and weights may vary with producer. Check availability with local manufacturers.

Figure 2.5.6 Section Properties – Normal Weight Concrete Ultra Span

Trade Name: Ultra Span ®
Licensing Organization: Ultra Span Technologies Inc., Winnipeg, Manitoba, Canada

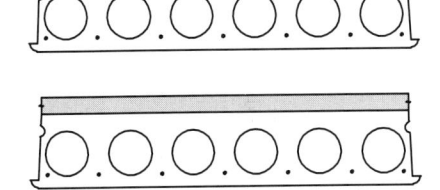

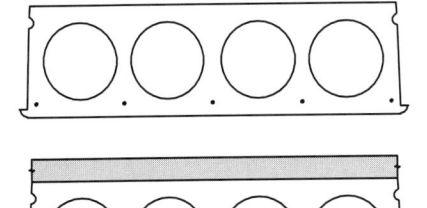

Section width x depth	Untopped				With 2 in. topping		
	A in.²	y_b in.	I in.⁴	wt psf	y_b in.	I in.⁴	wt psf
4'-0" x 4"	154	2.00	247	40	2.98	723	65
4'-0" x 6"	188	3.00	764	49	4.13	1,641	74
4'-0" x 8"	214	4.00	1,666	56	5.29	3,070	81
4'-0" x 10"	259	5.00	3,223	67	6.34	5,328	92
4'-0" x 12"	289	6.00	5,272	75	7.43	8,195	100

Note: All sections are not available from all producers. Check availability with local manufacturers.

HOLLOW-CORE SLABS

Figure 2.5.7 Section Properties – Normal Weight Concrete *Elematic*

Trade Name: Elematic ®
Equipment Manufacturer: Elematic Inc., Waukesha, Wisconsin

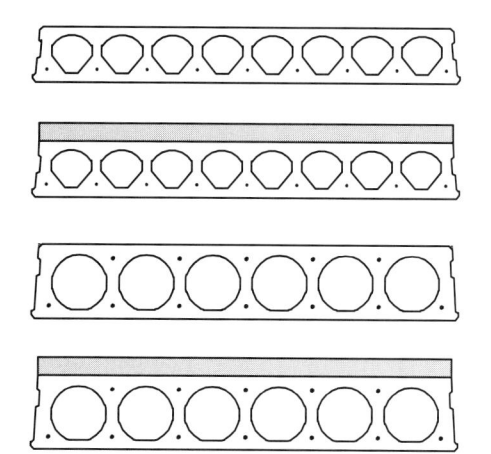

Section width x depth	Untopped A in.²	Untopped y_b in.	Untopped I in.⁴	Untopped wt psf	With 2 in. topping y_b in.	With 2 in. topping I in.⁴	With 2 in. topping wt psf
4'-0" x 6"	157	3.00	694	41	4.33	1,557	66
4'-0" x 8"	196	3.97	1,580	51	5.41	3,024	76
4'0" x 10"	249	5.00	3,108	65	6.44	5,280	90
4'-0" x 12"	279	6.20	5,104	74	7.90	8,406	99
4'-0" x 16"	346	8.30	11,339	91	10.20	16,883	116
8'-0" x 8"	404	4.00	3,219	54	5.60	6,475	79
8'-0" x 10"	549	5.00	6,642	73	6.50	11,827	98
8'-0" x 12"	620	6.10	10,588	82	7.60	17,915	107

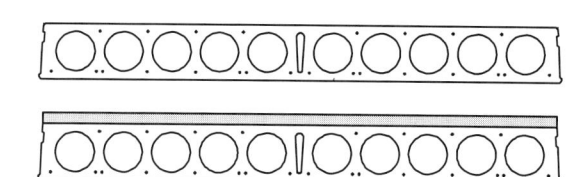

Note: All sections are not available from all producers. Check availability with local manufacturers.

Figure 2.5.8 Section Properties – Normal Weight Concrete *Roth*

Trade Name: Roth ®
Equipment Manufacturer: Elematic Inc., Waukesha, Wisconsin

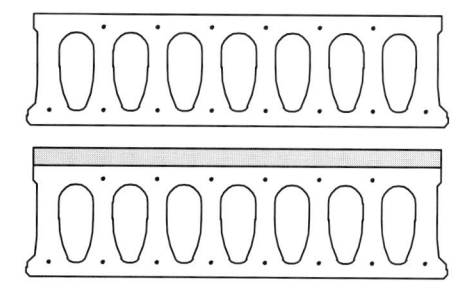

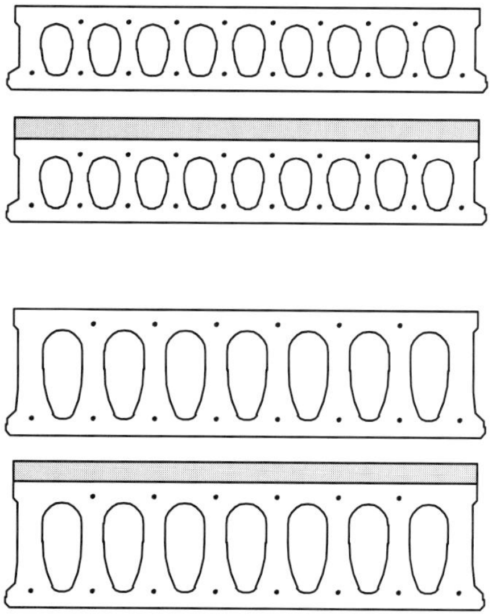

Section width x depth	Untopped A in.²	Untopped y_b in.	Untopped I in.⁴	Untopped wt psf	With 2 in. topping y_b in.	With 2 in. topping I in.⁴	With 2 in. topping wt psf
4'-0" x 6"	194	3.00	767	52	4.30	1,808	76
4'-0" x 8"	265	4.00	1,812	71	5.30	3,567	95
4'-0" x 10"	310	5.00	3,409	83	6.40	5,969	107
4'-0" x 12"	352	6.00	5,698	94	7.40	9,325	118
4'-0" x 16"	442	8.10	12,864	117	9.50	19,156	142
8'-0" x 8"	531	4.00	3,659	71	5.30	7,211	96

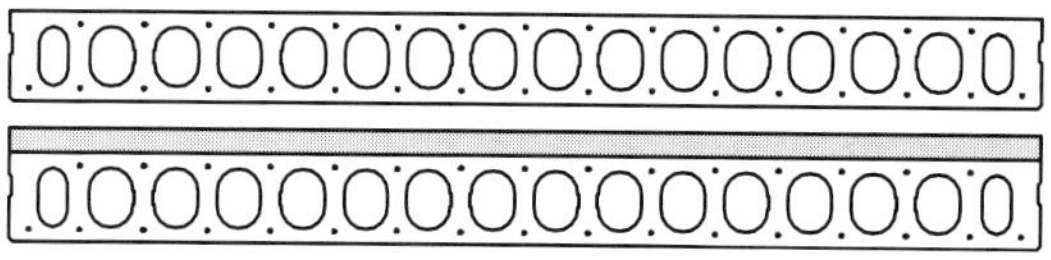

Note: All sections are not available from all producers. Check availability with local manufacturers.

Strand Pattern Designation
76-S

▲ ▲ ▲
└─ S = straight
└── Diameter of strand in 16ths
└─── No. of Strand (7)

Safe loads shown include dead load of 10 psf for untopped members and 15 psf for topped members. Remainder is live load. Long-time cambers include superimposed dead load but do not include live load.

Key
216 – Safe superimposed service load, psf
0.1 – Estimated camber at erection, in.
0.1 – Estimated long-time camber, in.

SOLID FLAT SLAB
4 in. Thick
Normal Weight Concrete

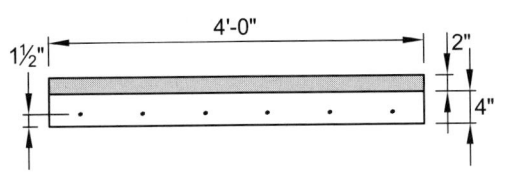

$f'_c = 5,000$ psi
$f_{pu} = 270,000$ psi

Section Properties

	Untopped	Topped
A =	192 in.²	288 in.²
I =	256 in.⁴	763 in.⁴
y_b =	2.00 in.	2.84 in.
y_t =	2.00 in.	3.16 in.
S_b =	128 in.³	269 in.³
S_t =	128 in.³	242 in.³
wt =	200 plf	300 plf
DL =	50 psf	75 psf
V/S =	1.85 in.	

FS4

Table of safe superimposed service load (psf) and cambers (in.) — No Topping

Strand Designation Code	10	11	12	13	14	15	16	17	18	19	20	21	22
66-S	216 0.1 0.1	182 0.1 0.1	147 0.1 0.0	118 0.0 0.0	95 0.0 −0.1	76 −0.1 −0.3	61 −0.2 −0.4	48 −0.3 −0.6	38 −0.4 −0.9	29 −0.6 −1.2			
76-S	250 0.1 0.1	208 0.1 0.1	167 0.1 0.1	135 0.1 0.0	110 0.0 −0.1	89 0.0 −0.2	72 −0.1 −0.3	58 −0.2 −0.5	47 −0.3 −0.7	37 −0.5 −1.0	28 −0.7 −1.4		
58-S	256 0.2 0.2	220 0.2 0.2	180 0.2 0.2	148 0.2 0.2	123 0.1 0.1	103 0.1 0.0	86 0.0 −0.1	72 0.0 −0.3	61 −0.1 −0.5	51 −0.3 −0.8	42 −0.4 −1.1	34 −0.7 −1.5	26 −0.9 −1.9
68-S	276 0.2 0.3	223 0.2 0.3	182 0.2 0.3	150 0.2 0.3	125 0.2 0.2	104 0.2 0.2	87 0.2 0.1	73 0.1 −0.1	62 0.0 −0.3	52 −0.1 −0.5	43 −0.3 −0.8	36 −0.4 −1.2	30 −0.7 −1.6

FS4 + 2

Table of safe superimposed service load (psf) and cambers (in.) — 2 in. Normal Weight Topping

Strand Designation Code	11	12	13	14	15	16	17	18	19	20	21	22
66-S	336 0.1 0.0	257 0.1 0.0	196 0.0 −0.1	147 0.0 −0.2	108 −0.1 −0.3	76 −0.2 −0.5	49 −0.3 −0.7	27 −0.4 −1.0				
76-S	385 0.1 0.0	298 0.1 0.0	231 0.1 −0.1	177 0.0 −0.1	134 0.0 −0.3	99 −0.1 −0.4	70 −0.2 −0.6	45 −0.3 −0.9				
58-S		356 0.2 0.1	296 0.2 0.0	239 0.1 0.0	187 0.1 −0.1	146 0.0 −0.3	111 0.0 −0.4	82 −0.1 −0.7	58 −0.3 −0.9	37 −0.4 −1.3		
68-S		360 0.2 0.1	289 0.2 0.1	232 0.2 0.0	185 0.2 0.0	146 0.1 −0.1	114 0.0 −0.3	87 −0.1 −0.5	64 −0.3 −0.7	43 −0.4 −1.0	25 −0.7 −1.4	−1.9

Strength is based on strain compatibility; bottom tension is limited to $7.5\sqrt{f'_c}$; see pages 2–7 through 2–10 for explanation.

Strand Pattern Designation
76-S

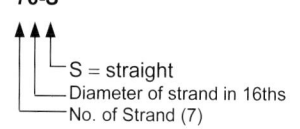

▲ ▲ ▲
- S = straight
- Diameter of strand in 16ths
- No. of Strand (7)

Safe loads shown include dead load of 10 psf for untopped members and 15 psf for topped members. Remainder is live load. Long-time cambers include superimposed dead load but do not include live load.

Key
- 361 – Safe superimposed service load, psf
- 0.1 – Estimated camber at erection, in.
- 0.2 – Estimated long-time camber, in.

SOLID FLAT SLAB
6 in. Thick
Normal Weight Concrete

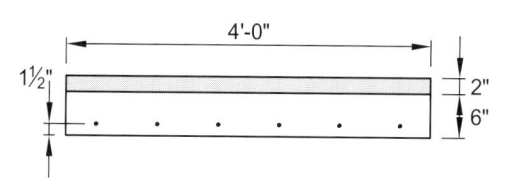

4'-0" wide, 1½" cover, 2" top, 6" total

$f'_c = 5,000$ psi
$f_{pu} = 270,000$ psi

Section Properties

	Untopped	Topped
A =	288 in.²	384 in.²
I =	864 in.⁴	1,834 in.⁴
y_b =	3.00 in.	3.82 in.
y_t =	3.00 in.	4.18 in.
S_b =	288 in.³	480 in.³
S_t =	288 in.³	439 in.³
wt =	300 plf	400 plf
DL =	75 psf	100 psf
V/S =	2.67 in.	

FS6 — No Topping

Table of safe superimposed service load (psf) and cambers (in.)

Strand Designation Code	Span, ft																						
	11	12	13	14	15	16	17	18	19	20	21	22	23	24	25	26	27	28	29	30	31	32
66-S	361	312	261	217	182	154	130	110	93	79	66	53	42	33								
	0.1	0.1	0.1	0.1	0.1	0.1	0.1	0.1	0.0	–0.1	–0.1	–0.2	–0.3	–0.5								
	0.2	0.2	0.2	0.2	0.1	0.1	0.1	0.0	–0.1	–0.2	–0.4	–0.5	–0.7	–1.0								
76-S		367	307	257	217	184	157	134	114	96	80	66	54	43	34	26						
		0.2	0.2	0.2	0.2	0.2	0.2	0.1	0.1	0.0	0.0	–0.1	–0.2	–0.3	–0.5	–0.7						
		0.2	0.2	0.2	0.2	0.2	0.2	0.1	0.0	–0.1	–0.2	–0.4	–0.5	–0.8	–1.0	–1.3						
58-S			350	308	273	239	203	173	148	126	107	91	77	65	54	44	35	28				
			0.3	0.3	0.3	0.3	0.3	0.3	0.3	0.2	0.2	0.1	0.0	–0.1	–0.2	–0.3	–0.5	–0.7				
			0.3	0.4	0.4	0.4	0.4	0.3	0.3	0.2	0.1	0.0	–0.2	–0.4	–0.6	–0.9	–1.2	–1.6				
68-S				371	329	282	241	207	178	154	132	114	98	84	71	60	50	42	34	27		
				0.4	0.4	0.4	0.4	0.4	0.4	0.4	0.4	0.3	0.3	0.2	0.1	0.0	–0.2	–0.4	–0.6	–0.9		
				0.5	0.5	0.5	0.5	0.5	0.5	0.5	0.4	0.3	0.2	0.0	–0.2	–0.4	–0.7	–1.1	–1.5	–1.9		
78-S				374	321	276	239	208	181	157	137	119	103	89	77	66	56	47	39	32	25	
				0.5	0.5	0.6	0.6	0.6	0.6	0.6	0.6	0.5	0.4	0.4	0.3	0.1	0.0	–0.2	–0.5	–0.7	–1.0	
				0.6	0.7	0.7	0.7	0.7	0.7	0.7	0.6	0.5	0.4	0.2	0.0	–0.3	–0.6	–0.9	–1.4	–1.8	–2.4	

FS6 + 2 — 2 in. Normal Weight Topping

Table of safe superimposed service load (psf) and cambers (in.)

Strand Designation Code	Span, ft																
	13	14	15	16	17	18	19	20	21	22	23	24	25	26	27	28	29
66-S	377	315	265	224	191	153	120	92	68	47	29						
	0.1	0.1	0.1	0.1	0.1	0.1	0.0	–0.1	–0.1	–0.2	–0.3						
	0.1	0.1	0.1	0.0	0.0	–0.1	–0.2	–0.3	–0.5	–0.7	–0.9						
76-S		371	314	267	227	184	148	118	91	68	48	31					
		0.2	0.2	0.2	0.2	0.1	0.1	0.0	0.0	–0.1	–0.2	–0.3					
		0.2	0.1	0.1	0.0	0.0	–0.1	–0.2	–0.4	–0.5	–0.7	–1.0					
58-S			392	347	297	247	205	169	137	110	87	66	48	32			
			0.3	0.3	0.3	0.3	0.3	0.2	0.2	0.1	0.0	–0.1	–0.2	–0.3			
			0.3	0.2	0.2	0.2	0.1	0.0	–0.1	–0.3	–0.4	–0.7	–0.9	–1.2			
68-S					357	304	256	214	179	148	121	98	77	59	42	28	
					0.4	0.4	0.4	0.4	0.4	0.3	0.3	0.2	0.1	0.0	–0.2	–0.4	
					0.4	0.3	0.3	0.2	0.1	0.0	–0.2	–0.3	–0.6	–0.9	–1.2	–1.5	
78-S						357	305	260	221	186	156	130	107	86	68	51	36
						0.6	0.6	0.6	0.6	0.6	0.5	0.4	0.4	0.3	0.1	0.0	–0.2
						0.5	0.5	0.4	0.4	0.3	0.1	0.0	–0.2	–0.5	–0.8	–1.1	–1.5

Strength is based on strain compatibility; bottom tension is limited to $7.5\sqrt{f'_c}$; see pages 2–7 through 2–10 for explanation.

Strand Pattern Designation
76-S

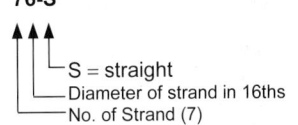

↑ ↑ ↑
- S = straight
- Diameter of strand in 16ths
- No. of Strand (7)

Safe loads shown include dead load of 10 psf for untopped members and 15 psf for topped members. Remainder is live load. Long-time cambers include superimposed dead load but do not include live load.

Key
- 361 – Safe superimposed service load, psf
- 0.1 – Estimated camber at erection, in.
- 0.2 – Estimated long-time camber, in.

SOLID FLAT SLAB
8 in. Thick
Normal Weight Concrete

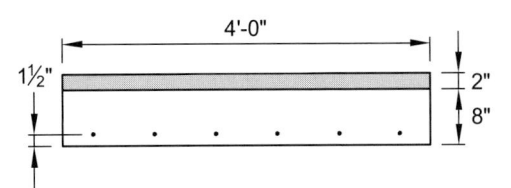

4'-0" wide, 1½" cover top, 2" topping shown, 8" thick

$f'_c = 5{,}000$ psi
$f_{pu} = 270{,}000$ psi

Section Properties

	Untopped	Topped
A =	384 in.²	480 in.²
I =	2,048 in.⁴	3,630 in.⁴
y_b =	4.00 in.	4.81 in.
y_t =	4.00 in.	5.19 in.
S_b =	512 in.³	755 in.³
S_t =	512 in.³	699 in.³
wt =	400 plf	500 plf
DL =	100 psf	125 psf
V/S =	3.43 in.	

FS8 — No Topping

Table of safe superimposed service load (psf) and cambers (in.)

Strand Designation Code	\multicolumn{24}{c}{Span, ft}
	13	14	15	16	17	18	19	20	21	22	23	24	25	26	27	28	29	30	31	32	33	34	35	36
66-S	361	330	278	236	201	171	146	125	107	91	77	65	54	44	36	28								
	0.1	0.1	0.1	0.1	0.1	0.1	0.1	0.1	0.0	0.0	−0.1	−0.2	−0.2	−0.3	−0.4	−0.6								
	0.2	0.2	0.1	0.1	0.1	0.1	0.1	0.0	0.0	−0.1	−0.2	−0.3	−0.5	−0.6	−0.8	−1.0								
76-S		393	333	284	243	209	180	155	134	116	100	86	73	62	52	41	32							
		0.1	0.2	0.2	0.2	0.2	0.1	0.1	0.1	0.1	0.0	−0.1	−0.1	−0.2	−0.3	−0.4	−0.6							
		0.2	0.2	0.2	0.2	0.2	0.2	0.1	0.1	0.0	−0.1	−0.2	−0.3	−0.5	−0.7	−0.9	−1.1							
58-S			372	324	281	245	214	187	164	144	125	107	92	78	65	54	44	35	27					
			0.2	0.3	0.3	0.3	0.3	0.2	0.2	0.2	0.1	0.1	0.0	−0.1	−0.2	−0.3	−0.4	−0.6	−0.8					
			0.3	0.3	0.3	0.3	0.3	0.3	0.2	0.2	0.1	0.0	−0.1	−0.3	−0.5	−0.7	−0.9	−1.2	−1.5					
68-S					394	344	301	265	232	203	177	154	134	117	101	87	74	63	52	43	34	27		
					0.3	0.4	0.4	0.4	0.4	0.4	0.4	0.3	0.3	0.2	0.2	0.1	0.0	−0.1	−0.3	−0.5	−0.6	−0.9		
					0.5	0.5	0.5	0.5	0.5	0.4	0.4	0.3	0.3	0.2	0.0	−0.1	−0.3	−0.5	−0.8	−1.0	−1.4	−1.7		
78-S							352	308	270	237	209	183	161	142	124	108	94	81	70	59	50	41	33	26
							0.5	0.5	0.5	0.5	0.5	0.5	0.5	0.4	0.4	0.3	0.2	0.1	0.0	−0.1	−0.3	−0.5	−0.7	−1.0
							0.6	0.7	0.7	0.7	0.6	0.6	0.5	0.5	0.3	0.2	0.1	−0.1	−0.3	−0.6	−0.9	−1.2	−1.6	−2.0

FS8 + 2 — 2 in. Normal Weight Topping

Table of safe superimposed service load (psf) and cambers (in.)

Strand Designation Code	\multicolumn{19}{c}{Span, ft}
	15	16	17	18	19	20	21	22	23	24	25	26	27	28	29	30	31	32	33
66-S	360	305	260	222	190	163	139	119	98	75	55	37							
	0.1	0.1	0.1	0.1	0.1	0.1	0.0	0.0	−0.1	−0.2	−0.2	−0.3							
	0.1	0.1	0.1	0.0	0.0	−0.1	−0.1	−0.2	−0.3	−0.4	−0.6	−0.8							
76-S		365	313	269	232	201	174	150	125	99	77	57	40						
		0.2	0.2	0.2	0.1	0.1	0.1	0.1	0.0	−0.1	−0.1	−0.2	−0.3						
		0.1	0.1	0.1	0.1	0.0	0.0	−0.1	−0.2	−0.3	−0.5	−0.6	−0.8						
58-S				359	313	274	240	210	177	147	121	98	78	59	43	28			
				0.3	0.3	0.3	0.2	0.2	0.2	0.1	0.1	0.0	−0.1	−0.2	−0.3	−0.4			
				0.2	0.2	0.2	0.2	0.1	0.0	−0.1	−0.2	−0.3	−0.5	−0.7	−0.9	−1.1			
68-S					381	335	296	261	224	190	161	135	112	91	72	56	40	27	
					0.4	0.4	0.4	0.4	0.4	0.3	0.3	0.2	0.2	0.1	0.0	−0.1	−0.3	−0.5	
					0.4	0.4	0.3	0.3	0.2	0.2	0.1	−0.1	−0.2	−0.4	−0.6	−0.8	−1.1	−1.4	
78-S						392	348	309	271	234	201	172	146	123	102	83	66	51	37
						0.5	0.5	0.5	0.5	0.5	0.5	0.4	0.4	0.3	0.2	0.1	0.0	−0.1	−0.3
						0.5	0.5	0.5	0.4	0.4	0.3	0.2	0.1	−0.1	−0.2	−0.5	−0.7	−1.0	−1.3

Strength is based on strain compatibility; bottom tension is limited to $7.5\sqrt{f'_c}$; see pages 2–7 through 2–10 for explanation.

RECTANGULAR BEAMS

Normal Weight Concrete

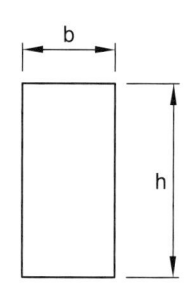

Designation	b in.	h in.	A in.²	I in.⁴	y_b in.	S in.³	wt plf
12RB16	12	16	192	4,096	8.00	512	200
12RB20	12	20	240	8,000	10.00	800	250
12RB24	12	24	288	13,824	12.00	1152	300
12RB28	12	28	336	21,952	14.00	1568	350
12RB32	12	32	384	32,768	16.00	2048	400
12RB36	12	36	432	46,656	18.00	2592	450
16RB24	16	24	384	18,432	12.00	1536	400
16RB28	16	28	448	29,269	14.00	2091	467
16RB32	16	32	512	43,691	16.00	2731	533
16RB36	16	36	576	62,208	18.00	3456	600
16RB40	16	40	640	85,333	20.00	4267	667

$f'_c = 5,000$ psi
$f_{pu} = 270,000$ psi
½ in. diameter
low-relaxation strand

1. Check local area for availability of other sizes.
2. Safe loads shown include 50% superimposed dead load and 50% live load. 800 psi top tension has been allowed, therefore, additional top reinforcement is required.
3. Safe loads can be significantly increased by use of structural composite topping.

Key
3553 – Safe superimposed service load, plf.
0.4 – Estimated camber at erection, in.
0.2 – Estimated long-time camber, in.

Table of safe superimposed service load (plf) and cambers (in.)

Desig-nation	No. Strand	y_s(end) in. y_s(center) in.	16	18	20	22	24	26	28	30	32	34	36	40	42	44	46	48	50	52
12RB16	58-S	3.00 / 3.00	3553 0.4 0.2	2772 0.5 0.2	2212 0.6 0.2	1799 0.8 0.2	1484 0.9 0.3	1239 1.0 0.3	1045 1.1 0.3											
12RB20	88-S	3.00 / 3.00	6163 0.4 0.2	4825 0.5 0.2	3867 0.6 0.3	3159 0.7 0.3	2620 0.9 0.4	2201 1.0 0.4	1868 1.1 0.4	1600 1.3 0.5	1380 1.4 0.5	1198 1.5 0.5	1046 1.7 0.5							
12RB24	108-S	3.60 / 3.60	8950 0.4 0.2	7018 0.4 0.2	5636 0.5 0.3	4613 0.7 0.3	3835 0.8 0.3	3230 0.9 0.4	2749 1.0 0.4	2362 1.1 0.5	2045 1.3 0.5	1782 1.4 0.6	1562 1.5 0.6	1375 1.6 0.6	1216 1.8 0.6	1079 1.9 0.7	960 2.0 0.6			
12RB28	128-S	4.00 / 4.00		9781 0.4 0.2	7866 0.5 0.2	6448 0.6 0.3	5370 0.7 0.3	4532 0.8 0.4	3866 0.9 0.4	3329 1.0 0.5	2890 1.2 0.5	2525 1.3 0.6	2220 1.4 0.6	1962 1.5 0.7	1741 1.7 0.7	1552 1.8 0.7	1387 1.9 0.8	1244 2.0 0.8	1118 2.1 0.8	1006 2.2 0.8
12RB32	138-S	4.77 / 4.77				8320 0.5 0.2	6936 0.6 0.3	5859 0.7 0.3	5005 0.8 0.3	4316 0.9 0.4	3752 1.0 0.4	3284 1.1 0.4	2892 1.2 0.5	2561 1.3 0.5	2278 1.4 0.5	2034 1.5 0.5	1823 1.6 0.6	1639 1.7 0.6	1477 1.8 0.6	1334 1.9 0.6
12RB36	158-S	5.07 / 5.07					9015 0.5 0.2	7624 0.6 0.3	6521 0.7 0.3	5631 0.8 0.4	4902 0.9 0.4	4298 1.0 0.4	3792 1.1 0.5	3364 1.2 0.5	2999 1.3 0.6	2684 1.4 0.6	2411 1.5 0.6	2173 1.6 0.6	1964 1.7 0.7	1780 1.8 0.7
16RB24	138-S	3.54 / 3.54	9397 0.4 0.2	7547 0.5 0.2	6177 0.6 0.3	5136 0.8 0.3	4325 0.9 0.4	3682 1.0 0.4	3164 1.1 0.5	2739 1.2 0.5	2387 1.4 0.5	2092 1.5 0.5	1843 1.6 0.6	1629 1.7 0.6	1446 1.8 0.6	1287 1.9 0.6	1149 2.0 0.6	1027 2.1 0.5		
16RB28	148-S	3.71 / 3.71			8730 0.5 0.2	7272 0.6 0.2	6137 0.7 0.3	5237 0.8 0.3	4510 0.9 0.3	3915 1.1 0.4	3423 1.2 0.4	3010 1.3 0.4	2660 1.4 0.4	2362 1.5 0.4	2105 1.6 0.4	1883 1.7 0.4	1688 1.8 0.4	1518 1.9 0.4	1368 1.9 0.3	
16RB32	188-S	4.67 / 4.67				9340 0.6 0.3	7891 0.7 0.3	6741 0.8 0.4	5813 0.9 0.4	5054 1.0 0.5	4425 1.1 0.5	3897 1.2 0.6	3451 1.3 0.6	3070 1.5 0.6	2742 1.6 0.7	2458 1.7 0.7	2210 1.8 0.7	1992 1.9 0.7	1800 2.0 0.7	
16RB36	208-S	5.40 / 5.40					9946 0.6 0.3	5805 0.7 0.3	7343 0.8 0.4	6391 0.9 0.4	5603 1.0 0.5	4942 1.1 0.5	4383 1.2 0.5	3905 1.3 0.6	3494 1.4 0.6	3138 1.5 0.6	2827 1.6 0.6	2555 1.7 0.6	2314 1.8 0.6	
16RB40	228-S	6.00 / 6.00							9122 0.7 0.3	7949 0.8 0.4	6976 0.9 0.4	6160 1.0 0.4	5470 1.1 0.5	4881 1.2 0.5	4374 1.3 0.5	3935 1.4 0.5	3552 1.5 0.6	3215 1.6 0.6	2918 1.7 0.6	

L-BEAMS

Normal Weight Concrete

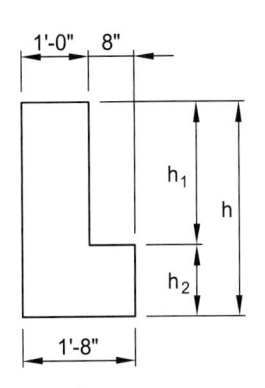

Designation	h in.	h_1/h_2 in./in.	A in.²	I in.⁴	y_b in.	S_b in.³	S_t in.³	wt plf
20LB20	20	12/8	304	10,160	8.74	1,163	902	317
20LB24	24	12/12	384	17,568	10.50	1,673	1,301	400
20LB28	28	16/12	432	27,883	12.22	2,282	1,767	450
20LB32	32	20/12	480	41,600	14.00	2,971	2,311	500
20LB36	36	24/12	528	59,119	15.82	3,737	2,930	550
20LB40	40	24/16	608	81,282	17.47	4,653	3,608	633
20LB44	44	28/16	656	108,107	19.27	5,610	4,372	683
20LB48	48	32/16	704	140,133	21.09	6,645	5,208	733
20LB52	52	36/16	752	177,752	22.94	7,749	6,117	783
20LB56	56	40/16	800	221,355	24.80	8,926	7,095	833
20LB60	60	44/16	848	271,332	26.68	10,170	8,143	883

1. Check local area for availability of other sizes.
2. Safe loads shown include 50% superimposed dead load and 50% live load. 800 psi top tension has been allowed, therefore, additional top reinforcement is required.
3. Safe loads can be significantly increased by use of structural composite topping.

f'_c = 5,000 psi
f_{pu} = 270,000 psi
½ in. diameter
low-relaxation strand

Key
6566 – Safe superimposed service load, plf.
0.3 – Estimated camber at erection, in.
0.1 – Estimated long-time camber, in.

Table of safe superimposed service load (plf) and cambers (in.)

Desig-nation	No. Strand	y_s(end) in. y_s(center) in.	16	18	20	22	24	26	28	30	32	34	36	38	40	42	44	46	48	50
20LB20	98-S	2.44 / 2.44	6566 0.3 0.1	5131 0.4 0.2	4105 0.5 0.2	3345 0.6 0.2	2768 0.7 0.2	2318 0.8 0.2	1961 0.9 0.3	1674 1.0 0.3	1438 1.0 0.3	1243 1.1 0.3	1079 1.2 0.2							
20LB24	108-S	2.80 / 2.80	9577 0.3 0.1	7495 0.3 0.1	6006 0.4 0.1	4904 0.5 0.1	4066 0.5 0.1	3414 0.6 0.2	2896 0.7 0.2	2479 0.8 0.2	2137 0.9 0.2	1854 0.9 0.2	1617 1.0 0.1	1416 1.0 0.1	1244 1.1 0.1	1097 1.1 0.0	969 1.2 0.0			
20LB28	128-S	3.33 / 3.33			8228 0.4 0.1	6733 0.4 0.1	5596 0.5 0.2	4711 0.6 0.2	4009 0.6 0.2	3443 0.7 0.2	2979 0.8 0.2	2595 0.9 0.2	2273 0.9 0.2	2000 1.0 0.2	1768 1.1 0.2	1567 1.1 0.2	1394 1.2 0.1	1243 1.2 0.1	1110 1.2 0.0	992 1.3 0.0
20LB32	148-S	3.71 / 3.71				8942 0.4 0.1	7446 0.5 0.2	6281 0.5 0.2	5356 0.6 0.2	4611 0.7 0.2	4001 0.7 0.2	3495 0.8 0.2	3071 0.9 0.3	2712 1.0 0.3	2406 1.0 0.3	2143 1.1 0.2	1914 1.2 0.2	1715 1.2 0.2	1540 1.3 0.2	1386 1.3 0.1
20LB36	168-S	4.25 / 4.25					9457 0.4 0.2	7988 0.5 0.2	6823 0.5 0.2	5883 0.6 0.2	5113 0.7 0.2	4476 0.8 0.3	3941 0.8 0.3	3489 0.9 0.3	3103 1.0 0.3	2771 1.1 0.3	2483 1.1 0.3	2231 1.2 0.3	2011 1.2 0.3	1816 1.3 0.2
20LB40	188-S	4.89 / 4.89						9812 0.4 0.2	8386 0.5 0.2	7235 0.6 0.2	6293 0.6 0.2	5513 0.7 0.2	4858 0.8 0.3	4305 0.8 0.3	3832 0.9 0.3	3425 1.0 0.3	3073 1.0 0.3	2765 1.1 0.3	2495 1.1 0.3	2257 1.2 0.3
20LB44	198-S	5.05 / 5.05							8959 0.5 0.2	7803 0.6 0.2	6845 0.6 0.2	6042 0.7 0.2	5363 0.8 0.2	4783 0.8 0.2	4284 0.9 0.2	3851 0.9 0.2	3474 1.0 0.2	3143 1.1 0.2	2850 1.1 0.2	
20LB48	218-S	5.81 / 5.81								9226 0.5 0.2	8100 0.6 0.2	7158 0.6 0.2	6360 0.7 0.2	5678 0.8 0.2	5092 0.8 0.2	4584 0.9 0.3	4140 0.9 0.3	3751 1.0 0.3	3408 1.1 0.3	
20LB52	238-S	6.17 / 6.17									9634 0.6 0.2	8521 0.6 0.2	7578 0.7 0.2	6774 0.7 0.3	6082 0.8 0.3	5482 0.9 0.3	4958 0.9 0.3	4499 1.0 0.3	4094 1.0 0.3	
20LB56	258-S	6.64 / 6.64										9954 0.6 0.2	8860 0.7 0.2	7927 0.7 0.3	7124 0.8 0.3	6427 0.8 0.3	5820 0.9 0.3	5287 1.0 0.3	4816 1.0 0.3	
20LB60	278-S	7.33 / 7.33											9089 0.7 0.3	8173 0.7 0.3	7380 0.8 0.3	6688 0.9 0.3	6080 0.9 0.3	5544 1.0 0.3		

L-BEAMS

Normal Weight Concrete

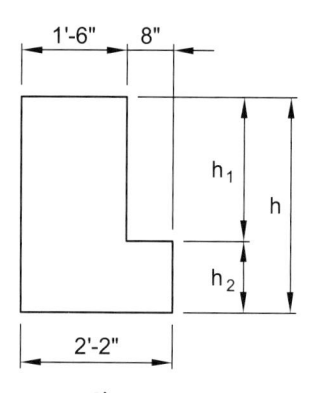

$f'_c = 5,000$ psi
$f_{pu} = 270,000$ psi
½ in. diameter
low-relaxation strand

Section Properties

Designation	h in.	h₁/h₂ in./in.	A in.²	I in.⁴	y_b in.	S_b in.³	S_t in.³	wt plf
26LB20	20	12/8	424	14,298	9.09	1,573	1,311	442
26LB24	24	12/12	528	24,716	10.91	2,265	1,888	550
26LB28	28	16/12	600	39,241	12.72	3,085	2,568	625
26LB32	32	20/12	672	58,533	14.57	4,017	3,358	700
26LB36	36	24/12	744	83,176	16.45	5,056	4,255	775
26LB40	40	24/16	848	114,381	18.19	6,288	5,244	883
26LB44	44	28/16	920	152,104	20.05	7,586	6,351	958
26LB48	48	32/16	992	197,159	21.94	8,986	7,566	1,033
26LB52	52	36/16	1,064	250,126	23.83	10,496	8,879	1,108
26LB56	56	40/16	1,136	311,586	25.75	12,100	10,300	1,183
26LB60	60	44/16	1,208	382,118	27.67	13,810	11,819	1,258

1. Check local area for availability of other sizes.
2. Safe loads shown include 50% superimposed dead load and 50% live load. 800 psi top tension has been allowed, therefore, additional top reinforcement is required.
3. Safe loads can be significantly increased by use of structural composite topping.

Key

9672 – Safe superimposed service load, plf.
0.4 – Estimated camber at erection, in.
0.2 – Estimated long-time camber, in.

Table of safe superimposed service load (plf) and cambers (in.)

Desig-nation	No. Strand	y_s(end) in. y_s(center) in.	16	18	20	22	24	26	28	30	32	34	36	38	40	42	44	46	48	50
26LB20	158-S	2.67 / 2.67	9672 0.4 0.2	7563 0.5 0.3	6054 0.6 0.3	4938 0.7 0.4	4089 0.8 0.4	3428 1.0 0.5	2903 1.1 0.5	2480 1.2 0.6	2134 1.4 0.6	1847 1.5 0.7	1607 1.6 0.7	1403 1.7 0.7	1230 1.8 0.7	1080 1.9 0.7	950 1.9 0.6			
26LB24	158-S	2.67 / 2.67			9165 0.5 0.2	7493 0.5 0.2	6221 0.6 0.2	5231 0.7 0.2	4445 0.8 0.3	3811 0.9 0.3	3293 1.0 0.3	2863 1.1 0.3	2503 1.2 0.3	2198 1.3 0.3	1938 1.3 0.3	1714 1.4 0.2	1520 1.5 0.2	1350 1.5 0.1	1202 1.5 0.1	1070 1.5 0.0
26LB28	188-S	3.33 / 3.33					8437 0.6 0.2	7170 0.6 0.2	6056 0.7 0.3	5207 0.8 0.3	4511 0.9 0.3	3935 1.0 0.3	3452 1.1 0.3	3043 1.2 0.3	2694 1.3 0.3	2394 1.3 0.3	2134 1.4 0.3	1907 1.5 0.3	1707 1.5 0.2	1532 1.6 0.2
26LB32	218-S	4.00 / 4.00							9265 0.6 0.2	7906 0.7 0.3	6809 0.7 0.3	5912 0.8 0.3	5169 0.9 0.3	4545 1.0 0.4	4018 1.1 0.4	3568 1.2 0.4	3180 1.2 0.4	2844 1.3 0.4	2551 1.4 0.4	2294 1.5 0.3	2067 1.5 0.3
26LB36	248-S	4.50 / 4.50								8722 0.7 0.3	7585 0.8 0.3	6643 0.9 0.3	5854 0.9 0.4	5186 1.0 0.4	4615 1.1 0.4	4125 1.2 0.4	3699 1.3 0.4	3328 1.3 0.4	3002 1.4 0.4	2715 1.5 0.4
26LB40	278-S	5.11 / 5.11									9372 0.7 0.3	8216 0.8 0.3	7246 0.9 0.3	6426 0.9 0.4	5726 1.0 0.4	5123 1.1 0.4	4601 1.2 0.4	4145 1.2 0.4	3745 1.3 0.4	3392 1.4 0.4
26LB44	288-S	5.29 / 5.29										8992 0.8 0.3	7986 0.8 0.3	7127 0.9 0.3	6388 1.0 0.3	5748 1.0 0.3	5189 1.1 0.3	4698 1.2 0.3	4266 1.2 0.3	
26LB48	328-S	5.75 / 5.75											9635 0.8 0.3	8609 0.9 0.4	7726 1.0 0.4	6961 1.0 0.4	6294 1.1 0.4	5708 1.2 0.4	5191 1.3 0.4	
26LB52	358-S	6.29 / 6.29													9137 0.9 0.4	8241 1.0 0.4	7459 1.1 0.4	6773 1.1 0.4	6167 1.2 0.5	
26LB56	378-S	7.00 / 7.00													9539 0.9 0.4	8641 1.0 0.4	7853 1.1 0.4	7158 1.1 0.4		
26LB60	388-S	7.68 / 7.68														9904 0.9 0.3	9008 0.9 0.3	8217 1.0 0.3		

2–44 PCI Design Handbook/Sixth Edition

INVERTED TEE BEAMS

Normal Weight Concrete

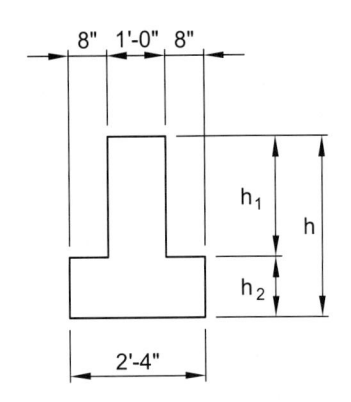

Section Properties								
Designation	h in.	h_1/h_2 in./in.	A in.^{2}	I in.^{4}	y_b in.	S_b in.^{3}	S_t in.^{3}	wt plf
28IT20	20	12/8	368	11,688	7.91	1,478	967	383
28IT24	24	12/12	480	20,275	9.60	2,112	1,408	500
28IT28	28	16/12	528	32,076	11.09	2,892	1,897	550
28IT32	32	20/12	576	47,872	12.67	3,778	2,477	600
28IT36	36	24/12	624	68,101	14.31	4,759	3,140	650
28IT40	40	24/16	736	93,503	15.83	5,907	3,869	767
28IT44	44	28/16	784	124,437	17.43	7,139	4,683	817
28IT48	48	32/16	832	161,424	19.08	8,460	5,582	867
28IT52	52	36/16	880	204,884	20.76	9,869	6,558	917
28IT56	56	40/16	928	255,229	22.48	11,354	7,614	967
28IT60	60	44/16	976	312,866	24.23	12,912	8,747	1,017

$f'_c = 5,000$ psi
$f_{pu} = 270,000$ psi
½ in. diameter
low-relaxation strand

1. Check local area for availability of other sizes.
2. Safe loads shown include 50% superimposed dead load and 50% live load. 800 psi top tension has been allowed, therefore, additional top reinforcement is required.
3. Safe loads can be significantly increased by use of structural composite topping.

Key
6511 – Safe superimposed service load, plf.
0.2 – Estimated camber at erection, in.
0.1 – Estimated long-time camber, in.

Table of safe superimposed service load (plf) and cambers (in.)

Designation	No. Strand	y_s(end) in. y_s(center) in.	Span, ft																	
			16	18	20	22	24	26	28	30	32	34	36	38	40	42	44	46	48	50
28IT20	98-S	2.44 / 2.44	6511 0.2 0.1	5076 0.3 0.1	4049 0.4 0.1	3289 0.4 0.1	2711 0.5 0.1	2262 0.5 0.1	1905 0.6 0.1	1617 0.7 0.0	1381 0.7 0.0	1186 0.7 0.0	1022 0.8 −0.1							
28IT24	188-S	2.73 / 2.73	9612 0.2 0.1	7504 0.3 0.1	5997 0.3 0.1	4882 0.4 0.1	4034 0.4 0.1	3374 0.5 0.1	2850 0.6 0.1	2427 0.6 0.1	2081 0.7 0.1	1795 0.7 0.1	1555 0.7 0.0	1351 0.8 0.0	1178 0.8 −0.1	1029 0.8 −0.2				
28IT28	138-S	3.08 / 3.08		8353 0.3 0.1	6822 0.3 0.1	5657 0.4 0.1	4750 0.5 0.1	4031 0.5 0.1	3451 0.6 0.1	2976 0.6 0.1	2582 0.7 0.1	2252 0.7 0.1	1973 0.8 0.1	1735 0.8 0.0	1530 0.8 0.0	1352 0.8 −0.1	1197 0.8 −0.2	1061 0.8 −0.2		
28IT32	158-S	3.47 / 3.47			9049 0.3 0.1	7521 0.4 0.1	5333 0.4 0.1	5389 0.5 0.1	4628 0.5 0.1	4006 0.6 0.1	3490 0.6 0.1	3057 0.7 0.1	2691 0.7 0.1	2379 0.8 0.1	2110 0.8 0.1	1876 0.9 0.0	1673 0.9 0.0	1495 0.9 0.0	1337 0.9 −0.1	
28IT36	168-S	3.50 / 3.50				9832 0.3 0.1	8295 0.4 0.1	7075 0.4 0.1	6092 0.5 0.1	5287 0.5 0.1	4619 0.6 0.1	4060 0.6 0.1	3587 0.7 0.1	3183 0.7 0.1	2835 0.8 0.1	2534 0.8 0.0	2271 0.9 0.0	2040 0.9 0.0	1836 0.9 −0.1	
28IT40	198-S	4.21 / 4.21						8638 0.4 0.1	7440 0.5 0.1	6460 0.5 0.1	5647 0.6 0.1	4966 0.6 0.1	4390 0.7 0.1	3898 0.7 0.1	3474 0.8 0.1	3107 0.8 0.1	2787 0.8 0.1	2506 0.9 0.1	2258 0.9 0.1	
28IT44	208-S	4.40 / 4.40							9186 0.4 0.1	7989 0.5 0.1	6997 0.5 0.1	6165 0.6 0.1	5462 0.6 0.1	4861 0.7 0.1	4344 0.7 0.1	3896 0.7 0.1	3505 0.8 0.1	3162 0.8 0.1	2859 0.8 0.0	
28IT48	228-S	4.55 / 4.55								9719 0.4 0.1	8525 0.5 0.1	7523 0.5 0.1	6676 0.6 0.1	5953 0.6 0.1	5330 0.7 0.1	4791 0.7 0.1	4320 0.8 0.1	3907 0.8 0.1	3542 0.9 0.1	
28IT52	248-S	5.17 / 5.17									9987 0.5 0.1	8823 0.5 0.1	7838 0.6 0.1	6998 0.6 0.1	6274 0.6 0.1	5647 0.7 0.1	4100 0.7 0.1	4619 0.8 0.1	4196 0.8 0.1	
28IT56	268-S	5.23 / 5.23											9307 0.5 0.2	8319 0.6 0.2	7469 0.6 0.2	6731 0.7 0.2	6088 0.7 0.2	5524 0.8 0.2	5026 0.8 0.2	
28IT60	288-S	5.57 / 5.57												9645 0.6 0.2	8668 0.6 0.2	7820 0.7 0.2	7081 0.7 0.2	6432 0.8 0.2	5859 0.8 0.2	

PCI Design Handbook/Sixth Edition

INVERTED TEE BEAMS

Normal Weight Concrete

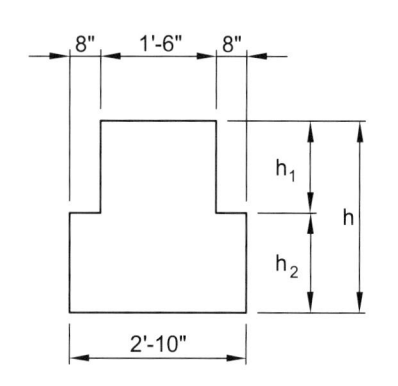

$f'_c = 5,000$ psi
$f_{pu} = 270,000$ psi
½ in. diameter
low-relaxation strand

Section Properties

Designation	h in.	h₁/h₂ in./in.	A in.²	I in.⁴	y_b in.	S_b in.³	S_t in.³	wt plf
34IT20	20	12/8	488	16,082	8.43	1,908	1,390	508
34IT24	24	12/12	624	27,825	10.15	2,741	2,009	650
34IT28	28	16/12	696	44,130	11.79	3,743	2,722	725
34IT32	32	20/12	768	65,856	13.50	4,878	3,560	800
34IT36	36	24/12	840	93,616	15.26	6,135	4,514	875
34IT40	40	24/16	976	128,656	16.85	7,635	5,558	1,017
34IT44	44	28/16	1,048	171,157	18.58	9,212	6,733	1,092
34IT48	48	23/16	1,120	221,906	20.34	10,910	8,023	1,167
34IT52	52	36/16	1,192	281,504	22.13	12,721	9,424	1,242
34IT60	60	44/16	1,336	439,623	25.78	17,053	12,847	1,392

1. Check local area for availability of other sizes.
2. Safe loads shown include 50% superimposed dead load and 50% live load. 800 psi top tension has been allowed, therefore, additional top reinforcement is required.
3. Safe loads can be significantly increased by use of structural composite topping.

Key
7822 – Safe superimposed service load, plf.
0.4 – Estimated camber at erection, in.
0.1 – Estimated long-time camber, in.

Table of safe superimposed service load (plf) and cambers (in.)

Desig-nation	No. Strand	y_s(end) in. y_s(center) in.	16	18	20	22	24	26	28	30	32	34	36	38	40	42	44	46	48	50
34IT20	148-S	2.29 / 2.29	7822 0.4 0.1	6253 0.5 0.2	5092 0.6 0.2	4209 0.7 0.2	3522 0.7 0.2	2977 0.8 0.2	2537 0.9 0.2	2177 1.0 0.2	1879 1.1 0.2	1629 1.1 0.2	1417 1.2 0.2	1237 1.2 0.1	1081 1.2 0.1					
34IT24	178-S	2.59 / 2.59		9221 0.4 0.2	7524 0.5 0.2	6233 0.6 0.2	5229 0.7 0.2	4432 0.7 0.3	3789 0.8 0.3	3262 0.9 0.3	2826 1.0 0.3	2461 1.1 0.3	2151 1.1 0.3	1887 1.2 0.2	1660 1.2 0.2	1463 1.2 0.2	1291 1.3 0.1	1140 1.3 0.0	1007 1.2 −0.1	
34IT28	208-S	3.00 / 3.00					8641 0.5 0.2	7271 0.6 0.2	6183 0.7 0.2	5306 0.7 0.3	4589 0.8 0.3	3994 0.9 0.3	3495 1.0 0.3	3073 1.0 0.3	2713 1.1 0.3	2403 1.2 0.3	2134 1.2 0.3	1900 1.3 0.2	1694 1.3 0.2	1513 1.3 0.1
34IT32	238-S	3.48 / 3.48							9589 0.5 0.2	8174 0.6 0.2	7032 0.7 0.3	6097 0.8 0.3	5323 0.8 0.3	4674 0.9 0.3	4124 1.0 0.3	3655 1.0 0.3	3252 1.1 0.3	2902 1.2 0.3	2597 1.2 0.3	2329 1.3 0.3	2093 1.3 0.2
34IT36	248-S	3.50 / 3.50								9223 0.6 0.2	8016 0.7 0.2	7015 0.7 0.2	6176 0.8 0.2	5466 0.9 0.2	4860 0.9 0.2	4338 1.0 0.2	3886 1.1 0.2	3492 1.1 0.2	3146 1.2 0.2	2840 1.2 0.2
34IT40	308-S	4.40 / 4.40									9720 0.6 0.3	8510 0.7 0.3	7497 0.8 0.4	6639 0.9 0.4	5907 0.9 0.4	5277 1.0 0.4	4731 1.1 0.4	4254 1.1 0.4	3836 1.2 0.4	3467 1.3 0.4
34IT44	308-S	4.40 / 4.40										9362 0.7 0.2	8307 0.7 0.2	7406 0.8 0.2	6630 0.9 0.2	5958 0.9 0.3	5372 1.0 0.3	4857 1.0 0.2	4403 1.1 0.2
34IT48	338-S	4.73 / 4.73												8963 0.8 0.3	8037 0.8 0.3	7234 0.9 0.3	6533 1.0 0.3	5919 1.0 0.3	5376 1.1 0.3
34IT52	368-S	5.22 / 5.22													9503 0.8 0.3	8564 0.9 0.3	7745 0.9 0.3	7026 1.0 0.3	6392 1.0 0.3
34IT56	398-S	5.59 / 5.59																	8269 1.0 0.3	7532 1.0 0.3
34IT60	408-S	6.00 / 6.00																	9564 0.8 0.3	8721 0.9 0.3

2–46 PCI Design Handbook/Sixth Edition

INVERTED TEE BEAMS

Normal Weight Concrete

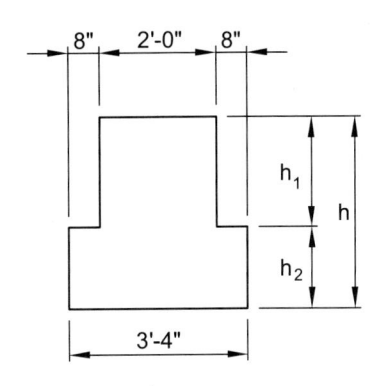

Section Properties

Designation	h in.	h₁/h₂ in./in.	A in.²	I in.⁴	y_b in.	S_b in.³	S_t in.³	wt plf
40IT20	20	12/8	608	20,321	8.74	2,325	1,805	633
40IT24	24	12/12	768	35,136	10.50	3,346	2,603	800
40IT28	28	16/12	864	55,765	12.22	4,563	3,534	900
40IT32	32	20/12	960	83,200	14.00	5,943	4,622	1,000
40IT36	36	24/12	1,056	118,237	15.82	7,474	5,859	1,100
40IT40	40	24/16	1,216	162,564	17.47	9,305	7,215	1,267
40IT44	44	28/16	1,312	216,215	19.27	11,220	8,743	1,367
40IT48	48	32/16	1,408	280,266	21.09	13,289	10,415	1,467
40IT52	52	36/16	1,504	355,503	22.94	15,497	12,233	1,567

$f'_c = 5,000$ psi
$f_{pu} = 270,000$ psi
½ in. diameter low-relaxation strand

1. Check local area for availability of other sizes.
2. Safe loads shown include 50% superimposed dead load and 50% live load. 800 psi top tension has been allowed, therefore, additional top reinforcement is required.
3. Safe loads can be significantly increased by use of structural composite topping.

Key
8427 – Safe superimposed service load, plf.
0.5 – Estimated camber at erection, in.
0.2 – Estimated long-time camber, in.

Table of safe superimposed service load (plf) and cambers (in.)

Designation	No. Strand	y_s(end) in. / y_s(center) in.	16	18	20	22	24	26	28	30	32	34	36	38	40	42	44	46	48	50
40IT20	188-S	2.22 / 2.22	8427 0.5 0.2	6870 0.6 0.2	5686 0.7 0.3	4764 0.8 0.3	4033 0.9 0.3	3444 1.0 0.3	2961 1.1 0.3	2561 1.2 0.3	2225 1.3 0.3	1942 1.3 0.3	1699 1.4 0.3	1491 1.4 0.2	1310 1.5 0.2	1153 1.5 0.1	1014 1.5 0.0		-0.1	
40IT24	228-S	2.67 / 2.67		9994 0.5 0.2	8288 0.6 0.2	6961 0.7 0.3	5907 0.8 0.3	5057 0.9 0.3	4362 1.0 0.3	3786 1.1 0.3	3303 1.2 0.3	2894 1.2 0.3	2545 1.3 0.3	2244 1.4 0.3	1984 1.4 0.3	1757 1.4 0.2	1558 1.5 0.1	1382 1.5 0.0		
40IT28	268-S	3.08 / 3.08				9672 0.6 0.3	8233 0.7 0.3	7073 0.8 0.3	6123 0.9 0.3	5336 1.0 0.3	4676 1.1 0.3	4118 1.2 0.4	3641 1.2 0.4	3231 1.3 0.4	2875 1.4 0.4	2565 1.4 0.3	2293 1.5 0.3	2052 1.5 0.3		
40IT32	308-S	3.33 / 3.33					9527 0.8 0.3	8269 0.8 0.3	7227 0.9 0.4	6354 1.0 0.4	5615 1.1 0.4	4984 1.2 0.4	4441 1.3 0.4	3970 1.3 0.4	3560 1.4 0.4	3199 1.5 0.4	2881 1.5 0.4			
40IT36	328-S	3.50 / 3.50						9410 0.8 0.3	8292 0.9 0.3	7345 1.0 0.3	6537 1.1 0.4	5842 1.1 0.4	5239 1.2 0.4	4713 1.3 0.4	4252 1.4 0.3	3844 1.4 0.3				
40IT40	388-S	4.32 / 4.32								8947 0.9 0.4	7969 1.0 0.4	7127 1.1 0.4	6398 1.2 0.4	5761 1.2 0.4	5202 1.3 0.4	4709 1.4 0.4				
40IT44	408-S	4.40 / 4.40									9950 0.9 0.3	8916 1.0 0.4	8020 1.1 0.4	7238 1.1 0.4	6552 1.2 0.4	5946 1.3 0.4				
40IT48	448-S	4.87 / 4.87											9652 1.0 0.4	8724 1.1 0.4	7910 1.2 0.4	7191 1.2 0.4				
40IT52	468-S	5.05 / 5.05												9494 1.1 0.4	8645 1.1 0.4					

PRECAST, PRESTRESSED COLUMNS

Figure 2.7.1 Design strength interaction curves for precast, prestressed concrete columns

CRITERIA
1. Minimum prestress = 225 psi
2. All strand assumed ½ in. diameter, f_{pu} = 270 ksi
3. Curves shown for partial development of strand near member end where $f_{pu} \approx f_{se}$
4. Horizontal portion of curve is the maximum for tied columns = $0.80\phi P_c$
5. Varies linearly from 0.9 for tension-controlled section to 0.65 for compression-controlled sections in accordance with ACI 318-02 Section 9.3.2

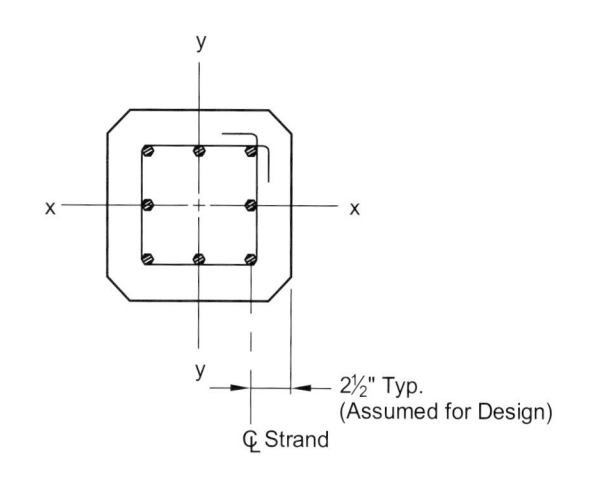

USE OF CURVES
1. Enter at left with applied factored axial load, P_u
2. Enter at bottom with applied magnified factored moment, δM_u
3. Intersection point must be to the left of curve indicating required concrete strength.

NOTATION
ϕP_n = Design axial strength
ϕM_n = Design flexural strength
ϕP_c = Design axial strength at zero eccentricity
A_g = Gross area of column
δ = Moment magnifier (Section 10.11–10.13 ACI 318-02)

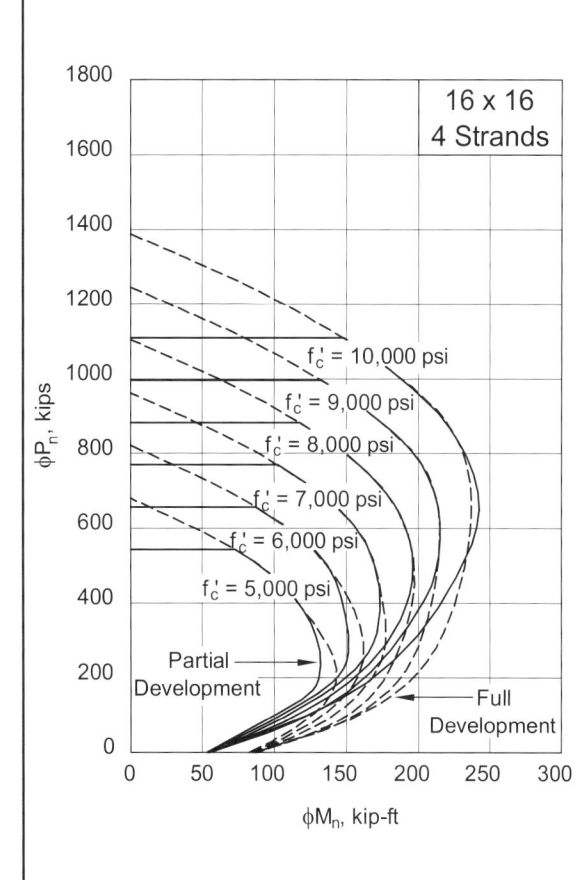

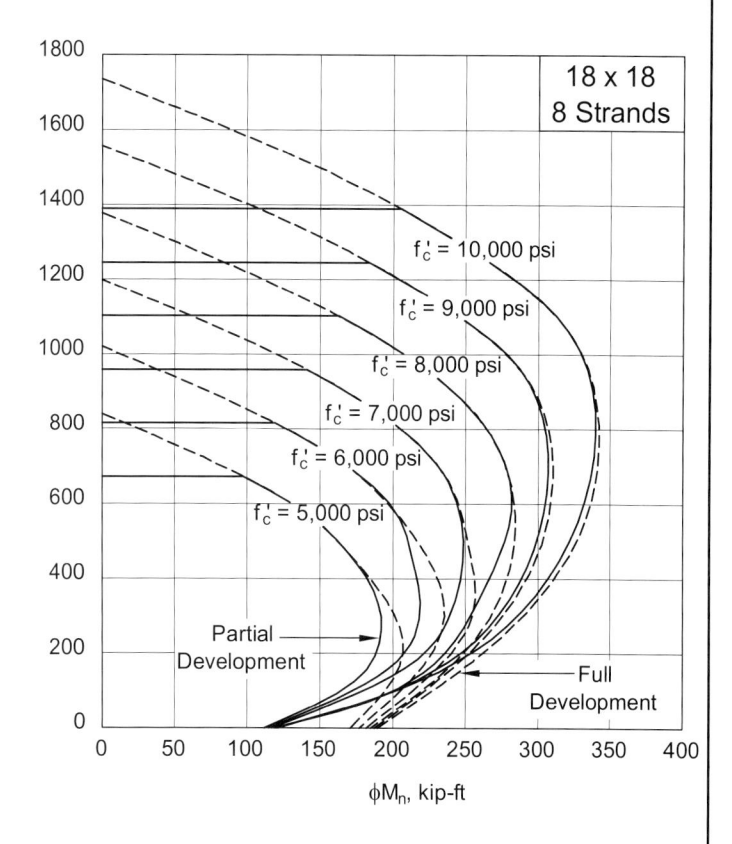

2½" Typ. (Assumed for Design)
℄ Strand

16 x 16
4 Strands

18 x 18
8 Strands

PRECAST, PRESTRESSED COLUMNS

Figure 2.7.1 Design strength interaction curves for precast, prestressed concrete columns (cont.)

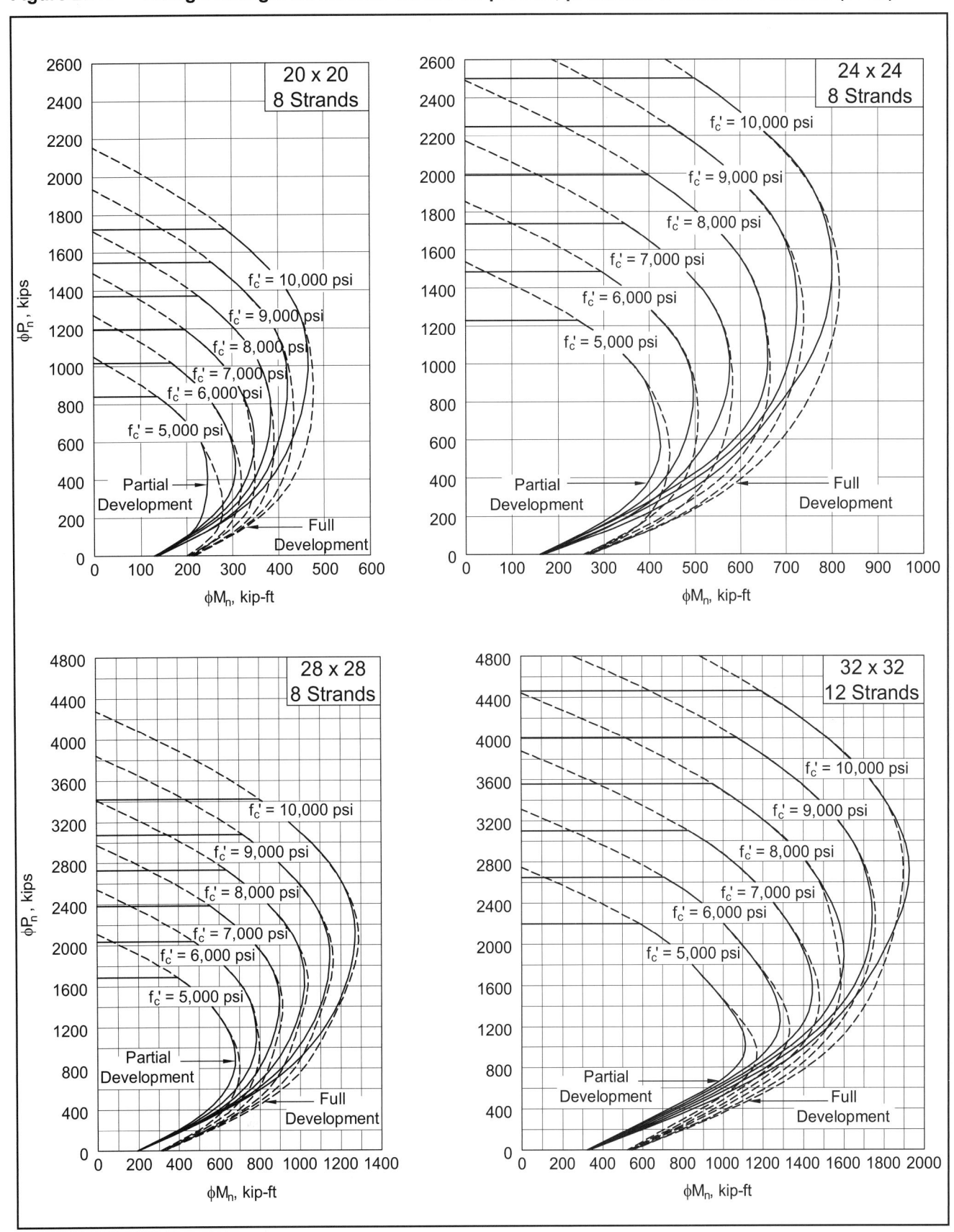

PRECAST, REINFORCED COLUMNS

Figure 2.7.2   Design strength interaction curves for precast, reinforced concrete columns

CRITERIA
1. Concrete f'_c = 5000 psi
2. Reinforcement f_y = 60,000 psi
3. Curves shown for full development of reinforcement
4. Horizontal portion of curve is the maximum for tied columns = $0.80\phi P_c$
5. Varies linearly from 0.9 for tension-controlled sections to 0.65 for compression-controlled sections in accordance with ACI 318-02 Section 9.3.2

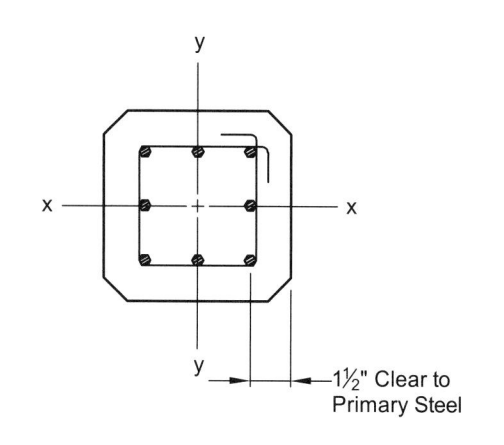

USE OF CURVES
1. Enter at left with applied factored axial load, P_u
2. Enter at bottom with applied magnified factored moment, δM_u
3. Intersection point must be to the left of curve indicating required reinforcement.

NOTATION
ϕP_n = Design axial strength
ϕM_n = Design flexural strength
ϕP_c = Design axial strength at zero eccentricity
A_g = Gross area of the column
δ = Moment magnifier (Section 10.11–10.13 ACI 318-02)

1½" Clear to Primary Steel

The interaction curves have been smoothed for plotting purposes. Exact calculated values may be slightly different.

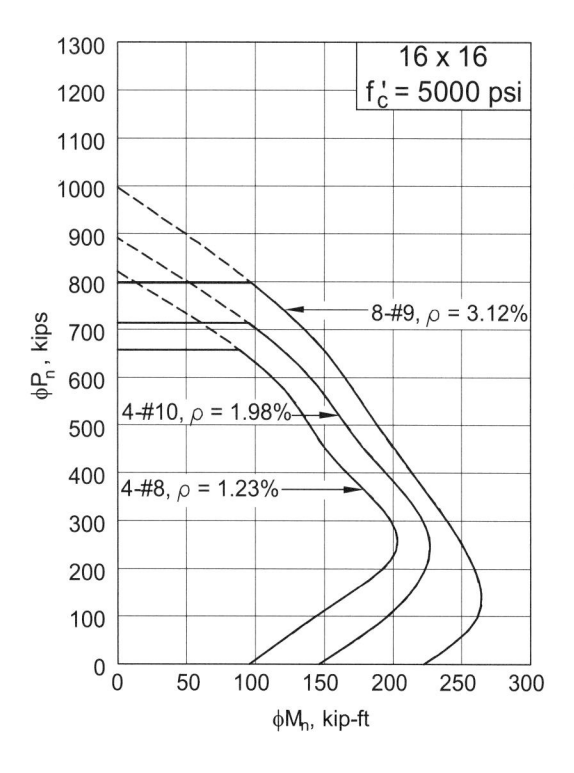

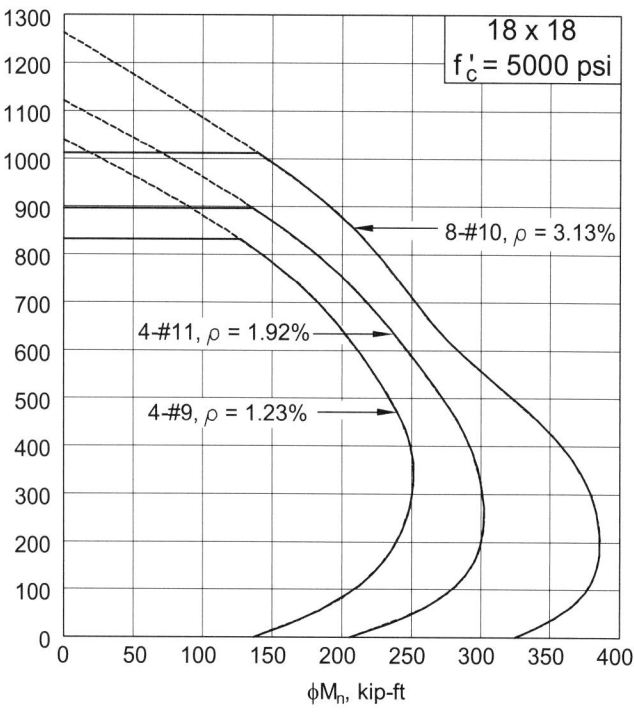

16 x 16, f'_c = 5000 psi
- 8-#9, ρ = 3.12%
- 4-#10, ρ = 1.98%
- 4-#8, ρ = 1.23%

18 x 18, f'_c = 5000 psi
- 8-#10, ρ = 3.13%
- 4-#11, ρ = 1.92%
- 4-#9, ρ = 1.23%

PRECAST, REINFORCED COLUMNS

Figure 2.7.2 Design strength interaction curves for precast, reinforced concrete columns (cont.)

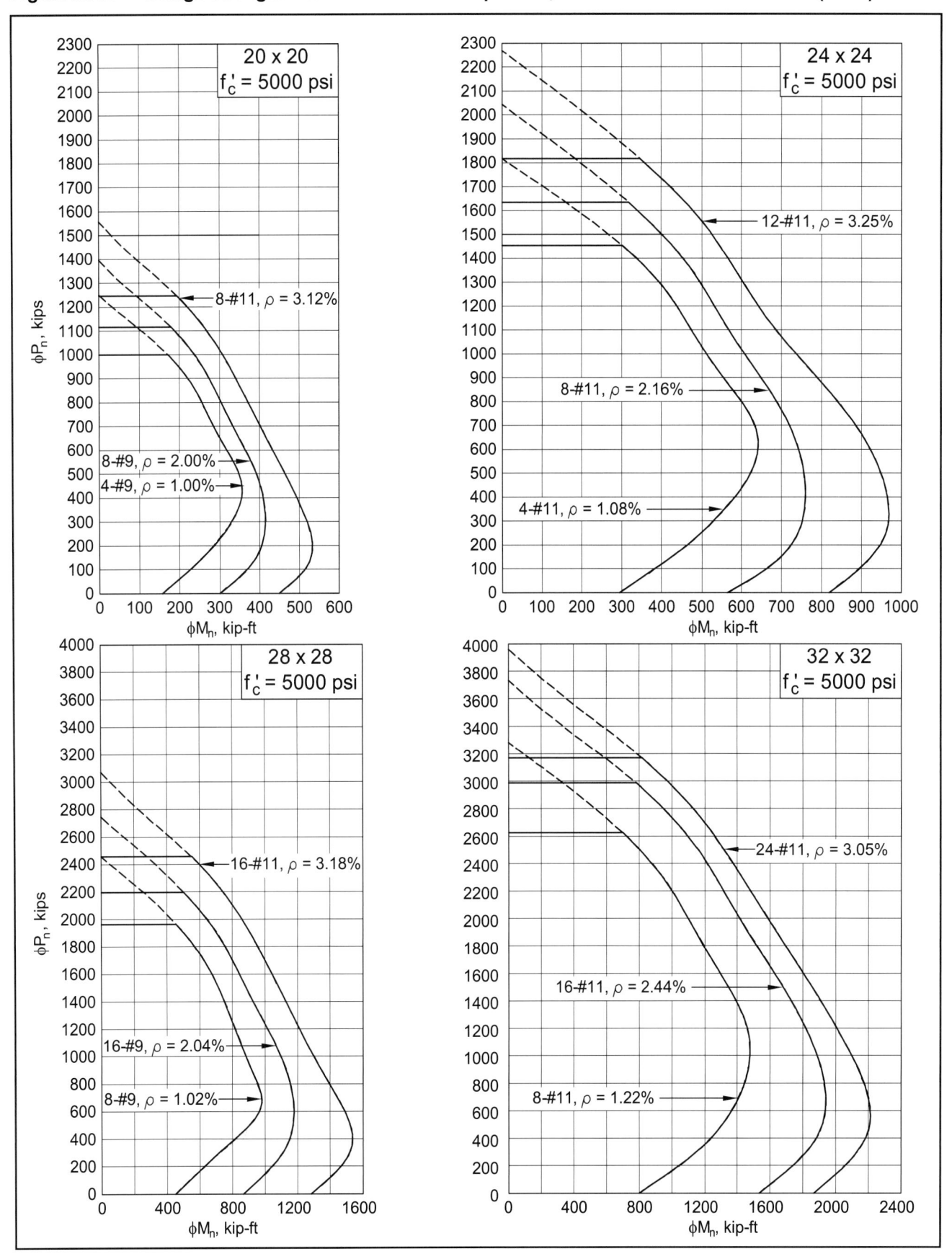

DOUBLE TEE WALL PANELS

Figure 2.7.3 Partial interaction curves for prestressed double tee wall panels

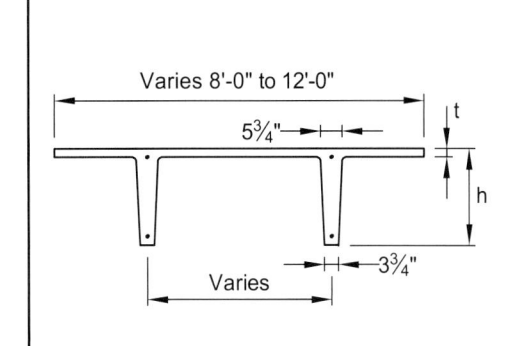

Mark	h, in.	t, in.	No. Strd.	ϕP_c	Partially Developed Strand Force ϕP_{nb}	ϕM_{nb}	ϕM_c	Fully Developed Strand Force ϕP_{nb}	ϕM_{nb}	ϕM_c
8DT16	16	2	4	871	547	164	57	513	179	95
8DT18	17	3	4	1136	775	196	64	743	213	107
10DT16	16	2	4	1003	667	175	57	634	192	96
10DT17	17	3	4	1335	962	208	64	929	226	108
12DT16	16	2	4	1136	789	185	57	757	202	96
12DT17	17	3	4	1534	1151	217	65	1118	236	108

f'_c = 5000 psi, Normal weight
Strand = ½ in. Diameter
f_{pu} = 270 ksi

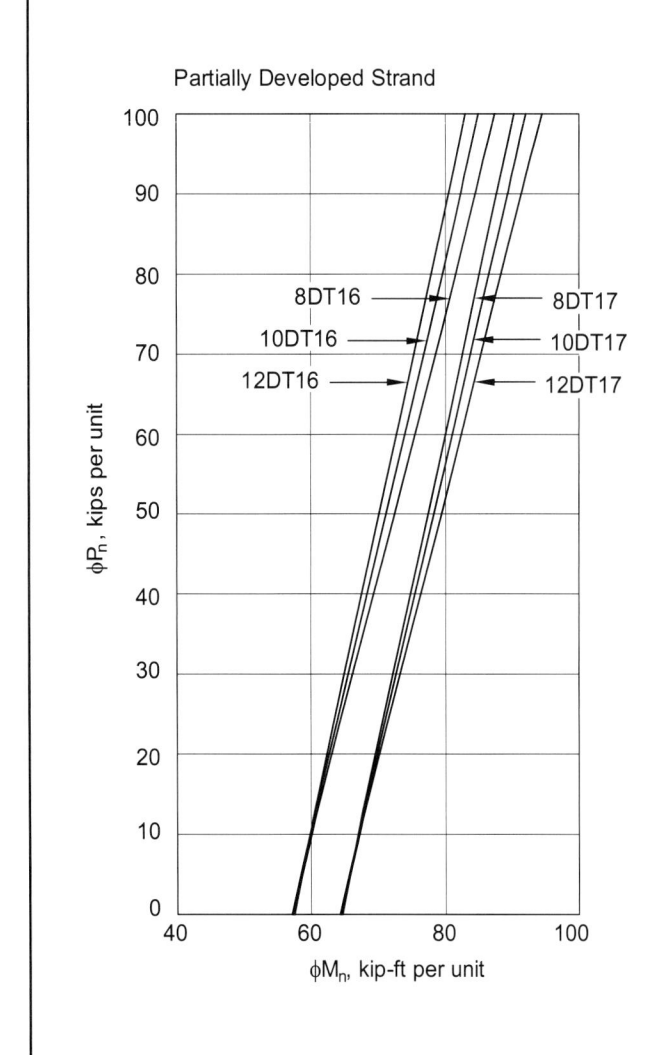

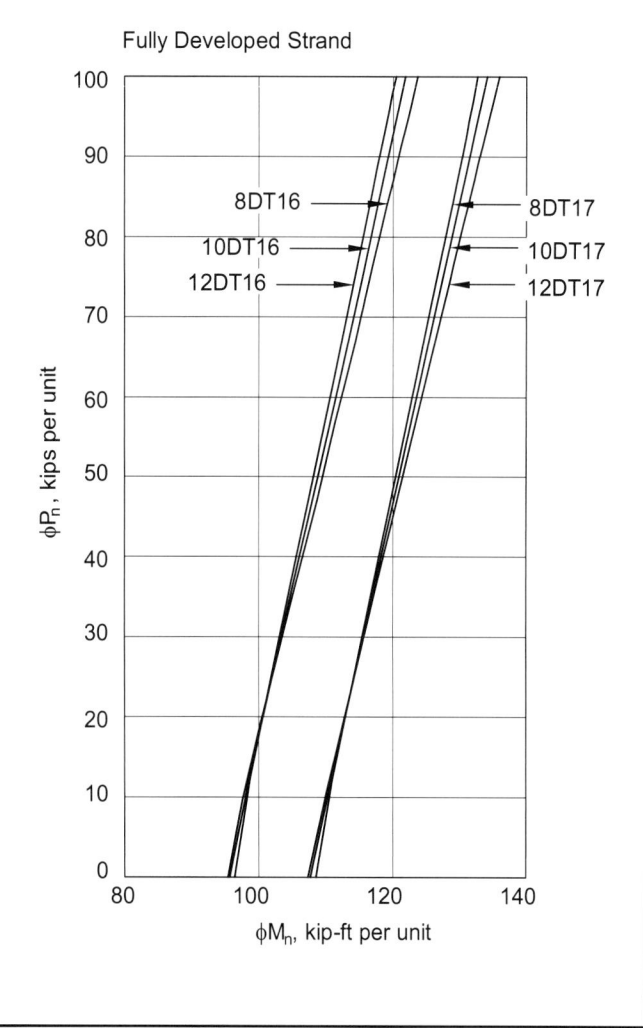

HOLLOW-CORE AND SANDWICH WALL PANELS

Figure 2.7.4   Partial interaction curves for prestressed hollow-core and sandwich wall panels

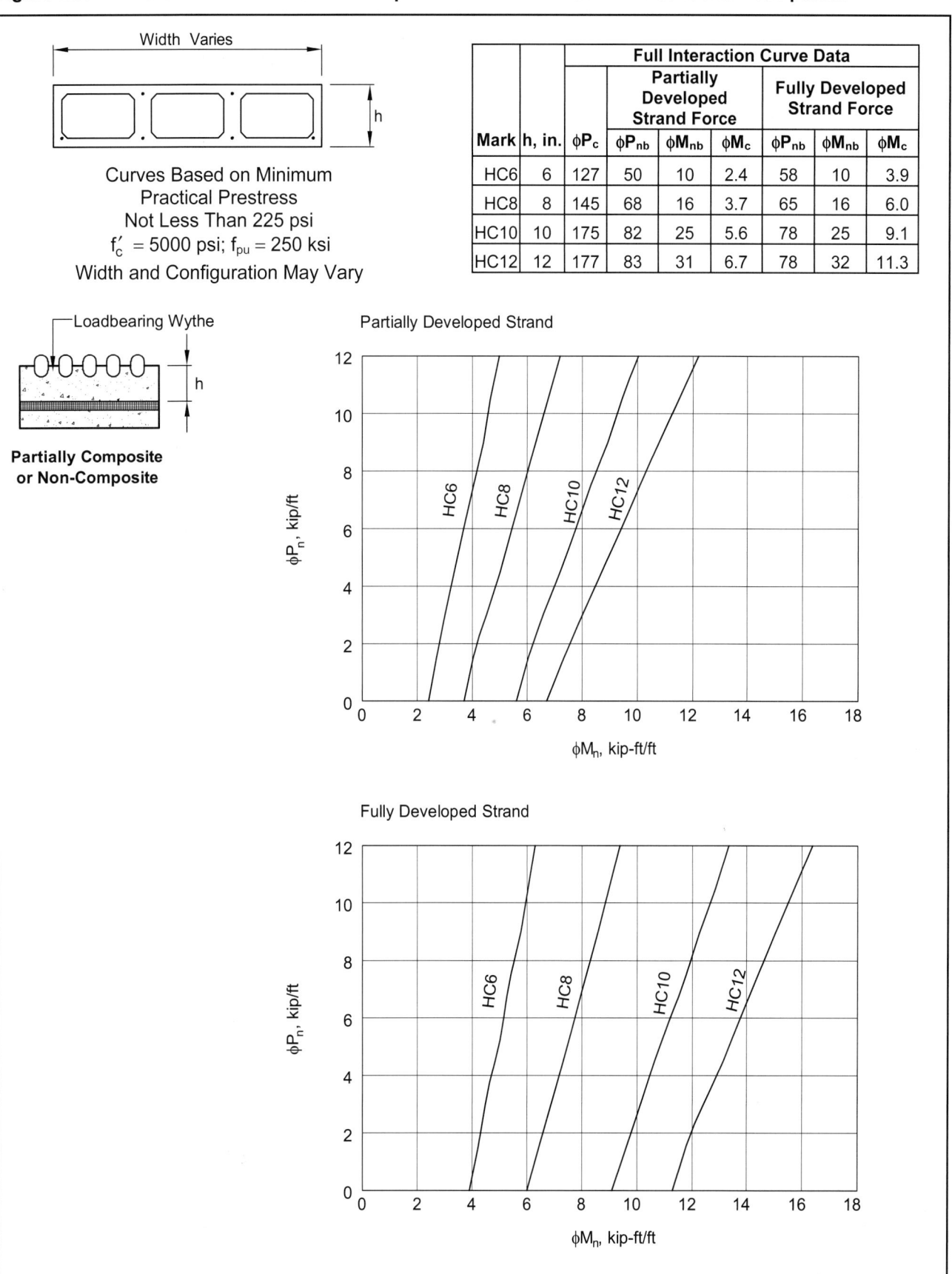

Curves Based on Minimum Practical Prestress Not Less Than 225 psi
$f'_c = 5000$ psi; $f_{pu} = 250$ ksi
Width and Configuration May Vary

Partially Composite or Non-Composite

Mark	h, in.	ϕP_c	Partially Developed Strand Force			Fully Developed Strand Force		
			ϕP_{nb}	ϕM_{nb}	ϕM_c	ϕP_{nb}	ϕM_{nb}	ϕM_c
HC6	6	127	50	10	2.4	58	10	3.9
HC8	8	145	68	16	3.7	65	16	6.0
HC10	10	175	82	25	5.6	78	25	9.1
HC12	12	177	83	31	6.7	78	32	11.3

Partially Developed Strand

Fully Developed Strand

PRECAST, PRESTRESSED SOLID AND SANDWICH WALL PANELS

Figure 2.7.5 Partial interaction curves for precast, prestressed solid and sandwich wall panels

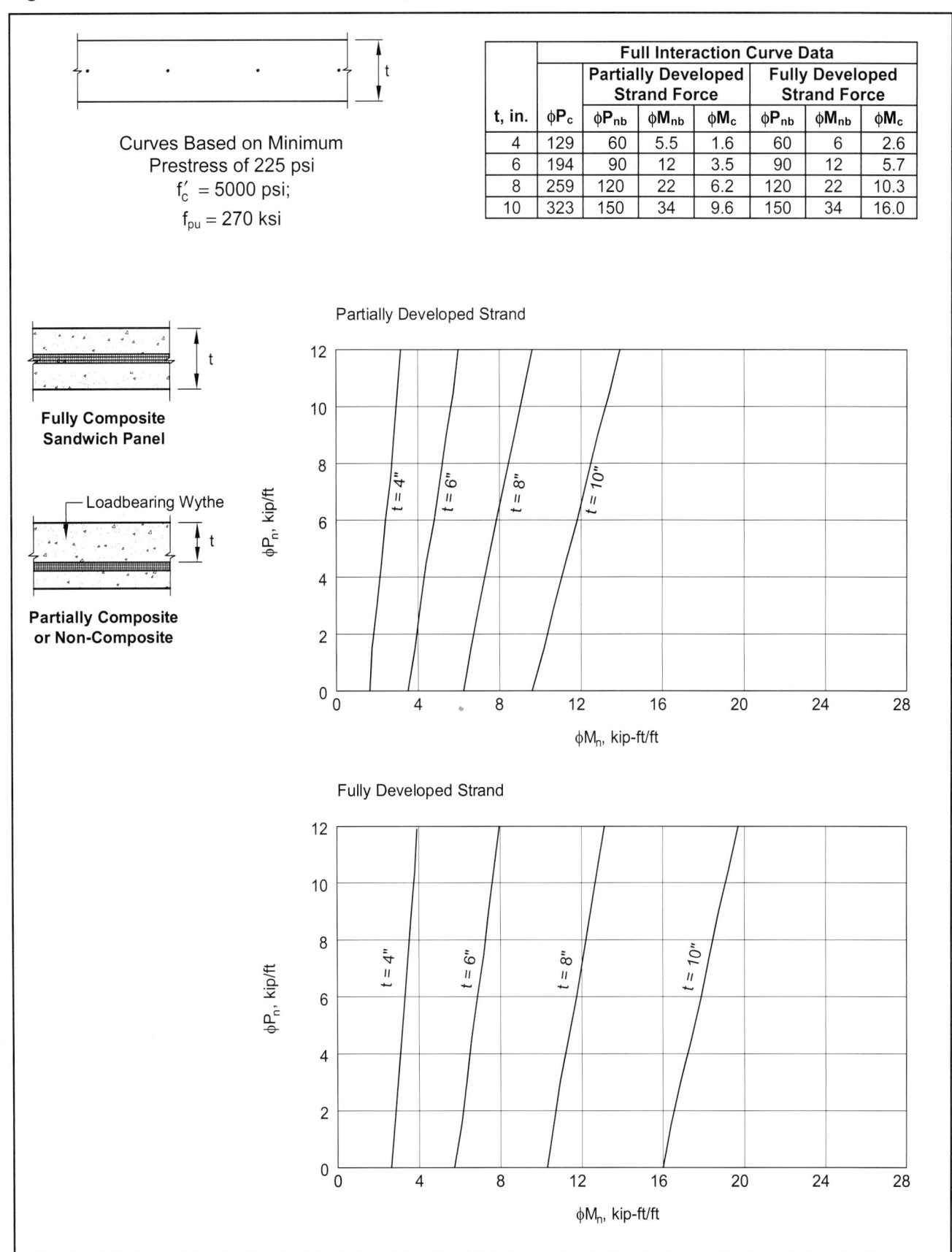

Curves Based on Minimum Prestress of 225 psi
$f'_c = 5000$ psi;
$f_{pu} = 270$ ksi

Fully Composite Sandwich Panel

Partially Composite or Non-Composite (Loadbearing Wythe)

		Full Interaction Curve Data					
		Partially Developed Strand Force			Fully Developed Strand Force		
t, in.	ϕP_c	ϕP_{nb}	ϕM_{nb}	ϕM_c	ϕP_{nb}	ϕM_{nb}	ϕM_c
4	129	60	5.5	1.6	60	6	2.6
6	194	90	12	3.5	90	12	5.7
8	259	120	22	6.2	120	22	10.3
10	323	150	34	9.6	150	34	16.0

Partially Developed Strand

Fully Developed Strand

2–54 PCI Design Handbook/Sixth Edition

PRECAST, REINFORCED SOLID AND SANDWICH WALL PANELS

Figure 2.7.6   Partial interaction curves for precast, reinforced concrete wall panels

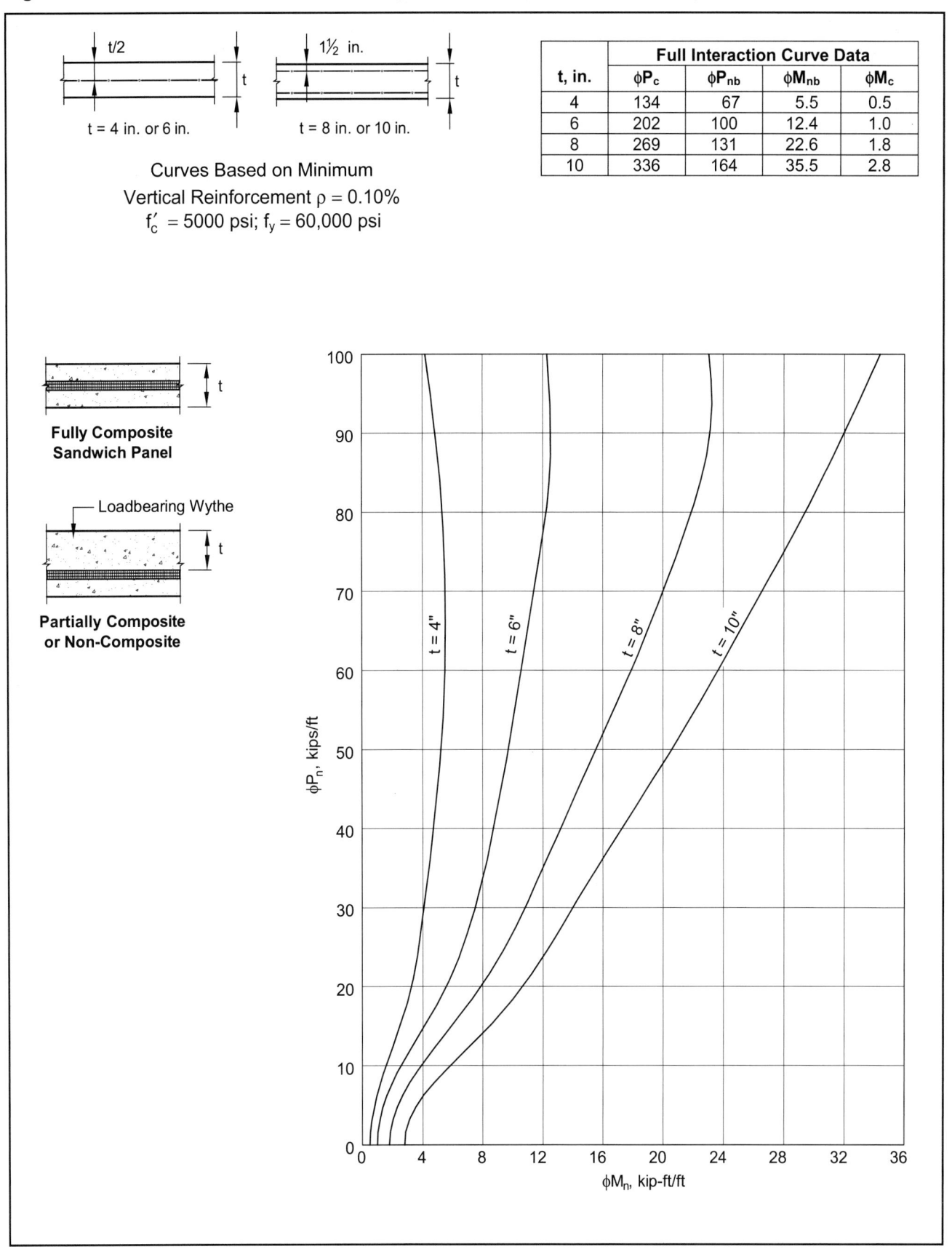

t = 4 in. or 6 in. (t/2)

t = 8 in. or 10 in. (1½ in.)

Curves Based on Minimum Vertical Reinforcement $\rho = 0.10\%$
$f'_c = 5000$ psi; $f_y = 60,000$ psi

t, in.	Full Interaction Curve Data			
	ϕP_c	ϕP_{nb}	ϕM_{nb}	ϕM_c
4	134	67	5.5	0.5
6	202	100	12.4	1.0
8	269	131	22.6	1.8
10	336	164	35.5	2.8

Fully Composite Sandwich Panel

Partially Composite or Non-Composite — Loadbearing Wythe

[Graph: ϕP_n, kips/ft (vertical axis, 0 to 100) vs. ϕM_n, kip-ft/ft (horizontal axis, 0 to 36), showing curves for t = 4", t = 6", t = 8", t = 10"]

PCI Design Handbook/Sixth Edition   2–55

PILES

Table 2.8.1 Section properties and allowable service loads for prestressed concrete piles

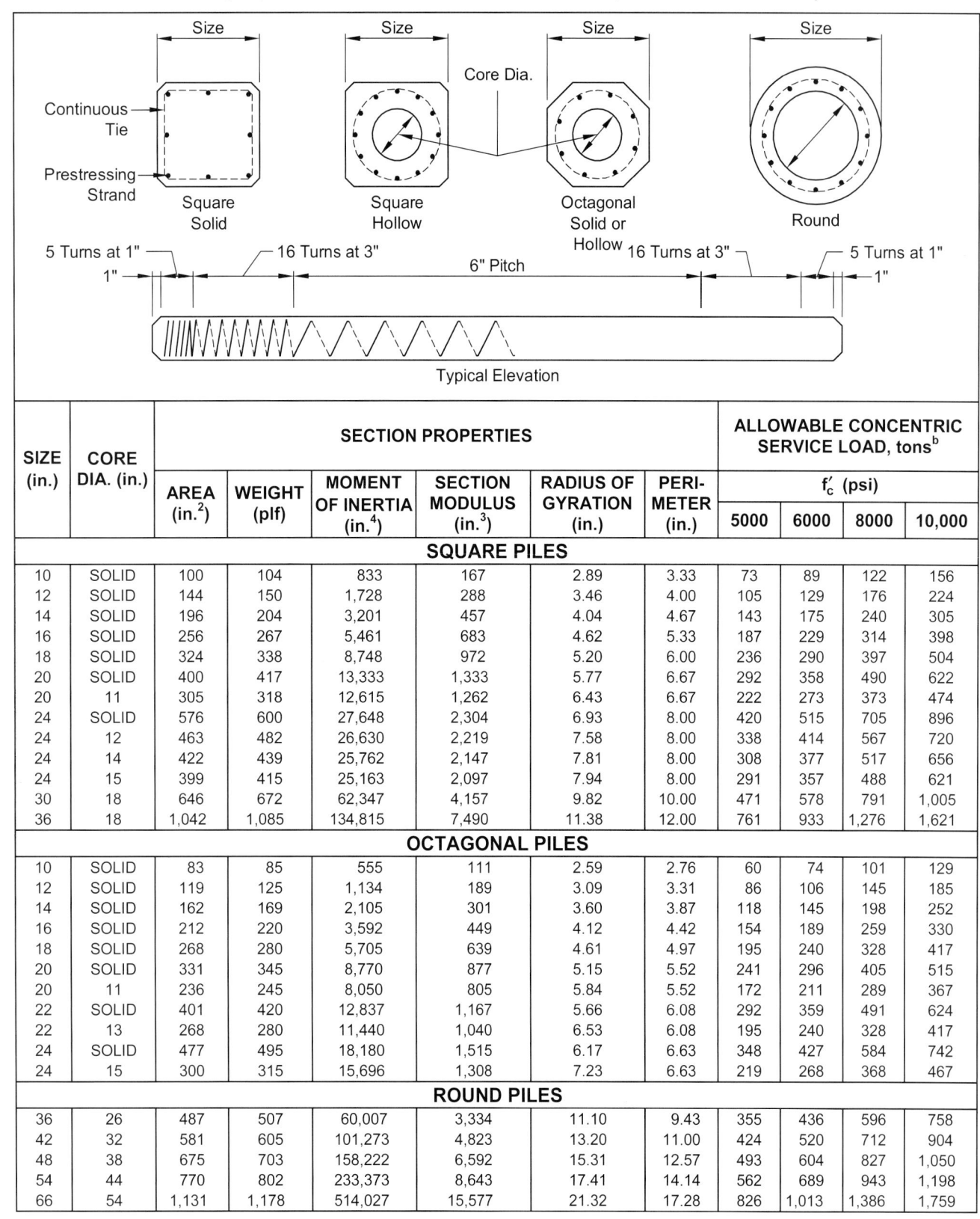

SIZE (in.)	CORE DIA. (in.)	AREA (in.²)	WEIGHT (plf)	MOMENT OF INERTIA (in.⁴)	SECTION MODULUS (in.³)	RADIUS OF GYRATION (in.)	PERIMETER (in.)	5000	6000	8000	10,000
\multicolumn{12}{c}{**SQUARE PILES**}											
10	SOLID	100	104	833	167	2.89	3.33	73	89	122	156
12	SOLID	144	150	1,728	288	3.46	4.00	105	129	176	224
14	SOLID	196	204	3,201	457	4.04	4.67	143	175	240	305
16	SOLID	256	267	5,461	683	4.62	5.33	187	229	314	398
18	SOLID	324	338	8,748	972	5.20	6.00	236	290	397	504
20	SOLID	400	417	13,333	1,333	5.77	6.67	292	358	490	622
20	11	305	318	12,615	1,262	6.43	6.67	222	273	373	474
24	SOLID	576	600	27,648	2,304	6.93	8.00	420	515	705	896
24	12	463	482	26,630	2,219	7.58	8.00	338	414	567	720
24	14	422	439	25,762	2,147	7.81	8.00	308	377	517	656
24	15	399	415	25,163	2,097	7.94	8.00	291	357	488	621
30	18	646	672	62,347	4,157	9.82	10.00	471	578	791	1,005
36	18	1,042	1,085	134,815	7,490	11.38	12.00	761	933	1,276	1,621
\multicolumn{12}{c}{**OCTAGONAL PILES**}											
10	SOLID	83	85	555	111	2.59	2.76	60	74	101	129
12	SOLID	119	125	1,134	189	3.09	3.31	86	106	145	185
14	SOLID	162	169	2,105	301	3.60	3.87	118	145	198	252
16	SOLID	212	220	3,592	449	4.12	4.42	154	189	259	330
18	SOLID	268	280	5,705	639	4.61	4.97	195	240	328	417
20	SOLID	331	345	8,770	877	5.15	5.52	241	296	405	515
20	11	236	245	8,050	805	5.84	5.52	172	211	289	367
22	SOLID	401	420	12,837	1,167	5.66	6.08	292	359	491	624
22	13	268	280	11,440	1,040	6.53	6.08	195	240	328	417
24	SOLID	477	495	18,180	1,515	6.17	6.63	348	427	584	742
24	15	300	315	15,696	1,308	7.23	6.63	219	268	368	467
\multicolumn{12}{c}{**ROUND PILES**}											
36	26	487	507	60,007	3,334	11.10	9.43	355	436	596	758
42	32	581	605	101,273	4,823	13.20	11.00	424	520	712	904
48	38	675	703	158,222	6,592	15.31	12.57	493	604	827	1,050
54	44	770	802	233,373	8,643	17.41	14.14	562	689	943	1,198
66	54	1,131	1,178	514,027	15,577	21.32	17.28	826	1,013	1,386	1,759

Section properties columns grouped under **SECTION PROPERTIES**; load columns grouped under **ALLOWABLE CONCENTRIC SERVICE LOAD, tons[b]** with f'_c (psi) sub-columns.

a. Form dimensions may vary with producers with corresponding variations in section properties.
b. Allowable loads based on $N = A_g(0.33 f'_c - 0.27 f_{pc})$; $f_{pc} = 700$ psi. Check local producer for available concrete strengths.

SHEET PILES

Table 2.8.2   Section properties and allowable service loads of prestressed sheet piles

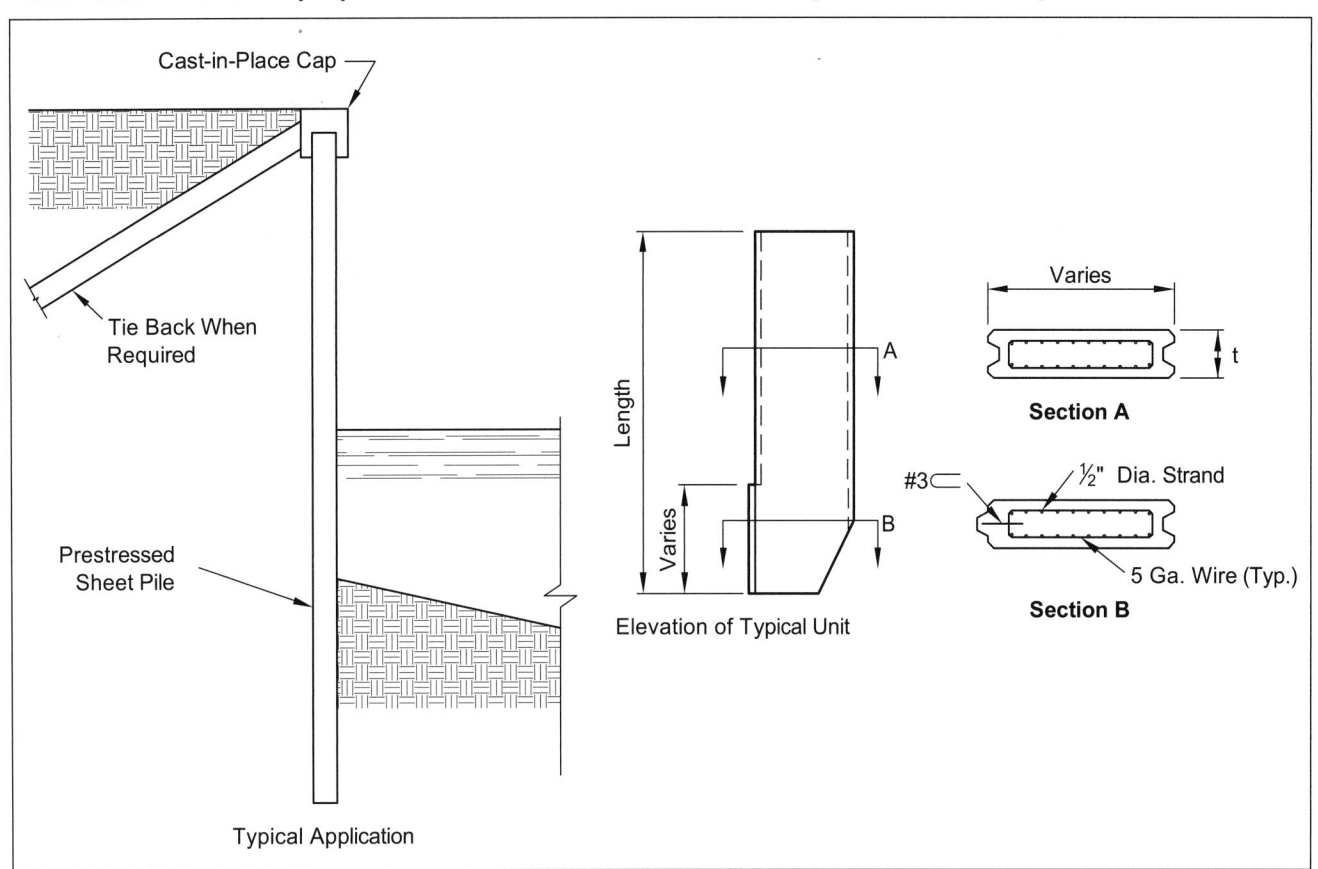

THICKNESS t (in.)	SECTION PROPERTIES PER FOOT OF WIDTH				MAXIMUM ALLOWABLE SERVICE LOAD MOMENT[b] kip-ft per foot	
	AREA (in.^{2})	WEIGHT[a] (psf)	MOMENT OF INERTIA (in.^{4})	SECTION MODULUS (in.^{3})	$f'_c = 5000$ (psi)	$f'_c = 6000$ (psi)
6[c]	72	75	216	72	6.0	7.2
8[c]	96	100	512	128	10.6	12.8
10	120	125	1,000	200	16.6	20.0
12	144	150	1,728	288	24.0	28.8
16	192	200	4,096	512	42.7	51.2
18	216	225	5,832	648	54.0	64.8
20	240	250	8,000	800	66.7	80.0
24	288	300	13,824	1,152	96.0	115.2

a. Normal weight concrete.
b. Based on zero tension and maximum $0.4f'_c$ compression.
c. Strand can be placed in a single layer in thin sections. Where site conditions require it, strand may be placed eccentrically.

CHAPTER 3
ANALYSIS AND DESIGN OF PRECAST, PRESTRESSED CONCRETE STRUCTURES

3.1 General 3–3
    3.1.1 Notation 3–3
    3.1.2 Precast Concrete Force Resisting Systems 3–6
        3.1.2.1 Emulation 3–6
        3.1.2.2 Non-Emulative Design 3–8

3.2 Code Requirements for Structural Loads 3–8
    3.2.1 Code References 3–8
        3.2.1.1 IBC 2003 3–8
        3.2.1.2 ASCE 7-2002 3–8
        3.2.1.3 NEHRP 2000 3–9
        3.2.1.4 NFPA 5000 3–9
        3.2.1.5 ACI 318-02 3–9
    3.2.2 Gravity Loads 3–9
        3.2.2.1 Dead Loads 3–9
        3.2.2.2 Live Loads 3–9
        3.2.2.3 Snow Loads 3–9
    3.2.3 Wind Loads 3–10
        3.2.3.1 ASCE 7 — Method 1 for Wind Design 3–10
    3.2.4 Earthquake Loads 3–13
        3.2.4.1 Base Shear
                (IBC 2003 — Section 1617.4, Equivalent Lateral Force Method) 3–13
        3.2.4.2 Vertical Distribution 3–16
        3.2.4.3 Lateral Distribution and Torsion 3–16
        3.2.4.4 Drift Effects 3–16
        3.2.4.5 Architectural Component Analysis 3–17
        3.2.4.6 Redundancy — Seismic Design Categories D, E, and F 3–22
    3.2.5 Lateral Soil Loads 3–22
    3.2.6 Load Combinations 3–22
        3.2.6.1 Load Factors for Diaphragms 3–22

3.3 Structural Integrity 3–23
    3.3.1 Introduction 3–23
    3.3.2 Precast Concrete Structures 3–23
    3.3.3 Large-Panel Bearing Wall Structures 3–24
    3.3.4 Hybrid Structures 3–25

3.4 Volume Changes 3–25
    3.4.1 Axial Volume Change Strains 3–28
    3.4.2 Expansion Joints 3–28
        3.4.2.1 Spacing of Expansion Joints 3–28
        3.4.2.2 Width of Expansion Joints 3–28
    3.4.3 Volume Change Effects in Moment-Resisting Frames 3–28
        3.4.3.1 Equivalent Volume Change 3–29
        3.4.3.2 Calculating Restraint Forces 3–29

3.5 Shear Wall Systems 3–33
    3.5.1 Introduction 3–33
    3.5.2 Principles of Shear Wall Buildings 3–33
    3.5.3 Code Requirements 3–34

  3.5.4 Design Guidelines for Shear Wall Structures 3–34
  3.5.5 Proportioning Shear Walls 3–35
  3.5.6 Stiffness Analysis 3–36
  3.5.7 Distribution of Lateral Loads 3–37
  3.5.8 Coupled Shear Walls 3–39
  3.5.9 Shear Walls with Large Openings 3–40
  3.5.10 Evaluation of Shear Wall Systems 3–40
  3.5.11 Example One-Story Building 3–41
  3.5.12 Example Three-Level Parking Structure 3–47

3.6 Moment-Resisting Building Frames 3–54
  3.6.1 Moment Resistance of Column Bases 3–54
  3.6.2 Fixity of Column Bases 3–59
  3.6.3 Computer Models for Frame Analysis 3–59
    3.6.3.1 Modeling Partially Fixed Bases 3–59
  3.6.4 Frame Classifications for Seismic Considerations 3–60
  3.6.5 Frame Analysis and Design — Ordinary Moment Frames............................. 3–61
  3.6.6 Special Moment Frames for Seismic Design Category C 3–69
  3.6.7 Special Moment Frames for High Seismic Design Categories 3–77

3.7 Shear Wall – Frame Interaction 3–84

3.8 Diaphragm Design 3–84
  3.8.1 Simple Diaphragm Design — The Horizontal Beam Analogy 3–84
    3.8.1.1 Shear Transfer Between Members 3–84
    3.8.1.2 Chord Forces 3–86
  3.8.2 Rigid and Flexible Diaphragms 3–87
    3.8.2.1 Defining Rigid or Flexible Diaphragms 3–87
    3.8.2.2 Behavior and Design Considerations 3–87
  3.8.3 Diaphragm Design Forces 3–88
    3.8.3.1 Code-Prescribed Forces.. 3–88
    3.8.3.2 Elastic Design of Diaphragms 3–90
    3.8.3.3 Performance-Based Design 3–97
  3.8.4 Diaphragm Detailing Considerations... 3–97
    3.8.4.1 General Detailing Requirements 3–98
    3.8.4.2 Wind and Low Seismic Hazard 3–98
    3.8.4.3 Moderate Seismic Hazard — Topped and Pretopped Systems......... 3–98
    3.8.4.4 Seismic Design Category D — Topped Systems............................. 3–99
    3.8.4.5 Untopped Systems for High Seismic Hazard 3–99
  3.8.5 Alternate Methods of Diaphragm Design 3–100
    3.8.5.1 Strut-and-Tie Modeling 3–100
    3.8.5.2 Finite Element Analysis 3–100

3.9 References 3–101

3.10 Design Aids... 3–103

ANALYSIS AND DESIGN OF PRECAST, PRESTRESSED CONCRETE STRUCTURES

3.1 General

3.1.1 Notation

a_p = amplification factor related to response of a system or component as affected by type of seismic attachment

A = area (with subscripts)

A_b = total area of anchor bolts which are in tension

A_{ps} = area of prestressing steel

A_{vf} = area of shear-friction reinforcement

A_w = area of shear wall

b = width of section or structure

C = coefficient of thermal expansion

C = compressive force

C_d = deflection amplification factor as given in Figure 3.10.8

C_e = exposure factor as determined from Figure 3.10.3

C_m = factor relating actual moment to equivalent uniform moment

C_p = external pressure coefficient to be used in determining wind loads for buildings

C_s = seismic response coefficient

C_t = thermal factor as determined from Figure 3.10.3

C_u = coefficient from Table 3.2.4.2

C_u = factored compressive force

D = dead load

e = eccentricity of axial load

E = effect of horizontal and vertical seismic forces

E = modulus of elasticity (with subscripts)

f'_c = concrete compressive strength

f'_{ci} = compressive strength of concrete at time of initial prestress

f_{pu} = specified tensile strength of prestressing steel

f_{ut} = factored tensile stress

f_y = yield strength of non-prestressed reinforcement

F = horizontal force in shear wall building and in moment-resisting frame

F_a = acceleration-based site coefficient (at 0.3-sec period)

F_b = degree of base fixity (decimal)

F_i = lateral force at Bay i or in Shear Wall i

F_i = restraining force in multistory columns at Level i

F_i, F_n, F_x = portion of seismic base shear, V, induced at Level i, n, or x, respectively

F_p = seismic force acting on component of structure

F_u = factored force

F_v = velocity-based site coefficient (at 1.0-sec period)

F_x, F_y = forces in x and y directions, respectively

g = assumed length over which elongation of anchor bolt takes place

g = acceleration due to gravity

h = height of shear wall

h = effective height of building

h_b = height of balanced snow load

h_c = clear height from top of balanced snow load to (1) closest point on adjacent upper roof, (2) top of parapet, or (3) top of projection on roof

h_d = height of snow drift

h_i, h_n, h_x = height above base Level i, n, or x, respectively

h_s = story height

I = importance factor as determined from Figure 3.10.1

I = moment of inertia (with subscripts)

I_b = moment of inertia of beam

I_{bp} = moment of inertia of base plate (vertical cross-sectional dimensions)

I_c = moment of inertia of column

I_{eq}	= approximate moment of inertia that results in flexural deflection equal to combined shear and flexural deflections of wall
I_f	= moment of inertia of footing (plan dimensions)
I_g	= uncracked moment of inertia
I_p	= polar moment of stiffness
k	= effective length factor
k_b, k_f, k_m	= coefficients used to determine forces and moments in beams and columns
k_s	= coefficient of subgrade reaction
K	= stiffnesses (with subscripts)
K'	= constant used for calculating equivalent creep and shrinkage shortening
K_d	= wind directionality factor
K_h	= velocity pressure exposure coefficient evaluated at height $z = h$
K_r	= relative stiffness
K_t	= constant used for calculation of equivalent temperature shortening
K_z	= velocity pressure exposure coefficient evaluated at Height z
K_{zt}	= topographic factor
ℓ	= distance between wall panel supports
ℓ	= length of span or structure
ℓ_u	= length of roof upwind of drift
ℓ_w	= length of weld
L	= live load
M	= unfactored moment
M_f	= foundation overturning design moment
M_j	= moment in multistory columns at Point j
M_R	= resisting moment
M_T	= torsional moment
M_u	= factored moment
M_x	= building overturning design moment at Level x as defined in Section 9.5.3.6
MWFRS	= main wind force resisting system
n	= number of panels in shear wall; number of bays in moment-resisting frame
p	= design pressure to be used in determining wind loads for buildings
p_d	= maximum intensity of drift surcharge load
p_f	= snow load on flat roofs ("flat" = roof slope ≤ 5 deg)
p_g	= ground snow load
p_L	= wind pressure acting on leeward face
p_{net30}	= net design wind pressure for Exposure B at $h = 30$ ft and $I = 1.0$
p_p	= combined net pressure on parapet
p_s	= combined windward and leeward net wind pressure
p_{s30}	= simplified design wind pressure for Exposure B at $h = 30$ ft and $I = 1.0$
p_w	= wind pressure acting on windward face
P	= applied axial load
P	= lateral force applied to shear wall
P	= vertical load acting at eccentricity e
P_o	= final prestress force in tendons
P_u	= factored axial load
P_x	= total unfactored vertical design load at, and above, Level x
PI	= plasticity index
q	= velocity pressure, in lb/ft^2 (N/m^2)
q_h	= velocity pressure evaluated at Height $z = h$
q_i	= velocity pressure for internal pressure determination
q_z	= velocity pressure evaluated at Height z
Q_E	= effect of horizontal seismic (earthquake-induced) forces
r	= rigidity
R	= response modification coefficient as given in Figure 3.10.8
R_p	= component response modification factor
S_1	= mapped maximum considered earthquake, 5% damped, spectral response acceleration at a period of 1 sec
S_{D1}	= design, 5% damped, spectral response acceleration at a period of 1 sec
S_{DS}	= design, 5% damped, spectral response acceleration at short periods
S_{M1}	= maximum considered earthquake, 5% damped, spectral response acceleration at a period of 1 sec. Adjusted for site class effects

S_{MS}	= maximum considered earthquake, 5% damped, spectral response acceleration at short periods adjusted for site class effects
S_S	= mapped maximum considered earthquake, 5% damped, spectral response acceleration at short periods
t	= thickness of member under consideration
T	= fundamental period of building
T	= tensile force
T_a	= approximate fundamental period of building
T_n	= nominal tensile strength
T_u	= factored tensile force
V	= basic wind speed obtained from Figure 3.10.5 in mph. The basic wind speed corresponds to a 3-sec gust speed at 33 ft above ground in Exposure Category C
V	= shear (with subscripts)
V	= total design lateral force or shear at base
V_n	= nominal shear strength
V_u	= factored shear force
V_x	= seismic design shear in Story x
w	= uniform load (with subscripts)
w	= width of snow drift, in ft (m)
w_i, w_n, w_x	= portion of W that is located at or assigned to Levels i, n, or x, respectively
W	= total lateral or gravity load (with subscripts)
x	= level under consideration
x_1	= distance from face of column to center of anchor bolts
x_2	= distance from face of column to base plate anchorage
x, y	= orthogonal distances of individual shear walls from center of rigidity
z	= height above ground level
β	= damping ratio, percent critical for buildings or other structures
β	= overstrength factor for seismic shear
β	= ratio of shear demand and shear capacity for story between Levels x and x–1
δ	= volume change shortening (with subscripts)
δ_e	= equivalent volume change shortening (with subscripts)
δ_{xe}	= deflection determined from elastic analysis that includes consideration of cracking
δ_x	= deflection of Level x at center of mass at and above Level x
Δ	= design story drift
Δ	= difference of deflections
Δ	= total equivalent shortening or column deflection
ϕ	= rotation (with subscripts)
ϕ	= strength reduction factor
ϕ_k	= stiffness reduction factor
γ	= flexibility coefficient (with subscripts)
γ	= snow density
λ	= adjustment factor for building height and exposure
μ	= static coefficient of friction
θ	= stability coefficient for P-delta effects
ρ	= reliability coefficient based on extent of structural redundancy present in building
Ω_0	= overstrength factor

3.1.2 Precast Concrete Force Resisting Systems

The design of precast/prestressed concrete structures involves the integration of many considerations. Most of these spatial, functional and architectural considerations are reviewed in Chapters 1 and 2. This chapter will focus on the design of the structural system.

Precast/prestressed concrete structures are the integration of the structural system as a whole, the connections and the individual components. Each aspect of design must consider the others as well as the functional requirements imposed by the building use. It is essential that the design loads be traced from their point of origin to the final support or foundation. Although not always required by code, it is desirable to design the members and their connections to achieve a ductile, rather than a brittle failure mode.

Resistance to gravity loads is largely a matter of component design, which is covered in Chapter 4 of this Handbook.

In addition to resisting gravity loads, a principal consideration in building design is the lateral force resisting system. Methods used to resist lateral forces, in the approximate order of economy, include:

1. **Cantilevered Columns or Wall Panels (Out of Plane):** This is usually only feasible in low-rise buildings. Base fixity can be attained through a moment couple between the footing and ground floor slab, or by fixing the wall or the column to the footing. In the latter case, a detailed analysis of the footing rotation can be made as described in Section 3.6.2.
2. **Shear Walls:** These can be precast concrete, cast-in-place concrete, or masonry. Shear walls are discussed in more detail in Section 3.5. When architectural or structural precast members are used for the exterior cladding, they can often be used as shear walls. Precast concrete box elements have been used effectively in low-rise to high-rise structures. The boxes are created as one complete unit, such as in a precast cell module, or can be created of individual precast walls connected together to create a box unit. Such box units have a much larger moment of inertia than individual walls and therefore can be important members in a lateral force resisting system.
3. **Steel or Concrete X-bracing:** This system has been used effectively in low and medium rise buildings. A related resistance system usually occurs naturally in parking structures with sloped ramps in the direction of traffic flow. The load path should be verified before the ramp of a parking structure is assumed as the stiffening element.
4. **Moment-Resisting Frames:** Building function may dictate the use of moment resisting frames. It is sometimes feasible to provide a moment connection at only one end of a member, or a connection that will resist moments with lateral forces in one direction but not in the other, in order to reduce the buildup of volume change restraint forces. To reduce the number of moment frames required, a combined shear wall-moment frame system may be used. Moment-resisting frames are discussed in more detail in Section 3.6.

All of the above systems depend on distribution of lateral loads through diaphragm action of the roof and floor systems (see Section 3.8).

The balance in system design is achieved not only by providing the strength, ductility and toughness to resist lateral forces. It is also important to consider the effects of concrete creep, shrinkage and temperature change. These effects are collectively known as volume changes. Details that result in over-restraint of volume changes can be as damaging as any externally applied force. Most buildings will never experience the design event that is represented by the requirements for wind or earthquakes, but they will experience the climatic temperature cycle every year.

In precast concrete structures, individual elements are connected at their joints with a variety of methods. These connections may include embedded steel shapes such as plates and angles, with headed stud or reinforcing bar anchorage; the steel embedments are field bolted or welded. These applications are "dry" connections. A "wet" connection consists of reinforcing bars protruding from the precast members, with the bars mechanically coupled or spliced. Cast-in-place concrete or grout at the joint completes this connection. Either dry or wet connections are used in both moment-resisting frame and shear wall systems.

3.1.2.1 Emulation

Precast concrete structures have been built in high seismic areas for many years. The building codes have allowed such construction provided that the precast members are connected so the unit will perform essentially the same as a cast-in-place concrete unit. These designs and details have become known as "Emulation" of cast-in-place concrete. The methods of detailing are described

Figure 3.1.2.1 Typical emulation details [6]

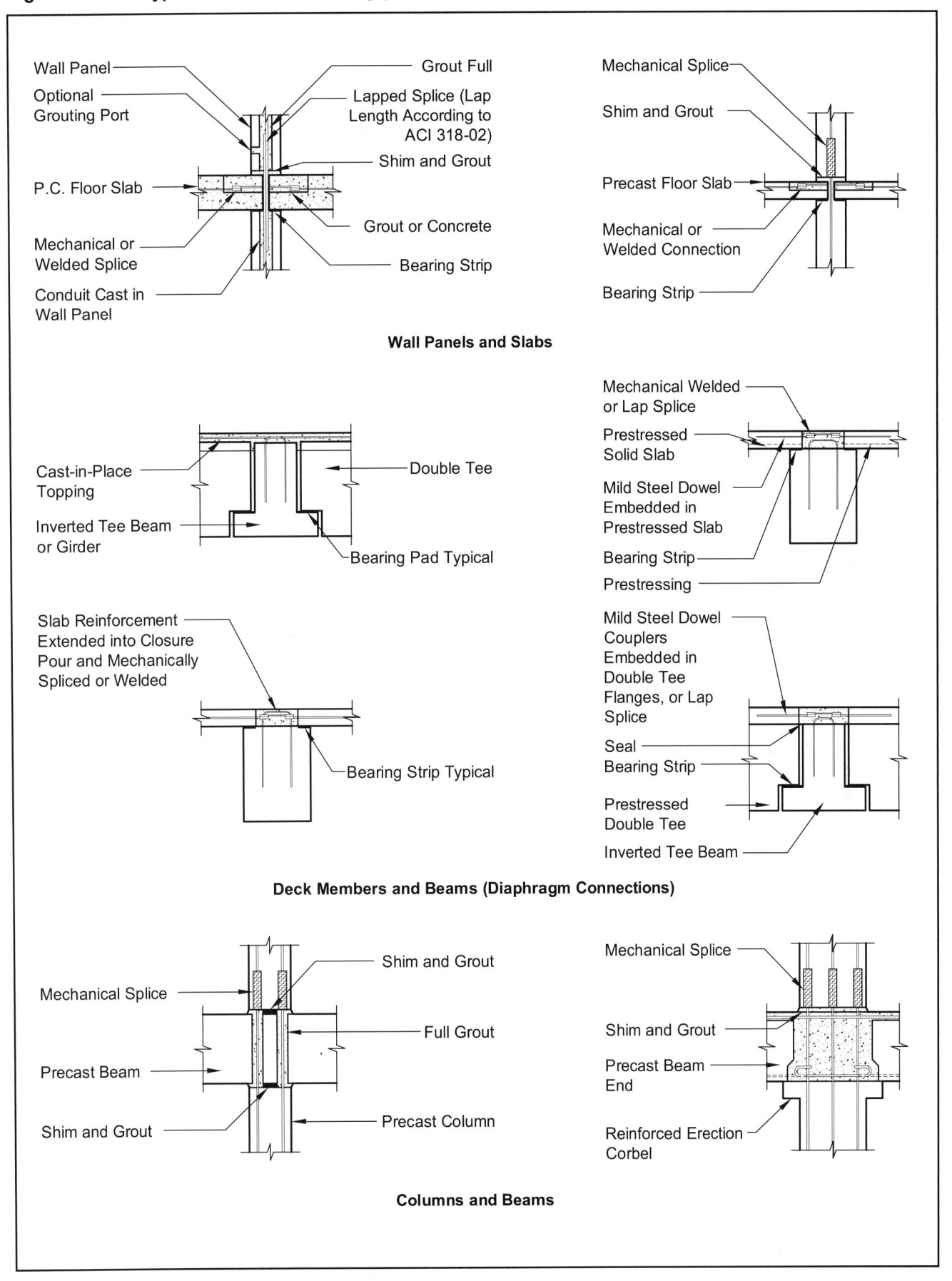

Wall Panels and Slabs

Deck Members and Beams (Diaphragm Connections)

Columns and Beams

and illustrated in the ACI Committee 550 report, "Emulating Cast-in-Place Detailing in Precast Concrete Structures." [6] Figure 3.1.2.1 shows several details excerpted from that report.

Emulation design creates construction that either is monolithic at the critical joint locations or provides connections that act as if they are monolithic at those locations. In general, emulative connections involve connecting reinforcing bars across a joint individually by lap splicing, welding or mechanical couplers. Concrete is made continuous by filling the joint with a high quality non-shrink or fiber-reinforced grout in horizontal joints or a cast-in-place closure pour in vertical joints. Engineering follows the rules for reinforced concrete design while taking into account the special circumstances for the connection system chosen, such as maintaining proper cover to mechanical couplers.

ACI 318-02, [5] while not using the word "emulation," has provisions for precast concrete special moment frames with "strong connections" (Section 21.6.2), and special structural (shear) walls (Sections 21.8 and 21.13).

3.1.2.2 Non-Emulative Design

Many advances have been made in the understanding of the seismic behavior of precast concrete frame structures. ACI 318-02 recognizes acceptance criteria for special moment frames based on validation testing, [7] in lieu of emulation. Similar criteria for special structural (shear) walls are in the 2000 NEHRP provisions. [3]

The PRESSS (Precast Seismic Structural Systems) program has researched systems which take advantage of the jointed nature of precast concrete, culminating with the testing of a five-story, 60%-scale building. [9] This test structure used precast frames with several different ductile joints in one direction, and precast concrete shear walls in the other direction. In 2002, a 39-story building using a hybrid frame system was completed in San Francisco. [10]

ACI 318-02 recognizes jointed precast construction for use in seismic areas in Sections 21.6.1, 21.6.3 and 21.13.2.

Regardless of the system used, it is imperative that lateral load paths and resisting elements be clearly defined. Where significant movement between adjacent elements is anticipated, ductile connections must be provided. One advantage of jointed construction is the ease of defining load paths through the connections. Connections can be designed for specific directional resistance while maintaining flexibility in one or more other directions. It is important in a precast structure to develop connections that tie all precast members into the lateral load resisting system.

Since wind and seismic ground motion are random in direction, a structure that is shaped so as to have sufficient resistance in any direction is necessary. Closed sections (boxes or tubes) have demonstrated markedly better behavior when compared to open sections because they provide more torsional resistance.

In structures where openness is important, such as parking structures, shear walls with large openings ("litewalls") have been successfully used. Attention must be paid to local flexure and shear in those elements which surround the openings.

To limit damage to non-structural elements, three options are open to the designer. First, the elements could be isolated from the structural system, so that these elements are not forced to undergo as much deformation as the supporting structure. Note, however, that if isolated from the structural system, these elements must maintain their own structural integrity. Second, the deflections of the system could be reduced in order to minimize deformation of the non-structural elements. This is typically attained through the use of shear walls. Third, the connection between individual elements and support elements could be designed to sustain large deformations and rotations without failure. Generally, the first or third approach is adopted for non-structural architectural wall panels (see Section 3.2.4.5).

3.2 Code Requirements for Structural Loads

3.2.1 Code References

3.2.1.1 IBC 2003 [1]

The International Building Code (IBC) is the result of many years of effort to combine the three most commonly adopted codes, the Uniform Building Code (UBC), the Standard Building Code (SBC), and the National Building Code (BOCA/NBC). To have the force of law, any code must be adopted by local jurisdictions (states, municipalities, etc.) Many jurisdictions are now adopting the IBC, while some remain governed by the former codes.

3.2.1.2 ASCE 7-2002 [2]

"Minimum Design Loads for Buildings and other Structures" (ASCE 7), published by the American Society of Civil Engineers, is a loading standard that is adopted by all the model codes. It specifies dead loads, live loads and load and resistance factors to be used for the strength design (LRFD) of buildings. It is usually not adopted directly by local jurisdictions, but may provide the basis for loads that vary with locality such as snow and wind. The earthquake provisions are those of NEHRP.

3.2.1.3 NEHRP 2000 [3]

The National Earthquake Hazards Reduction Program (NEHRP) provisions do not in themselves have the force of a code, but they are in a Federal Emergency Management Agency (FEMA) document, and are used as the basis for provisions of other codes and standards such as those produced by the IBC and ASCE 7. The 2000 edition contains major changes from prior editions that permit precast systems to be designed for high seismic regions, and provide acceptance criteria for non-emulative frames and walls.

3.2.1.4 NFPA 5000 [4]

The National Fire Protection Association (NFPA) has published a model building code that may be adopted by some jurisdictions. Load requirements are based on ASCE 7 and NEHRP. Concrete design follows ACI 318. Thus, structural designs shown in this Handbook will generally be in compliance with this code.

3.2.1.5 ACI 318-02 [5]

"Building Code Requirements for Structural Concrete" (ACI 318-02) and "Commentary" (ACI 318R-02) is the basis for concrete design throughout the United States. The 2002 edition has a great number of changes from previous editions, including extended provisions for the use of precast concrete in seismic regions, and the adoption of the ASCE-7 load and resistance factors. Since this Code was originally developed for cast-in-place concrete, provisions for precast and prestressed concrete have been slow to be adopted. Section 10.5 of this Handbook is a report of the PCI Building Code Committee that identifies provisions in ACI 318-02 which may need special interpretation when related to precast, prestressed concrete structures. Example problems in this Handbook typically use ACI 318-02 with the Section 10.5 modifications in the solutions.

3.2.2 Gravity Loads

3.2.2.1 Dead Loads

Dead loads include the self-weight of the structural components plus any materials or components that are attached to or permanently in place on the component or assembly. Since the dead loads are presumed to be determinable with a reasonable degree of accuracy, the load factor by ASCE 7 is a low 1.2. Weights of commonly used materials encountered in construction (taken from ASCE 7) are shown in Chapter 11, Design Aid 11.1.1.

3.2.2.2 Live Loads

Live loads are considered variable, transient and not accurately determinable, so the load factor is a higher 1.6. In some cases, a maximum live load may be calculated with a high degree of accuracy, for example fluid pressure, and a lower load factor is then used (see Section 3.2.6). Live loads recommended for use by ASCE 7 are given in Design Aid 11.1.2.

3.2.2.3 Snow Loads

Snow loads are treated differently from other live loads by ASCE 7 because they are very transient and vary by geographical location and terrain. The basic snow load for flat roofs is determined by:

$$p_f = 0.7 C_e C_t I p_g \qquad \text{(Eq. 3.2.2.1)}$$

limited by:

$p_f \geq I p_g$ where $p_g \leq 20$ psf
$p_f \geq 20I$ where $p_g > 20$ psf

where:

p_f = flat roof snow load (psf)
C_e = exposure factor from Figure 3.10.3(b)
C_t = thermal factor from Figure 3.10.3(c)
I = importance factor from Figure 3.10.1
p_g = ground snow load (psf) from Figures 3.10.2, 3.10.3(a), or specified by local authorities

Drift Effects: Drifting of snow due to roof projections such as penthouses, parapets and variable roof levels is also required to be taken into account. Drift loads are superimposed on the balanced loads. Figure 3.10.3(d) illustrates windward and leeward drifting. [2] Figure 3.10.4 illustrates the method specified by Ref. 2 for determining the loads caused by snow drifts:

1. Determine height of the balanced snow load:

$$h_b = \frac{p_f}{\gamma} \qquad \text{(Eq. 3.2.2.2)}$$

Example 3.2.2.1
Calculation of Snow Load

Given:
A flat roofed office building 450 ft long has a 50 ft long, 8 ft high penthouse centered along the length. The building is located in downtown Milwaukee, Wisconsin.

Problem:
Determine the roof snow load and drift load.

Solution:
From Figure 3.10.2: $p_g = 30$ psf

Assume that within the life of the structure, taller buildings may be built around it. Thus, from Figure 3.10.3(b), for Exposure B, sheltered, $C_e = 1.2$.

From Figure 3.10.3(c), for a heated building:
$C_t = 1.0$

Importance factor for office buildings, $I = 1.0$

From Eq. 3.2.2.1:

$p_f = 0.7(1.2)(1.0)(1.0)(30) = 25.2$ psf

Drifting:

1. $\gamma = 0.13p_g + 14 = 0.13(30) + 14$
   $= 17.9$ pcf
   $h_b = p_f/\gamma = 25.2/17.9 = 1.40$ ft

2. $\dfrac{h_c}{h_b} = \dfrac{8.0}{1.40} > 0.2$
   drifting must be considered

3. From Figure 3.10.4(b):
   For leeward wall, $h_d = 2.5$ ft
   For windward wall, $h_d = ¾(4.8) = 3.6$ ft
   Use $h_d = 3.6$ ft

4. $w = 4(3.6) = 14.4$ ft
   Snow drift load = $½\gamma h_d w$
   $= ½(17.9)(3.6)(14.4) = 464$ plf
   Applied as a line load at $14.4/3 = 4.8$ ft from the projection

where:
$\gamma = 0.13p_g + 14 \leq 30$ pcf (Eq. 3.2.2.3)

2. If $h_c/h_b \leq 0.2$, drift loads need not be applied

3. Determine h_d from Figure 3.10.4(b).
   - For leeward roofs, ℓ_u = length of the projection.
   - For windward roofs, ℓ_u = length of the lower roof on either side of the projection, and use three-quarters of the value from Figure 3.10.4(b) as h_d.
   - Use the larger of these two values.

4. $w = 4h_d$
   If $h_d > h_c$, $w = \dfrac{4h_d^2}{h_c} \leq 8h_c$ and $h_d = h_c$

   Snow drift load = $½\gamma w h_d$ (Eq. 3.2.2.4)
   applied at $w/3$ from the face of the projection.

3.2.3 Wind Loads

In most areas of the United States using IBC 2003, the earthquake loading will be more critical than wind, but wind loads should be checked. This section provides a method for such a check for most precast buildings. It is based on "Method 1 – Simplified Procedure" of ASCE 7, which is referenced by IBC 2003. The limitations for this method are:

1. Height ≤ 60 ft or least lateral dimension.
2. Enclosed building (includes parking structures).
3. Regular shaped.
4. No expansion joints.
5. Fundamental frequency ≥ 1 Hz. (Nearly all concrete buildings under 60 ft will qualify)
6. Flat or shallow pitched roof.
7. No unusual topography around the building.

ASCE 7 should also be reviewed for more details of these provisions, and for the other analysis methods that may be used for buildings that do not meet the listed limitations.

3.2.3.1 ASCE 7 – Method 1 for Wind Design

The following are the procedures required for this simplified analysis:

1. Determine the basic wind speed and directionality factor. The basic wind speed may be taken from Figure 3.10.5. The directionality factor is 0.85 for buildings; other structures

Figure 3.2.3.1 Wind pressure zones on typical building elevations

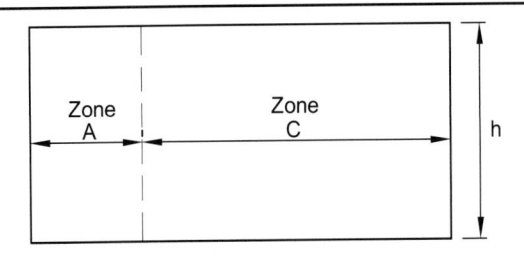

```
┌─────────────────────────────────────┐
│         ┊                           │
│  Zone   ┊        Zone            │ h
│   A    ←┼→        C                │
│         ┊                           │
└─────────────────────────────────────┘
```

Width of Zone A = the lesser of 20% of the least dimension of the building, or 80% of the mean roof height, but not less than 8% of the least dimension of the building, or 6 ft.

Zone A can be on either end, depending on wind direction.

such as chimneys and towers have slightly different factors. (Ref. 2, Table 6-4)

2. Determine importance factor from Figure 3.10.1.

3. Determine "Exposure Category" which applies to upwind direction (Note: Ref. 2 includes more detailed descriptions):

   - Exposure B: Urban and suburban areas, wooded areas
   - Exposure D: Flat, unobstructed areas outside hurricane-prone regions
   - Exposure C: All others

Example 3.2.3.1
Use of ASCE 7 Method 1 for Wind Load Determination

Given:
A 114 ft wide by 226 ft long by 54 ft tall hospital building in Memphis, TN. A section through a typical cladding panel is shown on the next page. Precast concrete wall panels are 7 ft tall by 28 ft long. A 6 ft high window is attached to the top of the panel, and an 8 ft high window is attached to the bottom.

Problem:
Determine the design wind load on the MWFRS and the wall panels.

Solution:
As this is an enclosed building under 60 ft high, Method 1 may be used.
From Figure 3.10.5, basic wind speed = 90 mph
From Section 3.2.3.1, Item 3, Exposure Category B

For MWFRS, use Eq. 3.2.3.1:

Interpolating from Figure 3.10.6(c): λ = 1.18
From Table 3.10.6(a): Zone A p_{s30} = 12.8 psf
                        Zone C p_{s30} = 8.5 psf

From Table 3.10.1:   I = 1.15

p_s (A) = $\lambda I p_{s30}$ = 1.18(1.15)(12.8) = 17.4 psf
p_s (C) = 1.18(1.15)(8.5) = 11.5 psf

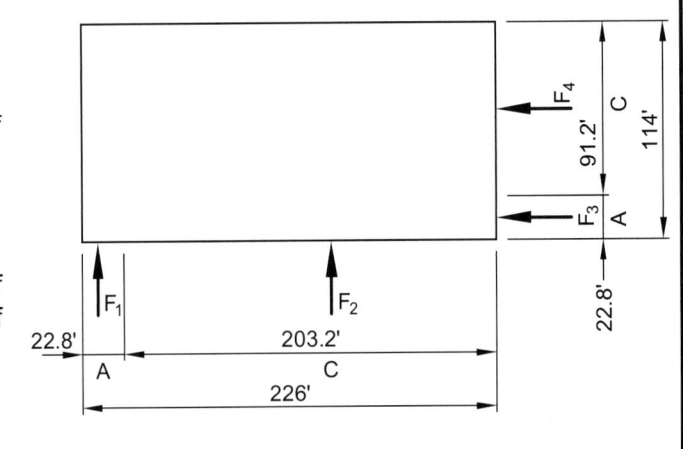

Plan

Example 3.2.3.1 (Cont.)
Use of ASCE 7 Method 1 for Wind Load Determination

From Figure 3.2.3.1:

For 226 ft length of building:
  Zone A width = b_A = Lesser of 0.2(114) or 0.8(54)
  Use b_A = 22.8 ft
  Zone C width = b_C = 226 − 22.8 = 203.2 ft
  $F_1 = b_A h[p_s(A)] = 22.8(54)(17.4)/1000 = 21.4$ kips
  $F_2 = b_B h[p_s(C)] = 203.2(54)(11.5)/1000 = 126.2$ kips
  Total force on length = 21.4 + 126.2 = 147.6 kips
  Resultant from left = $\dfrac{21.4(22.8/2) + 126.2(22.8 + 203.2/2)}{147.6} = 108$ ft
  Eccentricity = 226/2 − 108 = 5 ft

For 114 ft width of building:
  Calculate in a similar manner to above:
  Total force on width = 78 kips at eccentricity of 4.2 ft

Tributary area per panel = one-half of upper window + panel + one-half of lower window times the width
  = (6/2 + 7 + 8/2)28 = 392 ft²

Table 3.10.6(b) interpolating between 100 and 500 ft²:
$$\frac{392 - 100}{500 - 100} = 0.73$$

Inward pressure   = 12.4 − 0.73(12.4 − 10.9) = 11.3 psf
  $\lambda I p_{net30}$ = 1.18(1.15)(11.3)   = 15.3 psf
Outward pressure = 15.1 − 0.73(15.1 − 12.1) = 12.9 psf
  $\lambda I p_{net30}$ = 1.18(1.15)(12.9)   = 17.5 psf

Force on panel:
  Inward: 7.0(28)(15.3) = 2999 lb
  Outward: 7.0(28)(17.5) = 3430 lb

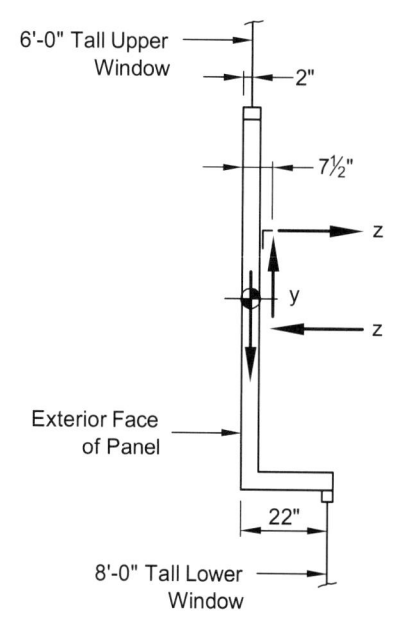

Force on panel from upper window:
  Inward: $\dfrac{6.0}{2}(28)(15.3) = 1285.2$ lb
  Outward: $\dfrac{6.0}{2}(28)(17.5) = 1470$ lb

Force on panel from lower window:
  Inward: $\dfrac{8.0}{2}(28)(15.3) = 1713.6$ lb
  Outward: $\dfrac{8.0}{2}(28)(17.5) = 1960$ lb

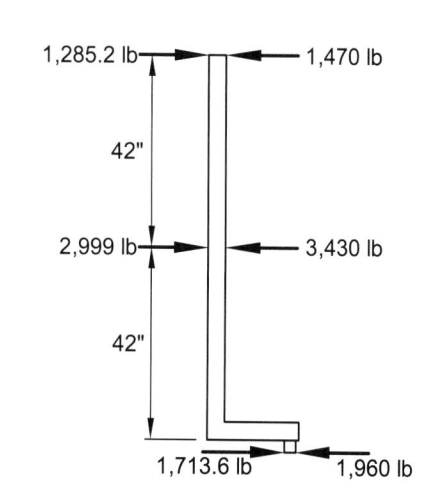

4. Determine the pressure zones on each side of the building from Figure 3.2.3.1:

5. The pressure on the Main Wind Force Resisting System (MWFRS) for each zone is then determined from:

$$p_s = \lambda I p_{s30} \quad \text{(Eq. 3.2.3.1)}$$

where:
- p_s = combined windward and leeward net pressures
- λ = coefficient from Figure 3.10.6(c)
- I = importance factor for wind from Figure 3.10.1
- p_{s30} = simplified design wind pressure from Figure 3.10.6(a)

6. The force on the MWFRS is then determined by multiplying the values of p_{s30} by their respective zone areas.

7. The pressure for cladding can be determined from:

$$p_{net} = \lambda I p_{net30} \quad \text{(Eq. 3.2.3.2)}$$

where:
- p_{net} = net design wind pressure on cladding
- λ = coefficient from Figure 3.10.6(c)
- p_{net30} = net design wind pressure from Figure 3.10.6(b). Note that there are two values for wind pressure to consider: inward (+) and outward (−)

3.2.4 Earthquake Loads

Earthquakes generate horizontal and vertical ground movement. When the seismic waves pass beneath a structure, the foundation will tend to move with the ground, while the superstructure will tend to remain in its original position. The lag between foundation and superstructure movement will cause distortions and develop forces in the structure. As the ground moves, changing distortions and forces are produced throughout the height of the structure.

The current philosophy for the design of earthquake resistant structures permits minor damage for moderate earthquakes, and accepts major damage for severe earthquakes, provided that complete collapse is prevented. The design details often require large, inelastic, deformations to occur in order to absorb the inertial forces. This is achieved by providing member and connection ductility. While this ductility can prevent total collapse, the resultant distortions may lead to significant damage to mechanical, electrical, and architectural elements. Seismic damage can be minimized by setting limitations on structural deflections, such as interstory drift.

The response of a structure to the ground motion of an earthquake depends on the structural system with its damping characteristics, and on the distribution of its mass. With mathematical idealization, a designer can determine the probable response of the structure to an imposed earthquake. IBC 2003 requires a dynamic analysis for structures that have highly irregular shapes or framing systems, or are particularly tall, and allows it for other structures. However, most buildings are not tall, have structural systems and shapes that are more or less regular, and most designers use the equivalent static force method for these structures.

There have been many recent seismic-related changes made to such documents as NEHRP and ACI 318. These changes are based on a significant amount of seismic research and observations from a number of recent earthquakes. Some of these changes are:

- Recognition of jointed panel construction as an alternative to emulation of monolithic construction.
- Achieving ductile structural behavior by using "strong" connections that remain elastic while nonlinear action (plastic hinging) occurs in the member away from the connection.
- Modification of drift computation and limiting drift.
- Deformation compatibility of structural elements and attached non-structural elements.
- Additional soil type classifications.
- Special considerations for building sites located near seismic faults.
- Special considerations for structures possessing redundancy.

3.2.4.1 Base Shear (IBC 2003 – Section 1617.4, Equivalent Lateral Force Method)

The procedure described here is applicable to all buildings in Seismic Design Categories A, B, C and to most precast structures in D. This method may not apply to buildings with irregularities in Seismic Design Categories D, E, or F, depending on the nature of the irregularity.

The seismic base shear, V, in a given direction is determined by:

$$V = C_s W \quad \text{(Eq. 3.2.4.1)}$$

where:

C_s = seismic response coefficient
W = total dead load of structure plus:

1. 25% of reduced floor live load in storage areas (live load in parking structures not included).
2. If partition load is included in gravity load include them here.
3. Total weight of permanent operating equipment.
4. 20 percent of flat roof snow load where flat roof snow load exceeds 30 psf.

The seismic response coefficient, C_s, is proportional to the design response spectrum. The design response spectrum has two segments. One is a short-period plateau and the other is a descending curve with lower values for longer building periods. Two coefficients, S_S and S_1, define these two segments, and vary with geographical location. Maps of the United States showing contours for S_S and S_1 are provided in IBC 2003. A CD-ROM is also available that enables determination of these values from longitude and latitude or zip code of the site.

To determine C_s:

1. Determine S_S and S_1 from the map, CD-ROM or from local building codes.
2. Determine site classification from soil reports or Figure 3.10.7(a). If site soils are not known, use Site Class D.
3. Calculate response accelerations:

$$S_{MS} = F_a S_S \qquad (Eq.\ 3.2.4.2)$$

$$S_{M1} = F_v S_1 \qquad (Eq.\ 3.2.4.3)$$

Figure 3.2.4.1 Design response spectrum

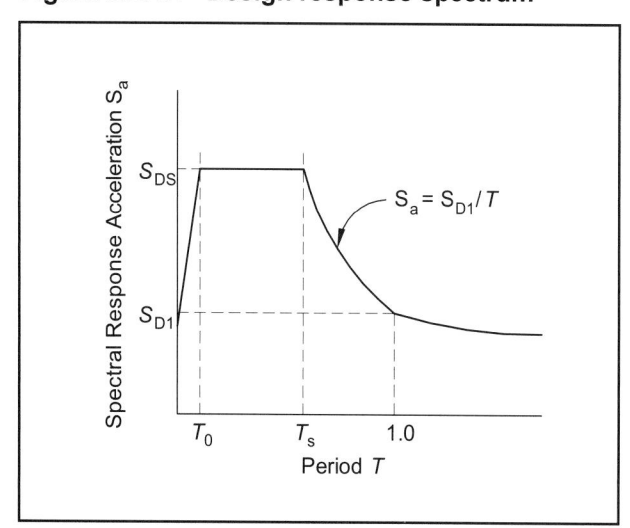

where:

F_a and F_v are site coefficients from Figure 3.10.7(b) and (c).

4. Calculate the 5%-damped design spectral response accelerations:

$$S_{DS} = (2/3)S_{MS} \qquad (Eq.\ 3.2.4.4)$$

$$S_{D1} = (2/3)S_{M1} \qquad (Eq.\ 3.2.4.5)$$

5. Determine the Seismic Design Category from Table 3.2.4.1. This will sometimes restrict the type of Seismic Force Resisting System (SFRS) used (see Figure 3.10.8).
6. Determine the fundamental period of the building from:

$$T_a = C_t h_n^x \qquad (Eq.\ 3.2.4.6)$$

where:

C_t = 0.016 for moment resisting frame systems of reinforced concrete in which the frames resist 100% of the required seismic forces and are not enclosed or adjoined by more rigid components that prevent that frame from deflecting when subjected to seismic forces
 = 0.020 for other concrete structural systems
h_n = distance from base to highest level (in feet)
x = 0.9 for concrete moment resisting frames
 = 0.75 for other concrete structural systems

Eq. 3.2.4.6 is considered to provide an "approximate period" by IBC 2003. In some cases, it may be useful to make a more accurate calculation of period from frame analysis. This can be done internally in some advanced stiffness analysis programs, but can be done relatively easily from the load inputs and the displacements using Rayleigh's formula:

$$T = 2\pi \sqrt{\frac{\sum_{i=1}^{n} w_i \delta_i^2}{g \sum_{i=1}^{n} F_i \delta_i}} \qquad (Eq.\ 3.2.4.7)$$

where:

w_i = dead load weight at Floor i
δ_i = elastic displacement at Floor i
F_i = lateral force at Floor i
g = acceleration of gravity, 386 in./sec^2

Table 3.2.4.1 Seismic design categories

(a) Based on short period response acceleration

Value of S_{DS}	Seismic Use Group[a]		
	I	II	III
$S_{DS} < 0.167g$	A	A	A
$0.167g \leq S_{DS} < 0.33g$	B	B	C
$0.33g \leq S_{DS} < 0.50g$	C	C	D
$0.50g \leq S_{DS}$	D[b]	D[b]	D[b]

(b) Based on 1-sec period response acceleration

Value of S_{D1}	Seismic Use Group[a]		
	I	II	III
$S_{D1} < 0.067g$	A	A	A
$0.067g \leq S_{D1} < 0.133g$	B	B	C
$0.133g \leq S_{D1} < 0.20g$	C	C	D
$0.20g \leq S_{D1}$	D[b]	D[b]	D[b]

a. "Category" in Figure 3.10.1.
b. Seismic Use Group I and II structures located on sites with mapped maximum considered earthquake spectral response acceleration at 1-sec period, S1, equal to or greater than the 0.75 g shall be assigned to Seismic Design Category E and Seismic Use Group III structures located on such sites shall be assigned to Seismic Design Category F.

Table 3.2.4.2 Coefficient for upper limit on calculated period

Design spectral response acceleration at 1-sec period, S_{D1}	Coefficient C_U
≥ 0.4	1.4
0.3	1.4
0.2	1.5
0.15	1.6
0.1	1.7
≤ 0.05	1.7

Example 3.2.4.1
Determination of Base Shear Coefficient, C_S

Given:
The hospital building in Memphis of Example 3.2.3.1. The building is a steel frame building with precast concrete shear walls designed by emulation as "Special reinforced concrete shear walls." A soils investigation has determined the site to be Class C.

Problem:
Determine the seismic response coefficient, C_S.

Solution:
1. From maps found in IBC 2003, $S_S = 1.5$, and $S_1 = 0.4$.
2. Site Class C
3. Response accelerations: From Figure 3.10.7(b), $F_a = 1.0$. From Figure 3.10.7 (c), $F_v = 1.4$.
   $S_{MS} = F_a S_s = 1.0(1.5) = 1.5$
   $S_{M1} = F_v S_1 = 1.4(0.4) = 0.56$
4. 5% damped design spectral response accelerations:
   $S_{DS} = (2/3)(1.5) = 1.0$
   $S_{D1} = (2/3)(0.56) = 0.37$
5. Approximate fundamental period:
   $C_T = 0.020$
   $h_n = 54$ ft
   $T_a = (0.020)(54)^{0.75} = 0.40$ sec
6. From Figure 3.10.8. Detail shear walls as "special walls." $R = 6$. From Figure 3.10.1, $I = 1.5$.
   Use lesser of Eqs. 3.2.4.9 and 3.2.4.10 for C_S.

From Eq. 3.2.4.9:
$$C_S = \frac{S_{DS}}{(R/I)} = \frac{1.0}{(6/1.5)} = 0.25$$

From Eq. 3.2.4.10:
$$C_S = \frac{S_{D1}}{(R/I)T} = \frac{0.37}{(6/1.5)(0.40)} = 0.23$$

Check Eq. 3.2.4.11: $0.044(1.0)(1.5)$
$= 0.066 < 0.23$ OK
Use $C_S = 0.23$

This is then multiplied by the dead weight of the building to determine the base shear.

However, the period cannot exceed:

$$T_{max} = C_u T_a \quad \text{(Eq. 3.2.4.8)}$$

where C_u is found from Table 3.2.4.2.

7. Determine C_s from the lesser of Eqs. 3.2.4.9 or 3.2.4.10:

$$C_s = \frac{S_{DS}}{(R/I)} \quad \text{(Eq. 3.2.4.9)}$$

where:
- R = response modification factor from Figure 3.10.8
- I = importance factor from Figure 3.10.1

$$C_s = \frac{S_{D1}}{(R/I)T} \quad \text{(Eq. 3.2.4.10)}$$

but cannot be less than:

$$C_s = 0.044 S_{DS} I \quad \text{(Eq. 3.2.4.11)}$$

In Seismic Design Categories E and F:

$$C_s = \frac{0.5 S_1}{(R/I)} \quad \text{(Eq. 3.2.4.12)}$$

3.2.4.2 Vertical Distribution

The lateral force at each level is calculated by:

$$F_x = C_{vx} V \quad \text{(Eq. 3.2.4.13)}$$

$$C_{vx} = \frac{w_x h_x^k}{\sum_{i=1}^{n} w_i h_i^k} \quad \text{(Eq. 3.2.4.14)}$$

where:
- C_{vx} = vertical distribution factor
- k = 1 for buildings having a period of 0.5 sec or less
- = 2 for buildings having a period of 2.5 sec or more

For buildings with a period between 0.5 and 2.5, determine k by linear interpolation.

- h_i and h_x = height from base to Level i or x
- V = base shear
- w_i and w_x = portion of total gravity load of building, W assigned to Level i or x

(Note: The denominator of Eq. 3.2.4.14 is the sum of the "$w_x h_x$" for all floors. Thus, for example, in a single-story building, the total base shear is assumed to be applied to the roof diaphragm.)

3.2.4.3 Lateral Distribution and Torsion

The force F_x as calculated in Section 3.2.4.2 is distributed to the resisting elements by the diaphragm as discussed in Section 3.8. Also, actual and accidental torsion must be considered in all buildings which do not have flexible diaphragms. (Note: It is conservative for application of this section to assume the diaphragm is not flexible.) The accidental torsion is calculated by assuming that the center of mass is located a distance of 5% of the plan dimension perpendicular to the applied load on either side of the actual center of mass. The total torsion is the sum of the actual torsion plus the accidental torsion.

3.2.4.4 Drift Effects

Seismic design includes requirements not only for strength, but also story drift, calculated as follows:

$$\delta_x = \frac{C_d \delta_{xe}}{I} \quad \text{(Eq. 3.2.4.15)}$$

where:
- δ_x = amplified deflection of Level x
- δ_{xe} = deflection of Level x determined from elastic analysis that includes consideration of cracking
- C_d = deflection amplification factor for structural system (Figure 3.10.8)
- I = seismic occupancy importance factor (Figure 3.10.1)

The C_d factor represents an approximation of the post-yield displacement.

In the evaluation of the building drift, the limit placed on building period by Eq. 3.2.4.8 and on the lower threshold limit Eq. 3.2.4.11 do not apply. A separate analysis using reduced lateral loads calculated without these constraints may be used to calculate the deflections used in the drift and stability evaluation.

Caution is needed when performing and interpreting the results of elastic analysis for precast moment frames. Designers using linear elastic analysis ordinarily calculate the load distribution and displacements using the load combinations as if the structure is instantly complete. This is usually not accurate. Erection of members and completion of connections is sequential. This sequential erection procedure can result in different load distributions and displacements than those one could calculate for a completed structure. For example, when deck members are placed on simple-span beams and moment connections later make the beam continuous, the deck load does not contribute to the negative moment at the connection. Calculated lateral forces and drifts can be similarly affected in unsymmetrical frames. Also, lateral sway may be taken out during erection as the frame is adjusted to plumb as erection proceeds.

Thus, load combinations for gravity dead loads as created by sequential erection should be analyzed so that these effects can be subtracted from the seismic load combinations to accurately reflect the effects of sequential erection. This can substantially reduce the design moments. Similarly, for drift derived from a linear elastic analysis, a load case or combination that includes only the seismic load should be used to separate the seismic deflections from the gravity dead load effects.

For drift derived from a linear elastic model, a load case or combination that includes only the seismic load can be used to separate the seismic deflections from the gravity effects.

A stability coefficient, θ, must be calculated:

$$\theta = \frac{P_x \Delta}{V_x h_{sx} C_d} \quad \text{(Eq. 3.2.4.16)}$$

where:
- P_x = total vertical unfactored load at and above Level x
- Δ = difference of deflections between Levels x and x–1
- V_x = seismic shear force acting between Levels x and x–1
- h_{sx} = story height below Level x
- C_d = deflection amplification factor (Figure 3.10.8)

The stability coefficient is limited to:

$$\theta_{max} = \frac{0.5}{\beta C_d} \leq 0.25 \quad \text{(Eq. 3.2.4.17)}$$

where β = ratio of shear demand to shear capacity between Levels x and x–1

P-Δ effects

In addition to the inelastic deformation increase, the design story drift must also be increased to account for P-Δ effects by a factor of $\frac{1}{1-\theta}$

If $\theta < 0.10$, P-Δ effects may be neglected.

The allowable story drift limits are shown in Table 3.10.9, which is taken from ASCE 7-02, Table 9.5.2.9.

3.2.4.5 Architectural Component Analysis

Non-structural architectural components, except those noted below, must resist seismic forces locally and be attached through a continuous load path to the seismic force resisting system.

Exceptions:
1. Seismic Design Category A.
2. Seismic Design Category B (other than parapets supported by bearing or shear walls), provided that the importance factor, I, is equal to 1.0.

For precast concrete cladding, the force required for these components and connections is given by Eq. 3.2.4.18, within the limits given by Eq. 3.2.4.19.

$$F_p = \frac{0.4 a_p S_{DS} W_p}{R_p}\left(1 + 2\frac{z}{h}\right) \quad \text{(Eq. 3.2.4.18)}$$

$$0.3 S_{DS} W_p \leq F_p \leq 1.6 S_{DS} W_p \quad \text{(Eq. 3.2.4.19)}$$

Note: Components with certain life-safety implications require a multiplier of 1.5 (not applicable to precast concrete panels.)

where:
- a_p = component amplification factor from Figure 3.10.10
- F_p = seismic design force centered at the component's center of gravity and distributed relative to component's mass distribution
- h = average roof height of structure
- R_p = component response modification factor from Figure 3.10.10
- S_{DS} = as previously defined
- W_p = component weight
- z = height in structure at attachment point $\leq$ h

Example 3.2.4.2
Architectural Precast Panel With Earthquake Loading

Given:
    The hospital in Memphis of Examples 3.2.3.1 and 3.2.4.1. Wall panels as shown.
    Concrete $f'_c = 5000$ psi (normal weight)
    Window weight = 10 psf

Problem:
Determine the seismic forces on the panel and compare with the wind forces of Example 3.2.3.1.

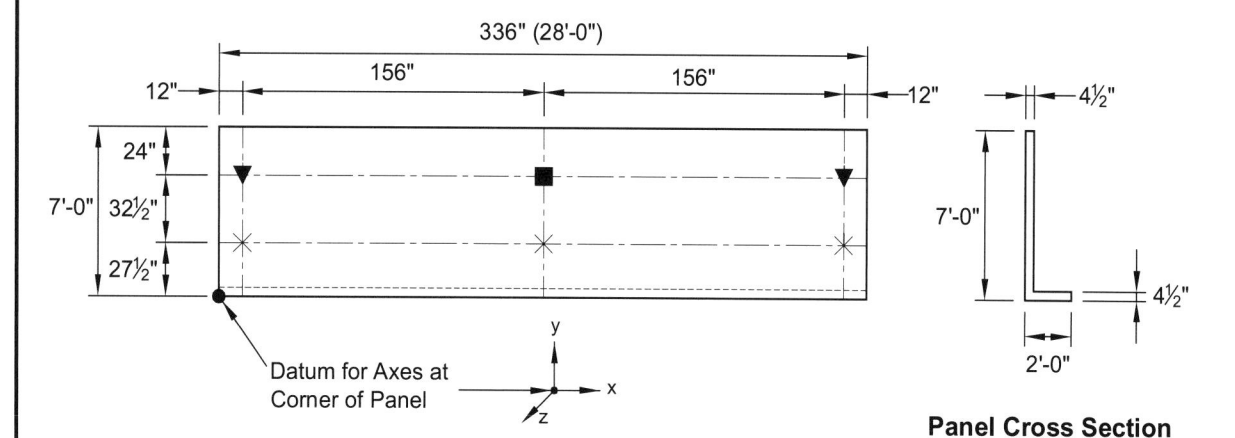

Panel Elevation
(As Viewed from Building Exterior)

Panel Cross Section

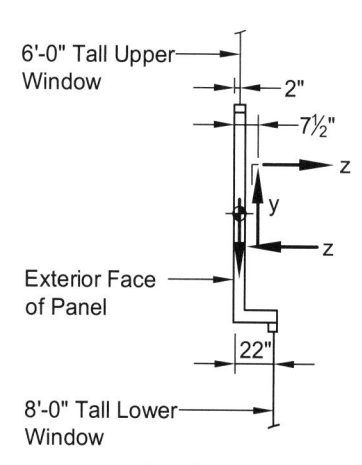

Section At Bearing
(Gravity Loading Reactions Shown)

▼ = Bearing: Resists Load in y and z Direction Only
■ = Lateral: Resists Load in x and z Direction Only
✕ = Tie-Back: Resists Load in z Direction Only

1. Connections Must Be Designed and Detailed to Ensure That Loads Do Not Occur in Directions Other Than Those Assumed.

2. Vertical Load Resistance at Bearing Connection is Located 7½ in. from the Exterior Face of the Panel.

3. Lateral Load (x-Direction) Resistance at Lateral Connection is Located 4½ in. from the Face of the Panel.

Solution:
    From Figure 3.10.10, Item 4:
    a. Wall element at top of building (z/h =1): $a_p = 1.0$, $R_p = 2.5$

$$F_p = \frac{0.4(1.0)(1.0)W_p}{2.5}(1+2) = 0.48W_p \quad \text{(Eq. 3.2.4.18)}$$

    b. Body of connections: $a_p = 1.0$, $R_p = 2.5$ (same as wall element)
       $F_p = 0.48W_p$

    c. Fasteners: $a_p = 1.25$, $R_p = 1.0$

$$F_p = \frac{0.4(1.25)(1.0)W_p}{1.0}(1+2) = 1.5W_p \quad \text{(Eq. 3.2.4.18)}$$

Example 3.2.4.2 (Cont.)
Architectural Precast Panel With Earthquake Loading

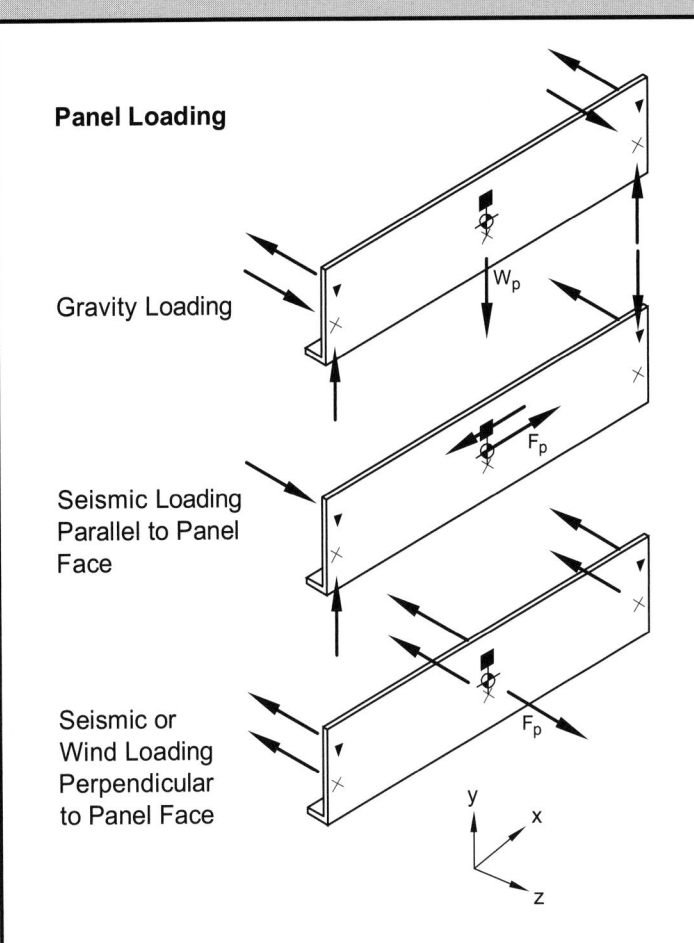

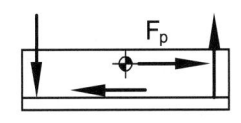

Panel Loading

Gravity Loading

Seismic Loading Parallel to Panel Face

Seismic or Wind Loading Perpendicular to Panel Face

Plan
(Vertical Reactions Not Shown for Clarity)

▼ = Bearing: Resists Load in y and z Direction Only
■ = Lateral: Resists Load in x and z Direction Only
✕ = Tie-Back: Resists Load in z Direction Only

Note: Load analysis for this case is done with F_p applied as a uniformly distributed load along the panel's length.

Cross-sectional area of panel = 465.75 in.^{2}
Center of gravity from datum (bottom outside corner): y = 34.5 in.; z = 4.5 in.

Panel wt = $\frac{465.75}{144}(150)$ = 485 lb/ft = 40.4 lb/in.

W_p = 485(28) = 13,580 lb
F_p = 0.48(13,580) = 6518 lb

Upper window height = 6 ft
  Total window weight on wall panel = 6(28)(10) = 1680 lb
  W_p of one-half of window = 3.0(10) = 30 plf
  F_p on panel = 0.48(30) = 14.4 plf
  14.4(28) = 403 lb, inward or outward

Lower window height = 8 ft
  (No weight on wall panel)
  W_p of one-half of window = 4.0(10) = 40 plf
  F_p on panel = 0.48(40) = 19.2 plf
  19.2(28) = 538 lb, inward or outward

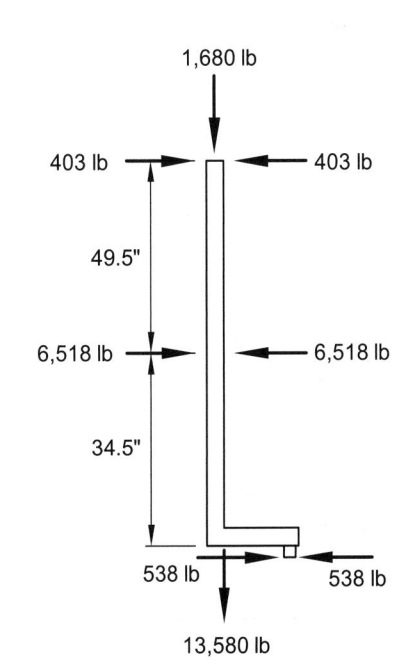

Example 3.2.4.2 (Cont.)
Architectural Precast Panel With Earthquake Loading

Determine center of dead load:

	W_p (lb)	z (in.)	$W_p z$ (lb-in.)
Panel Upper	13,580	4.5	61,110
Window Lower	1,680	2.0	3,360
Window	0	22.0	0
Total	15,260		64,470

Center of load from lower left: z = 64,470/15,260 = 4.2 in.

Dead loads to connections:
  Vertical = 15,260/2 = 7630 lb each connection
  Horizontal = 7630(7.5 − 4.2)/32.5 = 774.7/2 = 387 lb each connection
    Outward on top connection
    Inward on bottom connection

Determine center of seismic lateral force:

	F_p (lb)	y (in.)	z (in.)	$F_p y$ (lb-in.)	$F_p z$
Panel	6,518	34.5	4.5	224,871	29,331
Upper Window	403	84.0	2.0	33,852	806
Lower Window	538	0	22.0	0	11,836
Total	7,459			258,723	41,973

Center of force from lower left:
  y = 258,723/7459 = 34.7 in.
  z = 41,973/7459 = 5.6 in.

Determine center of outward wind lateral force:

	F_p (lb)	y (in.)	$F_p y$ (lb-in.)
Panel	3,430	42.0	144,060
Upper Window	1,470	84.0	123,480
Lower Window	1,960	0	0
	6,860		267,540

Center of force from lower left:
  y = 267,540/6860 = 39.0 in.

Center of inward wind lateral force is the same.
  F_p = 11.3/12.9(6860) = 6009 lb

For seismic in-out loads:
  y = 34.7 in.
  F_p = 7459 lb

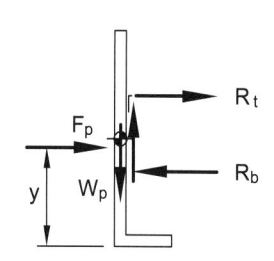

Moments about bottom connection:
  R_t = 7459(34.7 − 27.5)/32.5 = 1652 lb
  R_b = 7459 − 1652 = 5807 lb

For outward wind loads:
  y = 39.0 in.
  F_p = 6860 lb
  R_t = 6860(39.0 − 27.5)/32.5 = 2427 lb
  R_b = 6860 − 2427 = 4433 lb

For inward wind loads (proportional to pressure):
  R_t = (11.3/12.9)2427 = 2126 lb
  R_b = (11.3/12.9)4433 = 3883 lb

Example 3.2.4.2 (Cont.)
Architectural Precast Panel With Earthquake Loading

A continuous beam analysis shows that the center connection will take 58% of the load and each end connection will take 21%.

Seismic parallel to face:

(■) Parallel $= \pm 7459$ lb

(▼) Up-down $= \dfrac{7459(27.5 + 32.5 - 34.7)}{2(156)} = \pm 605$ lb

(▼) In-out $= \dfrac{7459(5.6 - 4.5)}{2(156)} = \pm 26$ lb

The appropriate load factor (Section 3.2.6) must be applied to loads to use strength design.

Summary of Factored Loads to Connections (lb)									
Connection	Dead Load[1]			Seismic[2]				Wind[3]	
	Vert (y)	In. (z)	Out (z)	Vert (y)	Horiz (x)	In. (z)	Out (z)	In. (z)	Out (z)
Top ctr ■					7,459	958	958	1,973	2,252
Top end ▼	9,156		484	605		347	347	714	815
Tie-b'k ctr X						3,368	3,368	3,603	4,114
Tie-b'k end X		484				1,219	1,219	1,305	1,489

1. Load factor of 1.2 applied.
2. Load factor of 1.0 applied.
3. Load factor of 1.6 applied.

Typical tie-back connection

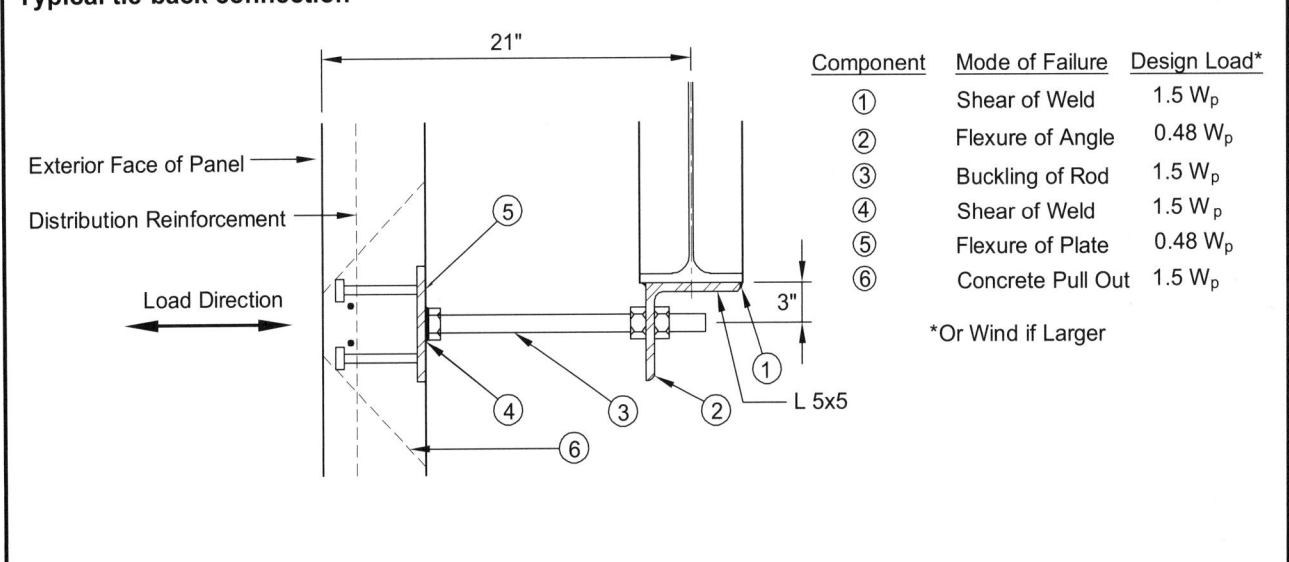

Component	Mode of Failure	Design Load*
①	Shear of Weld	1.5 W_p
②	Flexure of Angle	0.48 W_p
③	Buckling of Rod	1.5 W_p
④	Shear of Weld	1.5 W_p
⑤	Flexure of Plate	0.48 W_p
⑥	Concrete Pull Out	1.5 W_p

*Or Wind if Larger

3.2.4.6 Redundancy – Seismic Design Categories D, E, and F

In high seismic design categories, IBC-2003 and ASCE 7-02 require that a certain amount of redundancy be designed into the structure. This is done by increasing the earthquake force in high seismic areas (seismic design categories D, E, and F) by a reliability factor, ρ_i. ρ is the largest ρ_i over the height of the building, where ρ_i for the ith story is determined by:

$$\rho_i = 2 - \frac{20}{r_{maxi}\sqrt{A_i}} \qquad \text{(Eq. 3.2.4.20)}$$

ρ = 1.0 for structures in Seismic Design Categories A, B and C.

where, for each level:

r_{maxi} = For moment frames, the maximum of the sum of the shears in any two adjacent columns divided by the story shear. For columns common to two bays with moment-resisting connections on opposite sides, 70% of the shear in that column may be used in the column shear summary.

r_{maxi} = For shear walls, the maximum value of the product of the shear in the wall and $10/\ell_w$ divided by the story shear.

ℓ_w = length of wall
A_i = floor area

The value of ρ may not be less than 1.0, and need not exceed 1.5.

For structures with seismic-force-resisting systems in any direction comprised solely of special moment frames, the seismic-force-resisting system must be configured such that the value of ρ calculated in accordance with this section does not exceed 1.25 for structures assigned to Seismic Design Category D, and does not exceed 1.1 for structures assigned to Seismic Design Category E or F.

Example 3.6.7.1 illustrates the use of this factor.

(Note: For complex structures or combinations of seismic resisting systems, see IBC-2003, Section 1617.2.2.)

3.2.5 Lateral Soil Loads

Some precast structures require the consideration of lateral soil pressures. Design procedures involving soil loads are beyond the scope of this Handbook, and can be found in many texts and references.

3.2.6 Load Combinations

Refs. 1 through 3 specify the following load combinations:

$U = 1.4(D+F)$ (Eq. 3.2.6.1)
$U = 1.2(D+F+T) + 1.6(L+H)$ (Eq. 3.2.6.2)
$U = 1.2D + 1.6(L_r \text{ or } S \text{ or } R) + (1.0L \text{ or } 0.8W)$ (Eq. 3.2.6.3)
$U = 1.2D + 1.6W + 1.0L + 0.5(L_r \text{ or } S \text{ or } R)$ (Eq. 3.2.6.4)
$U = 1.2D + 1.0E + f_1L + 0.2S$ (Eq. 3.2.6.5)
$U = 0.9D + 1.6W + 1.6H$ (Eq. 3.2.6.6)
$U = 0.9D + 1.0E + 1.6H$ (Eq. 3.2.6.7)

where:
f_1 = 1.0 for floors in places of public assembly, for live loads in excess of 100 psf, and for parking garages, otherwise, $f_1 = 0.5$
D = Dead load
F = Pressure of fluids of known density and controlled depths
T = Effects of temperature, creep and shrinkage
L = Live load
H = Soil load
L_r = Roof live load
S = Snow load
R = Rain load
W = Wind load
E = Seismic load

IBC 2003 requires that the value of E in Eqs. 3.2.6.5 and 3.2.6.7 be defined by:

$$\rho Q_E \pm 0.2 S_{DS} D \qquad \text{(Eq. 3.2.6.8)}$$

This has the effect of changing those equations to:

$$U = (1.2 + 0.2 S_{DS})D + \rho Q_E + f_1 L + 0.2 S \qquad \text{(Eq. 3.2.6.5a)}$$

$$U = (0.9 - 0.2 S_{DS})D + \rho Q_E + 1.6H \qquad \text{(Eq. 3.2.6.7a)}$$

The coefficient, ρ, is 1.0 for buildings in Seismic Design Categories A, B, and C. See Section 3.2.4.6 for calculation of ρ in high Seismic Design Categories D, E, and F. Q_E is the horizontal seismic force on the component or system.

3.2.6.1 Load Factors for Diaphragms

Precast concrete diaphragms used in high seismic areas require special load combinations (see Section 3.8).

$$1.2D + f_1L + E_m \quad \text{(Eq. 3.2.6.9)}$$

$$0.9D + E_m \quad \text{(Eq. 3.2.6.10)}$$

where:
$$E_m = \Omega_o Q_E + 0.2S_{DS}D \quad \text{(Eq. 3.2.6.11)}$$

For the design of the diaphragm, which includes chord reinforcement and shear connections between precast elements, Section 3.8 recommends an overstrength factor $\Omega_o = 2.0$ in the above equation for Seismic Design Categories C, D, E and F. The value for the earthquake force for any Seismic Design Category should not be less than the F_{px} calculated by Eq. 3.8.3.2 or the largest level seismic force, whichever is greater. The same force should be used for diaphragm design for all levels of the building.

Since the overstrength factor is not used in Seismic Design Categories A and B, it is recommended, based on earlier versions of Ref. 3, that perimeter diaphragm reinforcement be designed based on strength reduction factors, ϕ, as follows: For continuous unspliced bars, $\phi = 0.9$. For continuous lap spliced or welded bars $\phi = 0.7$. For discontinuous bars, $\phi = 0.5$ (see also Section 3.8.3.3).

For connections from the diaphragms to the seismic force resisting system (SFRS)—shear walls or moment frames—Refs. 2, 3 and 4 require that the equations above be used with Ω_o the "system overstrength factor" (listed in Figure 3.10.8) for Seismic Design Categories C and higher.

3.3 Structural Integrity

3.3.1 Introduction

It is the intent of the structural integrity provisions of ACI 318-02 to improve the redundancy and ductility of structures and thereby reduce the risk of failure or collapse of parts or all of a building due to damage occurring to a relatively small area of a building. Code commentary emphasizes that the overall integrity of the structure can be substantially enhanced by minor changes in the detailing of reinforcement. In the event of damage to a beam, for example, it is important that displacement of its supporting member be minimized, so that other members will not be affected. For this reason, connection details which rely solely on friction caused by gravity loads are not permitted. Connections should also be detailed so as to minimize the potential for cracking due to restraint of volume change forces.

3.3.2 Precast Concrete Structures

For typical precast concrete structures, improved redundancy and ductility are achieved by connecting members into a load path to the lateral load resisting system. The load path in the lateral load resisting system must be continuous to the foundation.

Any individual member may be connected into this load path by several methods. For example, a loadbearing spandrel could be connected to a diaphragm (part of the lateral load resisting system) in more than one way. Structural integrity is typically achieved by connecting the spandrels into all or a portion of the deck members forming the diaphragm, which in turn would be connected to the supporting beams and the beams would be connected to their supporting columns. Alternatively, the spandrel could be connected only to its supporting columns, which in turn must then be connected to the diaphragm.

Vertical continuity is achieved by providing connections at horizontal joints of vertical members.

For precast concrete structures, the following provisions will satisfy the requirements of ACI 318-02, Sections 7.13.3 and 16.5. Example 3.3.2.1 demonstrates compliance with these requirements.

1. All members must be connected to the lateral force resisting system and their supporting members. Tension ties must be provided in the transverse, longitudinal, and vertical directions and around the perimeter of the structure.
2. The lateral force resisting system must be continuous to the foundation.
3. A diaphragm must be provided with connections between diaphragm elements, with tension ties around its perimeter and around openings that significantly interrupt diaphragm action. Section 16.5.2.4 of ACI 318-02 requires perimeter ties to provide a nominal strength of at least 16 kips and to be within 4 ft of the edge.
4. Column splices and column base connections must have a nominal tensile strength not less than $200A_g$ in lbs, where A_g is the gross area of the column in sq in. For a compression member with a larger cross section than required by consideration of loading, a reduced effective area, A_g, not less than one-half the total area, may be used.
5. Precast walls, other than cladding panels, must be connected across horizontal joints by a minimum of two connections per panel. Each connection is to have a nominal tensile strength of not less than 10 kips. When design forces result in no tension at the base, these

> **Example 3.3.2.1**
> **Compliance of a Precast Concrete Structure with the Structural Integrity Provisions**
>
> This example uses the design of the one-story building of Example 3.5.11.1 and some of the details shown in Chapter 6 to illustrate methods of compliance with the structural integrity provisions outlined in Section 3.3.2. The numbers used for the structural integrity provisions of this example refer to the same provision numbers used in Section 3.3.2.
>
> **Provision 1.   "Members must be connected..."**
> Compliance is provided by the connections between the roof diaphragm and the walls.
>
> **Provision 2.   "The lateral load resisting..."**
> Compliance is provided by the existence of the exterior shear walls and the connections of these walls to the roof diaphragm and the foundation. Example 6.14.4 shows a typical connection detail.
>
> **Provision 3.   "A diaphragm must be..."**
> Compliance is provided by the analysis, design and details of the roof diaphragm. Figure 3.8.1.2 and Examples 6.14.2 and 6.14.3 show example diaphragm connections.
>
> **Provision 4.   "Column splices and column..."**
> Section 6.12 shows column base plate and anchorage design.
>
> **Provision 5.   "Precast walls, other than..."**
> There are two shear wall-to-footing connections per panel. The connections designed to accommodate the loads determined for the wall panels to the foundation must have a minimum tensile strength of 10 kips.
>
> **Provision 6.   "Where precast elements..."**
> The connections between the exterior walls and the roof diaphragm in Example 3.5.11.1 are designed for seismic forces, which exceed these minimum requirements.
>
> **Provision 7.   "To accommodate volume..."**
> The details in Example 3.5.11.1 show how such connections are made.
>
> **Provision 8.   "Connections details that rely..."**
> All of the joints in the examples in this Handbook show positive connections.
>
> In conclusion, the structure of Example 3.5.11.1 with the clarifications noted in this section, satisfies all of the structural integrity provisions outlined in Section 3.3.2.

connections are permitted to be anchored into an appropriately reinforced slab on grade. If panels are too narrow to accommodate two connections, a single connection is satisfactory, as long as it is connected to adjacent panels.

6. Where precast elements form roof or floor diaphragms, the connections between the diaphragm and those members being laterally supported must have a nominal tensile strength not less than 300 lbs per linear ft.

7. To accommodate volume change strains (temperature and shrinkage) in supported beams, tie connections are typically located at the top of the member, with elastomeric pads used at the bottom-bearing surface. Such ties can be accomplished by welding, bolting, reinforcing steel in grout joints or bonded topping, or by doweling.

8. Connection details that rely solely on friction caused by gravity loads are not to be used. Exceptions may be permitted for heavy modular unit structures where resistance to overturning or sliding has a large factor of safety. Acceptance of such systems should be based on the provisions of ACI 318-02 Section 1.4.

3.3.3   Large-Panel Bearing Wall Structures

Large-panel bearing wall structures are a special category of precast concrete structures, with respect to structural integrity. Large panel structures are typically constructed with precast walls having a horizontal dimension greater than the vertical dimension, which is generally the height of one story. The panels are stacked for the height of the building and support the floor and roof decks. Criteria have been established for alternate load paths in such buildings three stories or more in height. [11]

Large-panel wall structures under three stories must meet the requirements of Section 3.3.2. For large-panel bearing wall structures three stories or more in height, minimum tie strength requirements are satisfied by the use of the following forces, as illustrated in Figure 3.3.3.1. It is not intended that these forces replace an analysis of the actual design forces required in the structure; these forces are not additive to the actual design forces.

The required forces are as follows:

T_1 = nominal tensile strength equal to 1500 lb per ft of floor or roof span. Ties may be encased in the floor units or in a topping, or may be concentrated at the wall. Spacing shall not exceed the spacing of bearing walls. Ties may be positioned in the walls within 2 ft of the plane of the floor or roof.

T_2 = nominal tensile strength sufficient to develop diaphragm action, but not less than 16,000 lbs, located within the depth of the floor or roof slab and within 4 ft of the edge. Ties may be reinforcing steel in a grout joint or in an edge beam; reinforced spandrels or wall anchored to the floor or roof may also be considered.

T_3 = nominal tensile strength not less than 1500 lb per linear ft of wall. Ties shall be spaced not greater than 10 ft on center. They may project from the precast element or be embedded in grout joints, with sufficient length and cover to develop the specified strength. At end walls, wall reinforcement shall extend into the floor, or mechanical anchorage between floor and wall shall be provided.

T_4 = nominal tensile strength not less than 3000 lb per horizontal ft of wall, with a minimum of two ties per wall. These ties shall be continuous from foundation to top of wall.

3.3.4 Hybrid Structures

The provisions of ACI 318-02 relate to concrete buildings only. Those connections which interface precast concrete components with other structural materials (e.g., masonry walls, steel or wood roofs) should provide the same load paths and follow the

Figure 3.3.3.1 Recommended forces in precast concrete bearing wall buildings

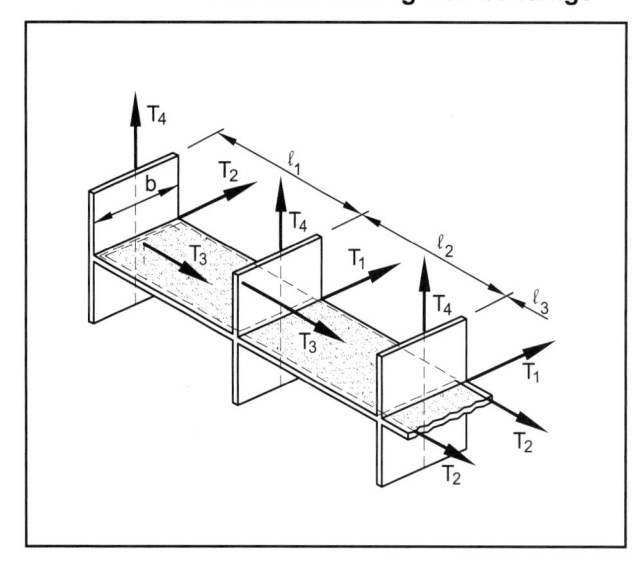

design philosophy described above. Since the precast concrete supplier rarely has control over such other materials, the Engineer of Record must design and detail the connections to satisfy structural integrity and include them in the contract drawings.

3.4 Volume Changes

Creep, shrinkage and temperature change and the forces caused by restraining these strains affect connections and service load behavior of precast concrete structures. Consequently, the effect of these strains and forces should not be ignored in the design and detailing. Volume change due to temperature variation can be positive (expansion) or negative (contraction) while volume changes from shrinkage and creep are only negative. Therefore, the critical combination is usually the contraction combination of creep, shrinkage and temperature drop. Figures 3.10.13 through 3.10.21 provide data for estimating the amount of shortening which may take place. Use of these aids is shown in Examples 3.4.1.1 and 3.4.1.2.

Properly detailed connections can minimize the effects of volume change strains. Connections should be detailed so that ductile deformation of one of the elements such as the connecting plate or connection bolt assembly can take place. Neglecting the effect of this connection deformation will produce unrealistically high computed restraint forces and can actually have a negative effect if connections are too strong, and inhibit necessary ductility.

Problems caused by volume change have appeared when prestressed members were welded to their bearings at both ends. When such members are connected only at the top using a ductile connection, experience has shown that volume changes are adequately accommodated. An excessively strong top connection may also attract high negative moment if compression resistance is encountered at the bearing location.

Vertical members such as loadbearing wall panels are also subject to volume change strains. The approximate magnitude can be calculated using Figures 3.10.11 through 3.10.17, adding the dead load strain to any prestress strain. The vertical member effects will only be significant in high-rise structures, and then only the differential movements between supporting columns and facade panels may significantly affect the performance of the members. Such effects can occur, for example, at the exterior of a building where precast panels are connected to a cast-in-place frame. The connection between the precast member and the supporting frames, and the connections between panels, must be detailed to accommodate volume changes.

Volume change forces are particularly important in parking structures because of their exposure to the elements. The primary volume change force to accommodate is due to temperature change. Again, connections between precast members and precast members to stiff elements such as shafts must accommodate these strains.

Example 3.4.1.1
Calculation of Volume Change Shortening

Given:
    Heated structure in Denver, Colorado
    Normal weight concrete beam – 12RB28
    8 – ½ in. diameter, 270 ksi, low relaxation strand
    Initial tension = $0.75f_{pu}$
    Assume initial prestress loss = 9%
    Release strength = 4500 psi (accelerated cure)
    Length = 24 ft

Problem:
    Determine the actual shortening that can be anticipated from:
    a. Casting to erection at 60 days.
    b. Erection to the end of service life.

Solution:
    From Figures 3.10.11 and 3.10.12:
    Design temperature change = 70°F
    Average ambient relative humidity = 55%

$A_{ps} = 8(0.153) = 1.224$ in.^{2}
$P_o = 1.224(270)(0.75)(1 - 0.09) = 225.6$ kips
$P_o/A = \dfrac{225.6(1000)}{12(28)} = 671$ psi

Volume-to-surface ratio = $\dfrac{12(28)}{2(12)+2(28)} = 4.2$ in.

a. At 60 days:

From Figure 3.10.13:
    Creep strain = 169×10^{-6} in./in.
    Shrinkage strain = 266×10^{-6} in./in.
From Figure 3.10.14:
    Creep correction factor = $0.88 + (71/200)(1.18 - 0.88) = 0.99$
From Figure 3.10.15:
    Relative humidity correction (creep) = $1.17 - 0.5(1.17 - 1.08) = 1.13$
    Relative humidity correction (shrinkage) = $1.29 - 0.5(1.29 - 1.14) = 1.22$

Example 3.4.1.1 (Cont.)
Calculation of Volume Change Shortening

From Figure 3.10.16:
Volume-to-surface ratio correction (creep) $= 0.48 - 0.2(0.48 - 0.36) = 0.46$
Volume-to-surface ratio correction (shrinkage) $= 0.46 - 0.2(0.46 - 0.31) = 0.43$

(Note: Temperature shortening is not significant for this calculation because the structure is in a controlled environment.)

Total strain:
Creep $= 169 \times 10^{-6}(0.99)(1.13)(0.46)$
$= 87 \times 10^{-6}$ in./in.
Shrinkage $= 266 \times 10^{-6}(1.22)(0.43)$
$= 140 \times 10^{-6}$ in./in.
Total strain $= 227 \times 10^{-6}$ in./in.
Total shortening $= 227 \times 10^{-6}(24)(12) = 0.065$ in.

b. At Final:
From Figure 3.10.13
Creep strain $= 315 \times 10^{-6}$ in./in.
Shrinkage strain $= 510 \times 10^{-6}$ in./in.

Factors from Figures 3.10.14 and 3.10.15 are the same as for 60 days.

From Figure 3.10.16:
Volume-to-surface ratio correction (creep):
$= 0.77 - 0.2(0.77 - 0.74) = 0.76$
Volume-to-surface ratio correction (shrinkage)
$= 0.75 - 0.2(0.75 - 0.64) = 0.73$
From Figure 3.10.17
Temperature strain $= 210 \times 10^{-6}$ in./in.
Total creep and shrinkage strain:
Creep $= 315 \times 10^{-6}(0.99)(1.13)(0.76)$
$= 268 \times 10^{-6}$ in./in.

Shrinkage $= 510 \times 10^{-6}(1.22)(0.73)$
$= 454 \times 10^{-6}$ in./in.
Total $= 722 \times 10^{-6}$ in./in.

Difference from 60 days to final
$= 722 - 227 = 495 \times 10^{-6}$ in./in.

Total strain $= 495 + 210 = 705 \times 10^{-6}$ in./in.

Total shortening $= 705 \times 10^{-6}(24)(12) = 0.203$ in./2 = 0.10 in. per end
This strain would then be accommodated in the detailing of the connections to provide for deformity of the connecting elements.

The behavior of actual structures indicates that reasonable estimates of volume change characteristics are satisfactory for the design of most structures even though test data relating volume changes to the variables shown in Figures 3.10.13 through 3.10.17 exhibit considerable scatter. Therefore, it is possible to reduce the variables and use approximate values as shown in Figures 3.10.18 and 3.10.19.

Example 3.4.1.2
Determine Volume Change Shortening by Figures 3.10.18 and 3.10.19

Given and Problem:
   Same as Example 3.4.1.1.

Solution:
   For prestressed, normal weight concrete, in a heated building, use Figure 3.10.18.
   For 55% relative humidity and temperature change of 70°F, interpolating from Figure 3.10.18:
   Actual strain = 687×10^{-6} in./in.
   This result is similar to the value of 705×10^{-6} calculated from Figures 3.10.13 through 3.10.17.

3.4.1 Axial Volume Change Strains

Figures 3.10.11 through 3.10.17 provide the data needed to determine volume change strains, as shown in Example 3.4.1.1.

3.4.2 Expansion Joints

Joints are placed in structures to limit the magnitude of forces, which result from volume change deformations (temperature changes, shrinkage and creep), and to permit movements (volume change, seismic) of structural elements.

Joints that permit contraction of the structure are needed to relieve the strains typically caused by temperature drop, creep and shrinkage which may be additive. Such joints are properly called contraction or control joints but are commonly referred to as expansion joints. Typically, the forces generated by a temperature difference are significantly greater than shrinkage and creep forces, and a true "expansion joint" may be needed in long buildings.

It is desirable to have as few expansion joints as possible. The purpose of this section is to present guidelines for determining the spacing and width of expansion joints.

3.4.2.1 Spacing of Expansion Joints

Recommended expansion joint spacing for precast concrete buildings is generally based on experience. Evaluation of joint spacing should consider the types of connections used, the column stiffnesses in simple span structures, the relative stiffness between beams and columns in framed structures, location of lateral load resisting elements and the weather exposure conditions. Non-heated structures, such as parking structures, are subjected to much greater temperature changes than enclosed structures, so the distance between expansion joints should consider the previously mentioned items.

Figure 3.10.20 shows joint spacing as recommended by the Federal Construction Council, and is adapted from Ref. 30. The spacings obtained from the graph in Figure 3.10.20 should be modified for various conditions as shown in the notes below the graph. Values for the design temperature change can be obtained from Figure 3.10.11.

When expansion joints are required in non-rectangular structures, such as T or L-shaped structures, they should be located at places where the plan or elevation dimensions change radically.

3.4.2.2 Width of Expansion Joints

The width of the joint can be calculated theoretically using a coefficient of expansion of 6×10^{-6} in./in./°F for normal weight concrete and 5×10^{-6} in./in./°F for sand-lightweight concrete. Since the primary problem in concrete buildings is contraction rather than expansion, joints that are too wide may result in problems with reduced bearing or loss of filler material. The joint width should consider the ambient temperature at time of erection. Seismic codes also stipulate joint widths to accommodate earthquake movement. It is typical to size seismic joints by summing the calculated drift of each building component and multiplying the amount by 0.60 to determine joint width.

3.4.3 Volume Change Effects in Moment-Resisting Frames

The restraint of volume changes in moment-resisting frames causes tension in the girders and deflections, shears and moments in the columns. The magnitude of the tension, shear, moment and deflection of a member is dependent on the distance from the center of stiffness of the frame or of several frames if they are connected.

The center of stiffness is that point of a frame, which is subject to a uniform unit shortening, at which no lateral movement will occur. For frames which are symmetrical with respect to bay sizes, story heights and member stiffnesses, the center of stiffness is located at the midpoint of the frame, as shown in Figure 3.10.21.

Tensions in girders are maximum in the bay nearest the center of stiffness. Deflections and moments in columns are maximum furthest from the center of stiffness. Thus, in Figure 3.10.21:

$$F_1 < F_2 < F_3$$
$$\Delta_1 > \Delta_2 > \Delta_3$$
$$M_1 > M_2 > M_3$$

The degree of fixity of the column base as described in Section 3.6.2 has a great effect on the magnitude of the forces and moments caused by volume change restraint. An assumption of a fully fixed base in the analysis of the structure may result in significant overestimation of the restraint forces, whereas assuming a pinned base may have the opposite effect. The degree of fixity used in the volume change analysis should be consistent with that used in the analysis of the column for other loadings, and the determination of slenderness effects. This will reduce the volumetric restraint force. To avoid a build-up of volume change forces, rather than connecting several frames, it is advisable to create flexible connections between sets of moment frames.

3.4.3.1 Equivalent Volume Change

If a horizontal framing member is connected at the ends such that the volume change shortening is restrained, a tensile force is built up in the member and transmitted to the supporting elements. However, since the shortening takes place gradually over a period of time, the effect of the shortening on the shears and moment of the support is lessened because of creep and microcracking of the member and its support.

For ease of design, the volume change shortenings can be treated in the same manner as short-term elastic deformations by using a concept of "equivalent" shortening.

Thus, the following relations can be assumed:

$$\delta_{ec} = \frac{\delta_c}{K_\ell} \qquad \text{(Eq. 3.4.3.1)}$$

$$\delta_{es} = \frac{\delta_s}{K_\ell} \qquad \text{(Eq. 3.4.3.2)}$$

where:

δ_{ec}, δ_{es} = equivalent creep and shrinkage shortenings, respectively
δ_c, δ_s = calculated creep and shrinkage shortenings, respectively
K_ℓ = a constant for design purposes which varies from 4 to 6

The value of K_ℓ will be near the lower end of the range when the members are heavily reinforced, and near the upper end when they are lightly reinforced. For most common structures, a value of $K_\ell = 5$ is sufficiently conservative.

Shortening due to temperature change will be similarly modified. However, the maximum temperature change will usually occur over a much shorter time, probably within 60 to 90 days.

Thus,

$$\delta_{et} = \delta_t / K_t \qquad \text{(Eq. 3.4.3.3)}$$

where:

δ_{et}, δ_t = the equivalent and calculated temperature shortening, respectively
K_t = a constant; recommended value = 1.5

Total equivalent shortening to be used for design:

$$\Delta = \delta_{ec} + \delta_{es} + \delta_{et} = \frac{\delta_c + \delta_s}{K_\ell} + \frac{\delta_t}{K_t}$$

(Eq. 3.4.3.4)

When the equivalent shortening is used in frame analysis for determining shears and moments in the supporting elements, the actual modulus of elasticity of the members is used, rather than a reduced modulus as used in other methods.

Figures 3.10.22 and 3.10.23 provide equivalent volume change strains for typical building frames.

3.4.3.2 Calculating Restraint Forces

Most "plane frame" computer analysis programs allow the input of shortening strains of members from volume changes. The equivalent strains as described in Section 3.4.2.1 can be input directly into such programs.

For frames that are approximately symmetrical, the coefficients from Figures 3.10.24 can be used with small error. The notation for this table is described in Figure 3.10.25.

Example 3.4.3.1
Calculation of Column Moment Caused by Volume Change Shortening of a Beam

Given:
The beam of Example 3.4.1.1 is supported and attached to two 16 x 16-in. columns as shown in the sketch. Use $E_c = 4.3 \times 10^6$ psi and f'_c (col.) = 5000 psi.

Problem:
Determine the horizontal force at the top of the column and the moment at the base of the column caused by volume change shortening of the beam.

Solution:

$$I_c = \frac{bh^3}{12} = \frac{16(16)^3}{12} = 5461 \text{ in.}^4$$

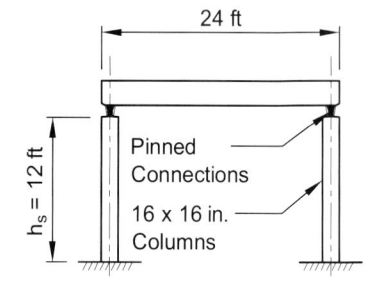

From Example 3.4.1.1:
Total volume change shortening from erection to final is 0.20 in., or 0.10 in. each end.

Calculate the equivalent shortening: $\Delta = \frac{\delta_c + \delta_s}{K_\ell} + \frac{\delta_t}{K_t}$

$$= \left(\frac{268 - 87 + 454 - 140}{5} + \frac{210}{1.5}\right)(10^{-6})(24)(12) = 239 \times 10^{-6}(288) = 0.07 \text{ in.}$$

$$\frac{\Delta}{2} = \frac{0.070}{2} = 0.035 \text{ in. each end}$$

$$\frac{\Delta}{2} = \frac{Fh_s^3}{3E_c I_c}$$

$$F = \frac{3E_c I_c \left(\frac{\Delta}{2}\right)}{h_s^3} = \frac{3(4.3 \times 10^6)(5461)(0.035)}{[12(12)]^3} = 826 \text{ lb}$$

$$M = Fh_s = 826(144) = 118{,}944 \text{ lb-in.} = 9.91 \text{ kip-ft}$$

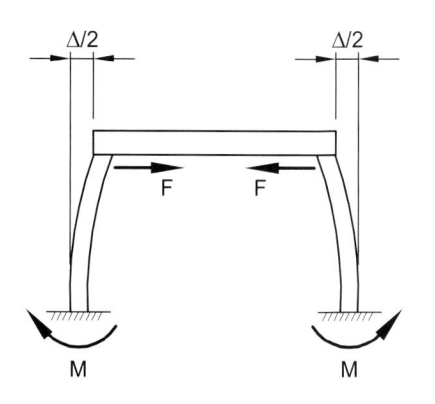

Example 3.4.3.2
Volume Change Restraint Forces

Given:
The four-bay, two-story frame shown below:
Beam modulus of elasticity = E_b = 4300 ksi
Column modulus of elasticity = E_c = 4700 ksi
Column bases 20% fixed (see Section 3.8.3)
Design R.H. = 70%
   Temperature zone is found from Figure 3.10.11.
   Building is heated.

Example 3.4.3.2 (Cont.)
Volume Change Restraint Forces

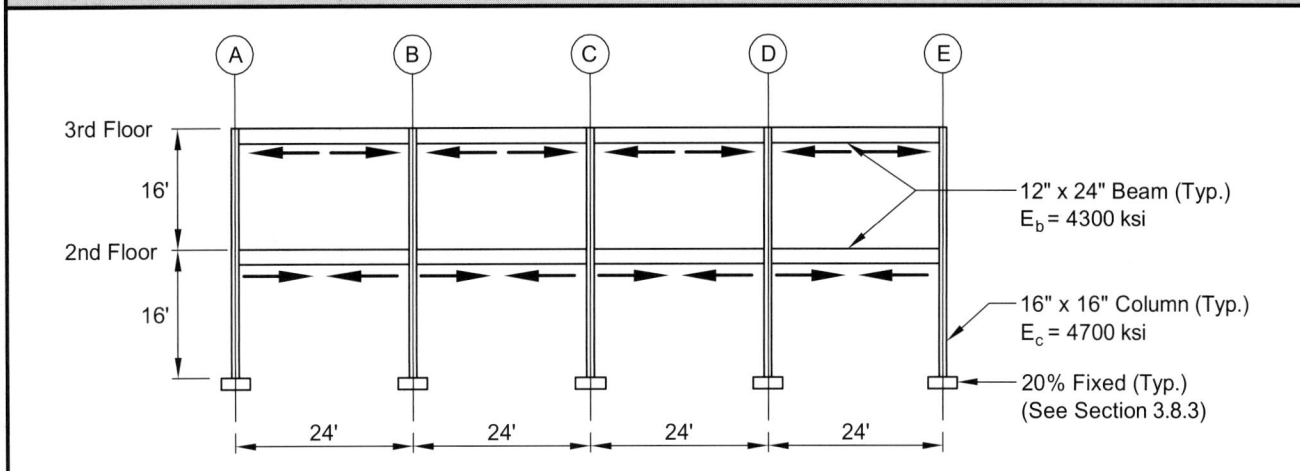

Problem:
  Determine the maximum tension in the beams and the maximum moment in the columns caused by volume change restraint.

Solution:
1. Determine relative stiffness between columns and beams:

$$I_b = \frac{12(24)^3}{12} = 13{,}824 \text{ in.}^4$$

$$\frac{E_b I_b}{\ell} = \frac{4300(13{,}824)}{24(12)} = 206{,}400$$

$$I_c = 16(16)^3/12 = 5461 \text{ in.}^4$$

$$\frac{E_c I_c}{h_s} = 4700(5461)/[16(12)] = 133{,}681$$

$$K_r = \frac{E_b I_b / \ell}{E_c I_c / h_s} = \frac{206{,}400}{133{,}381} = 1.5$$

2. Determine deflections:
From Figure 3.10.22:
  $\delta_e = 221 \times 10^{-6}$ in./in.
  $\Delta_B = \delta_e \ell = 0.000221(24)(12) = 0.064$ in.
  $\Delta_A = \delta_e (2\ell) = 0.128$ in.

3. Determine maximum beam tension:
Maximum tension is nearest the center of stiffness, i.e., Beams BC and CD, second floor.

From Figure 3.10.21:
  For n = 4 and i = 2:
  $k_b = 3.00$

Example 3.4.3.2 (Cont.)
Volume Change Restraint Forces

From Figure 3.10.24:
    For $K_r = 1.0$, fixed base; $k_f = 11.2$
    For $K_r = 2.0$, fixed base; $k_f = 11.6$

Therefore, for $K_r = 1.5$, $k_f = 11.4$

For pinned base: $k_f = 3.4$ (for $K_r = 1.0$ and 2.0)

For 20% fixed:

    $k_f = 3.4 + 0.20(11.4 - 3.4) = 5.0$

$$F_2 = \frac{k_i k_b \Delta_i E_c I_c}{h_s^3} = \frac{5.0(3.0)(0.064)(4700)(5461)}{[16(12)]^3} = 3.48 \text{ kips}$$

4. Determine maximum column moments:
For base moment, M_1:

From Figure 3.10.24, by interpolation similar to above:

    k_m (fixed)   = $(4.9 + 5.2)/2 = 5.05$

    k_m (pinned) = 0

    k_m (20% fixed) = $0 + 0.20(5.05) = 1.0$

$$M_1 = \frac{k_m \Delta_i E_c I_c}{h_s^2} = \frac{1.0(0.128)(4700)(5461)}{[16(12)]^2} = 89.1 \text{ kip-in.}$$

For second floor moment, M_{2L}:

    k_m (fixed)   = $(3.9 + 4.5)/2 = 4.2$

    k_m (pinned) = $(2.1 + 2.4)/2 = 2.25$

    k_m (20% fixed) = $2.25 + 0.20(4.20 - 2.25) = 2.64$

$$M_{2\ell} = \frac{2.64(0.128)(4700)(5461)}{[16(12)]^2} = 235 \text{ kip-in.}$$

These forces may be reduced by providing ductile connections at some locations.

3.5 Shear Wall Systems

3.5.1 Introduction

Buildings which use shear walls as the lateral force resisting system provide a safe, serviceable and economical solution for wind and earthquake resistance. Shear walls make up most common lateral force resisting systems in the precast, prestressed concrete industry. The excellent performance of shear wall buildings throughout the world, that have been subjected to earthquakes, testifies to the effectiveness of this system. Experience in earthquakes worldwide shows that, in many cases, shear wall buildings continue to be used with full functions after an earthquake. The design of these buildings has typically followed principles used for cast-in-place structures, with modifications made as appropriate for the jointed nature of a precast concrete structural system. Design methods used to achieve successful performance of precast shear wall structures have been largely left to the ingenuity and judgment of the design engineer. Observations of performance of structures in earthquakes show that where adequate strength and stiffness were provided to limit interstory drift to about 2%, the resulting displacements and damage were within acceptable levels. In regions of low and moderate seismicity, dry connections with small grout joints are generally used. In regions of high seismicity, connections to the foundation, and connections between precast walls, generally use details which emulate cast-in-place behavior (Section 3.1.2.1) and may include post-tensioning.

In the few cases where shear wall buildings have shown poor performance, the problems were generally related to details that were insufficiently ductile and thus resulted in brittle local failure (such as improper anchoring of embedments). Incomplete load paths such as improper diaphragm details were also problematic. Inadequate diaphragm behavior can be caused by inadequate reinforcement for shear and tension within the diaphragm, and insufficient ties between the diaphragm and the shear walls.

3.5.2 Principles of Shear Wall Buildings

Shear walls act as vertical cantilever beams, which transfer lateral forces acting parallel to the face of the wall, from the superstructure to the foundation. Shear walls should be oriented to resist lateral loads applied to the building along both principal axes of the building. Ideally, there should be at least two shear walls oriented to resist lateral loads along each principal axis of the building. If only one shear wall is oriented along one principal axis of the building, two shear walls should be provided along the orthogonal axis to resist diaphragm torsion. [see Figure 3.5.2.1(a)] Alternatively, it is acceptable to orient the three shear walls in any non-collinear position. Some codes require that lateral loads be applied in the direction of both principal axes simultaneously.

It is desirable to design shear walls as load-bearing panels, whenever possible. The increased dead load acting on the panel is an advantage because it increases the panel's resistance to uplift and overturning.

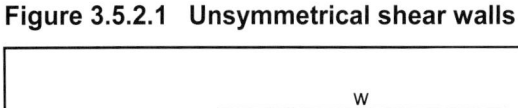

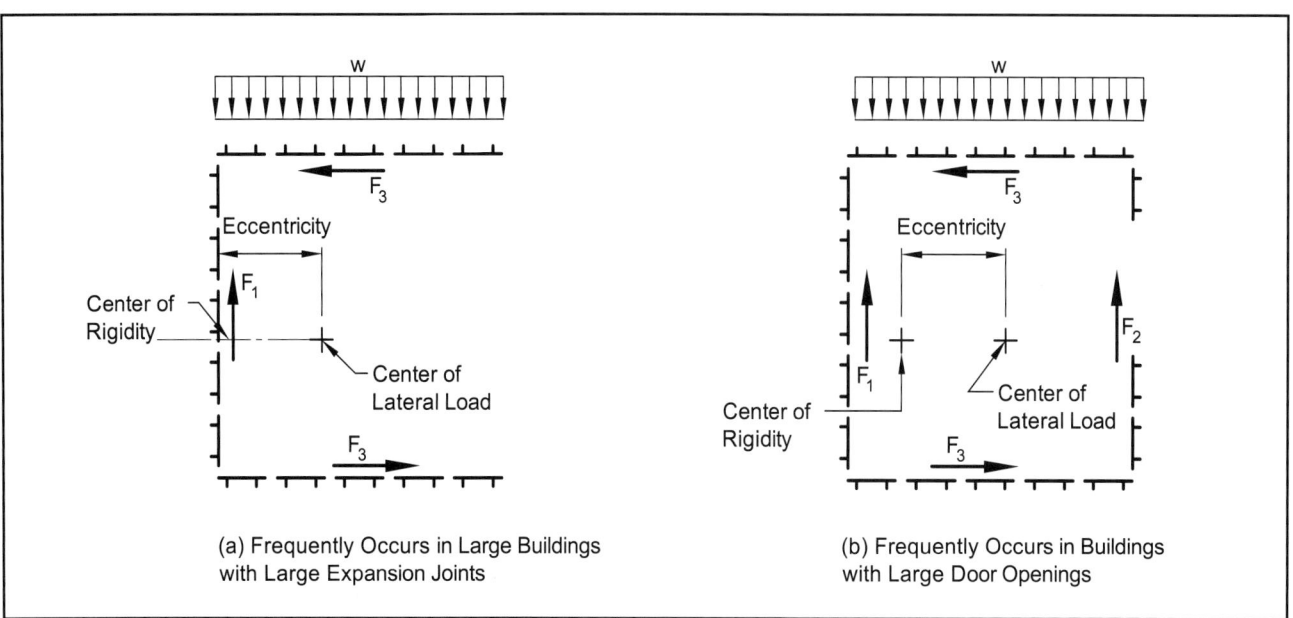

Figure 3.5.2.1 Unsymmetrical shear walls

(a) Frequently Occurs in Large Buildings with Large Expansion Joints

(b) Frequently Occurs in Buildings with Large Door Openings

The distribution of the total lateral force acting on a building to each individual shear wall is influenced by the following factors:

1. The supporting soil and footings.

2. The stiffness of the diaphragm.

3. The relative flexural and shear stiffnesses of the shear walls, and of connections.

4. The eccentricity of the lateral loads to the center of rigidity of the shear walls.

Generally, it is common practice to neglect the deformation of the soil and the footings when distributing shear forces among shear walls. If the depth-to-span ratio of a diaphragm is small, the diaphragm will be "flexible" and may deflect significantly when subjected to lateral loads. Flexible diaphragms distribute shears to each shear wall in proportion to the tributary width of diaphragm loading each shear wall.

If the depth-to-span ratio of a diaphragm is large, the diaphragm will be "rigid" and not deflect as significantly as a flexible diaphragm, when subjected to lateral loads. Rigid diaphragms distribute shears to each shear wall in proportion to the shear wall's relative stiffness. In precast concrete building design, it is common practice to assume that floor and roof diaphragms act as rigid diaphragms. See Section 3.8 for a more complete discussion of precast diaphragms.

3.5.3 Code Requirements

The International Building Code (IBC) provisions derived from the NEHRP recommendations will result in lateral forces that govern most precast building systems. The code now defines two categories of shear walls, ordinary and special. ACI 318-02 has created an additional intermediate wall category, but has assigned no distinct R, Ω_o and C_d factors to it (see Figure 3.10.8).

ACI 318-02 defines the various types of structural (shear) walls as follows (Note: ACI 318-02 uses the term "structural wall." In reference to lateral load resistance, this is the same as a shear wall. This Handbook uses the more traditional term "shear wall"):

Ordinary structural (shear) wall – "A wall complying with the requirements of Chapters 1 through 18," i.e., one with no special seismic detailing.

Intermediate precast structural (shear) wall – "A wall complying with all requirements of Chapters 1 through 18 in addition to 21.13." Section 21.13 requires that yielding of connections be in steel elements, or that the strength of the connection be 50% greater than required by analysis.

Special precast structural (shear) wall – "A precast wall complying with the requirements of 21.8. In addition, the requirements for ordinary reinforced concrete structural walls and the requirements of 21.2 shall be satisfied." This requires precast walls to be designed and detailed like cast-in-place walls, i. e., "emulative" design, in addition to having them meet the connection requirements of Section 21.13.

3.5.4 Design Guidelines for Shear Wall Structures

The following are suggested design guidelines for structures that have shear walls as the primary lateral load resisting elements:

1. Evaluation of building function and applicable precast frame
   a. In a warehouse type structure, it is common to include the exterior walls as part of the lateral load resisting system.
   b. In parking structures, shear walls can be located at stair and elevator towers, at the perimeter or ramped bays, at selected locations on the perimeter of the structure, or any in combination of the above locations.

2. Preliminary development of shear wall system
   a. Provide at least three non-collinear walls to ensure torsional as well as direct lateral resistance.
   b. Overturning will often be the governing criterion. Thus, the first choice is to use shear walls that also function as bearing walls.
   c. Arrange shear walls so that they minimize restraint due to volume changes.
   d. Consider whether the shear walls could be individual full height walls (vertical joints only).
   e. Consider the practicality of shipping and erection when selecting the size of wall panels.
   f. Balance the design requirements of the shear walls with the design requirements of the associated diaphragms.

3. Determination of vertical and lateral loads
   a. Determine the vertical gravity loads that are applicable to each of the shear walls.
   b. Use the applicable seismic design criteria to determine the magnitude of lateral load

at each floor, and compare with wind loading. Choose the critical condition for design.

4. Preliminary load analysis
   a. Determine the overturning moment, the lateral in plane shear and the axial load at the base of each of the shear walls.

5. Selection of shear walls
   a. Review the preliminary choice of shear wall size and location.
   b. Modify the number, location, and dimensions of shear walls as necessary to satisfy the requirements at the base of each. It is economically preferable that foundations not be subject to uplift.

6. Final load analysis
   a. Based on the final location and dimensions of shear walls, perform the final lateral load and vertical load analysis to determine the design load for each of the shear walls. Consider shear stiffness as well as flexural stiffness when distributing lateral loads to the shear walls.

7. Final shear wall design
   a. Design the shear wall reinforcement and the connections to the associated diaphragms.
   b. Where there is insufficient length of shear wall available to accommodate the necessary number of shear connectors, consider using an element in the plane of the diaphragm (drag strut) as an extension of the shear wall to pick up additional connectors to the diaphragm.
   c. Consider the additional requirements necessary to satisfy the structural integrity provisions of the code (see Section 3.3).

8. Diaphragm design
   a. Design the diaphragms to respond elastically to applied lateral loads in order to prevent formation of plastic regions in any diaphragm. See Section 3.8 for a more detailed discussion of diaphragm design.
   b. Design the diaphragms as beams, provide the necessary tensile reinforcement for each chord, and choose shear connectors using design procedures of Chapter 6, or shear reinforcement using shear-friction methods.
   c. Consider the additional requirements necessary to satisfy the structural integrity provisions of the code (see Section 3.3).

Figure 3.5.5.1 Translation and rotation of rigid diaphragms

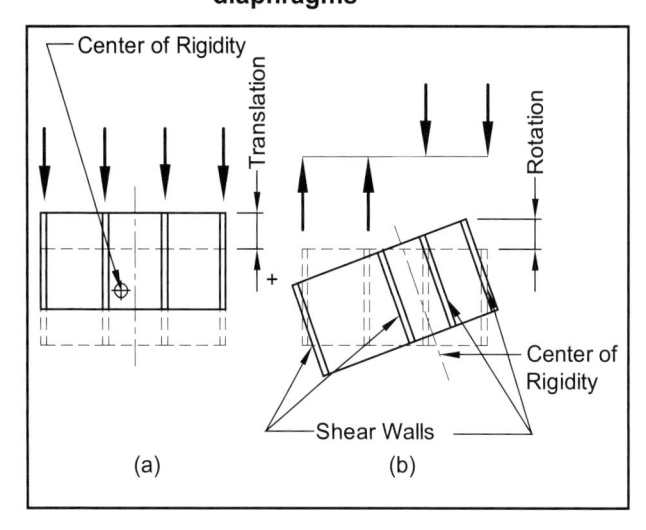

Figure 3.5.5.2 Effective width of wall perpendicular to shear walls [7]

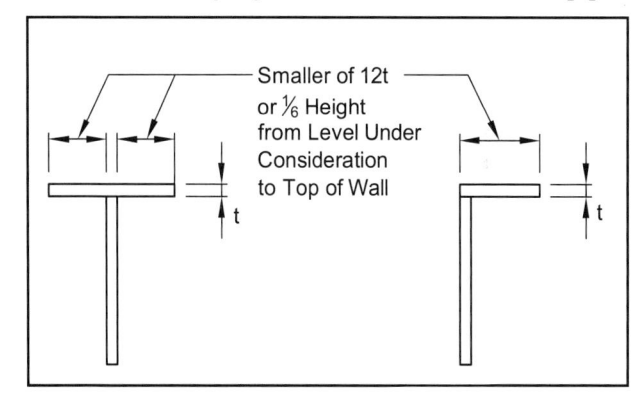

3.5.5 Proportioning Shear Walls

A rigid diaphragm subjected to lateral load will translate in the direction of the applied load [see Figure 3.5.5.1(a)]. The magnitude of the diaphragm translation is related to the sum of the rigidities of the resisting shear walls. If the center of rigidity is not coincident with the line of action of the applied loads, the diaphragm will tend to rotate about the center of rigidity [Figure 3.5.5.1(b)]. The location of the center of lateral load can be different for different load cases, such as wind loading and seismic loading. The dimension between the center of rigidity and center of lateral load is the eccentricity of the lateral load resisting system. IBC 2003 and ASCE 7-02 require the addition of extra eccentricity of 5% of the perpendicular plan dimension of the structure to provide for "accidental torsion."

It is desirable to design the wall panels as single uncoupled units. This reduces the cost of connections and the magnitude of volume change restraint forces that occur when many wall panels are connected together to form one large shear wall.

If the overturning moment results in excessive uplift to an individual, uncoupled shear wall panel, multiple wall panels can be connected together. Connecting individual vertical wall panels together greatly increases the shear resistance and reduces uplift of the shear wall. Locate the required panel-to-panel connections near mid-length of the wall to minimize volume change restraint forces. As previously stated, it is desirable to minimize the number of wall panels connected together to minimize volume change forces and number of connections required.

The stiffness of a shear wall may be increased by connecting it to perpendicular walls, which act as flanges. The effective flange width that can be assumed for such walls is illustrated in Figure 3.5.5.2, as long as the connections across the joint are sufficient to transfer forces as "strong" connections. Generally, the use of "flanges," increases the shear wall's flexural rigidity, but has little effect on its shear rigidity. In some structures, it may be desirable to connect perpendicular walls to a shear wall to increase its dead load, which increases the shear wall's resistance to overturning moment caused by lateral loads.

3.5.6 Stiffness Analysis

The distribution of lateral loads to shear walls in buildings with rigid diaphragms should be based on an analysis that includes relative rigidities of the shear walls. The total rigidity considers both flexural and shear stiffness. Rigidity of an individual wall is:

$$r = 1/\Delta \qquad \text{(Eq. 3.5.6.1)}$$

where:

Δ = sum of flexural and shear wall deflections.

For a structure with rectangular shear walls of the same material, with a wall height-to-length ratio of less than about 0.3, the flexural stiffness can be neglected, and the distribution made in accordance with the cross-sectional area of the walls. If the height-to-length ratio is greater than about 3.0, the shear stiffness can be neglected, and the distribution made in accordance with the moments of inertia, based on the plan dimensions of the wall.

When the height-to-length ratio is between 0.3 and 3.0, the effects of both shear and flexural deformations should be considered. In terms of stiffnesses:

$$\frac{1}{\Sigma K_i} = \frac{1}{\Sigma K_{si}} + \frac{1}{\Sigma K_{fi}} \qquad \text{(Eq. 3.5.6.2)}$$

where:

ΣK_i = summation of total stiffnesses at Level i
ΣK_{si} = summation of shear stiffnesses at Level i
ΣK_{fi} = summation of flexural stiffnesses at Level i

An equivalent moment of inertia, I_{eq}, can be derived for simplifying the calculation of wall rigidity. I_{eq} is an approximation of the moment of inertia that would result in a flexural deflection equal to the combined flexural and shear deflections of the wall. Figure 3.10.26 compares the deflections and I_{eq} for several load and restraint conditions, assuming the shear modulus, $G = 0.4E$. Note that the third case in Figure 3.10.26 assumes the wall is clamped rigidly at each floor. This rarely, if ever, occurs in precast construction.

The shear resisting area of the walls is not necessarily the total area of the walls. Just as only the web of a flanged beam really resists shear, wall groups perpendicular to the direction of load, or flange walls in wall groups will not contribute to shear stiffness, and their areas should be discounted. The configuration and assembly of walls or wall groups may, however, have an effect on the bending stiffness and the effective moment of inertia. It has been common practice to make calculations using the assumption that a wall or connected group acted as a combined unit as long as the connections in vertical joints between wall elements developed the required capacity to resist overturning or the VQ/I shear force. Codes now prescribe the evaluation of lateral drift using the deflection amplification factor, C_d (Figure 3.10.8) to determine P-Δ effects and for comparison to maximum limits. Therefore, there is a need to make a more accurate evaluation of the stiffness properties.

For precast shear walls that are stacked and connected across horizontal joints, the use of gross section to determine the moment of inertia is sufficiently accurate for determination of relative stiffness. There is, however, some question about the post-yield behavior at these joints which suggests that the displacement amplification factor, C_d, should be increased to more accurately reflect the reduction in stiffness.

Walls that are assembled with vertical joints, however, may behave differently. The use of ductile connections across vertical joints may have very beneficial effects as the location of clearly defined sites for inelastic behavior and energy dissipation without collateral damage to the major lateral force resisting elements. Research into the characteristics of these connections is found in Refs. 9 and 29. Obviously, the effective deformation characteristics of walls connected in this manner will be different from those of a comparable monolithic wall.

Reasonable assumptions based on connection detailing can be made to produce reliable results. Connections in vertical joints may be considered either "ductile" or "strong." Ductile connections are ones that are detailed for ductility and are intended to yield during the design event. These connections need to retain their load-carrying capacity at deformations consistent with the inelastic demand. They will provide the function of continuing to mobilize dead load resistance for overturning even after yielding has occurred. When these connections are used in vertical joints, the wall stiffness can conservatively be modeled as if each vertical panel acts independently. After yield, each vertical panel has only its own section for resistance, as a limit.

In some cases, it may be necessary or desirable to develop strong connections. These connections are proportioned to continue to carry loads elastically beyond the yielding of the primary yield mechanism in the walls. The design forces for this type of connection would be based on VQ/I multiplied by Ω_o to ensure these connections remain elastic through the design event, to account for the absence of the ductility factor. This kind of connection might be achieved with overlapping hairpins and longitudinal steel in a cast-in-place joint that develops strength in excess of a comparable monolithic wall section. For this condition, the stiffness of the resulting wall is equivalent to that of a comparable monolithic wall (see Section 3.1.2.1).

The type of connection, then, determines the design dimension of the wall that is modeled as a cantilever beam. The loads are applied level by level, and the effects of displacement from both shear and flexure are accumulated by superposition. Thus, story drift and total lateral displacement are derived. By accumulating the forces, the shear at each level and the overturning moments are calculated.

It is possible to use the selection of jointing and connection details to define the post-yield stiffnesses of groups of walls. With this technique, walls can be detailed to accept the loads for which they can develop overturning resistance or shed loads for which such capacity cannot be mobilized. In this way, the designer of precast wall systems can tune the stiffness through detailing.

3.5.7 Distribution of Lateral Loads

Lateral loads are distributed to each shear wall in proportion to its rigidity, if the diaphragm is rigid. A building is typically designed for lateral loads separately in two orthogonal directions.

When the shear walls are symmetrical with respect to the center of load application, the force resisted by any shear wall is:

$$F_i = (k_i/\Sigma r)/V_x \qquad \text{(Eq. 3.5.7.1)}$$

where:
  F_i = force resisted by an individual shear wall, i
  k_i = rigidity of Wall i
  Σr = sum of rigidities of all shear walls
  V_x = total lateral load

The analysis of structures which have shear walls placed asymmetrically with respect to the center of the lateral load, must consider the torsional effect in the analysis. Typical examples are shown in Figure 3.5.2.1. An approximate method of analysis based on a "polar moment of stiffness" is simple and direct:

Force in the y-direction is distributed to a given wall at a given level due to an applied force in the y-direction at that level:

$$F_y = \frac{V_y K_y}{\Sigma K_y} + \frac{TV_y(x)K_y}{\Sigma K_y(x^2) + \Sigma K_x(y^2)} \qquad \text{(Eq. 3.5.7.2)}$$

Force in the x-direction is distributed to a given wall at a given level due to an applied force in the y-direction at that level:

$$F_x = \frac{TV_y(y)K_x}{\Sigma K_y(x^2) + \Sigma K_x(y^2)} \qquad \text{(Eq. 3.5.7.3)}$$

where:
  V_y = lateral force at the level being considered
  K_x, K_y = rigidity in the x- and y-directions, respectively, of the wall under consideration
  $\Sigma K_x, \Sigma K_y$ = summation of rigidities of all walls at the level in the x- and y-directions, respectively
  x = distance of the wall from the center of stiffness in the x-direction
  y = distance of the wall from the center of stiffness in the y-direction

For most single-story buildings subjected to wind loads, a simplified, approximate analysis is commonly used to determine torsion in asymmetrically located shear walls. This type of analysis assumes a unit thickness for all shear walls, as described in Example 3.5.7.1.

Example 3.5.7.1
Design of Unsymmetrical Shear Walls

Given:
   The structure shown below. All walls are 8 ft high and 8 in. thick.

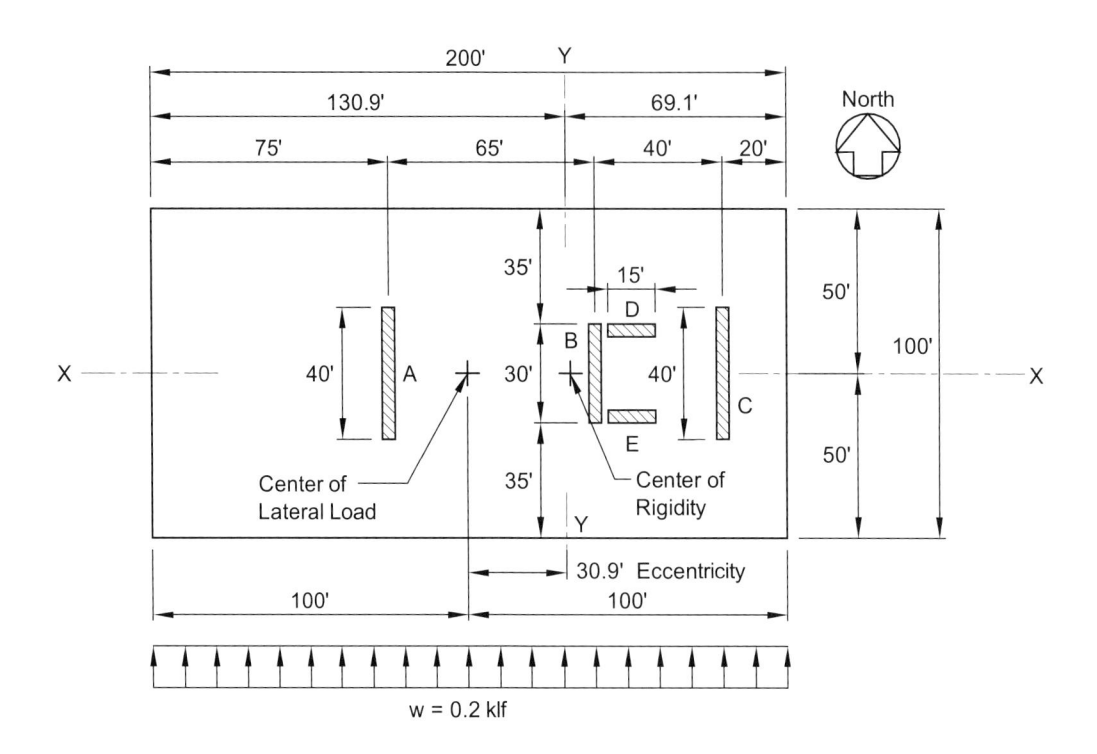

Problem:
   Determine the shear in each wall, assuming the floors and roof are rigid diaphragms. Walls D and E are not connected to Wall B.

Solution:
   Maximum height-to-length ratio of north-south walls = 8/30 < 0.3. Thus, for distribution of the direct wind shear, neglect flexural stiffness. Since walls are the same thickness and material, distribute in proportion to length.

Total lateral load, $V_x = 0.20 \times 200 = 40$ kips

Determine center of rigidity:

$$\bar{x} = \frac{40(75) + 30(140) + 40(180)}{40 + 30 + 40} = 130.9 \text{ ft from left}$$

$\bar{y}$ = center of building, since Walls D and E are placed symmetrically about the center of the building in the north-south direction.

Torsional moment, $M_T = 40(30.9) = 1236$ kip-ft

Example 3.5.7.1 (Cont.)
Design of Unsymmetrical Shear Walls

Determine the "polar moment of stiffness" of the shear wall group about the center of rigidity:

$I_p = I_{xx} + I_{yy}$
$I_{xx} = \Sigma \ell y^2$ of east-west walls $= 2(15)(15)^2 = 6750$ ft^3
$I_{yy} = \Sigma \ell x^2$ of north-south walls $= 40(130.9 - 75)^2 + 30(140 - 130.9)^2 + 40(180 - 130.9)^2 = 223,909$ ft^3
$I_p = 6750 + 223,909 = 230,659$ ft^3

Shear in north-south walls $= \dfrac{V_x \ell}{\Sigma \ell} + \dfrac{M_T x \ell}{I_p}$

Wall A $= \dfrac{40(40)}{110} + \dfrac{1236(130.9 - 75)(40)}{230,659} = 14.5 + 12.0 = 26.5$ kips

Wall B $= \dfrac{40(30)}{110} + \dfrac{1236(-9.1)(30)}{230,659} = 10.91 - 1.46 = 9.45$ kips

Wall C $= \dfrac{40(40)}{110} + \dfrac{1236(-49.1)(40)}{230,659} = 14.5 - 10.5 = 4.0$ kips

Shear in east-west walls $= \dfrac{M_T y \ell}{I_p} = \dfrac{1236(15)(15)}{230,659} = 1.21$ kips

3.5.8 Coupled Shear Walls

Two individual shear walls separated by large openings may be connected together with structural members that can resist axial and/or flexural loads. The combined stiffness of the two coupled shear walls is greater than the sum of their uncoupled stiffnesses. Coupling shear walls can reduce the lateral deflection (drift) in a building and reduce the magnitude of the moments for which a shear wall must be designed.

Figure 3.5.8.1 shows two examples of coupled shear walls. The effect of coupling is to increase the stiffness by transfer of shear and moment through the coupling beam. The wall curvatures are altered from that of a cantilever because of the frame action developed. Figure 3.5.8.2 shows how the deflected shapes differ in response to lateral loads.

Several approaches may be used to analyze the response of coupled shear walls. A simple approach is to ignore the coupling effect by considering the walls as independent cantilevers. This method results in a conservative wall design. However, if the coupling beams are rigidly connected, significant shears and moments will occur in the beams which are difficult to economically resist. To avoid the problem, the beam-to-panel connection can be detailed for little or no rigidity, or the beam can be designed to resist the actual shears and moments.

Finite element analysis may be used to determine the distribution of shears and moments within a coupled shear wall. The accuracy of such an analysis is a function of the finite element size used. This method may be suitable for coupled walls or walls with large openings.

A "plane frame" computer analysis will provide sufficiently accurate results for the great majority of structures. In modeling the coupled shear wall as a frame, the member dimensions must be considered, as a centerline basic analysis may yield inaccurate results. A suggested model is shown in Figure 3.5.8.3(a).

Either a finite element or frame analysis may be used to determine the deflection of a coupled shear wall, and hence its equivalent moment of inertia. This may then be used to determine the distribution of shears in a building which contains both solid and coupled shear walls. Some frame analysis programs do not calculate shear deformations; if significant, shear deformations may have to be calculated separately.

Figure 3.5.8.1 Coupled shear walls

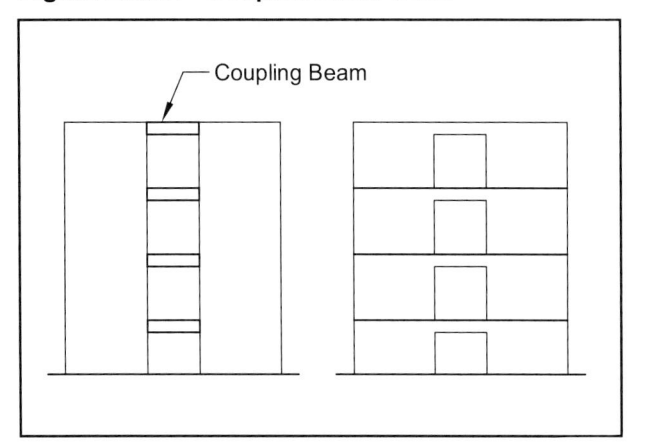

Figure 3.5.8.2 Response to lateral loads

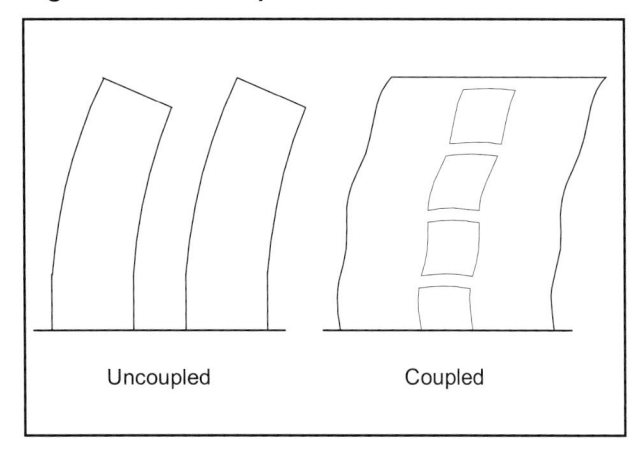

3.5.9 Shear Walls with Large Openings

As with coupled shear walls, the deflections yielded by computer analysis may be used to determine equivalent stiffness for determining lateral load distribution.

In very tall structures, vertical shear and axial deformations influence the rigidity of panels with large openings, so a more rigorous analysis may be required.

3.5.10 Evaluation of Shear Wall Systems

Once the forces on the shear walls have been determined, the walls and the structural system must be evaluated. Following are the design considerations:

1. Overturning. Resistance to overturning, M_{otr}, is compared to the overturning moment at the wall base by using Eq. 3.2.6.6 for wind loading and Eq. 3.2.6.7 for earthquake loads. If the resistance to overturning is exceeded by the overturning moment, it is usually possible to provide a positive hold-down from the wall to the foundation (e.g., vertical tie bars, post-tensioning, etc.). The foundation size may need to be increased, or in some cases, tension piles, ground anchors or other positive methods of resisting the overturning may need to be used. While the connections in vertical joints may reduce stiffness, they still may be adequate to mobilize overturning resistance. If none of these solutions work, additional shear walls or extension of existing walls may be necessary.

Figure 3.5.8.3 Computer models

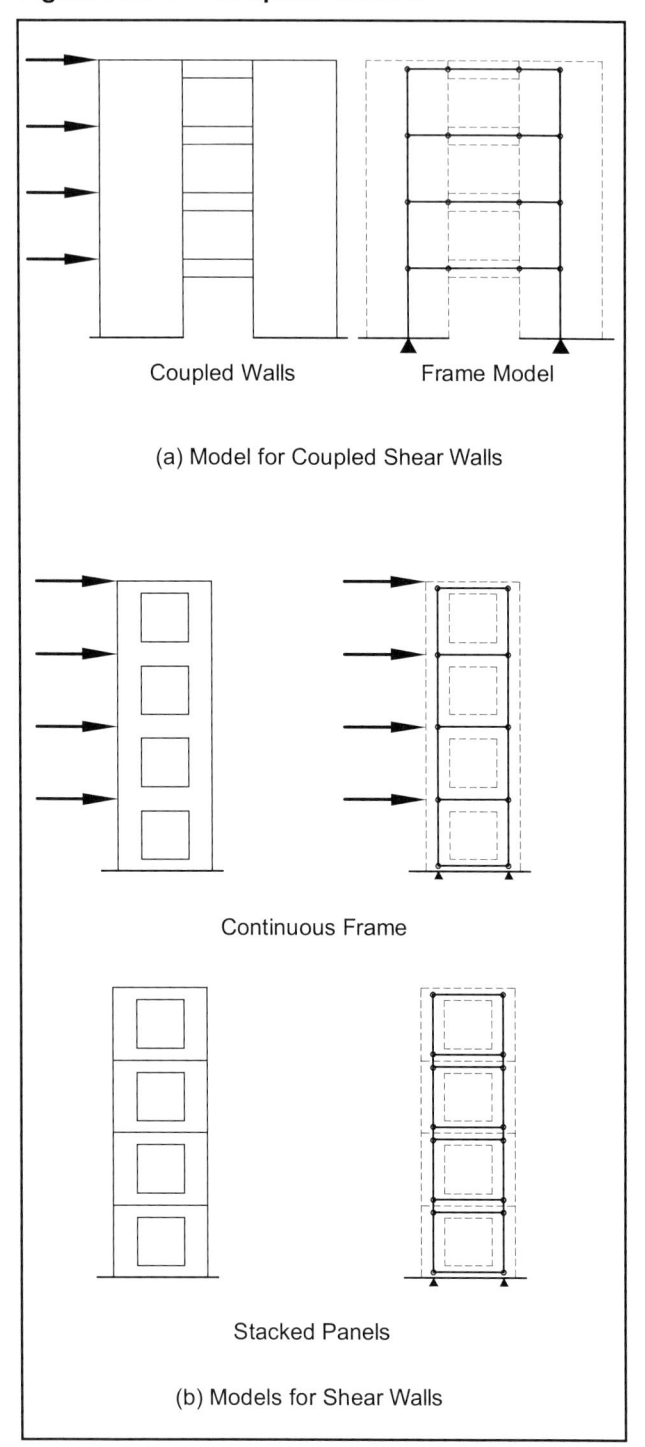

(a) Model for Coupled Shear Walls

(b) Models for Shear Walls

2. Base shear. The anchorage of the wall to the foundation also includes the transfer of shear. The principles of shear-friction may be applied (see Section 4.3.6). Although one side of the wall may have an opened joint due to flexure, with the anchorage acting in tension, there will be a region of compression that results from the combination of flexure and axial load on the wall. This compression force contributes to the shear transfer. Additional connections or mechanically spliced reinforcing bars may be required at the wall/foundation joint to provide additional shear capacity.
3. Drift. In addition to the limitations on drift in earthquakes, in some cases the walls must be designed for the P-Δ effects caused by the shifting of the point of application of gravity loads. If the stability coefficient, θ, exceeds 0.10, P-Δ effects must be considered. This is discussed in Section 3.2.4.4.
4. The drift used to evaluate the P-Δ effects is the drift from linear-elastic analysis drift, based on the strength design loads, multiplied by C_d (Figure 3.10.8). With shear walls, it is desirable to reduce the lateral displacement and avoid the need to consider P-Δ effects, but the drift must be calculated.

3.5.11   Example One-Story Building

By taking advantage of walls already present, one-story buildings usually can be designed to resist lateral loads (wind or earthquake) by shear wall and diaphragm action. If this is feasible, it is generally the most economical solution. Example 3.5.11.1 illustrates the procedures.

Because all loads must funnel through the connections, gravity and lateral loads must be considered together. Thus, this example shows both gravity and lateral load connections. The example emphasizes the concepts of free body diagrams and load paths.

This example illustrates one option for shear wall connections. Equilibrium is attained by balancing uplift tension in one panel with the matching compression couple force in an adjacent panel through simple (and ductile) shear connections across the vertical joints. In this system, tie-down connections may be required.

Analysis of a rigidly connected panel group is dependent upon a number of factors. Connection stiffness determines whether the group will act monolithically or act as a series of single shear walls. The aspect ratio of the rigidly connected group will affect the stress distribution at the base. While a panel group with a high height-to-length ratio may act nearly as a cantilever beam, a group with a medium to low ratio of height to length will act as a deep beam and exhibit nonlinear stress distributions. Determination of tie-down forces, vertical joint forces, and base shear distribution will be a function of the aspect ratio. The shear and flexural stiffness of the individual panels will have a similar effect on force distribution.

Example 3.5.11.1
Typical Single-Story Industrial Building

Given:
The single-story manufacturing building shown is located on Long Island, New York. The basic wind speed (Figure 3.10.5) is 120 mph, Exposure B. From seismic maps, $S_s = 0.30$ and $S_1 = 0.08$, Site Class D. 10-ft wide double tees are used on the roof, weighing 47 psf. Wall units are 10 ft wide sandwich panels with a 4 in. thick interior wythe, 2 in. insulation, and a 2 in. thick exterior wythe, weighing an average of 75 psf. A dead load of 10 psf is superimposed on the roof.

Problem:
Design members and connections for critical lateral load.

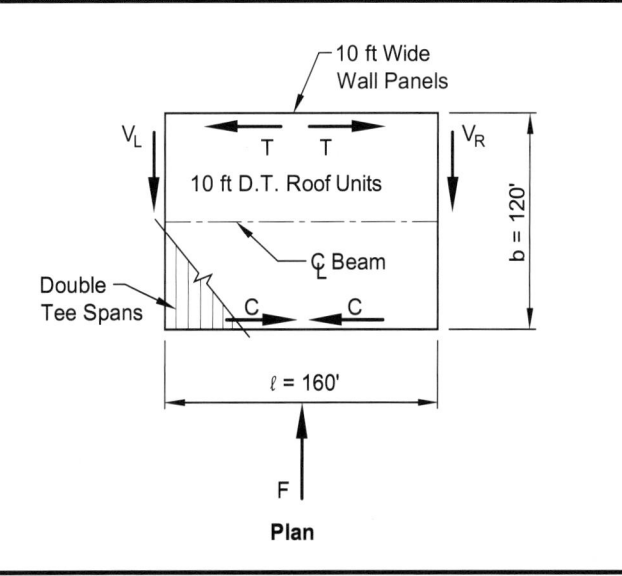

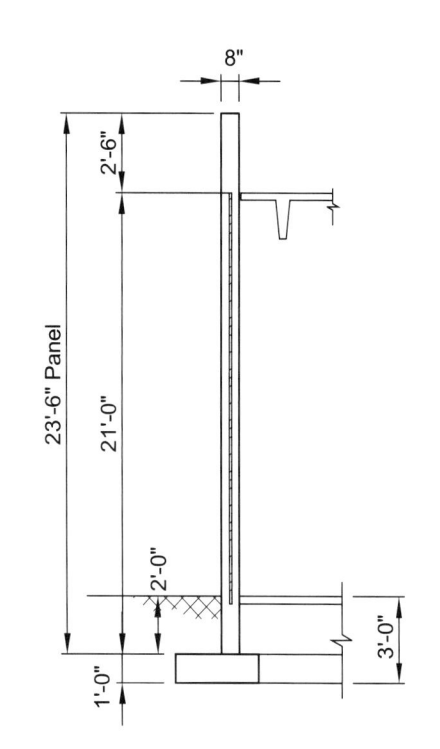

Example 3.5.11.1 (Cont.)
Typical Single-Story Industrial Building

Solution:

1. Using the "simplified procedure" of Section 3.2.3.1 for wind design, determine from Figures 3.10.6 and 3.10.1:
   $\lambda = 1.0$
   Zone A $p_{s30} = 22.8$ psf
   Zone C $p_{s30} = 15.1$ psf
   $I = 1.0$

   Mean roof height = 19.0 ft

   Zone A width = lesser of 0.2(120) = 24 ft or
   0.8(19.0) = 15.2 ft Use 15.2 ft

   Zone C width = 160 − (15.2) = 144.8 ft
   $p_s(A) = \lambda I p_{s30} = 1.0(1.0)(22.8) = 22.8$ psf
   $p_s(C) = 1.0(1.0)(15.1) = 15.1$ psf

   Wind force to roof:
   = [22.8(15.2) + (144.8)(15.1)](19/2 + 2.5)/1000 = 30.4 kips

2. Determine earthquake force: interpolating from Figure 3.10.7, $F_a = 1.6$, $F_v = 2.4$.
   $S_{MS} = F_a S_s = 1.6(0.30) = 0.48$
   $S_{M1} = F_v S_1 = 2.4(0.08) = 0.192$
   $S_{DS} = 2/3(S_{MS}) = 2(0.48)/3 = 0.320$
   $S_{D1} = 2/3(S_{M1}) = 2(0.192)/3 = 0.128$

   From Table 3.2.4.1: Seismic Design Category B

   Period, $T_a = C_T(h_n)^{3/4} = 0.020(21.0)^{3/4} = 0.196$

   From Figure 3.10.1, $I = 1.0$

   From Figure 3.10.8, for a bearing wall system with ordinary reinforced concrete shear walls: $R = 4$

   From Eq. 3.2.4.9:
   $$C_s = \frac{S_{DS}}{R/I} = \frac{0.320}{4/1.0} = 0.080$$

   From Eq. 3.2.4.10:
   $$C_s = \frac{S_{D1}}{(R/I)T} = \frac{0.128}{(4/1.0)0.196} = 0.163$$

   Check minimum:
   $C_s \geq 0.044 S_{DS} I = 0.044(0.320)(1.0) = 0.014 < 0.080$ Use $C_s = 0.080$

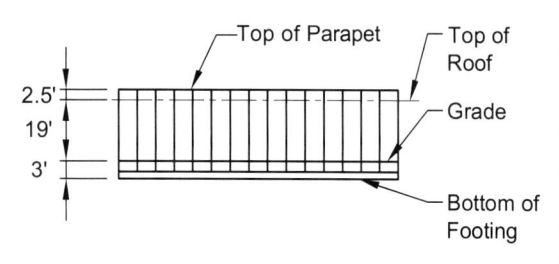

Example 3.5.11.1 (Cont.)
Typical Single-Story Industrial Building

Building weight:
    Walls = 75(23.5)[2(160) + 2(120)]/1000 = 987 kips (50% goes directly into foundation)
    Roof = 120(160)(47 + 10)/1000 = 1094 kips
    Beams and columns (estimated load contribution to roof) = 150 kips

$$W = \frac{987}{2} + 1094 + 150 = 1738 \text{ kips}$$

Base shear $V = C_sW = 0.080(1738) = 139.0$ kips

3. In a single-story building, the total base shear acts through the roof diaphragm. For a single-span diaphragm such as this, design is straightforward. For buildings with interior shear walls, it is more complex (see Section 3.8). For seismic forces, note the requirements for torsion analysis in Section 3.2.4.3. (No such requirement exists for wind load.) Assuming no substantial door openings in either shear wall, the center of mass is the center of the building. However, the 5% accidental torsion must be considered.

    Accidental torsion = 0.05(160) = 8.0 ft. Thus, assume the center of mass is 88 ft from the left wall and 72 ft from the right wall.

$$V_R = \frac{139.0(88)}{160} = 76.4 \text{ kips}$$

    As can be seen in Section 3.2.6, seismic forces thus calculated are factored (load factor = 1.0 for this seismic category), while a load factor of 1.6 is applied to the wind forces. The wind force is distributed equally to each shear wall:

1.6(32.2/2) = 25.8 kips
76.4 > 25.8 kips, seismic loading is critical.

    For the seismic diaphragm design (chord steel and shear connectors between roof tees) for this single-story building, use the F_{px} calculated by Eq. 3.8.3.1 or the total base shear, whichever is greater:

$$F_p = 0.2I_ES_{DS}w_P = 0.2(1.0)(0.312)(1738) = 108 \text{ kips} < 139.0 \text{ kips (governs)}$$

4. Check sliding resistance of the foundation:

Dead load on the footing:
Wall	= 75(23.5)(120)	= 211,500 lb
12 in. x 18 in. footing	= 1(1.5)(150)(120)	= 27,000 lb
Assume 2 ft backfill	= 100(1.5)(120)(2)	= 36,000 lb
	Total	= 274,500 lb

Assume coefficient of friction against granular soil, $\mu_s = 0.5$
Sliding resistance = $\mu_sN = 0.5(274.5) = 137.2$ kips say OK
(Note: This analysis is a close approximation. More detailed analysis may be required.)

Example 3.5.11.1 (Cont.)
Typical Single-Story Industrial Building

From Eq. 3.2.6.7a: $U = (0.9 - 0.2S_{DS})D - \rho Q_E = [0.9 - 0.2(0.312)]137.2 - 1.0(76.4) = 38.5 > 0$ OK

Determine the reinforcement and connection requirements for the diaphragm.

a. Connections from the roof to the walls:

   A connection similar to that shown in Example 6.14.3 may be used. That connection has a maximum capacity of 13.9 kips. On east and west walls, connections req'd = $\dfrac{V_R}{\text{conn. cap.}} = \dfrac{76.4}{13.9} = 5.5$

   Use minimum spacing of about 10 ft. Approx. 10 connections required per side.

b. Shear connections between double tees:
    The maximum shear is at first joint (10 ft) from left wall. The left wall is 88 ft from the center of force. Assuming a uniformly distributed lateral force:

$$\text{Shear} = \dfrac{88-10}{88}(76.4) = 67.7 \text{ kips}$$

A connection similar to one of those shown in Figure 3.8.1.2 may be used.

Note: Most engineers and precasters prefer a maximum connection spacing of about 8 to 15 ft for roof connections.

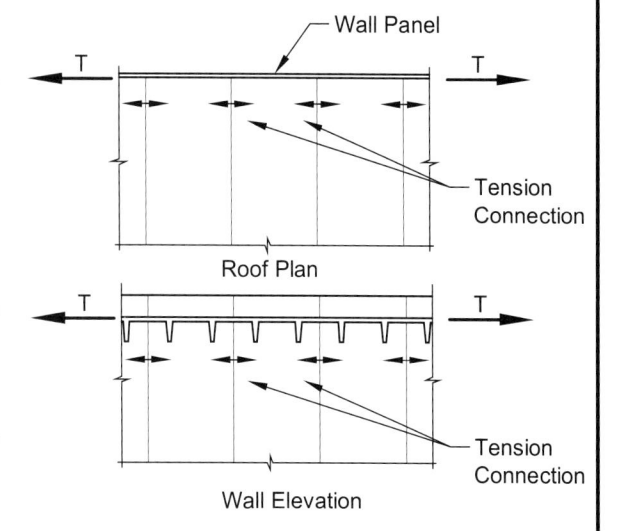

c. To determine chord reinforcement, the seismic force is assumed to be distributed uniformly across the building width, b:

Diaphragm moment:

$$\dfrac{F\ell}{8} = \dfrac{139.0(160)}{8} = 2780 \text{ ft-kips}$$

Chord force (see Plan):
Assume chord reinforcement is located 1 ft in from outside of wall. Since these are seismic forces, they can be considered factored:

$$T_u = C_u = \dfrac{M}{b-2} = \dfrac{2780}{120-2} = 23.6 \text{ kips}$$

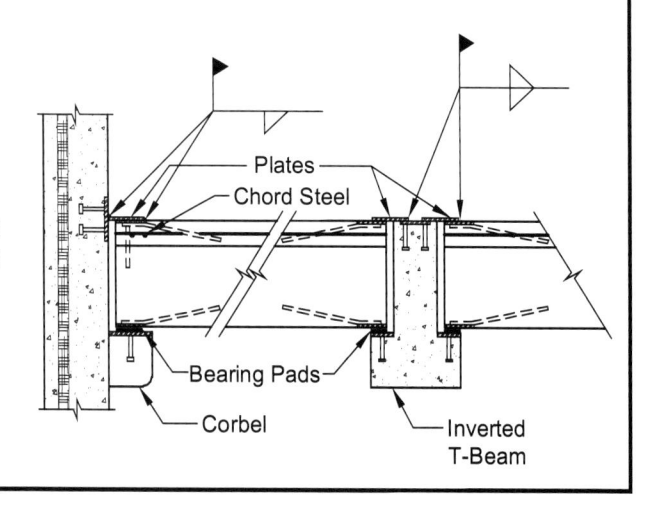

Alternative Placement of Chord Reinforcement

Example 3.5.11.1 (Cont.)
Typical Single-Story Industrial Building

Assume bars will be spliced, $\phi = 0.7$ (Section 3.2.6.1):

$$A_s = \frac{T_u}{\phi f_y} = \frac{23.6}{0.7(60)} = 0.56 \text{ in.}^2$$

This amount of reinforcement should be placed at the perimeter. The chord force can be transmitted between members by ties at the roof tees, wall panels or a combination, as illustrated. These ties and transmission of forces will usually provide the tie requirements for structural integrity outlined in Section 3.3.

d. Wall panel connections:

This shear wall may be designed to act as a series of independent units, without ties between the panels. The shear force is assumed to be distributed equally among the wall panels (see figure at right).

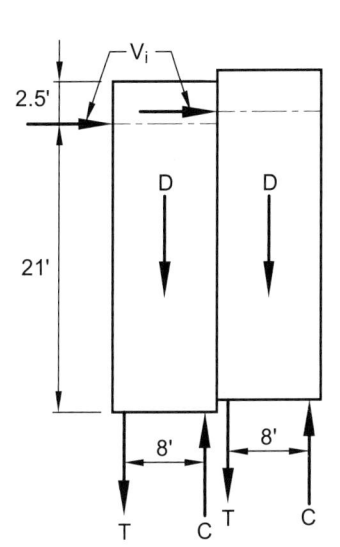

$n = 120/10 = 12$ panels
$V = V_R/n = 76.4/12 = 6.37$ kips/panel
$D = 75(10)(23.5)/1000 = 17.62$ kips

From Eq. 3.2.6.7a, D is multiplied by:

$(0.9 - 0.2S_{DS}) = [0.9 - 0.2(0.320)] = 0.836$
$D = 0.836 (17.62) = 14.73$ kips

Design base connection for $1.0E - (0.9 - 0.2S_{DS})D$

$$T_u = \frac{6.37(21) - 14.73(4)}{8} = 9.36 \text{ kips}$$

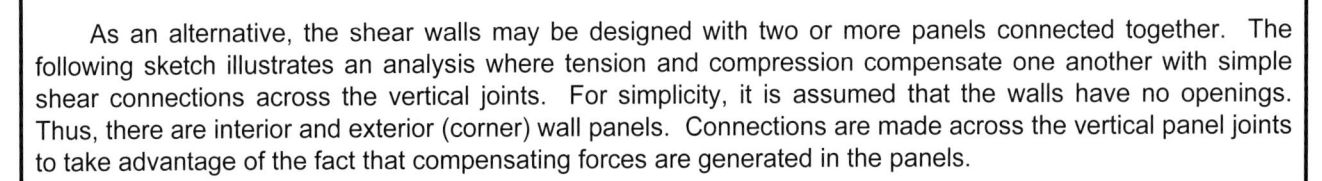

As an alternative, the shear walls may be designed with two or more panels connected together. The following sketch illustrates an analysis where tension and compression compensate one another with simple shear connections across the vertical joints. For simplicity, it is assumed that the walls have no openings. Thus, there are interior and exterior (corner) wall panels. Connections are made across the vertical panel joints to take advantage of the fact that compensating forces are generated in the panels.

Note: Determining connection forces requires solving classic equations of equilibrium. This will be done using factored loads that assume connection tension. Compression forces are assumed to be no problem, as the joint between the wall panel and the foundations is normally grouted.

Considering an interior panel:

ΣM about $C = 0$: $V(h) = V_1(b - a) + D(b/2 - a)$

$$V_1 = \frac{V(h) - D(b/2 - a)}{b - a}$$

$\Sigma V = 0$: $\quad C = D$

Example 3.5.11.1 (Cont.)
Typical Single-Story Industrial Building

Since this force system can exist for all interior panels, edge shears will balance to zero when all panels have the same dimensions and weight. The only requirement for the connections is a transfer of vertical shear. Therefore, connections which permit horizontal deformations can be used if volume change restraint is of concern. At the exterior panels, the edge shear V_1 from an exterior panel will be applied at one edge only. Because tension and compression base connections are not located at the panel edges, equilibrium may have to be satisfied with tension and compression connections to the foundation, or connections to the orthogonal panel that will allow the non-shear wall to contribute additional dead load at the corner.

At the tension side exterior panel, equilibrium can be determined by summing moments about the compression force, assuming the tension is taken by a tie-down into the foundation:

For this example, locate the foundation connections 1 ft from each side. The pertinent dimensions are:
  h = 21 ft; b = 10 ft; a = 1 ft; d = 8 ft
  V = 6.37 kips, D = 17.62 kips

Using Eq. 3.2.6.7:

$V_u = 1.0(6.37) = 6.37$ kips

Factored D.L. = 14.73 kips
  (see previous page)

For interior panels with factored loads:

$$V_1 = \frac{6.37(21) - 14.73(5-1)}{10-1} = 8.32 \text{ kips}$$

$C = D$

For the tension side exterior panel:

$$T_u = \frac{6.37(21) - 14.73(5-1) - 8.32(1)}{8}$$

$= 8.32$ kips

$C_u = T_u + D_u - V_1 = 8.32 + 14.73 - 8.32$

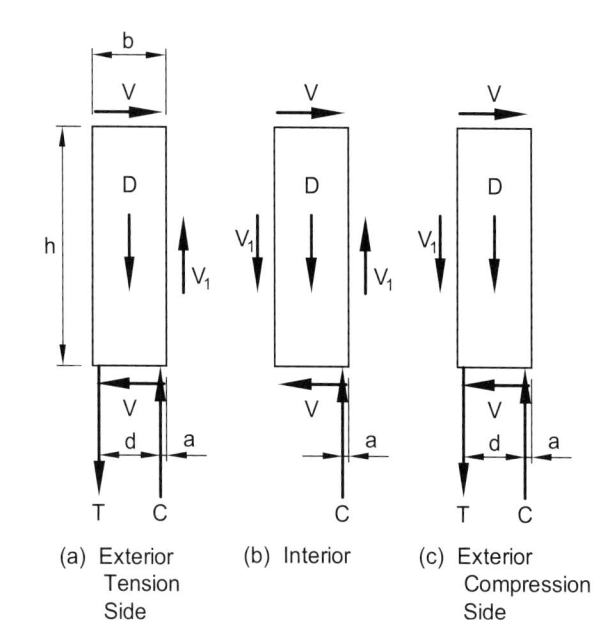

$= 14.73$ kips

(a) Exterior Tension Side
(b) Interior
(c) Exterior Compression Side

For the compression side exterior panel:

$$T_u = \frac{8.32(21) - 14.73(5-1) - 8.32(10-1)}{8} = 5.11 \text{ kips}$$

$C_u = T_u + D_u + V_1 = 5.11 + 14.73 + 8.32 = 28.2$ kips

3.5.12 Example Three-Level Parking Structure

Over the years, precast concrete parking structures have proven to be a reliable, economical means of providing attractive and secure parking for the public. Since prestressed concrete is capable of long spans, which accommodate ease of traffic circulation, columns are minimized. The resulting openness requires the designer to consider the

---

Example 3.5.12.1
Three-Level Parking Structure

Given:
The three-level parking structure shown is located near Savannah, GA. The basic wind speed (Figure 3.10.5) is determined to be 110 mph, Exposure B. From seismic maps, $S_s = 0.3$, $S_1 = 0.15$. A soil report indicates Site Class C.

Problem:
Determine the feasibility of a shear wall structure in this location.

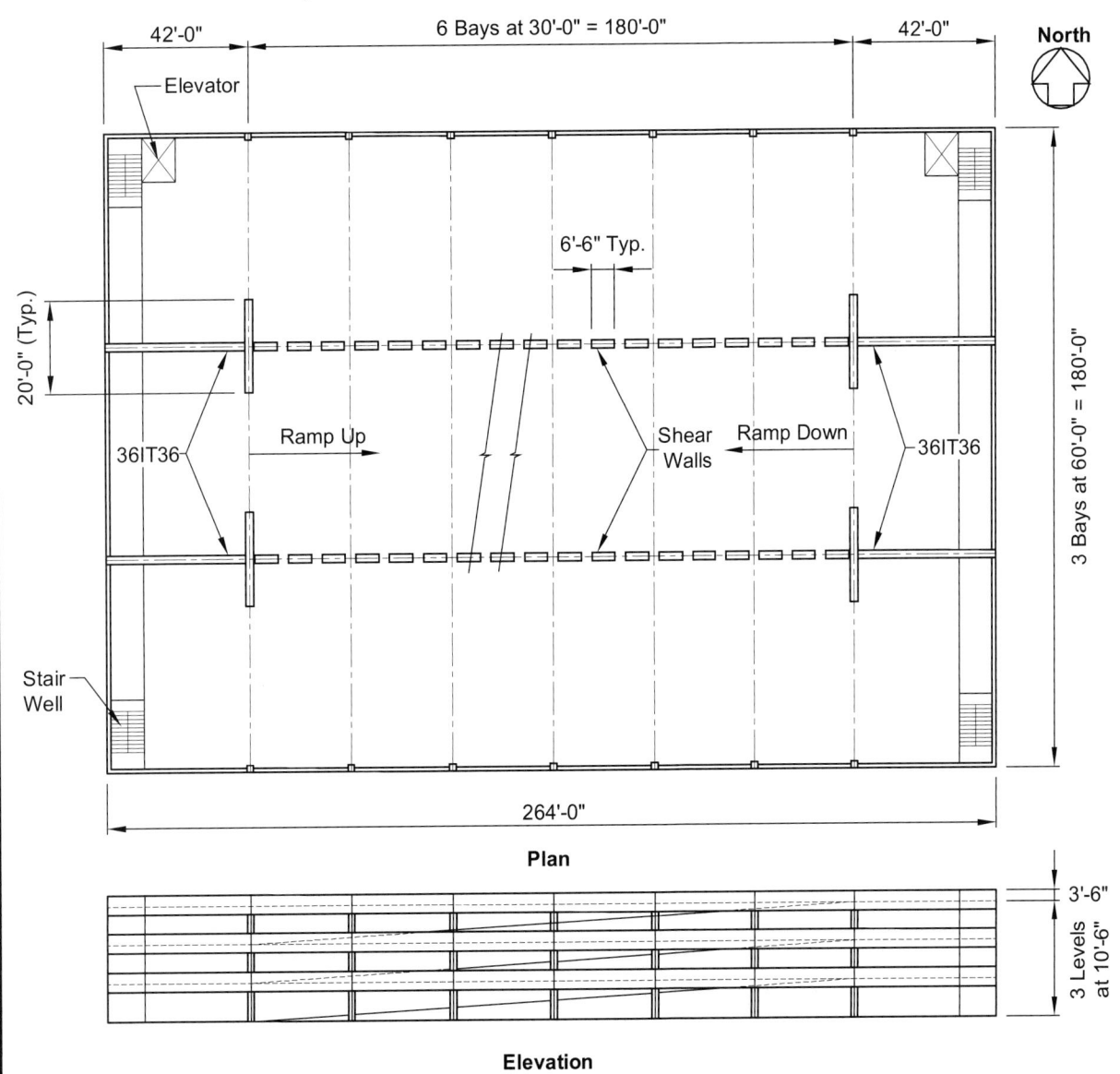

Plan

Elevation

PCI Design Handbook/Sixth Edition 3–47

Example 3.5.12.1 (Cont.)
Three-Level Parking Structure

Solution:
    Since the structure must resist moderate seismic forces and hurricane force winds as well as the gravity loads, the magnitude of the forces must first be determined to see which combinations of loads will control the design. By arranging support elements to resist both vertical and lateral loads, the goal of maintaining an open structure will be achieved.
    For gravity loads, 26 in. deep, 10 ft wide pretopped double tees will be used. The total weight of double tees, beams, columns, and curbs will be taken as 110 psf. The code specified live load is 50 psf. It is determined that for this magnitude of loading, 30 ft bays with 24 in. square columns, and a 36IT36 girder in the end bays, will support vertical loads.
    For lateral loads, a check is made to determine whether seismic or wind load will govern the design.

Wind analysis:
    The building height is 3(10.5) + 3.5 = 35 ft. Since it is less than 60 ft high, the ASCE 7 Method 1 of Section 3.2.3.1 may be used.

For a north-south wind:
$$p_s = \lambda I p_{s30}$$

Zone A width is the lesser of 0.2(180) = 36 ft or 0.8(31.5) = 25.2 ft
Zone A width = 25.2 ft
Zone C width = 264 − 25.2 = 238.8 ft

From Figure 3.10.6:
    Table (c): interpolate for $\lambda = 1.015$
    Table (a):
        p_{s30} (Zone A) = 19.2 psf
        p_{s30} (Zone C) = 12.7 psf

From Figure 3.10.1:
    I = 1.0
    p_s (A) = 1.015(1.0)(19.2) = 19.5 psf
    p_s (C) = 1.015(1.0)(12.7) = 12.9 psf
    (combined windward and leeward pressures)
    F_w = [19.5(25.2) + 12.9(238.8)]35/1000 = 125 kips

Factored wind load = 125(1.6) = 200 kips

Seismic analysis:

Determine F_a and F_v by interpolating from Figure 3.10.7(b) and (c).

For $S_s = 0.3$, $S_1 = 0.15$, Site Class C:
    F_a = 1.2, F_v = 1.65
    $S_{MS} = F_a S_s = 1.2(0.3)$ = 0.36
    $S_{M1} = F_v S_1 = 1.65(0.15)$ = 0.25
    $S_{DS} = (2/3)S_{MS} = (2/3)(0.36)$ = 0.24
    $S_{D1} = (2/3)S_{M1} = (2/3)(0.25)$ = 0.17

Example 3.5.12.1 (Cont.)
Three-Level Parking Structure

From Table 3.2.4.1 for Seismic Use Group 1, Seismic Design Category is "C."

Determine the period by Eq. 3.2.4.6:
Since this is a shear wall building, $C_T = 0.02$
Height to highest level = 3(10.5) = 31.5 ft

$$T_a = C_T h_n^{3/4} = 0.02(31.5)^{0.75} = 0.27 \text{ sec}$$

From Table 3.10.8: R = 4 or 5, depending on whether "ordinary" or "special" shear walls are used. For Seismic Design Category C, there are no restrictions on the type of shear walls that may be used. At this point in the analysis, assume "ordinary" shear walls. An economic analysis may be made later. Assume R = 4.

Use lesser of Eq. 3.2.4.9 or 3.2.4.10 for C_s:

Eq. 3.2.4.9:

$$C_s = \frac{S_{DS}}{R/I} = \frac{0.24}{4/1.0} = 0.06$$

Eq. 3.2.4.10:

$$C_s = \frac{S_{D1}}{(R/I)T} = \frac{0.17}{(4/1.0)0.27} = 0.16$$

Check minimum:
$C_s \geq 0.044 S_{DS} I = 0.044(0.24)(1.0) = 0.011$
Use $C_s = 0.06$

Weight of all components = 110 psf
Weight per level = 0.110(180)(264) = 5227 kips
Total weight, W = 3(5227) = 15,681 kips
Base shear, V = C_sW = 0.06(15,681) = 941 kips

Since this is much higher than the 200 kips wind load, seismic forces will control all of the design.

Substantial shear resisting elements are required. Loadbearing shear walls are chosen, primarily because the vertical gravity load will help resist the overturning moments due to applied lateral loads. However, by making the north-south shear walls loadbearing, the response modification factor is higher than for frames. It may be more economical to add two columns so that the lower factor for frames could be used. The walls could then be connected to the column to mobilize additional dead load. While the corner stairwells and elevator shafts could be used as part of the lateral load resisting system, this may result in high forces due to restraint of volumetric deformations; consequently, it is decided that the corners will be isolated from the main structure. Alternatively, it might have been decided to use these corner elements, and provide connections that are flexible in the direction of volumetric restraint.

The torsion is assumed to all be taken by the walls perpendicular to the direction of the lateral force bearing considered.

The distribution of seismic shears to each level using Eqs. 3.2.4.13 and 3.2.4.14 is shown in the table located on the following page.

Example 3.5.12.1 (Cont.)
Three-Level Parking Structure

Lateral Force Distribution Through Levels

(1) Level	(2) x	(3) h_x ft	(4) w_x kips	(5) $(3) \times (4)$ $w_x h_x$ kip-ft	(6) $(5)/\text{Total (5)}$ c_{vx}	(7) $(6) \times V_{base}$ F_x kips
3	3	31.5	5,227	164,650	0.500	471
2	2	21.0	5,227	109,767	0.333	313
1	1	10.5	5,227	54,883	0.167	157
Totals			15,681	329,300		941

For the north-south load resisting system, try two 8 in. thick loadbearing shear walls located at each end of the ramp. These walls support the 36IT36 girder, and may be as long as 30 ft without interfering with the traffic flow; a 20 ft length is used as a first iteration. The figure below illustrates the arrangement and loading.

The shear walls are located 90 ft from the center in the east-west direction of the structure, which is the center of the lateral force resisting system. However, accidental torsion must be considered (Section 3.2.4.3). The accidental eccentricity = 0.05(264) = 13.2 ft. Summing moments about the shear walls on one side, the force each pair of shear walls must resist is:

$$F_{2w} = \frac{941(180/2 + 13.2)}{180} = 540 \text{ kips or 270 kips to each wall}$$

The force at each level on the wall can be determined by the values of Column 6 in the table on previous page.

Level 3 $F_{1w} = 0.50(270) = 135$ kips
Level 2 $\quad\quad = 0.333(270) = 90$ kips
Level 1 $\quad\quad = 0.167(270) = 45$ kips

Overturning moment on the wall:
$\quad = 135(31.5) + 90(21) + 45(10.5) = 6615$ kip-ft

Dead load on each wall (includes all components)
$\quad = 3(21)(60)(0.110) = 416$ kips

Using Eq. 3.2.6.7a:
$\quad U = [0.9 - 0.2(0.24)]D + 1.0E$

$\quad = 0.85D + 1.0E$

$$T_u = \frac{6615 - 0.85(416)(10)}{18} = 171 \text{ kips}$$

$$A_s = \frac{T_u}{\phi f_y} = \frac{171}{0.9(60)} = 3.17 \text{ in.}^2$$

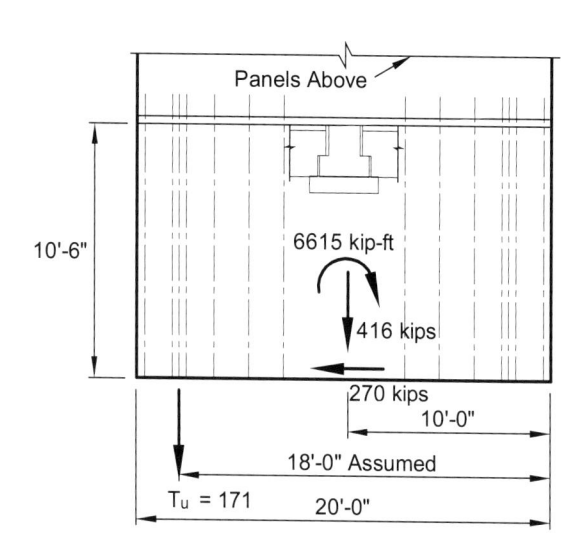

Example 3.5.12.1 (Cont.)
Three-Level Parking Structure

Use 4 – #8 bars = 3.17 in.²

The 4 – #8 bars will be centered 2 ft from each end of the wall. The force transfer between the precast shear wall and the foundation can be accomplished by reinforcing bars with grouted sleeves, rated mechanical couplers, or welding. Alternatively, post-tensioning bars could be chosen. The preliminary analysis is completed by examining the capacity of the foundation system to transfer this force to the supporting ground; that analysis is not shown here.

For resistance in the east-west direction, 18 individual loadbearing walls located along the length on each side of the interior ramped bay will be used. These 8-in. thick walls are spaced 10 ft on centers, supporting one 60-ft span double tee on each side of the wall as shown on the following page. Each wall is 6 ft 6 in. wide to accommodate the 5-ft stem spacing of the double tees, and to allow visibility between the wall units. As in the north-south direction, an accidental eccentricity of 5% must be considered:

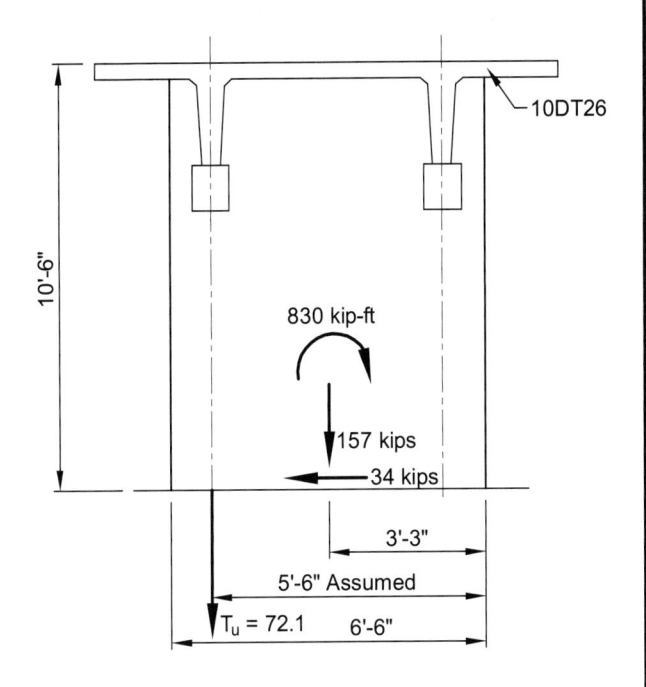

$e = 0.05(180) = 9$ ft

Total force in walls in one row:

$$F_{1ew} = \frac{941(60/2+9)}{60} = 612 \text{ kips, or } 612/18 = 34 \text{ kips to each wall. The force on each wall:}$$

Level 3 $F_w = 0.50(34) = 17$ kips
Level 2 $F_w = 0.333(34) = 11$ kips
Level 1 $F_w = 0.167(34) = 6$ kips

Overturning moment on the wall:
$= 17(31.5) + 11(21) + 6(10.5) = 830$ kip-ft

Dead load on each wall = one wall + one tee at each level (three levels):
$3[0.1(10.5)(6.5)+0.076(60)(10)] = 157$ kips

From Eq. 3.2.6.7a: $U = (0.9 - 0.2S_{DS})D + 1.0(Q_E)$
$= [0.9 - 0.2(0.24)]D + 1.0E = 0.85D + 1.0E$

$$T_u = \frac{830 - 0.85(157)(3.25)}{5.5} = 72.1 \text{ kips}$$

$$A_s = \frac{T_u}{\phi f_y} = \frac{72.1}{0.9(60)} = 1.34 \text{ in.}^2$$

Use 1 – #11 bar centered at 12 in. from each end of wall.

Example 3.5.12.1 (Cont.)
Three-Level Parking Structure

Diaphragm analysis

The diaphragm is modeled for north-south seismic forces as shown. As discussed in Section 3.8.3.1, the diaphragm force for Seismic Design Categories B and C may be calculated by Eq. 3.8.3.1:

$F_p = 0.2 I_E S_{DS} w_p + V_{px} = 0.2(1.0)(0.24)(5227) + 0 = 251$ kips, but not less than that shown in the table of lateral force distribution earlier in the example = 471 kips

It is common practice to assume the forces are distributed uniformly. To simplify the calculation, the force is divided among the three bays, and the flat and ramp areas are analyzed separately.

Total uniform load: $\dfrac{471}{264} = 1.784$ kip/ft

Uniform load on each bay = $w_1 = w_3 = \dfrac{1.784}{3} = 0.59$ kip/ft

In the flat area, half of the load of the center bay is assumed taken by each of the north and south bays.

$w_2 = 0.59 + 0.59/2 = 0.89$ kip/ft

Because the overhanging cantilevers will reduce the stresses in the level area, positive moments are calculated for the ramp, and the results conservatively used for the flat area. Negative moments also calculated.

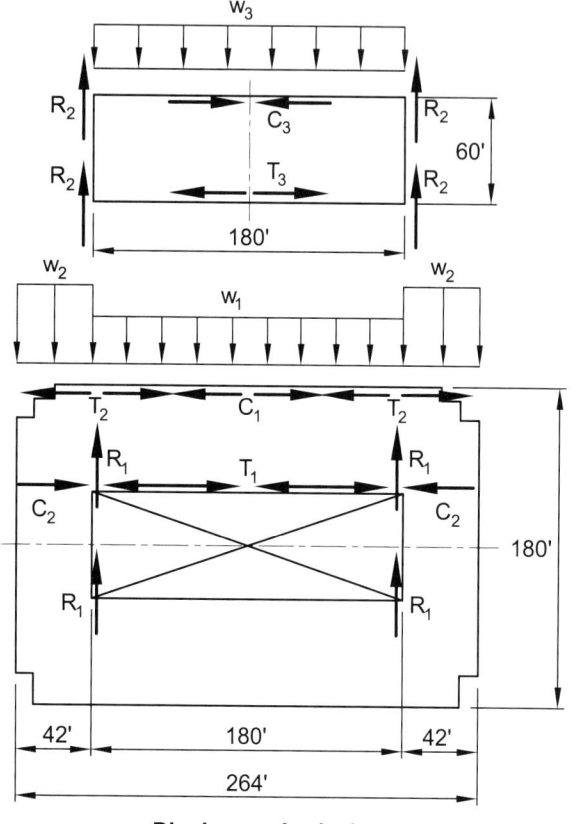

Diaphragm Analysis

Research [20] indicates that in a three-bay structure such as this one, the tee-to-beam joints at the end bays at the four inverted tee beams are particularly vulnerable. The pour strips over these beams should have transverse reinforcement across the joints to improve strength and ductility.

$+M_u = \dfrac{w_1(180)^2}{8} = \dfrac{0.59(180)^2}{8} = 2390$ kip-ft

$-M_u = \dfrac{w_2(42)^2}{2} = \dfrac{0.89(42)^2}{2} = -785$ kip-ft

$V_u = \dfrac{0.59(180)}{2} = 53$ kips

$R_2 = 53/2 = 27$ kips

Example 3.5.12.1 (Cont.)
Three-Level Parking Structure

Diaphragm Moment Design:
Assuming a 58 ft moment arm: $T_{3u} = 2390/58 = 41$ kips
This tensile force may be resisted by reinforcing bars placed into field applied concrete topping or curbs located at each end of the double tees, or by reinforcing steel shop welded to plates cast in the edges of the double tee flanges. These plates would be connected together in the field across the joint using splice plates and welds. Various suggestions for connections are provided in Chapter 6 of this Handbook and in Refs. 12 and 13.

From Section 3.2.6.1, $\phi = 0.7$:

$$A_s = \frac{T_{3u}}{\phi f_y} = \frac{41}{0.7(60)} = 0.98 \text{ in.}^2$$

Use 2 – #7 bars

Splice plate:
Required $A_{pl} = 0.98(60/36) = 1.63$ in.^{2}
Use a plate 6 in. x 6 in. x 5/16 in.
Provided $A_{pl} = 1.88$ in.^{2}

The required length of a 3/16 in. weld, from Figure 6.16.2:
$\ell_w = 41/4.18 = 9.8$ in.

use $\ell_w = 10$ in.

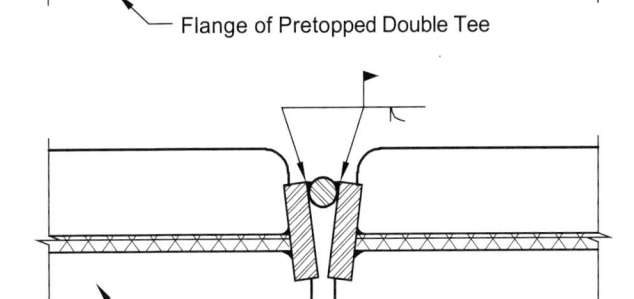

The arrangement of reinforcement is as shown.

Diaphragm Shear Design:
For the connection to the wall, the horizontal force is multiplied by $\Omega_o = 2.5$ (see Ref. 2, Section 9.5.2.6).
V_u to each wall $= (2.5)(24) = 60$ kips

10 ft of each wall is connected to a tee.
$V_u = 60/10 = 6.0$ klf

If flange-to-wall connectors are provided 5 ft on centers, required ϕV_n per connector $= 6.0(5) = 30$ kips.
Example 6.14.4 illustrates a connector design. A heavier connector than shown there would be required.

For the first interior tee-to-tee connection:
$V_u = 53 - 10(0.59) = 47.1$ kips
$V_u = 47.1/60 = 0.79$ klf

If flange connectors are provided 5 ft on centers, required ϕV_n per connector $= 0.79(5) = 4$ kips.
See Figure 3.8.1.2 for typical flange connections.

Conclusion:
The preliminary analysis indicates that the presumed sizes and arrangement of load resisting elements are reasonable. Refinements as may be architecturally required are then made, and the final analysis performed.

lateral load resisting system early on in the planning of the structure. This example shows that a sufficient number of walls can be provided without intruding on the parking and driving space, while allowing adequate security.

It should be noted that the structural concept illustrated in this example is also applicable for such other structures as office buildings.

3.6 Moment-Resisting Building Frames

Precast, prestressed concrete beams and deck members are usually most economical when they can be designed and connected into a structure as simple-span members. This is because:

1. Positive moment-resisting capacity is much easier and less expensive to achieve with pretensioned members than negative moment capacity at supports.

2. Connections which achieve continuity at the supports are usually complex and costly.

3. The restraint to volume changes that occurs in rigid connections may cause serious cracking and unsatisfactory performance or, in extreme cases, even structural failure.

Therefore, it is desirable to design precast, prestressed concrete structures with connections that allow lateral movement and rotation, and to design the structure to achieve lateral stability through the use of floor and roof diaphragms and shear walls.

However, in some structures, adequate shear walls interfere with the function of the building, or are more expensive than alternative solutions. In these cases, the lateral stability of the structure depends on the moment-resisting capacity of either the column bases, beam-to-column frames, or both.

When moment connections between beams and columns are required to resist lateral loads, it is desirable to make the moment connection after most of the dead loads have been applied. This requires careful detailing, specification of the construction process, and inspection. If such details are possible, the moment connections need only resist the negative moments from live load, lateral loads and volume changes, and will then be less costly.

3.6.1 Moment Resistance of Column Bases

Buildings without shear walls may depend on the fixity of the column base to resist lateral loads.

The ability of a spread footing to resist moments caused by lateral loads is dependent on the rotational characteristics of the base. The total rotation of the column base is a function of rotation between the footing and soil, bending in the base plate, and elongation of the anchor bolts, as shown in Figure 3.6.1.1. Further information on this type of analysis, as well as its use with other types of foundations, is given in Ref. 14.

The total rotation of the base is:

$$\phi_b = \phi_f + \phi_{bp} + \phi_{ab} \quad \text{(Eq. 3.6.1.1)}$$

If the axial load is large enough that there is no tension in the anchor bolts, ϕ_{bp} and ϕ_{ab} are zero, and:

$$\phi_b = \phi_f \quad \text{(Eq. 3.6.1.2)}$$

Rotational characteristics can be expressed in terms of flexibility or stiffness coefficients:

$$\phi = \gamma M = M/K \quad \text{(Eq. 3.6.1.3)}$$

where:

M = applied moment = Pe
e = eccentricity of applied load, P
γ = flexibility coefficient
K = stiffness coefficient = $\dfrac{1}{\gamma}$

If bending of the base plate and strain in the anchor bolts are assumed as shown in Figure 3.6.1.1, the flexibility coefficients for the base can be derived, and the total rotation of the base becomes:

$$\phi_b = M(\gamma_f + \gamma_{bp} + \gamma_{ab})$$

$$= Pe(\gamma_f + \gamma_{bp} + \gamma_{ab}) \quad \text{(Eq. 3.6.1.4)}$$

$$\gamma_f = \frac{1}{k_s I_f} \quad \text{(Eq. 3.6.1.5)}$$

$$\gamma_{bp} = \frac{(x_1 + x_2)^3 \left(\dfrac{2e}{h + 2x_1} - 1\right)}{6eE_s I_{bp}(h + x_1)} \quad \text{(Eq. 3.6.1.6)}$$

$$\gamma_{ab} = \frac{g\left(\dfrac{2e}{h + 2x_1} - 1\right)}{2eE_s A_b(h + x_1)} \quad \text{(Eq. 3.6.1.7)}$$

Figure 3.6.1.1 Assumptions used in derivation of rotational coefficients for column bases

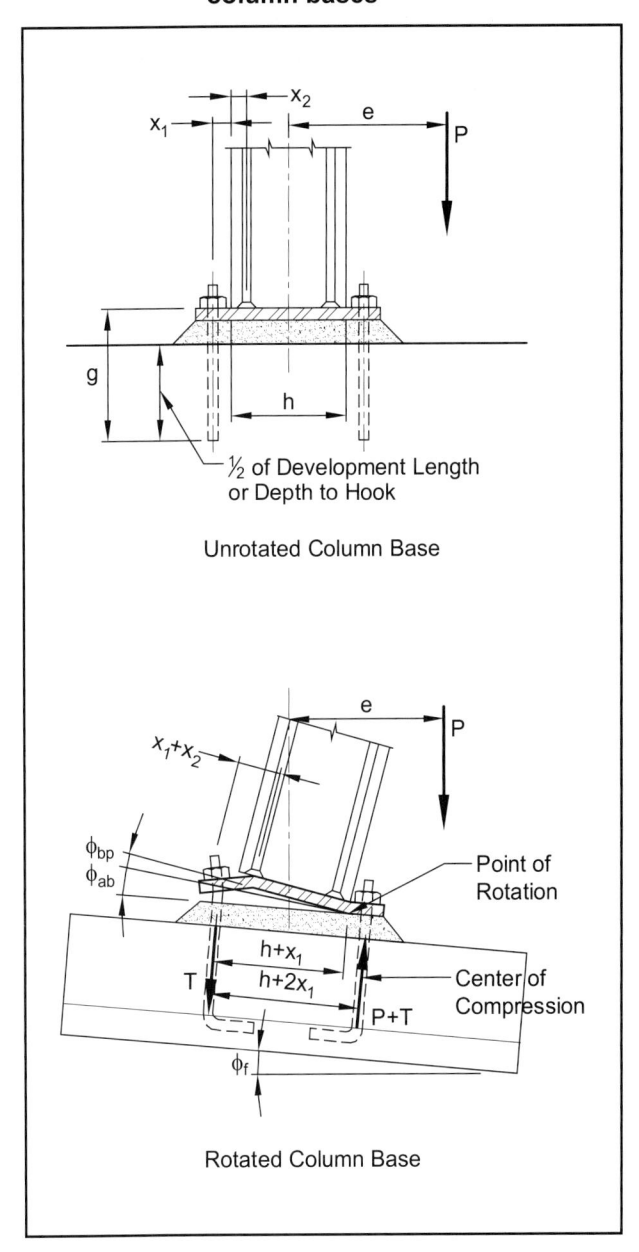

Figure 3.6.1.2 Approximate relation of allowable soil bearing value and coefficient of subgrade reaction, k_s

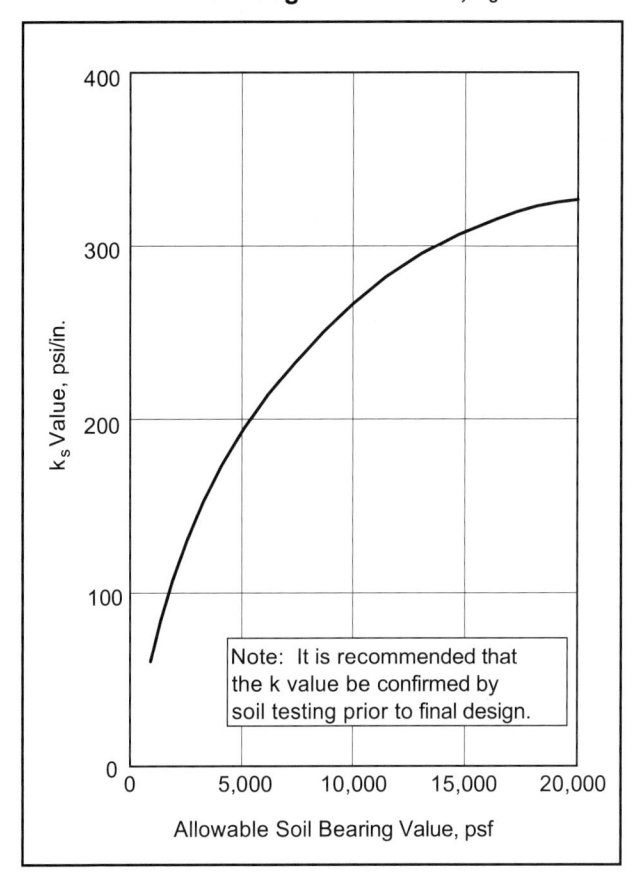

where:

γ_f = flexibility coefficient of footing/soil interaction

γ_{bp} = flexibility coefficient of base plate

γ_{ab} = flexibility coefficient of anchor bolts

k_s = coefficient of subgrade reaction from soil testing or Figure 3.6.1.2

I_f = moment of inertia of footing (plan dimensions)

E_s = modulus of elasticity of steel

I_{bp} = moment of inertia of base plate (vertical cross-section dimensions)

A_b = total area of anchor bolts in tension

h = width of column in direction of bending

x_1 = distance from face of column to the center of the anchor bolts, positive when anchor bolts are outside the column, and negative when anchor bolts are inside the column

x_2 = distance from the face of column to base plate anchorage

g = assumed length over which elongation of the anchor bolt takes place = one-half of development length + projection for anchor bolts made from reinforcing bars, or the length to the hook + projection for smooth anchor bolts (see Figure 3.6.1.1)

Rotation of the base may cause an additional eccentricity of the loads on the columns, causing moments which must be added to the moments induced by the lateral loads.

Note that in Eqs. 3.6.1.6 and 3.6.1.7, if the eccentricity, e, is less than $h/2 + x_1$ (inside the center of compression), γ_{bp} and γ_{ab} are less than zero, meaning that there is no rotation between the column and the footing, and only the rotation from soil deformation (Eq. 3.6.1.5) need be considered.

Example 3.6.1.1
Stability Analysis of an Unbraced Frame

Given:
    The column shown (see next page).
    Soil bearing capacity = 5000 psf
    P = 80 kips dead load, 30 kips snow load
    W = 2 kips wind load

Problem:
    Determine the column design loads and moments for stability as an unbraced frame.

Solution:
    ACI 318-02 requires that the column be designed for the following conditions (Section 3.2.6):
    Eq. 3.2.6.2: 1.2D + 1.6S
    Eq. 3.2.6.3: 1.2D + 1.6S + 0.8W
    Eq. 3.2.6.4: 1.2D + 1.6W + 0.5S
    Eq. 3.2.6.6: 0.9D + 1.6W

The maximum eccentricity occurs when Eq. 3.2.6.6 is applied. Moment at base of column:
    M = 2(16) = 32 kip-ft = 384 kip-in.
    P_u = 0.9D = 0.9(80) = 72 kips
    M_u = 1.6W = 1.6(384) = 614 kip-in.

Eccentricity due to wind load:

$$e = \frac{M_u}{P_u} = \frac{614}{72} = 8.53 \text{ in.}$$

To determine the moments caused by base rotation, an iterative procedure is required.

Estimate eccentricity due to rotation = 0.30 in.
    e = 8.53 + 0.30 = 8.83 in.

Check rotation between column and footing:
    h/2 + x_1 = 20/2 + (−2) = 8 in. < 8.53 in.

Thus, rotation between the column and footing must be checked:

Eq. 3.6.1.6: Since $x_1 + x_2$ = (−2)+2 = 0; γ_{bp} = 0

Eq. 3.6.1.7: Assume g = 15 in.

$$\gamma_{ab} = \frac{15\left(\frac{2(8.83)}{20+2(-2)} - 1\right)}{2(8.83)(29 \times 10^6)(2 \times 0.79)[20+(-2)]} = 0.12 \times 10^{-9}$$

Eq. 3.6.1.5:

$$I_f = \frac{[6(12)]^4}{12} = 2.24 \times 10^6 \text{ in.}^4$$

Example 3.6.1.1 (Cont.)
Stability Analysis of an Unbraced Frame

From Figure 3.6.1.2: $k_s \approx 200$ psi/in.

$$\gamma_f = \frac{1}{\left[200(2.24\times10^6)\right]} = 2.23\times10^{-9}$$

$M_u = 72(8.83) = 636$ kip-in.

$\phi_b = (\gamma_{ab} + \gamma_f)M_u$

$\quad = (0.12+2.23)(10^{-9})(636,000) = 0.00149$ radians

Eccentricity caused by rotation:
$\quad \phi_b h_s = 0.00149(16)(12) = 0.29 \approx 0.30$ in.

No further trial is required.

Check Eq. 3.2.6.3:
$\quad 1.2D = 1.2(80) = 96$ kips
$\quad 1.6S = 1.6(30) = 48$ kips
$\quad 0.8W = 0.8(384) = 307$ kip-in.

Eccentricity due to wind load:

$$e = \frac{M_u}{P_u} = \frac{307}{144} = 2.13 \text{ in.}$$

Estimate eccentricity due to rotation = 0.15 in.
$\quad e = 2.13 + 0.15 = 2.28$ in.

Check rotation between column and footing:
$$\frac{h}{2} + x_1 = \frac{20}{2} + (-2) = 8 \text{ in.} > 2.28$$

Thus, there is no movement between the column and footing.

$M_u = 144(2.28) = 328$ kip-in.

$\phi_b = \gamma_f M_u = 2.23(10^{-9})328,000$

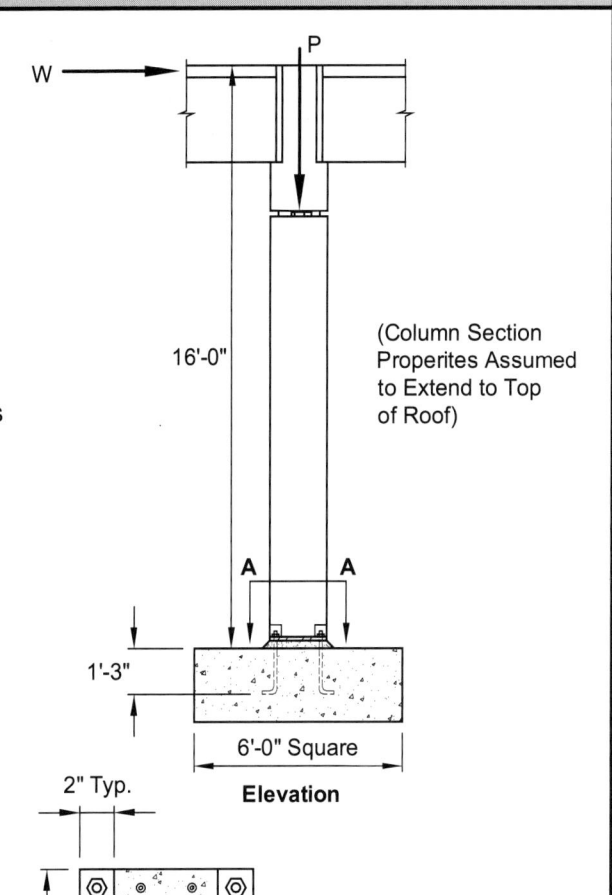

$\quad = 0.00073$ radians

Example 3.6.1.1 (Cont.)
Stability Analysis of an Unbraced Frame

Eccentricity caused by rotation:
$\phi_b h_s = 0.00073(16)(12) = 0.14$ in. ≈ 0.15 in.

No further trial is required.

Check Eq. 3.2.6.4:

$1.2D = 1.2(80) = 96$ kips
$0.5S = 0.5(30) = 15$ kips
$1.6W = 1.6(384) = 614$ kip-in.

Eccentricity due to wind load:
$e = \dfrac{M_u}{P_u} = \dfrac{614}{111} = 5.53$ in.

Estimate eccentricity due to rotation = 0.30 in.
$e = 5.53 + 0.30 = 5.83$ in.

Check rotation between column and footing:
$\dfrac{h}{2} + x_1 = \dfrac{20}{2} + (-2) = 8$ in. > 5.53 in.

Thus, there is no movement between the column and footing.
$M_u = 111(5.83) = 647$ kip-in.

$\phi_b = \gamma_f M_u = 2.23(10^{-9})647,000 = 0.00144$ radians

Eccentricity caused by rotation:
$\phi_b h_s = 0.00144(16)(12) = 0.28$ in. ≈ 0.30 in.

No further trial is required

Summary of column load cases:

Eq. 3.2.6.2: 1.2D + 1.6S: $P_u = 144$ kips, $M_u = 173$ kip-in. (minimum by Code)

Eq. 3.2.6.3: 1.2D + 1.6S + 0.8W: $P_u = 144$ kips, $M_u = 328$ kip-in.

Eq. 3.2.6.4: 1.2D + 1.6W + 0.5S: $P_u = 111$ kips, $M_u = 647$ kip-in.

Eq. 3.2.6.6: 0.9D + 1.6W: $P_u = 72$ kips, $M_u = 636$ kip-in.

3.6.2 Fixity of Column Bases

The degree of fixity of a column base is the ratio of the rotational stiffness of the base to the sum of the rotational stiffnesses of the column plus the base:

$$F_b = \frac{K_b}{K_b + K_c} \quad \text{(Eq. 3.6.2.1)}$$

where:
F_b = degree of base fixity, expressed as a decimal
$K_b = \dfrac{1}{\gamma_b}$
$K_c = \dfrac{4E_c I_c}{h_s}$
E_c = modulus of elasticity of the column concrete
I_c = moment of inertia of the column
h_s = column height

Example 3.6.2.1
Degree of Fixity

Determine the degree of fixity of the column base in Example 3.6.1.1:

For loads of Eq. 3.2.6.6:
E_c = 4300 ksi

$I_c = \dfrac{20^4}{12} = 13{,}333 \text{ in.}^4$

$K_c = \dfrac{4(4.3 \times 10^6)(13{,}333)}{16(12)} = 11.94 \times 10^8$

$\gamma_b = \gamma_{ab} + \gamma_f = 0.12 + 2.23 = 2.35$

$K_b = \dfrac{1}{\gamma_b} = \dfrac{1}{2.35 \times 10^{-9}} = 4.26 \times 10^8$

$F_b = \dfrac{4.26}{4.26 + 11.94} = 0.263$

For other load combinations:

$K_b = \dfrac{1}{2.23 \times 10^{-9}} = 4.48 \times 10^8$

$F_b = \dfrac{4.48}{4.48 + 11.94} = 0.273$

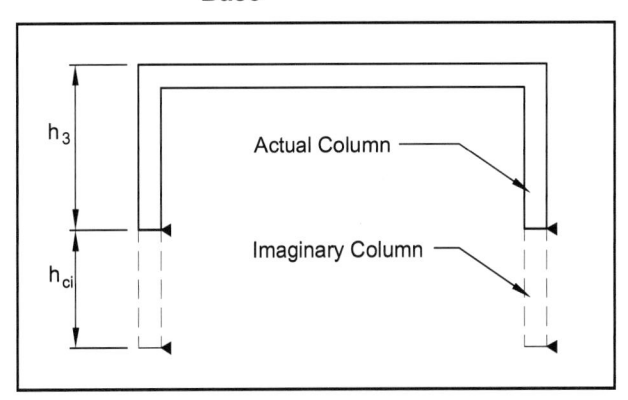

3.6.3 Computer Models for Frame Analysis

When precast frames are modeled as "sticks," as is usually done with steel frames, the results are often very misleading. For example, the structure as modeled in Figure 3.6.3.1(a) will indicate more flexibility than is actually true. Lateral drift will be overestimated, and the moments caused by axial shortening may be underestimated. Figure 3.6.3.1(b) shows a suggested model, which will better estimate the true condition.

3.6.3.1 Modeling Partially Fixed Bases

Many computer programs permit the direct modeling of various degrees of base fixity by the use of spring options. A simple way to model base fixity is to incorporate an imaginary column below the actual column base. If the bottom of the imaginary column is modeled as pinned, then the expression for its rotational stiffness is:

$$K_{ci} = \frac{3E_{ci} I_{ci}}{h_{ci}} \quad \text{(Eq. 3.6.3.1)}$$

where the subscript "ci" denotes the properties of the imaginary column (Figure 3.6.3.2), and $K_{ci} = K_b$ as calculated in Section 3.6.1, or, the degree of base fixity, F_b, can be determined or estimated, and K_b calculated from Eq. 3.6.2.1.

For the computer model, either I_{ci} or h_{ci} may be varied for different values of K_{ci}, with the other terms left constant for a given problem. It is usually preferable to use $E_{ci} = E_c$. For the assumptions of Figure 3.6.3.2:

$$h_{ci} = \frac{3E_{ci} I_{ci}}{K_{ci}} \quad \text{(Eq. 3.6.3.2)}$$

or

$$I_{ci} = \frac{K_{ci} h_{ci}}{3E_{ci}} \quad \text{(Eq. 3.6.3.3)}$$

Example 3.6.3.1 illustrates this procedure.

Figure 3.6.3.2 Model for Partially Fixed Column Base

Figure 3.6.3.1 Computer models

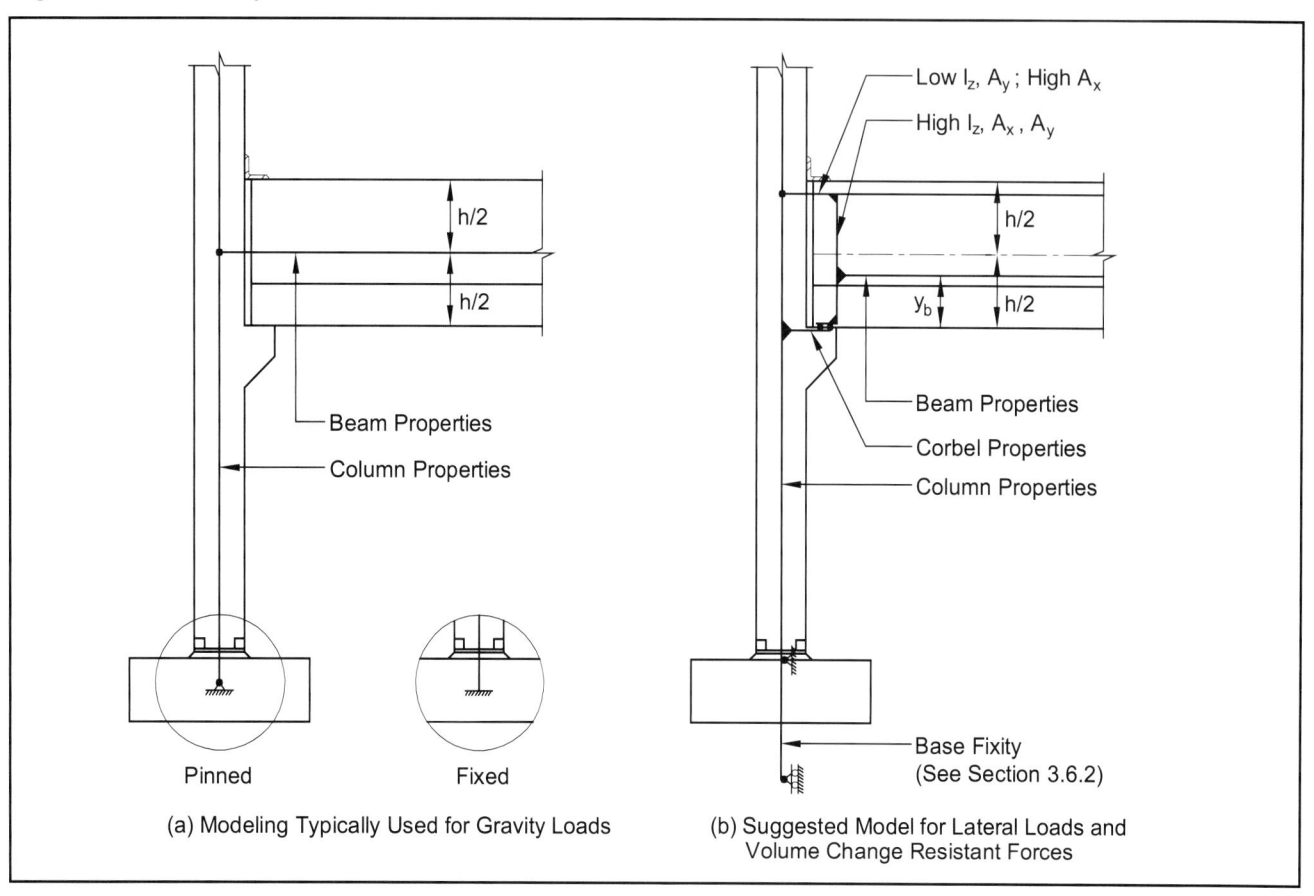

(a) Modeling Typically Used for Gravity Loads

(b) Suggested Model for Lateral Loads and Volume Change Resistant Forces

---

Example 3.6.3.1
Imaginary Column for Computer Model

Given:
 The column base of Examples 3.6.1.1 and 3.6.2.1.
 $f'_c = 5000$ psi $E_c = 4300$ ksi

Problem:
 Determine the length of an imaginary column to model the fixity of the column.

Solution:
$E_{ci} = E_c = 4.3 \times 10^6$ psi

$I_{ci} = I_c = 13{,}333$ in.^{4}

$K_{ci} = K_b = 4.48 \times 10^8$

$h_{ci} = \dfrac{3 E_{ci} I_{ci}}{K_{ci}} = \dfrac{3(4.3 \times 10^6)(13{,}333)}{4.48 \times 10^8}$
$= 383$ in. $= 31.9$ ft

3.6.4 Frame Classifications for Seismic Considerations

The evaluation of equivalent lateral forces for design for earthquake effects was reviewed in Section 3.2.4. The various reductions in equivalent lateral force from the full elastic response of a structure to seismic ground motion, Eqs. 3.2.4.2 to 3.2.4.8, reflect strength, ductility, toughness, redundancy and resonance. The implicit assumption of the code is that structures subject to higher levels of seismic risk should be designed to experience higher degrees of inelastic deformations without collapse. This is reflected by prescriptive requirements for detailing that become more stringent with higher seismic levels. These requirements are consistent with increases in the response modification factors that permit proportionally lower base shear relative to the ground acceleration.

The prescriptive detailing requirements are associated with seismic design categories determined in the model building codes from a combination of short periods and one-second period spectral acceleration and the soil conditions at the site. The lowest seismic design categories are A and B. Concrete frame structures assigned to these categories are permitted to use ordinary moment frames. These are defined as cast-in-place or precast frames meeting the requirements of Chapters 1 through 18 of ACI 318, and do not include special prescriptive requirements for seismic detailing. To reflect the low ductility and potential for unfavorable failure mechanisms, ordinary concrete moment frames are assigned a low response modification factor, R = 3 (see Figure 3.10.8).

The model codes do not permit ordinary moment frames to be used in the intermediate Seismic Design Category C structures. They require that concrete frames be at least intermediate moment frames. ACI 318-02 defines an intermediate moment frame as a cast-in-place frame meeting limited detailing requirements specified in Section 21.1.2. The rules for intermediate moment frames are intended to develop designs where the post-elastic behavior may result in significant damage, but that would avoid general collapse. Although an engineer might develop precast systems that emulate the monolithic cast-in-place concrete intermediate frame and use the assigned R value of 5, the code apparently intends that precast concrete frames in Category C structures should conform to the more stringent requirements for special moment frames.

The high seismic design categories are D, E and F. Structures in these categories are not permitted to use ordinary or intermediate moment frames, but must use special moment frames. A special moment frame of precast concrete must comply with ACI 318-02, Sections 21.2 through 21.6. The provisions for the design and detailing of these frames aim to produce structures with strong column-weak beam behavior. Yielding will be concentrated in the beam at the negative moment region of the beam/column interface. The remaining frame elements will be detailed with sufficient flexural and shear strength to maintain capacity through the post-elastic displacement of the frame. These frames are assigned an R = 8.

3.6.5 Frame Analysis and Design – Ordinary Moment Frames

When the engineer elects to use a frame and the seismic requirements permit an ordinary moment frame, it is usually most economical to develop that frame through connections made within the gravity load system. In this way, the components that are ordinarily required are extended to provide lateral resistance without additional components. The beams used for interior framing may not be suitable since they are most economically designed as prestressed with simple spans. Connecting these beams to induce negative moments at the columns is not favorable because these moments are additive to the effects of prestressing and put restraints on the elements with the highest creep and shrinkage movements. Deep precast spandrel beams on the building perimeter are an attractive option. These beams are often deeper than required to provide support for the floor framing for architectural reasons. With this depth, they are lightly prestressed. Connections can be made near the top and bottom of the beams with a larger tension/compression couple that moderates the forces. Example 3.6.5.1 illustrates such a design.

Example 3.6.5.1
Building with Ordinary Moment Frames

Given:

An eight-story office building is shown on the following pages. The example is a three-bay wide structure with two lines of interior framing of columns and inverted tee beams and two lines of exterior framing using columns and loadbearing spandrel beams that are also the architectural exterior cladding.

The floor-to-floor height is typically 13 ft, except that the first floor is 15 ft. The floor is framed with 24 in. deep, 10 ft wide tees with a minimum of 3 in. topping. The center bay on the elevation has a glass curtain wall so the exterior beams in this bay are mostly shallow L-beams.

There is a benefit in this separation of the frames on the perimeter because it reduces the length of the rigid assembly that is subject to restraint of volume change movements. The roof level is framed in the same way as the floor with a partial mechanical penthouse framed with light steel. For simplicity in this example, the loads at this level are considered comparable to the floor loads.

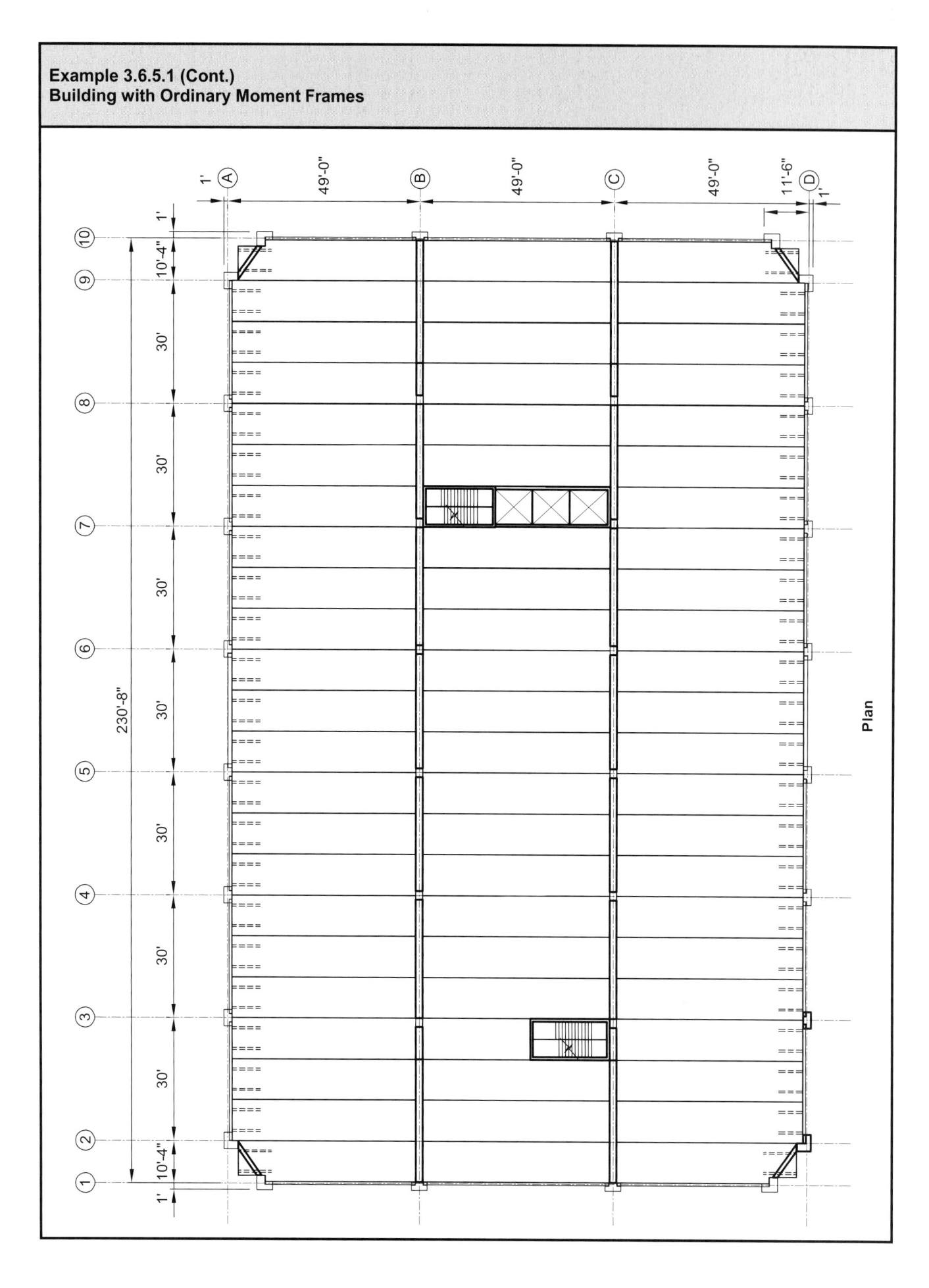

Example 3.6.5.1 (Cont.)
Building with Ordinary Moment Frames

Plan

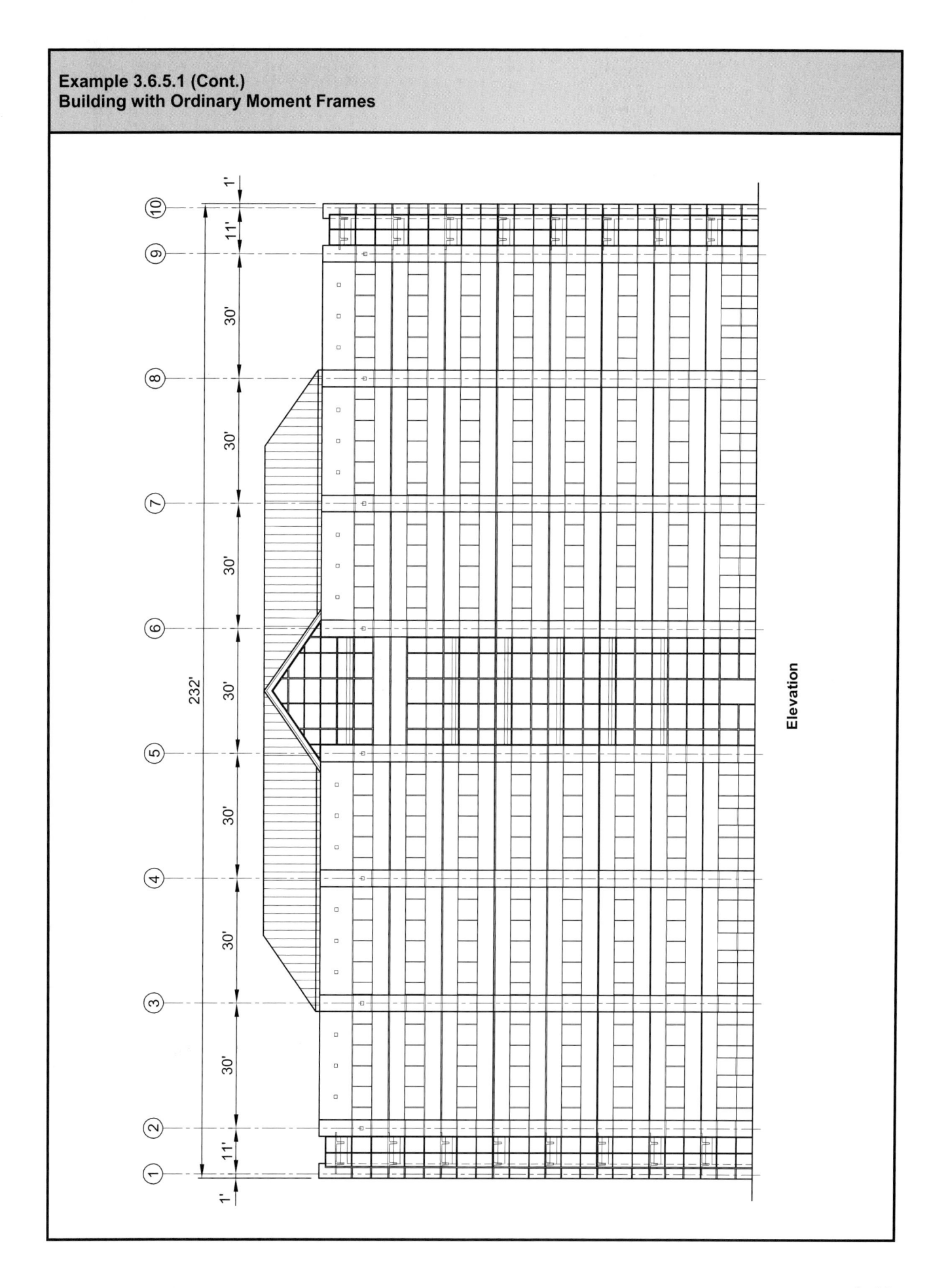

Example 3.6.5.1 (Cont.)
Building with Ordinary Moment Frames

Elevation

PCI Design Handbook/Sixth Edition 3–63

Example 3.6.5.1 (Cont.)
Building with Ordinary Moment Frames

Problem:
Design the lateral support system with frames in the long direction of the plan. (The core walls may provide support across the plan and are not evaluated in this example.)

Solution:
The design procedure for an ordinary moment frame building will generally follow these steps:

1. Determine the gravity loads for the structure. After calculating uniform dead and live loads and line loads, calculate the gravity loads level by level for each column and wall. This can be facilitated using a spreadsheet set up to collect and accumulate the loads to the foundations. Since one primary use of the total load is for column and foundation design, live load reductions should be calculated with multi-story tributary areas.

2. Calculate the seismic coefficients using the procedures shown in Section 3.2.4. Precast frame buildings often have calculated natural periods longer than the approximate period value and greater than the upper limit period for strength design, so it may save some iterative effort to use the upper limit period as an initial condition.

3. Calculate the lateral forces. For regions with low seismicity and high wind speeds, these forces might be governed by wind loads (Section 3.2.3). With the change to ASCE 7-02 load criteria and the larger mass in concrete structures, seismic loads will likely govern even in regions of low risk.

4. Define the configuration and structural models of frames for analysis.

5. Calculate a horizontal distribution of the forces to each frame (Section 3.2.4.3) considering the actual center of mass and the required allowance for accidental torsion.

6. Perform the stiffness analysis for the frames. This is usually done using a computer program. Several commercial programs are in common use by structural engineers.

7. Evaluate the frame displacements as discussed in Section 3.2.4.4. This may involve these steps:
   - Check the building period using the elastic displacements.
   - To reflect the post-elastic displacements, increase the elastic displacements by the C_d factor for the system. This value is 2½ for ordinary moment frames.
   - Check the interstory drift and verify that the results are within the limits prescribed by the code. (2% for seismic use Group I buildings that fall under the category of "all other buildings")
   - Check for the potential of P-Δ effects by calculating the value of θ from Eq. 3.2.4.16.
   - Evaluate component and connection forces for design feasibility.

If the analysis results are within code limits and acceptable component and connection capacity, proceed to the design of the components and connections. Otherwise, iterate on the structural model and repeat the analysis.

Stiffness analysis:
Frames constructed with precast components frequently require more sophisticated structural modeling for accurate stiffness analysis than simple lines for the beams and columns. The depths of interior beams and the widths of columns reduce unsupported lengths and increase effective stiffness. Some stiffness analysis computer programs include utilities to make these corrections. For frames made with deep beams and discrete connections, these adjustments are not sufficient to produce accurate models and to make the determination of connection forces easy. This is also discussed in Section 3.6.3.

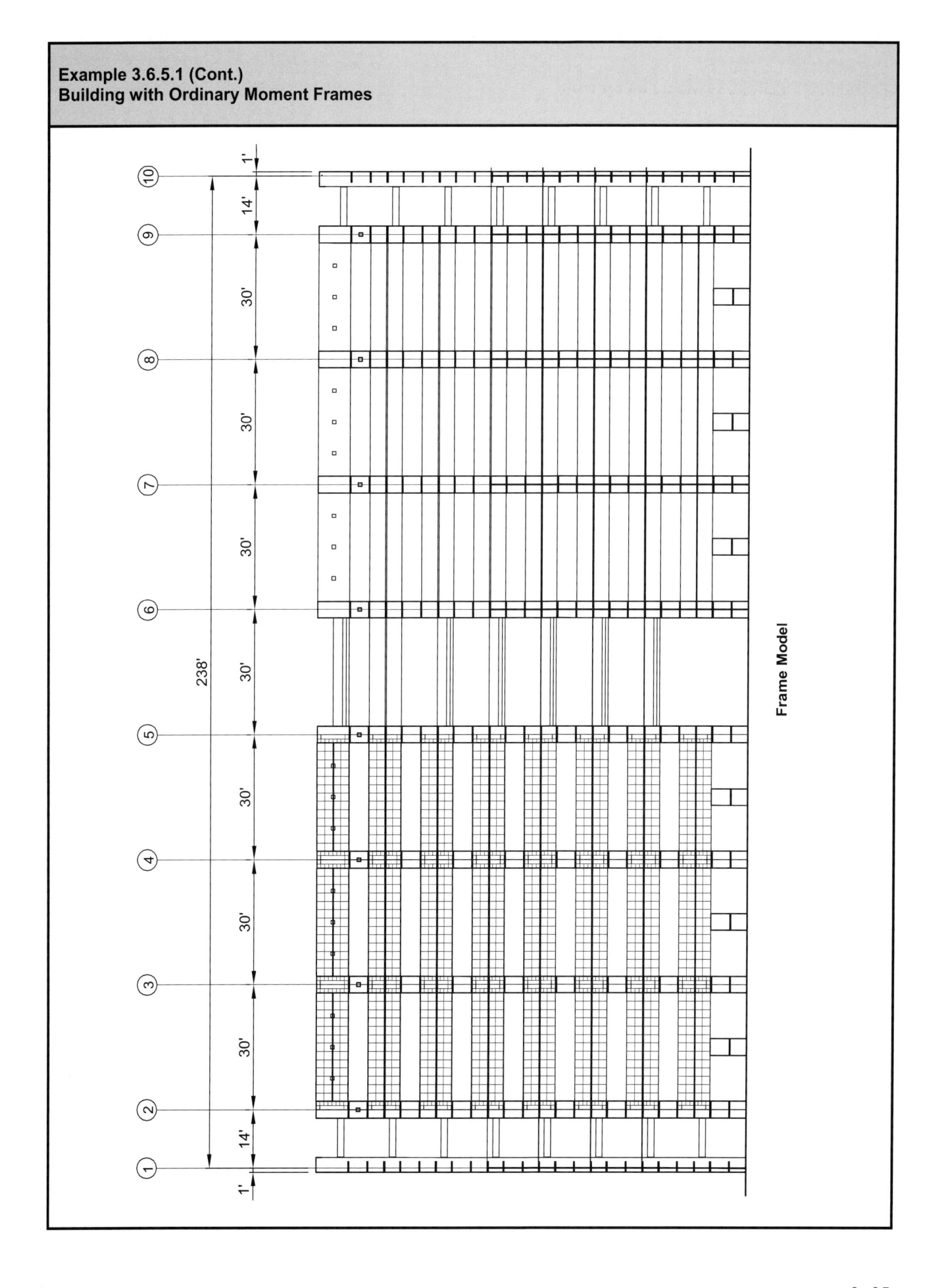

Example 3.6.5.1 (Cont.)
Building with Ordinary Moment Frames

Frame Model

Example 3.6.5.1 (Cont.)
Building with Ordinary Moment Frames

For the ordinary moment frame example, an improved structural model is used (see previous page). An example building elevation is shown with the curtain wall, glazing and light metal roof removed. On the right is a view of the spandrels and columns of one of four perimeter frames designed as the lateral force resisting system. On the left, the structural model is superimposed over the frame elevation to show the general relationship of the model to the structure (see page 3–65).

To meet the requirements of the code, the frame is analyzed as linear and elastic using the load combinations for gravity loads and combined gravity and seismic loads. The displacement amplification assumes this basic analysis and checks for potential P-Δ effects from the value of θ. For the gravity load cases, it may be necessary to increase moments derived from this linear analysis using the approximate method of moment magnification described in ACI 318-02 Sections 10.12 and 10.13. Many stiffness analysis programs today, however, have capabilities to make second order analysis and directly incorporate the P-Δ effects.

Gravity loads:
10DT24	= 47 psf
Field-placed topping (3 in. min., 3½ in. avg.)	= 44
Partition allowance	= 20 (Note: Partitions are sometimes considered live load.)
Total uniform dead load	=111 psf

Live load:
Office loading	= 50 psf
Corridors	= 80 psf
Design average	= 60 psf
Reduced live load = 60 × 0.4	= 24 psf

(The roof load includes snow load on the flat roof with sweeping limited by parapets as well as mechanical loads in the penthouse, light metal frame roof, and snow on the penthouse roof. For simplicity in the example, these are approximated by using the floor loads.)

Line dead loads:
Inverted tee beams	= 1330 plf
Spandrel beams	= 800 plf
Exterior columns (24 × 48)	= 1200 plf
Interior columns (24 × 30)	= 750 plf

A summary of the total dead load at each level is shown in Table 3.6.5.1(A).

Seismic Coefficients:
Follow the procedure outlined in Section 3.2.4.1 to determine C_s:

The mapped maximum considered earthquake (MCE) spectral response acceleration at short period and 1-sec period are determined from the spectral acceleration maps in IBC. For this example, a site in Richmond, Virginia, was assumed. The short period value, S_s, is 0.27 and the 1-sec period value, S_1, is 0.08.

Without a detailed geotechnical evaluation of the site, the default site class is D, which is used for the example.

The MCE spectral response accelerations must be adjusted for the site class effects in accordance with Figure 3.10.7. Values in the table are interpolated in the range of accelerations, yielding F_a = 1.568 and F_v = 2.4. Therefore:
$S_{MS} = F_a S_s = 1.568 \times 0.27 = 0.423$
$S_{M1} = F_v S_1 = 0.08 \times 2.4 = 0.192$

Example 3.6.5.1 (Cont.)
Building with Ordinary Moment Frames

For the 5% damped design spectral response accelerations:
$S_{DS} = 2/3 \times S_{ms} = 0.282$
$S_{D1} = 2/3 \times S_{m1} = 0.128$
From Table 3.2.4.1, building is in Seismic Design Category B.

The approximate building period is calculated from Eq. 3.2.4.6:
$T_a = C_T h_n^x = 0.016 \times (106)^{0.9} = 1.06$ sec

Calculate the seismic response coefficient, with the importance factor taken as 1.0:
From Eq. 3.2.4.9:
$C_s = \dfrac{S_{DS}}{(R/I)} = \dfrac{0.282}{(3/1)} = 0.094$

From Eq. 3.2.4.10:
$C_s = \dfrac{S_{D1}}{(R/I)T} = \dfrac{0.128}{(3/1)1.06} = 0.0401$

but not less than the value found from Eq. 3.2.4.11:
$C_s = 0.044 S_{DS} I = 0.044(0.282)(1.0)$
$\quad = 0.0124$
Use $C_s = 0.0401$

Calculate Base Shear from Eq. 3.2.4.1:
$V = C_s W = 0.0401(45,709) = 1833$ kips

The vertical distribution of the base shear is calculated from Eqs. 3.2.4.13 and 3.2.4.14:
$F_x = C_{vx} V$
where
$C_{vx} = \dfrac{w_x h_x^k}{\sum_{i=1}^{n} w_i h_i^k}$ and k varies with the period.

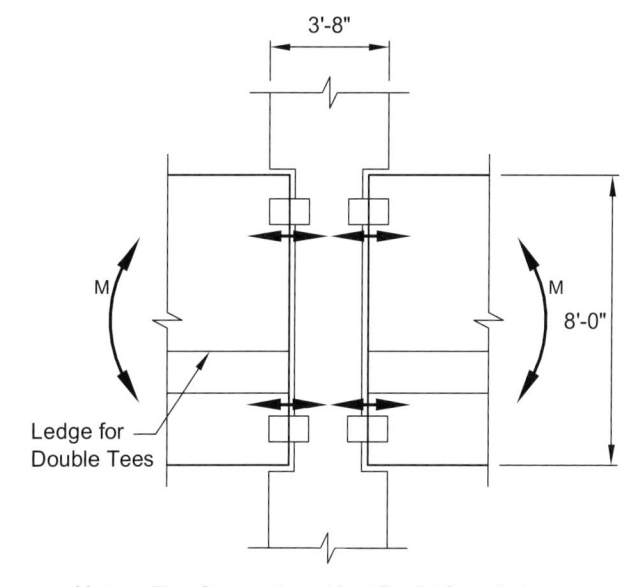

Note: Top Connections Must Resist Spandrel Beam Rotation as Well as in-Plane Seismic Tension and Compression.

Moment Connection

With the building period 1.06 sec, the interpolated value of k is 1.28. The resulting vertical distribution of the base shear is shown in Table 3.6.5.1(B).

The moment connection between the spandrel and the column is shown. The left side of the frame model shows the finite element model. With this model, the 8 ft deep spandrel beams are modeled with shell elements. The interior elements are 2 x 2 ft and the ends are modeled with 1 x 1 ft elements to improve the accuracy around the connections. The exterior columns are 22 x 48 in. for architectural appearance and for ample area to allow block-outs for spandrel beam bearing flush with the inside face. The columns are modeled as vertical line (beam) elements, with short stiff elements (corbels) used to model the eccentric support of the spandrel. Since precast spandrels rest on bearing pads without positive restraint at the bearing, this must be considered in the modeling. These "corbels" have had the axial reaction constraint and the moment constraint at the end shown released so that the element will accept only the vertical reaction from the beam. The frame connections are placed 1 ft above the base and 1 ft below the top of the beam as steel plates linking spandrel to column. The moments of inertia of the columns are reduced to 70% of values of gross section to

Example 3.6.5.1 (Cont.)
Building with Ordinary Moment Frames

approximate column cracking, but the spandrels in this configuration and with nominal prestressing are not likely to crack and are modeled with the full beam thickness. This model is also subject to the out-of-plane forces due to beam torsion and lateral support at the floor so that the frame analysis considers biaxial behavior in the columns. For simplicity in this example, the foundation is taken as a continuous strip under the frame line and the column-foundation stiffness interaction has not been evaluated.

Table 3.6.5.1(A) – Load summary

1	2	3	4	5	6	7	8	9
Level, x	w_x, kips	h_x, ft	h_x^k, ft	$w_x h_x^k$	C_{vx}	V, kips	F_x, kips	Story V_x
Penthouse	6,382	106	391	2,496,596	0.27	1,833	495	495
7	5,600	93	331	1,852,875	0.20	1,833	367	862
6	5,600	80	273	1,528,070	0.17	1,833	303	1,165
5	5,600	67	217	1,217,766	0.13	1,833	241	1,406
4	5,600	54	165	923,958	0.10	1,833	183	1,589
3	5,600	41	116	649,458	0.07	1,833	129	1,718
2	5,600	28	71	398,612	0.04	1,833	79	1,797
1	5,727	15	32	183,368	0.02	1,833	36	1,833
	45,709			9,250,703				

Calculate the horizontal distribution of forces to each frame.

As discussed in Section 3.2.4.3, with rigid diaphragms the design must include allowance for accidental torsion of 5% in addition to any torsion from the eccentricity of the center of mass from the center of stiffness. With the frames on the perimeter, this requirement is satisfied by using 0.55/2 times the story shear on each of the four frames.

The analysis of the frames should consider the load combinations required to complete the design as outlined in Section 3.2.6. For this case, using Eqs. 3.2.6.5a and 3.2.6.7a, with $S_{DS} = 0.282$, $\rho = 1.0$ (Seismic Design Category B) and $f_1 = 0.5$:

$$U = (1.2 + 0.2S_{DS})D + \rho Q_E + f_1 L + 0.2S = [1.2 + 0.2(0.282)]D + 1.0 Q_E + 0.5L + 0.2(0)$$
$$= 1.256D + Q_E + 0.5L$$
$$U = (0.9 - 0.2S_{DS})D + \rho Q_E + 1.6H = [0.9 - 0.2(0.282)]D1.0 Q_E + 1.6(0) = 0.844D + Q_E$$

Perform the stiffness analysis for the frames.

Section 10.11.1 of ACI 318-02 suggests reduced section properties to be used in structural analyses to account for "the influence of axial loads, the presence of cracked regions along the length of the member, and effects of duration of the loads." Since the beams are prestressed, cracking is unlikely, and gross section properties should be used. The columns, however, are assumed to be non-prestressed, and the recommendation of $I = 0.70 I_g$ may be followed.

As discussed in Section 3.2.4.4, separate analyses are performed for dead load and total load.

Evaluate the frame displacements.

For this example, the net lateral displacements for seismic loads are determined by the direct output of displacements for the seismic load only.

Determine the amplified displacement from Eq. 3.2.4.15. From Figure 3.10.8, C_d for ordinary moment frames is 2.5, From Figure 3.10.1, the importance factor, $I_E = 1.0$. Therefore, the amplifier is 2.5.

The table on this page provides the calculated displacements, the amplified displacements and the calculated story drift.

Example 3.6.5.1 (Cont.)
Building with Ordinary Moment Frames

The drift calculated is relatively small, indicating that the use of the spandrel beams and the architectural-sized columns on the building perimeter creates a relatively stiff structure even though it is jointed and completed with discrete welded connections.

Check the potential for P-Δ effects due to drift. P-Δ effects need not be considered if the stability coefficient, calculated by Eq. 3.2.4.16, is less than or equal to 0.10. For this example, the maximum calculated stability coefficient is 0.019, well below the threshold.

Evaluate component and connection forces for design feasibility.
The detailed design of the components and connections follow the procedures covered in Chapters 4 and 6.

Table 3.6.5.1(B) – Drift Calculations

Level	Net Displacement, in.	Amplifier C_d	Design Displacement, in.	Story Drift, in.
Penthouse	0.69	2.5	1.74	0.10
7	0.65	2.5	1.63	0.15
6	0.59	2.5	1.48	0.21
5	0.51	2.5	1.27	0.25
4	0.41	2.5	1.03	0.25
3	0.31	2.5	0.77	0.29
2	0.19	2.5	0.49	0.30
1	0.08	2.5	0.19	0.19

3.6.6 Special Moment Frames for Seismic Design Category C

As intermediate precast moment frames are not included in the ACI 318-02 Chapter 21 definition, precast frames in Category C should be special moment frames. It is possible to emulate a monolithic cast-in-place, intermediate moment frame system with precast components that meets all the requirements of Section 21.12, but the result would not be similar to the ordinary moment frame illustrated in Example 3.6.5.1. A precast frame emulating a monolithic cast-in-place frame will require special considerations for panelization, forming, transportation and erection. With these measures already taken, it is not a significant step to improve the detailing of these elements to achieve the emulation of special moment frames, and so achieve an R factor of 8 instead of 5. The reduction in base shear results in less demand for the special framing and may be the more economical alternative.

Emulation of cast-in-place concrete is discussed in Section 3.1.2.1.

The most economical framing for precast construction is simple spans. When continuity is introduced into a precast frame, it is important to:

- Retain as much simple span framing as the system can tolerate.
- Keep field connections easy and fast.
- Introduce those frames that are required, but no more.

In regions of moderate seismic risk, then, look at options that develop special frames with the monolithic joints fabricated in the plant and field joints with details that meet the Chapter 21 connection provisions. Use as few field connections as can be made consistent with limitations on shipping weights and sizes.

Example 3.6.6.1
Eight-Story Office Building in Seismic Design Category C.

Given:
This design example uses the same office building layout from Example 3.6.5.1 to show adaptation of the framing to special moment frames in regions of moderate seismic risk. The design develops support from six independent "H" frame stacks. The frames are located on the plan. These frames are made of elements with an inverted tee beam between two columns as shown below. The splices are located mid-height between the floors, using emulation detailing. [6] The outside frames include an overhung inverted tee beam toward the building ends. An overall building section is shown later.
The gravity loads for the structure are the same as for Example 3.6.5.1.

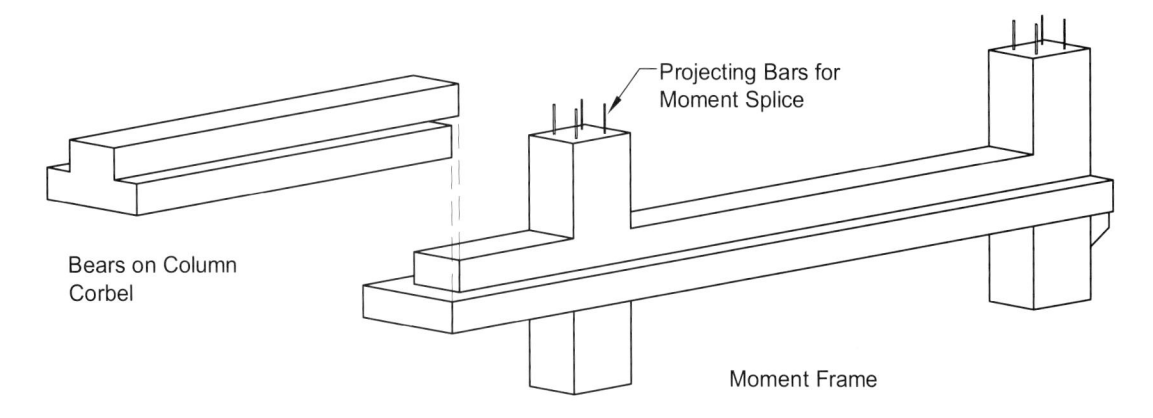

Problem:
Same as Example 3.6.5.1.

Solution:

Seismic Coefficients
1. For this example, a site in New York City was assumed to determine mapped spectral acceleration values. The short period value, S_s, is 0.43 and the 1-sec period value, S_1, is 0.095.
2. The default Site Class D is used for the example.
3. The MCE spectral response accelerations must be adjusted for the site class effects (see Figure 3.10.7). For the short period, $F_a = 1.456$ and $F_v = 2.4$. Therefore:
$S_{ms} = F_a S_s = 1.456(0.43) = 0.626$
$S_{m1} = F_v S_1 = 2.4(0.095) = 0.228$
4. The design values are two-thirds of the MCE values:
$S_{DS} = 2/3 \times S_{ms} = 0.418$
$S_{D1} = 2/3 \times S_{m1} = 0.152$
5. The approximate building period is the same as calculated in the previous example, $T = 1.06$ sec.
6. The seismic response coefficient from Eq. 3.2.4.10 is:
$$C_s = \frac{S_{D1}}{(R/I)T} = \frac{0.152}{(8/1)1.06} = 0.0179$$
The lower limit (Eq. 3.2.4.11) is:
$C_s = 0.044 S_{DS} I = 0.044(0.418) = 0.0184$ (governs)

It is significant to note that although the spectral response acceleration is greater for the moderate site, the response coefficient is less than half that of the previous example due to the high R factor assigned to special moment frames.

7. Calculate Lateral Forces:
$V = C_s W = 0.0184(45,709) = 841$ kips

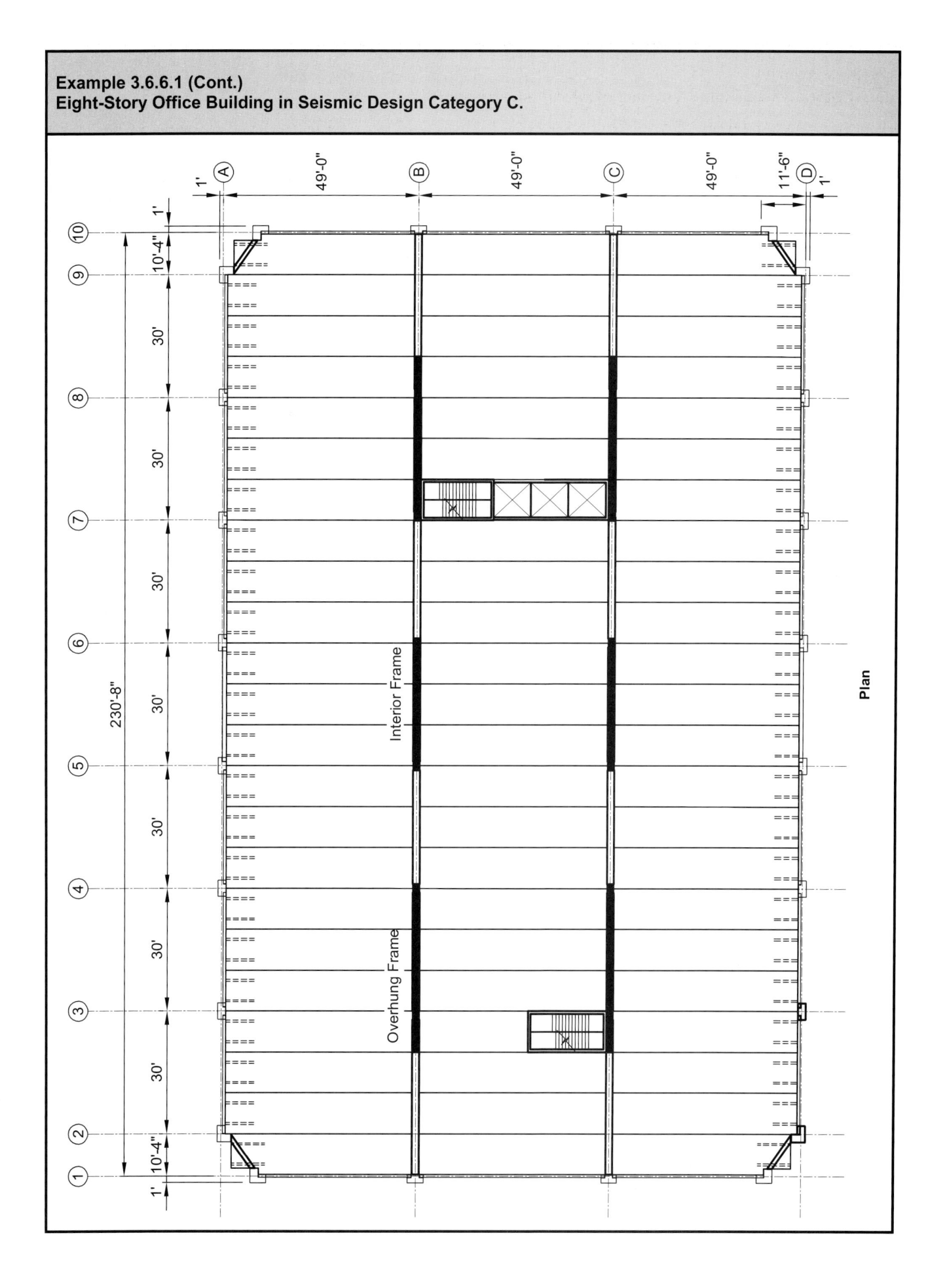

Example 3.6.6.1 (Cont.)
Eight-Story Office Building in Seismic Design Category C.

Plan

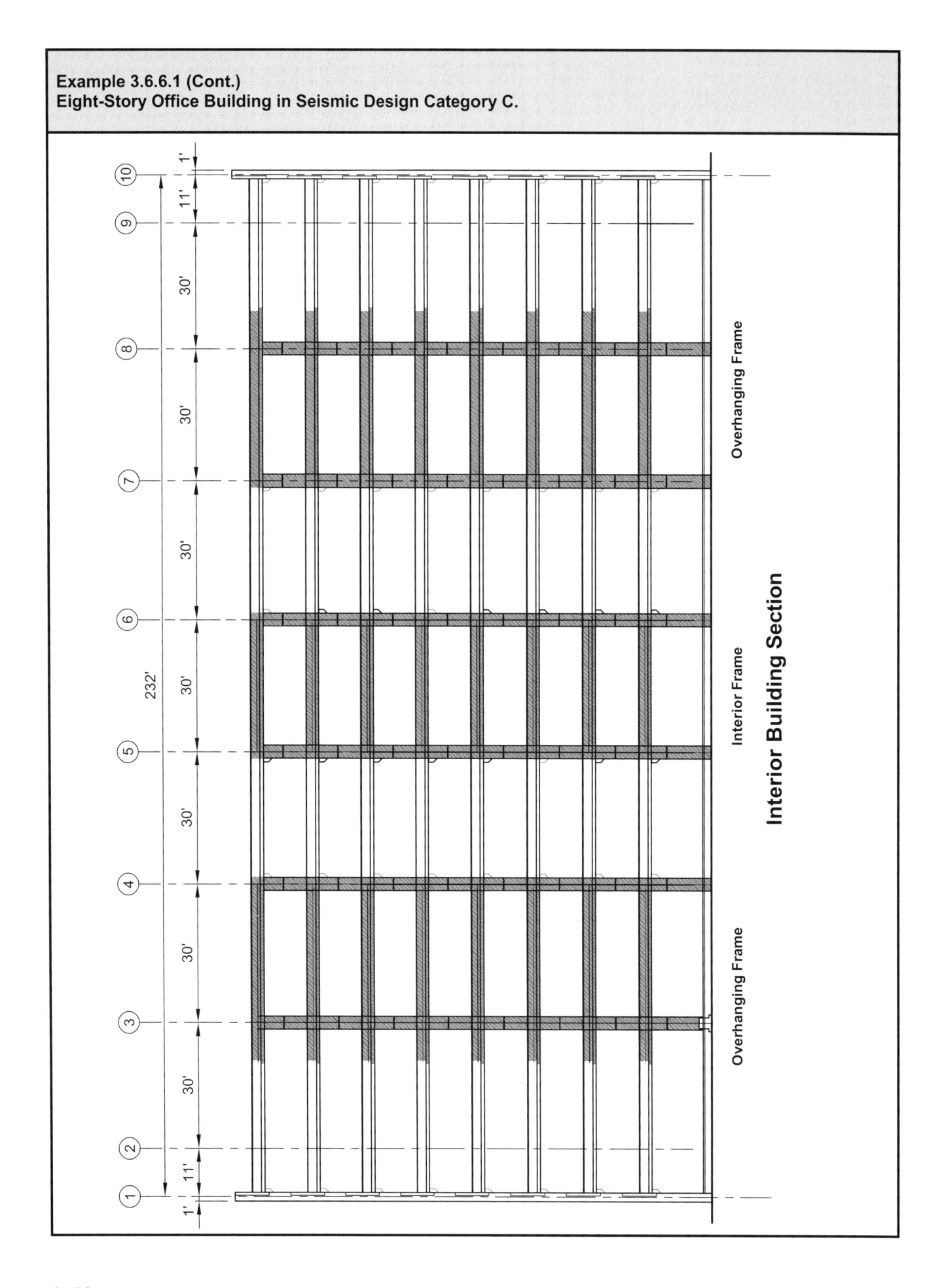

Example 3.6.6.1 (Cont.)
Eight-Story Office Building in Seismic Design Category C.

Interior Building Section

Example 3.6.6.1 (Cont.)
Eight-Story Office Building in Seismic Design Category C.

The vertical distribution is shown in the table below.

Level, x	W_x, kips	h_x, ft	h_x^k, ft	$w_x h_x^k$	C_{vx}	Applied force based on Strength Limits V, kips	Applied force based on Strength Limits F_x, kips	Applied force based on Drift limits V, kips	Applied force based on Drift limits F_x, kips
Penthouse	6,382	106	391	2,496,596	0.27	227	227	118	97
7	5,600	93	331	1,852,875	0.20	396	168	198	72
6	5,600	80	273	1,528,070	0.17	535	139	258	59
5	5,600	67	217	1,217,766	0.13	646	111	300	47
4	5,600	54	165	923,958	0.10	730	84	328	36
3	5,600	41	116	649,458	0.07	789	59	344	25
2	5,600	28	71	398,612	0.04	825	36	352	15
1	5,727	15	32	183,368	0.02	841	17	354	7
	45,709		0	9,250,703			841		359

Rayleigh's Formula used to calculate the period, T, is shown on following page.

Define the structural model of the frames of analysis.
  Detailed elevations of the outside and inside frames are shown on the previous pages.
  Although these six frames are independent, it is important in making the analysis to link at least the frames in a single plane together for analysis. Since the outside frames are not symmetric, they will tend to sway under gravity loads. To balance one end against the other and to reflect the rigid diaphragm between them, they are modeled together in one plane with link or truss elements representing the simple span beams between them. The simple beams rest on corbels without moment restraint.

Calculate the horizontal distribution of forces to each frame.
  The story shears are shared by six frames. With the frames on interior column lines for this example, it is assumed that the torsional effects will be taken by the orthogonal framing. Each line of frames, then, will resist half the story shear. The seismic load combinations, using Eqs. 3.2.6.5a and 3.2.6.7a with $S_{DS} = 0.418$, will be:

$$U = (1.2 + 0.2S_{DS})D + \rho Q_E + f_1 L + 0.2S = [1.2 + 0.2(0.418)]D + 1.0Q_E + 0.5L + 0.2(0)$$
$$= 1.28D + Q_E + 0.5L$$
$$U = (0.9 - 0.2S_{DS})D + \rho Q_E + 1.6H = [0.9 - 0.2(0.418)]D + 1.0Q_E + 1.6(0) = 0.816D + Q_E$$

Perform the stiffness analysis for the frames.
  The size of the columns is made 30 in. x 30 in. The inverted tee beams are 30 in. wide at the stem and 36 in. deep. To account for the effects of cracking, the effective moment of inertia of the beam is taken as $0.5I_g$. Based on an evaluation of gravity loads and approximate moments, the columns are uncracked except at the very top of the frames, and so the column section was modeled with the gross section properties. This can be checked with the analysis results. Since the columns and beams share common dimensions at the joints, the stiffness model includes rigid end sections or member end offsets to reflect the clear spans of the columns and the beams.

Evaluate frame displacements.
  For this special moment frame example, the near-symmetry of the system cause very small gravity load displacements. The following table provides the displacements and drifts from analysis using strength-level lateral loads. Design displacements are the elastic displacements times the C_d factor, 5.5.

Example 3.6.6.1 (Cont.)
Eight-Story Office Building in Seismic Design Category C.

Level	Calculated Displacement, in.	Amplifier C_d	Design Displacement, in.	Story Drift in.	Story Drift percent
Penthouse	1.75	5.5	9.63	0.55	0.31
7	1.65	5.5	9.08	0.88	0.56
6	1.49	5.5	8.20	1.10	0.71
5	1.29	5.5	7.10	1.32	0.85
4	1.05	5.5	5.78	1.43	0.92
3	0.79	5.5	4.35	1.54	0.99
2	0.51	5.5	2.81	1.49	0.95
1	0.24	5.5	1.32	1.32	0.85

Although the drift calculated is less than 1% at all levels, the drift for evaluating the stability can be determined from forces based on the calculated building period without the upper bound limit and without the lower bound force threshold. The displacements from this analysis can be used to calculate the period using Rayleigh's formula, Eq. 3.2.4.7. The following table provides that calculation:

Level, x	F_x	δ_x	δ_x^2	w_x	$w_x \delta_x^2$	$P_x \delta_x$
Penthouse	227	1.75	3.0625	6,382	19,544.88	397.25
7	169	1.65	2.7225	5,600	15,246.00	278.85
6	139	1.49	2.2201	5,600	12,432.56	207.11
5	111	1.29	1.6641	5,600	9,318.96	143.19
4	84	1.05	1.1025	5,600	6,174.00	88.20
3	59	0.79	0.6241	5,600	3,494.96	46.61
2	36	0.51	0.2601	5,600	1,456.56	18.36
1	17	0.24	0.0576	5,727	329.88	4.08
					67,998.00	1,184.00

g = acceleration due to gravity = 386 ft/sec²

$$T = 2\pi \sqrt{\frac{\sum_{i=1}^{n} w_x \delta_x^2}{g \sum_{i=n}^{n} P_x \delta_x}} = 2\pi \sqrt{\frac{67998}{386(1184)}} = 2.42 \text{ sec}$$

$$C_s = \frac{S_{D1}}{(R/I)T} = \frac{0.152}{(8/1)2.42} = 0.00785$$

$$V = C_s W = 0.00785(45709) = 359 \text{ kips}$$

Example 3.6.6.1 (Cont.)
Eight-Story Office Building in Seismic Design Category C.

This is the force used in the "Drift Limits" columns in the table on page 3–72.

Level	Calculated Displacement, in.	Amplifier C_d	Design Displacement, in.	Story Drift in.	Story Drift percent
Penthouse	1.04	5.5	5.72	0.39	0.25
7	0.97	5.5	5.34	0.55	0.35
6	0.87	5.5	4.79	0.72	0.46
5	0.74	5.5	4.07	0.83	0.53
4	0.59	5.5	3.25	0.83	0.53
3	0.44	5.5	2.42	0.88	0.56
2	0.28	5.5	1.54	0.88	0.56
1	0.12	5.5	0.66	0.66	0.38

Check the stability and P-Δ effects

It is required to consider P-Δ effects when the stability coefficient, θ, is greater than 0.10. The coefficient is defined in Eq. 3.2.4.16. The maximum of this coefficient is set by Eq. 3.2.4.17. For a special moment frame with $C_d = 5.5$, the limit is:

$$\theta = \frac{0.5}{\beta C_d} = \frac{0.5}{5.5} = 0.091 \text{ when } \beta = 1.0$$

β is the ratio of actual shear strength to the shear strength at any given level required by analysis and must be equal to or greater than 1.0. Since the moment in the beams at the face of the column govern the shear strength at a level, this is effectively the excess moment strength provided in the beam at this joint. For special moment frames, then, the P-Δ effect limit is greater than the system limit unless additional strength is provided in the beam-column joint.

The stability coefficient needs to be checked at each level. The following table provides this calculation from the initial trial analysis:

Level	P_x	Δ	V_x	h_{sx}	C_d	θ	Req'd β
Penthouse	7,190	0.39	118	156	5.5	0.028	0.30
7	13,600	0.55	198	156	5.5	0.044	0.48
6	20,010	0.72	258	156	5.5	0.065	0.72
5	26,410	0.83	300	156	5.5	0.085	0.94
4	32,830	0.83	328	156	5.5	0.097	1.06
3	39,240	0.88	344	156	5.5	0.117	1.29
2	45,650	0.88	352	156	5.5	0.133	1.46
1	52,190	0.66	354	174	5.5	0.102	1.12

From this check, it is found that even with low story drift (≈ 0.5%) the calculated θ exceeds the limit at the lower four levels unless additional strength is added to the beam-column joint. At the second level, this excess capacity requirement exceeds 40%. This indicates that even when the drift is low, if the lateral forces determined from actual period calculations are very low in comparison with the gravity loads at a level, additional strength or stiffness will be required to protect the structure.

Iteration of Design – Adjust Stiffness

The evaluation of the initial analysis indicates that the lower levels need additional stiffness to resolve the stability issue. This stiffness can be added by adjusting the columns, the beams or both. For this example, stiffness was added by increasing the depth of the inverted tee beams in the frame to 3 ft 6 in. at the lowest three levels. Again, the effective stiffness was taken as $0.5I_g$. The axial load and moments on the columns was checked and the uncracked section was verified, so the I_g remained the effective column section. From this revised model, new displacements were determined. The building period for the updated model was not

Example 3.6.6.1 (Cont.)
Eight-Story Office Building in Seismic Design Category C.

significantly different from the first trial. The following tables provide the updated calculation of drift and the updated evaluation of the stability coefficient:

Level	Calculated Displacement, in.	Amplifier C_d	Design Displacement, in.	Story Drift in.	Story Drift percent
Penthouse	0.853	5.5	4.69	0.38	0.25
7	0.784	5.5	4.31	0.54	0.35
6	0.685	5.5	3.77	0.67	0.43
5	0.563	5.5	3.10	0.74	0.48
4	0.428	5.5	2.35	0.69	0.44
3	0.302	5.5	1.66	0.62	0.39
2	0.190	5.5	1.05	0.58	0.37
1	0.085	5.5	0.47	0.47	0.27

Level	P_x	Δ	V_x	h_{sx}	C_d	θ	Req'd β
Penthouse	7,190	0.38	118	156	5.5	0.027	0.30
7	13,600	0.54	198	156	5.5	0.043	0.48
6	20,010	0.67	258	156	5.5	0.061	0.67
5	26,410	0.74	300	156	5.5	0.076	0.83
4	32,830	0.69	328	156	5.5	0.080	0.88
3	39,240	0.62	344	156	5.5	0.082	0.91
2	45,650	0.58	352	156	5.5	0.088	0.96
1	52,190	0.47	354	174	5.5	0.072	0.80

The revised analysis gives drifts at every level less than 0.5%. The values of θ are all less than 0.901 and there is no requirement for added story shear strength ($\beta < 1.0$).

Evaluate component and connection forces for design feasibility.
With the columns and beams sized to meet the drift and stability requirements, the model is re-run with the design lateral loads and load combinations. A brief review of the column base reaction forces and moments as well as typical column and beam forces finds values that are well within the capacity of the sections. The columns have base reactions that vary from 1500 to 3000 kips with moments that range up to 800 kip-ft, within the capacity of the 30 in. x 30 in. sections assumed. The beam moments in the deeper beam sections range up to 1,700 kip-ft, requiring 7 – #11 bars in the top of the 30 in. wide beam stem.

Detailed design of connections and components.
The detailed design of components and connections is covered in Chapters 4 and 6. The design of the frame elements of special moment frames requires detailing considerations that need to be applied to this system. Some of these requirements are:
- The negative moments in the beams at the face of the columns are expected to be the trigger for the primary yield mechanism. The reinforcement here is determined by the structural analysis.
- The reinforcement ratio in the flexural members may not exceed 0.025.
- At least two bars are required to be provided continuously top and bottom.
- The positive moment strength at the joint face must not be less than half the negative moment strength.
- Transverse reinforcement (stirrups and ties) is prescribed so that the inelastic regions are confined and shear strength is ensured. Column ties must be continued through the joint.
- The sum of the flexural strength of the columns above and below a joint must be 6/5 times the sum of the beam moment strength at the joint. (Here the "H" frame approach is favorable since there is a beam on only one side.)
- Ties and stirrups must be detailed with 135-degree hooks.

3.6.7 Special Moment Frames for High Seismic Design Categories

Precast moment frames in Seismic Design Categories D, E and F are required to be special moment frames. With the increasing lateral forces from higher accelerations, larger portions of the precast framing must be engaged than were needed in Example 3.6.6.1. The same office layout is used to illustrate strategies for resisting the increased loading. Two alternatives might be considered. First, the strength and stiffness of the interior frames can be increased by changing the interior simple span beams in the previous example to drop-in beams with strong connections made at the inflection points. That is, the "H" frames are panelized with overhanging beams projecting from both sides. Between the frames, simple-span inverted tee beams are erected for the immediate support of the floor. After the floors are set, connections are made at the ends of the drop-in beams to create continuity in the frame for lateral loads.

The second alternative is to develop moment frames on all four column lines across the structure. This alternative would place "H" frames along the outside column lines. The beams in these frames would be similar to the interior bays. The exterior would then be architectural precast concrete cladding that does not participate in the SFRS. It is not permissible to use the spandrel beams even if they are made monolithic with the columns, because ACI 318-02, Section 21.3.1.2 requires that the clear span of the beam be not less than four times the depth. The intent of this requirement is to provide the opportunity for plastic hinges to form in the beams.

Example 3.6.7.1
Eight-Story Office Building in High Seismic Design Category

Given:
An office building similar to that designed in Examples 3.6.5.1 and 3.6.6.1 is selected, and shown on the following pages.

Problem:
Adapt the framing used in the previous examples to a high seismic design category.

Solution:
For this example, the first alternative described in Section 3.6.7, and shown below, will be developed with frames on the interior column lines.

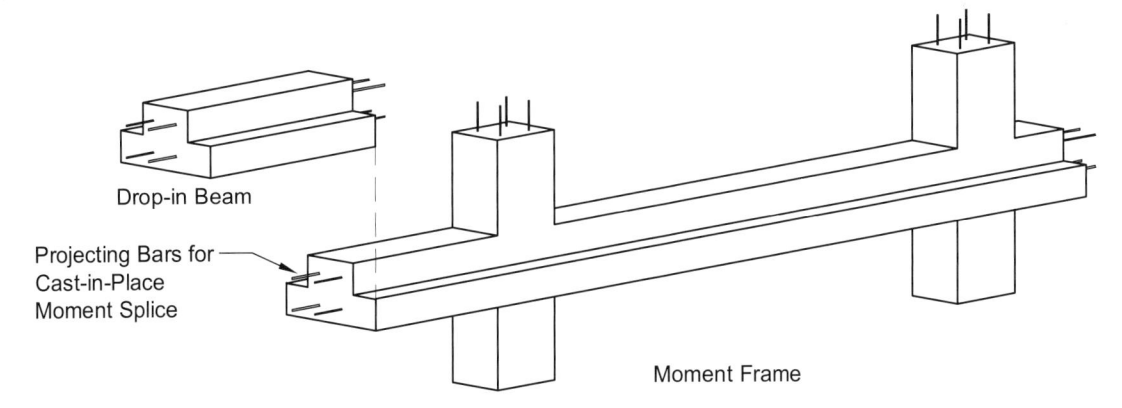

The gravity loads are the same as for Examples 3.6.5.1 and 3.6.6.1.

Seismic Coefficients.

1. For this example, a site in Seattle, WA, is assumed to determine mapped spectral acceleration values. The short period value, S_s, is 1.50 and the 1-sec period value, S_1, is 0.5.
2. The default Site Class D is used for the example.
3. The MCE spectral response accelerations must be adjusted for the site class effects. From Figure 3.10.7, $F_a = 1.0$ and $F_v = 1.5$. Therefore:

Example 3.6.7.1 (Cont.)
Eight-Story Office Building in High Seismic Design Category

$S_{MS} = F_a S_s = 1.5 \times 1.0 = 1.50$
$S_{M1} = F_v S_1 = 0.5 \times 1.5 = 0.75$

4. The design values are two-thirds of the MCE values:
   $S_{DS} = 2/3 (S_{ms}) = 1.00$
   $S_{D1} = 2/3 (S_{m1}) = 0.50$

5. The approximate building period is the same as calculated in the previous examples, T = 1.06 sec.

6. It was shown in Example 3.6.6.1 that where special precast moment resisting frames are used and limited to only those locations needed for the required lateral loads, the overall building stiffness is lower and the building period is longer than for comparable cast-in-place concrete structures. In the previous example, the building period exceeded the maximum period allowed for strength design by a large percentage. In an attempt to reduce the number of design iterations, it will be assumed that the period of this frame is longer than the maximum permitted for strength design. For $S_{D1} = 0.5 > 0.4$ in Table 3.2.4.2, the value of C_u is 1.4. $T = C_u T_a = 1.4 \times 1.06 = 1.49$ sec. With this period, the equation for the lower bound load threshold again governs the design, $C_s = 0.044$. The interpolated value of "k" (Section 3.2.4.2) = 1.49.

7. Calculate the lateral force from Eq. 3.2.4.1:

   $V = C_s W = 0.044(45,709) = 2011$ kips

   Since the structure is in Seismic Design Category D, the reliability or redundancy factor, ρ_x, must be determined by Eq. 3.2.4.20. There are 12 columns in the moment frames that resist east-west seismic forces, including two end columns.
   Thus, $r_{max} = 2/10(0.7) = 0.14$. The floor area, A_x, is $230.67(147) - 4(10.33^2)/2 \approx 33,500$ sq ft.

$$\rho_x = 2 - \frac{20}{r_{max}\sqrt{A_x}} = 2 - \frac{20}{(0.14)\sqrt{33,500}} = 1.22$$

Level	w_x, kips	h_x, ft	h_x^k, ft	$w_x h_x^k$	C_{vx}	Applied force based on Strength Limits V, kips	ρF_x, kips	Applied force based on Drift Limits* V, kips	ρF_x, kips
Penthouse	6,382	106	1,042	6,647,564	0.29	584	584	517	517
7	5,600	93	857	4,799,850	0.21	1,005	421	880	362
6	5,600	80	685	3,835,236	0.17	1,342	337	1,159	280
5	5,600	67	526	2,944,689	0.13	1,600	259	1,365	206
4	5,600	54	381	2,135,279	0.09	1,788	187	1,507	142
3	5,600	41	253	1,416,563	0.06	1,912	124	1,595	88
2	5,600	28	143	802,516	0.04	1,983	70	1,641	46
1	5,727	15	57	323,820	0.01	2,011	28	1,657	16
	45,709			22,905,517					

* Rayleigh's formula is used to calculate the period, T, shown on page 3–74.

The initial size of the columns is made generous for strength and control of drift, 30 in. x 48 in. The inverted tee beams are 30 in. wide at the stem and 42 in. deep. The moment of inertia of the columns is taken as $0.8I_g$ and of the beams, $0.50I_g$. Since the columns and beams share their dimensions at the joints, the stiffness model includes an adjustment (offset) that reflects the clear spans of the columns and beams.

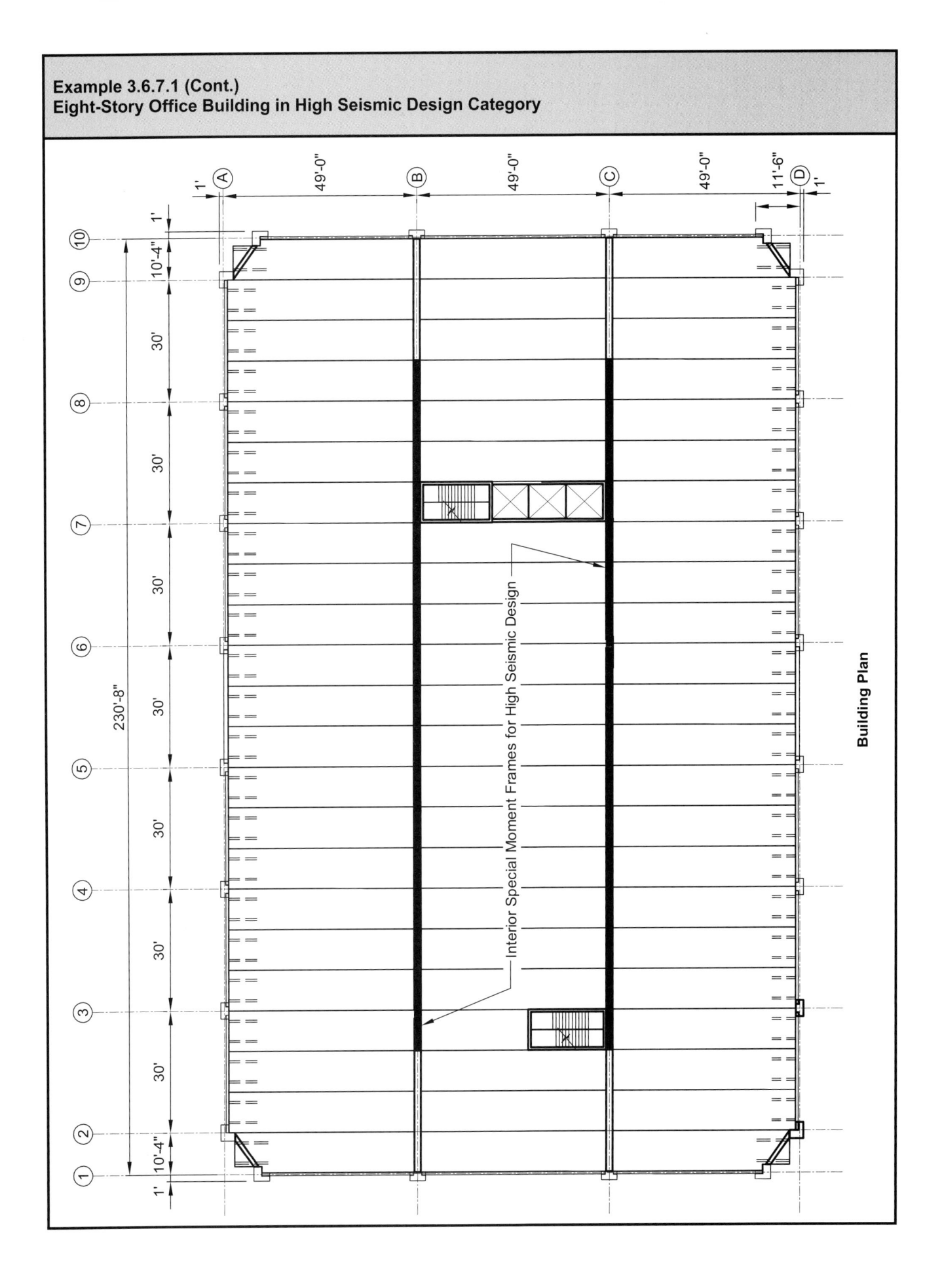

Example 3.6.7.1 (Cont.)
Eight-Story Office Building in High Seismic Design Category

Building Plan

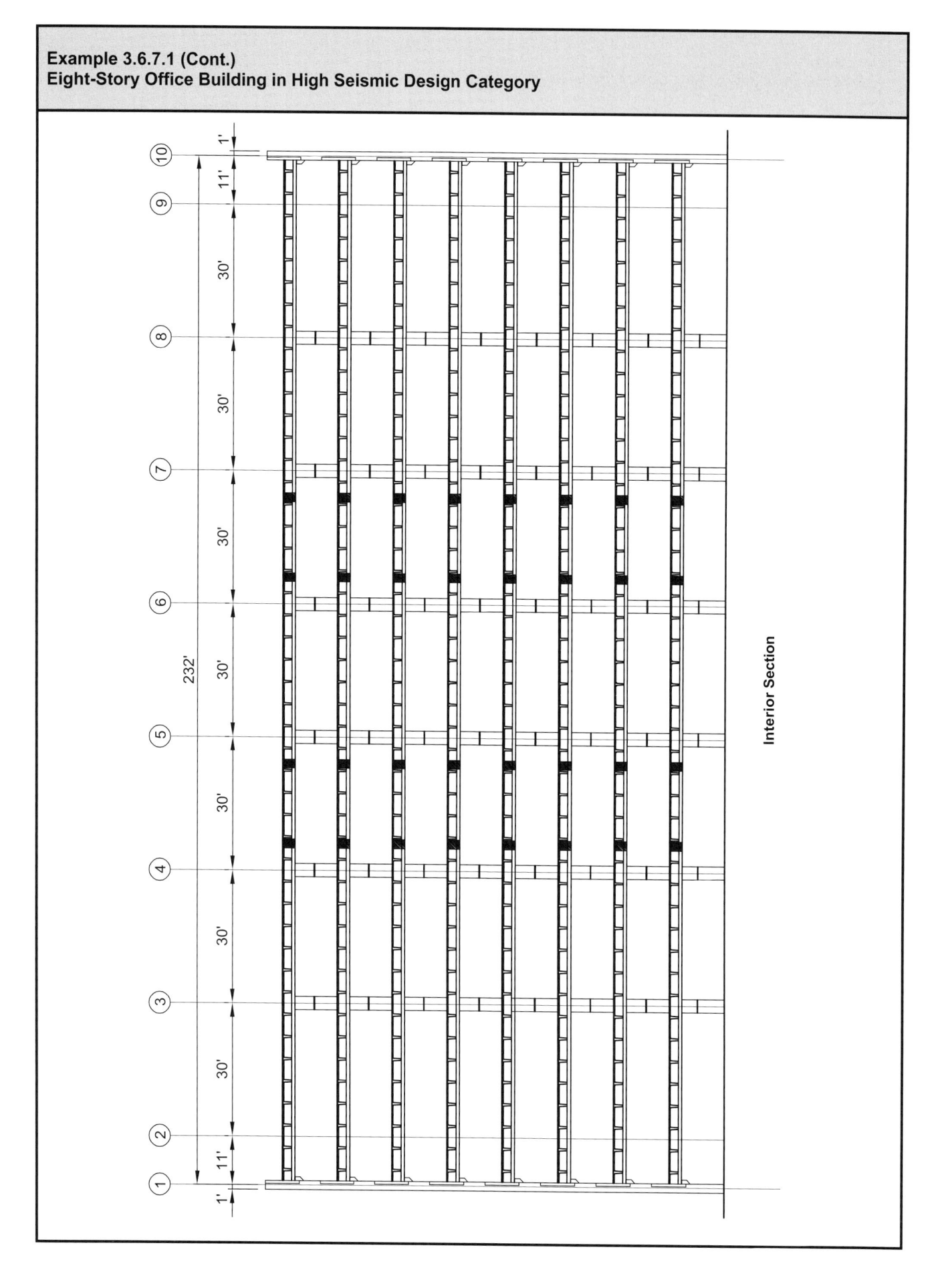

Example 3.6.7.1 (Cont.)
Eight-Story Office Building in High Seismic Design Category

Interior Section

Example 3.6.7.1 (Cont.)
Eight-Story Office Building in High Seismic Design Category

Calculate the horizontal distribution of forces to each frame.

The story shears are shared by the two interior frames. With the frames on interior column lines for this example, it is again assumed that the torsional effects will be taken by the orthogonal framing. Each line of frames, then, will take half the story shear. The load combinations are the same as reviewed before. With $S_{DS} = 1.0$, the net seismic combinations will be:

$$U = (1.2 + 0.2S_{DS})D + \rho Q_E + f_1 L + 0.2S = [1.2 + 0.2(1.0)]D + 1.0Q_E + 0.5L + 0.2(0)$$
$$= 1.4D + Q_E + 0.5L$$
$$U = (0.9 - 0.2S_{DS})D + \rho Q_E + 1.6H = [0.9 - 0.2(1.0)]D + 1.0Q_E + 1.6(0)$$
$$= 0.7D + Q_E$$

The computer analysis is run. With balance in the frames of this model, sway from the gravity loads is not significant.

Evaluate frame displacements.

Since the analysis was based on the C_s value calculated from the longest period allowed for strength design, the period needs to be checked. The floor displacements and applied lateral forces are used for this calculation. A table with calculations for period from Rayleigh's method follows:

Level, i	F_i	δ_i	δ_i^2	w_i	$w_i \delta_i^2$	$P_i \delta_i$
Penthouse	584	2.19	4.7961	6,382	30,608.71	1,278.96
7	421	2.05	4.2025	5,600	23,534.00	863.05
6	337	1.86	3.4596	5,600	19,373.76	626.82
5	259	1.60	2.5600	5,600	14,336.00	414.40
4	187	1.29	1.6641	5,600	9,318.96	241.23
3	124	0.95	0.9025	5,600	5,054.00	117.80
2	70	0.59	0.3481	5,600	1,949.36	41.30
1	28	0.24	0.0576	5,727	329.88	6.72
					104,505.00	3,590.00

g = acceleration due to gravity = 386 ft/sec^2

$$T = 2\pi \sqrt{\frac{\sum_{i=1}^{n} w_i \delta_i^2}{g \sum_{i=n}^{n} F_i \delta_i}} = 2\pi \sqrt{\frac{104,505}{386(3590)}} = 1.72 \text{ sec}$$

$$C_s = \frac{S_{D1}}{(R/I)T} = \frac{0.50}{(8/1)1.72} = 0.03634$$

$V = C_s W = 0.03634(45709) = 1661$ kips

This is the force used in the "Drift Limits" columns in the table on page 3–78.

This calculation shows that the calculated period is longer than the period used in the analysis, so the assumption was correct and the lateral forces used were appropriate.

A separate analysis is made using the calculated period to evaluate drift and stability limits. The force table above includes a column of the forces calculated from the C_s based on this longer period and the vertical distribution also reflects this period. The following table summarizes the displacements, amplified displacements and story drift from this analysis:

Example 3.6.7.1 (Cont.)
Eight-Story Office Building in High Seismic Design Category

Moment Splice in Beams

Moment Splices in Columns

Enlarged Elevation of Moment Frame

Example 3.6.7.1 (Cont.)
Eight-Story Office Building in High Seismic Design Category

Level	Net Displacement, in.	Amplifier C_d	Design Displacement, in.	Story Drift in.	Story Drift percent
Penthouse	1.85	5.5	10.17	0.66	0.42
7	1.73	5.5	9.53	0.94	0.60
6	1.56	5.5	8.59	1.21	0.78
5	1.34	5.5	7.38	1.43	0.92
4	1.08	5.5	5.92	1.60	1.02
3	0.79	5.5	4.38	1.65	1.06
2	0.49	5.5	2.68	1.60	1.02
1	0.20	5.5	1.12	1.10	0.63

The P-Δ and stability coefficient check is also calculated for this case.

Level	P_x	Δ	V_x	h_{sx}	C_d	θ	Req'd β
Penthouse	7,190	0.39	517	156	5.5	0.006	0.113
7	13,600	0.55	880	156	5.5	0.024	0.188
6	20,010	0.72	1,159	156	5.5	0.082	0.271
5	26,410	0.83	1,365	156	5.5	0.180	0.358
4	32,830	0.83	1,507	156	5.5	0.168	0.443
3	39,240	0.88	1,595	156	5.5	0.322	0.519
2	45,650	0.88	1,641	156	5.5	0.659	0.556
1	52,190	0.66	1,657	174	5.5	1.241	0.407

Evaluate component and connection forces for design feasibility.

A brief review of the column base reaction forces and moments as well as typical column and beam forces and moments finds that the values are within the design strength of the sections. The columns have base reactions varying from 1,500 to 3,000 kips with moments up to 2,500 kip-ft. These loads are within the design strength of the assumed sections. Beam end moments approach 2,300 kip-ft, requiring 10 – #11 bars in the 30 in. section. This may require that bars be bundled or be placed in two layers to fit, but the section is feasible. It indicates that the sections should not be reduced, even though the drift and stability values are low.

Detailed design of connections and components.

Some of the requirements of special moment frames were listed in Example 3.6.6.1. In that example, the connections between the precast "H" frames only required the use of splices in the columns at mid-height between floors. These splices can be made with commercially available hardware that has been tested to demonstrate Type 2 capability. The solution developed for this example requires additional connections to achieve continuity between the "H" frames. In ACI 318-02, a new Section 21.6 has been added to address the requirements for connections in special precast moment frames. This section has provisions for both ductile and strong connections.

The design developed here assumed strong connections would be used at the drop-in beam locations. These connections were located near the natural inflection points for continuous beams so that the moment demand on the connections would be low and they would be located away from the locations of ductile behavior. This requires that:

- The beam span for span-to-depth ratio limitations be the distance between locations of flexural yielding.
- The design strength of every strong connection be sufficient to develop the probable strength at the nearest location of flexural yielding.
- The primary longitudinal reinforcement must be made continuous across the connection so that it is developed away from the strong connection and the yield locations.

Column-to-column connections must develop 1.4 times the probable strength at the yield point. The connection moment strength must be at least 40% of the column moment strength within the story. This points to the use of Type 2 splices in the column-to-column connections.

3.7 Shear Wall – Frame Interaction

Rigid frames and shear walls exhibit different responses to lateral loads, which may be important, especially in high-rise structures. This difference is illustrated in Figure 3.7.1.

A frame bends predominantly in a shear mode as shown in Figure 3.7.1(a), while a shear wall deflects predominantly in a cantilever-bending mode as depicted in Figure 3.7.1(b). Elevator shafts, stairwells, and concrete walls normally exhibit this behavior.

It is not always easy to differentiate between modes of deformation. For example, a shear wall weakened by a row, or rows of openings may tend to act like a frame, and an infilled frame will tend to deflect in a bending mode. Also, the shear deformation of a shear wall can be more important than bending deformation if the height-to-length ratio is small, as discussed in Section 3.5.6.

If all vertical elements of a structure exhibit the same behavior under load, that is, if they are all frames or all shear walls, the load can be distributed to the units in proportion to their stiffnesses (see Section 3.5.7). However, because of the difference in bending modes, the load distribution in structures with both frames and shear walls is considerably more complex. Refs. 14 through 18 address this issue in detail.

3.8 Diaphragm Design

In structural systems, the framing of floor and roof components provides a function beyond the support of the gravity loads. The horizontal framing is designed as a diaphragm to collect and transmit lateral forces from wind or earthquakes to the vertical elements of the lateral force resisting system. It also provides lateral bracing to the vertical elements and protects them against abnormal loading.

Precast diaphragm systems are as varied as the components that form the horizontal framing. Hollow-core floors may be designed as diaphragms with perimeter reinforcing and grouted joints, or may be the form for reinforced cast-in-place topping. Double tee systems may include untopped or pretopped tees, pretopped tees with pour strips at the ends, or tees with cast-in-place topping.

In many precast structures, the configuration and behavior of the diaphragm may be very simple. Rectangular floors or roofs spanning between precast frames or walls provide connectivity and lateral load distribution and can easily be modeled as a deep horizontal beam. In other cases, the features of the structure may create more complex conditions. The features may include excessive horizontal spans between the vertical elements of the lateral force resisting system, large openings or discontinuities, large torsion effects from the eccentricity of the lateral force with respect to the center of stiffness, or lateral transfer requirements due to vertical discontinuities.

Figure 3.7.1 Deformation modes

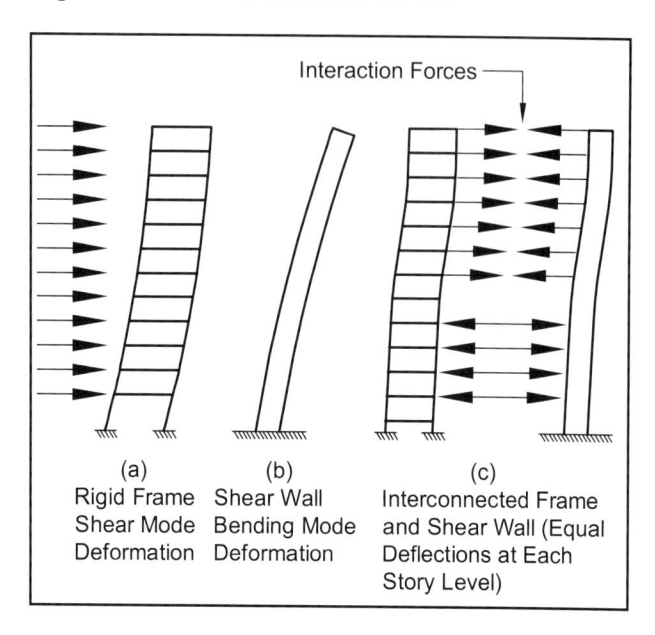

(a) Rigid Frame Shear Mode Deformation
(b) Shear Wall Bending Mode Deformation
(c) Interconnected Frame and Shear Wall (Equal Deflections at Each Story Level)

The performance of some precast parking structures during the 1994 Northridge earthquake prompted considerable research attention to the nature and behavior of diaphragms. Although this effort continues, considerable knowledge has been added to precast diaphragm technology in the past decade.

3.8.1 Simple Diaphragm Design – The Horizontal Beam Analogy

The diaphragm is analyzed by considering the roof or floor as a deep horizontal beam, analogous to a plate girder or I-beam. The shear walls or other lateral load resisting systems are the supports for this beam. As in a beam, tension and compression are induced in the chords or "flanges" of the analogous beam. The shear in the diaphragm is resisted by the "web" of the analogous beam. A diaphragm model using the analogous beam is shown in Figure 3.8.1.1.

3.8.1.1 Shear Transfer Between Members

In precast floors and roofs without composite topping, the individual components comprising a floor or roof diaphragm must be connected together to act as a single diaphragm. Joints between precast components, which are parallel to the lateral-force-

Figure 3.8.1.1 Analogous beam design of a diaphragm

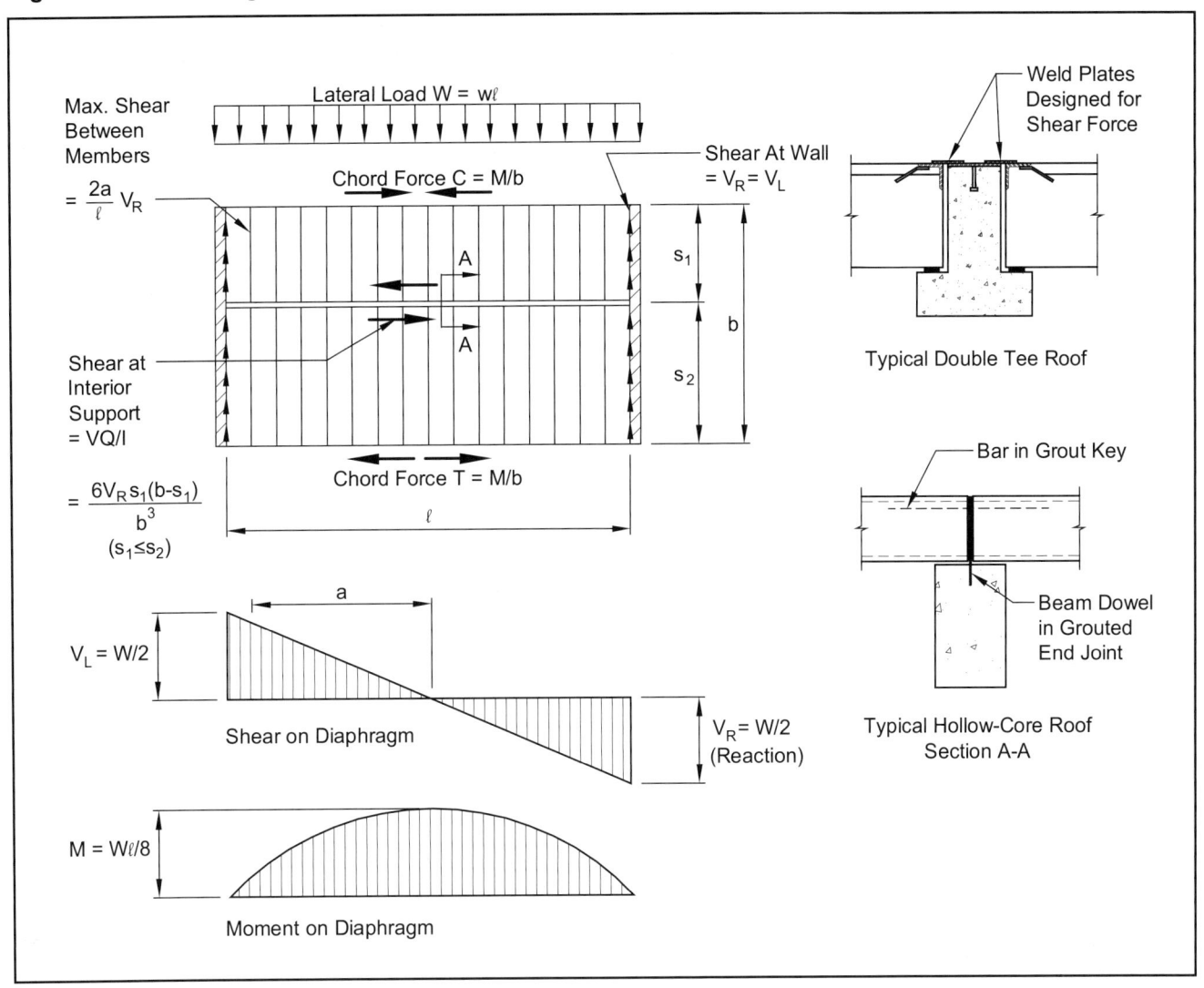

resisting system, must contain connections to resist the diaphragm shear forces as well as chord tension/compression forces at the edges of the diaphragm. Joints between the precast components, which are perpendicular to the lateral-force-resisting system, must contain connections to resist horizontal shear (VQ/I).

The types of connections used to connect precast components together to form diaphragms vary depending on the required connection strength, strain capacity to accommodate expected joint movement, and the preference of the precast supplier manufacturing and erecting the precast units. Three commonly used welded connections are shown in Figure 3.8.1.2.

Connections between members often serve functions in addition to the transfer of shear caused by lateral loads. For example, weld plates in flanged members are often used to adjust differential camber. Grout keys may be utilized in the joints to distribute concentrated loads.

Precast components may be fabricated with grout keys and connected by grouting the joint. For members connected by grout keys, a conservative value of 80 psi can be used for the design shear strength of the grouted key. If necessary, reinforcement placed as shown in Figure 3.8.1.3 can be used to transfer the shear. This steel is designed by the shear-friction principles discussed in Chapter 4.

In floors and roofs with composite topping, the topping itself, or in conjunction with the precast components, can act as the diaphragm, if adequately reinforced. Shear reinforcement can be determined by shear-friction analysis. Continuous reinforcement at the diaphragm boundaries is provided to resist chord tension forces.

Connections that transfer shear from the diaphragm to the shear walls or other lateral-force-resisting systems should be analyzed in the same manner as the connections between members. For

Figure 3.8.1.2 Typical flange weld plate details[1]

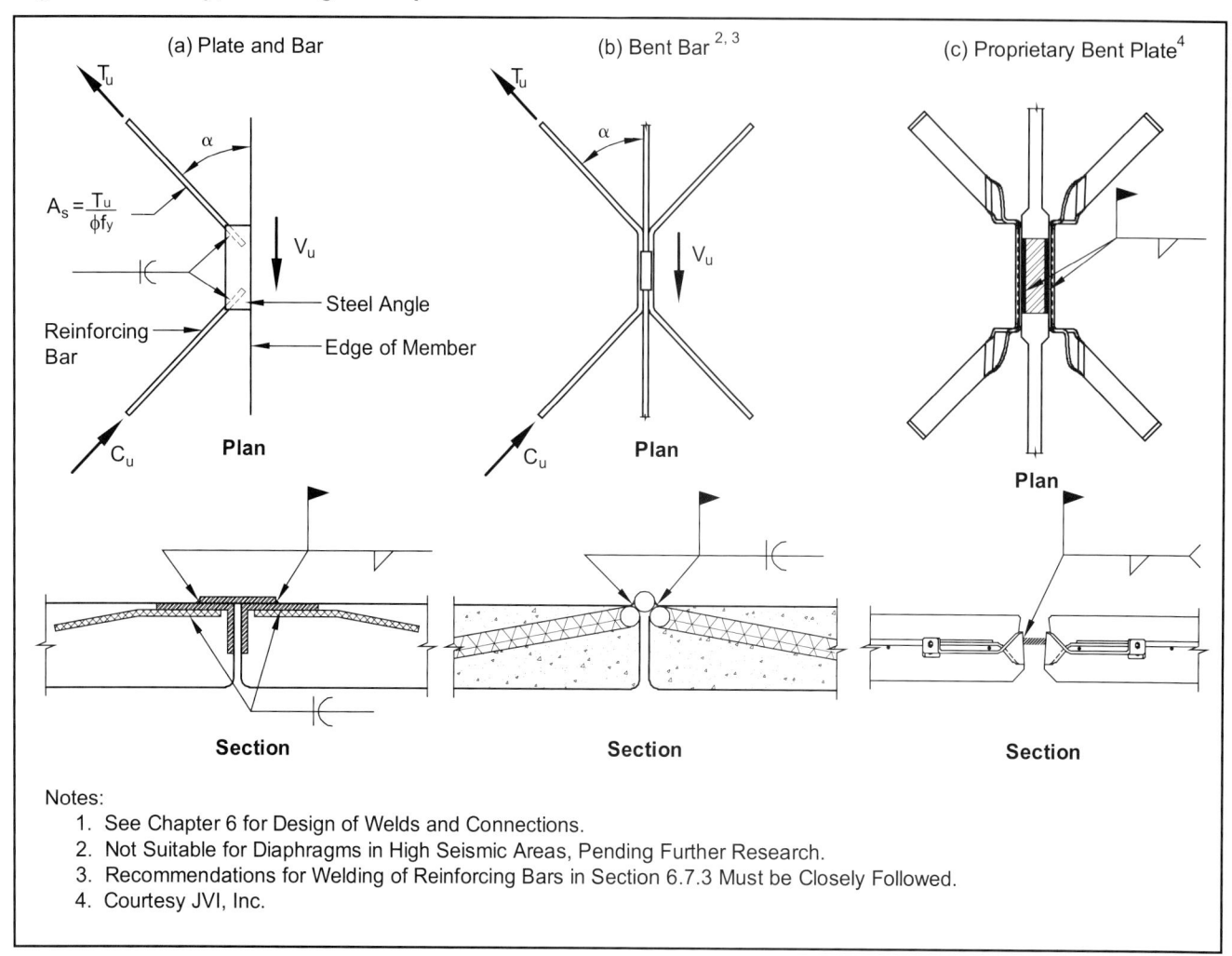

(a) Plate and Bar

$A_s = \dfrac{T_u}{\phi f_y}$

(b) Bent Bar [2,3]

(c) Proprietary Bent Plate[4]

Notes:
1. See Chapter 6 for Design of Welds and Connections.
2. Not Suitable for Diaphragms in High Seismic Areas, Pending Further Research.
3. Recommendations for Welding of Reinforcing Bars in Section 6.7.3 Must be Closely Followed.
4. Courtesy JVI, Inc.

Figure 3.8.1.3 Use of perimeter reinforcement as shear-friction steel

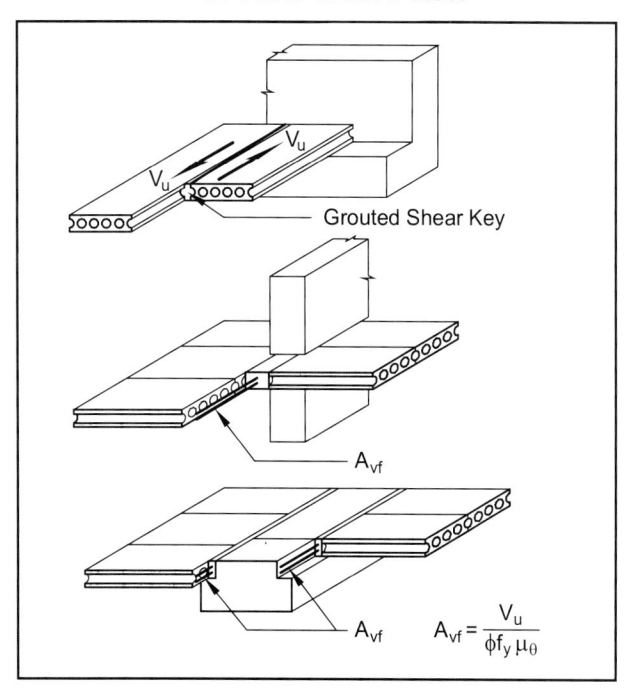

$A_{vf} = \dfrac{V_u}{\phi f_y \mu_e}$

rigid diaphragms, the reaction forces will be determined from the story shear with consideration of the maximum effects of torsion in the plane. For flexible diaphragms, the reactions are derived based on the tributary spans of the diaphragm between the vertical elements. Vertical elements that are within 5% of the length of the building perpendicular to their line of action to each other can be considered as the same line of resistance in this analysis.

3.8.1.2 Chord Forces

Chord forces are calculated as shown in Figure 3.8.1.1. For roofs with intermediate supports as shown, the shear stress is carried across the beam with weld plates or bars in grout keys as shown in Section A-A and Figure 3.8.1.1. Bars are designed by shear friction.

In decks consisting of flanged deck members, the chord tension at the perimeter of the building is usually transferred between members by reinforced topping or tension connections.

In all buildings, a minimum amount of perimeter reinforcement is required to satisfy structural integrity requirements (see Section 3.3). These minimum requirements may be more than enough to resist the chord tension.

3.8.2 Rigid and Flexible Diaphragms*

Since the inception of the precast, prestressed concrete industry, building structures have typically been designed using the assumption that the floor systems serve as rigid diaphragms between the vertical elements of the lateral force resisting system. A diaphragm is classified as rigid if it can distribute the horizontal forces to the vertical lateral load resisting elements in proportion to their relative stiffness. Close examination of the effective properties of diaphragms, along with long-span applications suggest that many precast diaphragms may in fact be flexible.

3.8.2.1 Defining Rigid or Flexible Diaphragms

A diaphragm is flexible for the purpose of distribution of story shear when the lateral deflection of the diaphragm under lateral load is more than twice the average story drift of adjoining vertical elements of the lateral force resisting system under equivalent tributary lateral loads. A rigid diaphragm is one that is not flexible. The distinction between rigid and flexible diaphragms is important not just for diaphragm design, but also for the design of the entire lateral force resisting system.

For structures with rigid diaphragms, the seismic design story shear is distributed to the various vertical elements of the lateral force resisting system based on the relative lateral stiffnesses of the vertical resisting elements. The general assumption is that the deformation within the diaphragm is not significant relative to that of the vertical system. This assumption implies that the diaphragm is capable of carrying loads to extreme points even when there are large differences in stiffness between individual vertical elements. It also implies that the deformation in the diaphragm does not have a significant effect on drift, so that gravity elements remote from the vertical lateral force resisting elements are not subject to significantly larger lateral displacements. These assumptions may not be conservative.

---

* This section was prepared by Ned M. Cleland of Blue Ridge Design. It relies heavily on the research work of Refs. 20 through 22 and others. He is indebted to Susanne Nakaki, Robert Fleischman, S.K. Ghosh, Jose Restrepo and Neil Hawkins for their contributions to the state-of-the-art of the subject. Research is continuing, and the reader is encouraged to keep abreast of developments.

For flexible diaphragms, the seismic design story shear is distributed to the various vertical elements based on the area of the diaphragm tributary to each line of resistance.

For systems with precast gravity load systems, some building code provisions have imposed an aspect ratio limit on precast diaphragms of 3 to 1 to ensure rigid behavior. It has been demonstrated in research studies that simply limiting the aspect ratio is not sufficient, since deflection is also a function of the actual span. Research by Fleischman et al. [20] has defined more precise parameters to identify flexible systems.

3.8.2.2 Behavior and Design Considerations

The behavior of diaphragms as rigid or flexible depends on many factors, including span, aspect ratio, jointing and connections. Consider a precast structure with shear walls at several lines of lateral support. Figure 3.8.3.1 shows a parking structure layout with stiff end walls and interior cruciform walls. The framing layout includes an interior ramp between Grid Lines 3 and 10, introducing a large interior discontinuity in the diaphragm.

It is not uncommon to use the walls as part of the gravity system so that the dead load of the structure helps resist overturning. Such a design could result in the interior cruciform walls being significantly less stiff than the end walls. For a rigid diaphragm assumption, the lateral forces can be applied as a combination of uniform and triangularly varying load in the plane to represent the effects of accidental torsional eccentricity.

Reactions can be calculated from the lateral distribution with proportional adjustments made for the magnitude of diaphragm loading. The determination of shears and bending in the statically indeterminate horizontal beams can be made by considering the interior walls as applying counteracting loads rather than as acting as supports. Since the interior wall forces in this example are low, due to lower relative stiffness, the moment diagram is close to one of a uniformly loaded beam spanning end to end. This is shown as an overlay to the framing on Figure 3.8.3.2. Since the ramp discontinuity separates each bay for most of the length, the bays share the loads as equal but separate beams. The required amount of chord reinforcement for flexure in such a system is excessive.

For the same system, the diaphragm may be designed as flexible. In this case, the equivalent beam is a continuous beam on four supports. The reactions to the walls are determined as continuous beam support reactions. The moment diagram for this assumption is quite different, as illustrated in

Figure 3.8.3.2. The maximum moments for this approach are considerably less than for the rigid diaphragm assumption. The chord reinforcement requirement is also much less.

Unfortunately, diaphragm design is not usually as simple as selecting one or the other approach. There is concern for local overloads. Redistribution of forces from overloaded elements requires some diaphragm rigidity. Flexibility can reduce the demand on the vertical elements and the shear and moments in the diaphragm, but actual rigidity inherent in the floor layout needed to meet the structure's function may make this reduction unsafe. If the design includes sufficient chord reinforcement to avoid yielding, the rigid model is safer because redistribution is ensured.

Unless a precast diaphragm clearly meets the conditions as rigid or flexible, design should consider the most severe effects of both assumptions as an envelope. For this to be economical, considerations will extend beyond the diaphragm design and include the vertical system. The best solution will be one where the stiffness of the vertical elements of the system produces reactions from a rigid diaphragm that are close to the reactions of a flexible model.

Steps in the design method should include:

1. An analysis with rigid diaphragm as a primary assumption.
2. A check of the distribution and diaphragm forces for a flexible assumption.
3. A comparison of the shear and moment results of the two approaches.
4. An evaluation of effective section properties and check of diaphragm deformation with respect to drift limits and the permissible deflection of the attached elements.
5. Adjustments in vertical element stiffness and placement to draw the results of analysis closer together and to limit drift to an acceptable magnitude.

3.8.3 Diaphragm Design Forces

Part of the challenge in the design of diaphragms is the selection of appropriate force criteria. For wind loads, this is straightforward. The combined windward and leeward wind pressures are considered a uniform load that must be collected at the building perimeter and distributed to the vertical elements of the lateral force resisting system. Code prescribed load factors are applied to ensure sufficient strength. When the design event is an earthquake, however, current code prescription for equivalent lateral forces may fail to adequately address dynamic and system effects under high seismic excitations. With the adoption of IBC 2003 [1] or NFPA 5000, [4] the lateral design of an overwhelming majority of precast structures in the United States will be governed by seismic criteria.

3.8.3.1 Code-Prescribed Forces

For the purpose of design, the lateral force resisting system is considered as the combination of vertical elements (walls or frames) and the horizontal diaphragms that distribute the loads to those vertical elements. The main wind force resisting system must be designed for vertical and horizontal loads based on the combined windward, leeward and uplift pressures calculated for appropriate wind speed, exposure, and gust response factors. The load used for diaphragm design is the same as used for the design of the vertical elements.

For seismic design, the current building codes require separate calculations of diaphragm design forces from those used in the design of the vertical element. The diaphragm design force is generally designated as F_p. This force cannot be less than the distributed force calculated in accordance with Section 3.2.4.2.

The requirements for diaphragm design forces for structures in Seismic Design Categories B or C from the IBC are in Section 1620.1.5:

$$F_p = 0.2I_E S_{DS} w_p + V_{px} \quad \text{(Eq. 3.8.3.1)}$$

where V_{px} is a term representing forces from above level that must be transferred through the diaphragm due to vertical system offsets or changes in stiffness. For Seismic Design Categories B and C, the diaphragm design force is not directly related to the base shear. The design force, related to the short-period acceleration coefficient S_{DS}, is reduced by the factor 0.2, reflecting consideration of ductility.

For Seismic Design Category D, IBC 2003 prescribes the diaphragm design forces:

$$F_{px} = \frac{\sum_{i=x}^{n} F_i}{\sum_{i=x}^{n} w_i} w_{px} \quad \text{(Eq. 3.8.3.2)}$$

$$0.2 S_{DS} I_E w_{px} \leq F_{px} \leq 0.4 S_{DS} I_E w_{px}$$

This equation follows the form of the UBC equation, but uses the NEHRP 97 forces as distributed vertically from the base shear. Again, these forces have been reduced by the response modification factor.

Figure 3.8.3.1 Typical parking structure plan

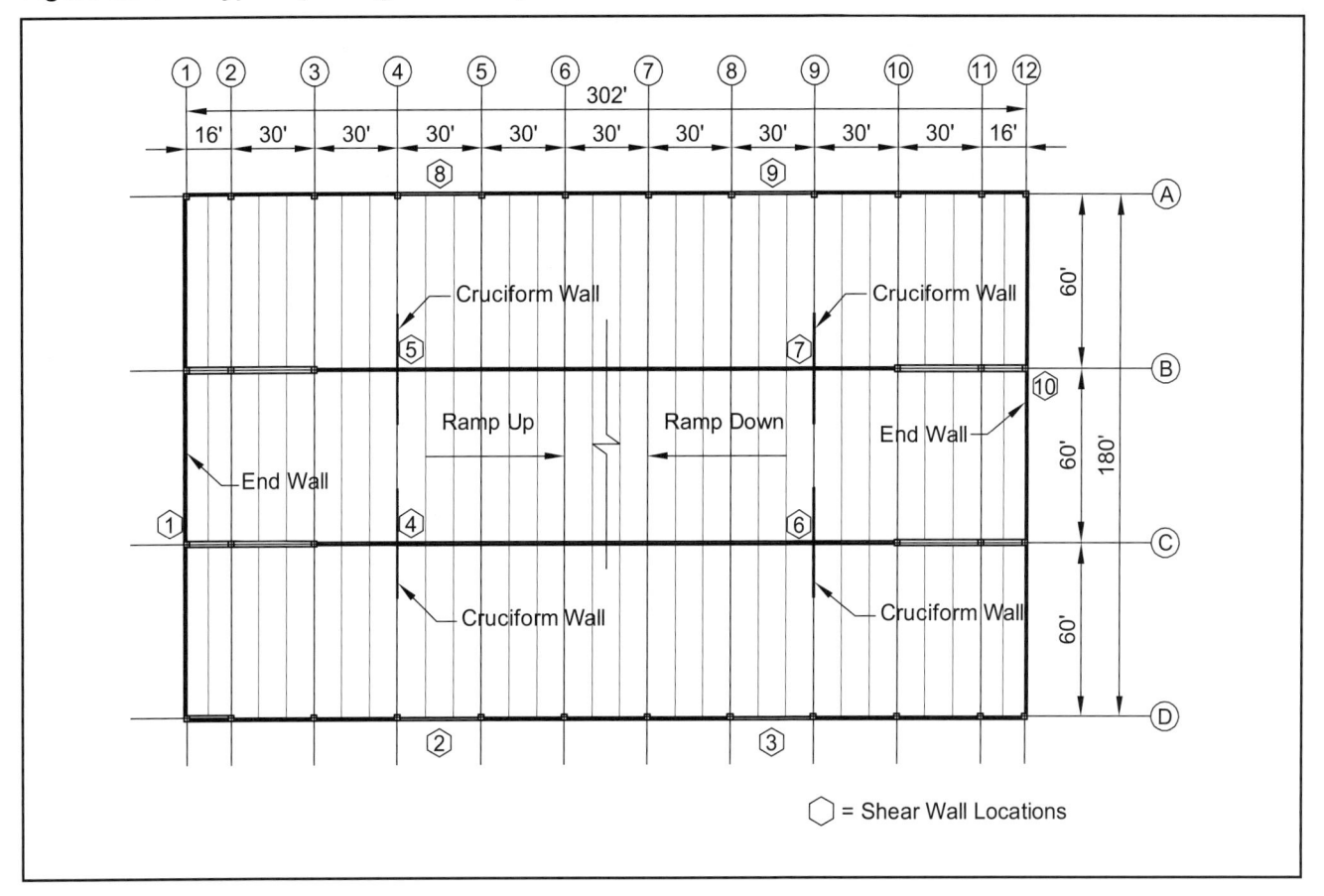

Figure 3.8.3.2 Comparison of rigid and flexible diaphragms

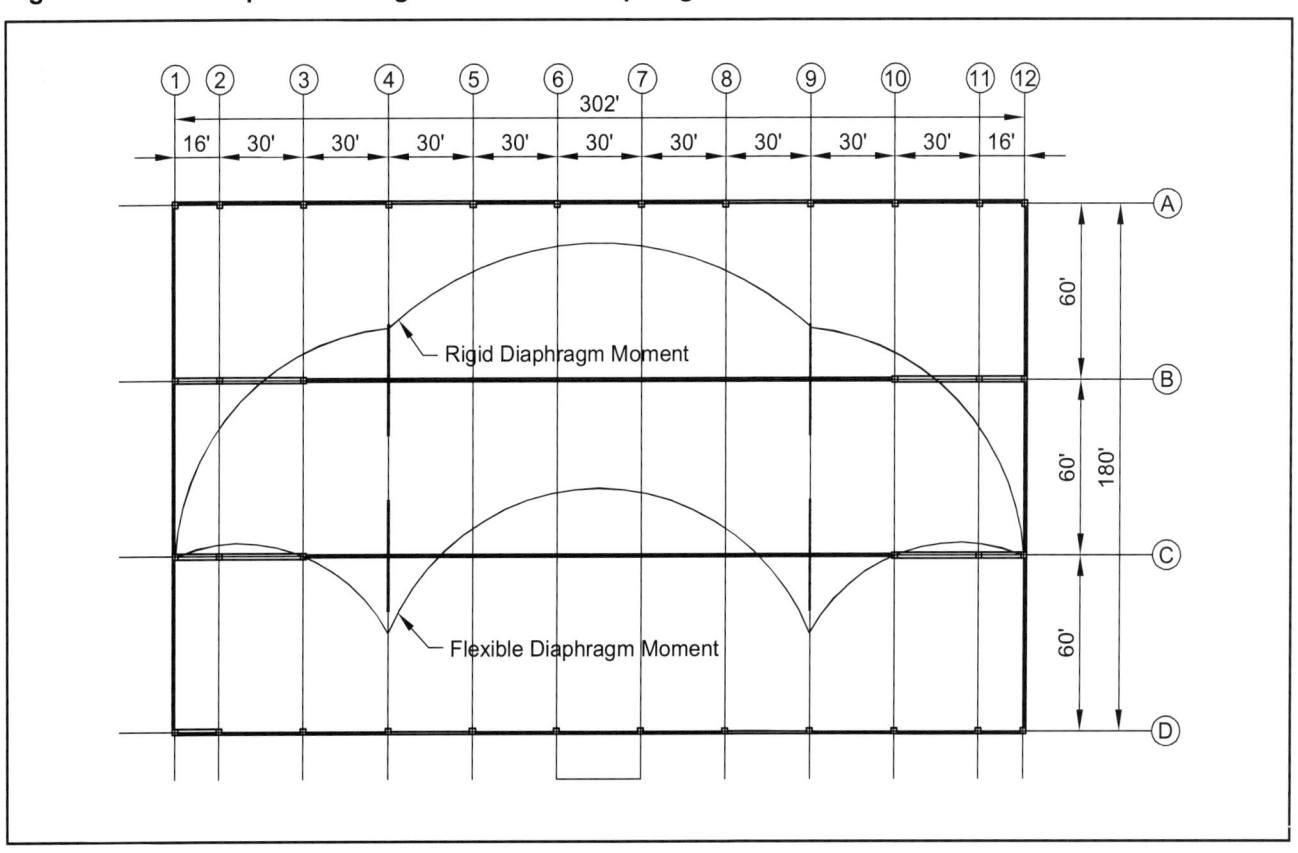

PCI Design Handbook/Sixth Edition

3–89

3.8.3.2 Elastic Design of Diaphragms

The forces used for design of the lateral force resisting system use reduction factors based on the inelastic capacity of the system. Each vertical system is assigned a response modification factor related to its ductility, strength and toughness. The assumption implicit in this approach is that the energy dissipation and post-yield deformation will be controlled by the characteristics of the vertical system. To be consistent, then, the diaphragm elements of the system need to have both the strength and the deformation capacity to ensure that the primary inelastic mechanism is developed. This means that the diaphragm may need to be designed to remain elastic as it develops and transfers forces to the vertical elements. For diaphragms assumed to act rigidly, there is a further reason to avoid yielding in the floor, as this may compromise the load paths that can redistribute loads to stiffer elements. The diaphragm must also continue to provide lateral support in the weak directions of vertical system elements and for elements not part of the lateral force resisting system.

The Code-prescribed diaphragm force provisions reviewed above do not require that elastic behavior in the diaphragm be generally maintained. Through IBC Section 1620.1.6, the special load combinations are required to be used in the design of collector elements and their splices, as discussed in Section 3.2.6.1.

In Eq. 3.2.6.11, the term Ω_o is the overstrength factor defined for each vertical lateral force resisting system shown in Figure 3.10.8, so this combination represents an approximation of the level of elastic forces. The code provisions appear to address the integrity of the diaphragm only through limits on deflection: "Permissible deflections shall be that deflection up to which the diaphragm and any attached distributing or resisting element will maintain its structural integrity under design load conditions, such that the resisting element will continue to support the design loads without danger to occupants of the structure." [1] It is not clear in the provisions whether these deflections include the inelastic behavior of the diaphragm. If so, no guidance on the deflection amplification is provided for any system. This provision also does not recognize that yielding in the diaphragm might create a failure mechanism without mobilizing the strength and ductility of the vertical elements of the LFRS. The code-prescribed design forces for diaphragms appear to contradict the implicit assumptions that are their basis.

To be consistent with the behavior inherently assumed in the prescribed seismic system parameters from the code, and to ensure the integrity of the system, simple diaphragms may be designed to remain elastic through the design event. Nakaki [22] lists additional benefits for providing for elastic design:

1. It eliminates the need to confine the concrete in the compression chord.
2. It prevents chord reinforcing from buckling when load is reversed and tension cracks close.
3. It improves bar splice performance without the need for lateral ties around the splice.
4. It eliminates uncontrolled inelastic diaphragm deformations.

Careful consideration of the function and details of diaphragms in precast concrete systems should reveal that design for elastic behavior is not simply a matter of using elastic design forces.

Floors and roofs composed of precast components are invariably jointed. These components may be double tees, hollow-core slabs or solid slabs. When the systems are dry (pretopped or untopped), the jointed nature is obvious. The components are connected to one another and to the supporting structure along the joints. When a cast-in-place topping is used, there is still a strong tendency for the diaphragm to behave in a jointed manner. The joints in precast members below the topping are planes of weakness similar to control joints in monolithic construction. Shrinkage strains tend to accumulate and form cracks in the topping above these joints. This is well recognized in parking structure design, so that joints are tooled in the wet topping above the precast joints to provide a regular line for the crack that can be sealed. These discrete joints, then, control the behavior of the diaphragm because the deformations consist almost entirely of the movements in these joints.

With precast diaphragms, it is common to differentiate between web and chord effects in the design of connections or topping reinforcement. With untopped double tees, the chord steel may be continuous reinforcing bars laid in pour strips at the ends of the tees, or welded connections that link continuous reinforcing bars embedded across the tee ends. The chord might use welded connections with tail bars that lap chord reinforcement in the flanges, but this approach has been penalized with a low capacity reduction factor ($\phi = 0.5$) in some codes based on older NEHRP recommended provisions. The web or flange connectors, then, are designed to transfer the diaphragm shear as well as to assist in the vertical alignment of the flanges. The most common design practice is that no shear capacity is attributed to the chord steel and no tension capacity is attributed to the flange connectors. As the diaphragm deforms, the opening of the joint is actually resisted by the tension capacity in both the chord and the web connections. This opening translates into tension strain concentrated in the small space of the joint that must be withstood by both types of connections. Web connections with high stiffness but low elastic strength may fail in tension and thus be unable to transfer loads in shear. Some connections have been designed and tested for their capacity to sustain much of their shear capacity in conjunction with the tension load and strain from joint opening. [24]

With hollow-core members, the chord is frequently made from continuous reinforcing bars placed in a cast-in-place strip beyond the ends of the slab. These bars also provide shear-friction steel to hold the joints together and permit shear transfer through the grouted joint.

When cast-in-place topping is used, the same discrete deformation at the precast joints will occur. The diaphragm reinforcement is usually provided as continuous reinforcing bars for the chords and welded wire reinforcement for the shrinkage and temperature steel that also acts as shear reinforcement. As the welded wire reinforcement crosses the joints, there is only the distance between the cross wires available to sustain the strain that results from the joint opening. In recognition of the vulnerability of this condition, ACI 318-02 includes the requirement that the spacing between the cross wires should not be less than 10 in. in order to allow more length to absorb the strain.

For the diaphragm to remain elastic, the design must consider the yield limits of web connections or reinforcement as well as chord connections and reinforcement. The chord steel may need to be increased beyond calculated moment requirements to limit joint strain to protect the web.

There is a growing consensus that the design for diaphragms to remain essentially elastic through the design earthquake while the vertical elements of the lateral force resisting system become inelastic is appropriate. The level of force above the code prescribed loads that is necessary to accomplish elastic performance is the subject of ongoing research. Preliminary indications from that research suggest that the overstrength factor for the system used (Ω_o) is more conservative than necessary. There is also concern that some common and practical configurations of diaphragms are particularly vulnerable at points of inelastic behavior. For these reasons, performance-based design may be more appropriate.

Example 3.8.3.1
Diaphragm Design

Given:
Four-level parking structure – Figure 3.8.3.1
302 ft long x 180 ft wide, three bays with central interior ramp
End walls – 48 ft high; Cruciform walls – 44 ft high

Seismic Design Category C

S_s = 0.43 S_1 = 0.095 S_{DS} = 0.418 S_{D1} = 0.152
C_s = 0.0836 V_{base} = 1910 kips

Floor weight	Force at floor level	Height above base
W_1 = 5850 kips	F_1 = 221 kips	h_1 = 12.5 ft
W_2 = 5740 kips	F_2 = 397 kips	h_2 = 23.0 ft
W_3 = 5740 kips	F_3 = 577 kips	h_3 = 33.5 ft
W_4 = 5410 kips	F_4 = 715 kips	h_4 = 44 ft

Problem:
Analyze diaphragm as both rigid and flexible, and compare the results. For this example, the accidental torsion for rigid diaphragms is not included to allow direct comparison of rigid and flexible diaphragm assumptions.

Solution:
Shear wall design – Initial:

Example 3.8.3.1 (Cont.)
Diaphragm Design

Total overturning moment (without torsion effect):
221(12.5) + 397(23) + 577(33.5) + 715(44) = 62,683 kip-ft
Resistance to overturning: Load factor = $0.9 - 0.2S_{DS} = 0.816$

End walls across center bay:	Full	Factored
Wall weight (10/12)(58)(48)(0.15)	= 348 kips	284 kips
Floor/Column loads	= 235 kips	191 kips

Moment = 284(30) + 191(60) = 19,980 kip-ft each end

Approximate additional resistance required = 62,683 kip-ft − (2)(19,980) = 22,723 kip-ft
Provided by four interior cruciform walls, 5,575 kip-ft/wall

Initial estimate for 24 ft long cross wall at cruciform

Dead load for wall group:	Full	Factored
Loadbearing wall (10/12)(20)(44)(0.15)	= 110 kips	90 kips
Non-loadbearing (est. initial)	= 132 kips	108 kips
Floor load (1200 sq ft × four levels)	= 408 kips	333 kips
Total	650 kips	531 kips

Moment = 531 × 12 = 6,372 kip-ft > 5,575 kip-ft: estimate ok.

Lateral Load Analysis:

Without torsional effect, lateral distribution for rigid diaphragm is simplified to direct distribution based on relative stiffness:

$$\frac{K_i}{E} = \frac{1}{\left(\dfrac{3h_j}{A_i} + \dfrac{h_j^3}{3I_i}\right)}$$

where:
h_j = height of Level j
A_i = shear area of Wall i
I_i = moment of inertia of Wall i

Take the stiffness of each end wall as that of a single wall from face to face of the columns, assuming that the connections to mobilize the dead load resistance to overturning are ductile (soft) connections. Take the stiffness of the cruciform wall as that of the flat wall in the direction of the loading only.
Distribution is done level by level:

	Shear	Overturning
End walls:	690.4 kips	23,244 kip-ft
Cruciform Walls:	124 kips	3,773 kip-ft

Lateral forces are distributed in proportion to the calculated dead load resistance to overturning. The interaction of vertical dead load and overturning moment may or may not result in tension at the base, and may require special tension connections. Wall-to-column connections can be made at the end walls that will mobilize the column dead load for increased resistance to overturning, without applying the column gravity loads to the wall. In this example, the solution is adequate for the diaphragm to be considered rigid.

Rigid Diaphragm Design:
The force required for diaphragm design is determined from Eq. 3.8.3.1:
$F_p = 0.2I_E S_{DS} w_p + V_{px}$

Example 3.8.3.1 (Cont.)
Diaphragm Design

For this example, $V_{px} = 0$, so
$F_p = 0.2(1.0)(0.418)5850 = 489$ kips

This is less than the maximum distributed horizontal force, 715 kips. In Seismic Design Category C, it is not necessary to apply the additional factor of 2.0 for overstrength. Thus:
$F_p = 715$ kips

A rigid diaphragm can be designed as a horizontal beam with reactions at the end walls and interior cruciform walls proportioned according to stiffness. The force was distributed to each level based on the stiffness of that level. The end walls at the roof include a parapet above the roof line, so they are somewhat stiffer than at the lower floors. Thus, the end reactions are greater at that level, placing a greater demand on the diaphragm. Separate analysis shows that 39.5% of the force is taken by each end wall and 5.25% by each of the four cruciform walls.

End wall resisting force: $715(0.395) = 282$ kips
Cruciform walls resist: $715(2)(0.0525) = 75$ kips per pair

Assuming a uniformly distributed force of $715/302 = 2.37$ klf, the resulting shear and moment diagrams are as shown below.

The lateral force effects can be divided equally into the three bays:
Moment per bay = $21,200/3 = 7067$ kip-ft

The moment arm of the ramped bay is approximately 55 ft
Chord force = $7067/55 = 128$ kips
$$A_s = \frac{F}{\phi F_y} = \frac{128}{0.9(60)} = 2.37 \text{ in.}^2. \text{ Use } 4 - \#7 \text{ bars.}$$

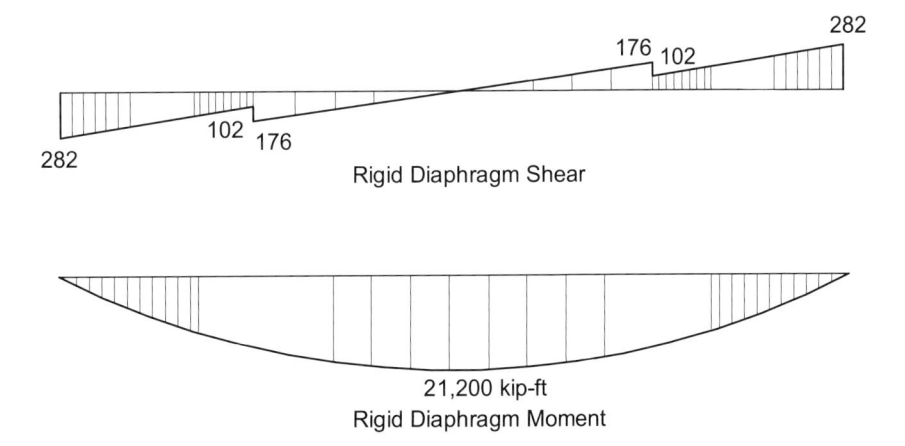

Rigid Diaphragm Shear

21,200 kip-ft
Rigid Diaphragm Moment

With this high moment, large chord forces and long effective span, there is a question about whether the diaphragm is truly a rigid diaphragm. For untopped diaphragms with the deformation strains concentrated at the joints and shear deformation from the connections, the effective stiffness may only be between 5% and 15% of the gross section. With heavily reinforced pour strips at the chord, 15% may be used. The deflection of the diaphragm calculated by assuming the properties vary from a three-bay section in the end bays to three separate sections at the ramp give a maximum mid-diaphragm displacement of almost 1.5 in. relative to the end walls. When the stiffness of the walls is used for estimating drift, it is appropriate to reduce the effective wall stiffness by 50% to 60% for the horizontal walls to account for joint opening at the horizontal joints. With

Example 3.8.3.1 (Cont.)
Diaphragm Design

the assumption of 50% of gross section, the calculated drift at the top of the walls is less than 0.1 in. The comparison of diaphragm deflection and wall drift shows that the diaphragm is flexible.

Flexible Diaphragm Analysis:
   A flexible diaphragm acts like a continuous beam across rigid supports. The diaphragm is not stiff enough to force the walls to have the same displacements. The analysis changes the shear and moment diagram for the diaphragm and the forces distributed to the shear walls.
   The continuous beam analysis includes the end bay sections that are the full width of the floor and interior bay sections that are the combination of the three bays. The section properties are taken as 15% of the gross section. The calculated displacement at the center of the garage is less than 0.1 in. The reactions on the walls at the ends are only 40.6 kips while the reactions at each interior support line (two cruciform walls) is 316 kips. The maximum shear in the diaphragm is 178 kips and the maximum moments are a positive 2,970 kip-ft at the middle and a negative 3,700 kip-ft at the interior wall supports, 1,235 kip-ft per bay. Shown below are the shear and moment diagrams. With a moment arm of 55 ft between the chord reinforcement or connections, the required chord force is 1235/55 = 22.5 kips and the steel area is 0.42 sq in., satisfied by 2 – #5 bars. Clearly, the design as a flexible diaphragm is more economical.

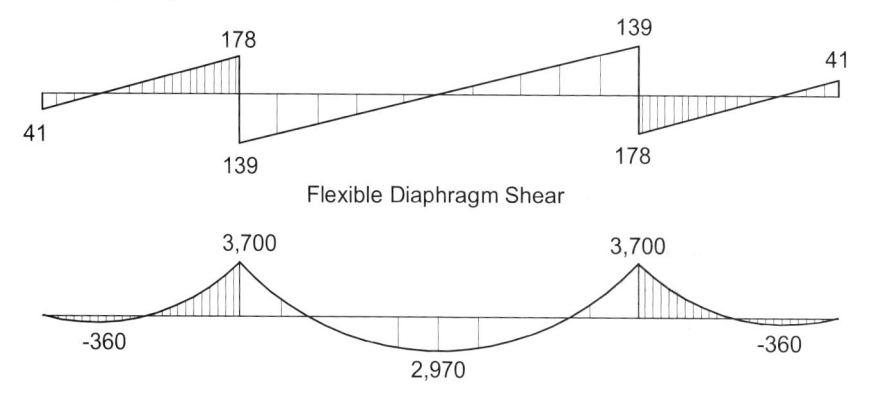

Flexible Diaphragm Shear

Flexible Diaphragm Moment

Flexible Diaphragm Forces and Shear Wall Analysis:
   The design of the diaphragm as flexible significantly reduces the reinforcement requirements, but it also changes the lateral load distribution to the walls. The analysis was made for the diaphragm force, but the wall forces can be derived by proportion from this analysis, as shown in the table:

		Proportion
Diaphragm Force	715	
End Wall Reaction	41	0.057
Cruciform Reaction	317	
Each Cruciform	159	0.222

Flexible Diaphragm Wall Force Distribution					
Level	1	2	3	4	Base
H	12.5	23.0	33.5	44.0	
Lateral Force at Level	221.0	397.0	577.0	715.0	
End Wall Force	12.6	22.6	32.8	40.7	108.6
End Wall Moment	157.0	519.0	1,099.0	1,789.0	3,563.7
Cruciform Force	49.0	88.1	128.0	158.6	423.7
Cruciform Moment	613.0	2,025.0	4,288.0	6,978.0	13,904.2

Example 3.8.3.1 (Cont.)
Diaphragm Design

The large change in wall loads and moments indicates the need for a revised wall design.

Shear Wall Design – Second Trial:
Based on the flexible diaphragm analysis, the end walls receive much less lateral force. This force may be easily resisted by the initial configuration. Connections from wall to column at the ends are not required to mobilize column dead load: 284(58/2) = 8,236 kip-ft > 3,564 kips. On the other hand, it should be more economical to reduce the end walls to match the lateral force requirement.

One solution is to keep the shear wall in the center of the middle bay and reduce its length. A 10 in. thick wall 36 ft long has a factored dead load of 162 kips and resistance to overturning of about 162 kips x 18 ft = 2,916 kip-ft. With non-loadbearing spandrel beams bearing on the wall ends, an additional resistance of about 620 kip-ft is developed, for a total of 3,536 kip-ft.

Another solution would be to add short walls on either side of the two end columns of the center bay with strong connections to act with each column. If the vertical walls were 4 ft wide, the total wall length would be 10 ft and the overturning resistance would be about 1,900 kip-ft each, 3,800 kip-ft total for the two columns.

The cruciform walls need to develop much greater resistance. The cross walls cannot project more than about 18 ft into the bay without interfering with the drive aisles of the parking layout. Walls 12 in. thick and 36 ft long connected to solid 10 in. loadbearing ramp walls have a dead load resistance to overturning of about 10,800 kip-ft, less than the 13,900 kip-ft demand. Additional capacity can be developed by connection to additional adjacent dead load or by positive reinforcement anchorage at the ends of the walls. Interaction design of the base indicates that 3 – #10 bars at the ends will develop sufficient capacity.

Lateral Force Analysis on Second Trial:
The assumptions of diaphragm stiffness are made to simplify complex behavior. They are generally considered conservative for considering the adverse effects of flexible conditions. On the other hand, there may also be problems if diaphragm behavior is more rigid than assumed. The diaphragm design, then, should be checked again for the rigid assumption. To do this, lateral force distribution to the revised walls must be updated. In the case where the central long wall is used at the ends, the stiffness of the 60 ft x 10 in. end wall is only a little less than each cruciform wall. Lateral distribution to the end walls is 276 kips shear and 9,040 kip-ft overturning moment. The distribution to each cruciform wall is 331 kips and 10,875 kip-ft. This distribution significantly exceeds the end wall capacity.

Where the end walls are designed as companion walls to the interior bay columns, the end wall stiffness is made small in comparison to the cruciform walls. Lateral distribution to the end walls is 17.9 kips shear and 503 kip-ft overturning moment to each wall. The distribution to each cruciform wall is 451 kips and 14,890 kip-ft. This distribution gives overturning 7% greater than the flexible distribution. With the top level wall forces used to proportion the diaphragm forces, the shear and moment diagrams are developed as shown below. The maximum moment, at the lines of the cruciform walls, is 5,800 kip-ft, a significant reduction from the first rigid diaphragm design. This will require chord reinforcement of 3 – #5 bars or 2 – #6 bars.

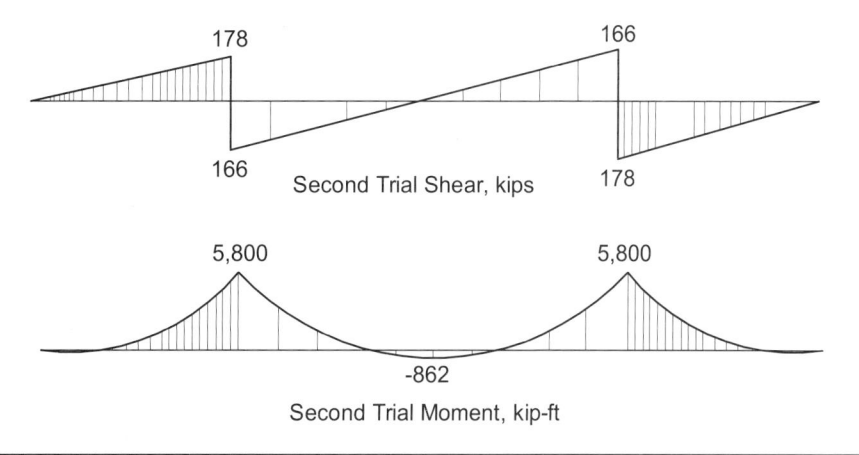

178 166
166 178
Second Trial Shear, kips

5,800 5,800
-862
Second Trial Moment, kip-ft

Example 3.8.3.1 (Cont.)
Diaphragm Design

This analysis indicates that the second option brings the rigid diaphragm solution closest to the flexible diaphragm condition. It also suggests that the design may be further refined by increasing the end wall stiffness beyond that required just for overturning.

Further Refinement – Trial 3:

To bring the rigid and flexible diaphragm solutions even closer together, the stiffness of the end walls should be increased. To increase the end wall stiffness, the total wall length is increased to 16 ft. With the stiffer walls, each end wall now has a base shear of 55.9 kips and an overturning moment of 1,666 kip-ft. Each cruciform wall has a base shear of 413 kips and an overturning moment of 13,728 kip-ft. For this solution, the shear and moment diagrams for the diaphragms are shown below. The maximum moment, at the cruciform walls, is now 3,600 kip-ft, slightly lower than the moment calculated for the flexible diaphragm. With this solution, the results from the flexible assumption and the rigid assumption are nearly identical.

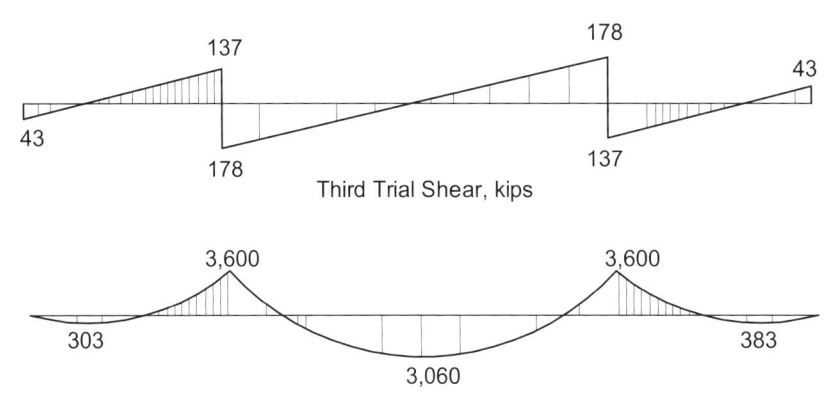

Third Trial Shear, kips

Third Trial Moment, kip-ft

Shear in the Joints and at the Walls:

The maximum shear from the analysis is 178 kips. This force is distributed across the three bays. With flange-to-flange connections at 5 ft spacing, there are thirty flange connections in each joint, discounting the space for the chord pour strips. This gives a requirement of about 6 kips per connection, well within the capacity of common commercial connection hardware.

The loads from the diaphragms must be transferred to the walls. The cruciform wall configuration is favorable because the highly loaded interior walls extend well into each of the bays. For these connections, IBC 2003 Section 1620.2.6 requires the application of the system overstrength factor, Ω_o, to the design forces. For shear walls, $\Omega_o = 2\frac{1}{2}$, so the maximum force to be carried to the cruciform walls in a line is 178 kips x 2.5 = 445 kips. For each walls in the line, 222 kips is required. Since the ramp bay is carried by walls on both sides, but the outside bays are connected only to walls on one side, the connections for the walls in the outside bays should carry two-thirds of the force or 148 kips. If common commercial inserts with 20 kips shear capacity are used, eight connections will be required on these walls, at a spacing of about 2 ft. Fewer connections are required at other locations.

Since the flange of the tee adjacent to the shear walls act as a collector, reinforcement in the flange in this area must be proportioned to transfer the forces from the diaphragm into the array of connections. For untopped diaphragms, the concrete at the potential failure surface is monolithic and the reinforcement requirement for flange can be determined by shear friction (Section 4.3.6). The potential shear crack is about 18 ft long and would extend around the flange embed plate anchorage bars, a total of approximately 21 ft. This steel area can be provided by a layer of 4x4 W4/W4 welded wire reinforcement.

3.8.3.3 Performance-Based Design

In an effort to find appropriate levels of force for precast diaphragm design, research has been conducted to evaluate their dynamic characteristics. [20, 25] These studies have produced important findings that bear directly on appropriate design.

An important observation is that the dynamic force distributions produced by the structures investigated can differ significantly from those provided by the equivalent lateral force patterns prescribed by the Codes. "For a rigid diaphragm structure following the formation of a base plastic hinge, the instantaneous effective centroid occurs well below the centroid implied by the equivalent lateral force pattern. This downward shift is due to the nature of higher modes in the instantaneous deformation state." [20]

The research has also identified vulnerabilities in common precast systems, particularly where there are large discontinuities, such as at parking structure ramps. The load paths across interior beams in these systems have particularly high force demands. Although design forces are applied in orthogonal directions, real accelerations can come from any direction. Oblique load trials also show some vulnerabilities where combined flexure and shear act on joints. For common precast systems with shear walls, the stiffness of the wall configurations can be significantly higher than the stiffness of the diaphragms. Under these circumstances, it may not be feasible to develop designs for diaphragms assured to remain elastic. It may be appropriate, instead, to establish performance criteria that include: "(1) the achievement of the diaphragm's elastic limit at life-safety; and (2) the exhaustion of the diaphragm's available ductility at collapse-prevention." [21]

The research papers referenced provide design guidelines that consider the specific parameters and characteristics of a building system and layout. Some general guidelines, however, can be derived from this work:

1. For structures of low and moderate seismic risk, the dynamic effects are less pronounced. If every floor diaphragm is designed for the force at the uppermost level derived from IBC Equation 16-65 (Eq. 3.8.3.2), additional load factors are not required for elastic diaphragm response under the design earthquake.

2. In regions of high seismic risk, special moment frames of reinforced concrete are sufficiently flexible to limit the direct transfer of ground acceleration to diaphragms at lower levels or the development of significant higher mode effects. Again, if every floor diaphragm is designed for the force at the uppermost level derived from Eq. 3.8.3.2, additional load factors are not required.

3. In regions of high seismic risk, shear wall buildings are most vulnerable to higher accelerations in diaphragms. As the buildings get taller, the effect becomes more severe. For most precast buildings, which are less than 80 ft tall, it is sufficient to apply a diaphragm load factor of 2 to the force at the uppermost level derived from Eq. 3.8.3.2, and to design each floor for that force.

4. In calculating diaphragm deflection, it is important to make a reasonable estimate of the effective section properties. A detailed analysis that considers the effects of joints, connections, and chord and web reinforcing may be made. A reasonable, but conservative, estimate can be made by taking the effective section as 10% to 15% of the gross section for topped systems and between 5% and 10% of the gross section for untopped systems.

5. Precast floors, topped or untopped, experience the strains of deformation by opening at the joints. These strains cannot be disregarded. Chord reinforcement should be proportioned to limit joint opening to the capacity of the shear reinforcement or connections as well as for strength. On the other hand, joint reinforcement or flange connections should have sufficient ductility to maintain required capacity while they undergo moderate joint strains.

3.8.4 Diaphragm Detailing Considerations

As the analysis of diaphragms for lateral forces most frequently follows the horizontal deep beam analogy, the design of the diaphragm connections and reinforcement also follow this model. "Diaphragms shall provide for both the shear and bending stresses resulting from these forces." [1] With large areas and depth relative to the spans, the characteristics of deep beams must often be considered in this design.

Diaphragms need to have elements to form ties or struts to distribute the anchorage forces into the diaphragm. The collector elements must be capable of transferring the seismic forces to the element providing the resistance to those forces.

3.8.4.1 General Detailing Requirements

The primary reinforcing steel or connection details that carry the diaphragm flexure through a tension/compression couple are in the chords. These may contain continuous reinforcing bars placed in a cast-in-place topping or strips or a continuous series of connections across the joints between components. These elements are placed near the boundaries of the diaphragm to increase the moment arm between the tension and compression forces for more efficient flexural resistance. Capacity-reduction factors, ϕ, have varied in past codes to reflect the reliability and ductility of chord details. Continuous reinforcement is treated as in other flexural elements with a ϕ of 0.90. For chords formed with connections welded at the joints, the ϕ factors might be treated more conservatively based on the reliability of the details.

Shear in the analogous beam is carried through shear capacity along or across the components. The transfer of shear across the joints is carried by connections or reinforcement specifically proportioned for the load. These connections may actually be weak in tension. It is important, however, that the shear capacity of a connection part or reinforcement is not reduced below the design requirement when the joint is subject to an opening. Even with cast-in-place topping, it is recognized that deformations in diaphragms are concentrated at the joints. The topping experiences reflection cracks above the joints in the precast member, and so is frequently tooled to control the cracking. Joint separation compromises the shear capacity in the concrete, so that all shear transfer is considered to be by shear-friction, dependent on the reinforcement crossing the joints.

Movement in the joints places a strain demand on the topping reinforcement or joint connections that must be considered in the detailing. The strain consideration may mean limiting that strain by higher strength and stiffness in the chord steel, accommodating the strain by providing connections or details that are tolerant, or a combination of the two.

In the design of long-span precast diaphragms, there are issues common to all levels of load. Long spans may be vulnerable to excessive deflections because details may result in a low effective stiffness. It is also important to consider stress concentrations at discontinuities, such as the separation points of ramps or at re-entrant corners near stairs or at the ends of spans. Additional strength or reinforcement may be needed in these locations.

Although the primary structural considerations for diaphragm design are strength and deflection control, the internal restraint that can develop must find some relief. Connections in diaphragms with strain capacity will permit small local deformations without distress, so that large restraint forces are not developed.

3.8.4.2 Wind and Low Seismic Hazard

For wind load and low seismic hazard (Seismic Design Category A), it is generally sufficient to design diaphragms for the strength requirements imposed by the applied forces. Openings and discontinuities may still require special attention for detailing, but these features are more sensitive to volume change movement than to lateral forces.

In addition, in Seismic Design Category B, the diaphragm design must also include the design of collector elements and their connections to the vertical supports to the higher strength criteria imposed by the special load combinations. At these locations, the design forces are calculated reactions increased by the overstrength factor, Ω_o.

3.8.4.3 Moderate Seismic Hazard – Topped and Pretopped Systems

Both topped and untopped diaphragms are commonly used in regions of moderate seismic risk (Seismic Design Category C). For systems that use concrete walls, including precast concrete, there are special requirements for diaphragm design in IBC 2003. In these systems, the diaphragm must include special continuous struts or ties between diaphragm chords for wall anchorage. When these diaphragms use sub-diaphragms, the aspect ratio of the sub-diaphragms is limited to 2½ to 1.

In moderate seismic risk areas or design categories, precast wall systems must conform to the requirements of intermediate precast walls. Again, the collector elements and their connections must be designed for the special load combinations with the overstrength factor. In addition, the interface with the wall must be with ductile connections including the reinforcing steel that would yield prior to crushing of the concrete. The body or mechanical parts of the connection (e.g., plates, welds, etc.) must have sufficient strength to permit development of $1.5f_y$ in the reinforcing steel.

"The design issues in a hollow-core diaphragm are the design of the connections to get the loads into the diaphragm, the strength and ductility of the slab system to transmit these loads to the lateral-resisting elements and the design of the connections required to unload the lateral forces from the diaphragm to the lateral-resisting elements." [26] For untopped hollow-core diaphragms, the chord steel may be added in the bearing areas in grout space outside the slab. This detailing requires some positive ties into grouted cores or keyways. Shear transfer can be designed through grouted keyways in the joints. Hollow-core slabs may also be topped, with the diaphragm reinforcement provided in the topping.

3.8.4.4 Seismic Design Category D – Topped Systems

ACI 318-02 includes specific provisions to address the requirements for cast-in-place topping as diaphragms. These provisions include systems with the topping composite with the precast components and systems with the topping non-composite part acting alone. The composite system requirements recognize that connections may be part of the design proportioned and detailed to transfer forces. They require that the interface be "clean, free of laitance, and intentionally roughened." [5]

The chord reinforcing steel design is determined from flexural analysis. Shear strength must be based entirely on reinforcement crossing the joint:

$$V_n = A_{cv}\rho_n f_y \qquad \text{(Eq. 3.8.4.1)}$$

where A_{cv} is based on the thickness of the topping slab and ρ_n is the steel ratio of the reinforcement crossing the precast joints. This is equivalent to shear-friction capacity with a μ of 1.0.

In seismic design for high seismic risk, research suggests that it may not be possible or practical to ensure complete elastic behavior of a diaphragm through the maximum considered earthquake. Some ductility needs to be provided to accommodate the potential for local overloads in the diaphragm. Some redundancy in the field-topped systems can be obtained by using mechanical connections between tees and between tees and beams for erection stability that stay in place and add to the capacity provided by reinforcement in the topping.

The potential for breaking welded wire reinforcement due to the strain demand across the joints must be considered in the detailing. Reinforcing steel needs to be compatible with the strain demand at the joint. Welded wire reinforcement with larger wire spacing or the use of reinforcing bars may be needed in some cases. Mesh in the topping must take the entire shear across the joint. Lapping in accordance with the code is necessary to maintain diaphragm integrity. In ACI 318-02, there is a minimum spacing requirement of 10 in. to promote strain capacity. In long spans with widely spaced joints, it may be necessary to consider the joint strain demand directly. The ϕ-factor for the shear design of the diaphragm must be no greater than that used in the shear design of the supporting vertical elements (columns or walls). This will sometimes result in $\phi = 0.6$, if the ϕ-factor for the shear design of shear walls is modified by Section 9.3.4 of ACI 318-02.

With the development of ductile yielding at joints in special moment frames or rocking from flexural yielding at the base of special walls, there are deformation demands placed on the diaphragm that are not addressed by the codes. Detailing of reinforcement in areas that are vulnerable to elongation of beams due to formation of plastic hinges should consider protection against loss of support, increased column displacements, and cracks in the diaphragm topping.

3.8.4.5 Untopped Systems for High Seismic Hazard

The provisions for precast diaphragms that are included in ACI 318-02 are for topped composite and topped non-composite diaphragms, so untopped diaphragms are implicitly not recognized. Section 21.2.1.5 of ACI 318-02, however, includes the general alternative clause that permits a system to be used if it is shown by experimental evidence and analysis to be equivalent in strength and toughness to comparable monolithic cast-in-place systems. Ref. 27 describes how these conditions can be met in a high Seismic Design Category building.

The "comparable monolithic reinforced concrete system" is the cast-in-place topping slab diaphragm of Section 21.9.3 of ACI 318-02. This is a cast-in-place topping acting alone proportioned and detailed to resist the design forces. The use of cast-in-place pour strips at the ends of the untopped tees designed for the tension and compression chord forces essentially provides this cast-in-place topping system for that part of the untopped diaphragm. The shear strength in the untopped system must be comparable to the shear strength required by the steel design from Eq. 21-11 of ACI 318-02.

The approach recommended here is to use design forces to achieve a performance level greater than prescribed by the code. The use of a diaphragm design load factor of 2 or higher will ensure that the resulting design will exceed the required strength.

To satisfy requirements for toughness, some attention needs to be paid to the strain at the joints and the capacity of the connections to provide their required strength through that strain. The joint spread can be checked by analysis. The total diaphragm deflection is calculated from effective section properties, and the associated strains distributed to the joints. This strain can be compared to the connection strain capacity. If necessary, the chord reinforcement may need to be increased to control this joint opening beyond the calculated requirements for chord strength.

Tests have been run on many prototype flange connectors to determine their strain capacity. [24] Although many of the common plant-fabricated connections designed with reinforcing bars butt-welded to the backs of plates failed to show sustained capacity or strain tolerance under reversed cyclic loading, some commercial flange connections, designed specifically to have improved strain capacity, showed that they have sustained shear capacity even with ¼ in. or more joint opening and under reversed cyclic loading. With the selection of tested welded connections as the replacement for steel reinforcement where the deformation reflected as joint opening is analyzed and controlled, it is possible to demonstrate equivalency of the untopped system and to provide effective design of untopped diaphragms in high seismic regions.

3.8.5 Alternate Methods of Diaphragm Design

The use of the horizontal deep beam analogy is a common and reasonably simple approach to diaphragm design. It can be seen, however, that there are some problems with this model in more complex and demanding configurations. In these cases, the designer may consider alternative analysis methods.

3.8.5.1 Strut-and-Tie Modeling

Appendix A of ACI 318-02 provides recommendations for the application of strut-and-tie modeling to concrete structures. The method considers the diaphragm as an idealized truss with compression struts, tension ties and intersection nodes. "Intermediate beams in the slab can act in a similar manner to stirrups in a beam, with diagonal compression forces being resisted by the concrete topping and the precast units where they are continuous." [28] With the jointed and rectilinear nature of precast layouts, some caution must be exercised in defining diagonal struts that may cross joints unless topping can be reinforced to form these elements. The method as applied to precast concrete is discussed in Ref. 28.

3.8.5.2 Finite Element Analysis

Although traditionally perceived as a research tool rather than as a design aid, Finite Element Analysis (FEA) has more recently found a place in the desktop computers in many engineering offices. In some cases, the power and simplicity of finite element modeling may become a practical and useful alternative when the diaphragm problem is not simple.

The key to successful diaphragm analysis using finite elements is in the modeling. When the focus of the analysis is the in-plane behavior, shell elements without transverse loads can be used to create the model. With precast systems, it is important to adequately define and model the joints between adjacent precast components. The size or spacing of the elements should match flange connection spacing so the nodes can be used as connection sites. The model is then copied or arrayed with spacing that leaves joints between the components. At beam lines, the depth of the beam might be represented by increasing the thickness of the shell element. It is equally important to add connections between the components that adequately model the connection characteristics. Some FEA programs provide member releases that can reduce or eliminate tension forces from beam elements used to model the shear connectors. Some programs also permit the definition of nonlinear springs so that test deformations can be modeled.

Simple modeling of planar behavior, as indicated above, can provide significant results and insight into areas with high load concentrations or deformations. In some cases, the prominence of shear deformation as a large component of deflection is evident. It may also be possible to identify deformation patterns that are more characteristic of tied arch behavior than of beam bending. Chord forces and connection forces are the primary results that can be used in design.

3.9 References

1. *International Building Code (IBC 2003)*, International Conference of Building Officials, Falls Church, VA, 2003.

2. *Minimum Design Loads for Buildings and Other Structures, Revision of ASCE 7-98 (SEI/ASCE 7-02)*, American Society of Civil Engineers, Reston, VA, 2003 (Co-sponsored by the Structural Engineering Institute).

3. *NEHRP (National Earthquake Hazards Reduction Program) Recommended Provisions for Seismic Regulations for New Buildings and Other Structures (NEHRP 2000)*, Building Seismic Safety Council, Washington, DC, 2000.

4. *NFPA 5000 Building Code*, National Fire Protection Association, Quincy, MA, 2002.

5. ACI Committee 318, "Building Code Requirements for Structural Concrete (ACI 318-02) and Commentary (ACI 318R-02)," American Concrete Institute, Farmington Hills, MI, 2002.

6. ACI Committee 550, "Emulating Cast-in-Place Detailing in Precast Concrete Structures (ACI 550.1R-01)," American Concrete Institute, Farmington Hills, MI, 2001.

7. *Acceptance Criteria for Moment Frames Based on Structural Testing (T1.1-01) and Commentary (T1.1R-01)*, Reported by ACI Innovation Task Group 1 and Collaborators, American Concrete Institute, Farmington Hills, MI, 2001.

8. Ghosh, S. K., and Hawkins, Neil M., "Seismic Design Provisions for Precast Concrete Structures in ACI 318," PCI JOURNAL, V. 46, No. 1, January-February 2001.

9. Priestley, M. J., Nigel, Sritharan, S. (Sri), Conley, James R., and Pampanin, Stefano, "Preliminary Results and Conclusions from the PRESSS Five-Story Precast Concrete Test Building," PCI JOURNAL, V. 44, No. 6, November-December 1999.

10. Englekirk, Robert E., "Design-Construction of The Paramount — A 39-Story Precast Pre-stressed Concrete Apartment Building," PCI JOURNAL, V. 47, No. 4, July-August 2002.

11. Speyer, Irwin J., "Considerations for the Design of Precast Concrete Bearing Wall Buildings to Withstand Abnormal Loads," PCI JOURNAL, V. 21, No. 2, March-April 1976.

12. *Precast/Prestressed Parking Structures: Recommended Practice for Design and Construction*, MNL-129-98, Precast/Prestressed Concrete Institute, Chicago, IL, 1998.

13. *Design and Typical Details of Connections for Precast and Prestressed Concrete*, Second Edition, MNL-123-88, Precast/Prestressed Concrete Institute, Chicago, IL, 1988.

14. Aristizabal-Ochoa, J. Dario, "Moment Restraint and Second-Order Analysis of a Cantilevered Precast Column Supported by an Isolated Footing," PCI JOURNAL, V. 47, No. 6, November-December 2002.

15. "Response of Multistory Concrete Structures to Lateral Forces," Special Publication SP-36, American Concrete Institute, Farmington Hills, MI, 1973.

16. ACI Committee 442, "Response of Buildings to Lateral Forces," *ACI Journal*, V. 68, No. 2, February 1971.

17. "Design of Combined Frames and Shear Walls," Advanced Engineering Bulletin No. 14, Portland Cement Association, Skokie, IL, 1965.

18. Fintel, Mark (Editor), *Handbook of Concrete Engineering*, Second Edition, Van Nostrand Reinhold Company, New York, NY, 1965.

19. MacLeod, I. A., *Shear Wall-Frame Interaction, A Design Aid*, Portland, Cement Association, Skokie, IL, April 1970.

20. Fleischman, Robert B., Farrow, Kenneth T., and Eastman, Kristin, "Seismic Performance of Perimeter Lateral-System Structures with Highly Flexible Diaphragms," *Earthquake Spectra*, V. 18, No. 2, May 2002, Earthquake Engineering Research Institute.

21. Naeim, Farzad, *The Seismic Design Handbook*, Second Edition, Kluwer Academic Publishers, Boston, MA, 2001.

22. Nakaki, S. D., Design Guidelines: Precast and Cast-in-Place Concrete Diaphragms, Earthquake Engineering Research Institute, Sacremento, CA, 2000.

23. Farrow, K. T., and Fleischman, R. B., "Seismic Design Recommendations for Precast Concrete Diaphragms in Long-Floor Span Construction," PCI JOURNAL, V. 48, No. 6, November-December 2003.

24. Oliva, M. G., "Testing of the JVI Flange Connectors for Precast Concrete Double-Tee Systems," University of Wisconsin, Madison, WI, 2000.

25. Rodriquez, M. E., Restrepo, J. I., and Carr, A. J., *Earthquake Engineering and Structural Dynamics,* Chapter 31 "Earthquake-Induced Floor Horizontal Accelerations in Buildings," pp. 693-718. Published by John Wiley & Sons, Ltd., New York, NY.

26. *Manual for the Design of Hollow-Core Slabs, Second Edition*, MNL-126-98, Precast/Prestressed Concrete Institute, Chicago, IL, 1998.

27. Cleland, N. M., and Ghosh, S. K., "Untopped Precast Concrete Diaphragms in High Seismic Applications," PCI JOURNAL, V. 47, No. 6, November-December 2002.

28. *Guidelines for the Use of Structural Precast Concrete in Buildings*, Second Edition, New Zealand Concrete Society and New Zealand Society for Earthquake Engineering, Centre for Advanced Engineering, University of Canterbury, Christchurch, New Zealand, 1999.

29. Schultz, A. E., and Magaña, R. A., "Seismic Behavior of Connections in Precast Concrete Shear Walls," Mete A. Sozen Symposium, ACI Special Publication SP-162, J. K. Wight and M. E., Kreger (Editors), American Concrete Institute, Farmington Hills, MI, 1996, pp. 273-311.

30. *Expansion Joints in Buildings*, Technical Report No. 65, Federal Construction Council, Building Research Advisory Board, Division of Engineering, National Research Council, National Academy of Sciences, Washington, DC, 1974.

3.10 DESIGN AIDS

Figure 3.10.1 Classification of building and other structures for importance factors, I[a]

CATEGORY[b]	NATURE OF OCCUPANCY	SEISMIC FACTOR	SNOW FACTOR	WIND FACTOR
I	Buildings and other structures that represent a low hazard to human life in the event of failure including, but not limited to: • Agricultural facilities • Certain temporary facilities • Minor storage facilities	1.00	0.8	0.87[c]
II	Buildings and other structures except those listed in Categories II, III, IV	1.00	1.0	1.00
III	Buildings and other structures that represent a substantial hazard to human life in the event of failure including, but not limited to: • Buildings and other structures where more than 300 people congregate in one area • Buildings and other structures with elementary school, secondary school or day-care facilities with a capacity greater than 250 • Buildings and other structures with a capacity greater than 500 for colleges or adult education facilities • Health care facilities with a capacity of 50 or more resident patients but not having surgery or emergency treatment facilities • Jails and detention facilities • Any other occupancy with an occupant load greater than 5,000 • Power-generating stations, water treatment for potable water, waste water treatment facilities and other public utility facilities not included in Category III • Buildings and other structures not included in Category III containing sufficient quantities of toxic or explosive substances to be dangerous to the public if released	1.25	1.1	1.15
IV	Buildings and other structures designated as essential facilities including, but not limited to: • Hospitals and other health care facilities having surgery or emergency treatment facilities • Fire, rescue and police stations and emergency vehicle garages • Designated earthquake, hurricane or other emergency shelters • Designated emergency preparedness, communication, and operation centers and other facilities required or emergency response • Power-generated stations and other public utility facilities required as emergency back-up facilities for Category III structures • Structures containing highly toxic materials as defined by Section 307 where the quantity of the material exceeds the maximum allowable quantity of Table 307.7(2) • Aviation control towers, air traffic control centers and emergency aircraft hangars • Buildings and other structures having critical national defense functions • Water treatment facilities required to maintain water pressure for fire suppression	1.50	1.2	1.15

a. IBC 2003, Table 1604.5 (Information same as SEI/ASCE 7-02, Table 1.1, 6.1, 7.4, 9.1.4).
b. For the purpose of Section 1616.2, Categories I and II are considered Seismic Use Group I, Category III is considered Seismic Use Group II and Category IV is equivalent to Seismic Use Group III.
c. In hurricane-prone regions with V > 100 mph, I = 0.77.

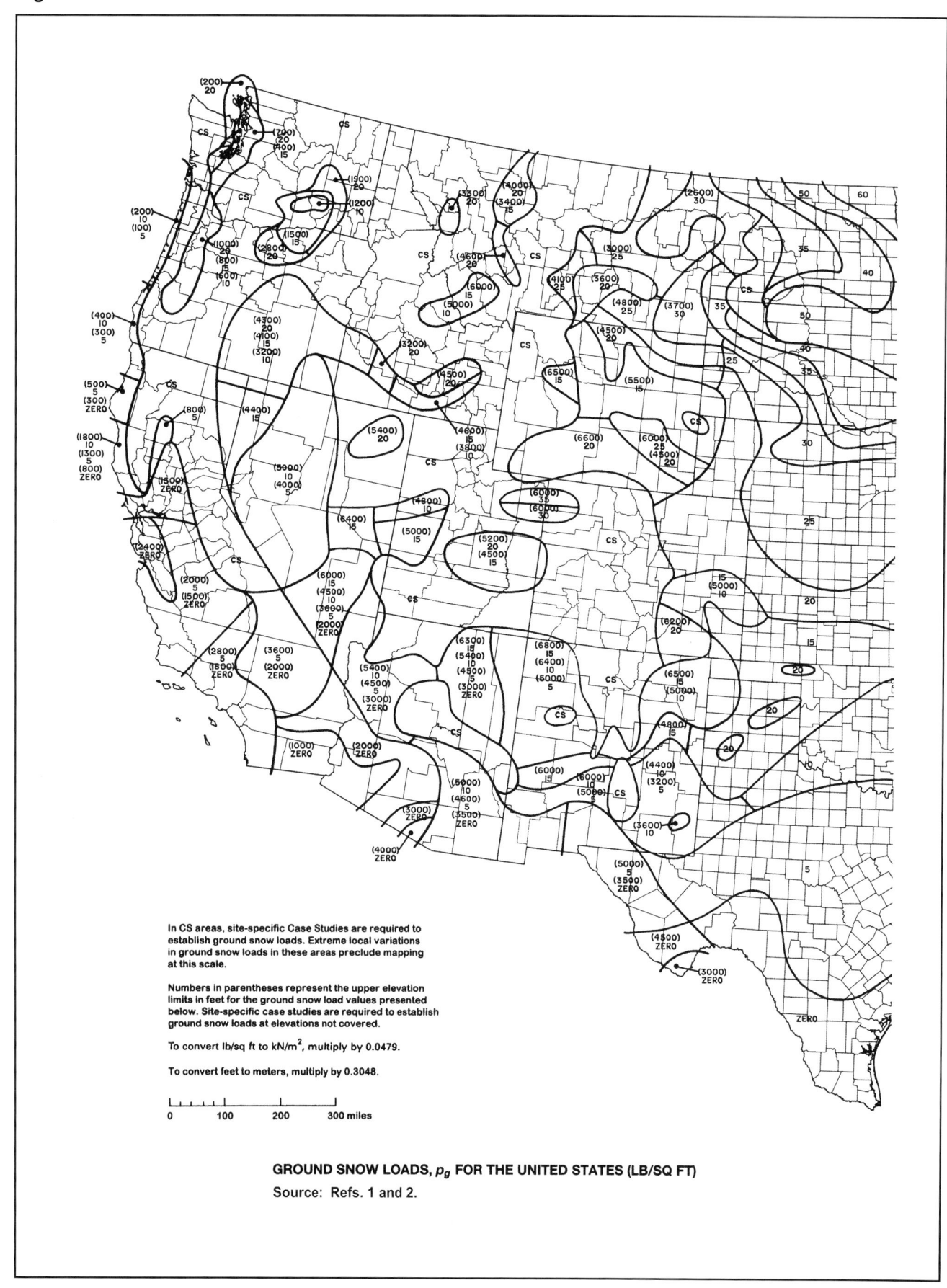

Figure 3.10.2 Ground snow loads[2]

GROUND SNOW LOADS, p_g FOR THE UNITED STATES (LB/SQ FT)
Source: Refs. 1 and 2.

In CS areas, site-specific Case Studies are required to establish ground snow loads. Extreme local variations in ground snow loads in these areas preclude mapping at this scale.

Numbers in parentheses represent the upper elevation limits in feet for the ground snow load values presented below. Site-specific case studies are required to establish ground snow loads at elevations not covered.

To convert lb/sq ft to kN/m^2, multiply by 0.0479.

To convert feet to meters, multiply by 0.3048.

Figure 3.10.2 Ground snow loads (Cont.)

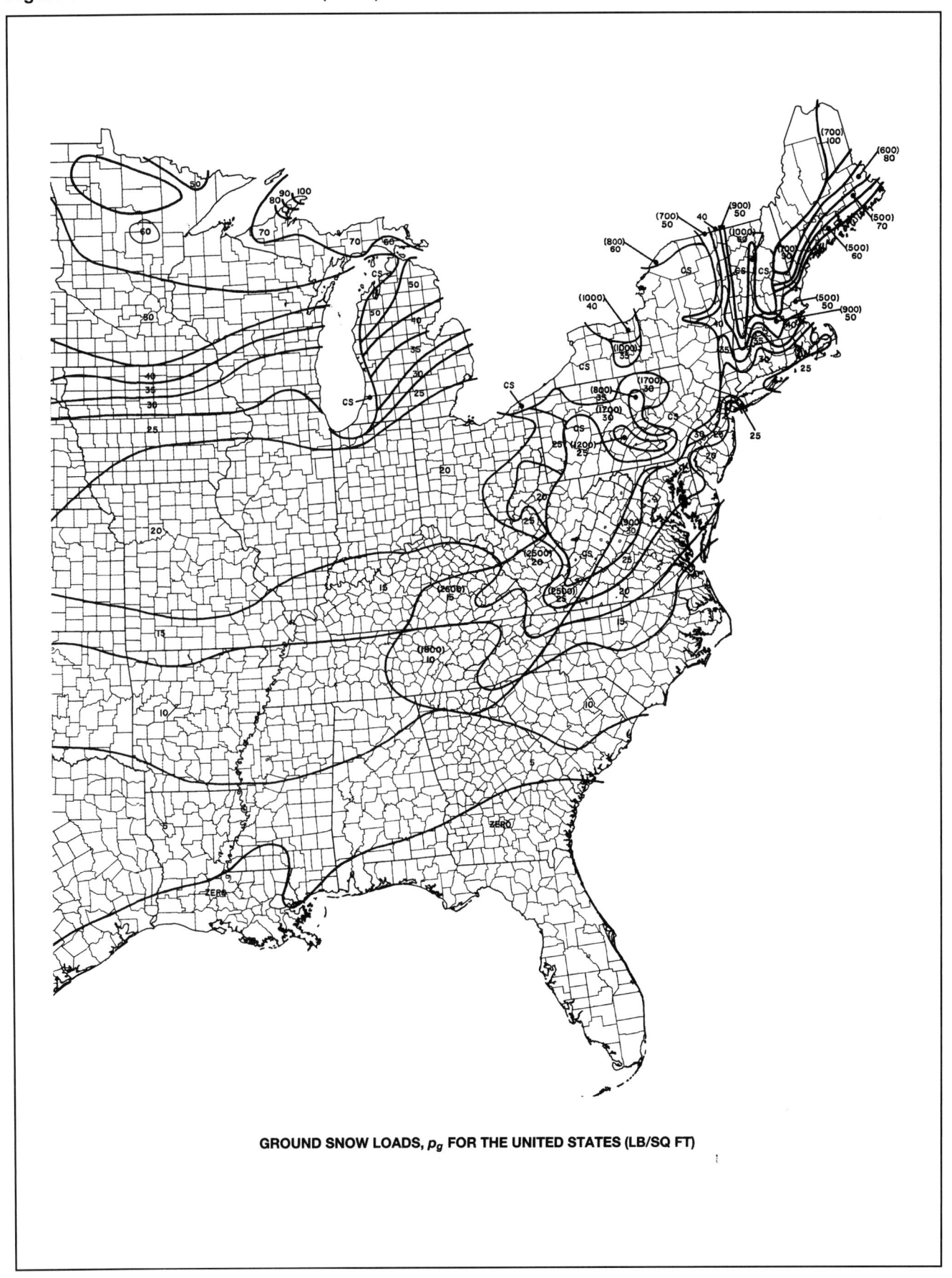

GROUND SNOW LOADS, p_g FOR THE UNITED STATES (LB/SQ FT)

Figure 3.10.3 Snow loading

(a) Ground snow load for Alaska locations

Location	P_g lb/ft²	(kN/m²)	Location	P_g lb/ft²	(kN/m²)	Location	P_g lb/ft²	(kN/m²)
Adak	30	(1.4)	Galena	60	(2.9)	Petersburg	150	(7.2)
Anchorage	50	(2.4)	Gulkana	70	(3.4)	St Paul Islands	40	(1.9)
Angoon	70	(3.4)	Homer	40	(1.9)	Seward	50	(2.4)
Barrow	25	(1.2)	Juneau	60	(2.9)	Shemya	25	(1.2)
Barter Island	35	(1.7)	Kenai	70	(3.4)	Sitka	50	(2.4)
Bethel	40	(1.9)	Kodiak	30	(1.4)	Talkeetna	120	(5.8)
Big Delta	50	(2.4)	Kotzebue	60	(2.9)	Unalakleet	50	(2.4)
Cold Bay	25	(1.2)	McGrath	70	(3.4)	Valdez	160	(7.7)
Cordova	100	(4.8)	Nenana	80	(3.8)	Whittier	300	(14.4)
Fairbanks	60	(2.9)	Nome	70	(3.4)	Wrangell	60	(2.9)
Fort Yukon	60	(2.9)	Palmer	50	(2.4)	Yakutat	150	(7.2)

(b) Exposure factor, C_e

Terrain Category	Fully Exposed	Exposure of Roof Partially Exposed	Sheltered
Exposure B: Urban and suburban areas, wooded areas	0.9	1.0	1.2
Exposure D: Flat, unobstructed areas outside hurricane-prone regions	0.9	1.0	1.1
Exposure C: All others	0.8	0.9	1.0
Above the treeline in windswept areas	0.7	0.8	N/A
In Alaska, in areas where trees do not exist within a 2-mile (3 km) radius of the site.	0.7	0.8	N/A

The terrain category and roof exposure condition chosen shall be representative of the anticipated conditions during the life of the structure. An exposure factor shall be determined for each roof of a structure.

(c) Thermal factor, C_t

Thermal Condition*	C_t
All structures except as indicated below	1.0
Structures kept just above freezing and others with cold, ventilated roofs in which the thermal resistance (R-value) between the ventilated space and the heated space exceeds 25°F hr sq ft/Btu (4.4 K m²/W)	1.1
Unheated structures and structures intentionally kept below freezing	1.2

*These conditions shall be representative of the anticipated conditions during winters for the life of the structure.

(d) Windward and Leeward Drifting

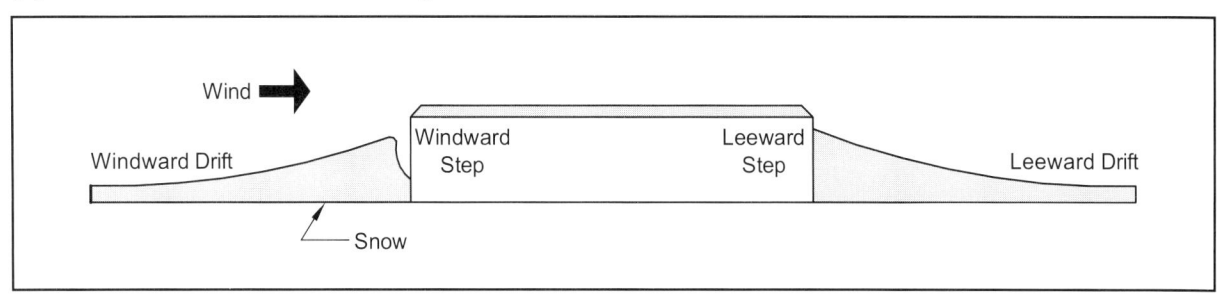

Source: Refs. 1 and 2.

Figure 3.10.4   Snow drifting

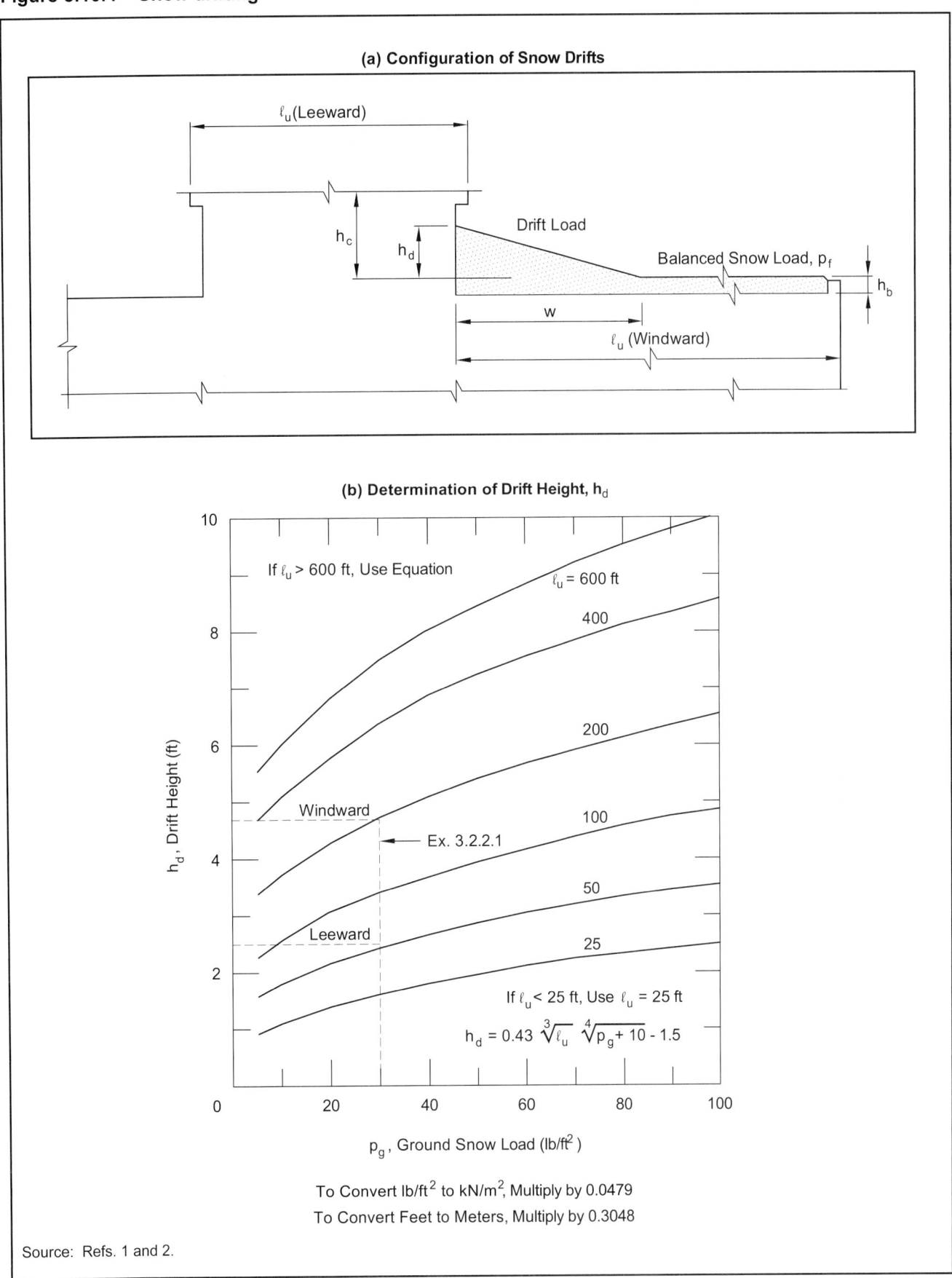

(a) Configuration of Snow Drifts

ℓ_u (Leeward), ℓ_u (Windward), h_c, h_d, h_b, w, Drift Load, Balanced Snow Load, p_f

(b) Determination of Drift Height, h_d

If ℓ_u > 600 ft, Use Equation

ℓ_u = 600 ft, 400, 200, 100, 50, 25

Windward
Leeward
← Ex. 3.2.2.1

If ℓ_u < 25 ft, Use ℓ_u = 25 ft

$$h_d = 0.43 \sqrt[3]{\ell_u} \sqrt[4]{p_g + 10} - 1.5$$

p_g, Ground Snow Load (lb/ft^2)

To Convert lb/ft^2 to kN/m^2, Multiply by 0.0479
To Convert Feet to Meters, Multiply by 0.3048

Source: Refs. 1 and 2.

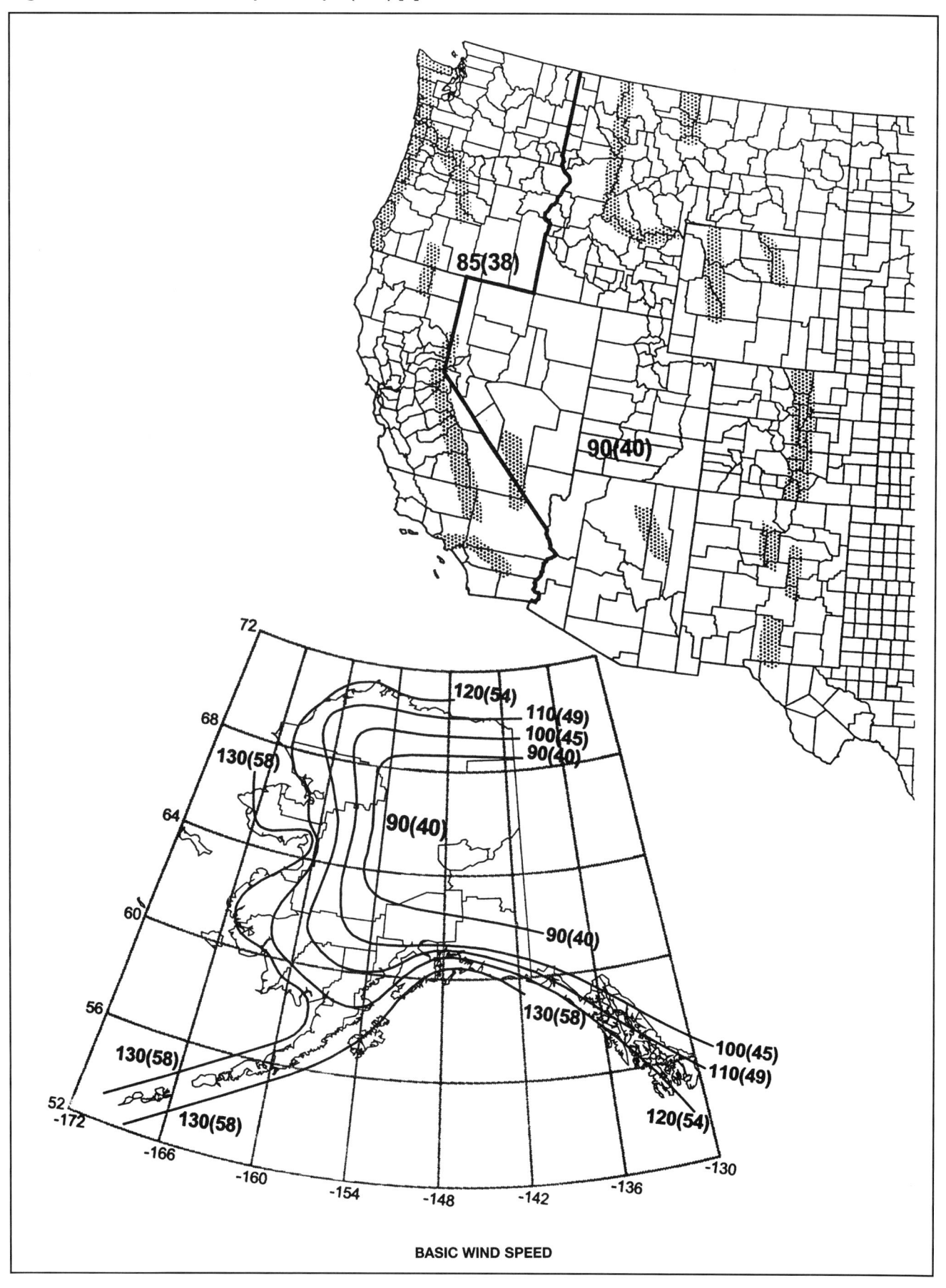

Figure 3.10.5 Basic wind speed, mph (m/s) [2]

Figure 3.10.5 Basic wind speed, mph (m/s) (Cont.) [2]

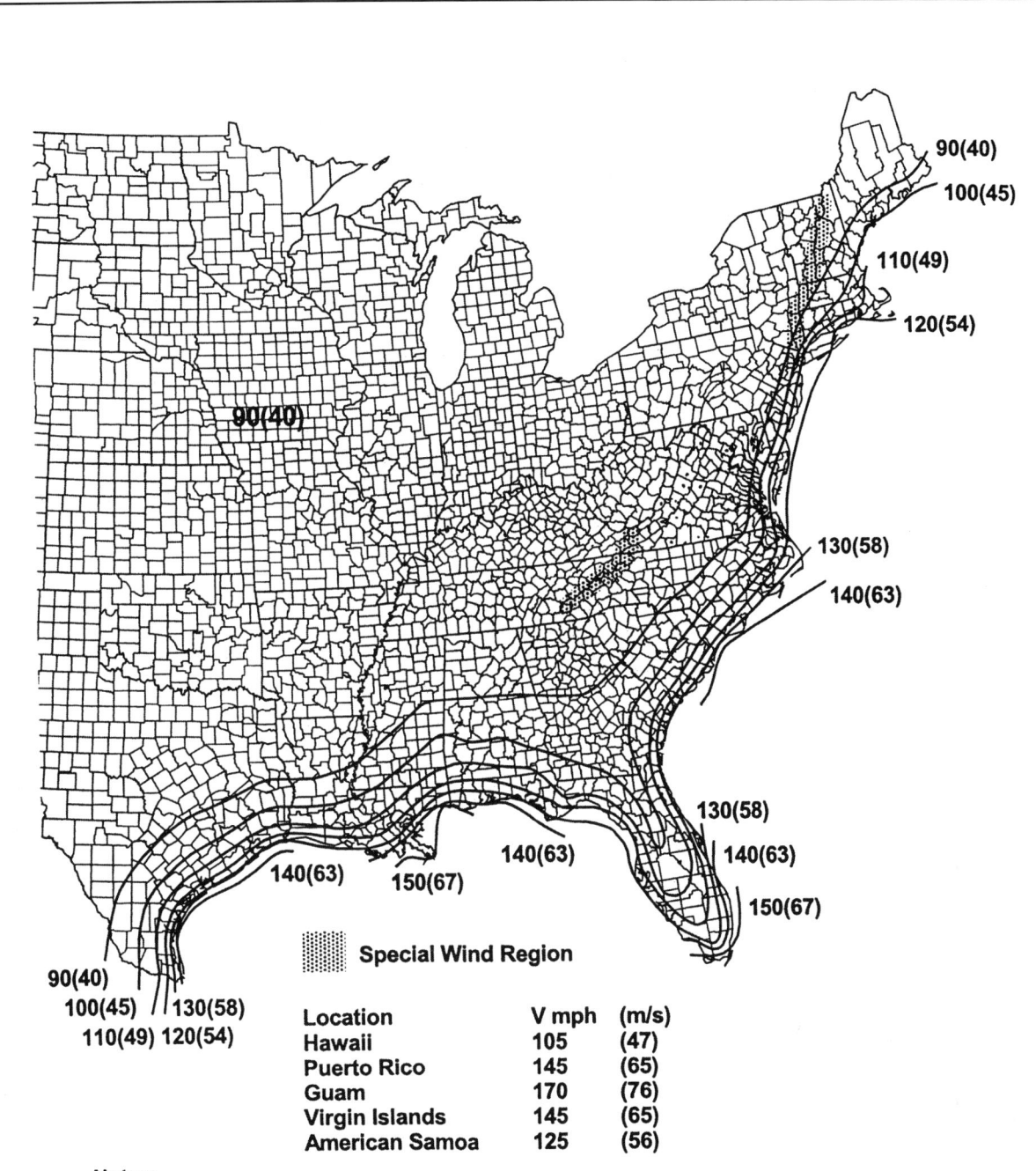

Location	V mph	(m/s)
Hawaii	105	(47)
Puerto Rico	145	(65)
Guam	170	(76)
Virgin Islands	145	(65)
American Samoa	125	(56)

Notes:
1. Values are nominal design 3-second gust wind speeds in miles per hour (m/s) at 33 ft (10 m) above ground for Exposure C category.
2. Linear interpolation between wind contours is permitted.
3. Islands and coastal areas outside the last contour shall use the last wind speed contour of the coastal area.
4. Mountainous terrain, gorges, ocean promontories, and special wind regions shall be examined for unusual wind conditions.

BASIC WIND SPEED

Figure 3.10.6 Factors for use with ASCE 7 Method 1 wind design [2]

(a) p_{s30} for horizontal walls for use in Eq. 3.2.3.1

Basic Wind Speed (mph)	p_{s30} Zone A	p_{s30} Zone C
85	11.5	7.6
90	12.8	8.5
100	15.9	10.5
110	19.2	12.7
120	22.8	15.1
130	26.8	17.8
140	31.1	20.6
150	35.7	23.7

(c) Adjustment factor, λ, for building height and exposure

Mean Roof Height (H)	Exposure B	Exposure C	Exposure D
15	1.00	1.21	1.47
20	1.00	1.29	1.55
25	1.00	1.35	1.61
30	1.00	1.40	1.66
35	1.05	1.45	1.70
40	1.09	1.49	1.74
45	1.12	1.53	1.78
50	1.16	1.56	1.81
55	1.19	1.59	1.84
60	1.22	1.62	1.87

(b) p_{net30} for use in components and cladding for use in Eq. 3.2.3.2

Zone	Basic Wind Speed (mph)	10	20	50	100	500
Zone A (Figure 3.2.3.1)	85	13.0/–17.4	12.4/–16.2	11.6/–14.7	11.1/–13.5	9.7/–10.8
	90	14.6/–19.5	13.9/–18.2	13.0/–16.5	12.4/–15.1	10.9/–12.1
	100	18.0/–24.1	17.2/–22.5	16.1/–20.3	15.3/–18.7	13.4/–14.9
	110	21.8/–29.1	20.8/–27.2	19.5/–24.6	18.5/–22.6	16.2/–18.1
	120	25.9/–34.7	24.7/–32.4	23.2/–29.3	22.0/–26.9	19.3/–21.5
	130	30.4/–40.7	29.0/–38.0	27.2/–34.3	25.9/–31.6	22.7/–25.2
	140	35.3/–47.2	33.7/–44.0	31.6/–39.8	30.0/–36.7	26.3/–29.3
	150	40.5/–54.2	38.7/–50.5	36.2/–45.7	34.4/–42.1	30.2/–33.6
Zone C (Figure 3.2.3.1)	85	13.0/–14.1	12.4/–13.5	11.6/–12.7	11.1/–12.2	9.7/–10.8
	90	14.6/–15.8	13.9/–15.1	13.0/–14.3	12.4/–13.6	10.9/–12.1
	100	18.0/–19.5	17.2/–18.7	16.1/–17.6	15.3/–16.8	13.4/–14.9
	110	21.8/–23.6	20.8/–22.6	19.5/–21.3	18.5/–20.4	16.2/–18.1
	120	25.9/–28.1	24.7/–26.9	23.2/–25.4	22.0/–24.2	19.3/–21.5
	130	30.4/–33.0	29.0/–31.6	27.2/–29.8	25.9/–28.4	22.7/–25.2
	140	35.3/–38.2	33.7/–36.7	31.6/–34.6	30.0/–33.0	26.3/–29.3
	150	40.5/–43.9	38.7/–42.1	36.2/–39.7	34.4/–37.8	30.2/–33.6

Effective Wind Area (sq ft)

Notes:
1. Method 1 applies only to enclosed buildings ≤ 60 ft in height. See Section 3.2.3.1 for other restrictions and design procedure.
2. Assume flat roof (pitch < 5 deg.).

From "Minimum Design Loads for Buildings and Other Structures," SEI/ASCE 7-02, Figures 6-2 and 6-3.

Figure 3.10.7 Site classifications and coefficients [1–3]

(a) Site class definitions

Site Class	Soil Profile Name	Average Properties in Top 100 ft, as per Section 1615.1.5		
		Soil Shear Wave Velocity, $\bar{v}_s$ (ft/s)	Standard Penetration Resistance, N	Soil Undrained Shear Strength, $\bar{s}_u$ (psf)
A	Hard rock	$\bar{v}_s > 5{,}000$	Not applicable	Not applicable
B	Rock	$2{,}500 \leq \bar{v}_s \leq 5{,}000$	Not applicable	Not applicable
C	Very dense soil and soft rock	$1{,}200 \leq \bar{v}_s \leq 2{,}500$	$\bar{N} > 50$	$\bar{s}_u \geq 2{,}000$
D	Stiff soil profile	$600 \leq \bar{v}_s \leq 1{,}200$	$15 \leq \bar{N} \leq 50$	$1{,}000 \leq \bar{s}_u \leq 2{,}000$
E	Soft soil profile	$\bar{v}_s < 600$	$\bar{N} < 15$	$\bar{s}_u < 1{,}000$
E	—	Any profile with more than 10 ft of soil having following characteristics: 1. Plasticity index PI > 20; 2. Moisture content w ≥ 40%, and 3. Undrained shear strength $\bar{s}_u < 500$ psf		
F	—	Any profile containing soils having one or more of the following characteristics: 1. Soils vulnerable to potential failure or collapse under seismic loading such as liquefiable soils, quick and highly organic clays, collapsible weakly cemented soils 2. Peats and/or highly organic clay (H > 10 ft of peat and/or highly organic clay where H = thickness soil) 3. Very high plasticity clays (H > 25 ft with plasticity index PI > 75) 4. Very thick soft/medium stiff clays (H > 120 ft)		

(b) Site coefficient F_a

Site Class	Mapped Spectral Response Acceleration at Short Periods[a]				
	$S_s \leq 0.25$	$S_s = 0.50$	$S_s = 0.75$	$S_s = 1.00$	$S_s = 1.25$
A	0.8	0.8	0.8	0.8	0.8
B	1.0	1.0	1.0	1.0	1.0
C	1.2	1.2	1.1	1.0	1.0
D	1.6	1.4	1.2	1.1	1.0
E	2.5	1.7	1.2	0.9	Note b
F	Note b	Note b	Note b	Note b	Note b

a. Use straight line interpolation for intermediate values of mapped spectral acceleration at short period, S_s.
b. Site-specific geotechnical investigation and dynamic site response analyses shall be performed to determine appropriate values.

(c) Site coefficient F_v

Site Class	Mapped Spectral Response Acceleration at short Periods[a]				
	$S_1 \leq 0.1$	$S_1 = 0.2$	$S_1 = 0.3$	$S_1 = 0.4$	$S_1 = 0.5$
A	0.8	0.8	0.8	0.8	0.8
B	1.0	1.0	1.0	1.0	1.0
C	1.7	1.6	1.5	1.4	1.3
D	2.4	2.0	1.8	1.6	1.5
E	3.5	3.2	2.8	2.4	Note b
F	Note b	Note b	Note b	Note b	Note b

a. Use straight line interpolation values of mapped spectral acceleration at 1-sec period, S_i.
b. Site-specific geotechnical investigation and dynamic site response analyses shall be performed to determine appropriate values.

Figure 3.10.8 Design coefficients and factors for concrete seismic force-resisting systems [1–3]

Basic Seismic Force-Resisting System	Response Modification Coefficeint, R[a]	System Over-strength Factor, Ω_0^g	Deflection Amplification Factor, C_d^b	Structural Systems Limitations and Building Height (ft) Limitations[c] Seismic Design Category				
				A & B	C	D[d]	E[e]	F[e]
Bearing Wall Systems								
Special reinforced concrete shear walls	5	2½	5	NL	NL	160	160	160
Ordinary reinforced concrete shear walls	4	2½	4	NL	NL	NP	NP	NP
Detailed plain concrete shear walls	2½	2½	2	NL	NP	NP	NP	NP
Ordinary plain concrete shear walls	1½	2½	1½	NL	NP	NP	NP	NP
Building Frame Systems								
Special reinforced concrete shear walls	6	2½	5	NL	NL	160	160	100
Ordinary reinforced concrete shear walls	5	2½	4½	NL	NL	NP	NP	NP
Detailed plain concrete shear walls	3	2½	2½	NL	NP	NP	NP	NP
Ordinary plain concrete shear walls	2	2½	2	NP	NP	NP	NP	NP
Moment Resisting Frame Systems								
Special reinforced concrete moment frames	8	3	5½	NL	NL	NL	NL	NL
Intermediate reinforced concrete moment frames[h]	5	3	4½	NL	NL	NP	NP	NP
Ordinary reinforced concrete moment frames	3	3	2½	NL	NP	NP	NP	NP
Dual System with Special Moment Frames Capable of Resisting at Least 25% of Prescribed Seismic Forces								
Special reinforced concrete shear walls	8	2½	6½	NL	NL	NL	NL	NL
Ordinary reinforced concrete shear walls	7	2½	6	NL	NL	NP	NP	NP
Dual Systems with Intermediate Moment Frames Capable of Resisting at Least 25% of Prescribed Seismic Forces[h]								
Special reinforced concrete shear walls	6	2½	5	NL	NL	160	100	100
Ordinary reinforced concrete shear walls	5½	2½	4½	NL	NL	NP	NP	NP
Inverted Pendulum Systems and Cantilevered Column Systems								
Special reinforced concrete moment frames	2½	2	1¼	NL	NL	NL	NL	NL

a. Response modification coefficient, *R*, for use throughout the standard. Note R reduces forces to a strength level, not an allowable stress level.
b. Deflection amplification factor, C_d, for use in Sections 9.5.3.7.1 and 9.5.3.7.2.
c. NL = Not Limited and NP = Not Permitted. For metric units use 30 m for 100 ft and use 50 m for 160 ft. Heights are measured from the base of the structure as defined in Section 9.2.1.
d. See Section 9.5.2.2.4.1 for a description of building systems limited to buildings with a height of 240 ft (75 m) or less.
e. See Sections 9.5.2.2.4 and 9.5.2.2.4.5 for building systems limited to buildings with a height of 160 ft (50 m) or less.
f. Ordinary moment frame is permitted to be used in lieu of intermediate moment frame in Seismic Design Categories B and C.
g. The tabulated value of the overstrength factor, Ω, may be reduced by subtracting ½ for structures with flexible diaphragms but shall not be taken as less than 2.0 for any structure.
h. ACI 318-02 does not recognize precast intermediate frames.

Figure 3.10.9 Allowable story drift, Δ_a, in. [1–3]

Building	Seismic Use Group		
	I	II	III
Buildings, other than masonry shear walls or masonry wall frame buildings, four stories or less in height with interior walls, partitions, ceilings, and exterior walls systems that have been designed to accommodate the story drifts	0.025 h_{sx}[b]	0.020 h_{sx}	0.015 h_{sx}
Masonry cantilever shear wall buildings[c]	0.010 h_{sx}	0.010 h_{sx}	0.010 h_{sx}
Other masonry shear wall buildings	0.007 h_{sx}	0.007 h_{sx}	0.007 h_{sx}
Masonry wall frame buildings	0.013 h_{sx}	0.013 h_{sx}	0.010 h_{sx}
All other buildings	0.020 h_{sx}	0.015 h_{sx}	0.010 h_{sx}

a. There shall be no drift limit for single-story buildings with interior walls, partitions, ceilings and exterior wall systems that have been designed to accommodate the story drifts.
b. h_{sx} is the story height below Level x.
c. Buildings in which the basic structural system consists of masonry shear walls designed as vertical elements cantilevered from their base or foundation support which are so constructed that moment transfer between shear walls (coupling) is negligible.

Figure 3.10.10 Architectural components coefficients [1–3]

Architectural Component or Element	Component Amplification Factor a_p	Component Response Modification Factor R_p
1. Interior non-structural walls and partitions (see also Section 1621.2.7, IBC 2003)		
a. Plain (unreinforced) masonry walls	1.0	1.5
b. Other walls and partitions	1.0	2.5
2. Cantilever elements (unbraced or braced to structural frame below its center of mass)		
a. Parapets and cantilever interior non-structural walls	2.5	2.5
b. Chimneys and stacks when laterally braced or supported by the structural frame	2.5	2.5
3. Cantilever elements (braced to structural frame above its center of mass)		
a. Parapets	1.0	2.5
b. Chimneys and stacks	1.0	2.5
c. Exterior non-structural walls	1.0	2.5
4. Exterior non-structural wall elements and connections		
a. Wall element	1.0	2.5
b. Body of wall panel connections	1.0	2.5
c. Fasteners of the connection system	1.25	1.0
5. Veneer		
a. Limited deformability elements and attachments	1.0	2.5
b. Low deformability elements or attachments	1.0	2.5
6. Penthouses (except when framed by an extension of the building frame)	2.5	3.5
7. Ceilings (see also Section 1621.2.5)	1.0	2.5

Figure 3.10.11 Maximum seasonal climatic temperature change, °F

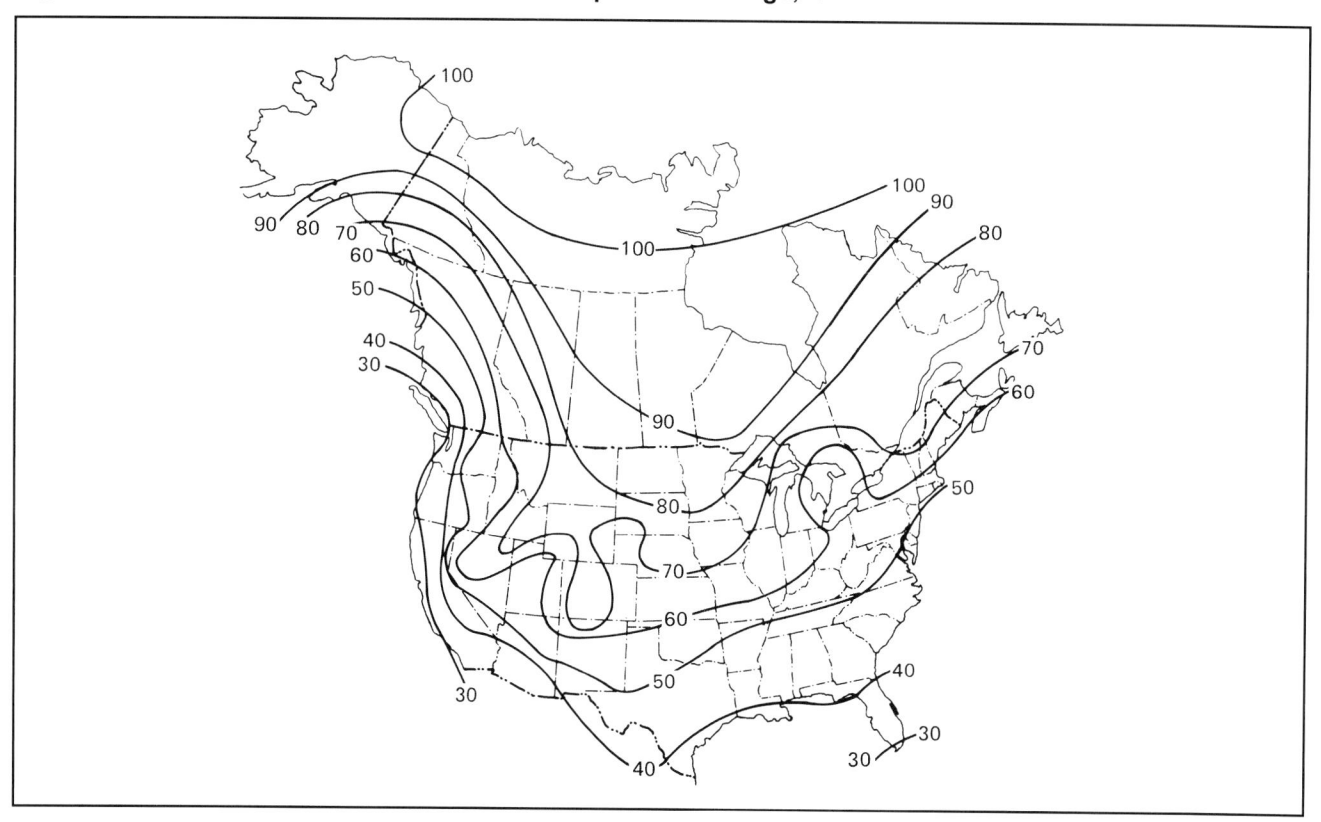

Figure 3.10.12 Annual average ambient relative humidity, percent

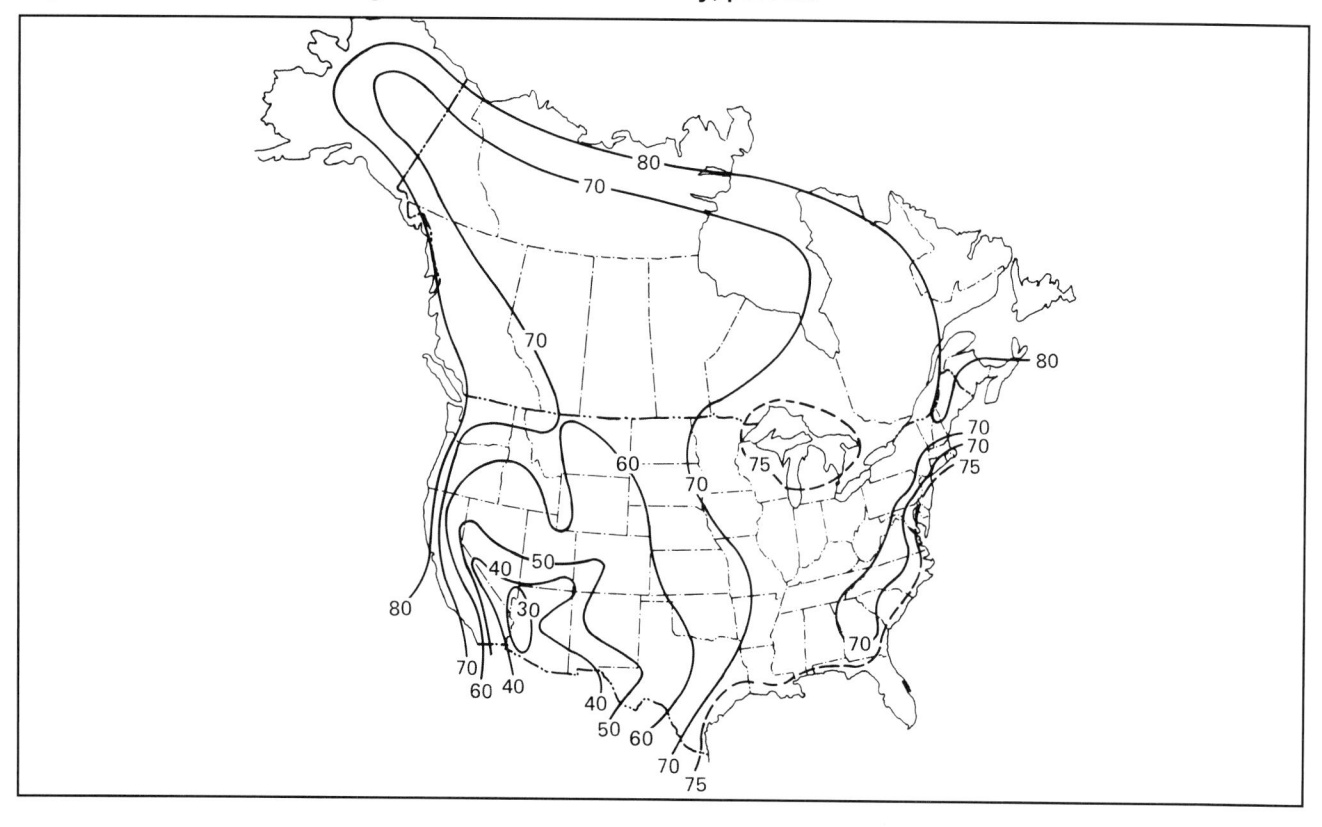

Figure 3.10.13 Creep and shrinkage strains (millionths)

	Concrete Release Strength = 3500 psi Average Prestress = 600 psi Relative Humidity = 70% Volume-to-Surface Ratio = 1.5 in.			
Time, days	**Creep**		**Shrinkage**	
	Normal weight	Lightweight	Moist Cure	Accelerated Curve
1	29	43	16	9
3	51	76	44	26
5	65	97	70	43
7	76	114	93	58
9	86	127	115	72
10	90	133	124	78
20	118	176	204	136
30	137	204	258	180
40	150	224	299	215
50	161	239	329	243
60	169	252	354	266
70	177	263	373	286
80	183	272	390	302
90	188	280	403	317
100	193	287	415	329
200	222	331	477	400
1 yr	244	363	511	443
3 yr	273	407	543	486
5 yr	283	422	549	495
Final	315	468	560	510

Figure 3.10.14 Correction factors for prestress and concrete strength (creep only)

Avg. P/A (psi)	Release Strength, f'_{ci} (psi)						
	2500	3000	3500	4000	4500	5000	6000
0	0.00	0.00	0.00	0.00	0.00	0.00	0.00
200	0.39	0.36	0.33	0.31	0.29	0.28	0.25
400	0.79	0.72	0.67	0.62	0.59	0.56	0.51
600	1.18	1.08	1.00	0.94	0.88	0.84	0.76
800	1.58	1.44	1.33	1.25	1.18	1.12	1.02
1000	1.97	1.80	1.67	1.56	1.47	1.39	1.27
1200	2.37	2.16	2.00	1.87	1.76	1.67	1.53
1400	2.76	2.52	2.33	2.18	2.06	1.95	1.78
1600		2.88	2.67	2.49	2.35	2.23	2.04
1800		3.24	3.00	2.81	2.65	2.51	2.29
2000			3.12	3.12	2.94	2.79	2.55
2200				3.43	3.23	3.07	2.80
2400				3.74	3.53	3.35	3.06
2600					3.82	3.63	3.31
2800						3.90	3.56
3000						4.18	3.82

Figure 3.10.15 Correction factors for relative humidity

Avg. ambient R.H. (from Figure 3.10.12)	Creep	Shrinkage
40	1.25	1.43
50	1.17	1.29
60	1.08	1.14
70	1.00	1.00
80	0.92	0.86
90	0.83	0.43
100	0.75	0.00

Figure 3.10.16 Correction factors for volume-to-surface ratio

Time, days	Creep V/S 1	2	3	4	5	6	Shrinkage V/S 1	2	3	4	5	6
1	1.30	0.78	0.49	0.32	0.21	0.15	1.25	0.80	0.50	0.31	0.19	0.11
3	1.29	0.78	0.50	0.33	0.22	0.15	1.24	0.80	0.51	0.31	0.19	0.11
5	1.28	0.79	0.51	0.33	0.23	0.16	1.23	0.81	0.52	0.32	0.20	0.12
7	1.28	0.79	0.51	0.34	0.23	0.16	1.23	0.81	0.52	0.33	0.20	0.12
9	1.27	0.80	0.52	0.35	0.24	0.17	1.22	0.82	0.53	0.34	0.21	0.12
10	1.26	0.80	0.52	0.35	0.24	0.17	1.21	0.82	0.53	0.34	0.21	0.13
20	1.23	0.82	0.56	0.39	0.27	0.19	1.19	0.84	0.57	0.37	0.23	0.14
30	1.21	0.83	0.58	0.41	0.30	0.21	1.17	0.85	0.59	0.40	0.26	0.16
40	1.20	0.84	0.60	0.44	0.32	0.23	1.15	0.86	0.62	0.42	0.28	0.17
50	1.19	0.85	0.62	0.46	0.34	0.25	1.14	0.87	0.63	0.44	0.29	0.19
60	1.18	0.86	0.64	0.48	0.36	0.26	1.13	0.88	0.65	0.46	0.31	0.20
70	1.17	0.86	0.65	0.49	0.37	0.28	1.12	0.88	0.66	0.48	0.32	0.21
80	1.16	0.87	0.66	0.51	0.39	0.29	1.12	0.89	0.67	0.49	0.34	0.22
90	1.16	0.87	0.67	0.52	0.40	0.31	1.11	0.89	0.68	0.50	0.35	0.23
100	1.15	0.87	0.68	0.53	0.42	0.32	1.11	0.89	0.69	0.51	0.36	0.24
200	1.13	0.90	0.74	0.61	0.51	0.42	1.08	0.92	0.75	0.59	0.44	0.31
1 yr	1.10	0.91	0.77	0.67	0.58	0.50	1.07	0.93	0.79	0.64	0.50	0.38
3 yr	1.10	0.92	0.81	0.73	0.67	0.62	1.06	0.94	0.82	0.71	0.59	0.47
5 yr	1.10	0.92	0.82	0.75	0.70	0.66	1.06	0.94	0.83	0.72	0.61	0.49
Final	1.09	0.93	0.83	0.77	0.74	0.72	1.05	0.95	0.85	0.75	0.64	0.54

Figure 3.10.17 Design temperature strains (millionths)

Temperature zone (from Figure 3.10.11)	Normal weight Heated	Unheated	Lightweight Heated	Unheated
10	30	45	25	38
20	60	90	50	75
30	90	135	75	113
40	120	180	100	150
50	150	225	125	188
60	180	270	150	225
70	210	315	175	263
80	240	360	200	300
90	270	405	225	338
100	300	450	250	375

Based on accepted coefficients of thermal expansion, reduced to account for thermal lag. [2]

Figure 3.10.18 Volume change strains for typical building elements (millionths)

Temp. zone (from map)	Prestressed members (P/A = 600 psi)									
	Normal weight concrete					Lightweight concrete				
	Avg. R.H. (from map)					Avg. R.H. (from map)				
	40	50	60	70	80	40	50	60	70	80
Heated buildings											
0	548	501	454	407	360	581	532	483	434	385
10	578	531	484	437	390	606	557	508	459	410
20	608	561	514	467	420	631	582	533	484	435
30	638	591	544	497	450	656	607	558	509	460
40	668	621	574	527	480	681	632	583	534	485
50	698	651	604	557	510	706	657	608	559	510
60	728	681	634	587	540	731	682	633	584	535
70	758	711	664	617	570	756	707	658	609	560
80	788	741	694	647	600	781	732	683	634	585
90	818	771	724	677	630	806	757	708	659	610
100	848	801	754	707	660	831	782	733	684	635
Unheated structures											
0	548	501	454	407	360	581	532	483	434	385
10	593	546	499	452	405	619	570	521	472	423
20	638	591	544	497	450	656	607	558	509	460
30	683	636	589	542	495	694	645	596	547	498
40	728	681	634	587	540	731	682	633	584	535
50	773	726	679	632	585	769	720	671	622	573
60	818	771	724	677	630	806	757	708	659	610
70	863	816	769	722	675	844	795	746	697	648
80	908	861	814	767	720	881	832	783	734	685
90	953	906	859	812	765	919	870	821	772	723
100	998	951	904	857	810	956	907	858	809	760

Figure 3.10.19 Volume change strains for typical building elements (millionths)

Temp. zone (from map)	Non-prestressed members									
	Normal weight concrete					Lightweight concrete				
	Avg. R.H. (from map)					Avg. R.H. (from map)				
	40	50	60	70	80	40	50	60	70	80
Heated buildings											
0	294	265	235	206	177	294	265	235	206	177
10	324	295	265	236	207	319	290	260	231	202
20	354	325	295	266	237	344	315	285	256	227
30	384	355	325	296	267	369	340	310	281	252
40	414	385	355	326	297	394	365	335	306	277
50	444	415	385	356	327	419	390	360	331	302
60	474	445	415	386	357	444	415	385	356	327
70	504	475	445	416	387	469	440	410	381	352
80	534	505	475	446	417	494	465	435	406	377
90	564	535	505	476	447	519	490	460	431	402
100	594	565	535	506	477	544	515	485	456	427
Unheated structures											
0	294	265	235	206	177	294	265	235	206	177
10	339	310	280	251	222	332	302	273	244	214
20	384	355	325	296	267	369	340	310	281	252
30	429	400	370	341	312	407	377	348	319	289
40	474	445	415	386	357	444	415	385	356	327
50	519	490	460	431	402	482	452	423	394	364
60	564	535	505	476	447	519	490	460	431	402
70	609	580	550	521	492	557	527	498	469	439
80	654	625	595	566	537	594	565	535	506	477
90	699	670	640	611	582	632	602	573	544	514
100	744	715	685	656	627	669	640	610	581	552

Figure 3.10.20 Length of structure without use of expansion joints [30]

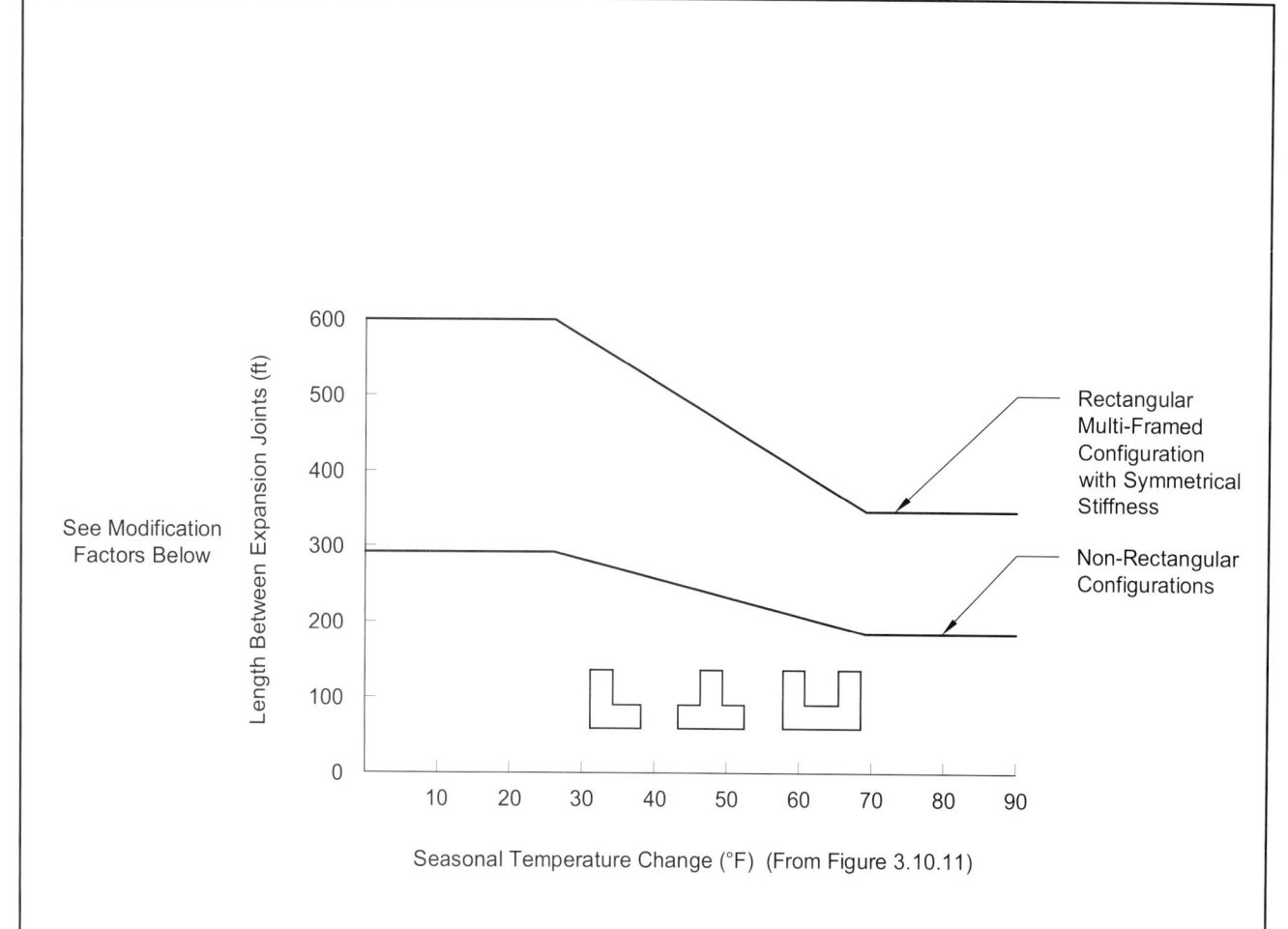

These curves are directly applicable to buildings of beam-and-column construction, hinged at the base, and with heated interiors. When other conditions prevail, the following rules are applicable:

(a) If the building will be heated only and will have hinged-column bases, use the length as specified;
(b) If the building will be air conditioned as well as heated, increase the length by 15% (provided the environmental control system will run continuously);
(c) If the building will be unheated, decrease the length by 33%;
(d) If the building will have fixed-column bases, decrease the length by 15%;
(e) If the building will have substantially greater stiffness against lateral displacement at one end of the plan dimension, decrease the length by 25%.

When more than one of these design conditions prevail in a building, the percentile factor to be applied should be the algebraic sum of the adjustment factors of all various applicable conditions.

Note: Figure 3.10.20 should be used as a preliminary guideline for the spacing of expansion joints. It should not be used as the absolute, authoritative, or sole means of determining the spacing of expansion joints. The spacing of expansion joints should be determined by the designer, based on analysis or experience, for the structure being considered.

Figure 3.10.21 Build-up of restraint forces in beams (k_b)

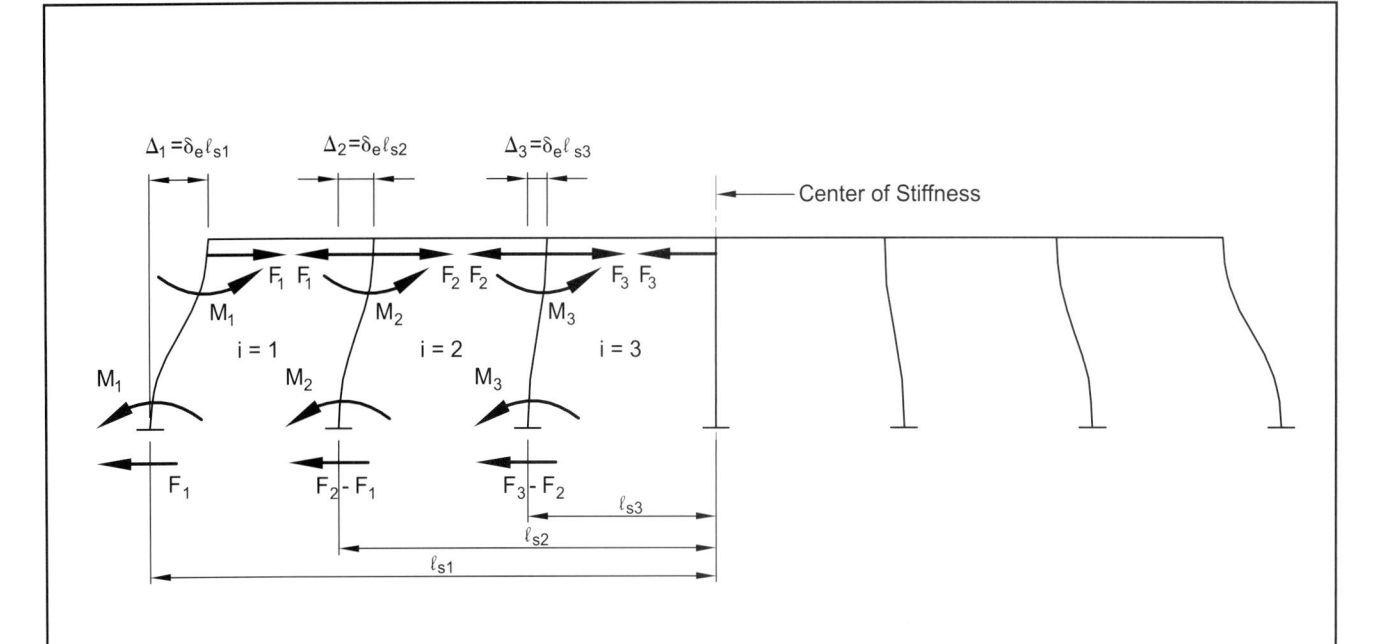

Values of k_b (see Figure 3.10.25)

Total number of bays (n)	Number of bays from end (i)							
	1	2	3	4	5	6	7	8
2	1.00							
3	1.00	4.00						
4	1.00	3.00						
5	1.00	2.67	9.00					
6	1.00	2.50	6.00					
7	1.00	2.40	5.00	16.00				
8	1.00	2.33	4.50	10.00				
9	1.00	2.29	4.20	8.00	25.00			
10	1.00	2.25	4.00	7.00	15.00			
11	1.00	2.22	3.86	6.40	11.67	36.00		
12	1.00	2.20	3.75	6.00	10.00	21.00		
13	1.00	2.18	3.67	5.71	9.00	16.00	49.00	
14	1.00	2.17	3.60	5.50	8.33	13.50	28.00	
15	1.00	2.15	3.55	5.33	7.86	12.00	21.00	64.00
16	1.00	2.14	3.50	5.20	7.50	11.00	17.50	36.00

Figure 3.10.22 Equivalent volume change strains for typical building elements (millionths)

Temp. zone (from map)	Prestressed members (P/A = 600 psi)									
	Normal weight concrete					Lightweight concrete				
	Avg. R.H. (from map)					Avg. R.H. (from map)				
	40	50	60	70	80	40	50	60	70	80
Heated buildings											
0	110	100	91	81	72	116	106	97	87	77
10	130	120	111	101	92	133	123	113	104	94
20	150	140	131	121	112	150	140	130	120	110
30	170	160	151	141	132	166	156	147	137	127
40	190	180	171	161	152	183	173	163	154	144
50	210	200	191	181	172	200	190	180	170	160
60	230	220	211	201	192	216	206	197	187	177
70	250	240	231	221	212	233	223	213	204	194
80	270	260	251	241	232	250	240	230	220	210
90	290	280	271	261	252	266	256	247	237	227
100	310	300	291	281	272	283	273	263	254	244
Unheated structures											
0	110	100	91	81	72	116	106	97	87	77
10	140	130	121	111	102	141	131	122	112	102
20	170	160	151	141	132	166	156	147	137	127
30	200	190	181	171	162	191	181	172	162	152
40	230	220	211	201	192	216	206	197	187	177
50	260	250	241	231	222	241	231	222	212	202
60	290	280	271	261	252	266	256	247	237	227
70	320	310	301	291	282	291	281	272	262	252
80	350	340	331	321	312	316	306	297	287	277
90	380	370	361	351	342	341	331	322	312	302
100	410	400	391	381	372	366	356	347	337	327

Figure 3.10.23 Equivalent volume change strains for typical building elements (millionths)

Temp. zone (from map)	Non-prestressed members									
	Normal weight concrete					Lightweight concrete				
	Avg. R.H. (from map)					Avg. R.H. (from map)				
	40	50	60	70	80	40	50	60	70	80
Heated buildings											
0	59	53	47	41	35	59	53	47	41	35
10	79	73	67	61	55	76	70	64	58	52
20	99	93	87	81	75	92	86	80	75	69
30	119	113	107	101	95	109	103	97	91	85
40	139	133	127	121	115	126	120	114	108	102
50	159	153	147	141	135	142	137	130	125	119
60	179	173	167	161	155	159	153	147	141	135
70	199	193	187	181	175	176	170	164	158	152
80	219	213	207	201	195	192	186	180	175	169
90	239	233	227	221	215	209	203	197	191	185
100	259	253	247	241	235	226	220	214	208	202
Unheated structures											
0	59	53	47	41	35	59	53	47	41	35
10	89	83	77	71	65	84	78	72	66	60
20	119	113	107	101	95	109	103	97	91	85
30	149	143	137	131	125	134	128	122	116	110
40	179	173	167	161	155	159	153	147	141	135
50	209	203	197	191	185	184	178	172	166	160
60	239	233	227	221	215	209	203	197	191	185
70	269	263	257	251	245	234	228	222	216	210
80	299	293	287	281	275	259	253	247	241	235
90	329	323	317	311	305	284	278	272	266	260
100	359	353	347	341	335	309	303	297	291	285

Figure 3.10.24 Coefficients k_f and k_m for forces and moments caused by volume change restraint forces

See Figure 3.10.25 for notation

No. of Stories	K_r $\dfrac{\sum E_b I_b / \ell}{\sum E_c I_c / h_s}$	Base Fixity	\multicolumn{4}{c	}{Values of k_f}	\multicolumn{6}{c	}{Values of k_m}
			F_1	F_2	F_3	F_4	Base M_1	2nd floor M_{2L}	2nd floor M_{2U}	3rd floor M_{3L}	3rd floor M_{3U}	4th M_4
1	0.0	Fixed	3.0	3.0			3.0	0.0				
		Pinned	0.0	0.0			0.0	0.0				
	0.5	Fixed	6.0	6.0			4.0	2.0				
		Pinned	1.2	1.2			0.0	1.2				
	1.0	Fixed	7.5	7.5			4.5	3.0				
		Pinned	1.7	1.7			0.0	1.7				
	2.0	Fixed	9.0	9.0			5.0	4.0				
		Pinned	2.2	2.2			0.0	2.2				
	4.0 or more	Fixed	10.1	10.1			5.4	4.7				
		Pinned	2.5	2.5			0.0	2.5				
2	0.0	Fixed	6.8	9.4	2.6		4.3	2.6	2.6	0.0		
		Pinned	0.0	3.0	1.5		0.0	1.5	1.5	0.0		
	0.5	Fixed	8.1	10.7	2.6		4.7	3.4	2.1	0.4		
		Pinned	1.9	3.4	1.4		0.0	1.9	1.2	0.2		
	1.0	Fixed	8.9	11.2	2.3		4.9	3.9	1.8	0.5		
		Pinned	2.1	3.4	1.3		0.0	2.1	1.0	0.3		
	2.0	Fixed	9.7	11.6	1.9		5.2	4.5	1.4	0.5		
		Pinned	2.4	3.4	1.0		0.0	2.4	0.8	0.3		
	4.0 or more	Fixed	10.4	11.9	1.4		5.5	5.0	1.0	0.4		
		Pinned	2.6	3.4	0.8		0.0	2.6	0.5	0.2		
3 or more	0.0	Fixed	7.1	10.6	4.1	0.70	4.4	2.8	2.8	0.7	0.70	0.00
		Pinned	1.6	3.6	2.4	0.40	0.0	1.6	1.6	0.4	0.40	0.00
	0.5	Fixed	8.2	11.1	3.5	0.50	4.7	3.5	2.2	0.7	0.40	0.09
		Pinned	1.9	3.6	1.9	0.30	0.0	1.9	1.2	0.4	0.20	0.05
	1.0	Fixed	8.9	11.4	2.9	0.40	5.0	3.9	1.9	0.7	0.30	0.09
		Pinned	2.2	3.5	1.6	0.20	0.0	2.2	1.0	0.4	0.20	0.05
	2.0	Fixed	9.7	11.7	2.2	0.20	5.2	4.7	1.4	0.6	0.20	0.06
		Pinned	2.4	3.5	1.2	0.10	0.0	2.4	0.8	0.3	0.10	0.03
	4.0 or more	Fixed	10.4	11.9	1.5	0.04	5.5	5.0	1.0	0.5	0.04	0.01
		Pinned	2.6	3.4	0.8	0.02	0.0	2.6	0.5	0.2	0.02	0.00

Figure 3.10.25 Use of Figure 3.10.24

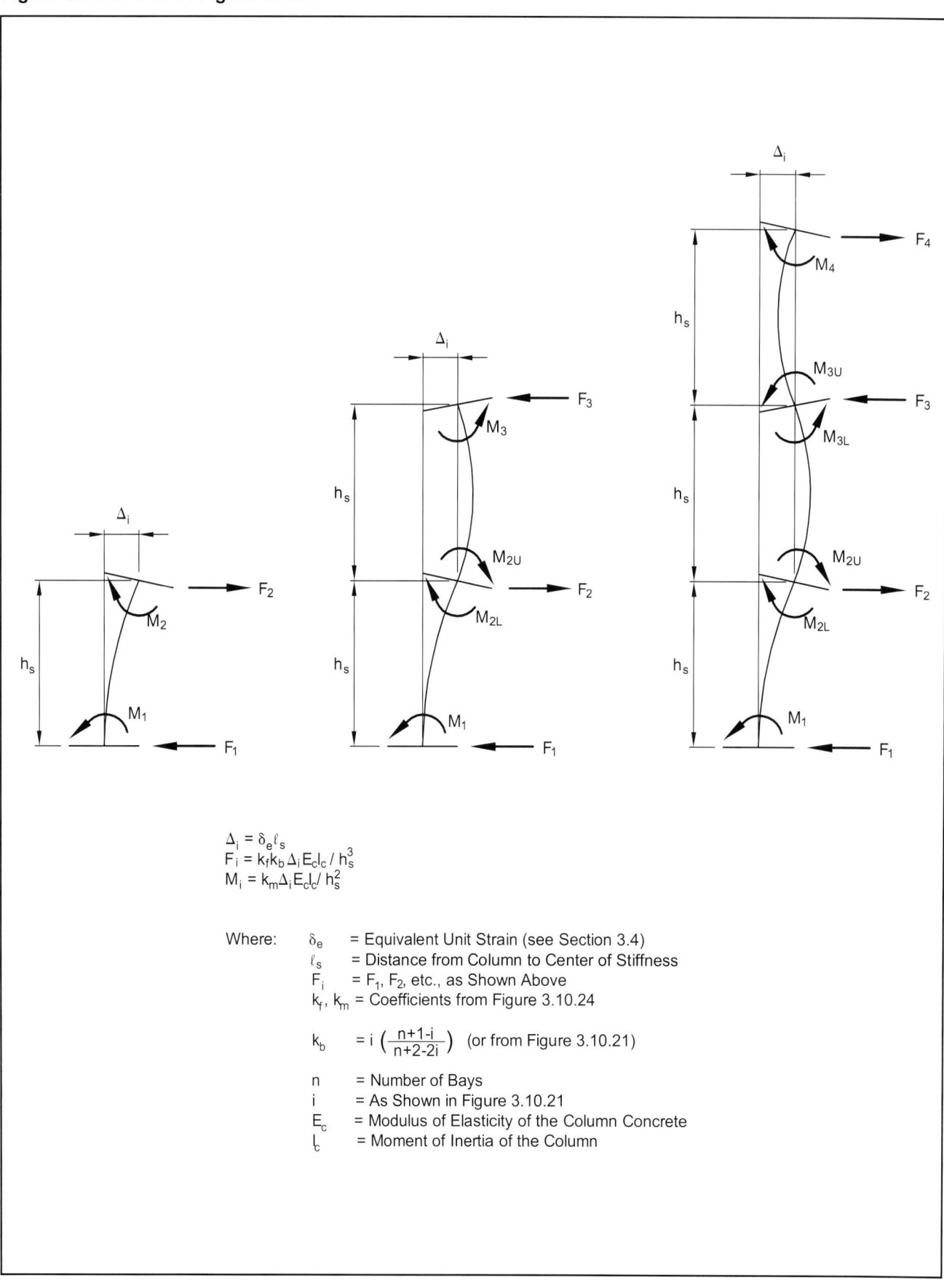

$\Delta_i = \delta_e \ell_s$
$F_i = k_f k_b \Delta_i E_c I_c / h_s^3$
$M_i = k_m \Delta_i E_c I_c / h_s^2$

Where:
δ_e = Equivalent Unit Strain (see Section 3.4)
ℓ_s = Distance from Column to Center of Stiffness
F_i = F_1, F_2, etc., as Shown Above
k_f, k_m = Coefficients from Figure 3.10.24

k_b = $i \left(\dfrac{n+1-i}{n+2-2i} \right)$ (or from Figure 3.10.21)

n = Number of Bays
i = As Shown in Figure 3.10.21
E_c = Modulus of Elasticity of the Column Concrete
I_c = Moment of Inertia of the Column

Figure 3.10.26 Shear wall deflection

Case	Deflection Due to Flexure	Deflection Due to Shear	Equivalent Moment of Inertia I_{eq} Single Story	Equivalent Moment of Inertia I_{eq} Multi-Story
(cantilever with point load P at top, height h, length ℓ, thickness t)	$\dfrac{Ph^3}{3EI}$	$\dfrac{2.78Ph}{A_w E}$ ($A_w = \ell t$)	$\dfrac{I}{1+\dfrac{8.34I}{A_w h^2}}$	$\dfrac{I}{1+\dfrac{13.4I}{A_w h^2}}$
(cantilever with uniform load w over height h)	$\dfrac{Wh^3}{8EI}$ W = wh	$\dfrac{1.39Wh}{A_w E}$ W = wh	NA	$\dfrac{I}{1+\dfrac{23.6I}{A_w h^2}}$
(fixed-fixed with point load P at top)	$\dfrac{Ph^3}{12EI}$	$\dfrac{2.78Ph}{A_w E}$	$\dfrac{I}{1+\dfrac{33.4I}{A_w h^2}}$	NA

CHAPTER 4
DESIGN OF PRECAST AND PRESTRESSED CONCRETE COMPONENTS

4.0	Notation	4–3
4.1	Introduction	4–8
4.2	Flexure	4–8
	4.2.1 Strength Design	4–8
	4.2.1.1 Depth of Stress Block	4–8
	4.2.1.2 Flanged Elements	4–8
	4.2.1.3 Strength Reduction Factor	4–10
	4.2.1.4 Limitations on Reinforcement	4–10
	4.2.1.5 Critical Section	4–11
	4.2.1.6 Analysis Using Code Equations	4–11
	4.2.1.7 Analysis Using Strain Compatibility	4–11
	4.2.1.8 Use of Design Aids	4–12
	4.2.2 Service Load Design	4–22
	4.2.2.1 Non-Prestressed Element Design	4–22
	4.2.2.2 Prestressed Element Design	4–22
	4.2.3 Prestress Transfer and Strand Development	4–41
	4.2.3.1 Strand Debonding	4–41
	4.2.4 End Stresses at Transfer	4–45
	4.2.5 Bending of Asymmetrical Sections	4–45
4.3	Shear	4–46
	4.3.1 Shear Resistance of Non-Prestressed Concrete	4–46
	4.3.2 Shear Resistance of Prestressed Concrete Members	4–46
	4.3.3 Design Using Design Aids	4–49
	4.3.4 Shear Reinforcement	4–51
	4.3.5 Horizontal Shear Transfer in Composite Members	4–51
	4.3.6 Shear-Friction	4–55
4.4	Torsion	4–57
4.5	Beams with Ledges	4–65
	4.5.1 Shear Strength of Ledge	4–65
	4.5.2 Transverse (Cantilever) Bending of Ledge	4–67
	4.5.3 Longitudinal Bending of Ledge	4–67
	4.5.4 Attachment of Ledge to Web	4–67
	4.5.5 Out-of-Plane Bending Near Beam End	4–67
	4.5.6 Pocketed Beams	4–73
4.6	Bearing	4–77
	4.6.1 Bearing on Plain Concrete	4–77
	4.6.2 Reinforced Concrete Bearing	4–77
	4.6.3 Dapped-End Bearing	4–79
	4.6.3.1 Flexure and Axial Tension in Extended End	4–80
	4.6.3.2 Direct Shear	4–80
	4.6.3.3 Diagonal Tension at Re-entrant Corner	4–81
	4.6.3.4 Diagonal Tension in Extended End	4–81
	4.6.3.5 Anchorage of Reinforcement	4–81
	4.6.3.6 Other Considerations	4–81

4.7	Loss of Prestress	4–84
	4.7.1 Sources of Stress Loss	4–84
	4.7.2 Range of Values for Total Loss	4–84
	4.7.3 Estimating Prestress Loss	4–84
	4.7.4 Critical Locations	4–85
4.8	Camber and Deflection	4–86
	4.8.1 Initial Camber	4–88
	4.8.2 Elastic Deflections	4–89
	4.8.3 Bilinear Behavior	4–89
	4.8.3.1 Cracked Section Analysis	4–90
	4.8.4 Long-Term Camber/Deflection	4–93
	4.8.5 Bowing	4–95
4.9	Compression Members	4–97
	4.9.1 Strength Design of Precast Concrete Compression Members	4–98
	4.9.2 Eccentrically Loaded Columns	4–106
	4.9.3 Slenderness Effects in Columns and Wall Panels	4–107
	4.9.3.1 Second-Order (P-Δ) Analysis	4–107
	4.9.4 Effective Width of Wall Panels	4–107
	4.9.5 Varying Section Properties of Compression Members	4–110
	4.9.6 Piles	4–111
	4.9.6.1 Service Load Design	4–111
	4.9.6.2 Strength Design	4–111
	4.9.6.3 Other Considerations	4–111
4.10	Special Considerations	4–114
	4.10.1 Load Distribution	4–115
	4.10.2 Openings through Decks	4–115
	4.10.3 Openings through Webs	4–115
	4.10.4 Continuity	4–117
	4.10.5 Cantilevers	4–117
4.11	References	4–118
4.12	Design Aids	4–120

DESIGN OF PRECAST AND PRESTRESSED CONCRETE COMPONENTS

4.0 Notation

a	=	depth of equivalent rectangular stress block
A	=	cross-sectional area
A_1	=	loaded area
A_2	=	area of lower base of largest frustum of a pyramid, cone or tapered wedge contained wholly within the support and having for its upper base the loaded area, and having side slopes of 1 vertical to 2 horizontal
A_{comp}	=	cross-sectional area of equivalent rectangular stress block
A_{cr}	=	area of crack interface
A_{cs}	=	area of horizontal shear ties
A_f	=	in a flanged section, area of flange outside web
A_g	=	gross area of concrete cross section
A_h	=	shear-friction steel across vertical crack at dapped ends
A_ℓ	=	total area of longitudinal reinforcement to resist torsion
A_ℓ	=	area of longitudinal reinforcement to resist bending in ledges
A_n	=	area of reinforcement required to resist axial tension
A_{ps}	=	area of prestressed reinforcement
A_{pw}	=	portion of prestressed reinforcement to develop web strength
A_s	=	area of non-prestressed tension reinforcement
A_s'	=	area of non-prestressed compression reinforcement
$A_{s,min}$	=	minimum amount of flexural reinforcement
A_{sh}	=	area of vertical reinforcement for horizontal or diagonal cracks
A_t	=	area of one leg of closed stirrup
A_t'	=	area of transformed cracked section
A_{top}	=	effective area of cast-in-place composite topping
A_v	=	area of diagonal tension reinforcement in dapped end
A_v	=	area of shear reinforcement
A_{vf}	=	area of shear-friction reinforcement
A_{vt}	=	required area of stirrups at end
A_w	=	area of web in a flanged section
$A_{w\ell}$	=	area of steel in vertical direction at beam end for torsional equilibrium
A_{wv}	=	area of steel in vertical direction at beam end for torsional equilibrium
b	=	width of compression or tension face of member
b_ℓ	=	L-beam width of web and one ledge
b_t	=	width of bearing area under concentrated loads on ledges
b_v	=	width of interface surface in a composite member (Section 4.3.5)
b_w	=	web width
c	=	distance from extreme compression fiber to neutral axis
c_c	=	clear cover to reinforcement from tension face
C	=	coefficient as defined in section used (with subscripts)
C	=	compressive force
C_c	=	compressive force capacity of composite topping
C_r	=	reduction coefficient (see Section 4.6.1)
CR	=	creep of concrete
d	=	distance from extreme compression fiber to centroid of longitudinal tension reinforcement, but need not be less than 0.8h for prestressed members (d_p is used for prestressed members when a distinction from d for non-prestressed reinforcement is relevant)
d'	=	distance from extreme compression fiber to centroid of compression reinforcement (d_p' is used for prestressed members when a distinction from non-prestressed reinforcement is relevant)

d_b	= nominal diameter of reinforcing bar or prestressing strand
d_c	= concrete thickness to center of reinforcement closest to tension face
d_e	= distance from center of load to beam end
d_ℓ	= depth of centroid of reinforcement in L-beam ledges
d_p, d_p'	= distance from extreme compression fiber to centroid of prestressing steel in tension and compression zones, respectively
d_t	= distance from extreme compression fiber to extreme tension steel
d_w	= depth of reinforcement from outside face of ledger beam
D	= distance from extreme compression fiber to centroid of tension reinforcement in a dapped end member
e	= eccentricity of design load or prestressing force parallel to axis measured from centroid of section
e'	= distance between c.g. of strand at end and c.g. of strand at lowest point $= e_c - e_e$
e_c	= eccentricity of prestressing force from centroid of section at center of span
e_e	= eccentricity of prestressing force from centroid of section at end of span
E_c	= modulus of elasticity of concrete
E_{ci}	= modulus of elasticity of concrete at time of initial prestress
E_{ps}	= modulus of elasticity of prestressed reinforcement
E_s	= modulus of elasticity of steel
ES	= elastic shortening
f_b	= stress in bottom fiber of cross section
f_c'	= specified compressive strength of concrete
f_{ct}	= specified compressive strength of composite topping
f_{cds}	= concrete stress at center of gravity of prestressing force due to all permanent (dead) loads not used in computing f_{cir}
f_{ci}'	= compressive strength of concrete at time of initial prestress
f_{cir}	= concrete stress at center of gravity of prestressing force immediately after transfer
f_d	= stress due to service dead load
f_e	= total load stress in excess of f_r
f_ℓ	= stress due to service live load
f_{pc}	= compressive stress in concrete at centroid of cross section due to prestress (after allowance for all prestress losses)
f_{pd}	= stress in prestressed reinforcement limited by strand development
f_{pe}	= compressive stress in concrete due to effective prestress forces only (after allowance for all prestress losses) at extreme fiber of section where tensile stress is caused by externally applied loads
f_{pi}	= compressive stress in concrete due to initial prestress force
f_{ps}	= stress in prestressed reinforcement at nominal strength of member
f_{pu}	= ultimate strength of prestressing steel
f_{py}	= specified yield strength of prestressing steel
f_r	= modulus of rupture of concrete
f_r	= allowable flexural tension
f_r'	= allowable flexural tension, computed using gross concrete section
f_s	= stress in non-prestressed reinforcement
f_s'	= stress in non-prestressed compression reinforcement
f_{se}	= effective stress in prestressing steel after losses
f_t	= stress in top fiber of cross section
f_t	= tensile stress at extreme tension fiber
$f_{t\ell}$	= final calculated stress in member
f_y	= specified yield strength of non-prestressed reinforcement
f_y'	= specified yield strength of compression reinforcement
$f_{y\ell}$	= yield strength of longitudinal torsional reinforcement, psi
f_{yv}	= yield strength of closed transverse torsional reinforcement, psi
F_h	= horizontal shear force
F.S.	= factor of safety
h	= depth of member above dap
h	= overall depth of non-dapped member
h	= unsupported length of pile
h_1	= distance from centroid of tensile reinforcement to neutral axis

h_2	= distance from extreme tension fiber to neutral axis
h_f	= depth of flange
h_ℓ	= ledger beam depth of ledge
h_p	= height of pocket in pocketed spandrel beam
h_s	= distance between torsional equilibrium reactions
H	= total depth of dapped end member
H_u	= factored torsional equilibrium reactions
I	= moment of inertia
I_e	= effective moment of inertia for computation of deflection
I_{equiv}	= equivalent constant moment of inertia in members with varying cross sections
I_g	= moment of inertia of gross section
I_t	= moment of inertia of transformed cracked section
J	= coefficient as defined in Section 4.7.4
j, k	= factors used in service load design (Section 4.2.2)
K	= coefficient as defined in Section 4.7.4 (with subscripts)
K_u	= design coefficient (see Figure 4.12.1)
K'_u	= design coefficient (see Figure 4.12.2)
ℓ	= span length
ℓ_d	= development length
ℓ_t	= strand transfer length
ℓ_{vh}	= horizontal shear length as defined in Figure 4.3.5.2
m	= modification factor for hanger steel calculation
M	= service load moment
M_a	= total moment at section
M_{cr}	= cracking moment
M_d	= moment due to service dead load (unfactored)
M_g	= moment due to weight of member (unfactored)
M_ℓ	= moment due to service live load (unfactored)
M_{max}	= maximum factored moment at section due to externally applied loads
M_n	= nominal moment strength of section
M_{nb}	= nominal moment strength under balanced conditions
M_{nc}	= moment due to beam self weight plus dead loads applied before composite action
M_o	= nominal moment strength of compression member with zero axial load
M_{sd}	= moment due to superimposed dead load plus sustained live load (unfactored)
M_{top}	= moment due to topping (unfactored)
M_u	= applied factored moment at section
n	= modular ratio
n	= number of reinforcing bars
N	= unfactored axial load
N_u	= factored horizontal or axial force
P	= prestress force after losses
P_e	= equivalent compression force
P_i	= initial prestress force
P_n	= axial load nominal strength of compression member at given eccentricity
P_{nb}	= axial load nominal strength under balanced conditions
P_o	= axial load nominal strength of compression member with zero eccentricity
P_o	= prestress force at transfer
r	= radius of gyration
RE	= relaxation of tendons
R.H.	= average ambient relative humidity
s	= reinforcement spacing
s	= spacing of concentrated loads on ledger beam ledge
S	= section modulus
S_b	= section modulus with respect to bottom fiber of precast section
S_{bc}	= section modulus with respect to bottom fiber of composite section
S_t	= section modulus with respect to top fiber of a cross section
SH	= shrinkage of concrete
t	= thickness (used for various parts of members with subscripts)
T	= tensile force
T	= torsion
T_c	= torsion strength of concrete
T_n	= nominal torsional moment strength

T_u	=	factored torsional moment on section
TL	=	total prestress loss
V_c	=	nominal shear strength provided by concrete
V_{ci}	=	nominal shear strength provided by concrete when diagonal cracking is result of combined shear and moment
V_{cw}	=	nominal shear strength provided by concrete when diagonal cracking is result of principal tensile stress in web in excess of cracking stress
V_d	=	dead load shear (unfactored)
V_i	=	factored shear force at section due to externally applied loads occurring simultaneously with M_{max}
V_ℓ	=	live load shear (unfactored)
V_n	=	nominal shear strength
V_p	=	vertical component of effective prestress force at section considered
V_s	=	nominal shear strength provided by shear reinforcement
V_u	=	factored shear force at section
V/S	=	volume-to-surface ratio
w	=	unfactored load per unit length of beam or per unit area of slab
w_c	=	unit weight of concrete
w_d	=	unfactored dead load per unit length
w_ℓ	=	unfactored live load per unit length
w_{sd}	=	dead load due to superimposed loading plus sustained live load
$w_{t\ell}$	=	unfactored total load per unit length
w_u	=	factored total load per unit length or area
x	=	distance from support to point being investigated
x	=	short side of component rectangle
x_1	=	short side of closed tie
y′	=	distance from top of c.g. to A_{comp}
y_1	=	long side of component rectangle
y_1	=	long side of closed tie
y_b	=	distance from bottom fiber to center of gravity of section
y_s	=	distance from centroid of prestressed reinforcement to bottom fiber
y_t	=	distance from centroid of gross section to extreme fiber in tension
y_t	=	distance from top fiber to center of gravity of section
α	=	distance from end of member to strand depression point (Figure 4.12.10)
α_t	=	torsion coefficient
β_1	=	factor defined in Section 4.2.1.1
Δ	=	deflection (with subscripts)
ε_c	=	concrete strain
ε_c	=	net tensile strain in extreme tension steel
ε_{cu}	=	ultimate concrete strain
$\varepsilon_{ps}, \varepsilon_{pd}, \varepsilon'_{ps}$	=	strain in prestressing steel corresponding to f_{ps}, f_{pd}, f'_{ps}
ε_s	=	net tensile strain in extreme tension steel
ε_s	=	strain in non-prestressed tension reinforcement
ε'_s	=	strain in non-prestressed compression reinforcement
ε_{sa}	=	strain in prestressing steel caused by external loads = $\varepsilon_{ps} - \varepsilon_{se}$
ε_{se}	=	strain in prestressing steel after losses
ε_y	=	maximum allowable strain in non-prestressed reinforcement
ϕ	=	strength reduction factor
γ	=	a factor dependent on level of prestress $= \sqrt{1+10\dfrac{f_{pc}}{f'_c}}$
γ_p	=	factor for type of prestressing tendon (see ACI Code Section 18.0 for values)
γ_t	=	factor used in designing hanger reinforcement (see Table 4.5.4.1)
λ	=	conversion factor for unit weight of concrete (Section 4.3.3)
μ	=	shear-friction coefficient (see Table 4.3.6.1)
μ_e	=	effective shear-friction coefficient
θ	=	angle of compression diagonals in truss analogy for torsion
ρ	=	A_s/bd = ratio of non-prestressed tension reinforcement
ρ'	=	A'_s/bd = ratio of non-prestressed compression reinforcement
ρ_{bal}	=	non-prestressed reinforcement ratio producing balanced strain conditions
ρ_{max}	=	maximum reinforcement ratio for non-prestressed members

ρ_p = A_{ps}/bd_p = ratio of prestressed reinforcement

ρ_w = $A_s/b_w d$ = ratio of non-prestressed tension reinforcement based on web width

ω = $\rho f_y / f'_c = A_s f_y / bd f'_c$

ω' = $\rho' f_y / f'_c = A'_s f_y / bd f'_c$

$\bar{\omega}$ = factor used in Figure 4.12.1

ω_{max} = maximum permissible ω

ω_p = $\rho_p f_{ps} / f'_c = A_{ps} f_{ps} / bd_p f'_c$

ω_{pu} = $\rho_p f_{pu} / f'_c = A_{ps} f_{pu} / bd_p f'_c$

ω_w, ω_{pw}, ω'_w = reinforcement indices for flanged-sections computed as for ω, ω_{pu}, and ω' except that b shall be the web width, and reinforcement area shall be that required to develop compressive strength of web only

ξ = time-dependent factor for sustained loads

4.1 Introduction

This chapter of the Handbook provides a summary of theory and procedures used in the design of precast and prestressed concrete components. Designs are based on the provisions of "Building Code Requirements for Structural Concrete (ACI 318-02)," [1] which is referred to as "the Code" or "ACI 318" in the Handbook. Occasionally, standard design practice may require interpretation of the Code. This has been done in Section 10.5 of the Handbook and is used where applicable in this chapter.

Two different phases must be considered in designing precast concrete components: (1) the manufacturing through erection phase and (2) the in-service conditions. The designer is referred to Chapter 5 for the first phase. This chapter is concerned with the in-service conditions.

The load tables in Chapter 2 will not provide all the design data necessary. In most cases, the engineer will select standard sections and typical details, with the detailed design carried out by the precast concrete producer or his consultant. The engineer of record should verify that the section selected is capable of satisfying both strength and serviceability criteria for the use intended. In cases where the engineer of record undertakes the complete design responsibility, consultation with producers in the area will ensure compatibility of design with the producer's production capability and will result in optimum quality and economy.

4.2 Flexure

Design for flexure in accordance with the Code requires that precast and prestressed concrete members be checked for both strength (Code Section 18.7) and serviceability requirements (Code Section 18.4). Members reinforced with non-prestressed steel require minimum reinforcement spacing in accordance with Code Section 10.6.4. Also see Section 4.2.2 of this Handbook.

4.2.1 Strength Design

Strength design is based on solution of the equations of equilibrium, normally using the rectangular stress block in accordance with Section 10.2.7 of the Code (see Figure 4.2.1.1). The stress in the prestressing steel at nominal strength, f_{ps}, can be determined by strain compatibility [2] or by the approximate equation given in the Code (Eq. 18-3). In carrying out strain compatibility analysis, the strand manufacturer's stress-strain relations may be used. Alternatively, the idealized stress-strain equa-

Figure 4.2.1.1 Nominal flexural resistance

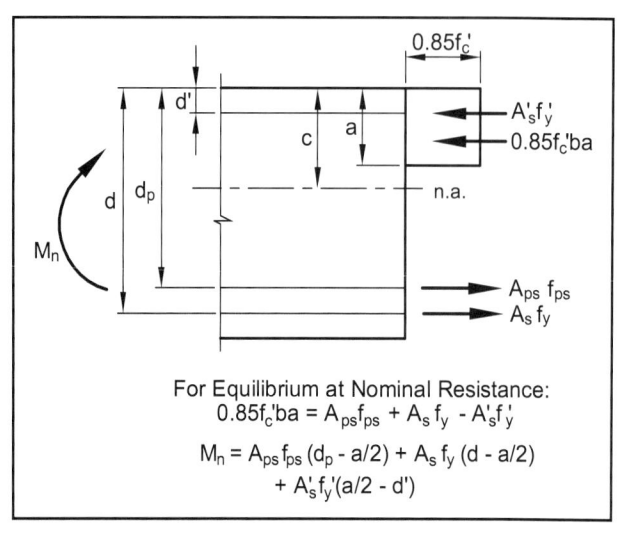

For Equilibrium at Nominal Resistance:
$0.85 f_c' ba = A_{ps} f_{ps} + A_s f_y - A_s' f_y'$

$M_n = A_{ps} f_{ps} (d_p - a/2) + A_s f_y (d - a/2) + A_s' f_y' (a/2 - d')$

tions given in Design Aid 11.2.5 may be used.

For elements with compression reinforcement, the nominal strength can be calculated by assuming that the compression reinforcement yields. This assumption should be subsequently verified from the strain diagram. The designer will normally choose a section and reinforcement and then determine if it meets the basic design strength requirement:

$\phi M_n \geq M_u$

A flowchart illustrating the calculation of the nominal strength of flexural elements is given in Figure 4.2.1.2.

4.2.1.1 Depth of Stress Block

The depth "a" of the rectangular stress block is related to the depth to the neutral axis "c" by the equation:

$a = \beta_1 c$

where, by Code Section 10.2.7.3:

β_1	f_c', psi
0.85	3000
0.85	4000
0.80	5000
0.75	6000
0.70	7000
0.65	8000 and higher

4.2.1.2 Flanged Elements

The equations for nominal strength given in Figure 4.2.1.1 apply to rectangular cross sections and flanged sections in which the stress block lies entirely within the depth of the flange h_f. The depth of the stress block "a", is obtained from the first equation of equilibrium in Figure 4.2.1.1:

Figure 4.2.1.2 Flowchart of nominal strength calculations for flexure

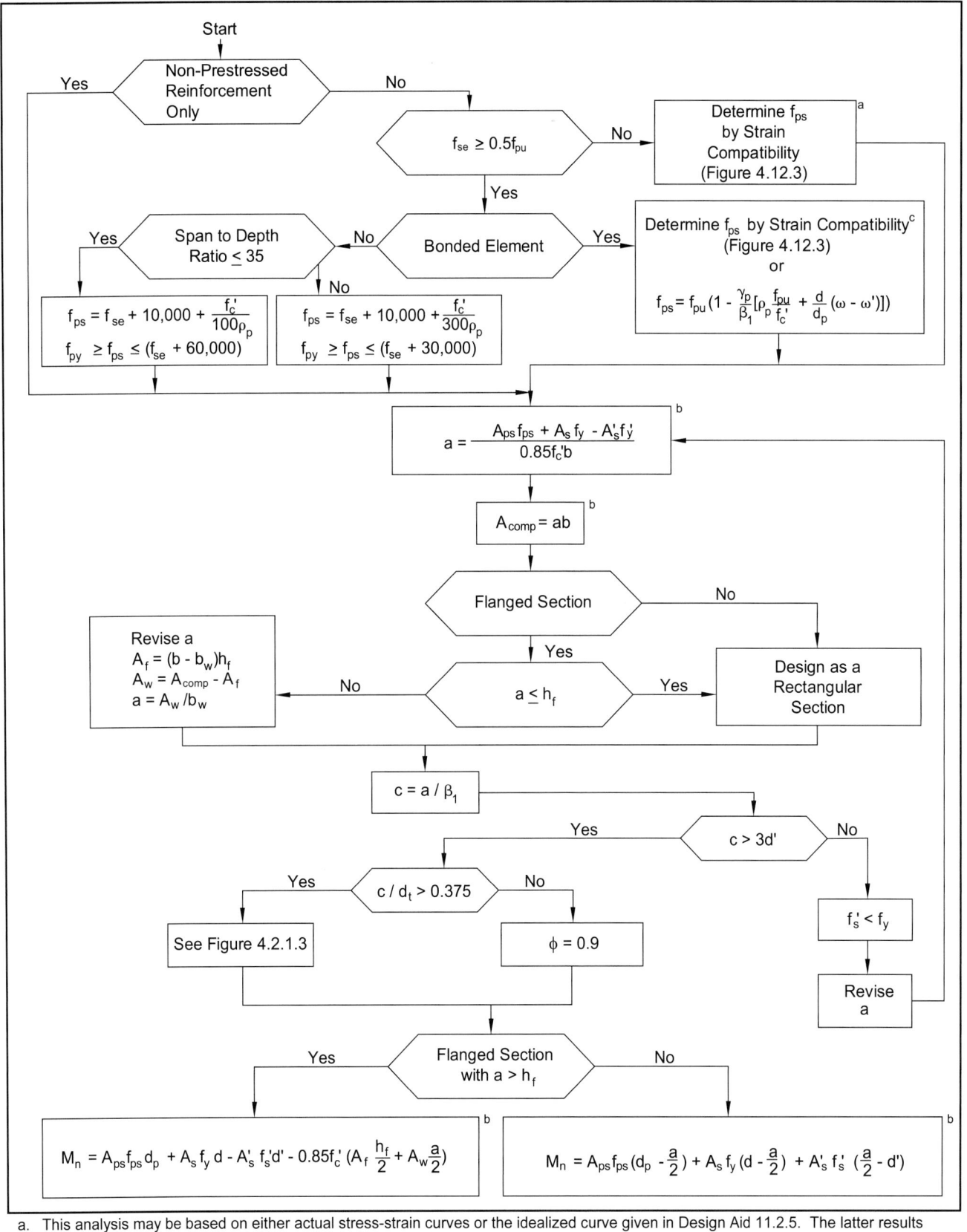

```
                          Start
                            │
                            ▼
              ┌─────────────────────────┐
   Yes ◄──────┤ Non-Prestressed         ├──── No
              │ Reinforcement Only      │
              └─────────────────────────┘
                                              │
                                              ▼
                                    ┌──────────────────┐      ┌──────────────────────┐
                                    │ f_se ≥ 0.5 f_pu  ├─No──►│ Determine f_ps by    │ [a]
                                    └──────────────────┘      │ Strain Compatibility │
                                              │               │ (Figure 4.12.3)      │
                                             Yes              └──────────────────────┘
                                              │
   ┌─────────────────┐    No    ┌──────────────────┐   Yes    ┌──────────────────────────────┐
   │ Span to Depth   │◄─────────┤ Bonded Element   ├─────────►│ Determine f_ps by Strain     │ [c]
   │ Ratio ≤ 35      │          └──────────────────┘          │ Compatibility (Figure 4.12.3)│
   └─────────────────┘                   │                    │ or                           │
       Yes │     │ No                                         │ equation for f_ps            │
```

$$f_{ps} = f_{se} + 10{,}000 + \frac{f'_c}{100\rho_p}$$
$$f_{py} \geq f_{ps} \leq (f_{se} + 60{,}000)$$

$$f_{ps} = f_{se} + 10{,}000 + \frac{f'_c}{300\rho_p}$$
$$f_{py} \geq f_{ps} \leq (f_{se} + 30{,}000)$$

$$f_{ps} = f_{pu}\left(1 - \frac{\gamma_p}{\beta_1}\left[\rho_p \frac{f_{pu}}{f'_c} + \frac{d}{d_p}(\omega - \omega')\right]\right)$$

$$a = \frac{A_{ps} f_{ps} + A_s f_y - A'_s f'_y}{0.85 f'_c b}$$ [b]

$$A_{comp} = ab$$ [b]

Flanged Section? — No → Design as a Rectangular Section

If Yes: $a \leq h_f$? — Yes → Design as a Rectangular Section

If No:
Revise a
$A_f = (b - b_w) h_f$
$A_w = A_{comp} - A_f$
$a = A_w / b_w$

$$c = a / \beta_1$$

$c > 3d'$? — No → $f'_s < f_y$ → Revise a
If Yes: $c/d_t > 0.375$?
 - Yes → See Figure 4.2.1.3
 - No → $\phi = 0.9$

Flanged Section with $a > h_f$?

Yes:
$$M_n = A_{ps} f_{ps} d_p + A_s f_y d - A'_s f'_s d' - 0.85 f'_c \left(A_f \frac{h_f}{2} + A_w \frac{a}{2}\right)$$ [b]

No:
$$M_n = A_{ps} f_{ps}\left(d_p - \frac{a}{2}\right) + A_s f_y\left(d - \frac{a}{2}\right) + A'_s f'_s\left(\frac{a}{2} - d'\right)$$ [b]

a. This analysis may be based on either actual stress-strain curves or the idealized curve given in Design Aid 11.2.5. The latter results in f_{ps} values given in Figure 4.12.3.
b. These equations assume a prismatic section. For irregular sections, reasonable approximations may be used. See also Figure 4.9.1.1
c. Check strand development length.

$$a = \frac{A_{ps}f_{ps} + A_s f_y - A'_s f'_y}{0.85 f'_c b} \quad \text{(Eq. 4.2.1.1)}$$

If $a > h_f$, and compression reinforcement is present, use of strain compatibility or reasonable approximations for it may be made to find the nominal strength. If compression reinforcement is not present, the nominal strength can be found using the Code equations shown in Figure 4.2.1.2 or by strain compatibility.

4.2.1.3 Strength Reduction Factor

ACI 318-02 classes prestressed and non-prestressed members as tension or compression controlled, and determines the strength reduction factor, ϕ, on that basis. A compression-controlled section is defined as one having a maximum net tensile strain in the extreme tension steel of 0.002 or less (for Grade 60 reinforcement and prestressing steel), and a tension-controlled section is defined as one having a maximum tensile strain in the steel of 0.005 or more. (In prestressed members, the strain limits are after the decompression.)

The net tensile strain limits of 0.002 and 0.005 may also be expressed in terms of c/d_t, where c is the depth to the neutral axis and d_t is the distance from the extreme compression fiber to the tension steel farthest from that fiber. Figure 4.2.1.3 shows ϕ as a function of ε_t and c/d_t.

For the rectangular sections with one layer of tension steel, the net tensile strain limit of 0.005 corresponds to $\omega = 0.32\beta_1$.

For flexural members, it is usually best to design sections of maximum moment to be tension controlled, adding compression steel, if necessary, to limit c/d_t to 0.375. In the transition zone shown in Figure 4.2.1.3, the reduction in ϕ cancels the benefits of higher amounts of reinforcement without balancing compression steel.

4.2.1.4 Limitations on Reinforcement

For non-prestressed flexural elements, except slabs of uniform thickness, the minimum reinforcement required, is:

$$A_{s,min} = 3\frac{\sqrt{f'_c}}{f_y} b_w d \ge \frac{200 b_w d}{f_y} \quad \text{(Eq. 4.2.1.2)}$$

For T-sections with flanges in tension, $A_{s,min}$ is the smaller of that determined by Eq. 4.2.1.2, where b_w is the width of the flange, or by:

$$A_{s,min} = \frac{6\sqrt{f'_c}}{f_y} b_w d \quad \text{(Eq. 4.2.1.3)}$$

Figure 4.2.1.3 Variation of ϕ with net tensile strain for Grade 60 reinforcement and for prestressing steel

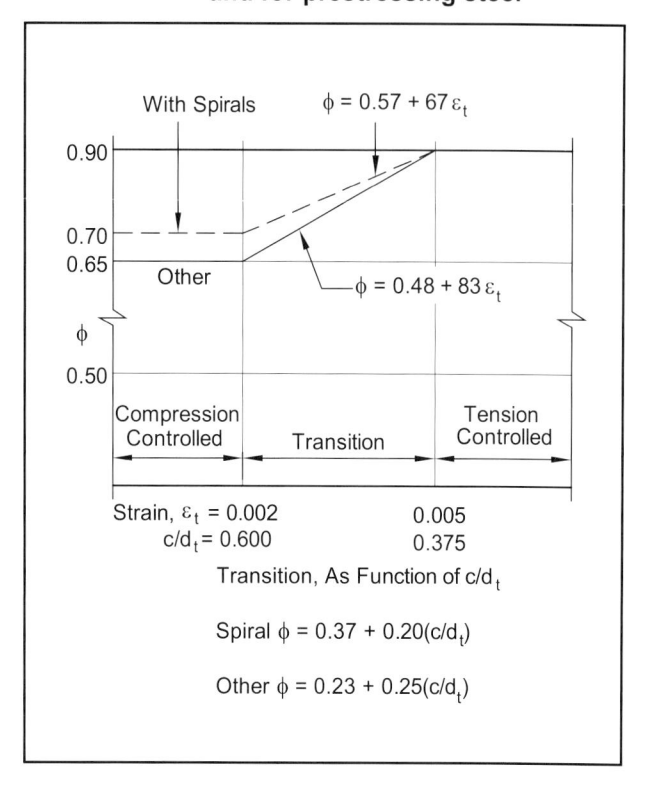

With Spirals $\phi = 0.57 + 67\varepsilon_t$
Other $\phi = 0.48 + 83\varepsilon_t$

Strain, $\varepsilon_t = 0.002$ 0.005
$c/d_t = 0.600$ 0.375

Transition, As Function of c/d_t

Spiral $\phi = 0.37 + 0.20(c/d_t)$

Other $\phi = 0.23 + 0.25(c/d_t)$

where:
  b_w = width of the web

The above limits do not apply if the area of tensile reinforcement provided is one-third greater than that required by analysis.

For slabs, the minimum flexural reinforcement is that amount required for shrinkage and temperature reinforcement, with a maximum spacing not exceeding three times the slab thickness or 18 in., whichever is smaller. Design Aid 11.2.13 shows minimum reinforcement requirements.

For prestressed flexural members to ensure adaquate ductility, the Code requires that the total prestressed and non-prestressed reinforcement be adequate to develop a design strength at least 1.2 times the cracking moment ($\phi M_n \ge 1.2 M_{cr}$). This provision is generally assumed to apply only at critical flexural sections, and is waived for members with shear and flexural strength at least twice that required by analysis. The cracking moment, M_{cr}, may be calculated by:

$$M_{cr} = S_{bc}\left[\frac{P}{A} + \frac{Pe}{S_b} + f_r\right] - M_{nc}\left(\frac{S_{bc}}{S_b} - 1\right)$$

(Eq. 4.2.1.4)

Figure 4.2.1.4 Critical section for flexural design

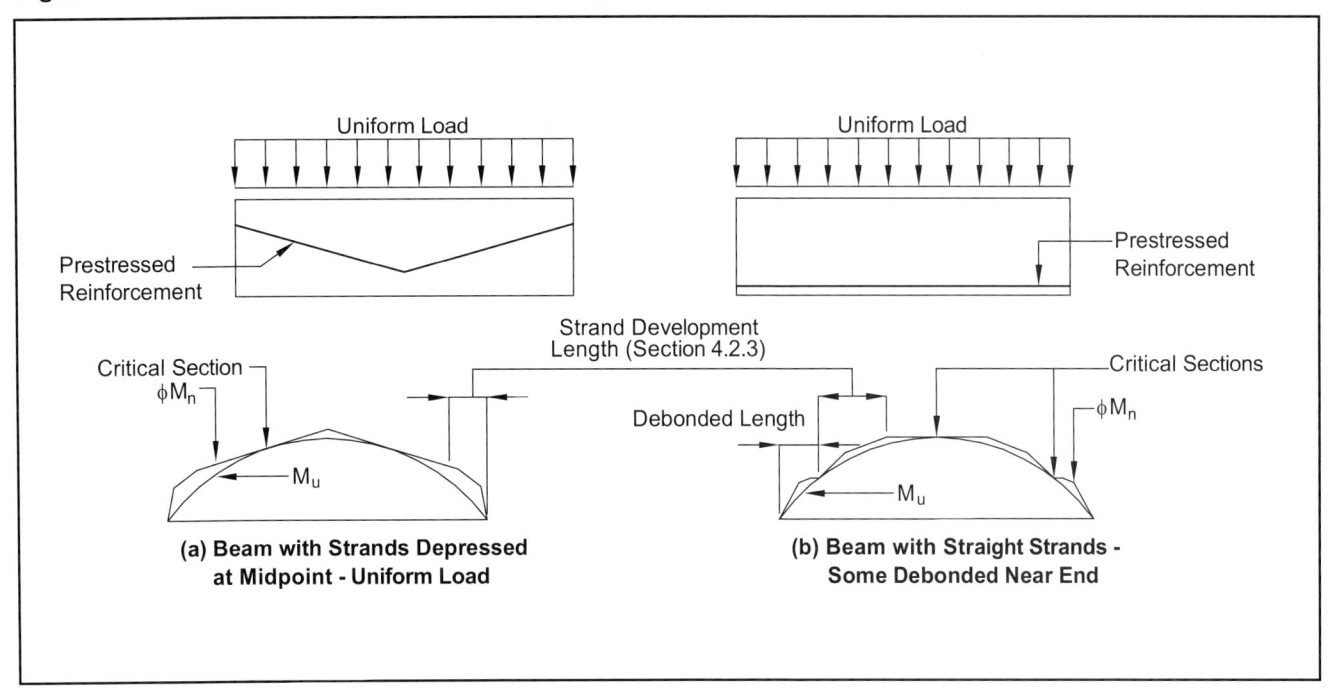

(a) Beam with Strands Depressed at Midpoint - Uniform Load

(b) Beam with Straight Strands - Some Debonded Near End

where:

M_{nc} = moment due to beam self-weight plus dead loads applied before composite action
S_b = section modulus with respect to the bottom fiber of the precast section
S_{bc} = section modulus with respect to the bottom fiber of a composite section
f_r = modulus of rupture, defined as $7.5\lambda\sqrt{f'_c}$

4.2.1.5 Critical Section

For simply supported, uniformly loaded, prismatic non-prestressed members, the critical section for flexural design will occur at midspan. For uniformly loaded prestressed members, in order to reduce the end stresses at release some strands are often depressed near midspan, or debonded for a length near the ends. For strands with a single-point depression, the critical section can usually be assumed at 0.4ℓ. For straight strands, the critical section will be at midspan, but if some strands are debonded near the end, an additional critical section may occur near the end of the debonded length as shown in Figure 4.2.1.4.

The presence of concentrated loads or non-prestressed reinforcement may alter the location of the critical section. In such cases, computer programs with the capability of checking the capacity at short intervals along the member length can be used to expedite the analysis.

4.2.1.6 Analysis Using Code Equations

Figure 4.2.1.2 essentially outlines the design procedures using the Code equations for prestressed and partially prestressed members. If compression reinforcement is present, the Code requires certain checks to ensure that the stress in the compression reinforcement is at its yield strength. In computing f_{ps}, if any compression reinforcement is taken into account, the term:

$$\left[\rho_p \frac{f_{pu}}{f'_c} + \frac{d}{d_p}(\omega - \omega')\right]$$

shall be taken not less than 0.17 and d' shall be no greater than $0.15d_p$. Alternatively, the yielding of the compression reinforcement is ensured if Eq. 4.2.1.5 is satisfied:

$$\frac{A_{ps}f_{ps} + A_s f_y - A'_s f'_s}{bd} \geq 0.85\beta_1 f'_c \frac{d'}{d}\left(\frac{87000}{87000 - f_y}\right)$$

(Eq. 4.2.1.5)

4.2.1.7 Analysis Using Strain Compatibility

Strain compatibility is an alternative, and more accurate method, to the Code equations. The procedure consists of assuming the location of the neutral axis, computing the strains in the prestressed and non-prestressed reinforcement, and establishing the depth of the stress block. Knowing the stress-strain relationship for the reinforcement, and assuming that the maximum strain in the concrete is 0.003, the forces in the reinforcement and in the concrete are determined and the sum of compression and tension forces is computed. If necessary, the neutral axis location is moved on a trial and error basis until the sum of forces is zero. The moment of these forces is then computed to obtain the nominal strength of the section.

4.2.1.8 Use of Design Aids

Figures 4.12.1 through 4.12.3 are provided to assist in the strength design of flexural members. Figure 4.12.1 can be used for members with prestressed or non-prestressed reinforcement, or combinations (partial prestressing). Note that to use this aid, it is necessary to determine f_{ps} from some other source, such as Eq. 18-3 of ACI 318-02 or Figure 4.12.3.

Figure 4.12.2 is for use only with fully prestressed members. The value of f_{ps} is determined by strain compatibility in a manner similar to that used in Figure 4.12.3. The reduction factor, ϕ, is included in the values of K'_u. Examples 4.2.1.1 through 4.2.1.4 illustrate the use of these design aids.

Example 4.2.1.1
Use of Figure 4.12.1 for Determination of Non-Prestressed Reinforcement

Given:
   The ledger beam shown.
   Applied factored moment, M_u = 1460 kip-ft
   f'_c = 5000 psi, normal weight concrete
   d = 72 in.
   f_y = 60 ksi

Problem:
   Find the amount of mild steel reinforcement, A_s, required.

Solution:
   Referring to Figure 4.12.1:
   $A_{ps} = 0$
   $A'_s = 0$

   $$K_u = \frac{\left(\frac{M_u}{\phi}\right)}{f'_c bd^2} = \frac{\frac{1460(12000)}{0.9}}{5000(8)(72)^2} = 0.0939$$

   For tension controlled sections, $\omega_{max} = 0.256$
   Required $\overline{\omega} = 0.10 < \omega_{max}$

   $$A_s = \frac{\overline{\omega} bdf'_c}{f_y} = \frac{0.10(8)(72)(5)}{60} = 4.80 \text{ in.}^2$$

   Use 5 – #9; $A_s = 5.00$ in.^{2}

   $$A_{s,min} = \left(\frac{3\sqrt{f'_c}}{f_y}b_w d = \frac{3\sqrt{5000}}{60,000}(8)(72)\right) = 2.04 \text{ in.}^2$$

   or $\frac{200 b_w d}{f_y} = 1.92$ in.^{2}

   2.04 in.^{2} < 5.00 in.^{2} OK

Note: Detailing of reinforcement must provide for adequate crack control (see Section 4.2.2.1).

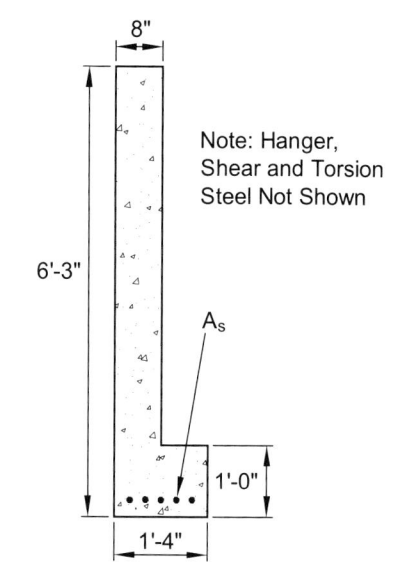

(Note: Hanger, Shear and Torsion Steel Not Shown)

Example 4.2.1.2
Use of Figure 4.12.2 for Determination of Prestressing Steel Requirements—Bonded Strand

Given:
 PCI standard rectangular beam 16RB24
 Applied factored moment, M_u = 600 kip-ft
 f'_c = 6000 psi normal weight concrete
 f_{pu} = 270 ksi, low-relaxation strand

Problem:
 Find the required amount of prestressing steel.

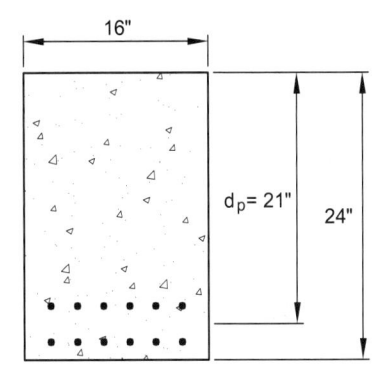

Solution:
Referring to Figure 4.12.2:

$$M_u \leq \phi M_n = \frac{K'_u bd_p^2}{12000} \text{ kip-ft}$$

Required $K'_u = \dfrac{M_u(12000)}{bd_p^2} = \dfrac{600(12000)}{16(21)^2} = 1020$

 for ω_{pu} = 0.22, K'_u = 1006
 for ω_{pu} = 0.23, K'_u = 1043

Therefore, by interpolation:

$$\omega_{pu} = 0.22 + \frac{1020 - 1006}{1043 - 1006}(0.01) = 0.224$$

$$A_{ps} = \frac{\omega_{pu} bd_p f'_c}{f_{pu}} = \frac{0.224(16)(21)(6)}{270} = 1.67 \text{ in.}^2$$

Use 12 – ½ in. diameter strands; A_{ps} = 1.84 in.²

Example 4.2.1.3
Use of Figure 4.12.3 — Values of f_{ps} by Stress-Strain Relationship — Bonded Strand

Given:
 3 ft - 4 in. x 8 in. hollow-core slab

Concrete:
 Use 5000 psi normal weight concrete

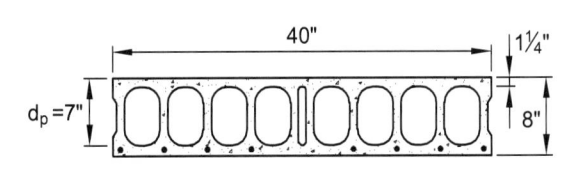

Prestressing Steel:
 8 – ⅜ in. diameter 270K low-relaxation strand
 A_{ps} = 8(0.085) = 0.68 in.²

Section Properties:
 A = 218 in.²
 S_b = 381 in.³
 y_b = 3.98 in.

Example 4.2.1.3 (Cont.)
Use of Figure 4.12.3 — Values of f_{ps} by Stress-Strain Relationship — Bonded Strand

Problem:
    Find design flexural strength, ϕM_n

Solution:
Determine $C\omega_{pu}$ for the section:

From table on Figure 4.12.3:
    for $f'_c = 5000$ psi, C = 1.06

$$C\omega_{pu} = C\frac{A_{ps}f_{pu}}{bd_pf'_c} + \frac{d}{d_p}(\omega - \omega')$$

Since $\omega = \omega' = 0$

$$C\omega_{pu} = \frac{1.06(0.68)(270)}{40(7)(5)} = 0.139$$

Entering Figure 4.12.3 with this parameter and an assumed effective stress, f_{se}, of 170 ksi gives a value of:
    $f_{ps} = 266$ ksi

Determine the flexural strength:

$$\phi M_n = \phi A_{ps} f_{ps}\left(d_p - \frac{a}{2}\right)$$

$$a = \frac{A_{ps}f_{ps}}{0.85 f'_c b}$$

$$= \frac{0.68(266)}{0.85(5)(40)} = 1.064$$

    a < flange thickness

$$c = \frac{a}{\beta_1} = \frac{1.064}{0.8} = 1.33$$

$$\frac{c}{d_p} = \frac{1.33}{7} = 0.19 < 0.375$$

tension controlled, $\phi = 0.9$

$$\phi M_n = 0.9(0.68)(266)\left[7 - \left(\frac{1.064}{2}\right)\right] = 1053 \text{ kip-in.} = 87.7 \text{ kip-ft}$$

Check the ductility requirement:
    $\phi M_n > 1.2 M_{cr}$
    $P = f_{se}A_{ps} = 170(0.68) = 116$ kips
    $e = 3$ in.

From Eq. 4.2.1.4:

$$1.2 M_{cr} = 1.2\left(\frac{P}{A} + \frac{Pe}{S_b} + 7.5\sqrt{f'_c}\right)S_b$$

$$1.2 M_{cr} = 1.2\left(\frac{116}{218} + \frac{116(3)}{381} + \frac{7.5\sqrt{5000}}{1000}\right)381 = 903 \text{ kip-in.} = 75.3 \text{ kip-ft} < 87.7 \text{ kip-ft} \quad \text{OK}$$

Example 4.2.1.4
Use of Figure 4.12.3 and Eq. 18-3 (ACI 318-02) for Partially Prestressed Member

Given:
 PCI standard double tee
 8DT24 + 2

Concrete:
 Precast: $f'_c = 5000$ psi
 Topping: $f'_c = 3000$ psi, normal weight

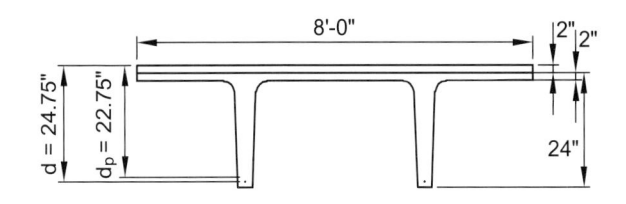

Reinforcement:
 12–½ in. diameter 270K low-relaxation strands (six each stem)
 $A_{ps} = 12(0.153) = 1.84$ in.^{2}
 A_s (2 – #6) $= 0.88$ in.^{2}

Problem:
 Find the design flexural strength of the composite section by the stress-strain relationship, Figure 4.12.3, and compare using Code Eq. 18-3.

Solution:
 Assume $f_{se} = 170$ ksi

$$C\omega_{pu} = C\frac{A_{ps}f_{pu}}{bd_p f'_c} + \frac{d}{d_p}(\omega - \omega')$$

$$\omega = \frac{A_s f_y}{bd f'_c} = \frac{0.88(60)}{96(24.75)(3)} = 0.0074; \; \omega' = 0$$

$$C\omega_{pu} = \frac{1.00(1.84)(270)}{96(22.75)(3)} + \frac{24.75}{22.75}(0.0074) = 0.084$$

From Figure 4.12.3:
 $f_{ps} = 268$ ksi

$$a = \frac{A_{ps}f_{ps} + A_s f_y}{0.85 f'_c b} = \frac{1.84(268) + 0.88(60)}{0.85(3)(96)} = 2.23 \text{ in.*}$$

$$M_n = A_{ps}f_{ps}\left(d_p - \frac{a}{2}\right) + A_s f_y\left(d - \frac{a}{2}\right) = 1.84(268)\left[22.75 - \left(\frac{2.23}{2}\right)\right] + 0.88(60)\left[24.75 - \left(\frac{2.23}{2}\right)\right]$$

 $= 11{,}917$ kip-in. $= 993$ kip-ft

$$c = \frac{a}{\beta_1} = \frac{2.23}{0.85} = 2.62$$

$$\frac{c}{d_t} = \frac{2.62}{24.5} = 0.107$$

 tension controlled, $\phi = 0.9$
 $\phi M_n = 0.9(993) = 894$ kip-ft

---

* Since a = 2.23 in. > 2.0 in. – the topping thickness, a more exact analysis requires a revised calculation to account for the higher strength concrete of the double tee flange. However, such a refinement is expected to produce a negligible difference in the results. In this case, the revised calculation yields a = 2.19 in. versus 2.23 in. and ϕM_n = 880 kip-ft versus 894 kip-ft.

Example 4.2.1.4 (Cont.)
Use of Figure 4.12.3 and Eq. 18-3 (ACI 318-02) for Partially Prestressed Member

Find f_{ps} using Eq. 18-3 (ACI 318-02):

$$f_{ps} = f_{pu}\left(1 - \frac{\gamma_p}{\beta_1}\left[\rho_p \frac{f_{pu}}{f'_c} + \frac{d}{d_p}(\omega - \omega')\right]\right)$$

$\omega' = 0$ in this example

$$f_{ps} = 270\left(1 - \frac{0.28}{0.85}\left[\frac{1.84(270)}{96(22.75)(3)} + \frac{24.75}{22.75}(0.0074)\right]\right) = 263 \text{ ksi (vs. 268 ksi from Figure 4.12.3)}$$

$\phi M_n = 879$ kip-ft

(Note: The ductility provisions [$\phi M_n \geq 1.2 M_{cr}$] should be checked as in Example 4.2.1.3)

Example 4.2.1.5
Flexural Strength of Double Tee Flange in Transverse Direction

For flanged sections, in addition to providing adequate flexural strength in the longitudinal direction, the flanges must be designed for bending in the transverse direction. This example illustrates a design for both uniformly distributed loads (Part A) and concentrated loads (Part B).

Fibers, which are sometimes used to control shrinkage cracks, do not transfer loads and, therefore, cannot be used to replace structural reinforcement such as welded wire reinforcement. This is particularly important for structural toppings over precast concrete decks. The reinforcement in these toppings cannot be replaced with fibers.

Part A

Given:
PCI standard double tee of Example 4.2.1.4. Topping is reinforced with 6x6–W1.4xW1.4 WWR, and the flange is reinforced with 6x6–W4.0xW4.0 WWR.

f'_c (topping) = 3000 psi
f'_c (precast) = 5000 psi
f_y = 60,000 psi*

Problem:
Find the uniform live load which the flange can support.

Solution:
From Design Aid 11.2.11
A_{s1} (WWR in precast) = W4.0 @ 6 in. = 0.080 in.²/ft
A_{s2} (WWR in topping) = W1.4 @ 6 in. = 0.028 in.²/ft

The cantilevered flange controls the design since the negative moment over the stem reduces the positive moment between the stems. Construct strain diagram:

Try c = 0.25 in.

$$\frac{\varepsilon_s + 0.003}{0.003} = \frac{d}{c}$$

---

* ACI 318-02, Section 3.5.3 allows f_y to exceed 60,000 psi if stress corresponds to a strain of 0.35%.

Example 4.2.1.5 (Cont.)
Flexural Strength of Double Tee Flange in Transverse Direction

Therefore:

$$\varepsilon_{s1} = \frac{0.003 d_1}{c} - 0.003$$

$$\varepsilon_{s2} = \frac{0.003 d_2}{c} - 0.003$$

$$\varepsilon_{s1} = \frac{0.003(1)}{0.25} - 0.003 = 0.009$$

$$\varepsilon_{s2} = \frac{0.003(3)}{0.25} - 0.003 = 0.033$$

$$\varepsilon_y = \frac{f_y}{E_s} = \frac{60}{29,000} = 0.0021 < 0.009$$

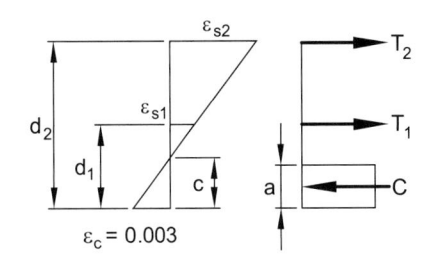

Therefore, reinforcement yields at both levels, and

$f_s = f_y = 60$ ksi
$a = \beta_1 c = 0.80(0.25) = 0.20$ in.
$C = 0.85 f'_c ba = 0.85(5)(12)(0.20) = 10.2$ kips/ft
$T_1 = A_{s1}f_y = 0.08(60) = 4.80$ kips/ft
$T_2 = A_{s2}f_y = 0.028(60) = 1.68$ kips/ft
$T_1 + T_2 = 6.48 < 10.2$ kips/ft

Therefore, try

$$a = \frac{(T_1 + T_2)}{(0.85 f'_c b)} = \frac{6.48}{[0.85(5)(12)]} = 0.13 \text{ in.}$$

$$c = \frac{0.13}{0.80} = 0.16$$

Check:

$$\varepsilon_{s1} = \frac{0.003(1)}{0.16} - 0.003 = 0.016 > 0.0021$$

Therefore, the reinforcement yields, and the analysis is valid.

Determine ϕ:

$$\frac{c}{d_t} = \frac{0.16}{1} = 0.16$$

From Figure 4.2.1.3, 0.16 < 0.375
Therefore, tension controls, and $\phi = 0.9$:

$$\phi M_n = \phi \left[T_1 \left(d_1 - \frac{a}{2} \right) + T_2 \left(d_2 - \frac{a}{2} \right) \right] = 0.9 \left[4.80 \left(1 - \frac{0.13}{2} \right) + 1.68 \left(3 - \frac{0.13}{2} \right) \right] = 8.48 \text{ kip-in./ft} = 707 \text{ lb-ft/ft}$$

Check minimum and maximum reinforcement:

$$d = \frac{0.028(3) + 0.08(1)}{0.028 + 0.080} = 1.52 \text{ in.}$$

Example 4.2.1.5 (Cont.)
Flexural Strength of Double Tee Flange in Transverse Direction

$b = 12$ in.

$$A_{s,min} = \frac{3\sqrt{f'_c}}{f_y} b_w d = \frac{3\sqrt{5000}(12)(1.52)}{60,000} = 0.064 \text{ in.}^2$$

A_s provided $= 0.028 + 0.080 = 0.108$ in.^{2} > 0.064 in.^{2} OK

Calculate allowable load:

w_d (flange self weight) = 50 psf

$$M_d = \frac{w_d \ell^2}{2} = \frac{50(1.2)(1.75)^2}{2} = 91.9 \text{ lb-ft/ft}$$

$M_\ell = 707 - 91.9 = 615.1$ lb-ft/ft

$$w_\ell = \frac{615.1(2)}{(1.75)^2(1.6)} = 251 \text{ psf}$$

Part B

Given:
    Pretopped double tee 10DT34 (see Chapter 2)
    $f'_c = 5,000$ psi
    $f_y = 65,000$ psi (WWR)*

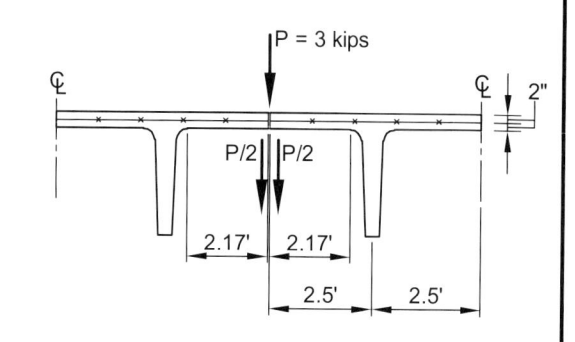

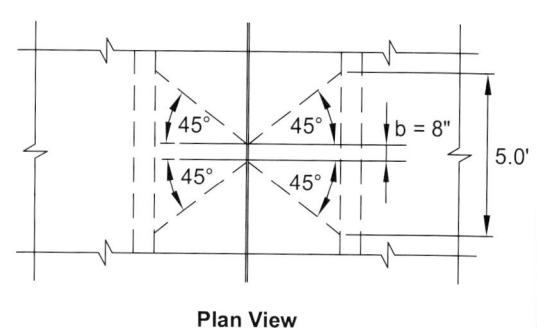

Problem:
    Design the flange for bending in the transverse direction for a concentrated live load of 3 kips (typical for parking structures) as shown.

Solution:
The following assumptions related to distribution of the concentrated load are typical:
1. Because of the flange-to-flange connection, the 3-kip load may be distributed to two adjacent double tees (1.5 kips per double tee). [32]
2. The load is considered applied over an area of about 20 in.^{2} with the dimension b equal to 6 to 10 in.
3. An angle of 45-deg is typically used for distribution of the concentrated load in each double tee flange.

Note: For this example, a 45-deg angle and the dimension b equal to 8 in. results in a distribution width at face of stem of 5 ft as shown.

Calculate factored moment per foot width:

w_d (self weight of flange) $= \frac{4}{12}(150) = 50$ psf

$$M_d \text{ (factored)} = \frac{1.2 w_d \ell^2}{2} = 1.2(50)(2.17)\left(\frac{2.17}{2}\right)\left(\frac{12}{1000}\right) = 1.69 \text{ kip-in./ft}$$

$$M_\ell \text{ (factored)} = 1.6\left(\frac{P}{2}\right)\ell = 1.6(1.5)(2.17)(12) = 62.50 \text{ kip-in. distributed over 5 ft. } \frac{62.50}{5} = 12.50 \text{ kip-in./ft}$$

$M_u = 1.69 + 12.50 = 14.19$ kip-in./ft

Calculate design moment strength with trial welded wire reinforcement that has W4 wire at 3 in. on centers.
    $A_s = 0.16$ in.^{2}/ft (see Design Aid 11.2.11)

---

* ACI 318-02, Section 3.5.3 allows f_y to exceed 60,000 psi if stress corresponds to a strain of 0.35%.

Example 4.2.1.5 (Cont.)
Flexural Strength of Double Tee Flange in Transverse Direction

$$a = \frac{A_s f_y}{0.85 f'_c b} = \frac{0.16(65)}{0.85(5)(12)} = 0.20 \text{ in.}$$

$$\phi M_n = \phi A_s f_y \left(d - \frac{a}{2}\right) = 0.9(0.16)(65)\left(2 - \frac{0.20}{2}\right)$$

$$= 17.78 \text{ kip-in./ft} > 14.19 \text{ kip-in./ft} \quad \text{OK}$$

Check ACI 318-02, Sections 10.5.4 and 7.12.2.1 for required minimum reinforcement:
A_s (shrinkage and temperature)

$$= \left[\frac{0.0018(60)}{f_y}\right] bd = 0.0018\left(\frac{60}{65}\right)(12)(4)$$

$$= 0.080 \text{ in.}^2/\text{ft} < 0.16 \text{ in.}^2/\text{ft} \quad \text{OK}$$

Use 12x3–W2.0xW4.0 WWR (one layer).
The unsupported corners of double tees should be reinforced a minimum of 1 – #4 L-bar.

Example 4.2.1.6
Design of a Partially Prestressed Flanged Section Using Strain Compatibility

Given:
    Inverted tee beam with 2 in. composite topping as shown.

Concrete:
    Precast: $f'_c = 5000$ psi
    Topping: $f'_c = 3000$ psi

Reinforcement:
    12–½ in. diameter 270K low-relaxation strand
    $A_{ps} = 12 (0.153) = 1.836$ in.^{2}
    $E_{ps} = 28,500$ ksi
    $E_s = 29,000$ ksi
    $A_s = 2 - \#7 = 1.2$ in.^{2}
    $A'_s = 2 - \#9 = 2.0$ in.^{2}

Problem:
    Find flexural strength, ϕM_n.

Solution:
    Determine effective flange width, b, from Section 8.10.2 of the Code; overhanging width = eight times thickness. (Note: May also be limited by one-quarter span.)
    $b = b_w + 2(8t) = 12 + 2(8)(2) = 44$ in.
    $d_p = 26 - 3 = 23$ in.
    $d = 26 - 2 = 24$ in.
    $d' = 1½$ in.
Assume 20% loss of prestress.

Example 4.2.1.6 (Cont.)
Design of a Partially Prestressed Flanged Section Using Strain Compatibility

Strand initially tensioned to 75% of f_{pu}:

$$f_{se} = (1 - 0.2)(0.75)(270) = 162 \text{ ksi}$$

$$\varepsilon_{se} = \frac{f_{se}}{E_{ps}} = \frac{162}{28,500} = 0.0057$$

Construct a strain diagram (see following page):
By proportional triangles:

$$\frac{\varepsilon_{sa} + 0.003}{d_p} = \frac{0.003}{c}$$

Therefore: $\dfrac{\varepsilon_{sa} + 0.003}{0.003} = \dfrac{d_p}{c}$ and

$$\varepsilon_{sa} = \frac{0.003 d_p}{c} - 0.003 = \frac{0.003(23)}{c} - 0.003 = \frac{0.069}{c} - 0.003$$

$$\varepsilon_{ps} = \varepsilon_{sa} + \varepsilon_{se} = \frac{0.069}{c} - 0.003 + 0.0057 = \frac{0.069}{c} + 0.0027$$

$$\varepsilon_s = \frac{0.003d}{c} - 0.003 = \frac{0.003(24)}{c} - 0.003 = \frac{0.072}{c} - 0.003$$

$$\varepsilon'_s = 0.003 - \frac{0.003 d'}{c} = 0.003 - \frac{0.003(1.5)}{c} = 0.003 - \frac{0.0045}{c}$$

$$\varepsilon_y = \frac{f_y}{E_s} = \frac{60}{29,000} = 0.0021$$

Try $c = 9$ in.

$$\varepsilon_{ps} = \frac{0.069}{c} + 0.0027 = \frac{0.069}{9} + 0.0027 = 0.0104$$

From the strand stress-strain curve equations in Design Aid 11.2.5:

$$f_{ps} = 270 - \frac{0.04}{\varepsilon_{ps} - 0.007} = 270 - \frac{0.04}{0.0104 - 0.007} = 258.2 \text{ ksi}$$

ε_s (from above) $= \dfrac{0.072}{9} - 0.003 = 0.0050 > 0.0021$; therefore, $f_s = 60$ ksi

ε'_s (from above) $= 0.003 - \dfrac{0.0045}{9} = 0.0025 > 0.0021$; therefore, $f'_s = 60$ ksi

$C_1 = 0.85 f'_{c(topping)} a_1 b = 0.85(3)(2)(44) = 224.4$ kips
$C_2 = 0.85 f'_{c(topping)} a_2 b_w = 0.85(3)(2)(12) = 61.2$ kips
$a_3 = \beta_1 c - 4 \text{ in.} = (0.85)(9) - 4 = 3.65$ in.
$C_3 = 0.85 f'_c a_3 b_w = 0.85(5)(3.65)(12) = 186.2$ kips
$C_4 = A'_s f'_s = 2.0(60) = 120$ kips
$C = C_1 + C_2 + C_3 + C_4 = 224.4 + 61.2 + 186.2 + 120 = 591.8$ kips
$T_1 = A_{ps} f_{ps} = 1.836(258.2) = 474.1$ kips
$T_2 = A_s f_s = 1.2(60) = 72$ kips
$T_1 + T_2 = 474.1 + 72 = 546.1$ kips $<<$ 591.8 kips

Example 4.2.1.6 (Cont.)
Design of a Partially Prestressed Flanged Section Using Strain Compatibility

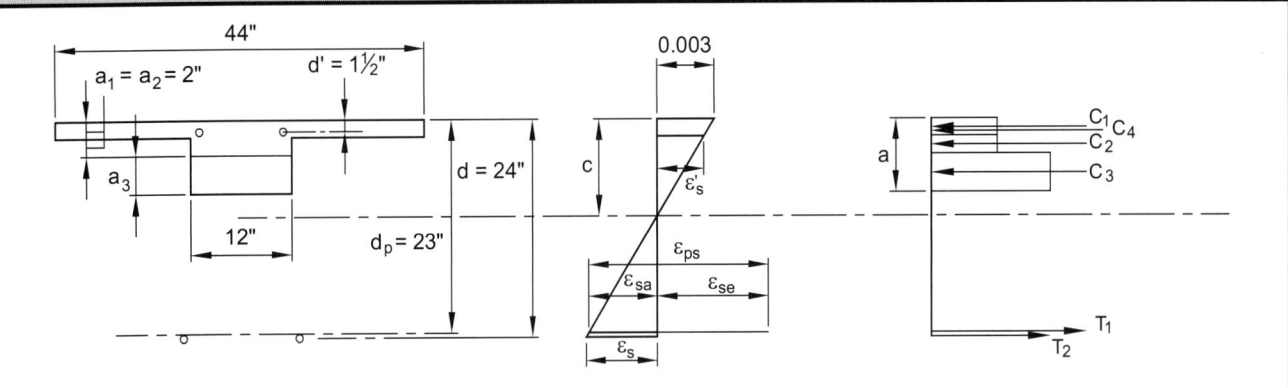

Try c = 8 in.

$$\varepsilon_{ps} = \frac{0.069}{c} + 0.0027 = \frac{0.069}{8} + 0.0027 = 0.0113$$

From the strand stress-strain curve equations in Design Aid 11.2.5:

$$f_{ps} = 270 - \frac{0.04}{\varepsilon_{ps} - 0.007} = 270 - \frac{0.04}{0.0113 - 0.007} = 260.7 \text{ ksi}$$

ε_s (from above) $= \frac{0.072}{8} - 0.003 = 0.0060 > 0.0021$; therefore, $f_s = 60$ ksi

ε'_s (from above) $= 0.003 - \frac{0.0045}{8} = 0.0024 > 0.0021$; therefore, $f'_s = 60$ ksi

$C_1 = 0.85 f'_{c(topping)} a_1 b = 0.85(3)(2)(44) = 224.4$ kips
$C_2 = 0.85 f'_{c(topping)} a_2 b_w = 0.85(3)(2)(12) = 61.2$ kips
$a_3 = \beta_1 c - 4$ in. $= (0.85)(8) - 4 = 2.8$ in.
$C_3 = 0.85 f'_c a_3 b_w = 0.85(5)(2.8)(12) = 142.8$ kips
$C_4 = A'_s f'_s = 2.0(60) = 120$ kips
$C = C_1 + C_2 + C_3 + C_4 = 224.4 + 61.2 + 142.8 + 120 = 548.4$ kips
$T_1 = A_{ps} f_{ps} = 1.836(260.7) = 478.6$ kips
$T_2 = A_s f_s = 1.2(60) = 72$ kips
$T_1 + T_2 = 478.6 + 72 = 550.6$ kips ≈ 548.4 kips

Therefore, use c = 8 in.

Check reinforcement limits by ACI 318-02:
    Net tensile strain of extreme depth:
    In this case, $\varepsilon_t = \varepsilon_s$
    $\varepsilon_t = 0.0060 > 0.005$

Therefore, it is a tension-controlled section and
    $\phi = 0.9$

Alternatively:
    Extreme depth $d_t = 24$ in.
$$\frac{c}{d_t} = \frac{8.0}{24} = 0.333 < 0.375$$

Example 4.2.1.6 (Cont.)
Design of a Partially Prestressed Flanged Section Using Strain Compatibility

From Figure 4.2.1.3, $\phi = 0.9$

$$M_n = 224.4(8-1) + 61.2(8-3) + 142.8\left(\frac{4}{2}\right) + 120(8-1.5) + 478.6(23-8) + 72(24-8)$$

$$= 1570.8 + 306.0 + 285.6 + 780.0 + 7179.0 + 1152.0 = 11{,}273 \text{ kip-in.} = 939 \text{ kip-ft}$$

$$\phi M_n = 0.9(939) = 845 \text{ kip-ft}$$

Notes:
1. This example shows a more precise method; approximate methods are satisfactory in most situations.
2. In this example, since the cast-in-place topping carries significantly more compression force than the precast member, $\beta_1 = 0.85$ corresponds to topping concrete. In other cases, where the compression is shared by the topping and precast member in different proportions, a reasonable average value for β_1 may be used.
3. In evaluating ω_{pu} and other similar factors, $f'_c = 3.5$ ksi is used to reflect the contribution of topping vs. precast member to the total compression force.
4. For over-reinforced members, nominal moment strength must be calculated based on the compression portion of the internal couple.

4.2.2 Service Load Design

Precast members are checked under service load, primarily for meeting serviceability criteria and to control cracking (see Section 9.8).

4.2.2.1 Non-Prestressed Element Design

Non-prestressed flexural elements are normally proportioned, and reinforcement selected, on the basis of the procedures described in Section 4.2.1. However, depending upon the application and exposure of the member, designers may want to control the degree of cracking. In some applications, such as architectural precast concrete panels, they may not want any discernible cracking. In other cases, minor cracking may be permitted. Code Section 10.6.4 limits the spacing of the reinforcement in the tension face, s, to:

$$s = \frac{540}{f_s} - 2.5c_c \leq 12\left(\frac{36}{f_s}\right) \quad \text{(Eq. 4.2.2.1)}$$

where:

f_s = calculated stress in reinforcement, ksi; may be assumed 60% of specified yield

c_c = clear cover to reinforcement from tension face

If no discernible cracking is the criterion, the flexural tensile stress level should be limited to the modulus of rupture divided by a safety factor of 1.5.

$$f'_r \leq \frac{7.5\lambda\sqrt{f'_c}}{1.5} = 5\lambda\sqrt{f'_c} \quad \text{(Eq. 4.2.2.2)}$$

where:

f'_r = allowable flexural tension, computed using gross concrete section
f'_c = concrete strength at time considered
λ = 1.0 for normal weight concrete
  = 0.85 for sand-lightweight concrete
  = 0.75 for all-lightweight concrete

4.2.2.2 Prestressed Element Design

To check serviceability of prestressed concrete members, stresses are calculated by the classical straight line theory as shown in Figure 4.2.2.1. While there are no theoretical limitations on stresses, Section 18.3.3 of ACI 318-02 classifies prestressed concrete flexural members as Class U (uncracked), Class C (cracked) or Class T (transition between uncracked and cracked). The classifications are based on the maximum tensile stress, f_t, at service loads in the precompressed tensile zone, as shown in Table 4.2.2.1.

The Code makes an exception to the use of Class C for "members not subject to fatigue or to aggressive exposure." The Commentary defines this as "an environment in which chemical attack (such as seawater, corrosive industrial atmosphere or sewer gas) is encountered." In such cases, members should be designed as Class U or T.

ACI 318-02 does, however, prescribe maximum stresses at the time of transfer of prestress. Research [3] has shown that the ACI 318-02 values are conservative, and the *PCI Standard Design Practice* (Handbook Section 10.5) suggests commonly used modifications. Both are shown in Table 4.2.2.2. The Section 10.5 modifications are used in the examples in this Chapter.

Where the tensile stresses exceed those in Table 4.2.2.2, bonded additional reinforcement is to be provided in the tensile zone to resist the total tensile force in the concrete. The Commentary to ACI-02 suggests that steel stress in this reinforcement be limited to 30 ksi, but Section 10.5 says that this is excessively conservative when the members are in tension at transfer, but in compression under service load and not exposed to weather (such as beams with field-applied topping or pour strips.) In such cases, f_s up to yield may be used. See Example 4.2.2.4.

Example 4.2.2.1
Non-Prestressed Panel Design

Given:
A precast wall panel exposed to view and to the weather, with dimensions as shown:
Concrete $f'_c = 5000$ psi
Service load moment, $M = 2.4$ kip-ft/ft

Problem:
Determine minimum thickness to result in "no discernible cracking" and the minimum reinforcement spacing requirements.

Solution:
For a 12 in. width:

From Eq. 4.2.2.2:
$$f'_r = 5\lambda\sqrt{f'_c} = 5(1.0)\sqrt{5000} = 353.6 \text{ psi}$$

Required section modulus:
$$\frac{M}{f'_r} = \frac{2.4(12)(1000)}{353.6} = 81.4 \text{ in.}^3$$

$$\frac{bt^2}{6} = 81.4$$

$$t = \sqrt{\frac{6(81.4)}{12}} = 6.38 \text{ in. say 7 in.}$$

Minimum reinforcement spacing:
Use $f_s = 0.6(60) = 36$ ksi
Assume #3 reinforcing bars.
Clear cover $1.5 - \frac{0.375}{2} = 1.31$ in.

From Eq. 4.2.2.1:
$$s = \frac{540}{36} - 2.5c_c = 15 - 2.5(1.31) = 11.7 \text{ in.}$$

Note: To complete the design, check strength as described in Section 4.2.1. In most cases, handling requirements will control the reinforcement selection of wall panels. See Chapter 5.

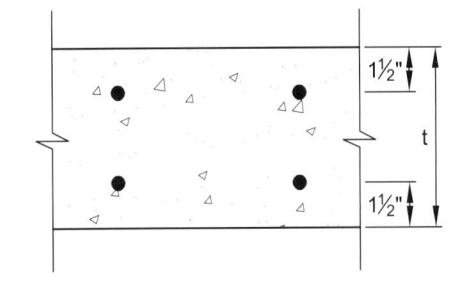

Figure 4.2.2.1 Calculation of service load stresses

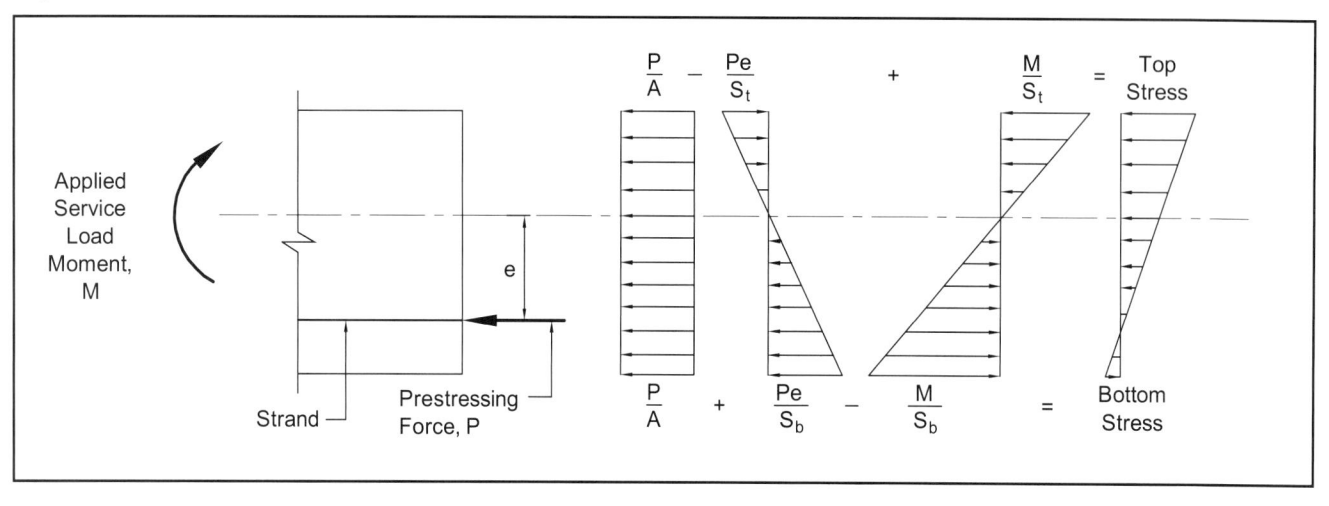

$$\frac{P}{A} - \frac{Pe}{S_t} + \frac{M}{S_t} = \text{Top Stress}$$

$$\frac{P}{A} + \frac{Pe}{S_b} - \frac{M}{S_b} = \text{Bottom Stress}$$

Table 4.2.2.1 Serviceability design requirements

	Prestressed			Non-prestressed
	Class U	**Class T**	**Class C**	
Assumed behavior	Uncracked	Transition between uncracked and cracked	Cracked	Cracked
Section properties for stress calculation at service loads	Gross section 18.3.4	Gross section 18.3.4	Cracked section 18.3.4	No requirement
Allowable stress at transfer	18.4.1	18.4.1	18.4.1	No requirement
Allowable compressive stress based on uncracked section properties	18.4.2	18.4.2	No requirement	No requirement
Tensile stress at service loads 18.3.3	$\leq 7.5\sqrt{f'_c}$	$7.5\sqrt{f'_c} < f_t \leq 12\sqrt{f'_c}$	No requirement	No requirement
Deflection calculation basis	9.5.4.1 Gross section	9.5.4.2 Cracked section, bilinear	9.5.4.2 Cracked section, bilinear	9.5.2, 9.5.2.3 Effective moment of inertia
Crack control	No requirement	No requirement	10.6.4 Modified by 18.4.4.1	10.6.4
Computation of Δf_s or f_s for crack control	—	—	Cracked section analysis	$\dfrac{M}{A_s \times \text{lever arm}}$ or $0.6 f_y$
Side skin reinforcement	No requirement	No requirement	10.6.7	10.6.7

Based on ACI 318-02 Table R18.3.3. Numbers in the table refer to ACI Code sections.

Composite Members

It is usually more economical to place cast-in-place composite topping without shoring the member, especially for deck members. This means that the weight of the topping and simultaneous construction live load must be carried by the precast member alone. Additional superimposed dead and live loads are carried by the composite section.

Example 4.2.2.3 through 4.2.2.4 illustrate a tabular form of superimposing the stresses caused by the prestress force and the dead and live load moments.

Table 4.2.2.2 Stress limits at release (transfer) of prestress (before time-dependent losses)

Concrete

a. Compression
   ACI 318-02 .... $0.60f'_{ci}$
   (see Section 10.5) .... $0.70f'_{ci}$

b. Tension (except at ends)
   ACI 318-02 .... $3\sqrt{f'_{ci}}$
   (see Section 10.5) .... $7.5\sqrt{f'_{ci}}$

c. Tension at ends* of simply supported members
   ACI 318-02 .... $6\sqrt{f'_c}$
   (see Section 10.5) .... $7.5\sqrt{f'_c}$

Prestressing steel

a. Tension due to tendon jacking
   force: .... $0.80f_{pu}$ or $0.94f_{py}$

b. Tension immediately after prestress transfer:
   Stress-relieved strand: .... $0.7f_{pu}$
   Low-relaxation strand: .... $0.74f_{pu}$

These values are commonly assumed to be after a small seating loss. Manufacturers may "fine tune" actual jacking force to compensate for temperature variation.

* At the point of prestress transfer.

Sign Convention

The customary sign convention used in the design of precast, prestressed concrete members for service load stresses is positive (+) for compression and negative (−) for tension. This convention is used throughout this Handbook.

Cracked Section Analysis

Note from Table 4.2.2.1 that Class C members are required to have service load stresses checked using cracked, transformed section properties. When this is done, there are no upper limits specified for tension. Deflections must be carefully checked (Section 4.8), and the spacing limitations for tensile reinforcement (prestressed or non-prestressed) of Code Section 10.6.4 must be observed – see Eq. 4.2.2.1.

When applying Eq. 4.2.2.1 to prestressed reinforcement, Δf_{ps} is substituted for f_s, where Δf_{ps} is the difference between the stress in the strands at service loads (based on cracked section) and the decompression stress, f_{dc}. The decompression stress may be taken as the effective prestress, f_{se}. Δf_{ps} cannot be taken greater than 36 ksi, and if it is 20 ksi or less spacing requirements need not be applied. See Examples 4.2.2.5 and 4.2.2.6.

When performing a cracked, transformed section analysis, a method described by Mast, [20] summarized below and illustrated in Examples 4.2.2.5 and 4.2.2.6 may be used.

In this method (see Figure 4.2.2.2), the prestress force is accounted for by calculating an internal moment, M_{int}, which is the applied moment, M_{ext}, reduced by the moment $P(y_p)$, where P is the prestress force acting at the level of the tendons and y_p is the distance from the applied P to the center of gravity of the cracked section. The force P is based on the stress that would exist in the tendons when the stress is zero in the adjacent concrete at the same level. This is called the "decompression stress," f_{dc}, and can be assumed as the effective prestress, f_{se}, with the elastic shortening loss, ES, added back.

A decompression stress also exists in non-prestressed reinforcement in the precompressed tensile zone. In the design of non-prestressed sections, this force is small and is usually neglected. In prestressed concrete sections, however, creep losses make this force more significant, and can be estimated as the sum of the creep loss, CR, and the shrinkage loss, SH.

Note that the decompression stress in non-prestressed reinforcement is caused by the concrete acting on the steel, rather than a direct loss of prestress, so the force is directionally opposite that of the decompression stress in the prestressing steel. If losses are calculated by the method shown in Section 4.7, the losses which contribute to the decompression stresses can be determined directly. If a lump sum or percentage loss is estimated, it is sufficiently precise to assume that elastic shortening loss, ES, is 40% of the total, and the combined creep and shrinkage losses, CR + SH, are 50% of the total loss.

For composite sections, the prestress and some dead load bending are applied to the precast section, which does not create stress in the cast-in-place composite slab. This causes a discontinuity in stress and strain at the interface, which remains while additional loads are applied to the composite beam. The method of Ref. 20 computes a fictitious force in the composite slab representing the interface stress caused by the moment on the bare precast section.

This is combined with the prestress force and the decompression forces and applied to the composite section. These forces are added into the iterative analysis of the composite section. Compression reinforcement contained within the topping is treated in a similar manner. Example 4.2.2.6 illustrates the above principles.

Example 4.2.2.2
Calculation of Critical Stresses — Straight Strands Class U Member

Given:
  4HC12 + 2 as shown
  Span = 36 ft

Section properties:

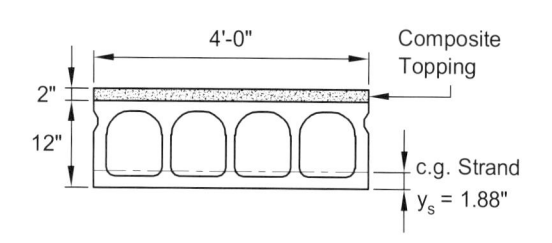

	Non-Composite	Composite
A =	265 in.^{2}	—
I =	4771 in.^{4}	7209 in.^{4}
y_b =	6.67 in.	8.10 in.
y_t =	5.33 in.	5.90 in.
S_b =	715.3 in.^{3}	890.0 in.^{3}
S_t =	895.1 in.^{3}	1221.9 in.^{3}
wt =	276 plf	376 plf
e =	4.79 in.	

Superimposed sustained load = 20 psf = 80 plf
Superimposed live load = 50 psf = 200 plf

Precast concrete (normal weight):
  f'_c = 6000 psi
  f'_{ci} = 4000 psi
  E_c = 4700 ksi

Topping concrete (normal weight):
  f'_c = 4000 psi
  E_c = 3800 ksi

Prestressing steel:
  5 – ½ in. diameter 270K low-relaxation strand
  A_{ps} = 5(0.153) = 0.765 in.^{2}
  Straight strands

Problem:
  Find critical service load stresses. Class U tension limit = $7.5\sqrt{f'_c}$ (Table 4.2.2.1).

Solution:
Prestress force:
  P_i = 0.75 $A_{ps}f_{pu}$ = 0.75(0.765)(270)   = 155 kips
  P_o (assume 10% initial loss) = 0.90(155) = 139 kips
  P (assume 18% total loss) = 0.82(155)   = 127 kips

Midspan service load moments:
  M_d = 0.276(36)^2(12)/8 = 537 kip-in.
  M_{top} = 0.100(36)^2(12)/8 = 194 kip-in.
  M_{sd} = 0.080(36)^2(12)/8 = 156 kip-in.
  M_ℓ = 0.200(36)^2(12)/8 = 389 kip-in.

Example 4.2.2.2 (Cont.)
Calculation of Critical Stresses — Straight Strands Class U Member

See the following table for service load stresses:

Load	Transfer Pt. at Release $P = P_o$		Midspan at Release $P = P_o$		Midspan at Service Load $P = P$		
	f_b	f_t	f_b	f_t	f_b	f_t (a)[a]	f_t (b)[a]
P/A	+525	+525	+525	+525	+479	+479	+479
Pe/S	+931	−744	+931	−744	+850	−680	−680
M_d/S	−164	+131	−751	+600	−751	+600	+600
M_{top}/S					−271	+217	+217
M_{sd}/S[b]					−175	+128	+128
M_ℓ/S[b]					−437		+318
Stresses	+1292	−88	+705	+381	−305	+744	+1062
Limiting Stresses	$0.70f'_{ci}$	$7.5\sqrt{f'_{ci}}$	$0.70f'_{ci}$[c]	$0.70f'_{ci}$[c]	$7.5\sqrt{f'_c}$	$0.45f'_c$	$0.6f'_c$
	+2800	−474	+2800	+2800	−581	+2700	+3600
	OK	OK	OK	OK	OK	OK	OK

a. (a) and (b) correspond to the (a) and (b) stress criteria in Code Section 18.4.2.
b. For all stresses, the composite section modulus is used. The stresses are calculated at the top of the precast section.
c. See Section 10.5.

Example 4.2.2.3
Calculation of Critical Stresses — Single-Point Depressed Strand Class T Member

Given:
   Span = 70 ft
   Superimposed sustained load = 10 psf = 80 plf
   Superimposed live load = 35 psf = 280 plf
   Select 8DT24 as shown

Concrete (normal weight):
   f'_c = 5000 psi
   f'_{ci} = 3500 psi

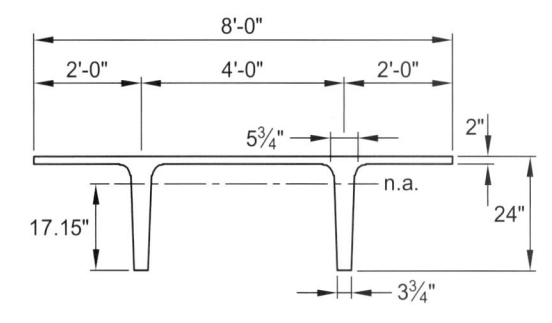

Prestressing steel:
   12 – ½ in. diameter 270K low-relaxation strand
   $A_{ps} = 12 \times 0.153 = 1.836$ in.²

Section properties:
   A = 401 in.²
   I = 20,985 in.⁴
   y_b = 17.15 in.
   y_t = 6.85 in.
   S_b = 1224 in.³
   S_t = 3063 in.³
   wt = 418 plf = 52 psf

Example 4.2.2.3 (Cont.)
Calculation of Critical Stresses — Single-Point Depressed Strand Class T Member

Eccentricities, single-point depression:
- e_e = 5.48 in.
- e_c = 13.90 in.
- e' = 13.90 − 5.48 = 8.42 in.
- e at $0.4\ell \left(\dfrac{0.8\ell}{2} \right)$ = 5.48 + 0.8(8.42) = 12.22 in.

Problem:
Find critical service load stresses. Class T tension limit = $12\sqrt{f'_c}$ (Table 4.2.2.1)

Solution:
Prestress force:
- P_i = 0.75$A_{ps}f_{pu}$ = 0.75(1.836)(270) = 372 kips
- P_o (assume 10% initial loss) = 0.90 (372) = 335 kips
- P (assume 20% total loss) = 0.80(372) = 298 kips

Service load moments at midspan:
- M_d = 0.418(70)²(12)/8 = 3072 kip-in.
- M_{sd} = 0.080(70)²(12)/8 = 588 kip-in.
- M_ℓ = 0.280(70)²(12)/8 = 2058 kip-in.

Service loads and moments at 0.4ℓ:
- M_d = 3072(0.96) = 2949 kip-in.
- M_{sd} = 588(0.96) = 564 kip-in.
- M_ℓ = 2058(0.96) = 1976 kip-in.

See the following table for service load stresses:

Load	Transfer Pt. at Release P = P₀		Midspan at Release P = P₀		0.4ℓ at Service Load P = P		
	f_b	f_t	f_b	f_t	f_b	f_t (a)[a]	f_t (b)[a]
P/A	+835	+835	+835	+835	+743	+743	+743
Pe/S	+1500	−599	+3804	−1520	+2975	−1189	−1189
M_d/S	−290	+116	−2510	+1003	−2409	+962	+962
M_sd/S					−461	+184	+184
M_ℓ/S					−1614		+645
Stresses	+2045	+352	+2129	+318	−766	+700	+1345
Limiting Stresses	0.70f'_{ci}[b]	0.70f'_{ci}[b]	0.70f'_{ci}[b]	0.70f'_{ci}[b]	$12\sqrt{f'_c}$	0.45f'_c	0.6f'_c
	+2450	+2450	+2450	+2450	−848	+2250	+3000
	OK	OK	OK	OK	OK	OK	OK

a. (a) and (b) correspond to the (a) and (b) stress criteria in Code Section 18.4.2.
b. See Section 10.5.

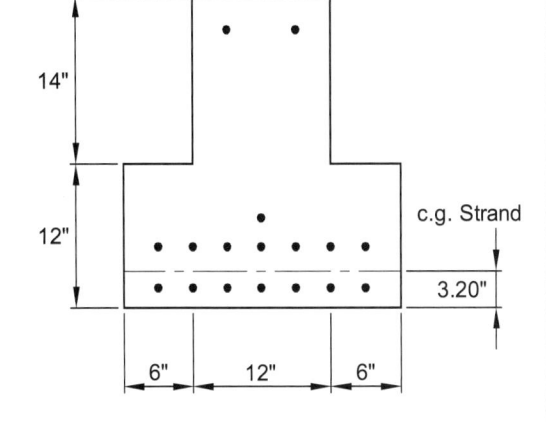

Example 4.2.2.4
Tensile Force to be Resisted by Top Reinforcement

Given:
  Span = 24 ft
  24IT26 as shown below

Concrete:
  $f'_c = 6000$ psi
  $f'_{ci} = 4000$ psi

Prestressing steel:
  15 – ½ in. diameter 270K low-relaxation strand
  $A_{ps} = 15(0.153) = 2.295$ in.^{2}

Section properties:
  A = 456 in.^{2}
  I = 24,132 in.^{4}
  y_b = 10.79 in.
  y_t = 15.21 in.
  S_b = 2237 in.^{3}
  S_t = 1587 in.^{3}
  Wt = 475 plf
  e = 7.59 in.

Problem:
  Find critical stresses at release.

Solution:
Prestress force:
  $P_i = 0.75 A_{ps} f_{pu} = 0.75(2.295)(270) = 465$ kips
  P_o (assume 10% initial loss) = 0.90(465) = 418 kips

Moment due to member weight:
At midspan:
  $M_d = 0.475(24)^2(12)/8 = 410$ kip-in.

Assume transfer length = $\left(\dfrac{f_{se}}{3}\right) d_b = (170/3)0.5 = 28.3$ in. = 2.36 ft (see Section 4.2.3)

At transfer point:
  $M_d = \dfrac{wx}{2}(\ell - x) = \dfrac{0.475(2.36)}{2}(24 - 2.36)(12) = 146$ kip-in.

Since the tensile stress exceeds the limits, reinforcement is required to resist the total tensile force, as follows:

  $c = \dfrac{f_t}{f_t + f_b}(h) = \dfrac{990}{990 + 2270}(26) = 7.90$ in.

  $T = \dfrac{c f_t b}{2} = \dfrac{7.90(990)(12)}{2} = 46{,}926$ lb

Example 4.2.2.4 (Cont.)
Tensile Force to be Resisted by Top Reinforcement

See the following table for stresses at release:

Load	Transfer Point at Release $P = P_o$ [a]		Midspan at Release $P = P_o$	
	f_b	f_t	f_b	f_t
P/A	+917	+917	+917	+917
Pe/S	+1418	−1999	+1418	−1999
M_d/S	−65	+92	−183	+258
Stresses	+2270	−990	+2152	−824
Limiting Stresses	$0.7f'_{ci}$	$7.5\sqrt{f'_{ci}}$	$0.7f'_{ci}$	$7.5\sqrt{f'_{ci}}$
	+2800	−474	+2800	−474
	OK	HIGH	OK	HIGH

a. See Section 10.5.

Similarly, the tension at midspan can be found as 35,591 lb.

The Commentary to the Code, Section 18.4.1 (b) and (c), suggests that reinforcement be proportioned to resist this tensile force at a stress of $0.6f_y$, but not more than 30 ksi. However, the "PCI Standard Design Practice" (Handbook Section 10.5) recommends that full yield strength be used. Using reinforcement with $f_y = 60$ ksi, and Section 10.5:

$$A_s(\text{end}) = \frac{46.9}{60} = 0.78 \text{ in.}^2$$

$$A_s(\text{midspan}) = \frac{35.6}{60} = 0.59 \text{ in.}^2$$

Top strands used as stirrup supports may also be used to carry this tensile force.

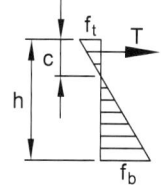

It should be noted that some cracking may occur even with reinforcement. Such cracking, however, has no structural significance, and is acceptable because the tension at the beam top will be reduced under service loads.

In summary, the steps for cracked section analysis are as follows:

1. Perform a gross section analysis and determine if the section is cracked at service load.
2. Estimate the decompression force P_{ps} in the prestressing steel. The decompression stress: $f_{dc} = (f_{se} - ES)$; $P_{ps} = f_{dc}A_{ps}$
3. Estimate the decompression force in the unstressed reinforcement: $P_s = (CR + SH)A'_s$, opposite in direction to the decompression in the prestressing steel. Combine this force with the decompression force P_{ps} to obtain a resultant decompression force P and a location for that resultant.
4. If the section is composite, compute the fictitious stresses and force in the composite slab and/or reinforcement created by extending the bare beam stress diagram through the composite slab. Combine this fictitious force with the decompression forces, $P_{ps} + P_s$, to obtain an equivalent force P_e, and its location.
5. Compute the combined transformed section properties (A, I, center of gravity) of all the steel elements using the modular ratio to transform the steel area to an equivalent concrete area. Similarly, compute the section properties of the composite slab.
6. Begin the iterative procedure by selecting a trial depth c to the neutral axis of the section. Compute the section properties (A, I, center of gravity) of the net concrete section between the compression face and the neutral axis, and combine the results with those of the transformed steel to obtain the combined properties of the transformed section.
7. Calculate the moment caused by applying the decompression force P (or the equivalent force P_e for composite sections) at a distance y_p from the calculated center of gravity of the transformed section. Combine this moment with the applied exterior moment M_{ext} to determine the interior moment M_{int}.

Figure 4.2.2.2 Forces and stresses on transformed cracked section

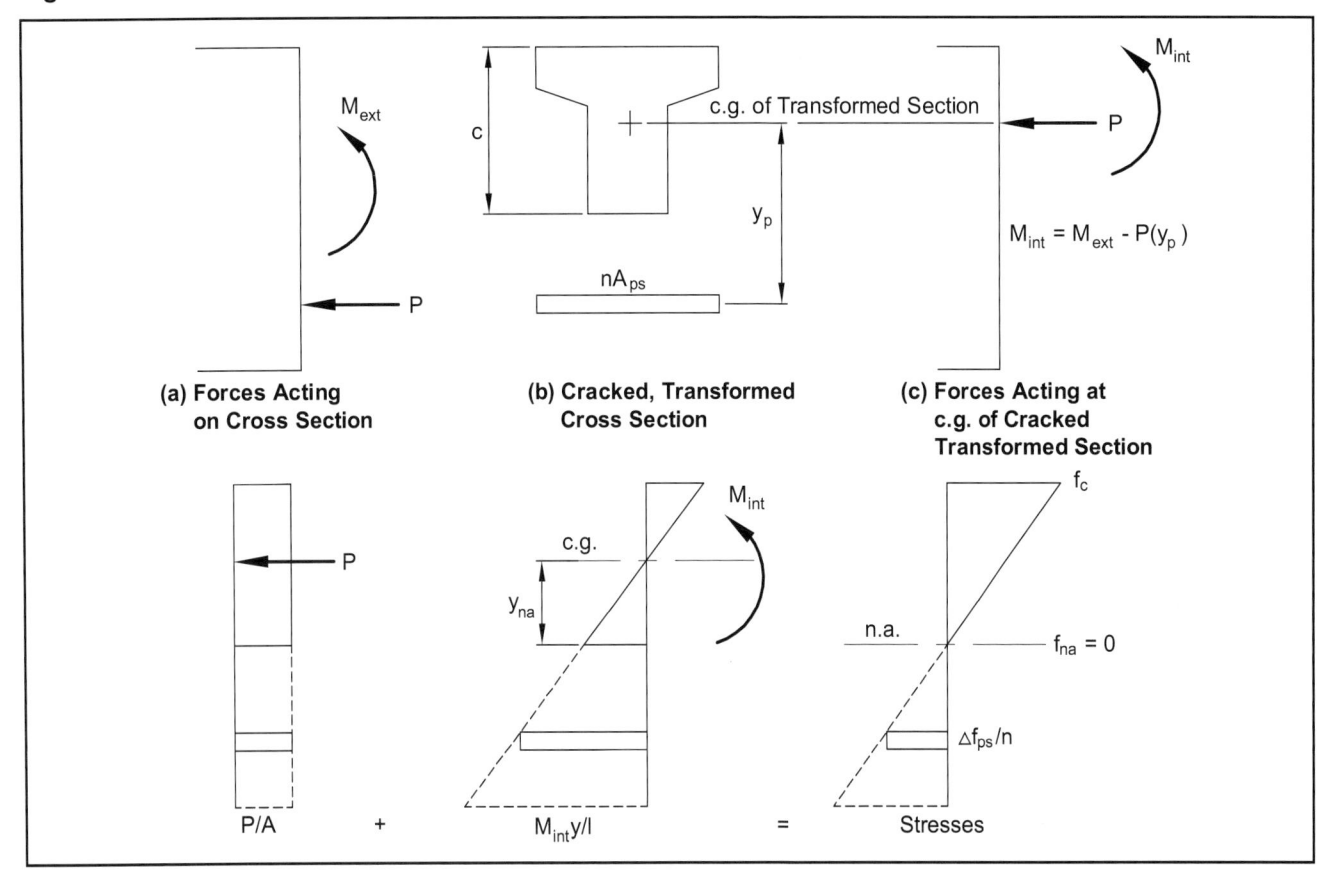

(a) Forces Acting on Cross Section

(b) Cracked, Transformed Cross Section

(c) Forces Acting at c.g. of Cracked Transformed Section

$M_{int} = M_{ext} - P(y_p)$

P/A + $M_{int}y/I$ = Stresses

8. Compute the stress at the assumed level of the neutral axis, c:

$$f_{na} = \frac{M_{int}}{\left(\frac{I_t}{y_{na}}\right)} \pm \frac{P_e}{A_t}$$

Adjust c with repetitive trials until the calculated stress is essentially zero.

9. Once the iterative procedure has converged: (a) Calculate stresses in the transformed section and compare with code compression limitations shown in the previous section. Also, use the calculated tensile stress in the steel to check for spacing requirements of Eq. 4.2.2.1, substituting Δf_{ps} for f_s. In composite sections, the fictitious stresses calculated in Step 4 are subtracted from the stresses calculated in the topping using transformed section properties. (b) To avoid error, it is advisable to check static equilibrium as illustrated in the examples.

10. The cracked section properties thus determined may also be used as described in Section 4.8 to determine the deflection of the member.

Example 4.2.2.5
Transformed Cracked Section Using Ref. 20: Example 1

Given:
Rectangular beam shown.
Gross section properties:
  A = 384 in.²
  I = 32,768 in.⁴
  $y_b = y_t = 16$ in.
  $S_b = S_t = 2048$ in.³
  Wt = 400 lb/ft

Example 4.2.2.5 (Cont.)
Transformed Cracked Section Using Ref. 20: Example 1

Concrete:
f'_c = 6000 psi (normal weight concrete)
E_c = 4696 ksi

Span = 40 ft
Loads and moments
Superimposed dead load = 1.00 kip/ft
Total dead load = 1.40 kips/ft
Live load = 1.25 kips/ft

$$M_d = \frac{1.40(40)^2(12)}{8} = 3360 \text{ kip-in.}$$

$$M_\ell = \frac{1.25(40)^2(12)}{8} = 3000 \text{ kip-in.}$$

M_{ext} = 3360 + 3000 = 6360 kip-in.

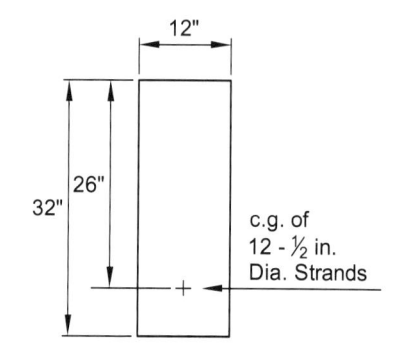

Prestress force:
A_{ps} = 12(0.153) = 1.836 in.^{2}
f_{pu} = 270 ksi
Initial prestress level = 0.75 f_{pu}
Estimated loss = 20%
f_{se} = 270 (0.75)(1 − 0.2) = 162 ksi
P = 1.836(162) = 297.4 kips
e = 26 − 16 = 10 in.

Solution:
1. Gross section analysis:

$$f_b = \frac{P}{A} + \frac{Pe}{S_b} - \frac{M_{ext}}{S_b}$$

$$= \frac{297.4(1000)}{384} + \frac{297.4(10)(1000)}{2048} - \frac{6360(1000)}{2048} = 774 + 1452 - 3105 = -879 \text{ psi}$$

Note: 879 psi is less than $12\sqrt{f'_c}$, so this is a Class T member, and cracked section analysis may not be required. The procedure is shown for illustration.

$f_r = 7.5\sqrt{f'_c} = 7.5\sqrt{6000} = 581$ psi < 879 psi
Section is cracked.

2. Estimate decompression force:
Elastic shortening loss (ES) is approximately 40% of total loss.
ES = 20 − 0.4(20) = 12%
f_{dc} = (1 − 0.12)(0.75)(270) = 178 ksi
P_e = 178(1.836) = 327 kips

3. Not applicable.

4. Not applicable.

Example 4.2.2.5 (Cont.)
Transformed Cracked Section Using Ref. 20: Example 1

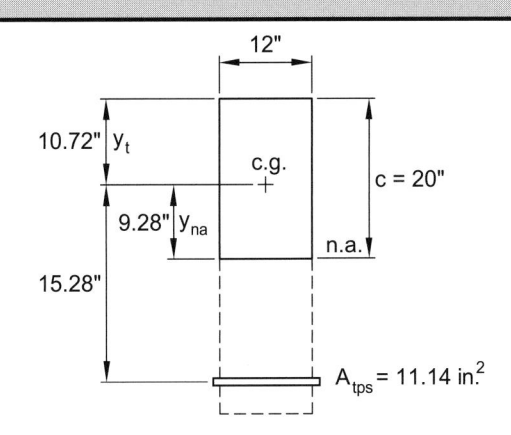

5. Transformed area of steel:

$$n = \frac{E_{ps}}{E_c} = \frac{28,500}{4696} = 6.07$$

$$A_{tps} = 1.836(6.07) = 11.14 \text{ in.}^2$$

6.(1) Select c = 20 in. for the first trial.

$$A_t = A_{tps} + bc = 11.14 + 20(12) = 251 \text{ in.}^2$$

$$y_t = \frac{12(20)(10) + 11.14(26)}{251} = 10.72 \text{ in.}$$

$$I_t = \frac{12(20)^3}{12} + 12(20)(10.72 - 10)^2 + 11.14(26 - 10.72)^2$$

$$= 8000 + 124 + 2601 = 10,725 \text{ in.}^4$$

$$\frac{I_t}{y_{na}} = \frac{I_t}{c - y_t} = \frac{10,725}{(20 - 10.72)} = 1156 \text{ in.}^3$$

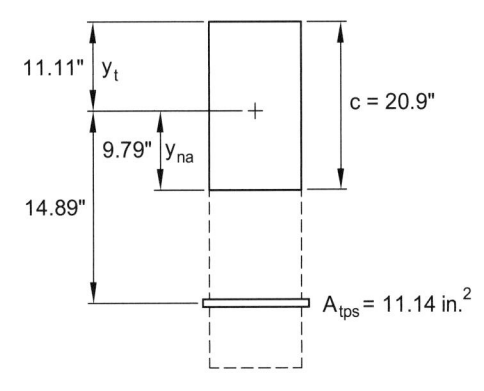

7.(1) $y_p = y_{ps} - y_t = 26 - 10.72 = 15.28$ in.

$M_{int} = M_{ext} - P_e y_p = 6360 - 327(15.28) = 1363$ kip-in.

8.(1) $\frac{P_e}{A_t} = \frac{327}{251} = +1.303$ ksi

$\frac{M_{int}}{I_t / y_{na}} = \frac{1363}{1156} = -1.179$ ksi

Difference = $+0.124$ ksi

For second trial, select c = 20.9 in.

6.(2) $A_t = 11.14 + 20.9(12) = 262 \text{ in.}^2$

$$y_t = \frac{bc(c/2) + A_{tps}d}{A_t} = \frac{12(20.9)(10.45) + 11.14(26)}{262} = 11.11 \text{ in.}$$

$$I_t = \frac{12(20.9)^3}{12} + 12(20.9)(11.11 - 10.45)^2 + 11.14(26 - 11.11)^2$$

$$= 9129 + 109 + 2470 = 11,708 \text{ in.}^4$$

$$\frac{I_t}{y_{na}} = \frac{11,708}{20.9 - 11.11} = 1196 \text{ in.}^3$$

7.(2) $y_p = 26 - 11.11 = 14.89$ in.

$M_{int} = 6360 - 327(14.89) = 1491$ lb-in.

Example 4.2.2.5 (Cont.)
Transformed Cracked Section Using Ref. 20: Example 1

8.(2) $\dfrac{P_e}{A_t} = \dfrac{327}{262} = +1.248$ ksi

$\dfrac{M_{int}}{I_t / y_{na}} = \dfrac{1491}{1196} = -1.247$ ksi

$\overline{+0.001 \text{ ksi}} \approx 0$ convergence

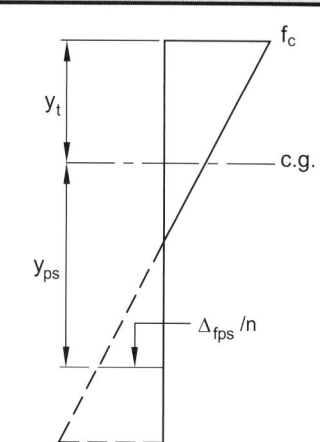

9.(a) Check stresses:

$\dfrac{I_t}{y_t} = \dfrac{11{,}708}{11.11} = 1054$ in.^{3}

$f_c = \dfrac{P}{A} + \dfrac{M_{int}}{I_t / y_p} = 1.248 + \dfrac{1491}{1054} = 2.663$ ksi $< 0.6 f'_c$

$\dfrac{I_t}{y_{ps}} = \dfrac{11{,}708}{14.89} = 786.3$ in.^{3}

$\Delta f_{ps} = \left[\dfrac{P}{A} + \dfrac{M_{int}}{I_t / y_{ps}}\right] n = \left[-1.248 + \dfrac{1491}{786.3}\right] 6.07 = 0.648(6.07)$

$= 3.935$ ksi < 20 ksi

No crack control spacing requirements.

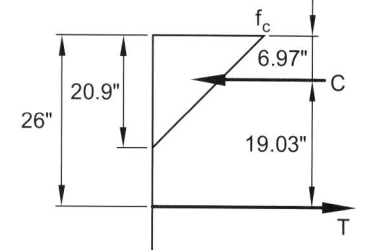

9.(b) Check equilibrium:

$C = \dfrac{f_c bc}{2} = \dfrac{2.663(12)(20.9)}{2} = 334$ kips

$T = P_e + \Delta f_{ps} A_{ps} = 327 + 3.935(1.836) = 334$ kips

$M = 334(19.03) = 6356$ kip-in. ≈ 6360 kip-in. (check)

Example 4.2.2.6
Transformed Cracked Section Using Ref. 20: Example 2

Given:
Inverted tee beam with 2 in. composite topping (Example 4.2.1.6).

Gross section properties:

	Precast	Composite
A =	408 in.^{2}	
I =	14,705 in.^{4}	31,912 in.^{4}
y_b =	9.24 in.	11.99 in.
y_t =	12.76 in.	14.01 in.
S_b =	1592 in.^{3}	2662 in.^{3}
S_t =	1152 in.^{3}	2277 in.^{3}
wt =	425 lb/ft	

Example 4.2.2.6 (Cont.)
Transformed Cracked Section Using Ref. 20: Example 2

Precast f'_c = 5000 psi
$\quad E_c$ = 4287 ksi
Topping f'_c = 3000 psi
$\quad E_c$ = 3321 ksi
n_{top} = 3321/4287 = 0.775

Span = 24 ft
Dead load (including beam weight)
$\quad$ = 4.15 kips/ft
Live load = 4.00 kips/ft

Dead load moment = $\dfrac{\omega \ell^3}{8} = \dfrac{4.15(24)^2}{8}$ = 298.8 kip-ft = 3586 kip-in.

Live load moment = $\dfrac{4.0(24)^2}{8}$ = 288.0 kip-ft = 3456 kip-in.

$\quad$ Total = M_{ext} = 586.8 kip-ft = 7042 kip-in.

Prestress force:
$\quad A_{ps}$ = 1.836 in.^{2}
$\quad f_{pu}$ = 270 ksi
$\quad E_{ps}$ = 28,500 ksi
$\quad$ Initial prestress level = 0.75f_{pu}
$\quad$ Estimated prestress loss = 20%
$\quad$ P $\;$ = 1.836(270)(0.75)(1 − 0.2) = 297.4 kips
$\quad$ e $\;$ = 9.24 − 3 = 6.24 in.
$\quad A_s$ = 2 − #7 = 1.2 in.^{2} (Grade 60)
$\quad E_s$ = 29,000 ksi
$\quad A'_s$ = 2 − #9 = 2.0 in.^{2} (Grade 60)
$\quad E_s$ = 29,000 ksi
$\quad n_{ps}$ = 28,500/4287 = 6.65
$\quad n_s$ = 29,000/4287 = 6.76

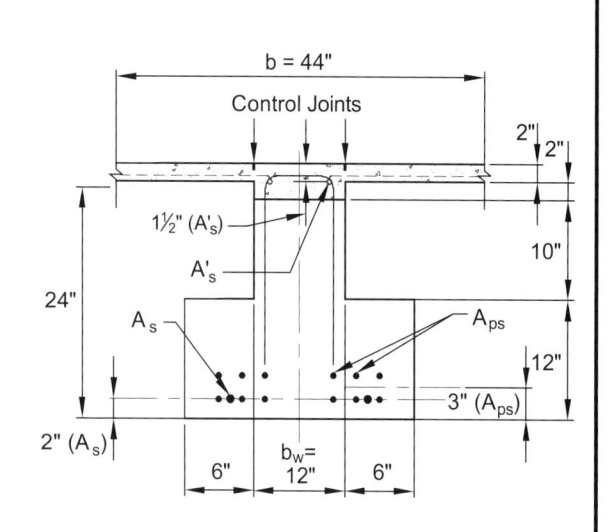

Solution:
1. Gross Section Analysis:

a. Stress on bare beam = Stress due to dead load moment

$f_b = \dfrac{P}{A} + \dfrac{Pe}{S_b} - \dfrac{M_d}{S_b} = \dfrac{297.4(1000)}{408} + \dfrac{297.4(1000)(6.24)}{1592} - \dfrac{3586(1000)}{1592} = 729 + 1166 - 2253$

$\quad$ = −358 psi (tension)

$f_t = \dfrac{P}{A} - \dfrac{Pe}{S_t} + \dfrac{M_d}{S_t} = \dfrac{297.4(1000)}{408} - \dfrac{297.4(1000)(6.24)}{1152} + \dfrac{3586(1000)}{1152} = 729 - 1611 + 3113$

$\quad$ = 2231 psi (compression)

b. Stress on composite section = Stress due to live load moment

$f_b = \dfrac{M_\ell}{S_{bc}} = \dfrac{3456(1000)}{2662} = -1298$ psi (tension)

Example 4.2.2.6 (Cont.)
Transformed Cracked Section Using Ref. 20: Example 2

$$f_t = \frac{M_\ell}{S_t} = \frac{3456(1000)}{2277} = 1518 \text{ psi (compression)}$$

$f_{t\ell} = -358 - 1298 = -1656$ psi (tension)

$\quad > 7.5\sqrt{5000}$ (section is cracked)

2. Estimate the decompression stress, f_{dc}.
   Estimate elastic shortening loss (ES) at 40% of TL
   Loss used for calculating f_{dc} = TL − ES:
   20 − 0.4(20) = 12%
   Strand stressed to $0.75 f_{pu}$
   $f_{dc} = (1 - 0.12)(0.75)(270) = 178$ ksi
   $P_{ps} = 178(1.836) = 327$ kips

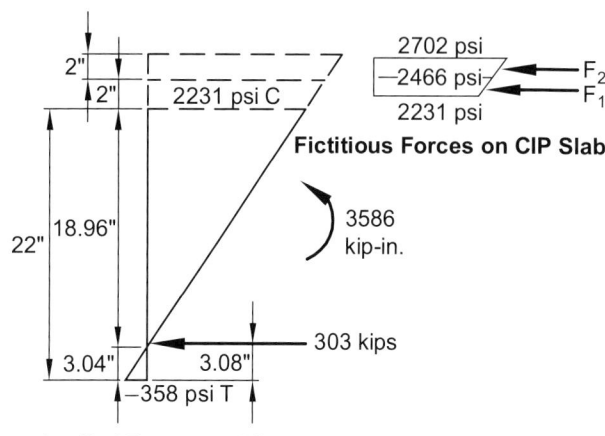

3. Estimate decompression stress in reinforcing bars:

Creep and shrinkage loss is approximately 50% of total loss:
   20 − 0.5(20) = 10%
   f_{dc} (bars) = 0.10(0.75)(270) = 20.25 ksi (compression)
   $P_s = -20.25(1.2) = -24$ kips
   Combine with P_{ps}: 327 − 24 = 303 kips
   Location: $\frac{327(3) - 24(2)}{303} = 3.08$ in. from bottom

4. Compute fictitious stresses and forces in composite slab and compression reinforcement:
   Combine with P_{ps} and P_s.
   Zero stress is at $\frac{2231}{358 + 2231}(22) = 18.96$ in. from top of precast member

Project stresses to top of composite section:
   Fictitious stress at interface of stem and flange:
   $= \frac{20.96}{18.96}(2231) = 2466$ psi

   Fictitious stress at top of flange:
   $= \frac{22.96}{18.96}(2231) = 2702$ psi

   Compression reinforcement is located 2.5 in. from top of precast member.
   Fictitious stress at level of A'_s
   $= \frac{18.96 + 2.5}{18.96}(2231) = 2525$ psi

Equivalent width of topping:
   $b_{te} = n_{top}(b) = 0.775(44) = 34.1$ in.
Equivalent width of CIP stem:
   $b_{se} = n_{top}(b_w) = 0.775(12) = 9.3$ in.

Transformed area of compression steel:
   $A'_{ts} = (n_s - 1)(A'_s) = 5.76(2.0) = 11.52$ in.^{2}

Example 4.2.2.6 (Cont.)
Transformed Cracked Section Using Ref. 20: Example 2

Note: Center of topping elements is assumed at center of thickness (error is small).

$$F_3 = \left(\frac{2525}{1000}\right)(11.52) = 29 \text{ kips (at 24.5 in. from bottom)}$$

$$F_2 = \left(\frac{2466+2702}{2(1000)}\right)(2)(34.1) = 176 \text{ kips (at 25 in. from bottom)}$$

$$F_1 = \left(\frac{2466+2231}{2(1000)}\right)(2)(9.3) = 44 \text{ kips (at 23 in. from bottom)}$$

$$P_{ps} + P_s = \underline{303} \text{ (at 3.08 in. from bottom)}$$
$$P_e = 552 \text{ kips}$$

Location:
$$y_1 = \frac{303(3.08)+44(23)+176(25)+29(24.5)}{552} = 12.78 \text{ in.}$$

5. Transformed section properties of steel elements and topping:

$$A_{tps} = n_{ps}(A_{ps}) = 6.65(1.836) = 12.21 \text{ in.}^2$$
$$A_{ts} = n_s(A_s) = 6.76(1.2) = 8.11 \text{ in.}^2$$
$$A'_{ts} = (n_s - 1)A'_s = 5.76(2.0) = 11.52 \text{ in.}^2$$

Equivalent area of topping slab:
$$= b_{te}(t) = 34.1(2) = 68.2 \text{ in.}^2$$
Equivalent area of CIP stem:
$$= b_{se}(t) = 9.3(2) = \underline{18.6 \text{ in.}^2}$$
$$\text{Total} = 86.8 \text{ in.}^2$$

$$A_{tt} = 86.8 + 11.52 = 98.3 \text{ in.}^2$$

c.g. of topping + $A'_s = \dfrac{68.2(1)+18.6(3)+11.52(1.5)}{98.3} = 1.44$ in. from top;

$$y_{tt} = 26 - 1.44 = 24.56 \text{ in.}$$

$$I_{top} = \frac{34.1(2)^3}{12} + 68.2(1.44-1.0)^2 + \frac{9.3(2)^3}{12} + 18.6(3.0-1.44)^2 + 11.52(1.5-1.44)^2 = 87.4 \text{ in.}^4$$

6.(1) Select c = 14 in. (d_1 = 10 in.) for first trial:

$$A_t = A_{tt} + b(d_1) + A_{tps} + A_{ts} = 98.3 + 12(10) + 12.21 + 8.11 = 238.6 \text{ in.}^2$$

$$y_t = \frac{A_{tt}y_{tt}+b(d_1)(y_3)+A_{tps}y_{tps}+A_{ts}y_{ts}}{A_t} = \frac{98.3(24.56)+12(10)(17)+12.21(3)+8.11(2)}{238.6} = 18.89 \text{ in.}$$

$$I_t = I_{top} + A_{tt}(y_{tt}-y_t)^2 + \frac{b(d_1)^3}{12} + (bd_1)(y_t-y_3)^2 + A_{tps}(y_t-y_{tps})^2 + A_{ts}(y_t-y_{ts})^2$$

$$= 87.4 + 98.3(24.56-18.89)^2 + \frac{12(10)^3}{12} + 12(10)(18.89-17)^2 + 12.21(18.89-3)^2 + 8.11(18.89-2)^2$$

$$= 87 + 3160 + 1000 + 429 + 3083 + 2314 = 10{,}073 \text{ in.}^4$$

$$y_{na} = 18.89 - 12 = 6.89 \text{ in.}$$

$$\frac{I_t}{y_{na}} = \frac{10073}{6.89} = 1462 \text{ in.}^3$$

Example 4.2.2.6 (Cont.)
Transformed Cracked Section Using Ref. 20: Example 2

7.(1) $y_p = y_t - y_1 = 18.89 - 12.78 = 6.11$ in.
$M_{int} = M_{ext} - P_e(y_p) = 7042 - 552(6.11) = 3669$ kip-in.

8.(1) $\dfrac{M_{int}}{I/y_{na}} = \dfrac{3669}{1462} = -2.510$ ksi

$\dfrac{P_e}{A_t} = \dfrac{552}{238.6} = \underline{+2.313}$ ksi

Difference = -0.197 ksi

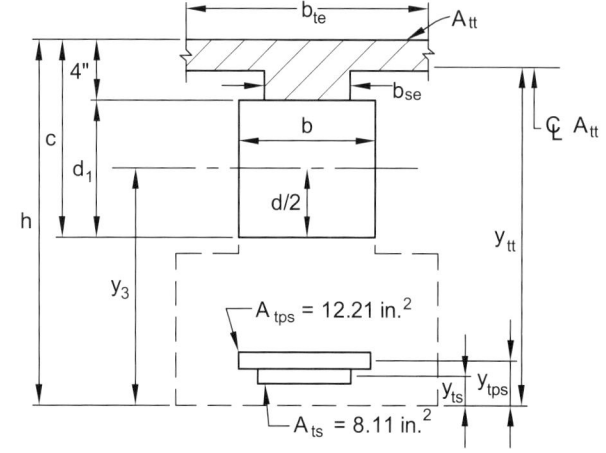

6.(2) Final solution at c = 13.44 in. (d_1 = 9.44 in.):
$A_t = A_{tt} + b(d_1) + A_{tps} + A_{ts}$
$= 98.3 + 12(9.44) + 12.21 + 8.11$
$= 231.9$ in.^{2}

$y_t = \dfrac{A_{tt}y_{tt} + b(d_1)(y_3) + A_{tps}y_{tps} + A_{ts}y_{ts}}{A_t}$

$= \dfrac{98.3(24.56) + 12(9.44)(17.28) + 12.21(3) + 8.11(2)}{231.9} = 19.08$ in.

$I_t = I_{top} + A_{tt}(y_{tt} - y_t)^2 + b\left(\dfrac{(d_1)^3}{12}\right) + (bd_1)(y_t - y_3)^2 + A_{tps}(y_t - y_{tps})^2 + A_{ts}(y_t - y_{ts})^2$

$= 87.4 + 98.3(24.56 - 19.08)^2 + \dfrac{12(9.44)^3}{12} + 12(9.44)(19.08 - 17.28)^2 + 12.21(19.08 - 3)^2$
$+ 8.11(19.08 - 2)^2 = 87 + 2952 + 841 + 367 + 3157 + 2366 = 9770$ in.^{4}

$y_{na} = 19.08 - 12.56 = 6.52$ in.

$\dfrac{I_t}{y_{na}} = \dfrac{9770}{6.52} = 1498$ in.^{3}

7.(2) $y_p = y_t - y_1 = 19.08 - 12.78 = 6.30$ in.
$M_{int} = M_{ext} - p_e(y_p) = 7042 - 552(6.30) = 3564$ kip-in.

8.(2) $\dfrac{M_{int}}{I/y_{na}} = \dfrac{3564}{1498} = -2.379$ ksi

$\dfrac{P_e}{A_t} = \dfrac{552}{231.9} = +2.380$ ksi

Difference = $\overline{+0.001}$ ksi

Use $I_{cr} = 9{,}770$ in.^{4}

9.(a) Check stresses:
Transformed stress at top of composite slab:
$\dfrac{P_e}{A_t} + \dfrac{M_{int}y_{top}}{I_{cr}} = \dfrac{552}{231.9} + \dfrac{3564(26 - 19.08)}{9770} = 4.905$ ksi = 4905 psi

Subtract fictitious stress calculated in Step 4: 4905 − 2702 = 2203 psi

Example 4.2.2.6 (Cont.)
Transformed Cracked Section Using Ref. 20: Example 2

Actual stress – multiply by n (topping)
= 0.775(2203) = 1707 psi (OK)
Transformed stress at top of precast member:
$$= \frac{13.44 - 4}{13.44}(4905) = 3445 \text{ psi}$$

(Note: This is slightly higher than $0.6f'_c = 3000$ psi allowed by Code, but at this location there would be no effect on performance. Higher concrete strength may be used.)

Less fictitious stress
= 3445 – 2231 = 1214 psi
Transformed stress at interface of CIP stem and slab:
= (4905 + 3445)/2 = 4175 psi

Less fictitious stress
= 4175 – 2466 = 1709 psi
Transformed stress at level of compression reinforcement:
$$= \frac{13.44 - 1.5}{13.44}(4905) = 4358 \text{ psi}$$

Less fictitious stress
= 4358 – 2525 = 1833 psi
Δf_{ps} = 3489 psi < 20 ksi

No crack control spacing requirements.

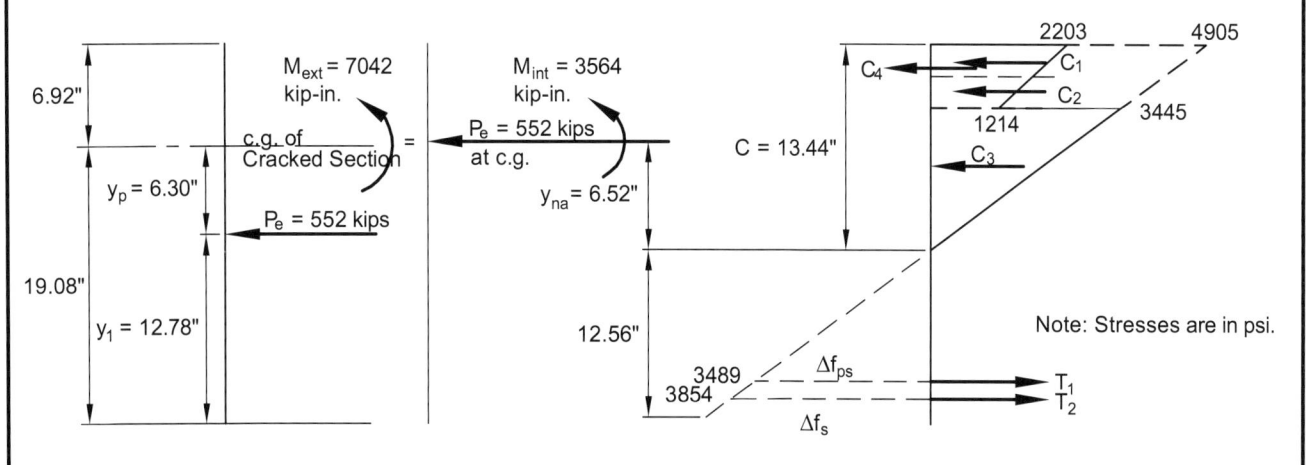

Final Solution

Stresses and Equilibrium Check

Note: Stresses are in psi.

9.(b) Check equilibrium:
Compression:
    CIP slab: b = 34.1 in.; t = 2 in.
$$C_1 = \frac{2203 + 1709}{2(1000)}(34.1)(2) = 133.4 \text{ kips}$$
    CIP stem: b = 9.3 in.; t = 2 in.

Example 4.2.2.6 (Cont.)
Transformed Cracked Section Using Ref. 20: Example 2

$$C_2 = \frac{1709+1214}{2(1000)}(9.3)(2) = 27.2 \text{ kips}$$

Precast member:
  b = 12 in.; t = 13.44 − 4 = 9.44 in.
$$C_3 = \frac{3445}{2(1000)}(12)(9.44) = 195.1 \text{ kips}$$

Compression reinforcement:
$$(n_s - 1)(A'_s) = 11.52 \text{ in.}^2$$
$$C_4 = \frac{1833}{1000}(11.52) = 21.1 \text{ kips}$$
$$C = 133.4 + 27.2 + 195.1 + 21.1 = 376.8 \text{ kips}$$

Center of compression:
$$\frac{133.4(1) + 27.2(3) + 195.1\left(4 + \frac{9.44}{3}\right) + 21.1(1.5)}{376.8} = 4.36 \text{ in. from top}$$

Transformed section tension stress:
By similar triangles:
  At level of A_{ps}: $\dfrac{23}{13.44} = \dfrac{\Delta f_{ps} + 4905}{4905}$

$$\Delta f_{ps} = \frac{23(4905)}{13.44} - 4905 = 3489 \text{ psi}$$

  Similarly,
$$\Delta f_s = \frac{24(4905)}{13.44} - 4905 = 3854 \text{ psi}$$

The decompression stresses calculated in Steps 2 and 3 must be added back:
  $T_1 = (n\Delta f_{ps} + f_{dc})A_{ps} = [(6.65)3.489 + 178](1.836) = 369.4$ kips
  $T_2 = (n\Delta f_{ps} + f_{dc})A_s = [6.76(3.854) - 20.25](1.2) = 7.0$ kips
  $T = 369.4 + 7.0 = 376.4$ kips ≈ 376.8 kips (Check)

Center of tension:
$$= \frac{369.4(3) + 7.0(2)}{376.4} = 2.98 \text{ in.}$$
  M = (26 − 2.98 − 4.36)(376.8) = 7031 kip-in. ≈ 7042 kip-in. (Check)

4.2.3 Prestress Transfer and Strand Development

In a pretensioned member, the prestress force is transferred to the concrete by bond. The length required to accomplish this transfer is called the "transfer length." It is given in the Commentary to the Code to be equal to $(f_{se}/3)d_b$.

The length required to develop the design strength of the strand, however, is much longer, and is specified in Code Section 12.9.1 by the equation:

$$\ell_d = \left(\frac{f_{se}}{3}\right)d_b + (f_{ps} - f_{se})d_b \qquad \text{(Eq. 4.2.3.1)}$$

Section 12.9.3 of the Code requires the development length to be doubled if:

1. Bonding of the strand does not extend to the end of the member (debonded or "shielded" strand), and
2. The member is designed such that tension will occur in the precompressed tensile zone under service loads.

In the Commentary to the Code, the variation of strand stress along the development length is given as shown in Figure 4.2.3.1. This is illustrated in Figure 4.12.4 for several strand sizes. The value of f_{se} is shown as 170 ksi, which is typical for low-relaxation strand tensioned to the maximum allowable. A more general design aid which includes all currently used strand sizes and other values of f_{se} is given in Chapter 11 (see Design Aid 11.2.6).

In short span flexural members, prestressing strands may not be developed at sections of high moment. In such cases, it is possible that a premature failure may occur in the concrete due to strand slip. [5] In such cases, the capacity of the section should be reduced to account for this changed failure mode, as illustrated in Example 4.2.3.1.

When a portion of the strands are debonded, zones are created where sections through the member will contain strands with unequal strains. In this case, calculation of nominal strength in the development region should be based on strain compatibility, or conservatively, the contribution of the debonded strand neglected until it is fully developed. Example 4.2.3.2 illustrates these principles.

Failures caused by bond slip are brittle, so a value of $\phi = 0.75$ is recommended to determine flexural capacity when this failure mode is possible. Also, quality control measures must be in place to ensure that prestressing strands will meet the transfer and development length requirements of the Code. [6]

Figure 4.2.3.1 Variation of strand stress along development length

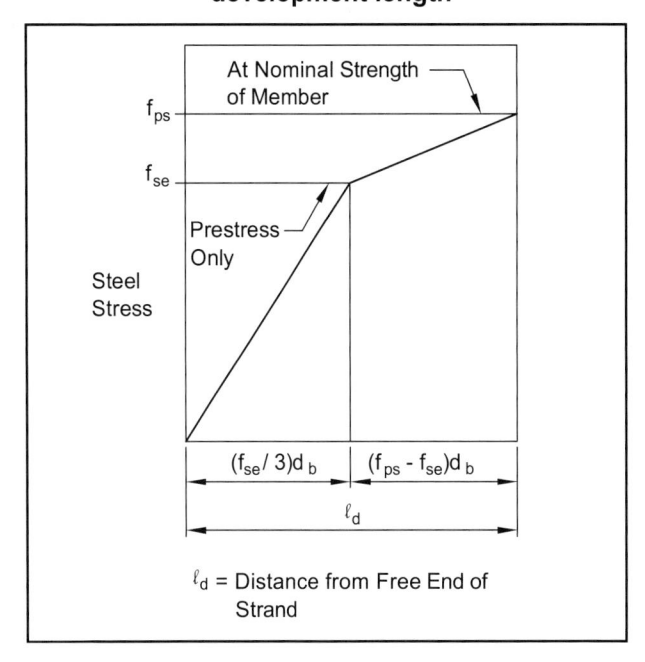

ℓ_d = Distance from Free End of Strand

4.2.3.1 Strand Debonding

For economic and production safety reasons, it is common practice to use all straight strands in a member rather than depressing at midspan. This may cause excessive end release stresses. Because of this and sometimes to improve camber control, producers of precast, prestressed concrete may choose to debond some of the strands at the end of the member. (This is sometimes called "shielding," because the debonding is usually done by placing a length of plastic tube or shield over the strand.)

Most engineers feel the practice of debonding should be avoided except when absolutely necessary, i.e., it is better to allow slightly higher release stresses than the Code prescribes. When debonding is needed, the following guidelines should be followed whenever possible (see Figure 4.2.3.2):

- Do not debond any strands in the bottom row.
- Stagger debonding at transfer length increments.
- Do not debond more than 50% of the strands below a dapped end.
- Avoid debonding adjacent strands.
- Provide vertical reinforcement (WWR, stirrups or bearing plate anchorage) at least equal to minimum shear reinforcement in the debonded area.

Example 4.2.3.1
Use of Figure 4.12.4—Design Stress for Underdeveloped Strand

Given:
  Span = 12 ft
  4HC8 as shown

Concrete:
  f'_c = 5000 psi, normal weight concrete

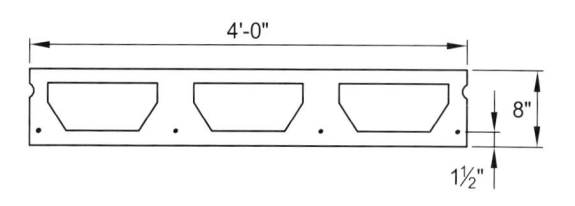

Prestressing steel:
  4 – ½ in. diameter 270K strands
  A_{ps} = 4 (0.153) = 0.612 in.^{2}

Solution:
If the strand is fully developed (see Figure 4.12.3):
$$C\omega_{pu} = \frac{CA_{ps}f_{pu}}{bd_pf'_c} = \frac{1.06(0.612)(270)}{48(6.5)(5)} = 0.11$$

From Figure 4.12.3, with f_{se} = 170 ksi
  f_{ps} = 268 ksi

The maximum development length available is:
  $\ell/2$ = 12(12/2) = 72 in.
From Figure 4.12.4, the maximum f_{pd} = 257 ksi

This value, rather than 268 ksi, should be used to calculate the design strength (M_n) of the member at midspan. Since failure would be by brittle bond slip, ϕ = 0.75.

Figure 4.2.3.2 Strand debonding limitations

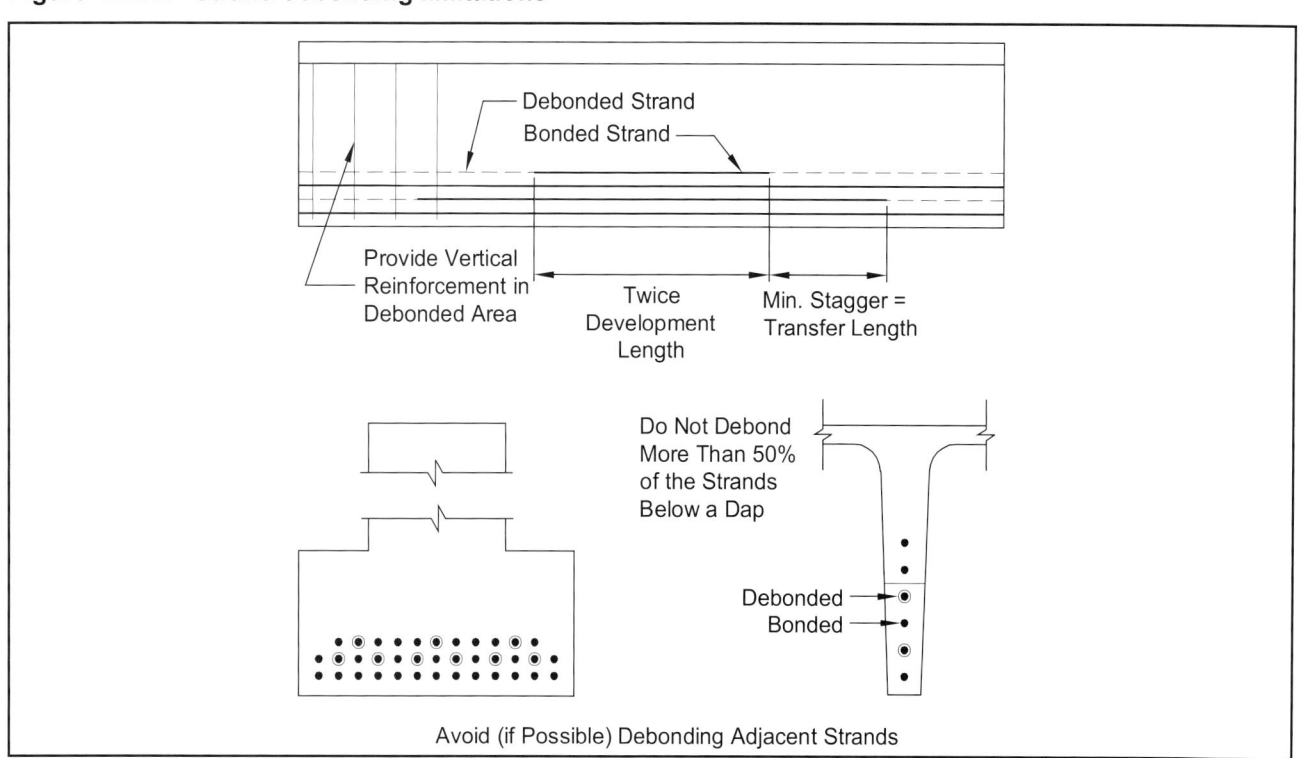

Example 4.2.3.2
Moment Capacity of Member with Debonded Strands in Development Region

Given:
  10-ft wide double tee, one stem shown above:
  10 – ½ in. diameter, 270 ksi strands (five each stem)
  $A_{ps} = 10(0.153) = 1.53$ in.^{2}

Concrete (normal weight):
  $f'_c = 5000$ psi
  $E_c = 4300$ ksi

Prestressing strands:
  $f_{pu} = 270$ ksi
  $f_{se} = 170$ ksi
  $E_{ps} = 28{,}500$ ksi
  $\varepsilon_{se} = 170/28500 = 0.0060$

Problem:
  Strand No. 3 is debonded for 5 ft from the end.
  Find M_n at 12 ft from the end.

Solution:

A. Determine capacity assuming the debonded strand does not slip:
    Maximum f_{ps} for fully bonded strand (from separate analysis) = 269 ksi.
    Transfer length = $(f_{se}/3)d_b = (170/3)(0.5) = 28.33$ in.

    For debonded strands, double the transfer length per ACI 318-02, Section 12.9.3.
    Transfer length for debonded strands = $2(28.33) = 56.7$ in.
    $\ell_d = [f_{ps} - (2/3)f_{se}]d_b = (269 - 113.3)(0.5) = 77.85$ in.
    ℓ_d for debonded strands = $2(77.85) = 155.7$ in.

   If the strand does not slip, the maximum strength the strand can develop at 12 ft from the end (7 ft or 84 in. from the point of debonding) is given by:
   $$f_{pd} = 170 + \frac{84 - 56.7}{155.7 - 56.7}(269 - 170) = 197.3 \text{ ksi}$$
   and the corresponding strain is:
   $$\varepsilon_{pd} = \frac{f_{pd}}{E_{ps}} = \frac{197.3}{28500} = 0.00692$$

   $\varepsilon_{sa} = \varepsilon_{pd} - \varepsilon_{se} = 0.00692 - 0.0060 = 0.00092$

(See the sketch on next page)

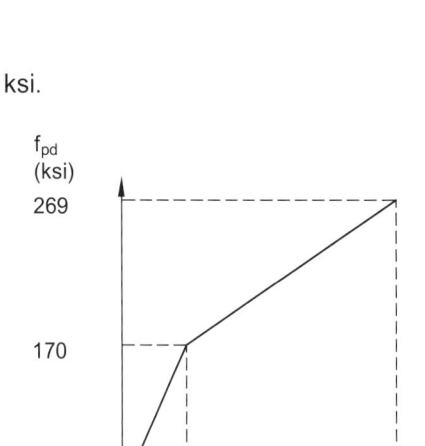

Bonded	28.3"	77.8"
Debonded	56.7"	155.7"

Distance from Point of Debonding
(End of Member for Fully Bonded Strains)

  Note: In this example, the spacing of the strands is such that the variation in strains is inconsequential. Thus, the stress in the strands may be assumed to be equal, and the centroid of the tension force, T, may be assumed to be at the centroid of the strand group.
  T = 197.3(1.53) = 301.9 kips

  Use iteration procedures, varying ε_c until the compression force, C, reasonably approximates the tension force, T.

Example 4.2.3.2 (Cont.)
Moment Capacity of Member with Debonded Strands in Development Region

Final trial:
$\varepsilon_c = 0.000265$

$c = \dfrac{0.000265}{0.000265 + 0.00092}(20) = 4.47$ in.

Compressive stress at the top of the flange
$= \varepsilon_c E_c = 0.000265(4300) = 1.14$ ksi

Compressive stress at the bottom of the flange
$= \dfrac{0.47}{4.47}(1.14) = 0.12$ ksi

Compression in the flange
$= \dfrac{1.14 + 0.12}{2}(120)(4) = 302.4$ kips

Compression in the webs
$= \dfrac{5.75(0.47)(0.12)}{2}(2 \text{ webs}) = 0.3$ kips

Total compression = 302.4 + 0.3
= 302.7 kips ≈ 301.9 kips

Use the sketch at the right to calculate moments:
$M_n = 301.9(15.84 + 2.70) - 0.3(2.70)$
= 5596 kip-in. = 466 kip-ft

Since the failure mode is brittle strand slip, by Code Section 9.3.2.7:
$\phi = 0.75$
$\phi M_n = 0.75(466) = 350$ kip-ft

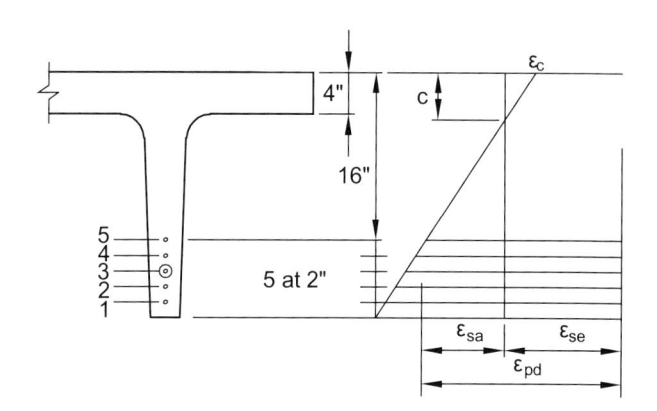

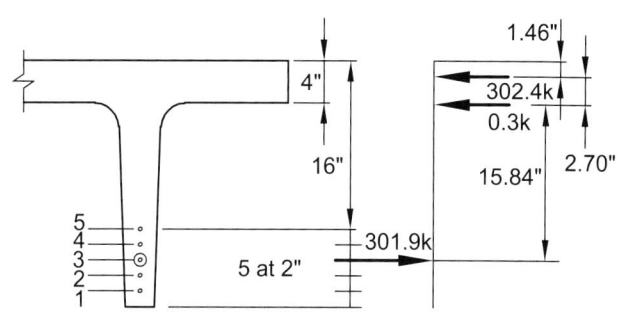

B. Assume debonded strand slips.

If additional load is applied causing higher strand stress than indicated in the above analysis, the strand will slip. The other strands, however, being fully bonded, are capable of much higher stress and strain. If additional load is applied, as in a load test, the stress and strains will redistribute, and load can be applied until the full ultimate strength of the strands is reached. Thus, a second, more conservative analysis would be to neglect the debonded strands and determine moment capacity with eight fully bonded strands.

$T = A_{ps}f_{ps} = 8(0.153)(269) = 329.3$ kips

$a = \dfrac{T}{0.85f'_c b} = \dfrac{329.3}{0.85(5)(120)} = 0.65$ in.

Since this is less than the flange thickness, the design is like a rectangular beam.

$M_n = T\left(d - \dfrac{a}{2}\right) = 329.3\left(20 - \dfrac{0.65}{2}\right) = 6480$ kip-in. = 540 kip-ft

In this case, the failure mode is ductile, $\phi = 0.9$.
$\phi M_n = 0.9(540) = 486$ kip-ft

Since the assumption of strand slip is more conservative, the design may be based on Method B.

Example 4.2.4.1
Calculation of End Reinforcement to Resist Bursting Stresses

Given:
Beam of Example 4.2.2.4
Transfer length = 28.3 in.

$$A_{vt} = \frac{0.021 P_o h}{f_s \ell_t} = \frac{0.021(418)(26)}{30(28.3)} = 0.27 \text{ in.}^2$$

h/5 = 26/5 = 5.2 say 6 in.

Provide at least 2 – #3 stirrups ($A_s = 0.44$ in.^{2}) within 6 in. of the end.

4.2.4 End Stresses at Transfer

At the time prestress force is transferred, tensile stresses perpendicular to the prestressing force (sometimes called "bursting" or "splitting" stresses) develop which may cause horizontal cracks near the end of the member. [7] These forces can be resisted by vertical reinforcement calculated by the following equation:

$$A_{vt} = \frac{0.021 P_o h}{f_s \ell_t} \quad \text{(Eq. 4.2.4.1)}$$

where:
- A_{vt} = required area of stirrups at the end of a member uniformly distributed over a length h/5 from the end
- P_o = prestress force at transfer
- h = depth of the member
- f_s = design stress in the stirrups, usually assumed to be 30 ksi
- ℓ_t = strand transfer length

Since this is a temporary stress at the time of prestress transfer, such reinforcement need not be in addition to shear and torsion reinforcement.

4.2.5 Bending of Asymmetrical Sections

Most precast, prestressed standard sections are symmetrical about their vertical axes. However, some sections, such as the stadium riser unit shown in Figure 4.2.5.1(a), are not, and it is necessary to calculate properties about the principal axes, U-U and V-V. Note that the principal axes may change when the unit is attached to the structure. Calculations of stresses of prestressed concrete stadium risers are illustrated in Ref. 8. Strength design, Figure 4.2.5.1(b), is more complex, and may be approximated or computed graphically.

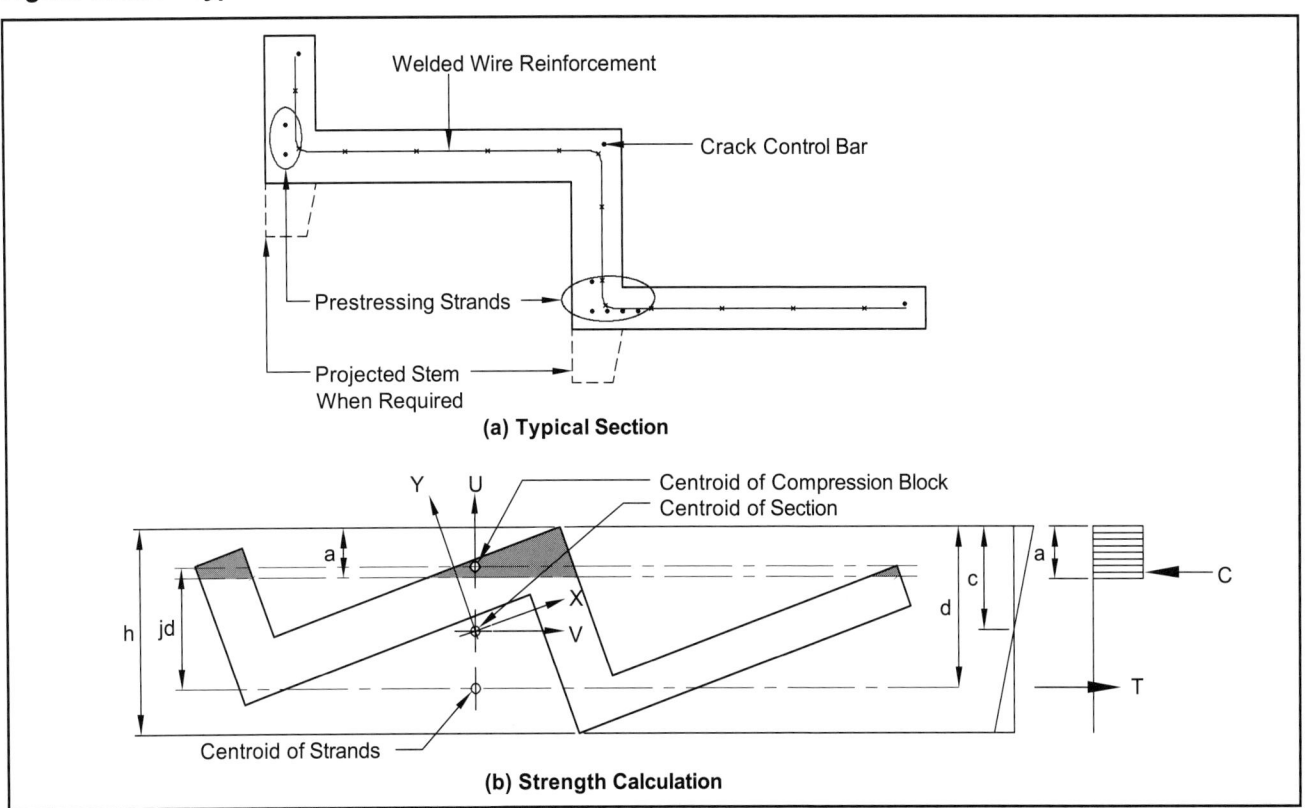

Figure 4.2.5.1 Typical stadium riser

(a) Typical Section

(b) Strength Calculation

4.3 Shear

The shear design of precast concrete members is covered in Chapter 11 of ACI 318-02. The shear resistance of precast concrete members must meet the requirement:

$$\phi V_n \geq V_u$$

where:
- $V_n = V_c + V_s$, and
- V_c = nominal shear strength of concrete
- V_s = nominal shear strength of shear reinforcement
- $\phi = 0.75$

If V_u is less than $\phi V_c/2$, no shear reinforcement is required. However, for flat deck members (hollow-core and solid slabs), and others proven by test, no shear reinforcement is required if the factored shear force, V_u, does not exceed the design shear strength of the concrete, ϕV_c. For other prestressed members, the minimum shear reinforcement is usually adequate.

The critical section for shear and torsion is indicated in the Code to be a distance "d" from the face of the support for non-prestressed members and "h/2" for prestressed members. However, precast concrete members on which the load is not applied at the top of the member, such as L-shaped beams, the distance "d" or "h" should be measured from the point of load application to the bottom, or, conservatively, the critical section taken at the face of the support. Also, if a concentrated load is applied near a support face, the critical section should be taken at the support face. See Code Section 11.1.3.

4.3.1 Shear Resistance of Non-Prestressed Concrete

In the absence of torsion and axial forces, the nominal shear resistance of concrete is given by:

$$V_c = 2\sqrt{f'_c}\, b_w d \quad \text{(Eq. 4.3.1.1)}$$

or if a more detailed analysis is performed:

$$V_c = \left(1.9\sqrt{f'_c} + 2500\rho_w \frac{V_u d}{M_u}\right) b_w d \leq 3.5\sqrt{f'_c}\, b_w d$$

(Eq. 4.3.1.2)

where:

$$\frac{V_u d}{M_u} \leq 1.0$$

See ACI Code for members subjected to significant axial forces in addition to shear. Torsion is presented in Section 4.4.

4.3.2 Shear Resistance of Prestressed Concrete Members

Shear design of prestressed concrete members is covered in ACI 318-02 by Eq. 11-9 through 11-12, reproduced below

Either Eq. 4.3.2.1 or the lesser of Eqs. 4.3.2.2 or 4.3.2.4 may be used; however, Eq. 4.3.2.1 is valid only if the effective prestress force is at least equal to 40% of the tensile strength of the prestressing strand. The Code places certain upper and lower limits on the use of these equations, which are shown in Figure 4.3.2.1.

Example 4.3.1.1
Design of Shear Reinforcement — Non-Prestressed Member

Given:
A spandrel beam as shown:
- b_w = 8 in.
- d = 87 in.
- Span, ℓ = 30 ft
- f'_c = 5 ksi, normal weight concrete
- f_y = 60,000 psi

Loading at top of member, w_u
- Dead load = 3.96(1.2) = 4.75 kips/ft
- Live load = 1.31(1.6) = 2.10 kips/ft
- Total = 6.85 kips/ft

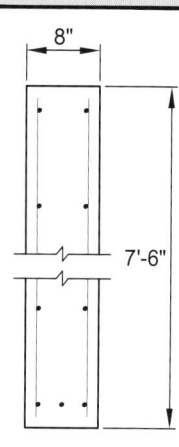

Example 4.3.1.1 (Cont.)
Design of Shear Reinforcement — Non-Prestressed Member

Problem:
Determine what size welded wire reinforcement will satisfy the shear requirements.

Solution:
Determine V_u at a distance d from support:

$V_u = \omega(\ell/2 - d/2) = 6.85(30/2 - 87/12) = 53.1$ kips

$V_c = 2\sqrt{f'_c}b_w d = \dfrac{2\sqrt{5000}}{1000}(8)(87) = 98.4$ kips

$\dfrac{\phi V_c}{2} = \dfrac{0.75(98.4)}{2} = 36.9$ kips

$V_c > V_u > \dfrac{\phi V_c}{2}$

Therefore, minimum shear reinforcement is required.
ACI 318-02, Eq. 11-13 requires a minimum amount of reinforcement be provided as follows:

$A_v = 0.75\sqrt{f'_c}\dfrac{b_w s}{f_y} = \dfrac{0.75(\sqrt{5000})(8)(12)}{60,000} = 0.08$ in.^{2}/ft

From Design Aid 11.2.11, select a WWR that has vertical wires: W4 at 6 in.
$A_v = 0.08$ in.^{2}/ft

See Section 12.13.2.4 of the Code for development of single leg WWR used as shear reinforcement.

$V_c = \left(0.6\sqrt{f'_c} + 700\dfrac{V_u d}{M_u}\right)b_w d$ (Eq. 4.3.2.1)

where: $\dfrac{V_u d}{M_u} \leq 1.0$

$V_{ci} = 0.6\sqrt{f'_c}b_w d + V_d + \dfrac{V_i M_{cr}}{M_{max}}$ (Eq. 4.3.2.2)

$M_{cr} = \left(\dfrac{I}{y_t}\right)(6\sqrt{f'_c} + f_{pe} - f_d)$ (Eq. 4.3.2.3)

$V_{cw} = (3.5\sqrt{f'_c} + 0.3f_{pc})b_w d + V_p$ (Eq. 4.3.2.4)

The value of d in the term $V_u d/M_u$ in Eq. 4.3.2.1 is the distance from the extreme compression fiber to the centroid of the prestressed reinforcement. In all other equations, d need not be less than 0.8h.

In unusual cases, such as members which carry heavy concentrated loads, or short spans with high superimposed loads, it may be necessary to construct a shear resistance diagram (V_c) and superimpose upon that a factored shear (V_u) diagram. The procedure is illustrated in Figure 4.3.2.1.

The steps for constructing the shear resistance diagram are as follows:

1. Draw a horizontal line at a value of $2\sqrt{f'_c}b_w d$
   (Note: The Code requires that this minimum be reduced to $1.7\sqrt{f'_c}b_w d$ when the stress in the strand after all losses is less than $0.4f_{pu}$. For precast, prestressed members, the value will generally be above $0.4 f_{pu}$.)

2. Construct the curved portion of the diagram. For this, either Eq. 4.3.2.2 or, more conservatively, Eq. 4.3.2.1 may be used. Usually, it is adequate to find three points on the curve.

Figure 4.3.2.1 Shear design

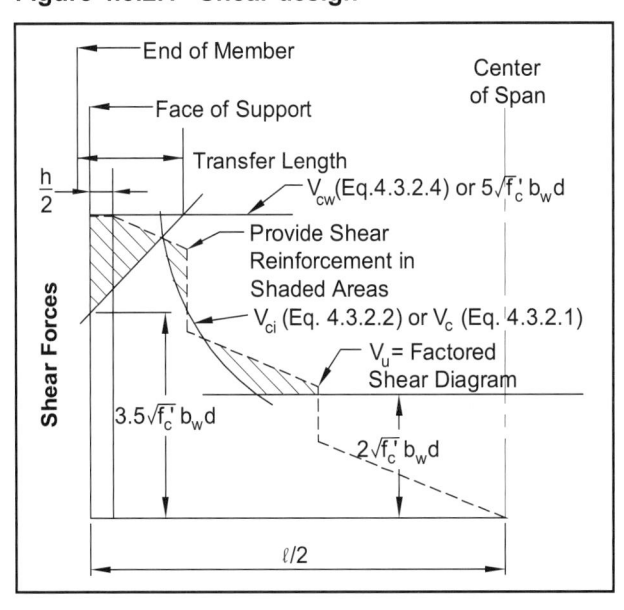

Example 4.3.2.1
Construction of Applied and Resisting Design Shear Diagrams

Given:
  2HC8 with span and loadings shown.

Section properties:
  $A = 110$ in.^{2}
  $I = 843$ in.^{4}
  $y_b = 4.0$ in.
  $b_w = 6.25$ in.
  $d = 5.875$ in.
  $h = 8.0$ in. ($0.8h = 6.4$ in.)
  wt = 57 psf = 114 plf

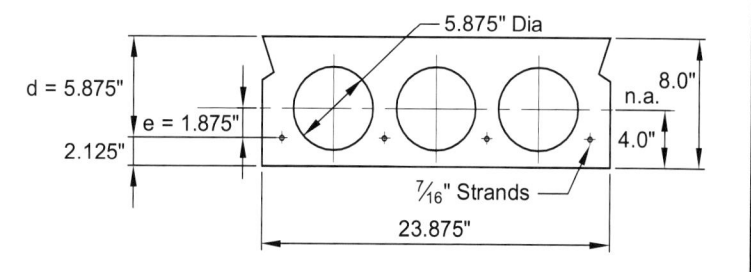

Concrete:
  $f'_c = 5000$ psi normal weight concrete

Problem:
Determine shear reinforcement requirement.

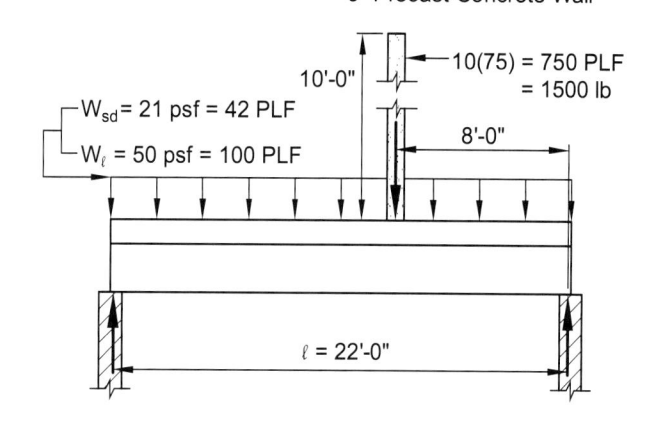

Solution:
1. Determine factored loads.
   Uniform dead load = 1.2(42 + 114) = 187 plf
   Uniform live load = 1.6(100) = 160 plf
   Concentrated dead load = 1.2(1500) = 1800 lb

2. Construct shear diagram.
   $\omega_u = 187 + 160 = 347$ plf
   $$R_L = \frac{\omega_u \ell(\ell/2) + P_x}{\ell} = \frac{0.347(22)(11) + 1.8(8)}{22}$$
   $= 4.47$ kips
   $R_R = 0.347(22) + 1.8 - 4.47 = 4.96$ kips

3. Construct the shear resistance diagram as described in the previous section.

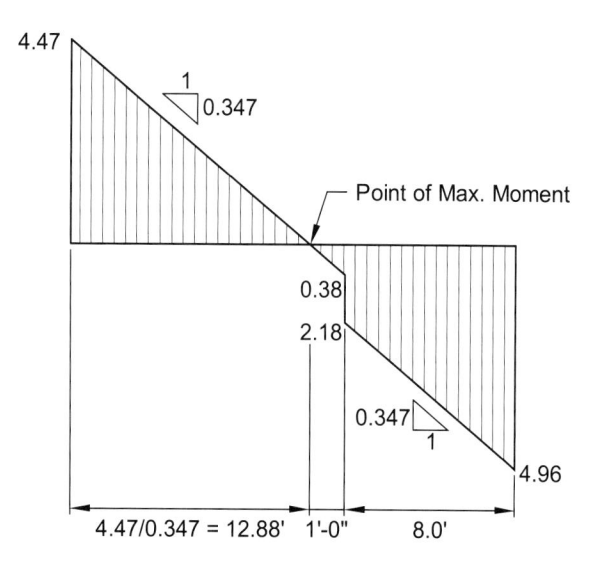

a. Construct line at $2\sqrt{f'_c}\, b_w d = 5.2$ kips
b. Construct V_c line by Eq. 4.3.2.1:
$$V_c = \left(0.6\sqrt{f'_c} + 700\frac{V_u d}{M_u}\right) b_w d = 1.56 + 25.7 \frac{V_u d}{M_u}$$

At 1, 2, and 4 ft from each end:
  V_u (left) $= 4.47 - 0.347x$
  M_u (left) $= \left(4.47x - \dfrac{0.347x^2}{2}\right)12$
  V_u (right) $= 4.96 - 0.347x$
  M_u (right) $= \left(4.96x - \dfrac{0.347x^2}{2}\right)12$

Example 4.3.2.1 (Cont.)
Construction of Applied and Resisting Design Shear Diagrams

See table at right.

c. Construct upper limit line.
Since Eq. 4.3.2.1 was used,

$$\text{upper limit} = 5\sqrt{f'_c}\,b_w d = \frac{5\sqrt{5000}(6.25)(5.875)}{1000}$$
$$= 13.0 \text{ kips}$$

d. Construct diagonal line at transfer zone from $3.5\sqrt{f'_c}\,b_w d = 9.1$ kips at end of member to 13.0 kips at $50d_b^* = 50(7/16) = 21.9$ in. $= 1.82$ ft.

e. Construct $\dfrac{V_u}{\phi}$ diagram:

4.47/0.75 = 5.96
0.38/0.75 = 0.51
2.18/0.75 = 2.91
4.96/0.75 = 6.61

It is apparent from these diagrams that no shear reinforcement is required.

Point	x (ft)	V_u (kips)	M_u (kip-in.)	$\dfrac{V_u d}{M_u}$	V_c (kips)
1	1	4.12	51.56	0.470	13.6
2	2	3.78	98.95	0.224	7.3
3	4	3.08	181.25	0.100	4.1
4	1	4.61	57.44	0.472	13.7
5	2	4.27	110.71	0.226	7.4
6	4	3.57	204.77	0.102	4.2

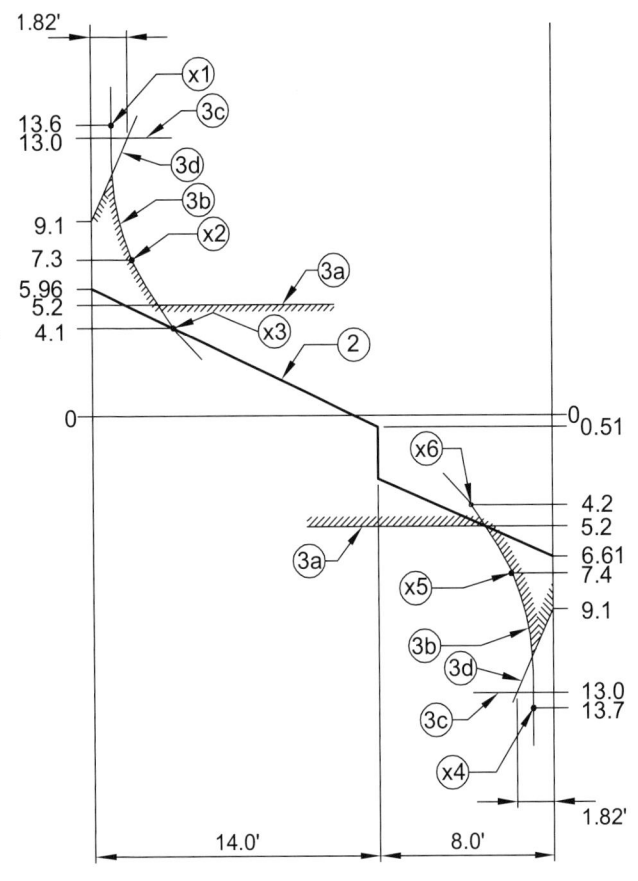

*Assumed transfer length. See ACI 318-02, Section 11.4.3.

---

3. Draw the upper limits line, V_{cw} from Eq. 4.3.2.4 if Eq. 4.3.2.2 has been used in Step 2, or $5\sqrt{f'_c}\,b_w d$ if Eq. 4.3.2.1 has been used.
4. The diagonal line at the upper left of Figure 4.3.2.1 delineates the upper limit of the shear resistance diagram in the prestress transfer zone. This line starts at a value of $3.5\sqrt{f'_c}\,b_w d$ at the end of the member, and intersects the V_{cw} line or $5\sqrt{f'_c}\,b_w d$ line at a transfer length from the end of the member. See Code Section 11.4.3.

4.3.3 Design Using Design Aids

Figures 4.12.5 through 4.12.9 are design aids to assist in determining the shear strength of precast, prestressed concrete members.

When lightweight concrete is used, the shear equations, Eqs. 4.3.1.1 through 4.3.2.4, are modified by substituting $\lambda\sqrt{f'_c}$ for $\sqrt{f'_c}$. The coefficient λ is defined as follows:

$$\lambda = \left(\frac{f_{ct}}{6.7}\right)\frac{1}{\sqrt{f'_c}} \leq 1.0$$

Example 4.3.3.1
Use of Figures 4.12.5 through 4.12.7 — Graphical Solution of Eq. 4.3.2.1 (Code Eq. 11-9)

Given:
- PCI standard double tee 10LDT24 + 2
- Span = 50 ft
- b_w = 9.5 in.
- d = 21 + 2 = 23 in.
- For depressed strand patterns $d \geq 0.8h$
- 0.8h = 20.8 in.
- d (near ends) = 16 in. < 20.8, use 20.8 in.
- d (near midspan) = 23 in. > 20.8, use 23 in.

Concrete:
- Precast: f'_c = 5000 psi, sand-lightweight, λ = 0.85
- Topping: f'_c = 4000 psi, normal weight

Reinforcement:
- Prestressing steel:
  - 270K strand, 108D1 pattern,
  - A_{ps} = 1.53 in.^{2}
  - e_e = 7.77 in.
  - e_c = 14.77 in.
- Shear reinforcement
  - f_y = 60 ksi

Loads:
- Dead load, w_d = 609 plf
- Live load, w = 800 plf

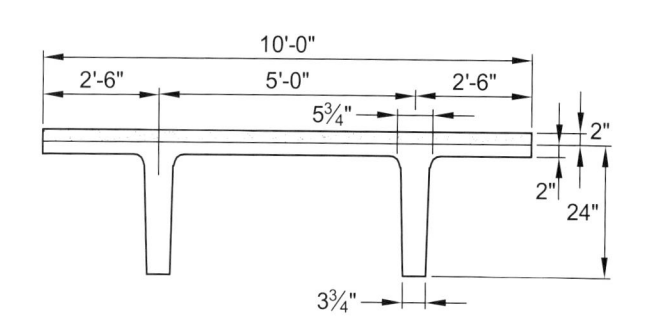

Problem: Find the value of excess shear force, $\frac{V_u}{\phi} - V_c$, along the span using Eq. 4.3.2.1 for V_c.

Solution (see figure at the end of this example):
The parameters needed for use of Figures 4.12.5 through 4.12.7 are:
Strand drape: 14.77 − 7.77 = 7 in. which is approximately equal to d/3, thus a shallow drape.

$$\frac{\ell}{d} = \frac{50 \times 12}{23} = 26.1$$

$\frac{V_u}{\phi}$ at support = $(1.2w_d + 1.6w_\ell)\left(\frac{\ell}{2}\right)\left(\frac{1}{\phi}\right)$ = (1.2 × 609 + 1.6 × 800)(50/2)(1/0.75)(1/1000) = 67.0 kips

The graphical solution (see figure at the end of this example) follows these steps:
(a) Draw a line from V_u/ϕ = 67.0 kips at support to V_u/ϕ = 0 at midspan.
(b) At end of member (conservative to assume support = end of member), d = 20.8 in., f_{pc} = V_p = 0
Prestress transfer length is assumed at $50d_b$ in ACI 318, Section 11.4.3. $50d_b$ = 25 in. = 0.042ℓ
At end by Eq. 4.3.2.4:
$V_{cw} = (3.5\lambda\sqrt{f'_c} + 0.3f_{pc})b_w d + V_p = [3.5(0.85)\sqrt{5000} + 0.3(0)](9.5)(20.8) + 0 = 41.6$ kips

At 50 d_b: $V_c = 5\lambda\sqrt{f'_c} b_w d = 5(0.85)\sqrt{5000}(9.5)(20.8) = 59.4$ kips

Draw line between these two values.
(c) Draw a curved line at ℓ/d = 26.1, where d = 23 in.
(d) Draw a vertical line at h/2 from face of support.
(e) The shaded area is the excess shear, $V_u/\phi - V_c$, for which shear reinforcement is required (see Example 4.3.4.2 for design of shear reinforcement).

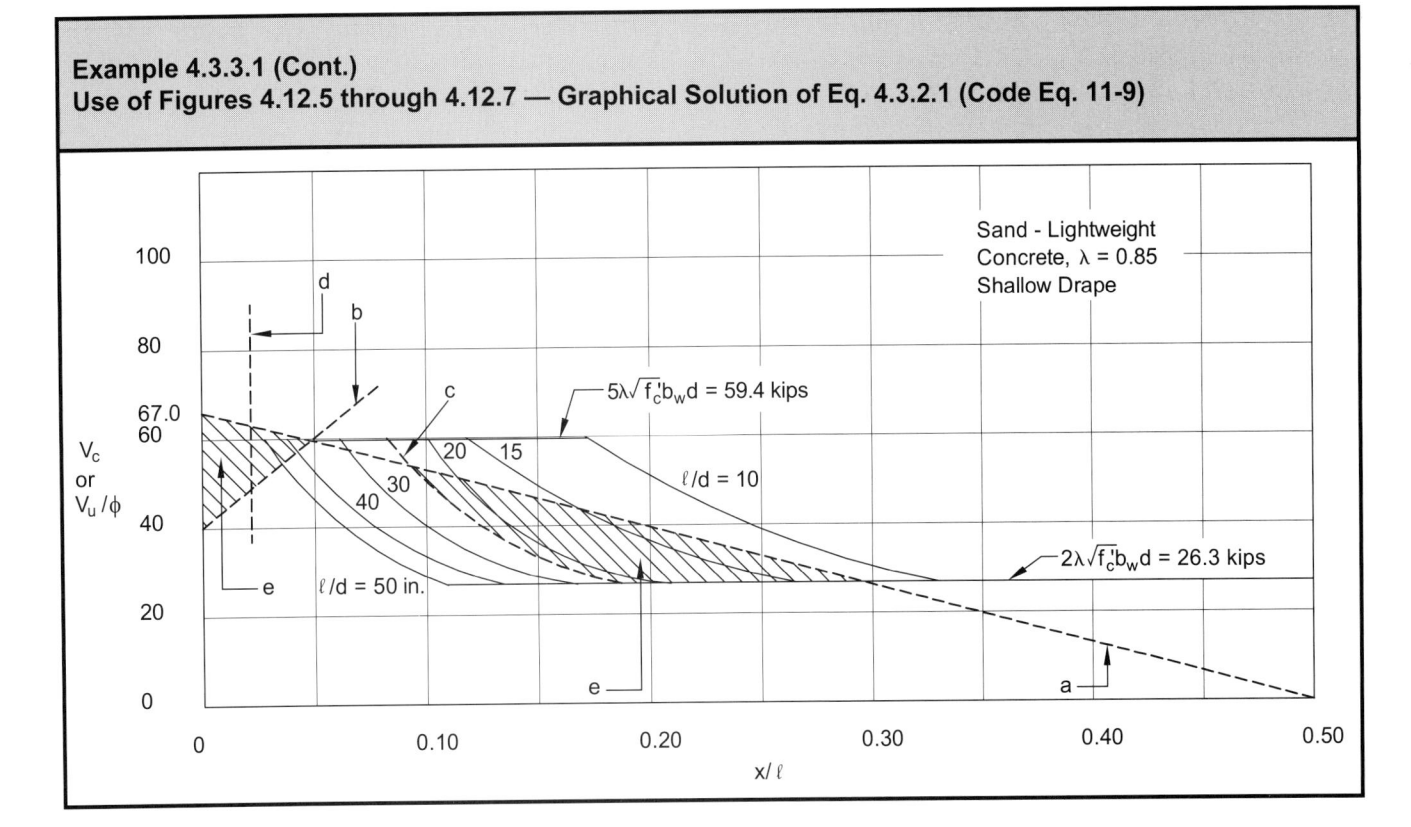

Example 4.3.3.1 (Cont.)
Use of Figures 4.12.5 through 4.12.7 — Graphical Solution of Eq. 4.3.2.1 (Code Eq. 11-9)

[Figure: Chart showing V_c or V_u/ϕ vs x/ℓ for Sand-Lightweight Concrete, $\lambda = 0.85$, Shallow Drape. Curves for $\ell/d = 10, 15, 20, 30, 40, 50$ in. Horizontal lines at $5\lambda\sqrt{f_c'}b_wd = 59.4$ kips and $2\lambda\sqrt{f_c'}b_wd = 26.3$ kips. Value 67.0 marked on vertical axis.]

In this equation, f_{ct} is the splitting tensile strength determined by test (ASTM C 496). For normal weight concrete, λ is equal to 1.0. If the value of f_{ct} is not known, $\lambda = 0.85$ for sand-lightweight concrete, and 0.75 for all-lightweight concrete is used. Figures 4.12.5 to 4.12.7 provide separate charts for normal weight and lightweight concrete. In these charts, it is assumed that f_{ct} is not known and the material is sand-lightweight, or $\lambda = 0.85$.

4.3.4 Shear Reinforcement

Shear reinforcement is required in all concrete members, except as noted in Section 11.5.5, ACI 318-02. The minimum area required by the ACI Code is determined using Eq. 11-13:

$$A_v = 0.75\sqrt{f_c'}\frac{b_ws}{f_y} \text{ but not less than } 50b_w\frac{s}{f_y}$$

(Eq. 4.3.4.1)

or, alternatively for prestressed members only, using Eq. 11-14:

$$A_v = \frac{A_{ps}f_{pu}s}{80f_yd}\sqrt{\frac{d}{b_w}}$$

(Eq. 4.3.4.2)

Figure 4.12.8 is a graphical solution for the minimum shear reinforcement by Eq. 4.3.4.2.

Shear reinforcement requirements are defined in ACI 318-02 by Eq. 11-15 which may be rewritten for vertical reinforcement as:

$$A_v = \frac{\left[\left(\frac{V_u}{\phi}\right) - V_c\right]s}{f_yd}$$

(Eq. 4.3.4.3)

Figure 4.12.9 may be used to design shear reinforcement by Eq. 4.3.4.3 for a given excess shear. Stirrup size, strength or spacing can be varied. Welded wire reinforcement may also be used for shear reinforcement in accordance with Section 11.5.1 of ACI 318-02; Sections 12.13.2.3 and 12.13.2.4 give development requirements.

4.3.5 Horizontal Shear Transfer in Composite Members

Cast-in-place concrete topping is often used on precast members to develop composite structures. The increased stiffness and strength may be required for gravity loads, for developing a diaphragm to transfer lateral loads, or to achieve a level floor.

In order for a precast member with topping to behave compositely, full transfer of horizontal shear forces must be ensured at the interface of the precast member and the cast-in-place topping. This requires that interface surfaces be clean and free of laitance. In addition, intentional roughening of surfaces and/or horizontal shear ties may also be required depending on the magnitude of shear force to be transferred. ACI 318-02 includes two methods for design of horizontal shear transfer. The procedure recommended and described below is in Code Section 17.5.3.

The horizontal shear force, F_h, which must be resisted is the total force in the topping; compression in positive moment regions and tension in negative moment regions are shown in Figure 4.3.5.1.

In a composite member which has an interface surface that is intentionally roughened but does not have horizontal shear ties, or where minimum ties are provided in accordance with Code Section 17.6 but the surface is not intentionally roughened, F_h should not exceed $\phi 80 b_v \ell_{vh}$, where b_v is the width of the interface surface and ℓ_{vh} is the horizontal shear length as defined in Figure 4.3.5.2. (Note: Experience and tests indicate that normal finishing methods used for precast concrete structural members will qualify as "intentionally roughened." Thus, horizontal shear strength of $\phi 80 b_v \ell_{vh}$ may be used for design.)

Code Section 17.5.2.3 requres that for an interface surface which is both intentionally roughened to a full amplitude of approximately ¼ in. and includes minimum horizontal shear ties per Code Section 17.6, F_h is limited to:

$$F_h = \phi(260 + 0.6\rho_v f_y)\lambda b_v \ell_{vh} \quad \text{(Eq. 4.3.5.1)}$$

$$F_h \leq \phi 500 b_v \ell_{vh}$$

For F_h exceeding $\phi 500 b_v \ell_{vh}$, the area of horizontal shear ties required in length ℓ_{vh} may be calculated by:

$$A_{cs} = \frac{F_h}{\phi \mu_e f_y} \quad \text{(Eq. 4.3.5.2)}$$

Example 4.3.4.1
Minimum Shear Reinforcement by Eq. 4.3.4.2 and Figure 4.12.8

Given:
    Double tee of Example 4.3.3.1.
    Shear reinforcement: Try W 2.9 wire each leg.
    $A_v = 2(0.029) = 0.058$ in.²

Problem:
    Determine the minimum amount of shear reinforcement required by Eq. 4.3.4.2 (Code Eq. 1114). Verify the result from Figure 4.12.8.

Solution:
    $b_w d = 9.5(22) = 209$ in.²
Note: As a simplification, d = 22 in. is used as an average value.

From Eq. 4.3.4.2:

$$A_v = 0.058 = \frac{1.53}{80}\left(\frac{270}{60}\right)\left(\frac{s}{22}\right)\sqrt{\frac{22}{9.5}}$$

thus:
    s = 9.7 in.

From Figure 4.12.8:
For:
    $A_{ps} = 1.53$ in.²
    $f_y = 60$ ksi
    $f_{pu} = 270$ ksi
    $b_w d = 209$ in.²
    $A_v = 0.075$ in.²/ft
corresponding s = 12(0.058)/0.075
    = 9.3 in. ≈ 9.7 in. OK

Per Code Section 11.5.4:
    s_{max} = ¾h ≤ 24 in.
    ¾h = 0.75(26) = 19.5 in. > 9.7 in. OK

Example 4.3.4.2
Use of Figure 4.12.9—Shear Reinforcement

Given:
    Double tee of Examples 4.3.3.1 and 4.3.4.1.

Problem:
    Design shear reinforcement for:
$$\left(\frac{V_u}{\phi}\right) - V_c = 10{,}000 \text{ lb}$$

Solution:
Excess shear:

$$= \left[\frac{\left(\frac{V_U}{\phi}\right) - V_c}{d}\right] = \left(\frac{10000}{23}\right) = 435 \text{ lb/in.}$$

From Figure 4.12.9:
Use two rows (one row per stem) of welded wire reinforcement W2.9, vertical wire spacing = 6 in.

Shear strength provided by reinforcement per stem:
    = 580 lb/in. > 435 lb/in. OK

where:

A_{cs} = area of horizontal shear ties, in.^{2}
F_h = horizontal shear force, lb
f_y = yield strength of horizontal shear ties, psi
μ_e = effective shear-friction coefficient; see Eq. 4.3.6.2.
ϕ = 0.75

For composite members, $\mu = 1.0\lambda$ and $A_{cr} = b_v \ell_{vh}$; therefore:

$$\mu_e = \frac{1000\lambda^2 b_v \ell_{vh}}{F_h} \leq 2.9 \quad \text{(Eq. 4.3.5.3)}$$

(See Table 4.3.6.1)

The value of F_h is limited to:

$$\left(\frac{F_h}{\phi}\right)(\text{max}) = 0.25\lambda^2 f'_c b_v \ell_{vh} \quad \text{(Eq. 4.3.5.4)}$$

$\leq 1000\lambda^2 b_v \ell_{vh}$

Section 17.5.3.1 of the Code requires that ties be placed along the member to "approximately reflect the distribution of shear forces in the member." Thus, the horizontal shear ties are usually either an extension of, or placed the same as shear reinforcement. Section 17.6.1 of ACI 318-02 also requires that ties, when required, be spaced no more than four times the least dimension of the supported element, nor 24 in., and meet the minimum shear reinforcement requirements of Section 11.5.5.3:

$$A_v = 0.75\sqrt{f'_c}\frac{b_w \ell_{vh}}{f_y} \text{ but not less than } \frac{50 b_w \ell_{vh}}{f_y}$$

(Eq. 4.3.5.5)

Anchorage of ties must satisfy Code Section 17.6.3. Research [4] has shown that this requirement can be satisfied by providing a minimum distance of 2.25, 2.75 and 3.25 in. between the shear transfer interface and the outside ends of standard hooks on No. 3, No. 4 and No. 5 ties, respectively.

(See also Section 10.5 for PCI Standard Practice regarding Code Section 17.6.3.)

Figure 4.3.5.1 Horizontal shear in composite section

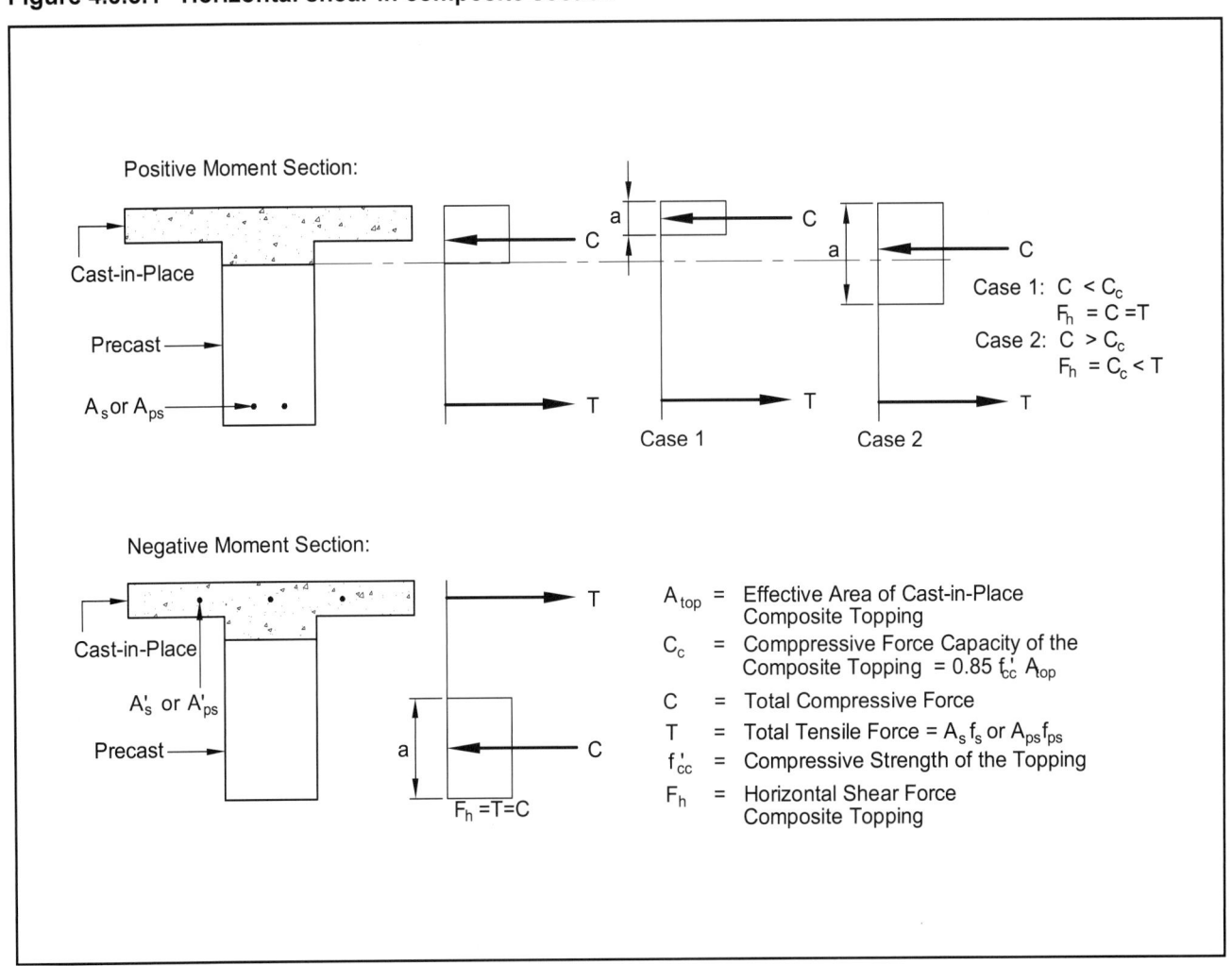

Case 1: $C < C_c$
$F_h = C = T$
Case 2: $C > C_c$
$F_h = C_c < T$

A_{top} = Effective Area of Cast-in-Place Composite Topping
C_c = Comppressive Force Capacity of the Composite Topping = $0.85 f'_{cc} A_{top}$
C = Total Compressive Force
T = Total Tensile Force = $A_s f_s$ or $A_{ps} f_{ps}$
f'_{cc} = Compressive Strength of the Topping
F_h = Horizontal Shear Force Composite Topping

Example 4.3.5.1
Horizontal Shear Design for Composite Beam

Given:
    Inverted tee beam with 2 in. composite topping
    Beam length = 24 ft

Concrete:
    Precast: f'_c = 5000 psi normal weight concrete
    Topping: f'_{cc} = 4000 psi normal weight concrete

Prestressing steel:
    12 – ½ in. diameter 270K strands
    A_{ps} = 12 × 0.153 = 1.836 in.²
    Tie steel: f_y = 60,000 psi

Problem:
    Determine the tie requirements to transfer horizontal shear force.

Solution:
$$A_{top} = b(2) + b_w(2) = 2(44) + 2(12) = 112 \text{ in.}^2$$
$$C = 0.85 f'_{cc} A_{top} + A'_s f'_s = 0.85(4)(112) + 2(60) = 380.8 + 120.0 = 500.8 \text{ kips}$$
$$\ell_{vh} = \frac{24(12)}{2} = 144 \text{ in.}$$
f_{ps} = 260.7 ksi (from Example 4.2.1.6)
$A_{ps} f_{ps}$ = 1.836(260.7) = 478.6 kips < 500.8 kips

Therefore, from Figure 4.3.5.1:
    F_h = 478.6 kips
$$\phi 80 b_v \ell_{vh} = \frac{0.75(80)(12)(144)}{1000} = 103.7 \text{ kips} < 478.6 \text{ kips}$$

Therefore, ties are required.
    λ = 1.0 (normal weight concrete)

Check maximum by Eq. 4.3.5.1:
$$\phi 500 b_v \ell_{vh} = \frac{0.75(500)(12)(144)}{1000} = 648 \text{ kips} > 478.6 \text{ kips}$$

Therefore, design by Eq. 4.3.5.1:
$$\phi(260 + 0.6 \rho_v f_y) \lambda b_v \ell_{vh} = 478,600 \text{ lb}$$
$$\phi \lambda b_v \ell_{vh} = 0.75(1.0)(12)(144) = 1296 \text{ in.}^2$$
$$0.6 \rho_v (60,000) = \frac{478,600 - 260(1296)}{1296} = 109.3$$
$$\rho_v = \frac{109.3}{0.6(60,000)} = 0.00304$$
$$A_{cs} = \rho_v b_w \ell_{vh} = 0.00304(12)(144) = 5.25 \text{ in.}^2$$

Example 4.3.5.1 (Cont.)
Horizontal Shear Design for Composite Beam

Check minimum requirements:
$$A_{cs,min} = 0.75\sqrt{f'_c}\frac{b_w \ell_{vh}}{f_y} = \frac{0.75(70.7)(12)(144)}{60000} = 1.52 \text{ in.}^2 < 5.25 \text{ in.}^2$$

Use #4 ties, $A_{cs} = 2(0.20) = 0.40$ in.²
Maximum tie spacing = 4(4) = 16 in. < 24 in.
For $A_{cs} = 5.25$ in.², number of ties in length $\ell_{vh} = 5.25/0.4 \approx 13$
Total number of ties in the beam = 26

Provide 5 – #4 ties at 6 in. spacing at each end and the remaining 16 – #4 at approximately 12 in. spacing in the middle portion of the beam.

In this example, it is assumed that the full flange width of 44 in. is part of the composite section. Control joints in topping, particularly in parking structures, are often located along the joints between ends of double tees and edges of tee beams for crack control as shown in the example figure. This raises the concern that the extended parts of the topping on each side of the tee beam web (16 in. for this example) may not behave compositely with the rest of the section. However, it can be shown that the usual steel provided in the topping, in most cases, generates sufficient shear-friction resistance (see Section 4.3.6) to ensure composite action. The following calculations support this observation.

Note: It is important that the control joints be tooled rather than saw-cut, as there is a danger of cutting the reinforcement.

The topping reinforcement is:
6x6–W2.9xW2.9
$A_s = 0.058$ in.²/ft = 0.58 in.² per half-span

The portion of the total compression force in the extended part of topping:

$$C'_c = \frac{16(2)}{112}(285.6) = 81.6 \text{ kips } (= V_u \text{ for use in calculations below})$$

$$\mu_e = \frac{1000\lambda A_{cr}\mu}{V_u} = \frac{1000(1.0)(2)(144)(1.4)}{81,600} = 4.94 > 3.4, \text{ use } 3.4$$

$$A'_{cs} = \frac{V_u}{\phi f_y \mu_e} = \frac{81.6}{0.75(60)(3.4)} = 0.53 \text{ in.}^2 < 0.58 \text{ in.}^2 \text{ OK}$$

$$A'_{cs,min} = \frac{50(2)(144)}{60,000} = 0.24 \text{ in.}^2 < 0.53 \text{ in.}^2 \text{ OK}$$

Maximum spacing = 4(2) = 8 in.

Thus, 6x6–W2.9xW2.9 is adequate.

4.3.6 Shear-Friction

Shear-friction is an extremely useful tool in the design of precast and prestressed concrete structures.

Use of the shear-friction theory is recognized by Section 11.7 of ACI 318-02, which states that shear friction is "to be applied where it is appropriate to consider shear transfer across a given plane, such as an existing or potential crack, an interface between dissimilar materials, or an interface between two concretes cast at different times."

A basic assumption used in applying the shear-friction concept is that concrete within the direct shear area of the connection will crack in the most undesirable manner. Ductility is achieved by placing reinforcement across this anticipated crack so that the tension developed by the reinforcement will provide a force normal to the crack. This normal force in combination with "friction" at the crack interface

Figure 4.3.5.2 Horizontal shear length

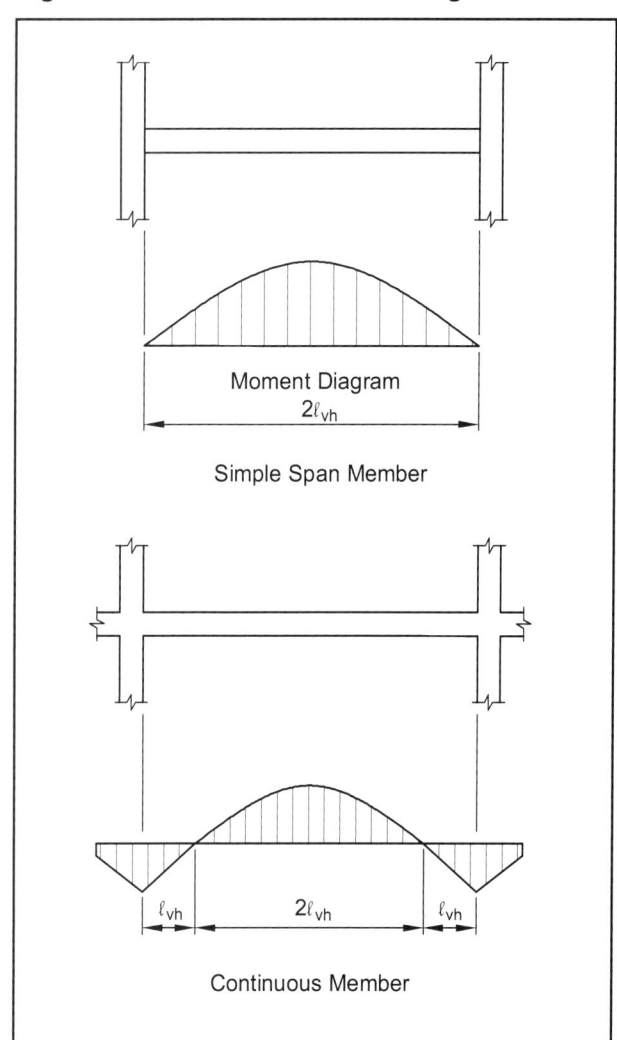

Simple Span Member

Moment Diagram
$2\ell_{vh}$

Continuous Member

ℓ_{vh} | $2\ell_{vh}$ | ℓ_{vh}

provides the shear resistance. The shear-friction analogy can be adapted to designs for reinforced concrete bearing, corbels, daps, composite sections, and other applications.

An "effective shear-friction coefficient," μ_e, may be used [9] when the concept is applied to precast concrete construction. Examples can be found in Sections 4.3.5 (Horizontal Shear Transfer in Composite Members), 4.6.3 (Dapped End Bearing) and 6.8 (Concrete Brackets and Corbels). Other uses may include connections of shear walls to foundations, shear connections in precast diaphragms, etc. The shear-friction reinforcement nominally perpendicular to the assumed crack plane can be determined by:

$$A_{vf} = \frac{V_u}{\phi f_y \mu_e} \quad \text{(Eq. 4.3.6.1)}$$

where:
- ϕ = 0.75
- A_{vf} = area of reinforcement nominally perpendicular to the assumed crack plane, in.²
- f_y = yield strength of A_{vf}, psi (equal to or less than 60,000 psi)
- V_u = applied factored shear force, parallel to the assumed crack plane, lb (limited by the values given in Table 4.3.6.1)

$$\mu_e = \frac{1000 \lambda A_{cr} \mu}{V_u} \quad \text{(Eq. 4.3.6.2)}$$

≤ values in Table 4.3.6.1
- λ = factor for use with lightweight concrete, see Section 4.3.3
- μ = shear-friction coefficient (values in Table 4.3.6.1)
- A_{cr} = area of the crack interface, in.²

When net axial tension is present, additional reinforcement area should be provided:

$$A_n = \frac{N_u}{\phi f_y} \quad \text{(Eq. 4.3.6.3)}$$

where:
- A_n = area of reinforcement required to resist axial tension, in.²
- N_u = applied factored tensile force nominally perpendicular to the assumed crack plane, lb. May be taken as 0.2 times the factored sustained parallel force. [14]
- ϕ = 0.75

Table 4.3.6.1 Recommended shear-friction coefficients

Crack interface condition	Recommended μ	Maximum μ_e	Maximum $V_u = \phi V_n$
1. Concrete to concrete, cast monolithically	1.4λ	3.4	$0.30\lambda^2 f'_c A_{cr} \leq 1000\lambda^2 A_{cr}$
2. Concrete to hardened concrete, with roughened surface	1.0λ	2.9	$0.25\lambda^2 f'_c A_{cr} \leq 1000\lambda^2 A_{cr}$
3. Concrete to concrete	0.6λ	2.2	$0.20\lambda^2 f'_c A_{cr} \leq 800\lambda^2 A_{cr}$
4. Concrete to steel	0.7λ	2.4	$0.20\lambda^2 f'_c A_{cr} \leq 800\lambda^2 A_{cr}$

All reinforcement, either side of the assumed crack plane, should be properly anchored to develop the design forces. The anchorage may be achieved by extending the reinforcement for the required development length (with or without hooks), or by welding to reinforcing bars, angles or plates.

4.4 Torsion

The torsion provisions in ACI 318-02 are based on models which are relatively compact, unlike most precast concrete sections subjected to torsion. Further, precast sections do not achieve the degree of fixity that is common in cast-in-place structures. The design procedure of ACI 318-02 (and previous editions since 1995) typically requires significantly greater reinforcement than previously used methods (see Section 10.5). This section is based on the Zia and McGee procedure, [10] updated by Zia and Hsu. [11] The Zia and McGee method is essentially the same as in pre-1995 Codes for non-prestressed concrete, but allows the effects of prestressing to be included.

The following is a step-by-step procedure for torsion design based on the above:

Step 1: Determine the design shear (V_u) and the torsional moment (T_u) at the critical section. The critical section for shear and torsion by ACI 318-02 (Section 11.1.3) is "d" from the face of the support for non-prestressed members, and "h/2" for prestressed members. If the load is not located at the top of the member, "d" is to be measured to the point of application of the load. If a concentrated load occurs within these dimensions, the critical section must be re-established to include the load.

Step 2: Determine if torsion can be neglected, i.e., is $T_u \leq T_{u(min)}$:

$$T_{u(min)} = \phi(0.5\lambda\sqrt{f'_c}\Sigma x^2 y)\gamma \qquad \text{(Eq. 4.4.1)}$$

where:
- T_u = factored torsional moment, lb-in.
- ϕ = 0.75
- λ = conversion factor for lightweight concrete
- f'_c = concrete compressive strength, psi
- x, y = short side and long side, respectively, of a component rectangle, in.
- γ = a factor dependent on the level of prestress
  $$= \sqrt{1 + 10\frac{f_{pc}}{f'_c}}$$
  = 1.0 for non-prestressed sections
- f_{pc} = average prestress after losses

In computing $\Sigma x^2 y$ for a flanged section, the section may be divided into component rectangles such that the quantity $\Sigma x^2 y$ would be the maximum, however, the overhanging flange width used in design should not be taken greater than three times the flange thickness.

If $T_u \leq T_{u(min)}$, no torsion reinforcement is required. Design is complete.

Step 3: Check to ensure that the required nominal torsional moment and shear strengths do not exceed the following maximum limits to avoid potential compression failures due to over-reinforcing:

$$T_{n(max)} = \frac{\left(\frac{1}{3}\right)K_t\lambda\sqrt{f'_c}\Sigma x^2 y}{\sqrt{1+\left(\frac{K_t V_u}{30 C_t T_u}\right)^2}} \geq \frac{T_u}{\phi} \qquad \text{(Eq. 4.4.2)}$$

$$V_{n(max)} = \frac{10\lambda\sqrt{f'_c}b_w d}{\sqrt{1+\left(\frac{30 C_t T_u}{K_t V_u}\right)^2}} \geq \frac{V_u}{\phi} \qquad \text{(Eq. 4.4.3)}$$

where:
- $K_t = \gamma\left(12 - 10\frac{f_{pc}}{f'_c}\right)$
- V_u = factored shear force, lb
- $C_t = \dfrac{b_w d}{\Sigma x^2 y}$

Step 4: If $T_u \leq T_{n(max)}$ and $V_u \leq V_{n(max)}$, design may be continued. Calculate torsion and shear carried by concrete:

$$T_c = \frac{T'_c}{\sqrt{1+\left(\frac{T'_c/T_u}{V'_c/V_u}\right)^2}} \qquad \text{(Eq. 4.4.4)}$$

$$V_c = \frac{V'_c}{\sqrt{1+\left(\frac{V'_c/V_u}{T'_c/T_u}\right)^2}} \qquad \text{(Eq. 4.4.5)}$$

where:
- T_c, T'_c = nominal torsional moment strength of concrete under combined shear and torsion and under pure torsion, respectively
- V_c, V'_c = nominal shear strength of concrete under combined shear and torsion and under pure shear, respectively
- $T'_c = 0.8\lambda\sqrt{f'_c}\Sigma x^2 y(2.5\gamma - 1.5)$
- $V'_c = V_c$ as calculated in Section 4.3.2

Step 5: Provide stirrups for torsional moment in excess of that carried by concrete. Stirrups must be closed and must be spaced not more than 12 in. or $(x_1 + y_1)/4$:

$$A_t = \frac{\left(\frac{T_u}{\phi} - T_c\right)s}{\alpha_t x_1 y_1 f_y} \qquad \text{(Eq. 4.4.6)}$$

where:
A_t = required area of one leg of closed tie, in.^{2}
x_1 = short side of closed tie, in.
y_1 = long side of closed tie, in.
s = tie spacing $\leq (x_1 + y_1)/4$ or 12 in.
$\alpha_t = [0.66 + 0.33\, y_1/x_1] < 1.5$
f_y = yield strength of closed tie, psi

Note: the required ties for torsion are in addition to those required for shear.

A minimum amount of web reinforcement should be provided to ensure reasonable ductility. The minimum area of closed stirrups which should be provided is:

$$(A_v + 2A_t)_{min} = 50\frac{b_w s}{f_y}(\gamma)^2 \leq 200\frac{b_w s}{f_y}$$

(Eq. 4.4.7)

where b_w is the web width of the beam.

Step 6: To resist the longitudinal component of the diagonal tension induced by torsion, longitudinal reinforcement approximately equal in volume to that of ties for torsion and effectively distributed around the perimeter (spaced $\leq$ 12 in.) should be provided.

Prestressing strand may be used to satisfy the longitudinal reinforcement requirement. However, to control diagonal crack width, the stress in the prestressing strand is limited to 60,000 psi (see Code Section 11.6.7.4) This longitudinal steel for torsion is in addition to that required for flexure. Thus:

$$A_\ell = \frac{2A_t(x_1 + y_1)}{s} \qquad \text{(Eq. 4.4.8)}$$

or

$$A_\ell = \left[\frac{400x}{f_y}\left(\frac{T_u}{T_u + \frac{V_u}{3C_t}}\right) - \frac{2A_t}{s}\right](x_1 + y_1)$$

(Eq. 4.4.9)

whichever is greater. The value of A_ℓ calculated from Eq. 4.4.9 need not exceed that obtained by substituting:

$$\frac{50b_w}{f_y}\left(1 + \frac{12f_{pc}}{f'_c}\right) \leq \frac{200b_w}{f_y} \text{ for } \frac{2A_t}{s}$$

In many cases, the bars or tendons may not be fully developed at the critical section. Thus, it may be necessary to provide U-bars at the ends of the members lapped with the longitudinal reinforcement.

Strands may be used as part of the longitudinal reinforcement. However, they must be developed, or the reduced capacity due to partial development used.

It should be noted that the design of the connections between the torsional member and its supports to provide adequate reactions is critical; also, the effects of torsion on the overall stability of the frame should be investigated. Guidelines for these design considerations are given in Chapters 5 and 6.

Example 4.4.1
Shear and Torsion Design of a Non-Prestressed Member

Given:
Precast loadbearing spandrel beam shown (see next page).
Span of spandrel beam = 30 ft clear
Tributary width of floor = 20 ft
f'_c = 5000 psi, normal weight concrete
Reinforcement f_y = 60,000 psi
d = 69 in.

Factored loads (kips/ft):
Dead load:
Precast floor 60 psf (20 ft) = 1.2(1.2) = 1.44
Topping (38 psf)(20 ft) = 0.76(1.2) = 0.91

Example 4.4.1 (Cont.)
Shear and Torsion Design of a Non-Prestressed Member

Superimposed (20 psf)(20 ft) = 0.4(1.2) = 0.48
Window = 0.05(1.2) = 0.06
Spandrel = 0.63(1.2) = 0.76
Live load: 50 psf × (20 ft) = 1.00(1.6) = 1.60 kips/ft
Total load = 5.25 kips/ft

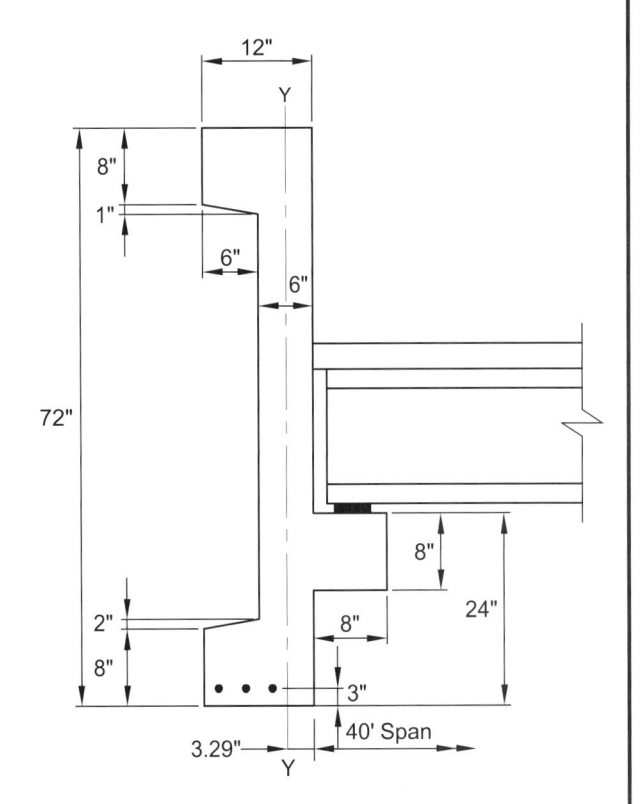

Problem:
Determine torsion reinforcement requirements.

Solution:
1. Compute shear (V_u) and torsion (T_u) at critical section, located at 24 in. − 3 in. = 1 ft - 9 in. from the face of the support, equal to 13.25 ft from midspan (see Section 4.3):
   w_u for shear = 5.25 kips/ft
   V_u = 5.25(13.25) = 69.6 kips
   w_u for torsion = 1.44 + 0.91 + 0.48 + 1.60
   = 4.43 kips/ft

Note: It is sufficiently accurate in this and most precast concrete sections to use the centroid rather than the shear center to calculate the eccentricity.

Eccentricity = ¾(8) + 3.29 = 9.29 in.
$T_u = w_u e(15 - 1.75) = 4.43(9.29)(13.25)$
= 545 kip-in.
$\Sigma x^2 y = 6^2(72) + 6^2(8)(2) + 8^2(8) = 3680$ in.^{3}
$\lambda \sqrt{f'_c} = \dfrac{1.0\sqrt{5000}}{1000} = 0.0707$ ksi

2. Determine if torsion can be neglected:
$T_{u(min)} = \phi(0.5\lambda\sqrt{f'_c}\Sigma x^2 y)\gamma = 0.75(0.5)(0.0707)(3680)(1.0) = 97.6$ kip-in. < 545 kip-in.
Consider torsion.

3. Check maximums by Eqs. 4.4.2 and 4.4.3 with:
$K_t = 12$
$C_t = \dfrac{b_w d}{\Sigma x^2 y} = \dfrac{6(69)}{3680} = 0.1125$

Eq. 4.4.2:

$$T_{n(max)} = \dfrac{\left(\dfrac{1}{3}\right) K_t \lambda \sqrt{f'_c} \Sigma x^2 y}{\sqrt{1+\left(\dfrac{K_t V_u}{30 C_t T_u}\right)^2}} \geq \dfrac{T_u}{\phi} = \dfrac{\left(\dfrac{1}{3}\right) 12(0.0707)(3680)}{\sqrt{1+\left[\dfrac{12(69.6)}{30(0.1125)(545)}\right]^2}} = 948 \text{ kip-in.} > \dfrac{545}{0.75} = 727 \text{ kip-in.}$$

Example 4.4.1 (Cont.)
Shear and Torsion Design of a Non-Prestressed Member

Eq. 4.4.3:
$$V_{n(max)} = \frac{10\lambda\sqrt{f'_c}b_w d}{\sqrt{1+\left(\frac{30C_t T_u}{K_t V_n}\right)^2}} \geq \frac{V_u}{\phi} = \frac{10(0.0707)(6)(69)}{\sqrt{1+\left[\frac{30(0.1125)(545)}{12(69.6)}\right]^2}} = 121 \text{ kips} > \frac{69.6}{0.75} = 92.8 \text{ kips}$$

4. Torsion and shear carried by concrete by Eqs. 4.4.4 and 4.4.5 with:
   $T'_c = 0.8(0.0707)(3680)(1) = 208$ kip-in.
Near the support, use Eq. 4.3.1.2:

From a separate analysis, determine:
   Flexural reinforcement = 3 – #8
   By Eq. 4.3.1.2, $V'_c = 73.9$ kips
   $T'_c / T_u = 208/545 = 0.382$
   $V'_c / V_u = 73.9/69.6 = 1.062$

Eq. 4.4.4:
$$T_c = \frac{T'_c}{\sqrt{1+\left(\frac{T'_c/T_u}{V'_c/V_u}\right)^2}} = \frac{208}{\sqrt{1+\left(\frac{0.382}{1.062}\right)^2}} = 195.7 \text{ kip-in.}$$

Eq. 4.4.5:
$$V_c = \frac{V'_c}{\sqrt{1+\left(\frac{V'_c/V_u}{T'_c/T_u}\right)^2}} = \frac{73.9}{\sqrt{1+\left(\frac{1.062}{0.382}\right)^2}} = 25.0 \text{ kips}$$

5. Determine transverse reinforcement by Eq. 4.4.6, assuming ¾ in. cover to reinforcement:
   $x_1 = 4$ in.; $y_1 = 70$ in.
   $$\alpha_t = \left(0.66 + 0.33\frac{y_1}{x_1}\right) = 0.66 + 0.33\left(\frac{70}{4}\right) = 6.44 \geq 1.5$$
   Use $\alpha_t = 1.5$.

$$\frac{A_t}{s} = \frac{\left(\frac{T_u}{\phi} - T_c\right)}{\alpha_t x_1 y_1 f_y} = \frac{\left(\frac{545}{0.75} - 195.7\right)}{1.5(4)(70)(60)} = 0.0211 \text{ in.}^2/\text{in.} = 0.253 \text{ in.}^2/\text{ft}$$

Stirrups required for shear (each side).

$$\frac{A_v}{2s} = \frac{\left(\frac{V_u}{\phi}\right) - V_c}{2f_y d} = \frac{\left(\frac{69.6}{0.75}\right) - 25.0}{2(60)(69)} = 0.0082 \text{ in.}^2/\text{in.} = 0.098 \text{ in.}^2/\text{ft}$$

0.098 + 0.253 = 0.351 in.² /ft (each side)

Example 4.4.1 (Cont.)
Shear and Torsion Design of a Non-Prestressed Member

The minimum area of closed stirrups is:
$$\frac{A_v + 2A_t}{s} = \frac{50b_w}{f_y} = \frac{50(6)}{60000}(12) = 0.06 < (0.351)(2) = 0.702$$

#4 @ 7 in. = 0.343 in.^{2}/ft say OK

Spacing may be increased toward midspan to a maximum of 12 in.

6. Determine longitudinal reinforcement requirements by Eqs. 4.4.8 or 4.4.9:

By Eq. 4.4.8:
$$A_\ell = \frac{2A_t(x_1 + y_1)}{s} = \frac{2(0.0208)(4 + 70)}{1}$$
$$= 3.08 \text{ in.}^2$$

or by Eq. 4.4.9:
$$\frac{2A_t}{s} = \frac{2(0.0208)}{1} = 0.042$$
$$\frac{50b_w}{f_y} = \frac{50(6)}{60000} = 0.005$$

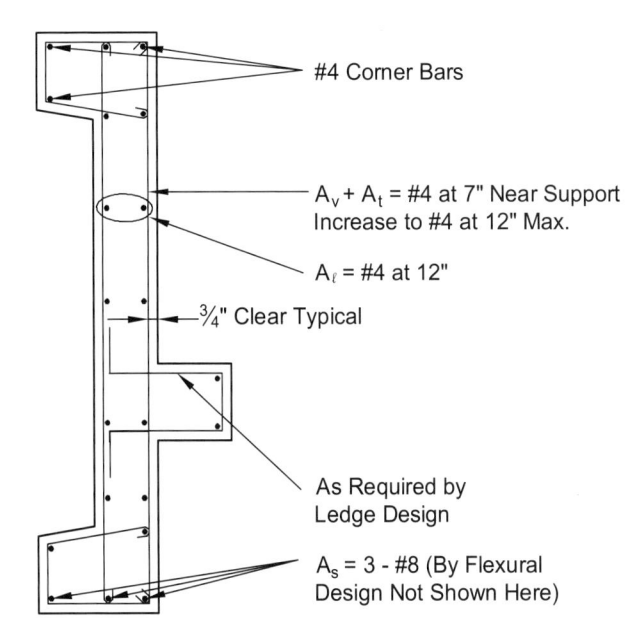

Use 0.042 in Eq. 4.4.9:
$$A_\ell = \left[\frac{400x}{f_y}\left(\frac{T_u}{T_u + \frac{V_u}{3C_t}}\right) - \frac{2A_t}{s}\right](x_1 + y_1) = \left[\frac{400(6)}{60000}\left(\frac{545}{545 + \frac{69.6}{3(0.1125)}}\right) - 0.042\right](4 + 70) = -1$$

Therefore, use A_ℓ = 3.08 in.^{2} distributed around perimeter. The maximum spacing is 12 in. Corner bars are required, and the flexural reinforcement not required for flexure can also be counted. Thus, additional #4 bars at 12 in. is adequate.

Figure labels:
- #4 Corner Bars
- $A_v + A_t$ = #4 at 7" Near Support, Increase to #4 at 12" Max.
- A_ℓ = #4 at 12"
- ¾" Clear Typical
- As Required by Ledge Design
- A_s = 3 - #8 (By Flexural Design Not Shown Here)

Example 4.4.2
Shear and Torsion Design in a Prestressed Concrete Member

Given:
  Precast, prestressed concrete spandrel beam shown on next page.
  Dead load of deck = 89.5 psf
  Live load = 50 psf

Beam Properties:
  A = 696 in.^{2}
  wt = 725 plf
  f'_c = 6000 psi, normal weight concrete
  f_y = 60 ksi

Example 4.4.2 (Cont.)
Shear and Torsion Design in a Prestressed Concrete Member

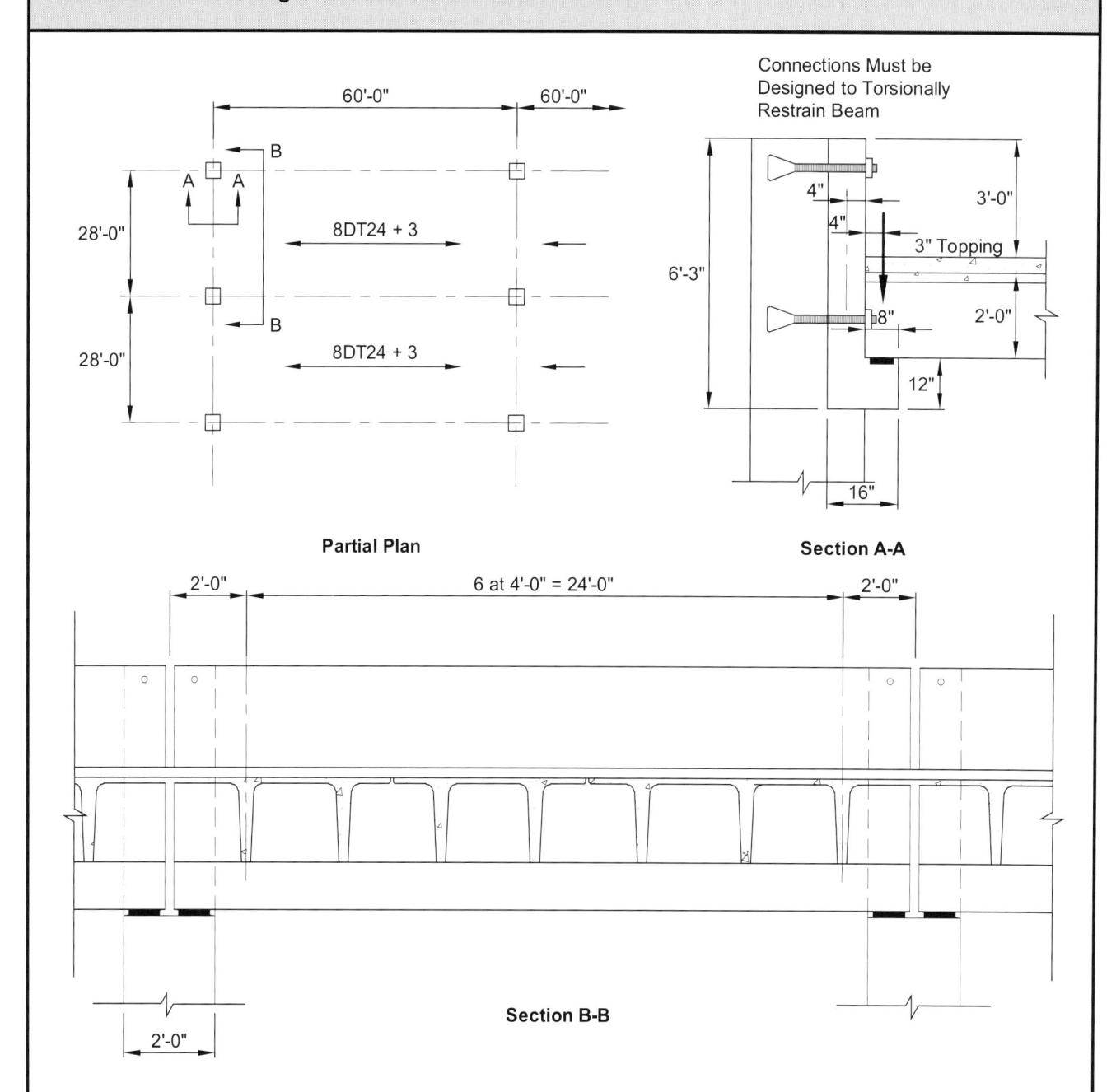

Partial Plan

Section A-A

Section B-B

Prestressing:
    6 – ½ in. diameter, 270 ksi strands
    $A_{ps} = 6(0.153) = 0.918$ in.^{2}
    d = 72 in.

Problem:
    Determine shear and torsion reinforcement for the spandrel beam at critical section.

Solution:
1.   Calculate V_u and T_u:

Example 4.4.2 (Cont.)
Shear and Torsion Design in a Prestressed Concrete Member

Determine factored loads:
    Dead load of beam = 1.2(0.725) = 0.87 kips per ft
    Dead load of deck = $1.2(0.0895)\dfrac{60}{2}(4)$ = 12.89 kips per stem
    Live load of deck = 1.6(0.050)(30)(4) = 9.6 kips per stem
V_u at center of column:
    = [0.87(28) + 7(12.89 + 9.60)](½) = 90.9 kips

Center of support to h/2
$$= 0.5 + \frac{6.25}{2} = 3.625 \text{ ft}$$

Since a concentrated load occurs within h/2, the critical section is at face of support. Note: the critical section may also be limited by the height at the ledge.

V_u at face of support:
    = 90.9 − 0.87(1.0) = 90.0 kips
T_u at support [torsion arm = 4 in. + $\dfrac{3}{4}$(8) = 10 in.]:
    = (12.89 + 9.6)(7)(½)(10) = 787 kip-in. (same at face of support)

2. Determine if $T_u \leq T_{u(min)}$ by Eq. 4.4.1:

$T_{u(min)}$ = $\phi(0.5\lambda\sqrt{f'_c}\Sigma x^2 y)\gamma$
ϕ = 0.75
$\lambda\sqrt{f'_c}$ = $\dfrac{1.0\sqrt{6000}}{1000}$ = 0.0775
$\Sigma x^2 y$ = $[(8^2)(63) + (12^2)(16)]$ = 6336 in.³

Determine f_{pc} at critical section approximately 11 in. from end of member. From Figure 4.12.4, strand develops 170 ksi at 28 in. Use approximate linear interpolation:
f_{pd} = (11/28)(170) = 66.8 ksi
P_{pd} = 66.8(0.918) = 61.3 kips
f_{pc} = $\dfrac{P_{pd}}{A} = \dfrac{61.3}{696}$ = 0.088 ksi
γ = $\sqrt{1 + \dfrac{10 f_{pc}}{f'_c}} = \sqrt{1 + \dfrac{10(0.088)}{6}} = 1.07$

$T_{u(min)}$ = $\phi(0.5\lambda\sqrt{f'_c}\Sigma x^2 y)\gamma$ = 0.75(0.5)(0.0775)(6336)(1.07) = 197 kip-in. < 787 kip-in.
Consider torsion.

3. Check maximums by Eq. 4.4.2 and 4.4.3:

K_t = $\gamma\left(12 - \dfrac{10 f_{pc}}{f'_c}\right) = 1.07\left(12 - \dfrac{10(0.088)}{6}\right) = 12.7$

C_t = $\dfrac{b_w d}{\Sigma x^2 y} = \dfrac{8(69)}{6336} = 0.087$

Example 4.4.2 (Cont.)
Shear and Torsion Design in a Prestressed Concrete Member

Eq. 4.4.2:

$$T_{n(max)} = \frac{\left(\frac{1}{3}\right)K_t\lambda\sqrt{f'_c}\Sigma x^2 y}{\sqrt{1+\left(\frac{K_t V_u}{30 C_t T_u}\right)^2}} = \frac{\left(\frac{1}{3}\right)(12.7)(0.0775)(6336)}{\sqrt{1+\left(\frac{12.7(90.0)}{30(0.087)(787)}\right)^2}} = 1816 \text{ kip-in.} > \frac{787}{0.75} = 1049 \text{ kip-in.}$$

Eq. 4.4.3:

$$V_{n(max)} = \frac{10\lambda\sqrt{f'_c}b_w d}{\sqrt{1+\left(\frac{30 C_t T_u}{K_t V_u}\right)^2}} = \frac{10(0.0775)(8)(72)}{\sqrt{1+\left(\frac{30(0.087)(787)}{12.7(90.0)}\right)^2}} = 217.1 \text{ kips} > \frac{90.0}{0.75} = 120.0 \text{ kips}$$

4. Calculate torsion and shear carried by concrete:

$$T'_c = 0.8\lambda\sqrt{f'_c}\Sigma x^2 y(2.5\gamma - 1.5) = 0.8(0.0775)(6336)[2.5(1.07) - 1.5] = 462 \text{ kip-in.}$$

Shear near the support, use Eq. 4.3.2.4:

$$V'_c = (3.5\lambda\sqrt{f'_c} + 0.3 f_{pc})b_w d = [3.5(0.0775) + 0.3(0.088)](8)(72) = 171.4 \text{ kips}$$

$T'_c / T_u = 462/787 = 0.59$
$V'_c / V_u = 171.4/90.0 = 1.90$

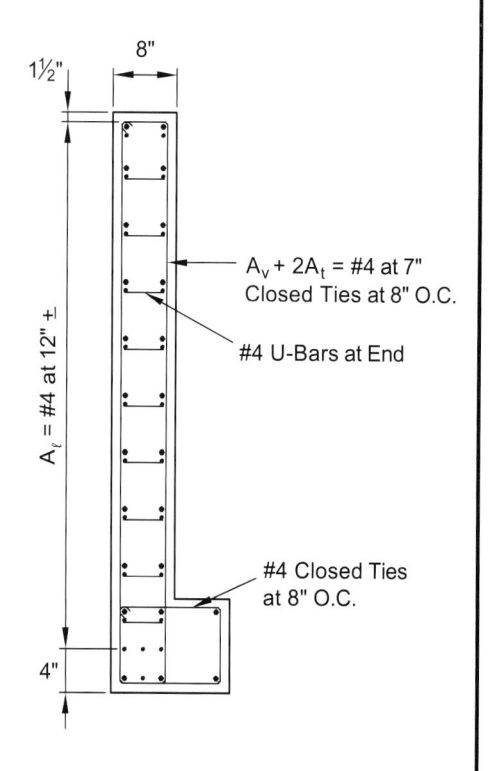

Eq. 4.4.4:

$$T_c = \frac{T'_c}{\sqrt{1+\left(\frac{T'_c / T_u}{V'_c / V_u}\right)^2}} = \frac{462}{\sqrt{1+\left(\frac{0.59}{1.90}\right)^2}} = 441 \text{ kip-in.}$$

Eq. 4.4.5:

$$V_c = \frac{V'_c}{\sqrt{1+\left(\frac{V'_c / V_u}{T'_c / T_u}\right)^2}} = \frac{171.4}{\sqrt{1+\left(\frac{1.90}{0.59}\right)^2}} = 50.8 \text{ kips}$$

5. Determine transverse reinforcement by Eq. 4.4.6:
   With 1 in. cover (1.25 in. to center of steel):
   $x_1 = 5.5$ in.; $y_1 = 72.5$ in.

$$\alpha_t = \left(0.66 + 0.33\frac{y_1}{x_1}\right) \leq 1.5 = 0.66 + 0.33(72.5/5.5)$$

$$= 5.01$$

Use $\alpha = 1.5$:

$$\frac{A_t}{s} = \frac{\left(\frac{T_u}{\phi} - T_c\right)}{\alpha_t x_1 y_1 f_y} = \frac{\left(\frac{787}{0.75} - 441\right)}{1.5(5.5)(72.5)(60)} = 0.0170 \text{ in.}^2/\text{in.} = 0.20 \text{ in.}^2/\text{ft}$$

Stirrups required for shear (each side).

Example 4.4.2 (Cont.)
Shear and Torsion Design in a Prestressed Concrete Member

$$\frac{A_v}{2s} = \frac{\left(\frac{V_u}{\phi} - V_c\right)}{2f_y d} = \frac{\left(\frac{90.0}{0.75} - 50.8\right)}{2(60)(72)} = 0.0080 \text{ in.}^2/\text{in.} = 0.10 \text{ in.}^2/\text{ft}$$

$$0.10 + 0.20 = 0.30 \text{ in.}^2/\text{ft per leg}$$

The minimum area of closed stirrups is:

$$\frac{A_v + 2A_t}{s} = \frac{50b_w}{f_y} = \frac{50(8)(12)}{60000} = 0.08 \text{ in.}^2/\text{ft} < 0.30(2) = 0.60$$

#4 @ 8 in. = 0.30 in.²/ft

Spacing may be increased toward midspan to a maximum of 12 in.

6. Determine longitudinal reinforcement requirements by Eqs. 4.4.8 or 4.4.9:

By Eq. 4.4.8:
$$A_\ell = \frac{2A_t(x_1 + y_1)}{s}$$

$$\frac{2(0.0170)(5.5 + 72.5)}{1} = 2.65 \text{ in.}^2$$

or by Eq. 4.4.9:

$$\frac{2A_t}{s} = \frac{2(0.0170)}{1} = 0.034$$

$$\frac{50b_w}{f_y}\left[1 + 12\left(\frac{f_{pc}}{f'_c}\right)\right] = \frac{50(8)}{60000}\left[1 + 12\left(\frac{0.088}{6}\right)\right] = 0.008$$

Use 0.034 in Eq. 4.4.9:

$$A_\ell = \left[\frac{400x}{f_y}\left[\frac{T_u}{T_u + \frac{V_u}{3C_t}}\right] - \frac{2A_t}{s}\right](x_1 + y_1) = \left[\frac{400(8)}{60000}\left(\frac{787}{787 + \frac{90.0}{3(0.087)}}\right) - 0.034\right](5.5 + 72.5) = 0.24 \text{ in.}^2$$

Use A_ℓ = 2.65 in.² distributed around perimeter. The maximum spacing is 12 in. Prestressing strands not required for flexure can also be counted. Thus, additional #4 at 12 in. is adequate.

The transverse and longitudinal reinforcement can be reduced to the interior span side of the first double tee stem.

4.5 Beams with Ledges

As shown in the previous example, one of the most common occurrences of torsion in precast members is in L-shaped beams (beams with a ledge on one side). Significant torsion may also occur in inverted tee beams with severely unbalanced loads. This section covers additional design items related to the beam end and the ledge and its attachment to the web. These items are discussed in Refs. 10 through 13 and are covered as shown in Figure 4.5.1.

4.5.1 Shear Strength of Ledge

The design shear strength of continuous beam ledges, supporting concentrated loads, can be determined by the lesser of Eqs. 4.5.1.1 and 4.5.1.2:

for s > b_t + h_ℓ

$$\phi V_n = 3\phi\lambda\sqrt{f'_c}h_\ell[2(b_\ell-b)+b_t+h_\ell]$$
(Eq. 4.5.1.1)

$$\phi V_n = \phi\lambda\sqrt{f'_c}h_\ell[2(b_\ell-b)+b_t+h_\ell+2d_e]$$
(Eq. 4.5.1.2)

For s < b_t + h_ℓ, and equal concentrated loads, use the lesser of Eqs. 4.5.1.1a, 4.5.1.2a or 4.5.1.3.

$$\phi V_n = 1.5\phi\lambda\sqrt{f'_c}h_\ell[2(b_\ell-b)+b_t+h_\ell+s]$$
(Eq. 4.5.1.1a)

$$\phi V_n = \phi\lambda\sqrt{f'_c}h_\ell\left[(b_\ell-b)+\left(\frac{b_t+h_\ell}{2}\right)+d_e+s\right]$$
(Eq. 4.5.1.2a)

where:
- ϕ = 0.75
- h_ℓ = depth of beam ledge, in.
- b = beam web width
- b_ℓ = width of web and one ledge, in.
- b_t = width of bearing area, in.
- s = spacing of concentrated loads, in.
- d_e = distance from center of load to end of beam, in.

If the ledge supports a continuous load or closely spaced concentrated loads, the design shear strength is:

$$\phi V_n = 24\phi h_\ell\lambda\sqrt{f'_c}$$
(Eq. 4.5.1.3)

where:
- ϕV_n = design shear strength, lb/ft

If the applied factored load exceeds the strength as determined by Eqs. 4.5.1.1, 4.5.1.2 or

Figure 4.5.1   Design of beam ledges

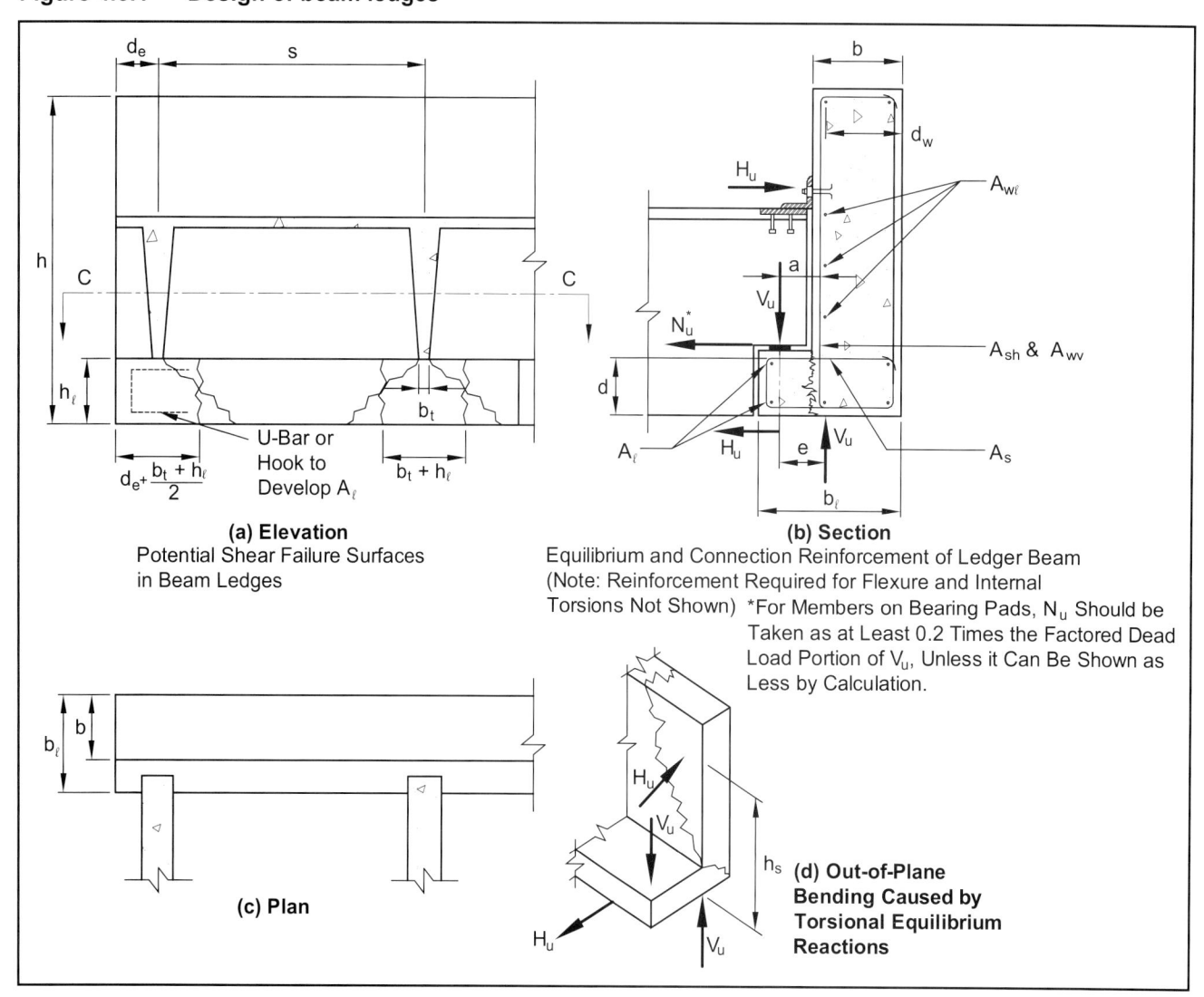

(a) Elevation
Potential Shear Failure Surfaces in Beam Ledges

(b) Section
Equilibrium and Connection Reinforcement of Ledger Beam
(Note: Reinforcement Required for Flexure and Internal Torsions Not Shown) *For Members on Bearing Pads, N_u Should be Taken as at Least 0.2 Times the Factored Dead Load Portion of V_u, Unless it Can Be Shown as Less by Calculation.

(c) Plan

(d) Out-of-Plane Bending Caused by Torsional Equilibrium Reactions

4.5.1.3, the ledge should be designed for shear transfer and diagonal tension in accordance with Sections 4.6.3.2 through 4.6.3.4.

4.5.2 Transverse (Cantilever) Bending of Ledge

Transverse (cantilever) bending of the ledge requires flexural reinforcement, A_s, which is computed by Eq. 4.5.2.1. Such reinforcement may be uniformly spaced over a width of $6h_\ell$ on either side of the bearing, but not to exceed half the distance to the next load. Bar spacing should not exceed the ledge depth, h_ℓ, or 18 in.

$$A_s = \frac{1}{\phi f_y}\left[V_u\left(\frac{a}{d}\right) + N_u\left(\frac{h_\ell}{d}\right)\right] \quad \text{(Eq. 4.5.2.1)}$$

(See Figure 4.5.1 for definitions.)

4.5.3 Longitudinal Bending of Ledge

Longitudinal reinforcement, calculated by Eq. 4.5.3.1, should be placed in both the top and bottom of the ledge portion of the beam:

$$A_\ell = 200(b_\ell - b)d_\ell / f_y \quad \text{(Eq. 4.5.3.1)}$$

where:
d_ℓ = design depth of A_ℓ reinforcement
(See Figure 4.5.1 for other definitions.)

4.5.4 Attachment of Ledge to Web

Hanger steel, A_{sh}, computed by Eq. 4.5.4.1 is required for attachment of the ledge to the web. Distribution and spacing of A_{sh} reinforcement should follow the same guidelines as for A_s reinforcement in Section 4.5.2. A_{sh} is not additive to shear and torsion reinforcement designed in accordance with Sections 4.3 and 4.4.

$$A_{sh} = \frac{V_u}{\phi f_y}(m) \quad \text{(Eq. 4.5.4.1)}$$

where:
V_u = applied factored load
ϕ = 0.75
f_y = yield strength of A_{sh} reinforcement
m = a modification factor which can be derived from Eqs. 6 through 9 of Ref. 15 and is dependent on beam section geometry (see Table 4.5.4.1), and a factor γ_t which accounts for proportioning of applied torsion between the ledge and the web. If closed stirrups are provided in the ledge, γ_t may conservatively be taken as 1.0. If closed stirrups are not provided in the ledge, γ_t may be taken as zero.

Figure 4.5.4.1 Unbalanced loads on an inverted tee beam

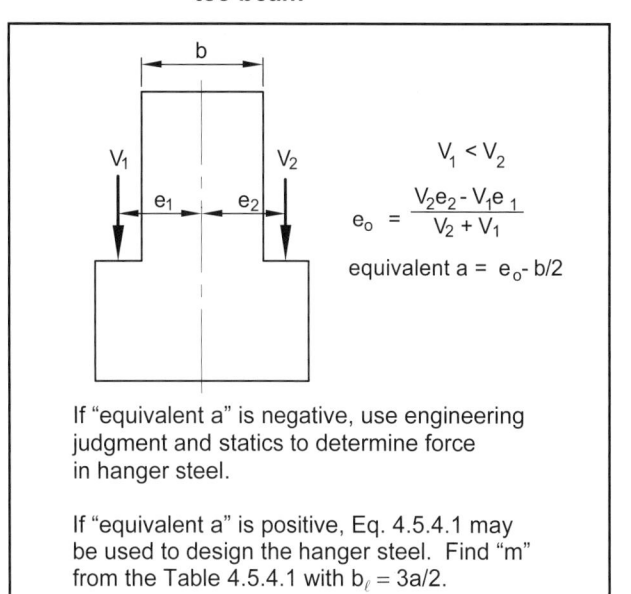

$V_1 < V_2$

$$e_o = \frac{V_2 e_2 - V_1 e_1}{V_2 + V_1}$$

equivalent $a = e_o - b/2$

If "equivalent a" is negative, use engineering judgment and statics to determine force in hanger steel.

If "equivalent a" is positive, Eq. 4.5.4.1 may be used to design the hanger steel. Find "m" from the Table 4.5.4.1 with $b_\ell = 3a/2$.

In the case of an inverted tee beam with unbalanced loads, an "equivalent a" can be calculated as shown in Figure 4.5.4.1. Use the values of e_o and b shown in Figure 4.5.4.1.

4.5.5 Out-of-Plane Bending Near Beam End

In the Ref. 15 study, it was found that when the reaction is not colinear with applied loads, as illustrated in Figure 4.5.1, the resulting out-of-plane bending may require additional vertical and horizontal reinforcement. These are computed by Eq. 4.5.5.1 and provided on the inside face of the beam. This reinforcement is not additive to the reinforcement for internal torsion. The $A_{w\ell}$ and A_{wv} bars should be evenly distributed over a height and width equal to h_s. See Figure 4.5.1(d).

$$A_{wv} = A_{w\ell} = \frac{V_u e}{2\phi f_y d_w} \quad \text{(Eq. 4.5.5.1)}$$

Terms are as shown in Figure 4.5.1(b).

where:
V_u = factored shear force at critical section
e = eccentricity, see Figure 4.5.1(b)
ϕ = 0.75 (Note: The use of ϕ = 0.75 instead of 0.90 (flexure) compensates for the use of d_w in place of the actual, somewhat smaller, lever arm.)
f_y = yield strength
d_w = depth of A_{wv} and $A_{w\ell}$ reinforcement from outside face of beam

Note that if the out-of-plane eccentricity, e, is very small, then the additional reinforcement may not be required.

Example 4.5.1
L-Beam End and Ledge Design

Given:
 8 ft wide double tees resting on a standard L-beam similar to that shown in Figure 4.5.1. Layout of tees is irregular so that a stem can be placed at any point on the ledge.
 V_u = 18 kips per stem
 N_u = 3 kips per stem
 h_ℓ = 12 in.
 d = 11 in.
 d_ℓ = 11 in.
 b_t = 3 in.
 b_ℓ = 14 in.
 h = 36 in.
 b = 8 in.
 s = 48 in.
 f'_c = 5000 psi (normal weight concrete)
 f_y = 60 ksi

Problem:
 Investigate the strength of the ledge and its attachment to the web. Determine required reinforcement.

Solution:
 Minimum $d_e = b_t/2$ = 1.5 in.
 Since $s > b_t + h_\ell$ and $d_e < 2(b_\ell - b) + b_t + h_\ell$
Use Eq. 4.5.1.2:
$$\phi V_n = \phi \lambda \sqrt{f'_c} h_\ell [2(b_\ell - b) + b_t + h_\ell + 2d_e] = 0.75(1)\sqrt{5000}(12)[2(6) + 3 + 12 + 2(1.5)] = 19{,}100 \text{ lb}$$
$$= 19.1 \text{ kips} > 18 \text{ kips} \quad \text{OK}$$

Determine reinforcement for flexure and axial tension:
 Shear span, $a = \frac{3}{4}(b_\ell - b) + 1.5 = 6$ in.

By Eq. 4.5.2.1:
$$A_s = \frac{1}{\phi f_y}\left[V_u\left(\frac{a}{d}\right) + N_u\left(\frac{h_\ell}{d}\right)\right] = \frac{1}{0.75(60)}\left[18\left(\frac{6}{11}\right) + 3\left(\frac{12}{11}\right)\right] = 0.29 \text{ in.}^2$$
 $6h_\ell$ = 6 ft > s/2 = 2 ft

Therefore, distribute reinforcement over s/2 each side of the load.
 (s/2)(2) = 4 ft
 Maximum bar spacing = h_ℓ = 12 in.
 Use #3 bars at 12 in. = 0.44 in.² for each 4 ft
Place two additional bars at the beam end to provide equivalent reinforcement for stem placed near the end.

By Eq. 4.5.4.1:
$$A_{sh} = \frac{V_u}{\phi f_y}(m) = \frac{18}{0.75(60)}(1.33) = 0.53 \text{ in.}^2$$

Example 4.5.1 (Cont.)
L-Beam End and Ledge Design

Note: m = 1.33 is obtained from Table 4.5.4.1 corresponding to: $b_\ell/b = 14/8 = 1.75$, $h_\ell/h = 12/36 = 0.33$, b = 8, and $\gamma_t = 1.0$ (closed stirrups in the ledge).

$$A_{sh} = \frac{0.53}{4} = 0.13 \text{ in.}^2/\text{ft}$$

Maximum bar spacing = h_ℓ = 12 in.
A_{sh} = #3 at 10 in. = 0.132 in.²/ft

By Eq. 4.5.3.1:
$$A_\ell = \frac{200(b_\ell - b)d_\ell}{f_y} = \frac{200(14-8)(11)}{60000} = 0.22 \text{ in.}^2$$

Use 1 – #3 top and 1 – #3 bottom = 0.22 in.²
By Eq. 4.5.5.1:
$$A_{wv} = A_{w\ell} = \frac{V_u e}{2\phi f_y d_w}$$

Assume:
$V_u e$ = 820 kip-in.
d_w = b – 1.5 = 6.5 in.
h_s = 18 in.

$$A_{wv} = A_{w\ell} = \frac{820}{2(0.75)(60)(6.5)} = 1.40 \text{ in.}^2$$

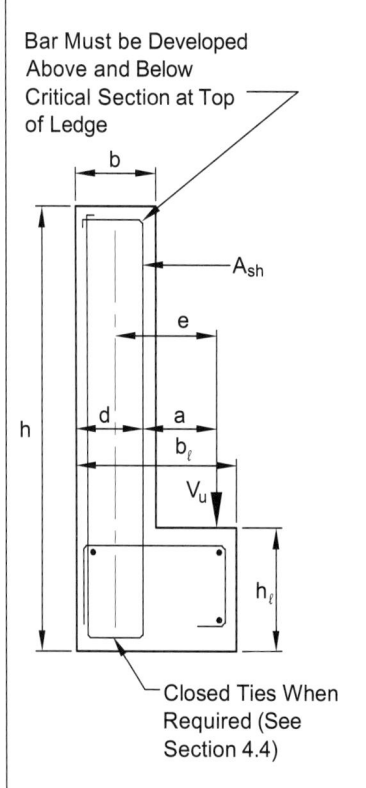

Table 4.5.4.1   Modification factor, m, for design of hanger steel

$$A_{sh} = \frac{V_u}{\phi f_y}(m) \quad \text{(Eq. 4.5.4.1)}$$

where:

$$m = \frac{\left[(d+a) - \left(3 - 2\frac{h_\ell}{h}\right)\left(\frac{h_\ell}{h}\right)^2 \left(\frac{b_\ell}{2}\right) - e\gamma_t \frac{(x^2 y)_\ell}{\Sigma x^2 y}\right]}{d}$$

x, y = shorter and longer sides, respectively, of the component rectangles forming the ledge and the web parts of the beam.
γ_t = 0, when closed ties are not used in the ledge.
γ_t = 1.0 when closed ties are used in the ledge.

The table values are based on the following assumptions:
(1)   d = b – 1.25
(2) V_u is applied at a distance equal to three-quarters of the ledge projection from the inside face of web.

(3) $\dfrac{(x^2 y)_\ell}{\Sigma x^2 y} = \dfrac{(b_\ell - b)^2 \dfrac{h_\ell}{h}}{(b_\ell - b)^2 \dfrac{h_\ell}{h} + b^2}$

This is a conservative assumption for commonly used ledger beam sizes.

Note: A lower limit of m = 0.6 is suggested to account for the limited variability in beam sizes tested. Values of m smaller than 0.6 are flagged with an asterisk in the table.

Table 4.5.4.1 (Cont.) Modification factor, m, for design of hanger steel

		\multicolumn{2}{c	}{b = 6 in.}	\multicolumn{2}{c	}{b = 8 in.}	\multicolumn{2}{c	}{b = 10 in.}
b_ℓ/b	γ_t h_ℓ/h	0.00	1.00	0.00	1.00	0.00	1.00
1.25	0.10	1.45	1.45	1.36	1.36	1.31	1.31
	0.15	1.43	1.42	1.34	1.33	1.29	1.28
	0.20	1.39	1.38	1.31	1.30	1.26	1.25
	0.25	1.35	1.34	1.27	1.26	1.22	1.21
	0.30	1.30	1.29	1.22	1.21	1.18	1.17
	0.35	1.25	1.23	1.18	1.16	1.13	1.12
	0.40	1.20	1.18	1.12	1.10	1.05	1.06
	0.45	1.14	1.12	1.07	1.05	1.03	1.01
	0.50	1.08	1.05	1.01	0.99	0.98	0.95
	0.55	1.02	0.99	0.96	0.93	0.92	0.90
	0.60	0.96	0.93	0.90	0.87	0.87	0.84
	0.65	0.91	0.87	0.85	0.82	0.82	0.79
	0.70	0.86	0.82	0.80	0.77	0.77	0.74
	0.75	0.81	0.77	0.76	0.72	0.73	0.70
1.50	0.10	1.66	1.63	1.56	1.53	1.50	1.48
	0.15	1.63	1.59	1.53	1.49	1.47	1.44
	0.20	1.59	1.54	1.49	1.44	1.44	1.39
	0.25	1.54	1.48	1.44	1.38	1.39	1.34
	0.30	1.48	1.41	1.39	1.32	1.34	1.27
	0.35	1.42	1.33	1.33	1.25	1.28	1.21
	0.40	1.35	1.26	1.27	1.18	1.22	1.14
	0.45	1.28	1.18	1.20	1.10	1.16	1.06
	0.50	1.21	1.10	1.14	1.03	1.10	0.99
	0.55	1.14	1.01	1.07	0.85	1.03	0.92
	0.60	1.07	0.93	1.01	0.88	0.97	0.85
	0.65	1.01	0.86	0.94	0.80	0.91	0.78
	0.70	0.94	0.79	0.89	0.74	0.85	0.71
	0.75	0.89	0.72	0.83	0.68	0.80	0.65
1.75	0.10	1.87	1.80	1.75	1.69	1.69	1.63
	0.15	1.83	1.73	1.72	1.63	1.66	1.57
	0.20	1.78	1.65	1.67	1.55	1.61	1.50
	0.25	1.73	1.57	1.62	1.47	1.56	1.42
	0.30	1.66	1.48	1.56	1.38	1.50	1.34
	0.35	1.59	1.38	1.49	1.29	1.44	1.25
	0.40	1.51	1.28	1.42	1.20	1.37	1.15
	0.45	1.43	1.17	1.34	1.10	1.29	1.06
	0.50	1.35	1.07	1.26	1.00	1.22	0.96
	0.55	1.26	0.96	1.18	0.90	1.14	0.87
	0.60	1.18	0.86	1.11	0.81	1.07	0.78
	0.65	1.10	0.76	1.04	0.72	1.00	0.69
	0.70	1.03	0.67	0.97	0.63	0.93	0.61
	0.75	0.97	0.59*	0.91	0.55*	0.87	0.53*

* Values of m smaller than 0.6.

Table 4.5.4.1 (Cont.) Modification factor, m, for design of hanger steel

b_ℓ/b	γ_t h_ℓ/h	b = 6 in. 0.00	1.00	b = 8 in. 0.00	1.00	b = 10 in. 0.00	1.00
2.00	0.10	2.07	1.94	1.95	1.82	1.88	1.75
	0.15	2.03	1.84	1.91	1.73	1.84	1.66
	0.20	1.98	1.73	1.86	1.62	1.79	1.57
	0.25	1.91	1.62	1.79	1.52	1.73	1.46
	0.30	1.84	1.49	1.72	1.40	1.66	1.35
	0.35	1.75	1.37	1.65	1.28	1.59	1.24
	0.40	1.66	1.24	1.56	1.16	1.51	1.12
	0.45	1.57	1.11	1.48	1.04	1.42	1.01
	0.50	1.48	0.98	1.39	0.92	1.34	0.89
	0.55	1.38	0.86	1.30	0.80	1.25	0.78
	0.60	1.29	0.74	1.21	0.69	1.17	0.67
	0.65	1.20	0.62	1.13	0.58*	1.09	0.56*
	0.70	1.12	0.51*	1.05	0.48*	1.01	0.46*
	0.75	1.04	0.41*	0.98	0.38*	0.94	0.37*
2.25	0.10	2.28	2.05	2.14	1.93	2.06	1.86
	0.15	2.23	1.91	2.10	1.80	2.02	1.73
	0.20	2.17	1.77	2.04	1.66	1.97	1.60
	0.25	2.10	1.62	1.97	1.52	1.90	1.47
	0.30	2.01	1.47	1.89	1.38	1.82	1.33
	0.35	1.92	1.32	1.80	1.24	1.74	1.20
	0.40	1.82	1.17	1.71	1.10	1.65	1.06
	0.45	1.72	1.02	1.61	0.95	1.55	0.92
	0.50	1.61	0.87	1.51	0.81	1.46	0.79
	0.55	1.50	0.72	1.41	0.68	1.36	0.65
	0.60	1.40	0.58*	1.31	0.55*	1.27	0.53*
	0.65	1.30	0.45*	1.22	0.42*	1.18	0.40*
	0.70	1.21	0.32*	1.13	0.30*	1.09	0.29*
	0.75	1.12	0.21*	1.05	0.20*	1.02	0.19*

b_ℓ/b	γ_t h_ℓ/h	b = 12 in. 0.00	1.00	b = 14 in. 0.00	1.00	b = 16 in. 0.00	1.00
1.25	0.10	1.28	1.28	1.26	1.26	1.25	1.24
	0.15	1.26	1.25	1.24	1.23	1.23	1.22
	0.20	1.23	1.22	1.21	1.20	1.20	1.19
	0.25	1.19	1.18	1.17	1.16	1.16	1.15
	0.30	1.15	1.14	1.13	1.12	1.12	1.11
	0.40	1.06	1.04	1.04	1.02	1.03	1.01
	0.45	1.01	0.99	0.99	0.97	0.98	0.96
	0.50	0.85	0.93	0.94	0.92	0.93	0.91
	0.55	0.90	0.88	0.89	0.86	0.88	0.85
	0.60	0.85	0.82	0.84	0.81	0.83	0.80
	0.65	0.80	0.77	0.79	0.76	0.78	0.75
	0.70	0.76	0.72	0.74	0.71	0.73	0.70
	0.75	0.71	0.68	0.70	0.67	0.69	0.66

* Values of m smaller than 0.6.

Table 4.5.4.1 (Cont.)   Modification factor, m, for design of hanger steel

b_ℓ/b	γ_t h_ℓ/h	b = 12 in. 0.00	b = 12 in. 1.00	b = 14 in. 0.00	b = 14 in. 1.00	b = 16 in. 0.00	b = 16 in. 1.00
1.50	0.10	1.47	1.44	1.44	1.42	1.43	1.40
	0.15	1.44	1.41	1.42	1.38	1.40	1.37
	0.20	1.40	1.36	1.38	1.34	1.36	1.32
	0.25	1.36	1.30	1.34	1.28	1.32	1.27
	0.30	1.31	1.24	1.29	1.22	1.27	1.21
	0.35	1.25	1.18	1.23	1.16	1.22	1.15
	0.40	1.20	1.11	1.18	1.09	1.16	1.08
	0.45	1.13	1.04	1.12	1.02	1.10	1.01
	0.50	1.07	0.97	1.05	0.95	1.04	0.94
	0.55	1.01	0.90	0.99	0.88	0.98	0.87
	0.60	0.95	0.83	0.93	0.81	0.92	0.80
	0.65	0.89	0.76	0.87	0.75	0.86	0.74
	0.70	0.83	0.69	0.82	0.68	0.81	0.67
	0.75	0.78	0.64	0.77	0.63	0.76	0.62
1.75	0.10	1.65	1.59	1.62	1.56	1.60	1.55
	0.15	1.62	1.53	1.59	1.51	1.57	1.49
	0.20	1.58	1.46	1.55	1.44	1.53	1.42
	0.25	1.52	1.39	1.50	1.36	1.48	1.35
	0.30	1.47	0.30	1.44	1.28	1.52	1.27
	0.35	1.40	1.22	1.38	1.20	1.36	1.18
	0.40	1.33	1.13	1.31	1.11	1.30	1.10
	0.45	1.26	1.04	1.24	1.02	1.23	1.01
	0.50	1.19	0.94	1.17	0.93	1.16	0.92
	0.55	1.12	0.85	1.10	0.84	1.08	0.83
	0.60	1.04	0.76	1.03	0.75	1.01	0.74
	0.65	0.98	0.68	0.96	0.66	0.95	0.66
	0.70	0.91	0.59*	0.90	0.58*	0.89	0.58*
	0.75	0.85	0.52*	0.84	0.51*	0.83	0.51*
2.00	0.10	1.83	1.71	1.80	1.69	1.78	1.67
	0.15	1.80	1.63	1.77	1.60	1.75	1.58
	0.20	1.75	1.53	1.72	1.50	1.70	1.49
	0.25	1.69	1.43	1.66	1.40	1.64	1.39
	0.30	1.62	1.32	1.60	1.30	1.58	1.28
	0.35	1.55	1.21	1.52	1.19	1.51	1.28
	0.40	1.47	1.10	1.45	1.08	1.43	1.07
	0.45	1.39	0.98	1.37	0.97	1.35	0.96
	0.50	1.31	0.87	1.28	0.86	1.27	0.85
	0.55	1.22	0.76	1.20	0.75	1.19	0.74
	0.60	1.14	0.65	1.12	0.64	1.11	0.63
	0.65	1.06	0.55*	1.05	0.54*	1.03	0.63
	0.70	0.99	0.45*	0.97	0.44*	0.96	0.44*
	0.75	0.92	0.36*	0.91	0.36*	0.90	0.35*

* Values of m smaller than 0.6.

Table 4.5.4.1 (Cont.) Modification factor, m, for design of hanger steel

b_ℓ/b	γ_t h_ℓ/h	b = 12 in. 0.00	b = 12 in. 1.00	b = 14 in. 0.00	b = 14 in. 1.00	b = 16 in. 0.00	b = 16 in. 1.00
2.25	0.10	1.98	1.78	1.96	1.76	1.94	1.75
	0.15	1.94	1.66	1.92	1.64	1.90	1.63
	0.20	1.89	1.54	1.87	1.52	1.85	1.51
	0.25	1.82	1.41	1.80	0.39	1.79	1.38
	0.30	1.75	1.28	1.73	1.27	1.71	1.25
	0.35	1.67	1.15	1.65	1.14	1.63	1.12
	0.40	1.58	1.02	1.56	1.00	1.55	1.00
	0.45	1.79	0.88	1.47	0.87	1.46	0.87
	0.50	1.40	0.75	1.38	0.75	1.37	0.74
	0.55	1.31	0.63	1.29	0.62	1.28	0.61
	0.60	1.22	0.50*	1.20	0.50*	1.19	0.49*
	0.65	1.13	0.39*	1.12	0.38*	1.11	0.38*
	0.70	1.05	0.28*	1.04	0.28*	1.03	0.27*
	0.75	0.98	0.18*	0.96	0.18*	0.95	0.18*

* Values of m smaller than 0.6.

4.5.6 Pocketed Beams

As an alternative to beams with ledges, pocketed beams may be used to provide support for stemmed members. Because of double tee erection constraints, pocketed beams can only be used to support one end of the member. They are frequently used on the exterior line of columns in parking structures. Pocketed beams have the advantages of minimal torsion, more simplified forming, economical production and a flush interior face.

Example 4.5.6.1
Design of a Prestressed Pocketed Spandrel Beam

Given:
  Precast pocketed spandrel beam shown on next page.
  Dead load of deck = 68 psf (10DT34 lightweight)
  Live load of deck (reduced by IBC 1607.9.2) = 30 psf

Beam Properties:
  A = 624 in.^{2}
  wt = 650 plf
  f'_c = 6000 psi
  f_y = 60 ksi

Prestressing:
  6 – ½ in. diameter, 270 ksi strands
  A_{ps} = 6(0.153) = 0.918 in.^{2}
  f_{se} = 170 ksi

Problem:
  Determine shear, torsion, dap, and pocket reinforcement requirements. The differences between this and the similarly loaded beam of Example 4.4.2 is illustrated.

Example 4.5.6.1 (Cont.)
Design of a Prestressed Pocketed Spandrel Beam

Solution:
1. Calculate V_u and T_u:
Determine factored loads:
Tee Stem:
$$P_u = V_u = \frac{60}{2}\left(\frac{10}{2}\right)\frac{[68(1.2) + 30(1.6)]}{1000}$$
$$= 19.4 \text{ kips/stem}$$

Spandrel (8 × 78 in. normal weight concrete):
$$\frac{650}{1000}(1.2) = 0.78 \text{ klf}$$

Determine eccentricity, e. Stem load is applied at one-quarter of the pocket depth from the outside face of the beam:
$$e = \frac{b}{2} - 6\left(\frac{1}{4}\right) = 2.5 \text{ in.}$$

Reaction at column:
19.4(3) + 0.78(14.917) = 69.8 kips

At support face:
V_u = 69.8 − 0.78(11/12) = 69.1 kips
T_u = 19.4(3)(2.5) = 145.5 kip-in.

2. Determine if $T_u \leq T_{u(min)}$ by Eq. 4.4.1:
$T_{u(min)} = \phi(0.5\lambda\sqrt{f'_c}\Sigma x^2 y)\gamma$
$\phi = 0.75$
$\lambda\sqrt{f'_c} = \dfrac{1.0\sqrt{6000}}{1000} = 0.0775$
$\Sigma x^2 y = (8^2 \times 78) = 4992 \text{ in.}^3$

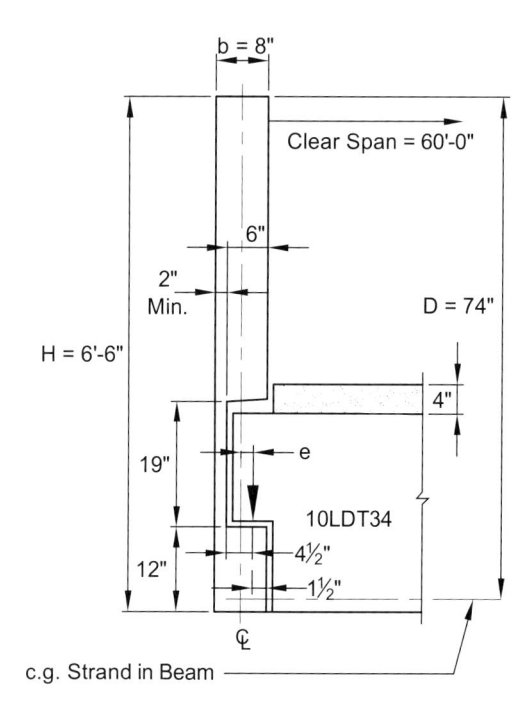

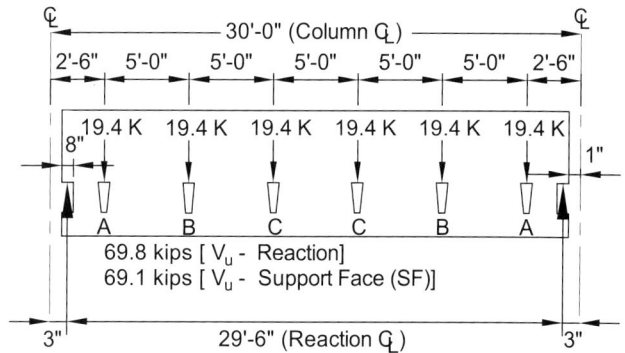

Determine f_{pc} at critical section approximately 23 in. from end of member. From Figure 4.12.4, strand develops 170 ksi (f_{se}) at $\ell_d = 28$ in.

f_{pd} = (23/28)(170) = 139.6 ksi
P_{pd} = 139.6(0.918) = 128.2 kips
$f_{pc} = \dfrac{P_{pd}}{A} = \dfrac{128.2}{624} = 0.205 \text{ ksi}$
$\gamma = \sqrt{1 + \dfrac{10 f_{pc}}{f'_c}} = \sqrt{1 + \dfrac{10(0.205)}{6}} = 1.16$

$T_{u(min)} = \phi(0.5\lambda\sqrt{f'_c}\Sigma x^2 y)\gamma$
= 0.75(0.5)(0.0775)(4992)(1.16)
= 168.3 kip-in. > 145.5 kip-in.

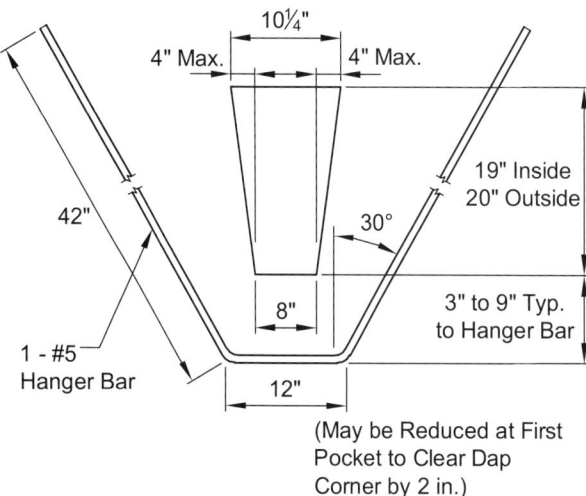

Example 4.5.6.1 (Cont.)
Design of a Prestressed Pocketed Spandrel Beam

Torsion need not be considered
Therefore, design transverse reinforcement for shear only.

V_c near end of member = $3.5\lambda\sqrt{f'_c}b_w d = \dfrac{3.5(1.0\sqrt{6000})(8.0)(74)}{1000} = 160.6$ kips

At edge of Pocket A (approx. 29 in. from end)
d above pocket ≈ 45 in.

$f_{pc} = \dfrac{170}{624} = 0.272$ ksi

$V_c = (3.5\lambda\sqrt{f'_c} + 0.3f_{pc})b_w d = [3.5(0.0775) + 0.3(0.272)](8)(45) = 127.0$ kips

$V_u = 69.8 - 0.78(29/12) = 67.9$ kips < 127.0 kips

(Note: A more complete shear analysis is shown in Section 4.3)

$V_c > V_u > V_c/2$ use minimum shear reinforcement.

From Section 4.3.4, calculate minimum shear reinforcement:

$\dfrac{A_v}{s} = 0.75\lambda\sqrt{f'_c}\dfrac{b_w}{f_y} = 0.75(1)\sqrt{6000}\dfrac{8}{60000} = 0.00775$ in.²/in. = 0.093 in.²/ft

Alternative:
At end:

$\dfrac{A_v}{s} = \dfrac{A_{ps}f_{pu}}{80f_y d}\sqrt{\dfrac{d}{b_w}} = \dfrac{0.918(270)}{80(60)(74)}\sqrt{\dfrac{74}{8}} = 0.00212$ in.²/in. = 0.025 in.²/ft

Over pocket:

$= \dfrac{0.918(270)}{80(60)(45)}\sqrt{\dfrac{45}{8}} = 0.00272$ in.²/in. = 0.033 in.²/ft

These requirements are easily met with welded wire reinforcement in each face of the beam.

Logitudinal reinforcement: The Code does not require "skin reinforcement" for prestressed members, unless it is a "Class C" member (see Code Section 18.4.4.4); however, it is generally considered good practice to provide some horizontal steel. The requirements for out-of-plane bending near the end (Section 4.5.5) must be met (only tee stem loads are eccentric):

$A_{wv} = A_{w\ell} = \dfrac{V_u e}{2\phi f_y d_w} = \dfrac{3(19.4)(2.5)}{2(0.90)(60)(6)}$

$= 0.22$ in.² on inside face

6x6W2.9xW2.9 in each face (cut out around pockets would meet all of the requirements.)

Determine hanger reinforcement at beam pockets:
Vertical component of each leg of hanger bar

$= \dfrac{19.4}{2} = 9.7$ kips

V_u (each leg) $= \dfrac{9.7}{\cos 30°} = 11.2$ kips

(May be Reduced at First Pocket to Clear Dap Corner by 2 in.)

Example 4.5.6.1 (Cont.)
Design of a Prestressed Pocketed Spandrel Beam

$$A_{sh} \text{ (each leg)} = \frac{V_u}{\phi f_y} = \frac{11.2}{(0.75)(60)} = 0.25 \text{ in.}^2$$

Use 1 – #5 bar; $A_{sh} = 0.31$ in.^{2} > 0.25 in.^{2}
From Design Aid 11.2.8, $\ell_d = 19$ in. above pocket top
Select #5 with verical leg length of 42 in. minimum.

Reinforcement at the end of the member should now be designed in accordance with the step-by-step procedure shown in Section 4.6.3. The result is reinforcement similar to that shown below.

3. Other considerations.
a. For parking structures the spandrel beam must be checked for vehicle barrier load criteria. Most codes require the application of 6000 lb (unfactored) applied horizontally 18 in. above the floor.
b. Local code requirements for snow loading, including drifting, should be checked in addition to the 30 psf live load on the roof decks of parking structures.

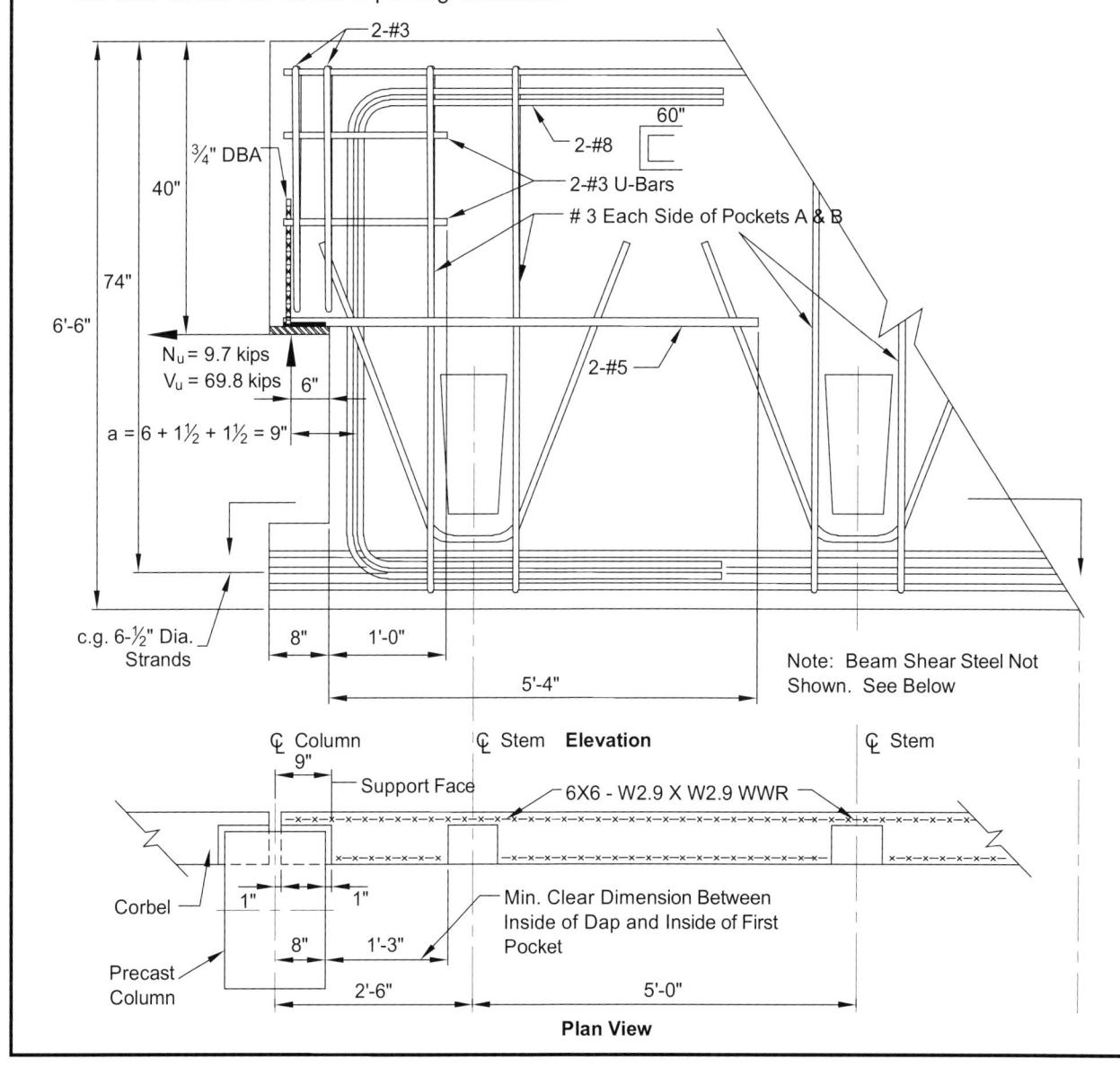

4.6 Bearing

4.6.1 Bearing on Plain Concrete

Plain concrete bearing may be used in situations where the bearing is uniform and the bearing stresses are low as is typical, for example, in hollow-core and solid slabs. In other situations, and in thin stemmed members where the bearing area is smaller than 20 in.², a minimum reinforcement equal to $N_u/\phi f_y$ (but not less than one No. 3 bar) is recommended. It is recommened that, when bearing pads are used that N_u be taken as 0.2 times the sustained load portion of V_u unless otherwise calculated.

The design bearing strength of plain concrete may be calculated as:

$$\phi V_n = \phi C_r (0.85 f'_c A_1) \sqrt{\frac{A_2}{A_1}} \leq 1.1 f'_c A_1$$

(Eq. 4.6.1.1)

where:

ϕV_n = design bearing strength
ϕ = 0.65
$C_r = \left(\dfrac{sw}{200}\right)^{\frac{N_u}{V_u}}$

= 1.0 when reinforcement is provided in direction of N_u, in accordance with Section 4.6.2 or when N_u is zero. The product sw should not be taken greater than 9.0 in.²

s = distance from edge to load point
A_1 = loaded area = b(w) (see Figure 4.6.1)
A_2 = the area of the lower base of the largest frustum of a pyramid, cone, or tapered wedge contained wholly within the support and having for its upper base the loaded area, and having side slopes of 1 vertical to 2 horizontal (see Figure 4.6.1)

4.6.2 Reinforced Concrete Bearing

If the applied load, V_u, exceeds the design bearing strength, ϕV_n, as calculated by Eq. 4.6.1.1, reinforcement is required in the bearing area. This reinforcement can be designed by shear-friction as discussed in Section 4.3.6. Referring to Figure 4.6.2, the reinforcement $A_{vf} + A_n$ nominally parallel to the direction of the axial load, N_u, is determined by Eqs. 4.3.6.1 through 4.3.6.3 with A_{cr} equal to b × h (i.e., $\theta \approx 0°$, see Ref. 14). This reinforcement must be appropriately developed by hooks or by welding to an anchor bar, bearing plate or angle.

Vertical reinforcement across potential horizontal cracks can be calculated by:

$$A_{sh} = \frac{(A_{vf} + A_n) f_y}{\mu_e f_{ys}}$$

(Eq. 4.6.2.1)

where:

$\mu_e = \dfrac{1000 \lambda A_{cr} \mu}{(A_{vf} + A_n) f_{ys}} \leq$ values in Table 4.3.6.1

f_y = yield strength of $A_{vf} + A_n$
f_{ys} = yield strength of A_{sh}, psi
A_{cr} = $\ell_d b$, in.² (see Figure 4.6.2)
b = average member width, in.
ℓ_d = development length of $A_{vf} + A_n$ bars, in.
μ_e = shear-friction coefficient (see Table 4.3.6.1)
λ = conversion factor for lightweight concrete (see Section 4.3.3)

Stirrups or mesh used for diagonal tension reinforcement can be considered to act as A_{sh} reinforcement.

When members are subjected to bearing force in excess of $1.1 f'_c A_1$, confinement reinforcement in all directions may be necessary.

Figure 4.6.1   Bearing on plain concrete

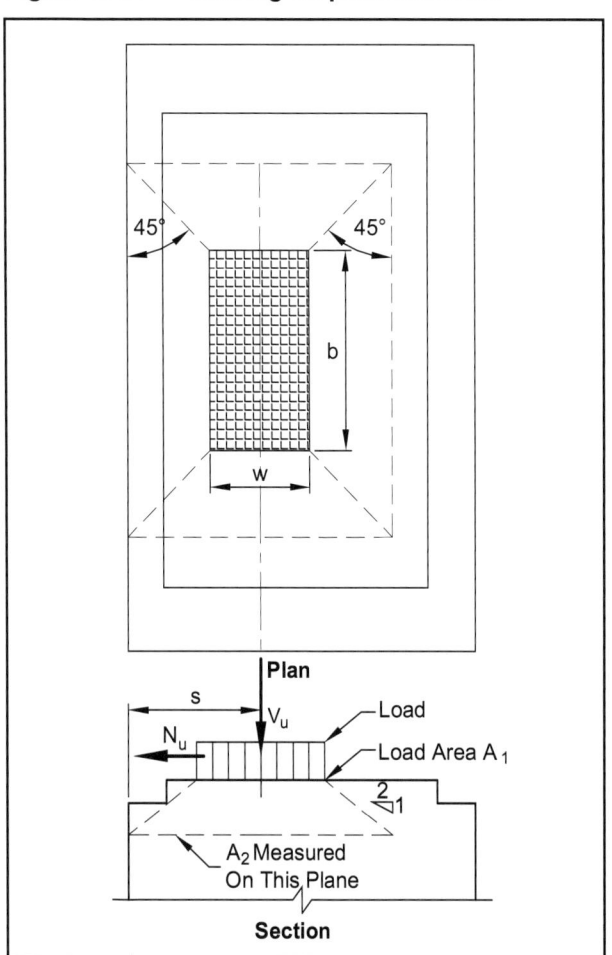

Example 4.6.1
Reinforced Bearing for a Rectangular Beam

Given:
  PCI standard rectangular beam 12RB28
  V_u = 94 kips (includes all load factors)
  N_u = 15 kips (includes all load factors)
  Bearing pad = 4 in. x 10 in.
  f_y = 60,000 psi (all reinforcement)
  f'_c = 5000 psi (normal weight concrete)

Problem:
  Determine reinforcement requirements for the end of the member.

Solution:
  $A_{cr} = b(h) = 12(28) = 336$ in.^{2}
  Check max V_n from Table 4.3.6.1.
  $1000\lambda^2 A_{cr} = 1000(1.0)^2(336)/1000 = 336$ kips
  max $V_u = 0.75(336) = 252$ kips > 94 kips OK

By Eq. 4.3.6.2:
$$\mu_e = \frac{1000\lambda A_{cr}\mu}{V_u} = \frac{1000(1)(336)(1.4)}{94,000} = 5.0 > 3.4, \text{ use } 3.4$$

By Eq. 4.3.6.1:
$$A_{vf} = \frac{V_u}{\phi f_y \mu_e} = \frac{94000}{0.75(60000)(3.4)} = 0.61 \text{ in.}^2$$

By Eq. 4.3.6.3:
$$A_n = \frac{N_u}{\phi f_y} = \frac{15,000}{0.75(60000)} = 0.33 \text{ in.}^2$$
$$A_{vf} + A_n = 0.61 + 0.33 = 0.94 \text{ in.}^2$$

Use 3 – #5 = 0.93 in.^{2}
Determine ℓ_d from Design Aid 11.2.8.
  for $f'_c = 5000$ psi:
  ℓ_d = 21 in.
  $A_{cr} = \ell_d b = (21)(12) = 252$ in.^{2}
  $$\mu_e = \frac{1000(1)(252)(1.4)(1)}{0.94(60,000)} = 6.3 > 3.4$$

Use $\mu_e = 3.4$:
$$A_{sh} = \frac{(A_{vf} + A_n)f_y}{\mu_e f_{ys}} = \frac{0.94(60,000)}{3.4(60,000)} = 0.28 \text{ in.}^2$$
Use 2 – #3 stirrups = 0.44 in.^{2}

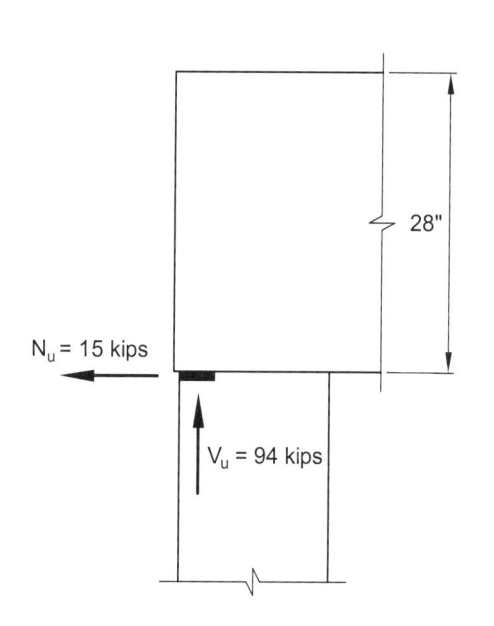

4.6.3 Dapped-End Bearing

Design of bearing areas which are recessed or dapped into the end of the member requires the investigation of several potential failure modes. These are numbered and shown in Figure 4.6.3.1 and listed in this section along with the reinforcement required for each consideration. The design equations given in this section are based primarily on Refs. 15 and 16, and are appropriate for cases where shear span-to-depth ratio (a/d in Figure 4.6.3.1) is not more than 1.0.

Dap reinforcement (Eqs. 4.6.3.1 through 4.6.3.7) should be provided in all cases where any one or more of the following conditions occurs:
1. The depth of the recess exceeds 0.2H or 8 in.
2. The width of the recess (ℓ_p in Figure 4.6.3.1) exceeds 12 in.
3a. For members less than 8 in. wide, less than one-half of the main flexural reinforcement extends to the end of the member above the dap.
3b. For members 8 in. or more wide, less than one-third of the main flexural reinforcement extends to the end of the member above the dap.

These criteria indicate that only short, shallow recesses, having the minimum amount of main reinforcement or more extending into the nib above the dap, do not require all the dap reinforcement.

Experience and some unpublished tests verify that, for short, shallow recesses, the hanger reinforcement, A_{sh} and A'_{sh}, is not necessary. However, in these cases it is recommended that confinement reinforcement, A_v, and flexural reinforcement, $A_{vf} + A_n$, in accordance with Section 4.6.2, be provided.

The potential failure modes are as follows:
1. Flexure (cantilever bending) and axial tension in the extended end. Provide flexural reinforcement, A_f, plus axial tension reinforcement, A_n.
2. Direct shear at the junction of the dap and the main body of the member. Provide shear-friction reinforcement composed of A_{vf} and A_h, plus axial tension reinforcement, A_n.
3. Diagonal tension emanating from the re-entrant corner. Provide shear reinforcement, A_{sh}.
4. Diagonal tension in the extended end. Provide shear reinforcement composed of A_h and A_v.
5. Diagonal tension in the undapped portion. This is resisted by providing a full development length for A_s beyond the potential crack. [15]

Each of these potential failure modes should be investigated separately. The reinforcement requirements are not cumulative, that is, A_s is the greater of that required by 1 or 2, not the sum. A_h is the greater of that required by 2 or 4, not the sum.

Figure 4.6.2   Reinforced concrete bearing

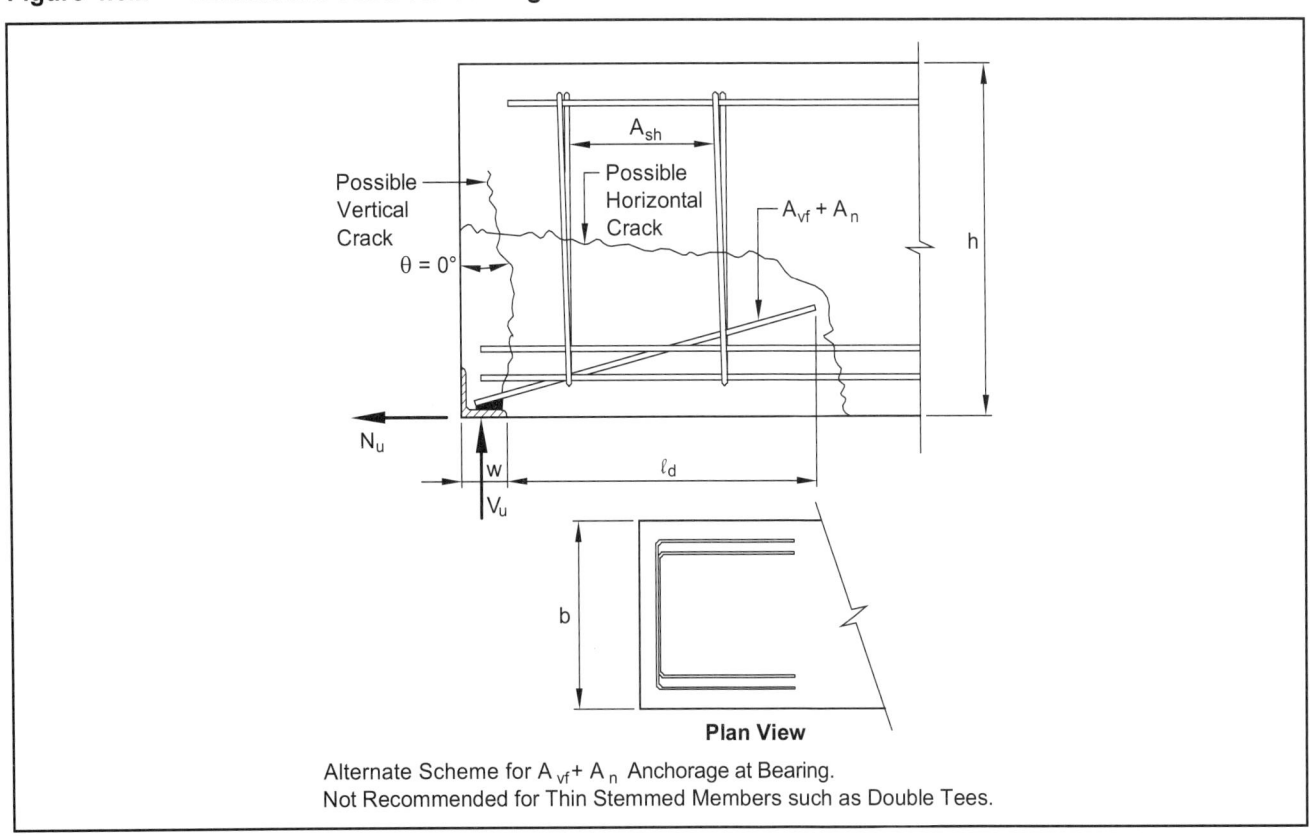

Figure 4.6.3.1 Potential failure modes and required reinforcement in dapped-end connections

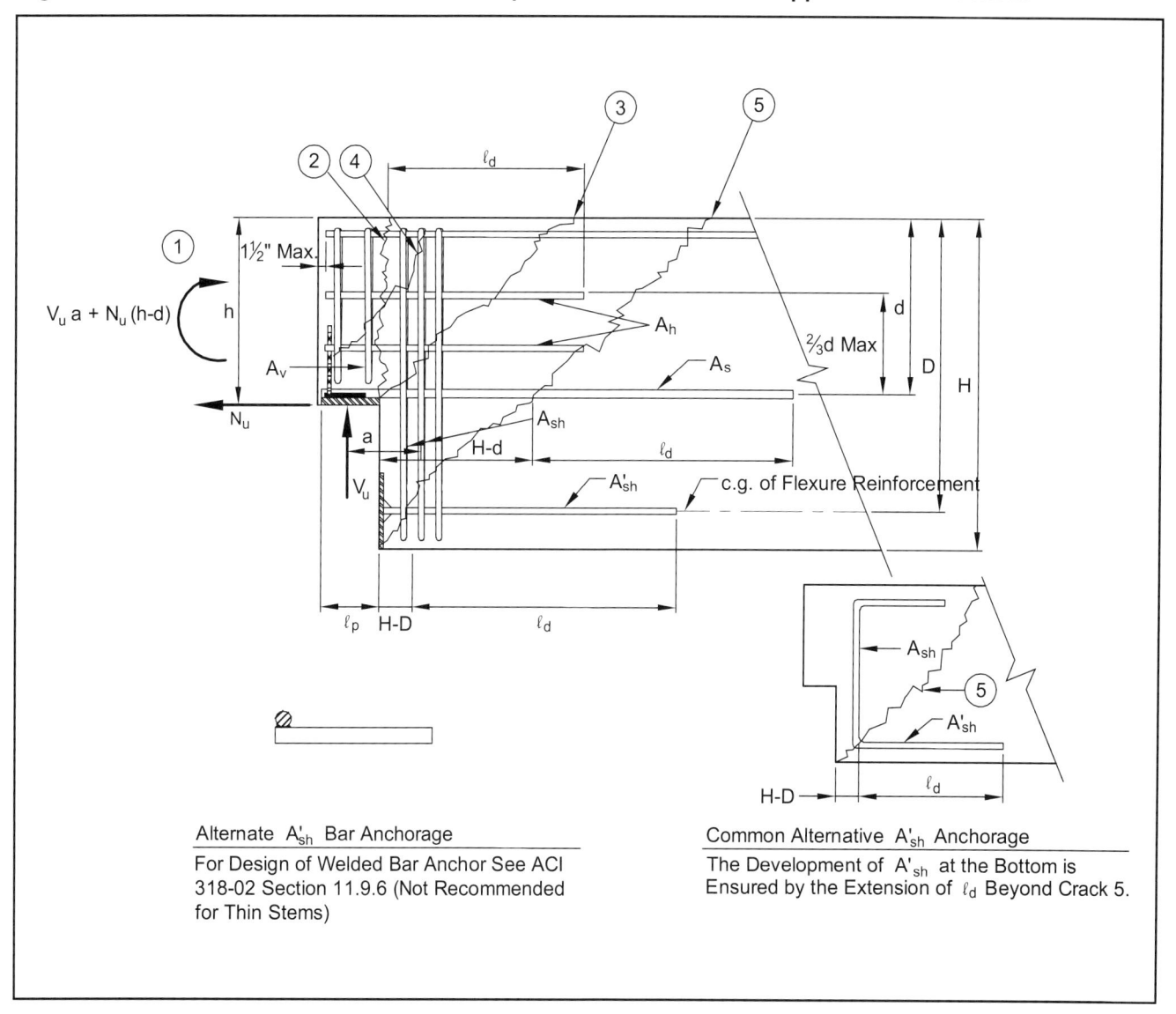

Alternate A'_{sh} Bar Anchorage
For Design of Welded Bar Anchor See ACI 318-02 Section 11.9.6 (Not Recommended for Thin Stems)

Common Alternative A'_{sh} Anchorage
The Development of A'_{sh} at the Bottom is Ensured by the Extension of ℓ_d Beyond Crack 5.

4.6.3.1 Flexure and Axial Tension in Extended End

The horizontal reinforcement is determined in a manner similar to that for column corbels, Section 6.8. Thus:

$$A_s = A_f + A_n$$
$$= \frac{1}{\phi f_y}\left[V_u\left(\frac{a}{d}\right) + N_u\left(\frac{h}{d}\right)\right] \quad \text{(Eq. 4.6.3.1)}$$

where:
$\phi = 0.75^*$

---

* To be exact, Eq. 4.6.3.1 should have $j_u d$ in the denominator. The use of $\phi = 0.75$ instead of 0.90 (flexure) compensates for this approximation. Also, for uniformity $\phi = 0.75$ is used throughout Section 4.6 even though $\phi = 0.90$ would be more appropriate and may be used where flexure and direct tension are addressed.

a = shear span, in., measured from load to center of A_{sh}
h = depth of member above dap, in.
d = distance from top to center of reinforcement, A_s, in.
f_y = yield strength of flexural reinforcement, psi
N_u = 0.2 times sustained load portion of V_u unless otherwise calculated (when bearing pads are used). [14]

4.6.3.2 Direct Shear

The potential vertical crack shown in Figure 4.6.3.1 is resisted by a combination of A_s and A_h. This reinforcement can be calculated by Eqs. 4.6.3.2 through 4.6.3.4:

$$A_s = \frac{2V_u}{3\phi f_y \mu_e} + A_n \quad \text{(Eq. 4.6.3.2)}$$

$$A_n = \frac{N_u}{\phi f_y} \quad \text{(Eq. 4.6.3.3)}$$

$$A_h = 0.5(A_s - A_n) \quad \text{(Eq. 4.6.3.4)}$$

where:
$\phi = 0.75$
f_y = yield strength of A_s, A_n, A_h, psi
$\mu_e = \dfrac{1000\lambda b h_\mu}{V_u} \le$ values in Table 4.3.6.1

The shear strength of the extended end is limited by the maximum values given in Table 4.3.6.1.

4.6.3.3 Diagonal Tension at Re-entrant Corner

The reinforcement required to resist diagonal tension cracking starting from the re-entrant corner, shown as 3 in Figure 4.6.3.1, can be calculated from:

$$A_{sh} = \frac{V_u}{\phi f_y} \quad \text{(Eq. 4.6.3.5)}$$

where:
$\phi = 0.75$
V_u = applied factored load
A_{sh} = vertical or diagonal bars across potential diagonal tension crack, in.^{2}
f_y = yield strength of A_{sh}

4.6.3.4 Diagonal Tension in Extended End

Additional reinforcement for Crack 4 in Figure 4.6.3.1 is required in the extended end, such that:

$$\phi V_n = \phi(A_v f_y + A_h f_y + 2bd\lambda\sqrt{f'_c}) \quad \text{(Eq. 4.6.3.6)}$$

At least one-half of the reinforcement required in this area should be placed vertically. Thus:

$$\min A_v = \frac{1}{2f_y}\left(\frac{V_u}{\phi} - 2bd\lambda\sqrt{f'_c}\right) \quad \text{(Eq. 4.6.3.7)}$$

4.6.3.5 Anchorage of Reinforcement

With reference to Figure 4.6.3.1:

1. Horizontal bars A_s should be extended a minimum of ℓ_d past Crack 5, and anchored at the end of the beam by welding to cross bars, plates or angles.
2. Horizontal bars A_h should be extended a minimum of ℓ_d past Crack 2 and anchored at the end of the beam by hooks or other suitable means.
3. To ensure development of hanger reinforcement, A_{sh}, it may be bent and continued parallel to the beam bottom, or separate horizontal reinforcement, $A'_{sh} \ge A_{sh}$, must be provided. The extension of reinforcement at beam bottom must be at least ℓ_d beyond Crack 5. The A'_{sh} reinforcement may be anchored on the dap side by welding it to a plate (as shown in Figure 4.6.3.1), angle or cross bar. The beam flexure reinforcement may also be used to ensure development of A_{sh} reinforcement provided that the flexure reinforcement is adequately anchored on the dap side.
4. Vertical reinforcement, A_v, should be properly anchored by hooks as required by ACI 318-02.
5. Welded wire reinforcement in place of bars may be used for reinforcement. It should be anchored in accordance with ACI 318-02.

4.6.3.6 Other Considerations

1. The depth of the extended end should not be less than about one-half the depth of the beam, unless the beam is significantly deeper than necessary for other than structural reasons.
2. The hanger reinforcement, A_{sh}, should be placed as close as practical to the re-entrant corner. This reinforcement requirement is not additive to other shear reinforcement requirements, but should be in addition to torsion steel.
3. If the flexural stress in the full depth section immediately beyond the dap, using factored loads and gross section properties, exceeds $7.5\sqrt{f'_c}$, longitudinal reinforcement should be placed in the beam to develop the required flexural strength.
4. The Ref. 18 study found that, due to formation of the critical diagonal tension crack (Crack 5 in Figure 4.6.3.1), it was not possible to develop a full depth beam shear strength greater than the diagonal tension cracking shear in the vicinity of the dap. It is, therefore, suggested that, for a length of the beam equal to the overall depth, H, of the beam, the nominal shear strength of concrete, V_c, be taken as the lesser of V_{ci} and V_{cw} calculated at H/2 from the end of the full depth web.

Example 4.6.3.1
Reinforcement for Dapped-End Beam

Given:
The 16RB28 beam with a dapped end is shown below.
V_u = 100 kips (includes all load factors)
N_u = 15 kips (includes all load factors)
f'_c = 5000 psi (normal weight)
f_y = 60 ksi (for all reinforcement)

Problem:
Determine the required reinforcements A_s, A_h, A_{sh}, and A_v shown in Figure 4.6.3.1.

Solution:
1. Flexure in extended end:
By Eq. 4.6.3.1:
Assume: Shear span, a = 6 in., d = 15 in.

$$A_s = \frac{1}{\phi f_y}\left[V_u\left(\frac{a}{d}\right) + N_u\left(\frac{h}{d}\right)\right] = \frac{1}{0.75(60)}\left[100\left(\frac{6}{15}\right) + 15\left(\frac{16}{15}\right)\right] = 1.24 \text{ in.}^2$$

2. Direct shear:

$$\mu_e = \frac{1000\lambda bh\mu}{V_u} = \frac{1000(1)(16)(16)(1.4)}{100000} = 3.58 > 3.4 \text{ Use } 3.4$$

By Eqs. 4.6.3.2 and 4.6.3.3:

$$A_s = \frac{2V_u}{3\phi f_y \mu_e} + \frac{N_u}{\phi f_y} = \frac{2(100)}{3(0.75)(60)(3.4)} + \frac{15}{0.75(60)} = 0.44 + 0.33 = 0.77 \text{ in.}^2 < 1.24 \text{ in.}^2$$

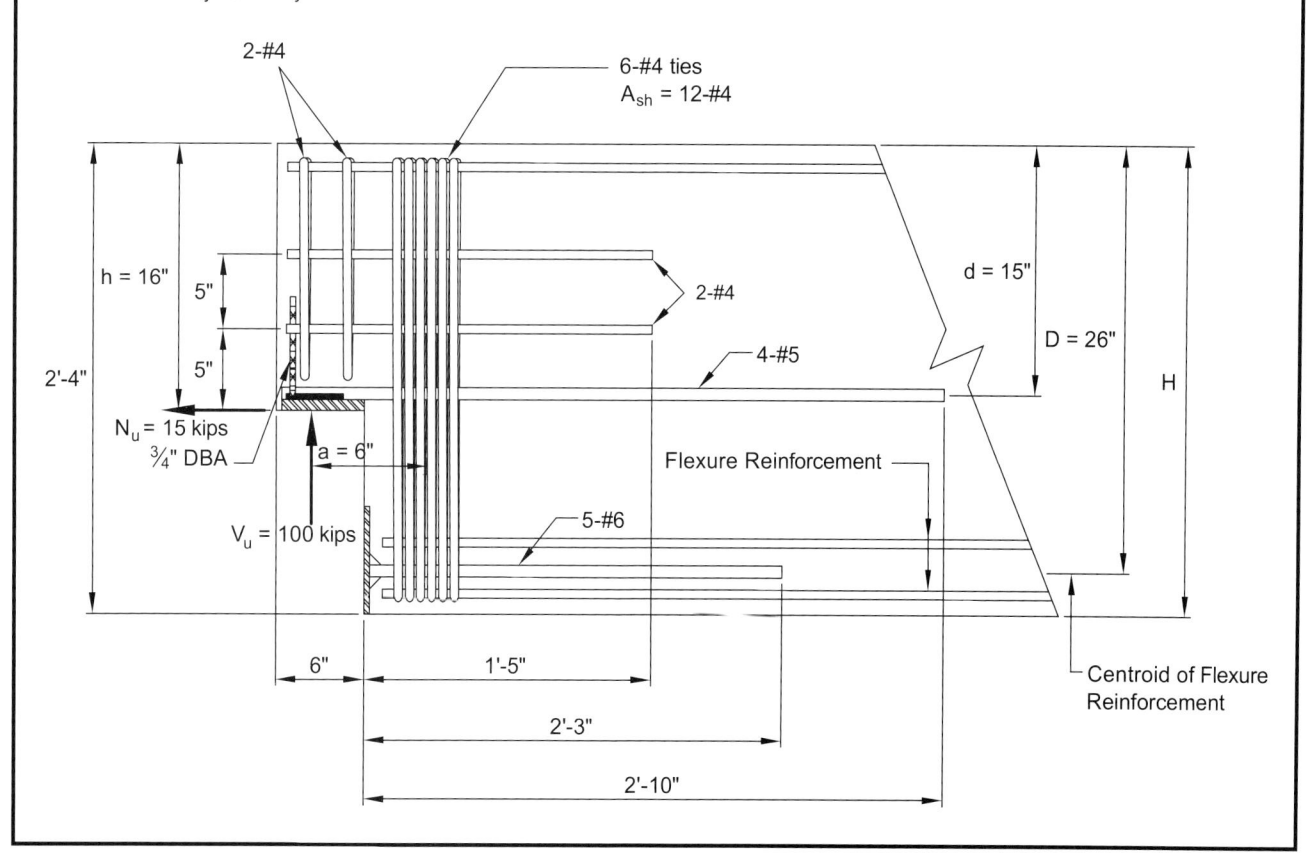

Example 4.6.3.1 (Cont.)
Reinforcement for Dapped-End Beam

Therefore $A_s = 1.24$ in.^{2}
    Use 4 – #5, $A_s = 1.24$ in.^{2}

By Eq. 4.6.3.4:
    $A_h = 0.5(A_s - A_n) = 0.5(1.24 - 0.33) = 0.45$ in.^{2}

Check shear strength, Table 4.3.6.1:
    $\phi V_n = \phi(1000\lambda^2 bd) = 0.75(1000)(1)^2(16)(15)/1000 = 180$ kips > 100 kips   OK
    Use 2 – #3 U-bars. $A_h = 0.44$ in.^{2} (see also Step 4 below)

3. Diagonal tension at re-entrant corner:
By Eq. 4.6.3.5:
    $A_{sh} = \dfrac{V_u}{\phi f_y} = \dfrac{100}{0.75(60)} = 2.22$ in.^{2}
    Use 6 – #4 closed ties = 12 – #4 = 2.40 in.^{2}
    For A'_{sh} (minimum area = A_{sh})
    Use 5 – #6 = 2.20 in.^{2}

4. Diagonal tension in the extended end:
    Concrete capacity = $2\lambda\sqrt{f'_c}\,bd = \dfrac{2(1)\sqrt{5000}(16)(15)}{1000} = 33.9$ kips

By Eq. 4.6.3.7:
    $A_v = \dfrac{1}{2f_y}\left[\dfrac{V_u}{\phi} - 2bd\lambda\sqrt{f'_c}\right] = \dfrac{1}{2(60)}\left(\dfrac{100}{0.75} - 33.9\right) = 0.83$ in.^{2}
    Try 2 – #4 stirrups = 0.80 in.^{2} say OK
    Check Eq. 4.6.3.6:
    $\phi V_n = \phi(A_v f_y + A_h f_y + 2\lambda\sqrt{f'_c}\,bd)$
    $\phi V_n = 0.75[0.80(60) + 0.44(60) + 33.9] = 81.2$ kips < 100 kips
    Change A_h to 2 – #4
    $\phi V_n = 97.4$ kips $\approx$ 100 kips   say OK

Check anchorage requirements:

A_s bars:
From Design Aid 11.2.9:
    $f_y = 60,000$ psi
    $f'_c = 5000$ psi
    #5 bars
    $\ell_d = 21$ in.
Extension past dap = $H - d + \ell_d = 28 - 15 + 21 = 34$ in.

A_h bars:
From Design Aid 11.2.9, for #4 bars:
    $\ell_d = 17$ in.

A'_{sh} bars:
From Design Aid 11.2.9, for #6 bars:
    $\ell_d = 25$ in.
    Bar length = $H - D + \ell_d = 28 - 26 + 25 = 27$ in.

4.7 Loss of Prestress

Loss of prestress is the reduction of tensile stress in prestressing tendons due to shortening of the concrete around the tendons, relaxation of stress within the tendons and external factors which reduce the total initial force before it is applied to the concrete. ACI 318-02 identifies the sources of loss of prestress listed in Section 4.7.1.

Accurate determination of losses is more important in some prestressed concrete members than in others. Losses have no effect on the ultimate strength of a flexural member unless the tendons are unbonded or if the final stress after losses is less than $0.50f_{pu}$. Underestimation or overestimation of losses can affect service conditions such as camber, deflection and cracking.

4.7.1 Sources of Stress Loss

1. **Anchorage Seating Loss and Friction**

   Anchorage seating loss and friction loss due to intended or unintended curvature in post-tensioning tendons are two mechanical sources of loss. They represent the difference between the tension applied to the tendon by the jacking unit and the initial tension available for application to the concrete by the tendon. Their magnitude can be determined with reasonable accuracy and, in many cases, they are fully or partially compensated for by overjacking.

2. **Elastic Shortening of Concrete**

   The concrete around the tendons shortens as the prestressing force is applied to it. Tendons which are already bonded to the concrete shorten with it.

3. **Shrinkage of Concrete**

   Loss of stress in the tendon due to shrinkage of the concrete surrounding it is proportional to that part of the shrinkage that takes place after the transfer of prestress force to the concrete.

4. **Creep of Concrete and Relaxation of Tendons**

   Creep of concrete and relaxation of tendons complicate stress loss calculations. The rate of loss due to each of these factors changes when the stress level changes and the stress level is changing constantly throughout the life of the structure. Therefore, the rates of loss due to creep and relaxation are constantly changing.

4.7.2 Range of Values for Total Loss

Total loss of prestress in typical members will range from about 25,000 to 50,000 psi for normal weight concrete members, and from about 30,000 to 55,000 psi for sand-lightweight members.

The load tables in Chapter 2 have a lower limit on loss of 30,000 psi.

4.7.3 Estimating Prestress Loss

This section is based on the report of a task group sponsored by ACI-ASCE Committee 423, Prestressed Concrete. [19] That report gives simple equations for estimating losses of prestress which would enable the designer to estimate the various types of prestress loss rather than using a lump sum value. It is believed that these equations, intended for practical design applications, provide fairly realistic values for normal design conditions. For unusual design situations and special structures, more detailed analyses may be warranted.

$$TL = ES + CR + SH + RE \quad \text{(Eq. 4.7.3.1)}$$

where:
- TL = total loss (psi), and other terms are losses due to:
- ES = elastic shortening
- CR = creep of concrete
- SH = shrinkage of concrete
- RE = relaxation of tendons

$$ES = K_{es}E_{ps}f_{cir}/E_{ci} \quad \text{(Eq. 4.7.3.2)}$$

where:
- K_{es} = 1.0 for pretensioned members
- E_{ps} = modulus of elasticity of prestressing tendons (about 28.5×10^6 psi)
- E_{ci} = modulus of elasticity of concrete at time prestress is applied
- f_{cir} = net compressive stress in concrete at center of gravity of prestressing force immediately after the prestress has been applied to the concrete:

$$f_{cir} = K_{cir}\left(\frac{P_i}{A_g} + \frac{P_i e^2}{I_g}\right) - \frac{M_g e}{I_g} \quad \text{(Eq. 4.7.3.3)}$$

where:
- K_{cir} = 0.9 for pretensioned members
- P_i = initial prestress force (after anchorage seating loss)
- e = eccentricity of center of gravity of tendons with respect to center of gravity of concrete at the cross section considered
- A_g = area of gross concrete section at the cross section considered
- I_g = moment of inertia of gross concrete section at the cross section considered
- M_g = bending moment due to dead weight of prestressed member and any other permanent loads in place at time of prestressing

$$CR = K_{cr}(E_{ps}/E_c)(f_{cir} - f_{cds}) \quad \text{(Eq. 4.7.3.4)}$$

where:
- K_{cr} = 2.0 normal weight concrete
- = 1.6 sand-lightweight concrete
- f_{cds} = stress in concrete at center of gravity of prestressing force due to all superimposed permanent dead loads that are applied to the member after it has been prestressed
- E_c = modulus of elasticity of concrete at 28 days

$$f_{cds} = M_{sd}(e)/I_g \quad \text{(Eq. 4.7.3.5)}$$

where:
- M_{sd} = moment due to all superimposed permanent dead and sustained loads applied after prestressing

$$SH = (8.2 \times 10^{-6})K_{sh}E_{ps}(1 - 0.06V/S)(100 - R.H.)$$
$$\text{(Eq. 4.7.3.6)}$$

where:
- K_{sh} = 1.0 for pretensioned members
- V/S = volume-to-surface ratio
- R.H. = average ambient relative humidity (see Figure 3.12.2)

$$RE = [K_{re} - J(SH + CR + ES)]C \quad \text{(Eq. 4.7.3.7)}$$

where:

Values for K_{re} and J are taken from Table 4.7.3.1.

For values of coefficient, C, see Table 4.7.3.2 or calculate using Eqs. 4.7.3.8. through 4.7.3.12.

For stress-relieved strand:

$0.75 \geq \dfrac{f_{pi}}{f_{pu}} \geq 0.70$:

$$C = 1 + 9\left(\dfrac{f_{pi}}{f_{pu}} - 0.7\right) \quad \text{(Eq. 4.7.3.8)}$$

$0.70 > \dfrac{f_{pi}}{f_{pu}} \geq 0.51$:

$$C = \dfrac{\left(\dfrac{f_{pi}}{f_{pu}}\right)\left(\dfrac{f_{pi}}{f_{pu}}\right)}{0.19 \quad 0.85} - 0.55 \quad \text{(Eq. 4.7.3.9)}$$

$\left(\dfrac{f_{pi}}{f_{pu}}\right) < 0.51$:

$$C = \dfrac{\left(\dfrac{f_{pi}}{f_{pu}}\right)}{3.83} \quad \text{(Eq. 4.7.3.10)}$$

For low-relaxation strand:

$\left(\dfrac{f_{pi}}{f_{pu}}\right) \geq 0.54$:

$$C = \dfrac{\left(\dfrac{f_{pi}}{f_{pu}}\right)\left(\dfrac{f_{pi}}{f_{pu}}\right)}{0.21 \quad 0.9} - 0.55 \quad \text{(Eq. 4.7.3.11)}$$

$\left(\dfrac{f_{pi}}{f_{pu}}\right) \leq 0.54$:

$$C = \dfrac{\left(\dfrac{f_{pi}}{f_{pu}}\right)}{4.25} \quad \text{(Eq. 4.7.3.12)}$$

where:
$$f_{pi} = \dfrac{P_i}{A_{ps}}$$

f_{pu} = ultimate strength of prestressing steel

4.7.4 Critical Locations

Computations for stress losses due to elastic shortening and creep of concrete are based on the compressive stress in the concrete at the center of gravity (cgs) of the prestressing force.

For bonded tendons, stress losses are computed at that point on the span where flexural tensile stresses are most critical. In members with straight, parabolic or approximately parabolic tendons this is usually midspan. In members with

Table 4.7.3.1 Values of K_{re} and J

Type of tendon	K_{re}	J
270 Grade stress-relieved strand or wire	20,000	0.15
250 Grade stress-relieved strand or wire	18,500	0.14
240 or 235 Grade stress-relieved wire	17,600	0.13
270 Grade low-relaxation strand	5,000	0.040
250 Grade low-relaxation wire	4,630	0.037
240 or 235 Grade low-relaxation wire	4,400	0.035
145 or 160 Grade stress-relieved bar	6,000	0.05

Table 4.7.3.2 Values of C

f_{pi}/f_{pu}	Stress-relieved strand or wire	Stress-relieved bar or low-relaxation strand or wire
0.80		1.28
0.79		1.22
0.78		1.16
0.77		1.11
0.76		1.05
0.75	1.45	1.00
0.74	1.36	0.95
0.73	1.27	0.90
0.72	1.18	0.85
0.71	1.09	0.80
0.70	1.00	0.75
0.69	0.94	0.70
0.68	0.89	0.66
0.67	0.83	0.61
0.66	0.78	0.57
0.65	0.73	0.53
0.64	0.68	0.49
0.63	0.63	0.45
0.62	0.58	0.41
0.61	0.53	0.37
0.60	0.49	0.33

tendons deflected at midspan only, the critical point is generally near the 0.4 point of the span. Since the tendons are bonded, only the stresses at the critical point need to be considered. (Unless critical sections occur within the development length – see Section 4.2.3.) Stresses or stress changes at other points along the member do not affect the stresses or stress losses at the critical point.

4.8 Camber and Deflection

Most precast, prestressed concrete flexural members will have a net positive (upward) camber at the time of transfer of prestress, caused by the eccentricity of the prestressing force. This camber may increase or decrease with time, depending on the stress distribution across the member under sustained loads. Camber tolerances are suggested in Chapter 8 of this Handbook.

Limitations on instantaneous deflections and time-dependent cambers and deflections are specified in the ACI Code. Table 9.5(B) of the Code is reprinted for reference (see Table 4.8.1).

The following sections contain suggested methods for computing cambers and deflections. There are many inherent variables that affect camber and deflection, such as concrete mix, storage method, time of release of prestress, time of erection and

---

Example 4.7.1
Loss of Prestress

Given:
10LDT 32 + 2 as shown:
  Span = 70 ft
No superimposed dead load except topping
  R.H. = 75%

Section properties (untopped):
  A   = 615 in.²
  I   = 59,720 in.⁴
  S_b = 2717 in.³
  V/S = 615/364 = 1.69 in.
  Wt = 491 plf
  Wt of topping = 250 plf

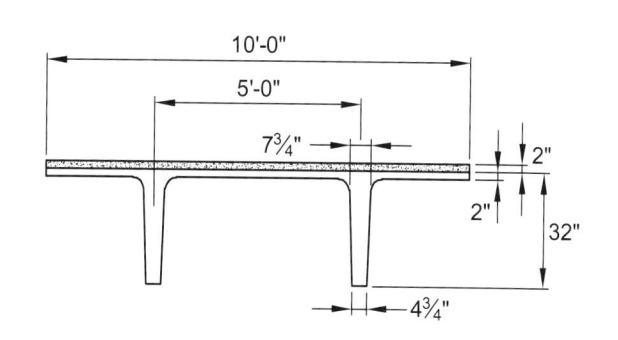

Concrete:
Precast: Sand-lightweight
  f'_c = 5000 psi
  E_c = 3.0 × 10⁶ psi
  f'_{ci} = 3500 psi
  E_{ci} = 2.5 × 10⁶ psi
Topping: Normal weight concrete

Example 4.7.1 (Cont.)
Loss of Prestress

Prestressing steel:
  12 – ½ in. diameter 270 ksi low-relaxation strands
  $A_{ps} = 12(0.153) = 1.836$ in.^{2}
  $E_{ps} = 28.5 \times 10^6$ psi

Depressed at midspan:
  $e_e = 12.81$ in.
  $e_c = 18.73$ in.

Problem:
  Determine total loss of prestress.

Solution:
For depressed strand, critical section is at 0.4ℓ. Determine moments, eccentricity, and prestress force.

$$M \text{ at } 0.4\ell = \frac{wx}{2}(\ell - x) = \frac{w(0.4\ell)}{2}(\ell - 0.4\ell) = 0.12\,w\ell^2$$

  $M_g = 0.12(0.491)(70)^2 = 289$ kip-ft
  $M_{sd} = 0.12(0.250)(70)^2 = 147$ kip-ft
  e at $0.4\ell = 12.81 + 0.8(18.73 - 12.81) = 17.55$ in.

Assume compensation for anchorage seating loss during prestressing.
  $P_i = 0.75\,A_{ps}f_{pu} = 0.75(1.836)(270) = 371.8$ kips

Determine f_{cir} and f_{cds}:

$$f_{cir} = K_{cir}\left(\frac{P_i}{A_g} + \frac{P_i e^2}{I_g}\right) - \frac{M_g e}{I_g} = 0.9\left(\frac{371.8}{615} + \frac{371.8(17.55)^2}{59720}\right) - \frac{(289)(12)(17.55)}{59720} = 1.251 \text{ ksi} = 1251 \text{ psi}$$

$$f_{cds} = \frac{M_{sd}(e)}{I_g} = \frac{147(12)(17.55)}{59720} = 0.518 \text{ ksi} = 518 \text{ psi}$$

$$ES = \frac{K_{es}E_{ps}f_{cir}}{E_{ci}} = \frac{(1)(28.5\times 10^6)(1251)}{2.5\times 10^6} = 14{,}261 \text{ psi}$$

$$CR = K_{cr}\frac{E_{ps}}{E_c}(f_{cir} - f_{cds}) = (1.6)\frac{28.5\times 10^6}{3.0\times 10^6}(1251 - 518) = 11{,}142 \text{ psi}$$

$$SH = (8.2\times 10^{-6})K_{sh}E_{ps}\left(1 - 0.06\left(\frac{V}{S}\right)\right)(100 - \text{R.H.})$$
$$= (8.2\times 10^{-6})(1)(28.5\times 10^6) \times [1 - 0.06(1.69)](100 - 75) = 5250 \text{ psi}$$

  $RE = [K_{re} - J(SH + CR + ES)]C$

From Table 4.7.3.1:
  $K_{re} = 5000$
  $J = 0.04$
  $f_{pi}/f_{pu} = 0.75$

From Table 4.7.3.2:
  $C = 1.0$
  $RE = [5000 - 0.04(5250 + 11{,}142 + 14{,}261)](1) = 3774$ psi
  $T.L. = ES + CR + SH + RE = 14{,}261 + 11{,}142 + 5250 + 3774 = 34{,}427$ psi $= 34.4$ ksi
Final prestress force $= 371.8 - 34.5(1.836) = 308$ kips

placement of superimposed loads, relative humidity, etc. Because of this, calculated long-term values should never be considered any better than estimates. Non-structural components attached to members which could be affected by camber variations, such as partitions or folding doors, should be placed with adequate allowance for variation. Calculation of topping quantities should also recognize the imprecision of camber calculations.

It should also be recognized that camber of precast, prestressed members is a result of the placement of the strands needed to resist the design moments and service load stresses. It is not practical to alter the forms of the members to produce a desired camber. Therefore, cambers should not be specified, but their inherent existence should be recognized.

4.8.1 Initial Camber

Initial camber can be calculated using conventional moment-area equations. Figure 4.12.10 has equations for the camber caused by prestress force for the most common strand patterns used in precast, prestressed members. Design Aids 11.1.3 and 11.1.4 provide deflection equations for typical loading conditions and more general camber equations.

Table 4.8.1 Maximum permissible computed deflections

Type of member	Deflection to be considered	Deflection limitation
Flat roofs not supporting or attached to non-structural elements likely to be damaged by large deflections	Immediate deflection due to live load	$\ell/180$[a]
Floors not supporting or attached to non-structural elements likely to be damaged by large deflections	Immediate deflection due to live load	$\ell/360$
Roof or floor construction supporting or attached to non-structural elements likely to be damaged by large deflections	That part of the total deflection occurring after attachment of non-structural elements (sum of the long-term deflection due to all sustained loads and the immediate deflection due to any additional live load)[c]	$\ell/480$[b]
Roof or floor construction supporting or attached to non-structural elements not likely to be damaged by large deflections		$\ell/240$[d]

a. Limit not intended to safeguard against ponding. Ponding should be checked by suitable calculations of deflections, including added deflections due to ponded water, and considering long-term effects of all sustained loads, camber, construction tolerances, and reliability of provisions for drainage.
b. Limit may be exceeded if adequate measures are taken to prevent damage to supported or attached elements.
c. Long-term deflection shall be determined in accordance with Sections 9.5.2.5 or 9.5.4.3, ACI 318-02, but may be reduced by amount of deflection calculated to occur before attachment of nonstructural elements. This amount shall be determined on the basis of accepted engineering data relating to time-deflection characteristics of members similar to those being considered.
d. But not greater than tolerance provided for non-structural elements. Limit may be exceeded if camber is provided so that total deflection minus camber does not exceed limit.

Example 4.8.1.1
Calculation of Initial Camber

Given:
    8DT24 of Example 4.2.2.3.

Section Properties:
    $A = 401$ in.^{2} $S_b = 1224$ in.^{3}
    $I = 20{,}985$ in.^{4} $S_t = 3063$ in.^{3}
    $y_b = 17.15$ in.   wt' = 418 plf
    $y_t = 6.85$ in.   DL = 52 psf

Example 4.8.1.1 (Cont.)
Calculation of Initial Camber

Concrete:
f'_c = 5000 psi   Normal weight concrete (150 pcf)

$E_c = 33w^{1.5}\sqrt{f'_c} = 33(150)^{1.5}\dfrac{\sqrt{5000}}{1000} = 4287$ ksi

f'_{ci} = 3500 psi

$E_{ci} = 33(150)^{1.5}\dfrac{\sqrt{3500}}{1000} = 3587$ ksi

(Note: The values of E_c and E_{ci} could also be read from Design Aid 11.2.2.)

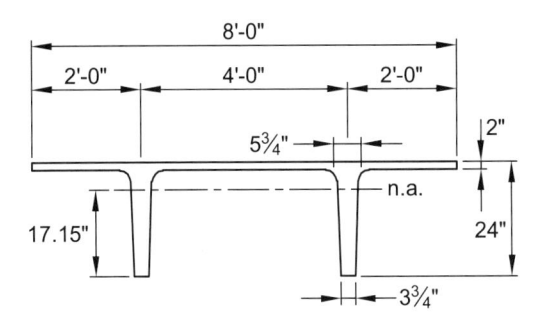

Problem:
Find the initial camber at time of transfer of prestress.

Solution:
The prestress force at transfer and strand eccentricities are calculated in Example 4.2.2.3 and are shown on the drawing.

Calculate the upward component using equations given in Figure 4.12.10.

$$\Delta \uparrow = \frac{P_o e_e \ell^2}{8E_{ci}I} + \frac{P_o e' \ell^2}{12E_{ci}I} = \frac{335(5.48)\left[(70)(12)\right]^2}{8(3587)(20,985)} + \frac{335(8.42)\left[(70)(12)\right]^2}{12(3587)(20,985)} = 2.15 + 2.20 = 4.35 \text{ in.} \uparrow$$

Deduct deflection caused by weight of member:

$$\Delta \downarrow = \frac{5w\ell^4}{384E_{ci}I} = \frac{5\left(\dfrac{0.418}{12}\right)\left[(70)(12)\right]^4}{384(3587)(20,985)} = 3.00 \text{ in.} \downarrow$$

Net camber at release = 4.35 − 3.00 = 1.35 in. ↑

4.8.2 Elastic Deflections

Calculation of instantaneous deflections of both prestressed and non-prestressed members caused by superimposed service loads follows classical methods of mechanics. Design equations for various load conditions are given in Chapter 11 of this Handbook. If the bottom tension in a simple span member does not exceed the modulus of rupture, the deflection is calculated using the uncracked moment of inertia of the section. The modulus of rupture of concrete is defined in Chapter 9 of the Code as:

$f_r = 7.5\lambda\sqrt{f'_c}$ (Eq. 4.8.2.1)

where λ is a coefficient related to the unit weight of concrete; see Section 4.3.3 for definition.

4.8.3 Bilinear Behavior

Section 9.5.4.2 of the Code requires that for prestressed concrete flexural members which fall into Class T or Class C (see Section 4.2.2.2), deflection calculations be based on transformed cracked section analysis, and permits the use of "bilinear moment-deflection relationships." This means that the deflection before the member has cracked is calculated using the gross (uncracked) moment of inertia, I_g, and the additional deflection after cracking is calculated using the moment of inertia of the cracked section. Examples 4.8.3.1 and 4.8.3.2 illustrate the direct application of this principle. As an alternative, the Code allows the determination of an "effective moment of inertia," I_e, and the deflection then calculated by substituting I_e for I_g in the deflection calculation, as illustrated in Examples 4.8.3.2 and 4.8.3.4. Figure 4.8.3.1 Illustrates the differences between the two methods.

The equation for effective moment of inertia is:

$$I_e = \left(\frac{M_{cr}}{M_a}\right)^3 I_g + \left[1 - \left(\frac{M_{cr}}{M_a}\right)^3\right] I_{cr} \quad \text{(Eq. 4.8.3.1)}$$

$$\frac{M_{cr}}{M_a} = 1 - \left(\frac{f_{t\ell} - f_r}{f_\ell}\right) \quad \text{(Eq. 4.8.3.2)}$$

where:
$f_{t\ell}$ = the final calculated total stress in the member
f_ℓ = calculated stress due to live load

4.8.3.1 Cracked Section Analysis

For many applications, particularly Class T members, the conservative empirical relationship:

$$I_{cr} = nA_{ps}d_p^2\left(1 - 1.6\sqrt{n\rho_p}\right) \quad \text{(Eq. 4.8.3.3)}$$

may be used to determine the cracked moment of inertia. Figure 4.12.11 gives coefficients for use in solving this equation. Iterative procedures used for non-prestressed members which neglect the effects of prestressing are also conservative, but may be adequate for Class T members.

For Class C members, however, and for other applications where a more exact analysis may be desired, a method described by Mast, [20] summarized in Section 4.2.2.2 may be used.

Figure 4.8.3.1 Bilinear moment-deflection: effective moment of inertia

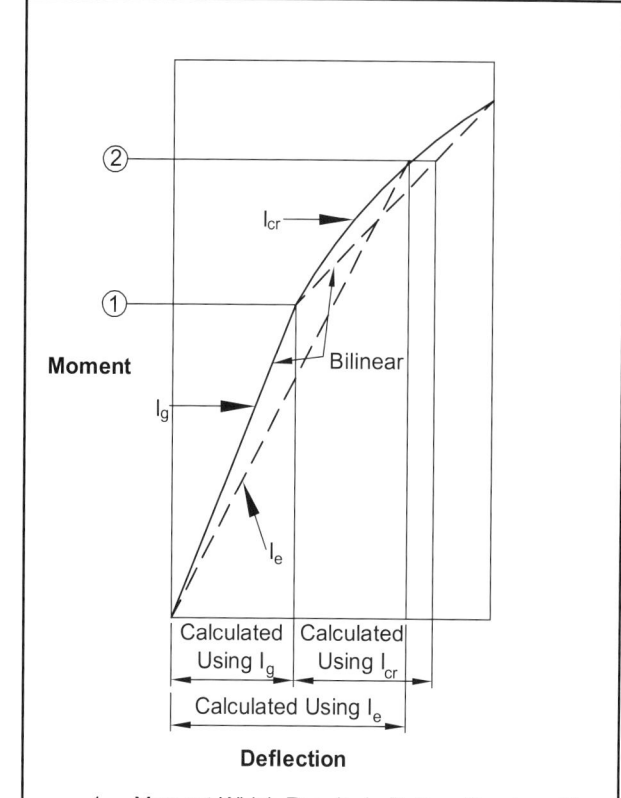

1. Moment Which Results in Bottom Tension of f_r
2. Moment Corresponding to Final Bottom Tension
(Note: Prestress Effects Not Included in Illustration)

Example 4.8.3.1
Deflection Calculation Using Bilinear Moment-Deflection Relationship – Example 1

Given:
8DT24 from Examples 4.2.2.3 and 4.8.1.1.

Problem:
Determine the total deflection caused by the specified uniform live load.

Solution:
Determine $f_r = 7.5\sqrt{f'_c} = 530$ psi

From Example 4.2.2.3, the final tensile stress is 766 psi, which is more than 530 psi, but less than $12\sqrt{f'_c}$.

Therefore, it is a Class T member.

Example 4.8.3.1 (Cont.)
Deflection Calculation Using Bilinear Moment-Deflection Relationship – Example 1

Determine I_{cr} from Figure 4.12.11:

A_{ps} = 1.836 in.^{2} (see Example 4.2.2.3)
d_p at midspan = $e_c + y_t$ = 13.90 + 6.85 = 20.75 in.

(Note: It is within the precision of the calculation method and observed behavior to use midspan d_p and to calculate the deflection at midspan, although the maximum tensile stress in this case is assumed at 0.4ℓ.)

$$\rho_p = \frac{A_{ps}}{bd_p} = \frac{1.836}{(96)(20.75)} = 0.00092$$

C = 0.0056

$I_{cr} = Cbd_p^3 = 0.0056(96)(20.75)^3 = 4803$ in.^{4}

Determine the portion of the live load that would result in a bottom tension of 530 psi.

766 − 530 = 236 psi

The tension caused by live load alone is 1614 psi; therefore, the portion of the live load that would result in a bottom tension of 530 psi is:

$$\frac{1614 - 236}{1614}(0.280) = 0.239 \text{ kips/ft}$$

0.280 − 0.239 = 0.041 kips/ft

$$\Delta_g = \frac{5w\ell^4}{384E_cI_g} = \frac{5\left(\frac{0.239}{12}\right)[70(12)]^4}{384(4287)(20,985)} = 1.44 \text{ in.}$$

$$\Delta_{cr} = \frac{5\left(\frac{0.041}{12}\right)[70(12)]^4}{384(4287)(4803)} = 1.08 \text{ in.}$$

Total live load deflection = Δ_ℓ = 1.44 + 1.08 = 2.52 in.

Example 4.8.3.2
Deflection Calculation Using Bilinear Moment-Deflection Relationship – Example 2

Given:
    Rectangular beam from Example 4.2.2.5.

Problem:
    Determine deflection caused by the specified live load.

Example 4.8.3.2 (Cont.)
Deflection Calculation Using Bilinear Moment-Deflection Relationship – Example 2

Solution:
From Example 4.2.2.5:

$$\text{Tension caused by live load} = \frac{M}{S} = \frac{3000(1000)}{2048} = 1465 \text{ psi}$$

Portion of live load that will result in cracking (581 psi)
$$879 - 581 = 298 \text{ psi}$$
$$\frac{1465 - 298}{1465}(1.25) = 0.996 \text{ kips/ft}$$

$$\Delta_g = \frac{5wl^4}{384 E_c I_g} = \frac{5(0.996/12)\left[(40)(12)\right]^4}{384(4696)(32768)} = 0.37 \text{ in.}$$

$$\Delta_{cr} = \frac{5wl^4}{384 E_c I_{cr}} = \frac{5\left[(1.25 - 0.996)/12\right]\left[(40)(12)\right]^4}{384(4696)(11{,}708)} = 0.27 \text{ in.}$$

Total live load deflection = 0.37 + 0.27 = 0.64 in.
From Table 4.8.1, $\ell/480 = 40(12)/480 = 1$ in. Section is satisfactory for all applications.

Example 4.8.3.3
Deflection Calculation Using Effective Moment of Inertia – Example 1

Given:
Same section and loading conditions of Examples 4.2.2.3, 4.8.1.1 and 4.8.3.1.

Problem:
Determine the deflection caused by live load using the I_e method.

Solution:
From the table of stresses in Example 4.2.2.3:
- $f_{t\ell}$ = 766 psi (tension)
- f_ℓ = 1614 psi (tension)
- f_r = $7.5\sqrt{f'_c}$ = 530 psi (tension)

$$\frac{M_{cr}}{M_a} = 1 - \left(\frac{766 - 530}{1614}\right) = 0.854$$

$$\left(\frac{M_{cr}}{M_a}\right)^3 = (0.854)^3 = 0.623$$

$$1 - \left(\frac{M_{cr}}{M_a}\right)^3 = 1 - 0.623 = 0.377$$

$$I_e = 0.623(20{,}985) + 0.377(4803) = 14{,}884 \text{ in.}^4$$

$$\Delta_\ell = \frac{5wl^4}{384 E_c I_e} = \frac{5\left(\dfrac{0.280}{12}\right)\left[70(12)\right]^4}{384(4287)(14884)} = 2.37 \text{ in.}$$

Example 4.8.3.4
Deflection Calculation Using Effective Moment of Inertia – Example 2

Given:
The composite section of Example 4.2.2.6:
- Span = 24 ft
- E_c = 4287 ksi
- Live Load = 4.0 kips/ft
- I_g = 31,984 in.^{4}
- f_ℓ = −1298 psi tension
- I_{cr} = 9770 in.^{4}
- $f_{t\ell}$ = −1656 psi tension
- $f_r = -7.5\sqrt{5000} = -530$ psi

Problem:
Determine live load deflection and compare with Table 4.8.1.

Solution:
From Eq. 4.8.3.2:
$$\frac{M_{cr}}{M_a} = 1 - \left(\frac{f_{t\ell} - f_r}{f_\ell}\right) = 1 - \left(\frac{1656 - 530}{1298}\right) = 0.133$$

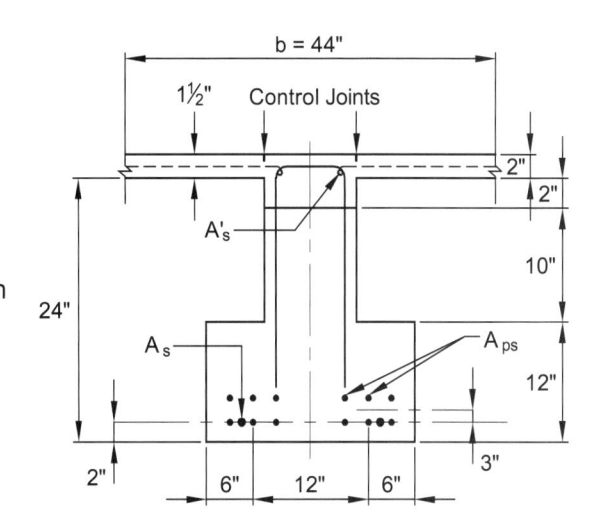

From Eq. 4.8.3.1:
$$I_e = \left(\frac{M_{cr}}{M_a}\right)^3 I_g + \left[1 - \left(\frac{M_{cr}}{M_a}\right)^3\right] I_{cr} = (0.133)^3(31912) + \left[1 - (0.133)^3\right](9770) = 9822 \text{ in.}^4$$

Check conformance with Table 4.8.1 assuming beam is not supporting or attached to non-structural elements likely to be damaged by large deflections:

Permissible live load deflection: $= \dfrac{\ell}{360} = \dfrac{24(12)}{360} = 0.80$ in.

Live load deflection $= \dfrac{5w\ell^4}{384 E_c I_e} = \dfrac{5(4.0)(24)^4(1728)}{384(4287)(9822)} = 0.71$ in. < 0.80 in. OK

4.8.4 Long-Term Camber/Deflection

ACI 318-02 provides a multiplier, λ, applied to initial deflection for estimating the long-term deflection of non-prestressed reinforced concrete members (Code Section 9.5.2.5):

$$\lambda = \frac{\xi}{1 + 50\rho'} \quad \text{(Eq. 4.8.4.1)}$$

where ξ is a factor related to length of time, and ρ' is the ratio of compressive reinforcement. (There is no corresponding ratio for prestressed concrete in ACI 318-02.)

The determination of long-term cambers and deflections in precast, prestressed members is somewhat more complex because of (1) the effect of prestress and the loss of prestress over time, (2) the strength gain of concrete after release of prestress, and because (3) the camber or deflection is important not only at the "initial" and "final" stages, but also at erection, which occurs at some intermediate stage, usually from 30 to 60 days after casting.

It has been customary in the design of precast, prestressed concrete members to estimate the camber of a member after a period of time by multiplying the initial calculated camber by some factor, usually based on the experience of the designer. To properly use these "multipliers," the upward and downward components of the initial calculated camber should be separated in order to take into account the effects of loss of prestress, which only affects the upward component.

Table 4.8.4.1 provides suggested multipliers which can be used as a guide in estimating long-term cambers and deflections for typical members, i.e., those members which are within the span-to-depth ratios recommended in this Handbook (see Section 2.2.2). Derivation of these multipliers is contained in Ref. 21.

For members which are made continuous for superimposed dead load and live load, the final multipliers, (3) through (6) Table 4.8.4.1, may be reduced.

Long-term effects can be substantially reduced by adding non-prestressed reinforcement in prestressed concrete members. The reduction effects proposed in Ref. 22 can be applied to the approximate multipliers found in Table 4.8.4.1 as follows:

$$C_2 = \frac{C_1 + \frac{A_s}{A_{ps}}}{1 + \frac{A_s}{A_{ps}}} \quad \text{(Eq. 4.8.4.2)}$$

where:
C_1 = multiplier from Table 4.8.4.1
C_2 = revised multiplier
A_s = area of non-prestressed reinforcement
A_{ps} = area of prestressed steel

Table 4.8.4.1 Suggested simple span multipliers to be used as a guide in estimating long-term cambers and deflections for typical prestressed members

	Without Composite Topping	With Composite Topping
At erection:		
(1) Deflection (downward) component—apply to the elastic deflection due to the member weight at release of prestress.	1.85	1.85
(2) Camber (upward) component—apply to the elastic camber due to prestress at the time of release of prestress.	1.80	1.80
Final:		
(3) Deflection (downward) component—apply to the elastic deflection due to the member weight at release of prestress.	2.70	2.40
(4) Camber (upward) component—apply to the elastic camber due to prestress at the time of release of prestress.	2.45	2.20
(5) Deflection (downward)—apply to elastic deflection due to superimposed dead load only.	3.00	3.00
(6) Deflection (downward)—apply to elastic deflection caused by the composite topping.	—	2.30

Example 4.8.4.1
Use of Multipliers for Determining Long-Term Cambers and Deflections

Given:
8DT24 for Examples 4.2.2.3, 4.8.1.1, and 4.8.3.1. Non-structural elements are attached, but not likely to be damaged by deflections (light fixtures, etc.).

Problem:
Estimate the camber and deflection and determine if they meet the requirements of Table 9.5(b) of the Code (see Table 4.8.1).

Solution:
Calculate the instantaneous deflections caused by the superimposed dead and live loads:

$$\Delta_{sd} = \frac{5w\ell^4}{(384)E_c I} = \frac{5\left(\frac{0.080}{12}\right)[70(12)]^4}{384(4287)(20985)} = 0.48 \text{ in.} \downarrow$$

Example 4.8.4.1 (Cont.)
Use of Multipliers for Determining Long-Term Cambers and Deflections

$\Delta_\ell = 2.52$ in.↓ (Example 4.8.3.1)

For convenience, a tabular format is used.

The estimated critical cambers would then be:
At erection of the member:
After w_{sd} is applied = 1.80 in.
"Final" long-term camber = 1.12 in.
The deflection limitation of Table 4.8.1 for the above condition is $\ell/240$.

$$\frac{70(12)}{240} = 3.50 \text{ in.}$$

Total deflection occurring after attachment of non-structural elements:
$\Delta_t = (1.80 - 1.12) + 2.52 = 3.20$ in. < 3.50 OK

	(1) Release	Multiplier	(2) Erection	Multiplier	(3) Final
Prestress	4.35↑	1.80 × (1)	7.83↑	2.45 × (1)	10.66↑
w_d	3.00↓	1.85 × (1)	5.55↓	2.7 × (1)	8.10↓
	1.35↑		**2.28↑**		**2.56↑**
w_{sd}			0.48↓	3.0 × (2)	1.44↓
			1.80↑	final camber =	**1.12↑**
w_ℓ					2.52↓
					1.40↓

4.8.5 Bowing

A strain gradient between the inside and outside of a wall panel, or between the top and underside of an uninsulated roof member, can cause the member to bow. The theoretical magnitude of bowing (see Figure 4.8.5.1) can be determined by:

$$\Delta = \alpha \frac{\ell^2}{8h} \quad \text{(Eq. 4.8.5.1)}$$

where:
α = strain gradient across panel thickness
ℓ = distance between supports
h = member thickness

For a temperature difference between inside and outside of a panel:

$$\alpha = C(T_1 - T_2) \quad \text{(Eq. 4.8.5.2)}$$

where:
C = coefficient of thermal expansion
T_1, T_2 = inside and outside temperatures, respectively

Limited records of temperature measurements indicate that in open structures, such as the roofs of

Figure 4.8.5.1 Thermal bow of wall panel

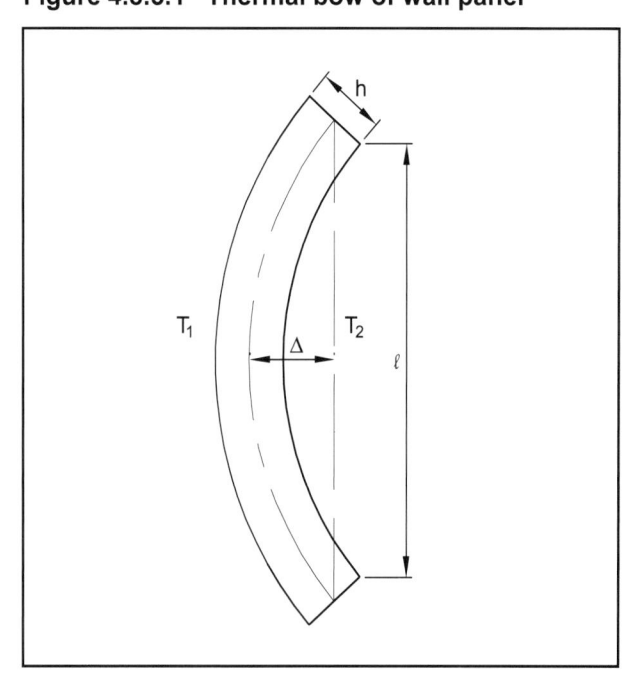

parking decks, the temperature differential seldom exceeds 30 to 40°F. In an insulated sandwich wall panel, the theoretical difference can be higher, but this is tempered by "thermal lag" due to the mass of the concrete (see Section 9.1).

Moisture differences between the inside and outside of an enclosed building can also cause bowing; however, the calculation is much less precise and involves more variables. The exterior layer of the concrete panel absorbs moisture from the atmosphere and periodic precipitation, while the interior layer is relatively dry, especially when the building is heated. This causes the inside layer to shrink more than the outside, causing an outward bow. The outward shrinkage bowing would tend to balance the theoretical inward thermal bowing in cold weather, which is believed to explain the observation that "wall panels always bow out."

While the magnitude of bowing is usually not very significant, in the case of wall panels it may cause unacceptable separation at the corners (see Figure 4.8.5.2), and possible damage to joint sealants. It may, therefore, be desirable to restrain bowing with one or more connectors between panels. As illustrated in the previous example, Figure 4.12.12 gives equations for calculating the required restraint and the moments this restraint would cause in the panel.

The mid-height panel restraining force illustrated in Example 4.8.5.1 can occur in a multi-story loadbearing wall panel. For example, if the wall panel is two stories tall and if the intermediate floor is laterally braced against sidesway, a restraining force resisting thermal bowing can develop in a connection between the wall panel and the intermediate floor. The wall panel should be connected to the intermediate floor, to avoid loss of floor support when the wall panel bows outward. If the intermediate floor is not braced against sidesway, the P-Δ wall analysis should consider the secondary moment due to the intermediate floor

Figure 4.8.5.2 Corner separation due to thermal bow

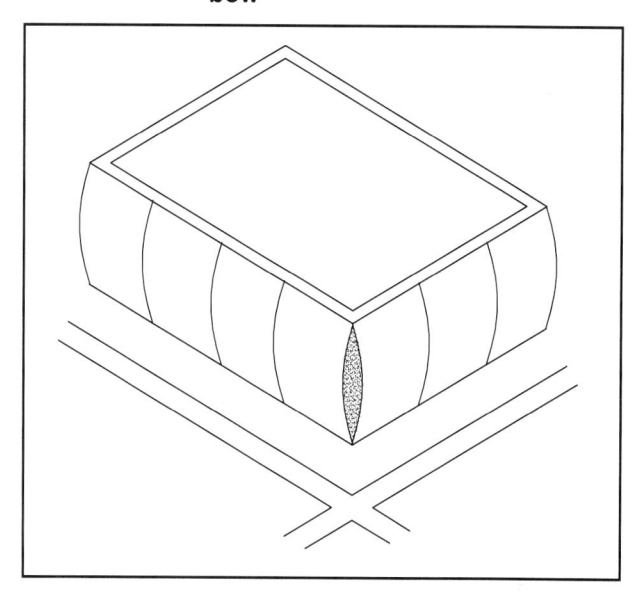

Example 4.8.5.1
Thermal Bow in a Wall Panel

Given:
A 20 ft high, 6 in. thick wall panel as shown on next page. Assume a coefficient of thermal expansion
$C = 6 \times 10^{-6}$ in./in./°F

Temperature differential
$T_1 - T_2 = 35°F$
$E_c = 4300$ ksi

Problem:
Determine the potential thermal bow, Δ_1, the force, P, required at mid-height to restrain the bowing, the stress in the panel caused by the restraint, and the residual bow, Δ_2.

Solution:
From Eqs. 4.8.5.1 and 4.8.5.2:

$$\Delta_1 = \frac{(6 \times 10^{-6})(35)[20(12)]^2}{8(6)} = 0.252 \text{ in.}$$

Example 4.8.5.1 (Cont.)
Thermal Bow in a Wall Panel

From Figure 4.12.12 for daily temperature change:
$E_t = 0.75(4300) = 3225$ ksi

$$I = \frac{(bh^3)}{12} = \frac{[12(6)^3]}{12} = 216 \text{ in.}^4/\text{ft}$$

From Figure 4.12.12, Case 1:

$$P = \frac{48E_t I \Delta}{\ell^3} = \frac{[48(3225)(216)(0.25)]}{[20(12)]^3} = 0.605 \text{ kip/ft of width}$$

$$M = \frac{P\ell}{4} = \frac{[0.605(20)]}{4} = 3.02 \frac{\text{kip-ft}}{\text{ft}} \text{ of width}$$

Panel stress $= My/I$
$$= \frac{[3.02(12,000)(3)]}{216} = 503 \text{ psi}$$

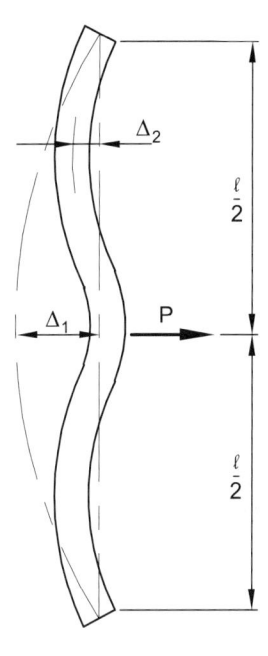

The residual bow can be calculated by adjusting the equation in Figure 4.12.12, Case 5, to read:

$$\Delta_2 = \frac{M\ell^2}{(16 \times E_t I)}; \text{ and substituting}$$

$\ell/2$ for ℓ:

$$\Delta_2 = \frac{3.02(12)[10(12)]^2}{16(3225)(216)} = 0.047 \text{ in.}$$

gravity reaction loading the wall panel near, or at the point of its maximum bow.

If non-structural elements, such as drywall ceilings, or interior drywall or masonry partitions, are attached to wall panels unrestrained against bowing, those items should be attached with "soft" or flexible joints. Differential temperature can cause upward bowing in roof members, especially in open structures such as parking decks. If these members are restrained from rotations at the ends, positive moments (bottom tension) can develop at the support, as shown in Cases 4 and 5, Figure 4.12.12. The bottom tension can cause severe cracking. Examination of the equations shows that the thermal induced positive moments are independent of the span length. (For example, substitute Eq. 4.8.5.1 for Δ in Figure 4.12.12, Case 4.) Note from Figure 4.12.12 that if only one end is restrained, as is sometimes done to relieve axial volume change force, the restraint moment is doubled. Also note that, since thermal bow occurs with daily temperature changes, the cyclical effects could magnify the potential damage.

4.9 Compression Members

Precast and prestressed concrete columns and loadbearing wall panels are usually proportioned on the basis of strength design. Stresses under service conditions, particularly during handling and erection (especially of wall panels) must also be considered. The procedures in this section are based on Chapter

Example 4.8.5.2
Thermal Force and Bow in a Roof Member

Given:

A 30IT24 inverted tee beam supporting the double tees of the upper level of a parking deck, as shown below, is welded at each end at the bearing to the column, and subject to a thermal gradient of 35°F.
Note: Welding at the bearings of precast members is <u>highly discouraged</u> – see note at the end of the example.

Coefficient of thermal expansion:
$C = 6 \times 10^{-6}$ in./in./°F
$T_1 - T_2 = 35°F$
$E_c = 4300$ ksi, for the composite section
$I = 51,725$ in.^{4}

Problem:
Find the tensile force developed at the support.

Solution:
From Eqs. 4.8.5.1 and 4.8.5.2:

$$\Delta = \frac{C(T_1 - T_2)\ell^2}{8h} = \frac{(6 \times 10^6)(35)[24(12)]^2}{8(27)} = 0.081 \text{ in.}$$

From Figure 4.12.12, Case 4:
$E_t = 0.75(4300) = 3225$ ksi

$$M = \frac{8E_t I \Delta}{\ell^2} = \frac{8(3225)(51,725)(0.081)}{[24(12)]^2} = 1303 \text{ kip-in.}$$

$$T = \frac{M}{d} = \frac{1303}{(24+1.5)} = 51.1 \text{ kips}$$

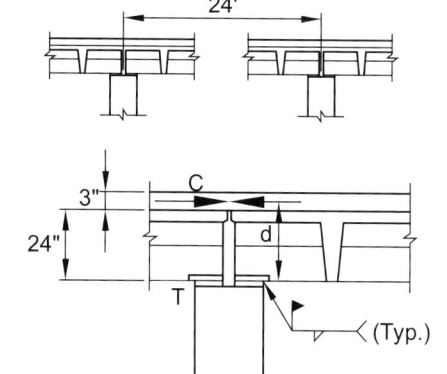

Note: This weld is highly discouraged - shown here for illustrative purposes only.

This example illustrates that large forces that can occur when roof members are welded at the bearings, and why cracking occurs. The force calculated is an upper bound value, and does not consider the relieving effects of connection extension, microcracking and column flexibility. Nevertheless, an alternative method which does not use welding at the bearings is strongly recommended.

---

10 of the Code and on the recommendations of the PCI Committee on Prestressed Concrete Columns (referred to in this section as "the Recommended Practice"). [23]

4.9.1 Strength Design of Precast Concrete Compression Members

The capacity of a reinforced concrete compression member with eccentric loads is most easily determined by constructing a capacity interaction curve. Points on this curve are calculated using the compatibility of strains and solving the equations of equilibrium as prescribed in Chapter 10 of the Code.

Solution of these equations is illustrated in Figure 4.9.1.1.

ACI 318-02 waives the minimum vertical reinforcement requirements for compression members if the concrete is prestressed to at least an average of 225 psi after all losses. In addition, the PCI Recommended Practice permits the elimination of lateral ties if:

- It is a compression-controlled section.
- Non-prestressed reinforcement is not considered in the calculation of P_n. Non-prestressed reinforcement which is added for tension (e.g., for handling) is not considered in the calculation of P_n.

- The nominal capacity is multiplied by 0.85.

Interaction curves for typical prestressed square columns and wall panels are provided in Chapter 2.

Construction of an interaction curve usually follows these steps:

Step 1:
Determine P_o for $M_n = 0$. See Figure 4.9.1.1(c).

Step 2:
Determine the point of maximum moment. For members with non-prestressed reinforcement, this is the balance point, which occurs when the net tensile strain in the extreme tension steel is equal to f_y/E_s. For symmetrical prestressed members, it is sufficiently precise to assume that the point occurs when the compression block, a, is one-half the member depth.

Step 3:
Determine M_o for $P_n = 0$. This is normally done by neglecting the reinforcement above the neutral axis and determining the moment capacity by one of the methods described in Section 4.2.1.

Step 4:
For each additional point on the interaction curve, proceed as follows:
a. Select a value of "c" and calculate $a = \beta_1 c$.
b. Determine the value of A_{comp} from the geometry of the section. See Figure 4.9.1.1(a).
c. Determine the strain in the reinforcement assuming that $\varepsilon_c = 0.003$ at the compression face of the column. For prestressed reinforcement, add the strain due to the effective prestress $\varepsilon_{se} = f_{se}/E_{ps}$.
d. Determine the stress in the reinforcement. For non-prestressed reinforcement, $f_s = \varepsilon_s E_s \leq f_y$. For prestressed reinforcement, the stress is determined from a stress-strain relationship (see Design Aid 11.2.5). If the maximum factored moment occurs near the end of a prestressed element, where the strand is not fully developed, an appropriate reduction in the value of f_{ps} should be made as described in Section 4.2.3.
e. Calculate P_n and M_n by statics.
f. Calculate ϕP_n and ϕM_n. The Code describes that for compression-controlled sections (without spiral reinforcement) with net tensile strain ε_t in the extreme tension steel less than or equal to that at the balance point, $\phi = 0.65$. For Grade 60 reinforcement and for prestressed steel, this occurs when $\varepsilon_t \leq 0.002$. For tension-controlled sections in which $\varepsilon_t \geq 0.005$, $\phi = 0.9$. For sections in which ε_t is between these limits, $\phi = 0.48 + 83\varepsilon_t$. For each point plotted on the nominal strength curve, multiply P_n and M_n by ϕ to obtain the design strength curve.

Step 5:
Calculate the maximum factored axial resistance specified by the Code as:
$0.80\phi P_o$ for tied columns
$0.85\phi P_o$ for spiral columns
(See also Section 4.9.1)

For cross sections which are not rectangular, it is necessary to determine separate curves for each direction of the applied moment. Further, since most architectural precast column cross sections are not rectangular, the "a" distance only defines the depth of the rectangular concrete stress distribution. Instead of using a/2, as for a rectangular cross section, it is necessary to calculate the actual centroid of the compression area which is indicated as y'.

As noted in Step 4d, the flexural resistance is reduced for prestressed elements at locations within a distance equal to the strand development length from each end. The flexural resistance of the prestressed reinforcement in this zone can be supplemented by non-prestressed reinforcement that is anchored to end plates, or otherwise developed.

The interaction curves in Chapter 2 are based on a maximum value of $f_{ps} = f_{se}$. This can be assumed to be at the point where the prestressing force is transformed to the concrete. Additional anchored end reinforcement should be supplied which is equivalent to a development length equal to the assumed transfer length. The required area of end reinforcement can be determined by matching interaction curves, or can be approximated by the following equation if the bar locations approximately match the strand locations:

$$A_s = \frac{A_{ps} f_{se}}{f_y} \quad \text{(Eq. 4.9.1.1)}$$

where:
A_s = required area of bars
A_{ps} = area of prestressing steel
f_{se} = strand stress after losses
f_y = yield strength of bars

The effects of adding end reinforcement to a 24 x 24 in. prestressed concrete column, thus improving moment capacity in the end 2 ft, are shown in Figure 4.9.1.2.

Figure 4.9.1.1 Equilibrium equations for prestressed and non-prestressed compression members

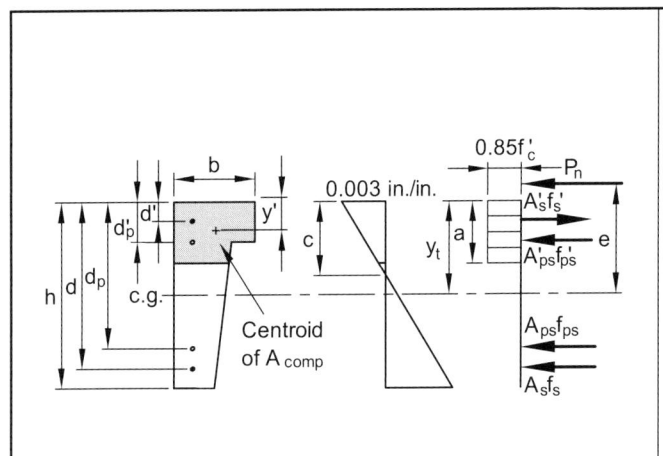

(a) Basic relationships

$\varepsilon'_s = (0.003/c)(c-d')$

$\varepsilon_s = (0.003/c)(d-c)$

$\varepsilon'_{ps} = f_{se}/E_{ps} - (0.003/c)(c-d'_p) \leq 0.035$

$\varepsilon_{ps} = f_{se}/E_{ps} + (0.003/c)(d_p-c) \leq 0.035$

$f'_s = \varepsilon'_s E_s \leq f_y$

$f_s = \varepsilon_s E_s \leq f_y$

f'_{ps} from stress-strain diagram $\leq f_{pu}$

f_{ps} from stress-strain diagram $\geq f_{pu}$

$P_n^* = (A_{comp} - A'_s - A'_{ps})(0.85f'_c)$
$\quad + A'_s f'_s - A_s f_s - A'_{ps} f'_{ps} - A_{ps} f_{ps}$

$M_n = P_n e = (A_{comp} - A'_s - A'_{ps})(y_t - y')(0.85f'_c)$
$\quad + A'_s f'_s (y_t - d') + A_s f_s (d - y_t)$
$\quad - A'_{ps} f'_{ps} (y_t - d'_p) + A_{ps} f_{ps} (d - y_t)$

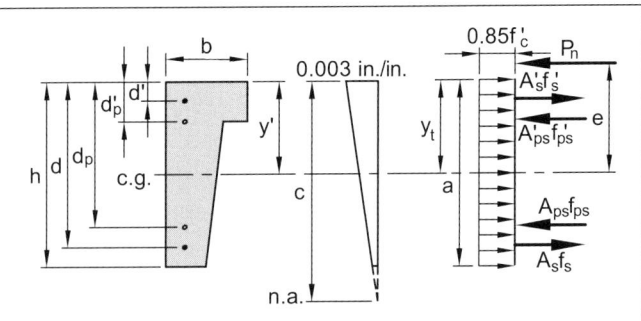

(b) Special case with neutral axis outside of the section

$A_{comp} = A$ if $a > h$

$\varepsilon'_s = (0.003/c)(c-d')$

$\varepsilon_s = (0.003/c)(c-d)$

$\varepsilon'_{ps} = f_{se}/E_{ps} - (0.003/c)(c-d'_p) \leq 0.035$

$\varepsilon_{ps} = f_{se}/E_{ps} + (0.003/c)(d_p-c) \leq 0.035$

Remaining equations same as above.

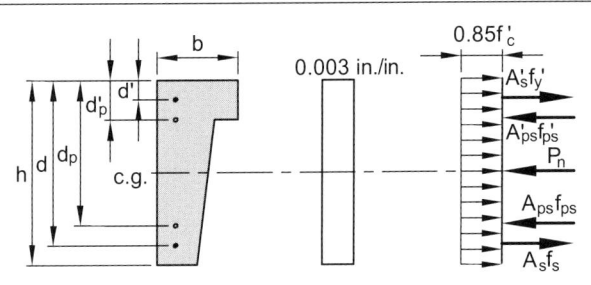

(c) Special case when: $M_n = 0$, $P_n = P_o$

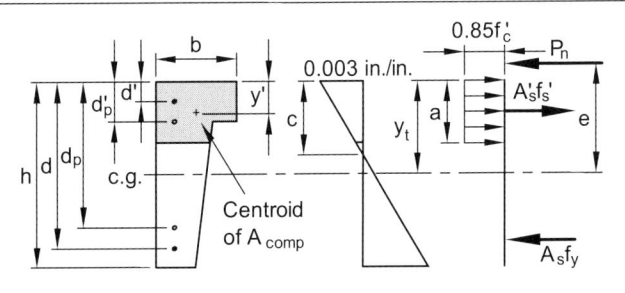

$P_n^* = 0.85f'_c(A - A'_s - A'_{ps} - A_{ps} - A_s) -$
$\quad (A'_{ps} + A_{ps})(f_{se} - 0.003E_{ps}) + (A'_s + A_s)f_y$

(d) Special case at compression controlled limit for members reinforced with prestressing strand and/or Grade 60 reinforcement

d_t = the greater of d and d_p

$c = \dfrac{0.003 d_t}{0.003 + 0.002} = 0.6 d_t$

Calculate P_n and M_n per Case (a) above

*Ties may be omitted if:
- It is a compression-controlled section.
- Non-prestressed reinforcement is not considered in the calculation of P_n. Non-prestressed reinforcement which is added for tension (e.g., for handling) is not considered in the calculation of P_n.
- The nominal capacity is multiplied by 0.85.

Figure 4.9.1.2 End reinforcement in a precast, prestressed concrete column

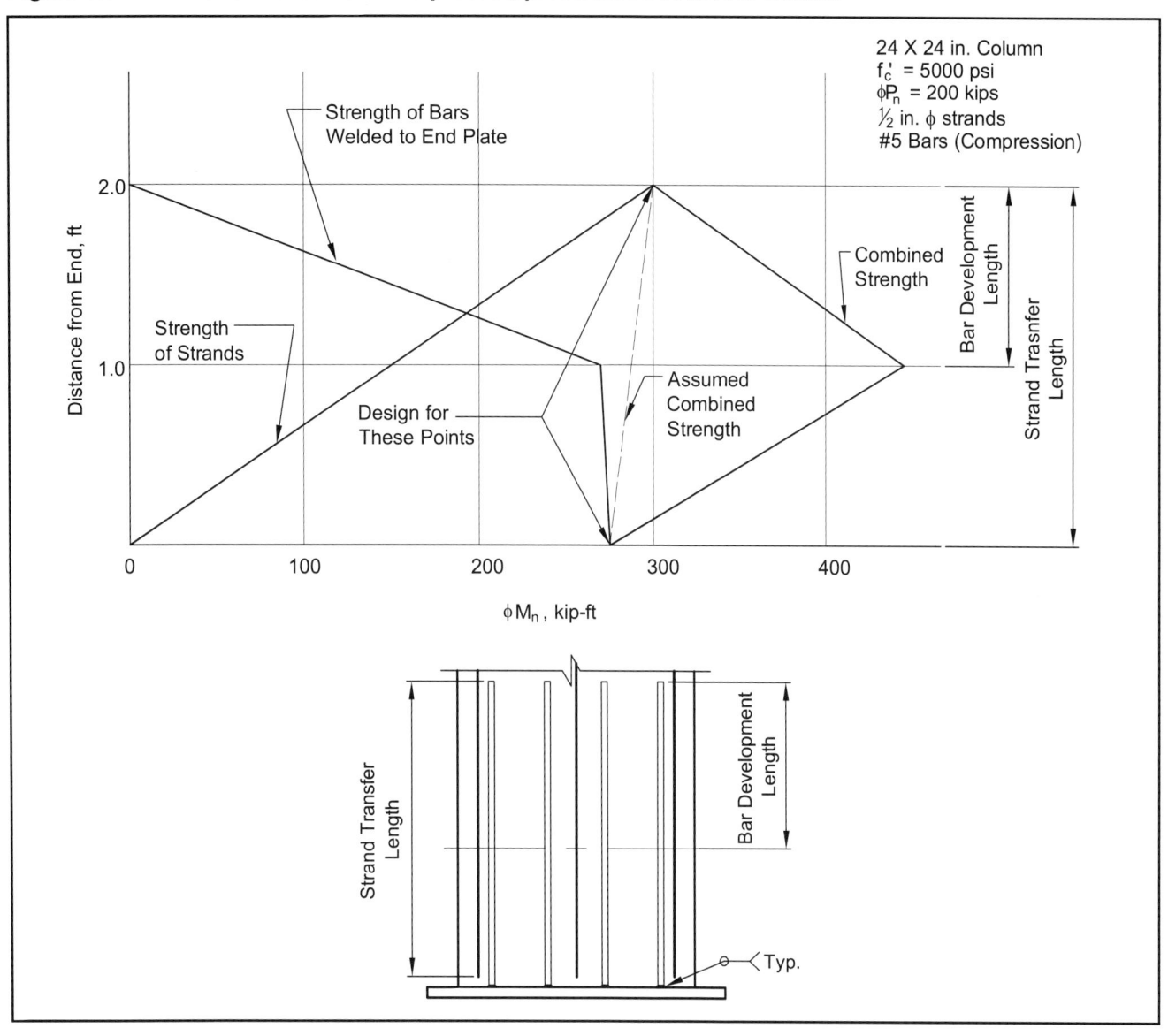

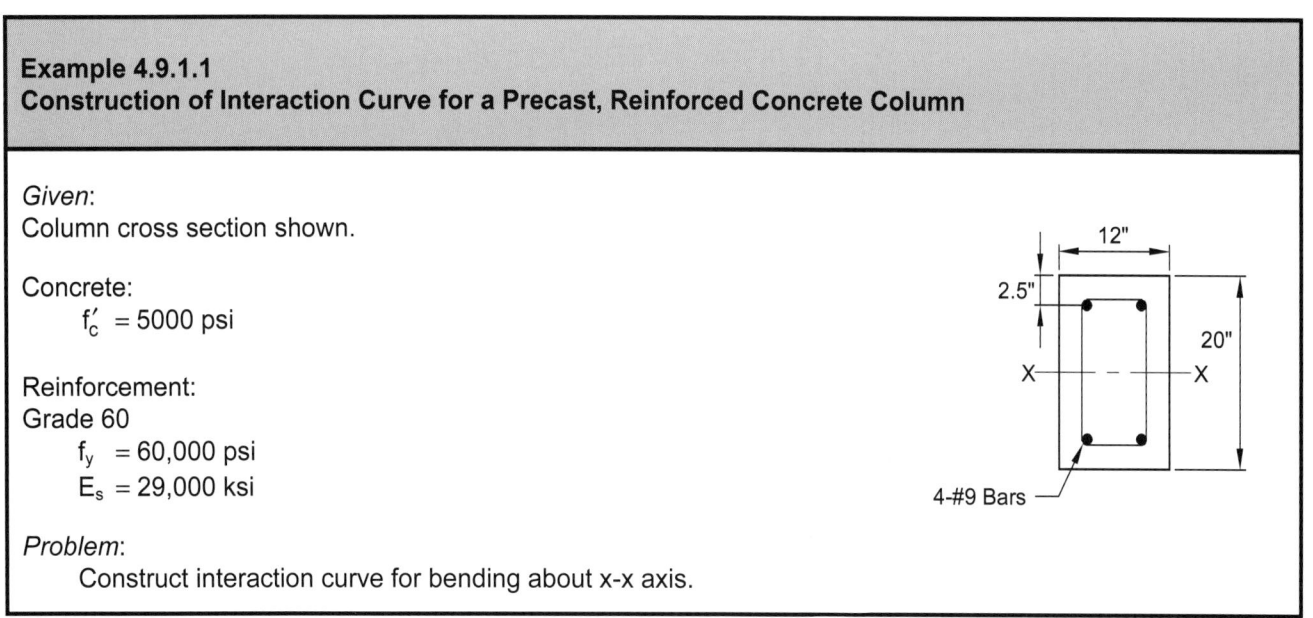

24 X 24 in. Column
f'_c = 5000 psi
ϕP_n = 200 kips
½ in. φ strands
#5 Bars (Compression)

Example 4.9.1.1
Construction of Interaction Curve for a Precast, Reinforced Concrete Column

Given:
Column cross section shown.

Concrete:
    f'_c = 5000 psi

Reinforcement:
Grade 60
    f_y = 60,000 psi
    E_s = 29,000 ksi

Problem:
    Construct interaction curve for bending about x-x axis.

12"
2.5"
20"
4-#9 Bars

Example 4.9.1.1 (Cont.)
Construction of Interaction Curve for a Precast, Reinforced Concrete Column

Solution:
Determine following parameters:

$\beta_1 = 0.85 - 0.05 = 0.80$

$d = 20 - 2.5 = 17.5$ in.

$d' = 2.5$ in.

$y_t = 10$ in.

$0.85f'_c = 0.85(5) = 4.25$ ksi

$A_g = 12(20) = 240$ in.^{2}

$A_s = A'_s = 2.00$ in.^{2}

Step 1. Determine P_o from Figure 4.9.1.1(c):

With no prestressing steel, the equation reduces to:

$P_o = P_n = 0.85f'_c(A - A'_s - A_s) + (A'_s + A_s)f_y = 4.25(240 - 2 - 2) + (2 + 2)60 = 1243$ kips

$\phi P_o = 0.65(1243) = 808$ kips

Step 2. Determine P_{nb} and M_{nb} from Figure 4.9.1.1(d) and (a):

$d_t = d = 17.5$ in.; $c = 0.6d_t = 0.6(17.5) = 10.5$ in.; $a = \beta_1 c = 0.80(10.5) = 8.40$ in.

$f'_s = \varepsilon'_s E_s = \left[\dfrac{0.003}{c}(c - d')\right]E_s = \left[\dfrac{0.003}{10.5}(10.5 - 2.5)\right]29{,}000 = 66.3$ ksi > 60 ksi

Therefore, $f'_s = f_y = 60$ ksi

$A_{comp} = ab = 8.40(12) = 100.8$ in.^{2}

With no prestressing steel:

$P_{nb} = (A_{comp} - A'_s)(0.85f'_c) + A'_s f'_s = A_s f_s = (100.8 - 2)4.25 + 2(60) - 2(60) = 419.9$ kips

$\phi P_{nb} = 0.65(419.9) = 273$ kips

$y' = a/2 = 4.20$ in.

$M_{nb} = (A_{comp} - A'_s)(y_t - y')(0.85f'_c) + A'_s f'_s(y_t - d') + A_s f_s(d - y_t)$

$= (100.8 - 2)(10 - 4.20)(4.25) + 2(60)(10 - 2.5) + 2(60)(17.5 - 10)$

$= 2435 + 900 + 900 = 4235$ kip-in.

$\phi M_{nb} = 0.65(4235) = 2752$ kip-in. $= 229$ kip-ft

Step 3. Determine M_o; use conservative solution neglecting compressive reinforcement:

$a = \dfrac{A_s f_y}{0.85f'_c b} = \dfrac{2.0(60)}{4.25(12)} = 2.35$ in.

$M_o = A_s f_y\left(d - \dfrac{a}{2}\right) = (2.0)(60)\left(17.5 - \dfrac{2.35}{2}\right) = 1959$ kip-in.

Example 4.9.1.1 (Cont.)
Construction of Interaction Curve for a Precast, Reinforced Concrete Column

$$c = \frac{a}{\beta_1} = \frac{2.35}{0.8} = 2.94$$

$$\varepsilon_t = \frac{0.003(d-c)}{c} = 0.0149$$

$$\varepsilon_t > 0.005; \phi = 0.9$$

For $\phi = 0.9$, $\phi M_o = 1763$ kip-in. $= 147$ kip-ft

To determine intermediate points on the curve:

Step 4(a).
Set $a = 6$ in., $c = \dfrac{6}{0.80} = 7.5$ in.

Step 4(b).
$A_{comp} = 6(12) = 72$ in.^{2}

Step 4(c).
Use Figure 4.9.1.1(a):

$$f'_s = 29,000\left[\frac{0.003}{7.5}(7.5-2.5)\right] = 58.0 \text{ ksi} < f_y$$

$$f_s = 29,000\left[\frac{0.003}{7.5}(17.5-7.5)\right] = 116 \text{ ksi} > f_y$$

Use $f_s = f_y = 60$ ksi

Steps 4(d) and 4(e).

$$P_n = (72-2)4.25 + 2.0(58) - 2.0(60) = 293.5 \text{ kips}$$

$$\varepsilon_t = \frac{0.003(17.75-7.5)}{7.5} = 0.004$$

$$\phi = (0.48 + 83\varepsilon_t) = 0.81$$

$$\phi P_n = 0.81(293.5) = 238 \text{ kips}$$

$$\phi M_n = 0.81[(72-2)(10-3)4.25 + 2.0(60)(17.5-10) + 2.0(58)(10-2.5)] = 0.81(2082 + 900 + 870)$$
$$= 3120 \text{ kip-in.} = 260 \text{ kip-ft}$$

(Note: Steps 4a to 4e can be repeated for as many points as desired.)
A plot of these points is shown above.

Step 5.
Calculate maximum design load:
$$= 0.80 \phi P_o = 0.80(808) = 646 \text{ kips}$$

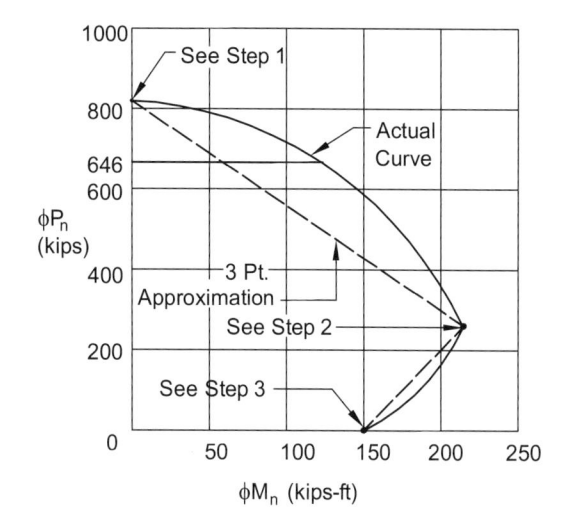

Example 4.9.1.2
Calculation of Interaction Points for a Prestressed Concrete Compression Member

Given:
Hollow-core wall panel shown:

Concrete:
$f'_c = 6000$ psi
$A = 204$ in.^{2}

Prestressing steel:
$f_{pu} = 270$ ksi
$E_{ps} = 28,500$ ksi
$f_{se} = 150$ ksi
5 – ⅜ in. diameter 270K strands
A_{ps} (bott) = 3(0.085) = 0.255 in.^{2}
A'_{ps} (top) = 2(0.085) = 0.170 in.^{2}

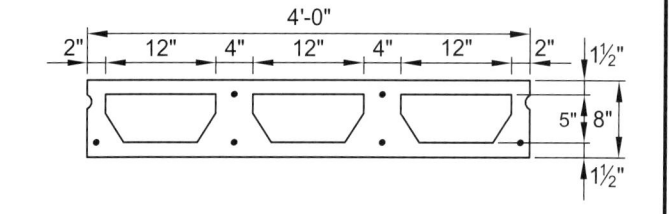

Problem:
Calculate a point on the design interaction curve for a = 2 in.

Solution:
Step 1:
$\beta_1 = 0.85 - 2(.05) = 0.75$
a = 2 in., c = $\frac{2}{0.75}$ = 2.67 in.

Step 2:
$A_{comp} = 48(1.5) + 12(2 - 1.5) = 78$ in.^{2}
$y' = \frac{48(1.5)(1.5/2) + 12(0.5)(1.5 + 0.5/2)}{78} = 0.83$ in.

Step 3:
$\varepsilon_{se} = \frac{f_{se}}{E_{ps}} = \frac{150}{28,500} = 0.00526$ in./in.

Step 4:
From Figure 4.9.1.1(a):

$\varepsilon'_{ps} = 0.00526 - \frac{0.003}{2.67}(2.67 - 1.5) = 0.00526 - 0.00131 = 0.00395$ in./in.

From Design Aid 11.2.5, this strain is on the linear part of the curve:
$f'_{ps} = \varepsilon'_{ps} E_s = 0.00395(28,500) = 113$ ksi
$\varepsilon_t = \frac{0.003}{2.67}(6.5 - 2.67) = 0.00430$
$\varepsilon_{ps} = 0.00526 + 0.00430 = 0.00956$
$\phi = 0.48 + 83\varepsilon_t = 0.84$

Example 4.9.1.2 (Cont.)
Calculation of Interaction Points for a Prestressed Concrete Compression Member

From Design Aid 11.2.5, f_{ps} = 252 ksi
From Figure 4.9.1.1(a):
P_n = $(A_{comp})0.85f'_c - A'_{ps}f'_{ps} - A_{ps}f_{ps}$ = 78(0.85)(6) − 0.170(113) − 0.255(252) = 397.8 − 19.2 − 64.3
    = 314.3 kips
ϕP_n = 0.84(314.3) = 264 kips
M_n = 397.8(4 − 0.83) − 19.2(4 − 1.5) + 64.3(6.5 − 4) = 1261.0 − 48.0 + 160.8 = 1373.8 kip-in.
    = 114.5 kip-ft
ϕM_n = 0.84(1373.8) = 1154 kip-in. = 96.2 kip-ft

Since no lateral ties are used in this member, the values should be multiplied by 0.85.
    ϕP_n = 0.85(264) = 224 kips
    ϕM_n = 0.85(96.2) = 81.8 kip-ft

Note: This is for fully developed strand. If the capacity at a point near the end of the transfer zone is desired, then $f_{ps} \leq f_{se}$ = 150 ksi, ϕ = 0.65 because $\varepsilon_t < 0.002$.

    ϕP_n = 0.85(0.65)[397.8 − 19.2 − 0.255(150)] = 188.0 kips
    ϕM_n = 0.85(0.65)[1261.0 − 48.0 + 0.255(150)(6.5 − 4)] = 723.0 kip-in. = 60.3 kip-ft

Note: For compliance with ACI Code Section 18.8.3, cracking moment, M_{cr}, must be calculated to verify $\phi M_n \geq 1.2 M_{cr}$ for both faces of the wall panel, except for flexural members with shear and flexural strength at least twice that required by ACI 318-02 Section 9.2. The effects of prestressing and its eccentricity should be included in calculating M_{cr} and other aspects of wall panel behavior, such as deflections.

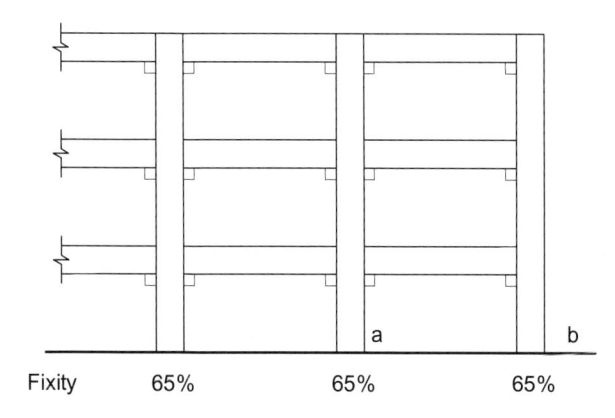

Example 4.9.2.1
Use of Figures 4.12.13 and 4.12.14

Problem:
    Using Figure 4.12.14, determine the maximum restraining force and moment in the lowest story of a three-story frame for:
a. An interior column in a multi-bay frame
b. An exterior column

Given:
Beam reactions to column haunch at each level:
    Dead load = 50 kips
    Live load = 20 kips
    Eccentricity e = 14 in.
    Story height h_s = 16 ft
Column base is determined to be 65% fixed.

Solution:
Factored loads:
    Dead load = 1.2(50) = 60.0 kips
    Live load = 1.6(20) = <u>32.0 kips</u>
                            92.0 kips

Example 4.9.2.1 (Cont.)
Use of Figures 4.12.13 and 4.12.14

1. For the interior column, the dead load reaction would be the same on either side, thus, no moment results. The live load could occur on any one side at any floor, hence, use of the coefficients in the "Σ Max" line:
   $P_u e = 32.0(14) = 448$ kip-in. $= 37.3$ kip-ft

   To determine the maximum moment at Point B:
   For a pinned base: $k_m = 0.67$
   For a fixed base: $k_m = 0.77$
   For 65% fixed: $k_m = 0.67 + 0.65(0.77 - 0.67)$
   $= 0.74$
   $M_u = k_m P_u e = 0.74(37.3) = 27.6$ kip-ft

   Maximum restraining force at Level 2:
   $F_u = k_f P_u e / h_s$
   $k_f = 1.40 + 0.65(1.62 - 1.40) = 1.54$
   $F_u = 1.54(37.3)/16 = 3.59$ kips
   (tension or compression)

2. For the exterior column, the total load is eccentric on the same side of the column, hence use the coefficients in the "Σ One Side" line:
   $P_u e = 92.0(14) = 1288$ kip-in. $= 107.3$ kip-ft

   To determine the maximum moment at Point B:
   For a pinned base: $k_m = 0.40$
   For a fixed base: $k_m = 0.46$
   For 65% fixed: $k_m = 0.40 + 0.65(0.46 - 0.40)$
   $= 0.44$
   $M_u = k_m P_u e = 0.44(107.3) = 47.2$ kip-ft

   Maximum restraining force at Level 2:
   $F_u = k_f P_u e / h_s$
   $k_f = -0.60 - 0.65(-0.60 + 0.22) = -0.35$
   $F_u = -0.35(107.3)/16 = -2.34$ kips (tension)

4.9.2 Eccentrically Loaded Columns

Many precast concrete structures utilize multi-story columns with simple span beams resting on haunches. Figures 4.12.13 and 4.12.14 are provided as aids for determining the various combinations of load and moment that can occur with such columns.

The following conditions and limitations apply to Figures 4.12.13 and 4.12.14.

1. The coefficients are only valid for columns braced against sidesway.
2. For partially fixed column bases (see Section 3.8.3), a straight line interpolation between the coefficients for pinned and fixed bases can be used with small error.
3. For taller columns, the coefficients for the four-story columns can be used with small error.
4. The coefficients in the "Σ Max" line will yield the maximum required restraining force, F_b, and column moments caused by loads (equal at each level) which can occur on either side of the column, for example, live loads on interior columns. The maximum force will not necessarily occur with the same loading pattern that causes the maximum moment.
5. The coefficients in the "Σ One Side" line will yield the maximum moments which can occur if the column is loaded on only one side, such as the end column in a bay.

4.9.3 Slenderness Effects in Columns and Wall Panels

Sections 10.10 through 10.13 of ACI 318-02 contain provisions for evaluating slenderness effects (buckling) of columns. Additional recommendations are given in the Recommended Practice. [23]

The term "slenderness effects," can be described as the moments in a member produced when the line of action of the axial force is not coincident with the displaced centroid of the member. These moments, which are not accounted for in the primary analysis, are thus termed "secondary moments." These secondary moments arise from changes in the geometry of the structure, and may be caused by one or more of the following:

1. Relative displacement of the ends of the member due to:
   a. Lateral or unbalanced vertical loads in an unbraced frame, usually labeled "translation" or "sidesway."
   b. Manufacturing and erection tolerances.
2. Deflections away from the end of the member due to:
   a. End moment due to eccentricity of the axial load.
   b. End moments due to frame action-continuity, fixity or partial fixity of the ends.
   c. Applied lateral loads, such as wind.
   d. Thermal bowing from differential temperature (see Section 4.8.5).
   e. Manufacturing tolerances.
   f. Camber due to prestressing.

Section 10.10.1 of ACI 318-02 describes the use of second-order analysis for slenderness effects. Section 10.10.2 also allows the use of an approximate procedure termed "Moment Magnification." Precast frames with columns reinforced with non-prestressed reinforcement which meet the requirements for minimum reinforcement, may be designed using Moment Magnification, but prestressed compression members usually have less than the minimum 1% vertical reinforcement. The Recommended Practice [23] suggests ways to modify the Code equations used in Moment Magnification, but the second-order, or "P-Δ" analysis is preferred.

4.9.3.1 Second-Order (P-Δ) Analysis

The usual procedure for this analysis is to perform an elastic type analysis using factored loads. While deflections are usually only a concern under service load, the deflections calculated for this purpose are to avoid a stability failure, so it is logical to provide the same safety factor as for strength design.

Out-of-plumbness (Items 1b and 2e above) are initially assumed based on experience and/or specified tolerances. The thermal bowing effect is usually neglected in columns, but may be significant in exterior wall panels (see Section 4.8.5).

At each iteration, the lateral deflection is calculated, and the moments caused by the axial load acting at that deflection are accumulated. After three or four iterations, the increase in deflection should be negligible (convergence). If it is not, the member may be approaching stability failure, and the section dimensions should be re-evaluated.

If the calculated moments indicate that cracking will occur, this needs to be taken into account in the deflection calculations. The stiffness used in the second order analysis should represent the stiffness of the members immediately before failure. This may involve iterations within iterations, greatly complicating the procedure, although approximations of cracked section properties are usually satisfactory. Section 10.11.1 of ACI 318-02 has cracked member properties for different member types for use in second-order analysis of frames. These values are the lower bound of what can be expected for equivalent moments of inertia of cracked members and include a stiffness reduction factor ϕ_K to account for variability of second-order deflections.

Effects of creep should also be included. The most common method is to divide the stiffness (EI) by the factor $1 + \beta_d$ as specified in the ACI moment magnification method.

In unbraced continuous frames, the joint translations effect and the frame response are interdependent. Thus, a practical analysis, especially when potential cracking is a parameter, usually will require a more sophisticated approach with the aid of a computer.

A good review of second-order analysis, along with an extensive bibliography and an outline of a complete program, is contained in Ref. 24.

An example of P-Δ calculations for a simple yet frequently encountered problem is shown in Example 4.9.3.1. In the following example only two of the ACI load combinations are checked, presuming by engineering judgment that these were the controlling cases.

4.9.4 Effective Width of Wall Panels

The Recommended Practice specifies that the portion of a wall considered as effective for supporting concentrated loads or for determining the effects of slenderness shall be the least of the following:

Example 4.9.3.1
Second-Order Analysis of an Uncracked Member

Given:
An 8 in. thick, 8 ft wide prestressed wall panel as shown.
Loading assumptions are as follows:
1. Axial load eccentricity = 1 in. (at one end)
2. Assume midspan bowing = 1.0 in. outward. (Note: Production tolerances allow $\ell/360$, see Chapter 8.)
3. Wind load = 30 psf.
4. Non-sway frame – no joint translation.
5. Joints assumed pinned top and bottom.

Concrete: f'_c = 5000 psi
E_c = 4300 ksi

D.L. = 10 kips
L.L. = 6 kips (Roof)

30'-0" = 360"

8"

Problem:
Determine if standard panel (Figure 2.6.5) is adequate.

Solution:
Case 1: No wind
$P_u = 1.2D + 1.6L_r = 1.2(10) + 1.6(6) = 21.6$ kips
$I_g = bh^3/12 = 96(8)^3/12 = 4096$ in.^{4}

Note: The bending stiffness EI is multiplied by 0.70 per Section 10.11.1, ACI 318-02, for uncracked wall sections. This includes a stiffness reduction factor ϕ_K to cover the variability of computed deflections. Once the moments are established, the ϕ factor from Section 9.3.2.2 of ACI 318-02 is used to determine the strength of the cross section.

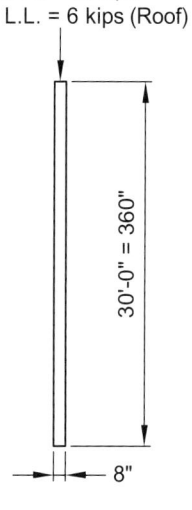

$$\beta_d = \text{factored} \frac{DL}{TL} = \frac{12}{21.6} = 0.56$$

The β_d factor is used to account for sustained load effects by modifying the stiffness. It is reasonable to use:

$$EI_{eff} = \frac{0.70 E_c I_g}{1+\beta_d} = \frac{0.70(4300)(4096)}{1.56} = 7.90 \times 10^6 \text{ kip-in.}^2$$

Check P-critical using Euler's formula:

$$P_c = \frac{\pi^2 EI_{eff}}{\ell^2} = \frac{\pi^2 (7.90 \times 10^6)}{(360)^2} = 602 \text{ kips} > 21.6 \text{ kips} \quad OK$$

Moments on panel:
Deflection at midspan due to $P_u e$:

$$\Delta_i = \frac{P_u e \ell^2}{16 EI} = \frac{21.6(1)(360)^2}{16(7.90 \times 10^6)} = 0.0221 \text{ in.}$$

Total midspan deflection including initial bow:
= 1.0 + 0.0221 = 1.022 in.
Deflection due to P-Δ moment at midspan:

$$\Delta = \frac{P_u e \ell^2}{8 EI} = \frac{21.6 e (360)^2}{8(7.90 \times 10^6)} = 0.044 e$$

Example 4.9.3.1 (Cont.)
Second-Order Analysis of an Uncracked Member

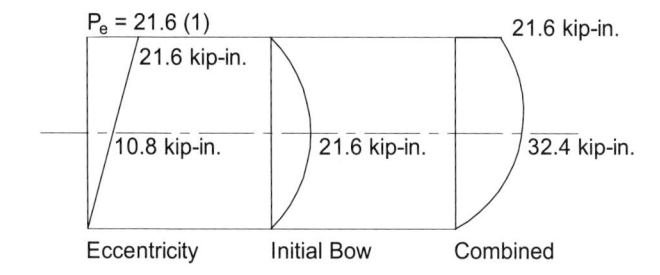

First iteration:
$\Delta = 0.044(1.022) = 0.045$ in.

Second iteration:
$e = 1.022 + 0.045 = 1.067$ in.
$\Delta = 0.044(1.067) = 0.047$ in.

Third iteration:
$e = 1.022 + 0.047 = 1.069$ in.
$\Delta = 0.044(1.069)$
$= 0.047$ in. (convergence)

or using a geometric series to calculate the midspan deflection in one step:

$$e = \frac{e_0}{1 - \frac{\Delta_i}{e}} = \frac{1.022}{1 - \frac{0.0221}{0.5}} = 1.069 \text{ in.}$$

M_u at midheight $= 10.8 + 21.6(1.069) = 33.9$ kip-in.

Check for cracking at midheight:
$\frac{M_u y}{I} = 33.9(4)(1000)/4096 \quad = -33$ psi
Half Panel wt.: $[100(15)/(8(12))](1.2) = 19$ psi
Prestress $\quad = 250$ psi
Dead load $= 10(1000)(1.2)/[8(96)] \quad = \underline{16 \text{ psi}}$
Net stress (compression) $\quad = 252$ psi

Therefore, the analysis is valid.
Using interaction curve Figure 2.6.5:
$P_u = 21.6/8 + 1.2(15)0.1 = 4.50$ kips/ft
$M_u = 33.9/[12(8)] = 0.35$ kip-ft/ft

Point is below curve, OK.

Case 2: Include wind

Axial load without wind is:
$P_u = 1.2D + 1.6L + 0.8W = [1.2(10) + 1.6(6)] = 21.6$ kips
Deflection due to $P_u e$ (see Case 1)
$= 0.0221$ in.
Additive wind load would be suction $= 30$ psf
$w_u = 30(8)(0.8) = 192$ lb/ft

When considering wind, $\beta_d = 0$

$$EI_{eff} = \frac{0.70 E_c I_g}{1 + \beta_d} = \frac{0.70(4300)(4096)}{1.0} = 1.23 \times 10^7 \text{ kip-in.}^2$$

Example 4.9.3.1 (Cont.)
Second-Order Analysis of an Uncracked Member

Deflection due to wind:

$$\frac{5w_u\ell^4}{384EI} = \frac{5\left(\frac{0.192}{12}\right)(360)^4}{384(1.23 \times 10^7)} = 0.28 \text{ in.}$$

Total initial midspan bow including eccentricity and wind:
  e = 1.0 + 0.022 + 0.28 = 1.302 in.

Deflection due to P-Δ moment at midspan:

$$\Delta = \frac{Pe\ell^2}{8EI} = \frac{21.6e(360)^2}{8(1.23 \times 10^7)} = 0.028e$$

First iteration:
  Δ = 0.028(1.302) = 0.036 in.

Second iteration:
  e = 1.302 + 0.036 = 1.338 in.
  Δ = 0.028(1.338) = 0.037 in.

Third iteration:
  e = 1.302 + 0.037 = 1.339 in.
  Δ = 0.028(1.339) = 0.037 in. (convergence)

$$M_u = \frac{21.6(1)}{2} + 21.6(1.339) + \frac{\left(\frac{0.192}{12}\right)(360)^2}{8} = 299 \text{ kip-in.}$$

  $M_u y/I$ = 299(4)(1000)/4096 = −292 psi

Panel weight	= 19 psi
Prestress	= 250 psi
Dead load	= 16 psi
Net stress (tension)	= −7 psi

  $f_r = 7.5\sqrt{f'_c}$ = 530 psi > 7 psi

Therefore, the analysis is valid.
  P_u = 21.6/8 = 2.7 kips/ft
  M_u = 299/[(12)(8)] = 3.1 kip-ft/ft

By comparison with Figure 2.6.5, OK.

---

a. The center-to-center distance between loads.
b. The width of the loaded portion plus six times the wall thickness on either side (Figure 4.9.4.1).
c. The width of the rib (in ribbed wall panels) plus six times the thickness of the wall between ribs on either side of the rib (Figure 4.9.4.1).
d. 0.4 times the actual height of the wall.

4.9.5 Varying Section Properties of Compression Members

Architectural wall panels will frequently be of a configuration that varies over the unsupported height of the panel. While there are precise methods of determining the effects of slenderness for such

Figure 4.9.4.1 Effective width of wall panels

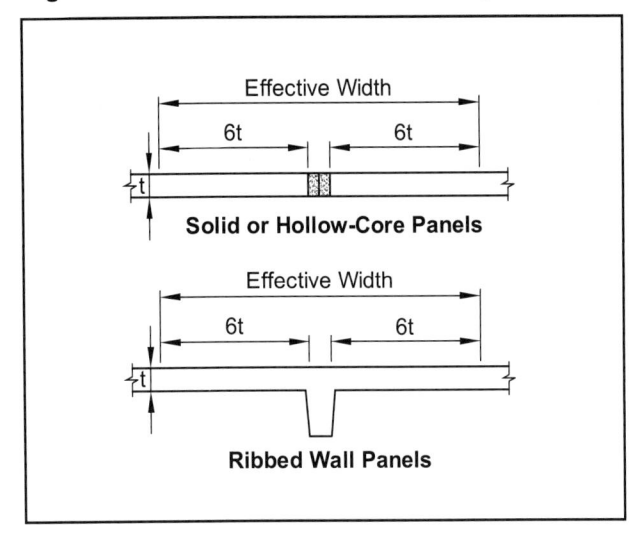

Figure 4.9.6.1 Typical pile interaction curve

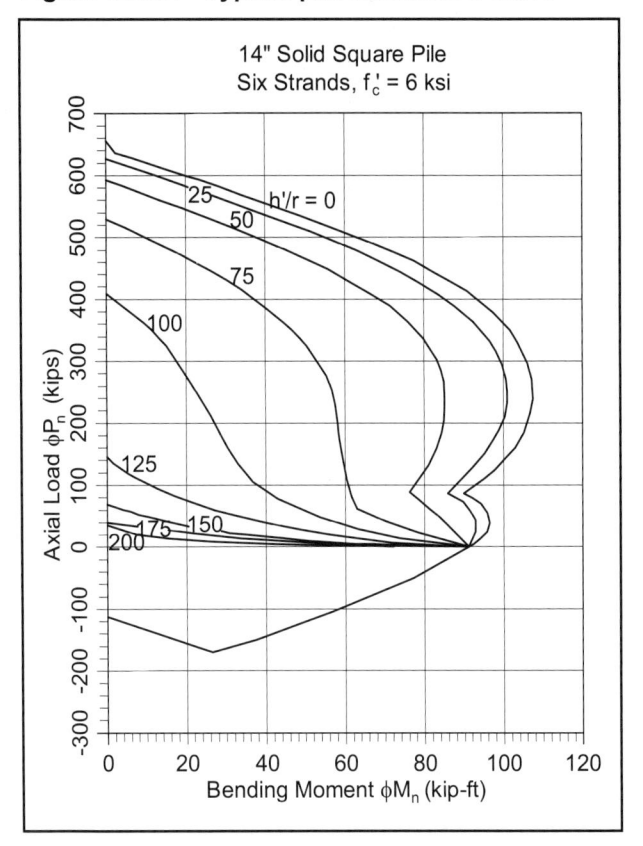

members, the approximate nature of the analysis procedures used do not warrant such precision. Example 4.9.5.1 illustrates approximate methods for determining section properties used in evaluating slenderness effects.

4.9.6 Piles

In this section, procedures are discussed to determine the structural capacity of prestressed concrete piles. It is assumed that the soil is adequate to allow the pile to reach its design strength. The soil capacity must be evaluated by a geotechnical engineer. Experience has shown that in piles subjected to concentric axial loads, the frictional and bearing resistance of the soil may control the design more often than the service load stresses and the strength of the pile itself. Refs. 33, 34 and 35 provide additional information.

4.9.6.1 Service Load Design

Except when load factors are much higher than normal, service load stresses are usually more critical than strength. Stresses are computed by:

$$f = f_{pc} \pm \frac{N}{A} \pm \frac{M}{S} \quad \text{(Eq. 4.9.6.1)}$$

where:
- f_{pc} = average prestress
- N = unfactored axial load
- A = cross-sectional area
- M = service load moment
- S = section modulus

The allowable stress varies with the code, specification, or guide being used, and the type of loading being investigated.

4.9.6.2 Strength Design

Design strength may be obtained from a strain compatibility analysis similar to that used for columns in Section 4.9.1, and as illustrated in Example 4.9.6.1, or from interaction diagrams based on such an analysis. A typical pile interaction diagram is shown in Figure 4.9.6.1. Note that piles may be in tension, hence the lower part of the diagram. A series of interaction curves for most common pile sections has been published by PCI. [36]

Shear reinforcement is determined by strength design (Section 4.3). Wire spirals are often used for shear reinforcement (see illustration on Table 2.8.1.)

The most severe conditions a pile will endure usually occur during driving. This is discussed in Ref. 35.

4.9.6.3 Other Considerations

The stress in the reinforcing steel used to transfer the tension load from the foundation or pile cap to the pile is usually limited to 30 ksi service load tension.

When non-prestressed reinforcement is used to transfer tension loads to the pile, it should extend a sufficient length into the pile to develop both the pre-stressed and non-prestressed reinforcement (see Section 4.2.3).

Example 4.9.5.1
Approximate Section Properties of an Architectural Mullion Panel

Given:
The loadbearing architectural wall panel is shown at right.
$f'_c = 5000$ psi
$E_c = 4300$ ksi

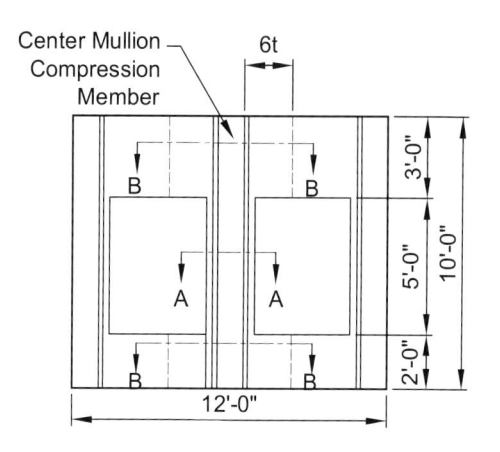

Problem:
Determine an approximate moment of inertia for slenderness analysis.

Solution:
One method is to determine a simple span deflection with a unit uniform load as follows:
Using the moment area method, the center or mid-height deflection is:

$\Delta_o = 0.013$ in.

$$\Delta_o = \frac{5w(\ell)^4}{384EI_{equiv}}$$

$$I_{equiv} = \frac{5w(\ell)^4}{\Delta_o 384E} = \frac{5(1/12)[10(12)]^4}{0.013(384)(4300)} = 4025 \text{ in.}^4$$

A second, more approximate method would be to use a "weighted average" of the moments of inertia of the two sections. In this case:

$$I_{equiv} = \frac{I_1 h_1 + I_2 h_2}{h_1 + h_2}$$

$$= \frac{5638(3+2) + 3891(5)}{10} = 4764 \text{ in.}^4$$

Once an equivalent moment of inertia is determined, slenderness effects are evaluated as described in Section 4.9.3.

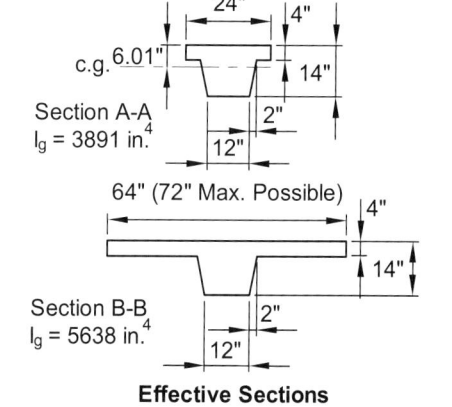

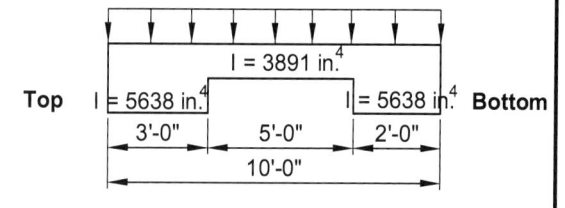

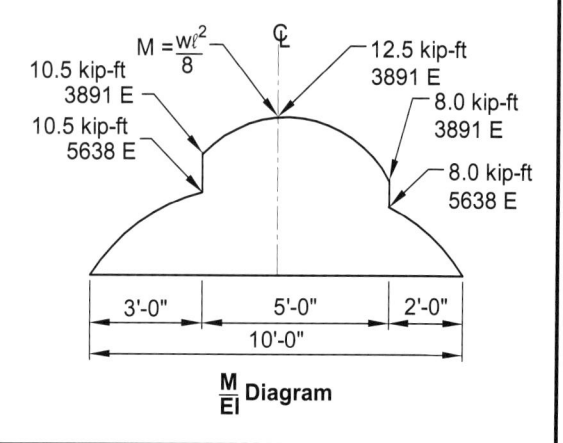

Piles subject to large tension loads in combination with bending may have failure of the pile initiated by rupture of the prestressing strands before failure of the concrete in compression. This can be noted in the break of the smooth curve of the interaction diagrams in the diagram tension region.

In piles that are subject to tension, with or without bending, the nominal strength of the prestressing strands should be equal to or greater than 1.2 times the cracking load of the pile, based on a modulus of rupture, $f_r = 7.5\sqrt{f'_c}$. This requirement may be waived if the available steel strength is greater than twice the factored tension plus bending loads.

Piles in compression should be designed for the actual calculated moment, but not less than the moment induced by a load eccentricity of 0.05 times the pile diameter or width.

Example 4.9.6.1
Sheet Pile

Given:
- A = 432 in.^{2}
- I = 5184 in.^{4}
- S = 5184/6 = 864 in.^{3}
- f'_c = 6000 psi

Stress to 75% of f_{pu}; assume 18% loss

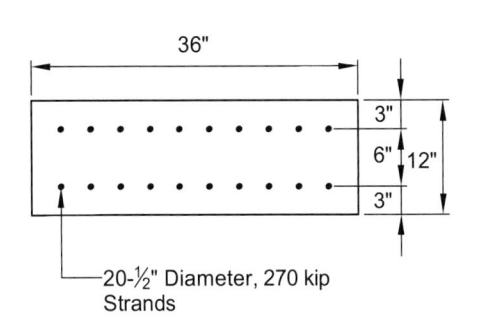

Problem:
For the 12 in. x 36 in. sheet pile shown, compute allowable moment.

Solution:
$$f_{se} = 0.75(0.82)(270) = 166 \text{ ksi}$$
$$A_{ps} = 20(0.153) = 3.06 \text{ in.}^2$$
$$f_{pc} = \frac{166(3.06)}{432} = 1.176 \text{ ksi}$$

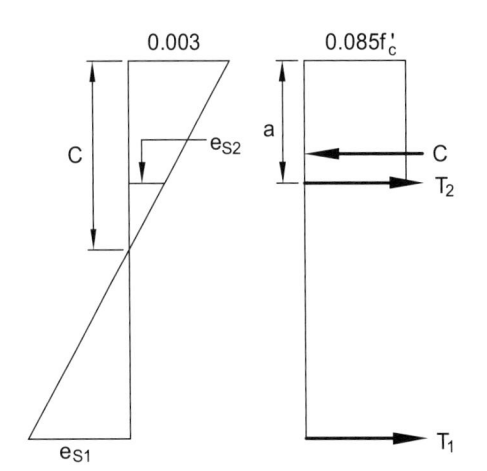

Moment Capacity:
(a) Service loads
   For allowable tension of $4\sqrt{f'_c}$ or 0.310 ksi:

   Allowable moment:
   $$= (4\sqrt{f'_c} + f_{pc})S = (0.310 + 1.176)864 = 1284 \text{ kip-in.}$$
   per pile = 428 kip-in. per ft

(b) Strength design
   Use strain compatibility:
   Using an iteration procedure, determine the depth of neutral axis, c, to be 4.33 in. (see Example 4.2.1.6.)
   $$\varepsilon_{se} = f_{se}E_{ps} = 166/28{,}500 = 0.00582 \text{ in./in.}$$

Last iteration:
$$\varepsilon_{s1} = \frac{9.00 - 4.33}{4.33}(0.003) + 0.00582 = 0.00906$$

$$f_{s1} = 270 - \frac{0.04}{0.00906 - 0.007} = 250.6 \text{ ksi}$$

Example 4.9.6.1 (Cont.)
Sheet Pile

$T_1 = 250.6(0.153)(10) = 383.4$ kips
$\varepsilon_{s2} = \dfrac{4.33 - 3.00}{4.33}(-0.003) + 0.00582 = 0.00490$
$f_{s2} = 0.00490 \ (28,500) = 139.7$ ksi
$T_2 = 139.7(0.153)(10) = 213.7$ kips
$T_1 + T_2 = 383.4 + 213.7 = 597.1$ kips
$a = \beta_1 c = 0.75(4.33) = 3.25$ in.
$C = 0.85 f'_c ba = 0.85(6)(36)(3.25)$
$\quad = 596.7$ kips (close to $T_1 + T_2$) OK
$M_n = -596.7(3.25/2) + 213.7(3) + 383.4(9) = 3122$ kip-in.
$\varepsilon_t = \varepsilon_{s1} - \varepsilon_{se} = 0.00906 - 0.00582 = 0.00324$
$\phi = 0.48 + 83\varepsilon_t = 0.75$
$\phi M_n = 0.75(3123) = 2342$ kip-in. $= 195$ kip-ft

4.10 Special Considerations

This section outlines solutions of special situations which may arise in the design of a precast floor or roof system. Since production methods of products vary greatly, local producers should be consulted. Also, test data may indicate that the conservative guidelines presented here may be exceeded for specific applications with engineering judgment considered.

Figure 4.10.1.1 Assumed load distribution for hollow-core slabs

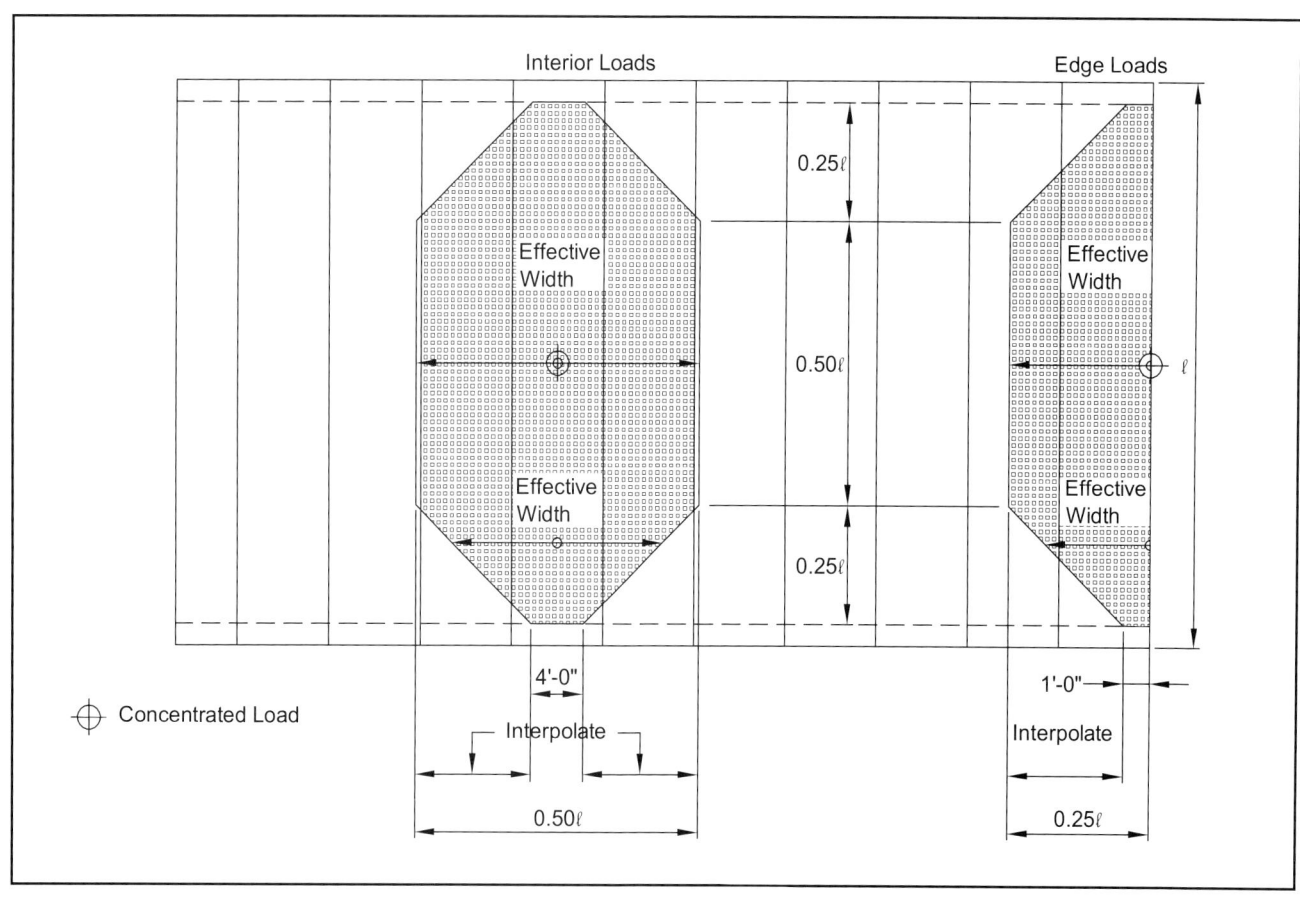

4.10.1 Load Distribution

Frequently, floors and roofs are subjected to line loads, for example from walls, and concentrated loads. The ability of hollow-core systems to transfer or distribute loads laterally through grouted shear keys has been demonstrated in several published reports, [25–27] and many unpublished tests.

Based on tests, analysis and experience, the PCI Hollow-Core Slab Producers Committee recommends [28] that line and concentrated loads be resisted by an effective section as described in Figure 4.10.1.1. Exception: If the total deck width, perpendicular to the span, is less than the span, modification may be required. Contact local producers for recommendations.

Load distribution in stemmed members may not necessarily follow the same pattern, because of different torsional resistance properties.

4.10.2 Openings through Decks

Openings may be provided in precast decks by: (1) saw cutting after the deck is installed and grouted, (2) forming (blocking out) or sawing in the plant, or (3) using short units with steel headers or other connections. In hollow-core or solid slabs, structural capacity is least affected by orienting the longest dimension of an opening parallel to a span, or by coring small holes to cut the fewest strands. Openings in stemmed members must not cut through the stem, and for double tees, should be narrower than the stem spacing less the top stem width less 2 in. (see Figure 9.10.4.1).

The following are reasonable guidelines regarding the design of hollow-core slabs around openings. Some producers may have data to support a different procedure:

1. An opening located near the end of the span and extending into the span less than the lesser of 0.125ℓ or 4 ft may be neglected when designing for flexure in the midspan region.
2. Strand development must be considered on each side of an opening which cuts strand (see Section 4.2.3).
3. Slabs which are adjacent to long openings ($\ell/4$ or more), or occur near midspan, may be considered to have a free edge for flexural design.
4. Slabs which are adjacent to openings closer to the end than $3\ell/8$ may be considered to have a free edge for shear design.

4.10.3 Openings through Webs

In special situations, horizontal openings may be required in stems of deck members. These openings will add significantly to the cost of the product and the constructability of such modifications should be carefully considered. Refs. 29–31 summarize research on the subject including the results of full-scale testing and design recommendations. The principal recommendations from those references are:

1. Web openings should be placed outside the strand development area. See Figure 4.10.3.
2. Vertical stirrups should be placed on each side of an opening to control cracks in the vicinity of the openings.
3. The openings should be located in regions of the member with low shear and below the compression block.
4. The member should be subjected to primarily uniformly distributed loads. Significant concentrated loads may be placed directly above the solid area between openings, or otherwise carefully designed for.
5. The minimum distance between the openings should be equal to the height of the opening but not less than 10 in.
6. Members should be designed so that tensile stresses do not exceed the modulus of rupture.

Example 4.10.1
Load Distribution

Given:
An untopped hollow-core floor with 4 ft wide 8 in. deep slabs, and supporting a loadbearing wall and concentrated loads as shown on the next page.

Problem:
Determine the design loads for the slab supporting the wall and concentrated loads.

Example 4.10.1 (Cont.)
Load Distribution

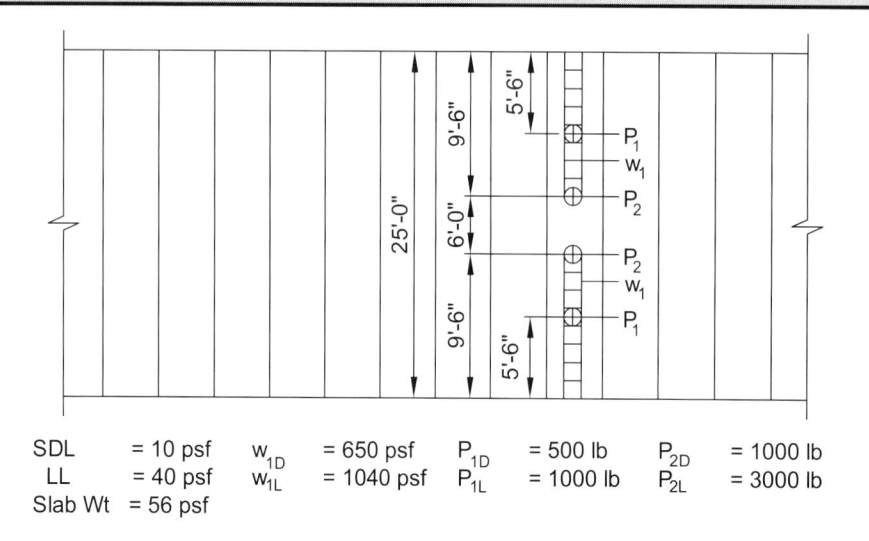

SDL = 10 psf w_{1D} = 650 psf P_{1D} = 500 lb P_{2D} = 1000 lb
LL = 40 psf w_{1L} = 1040 psf P_{1L} = 1000 lb P_{2L} = 3000 lb
Slab Wt = 56 psf

Solution:
(Note: Each step corresponds to a line number in the table shown below)
1. Calculate the shears and moments for the non-distributable (uniform) loads:
   w_u = 1.2(56 + 10) + 1.6(40) = 143 psf
2. Calculate the shears and moments for the distributable (concentrated and line) loads:
   w_u = 1.2(650) + 1.6(1040) = 2444 lb/ft
   P_{1u} = 1.2(500) + 1.6(1000) = 2200 lb
   P_{2u} = 1.2(1000) + 1.6(3000) = 6000 lb
3. Calculate the effective width along the span:
   At the support, width = 4.0 ft
   At 0.25ℓ (6.25 ft), width = 0.5ℓ = 12.5 ft
   Between x = 0 and x = 6.25 ft
   Width = 4 + (x/6.25)(12.5 − 4) = 4 + 1.36x
4. Divide distributable shears and moments from Step 2 by the effective widths from Step 3.
5. Add the distributed shear and moment to the non-distributable shear and moment from Step 1.

Once the moments and shears are determined, the slabs are designed as described in Sections 4.2, 4.3, and 4.6. This method is suitable for computer solution. For manual calculations, the procedure can be simplified by investigating only critical sections. For example, shear may be determined by dividing all distributable loads by 4 ft, and flexure at midspan can be checked by dividing the distributable loads by 0.5ℓ.

Shear and Moments											
Distance from support		0	0.333'	1'	2'	3'	4'	5'	7.5'	10'	12.5'
1. Non-distri- butable loads	V_u	1.79	1.74	1.64	1.50	1.36	1.22	1.07	0.72	0.36	0.00
	M_u	0.00	0.59	1.72	3.29	4.72	6.01	7.15	9.38	10.73	11.17
2. Distributable loads	V_u	31.42	30.60	28.97	26.53	24.09	21.64	19.20	10.89	0.00	0.00
	M_u	0.00	10.33	30.20	57.95	83.26	106.12	126.54	162.50	179.39	179.39
3. Effective width	Eff, W	4.00	4.45	5.36	6.72	8.08	9.44	10.80	12.50	12.50	12.50
4. Distributed shears & moments	V_u	7.85	6.88	5.41	3.95	2.98	2.29	1.78	0.87	0.00	0.00
	M_u	0.00	2.32	5.63	8.62	10.30	11.24	11.72	13.00	14.35	14.35
5. Design shears & moments	V_u	9.64	8.62	7.05	5.45	4.34	3.51	2.85	1.59	0.36	0.00
	M_u	0.00	2.91	7.35	11.91	15.02	17.25	18.87	22.38	25.08	25.52
V_u in kip/ft, M_u in kip-ft/ft											

Figure 4.10.3 Openings through webs

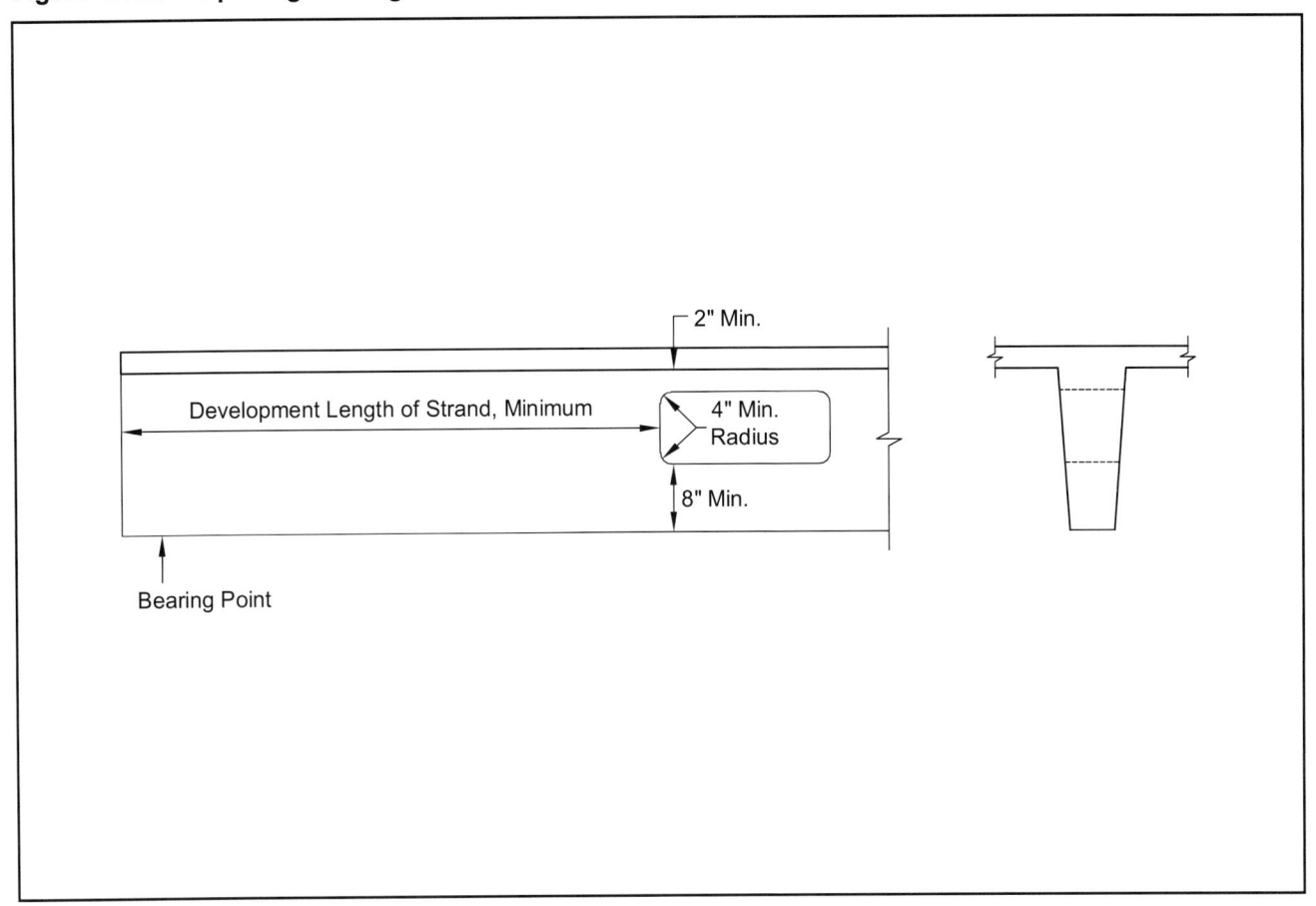

Refs. 28 and 29 provide design examples and further recommendations. These references should be reviewed before designing or specifying horizontal openings in webs.

4.10.4 Continuity

In most applications, precast members are used as part of a simple span system. In some cases, continuous frames are designed to be part or all of the lateral force resistant system (see Chapter 3). In other cases, reinforcement is added in topping or pour strips over deck supports, and a positive compression block constructed at the bottoms of the deck members. This can make an effective continuous structure with the resulting structural efficiency. Such a system is, however, highly dependent on proper design, detailing and construction, and local precasters should be contacted before such a system is specified.

4.10.5 Cantilevers

The method by which precast, prestressed members resist cantilever moments depends on (1) the method of production, (2) the length and loading requirements of the cantilever, and (3) the size of the project (amount of repetition).

If a cantilever is not long enough to fully develop top strands, a reduced value of f_{ps} must be used, as discussed in Section 4.2.3. As with reinforcing bars, due to settlement of concrete, top strands often do not bond as well as the ACI development equation indicates, especially in dry cast systems. It is often necessary, or at least desirable, to debond the top strands in positive moment regions, and the bottom strands in the cantilever.

In some cases, it is preferable to design cantilevers as reinforced concrete members using deformed reinforcing bars to provide the negative moment resistance. In machine-made products, the steel can be placed in grout keys, composite topping, or concreted into cores.

Top tension under service loads should be limited to $7.5\sqrt{f'_c}$, including prestress, so that the section remains uncracked, allowing better prediction of service load deflections.

Consultation with local producers is recommended before choosing a method of reinforcement for cantilevers. Some have developed standard methods that work best with their particular system, and have proven them with tests or experience.

4.11 References

1. ACI Committee 318, "Building Code Requirements for Structural Concrete, ACI 318-02, and Commentary, ACI 318R-02," American Concrete Institute, Farmington Hills, MI, 1995.

2. Naaman, A. E., "Ultimate Analysis of Prestressed and Partially Prestressed Sections by Strain Compatibility," PCI JOURNAL, V. 22, No. 1, January-February 1977.

3. Noppakunwijai, P., Tadros, M., Ma, Z., and Mast, R., "Strength Design of Pretensioned Flexural Concrete Members at Prestress Transfer," PCI JOURNAL, V. 46, No. 1, January-February 2001.

4. Mattock, Alan H., "Anchorage of Stirrups in a Thin Cast-in-Place Topping," PCI JOURNAL, V. 32, No. 6, November-December 1987.

5. Martin, L., and Korkosz, W., "Strength of Prestressed Concrete Members at Sections Where Strands Are Not Fully Developed," PCI JOURNAL, V. 40, No. 5, September-October 1995.

6. Logan, Donald R., "Acceptance Criteria for Bond Quality of Strand for Pretensioned Prestressed Concrete Applications," PCI JOURNAL, V. 42, No. 2, March-April 1997.

7. Marshal, W. T., and Mattock, A. H., "Control of Horizontal Cracking in the Ends of Pretensioned Prestressed Concrete Girders," PCI JOURNAL, V. 7, No. 5, October 1962.

8. Kelly, John B., and Pike, Kenneth J., "Design and Production of Prestressed L-Shaped Bleacher Seat Units," PCI JOURNAL, V. 18, No. 5, September-October 1973.

9. Shaikh, A. F., "Proposed Revisions to Shear-Friction Provisions," PCI JOURNAL, V. 23, No. 2, March-April 1978.

10. Zia, Paul and McGee, W. D., "Torsion Design of Prestressed Concrete," PCI JOURNAL, V. 19, No. 2, March-April 1974.

11. Zia, Paul and Hsu, T.C., "Design for Torsion and Shear in Prestressed Concrete," Preprint 3424, American Society of Civil Engineers, October, 1978. Reprinted in revised form in PCI JOURNAL, V. 49, No. 3, May-June 2004.

12. Mirza, S. A., and Furlong, R. W., "Serviceability Behavior and Failure Mechanisms of Concrete Inverted T-Beam Bridge Bentcaps," Journal of the American Concrete Institute, V. 80, No.4, July-August 1983.

13. Mirza, S. A., and Furlong, R. W., "Strength Criteria for Concrete Inverted T-Girders," ASCE Journal of Structural Engineering, V. 109, No. 8, August 1983.

14. Raths, Charles H., "Spandrel Beam Behavior and Design," PCI JOURNAL, V. 29, No. 2, March-April 1984.

15. Klein, G. J., "Design of Spandrel Beams," Research Project No. 5, Precast/Prestressed Concrete Institute, Chicago, IL, 1986; Summary Paper in PCI JOURNAL, V. 31, No. 5, September-October 1986.

16. *Design and Typical Details of Connections for Precast and Prestressed Concrete,* Second Edition, MNL-123-88, Precast/Prestressed Concrete Institute, Chicago, IL, 1988.

17. Mattock, A. H., and Chan, T. C., "Design and Behavior of Dapped-End Beams," PCI JOURNAL, V. 24, No. 6, November-December 1979.

18. Mattock, A. H., and Theryo, T. S., "Strength of Precast Prestressed Concrete Members with Dapped Ends," Research Project No. 6, Precast/Prestressed Concrete Institute, Chicago, IL, 1986; Summary Paper in PCI JOURNAL, V. 31, No. 5, September-October 1986.

19. Zia, Paul, Preston, H. K., Scott, N. L., and Workman, E. B., "Estimating Prestress Losses," Concrete International, V. 1, No. 6, June 1979.

20. Mast, Robert F., "Analysis of Cracked Prestressed Sections: A Practical Approach," PCI JOURNAL, V. 43, No. 4, July-August 1998.

21. Martin, L. D., "A Rational Method for Estimating Camber and Deflection of Precast Prestressed Members," PCI JOURNAL, V. 22, No. 1, January-February 1977.

22. Shaikh, A. F., and Branson, D. E., "Non-Tensioned Steel in Prestressed Concrete Beams," PCI JOURNAL, V. 15, No. 1, February 1970.

23. PCI Committee on Prestressed Concrete Columns, "Recommended Practice for the Design of Prestressed Concrete Columns and Walls," PCI JOURNAL, V. 33, No. 4, July-August 1988.

24. Nathan, Noel D., "Rational Analysis and Design of Prestressed Concrete Beam Columns and Wall Panels," PCI JOURNAL, V. 30, No. 3, May-June 1985.

25. LaGue, David J., "Load Distribution Tests for Precast Prestressed Hollow-Core Slab Construction," PCI JOURNAL, V. 16, No. 6, November-December 1971.

26. Johnson, Ted, and Ghadiali, Zohair, "Load Distribution Test on Precast Hollow-Core Slabs with Openings," PCI JOURNAL, V. 17, No. 5, September-October 1972.

27. Pfeifer, Donald W., and Nelson, Theodore A., "Tests to Determine the Lateral Distribution of Vertical Loads in a Long-Span Hollow-Core Floor Assembly," PCI JOURNAL, V. 28, No. 6, November-December 1983.

28. *PCI Manual for the Design of Hollow-Core Slabs*, MNL-126-85, Precast/Prestressed Concrete Institute, Chicago, IL, 1985.

29. Savage, J. M., Tadros, M. K., Arumugasaamy, P., and Fisher, L. G., "Behavior and Design of Double Tees with Web Openings," PCI JOURNAL, V. 41, No. 1, January-February 1996.

30. Saleh, M. A., "Optimization of Prefabricated Joists," Ph.D. Dissertation, University of Nebraska-Lincoln, NE, December 1996.

31. Saleh, M. A., Brady, P. A., Einea, A., and Tadros, M. K., "Design and Performance of Prestressed Precast Reinforced Concrete Double-Tee Beams with Web Openings," U.S. Army Corps of Engineers, USACERL Technical Report 97, April 1997.

32. Aswad, Alex and Burnley, George, "Point Load Tests of Double Tee Flanges," PCI JOURNAL, V. 36, No. 4, July-August 1991.

33. ACI Committee 543, "Design, Manufacture and Installation of Concrete Piles (ACI 543R-00)," Farmington Hills, MI.

34. "Recommended Practice for Design, Manufacture and Installation of Prestressed Concrete Piling," PCI JOURNAL, V. 38, No. 2, March-April 1993.

35. *Bridge Design Manual*, Second Edition, MNL-133-97, Precast/Prestressed Concrete Institute, Chicago, IL, 1997.

36. "Prestressed Concrete Piling Interaction Diagrams" (Explained by and to be used in conjunction with Ref. 34). Available from Precast/Prestressed Concrete Institute, Chicago, IL.

37. Wan, B., Harries, K. A., and Petrou, M. F., "Transfer Length of Strands in Prestressed Concrete Piles" *ACI Structural Journal*, V. 99, No. 5, September-October 2002.

4.12 DESIGN AIDS

Figure 4.12.1 Flexural resistance coefficients for elements with non-prestressed, prestressed reinforcement, or combinations of both

Procedure:
Design:

1. Determine $K_u = \dfrac{M_u/\phi - A'_s f'_y (d - d')}{f'_c b d^2}$

2. Find $\bar{\omega}$ from table.
3. For prestressed reinforcement, estimate f_{ps}.
4. Select A_{ps}, A_s and A'_s from:
$$A_{ps}f_{ps} + A_s f_y - A'_s f'_y = \bar{\omega} b d f'_c$$
5. Check assumed value of f_{ps}.

Analysis:

1. Determine $\bar{\omega} = \dfrac{A_{ps}f_{ps} + A_s f_y - A'_s f'_y}{b d f'_c}$

2. Find K_u from table.
3. Determine:
$$\phi M_n = \phi [K_u f'_c b d^2 + A'_s f'_y (d - d')]$$

Basis:
$K_u = \bar{\omega}(1 - 0.59\bar{\omega})$
M_u in units of lb-in.
$\omega_{max} = 0.32\beta_1$

ω_{max} for tension controlled section

f'_c (psi)	3000	4000	5000	6000	7000	8000 & up
ω_{max}	0.272	0.272	0.256	0.240	0.224	0.208

Values of K_u

$\bar{\omega}$	0.000	0.001	0.002	0.003	0.004	0.005	0.006	0.007	0.008	0.009
0.00	0.0000	0.0010	0.0020	0.0030	0.0040	0.0050	0.0060	0.0070	0.0080	0.0090
0.01	0.0099	0.0109	0.0119	0.0129	0.0139	0.0149	0.0158	0.0168	0.0178	0.0188
0.02	0.0198	0.0207	0.0217	0.0227	0.0237	0.0246	0.0256	0.0266	0.0275	0.0285
0.03	0.0295	0.0304	0.0314	0.0324	0.0333	0.0343	0.0352	0.0362	0.0371	0.0381
0.04	0.0391	0.0400	0.0410	0.0419	0.0429	0.0438	0.0448	0.0457	0.0466	0.0476
0.05	0.0485	0.0495	0.0504	0.0513	0.0523	0.0532	0.0541	0.0551	0.0560	0.0569
0.06	0.0579	0.0588	0.0597	0.0607	0.0616	0.0625	0.0634	0.0644	0.0653	0.0662
0.07	0.0671	0.0680	0.0689	0.0699	0.0708	0.0717	0.0726	0.0735	0.0744	0.0753
0.08	0.0762	0.0771	0.0780	0.0789	0.0798	0.0807	0.0816	0.0825	0.0834	0.0843
0.09	0.0852	0.0861	0.0870	0.0879	0.0888	0.0897	0.0906	0.0914	0.0923	0.0932
0.10	0.0941	0.0950	0.0959	0.0967	0.0976	0.0985	0.0994	0.1002	0.1011	0.1020
0.11	0.1029	0.1037	0.1046	0.1055	0.1063	0.1072	0.1081	0.1089	0.1098	0.1106
0.12	0.1115	0.1124	0.1132	0.1141	0.1149	0.1158	0.1166	0.1175	0.1183	0.1192
0.13	0.1200	0.1209	0.1217	0.1226	0.1234	0.1242	0.1251	0.1259	0.1268	0.1276
0.14	0.1284	0.1293	0.1301	0.1309	0.1318	0.1326	0.1334	0.1343	0.1351	0.1359
0.15	0.1367	0.1375	0.1384	0.1392	0.1400	0.1408	0.1416	0.1425	0.1433	0.1441
0.16	0.1449	0.1457	0.1465	0.1473	0.1481	0.1489	0.1497	0.1505	0.1513	0.1521
0.17	0.1529	0.1537	0.1545	0.1553	0.1561	0.1569	0.1577	0.1585	0.1593	0.1601
0.18	0.1609	0.1617	0.1625	0.1632	0.1640	0.1648	0.1656	0.1664	0.1671	0.1679
0.19	0.1687	0.1695	0.1703	0.1710	0.1718	0.1726	0.1733	0.1741	0.1749	0.1756
0.20	0.1764	0.1772	0.1779	0.1787	0.1794	0.1802	0.1810	0.1817	0.1825	0.1832
0.21	0.1840	0.1847	0.1855	0.1862	0.1870	0.1877	0.1885	0.1892	0.1900	0.1907
0.22	0.1914	0.1922	0.1929	0.1937	0.1944	0.1951	0.1959	0.1966	0.1973	0.1981
0.23	0.1988	0.1995	0.2002	02010	0.2017	0.2024	0.2031	0.2039	0.2046	0.2053
0.24	0.2060	0.2067	0.2074	02082	0.2089	0.2096	0.2103	0.2110	0.2117	0.2124
0.25	0.2131	0.2138	0.2145	0.2152	0.2159	0.2166	0.2173	0.2180	0.2187	0.2194
0.26	0.2201	0.2208	0.2215	0.2222	0.2229	0.2366	0.2243	0.2249	0.2256	0.2263
0.27	0.2270	0.2277	0.2283	0.2290	0.2297	0.2304	0.2311	0.2317	0.2324	0.2331
0.28	0.2337	0.2344	0.2351	0.2357	0.2364	0.2371	0.2377	0.2384	0.2391	0.2397
0.29	0.2404	0.2410	0.2417	0.2423	0.2430	0.2437	0.2443	0.2450	0.2456	0.2463
0.30	0.2469									

Figure 4.12.2 Coefficients K'_u for determining flexural design strength-bonded prestressing steel

Procedure:

1. Determine $\omega_{pu} = \left(\dfrac{A_{ps}}{bd_p}\right)\left(\dfrac{f_{pu}}{f'_c}\right)$

2. Find K'_u from table.

3. Determine $\phi M_n = K'_u \dfrac{bd_p^2}{12000}$ (kip-ft)

Basis:

$$K'_u = \dfrac{\phi f_{ps} f'_c}{f_{pu}}(\omega_{pu})\left[1-(0.59\omega_{pu})\left(\dfrac{f_{ps}}{f_{pu}}\right)\right]$$

Note:
K'_u from this table is approximately equivalent to $\phi K_u f'_c$ from Figure 4.12.1.

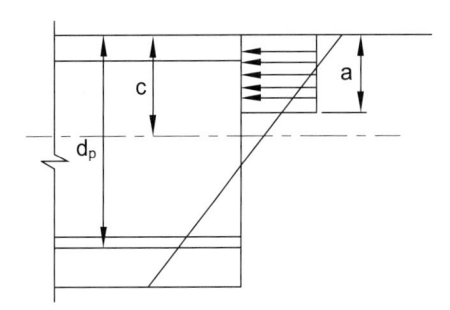

Table values are based on a strain compatibility analysis, using a stress-strain curve for prestressing strand similar to that in Design Aid 11.2.5.

Values of K'_u

f'_c	ω_{pu}	0.00	0.01	0.02	0.03	0.04	0.05	0.06	0.07	0.08	0.09
3000	0.0	0	27	53	79	105	131	156	180	205	229
	0.1	252	275	298	321	343	364	386	407	427	448
	0.2	467	487	506	524	543	560	578	595	611	613
	0.3	614	615	616	617	618	619	620	621	621	622
4000	0.0	0	36	71	106	140	174	207	240	273	305
	0.1	336	367	398	428	457	486	514	542	570	597
	0.2	623	649	674	699	724	747	770	793	815	817
	0.3	818	820	821	823	824	825	826	827	828	829
5000	0.0	0	45	89	132	175	218	259	300	341	381
	0.1	420	459	497	534	571	607	642	677	711	745
	0.2	777	810	841	872	902	931	960	971	974	976
	0.3	979	981	983	985	986	988	990	991	992	993
6000	0.0	0	54	107	159	210	261	311	360	409	457
	0.1	504	550	595	640	684	727	769	811	852	892
	0.2	931	969	1,006	1,043	1,078	1,104	1,107	1,111	1,114	1,117
	0.3	1,120	1,123	1,126	1,128	1,131	1,133	1,135	1,136	1,138	1,139
7000	0.0	0	63	124	185	245	304	363	420	477	532
	0.1	587	641	694	746	797	847	896	944	991	1,037
	0.2	1,082	1,126	1,169	1,211	1,218	1,223	1,228	1,232	1,236	1,240
	0.3	1,244	1,247	1,251	1,254	1,256	1,258	1,260	1,263	1,265	1,267
8000	0.0	0	72	142	212	280	348	414	480	544	608
	0.1	670	731	792	851	909	966	1,022	1,076	1,129	1,181
	0.2	1,232	1,282	1,305	1,312	1,318	1,324	1,329	1,334	1,339	1,344
	0.3	1,348	1,352	1,355	1,358	1,361	1,364	1,367	1,369	1,372	1,375

Figure 4.12.3 Values of f_{ps} by stress-strain relationship – Bonded strand

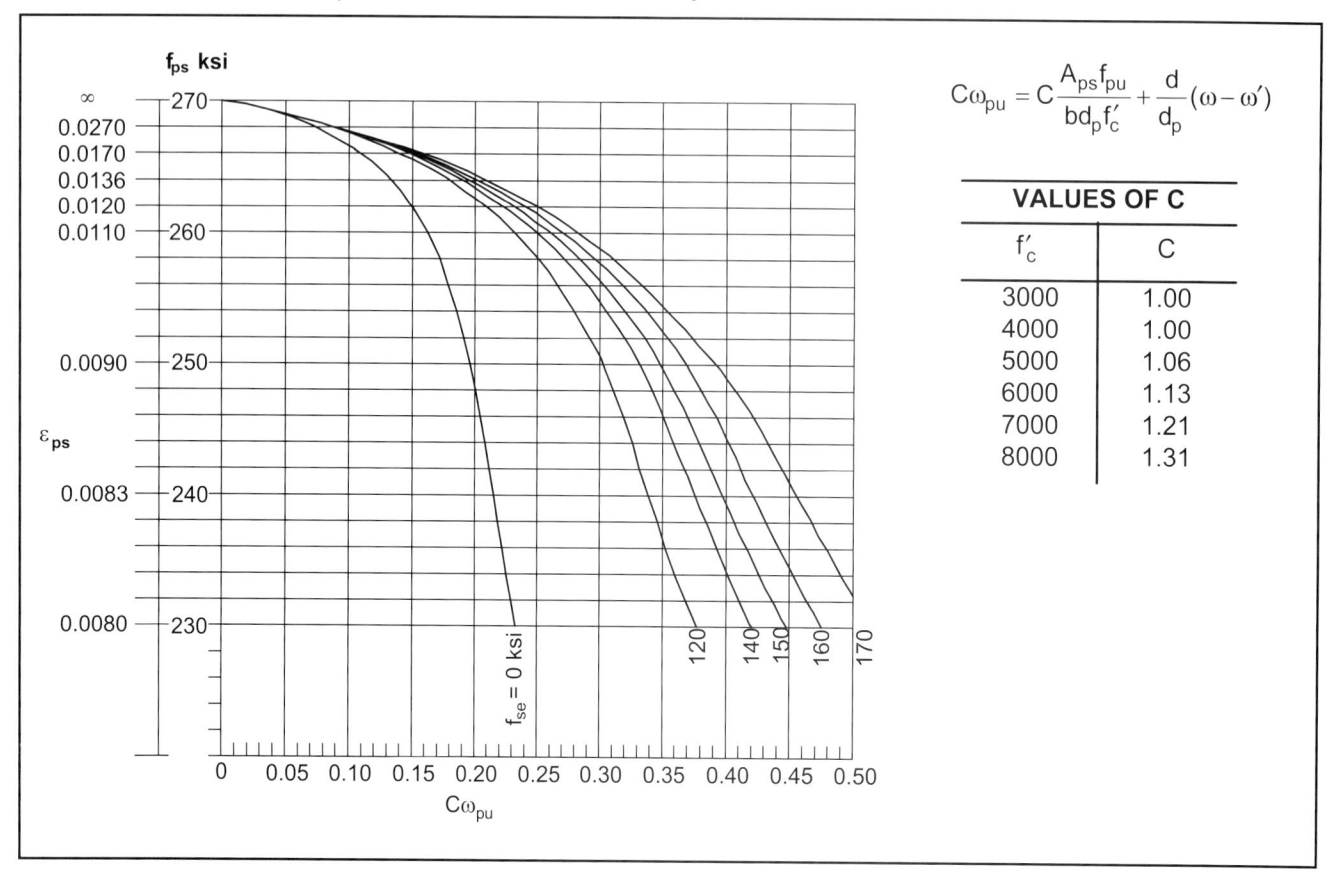

$$C\omega_{pu} = C\frac{A_{ps}f_{pu}}{bd_pf'_c} + \frac{d}{d_p}(\omega - \omega')$$

VALUES OF C

f'_c	C
3000	1.00
4000	1.00
5000	1.06
6000	1.13
7000	1.21
8000	1.31

Figure 4.12.4 Design stress for underdeveloped strand

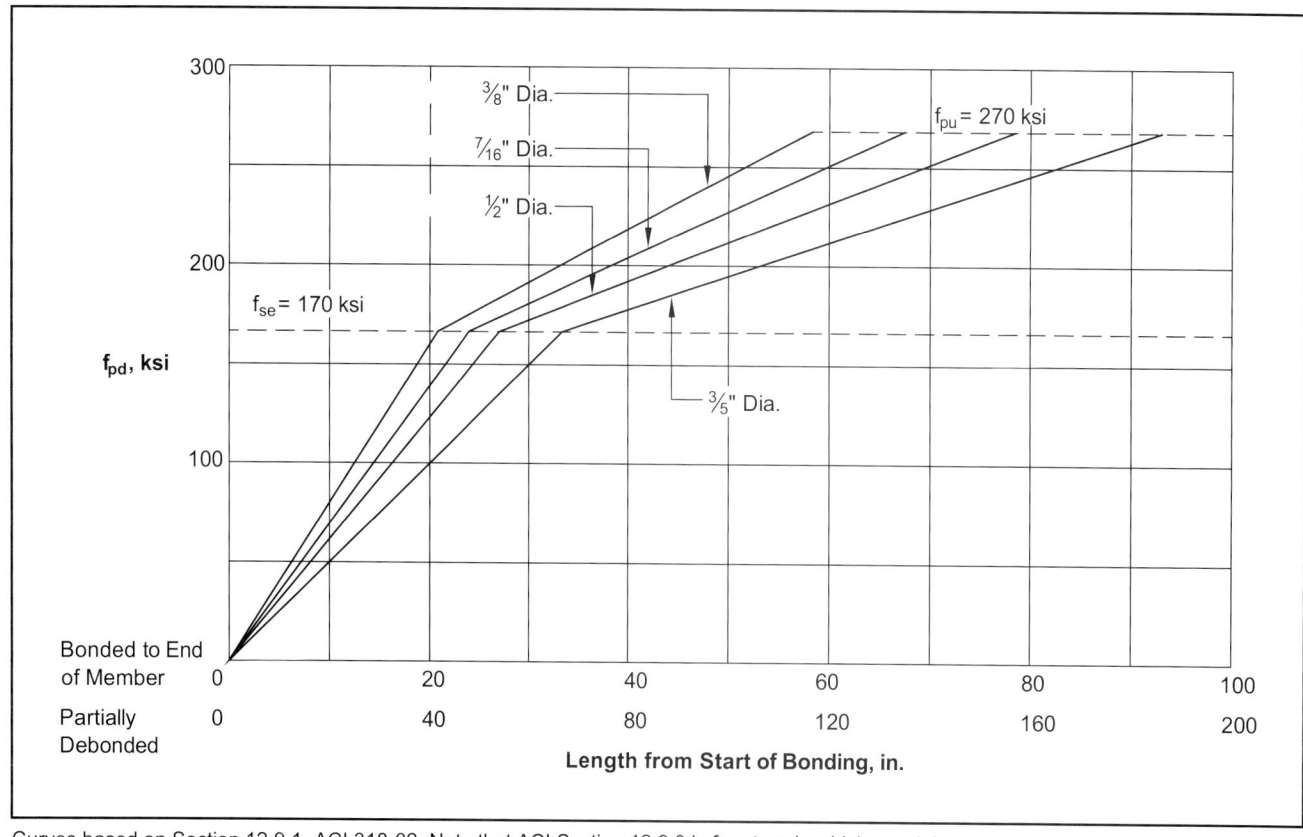

Curves based on Section 12.9.1, ACI 318-02. Note that ACI Section 12.9.3 is for strands which are debonded near member ends.

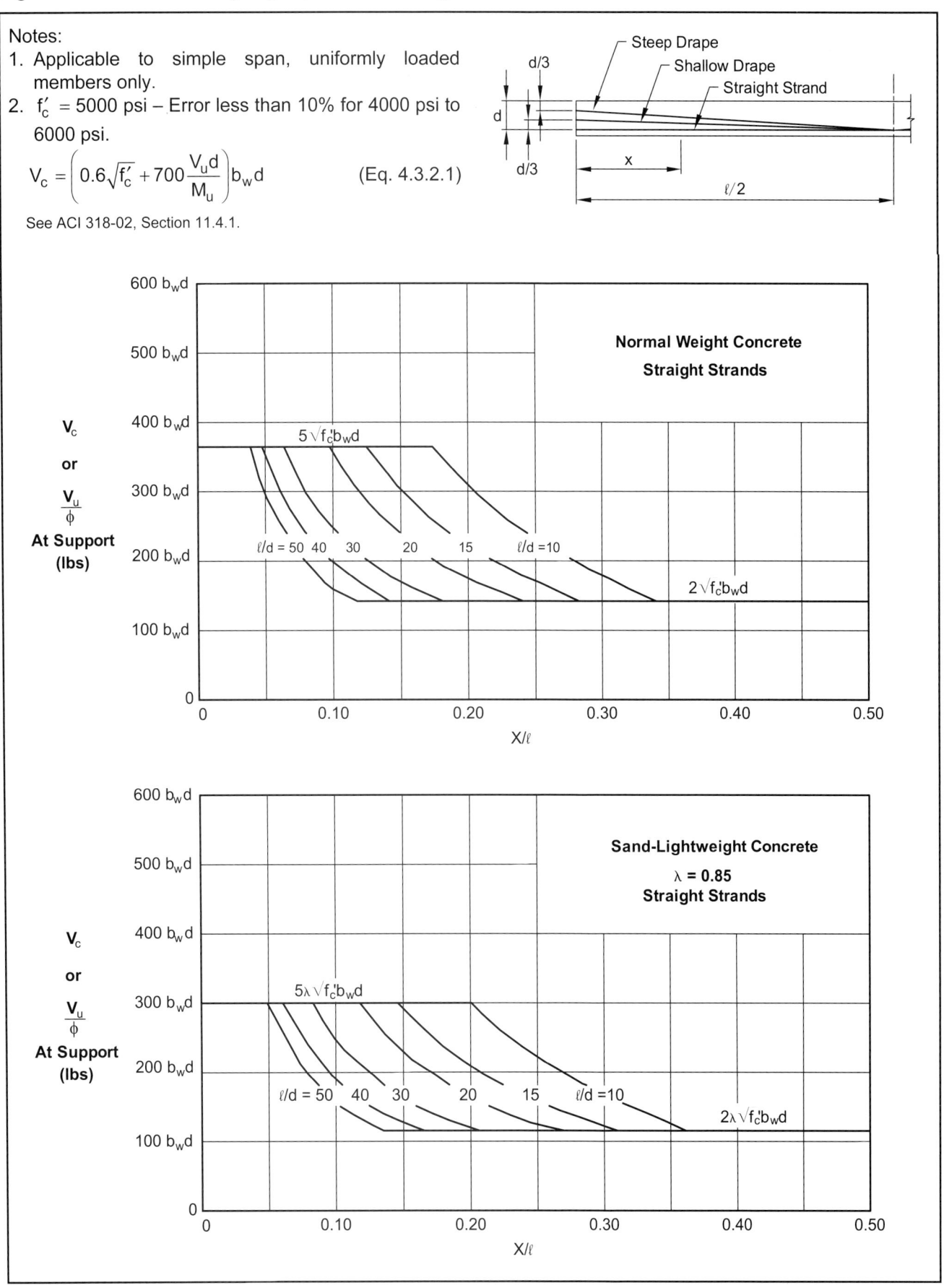

Figure 4.12.5 Shear design by Eq. 4.3.2.1 – Straight strands

Notes:
1. Applicable to simple span, uniformly loaded members only.
2. $f'_c = 5000$ psi – Error less than 10% for 4000 psi to 6000 psi.

$$V_c = \left(0.6\sqrt{f'_c} + 700\frac{V_u d}{M_u}\right)b_w d \qquad \text{(Eq. 4.3.2.1)}$$

See ACI 318-02, Section 11.4.1.

Normal Weight Concrete
Straight Strands

V_c or $\dfrac{V_u}{\phi}$ At Support (lbs) vs X/ℓ

Upper bound: $5\sqrt{f'_c}b_w d$; Lower bound: $2\sqrt{f'_c}b_w d$; Curves for $\ell/d = 50, 40, 30, 20, 15, 10$.

Sand-Lightweight Concrete
λ = 0.85
Straight Strands

Upper bound: $5\lambda\sqrt{f'_c}b_w d$; Lower bound: $2\lambda\sqrt{f'_c}b_w d$; Curves for $\ell/d = 50, 40, 30, 20, 15, 10$.

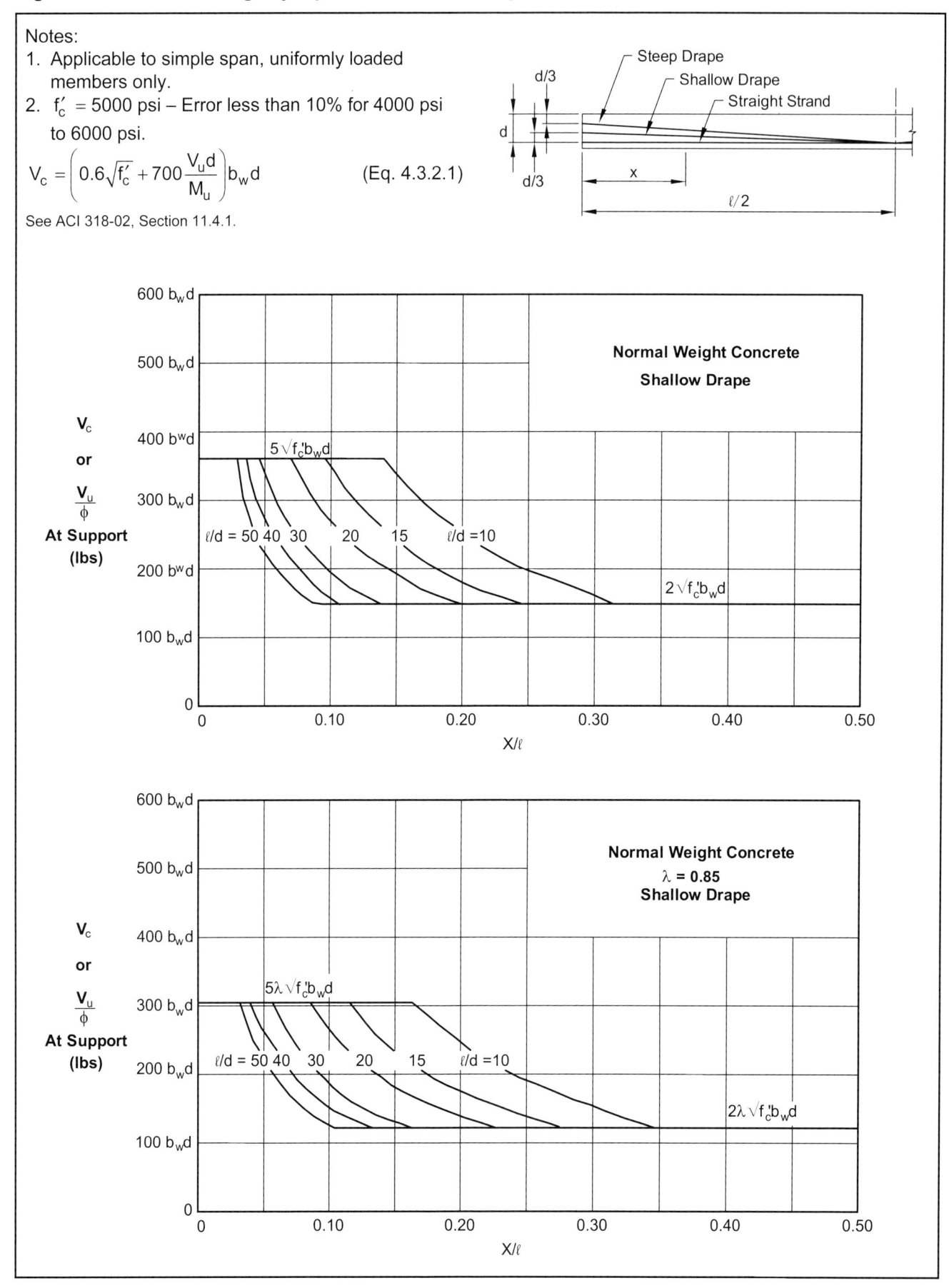

Figure 4.12.6 Shear design by Eq. 4.3.2.1 – Shallow drape

Notes:
1. Applicable to simple span, uniformly loaded members only.
2. $f'_c = 5000$ psi – Error less than 10% for 4000 psi to 6000 psi.

$$V_c = \left(0.6\sqrt{f'_c} + 700\frac{V_u d}{M_u}\right)b_w d \qquad \text{(Eq. 4.3.2.1)}$$

See ACI 318-02, Section 11.4.1.

Normal Weight Concrete — Shallow Drape

Curves for $\ell/d = 50, 40, 30, 20, 15, 10$, bounded by $5\sqrt{f'_c}b_w d$ upper limit and $2\sqrt{f'_c}b_w d$ lower limit. Vertical axis: V_c or V_u/ϕ at Support (lbs). Horizontal axis: X/ℓ.

Normal Weight Concrete — $\lambda = 0.85$ — Shallow Drape

Curves for $\ell/d = 50, 40, 30, 20, 15, 10$, bounded by $5\lambda\sqrt{f'_c}b_w d$ upper limit and $2\lambda\sqrt{f'_c}b_w d$ lower limit. Vertical axis: V_c or V_u/ϕ at Support (lbs). Horizontal axis: X/ℓ.

Figure 4.12.7 Shear design by Eq. 4.3.2.1 – Steep drape

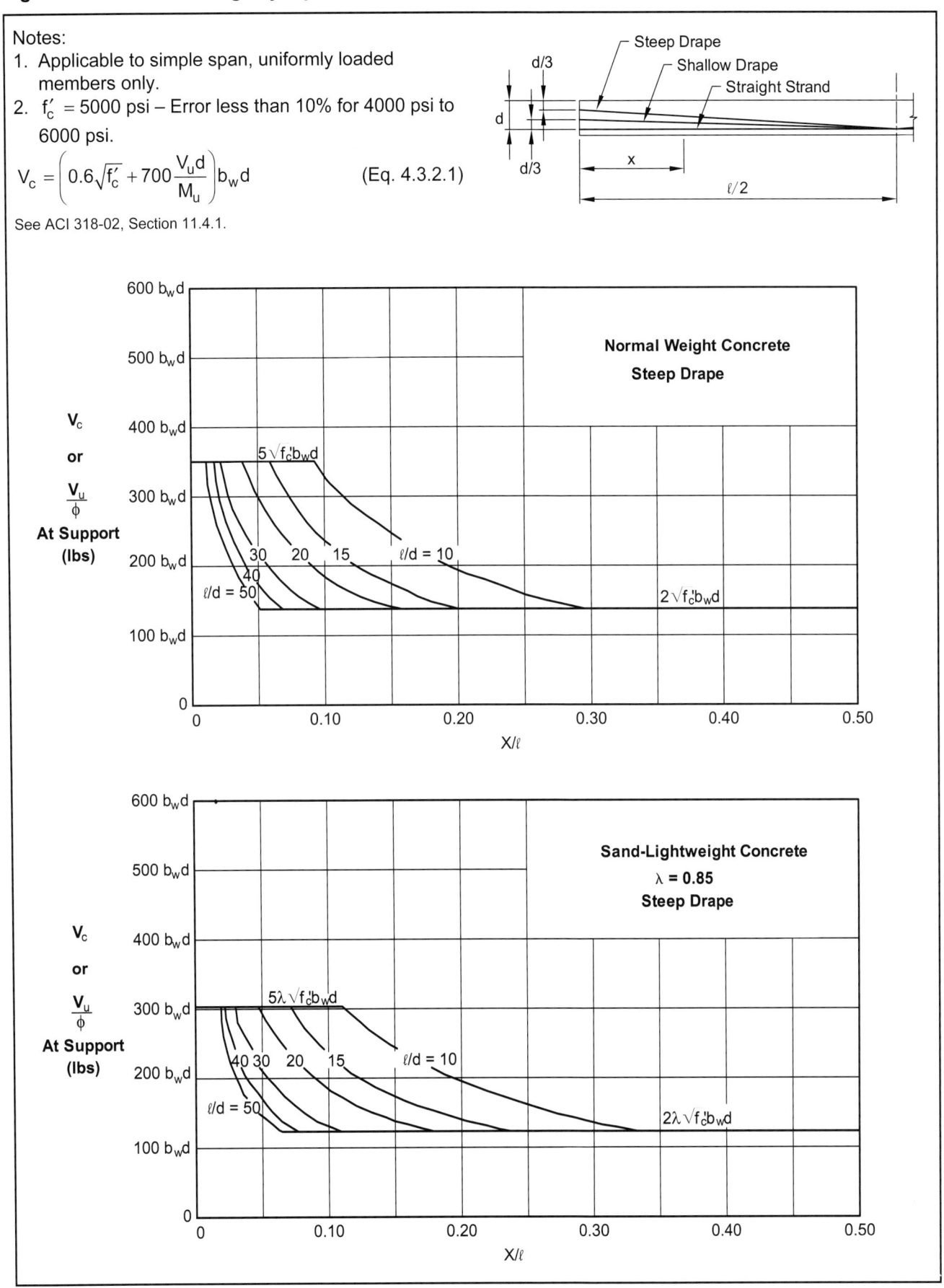

Notes:
1. Applicable to simple span, uniformly loaded members only.
2. $f'_c = 5000$ psi – Error less than 10% for 4000 psi to 6000 psi.

$$V_c = \left(0.6\sqrt{f'_c} + 700\frac{V_u d}{M_u}\right)b_w d \qquad \text{(Eq. 4.3.2.1)}$$

See ACI 318-02, Section 11.4.1.

Normal Weight Concrete — Steep Drape

V_c or $\dfrac{V_u}{\phi}$ At Support (lbs) vs. X/ℓ

Sand-Lightweight Concrete — $\lambda = 0.85$ — Steep Drape

V_c or $\dfrac{V_u}{\phi}$ At Support (lbs) vs. X/ℓ

PCI Design Handbook/Sixth Edition

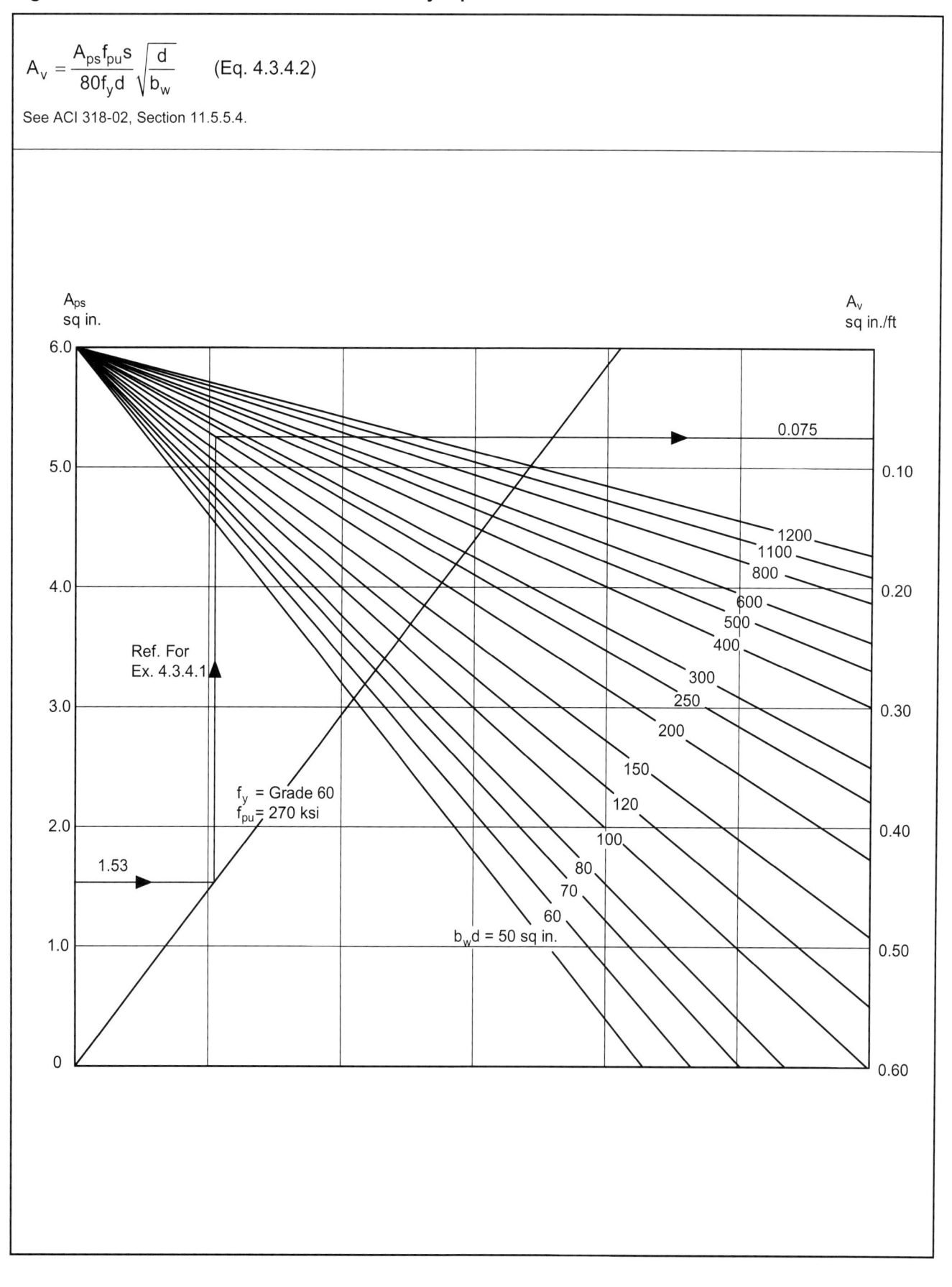

Figure 4.12.8 Minimum shear reinforcement by Eq. 4.3.4.2

$$A_v = \frac{A_{ps} f_{pu} s}{80 f_y d} \sqrt{\frac{d}{b_w}} \quad \text{(Eq. 4.3.4.2)}$$

See ACI 318-02, Section 11.5.5.4.

4–126   PCI Design Handbook/Sixth Edition

Figure 4.12.9 Shear reinforcement

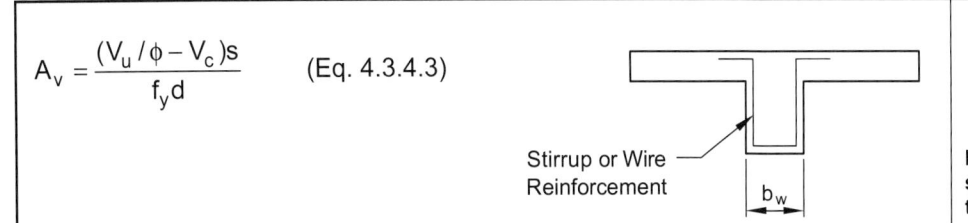

$$A_v = \frac{(V_u/\phi - V_c)s}{f_y d} \quad \text{(Eq. 4.3.4.3)}$$

Note: Other configurations of shear reinforcement may be used to simplify production of precast members.

Vertical Deformed Bar Stirrups

Maximum values of $\frac{(V_u/\phi - V_c)}{d}$ (lb/in.)

$f_y = 60{,}000$ psi

Stirrup Spacing (in.)	No 3 $A_v = 0.22$	No. 4 $A_v = 0.40$	No. 5 $A_v = 0.62$	No 3 $A_v = 0.22$	No. 4 $A_v = 0.40$	No. 5 $A_v = 0.62$	Stirrup Spacing (in.)
2.0	6600	12000	18600	1320	2400	3720	10.0
2.5	5280	9600	14880	1200	2182	3382	11.0
3.0	4400	8000	12400	1100	2000	3100	12.0
3.5	3771	6857	10629	1015	1846	2862	13.0
4.0	3300	6000	9300	943	1714	2657	14.0
4.5	2933	5333	8267	880	1600	2480	15.0
5.0	2640	4800	7440	825	1500	2325	16.0
5.5	2400	4364	6764	776	1412	2188	17.0
6.0	2200	4000	6200	733	1333	2067	18.0
7.0	1886	3429	5314	660	1200	1860	20.0
8.0	1650	3000	4650	600	1091	1691	22.0
9.0	1467	2667	4133	550	1000	1550	24.0

Welded Wire Reinforcement (WWR) as Shear Reinforcement ($f_y = 60{,}000$ psi)

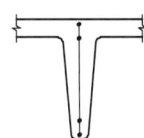

Maximum values of $\frac{(V_u/\phi - V_c)}{d}$ (lb/in.)

Spacing of Vertical Wire (in.)	One Row Vertical Wire W7.5 $A_v = 0.075$	W5.5 $A_v = 0.055$	W4 $A_v = 0.040$	W2.9 $A_v = 0.029$	Two Rows Vertical Wire W7.5 $A_v = 0.150$	W5.5 $A_v = 0.110$	W4 $A_v = 0.080$	W2.9 $A_v = 0.058$	Spacing of Vertical Wire (in.)
2	2250	1650	1200	870	4500	3300	2400	1740	2
3	1500	1100	800	580	3000	2200	1600	1160	3
4	1125	825	600	435	2250	1650	1200	870	4
6	750	550	400	290	1500	1100	800	580	6

Figure 4.12.10 Camber equations for typical strand profiles

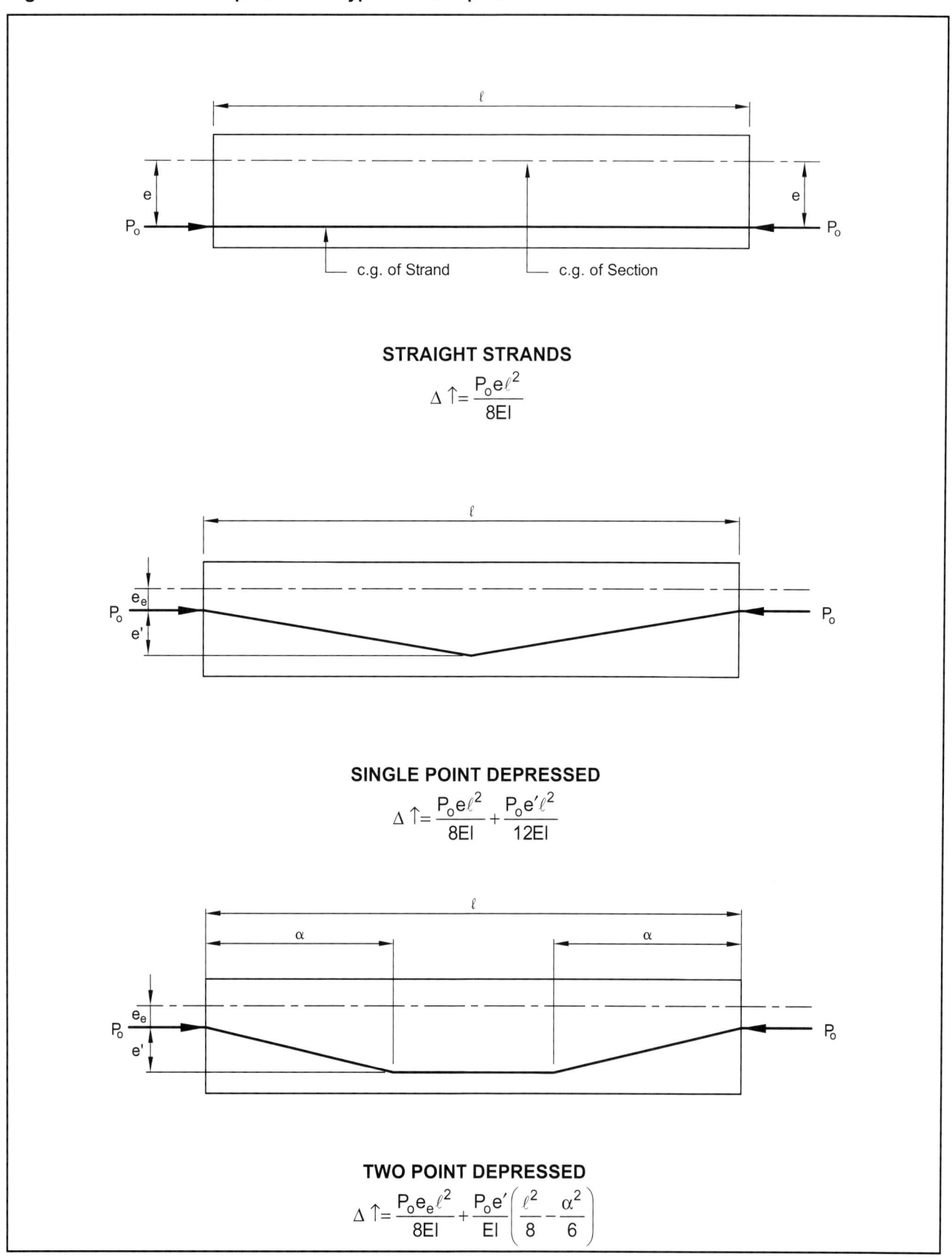

STRAIGHT STRANDS

$$\Delta \uparrow = \frac{P_o e \ell^2}{8EI}$$

SINGLE POINT DEPRESSED

$$\Delta \uparrow = \frac{P_o e \ell^2}{8EI} + \frac{P_o e' \ell^2}{12EI}$$

TWO POINT DEPRESSED

$$\Delta \uparrow = \frac{P_o e_e \ell^2}{8EI} + \frac{P_o e'}{EI}\left(\frac{\ell^2}{8} - \frac{\alpha^2}{6}\right)$$

Figure 4.12.11 Moment of inertia of transformed section – Prestressed members

$I_{cr} = nA_{ps}d_p^2(1-1.6\sqrt{n\rho_p})$
$= n\rho_p(1-1.6\sqrt{n\rho_p}) \times bd_p^3$
$= C(\text{from table}) \times bd_p^3$

where: $\rho_p = \dfrac{A_{ps}}{bd_p}$, $n = \dfrac{E_s}{E_c}$

$E_s = 28.5 \times 10^6$ psi
$E_c = 33w_c^{1.5}\sqrt{f_c'}$ (ACI Section 8.5.1)
$w_c = 145$ lb/ft^3 normal weight concrete
$w_c = 115$ lb/ft^3 sand-lightweight concrete

Values of Coefficient, C

| | ρ_p | \multicolumn{6}{c}{f_c', psi} |
		3000	4000	5000	6000	7000	8000
Normal Weight Concrete	0.0005	0.0041	0.0036	0.0032	0.0029	0.0027	0.0026
	0.0010	0.0077	0.0068	0.0061	0.0056	0.0052	0.0049
	0.0015	0.0111	0.0098	0.0089	0.0082	0.0076	0.0072
	0.0020	0.0143	0.0126	0.0115	0.0106	0.0099	0.0093
	0.0025	0.0173	0.0153	0.0139	0.0129	0.0120	0.0113
	0.0030	0.0201	0.0179	0.0163	0.0151	0.0141	0.0133
	0.0035	0.0228	0.0203	0.0185	0.0172	0.0161	0.0152
	0.0040	0.0254	0.0226	0.0207	0.0192	0.0180	0.0170
	0.0045	0.0278	0.0248	0.0227	0.0211	0.0198	0.0188
	0.0050	0.0300	0.0270	0.0247	0.0230	0.0216	0.0205
	0.0055	0.0322	0.0290	0.0266	0.0248	0.0233	0.0221
	0.0060	0.0343	0.0309	0.0284	0.0265	0.0250	0.0237
	0.0065	0.0362	0.0327	0.0302	0.0282	0.0266	0.0253
	0.0070	0.0381	0.0345	0.0319	0.0298	0.0281	0.0267
	0.0075	0.0398	0.0362	0.0335	0.0314	0.0296	0.0282
	0.0080	0.0415	0.0378	0.0350	0.0329	0.0311	0.0296
	0.0085	0.0430	0.0393	0.0365	0.0343	0.0325	0.0309
	0.0090	0.0445	0.0408	0.0380	0.0357	0.0338	0.0323
	0.0095	0.0459	0.0422	0.0393	0.0370	0.0351	0.0335
	0.0100	0.0472	0.0435	0.0406	0.0383	0.0364	0.0348
Sand-Lightweight Concrete	0.0005	0.0056	0.0049	0.0044	0.0040	0.0029	0.0036
	0.0010	0.0105	0.0092	0.0083	0.0077	0.0071	0.0067
	0.0015	0.0149	0.0132	0.0120	0.0110	0.0103	0.0097
	0.0020	0.0190	0.0169	0.0153	0.0142	0.0133	0.0125
	0.0025	0.0228	0.0203	0.0185	0.0172	0.0161	0.0152
	0.0030	0.0263	0.0235	0.0215	0.0200	0.0187	0.0177
	0.0035	0.0296	0.0265	0.0243	0.0226	0.0213	0.0201
	0.0040	0.0326	0.0294	0.0270	0.0252	0.0237	0.0224
	0.0045	0.0354	0.0320	0.0295	0.0275	0.0260	0.0246
	0.0050	0.0381	0.0345	0.0319	0.0298	0.0281	0.0267
	0.0055	0.0405	0.0368	0.0341	0.0320	0.0302	0.0288
	0.0060	0.0427	0.0390	0.0362	0.0340	0.0322	0.0307
	0.0065	0.0448	0.0411	0.0382	0.0360	0.0341	0.0325
	0.0070	0.0466	0.0430	0.0401	0.0378	0.0359	0.0343
	0.0075	0.0484	0.0447	0.0419	0.0395	0.0376	0.0359
	0.0080	0.0499	0.0464	0.0435	0.0412	0.0392	0.0375
	0.0085	0.0513	0.0478	0.0451	0.0428	0.0408	0.0391
	0.0090	0.0526	0.0493	0.0465	0.0442	0.0422	0.0405
	0.0095	0.0537	0.0506	0.0479	0.0456	0.0436	0.0419
	0.0100	0.0547	0.0518	0.0492	0.0469	0.0449	0.0432

Figure 4.12.12 Forces required to restrain bowing

Intermediate Restraint (Ends Free to Rotate)	End Restraint
Case 1: Single Restraint at Midspan $P = \dfrac{48E_t I \Delta}{\ell^3}$ Moment in panel $= \dfrac{P\ell}{4}$	**Case 4: Both Ends Restrained** $M = \dfrac{8E_t I \Delta}{\ell^2}$
Case 2: Two Restraint Points $P = \dfrac{24E_t I \Delta}{3a\ell^2 - 4a^3}$ Moment in panel $= Pa$	**Case 5: One End Restrained** $M = \dfrac{16E_t I \Delta}{\ell^2}$
Case 3: Three or More Restraint Points (Approximate Uniform Continuous Restraint) $\Sigma P = w\ell = \dfrac{77E_t I \Delta}{\ell^3}$ Moment in panel $= \dfrac{w\ell^2}{8} = \Sigma P \left(\dfrac{\ell}{8}\right)$	For Daily Temperature Change, Use $E_t = 0.75\, E_c$ For Season Changes, Use $E_t = 0.50\, E_c$

Figure 4.12.13 Use of Figure 4.12.14

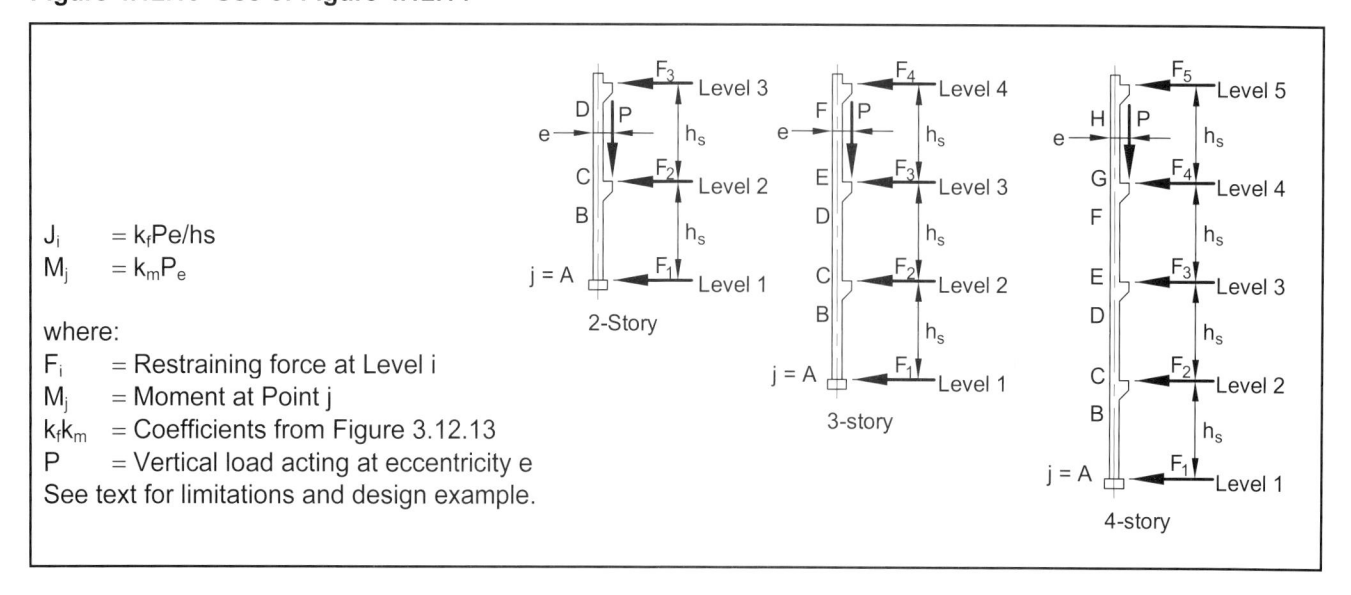

$J_i = k_f Pe/h_s$
$M_j = k_m P_e$

where:
F_i = Restraining force at Level i
M_j = Moment at Point j
k_f, k_m = Coefficients from Figure 3.12.13
P = Vertical load acting at eccentricity e
See text for limitations and design example.

Figure 4.12.14 Coefficients k_f and k_m for determining moments and restraining forces on eccentrically loaded columns braced against sidesway

See Figure 4.12.13 for explanation of terms
+ Indicates clockwise moments on the columns and compression in the restraining beam

No. of stories	Base fixity	Packing at level	k_f at level 1	2	3	4	5	k_m at point A	B	C	D	E	F	G	H
2	PINNED	3	+0.25	−1.50	+1.25			0	−0.25	+0.25	+1.00				
		2	−0.50	0	+0.50			0	+0.50	+0.50	0				
		ΣMax	±0.75	±1.50	±1.75			0	±0.75	±0.75	±1.00				
		ΣOne Side	−0.25	−1.50	+1.75			0	+0.25	+0.75	+1.00				
	FIXED	3	+0.43	−1.72	+1.29			−0.14	−0.29	+0.29	+1.00				
		2	−0.86	+0.43	+0.43			+0.29	+0.57	+0.43	0				
		ΣMax	±1.29	±2.15	±1.72			±0.43	±0.86	±0.72	±1.00				
		ΣOne Side	−0.43	−1.29	+1.72			+0.15	+0.28	+0.72	+1.00				
3	PINNED	4	−0.07	+0.40	−1.60	+1.27		0	+0.07	−0.07	−0.27	+0.27	+1.00		
		3	+0.13	−0.80	+0.20	+0.47		0	−0.13	+0.13	+0.53	+0.47	0		
		2	−0.47	−0.20	+0.80	−0.13		0	+0.47	+0.53	+0.13	−0.13	0		
		ΣMax	±0.67	±1.40	±2.60	±1.87		0	±0.67	±0.73	±0.93	±0.87	±1.00		
		ΣOne Side	−0.41	−0.60	−0.60	+1.61		0	+0.40	+0.60	+0.40	+0.60	+1.00		
	FIXED	4	−0.12	+0.47	−1.62	+1.27		+0.04	+0.08	−0.08	−0.27	+0.27	+1.00		
		3	+0.23	−0.92	+0.23	+0.46		−0.08	−0.15	+0.15	+0.54	+0.46	0		
		2	−0.81	+0.23	+0.70	−0.12		+0.27	+0.54	+0.46	+0.12	−0.12	0		
		ΣMax	±1.16	±1.62	±2.55	±1.85		±0.38	±0.77	±0.69	±0.92	±0.85	±1.00		
		ΣOne Side	−0.70	−0.22	−0.69	+1.61		+0.23	+0.46	+0.54	+0.38	+0.62	+1.00		
4	PINNED	5	+0.02	−0.11	+0.43	−1.61	+1.27	0	−0.02	+0.02	+0.07	−0.07	−0.27	+0.27	+1.00
		4	−0.04	+0.22	−0.86	+0.22	+0.46	0	+0.04	−0.04	−0.14	+0.14	+0.54	+0.46	0
		3	+0.13	−0.75	0	+0.75	−0.12	0	−0.13	+0.13	+0.50	+0.50	+0.12	−0.12	0
		2	−0.46	−0.22	+0.86	−0.22	+0.04	0	+0.46	+0.54	+0.14	−0.14	−0.04	+0.04	0
		ΣMax	±0.65	±1.30	±2.15	±2.80	±1.89	0	±0.64	±0.72	±0.86	±0.86	±0.97	±0.89	±1.00
		ΣOne Side	−0.35	−0.86	+0.43	−0.86	+1.65	0	+0.35	+0.65	+0.57	+0.43	+0.35	+0.65	+1.00
	FIXED	5	+0.03	−0.12	+0.43	−1.61	+1.27	−0.01	−0.02	+0.02	+0.07	−0.07	−0.27	+0.27	+1.00
		4	−0.06	+0.25	−0.87	+0.22	+0.46	+0.02	+0.04	−0.04	−0.14	+0.14	+0.54	+0.46	0
		3	+0.22	−0.87	+0.03	+0.74	−0.12	−0.07	−0.14	+0.14	+0.51	+0.50	+0.12	−0.12	0
		2	−0.80	+0.21	+0.74	−0.18	+0.03	+0.27	+0.54	+0.46	+0.12	−0.12	−0.03	+0.03	0
		ΣMax	±1.11	±1.45	±2.07	±2.75	±1.88	±0.37	±0.74	±0.67	±0.84	±0.83	±0.93	±0.88	±1.00
		ΣOne Side	−0.61	−0.53	+0.33	−0.83	+1.64	+0.21	+0.41	+0.59	+0.56	+0.44	+0.36	+0.64	+1.00

CHAPTER 5
PRODUCT HANDLING AND ERECTION BRACING

5.1	General	5-2
	5.1.1 Notation	5-2
5.2	Product Handling	5-3
	5.2.1 Introduction	5-3
	5.2.2 Planning and Setup	5-3
5.3	Stripping	5-3
	5.3.1 General	5-3
	5.3.2 Rigging Configurations	5-7
	5.3.3 Stripping Design	5-9
	5.3.3.1 Form Suction and Impact Factors	5-9
	5.3.3.2 Factors of Safety	5-9
	5.3.3.3 Stress Limits and Crack Control	5-9
	5.3.3.4 Benefits of Prestressing	5-10
	5.3.4 Handling Devices	5-10
	5.3.4.1 Aircraft Cable Loops	5-11
	5.3.4.2 Prestressing Strand Loops	5-11
	5.3.4.3 Threaded Inserts	5-11
	5.3.4.4 Proprietary Devices	5-12
5.4	Yarding and Storage	5-19
	5.4.1 Lateral Stability	5-19
	5.4.2 Storage	5-19
5.5	Transportation	5-20
5.6	Erection	5-22
5.7	Erection Bracing	5-26
	5.7.1 Introduction	5-26
	5.7.1.1 Responsibilities	5-26
	5.7.1.2 Handling Equipment	5-26
	5.7.1.3 Surveying and Layout	5-26
	5.7.2 Loads	5-26
	5.7.3 Factors of Safety	5-28
	5.7.4 Bracing Equipment and Materials	5-28
	5.7.5 Erection Analysis and Associated Bracing Requirements	5-28
5.8	References	5-35

PRODUCT HANDLING AND ERECTION BRACING

5.1 General

5.1.1 Notation

a	=	dimension defined in section used
A	=	area; average effective area around one reinforcing bar
A_s	=	area of reinforcement
A'_s	=	area of compression reinforcement
b	=	dimension defined in section used
d	=	depth from extreme compression fiber to centroid of tension reinforcement
e	=	eccentricity of force about center of gravity
E_{ci}	=	modulus of elasticity of concrete at age other than final
f_b, f_t	=	stress in bottom and top fiber, respectively
f'_c	=	specified concrete compressive strength
f'_{ci}	=	concrete compressive strength at time considered
f_{ct}	=	splitting tensile strength
f_r	=	allowable flexural tensile stress computed using gross concrete section
f_y	=	yield strength of steel
F	=	multiplication factor (see Figure 5.3.2.1)
I	=	moment of inertia (with subscripts)
ℓ	=	span; length of precast unit
ℓ_e	=	embedment length
M	=	bending moment (with subscripts)
M_x, M_y	=	see Figures 5.3.1.1 and 5.3.1.2
P	=	axial load
P, P_H, P_V	=	see Figure 5.3.2.2
R	=	reaction (with subscripts)
S_b, S_t	=	section modulus with respect to bottom and top, respectively
t	=	thickness of panel
T	=	tension load on sling
w	=	weight per unit length or area
w_b	=	wind load on beam
w_c	=	wind load on column
W	=	total load
y_b, y_c, y_t	=	see Figure 5.3.2.2
y_{max}	=	instantaneous maximum displacement
y_t	=	time dependent displacement; height of roll axis above center of gravity of beam
ϕ	=	angle of lift line from vertical in transverse direction; strength reduction factor
λ	=	correction factor related to unit weight of concrete; deflection amplification factor
μ	=	coefficient of curvature friction for unbonded tendons
θ	=	angle of lift lines in longitudinal directions (sling angle); angle of roll in long slender members
θ_{max}	=	maximum permissible tilt angle
ρ	=	reinforcement ratio for non-prestressed tension reinforcement
ρ'	=	reinforcement ratio for non-prestressed compression reinforcement

5.2 Product Handling

5.2.1 Introduction

The loads and forces on precast and prestressed concrete members during production, transportation or erection will frequently require a separate analysis because concrete strengths are lower and support points and orientation are usually different from members in their final load carrying position.

Most structural products are manufactured in long-line steel forms with fixed cross section dimensions. Standard structural product dimensions are shown in Chapter 2 and in manufacturers' catalogs. Architectural wall panels are formed in a variety of sizes and shapes. The project architect usually designs these characteristics and the members are then manufactured in molds fabricated specifically for the project.

The most economical piece size for a project is usually the largest, considering the following factors:

- Stability and stresses on the element during handling.
- Transportation size and weight regulations and equipment restrictions.
- Available crane capacity at both the plant and the project site. Position of the crane must be considered, since capacity is a function of reach.
- Storage space, truck turning radius, and other site restrictions.

5.2.2 Planning and Setup

Proper planning of production methods is essential. Once a piece has been fabricated, it is necessary to remove it from the mold without being damaged.

Positive drafts or breakaway forms should be used to allow a member to lift away from the casting bed without becoming wedged within the form (see Figure 5.2.2.1). Adequate draft also serves to reduce trapped air bubbles.

Lifting points must be located to keep member stresses within limits and to ensure proper alignment of the piece as it is being lifted. Common lifting configurations and analysis of these arrangements are discussed later in this chapter. Members with unsymmetrical geometry or projecting sections may require supplemental lifting points and auxiliary lifting lines to achieve even support during handling. "Come-alongs" or "chain-falls" are frequently used for these auxiliary lines (see Figure 5.2.2.2). When

Figure 5.2.2.1 Positive drafts and breakaway forms

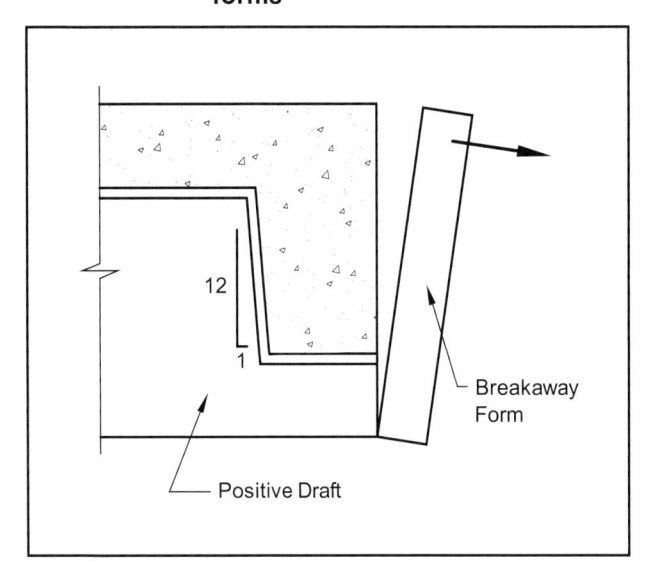

the member has areas of small cross section or large cantilevers, it may be necessary to add a structural steel "strongback" to the piece to provide added strength (see Figures 5.2.2.3 and 5.2.2.4).

Members that require a secondary process prior to shipment, such as sandblasting or attachment of haunches, may need to be rotated at the production facility. In these cases, it may be necessary to cast in extra lifting devices to facilitate these maneuvers.

When developing member shapes, the designer should consider the extra costs associated with special rigging or forming, and pieces requiring multiple handling.

5.3 Stripping

5.3.1 General

Orientation of members during storage, shipping and final in-place position is critical in determining stripping requirements. They can be horizontal, vertical or some angle in between.

The number and location of lifting devices are chosen to keep stresses within the allowable limits, which depends on whether the "no cracking" or "controlled cracking" criteria is to be used.* It is desirable to use the same lifting devices for both stripping and erection; however, additional devices may be required to rotate the member to its final position.

Panels that are stripped by rotating about one edge with lifting devices at the opposite edge will develop moments as shown in Figure 5.3.1.1. When panels are stripped this way, care should be taken to

---

* See Section 4.2.2.1 for discussion of crack control.

Figure 5.2.2.2 Supplemental lifting points

Supplemental Line with "Come-Along"

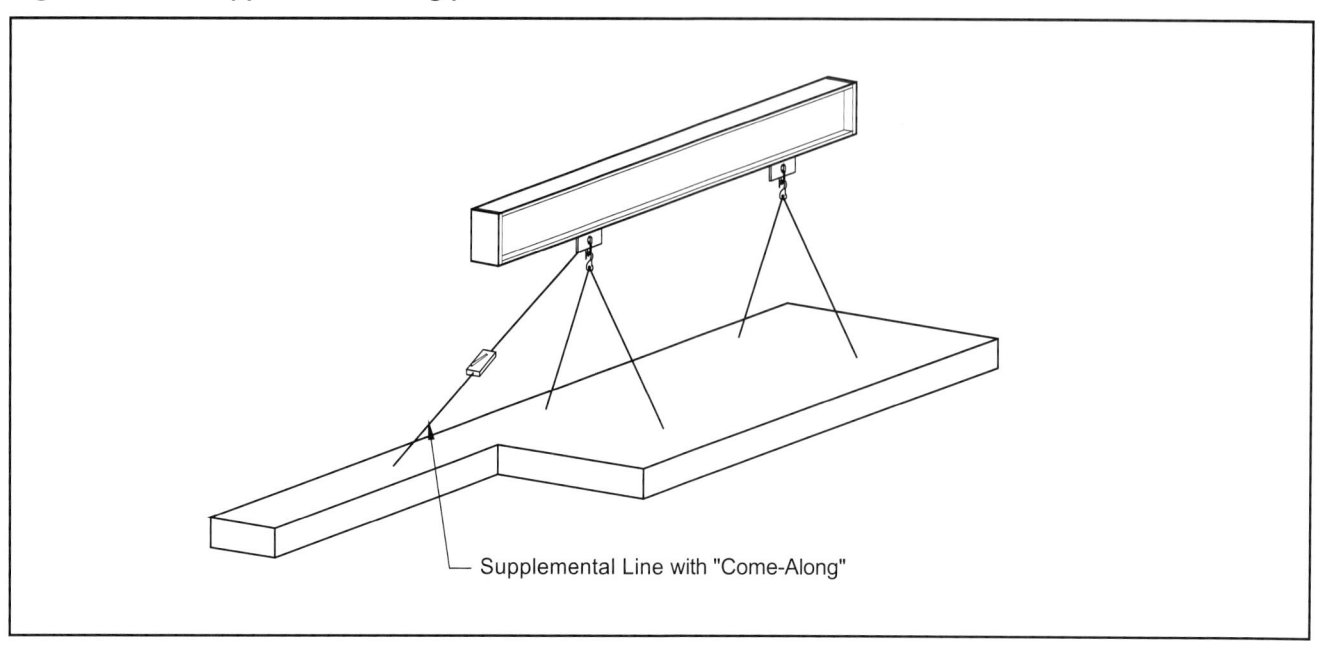

Figure 5.2.2.3 Strongback for slender section

Steel Angle Bolted to Slender Section

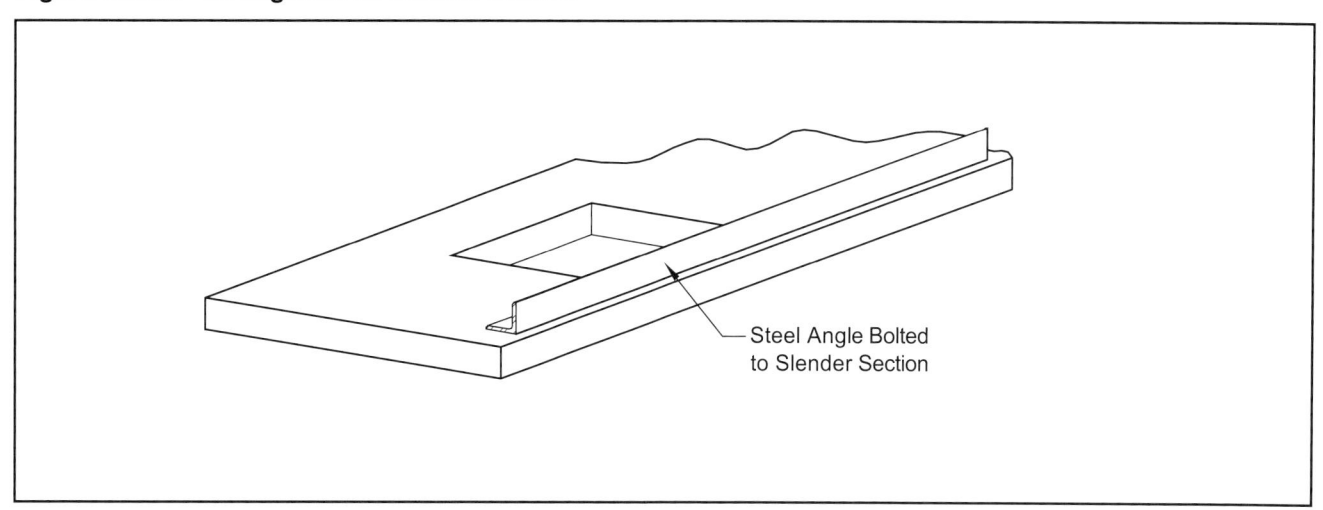

Figure 5.2.2.4 Strongback for cantilevers

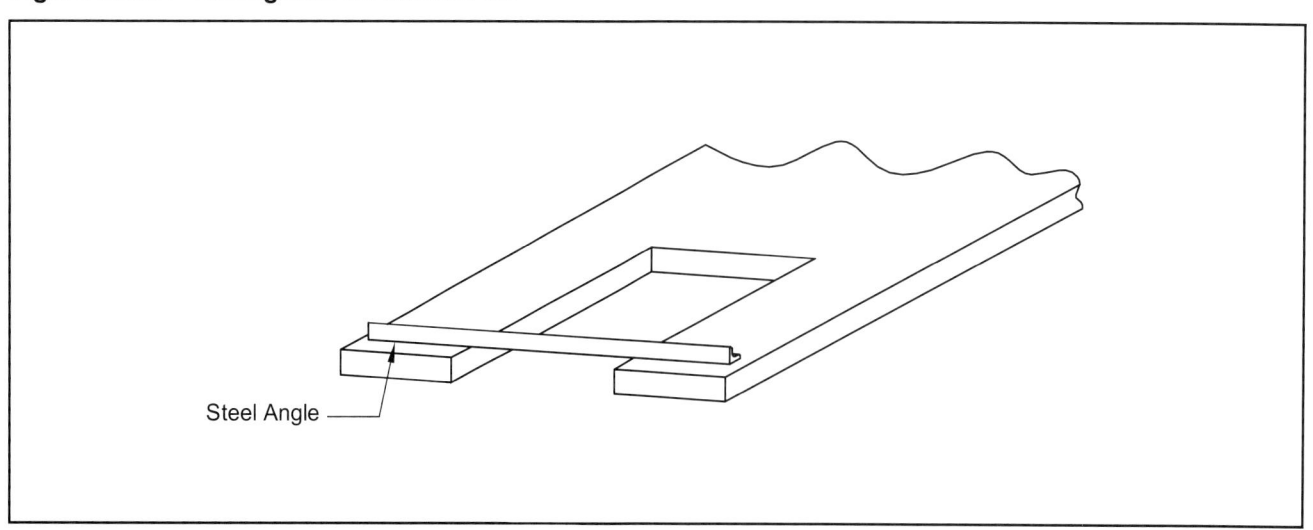

Steel Angle

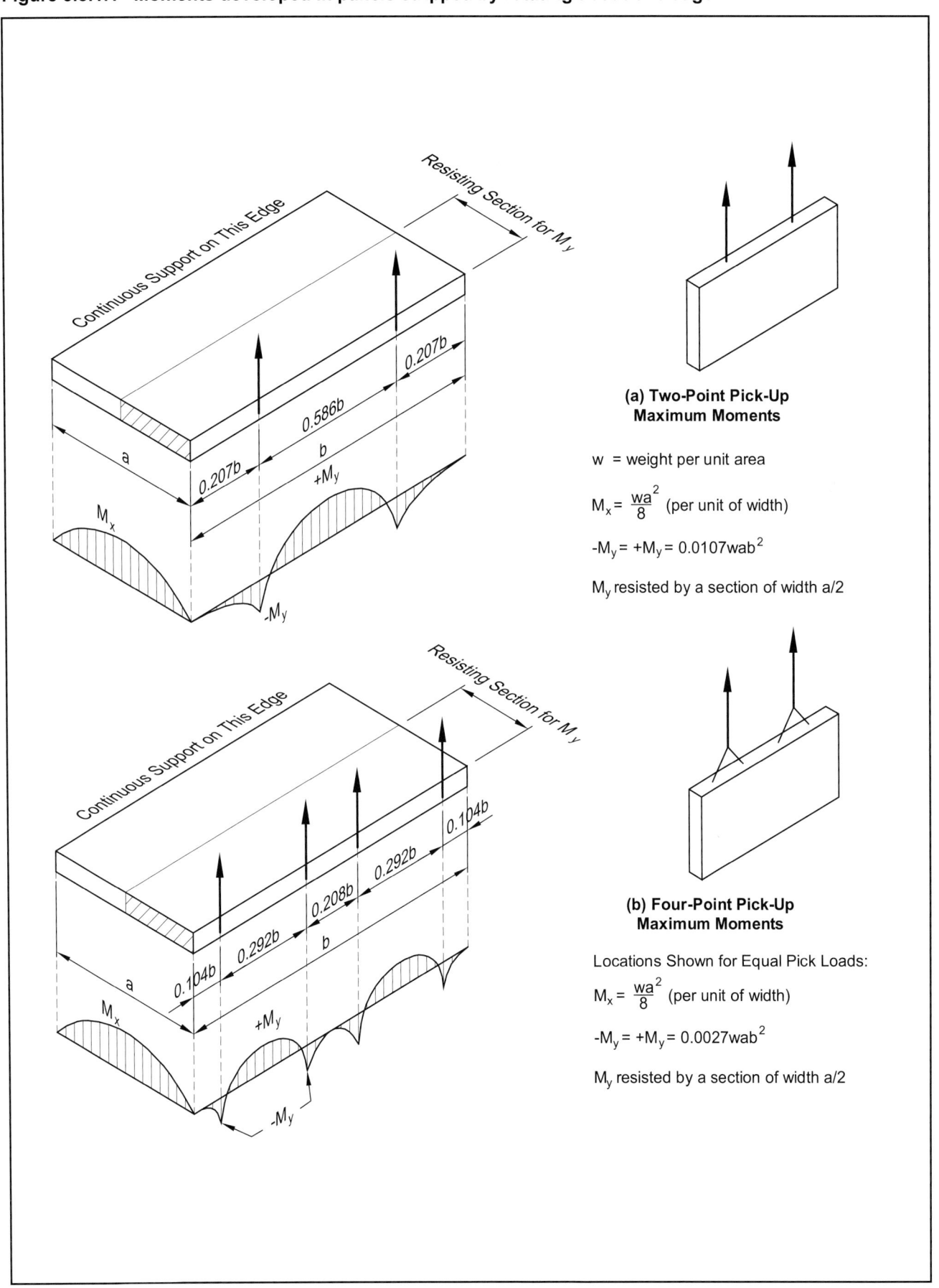

Figure 5.3.1.1 Moments developed in panels stripped by rotating about one edge

(a) Two-Point Pick-Up Maximum Moments

w = weight per unit area

$M_x = \dfrac{wa^2}{8}$ (per unit of width)

$-M_y = +M_y = 0.0107wab^2$

M_y resisted by a section of width $a/2$

(b) Four-Point Pick-Up Maximum Moments

Locations Shown for Equal Pick Loads:

$M_x = \dfrac{wa^2}{8}$ (per unit of width)

$-M_y = +M_y = 0.0027wab^2$

M_y resisted by a section of width $a/2$

Figure 5.3.1.2 Moments developed in panels stripped flat

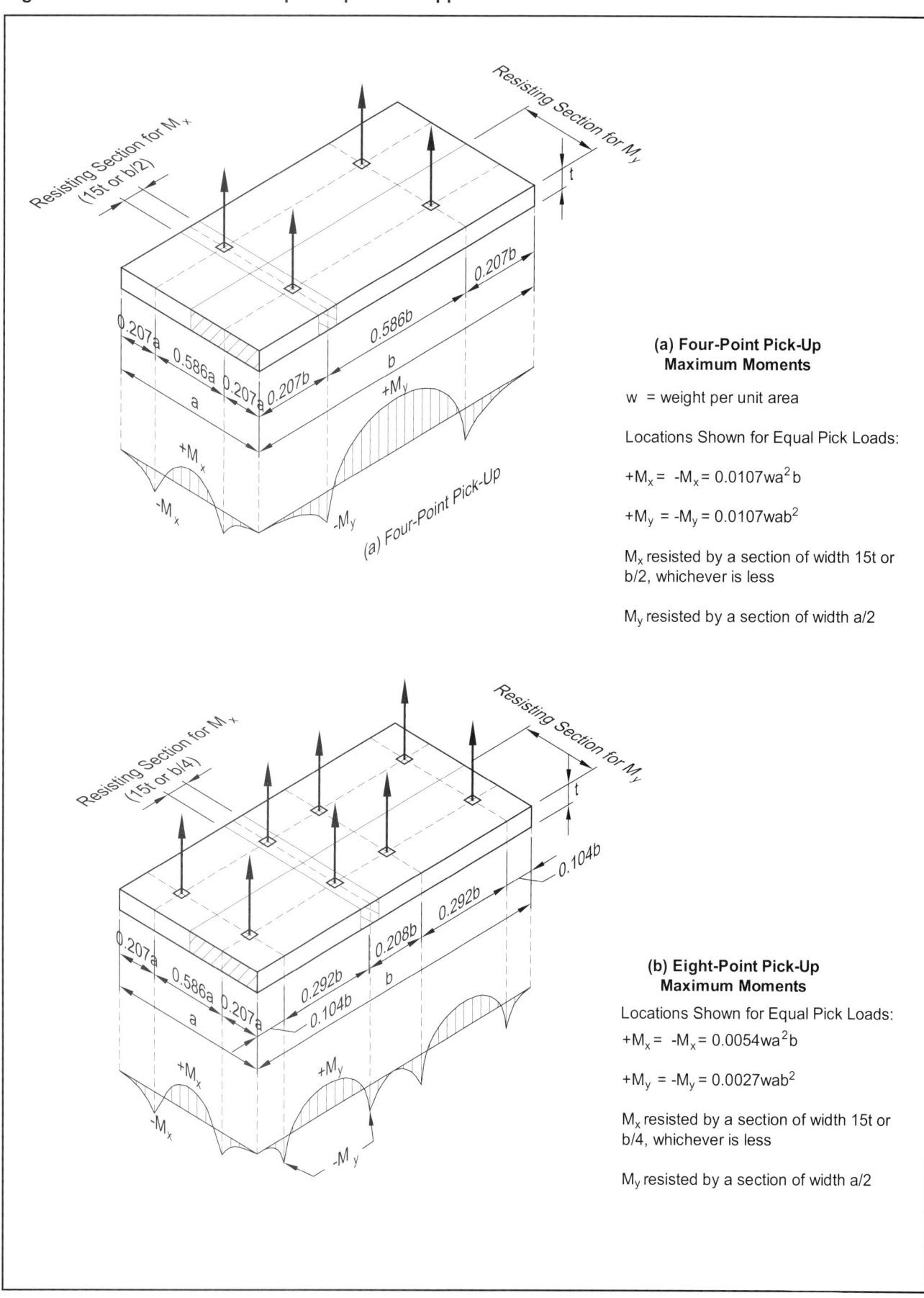

(a) Four-Point Pick-Up Maximum Moments

w = weight per unit area

Locations Shown for Equal Pick Loads:

$+M_x = -M_x = 0.0107wa^2b$

$+M_y = -M_y = 0.0107wab^2$

M_x resisted by a section of width 15t or b/2, whichever is less

M_y resisted by a section of width a/2

(b) Eight-Point Pick-Up Maximum Moments

Locations Shown for Equal Pick Loads:

$+M_x = -M_x = 0.0054wa^2b$

$+M_y = -M_y = 0.0027wab^2$

M_x resisted by a section of width 15t or b/4, whichever is less

M_y resisted by a section of width a/2

Figure 5.3.1.3 Stripping from a tilt table

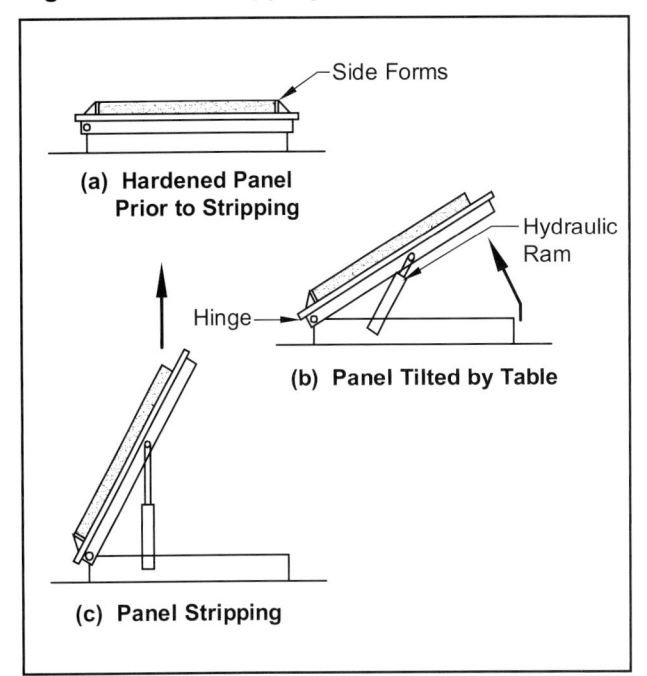

Figure 5.3.2.1 Force in lift lines

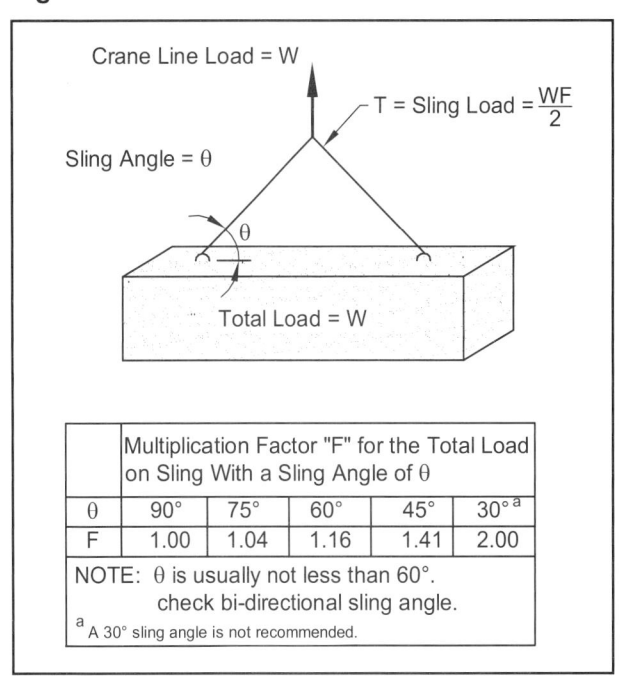

Multiplication Factor "F" for the Total Load on Sling With a Sling Angle of θ						
θ	90°	75°	60°	45°	30°[a]
F	1.00	1.04	1.16	1.41	2.00

NOTE: θ is usually not less than 60°.
check bi-directional sling angle.
[a] A 30° sling angle is not recommended.

prevent spalling of the edge along which the rotation occurs. A compressible material or sand bed will help protect this edge. Members that are stripped flat from the mold will develop the moments shown in Figure 5.3.1.2.

To determine stresses in flat panels for either rotation or flat stripping, the calculated moment may be assumed to be resisted by the effective widths shown.

In some plants, tilt tables or turning rigs are used to reduce stripping stresses, as shown in Figure 5.3.1.3.

When a panel is ribbed or is of a configuration or size such that stripping by rotation or tilting is not practical, vertical pick-up points on the top surface can be used. These lifting points should be located to minimize the tensile stresses on the exposed face.

Since the section modulus with respect to the top and bottom faces may not be the same, the designer must select the controlling design limitation:

1. Tensile stresses on both faces to be less than that which would cause cracking.
2. Tensile stress on one face to be less than that which would cause cracking, with controlled cracking permitted on the unexposed face.
3. Controlled cracking permitted on both faces.

If only one of the faces is exposed to view, the exposed face will generally control the stripping method.

5.3.2 Rigging Configurations

Stresses and forces occurring during handling are also influenced by the type of rigging used to hook up to the member.

Lift line forces for a two-point lift using inclined lines are shown in Figure 5.3.2.1. When the sling angle is small, the components of force parallel to the longitudinal axis of the member may generate a significant moment due to P-Δ effects. While this effect can and should be accounted for, it is not recommended that it be allowed to dominate design moments. Rather, consideration should be given to using spreader beams, two cranes or other mechanisms to increase the sling angle. Any such special handling required by the design should be clearly shown on the shop drawings.

In addition to longitudinal bending moments, a transverse bending moment may be caused by the orientation of the pick-up points with respect to the transverse dimension (see Figure 5.3.2.2). For the section shown, a critical moment could occur between the ribs because of the thin cross section.

The design guidelines listed above apply to elements of constant cross section. For elements of varying cross section, the location of lift points can be calculated and adjusted. Rolling blocks and spreader beams can be used on long elements of varying section (Figure 5.3.2.3), which makes the forces in the lifting lines equal. The member can then be analyzed as a beam with varying load supported by equal reactions. The force in the lift lines can be determined as shown in Figure 5.3.2.1.

Figure 5.3.2.2 Moments caused by eccentric lifting

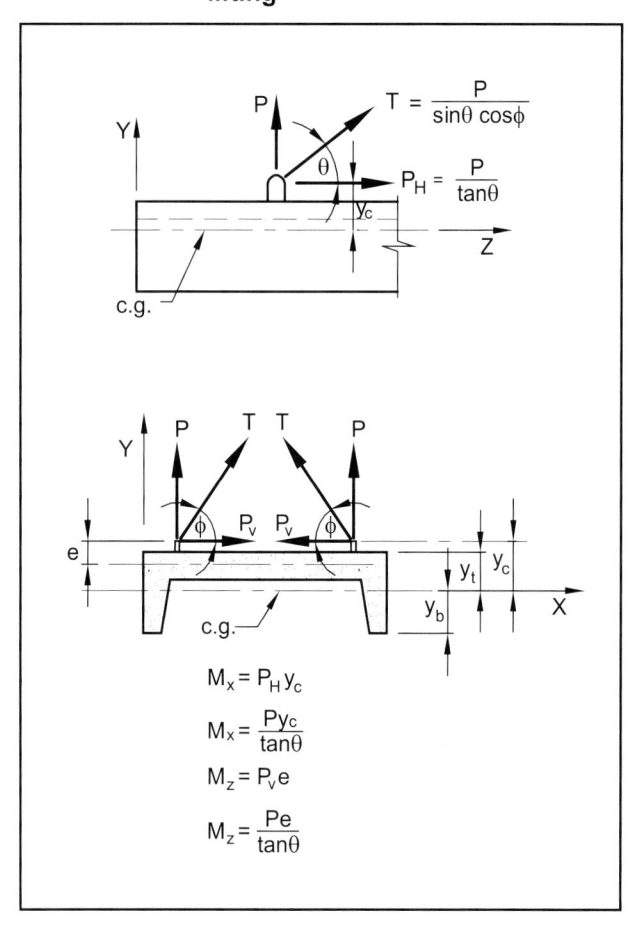

$$T = \frac{P}{\sin\theta \cos\phi}$$

$$P_H = \frac{P}{\tan\theta}$$

$$M_x = P_H y_c$$

$$M_x = \frac{P y_c}{\tan\theta}$$

$$M_z = P_v e$$

$$M_z = \frac{P e}{\tan\theta}$$

Figure 5.3.2.3 Arrangement for equalizing lifting loads

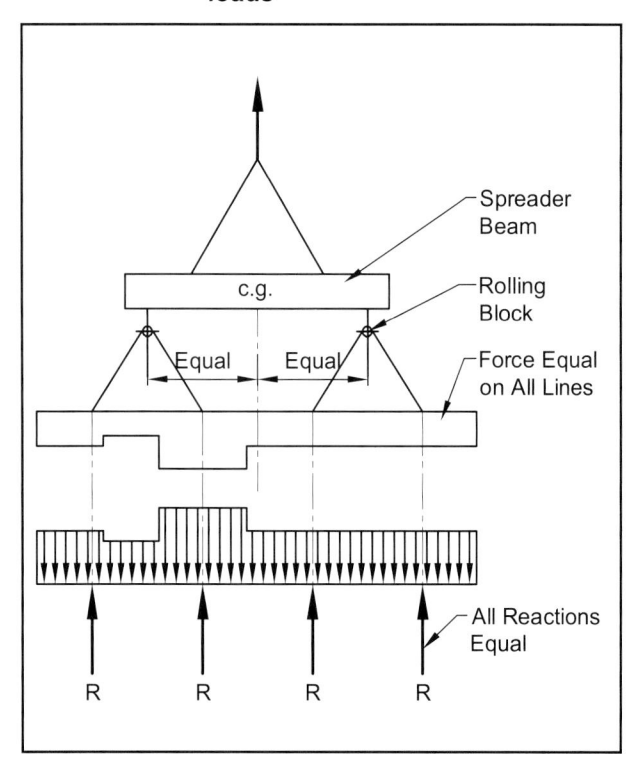

As an alternate to rolling blocks, multiple crane lines or spreader beams can be used. Availability of such equipment and preferred methods will vary among precasters.

Advantages of multiple crane lines or a straddle leg yard crane are that lift line forces are straight up and different forces can be applied to each crane line (Figure 5.3.2.4). For elements of varying cross section, panels with large openings or panels with an extended narrow leg, the different lift line forces allow for the larger or stiffer cross sections to resist the heavier lift line forces.

Using a spreader beam (Figure 5.3.2.5) can also eliminate the use of rolling blocks. Note that the spreader beam must be sufficiently stiffer than the concrete panel to limit panel deflections and stress cracking. Also, lifting hook locations, hook heights, and sling lengths are critical to ensure even lifting of the member. For analysis, rather than equal forces at each lifting hook location, the panel acts as a continuous beam over multiple supports.

Figure 5.3.2.4 Hook lifting

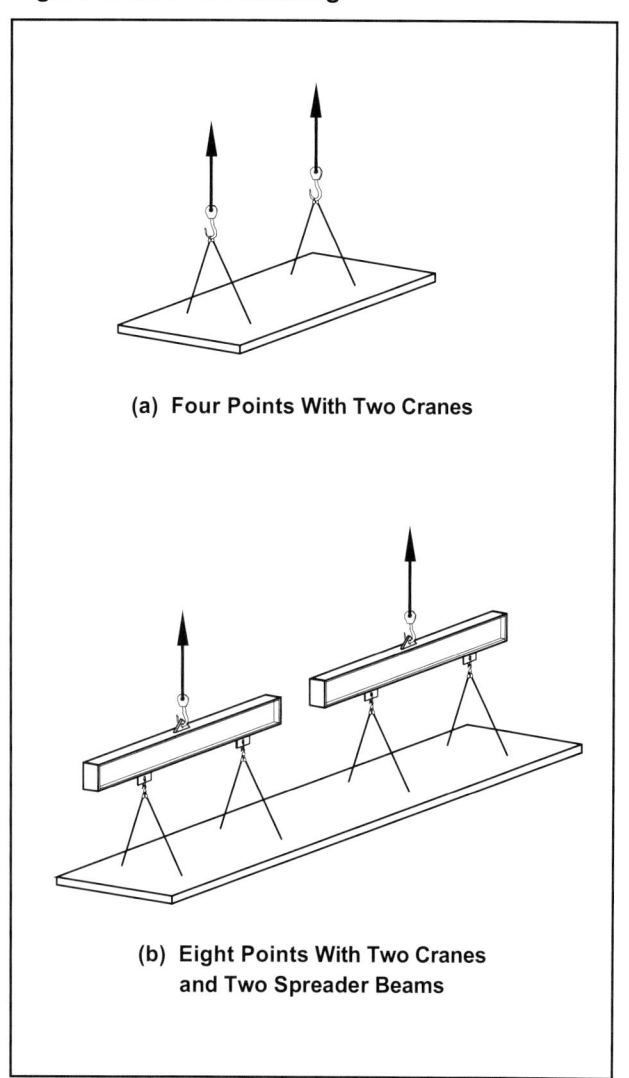

(a) Four Points With Two Cranes

(b) Eight Points With Two Cranes and Two Spreader Beams

Figure 5.3.2.5 Use of spreader beams

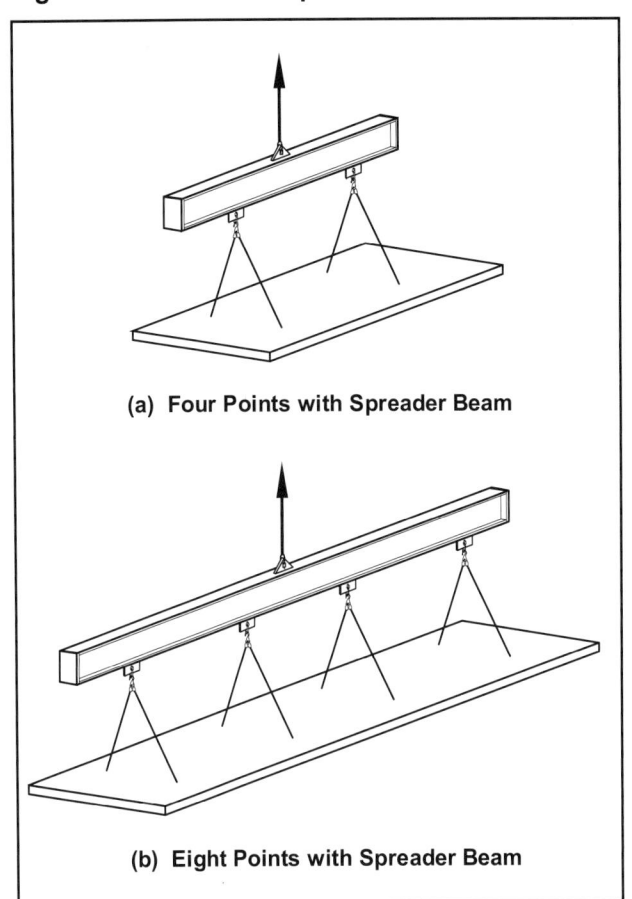

(a) Four Points with Spreader Beam

(b) Eight Points with Spreader Beam

Figure 5.3.3.1 Post-tensioned wall panel

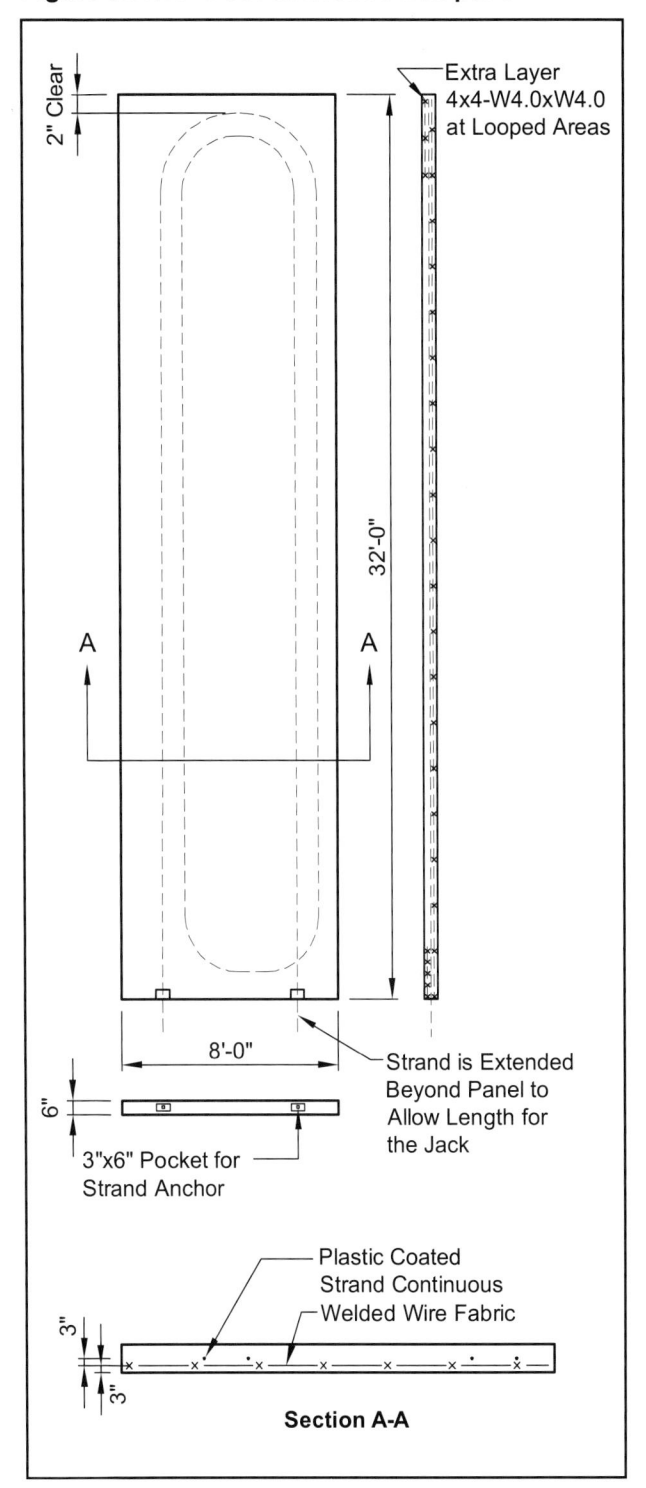

5.3.3 Stripping Design

5.3.3.1 Form Suction and Impact Factors

To account for the forces on the member caused by form suction and impact, it is common practice to apply a multiplier to the member weight and treat the resulting force as an equivalent static service load. The multipliers cannot be quantitatively derived, so they are based on experience. Table 5.3.3.1 provides typical values.

5.3.3.2 Factors of Safety

When designing for stripping and handling, the following safety factors are recommended:

1. Use embedded inserts and erection devices with a pullout strength at least equal to four times the calculated load on the device.
2. For members designed "without cracking," the modulus of rupture $(7.5\sqrt{f'_c})$, is divided by a safety factor of 1.5.

$$\frac{7.5\lambda\sqrt{f'_c}}{1.5} = 5\lambda\sqrt{f'_c} \qquad \text{(Eq. 5.3.3.1)}$$

5.3.3.3 Stress Limits and Crack Control

Stress limits for prestressed members during production are discussed in Section 4.2.2.2 of this Handbook. ACI 318-02 does not restrict stresses on non-prestressed members, but does specify minimum reinforcement spacing, as discussed in Section 4.2.2.1. Members which are exposed to view will

Table 5.3.3.1 Equivalent static load multipliers to account for stripping and dynamic forces[a,c]

Product Type	Finish	
	Exposed aggregate with retarder	Smooth mold (form oil only)
Flat, with removable side forms, no false joints or reveals	1.2	1.3
Flat, with false joints and/or reveals	1.3	1.4
Fluted, with proper draft[d]	1.4	1.6
Sculptured	1.5	1.7
Yard handling[b] and erection[c]		
All products	1.2	
Travel[b]		
All products	1.5	

a. These factors are used in flexural design of panels and are not to be applied to required safety factors on lifting devices. At stripping, suction between product and form introduces forces, which are treated here by introducing a multiplier on product weight. It would be more accurate to establish these multipliers based on the actual contact area and a suction factor independent of product weight.
b. Certain unfavorable conditions in road surface, equipment, etc., may require use of higher values.
c. Under certain circumstances may be higher.
d. For example, tee, channels and fluted panels.

Table 5.3.3.2 Suggested maximum strand diameter

Concrete thickness, (in.)	Strand diameter, (in.)
less than 3	3/8
3 or more	1/2

generally be designed for the "no discernible cracking criteria" (see Eq. 4.2.2.2), which limits the stress to $5\lambda\sqrt{f'_c}$. In the case of stripping stresses, f'_{ci} should be substituted for f'_c. Whether or not the members are exposed to view, the strength design and crack control requirements of ACI 318-02, as discussed in Chapter 4 of this Handbook, must be followed.

5.3.3.4 Benefits of Prestressing

Panels can be prestressed, using either pretensioning or post-tensioning. Design is based on Chapter 18 of ACI 318-02, as described in Chapter 4 of this Handbook. Further, tensile stresses should be restricted to less than $5\lambda\sqrt{f'_{ci}}$.

It is recommended that the average stress due to prestressing, after losses, be within a range of 125 to 800 psi. The prestressing force should be concentric with the effective cross section in order to minimize camber, although some manufacturers prefer to have a slight inward bow in the in-place position to counteract thermal bow (see Section 4.8.5). It should be noted that concentrically prestressed members do not camber, hence the form adhesion may be larger than with members that do camber.

In order to minimize the possibility of splitting cracks in thin pretensioned members, the strand diameter should not exceed that shown in Table 5.3.3.2. Additional light transverse reinforcement may be required to control longitudinal cracking.

When wall panels are post-tensioned, care must be taken to ensure proper transfer of force at the anchorage and protection of anchors and tendons against corrosion. Straight strands or bars may be used, or, to reduce the number of anchors, the method shown in Figure 5.3.3.1 may be used. Plastic coated tendons with a low coefficient of curvature friction ($\mu = 0.03$ to 0.05) are looped within the panel and anchors installed at only one end. The tendons remain unbonded.

It should be noted that if an unbonded tendon is cut, the prestress is lost. This can sometimes happen if an unplanned opening is cut in at a later date.

5.3.4 Handling Devices

The most common lifting devices are prestressing strand or aircraft cable loops projecting from the concrete, threaded inserts, and proprietary devices.

Since lifting devices are subject to dynamic loads, ductility of the material is a requirement. Deformed reinforcing bars should not be used as the deformations result in stress concentrations from the shackle pin. Also, reinforcing bars may be hard grade or re-rolled rail steel with little ductility and low impact strength at cold temperatures. Strain hardening from bending the bars may also cause embrittlement. Smooth bars of a known steel grade may be used if adequate embedment or mechanical anchorage is provided. The diameter must be such that localized failure will not occur by bearing on the shackle pin.

Table 5.3.4.1 Safe working load of 7x19 aircraft cable used as lifting loops[a]

Diameter (in.)	Safe Load (kips)[b]
3/8	3.6
7/16	4.4
1/2	5.7

a. Seven strands with 19 wires each.
b. Based on a single strand with a factor of safety of 4 applied to the minimum breaking strength of galvanized cable. The user should consider embedment, loop diameter and other factors discussed for strand loops. Aircraft cables are usually wrapped around reinforcement in the precast product.

5.3.4.1 Aircraft Cable Loops

For smaller precast members, aircraft cable can be used for stripping and erection purposes. Aircraft cable comes in several sizes (Table 5.3.4.1) with different capacities. The flexible cable is easier to handle and will not leave rust stains on precast concrete. For some small precast members such as coping, the flexible loops can be cast in ends of members and tucked back in the joints after erection. Aircraft cable loops should not be used as multiple loops in a single location, as even pull on multiple cables in a single hook is extremely difficult to achieve. User should ensure that the cable is clean and that each leg of the loop is embedded a minimum of 48 in.

5.3.4.2 Prestressing Strand Loops

Prestressing strand, both new and used, may be used for lifting loops. The capacity of a lifting loop embedded in concrete is dependent upon the strength of the strand, length of embedment, the condition of the strand, the diameter of the loop, and the strength of the concrete.

As a result of observations of lift loop behavior during the past few years, it is important that certain procedures be followed to prevent both strand slippage and strand failure. Precast producers' tests and/or experience offer the best guidelines for the load capacity to use. A safety factor of 4 against slippage or breakage should be used. In lieu of test data, the recommendations listed below should be considered when using strand as lifting loops.

1. Minimum embedment for each leg of the loop should be 24 in.
2. The strand surface must be free of contaminants, such as form oil, grease, mud, or loose rust, which could reduce the bond of the strand to the concrete.
3. The diameter of the hook or fitting around which the strand lifting eye will be placed should be at least four times the diameter of the strand being used.
4. Do not use heavily corroded strand or strand of unknown size and strength.

In the absence of test or experience, it is recommended that the safe load on a single 1/2 in. diameter 270 ksi strand loop satisfying the above recommendations not exceed 8 kips. The safe working load of multiple loops may be conservatively obtained by multiplying the safe load for one loop by 1.7 for double loops and 2.2 for triple loops. To avoid overstress in one loop when using multiple loops, care should be taken in the fabrication to ensure that all strands are bent the same. Thin wall conduit over the strands in the region of the bend has been used to reduce the potential for overstress.

When using double or triple loops, the embedded ends may need to be spread apart for concrete consolidation around embedded ends without voids being formed by bundled strand.

5.3.4.3 Threaded Inserts

Lifting heavy members with threaded inserts should be carefully assessed. Threaded inserts can have NC (National Course) or coil threads. Anchorage is provided by loop, strut or reinforcing bar (Figure 5.3.4.1). Inserts must be placed accurately because their safe working load decreases sharply if they are not perpendicular to the bearing surface, or if they are not in a straight line with the applied force. Embedment of inserts close to an edge will greatly reduce the effective area of the resisting concrete shear cone and thus reduce the tension safe working load of the embedded insert.

When properly designed for both insert and concrete capacities, threaded inserts have many advantages. However, correct usage is sometimes difficult to inspect during handling operations. In order to ensure that an embedded insert acts primarily in tension, a swivel plate as indicated in Figure 5.3.4.2 should be used. It is extremely important that sufficient threads be engaged to develop the strength of the bolt. Some manufacturers use a long bolt with a nut between the bolt head and the swivel plate. After the bolt has "bottomed out," the nut is turned against the swivel plate.

For straight tension loads only, eye bolts or wire rope loops (Figure 5.3.4.1) provide a fast method for handling precast members. Do not use either device if shear loading conditions exist.

Connection hardware designed for in-service loads should not be used for lifting or handling any member except the lightest units, unless approved by the designer. In order to prevent field error, inserts used for lifting should be of a different type or size than those used for the final connections.

5.3.4.4 Proprietary Devices

A variety of castings or stock steel devices, machined to accept specialized lifting assemblies are used in the precast industry (Figure 5.3.4.3). These proprietary devices are usually recessed (using a "pocket former") to provide access to the lifting unit. The recess allows one panel to be placed against another without cutting off the lifting device, and also helps prevent spalling around the device. Longer devices are used for edge lifting or deep precast concrete members. Shallow devices are available for thin precast concrete members.

The longer devices usually engage a reinforcing bar to provide greater pullout capacity, and often have holes for the bar to pass through as shown in Figure 5.3.4.3(a). These units have a rated capacity as high as 22 tons, with reductions for thin panels or close edge distances. Supplemental reinforcement may be required to achieve these values.

Shallow units usually have a spread foot or base to increase pullout capacity. Reinforcing bars are required in two directions over the base to fully develop the lifting unit, as shown in Figure 5.3.4.3(b). They are rated up to 8 tons. Some lifting eyes do not swivel, so rotation may be a concern.

In all cases manufacturer recommendations should be rigorously followed when using any of these devices.

Figure 5.3.4.1 Threaded inserts

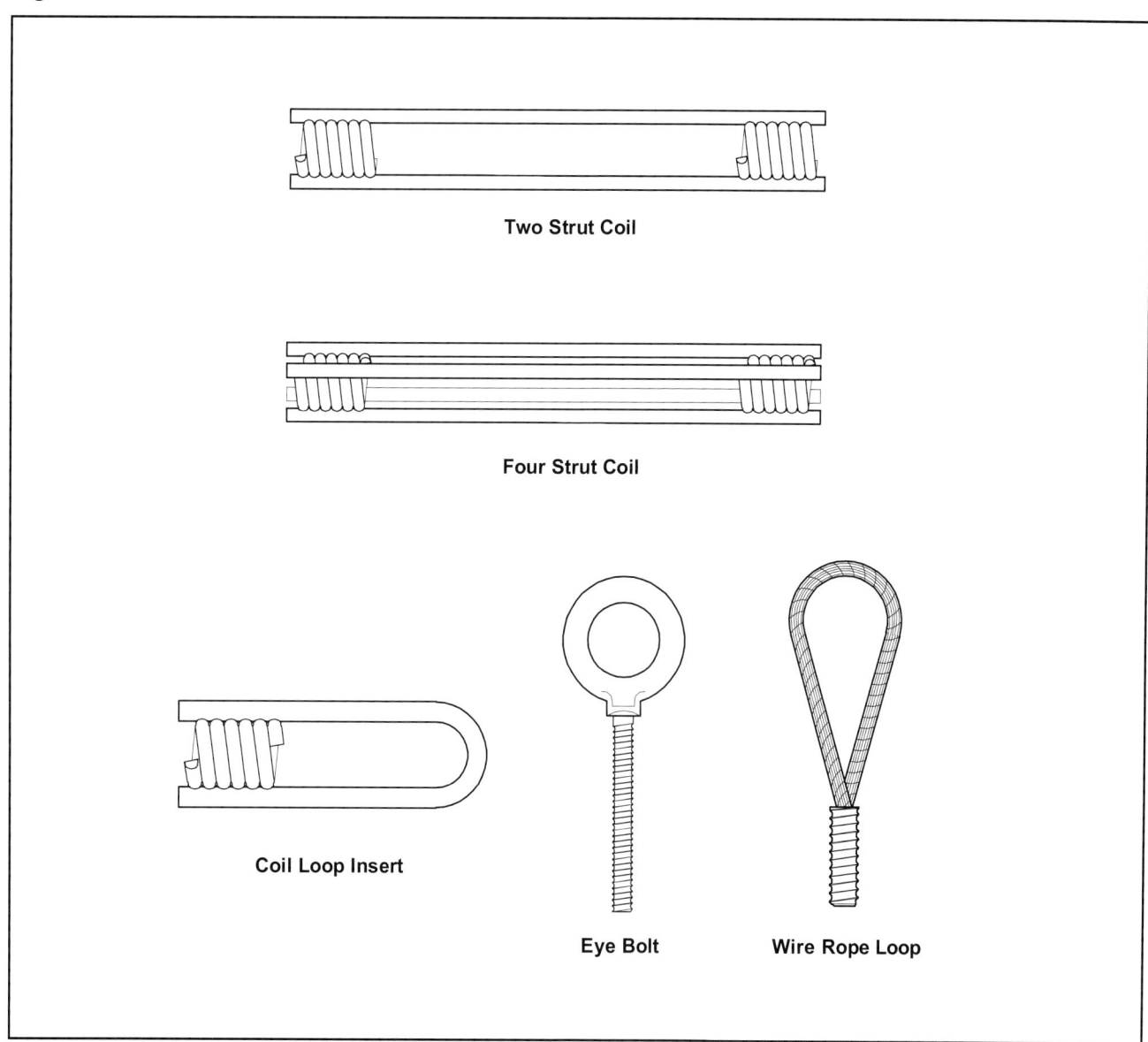

Two Strut Coil

Four Strut Coil

Coil Loop Insert

Eye Bolt

Wire Rope Loop

Figure 5.3.4.2 Swivel plate

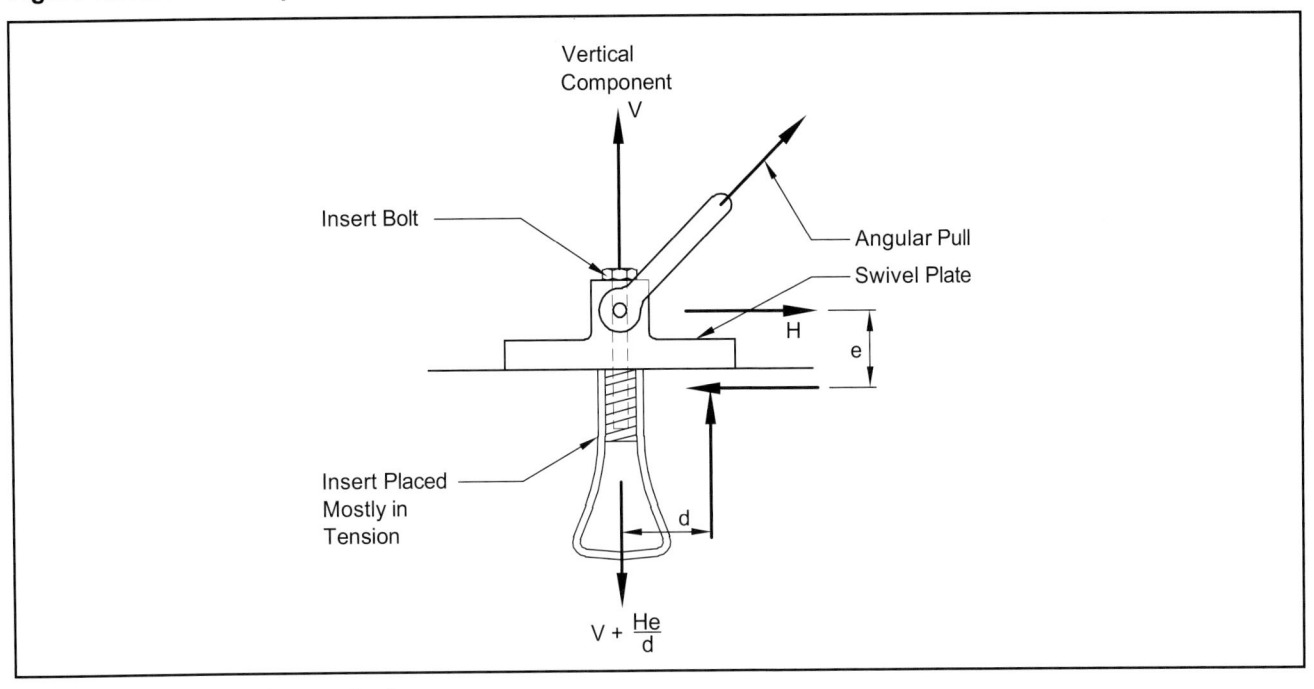

Figure 5.3.4.3 Proprietary devices

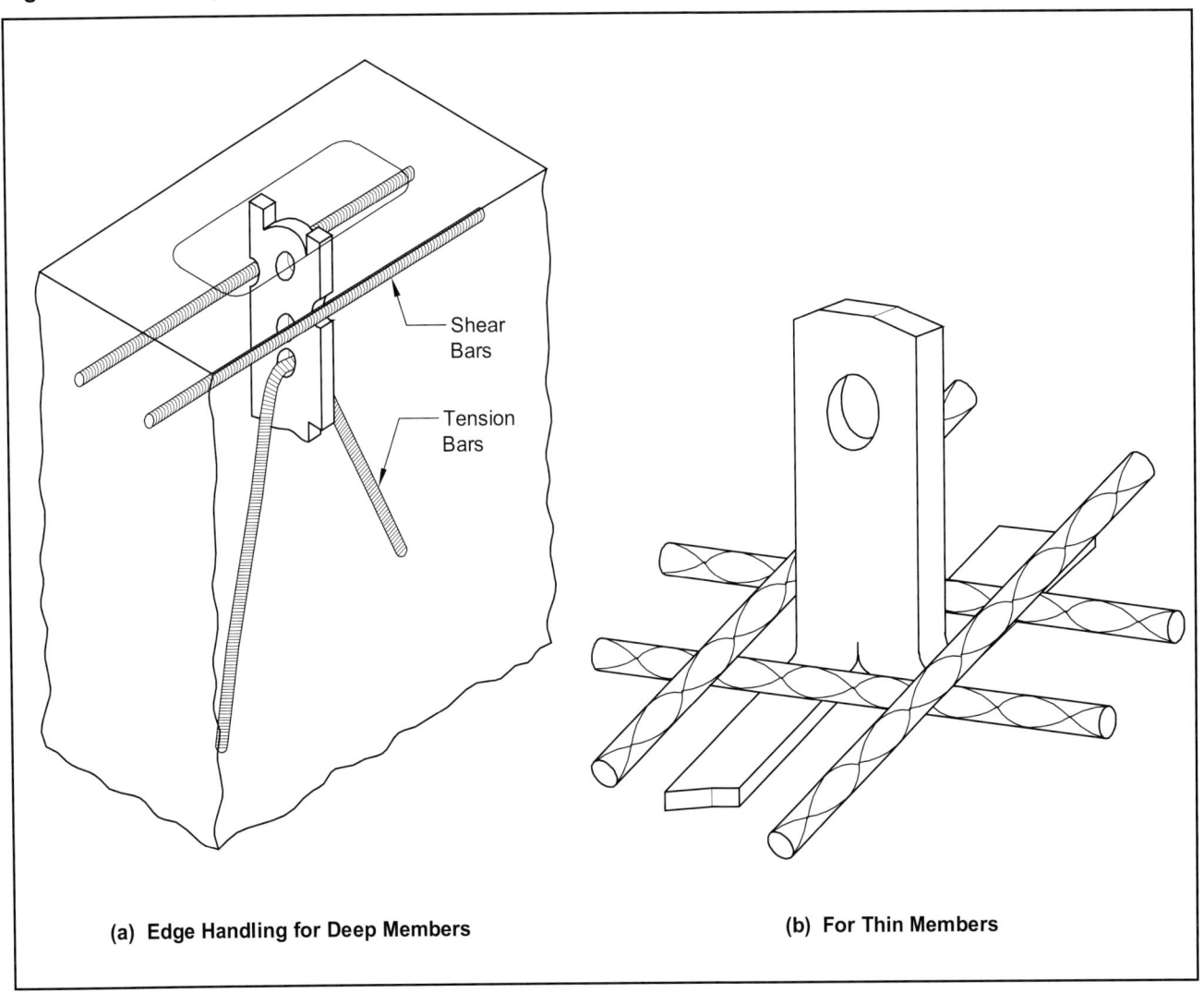

(a) Edge Handling for Deep Members

(b) For Thin Members

Example 5.3.1
Design of Wall Panel for Stripping

This example and others in Chapter 5 illustrate the use of many of the recommendations in this chapter. They are intended to be illustrative and general only. Each manufacturer will have its own preferred methods of handling.

Given:
A flat panel used as a loadbearing wall on a two-story structure, as shown below.
Section properties (nominal dimensions are used for design):

Solid panel
$A = 960$ in.^{2}
$S_b = S_t = 1280$ in.^{3}
$I_x = 5120$ in.^{4}

Panel with openings
$A = 480$ in.^{2}
$S_b = S_t = 640$ in.^{3}
$I_x = 2560$ in.^{4}

Unit weight @ 150 pcf = 100 psf = 0.100 ksf
Total weight = 35.2 kips (solid panel)
= 29.2 kips (panel with openings)

Stripping method:
Inside crane height prevents panel from being turned on edge directly in mold, therefore, strip flat.

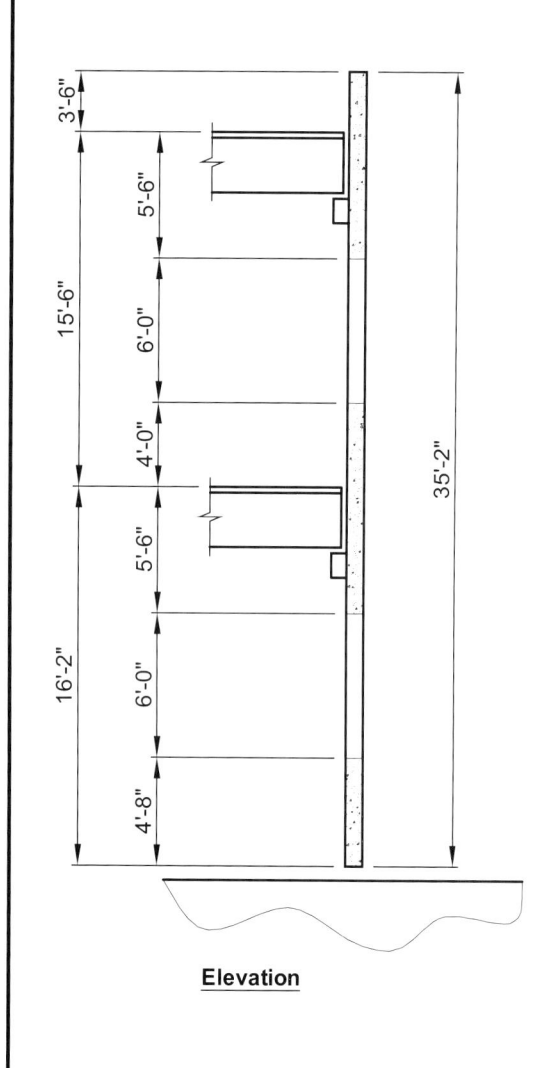

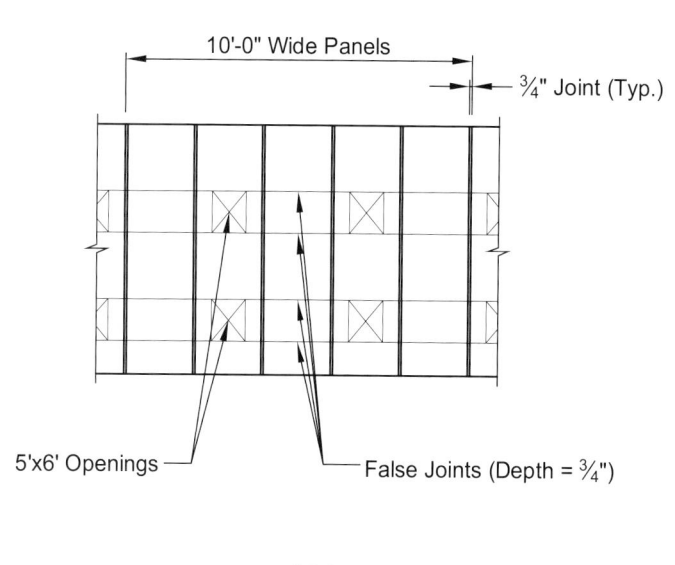

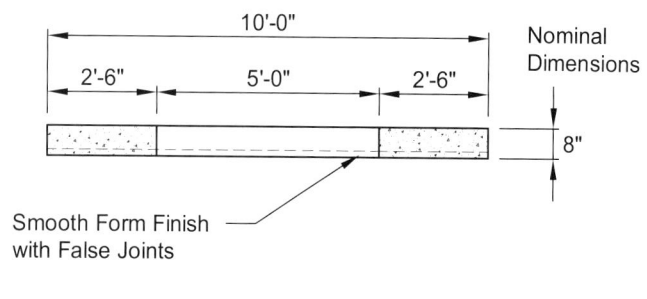

Elevation

Typical Panel Section

$f'_c = 5000$ psi $f'_{ci} = 3000$ psi

Example 5.3.1 (Cont.)
Design of Wall Panel for Stripping

Handling multipliers:
Exposed flat surface has a smooth form finish with false joints. Side rails are removable. Use multiplier of 1.4 (Table 5.3.3.1).

f'_{ci} at stripping = 3000 psi

Allowable tensile stresses at stripping and lifting: $5\lambda\sqrt{f'_{ci}} = \dfrac{5(1.0)\sqrt{3000}}{1000} = 0.274$ ksi

Problem:
Check critical stresses involved with stripping. Limit stresses to 0.274 ksi.

Solution:

(a) Solid panel
From Figure 5.3.1.2: a = 10 ft, b = 35.2 ft, a/2 = 5 ft = 60 in.
S for resisting section (half of panel width) = $60(8)^2/6 = 640$ in.^{3}

Try 4-point pick, Figure 5.3.1.2(a):
$M_y = 0.0107\,wab^2 = 0.0107(0.100)(10)(35.2)^2(12)(1.4) = 223$ kip-in.
$f_t = f_b = 223/640 = 0.348$ ksi > 0.274 ksi <u>four-point pick not adequate</u>

Try 8-point pick, Figure 5.3.1.2(b):
$M_y = 0.0027\,wab^2 = 0.0027(0.100)(10)(35.2)^2(12)(1.4)$
$= 56.2$ kip-in.
$f_t = f_b = 56.2/640 = 0.088$ ksi < 0.274 ksi OK
W = 35.2 kips
W/2 = 17.6 kips
W/4 = 8.8 kips

Using mechanics of materials:
$-M_{max} = \dfrac{-w\ell^2}{2} = \dfrac{-1.0(3.65)^2}{2}$
$= -6.7$ kip-ft
$+M_{max}$ = area under shear diagram
$= -6.7 + \dfrac{1}{2}(5.15)\left(\dfrac{10.3}{2}\right)$ kip-ft
$= +6.7$ kip-ft

$f_t = \dfrac{(6.7)(1.4)(12)}{1280}$
$= 0.088$ ksi < 0.274 ksi

(Note: If maximum moment occurs near a false joint, reduced section properties should be used.)

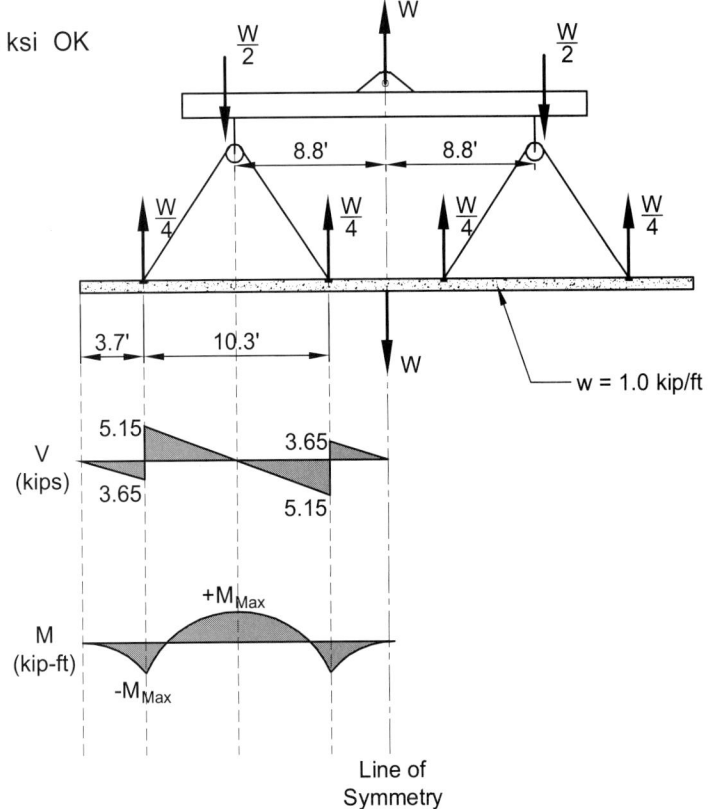

Example 5.3.1 (Cont.)
Design of Wall Panel for Stripping

(b) Panel with openings

Analyzing as a continuous beam with rigid supports:

$$f_t = \frac{(6.8)(1.4)(12)}{1280}$$
$$= 0.089 \text{ ksi}$$
or $f_t = \frac{(5.9)(1.4)(12)}{640}$
$$= 0.155 \text{ ksi}$$
0.155 ksi < 0.274 ksi   OK

Because there is a false joint near the point of maximum stress, recalculate stress using reduced cross section.

S_t at false joint:

$$= \frac{(120-60)\left(8-\frac{3}{4}\right)^2}{6}$$
$$= 526 \text{ in.}^3$$
$$f_t = 0.155\left(\frac{640}{526}\right)$$
$$= 0.189 \text{ ksi} < 0.274 \text{ ksi} \text{ OK}$$

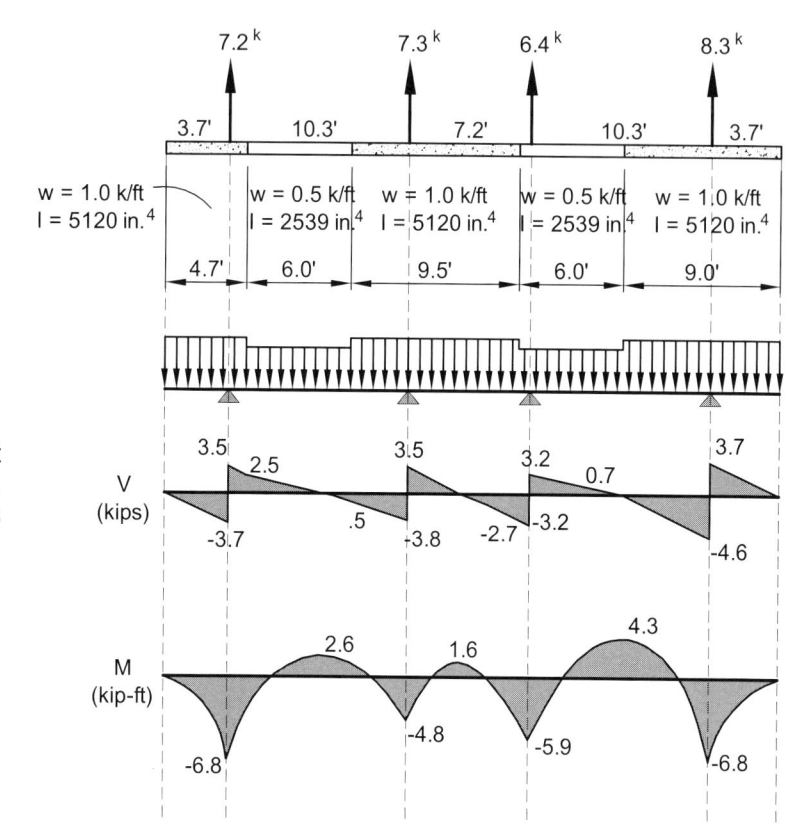

Alternatively, as a conservative approximation, use $\pm M_{max}$ based on a solid wall panel, but determine stresses using section properties of panel with openings:

M_y = 56.2 kip-in. for half of panel width,
or 112.4 kip-in. for full panel width

$$f_t = \frac{112.4 \text{ kip-in.}}{526 \text{ in.}^3}$$
$$= 0.214 \text{ ksi} < 0.274 \text{ ksi} \text{ OK}$$

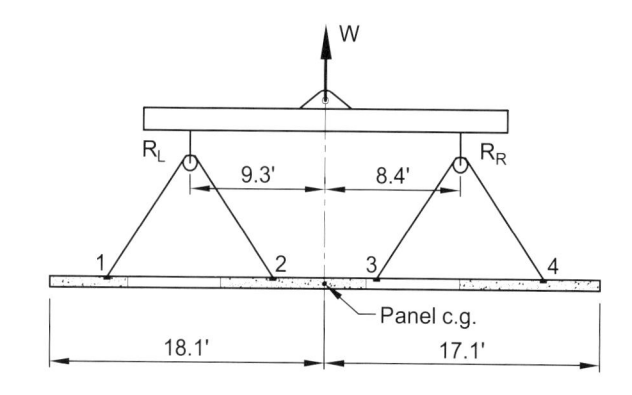

If using a rolling block for handling as shown in Figure 5.3.2.3, the panel cannot be analyzed as shown in Figure 5.3.1.2. Since each leg of a continuous cable running over a rolling block must carry equal loads, the panel can be analyzed as follows. Using the diagram above:

Example 5.3.1 (Cont.)
Design of Wall Panel for Stripping

Check panel with openings using rolling block handling:

$w = 29.2$ kips

$R_L = \dfrac{29.2(8.4)}{9.3+8.4} = 13.9$ kips

$R_1 = R_2 = \dfrac{13.9}{2} = 7.0$ kips

$R_R = 29.2 - 13.9 = 15.3$ kips

$R_3 = R_4 = \dfrac{15.3}{2} = 7.6$ kips

$M_{max} = 13.5$ kip-in.

$f_t = \dfrac{(13.5)(1.4)(12)}{526}$

$= 0.431$ ksi > 0.274 ksi

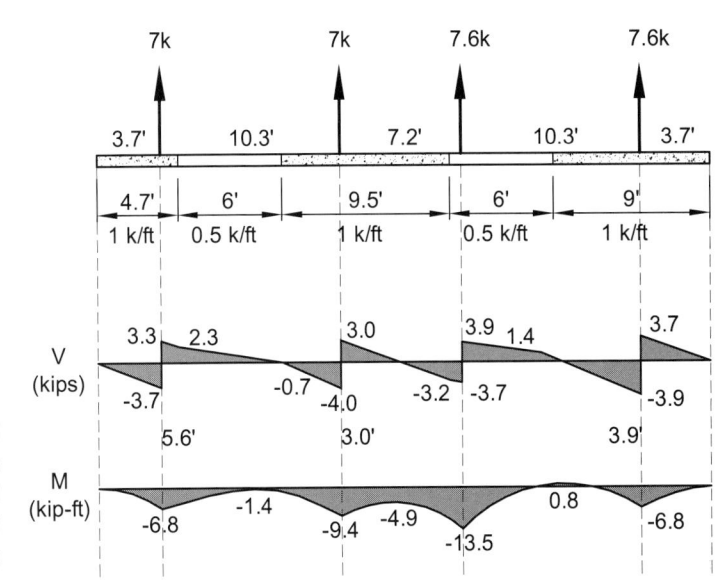

Use of rolling block with standard eight-point pick locations for these panels is not acceptable under the prescribed design criteria. The designer should choose to adjust the anchor locations in an effort to minimize stresses. Alternatively, introduce prestressing to lower or eliminate net tensile stresses (see Example 5.3.2).

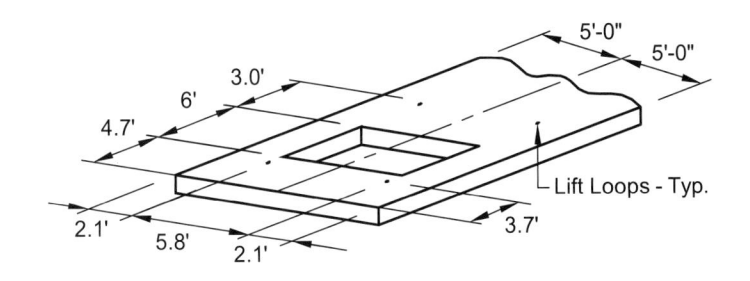

Check transverse bending:
Consider lower portion of panel with openings. Note that Figure 5.3.1.2 is based on a solid panel with uniformly distributed weight and cross section. Without the concrete in the area of the opening, the weight is reduced and unevenly distributed. Also, the resisting section is limited to a width of 4.7 ft.

Section through lifters:
$w_2 = (4.7)(0.67)(0.15) = 0.47$ kips/ft

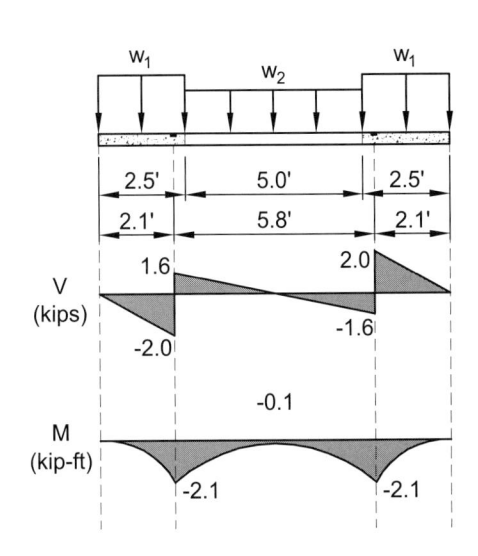

From continuous beam analysis, load carried by bottom two anchors is 7.2 kips, therefore:

$w_1 = \dfrac{7.2 - (0.47)(5)}{2(2.5)} = 0.97$ kips/ft

$M_{max} = 2.1$ ft-kips $= 25.2$ kip-in.

Check added moment due to sling angle (Figure 5.3.2.2):
Using recessed proprietary lifting anchor, $e = 3.5$ in.

Sling angle $\theta = 60$ degrees

$P = \dfrac{7.2}{2} = 3.6$ kips

Example 5.3.1 (Cont.)
Design of Wall Panel for Stripping

$$M_z = 3.6\left(\frac{3.5}{\tan 60}\right) = 7.3 \text{ kip-in.}$$

$$M_{tot} = 25.2 + 7.3 = 32.5 \text{ kip-in.}$$

Resisting section:
- b = 4.7 ft = 56 in.
- t = 8 in.

$$S = \frac{bt^2}{6} = \frac{(56)(8)^2}{6} = 597 \text{ in.}^3$$

$$f = \frac{M_{tot}(\text{handling multiplier})}{S} = \frac{(32.5)(1.4)}{597} = 0.08 \text{ ksi} < 0.274 \text{ ksi} \quad \text{OK}$$

Example 5.3.2
Use of Prestressing

Given:
The panel of Example 5.3.1.

Problem:
Determine required number of ½ in. diameter, 270 ksi strands pulled to 28.9 kips to prevent cracking in window panel. Assume 10% loss of prestress.

From Example 5.3.1, tensile stress is 0.431 ksi. The desired level of tensile stress is $5\sqrt{3000}$ or 0.274 ksi.

Therefore:
Compressive stress required = 0.431 − 0.274 = 0.157 ksi

Solution:
From Example 5.3.1, max moment/stress occurs at lifting points (-M). This results in tensile stresses on the top face.

$$A = (2)(30)(7.25) = 435 \text{ in.}^2$$

$$S_t = \frac{(2)(30)(7.25)^2}{6} = 526 \text{ in.}^3$$

$$\frac{P}{A} = \frac{(\text{no. of strands})(28.9)(0.90)}{435} = 0.060(\text{no. of strands}) \text{ (ksi)}$$

$$\frac{M}{S} = \frac{(\text{no. of strands})(28.9)(0.90)\left(\frac{7.25}{2} - 4\right)}{526} = -0.019(\text{no. of strands}) \text{ (ksi)}$$

Therefore:
0.060(no. of strands) − 0.019(no. of strands) = 0.157 ksi
No. of strands = 3.8
Add four strands to panel (two on each side of opening)

Figure 5.4.1.1 Equilibrium of beam in tilted position

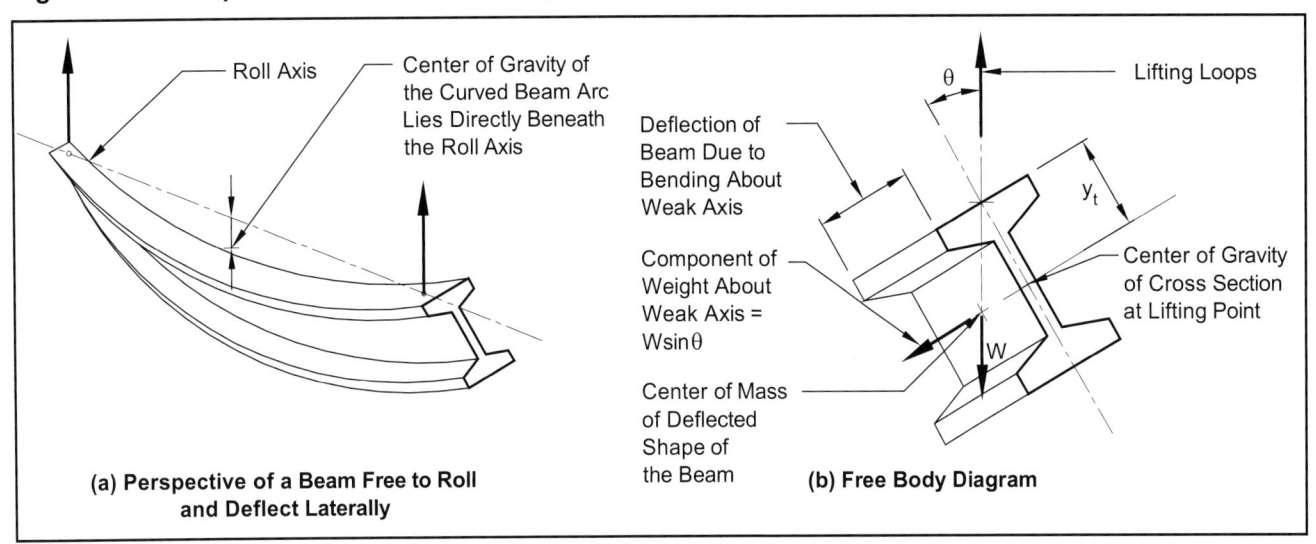

5.4 Yarding and Storage

5.4.1 Lateral Stability

Prestressed concrete members generally are sufficiently stiff to prevent lateral buckling. However, during handling and transportation, support flexibility may result in lateral roll of the beam, thus producing lateral bending. This is of particular concern with long-span bridge beams and spandrels in parking structures.

The equilibrium conditions for a hanging beam are shown in Figure 5.4.1.1. When a beam hangs from lifting points, it may roll about an axis through the lifting points. The safety and stability of long beams subject to roll is discussed in Refs. 5, 6, 7 and 8.

5.4.2 Storage

Wherever possible, a member should be stored on points of support located at or near those used for stripping and handling. Thus, the design for stripping and handling will usually control. Where points other than those used for stripping or handling are used for storage, the storage condition must be checked.

If support is provided at more than two points, and the design is based on more than two supports, precautions must be taken so that the element does not bridge over one of the supports due to differential support settlement. Particular care must be taken for prestressed members, with consideration made for the effect of prestressing. Designing for equal stresses on both faces of the member will help to minimize deformations in storage.

Warpage in storage may be caused by a temperature or shrinkage differential between surfaces, creep and storage conditions. Warpage cannot be totally eliminated, although it can be minimized by providing blocking so that the panel remains plane. Where feasible, the member should be oriented in the yard so that the sun does not overheat one side (see Section 4.8.5 for a discussion of thermal bowing). Storing members so that flexure is resisted about the strong axis will minimize stresses and deformations.

For the support condition shown in Figure 5.4.2.1, warping can occur in both directions. By superposition, the total instantaneous deflection, y_{max}, at the maximum point can be estimated by:

$$y_{max} = \frac{1.875(w)\sin\theta}{E_{ci}}\left[\frac{a^4}{I_c} + \frac{\ell^4}{I_b}\right] \quad \text{(Eq. 5.4.2.1)}$$

where:

- w = panel weight, psf
- E_{ci} = modulus of elasticity of concrete at age other than final, psi
- a = panel support height, in.
- ℓ = horizontal distance between supports, in.
- I_c, I_b = moment of inertia of uncracked section in the respective directions for 1 in. width of panel, in.[4]

This instantaneous deflection should be modified by a factor to account for the time-dependent effects of creep and shrinkage. ACI 318-02 suggests the total deformation y_t, at any time can be estimated as:

$$y_t = y_{max}(1 + \lambda) \quad \text{(Eq. 5.4.2.2)}$$

where:

- y_t = time-dependent displacement
- y_{max} = instantaneous displacement

Figure 5.4.2.1 Panel warpage in storage

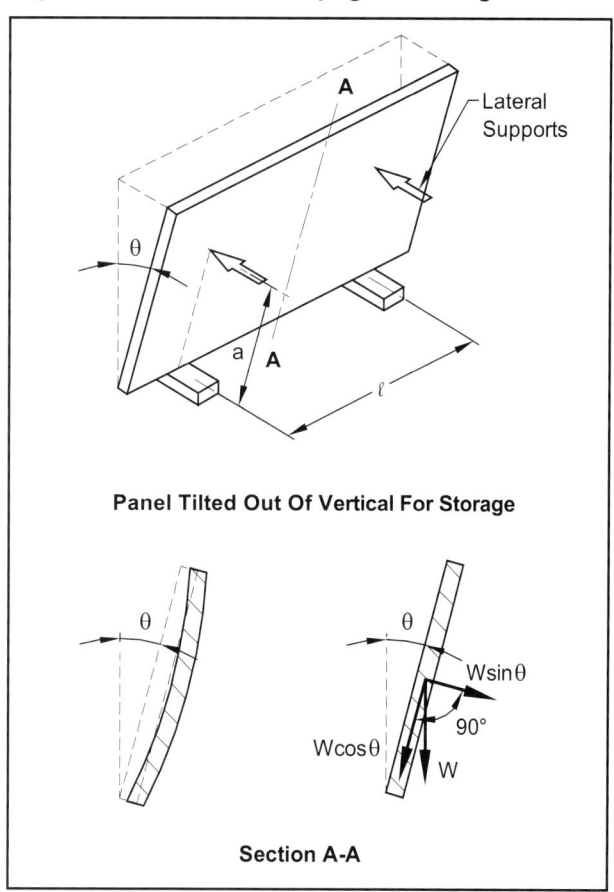

Panel Tilted Out Of Vertical For Storage

Section A-A

Figure 5.4.2.2 Effect of compression reinforcement on creep

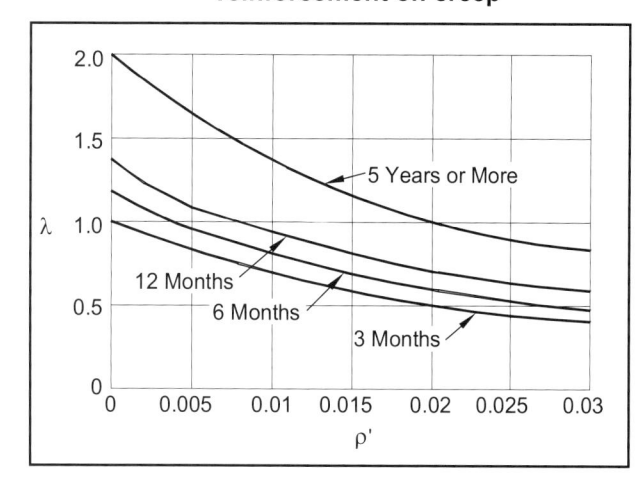

Figure 5.4.2.3 Area considered to compute ρ'

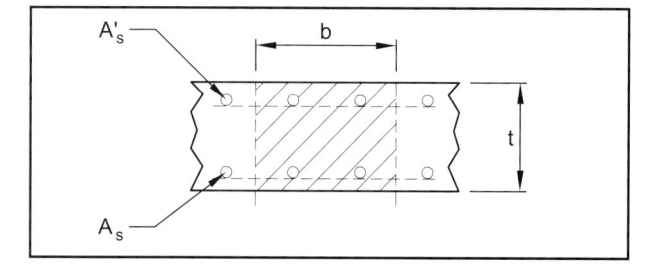

λ = amplification due to creep and shrinkage (Figure 5.4.2.2)
ρ' = reinforcement ratio for non-prestressed compression reinforcement, A_s'/bt (Figure 5.4.2.3)

5.5   Transportation

The method used for transporting precast concrete products can affect the structural design because of size and weight limitations and the dynamic effects imposed by road conditions.

Except for long prestressed deck members, most products are transported on either flatbed or low-boy trailers. These trailers deform during hauling. Thus, support at more than two points can be achieved only after considerable modification of the trailer, and even then results may be doubtful.

Size and weight limitations vary from one state to another, so a check of local regulations is necessary when large units are transported. Loads are further restricted on some secondary roads during spring thaws.

The common payload for standard trailers without special permits is 20 tons (may vary with state law) with width and height restricted to 8 ft and length to 40 ft. Low-boy trailers permit the height to be increased to about 10 to 12 ft. However, low-boys cost more to operate and have a shorter bed length. In some states, a total height (roadbed to top of load) of 13 ft 6 in. is allowed without special permit. This height may require special routing to avoid low overpasses and overhead wires.

Maximum width with permit varies among states, and even among cities – from 10 to 14 ft. Some states allow lengths over 70 ft with only a simple permit, while others require, for any load over 55 ft, a special permit, escorts front and rear and travel limited to certain times of the day. In some states, weights of up to 100 tons are allowed with permit, while in other states there are very severe restrictions on loads over 25 tons.

These restrictions add to the cost of precast concrete units, and should be compared with savings realized by combining smaller units into one large unit. When possible, a precast unit, or several units combined, should approximate the usual payload of 20 tons. For example, an 11-ton unit may not be economical, because only one member can be shipped on one load.

Erection is simplified when members are transported in the same orientation they will have in the structure. For example, single-story wall panels can be transported on A-frames with the panels upright (see Figure 5.5.1). A-frames also provide good lateral support and the desired two points of vertical support. Longer units can be transported on their sides to take advantage of the increased stiffness compared with flat shipment (see Figure 5.5.2). In all cases, the panel support locations should be consistent with the panel design. Panels with large openings sometimes require strongbacks, braces or ties to keep stresses within the design values (see Figures 5.2.2.3 and 5.2.2.4).

During transportation, units are usually supported with one or both ends cantilevered. For members not symmetrical with respect to the bending axis, the expressions given in Figure 5.5.3 can be used for determining the location of supports to give equal tensile stresses for positive and negative bending moments.

Figure 5.5.1 Transporting single-story panels

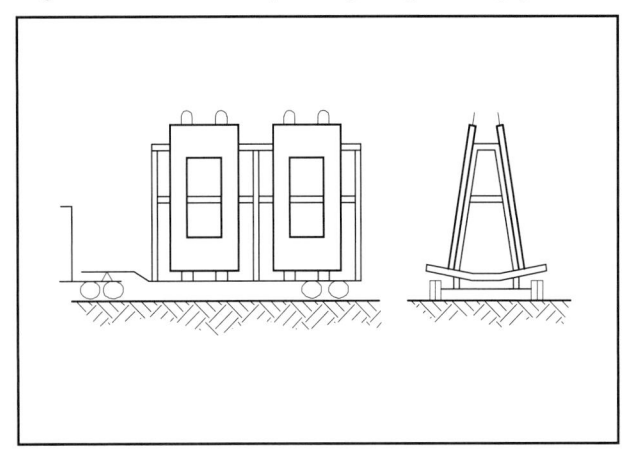

Figure 5.5.2 Transporting long panels

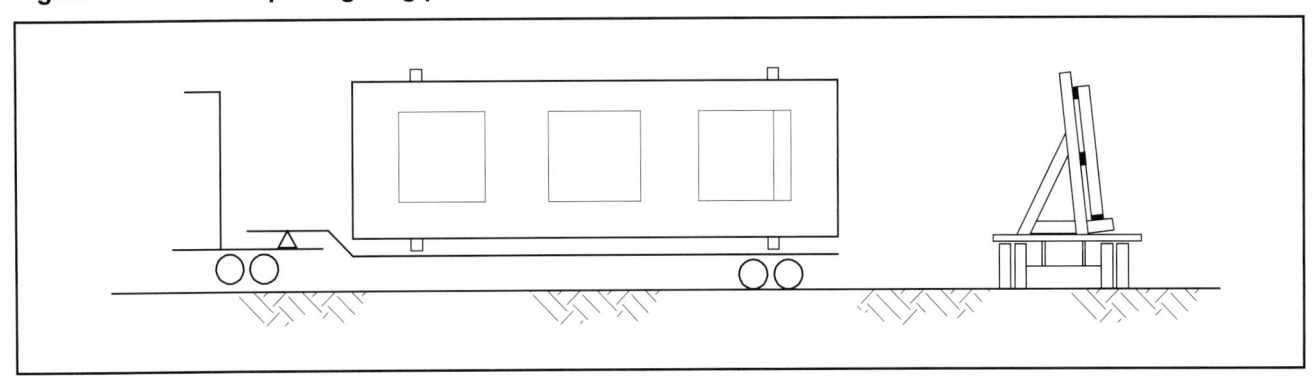

Figure 5.5.3 Equations for equal tensile stresses at top and bottom of member

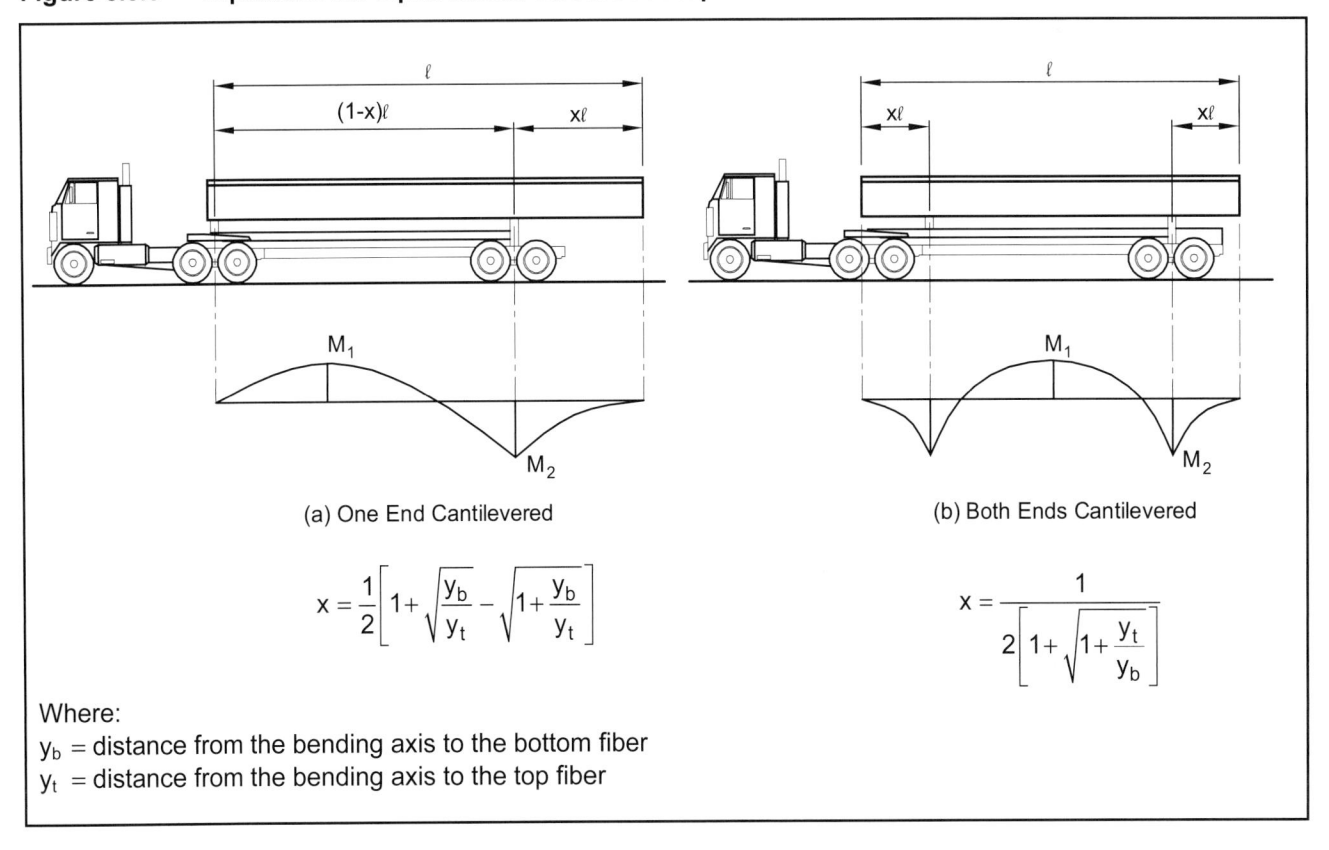

(a) One End Cantilevered

$$x = \frac{1}{2}\left[1 + \sqrt{\frac{y_b}{y_t}} - \sqrt{1 + \frac{y_b}{y_t}}\right]$$

(b) Both Ends Cantilevered

$$x = \frac{1}{2\left[1 + \sqrt{1 + \frac{y_t}{y_b}}\right]}$$

Where:
y_b = distance from the bending axis to the bottom fiber
y_t = distance from the bending axis to the top fiber

Example 5.5.1
Panel Shipping

Check the panels from Example 5.3.1 for proper method of transporting panels to the jobsite. Assume concrete has achieved 28-day strength. Use transportation load multiplier of 1.5.

Try shipping flat with supports at four locations along the length of the panel. It is very likely that the trailers will flex during transport. This flexing action will raise or lower support points thus causing bridging and unanticipated stresses. Try adjusting support points to create pairs of support points, 5 ft apart from one another, at each end of the panel. This configuration should ensure that all four supports remain in contact at all times during transportation.

As an alternative, panels may be shipped on edge, resting against A-frames, if height permits.

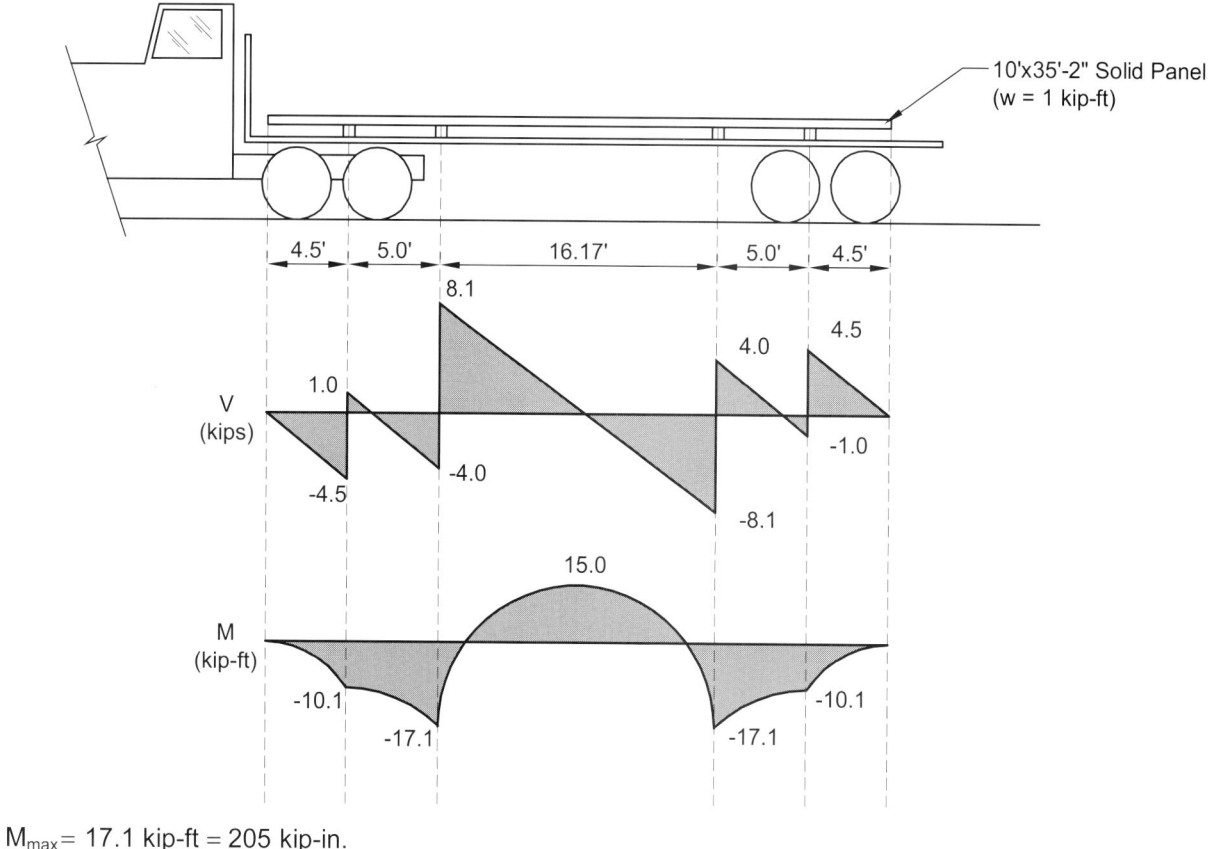

$M_{max} = 17.1$ kip-ft = 205 kip-in.

$$f = \frac{205(1.5)}{1280} = 0.240 \text{ ksi} < 5\sqrt{5000} = 0.354 \text{ ksi} \quad \text{OK}$$

5.6 Erection

Precast concrete members frequently must be reoriented from the position used to transport to its final construction position. The analysis for this "tripping" (rotating) operation is similar to that used during other handling stages. Figure 5.6.1 shows maximum moments for several commonly used tripping techniques.

When using two crane lines, the center of gravity must be between them in order to prevent a sudden shifting of the load while it is being rotated. To ensure that this is avoided, the stability condition shown in Figure 5.6.2 must be met. The capacities of lifting devices must be checked for the forces imposed during the tripping operation, since the directions vary.

When rotating a panel with two crane lines, the pick points should be located to prevent the panel from an uncontrolled roll on the roller blocks. This can be done by slightly offsetting the pick point locations to shift the weight toward the upper crane line lift points, or by using chain drags on the rolling block.

Figure 5.6.1 Typical tripping (rotating) positions for erection of wall panels

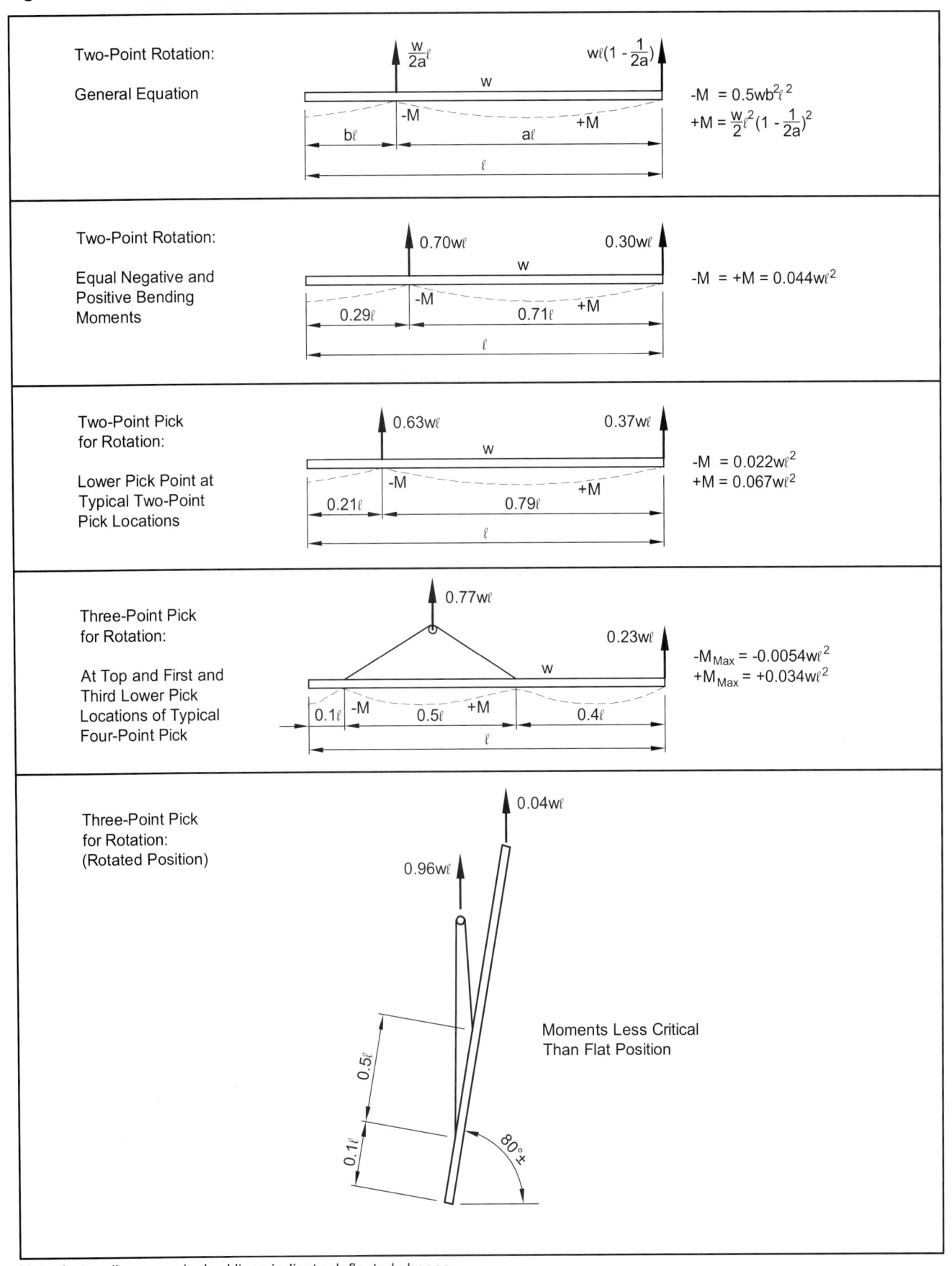

Two-Point Rotation:

General Equation

Reactions: $\frac{w}{2a}\ell$ and $w\ell(1 - \frac{1}{2a})$

Dimensions: $b\ell$, $a\ell$, total ℓ

$-M = 0.5wb^2\ell^2$
$+M = \frac{w}{2}\ell^2(1 - \frac{1}{2a})^2$

Two-Point Rotation:

Equal Negative and Positive Bending Moments

Reactions: $0.70w\ell$ and $0.30w\ell$

Dimensions: 0.29ℓ, 0.71ℓ, total ℓ

$-M = +M = 0.044w\ell^2$

Two-Point Pick for Rotation:

Lower Pick Point at Typical Two-Point Pick Locations

Reactions: $0.63w\ell$ and $0.37w\ell$

Dimensions: 0.21ℓ, 0.79ℓ, total ℓ

$-M = 0.022w\ell^2$
$+M = 0.067w\ell^2$

Three-Point Pick for Rotation:

At Top and First and Third Lower Pick Locations of Typical Four-Point Pick

Reactions: $0.77w\ell$ and $0.23w\ell$

Dimensions: 0.1ℓ, 0.5ℓ, 0.4ℓ, total ℓ

$-M_{Max} = -0.0054w\ell^2$
$+M_{Max} = +0.034w\ell^2$

Three-Point Pick for Rotation: (Rotated Position)

Reactions: $0.96w\ell$ and $0.04w\ell$

Dimensions: 0.5ℓ, 0.1ℓ, angle $80° \pm$

Moments Less Critical Than Flat Position

Note: In top diagrams, dashed lines indicate deflected shapes.

Figure 5.6.2 Stability during erection

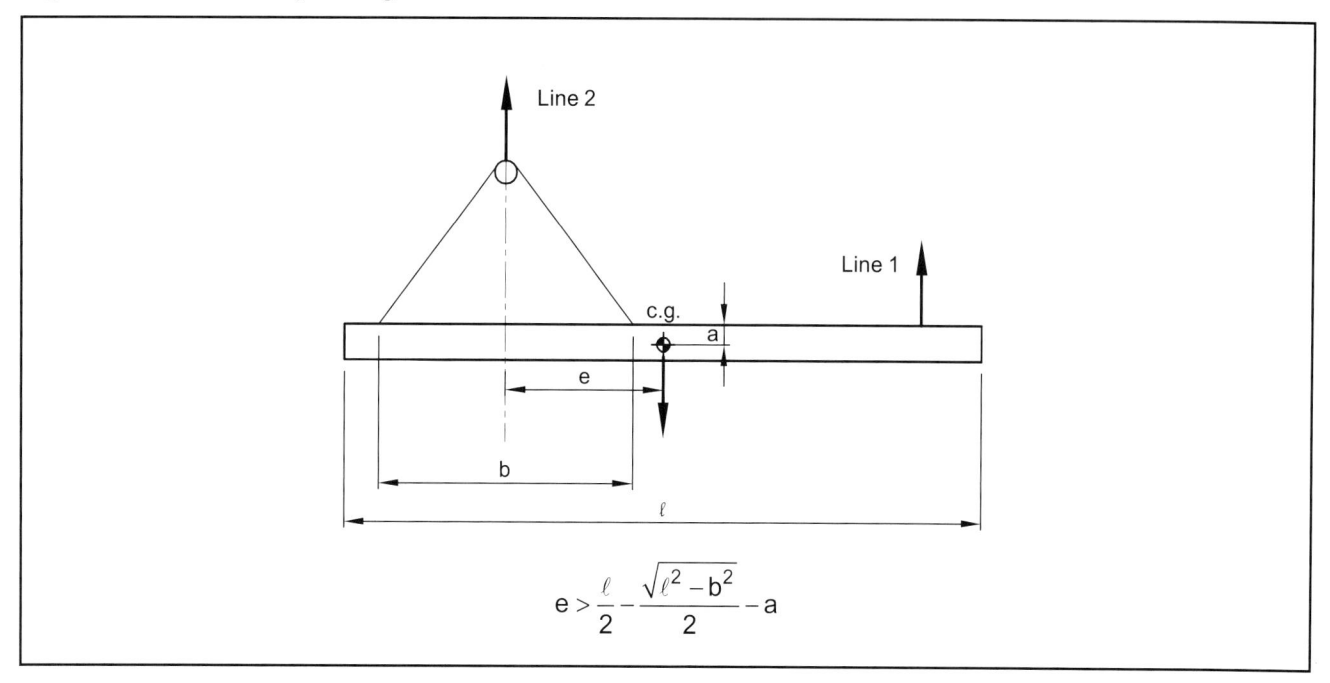

$$e > \frac{\ell}{2} - \frac{\sqrt{\ell^2 - b^2}}{2} - a$$

Example 5.6.1
Erecting Wall Panels

Given:
   The wall panels with openings of Examples 5.3.1 and 5.5.1.

Problem:
   Determine appropriate procedures for erecting the wall panels with openings.

From Example 5.5.1, panel will be shipped flat.
Limit stresses to $5\sqrt{f'_c}$ (0.354 ksi).
Crane has main and auxiliary lines.
A telescoping man lift is available on site.

Solution:
   Try three-point rotation up using stripping inserts and rolling block:
   To simplify, conservatively use solid panel (no openings) to determine moments.

$$W = 1.0 \text{ kip/ft}$$

$$R_B = \frac{(35.17)(1)\left(\frac{35.17}{2}\right)}{35.17\left(0.604 + \frac{0.292}{2}\right)} = 23.4 \text{ kips}$$

$$R_{B_1} = R_{B_2} = \frac{23.4}{2} = 11.7 \text{ kips}$$

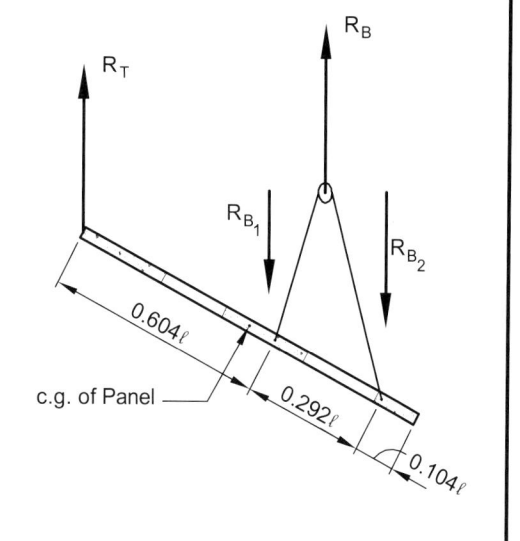

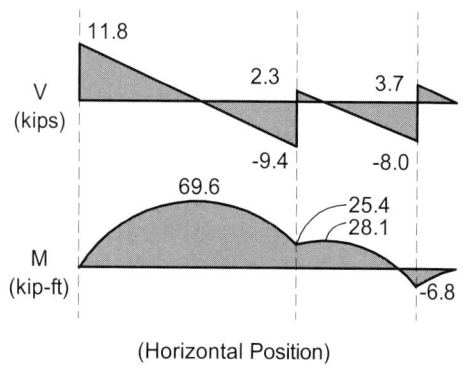

V (kips): 11.8, 2.3, 3.7, −9.4, −8.0

M (kip-ft): 69.6, 25.4, 28.1, −6.8

(Horizontal Position)

5–24 PCI Design Handbook/Sixth Edition

Example 5.6.1 (Cont.)
Erecting Wall Panels

$R_T = R_T = (35.17)(1) - 23.4 = 11.8$ kips

In horizontal position:

$M_{MAX} = 69.6$ kip-ft $= 835.2$ kip-in.

$f = \dfrac{1.2(835.2)}{635} = 1.58$ ksi > 0.354 ksi

Three-point pick not adequate.

Knowing from the stripping analysis (Example 5.3.1) that a eight-point pick can be used, the configurations shown here may be used. However, this rigging may become unstable at some point during tripping, i.e., continued rotation without tension in Line A. Therefore, the lower end of the panel must stay within inches of the ground to maintain control.

(a) Because the above configuration requires six rolling blocks and can be cumbersome, the method shown below may be an alternative.

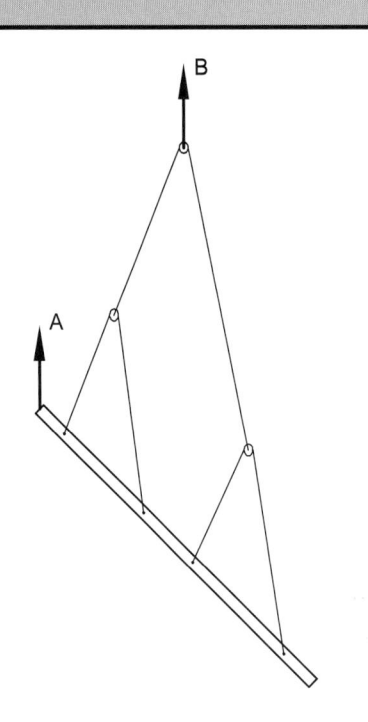

Note: In Figures (c) and (d), both "A" and "B" must be designed to carry the full weight of the panel.

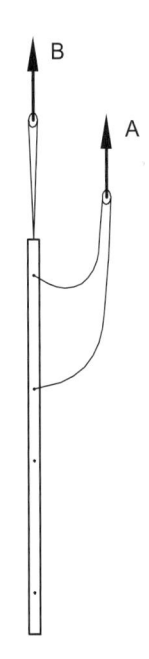

(b) Hook both crane lines to stripping inserts.

(c) With panel essentially vertical, disengage Crane Line B from bottom insert.

(d) Rehook Line B to lifting inserts in top edge of panel from manlift, lift panel from Line B.

5.7 Erection Bracing

5.7.1 Introduction

This section deals with the temporary bracing which may be necessary to maintain structural stability of a precast structure during construction. When possible, the final connections should be used to provide at least part of the erection bracing, but additional bracing apparatus is sometimes required to resist all of the temporary loads. These temporary loads may include wind, seismic, eccentric dead loads including construction loads, unbalanced conditions due to erection sequence and incomplete connections. Due to the low probability of design loads occurring during erection, engineering judgment should be used to establish a reasonable design load, Section 5.7.2. has suggested guidelines.

5.7.1.1 Responsibilities

Proper planning of the construction process is essential for efficient and safe erection. Sequence of erection must be established early, and the effects accounted for in the bracing analysis and the preparation of shop drawings.

The responsibility for the erection of precast concrete may vary as follows (see also Section 10.3):

1. The precast concrete manufacturer supplies the product erected, either with his own forces, or by an independent erector.
2. The manufacturer is responsible only for supplying the product, F.O.B. plant or jobsite. Erection is done either by the general contractor or by an independent erector under a separate agreement.
3. The products are purchased by an independent erector who has a contract to furnish the complete precast concrete package.

Responsibility for stability during erection must be clearly understood. Design for erection conditions must be in accordance with all local, state and federal regulations. It is desirable that this design be directed or approved by a Professional Engineer. Erection drawings define the procedure on how to assemble the components into the final structure. The erection drawings should also address the stability of the structure during construction and include temporary connections. When necessary, special drawings may be required to include shoring, guying, bracing and specific erection sequences. Additional guidelines are available in Ref. 3. For large and/or complex projects, a pre-job conference prior to the preparation of erection drawings may be warranted, in order to discuss erection methods and to coordinate with other trades.

5.7.1.2 Handling Equipment

The type of jobsite handling equipment selected may influence the erection sequence, and hence affect the temporary bracing requirements. Several types of erection equipment are available, including truck-mounted and crawler mobile cranes, hydraulic cranes, tower cranes, monorail systems, derricks and others. The *PCI Recommended Practice for Erection of Precast Concrete* [3] provides more information on the uses of each.

5.7.1.3 Surveying and Layout

Before products are shipped to the jobsite, a field check of the project is recommended to ensure that prior construction is suitable to accept the precast units. This check should include location, line and grade of bearing surfaces, notches, blockouts, anchor bolts, cast-in hardware, and dimensional deviations. Site conditions such as access ramps, overhead electrical lines, truck access, etc., should also be checked. Any discrepancies between actual conditions and those shown on drawings should be addressed before erection is started.

Surveys should be required before, during and after erection:

1. Before, so that the starting point is clearly established and any potential difficulties with the support structure are determined early.
2. During, to maintain alignment.
3. After, to ensure that the products have been erected within tolerances.

5.7.2 Loads

The publication *Design Loads on Structures During Construction* (SEI/ASCE 37-02) [9] provides minimum design loads, including wind, earthquake and construction loads and load combinations for partially completed structures and structures used during construction.

In addition to working stress or strength design using loads from the above publication, the designer must consider the effect of temporary loading on stability and bracing design. Examples are shown in Figure 5.7.2.1.

a. Columns with eccentric loads from other framing members produce sidesway which means the columns lean out of plumb. A similar condition can exist when cladding panels are erected on one side of a multistory structure.

b. Unbalanced loads due to partially complete erection may result in beam rotation. The erection drawings should address these conditions. Some solutions are:

- Install wood wedges between flange of tee and top of beam.
- Use connection to columns that prevent rotation.
- Erect tees on both sides of beam.
- Prop tees to level below.

c. Rotations and deflections of framing members may be caused by cladding panels. This may result in alignment problems and require connections that allow for alignment adjustment after all panels are erected.

If construction equipment such as concrete buggies, man-lifts, etc., are to be used, information such as wheel loads and spacing should be conveyed to the designer of the precast members and the designer of the erection bracing.

Figure 5.7.2.1 Temporary loading conditions that affect stability

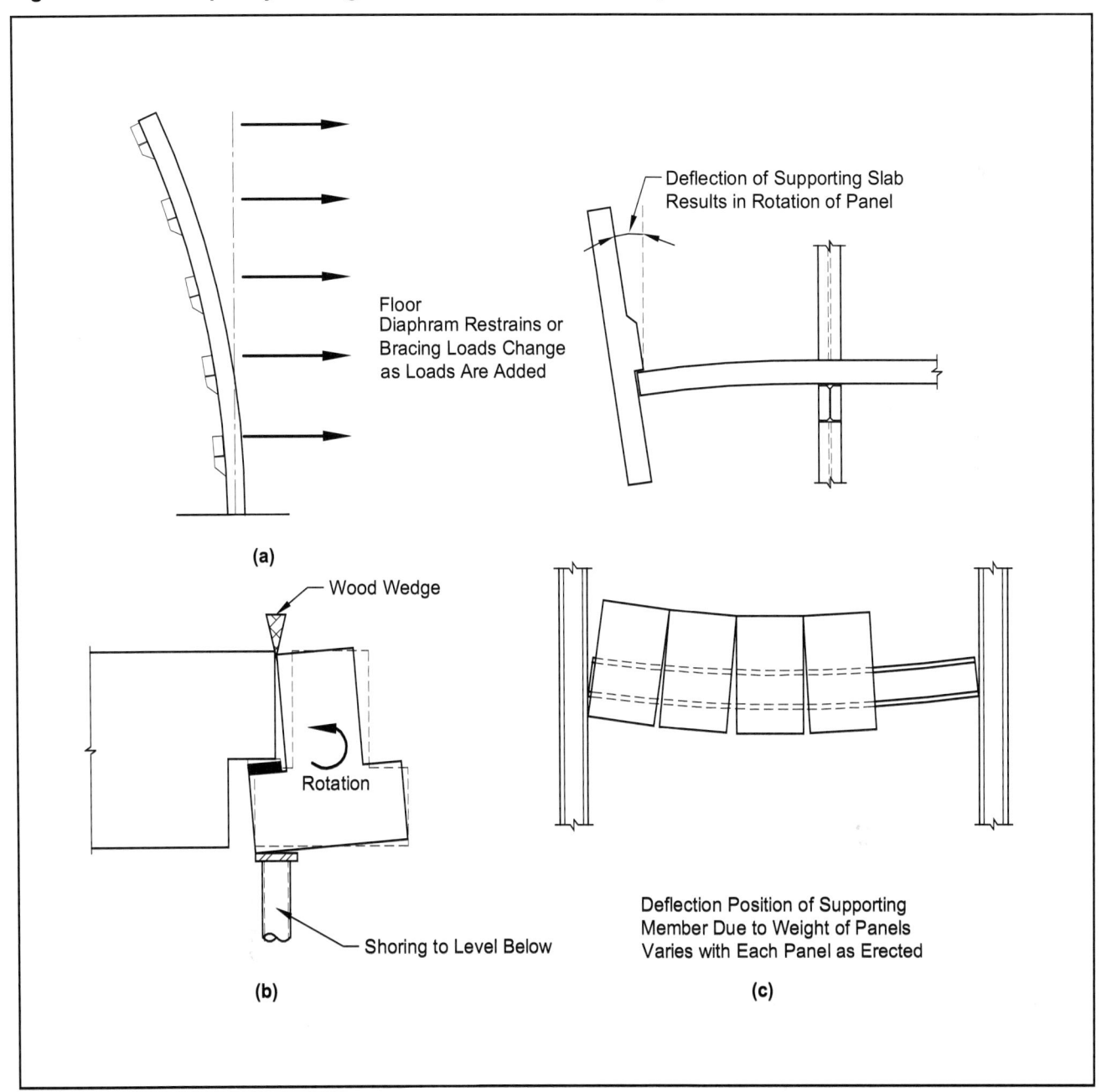

Table 5.7.3.1 Suggested factors of safety for construction loads

Bracing inserts cast into precast members	3
Reusable hardware	5
Lifting inserts	4

5.7.3 Factors of Safety

Safety factors used for temporary loading conditions are a matter of engineering judgment, and should consider failure mode (brittle or ductile), predictability of loads, quality control of products and construction, opportunity for human error, and economics. The total factor of safety also depends on load factors and capacity reduction factors used in the design of the entire bracing system. These must be consistent with applicable code requirements. Suggested safety factors are shown in Table 5.7.3.1.

5.7.4 Bracing Equipment and Materials

For most one-story and two-story high components that require bracing, steel pipe braces similar to those shown in Figure 5.7.4.1 are used.

A wide range of bracing types are available from a number of suppliers who should be consulted for dimensions and capacities. Pipe braces resist both tension and compression. When long braces are used in compression, it may be necessary to provide lateral restraint to the brace to prevent buckling. Proper anchoring of the braces to the precast members and deadmen must be considered. When the pipe braces are in tension, there may be significant shear and tension loads applied to the deadmen. Properly designed deadmen are a requirement for safe bracing.

Cable guys with turnbuckles are normally used for taller structures. Since wire rope used in cable guys can resist only tension, they are usually used in combination with other cable guys in an opposite direction. Compression struts, which may be the precast concrete components, are needed to complete truss action of the bracing system. A number of wire rope types are available. [4] Note that capacity of these systems is often governed by the turnbuckle capacity.

5.7.5 Erection Analysis and Associated Bracing Requirements

The following examples demonstrate suggested procedures to ensure structural stability and safety during various stages of construction. Actual loads, factors of safety, equipment used, etc., must be evaluated for each project.

General Considerations

The examples in this section demonstrate that careful planning of the erection sequence is important. This plan is usually developed by a coordinated effort involving the general contractor, precast erector, precaster production and shipping departments and a structural engineer. A properly planned erection sequence can reduce bracing requirements. For example, with wall panel systems a corner can first be erected so that immediate stability can be achieved. Similar considerations for shear wall structures can also reduce bracing requirements. All parties should be made aware of the necessity of closely following erection with the welded diaphragm connections. This includes the diaphragm to shear wall connections.

In order for precast erection to flow smoothly:

1. The site access and preparation must be ready.

2. The to-be-erected products must be ready.

3. Precast shipping must be planned.

4. The erection equipment must be ready.

5. Bracing equipment and deadmen must be ready.

Figure 5.7.4.1 Typical pipe braces

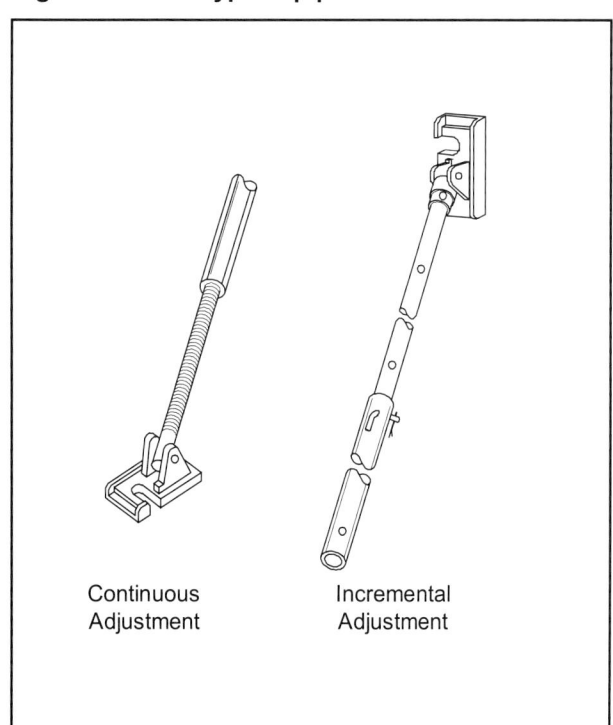

Continuous Adjustment   Incremental Adjustment

Example 5.7.5.1
Erecting a Single-Story Industrial Building

Given:
A single-story building with the plan shown schematically below. Totally precast structure, i.e., columns, inverted tee beams, double tee roof, loadbearing and non-loadbearing double tee wall panels. The structure is located in a 90 mph wind zone with Exposure Category C. The expected duration of the unbraced members is only a few days. The final stability of the structure is provided by the exterior walls acting as shear walls. There is no expansion joint in the structure.

Planned erection sequence: With a crane in the center bay, start at Column Line 8 and move toward Line 1, erecting all three bays progressively.

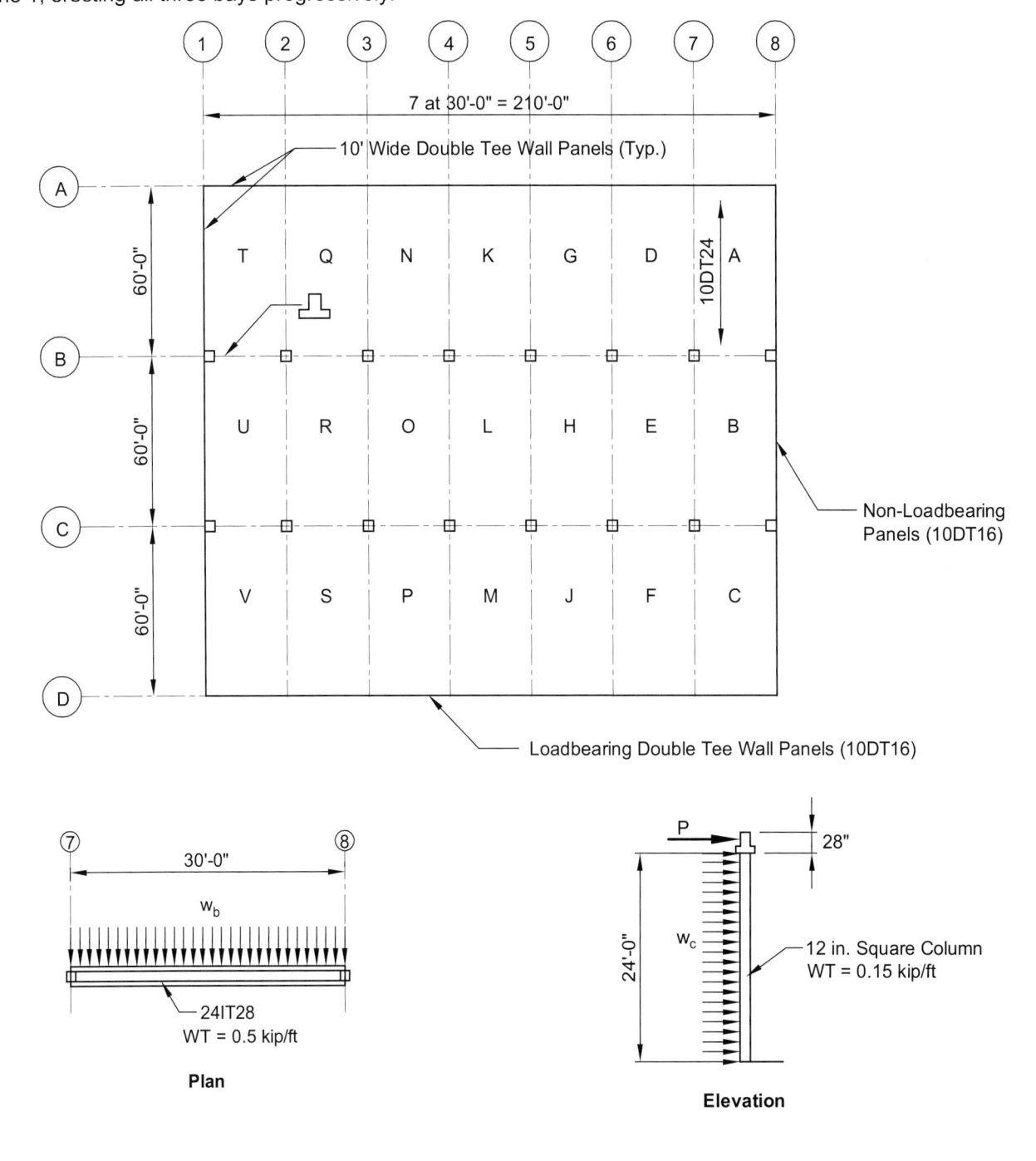

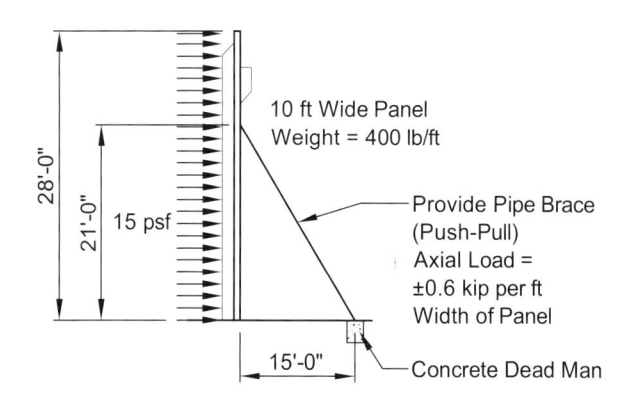

Example 5.7.5.1 (Cont.)
Erecting a Single-Story Industrial Building

Problem:
Determine temporary erection bracing systems.

Solution:
Calculation of Temporary Wind Load

ASCE 37 prescribes the use of ASCE 7-02 [10] for wind loads during construction. From ASCE 37 the duration factor for less than six weeks is 0.75 and the Importance Factor I is 1.0. The exposed members are assumed to resist wind loads in a manner similar to signs; therefore, the wind loads on the members are determined using the methods applicable for signs. (Refer to Section 6.5.13 of ASCE 7-02.)

An analysis based on the above shows that the appropriate wind pressure is 15 psf for this example.

To demonstrate the design procedure, the following temporary conditions will be considered:

1. Erect Columns B8 and B7 with inverted tee beam between them and check as free standing on the base plate.

w_b = 0.015(2.33) = 0.035 kips/ft

P = 0.035(30)/2 = 0.53 kips

w_c = 0.015(1) = 0.015 kips/ft

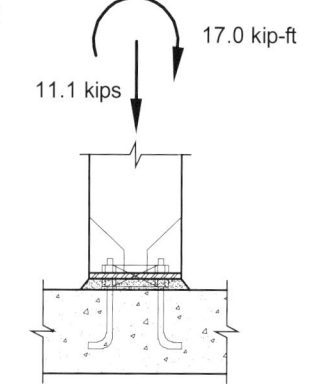

Moment at column base:
= 0.53(24)+0.015(24)²/2 = 17.0 kip-ft

Dead load at column base:
= 0.5(30)/2+0.15(24) = 11.1 kips

Design base plate and anchor bolts as described in Chapter 6.

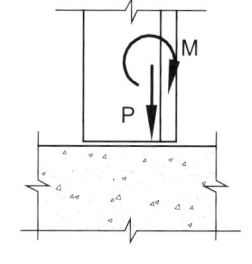

2. Erect wall panel on Line A from 7 to 8. There are two cases to consider:

Case a: If the base of the wall has some moment resisting capacity, the base should be designed for:
M = 0.015(10)(0.5)(28)²/2 = 29.4 kip-ft/stem
P = 0.4(28)/2 = 5.6 kips/stem

Example 5.7.5.1 (Cont.)
Erecting a Single-Story Industrial Building

Case b: If the wall panel has no moment resisting capacity, a brace must be provided, as shown below. Check the unsupported length of the brace to prevent buckling under compressive loads. Also check the horizontal reaction at the base of the panel.

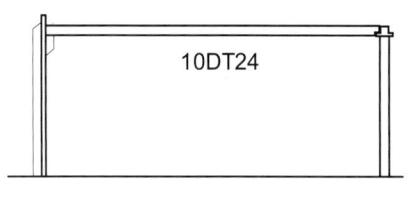

Section Through Bay A

3. Erect double tee roof members in Bay A. Assuming the wall panel has moment resisting capacity at the base, two loading conditions must be checked:

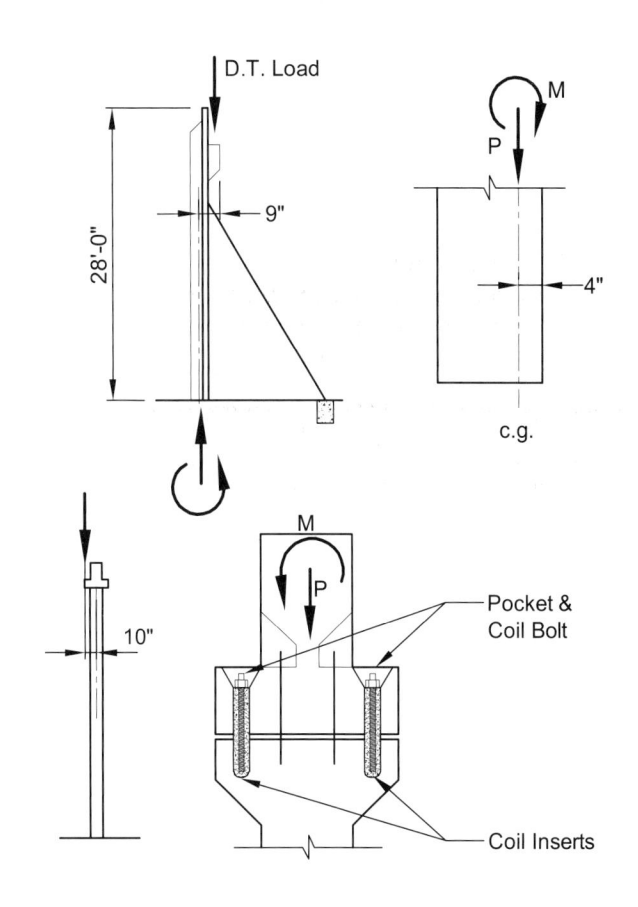

Case a: Dead Loads Only:

Load from roof tee = 0.047(5)(60/2)
 = 7.0 kips/stem
M = 7.0(9/12) = 5.25 kip-ft/stem
P = 7.0 + 5.6 = 12.6 kips/stem

Check beam-to-column connection. This will be critical when all three roof tees in Bay A are in place.

Load from double tees
 = 7.0 kips/stem times three stems = 21 kips
M = 21.0 (10/12) = 17.5 kip-ft
P = 21.0 + 0.5(30/2) = 28.5 kips

Check column base connection:

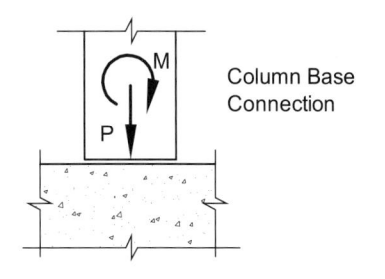

M = 17.5 kip-ft (from above)
P = 28.5 + 0.15(24) = 32.1 kips

Column Base Connection

Example 5.7.5.1 (Cont.)
Erecting a Single-Story Industrial Building

Case b: Dead Load plus Wind Load:

In this example, it is apparent that the loading for 1 and 2 above will be more critical than this condition.
The designer should always try to use the permanent connections for erection stability. In this example, the final connections between the wall and roof deck and the beam and roof deck will provide some degree of moment resistant capacity. This will reduce the moments on some of the connections considered above.

4. Erect wall panels on Line 8 from A to B.

(Note: If the crane reach is limited, these may be erected immediately after the first roof tee is placed.) These will be connected to the roof with permanent diaphragm transfer connections. This provides a rigid corner in one direction, which will provide stability to the remainder of the structure in that direction.
Continue erection working from the rigid corner in Bay A. It is unlikely that other temporary conditions will be more critical than those encountered in Bay A. However, each case should be considered and analyzed where appropriate.

Example 5.7.5.2
Erecting a Multi-Level Parking Structure

Given:

A typical four-story parking structure is shown schematically on the following page. The structural system is pretopped double tees on inverted tee beams, interior litewalls and loadbearing spandrel panels. Multi-level columns are one piece. No expansion joint.
Loads: Wind – 10 psf. Gravity construction loads – 5 psf. This is lower than recommended in ASCE 37-02 because there are virtually no interior finishing materials, such as masonry or drywall in this construction.
Final stability will be provided by the interior litewalls in the long direction and the shear walls shown in the short direction.

Problem:
Outline critical design conditions during erection and show a detailed erection sequence.

Solution:

Outline of critical erection design conditions:
1. Free standing columns. Design columns and base plates in accordance with Chapters 3, 4 and 6.
2. Determine bracing forces for wind loads from either direction.
3. Determine bracing and deadman requirements.
4. Determine forces on inserts used for bracing and select anchorages in accordance with Chapter 6.
5. Check column designs with temporary loading and bracing.
6. Check diaphragm design and determine which permanent connections must be made during erection. Determine need for temporary connections.
7. Check inverted tee beams and their connections for loading on one side only.

Example 5.7.5.2 (Cont.)
Erecting a Multi-Level Parking Structure

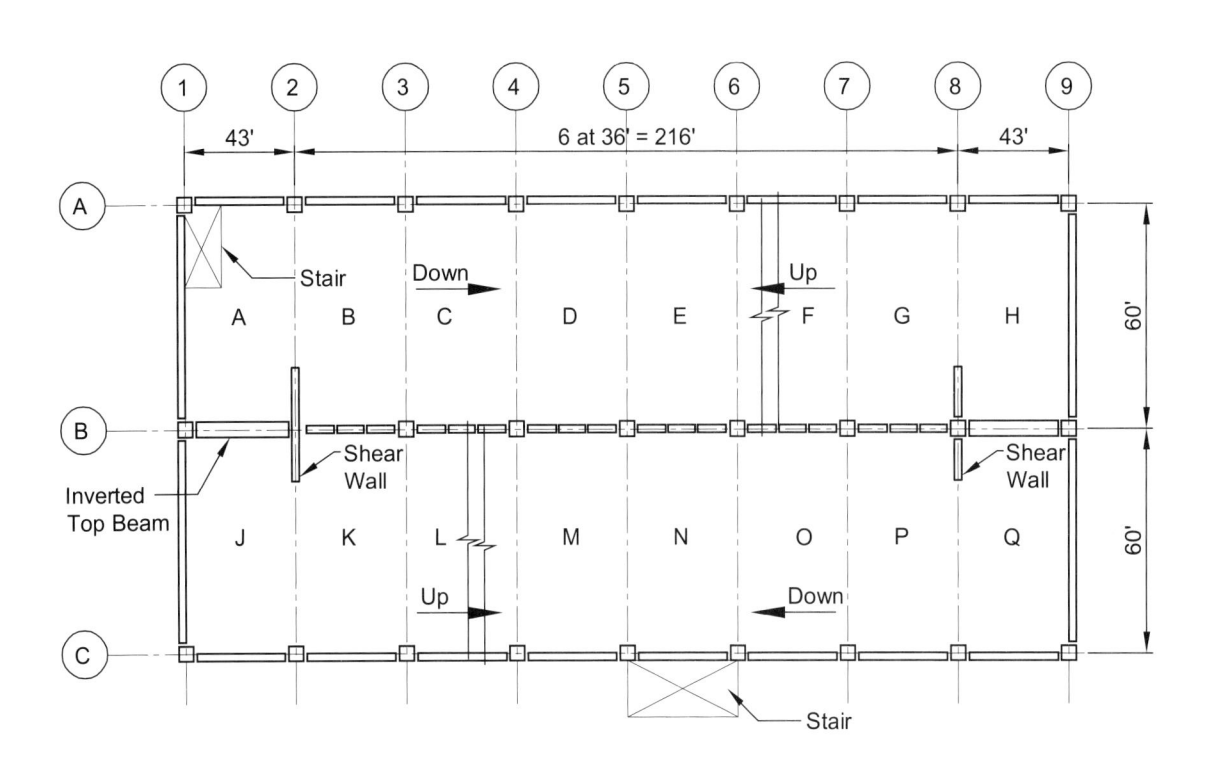

Typical Floor Plan

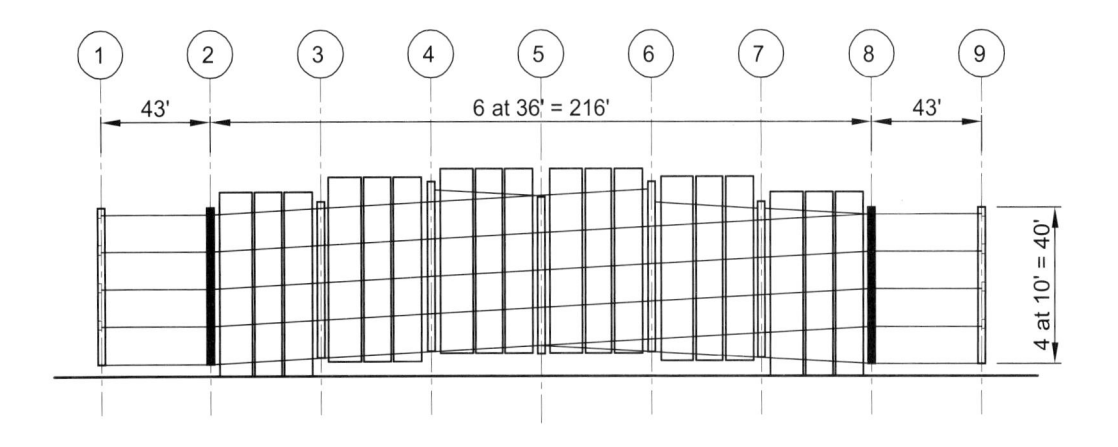

Section

Example 5.7.5.2 (Cont.)
Erecting a Multi-Level Parking Structure

Sequence of erection:
Start at column line 1 and, with crane in north bay, erect vertically and back out of structure at Line 9 in the following sequence:

1. Erect columns and shear walls A1, B1, A2 and B2 and shear wall. Check as free standing columns with 10 psf wind on surface. Design pipe braces if necessary.
2. Erect all levels in Bay A.
3. Install diaphragm connection to shear wall at 2. Check capacity required with full wind load on exposed surfaces.
4. Erect Column A3 and litewall nearest 2.
5. Erect all levels in Bay B.
6. Check litewall capacity for full wind load on exposed surfaces.
7. Erect Columns C1 and C2.
8. Erect all levels in Bay J.
9. Install diaphragm connections to shear wall at 2. Check capacity required with full load on exposed surfaces.
10. Erect Column A4 and B4 litewalls between 3 and 4.
11. Erect all levels in Bay C.
12. Erect Column C3.
13. Erect all levels in Bay K.
14. Erect Columns A5 and B5 and litewalls between 4 and 5.
15. Erect all levels in Bay D.
16. Install X-bracing from A5 to B5. Check capacity (include turnbuckle). This will help keep columns in plumb.
17. Erect Column C4.
18. Erect all levels of Bay L.
19. Continue in same sequence.

Additional Notes:
1. After each level is erected weld loose plate from spandrel and litewall to plate in double tee. (see above drawing)
2. Install additional X-bracing on Line 5 from B to C after framing is erected.
3. Install cables to columns as required to keep columns plumb.
   Maintain tightness of cables.

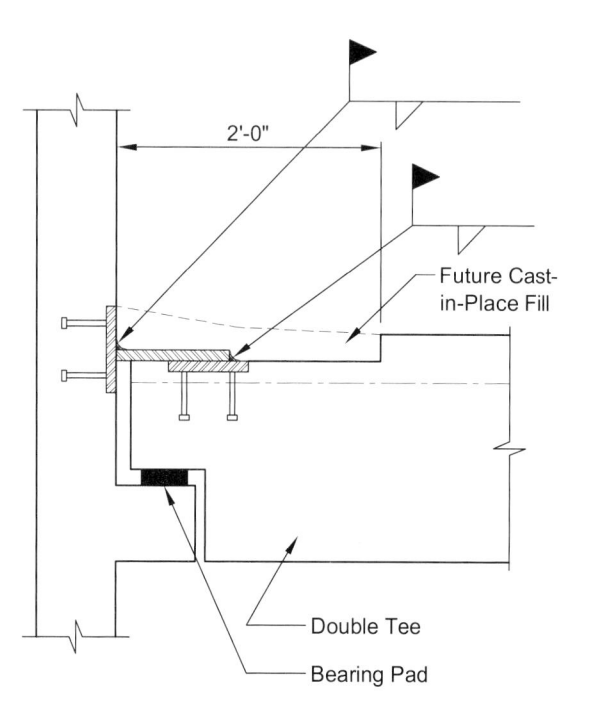

5.8 References

1. Tanner, John, "Architectural Panel Design and Production Using Post-Tensioning," PCI JOURNAL, V. 22, No. 3, May-June 1977.

2. Anderson, A. R., "Lateral Stability of Long Prestressed Concrete Beams," PCI JOURNAL, V. 16, No. 3, May-June 1971.

3. *Erectors Manual – Standards and Guideline for the Erection of Precast Concrete Products*, MNL-127-99, Precast/Prestressed Concrete Institute, Chicago, IL, 1999.

4. *Wire Rope Users Manual*, Committee of Wire Rope Producers, American Iron and Steel Institute, Washington, DC, 1985.

5. Mast, Robert F., "Lateral Stability of Long Prestressed Concrete Beams – Part 1," PCI JOURNAL, V. 34, No.1, January-February 1989.

6. Mast, Robert F., "Lateral Stability of Long Prestressed Concrete Beams – Part 2," PCI JOURNAL, V. 38, No. 1, January-February 1993.

7. *PCI Bridge Design Manual*, MNL-133-97, Precast/Prestressed Concrete Institute, Chicago, IL, 1997.

8. Mast, Robert F., "Lateral Bending Test to Destruction of a 149 ft Prestressed Concrete I-Beam," PCI JOURNAL, V. 39, No. 4, July-August 1994.

9. *Design Loads on Structures During Construction (SEI/ASCE 37-02)*, American Society of Civil Engineers, Reston, VA, 2002 (Co-sponsored by the Structural Engineering Institute).

10. *Minimum Design Loads for Buildings and Other Structures, Revision of ASCE 7-98 (SEI/ASCE 7-02)*, American Society of Civil Engineers, Reston, VA, 2003 (Co-sponsored by the Structural Engineering Institute).

CHAPTER 6
DESIGN OF CONNECTIONS

6.1 General 6–3
    6.1.1 Notation 6–3
    6.1.2 Introduction 6–6

6.2 Loads and Load Factors 6–6
    6.2.1 Strength Reduction Factors (ϕ-factors) 6–6
        6.2.1.1 Concrete Flexure and Axial Load 6–6
        6.2.1.2 Concrete Corbels 6–6
        6.2.1.3 Anchorages 6–6
        6.2.1.4 Steel 6–6
    6.2.2 Over Load Factors (OLF) 6–7

6.3 Connection Design Criteria 6–7

6.4 Connection Hardware and Load Transfer Devices 6–8
    6.4.1 Headed Concrete Anchors or Studs 6–8
    6.4.2 Steel Shapes 6–8
    6.4.3 Reinforcing Bars 6–8
    6.4.4 Reinforcing Bar Couplers 6–9
        6.4.4.1 Reinforcing Bar Splicing Systems 6–9
    6.4.5 Deformed Bar Anchors (DBA) 6–9
    6.4.6 Bolts and Threaded Connectors 6–9
    6.4.7 High Strength Threaded Rods 6–10
    6.4.8 Specialty Inserts 6–10
    6.4.9 Post-Installed Anchors 6–10
        6.4.9.1 Expansion Anchors 6–10
        6.4.9.2 Adhesive Anchors 6–10

6.5 Headed Concrete Anchor Design 6–11
    6.5.1 Steel Materials 6–11
        6.5.1.1 Steel Plates and Shapes 6–11
        6.5.1.2 Headed Concrete Anchors (HCA) or Studs 6–11
    6.5.2 HCA – Steel Strength 6–12
    6.5.3 HCA – Concrete Strength 6–12
        6.5.3.1 Supplemental Reinforcement 6–12
        6.5.3.2 Cracked Concrete 6–12
        6.5.3.3 The 5 Percent Fractile 6–13
        6.5.3.4 Strength Reduction Factor, ϕ 6–13
    6.5.4 Concrete Tension Strength 6–13
        6.5.4.1 Breakout Strength 6–13
        6.5.4.2 Pullout Strength 6–17
        6.5.4.3 Side-Face Blowout 6–17
    6.5.5 Concrete Shear Strength 6–17
        6.5.5.1 Front Edge (d_{e3}) 6–17
        6.5.5.2 Corners 6–20
        6.5.5.3 Side Edge (d_{e1} or d_{e2}) 6–22
    6.5.6 Back Edge (d_{e4}) 6–25
    6.5.7 In-the-Field 6–25
    6.5.8 Interaction of Tension and Shear 6–26

6.6 Structural Steel Design 6–28
    6.6.1 Design Flexural Strength 6–28

		6.6.1.1 Built-up Steel Sections	6–28
	6.6.2	Design Shear Strength	6–28
	6.6.3	Design Torsional Strength	6–30
		6.6.3.1 Rectangular Solid Plates in Torsion	6–30
		6.6.3.2 Rectangular Hollow Sections (Structural Tubing) in Torsion	6–31
	6.6.4	Combined Loading	6–32
	6.6.5	Unstiffened Connection Angles	6–32
	6.6.6	Triangular Stiffener Design	6–34
	6.6.7	Non-Triangular Stiffener Design	6–35

6.7	Welding		6–36
	6.7.1	Corrosion Resistance of Welded Connections	6–36
		6.7.1.1 Hot Dip Galvanized	6–36
		6.7.1.2 Stainless Steel	6–37
	6.7.2	Weld Design	6–38
		6.7.2.1 Full Penetration Welds	6–38
		6.7.2.2 Fillet Welds	6–38
		6.7.2.3 Partial Penetration Groove Welds	6–38
	6.7.3	Reinforcing Bar Welding	6–38
		6.7.3.1 Carbon Equivalent	6–38
		6.7.3.2 Strength of Reinforcing Bar Weld	6–40
	6.7.4	Cracking in Concrete Around Welded Connections	6–41
	6.7.5	Design of Group Welds	6–41
		6.7.5.1 Elastic Vector Method	6–42
		6.7.5.2 Instantaneous Center Method (ICM)	6–43

6.8	Concrete Brackets or Corbels		6–47
	6.8.1	Cantilever Beam Design Method	6–47
	6.8.2	Strut-and-Tie Design Method	6–50
		6.8.2.1 Bearing Area	6–50
		6.8.2.2 Truss Geometry	6–50
	6.8.3	Comparison of Corbel Design Methods	6–50
	6.8.4	Development of Corbel Reinforcement	6–54

6.9	Structural Steel Corbels	6–54

6.10	Hanger Connections		6–59
	6.10.1	Cazaly Hanger	6–59
	6.10.2	Loov Hanger	6–61

6.11	Bearing Pads		6–62
	6.11.1	Design Recommendations	6–64

6.12	Column Bases		6–66
	6.12.1	Base Plates	6–66
	6.12.2	Anchor Bolts	6–66
		6.12.2.1 Embedment Strength of Anchor Bolts in Tension	6–66

6.13	Moment Connections	6–71

6.14	Typical Connection Designs for Lateral Load Resisting Systems	6–71

6.15	References	6–83

6.16	Design Aids	6–85

DESIGN OF CONNECTIONS

6.1 General

6.1.1 Notation

a = center of exterior cantilever bearing to center of Cazaly hanger strap

a = depth of equivalent rectangular compression stress block

a = leg size of fillet weld

a = shear span

A_b = area of bar or stud

A_c = effective cross-sectional area at one end of strut in a strut-and-tie model, taken perpendicular to axis of strut

A_{cr} = area of crack interface

A_f = area of flexural reinforcement in corbel

A_h = area of shear reinforcement parallel to flexural tension reinforcement in corbel

A_n = area of reinforcement required to resist axial tension

A_n = area of face of nodal zone or section through nodal zone

A_N = projected surface area for stud or group of studs

A_{No} = projected surface area of one anchor when not limited by edge distance

A_s = area of steel

A_{se} = nominal area of headed stud shank

A_{vf} = area of shear-friction reinforcement

A_w = area of weld

b = dimension (see specific application)

b = vertical dimension of stiffener

b = width of compression stress block

b_n = net length of angle

BED = distance from back row of studs to front edge

c_c = concrete cover

C = compressive force

C_{bs} = breakout strength coefficient

C_{crb} = cracking coefficient (breakout)

C_{crp} = cracking coefficient (pullout)

$C_x, C_y, C_h, C_{ev}, C_{vcr}$ = coefficients used in design of shear strength of studs

CE = carbon equivalent

d = depth of welds in weld group

d = depth to centroid of reinforcement

d = diameter of stud

d_b = bar diameter

$d_{e1}, d_{e2}, d_{e3}, d_{e4}$ = edge distance from stud group

d_{hs} = head diameter of stud

d_o = shank diameter of stud

D = durometer (shore A hardness)

e = eccentricity of load

e_i = center of bolt to horizontal reaction

e_v = eccentricity of vertical load

e'_v = eccentricity of shear force on a group of anchors

e_{v1} = eccentricity from shear load to anchorage centroid

e_x, e_y = eccentricity of load in x,y directions

f'_c = specified compressive strength of concrete

f_{cu} = effective compressive strength of concrete in strut or nodal zone

f_r = resultant stress on weld

f_{un} = maximum normal stress

f_{ut} = minimum tensile strength of steel

f_{uv} = maximum shear stress

f_x = combined shear and torsion stress in horizontal direction

f_y = combined shear and torsion stress in vertical direction

f_y = yield strength of reinforcement or headed stud

ΣF = greatest sum of factored anchor bolt forces on one side of column

F_{exx} = classification strength of weld metal

F_{ns} = nominal strength of strut

F_{nt} = nominal strength of tie

F_y = yield strength of structural steel

F_u = tensile strength of base metal

g = gage of angle

G = shear modulus

h = horizontal dimension of stiffener

h = total depth

PCI Design Handbook/Sixth Edition

h_c	=	column width
h_{ef}	=	effective stud embedment length
h'_{ef}	=	design effective stud embedment length for narrow section
HCA	=	headed concrete anchors
I_p	=	polar moment of inertia
I_{xx}, I_{yy}	=	moment of inertia of weld segment with respect to its own axes
J_t	=	torsional constant
k	=	distance from back face of angle to web fillet toe
ℓ_b	=	bearing length
ℓ_d	=	development length
ℓ_e	=	embedment length
ℓ_ℓ	=	angle leg length
ℓ_p	=	bearing length of exterior cantilever in hanger connections
ℓ_w	=	length of weld
L	=	nominal length of stud
M_p	=	moment capacity of steel section
M_u	=	factored moment
n	=	number of studs in group
n_s	=	number of studs in back row of group
n_x, n_y	=	number of studs in x and y row
N_b	=	basic concrete breakout strength in tension
N_{cb}	=	nominal concrete breakout strength
N_{cbg}	=	nominal concrete breakout strength of stud group
N_n	=	nominal tensile strength of stud
N_{ph}	=	nominal concrete pullout strength
N_s	=	nominal tensile strength based on steel capacity
N_t	=	applied tensile force
N_u	=	factored horizontal or axial force
N_{sb}	=	nominal side face blowout strength of single stud
N_{sbg}	=	nominal side face blowout strength of group of studs
P	=	applied load
P_n	=	nominal strength
P_u	=	applied factored load
P_x, P_y	=	applied force in x, y direction
s	=	spacing of reinforcing (see specific application)
s	=	width of Cazaly hanger strap
S	=	section modulus
S	=	shape factor
SED	=	critical distance perpendicular to load
t	=	(with subscripts) thickness
T	=	tensile force
T_n	=	nominal torsional strength
T_u	=	factored tensile force
V_c	=	nominal strength of section controlled by concrete
V_{cp}	=	nominal shear strength of stud group "in the field"
V_n	=	nominal bearing or shear strength of element
V_s	=	nominal shear strength based on steel capacity
V_u	=	factored shear force
w	=	dimension (see specific application)
w	=	uniform load
w_s	=	effective width of strut
W	=	bearing strip width
x_c	=	distance from centerline of bolt to face of column
x_o	=	base plate projection
x_t	=	distance from centerline of bolt to centerline of column reinforcement
x, y	=	horizontal and vertical distance, respectively
$\bar{x}, \bar{y}$	=	distance of weld to center of gravity to weld group
X, Y	=	overall dimensions (width and length) of stud group
z	=	ratio used in stiffener design
Z, Z_s	=	plastic section modulus of structural steel section
α	=	hanger bar angle
β_n	=	factor to account for effect of anchorage ties on effective compressive strength of a nodal zone
β_s	=	factor to account for effect of cracking and confining reinforcement of effective compressive strength of concrete in strut

Δ = design horizontal movement at end of member

ϕ = strength reduction factor

λ = coefficient for use with lightweight concrete

μ = shear-friction coefficient

μ_e = effective shear-friction coefficient

μ_s = static coefficient of friction

τ = torsional shear strength on structural steel

$\psi_{ec,N}$ = factor used to modify tensile strength of anchors based on eccentricity of applied loads (from ACI 318-05, same as ψ_3 in ACI 318-02)

$\psi_{ed,N}$ = factor used to modify tensile strength of anchors based on proximity to edges of concrete member (from ACI 318-05, same as ψ_2 in ACI 318-02)

6.1.2 Introduction

The design procedures presented in this chapter follow ACI 318-02 [1] and other relevant national model building code requirements except where modified to reflect current industry practice (see also Section 10.5 of this Handbook).

The design of connections is one of the most important considerations in the structural design of a precast concrete structure. There may be several successful solutions to each connection problem, and the design methods and examples included in this chapter are not the only acceptable ones. Information is included on the design of common precast concrete connections. It is intended for use by those with an understanding of engineering mechanics and structural design, and in no case should it replace good engineering judgment.

The purpose of a connection is to transfer load, restrain movement and/or provide stability. Within any one connection, there may be several load transfers; each one must be designed for adequate strength, ductility, and be appropriately detailed. The detailing should take into account allowable tolerances, provide for a good fit between the selected materials and avoid interference between strand or reinforcing steel and the connection components, such as headed studs or deformed bar anchors. In the sections that follow, different methods of transferring load will be examined separately, then it will be shown how some of these are combined in typical connection situations. More information on some aspects of connection design can be found in Ref. 2.

6.2 Loads and Load Factors

With noted exceptions, such as bearing pads, the design methods in this chapter are based on strength design relationships.

Connections are usually designed with factored loads determined from the building analysis. Starting with the 2002 Edition, ACI 318 [1] has adopted load factors from ASCE-7 (see Chapter 3), as shown in Section 3.2.6.

In addition to gravity, wind and seismic loads, forces resulting from restraint of volume changes as well as those required for compatibility of deformations must be considered. For flexural members, it is recommended that bearing connections be designed for a minimum horizontal tensile force, acting parallel to the span, of 0.2 times the factored dead load transferred at the bearing unless a smaller value can be justified by using properly designed bearing pads (see Section 6.11).

6.2.1 Strength Reduction Factors (ϕ-factors)

6.2.1.1 Concrete Flexure and Axial Load

Except for concrete anchorages and concrete corbels, discussed below, ϕ-factors from ACI 318-02, Section 9.3 are used:

- Members designed as tension-controlled flexural members .... $\phi = 0.90$
- Shear and torsion .... $\phi = 0.75$
- Bearing .... $\phi = 0.65$
- Strut-and-tie models (Section 6.8.2) and struts, ties, nodal zones, and bearing areas in such models .... $\phi = 0.75$

6.2.1.2 Concrete Corbels

The ϕ-factor for concrete corbels designed by the cantilever beam method is specified in ACI 318-02, Section 11.9.3.1 .... $\phi = 0.75$

6.2.1.3 Anchorages

The ϕ-factors for concrete anchorages are specified in Appendix D, Section D.4.4 of ACI 318-02. This section describes two conditions governing concrete strength:

- "Condition A" applies where the potential concrete failure surfaces are crossed by supplementary reinforcement (see Example 6.5.4.1 for design of such reinforcement) tension or shear loads on cast-in headed studs, headed bolts, or hooked bolts .... $\phi = 0.75$
- "Condition B" applies where such supplementary reinforcement is not provided, or where pullout or pryout strength governs .... $\phi = 0.70$

Headed studs governed by steel strength:
  Tension .... $\phi = 0.75$
  Shear .... $\phi = 0.65$

Post-installed anchors (Section 6.4.9) are required to be designed with ϕ-factors which range from 0.45 to 0.75, depending on sensitivity to installation and reliability.

6.2.1.4 Steel

The ϕ-factors for steel construction are specified in AISC Manual of Steel Construction – Load and Resistance Factor Design [6].

Structural steel sections:

- Tension and shear .... $\phi = 0.90$
- Welds .... $\phi = 0.75$
- Bolts and threaded fasteners .... $\phi = 0.75$

6.2.2 Over Load Factors (OLF)

To ensure that the overall safety of the connection is adequate, the use of an additional load factor (Over Load Factor, OLF) in the range 1.0 to 1.33 has historically been used by the industry. The need and the magnitude of this additional load factor for a particular connection must depend on the Engineer's judgment and consideration of the following:

Mode of Failure. For conditions where the predicted failure mode is non-ductile, such as due to the use of very short studs, it is appropriate to use larger overload factors.

Consequences of Failure. If failure of a connection is likely to produce catastrophic results (non-redundancy), the connection should have a larger overall factor of safety.

Sensitivity of Connection to Tolerances. Production and erection tolerances, interface tolerances with other trades, as well as movements due to volume changes and applied loads may produce changes in load transfer positions on the connection. The magnitude of the additional load factor should be consistent with this sensitivity as well as the load transfer position considered in design. For example, if the most adverse combination of tolerances is used to establish the load transfer location, an additional load factor may be unnecessary.

Requirements of Local Codes: To prevent a connection from being the weak link in a precast concrete structure, some codes require larger load factors to be used in the analysis and design of the connections. When these factors are used, it is not necessary to use additional load factors. When using Appendix D of ACI 318-02, the reduced ϕ-factors make the use of an OLF unnecessary.

6.3 Connection Design Criteria

Precast concrete connections must meet a variety of design and performance criteria, and not all connections are required to meet the same criteria. These criteria include:

Strength. A connection must have the strength to transfer the forces to which it will be subjected during its lifetime, including those caused by volume change restraint and those required to maintain stability.

Ductility. This is the ability to undergo relatively large inelastic deformations without failure. In connections, ductility is achieved by designing and detailing so that steel devices yield prior to weld failure or concrete failure. Concrete typically fails in a brittle manner unless it is confined.

Volume Change Accommodation. Restraint of creep, shrinkage and temperature change strains can cause large stresses in precast concrete members and their connections. If these stresses develop due to rigid connections, they must be considered in the design. It is usually far better if the connection allows some movement to take place, thus relieving these stresses induced by volume change strain.

Durability. When a connection is exposed to weather, or used in a corrosive environment, steel elements should be adequately covered by concrete, painted, epoxy coated or galvanized. Stainless steel may also be used.

Fire Resistance. Connections, which could jeopardize the structure's stability if weakened by fire, should be protected to the same degree as that required for the members that they connect.

Constructibility. The following items should be considered when designing connections:

- Standardize products and connections.
- Avoid reinforcement and hardware congestion.
- Check material and size availability.
- Avoid penetration of forms, where possible.
- Reduce post-stripping work.
- Be aware of material sizes and limitations.
- Consider clearances and tolerances.
- Avoid non-standard production and erection tolerances.
- Use standard hardware items and as few different sizes as possible.
- Use repetitive details.
- Plan for the shortest possible hoist hook-up time.
- Provide for field adjustment.
- Providing accessibility while the piece is being placed is critical.
- Use connections that are not susceptible to damage in handling.
- Allow for adjustment after the product is un-hooked from the crane.
- Minimize weld heat buildup in the surrounding concrete, or allow for expansion.
- Determine if special inspection is required for the material and welding process (refer to Building Code).

6.4 Connection Hardware and Load Transfer Devices

A wide variety of hardware including reinforcing bars, headed studs, coil inserts, structural steel shapes, bolts, threaded rods, and other materials are used in connections. These devices provide load transfer to concrete via anchorage in the concrete by bond, bearing or by a shear cone resistance mechanism. It is preferable to have a steel material failure, typically defined by yielding, to govern the connection strength because such failures are more predictable and ductile. In seismic regions, this ductile behavior may be a code requirement. Load transfer should be as direct as possible to reduce the complexity and increase the efficiency of the connection.

6.4.1 Headed Concrete Anchors or Studs

A headed concrete anchor (stud) is a smooth shaft with an integral head. These devices are typically welded to a plate or another structural steel shape. The studs may be hand welded, but are more often attached using a stud gun. The design of these devices can be found in Section 6.5.

6.4.2 Steel Shapes

Steel shapes can be used for various load transfer configurations; i.e., wide flange sections, structural tubes, channels, and angles. They can act independently or with other load transfer devices. The design of these devices can be found in Section 6.6.

6.4.3 Reinforcing Bars

Reinforcing bars are usually anchored by bonding to the concrete. When insufficient length is available to anchor the bars by bond alone, supplemental mechanical anchorage is required. This can be accomplished by hooks, bolts, washers, threaded ends with heads, or welded cross-bars.

Load transfer between bars may be achieved by welds, lap splices or mechanical couplers. Required development lengths and standard hook dimensions are given in Chapter 11.

Reinforcing bars have been shown to develop yield strength when they are anchored by embedment in flexible metallic interlocking conduit using grout as shown in Figure 6.4.3.1. [25] The conduit must have sufficient concrete around it for adequate confinement. This scheme can be used to transfer tension, compression or shear forces and is convenient for certain connections, such as column to footing, column to column, and shear wall connections. Note that alternate systems are also available including a system which uses a plastic sleeve.

The following limitations are recommended:

1. The minimum concrete side cover over the conduit should be 3 in.
2. The conduit should have a minimum thickness of 0.023 in. (24 gage) and there should be a minimum annular space of ½ in. around the bar.
3. The grout strength should not be less than the specified concrete strength or 5000 psi.
4. The specified grout should be non-shrink.
5. For tension, the minimum embedment length for any diameter bar is 12 in.
6. Reinforcing bar ductility requirements should be met for seismic applications.
7. Confinement steel is required in most applications.
8. Care should be taken to prevent water from entering the conduit before grouting, especially during freezing weather.

Figure 6.4.3.1 Anchorage in grouted conduit [25]

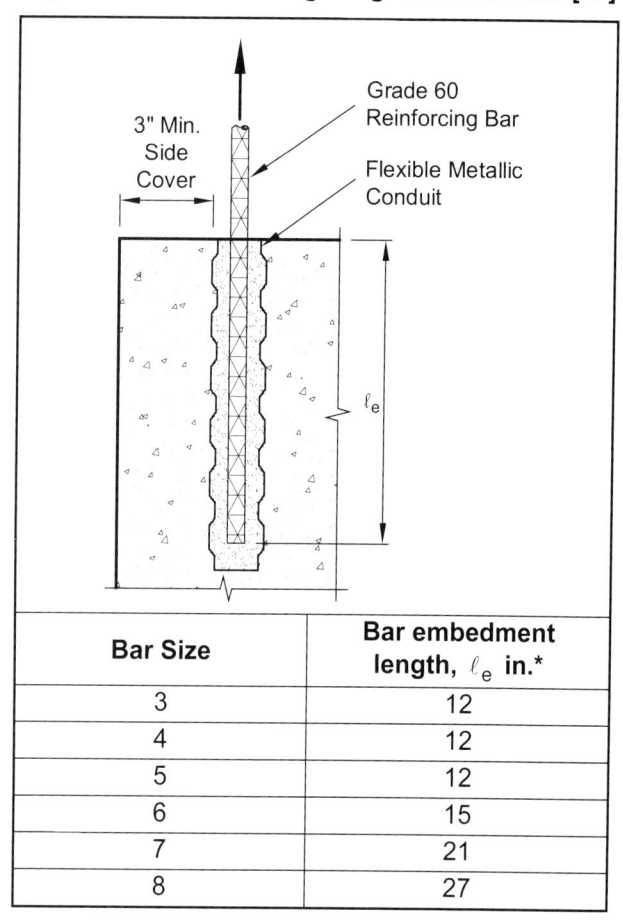

Bar Size	Bar embedment length, ℓ_e in.*
3	12
4	12
5	12
6	15
7	21
8	27

* For grout strengths higher than 5000 psi, multiply table values by $\sqrt{5000/f'_c}$.

6.4.4 Reinforcing Bar Couplers

Proprietary bar coupling devices are available as an alternative to lap splices or welding.

Manufacturers of these devices furnish design information and test data. Refs. 3 and 4 contain more detailed information.

6.4.4.1 Reinforcing Bar Splicing Systems

Several methods are available to splice reinforcing bars when lapping is not an option. This includes clamping, welding, threaded sleeves and a variety of proprietary grouted systems. Some grouted systems used extensively in precast concrete are capable of providing reinforcing bar tension and compression splice capacities equal to 150 percent of the specified yield of the bars. Therefore, they are especially applicable for moment-resistant connections for columns, shear walls, and beams, and are accepted as an emulation of a monolithically cast member (see Section 3.1.2.1). Some codes may require special inspection when using splicing systems.

Depending on the system used, splices are available for bar sizes ranging from #4 through #18. The splicing mechanism is typically a cylindrical shaped steel sleeve with open ends for insertion of the reinforced bar. Once the bar is in place, a specially formulated high early strength, non-shrink grout is pumped into the sleeve by way of PVC tubes attached to inlet and outlet grouting ports on the sleeves. A variation of the system is a sleeve with internal threads at one end, which allows a threaded reinforcing bar to be securely connected to the sleeve. The opposite end of the steel sleeve is open for bar insertion and grouting.

Two grouting methods: pre-grouting and post-grouting are commonly used:

Pre-grouting. The grout sleeve is located at the top of the lower precast member to be spliced. The grout is gravity fed into the splice sleeves and the upper member with projecting bar is erected into position.

Post-grouting. The grout sleeves are located in the bottom of the upper member with reinforcing bars projecting out of the top of the lower member. Following the placement of the upper member, grouting is achieved by low pressure pumping as mentioned above.

The following items require special attention when using reinforcing bar grout sleeves and grouted anchorages:

1. Use only the grout recommended by the grout sleeve manufacturer.
2. Refer to the manufacturer's literature for cold and hot weather precautions during grouting.
3. Placement of bars projecting from cast-in-place foundations is critical. These bars must be held securely in position during casting to within very close tolerances (see manufacturer's recommendations). When multiple bars are used, a template is recommended for field placement.
4. Add approximately 6 in. to the required projection of dowels cast into cast-in-place foundations and trim the bars to the required elevation prior to starting erection.
5. When grout sleeves are not vertical in the grouting condition, care must be taken to avoid air pockets within the sleeve.
6. Care should be taken to prevent water from entering the sleeve before grouting, especially during freezing weather.
7. The bar embedment requirements may be different for different systems.

Figure 3.1.2.1 shows several uses of these systems.

6.4.5 Deformed Bar Anchors (DBA)

Deformed bar anchors (DBA) are usually automatically stud welded to steel plates, similar to headed studs. They are anchored to the concrete by bond. DBAs conform to AWS D1.1-02, Table 7.1, Type C studs with an ultimate tensile strength (f_{ut}) of 80,000 psi and the yield strength (f_y) of 70,000 psi. Type C studs are cold-worked, deformed steel bars manufactured in conformance with ASTM A496. The development lengths are the same as for Grade 60 reinforcing bars per ACI 318-02. For these bars, which have a specified yield strength f_y exceeding 60,000 psi, ACI 318-02, Section R3.5.3.6 permits a higher f_y. As such, the f_y used should be the stress corresponding to a strain of 0.35 percent if the yield strength specified in the design exceeds 60,000 psi. In addition, the development length calculation should recognize this increase in strength.

6.4.6 Bolts and Threaded Connectors

In most connections, bolts are shipped loose and threaded into inserts. The design of embedded inserts for concrete strength is similar to that for studs. Occasionally a precast concrete member will be cast with a threaded connector projecting from the face. This is usually undesirable because of possible damage during handling.

High strength bolts are used infrequently in precast concrete connections because it is questionable as to whether the bolt pretension can be maintained when tightened against concrete.

When used, AISC recommendations should be followed. [5, 6] Allowable loads on bolts are shown in Figure 6.16.6.

6.4.7 High Strength Threaded Rods

Rods with threads and specially designed nuts and couplers are frequently used for erection (Chapter 5), and sometimes used for permanent connections. Working loads are shown in Figure 6.16.7.

6.4.8 Specialty Inserts

Many specialty inserts are used in the precast concrete industry for both permanent connections and for lifting, handling and bracing conditions. Several of these inserts have become accepted as industry standards due to their history of satisfactory performance, ease of use, and unique solutions for their particular application. The various threaded coil inserts for lifting and permanent connections and the slotted insert for lateral tie back connections are just a few of many examples. When used as part of a connection assembly designed for factored loads, the ϕ-factors for headed studs (Section 6.2.1) are commonly used. Designers should refer to the valuable technical resources provided by the manufacturers of such specialty products. Because of the proprietary nature of such inserts, manufacturers' catalogs should be consulted for capacities and design information. Most companies will provide test data confirming their published capacity information. These inserts commonly carry a safety factor of 3 or 4 to 1 against working (unfactored) loads when used for lifting and handling.

6.4.9 Post-Installed Anchors

Anchors used in concrete construction fall in two broad categories; cast-in-place anchors and post-installed anchors.

With increasing demand for more flexibility in the planning and construction, and for repair and retrofit applications, the post-installed anchors have seen increased use.

6.4.9.1 Expansion Anchors

All expansion inserts are proprietary, and design values should be taken from manufacturers' catalogs. The concrete capacity design (CCD) method provides guidance for calculating these

Figure 6.4.9.1 Anchorage of post-installed expansion inserts

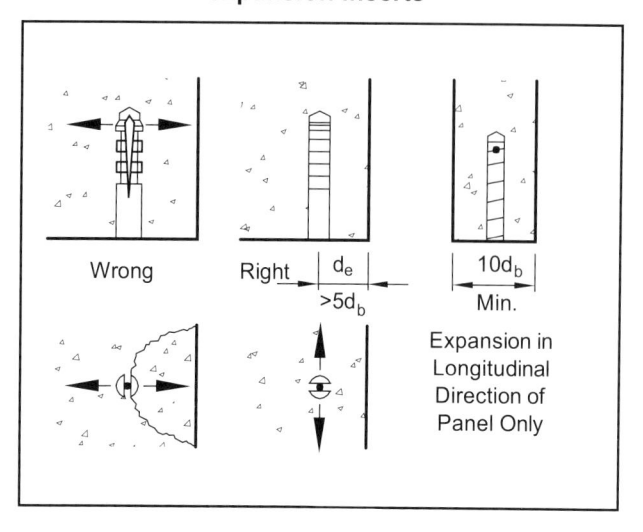

design values and is located in ACI 318-02, Appendix D. Edge distances for expansion inserts are more critical than for cast-in inserts. Expanding the insert in the direction of the edge should be avoided (see Figure 6.4.9.1).

The performance of expansion inserts when subjected to stress reversal, vibrations or earthquake loading is not sufficiently known. Their use for these load conditions should be carefully considered by the designer. The designer should note that the anchorage strength depends entirely on the lateral force (wedge action) on the concrete. Thus, the hole size must be carefully controlled and the hole must be very clean to achieve the published values of shear and tension.

6.4.9.2 Adhesive Anchors

An adhesive anchor is a threaded rod or reinforcing bar inserted into a hole drilled in hardened concrete. The diameter of the hole is typically 15 to 25 percent larger than the anchor diameter. It is filled with an adhesive that bonds the steel anchor to the concrete.

While there are generally accepted and, more recently, codified procedures, for the design of cast-in-place anchors such as headed studs, comparable information is not available on adhesive anchors. Thus, the designer must rely on manufacturers' recommendations to estimate the strength of these anchors.

Refs. 22 and 23 relate to the calculation of tensile and shear strength of adhesive anchors. While this information is limited in scope, the designer should find it useful to correlate with manufacturers' listed capacities to get a better assessment of actual factor of safety associated with adhesive anchors. The manufacturer installation procedures must be followed. Fire rating may be reduced when adhesive anchors are used, and they may have limited use in other elevated temperature or highly corrosive atmospheres. Welding near the anchor after it is in place should be done with caution. Because adhesive anchors rely on careful mixing and proportioning of adhesives, varying installation temperatures and shelf life of adhesives, a certain number should be proof tested prior to acceptance.

6.5 Headed Concrete Anchor Design

Anchorage to concrete and the design of welded headed studs has undergone a significant transformation since the Fifth Edition of the Handbook. The "Concrete Capacity Design" (CCD) approach has been incorporated into ACI 318-02 Building Code, Appendix D. [1] This method is primarily based on post-installed anchor tests with a limited amount of data applicable to headed studs used in the precast concrete industry. While there have been no failures or serious problems reported that are attributable to the use of design methods published in previous editions of the PCI Design Handbook, to be responsive to new data, modifications in welded headed stud design have been made. PCI sponsored an extensive research project, conducted by Wiss, Janney, Elstner Associates, Inc., (WJE), to study design criteria of headed stud groups loaded in shear and the combined effects of shear and tension. The shear provisions contained herein are the culmination of this research completed in 2002. Section D.4.2 of ACI 318-02 specifically permits alternate procedures, providing certain criteria are met. The WJE tests meet that criteria, and thus meet the Code.

An important factor in the performance of headed studs when controlled by concrete capacity is the location of the stud relative to the member edges. In shear, design capacity can be increased with confinement reinforcement. In tension, ductility can be provided by reinforcement that crosses the potential failure surfaces.

Welded headed studs are designed to resist direct tension, shear, or a combination of the two. The design equations given in this section are applicable to studs which are welded to steel plates or other structural members, and embedded in unconfined concrete.

Where feasible, headed stud connections should be designed and detailed such that the connection failure is precipitated by failure (typically defined as yielding) of the stud material rather than failure of the surrounding concrete. The in-place strength should be taken as the smaller of the values based on concrete and steel.

6.5.1 Steel Materials

6.5.1.1 Steel Plates and Shapes

Steel plates and shapes used with studs may be any of the structural steels used for welded steel structures. By far the most common conforms to ASTM A-36. Stainless steel usage is discussed in Chapter 1 and Section 6.7.1.2.

The minimum plate thickness to which studs are attached should be:

$$t_p(min) = \tfrac{1}{2}d_o \qquad \text{(Eq. 6.5.1.1)}$$

Thicker plates may be required for bending resistance or to ensure a more uniform load distribution to the attached studs.

6.5.1.2 Headed Concrete Anchors (HCA) or Studs

Table 6.5.1.1, adapted from AWS D1.1-02 Table 7.1, shows the current minimum tensile and yield strengths for headed studs. Type B studs are the most commonly used in precast concrete construction, although the smaller studs, Type A, may occasionally be appropriate for small loads.

Stainless steel studs can be welded to either stainless steel or mild carbon steel. Fully annealed stainless steel studs are recommended when welding stainless steel studs to a mild carbon steel base metal. Annealed stud use has been shown to be imperative for stainless steel studs welded to carbon steel plates subject to repetitive or cyclic loads. In such cases, stress corrosion failure in the weld can occur, and annealed stud use minimizes

Table 6.5.1.1 Minimum mechanical property requirements for headed studs*

Property (diameters)	Type A (¼ and ⅜ in.)	Type B (½ to 1 in.)
Tensile strength (min.) f_{ut}	61,000 psi	65,000 psi
Yield strength f_y (0.2% offset)	49,000 psi	51,000 psi
Elongation (min. % in 2 in.)	17%	20%
Reduction of area (min.)	50%	50%

* From AWS D1.1-02, [8] Table 7.1.

Table 6.5.1.2 Dimensions of headed studs

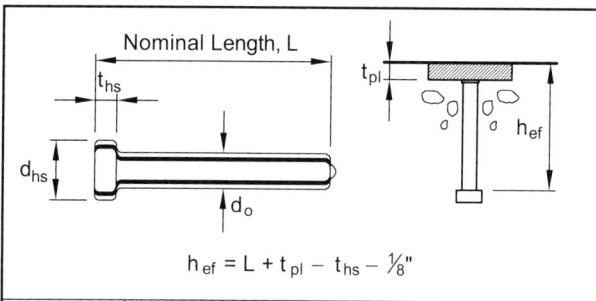

$h_{ef} = L + t_{pl} - t_{hs} - 1/8"$

Shank Diameter, d_o (in.)	Shank Area, A_{se} (in.^{2})	Head Diameter, d_{hs} (in.)	Head Thickness, t_{hs} (in.)	Bearing Area, A_{brg} (in.^{2})
1/4	0.05	1/2	3/16	0.15
3/8	0.11	3/4	9/32	0.33
1/2	0.20	1	5/16	0.59
5/8	0.31	1 1/4	5/16	0.92
3/4	0.44	1 1/4	3/8	0.79
7/8	0.60	1 3/8	3/8	0.88

the chance of weld cracking and failure. Consult the supplier of headed studs to obtain additional information on stainless steel stud use and availability.

6.5.2   HCA – Steel Strength

The design tensile or shear strength governed by steel failure is given by:

$$\phi V_s = \phi N_s = \phi(n)(A_{se}f_{ut}) \quad \text{(Eq. 6.5.2.1)}$$

where:
- ϕ = steel strength reduction factor
  - = 0.65 (shear)
  - = 0.75 (tension)
- V_s = nominal shear strength of an anchorage based on steel capacity
- N_s = nominal tensile strength of an anchorage based on steel capacity
- n = number of headed studs in the anchorage assembly
- A_{se} = nominal area of the headed stud shank
- f_{ut} = specified tensile strength of the stud steel
  - = 65 ksi for Type B headed studs normally used in precast anchorages (see Table 6.5.1.1)

6.5.3   HCA – Concrete Strength

ACI 318-02 includes a new section, Appendix D, titled "Anchoring to Concrete." This section covers all types of concrete anchors, including headed anchors, or studs, commonly used in precast concrete connections. These provisions, in general, have resulted in more conservative designs than those shown in previous editions of this Handbook. While there have been no reported incidences of structural failure resulting from the use of the previous design methods, this edition of the Handbook is, in general, based on the Code provisions. Several items regarding Appendix D are noted:

6.5.3.1   Supplemental Reinforcement

Section D.4.2.1 of ACI 318-02 discusses the use of supplemental reinforcement to enhance the strength of a concrete anchor. The Commentary, Section D.4.2.1, states: "The addition of supplementary reinforcement in the direction of load, confining reinforcement, or both, can greatly enhance the strength and ductility of the anchor connection. Such enhancement is practical with cast-in anchors such as those used in precast sections…reinforcement oriented in the direction of load and proportioned to resist the total load within the breakout prism, and fully anchored on both sides of the breakout planes, may be provided instead of calculating breakout capacity."

6.5.3.2   Cracked Concrete

The Appendix D equations presume that any anchorage device, including headed studs, will be embedded into cracked concrete. They are written so that the reduction caused by embedding in a crack is included, and adjustment factors are applied if the member is presumed not cracked. The uncracked factors are dependent on the anchor type, reinforcement, and crack width assumption. By providing design equations for cracked concrete, the Code assumes the anchorage is located where concrete cracking is likely. If an anchorage is embedded in a flexural compression zone, a column or other compression member, or other locations where cracking is not likely, as is the usual case when used in precast concrete, then uncracked concrete design is appropriate. Examples of where cracking should be considered include the tension zones of non-prestressed members and regions away from the ends of precast wall panels that may crack during handling.

This Handbook assumes that the majority of the precast member anchorages are in uncracked regions. This is reasonable because many precast members are prestressed and most of the anchorages designed for precast concrete are located where cracking is unlikely. Therefore, all of the tension and shear design equations in this Section are written as "uncracked" equations. There is thus an additional cracked concrete factor, C_{crb} for concrete breakout, or C_{crp} for pullout, which reduces the headed stud capacity due to cracking. Application of this coefficient to the Handbook equations will produce the same result as the ACI 318 Appendix D equations for tension.

6.5.3.3 The 5 Percent Fractile

ACI 318-02, Section D.4.2 states, in part: "...The nominal strength shall be based on the 5 percent fractile of the basic individual anchor strength..." This is a statistical concept that, simply stated, means that if a design equation is based on tests, 5 percent of the tests are allowed to fail. This allows us to say with 90 percent confidence that 95 percent of the test actual strengths exceed the equation thus derived. As described in the Appendix D Commentary, the determination of the coefficient κ, associated with the 5 percent fractile ($\kappa\sigma$) depends on the number of tests (sample population), n, used to compute $\bar{x}$ and σ. The term $\bar{x}$ is the sample mean and σ is the common standard deviation from data analysis. Example values of κ are:

$n = \infty$ $\kappa = 1.645$
$n = 40$ $\kappa = 2.010$
$n = 10$ $\kappa = 2.568$

ACI 318, Appendix D, Section D4.2, and the related commentary, RD.4.2 and RD.4.3 clearly permit alternative designs, providing the "5% fractile" criterion is met.

As a result of extensive research and testing sponsored by PCI [7], the equations in this section are based on:

- **Tension.** The equations are as shown in Appendix D, ACI 318-02, except for the cracking coefficients as previously stated.
- **Shear.** The shear provisions are based on the Ref. 7 tests. Equations derived from those tests meet the "5% fractile" criterion discussed above.
- **Combined shear and tension** (see Section 6.5.8). The Ref. 7 tests did not find significant variance from the Appendix D recommendations.

6.5.3.4 Strength Reduction Factor, ϕ

Appendix D requires that, for either tension or shear, the ϕ-factor be 0.75 when confinement reinforcement is present and 0.70 when it is not. It is highly recommended that such confinement be provided around all concrete controlled headed stud connections. Example 6.5.4.1 illustrates the use of shear-friction to design such confinement reinforcement.

6.5.4 Concrete Tension Strength*

The design tensile strength governed by concrete failure is the minimum of the following three modes:

- **Breakout ϕN_{cb}.** This is usually the most critical failure mode, and is a function of the failure planes shown in Figure 6.5.4.1.
- **Pullout ϕN_{ph}.** This is a function of bearing on the head of the stud.
- **Side-face blowout ϕN_{sb}.** To avoid this failure mode, studs cannot be closer to an edge than $0.4h_{ef}$. For studs closer than that, see Section D.5.4 of Appendix D, ACI 318-02. Note: This situation is not common practice.

Each of these possible modes must be checked when a stud connection is designed.

6.5.4.1 Breakout Strength

The equation for nominal concrete breakout strength for a stud or group of studs loaded concentrically is:

$$N_{cb} = N_{cbg} = C_{bs} A_N C_{crb} \psi_{ed,N} \quad \text{(Eq. 6.5.4.1)}$$

where:

C_{bs} = breakout strength coefficient

$$= 3.33\lambda \sqrt{\frac{f'_c}{h_{ef}}} \quad \text{(Eq. 6.5.4.2)}$$

(Note: This is equivalent to $\dfrac{N_b}{A_{No}}\psi_3$ of ACI 318-02.)

A_N = projected surface area for a stud or group of studs (sq in.) (see Figure 6.5.4.1).

h_{ef} = effective embedment depth (in.). For headed studs welded to a plate flush with the surface, it is the nominal length less the head thickness, plus the plate thickness, deducting the stud burnoff lost during the welding process (may be assumed as ≈ ⅛ in. – see Table 6.5.1.2).

---

* The tension provisions are shown as applied to headed studs. They are based on Appendix D of ACI 318-02, and thus are applicable to most anchorages. See also Section 6.12 of this Handbook.

In some cases, anchors may be affected by three or four edges. For example, anchors may be located in an edge of a wall, near the end. If the largest edge distance, $d_{e,max} \leq 1.5h_{ef}$, then an embedment depth, $h'_{ef} \leq d_{e,max}/1.5$, should be used in Eqs. 6.5.4.1 through 6.5.4.4 (see Figure 6.5.4.2, Cases 5 and 6). For further explanation, see Section RD.5.2.3 of ACI 318.02.

Cracking Factor

Cracking can reduce the capacity of the anchorage, and thus a cracking coefficient (breakout) described in Section 6.5.3 is applied as appropriate:

C_{crb} = 1.0 for concrete assumed uncracked (most common)
= 0.8 for locations likely to become cracked

Edge Distance Factor

$\psi_{ed,N}$ = modification for edge distance

$$= 0.7 + 0.3\left(\frac{d_{e,min}}{1.5h_{ef}}\right) \leq 1.0 \quad \text{(Eq. 6.5.4.3)}$$

where:

$d_{e,min}$ = minimum dimension of d_{e1}, d_{e2}, d_{e3}, or d_{e4} (in.), Figure 6.5.4.2.

Figure 6.5.4.1 Calculation of surface projection for anchors, A_N *

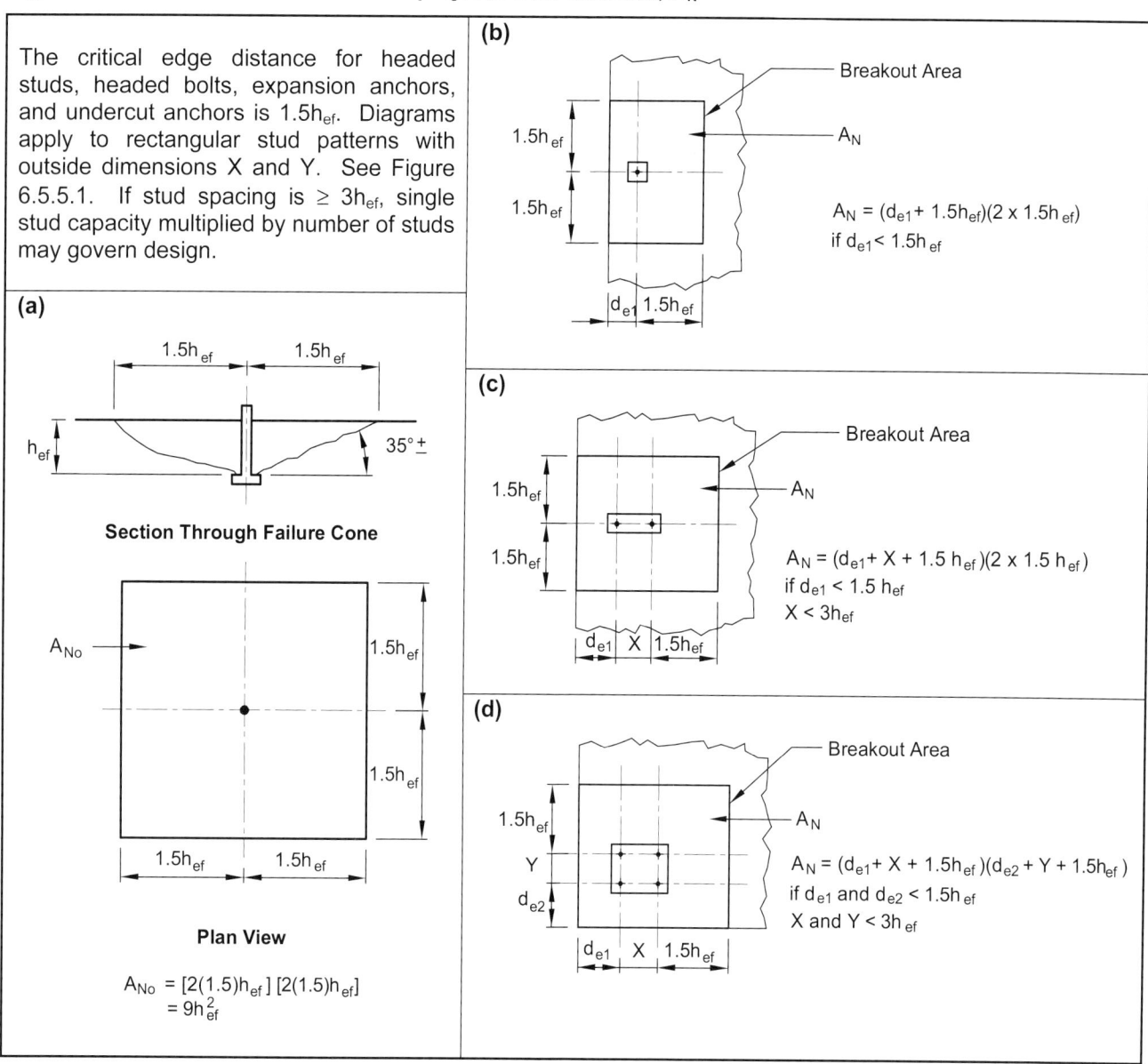

The critical edge distance for headed studs, headed bolts, expansion anchors, and undercut anchors is $1.5h_{ef}$. Diagrams apply to rectangular stud patterns with outside dimensions X and Y. See Figure 6.5.5.1. If stud spacing is $\geq 3h_{ef}$, single stud capacity multiplied by number of studs may govern design.

(a) Section Through Failure Cone

Plan View

$A_{No} = [2(1.5)h_{ef}][2(1.5)h_{ef}]$
$= 9h_{ef}^2$

(b) $A_N = (d_{e1} + 1.5h_{ef})(2 \times 1.5h_{ef})$
if $d_{e1} < 1.5h_{ef}$

(c) $A_N = (d_{e1} + X + 1.5h_{ef})(2 \times 1.5h_{ef})$
if $d_{e1} < 1.5h_{ef}$
$X < 3h_{ef}$

(d) $A_N = (d_{e1} + X + 1.5h_{ef})(d_{e2} + Y + 1.5h_{ef})$
if d_{e1} and $d_{e2} < 1.5h_{ef}$
X and Y $< 3h_{ef}$

* From Figure RD.5.2.1, ACI 318-02.

Figure 6.5.4.2 Design tensile strength of stud groups (breakout strength)

Case 1: Not near a free edge

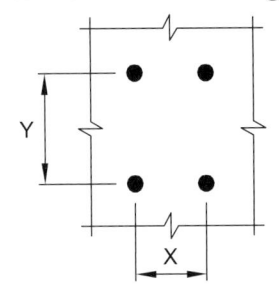

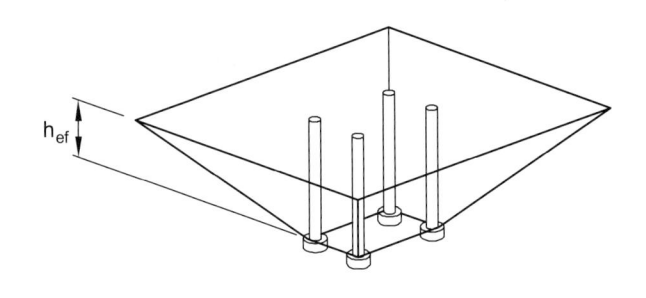

$$\phi N_{cb} = \phi C_{bs}(X + 3h_{ef})(Y + 3h_{ef})C_{crb}$$

Case 2: Free edge on one side

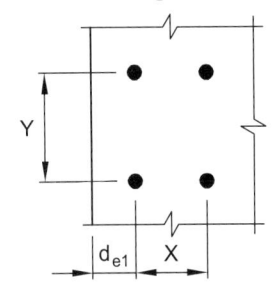

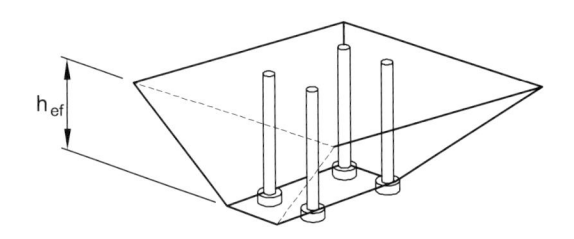

$$\phi N_{cb} = \phi C_{bs}(d_{e1} + X + 1.5h_{ef})(Y + 3h_{ef})(\psi_{ed,N})C_{crb}$$

Case 3: Free edges on two opposite sides

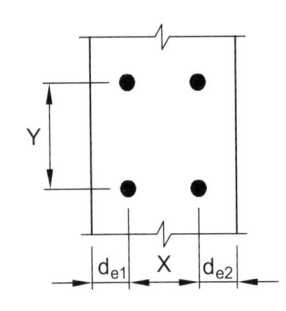

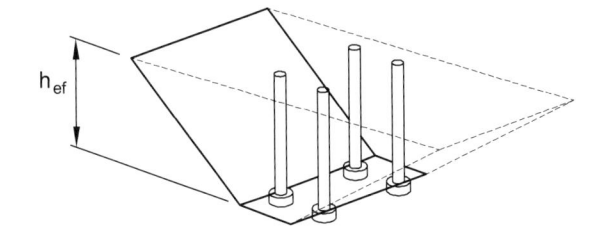

$$\phi N_{cb} = \phi C_{bs}(d_{e1} + X + d_{e2})(Y + 3h_{ef})(\psi_{ed,N})C_{crb}$$

Notes:

Diagrams apply to rectangular stud patterns with outside dimensions X and Y. See Figure 6.5.5.1. If stud spacing is $\geq 3h_{ef}$, single stud capacity multiplied by number of studs may govern design.

Near a free edge implies d_{e1}, d_{e2}, d_{e3} or $d_{e4} < 1.5 h_{ef}$

ϕ = 0.75 with confinement reinforcement
  = 0.70 without confinement reinforcement

$$C_{bs} = 3.33\lambda\sqrt{\frac{f'_c}{h_{ef}}}$$

For $d_{e,min}$ = least of d_{e1}, d_{e2}, d_{e3}, d_{e4} for each case: $\psi_{ed,N} = 0.7 + 0.3\dfrac{d_{e,min}}{1.5h_{ef}}$

For studs to behave as a group rather than separate shear cones, stud spacing $\leq 3h_{ef}$
(Notes continued on next page)

Figure 6.5.4.2 (Cont.) Design tensile strength of stud groups

Case 4: Free edges on two adjacent sides

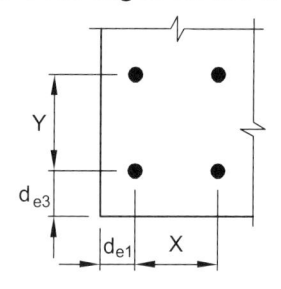

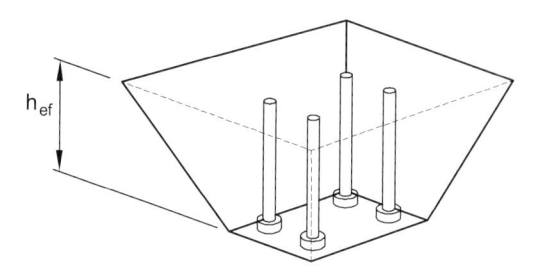

$$\phi N_{cb} = \phi C_{bs}(d_{e1} + X + 1.5h_{ef})(d_{e3} + Y + 1.5h_{ef})(\psi_{ed,N})C_{crb}$$

Case 5: Free edges on three sides

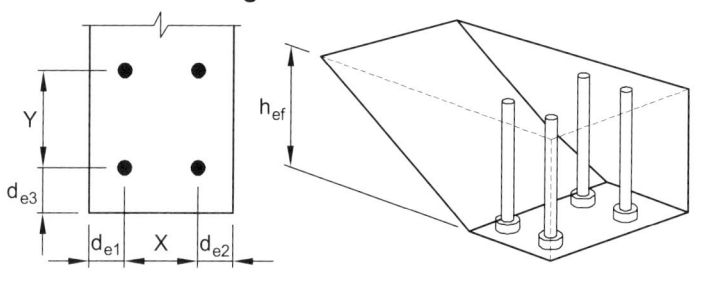

$$\phi N_{cb} = \phi C_{bs}(d_{e1} + X + d_{e2})(d_{e3} + Y + 1.5h_{ef})(\psi_{ed,N})C_{crb}$$

Case 6: Free edges on four sides

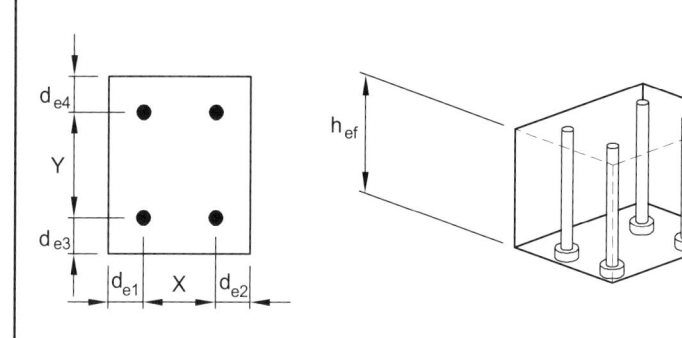

$$\phi N_{cb} = \phi C_{bs}(d_{e1} + X + d_{e2})(d_{e3} + Y + d_{e4})(\psi_{ed,N})C_{crb}$$

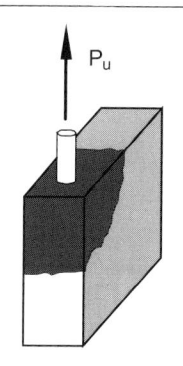

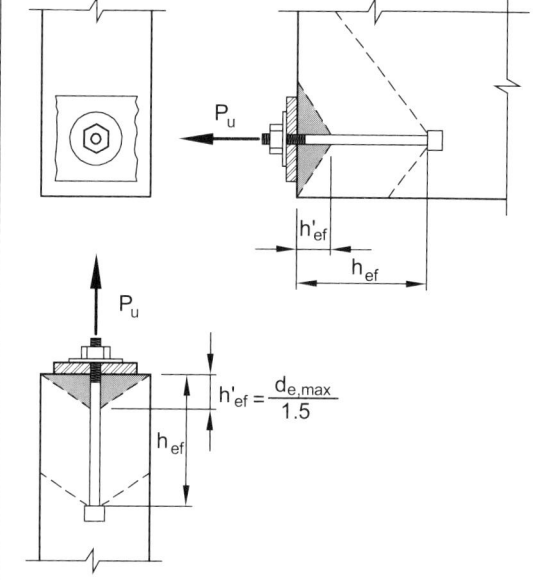

Applicable to Cases 5 and 6
From Figure RD.5.2.3, ACI 318-02

(See also notes on previous page.)

For studs to behave as a group, rather than as separate shear cones, stud spacing $\leq 3h_{ef}$

$\phi = 0.75$ with confinement reinforcement
$\quad = 0.70$ without confinement reinforcement

$$C_{bs} = 3.33\lambda\sqrt{\frac{f'_c}{h_{ef}}}$$

For $d_{e,min}$ = least of $d_{e1}, d_{e2}, d_{e3}, d_{e4}$ for each case: $\psi_{ed,N} = 0.7 + 0.3\dfrac{d_{e,min}}{1.5h_{ef}}$

Eccentricity Factor

It is usually possible to locate headed stud connections so that the load application can logically be assumed concentric. If not, the following modification should be applied to the breakout strength, Eq. 6.5.4.1:

$$\psi_{ec,N} = \frac{1}{\left(1 + \frac{2e'_N}{3h_{ef}}\right)} \leq 1 \quad \text{(Eq. 6.5.4.4)}$$

where:
e'_N = eccentricity of the tensile force relative to the center of the stud group (in.)
$\leq s/2$

6.5.4.2 Pullout Strength

The nominal pullout strength is a function of the bearing of the stud head against the concrete. The strength is:

$$N_{pn} = 11.2 A_{brg} f'_c C_{crp} \quad \text{(Eq. 6.5.4.5)}$$

where:
A_{brg} = bearing area of the stud head in tension (sq in.)
 = area of the head − area of the shank (values are shown in Table 6.5.1.2)
C_{crp} = cracking coefficient (pullout)
 = 1.0 for concrete assumed uncracked (most common)
 = 0.7 for locations likely to become cracked

6.5.4.3 Side-Face Blowout

For a single headed stud located close to an edge ($d_{e1} < 0.4 h_{ef}$):

$$N_{sb} = 160 d_{e1} \sqrt{A_{brg}} \sqrt{f'_c} \quad \text{(Eq. 6.5.4.6)}$$

where:
N_{sb} = nominal side-face blowout strength
d_{e1} = distance to closest edge
A_{brg} = area of head (Table 6.5.1.2)

If the single headed stud is located at a perpendicular distance, d_{e2}, less then $3d_{e1}$ from an edge, N_{sb}, is multiplied by:

$$\frac{\left(1 + \frac{d_{e2}}{d_{e1}}\right)}{4} \quad \text{(Eq. 6.5.4.7)}$$

where:
$1 \leq d_{e2}/d_{e1} \leq 3$

For multiple headed anchors located close to an edge ($d_{e1} < 0.4 h_{ef}$):

$$N_{sbg} = \left(1 + \frac{s_o}{6 d_{e1}}\right) N_{sb} \quad \text{(Eq. 6.5.4.8)}$$

where:
s_o = spacing of the outer anchors along the edge in the group
N_{sb} = value from Eq. 6.5.4.6

6.5.5 Concrete Shear Strength

The design shear strength governed by concrete failure is based on the testing described in Ref. 7. The in-place strength should be taken as the minimum value based on computing both the concrete and steel strength.

6.5.5.1 Front Edge (d_{e3})

This condition represents a majority of shear loaded connections. The shear force is applied perpendicular to the front edge d_{e3}.

Basic Strength

$$\phi V_{c3} = \phi V_{co3}(C_{x3})(C_{h3})(C_{ev3})(C_{vcr}) \quad \text{(Eq. 6.5.5.1)}$$

where:
ϕ = concrete strength reduction factor
 = 0.70 without confinement steel
 = 0.75 with confinement steel
V_{c3} = nominal concrete breakout strength for a single or multiple stud connection, accounting for member and connection geometry (lbs)
V_{co3} = concrete breakout strength for a single stud connection unaffected by connection or member geometry (lbs)
C_{X3} = coefficient for overall X spacing of a connection with two or more X rows for a d_{e3} type anchorage
C_{h3} = coefficient for member thickness (h) for a d_{e3} type anchorage
C_{ev3} = coefficient for eccentric shear force influences for a d_{e3} type anchorage
C_{vcr} = coefficient for cracking in a member, loaded in shear

Example 6.5.4.1
Tension Strength of Stud Groups

Given:
A flush-mounted base plate with four headed studs embedded in a corner of a 24 in. thick foundation slab.
4 – ¾ in. diameter headed studs welded to ½ in. thick plate.
Nominal stud length = 8 in.
f'_c = 4000 psi (normal weight concrete)
f_y = 60,000 psi

Problem:
Determine the design tension strength of the stud group.

Solution:

1. Depth of embedment from Table 6.5.1.2:

   h_{ef} = L + t_{pl} − t_{hs} − ⅛ in. = 8 + ½ − ⅜ − ⅛ = 8 in.

2. Check for edge effect:

   For the given problem (see Figure 6.5.4.2, Case 4)
   X = 16 in., Y = 8 in., d_{e1} = 4 in., d_{e3} = 6 in.
   d_{e1} and d_{e3} are both less than 1.5h_{ef} = 12 in., edge effects apply
   $d_{e,min}$ = 4 in.
   From Eq. 6.5.4.3:
   $$\psi_{ed,N} = 0.7 + 0.3\frac{d_{e,min}}{1.5h_{ef}} = 0.7 + 0.3\left[\frac{4}{1.5(8)}\right] = 0.8$$

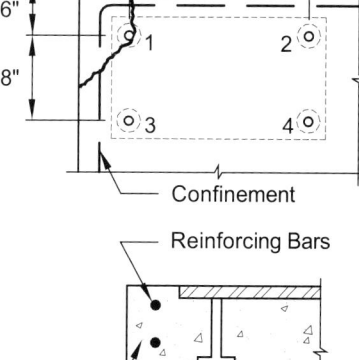

3. Check strength based on concrete breakout:

   From Eq. 6.5.4.2: $C_{bs} = 3.33\sqrt{\frac{f'_c}{h_{ef}}} = 3.33\sqrt{\frac{4000}{8}} = 74.5$

   From Eq. 6.5.4.1: $N_{cbg} = \phi C_{bs}(d_{e1} + X + 1.5h_{ef})(d_{e3} + Y + 1.5h_{ef})(\psi_{ed,N})C_{crb}$

   $= (0.75)(74.5)(4 + 16 + 12)(6 + 8 + 12)\left(\frac{0.8}{1000}\right)(1.0) = 37.2$ kips

Use 2 – #6 L-bar around stud group. These bars should extend ℓ_d past the breakout surface.

4. Check steel strength.
   From Eq. 6.5.2.1 and Tables 6.5.1.1 and 6.5.1.2: $\phi N_s = \phi n A_{se} f_{ut}$
   ϕN_s = 0.75(4)(0.44)(65) = 85.8 kips (ϕ = 0.75, see Section 6.2.1)

5. Check pullout strength. From Eq. 6.5.4.5 (ϕ = 0.70, see Section 6.2.1)
   From Table 6.5.1.2: A_{brg} = 0.79 in.² x 4 studs = 3.16 in.²
   From Eq. 6.5.4.5: $\phi N_{pn} = \phi(11.2)(A_{brg})f'_c C_{crp} = 0.70(11.2)(3.16)(4)(1.0) = 99.1$ kips

6. Check side-face blowout strength:
   $d_{e,min}$ = 4 in. > 0.4h_{ef} = 0.4(8) = 3.2 in. Therefore, it is not critical.

Example 6.5.4.1 (Cont.)
Tension Strength of Stud Groups

Tension strength of the group is 37.2 kips.

Design confinement reinforcement by shear-friction, Section 4.3.6:
Area of the crack plane is conservatively 4 in. by 8 in. = 32 in.^{2}

$$\mu_e = \frac{1000 A_{cr} \mu}{V_u} = \frac{1000(32)1.4}{37,200} = 1.20 < 3.4$$

$$A_{vf} = \frac{V_u}{\phi f_y \mu_e} = \frac{37.2}{0.75(60)(1.2)} = 0.68 \text{ in.}^2$$

Use 2 – #6 L-bar around stud group. These bars should extend ℓ_d past the breakout surface.

Figure 6.5.5.1 Headed stud plate edge variables

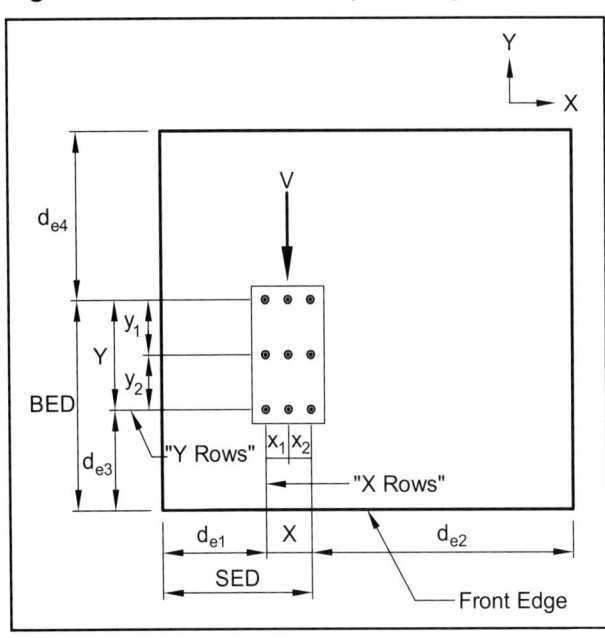

Single Anchor Strength

$$V_{co3} = 16.5 \lambda \sqrt{f'_c} (BED)^{1.33} \quad \text{(Eq. 6.5.5.2)}$$

where:
- λ = lightweight concrete factor (see Section 4.3.3)
- BED = distance from back row of studs to front edge
  $= d_{e3} + \sum y_i = d_{e3} + Y$ (in.) (Eq. 6.5.5.3)
- d_{e3} = distance from front stud to front edge (in.)
- $y_1, y_2, \ldots$ = center-to-center spacing of stud rows in Y direction (in.)
- Y = total out to out dimension of stud rows (in.)

X-Spacing Factor

$$C_{X3} = 0.85 + \frac{X}{3BED} \leq n_{studs\text{-}back} \quad \text{(Eq. 6.5.5.4)}$$
$$= 1.0, \text{ when } X = 0$$

where:
- $x_1, x_2, \ldots$ = center-to-center spacing of stud rows in X direction (in.)
- X = overall, out-to-out dimension of outermost studs in back row of anchorage = Σx (in.)
- $n_{studs\text{-}back}$ = number of studs in back row

Thickness Factor

$$C_{h3} = 1.0 \quad \text{for } h > 1.75 \text{ BED}$$
$$= 0.75 \sqrt{\frac{h}{BED}} \quad \text{for } h \leq 1.75 \text{ BED}$$
(Eq. 6.5.5.5)

where:
- h = member thickness (in.)

Eccentricity Factor

for $e'_v \leq \dfrac{X}{2}$

$$C_{ev3} = \frac{1}{1 + 0.67 \left(\dfrac{e'_v}{BED} \right)} \leq 1.0 \quad \text{(Eq. 6.5.5.6)}$$

where:
- e'_v = eccentricity of shear force on a group of anchors; distance between point of shear force application and geometric centroid of group of anchors resisting shear in direction of applied shear (in.)

Cracking Factor (see Figure 6.5.5.2)

$C_{vcr} = 1.0$ for uncracked concrete

For cracked concrete:

C_{vcr} = 0.70 no edge reinforcement or reinforcement < No. 4 bar
  = 0.85 edge reinforcement ≥ No. 4 bar
  = 1.0 edge reinforcement. ≥ No. 4 bar and confined within stirrups with a spacing ≤ 4 in.

6.5.5.2 Corners

The corner is considered to be a special case of the front edge loaded anchorage. If the shear force is applied perpendicular to the front edge, and the anchorage is located close to the corner, a different concrete breakout mode occurs. A corner condition should be considered when:

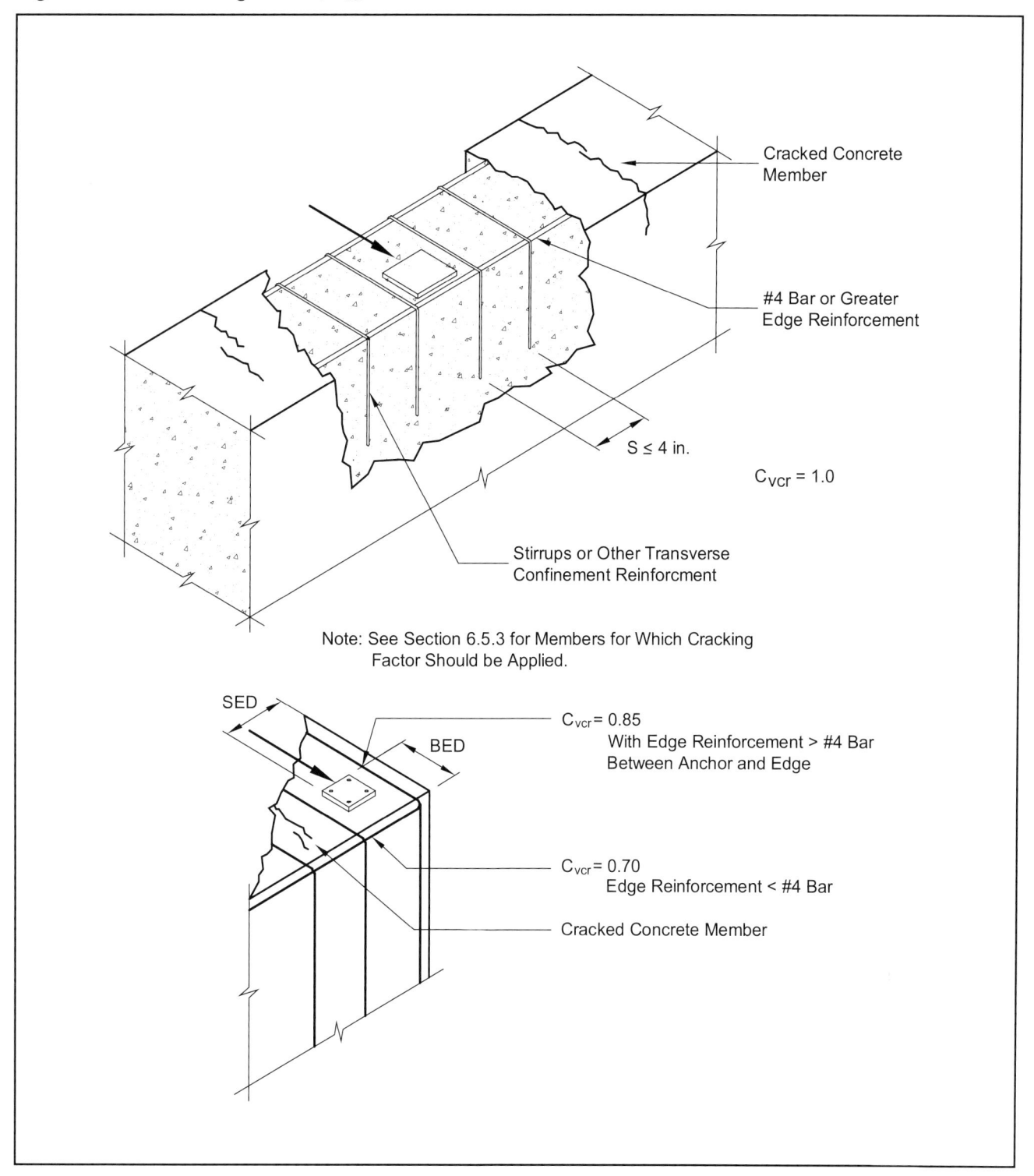

Figure 6.5.5.2 Cracking factors, C_{vcr}

- C_{vcr} = 1.0, S ≤ 4 in., #4 Bar or Greater Edge Reinforcement, Stirrups or Other Transverse Confinement Reinforcement, Cracked Concrete Member

Note: See Section 6.5.3 for Members for Which Cracking Factor Should be Applied.

- C_{vcr} = 0.85 With Edge Reinforcement > #4 Bar Between Anchor and Edge
- C_{vcr} = 0.70 Edge Reinforcement < #4 Bar
- Cracked Concrete Member

(SED, BED)

Example 6.5.5.1
Headed Concrete Anchor Front Edge Failure Mode

Given:
Plate with headed studs as shown, placed in a position where cracking is unlikely. The 8 in. thick panel has a 28-day concrete strength of 5000 psi. The plate is loaded with an eccentricity of 1½ in. from the centerline of the stud group. The panel has #5 confinement bars around the perimeter of the panel.

Problem:
Determine the design shear strength of the stud group.

Check for Corner Condition

$$\frac{SED}{BED} \geq 3 \qquad \frac{48+4}{12+4} = 3.25 \geq 3$$

Not a Corner Condition

Solution:

Steel Capacity:
From Table 6.5.1.1: $f_{ut} = 65$ ksi
From Eq. 6.5.2.1: $\phi V_s = \phi(n_s)(A_s f_{ut}) = 0.65(4)(0.20)(65) = 33.8$ kips

Concrete Capacity:
From Eq. 6.5.5.1: $\phi V_{c3} = \phi V_{co3}(C_{x3})(C_{h3})(C_{ev3})(C_{vcr})$
From Eq. 6.5.5.3: BED $= d_{e3} + Y = 12 + 4 = 16$ in.
From Eq. 6.5.5.2: $V_{co3} = 16.5\lambda\sqrt{f'_c}(BED)^{1.33} = \frac{16.5(1.0)\sqrt{5000}(16)^{1.33}}{1000} = 47.0$ kips

X-Spacing Factor:
From Eq. 6.5.5.4: $C_{X3} = 0.85 + \frac{X}{3BED} = 0.85 + \frac{4}{3(16)} = 0.93 \leq n_{studs\text{-}back}$

Thickness Factor:
From Eq. 6.5.5.5: $C_{h3} = 0.75\sqrt{\frac{h}{BED}} = 0.75\sqrt{\frac{8}{16}} = 0.53$

Eccentricity Factor:
From Eq. 6.5.5.6: $C_{ev3} = \dfrac{1}{1+0.67\left(\dfrac{e'_v}{BED}\right)} = \dfrac{1}{1+0.67\left(\dfrac{1.5}{16}\right)} = 0.94 < 1.0$

Cracked Concrete Factor:
$C_{vcr} = 1.0$

With confinement steel $\phi = 0.75$

$\phi V_{c3} = \phi V_{co3}(C_{x3})(C_{h3})(C_{ev3})(C_{vcr}) = 0.75(47.0)(0.93)(0.53)(0.94)(1.0) = 16.3$ kips < 33.8

Use $\phi V_{c3} = 16.3$ kips
If higher capacity is desired, use tail bars welded to plate.
For example, if 2 – #5 tail bars are used:
$\phi V_{c3} = \phi A_s f_y = 0.9(2)(0.31)(60) = 33.5$ kips

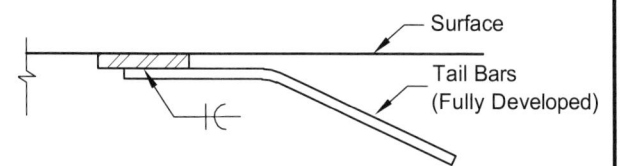

$$0.2 \leq \frac{SED}{BED} \leq 3.0 \quad \text{(Eq. 6.5.5.7)}$$

where the Side Edge Distance (SED) as shown in Figure 6.5.5.1, is defined as:

$$SED = d_{e1} + \Sigma x = d_{e1} + X \text{ (in.)} \quad \text{(Eq. 6.5.5.8)}$$

Basic Strength

The strength governed by concrete breakout at the corner is thus given by:

$$\phi V_{c3} = \phi V_{co3}(C_{c3})(C_{h3})(C_{ev3})(C_{vcr}) \quad \text{(Eq. 6.5.5.9)}$$

where:

ϕ = strength reduction factor
  = 0.70 without confinement steel
  = 0.75 with confinement steel
C_{c3} = coefficient for corner influence for a d_{e3} type anchorage
C_{h3}, C_{vcr}, C_{ev3} are as previously defined

Corner Factor

$$C_{c3} = 0.7 \sqrt[3]{\frac{SED}{BED}} \leq 1.0 \quad \text{(Eq. 6.5.5.10)}$$

Note that there is no C_{x3} factor when computing a corner capacity.

For the special case of a large X-spacing stud anchorage located near a corner, such that SED/BED > 3, a corner failure may still result, if $d_{e1} \leq 2.5$ BED (see Figure 6.5.5.3).

6.5.5.3 Side Edge (d_{e1} or d_{e2})

A connection loaded in shear parallel to a side edge results in a concrete breakout failure different from the front edge breakout mode. In this case, the shear force is applied parallel to the side edge (d_{e1} in Figure 6.5.5.1). The anchorage will likely behave in a side edge mode if:

$$\frac{SED}{BED} \leq 0.2 \quad \text{(Eq. 6.5.5.11)}$$

The research (Ref. 7) determined that the corner influence can be quite large, especially in thin panels. If the above ratio is close to the 0.2 value, it is recommended that a corner breakout condition be investigated, as it may still control for large BED values. Note also that minimum distribution reinforcement is required.

Basic Strength

The strength governed by concrete breakout at the side edge is given by:

$$\phi V_{c1} = \phi V_{co1}(C_{X1})(C_{Y1})(C_{ev1})(C_{vcr}) \quad \text{(Eq. 6.5.5.12)}$$

where:

ϕ = strength reduction factor
  = 0.70 without confinement steel
  = 0.75 with confinement reinforcement

Figure 6.5.5.3 Corner transition to a front edge breakout

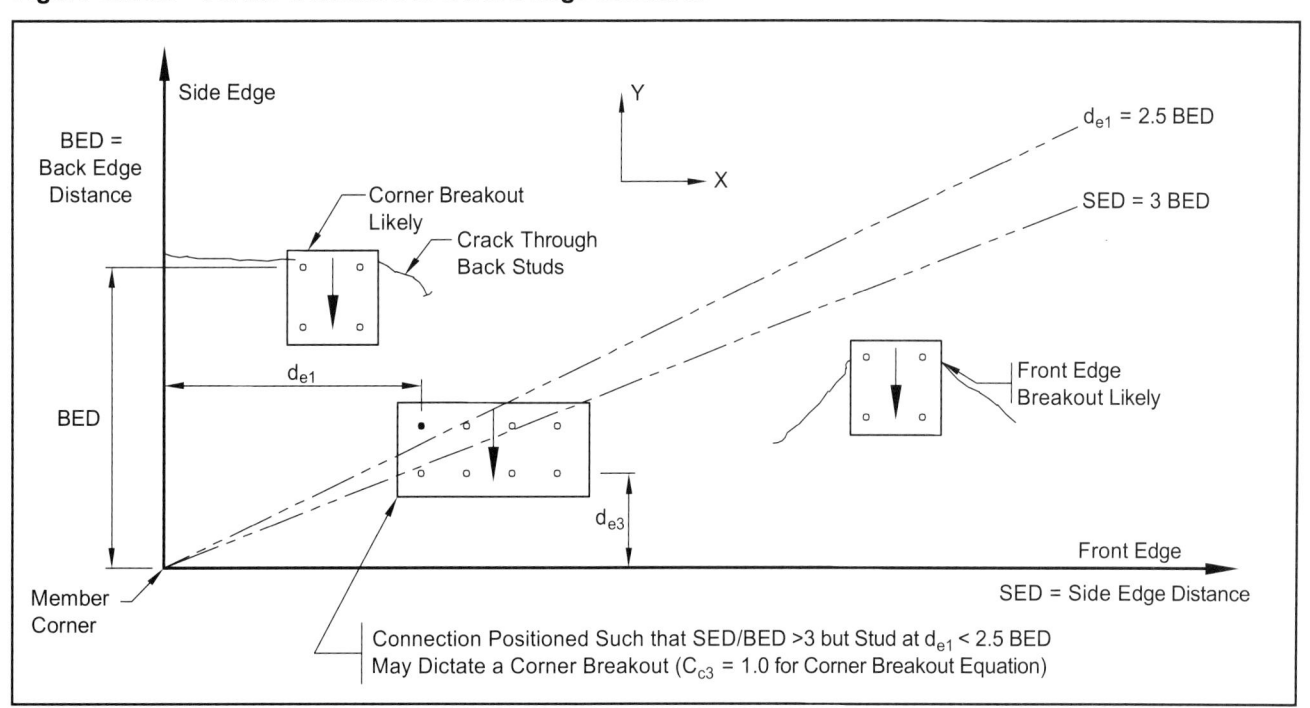

V_{co1} = nominal concrete breakout strength for a single stud connection unaffected by connection or member geometry (lbs)

C_{X1} = coefficient for overall X spacing of a connection with two or more X rows for a d_{e1} type anchorage

C_{Y1} = coefficient for overall Y spacing of a connection with two or more Y rows for a d_{e1} type anchorage

C_{ev1} = coefficient for in-plane, eccentric shear load for a d_{e1} type anchorage

C_{vcr} = coefficient for cracking in a member, loaded in shear (see previous section)

Single Anchor Strength

$$V_{co1} = 87\lambda\sqrt{f'_c}(d_{e1})^{1.33}(d_o)^{0.75} \quad \text{(Eq. 6.5.5.13)}$$

where:
λ = lightweight concrete factor
d_{e1} = distance from side stud to side edge (in.)
d_o = stud diameter (in.)

X-Spacing Factor

For a one edge connection condition or a single y-row of studs in a two, parallel edge condition (see Figure 6.5.5.4):

$$C_{X1} = \frac{n_x x}{2.5 d_{e1}} + 2 - n_{sides} \quad \text{(Eq. 6.5.5.14)}$$

where $1 \leq C_{X1} \leq n_x$

$C_{X1} = 1.0$, when $x = 0$

where:
n_x = number of X-rows
x = individual X-row spacing (in.)
n_{sides} = number of edges or sides that influence the X direction (1 or 2, i.e., 2 for a column in which connection is placed equidistant from each side)

For all multiple Y-row anchorages located adjacent to two parallel edges, such as a column corbel connection, the X-spacing for two or more studs in the row:

$$C_{x1} = n_x \quad \text{(Eq. 6.5.5.15)}$$

Y-Spacing Factor

$C_{Y1} = 1.0$ for $n_y = 1$ (one Y-row)

$$C_{Y1} = \frac{(n_y Y)^{0.25}}{0.6 d_{e1}} + 0.15 \leq 1.0$$

for $n_y > 1.0$ (Eq. 6.5.5.16)

where:
n_y = number of Y-row stud lines
Y = out-to-out Y-row spacing = Σy (in.)

Eccentricity Factor

$$C_{ev1} = 1.0 - \left(\frac{e_{v1}}{4 d_{e1}}\right) \leq 1.0 \quad \text{(Eq. 6.5.5.17)}$$

where:
e_{v1} = eccentricity from shear load to anchorage centroid (in.)

Figure 6.5.5.4 Conditions for calculations of C_{x1}, for side edges

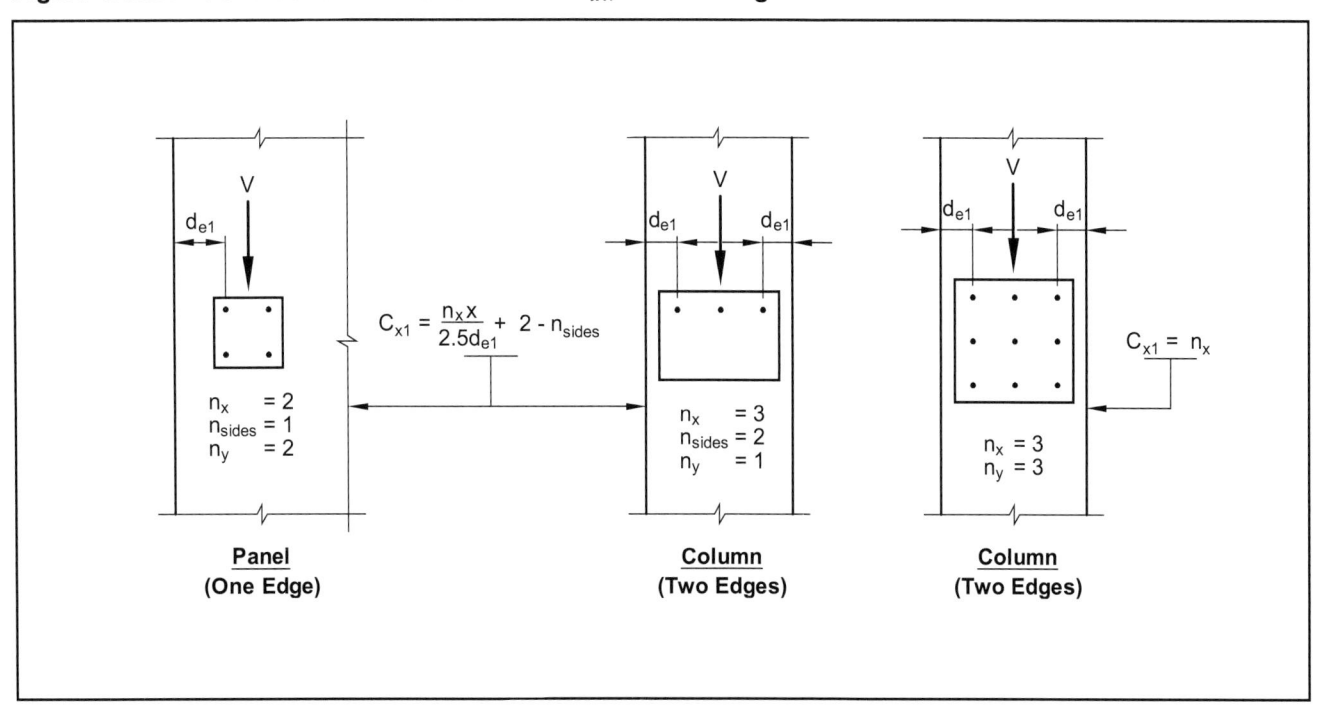

Panel (One Edge)
$n_x = 2$
$n_{sides} = 1$
$n_y = 2$

$C_{x1} = \frac{n_x x}{2.5 d_{e1}} + 2 - n_{sides}$

Column (Two Edges)
$n_x = 3$
$n_{sides} = 2$
$n_y = 1$

Column (Two Edges)
$C_{x1} = n_x$
$n_x = 3$
$n_y = 3$

Example 6.5.5.2
Headed Concrete Anchor Corner Failure Mode

Given:
 Plate with headed studs as shown, placed in a position where cracking is unlikely. The 8 in. thick panel has a 28-day concrete strength of 5000 psi. The panel has #5 confinement bars around the perimeter. The plate is loaded with an eccentricity of 1½ in. from the centerline of the stud group.

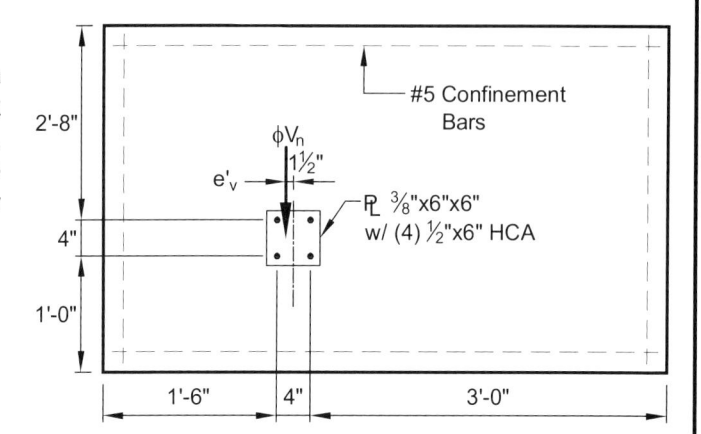

Problem:
 Determine the nominal shear strength of plate.
Check for corner condition by Eq. 6.5.5.7:
$$0.2 \le \frac{SED}{BED} \le 3; \quad \frac{18+4}{12+4} = 1.375$$
Thus, corner breakout is likely.

Solution:
Steel strength (same as Example 6.5.5.1): $\phi V_s = 33.8$ kips
Concrete strength from Eq. 6.5.5.9: $\phi V_{c3} = \phi V_{co3} (C_{vcr})(C_{ev3})(C_{h3})(C_{c3})$
 d_{e3} Single Anchor Capacity (same as Example 6.5.5.1): $V_{co3} = 47$ kips

Thickness Factor (same as Example 6.5.5.1): $C_{h3} = 0.53$

Corner-Spacing Factor from Eq. 6.5.5.10: $C_{c3} = 0.7 \sqrt[3]{\frac{SED}{BED}} = 0.7 \sqrt[3]{\frac{18+4}{12+4}} = 0.78$

Eccentricity Factor (same as Example 6.5.5.1): $C_{ev3} = 0.94 < 1.0$

Cracked Concrete Factor: $C_{vcr} = 1.0$
With confinement reinforcement, $\phi = 0.75$
 $\phi V_{c3} = \phi V_{co3}(C_{h3})(C_{c3})(C_{vcr})(C_{ev3}) = 0.75(47)(0.53)(0.78)(1.0)(0.94) = 13.7$ kips
 Use $\phi V_n = 13.7$ kips

Example 6.5.5.3
Headed Concrete Anchor Side Edge Failure Mode

Given:
 Headed Stud Plate as shown. The 8 in. thick reinforced concrete panel has a 28-day concrete strength of 5000 psi. The panel has #5 confinement bars around the perimeter. It is placed in a position where cracking is unlikely.

Problem:
Determine the design shear strength of the stud group.
Check for Corner Condition from Eq. 6.5.5.7.
$$\frac{SED}{BED} \le 0.2 \quad \frac{6+4}{72+4} = 0.13 \le 0.2$$

Example 6.5.5.3 (Cont.)
Headed Concrete Anchor Side Edge Failure Mode

Not a Corner Condition – Solve as Side-Edge condition.

Solution:

Steel strength (same as Example 6.5.5.1):
$\phi V_s = 33.8$ kips

Concrete strength: $\phi V_{c1} = \phi V_{co1}(C_{X1})(C_{Y1})(C_{ev1})(C_{vcr})$

d_{e1} *Single Anchor strength:*

$$V_{co1} = 87\lambda\sqrt{f'_c}(d_{e1})^{1.33}(d_o)^{0.75} = \frac{87(1.0)\sqrt{5000}(6)^{1.33}(0.5)^{0.75}}{1000}$$
$= 39.6$ kips

X-Spacing Factor:
$$C_{X1} = \frac{n_x X}{2.5d_{e1}} + 2 - n_{sides} = \frac{2(4)}{2.5(6)} + 2 - 1 = 1.53$$

Y-Spacing Factor:
$$C_{Y1} = \frac{(n_y Y)^{0.25}}{0.6d_{e1}} + 0.15 = \frac{[2(4)]^{0.25}}{0.6(6)} + 0.15 = 0.62$$

Eccentricity Factor: $C_{ev1} = 1.0$

Cracked Concrete Factor: $C_{vcr} = 1.0$
$\phi V_{c1} = \phi V_{co1}(C_{X1})(C_{Y1})(C_{ev1})(C_{vcr}) = 0.75(39.6)(1.53)(0.62)(1.0)(1.0) = 28.2$ kips
Use $\phi V_n = 28.2$ kips

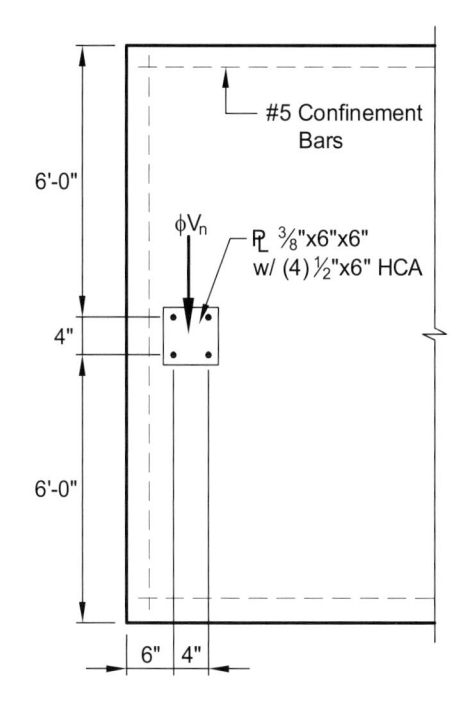

6.5.6 Back Edge (d_{e4})

The shear force is applied perpendicular to the back edge illustrated in Figure 6.5.5.1. Under a condition of pure shear, the back edge has been found through testing to have no influence on the connection capacity. Proper concrete clear cover from the studs to the edge must be maintained.

6.5.7 In-the-Field

When a headed stud anchorage is sufficiently away from all edges, termed "in-the-field" of the member, the anchorage strength will normally be governed by the steel strength. Pry-out failure is a concrete breakout failure that may occur when short, stocky studs are used.

For $h_{ef}/d_o \leq 4.5$ (in normal weight concrete):

$\phi V_{cp} = \phi 215 \psi_y n\sqrt{f'_c}(d_o)^{1.5}(h_{ef})^{0.5}$ (Eq. 6.5.7.1)

where:

$\psi_y = \dfrac{\sqrt{y}}{4d_o}$ for $\dfrac{y}{d} \leq 20 = 1.0$ for $y = 0$

y = center of center spacing of studs in direction of load
h_{ef} = effective stud embedment length
ϕ = strength reduction factor
  = 0.70 without confinement reinforcement
  = 0.75 with confinement reinforcement
V_{cp} = nominal pry-out shear strength (lbs)
f'_c = specified compressive strength of concrete (psi)
λ = concrete unit weight factor
f_{ut} = tensile strength of stud (psi) (see Table 6.5.1.1)
d_o = shank diameter of stud

For short studs in lightweight aggregate concrete, the h_{ef}/d_o ratio limit is greater than 4.5. Therefore, this capacity equation should be checked for short studs embedded in lightweight aggregate concrete and in-the-field.

6.5.8 Interaction of Tension and Shear

ACI 318-02, Section D.7, prescribes a tri-linear interaction as shown in Figure 6.5.8.1. This drawing shows that when both tension and shear are applied to a connection with anchors:

- If the applied shear, V_u, is less than or equal to 20% of the shear strength, ϕV_n, the shear can be neglected, and the connection designed for tension alone.
- If the applied tension, N_u, is less than or equal to 20% of the tensile strength, ϕN_n, the tension can be neglected, and the connection designed for shear alone.
- If $V_u > 0.2\phi V_n$ or $N_u > 0.2\phi N_n$, then:

$$\frac{N_u}{\phi N_n} + \frac{V_u}{\phi V_n} \leq 1.2 \qquad \text{(Eq. 6.5.8.1)}$$

ACI 318-02, Section RD.7 (Commentary), also suggests the use of a unity interaction based on exponents of 5/3, which has been used in previous editions of the Handbook. The interaction curves, shown in Figure 6.5.8.1, illustrate that checking both interaction equations may be beneficial, depending on the tension/shear ratios present on the given connection:

$$\left(\frac{N_u}{\phi N_n}\right)^{5/3} + \left(\frac{V_u}{\phi V_n}\right)^{5/3} \leq 1.0 \qquad \text{(Eq. 6.5.8.2)}$$

Note: it is only required to meet either Eq. 6.5.8.1 or 6.5.8.2, not both.

Figure 6.5.8.1 Tension-shear interaction

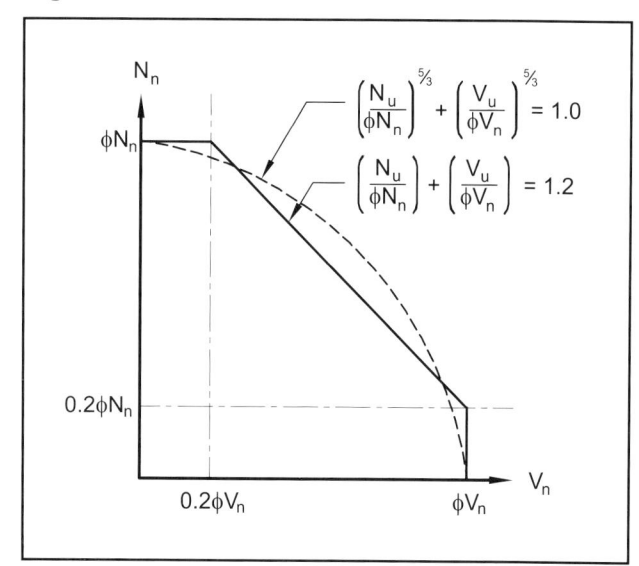

Example 6.5.8.1
Design of Welded Headed Studs for Combined Loads

Given:
A ½ in. thick plate with headed studs for attachment of a steel bracket to a column as shown at the right.

f'_c = 6000 psi normal weight concrete
λ = 1.0
8 – ½ in. diameter studs
A_{se} = 0.20 in.² (Table 6.5.1.2)

Nominal stud length = 6 in.
f_{ut} = 65,000 psi (Table 6.5.1.1)
V_u = 25 kips
N_u = 4 kips (all load factors and overloaded factors included)
Column size: 18 in. x 18 in.

Problem:
Determine if the studs are adequate for the connection. See Section 6.6 for bracket design.

Example 6.5.8.1 (Cont.)
Design of Welded Headed Studs for Combined Loads

Solution:
The eccentric shear force, V_u, is resolved by the force couple shown in the sketch at the right, assuming that the tensile force is resisted by the top two rows of studs, with breakout planes as shown. Note: assumptions for load distribution are a matter of engineering judgment.

$$N_{hu} = \frac{V_u e}{d_c} + N_u = \frac{25(6)}{10} + 4 = 19.0 \text{ kips}$$

For tension on the top group of studs:

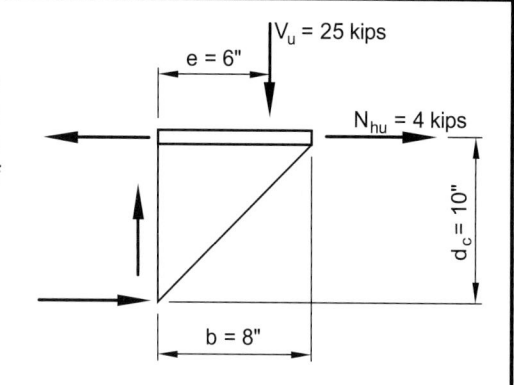

Concrete:
Provide ties around vertical bars in the column to ensure confinement: $\phi = 0.75$
From Table 6.5.1.2, $h_{ef} = L + t_{pl} - t_{hs} - \frac{1}{8}$ in.
$= 6 + 0.5 - 0.3125 - 0.125 = 6.06$ in.

From Eq. 6.5.4.2: $C_{bs} = 3.33\lambda \sqrt{\frac{f'_c}{h_{ef}}} = 3.33(1.0)\sqrt{\frac{6000}{6.06}} = 104.8$

From Figure 6.5.4.2, Case 3: $d_{e1} = d_{e2} = 6$ in.; $X = 6$ in.; $Y = 3$ in.

$\phi N_{cb} = \phi C_{bs}(d_{e1} + X + d_{e2})(Y + 3h_{ef})\psi_{ed,N}(C_{crb})$

$\psi_{ec,N} = 0.7 + 0.3\frac{d_{e,min}}{1.5h_{ef}} = 0.7 + (0.3)\frac{6}{1.5(6.06)} = 0.898$

$\phi N_{cb} = 0.75(104.8)(6 + 6 + 6)(3 + 18.2)(0.898)(1.0)/1000 = 26.9$ kips

Steel:
From Eq. 6.5.2.1: $\phi N_s = \phi n A_{se} f_{ut} = 0.75(4)(0.2)(65) = 39.0$ kips
$\phi N_n = 26.9$ kips

Assume the shear force is distributed equally between the top and bottom shear groups (engineering judgment): $V_u/2 = 25/2 = 12.5$ kips. Evaluate the top group of studs for combined shear and tension:

For concrete shear strength, it is apparent that "side edge" breakout will be critical:

From Figure 6.5.5.1: $d_{e1} = d_{e2} = 6$ in.; $X = 6$ in.; $Y = 3$ in.
From Eq. 6.5.5.12: $\phi V_{c1} = \phi V_{co1}(C_{X1})(C_{Y1})(C_{ev1})(C_{vcr})$
From Eq. 6.5.5.13: $V_{co1} = 87\lambda\sqrt{f'_c}(d_{e1})^{1.33}(d_o)^{0.75} = 87(1.0)(\sqrt{6000})(6)^{1.33}(0.5)^{0.75}/1000 = 43.7$ kips
From Eq. 6.5.5.15: $C_{X1} = n_{x1} = 2$
From Eq. 6.5.5.16: $C_{Y1} = \frac{(n_y Y)^{0.25}}{0.6 d_{e1}} + 0.15$; $n_Y = 2$, $Y = 3$ in., $C_{Y1} = \frac{[2(3)]^{0.25}}{0.6(6)} + 0.15 = 0.58$

$C_{ev1} = 1.0$; $C_{vcr} = 1.0$
$\phi V_c = \phi V_{c1} = 0.75(43.7)(2.0)(0.58)(1.0)(1.0) = 38.0$ kips

For steel shear strength:
From Eq. 6.5.2.1: $\phi V_s = \phi n A_{se} f_{ut} = 0.65(4)(0.2)(65) = 33.8$ kips

Use $\phi V_n = 33.8$ kips

Example 6.5.8.1 (Cont.)
Design of Welded Headed Studs for Combined Loads

Combined loading:

$N_{hu} = 19.0$ kips; $\phi N_n = 26.9$ kips; $\dfrac{N_{hu}}{\phi N_n} = \dfrac{19.0}{26.9} = 0.71 > 0.2$

$V_u = 12.5$ kips; $\phi V_n = 33.8$ kips; $\dfrac{V_u}{\phi V_n} = \dfrac{12.5}{33.8} = 0.37 > 0.2$

$\dfrac{N_{hu}}{\phi N_n} + \dfrac{V_u}{\phi V_n} = 0.71 + 0.37 = 1.08 < 1.2$

or $\left(\dfrac{N_{hu}}{\phi N_n}\right)^{5/3} + \left(\dfrac{V_u}{\phi V_n}\right)^{5/3} = 0.56 + 0.19 = 0.75 < 1.0$

Thus, the stud design is satisfactory by either Eq. 6.5.8.1 or 6.5.8.2.

Check embedded plate for bending between studs:

Plate $M_u = \dfrac{P\ell}{4} = \dfrac{19(3)}{4} = 14.25$ kip-in.

$\phi M_n = \phi F_y Z = 0.9(36)\left(\dfrac{bt^2}{4}\right)$, where b = 10 in.

For $\phi M_n \geq M_u$, calculate minimum plate thickness:

$M_u = \phi F_y Z = 0.9(36)\dfrac{10t^2}{4} = 14.25$

$t = \sqrt{\dfrac{14.25(4)}{0.9(36)(10)}} = 0.42$ in. Use ½ in. thick plate

Triangular stiffener design should be done in accordance with Section 6.6.6. Use ⅜ in. stiffener, see Example 6.6.6.1.

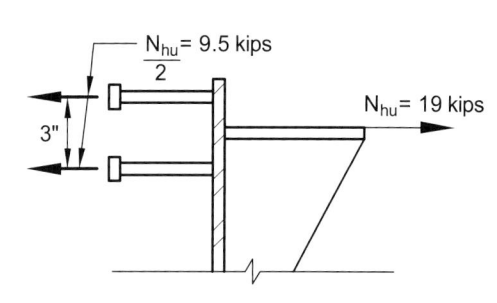

6.6 Structural Steel Design

Structural steel plates, angles, wide flange beams, channels, tubes, etc., are often used in connections. When designed using factored loads, the AISC LRFD Manual [6] is the appropriate design reference. For typical steel sections used in connections, plastic section properties and yield strengths with appropriate strength reduction factors are used.

6.6.1 Design Flexural Strength

The design flexural strength of a steel section is:

$\phi M_p = \phi F_y Z_s$ (Eq. 6.6.1.1)

where:
$\phi = 0.9$
F_y = material yield strength
Z_s = plastic section modulus

For thin plates bent about their major axis, buckling provisions may govern the design. Typical plastic section properties for various shapes are listed in Chapter 11.

6.6.1.1 Built-up Steel Sections

Sections built-up from steel plates are often used in precast structures for such things as beam seats. The design of a typical beam seat is illustrated in Example 6.6.1.1.

6.6.2 Design Shear Strength

The design shear strength of a section is determined by the following equation: [6]

$\phi V_n = \phi(0.6 F_y) A_w$ (Eq. 6.6.2.1)

Example 6.6.1.1
Plastic Section Modulus of a Built-up Steel Member

Given:

The built-up steel member shown below. The stiffeners have adequate welds so that the assembly acts monolithically. (Note: this small assembly could be used for the bearing seat of a hanger assembly, spanning just a few inches)

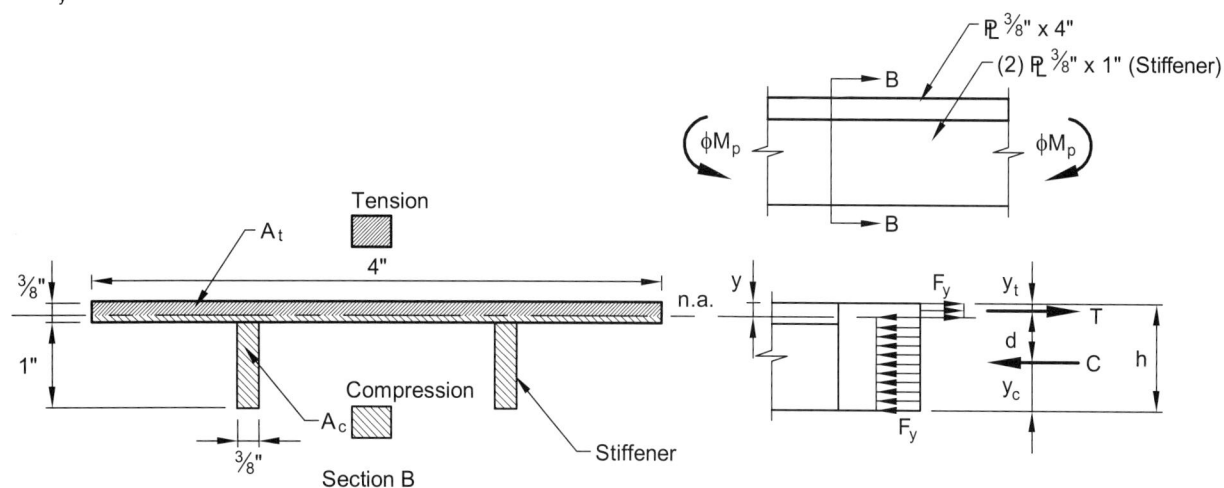

$F_y = 36$ ksi

Problem:

Determine the nominal plastic moment strength of the plate assembly.

Solution:

Let y = distance from the top of the member to the centroid.
Assume the tension area is within the flange thickness, i.e., $y \leq 3/8$ in.

$\Sigma F_x = 0$: T = C; $A_t F_y = A_c F_y$; $A_t = A_c$

$A_t = 4y$
$A_c = (3/8 - y)(4) + 2(3/8)(1) = 18/8 - 4y$
$4y = 18/8 - 4y$
$8y = 18/8$
$y = 18/64 = 0.281$ in.

Centroids of each area based on "y":

$y_t = \dfrac{y}{2} = \dfrac{0.281}{2} = 0.14$ in.

$y_c = \dfrac{\Sigma A \bar{y}}{\Sigma A} = \dfrac{2\left[0.375(1)\dfrac{1}{2}\right] + (0.375 - y)(4)\left(1 + \dfrac{0.375 - y}{2}\right)}{2[0.375(1)] + (0.375 - y)4}$

$= \dfrac{2\left[0.375(1)\dfrac{1}{2}\right] + (0.375 - 0.281)(4)\left(1 + \dfrac{0.375 - 0.281}{2}\right)}{2[0.375(1)] + (0.375 - 0.281)4} = 0.683$ in.

Plastic Section Modulus:

$Z_s = A_t d = A_t(h - y_t - y_c) = [4(0.281)](1.375 - 0.14 - 0.683) = 0.62$ in.^{3}

Design Strength:

$\phi M_p = \phi F_y Z_s = 0.9(36)(0.62) = 20.1$ kip-in.

where:
- $\phi = 0.9$
- A_w = area subjected to shear, typically taken as the web area (in.²)

This equation is only valid for sections that conform to the following limits:

$$\frac{h}{t_w} \leq \frac{418}{\sqrt{F_y}} \quad \text{(Eq. 6.6.2.2)}$$

where:
- h = height of web or clear distance between flanges (in.)
- t_w = thickness of shear web (in.)
- F_y = yield stress of web material (ksi)

For shear members with geometry outside this range, see Ref. 6.

6.6.3 Design Torsional Strength

Torsional strength of a steel member is related to the shear strength, as the torsional shear stresses (as calculated by Eq. 6.6.3.1) are numerically additive to the direct shear stresses. The design of a section subject to torsion must also consider the torsional rotation. Equations for torsional rotation of members with various support systems are shown in Design Aid 11.1.6.

6.6.3.1 Rectangular Solid Plates in Torsion

(See Figure 6.6.3.1) The torsional shear stress is:

$$\tau_u = \frac{T_u}{\alpha h t^2} \quad \text{(Eq. 6.6.3.1)}$$

where:
- τ = torsional shear stress
- t = total material thickness
- h = total material height
- α = constant from Table 6.6.3.1
- T_u = applied torsional moment about the Z axis

The maximum shear stress based on Ref. 6 is:

$$\tau_{max} = \phi(0.6F_y) \quad \text{(Eq. 6.6.3.2)}$$

where:
- $\phi = 0.9$
- F_y = yield strength of plate

Setting the maximum shear stress equal to the allowable limits yields:

$$\phi T_n = \phi(0.6F_y)\alpha h t^2 \quad \text{(Eq. 6.6.3.3)}$$

Check torsional rotation which is dependent on loading and the end support conditions. Equations are given in Design Aid 11.1.6. For example, for a single torsion moment applied to a cantilever:

$$\theta = \frac{t\ell}{GJ_T}$$

where:
- θ = rotation angle in radians
- ℓ = length of cantilever
- G = shear modulus (11,600 ksi for steel)
- $J_T = \beta h t^3$ (torsional constant)
- β = value based on h/t (see Table 6.6.3.1)
- h = total material height
- t = total material thickness

Figure 6.6.3.1 Rectangular section subjected to torsion about the Z axis

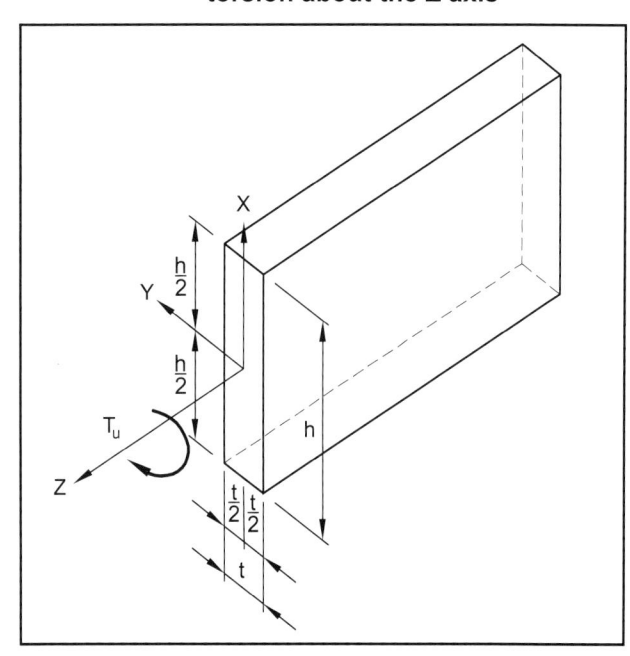

Table 6.6.3.1 Torsion Constants α and β

For solid rectangular sections (Figure 6.6.3.1)	h/t	1.00	1.50	1.75	2.00	2.50	3.00	4.00	6.00	8.00	10.00	∞
	α	0.208	0.231	0.239	0.246	0.258	0.267	0.282	0.299	0.307	0.313	0.333
	β	0.141	0.196	0.214	0.229	0.249	0.263	0.281	0.299	0.307	0.313	0.333

6.6.3.2 Rectangular Hollow Sections (Structural Tubing) in Torsion

$$\tau_u = \frac{T_u}{2\overline{A}t} \quad \text{(Eq. 6.6.3.4)}$$

where:
- t = thickness of tube wall
- $\overline{A}$ = area enclosed by centerline of tube walls
- = wd (see sketch at right)

Setting the maximum shear stress equal to the allowable limits yields:

$$\phi T_n = 2\phi(0.6F_y)\overline{A}t \quad \text{(Eq. 6.6.3.5)}$$

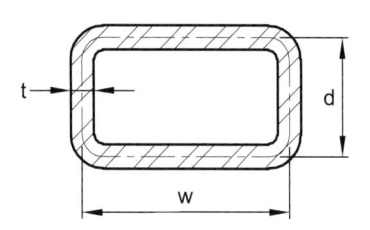

Setting the maximum shear stress equal to the allowable limits yields:

$$\phi T_n = 2\phi(0.6F_y)\overline{A}t \quad \text{(Eq. 6.6.3.5)}$$

Torsional rotation must also be checked (see Example 6.6.3.1).

Example 6.6.3.1
Design of Steel Structural Tube in Torsion

Given:

A beam made of steel structural tube HSS 8 x 6 x ½ spans 8.5 ft, and is welded to plates embedded in walls at each end. The beam supports a resultant uniform factored load, w_u, of 3 kips/ft, including overload factors, with a 4 in. resultant eccentricity, e.

- w_u = 3 kips/ft
- w = 7.5 in., d = 5.5 in., t = 0.465 in.
- F_y = 46 ksi
- ϕ = 0.9 (see Section 6.2.1)

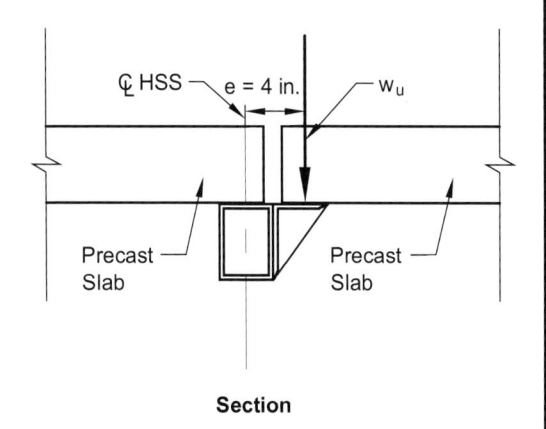

Problem:
Check shear stress and rotation of the beam.

Solution:
Use Design Aid 11.1.6(6):
Uniform torque, $T_u = w_u e = 3.0(4) = 12$ kip-in./ft

$$T_u \text{ at support} = \frac{T_u \ell}{2} = \frac{12(8.5)}{2} = 51 \text{ kip-in.}$$

Combined direct and torsional shear stresses:

$$A_w = 2(8)(0.5) = 8 \text{ in.}^2$$

$$\overline{A} = (5.5)(7.5) = 41.25 \text{ in.}^2$$

$$\frac{w_u}{2A_w} + \frac{T_u}{2\overline{A}t} = \frac{3.0(8.5)}{2(8)} + \frac{51}{(2)41.25} = 1.59 + 0.62 = 2.21 \text{ ksi}$$

$\phi(0.6)F_y = 0.9(0.6)(46) = 24.8$ ksi > 2.21 ksi OK

Example 6.6.3.1 (Cont.)
Design of Steel Structural Tube in Torsion

Check rotation from Design Aid 11.1.6(6).

Assume uniform service load torque = $T = T_u/2 = 12/2 = 6$ kip-in./ft = 0.5 kip-in./in.

Torsional Constant, $J_T = \dfrac{2tw^2d^2}{w+d} = \dfrac{2(0.465)(7.5)^2(5.5)^2}{7.5+5.5} = 121.7$ in.^{4}

Shear Modulus, $G = 11,600$ ksi

$\theta = \dfrac{T\ell^2}{8GJ_T} = \dfrac{0.5(8.5 \times 12)^2}{8(11,600)(121.7)} = 0.00046$ radians = 0.026 deg

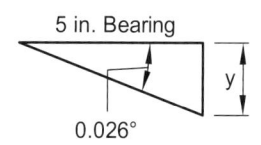

An angle of 0.026 deg over a 5 in. bearing distance results in a vertical displacement:

y = tan(0.026)5 = 0.0023 in.
Therefore, rotation is not significant.

6.6.4 Combined Loading

The stresses under factored loads are calculated using elastic analysis and the stresses due to individual loads are added appropriately to obtain resultant stresses (normal stress due to axial load and bending moment or shear stress due to shear force and torsion).

These resultant stresses may be combined to produce maximum (principal) stresses; however, for flexural members (such as wide flange members, channels or tube sections) such refinement is not necessary and the resultant stresses are checked with the LRFD limit states for yielding.

For yielding under normal stress:

$\phi f_{un} = \phi F_y$ (Eq. 6.6.4.1)

where:
    $\phi = 0.9$

For yielding under shear stress:

$\phi f_{uv} = \phi(0.6F_y)$ (Eq. 6.6.4.2)

where:
    $\phi = 0.9$

6.6.5 Unstiffened Connection Angles

Connection angles used to support small precast members can be designed by principles of mechanics as shown in Figure 6.6.5.1. In addition to the applied vertical and horizontal loads, the design should include all loads induced by restraint of relative movement between the precast member and the supporting member.

It is assumed the centerline of the compression force is at $0.1(\ell_\ell - g)$ for bolted connections and at $0.1\ell_\ell$ from the top of the angle for welded connections (see Figure 6.6.5.1). The compression force is determined using this assumption, then it is verified that there is adequate bearing area for this assumption. (Note: This is a simplified, conservative method of analysis. Other methods may be used.)

For bolted connection:
    $P_u(comp) \leq \phi P_n$

$\phi P_n \leq 0.2(\ell_\ell - g)\phi\, 0.85\, f'_c\, b$ (Eq. 6.6.5.1)

For welded connection:
    $P_u(comp) \leq \phi P_n$

$\phi P_n \leq 0.2(\ell_\ell)\phi\, 0.85\, f'_c\, b$ (Eq. 6.6.5.2)

where:
    $\phi = 0.65$
    f'_c = concrete strength
    b = width of angle leg

Welds should be designed in accordance with Section 6.7.

The design strength of non-stiffened angles loaded as shown in Figure 6.6.5.1 can be determined by:

$$\phi V_n = \frac{\phi F_y b_n t^2}{4e_v} \quad \text{(Eq. 6.6.5.3)}$$

where:
- $\phi = 0.9$
- b_n = (length of angle) − (hole diameter) −1/16 in.
- e_v = applied eccentricity and construction tolerances
- t = thickness of angle

Note: For welded angles, Figure 6.6.5.1 (b), e_v may be taken equal to actual eccentricity minus k, where k is the distance from the back face of angle to web toe of fillet. The tension requirement on the bolt and insert (Figure 6.6.5.2) can be calculated by:

$$P_u = V_u \frac{e_v}{e_i} \quad \text{(Eq. 6.6.5.4)}$$

For angles loaded axially, N_u, Figure 6.6.5.3, the design strength of non-gusseted angles can be calculated by:

$$\phi N_n = \frac{\phi F_y t^2 b_n}{4g} \geq N_u \quad \text{(Eq. 6.6.5.5)}$$

Figure 6.6.5.1 Connection angle design parameters

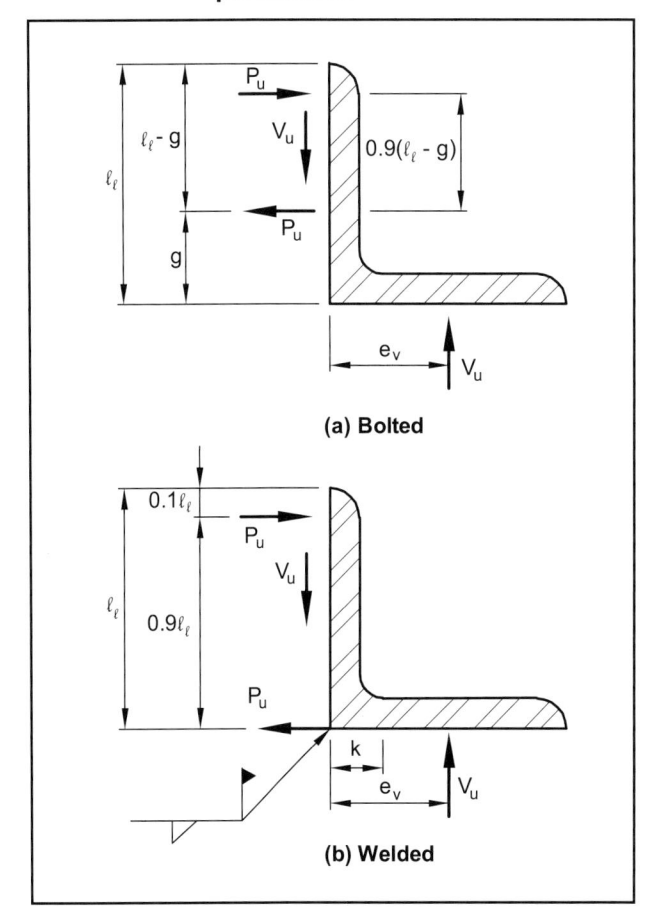

(a) Bolted

(b) Welded

and the design tension requirement for the bolt and insert can be calculated by:

$$\phi P_n = \phi N_n \left(1 + \frac{g}{e_i}\right) \geq P_u \quad \text{(Eq. 6.6.5.6)}$$

where:
- $\phi = 0.9$
- g = gage of angle

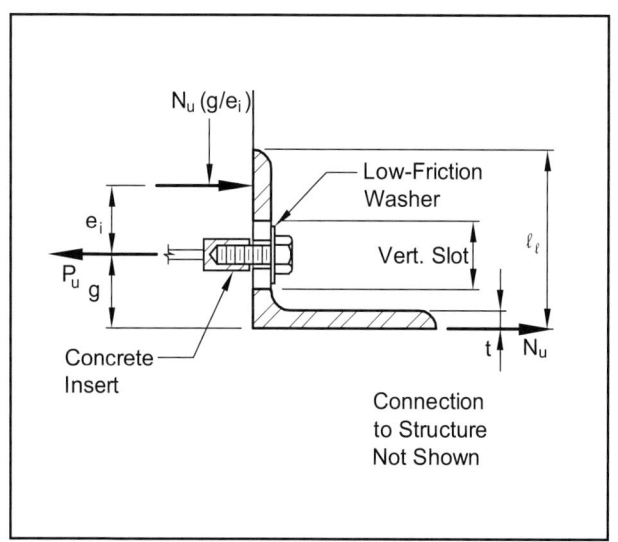

The minimum edge distance for bolt holes is shown in Table 6.6.5.1. These holes may be punched, reamed or drilled holes. The edge distance is measured from the center of the holes.

For connections made by welding instead of bolting, the welds should be designed in accordance with Section 6.7.

Capacity of angles based on Eqs. 6.6.5.3 and 6.6.5.5 can be read from Figures 6.16.8 and 6.16.9.

Figure 6.6.5.2 Shear force on unstiffened angle

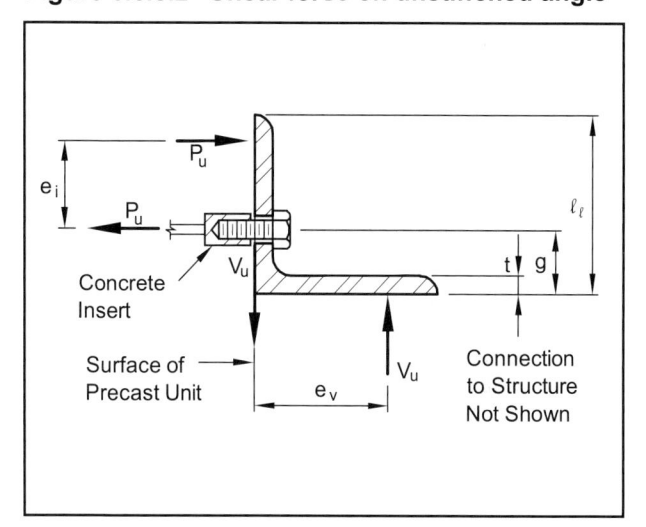

Figure 6.6.5.3 Normal force on unstiffened angle

Table 6.6.5.1   Minimum edge distance

Bolt diameter[a] (in.)	At sheared edges	At rolled edges of plates, shapes or bars or gas-cut edges[b]
½	⅞	¾
⅝	1⅛	⅞
¾	1¼	1
⅞	1½ [c]	1⅛
1	1¾ [c]	1¼
1¼	2¼	1⅝

a. When oversized or slotted holes are used, edge distances should be increased to provide the same distance from edge of hole to free edge as given by the tabulated edge distances. Use ASTM F436 washers, or 5/16 in. thick plate washer. [6]
b. All edge distances in this column may be reduced ⅛ in. when the hole is at a point where stress does not exceed 25% of the maximum allowable stress in the element.
c. These may be 1¼ in. at the ends of beam connection angles.

Table 6.6.5.2   Normal gages for angles

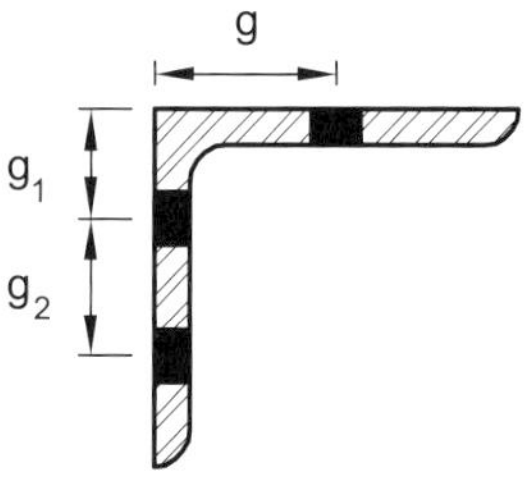

Angle Leg Length (in.)	Normal Angle Gage, g (in.)	Two-Hole Gage (in.)	
		g_1	g_2
2	1⅛		
2½	1⅜		
3	1¾		
3½	2		
4	2½		
5	3	2	1¾
6	3½	2¼	2½
7	4	2½	3
8	4½	3	3

6.6.6   Triangular Stiffener Design

Yielding along the free edge frequently occurs prior to buckling and stress redistribution occurs within the system. [11]

The design normal force, ϕN_n, is assumed to be resisted by the top line of weld of the bearing seat and has no impact on the design of the stiffener.

A ratio z has been established for triangular stiffeners that relates average stress, $\frac{V_u}{bt}$, to the maximum stress f_{max}:

$$z = 1.39 - 2.2\left(\frac{b}{a}\right) + 1.27\left(\frac{b}{a}\right)^2 - 0.25\left(\frac{b}{a}\right)^3$$

(Eq. 6.6.6.1)

Figure 6.6.6.1   Triangular stiffener

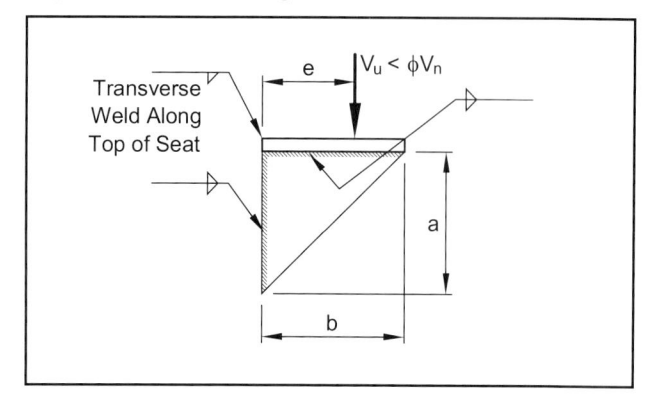

Figure 6.6.6.2   Triangular stiffener design limits

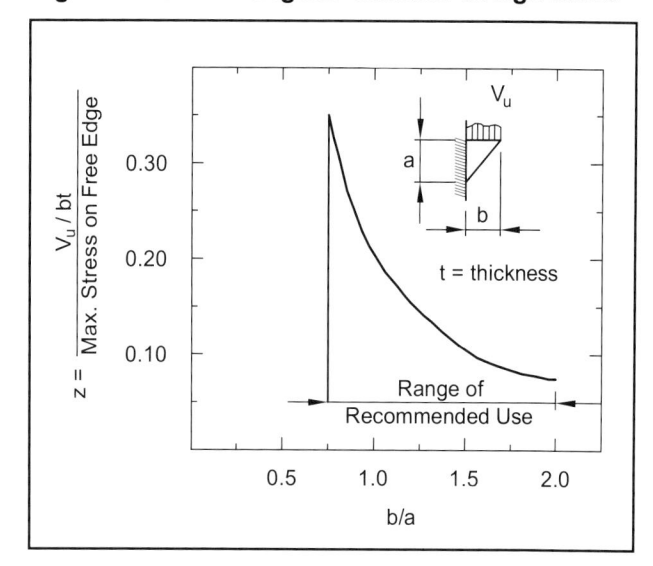

where:
b = projection of stiffener
a = height of stiffener

The design strength is limited when the free edge reaches the material's yield strength:

$$\phi V_n = \phi F_y z b t \quad \text{(Eq. 6.6.6.2)}$$

where:
ϕ = 0.85
F_y = yield strength of the stiffener
t = stiffener thickness

To ensure yielding along the free edge, the following limits should be satisfied:

If $0.75 \leq \frac{b}{a} \leq 1.0$, then $\frac{b}{t} \leq \frac{250}{\sqrt{F_y}}$ (Eq. 6.6.6.3)

If $1.0 \leq \frac{b}{a} \leq 2.0$, then $\frac{b}{t} \leq \frac{250\left(\frac{b}{a}\right)}{\sqrt{F_y}}$ (Eq. 6.6.6.4)

6.6.7 Non-Triangular Stiffener Design

The non-triangular stiffened beam seat is designed based on the recommendations of Ref. 11.

The minimum stiffener thickness to ensure yielding along the leading free edge is based on Ref. 6, Section B5, Local Buckling Criteria:

$$t_s = \frac{b\sqrt{F_y}}{95} \qquad \text{(Eq. 6.6.7.1)}$$

Using the minimum thickness, the nominal strength of the stiffener may be determined using a combined load analogy. The nominal normal force, ϕN_n, shown in Figure 6.6.7.1, is assumed to be resisted by the top line of weld on the bearing seat and has no impact on the design of the stiffener.

According to Ref. 6, bearing stress, f_b, at outer edge of stiffener must satisfy:

$$f_b = \frac{P}{A} + \frac{Mc}{I} \leq \phi 1.8 F_y$$

Therefore:

$$\phi(1.8F_y) = \frac{\phi V_n}{t_s b} + \frac{\left[\phi V_n \left(e - \frac{b}{2}\right)\right]\frac{b}{2}}{t_s b^3 / 12}$$

Therefore:

$$\phi(1.8F_y) = \frac{\phi V_n}{t_s b} + \frac{\left[\phi V_n \left(e - \frac{b}{2}\right)\right]\frac{b}{2}}{t_s b^3 / 12}$$

$$\phi V_n = \frac{\phi(1.8F_y)}{\dfrac{1}{t_s b} + \dfrac{\left(e - \dfrac{b}{2}\right)\dfrac{b}{2}}{t_s b^3 / 12}} \qquad \text{(Eq. 6.6.7.2)}$$

Figure 6.6.7.1 Non-triangular stiffener

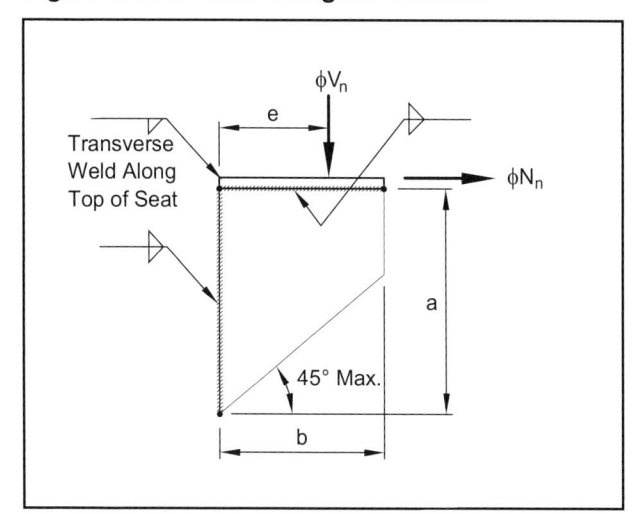

Example 6.6.6.1
Triangular Stiffener Analysis

Given:
The stiffened seat connection is shown at right.
Stiffener thickness, $t_s = 3/8$ in. $F_y = 36$ ksi

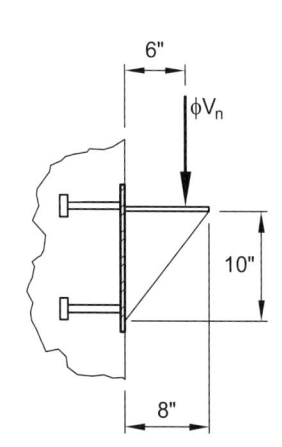

Problem:
Determine the design shear resistance of the stiffener.

Solution:

$$\frac{b}{a} = \frac{8}{10} = 0.80 \text{ which is} > 0.75 \text{ and} < 1.0$$

Therefore, $\dfrac{b}{t}$ must be $\leq \dfrac{250}{\sqrt{F_y}} = \dfrac{250}{\sqrt{36}} = 41.7$

$$\frac{b}{t} = \frac{8}{0.375} = 21.3 < 41.7 \quad \text{OK}$$

$$z = 1.39 - 2.2\left(\frac{b}{a}\right) + 1.27\left(\frac{b}{a}\right)^2 - 0.25\left(\frac{b}{a}\right)^3 = 1.39 - 2.2(0.80) + 1.27(0.80)^2 - 0.25(0.80)^3 = 0.315$$

$$\phi V_n = \phi F_y z b t = 0.85(36)(0.315)(8)(0.375) = 28.9 \text{ kips}$$

PCI Design Handbook/Sixth Edition

Example 6.6.6.2
Triangular Stiffener Design

Given:
The stiffened seat connection of Example 6.6.6.1.

Applied load V_u = 24 kips, F_y = 36 ksi (including overload factors)

Problem:
Determine the required stiffener thickness.

$$\frac{b}{a} = \frac{8}{10} = 0.80 \text{ which is} > 0.75 \text{ and} < 1.0$$

From Example 6.6.6.1:
$$z = 0.315$$

$$\phi V_n = \phi F_y z b t$$

$$t = \frac{V_u}{\phi F_y z b} = \frac{24}{0.85(36)(0.315)(8)} = 0.31 \text{ in.}$$

$$\frac{b}{t_{min}} \leq \frac{250}{\sqrt{F_y}}$$

$$t_{min} = \frac{b\sqrt{F_y}}{250} = \frac{8\sqrt{36}}{250}$$
$$= 0.192 \text{ in.} < 0.31 \text{ in.} \quad \text{OK}$$

where:
$\phi = 0.75$

The stiffener thickness required based on an applied load is:

$$t_s = V_u \left[\frac{1}{b} + \frac{6\left(e - \frac{b}{2}\right)}{b^2} \right] \frac{1}{\phi(1.8 F_y)} \quad \text{(Eq. 6.6.7.3)}$$

6.7 Welding

Nearly all structural steel used in precast concrete connections conforms with ASTM A36 steel. Thus, it is readily weldable with standard equipment [8] and standard welding procedures.

6.7.1 Corrosion Resistance of Welded Connections

For corrosion resistance, exposed components of connections are sometimes hot-dip galvanized or stainless steel. Connections that are covered, patched, or painted and/or not exposed to weather are typically unpainted steel.

The requirements and preparation for welding varies for these two types of steel.

6.7.1.1 Hot Dip Galvanized

Welding of hot-dip galvanized steel requires special care, such as thorough removal of galvanizing material and following qualified welding procedures. [2, 10]

Example 6.6.7.1
Non-Triangular Stiffener Analysis

Given:
The stiffened beam seat shown.
Stiffener thickness, $t_s = 3/8$ in.
F_y = 36 ksi

Problem:
Determine the design shear resistance of the stiffener.

Solution:
From Eq. 6.6.7.2:

$$\phi V_n = \frac{\phi(1.8 F_y)}{\dfrac{1}{b(t_s)} + \dfrac{\left(e - \dfrac{b}{2}\right)\dfrac{b}{2}}{t_s b^3 / 12}} = \frac{0.75(1.8(36))}{\dfrac{1}{6(0.375)} + \dfrac{\left(4.5 - \dfrac{6}{2}\right)\left(\dfrac{6}{2}\right)}{(0.375)(6)^3 / 12}} = 43.7 \text{ kips}$$

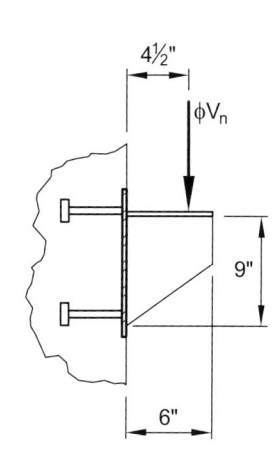

Table 6.7.1.1 Average Stainless Steel Properties [24]

Type	Tensile Strength	Yield Strength @ 0.2% Offset	Elongation in 2 in.
	(ksi)	(ksi)	(%)
304	84	42	55
308	115	80	40
309	90	45	45
310	95	45	45
316	84	42	50

Modulus of Elasticity = 28,000 ksi
Shear Modulus = 12,500 ksi
Coef. of Thermal Expansion @ 600°F
 = 9.9×10^{-6} in./in./°F
Density = 0.29 lbs/in.^{3}

Commonly, the galvanized material needs to be removed in the weld area prior to the welding procedure or a prequalified welding procedure needs to be submitted. Upon completion and cooling, the portion of the galvanic material that was removed is reapplied with a zinc rich paint.

6.7.1.2   Stainless Steel

Stainless steel is often used in connections because of its excellent resistance to corrosion. AISI designates three main types of stainless steels: austenitic, martensitic, and ferritic. The 200 Series is considered an austenitic stainless steel containing chromium, nickel, and maganese. Chromium-nickel stainless steels constitute the 300 Series and are also classified as austenitic. Hardenable martensitic and non-hardenable ferritic stainless steels comprise the 400 Series. Both of these 400 Series stainless steels are highly magnetic and classified as straight chrome stainless steels.

AISI Type 304 and 316 stainless steels are most commonly used in connections and structural applications. Some caution must be exercised when either of these two alloys are welded. These chrome-nickel stainless steels contain small amounts of carbon, which combines with chromium to produce chromium carbide; the chromium carbide by-product is not corrosion resistant, and is thus undesirable. The chemical reaction that forms chromium carbide is known as carbide precipitation, which can occur from the high heat of the welding process followed by slow cooling. Both Types 304 and 316 have low carbon alloys (304L and 316L), that should be considered to eliminate the formation of chromium carbides. Local availability should be checked for these alloys. It should be noted that stainless steel should not be cut with an acetylene torch.

Plate Welding. Stainless steel has slightly different physical properties than conventional mild carbon steel, making it slightly more difficult to weld. Specifically, stainless steel has a:
- Lower melting temperature
- Lower coefficient of thermal conductivity
- Higher electrical resistance
- Higher coefficient of thermal expansion

Slightly faster travel speeds are suggested for stainless steel welding, which will reduce the heat input and thus minimize joint distortion and cracking in the surrounding concrete. The lower melting temperature usually dictates lower weld current input. The PCI Connections Manual provides more specific information on the mechanics of stainless steel welding. [2]

Stainless Steel Welding. Type 304 (or 304L) stainless is usually welded with a Type 308 (or 308L) filler metal. Type 316 (or 316L) stainless typically uses a Type 316 (316L) filler metal. Other filler metals are available for the other stainless types. Consult with welding suppliers or manufacturers for the most suitable filler.

Dissimilar Metal Welding. Stainless steel is often welded to mild carbon steel, or vice versa; however, precautions must be followed to ensure a good quality weld is obtained. The key to this weld process is selecting a filler metal that can successfully combine with both base metals while avoiding the formation of internal joint flaws. Stresses at the weld interface of a dissimilar metal joint may be high because of differences in coefficients of thermal expansion.

When joining austenitic stainless steels to carbon steels or low-alloy steels, a stainless steel welding rod sufficiently high in total alloy content should be used, such as Type 309 or 309L, or other highly alloyed electrodes. For quality control, a longitudinal bend specimen is considered superior to a transverse bend specimen for dissimilar metal joining, because all regions of the welded joint participate in the bend. [12] Refer to AWS standards for more information on bend test specimens.

Carbon steels with less than 0.20 percent carbon can normally be welded with austenitic fillers without preheat. When the carbon is greater than 0.30 percent, temperature control is necessary. As alloy content increases, i.e., in the case of low-alloy steels, preheat control is usually essential. The deposition of carbon steel or low alloy steel weld metal on stainless steel can result in hard, brittle weld deposits that frequently crack when deposited and likely will fail in-service. Weld procedure qualifications should be conducted on all proposed combinations of base metal and electrodes to determine and identify any potential problems.

Stud Welding. The austenitic stainless steels (300 Series) are usually recommended for arc stud welding. Stainless steel studs can be welded to either stainless steel or mild carbon steel. For the latter case, it is generally recommended that the base metal carbon content not exceed 0.20 percent. Otherwise, specific types of stainless steel studs must be used (e.g., Types 308, 309 or 310).

It is also recommended that fully annealed stainless steel studs be used when welding stainless steel studs to a mild carbon steel base metal. Annealed studs have been shown to be imperative for stainless steel studs welded to carbon steel plates subject to repetitive or cyclic loads. In such cases, stress corrosion failure in the weld can occur, and annealed stud use minimizes the chance of weld cracking and failure. [13]

6.7.2 Weld Design

The limits on weld stress are based on ASW D1.1-02 [8] for steel plates and structural steel shapes and AWS D1.4-02 [9] for reinforcing bars. The design strength of welds for use with factored loads can be determined from the provisions of the AISC LRFD Manual. [6] The most commonly used welds in connections are full penetration welds, partial penetration groove welds, and fillet welds. These are illustrated in Chapter 11, Design Aid 11.4.2.

6.7.2.1 Full Penetration Welds

Full penetration strength is typically taken as the strength of the base metal. At least the "matching" weld metal specified by AWS D1.1 must be used. If a higher weld strength than the "matching" is used, the strength of the weld is limited to the strength of the base material.

The base metal design stress is limited to:

$$\phi f_n = \phi(0.6F_u) \qquad \text{(Eq. 6.7.2.1)}$$

where:
$\phi = 0.75$
F_u = ultimate tensile strength of base metal

Allowable stresses and design strengths of weld are given in Figure 6.16.1.

6.7.2.2 Fillet Welds

Fillet welds are the most commonly used welds in connections. The design strength may be limited by either the weld strength or the base metal strength. The weld design stress is limited to:

$$\phi f_n = \phi(0.6F_{exx}) \qquad \text{(Eq. 6.7.2.2)}$$

where:
$\phi = 0.75$

Strengths of fillet welds are given in Figure 6.16.2.

6.7.2.3 Partial Penetration Groove Welds

Partial penetration groove welds have various design strength limits. The most typical application of this weld is in lap splices of reinforcing bars or for attachment of reinforcement to connection plates.

When loaded in shear, the design stress is:

$$\phi f_n = \phi(0.6F_{exx}) \qquad \text{(Eq. 6.7.2.3)}$$

where:
$\phi = 0.75$

6.7.3 Reinforcing Bar Welding

Reinforcing bars are often welded to plate material for connection ductility, development length, anchorage, and placement. The weldability of reinforcement is based on the chemical composition of the bar.

Welding of reinforcing bars is covered by ANSI/AWS D1.4-02, "Structural Welding Code-Reinforcing Steel," [9] and by ANSI/AWS D1.1, "Structural Welding Code-Steel," [8] by the American Welding Society.

6.7.3.1 Carbon Equivalent

Weldability is defined by AWS as a function of the chemical composition of steel as shown in the mill report by the following formula:

$$C.E. = \%C + \frac{\%Mn}{6} + \frac{\%Cu}{40} + \frac{\%Ni}{20} \qquad \text{(Eq. 6.7.3.1)}$$

where:
C.E. = carbon equivalent

Figure 6.7.3.1 Typical reinforcing bar welds

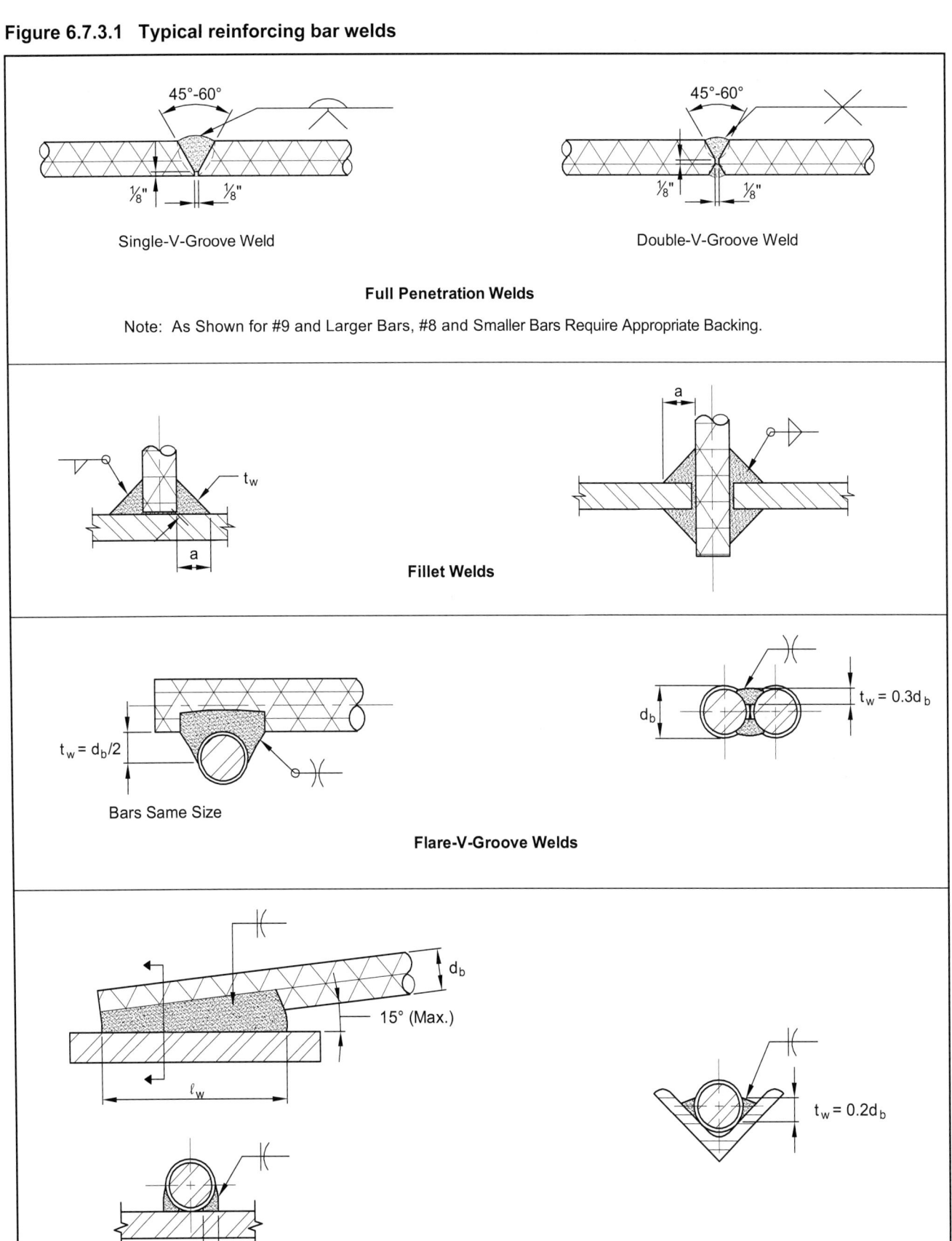

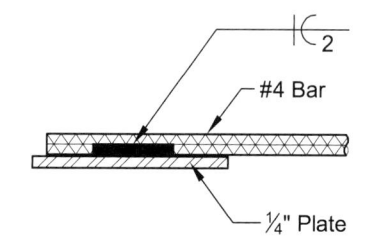

> **Example 6.7.3.1**
> **Reinforcing Bar Weld Analysis**
>
> *Given:*
> A #4 reinforcing bar welded to a ¼ in. thick plate as shown.
>   Plate F_y = 36 ksi
>   Weld electrode = E70
>
> The mill certificate shows the following:
> F_y = 62 ksi – Assume 60 ksi
>   %C = 0.24%
>   %Mn = 1.18%
>   %Cu = 0.27%
>   %Ni = 0.07%
>
> *Problem:*
> (1) Check the carbon equivalent of the bar and (2) determine the required weld length to develop the full strength of the bar.
>
> *Solution:*
> (1) Carbon Equivalent (Section 6.7.3.1)
> $$\text{C.E.} = \%C + \frac{\%Mn}{6} + \frac{\%Cu}{40} + \frac{\%Ni}{20} = 0.24 + \frac{1.18}{6} + \frac{0.27}{40} + \frac{0.07}{20} = 0.45\% < 0.55\% \quad \text{OK}$$
>
> (2) Required weld length:
> Strength of bar: $T_n = A_s f_y = (0.20)(60) = 12.0$ kips
> The weld on each side of the bar must resist 12/2 = 6 kips
> From Figure 6.7.3.1: The effective throat of a single side of a flare-bevel-groove weld = $0.2d_b$
>
> From Eq. 6.7.3.2:
>
> $\phi T_n = \phi(0.6 F_{exx}) t_w \ell_w = 0.75(0.6)(70)(0.2)(0.5) \ell_w = 3.15 \ell_w$
> $\ell_w = 6/3.15 = 1.9$ in. Use $\ell_w = 2$ in. each side
>
> Check plate shear:
> $\phi V_n = \phi(0.6 F_y) t_{pl} \ell_w = 0.9(0.6)(36)(0.25)(2) = 9.7$ kips > 6 kips OK
>
> Note: The solution may be read directly from Figure 6.16.3.

Three other elements, Cr, Mo and V usually appear only as trace elements, so they are often not included in the mill report. For reinforcing bars that are to be welded, the carbon equivalent should be requested with the order from the mill.

ANSI/AWS D1.4-02 indicates that most reinforcing bars can be welded. However, stringent preheat and other quality control measures are required for bars with high carbon equivalents. Except for welding shops with proven quality control procedures that meet ANSI/AWS D1.4-02, it is recommended that carbon equivalents be less than 0.45% for No. 7 and larger bars, and 0.55% for No. 6 and smaller bars. Bars which meet ASTM A 706 are specially formulated to meet the above requirements. Preheat is still required for grade A706 but does not require as much as A615.

Most reinforcing bars which meet ASTM 615 will not meet the above chemistry requirements. Preheat, as discussed above, will therefore, be required, based on the mill reports of the bars to be welded.

6.7.3.2   Strength of Reinforcing Bar Weld

Figure 6.7.3.1 shows the most common welds used with reinforcing bars. Full penetration groove welds can be considered to have the same nominal strength as the bar when matching weld metal is used. The design strength of the other weld types can be calculated using the values from Figures 6.16.1 and 6.16.2.

The design strength of a weld is based on the shear through the effective throat:

$$\phi T_n = \phi(0.6 F_{exx}) t_w \ell_w \qquad \text{(Eq. 6.7.3.2)}$$

where:
ϕ = 0.75
F_{exx} = electrode weld strength
ℓ_w = length of weld
t_w = effective throat of weld

For a fillet weld, the effective throat is:

$$t_w = \frac{a}{\sqrt{2}} \qquad \text{(Eq. 6.7.3.3)}$$

where:
a = weld size

Figures 6.16.3 through 6.16.5 show welding required to develop the full strength of reinforcing bars.

The welded cross-bar detail shown in Figure 6.7.3.1 is not included in ANSI/AWS D1.4. However, it has been used in numerous structures and verified by tests [17] and, when the diameter of the cross bar is at least the same size as the main bar, the full strength of the main bar has been shown to be developed.

Reinforcing bars should not be welded within 2 in. or two bar diameters of a bend to avoid potential crystallization and the associated brittle behavior.

ANSI/AWS D1.4 requires that tack welds be made using the same preheat and quality control requirements as permanent welds, and prohibits them unless authorized by the engineer.

Typical weld symbols are shown in Chapter 11 of this Handbook.

6.7.4 Cracking in Concrete Around Welded Connections

When welding is performed on components that are embedded in concrete, thermal expansion and distortion of the steel may destroy bond between the steel and concrete or induce cracking or spalling in the surrounding concrete.

The extent of cracking and plate distortion is dependent on the amount of heat generated during welding, the stiffness of the steel member, and restraint caused by plate anchors.

Heat may be reduced or monitored by:
- Use of low-heat welding rods of small size.
- Use of intermittent rather than continuous welds.
- Use of smaller welds in multiple passes.
- Use of heat indicator.

Concrete cracking may be minimized by providing a space around the metal by forming, taping, or using sealing foam. Weather stripping may also reduce risk of damage.

6.7.5 Design of Group Welds

There are two types of group weld designs as described below:

The Elastic Vector Method (EVM) calculates the resultant stress on a weld group due to applied loads. Typically, three-dimensional loading configurations are evaluated against two-dimensional weld geometry. The stress is calculated using a traditional mechanics of materials approach. This analysis procedure is typically used for out-of-plane loading. The results are conservative and do not take into account the deformation capability of the weld group.

The Instantaneous Center Method (ICM) calculates the resultant load based on strength design. The weld stresses are functions of the rotational displacement about an instantaneous center. [6, 11] This analysis procedure is typically used for in-plane loading. Where deformations are a concern, such as bearing, care should be used due to the nature of the analysis procedure. Depending on the geometry, the calculated capacity can be significantly greater than the Elastic Method.

In addition to the method presented, AISC has published design tables for various weld geometries. [6]

Figure 6.7.5.1 Typical eccentric loading and group welds

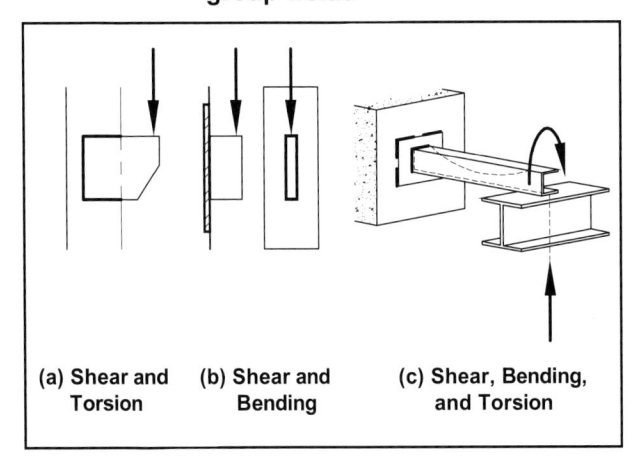

(a) Shear and Torsion (b) Shear and Bending (c) Shear, Bending, and Torsion

6.7.5.1 Elastic Vector Method

Consider the geometry in Figure 6.7.5.2.

The applied loads P_x, P_y, and P_z, each act at their respective eccentricities. The stress at each point (1, 2, 3, and 4) can be calculated using basic mechanics of materials.

The weld group has the following properties using a unit thickness:

A_W = total area of weld group
I_{xx} = moment of inertia about X-centroid axis
I_{yy} = moment of inertia about Y-centroid axis
I_p = polar moment of inertia $I_{xx} + I_{yy}$
$\bar{X}$ = dimension of weld groups centroid in X direction
$\bar{Y}$ = dimension of weld groups centroid in Y direction

The state of stress at each point is calculated with the following assumptions:

- Isotropic elastic material.
- Plane sections remain plane.

$$f_x = \frac{P_x}{A_w} + \frac{M_z y}{I_p} \quad \text{(Eq. 6.7.5.1)}$$

$$f_y = \frac{P_y}{A_w} + \frac{M_z x}{I_p} \quad \text{(Eq. 6.7.5.2)}$$

$$f_z = \frac{P_z}{A_w} + \frac{M_x y}{I_{xx}} + \frac{M_y x}{I_{yy}} \quad \text{(Eq. 6.7.5.3)}$$

$$f_r = \sqrt{f_x^2 + f_y^2 + f_z^2} \quad \text{(Eq. 6.7.5.4)}$$

where:
A_w = the total area for the weld group. The area of weld to be considered for steel sections, such as wide flange beams and channels, is that along the web of the member. [26]

$$A_w = \left(\frac{a}{\sqrt{2}}\right)\ell_w \quad \text{(Eq. 6.7.5.5)}$$

a = leg size of weld
M_z = total torsional moment about z axis
M_x = total bending moment about x axis
M_y = total bending moment about y axis
y and x = distances from c.g. of weld group to point in question

Figure 6.7.5.2 Elastic vector method – typical geometry

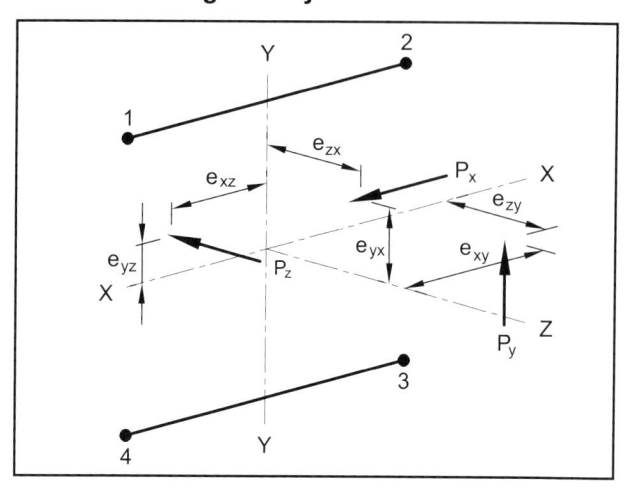

Figure 6.7.5.3 Instantaneous Center Method – differential free body diagram

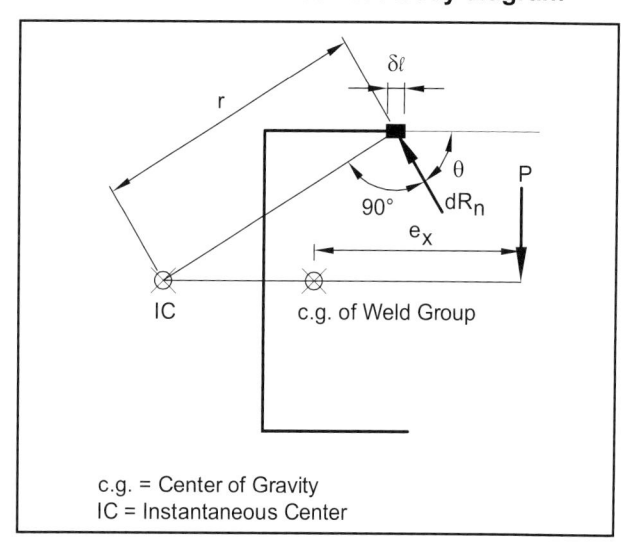

c.g. = Center of Gravity
IC = Instantaneous Center

Figure 6.7.5.4 Instantaneous Center Method – discrete free body diagram

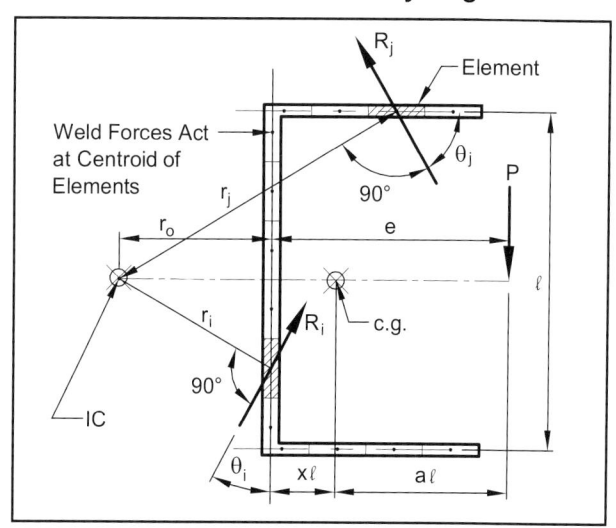

The resultant state of stress is compared to the following:

$$\phi f_n = \phi(0.6 F_{exx}) \quad \text{(Eq. 6.7.5.6)}$$

where:
$\phi = 0.75$
F_{exx} = weld electrode used

$$\phi f_n \geq f_v \quad \text{(Eq. 6.7.5.7)}$$

6.7.5.2 Instantaneous Center Method (ICM)

Consider the geometry in Figure 6.7.5.3. Because it is displacement based, the Instantaneous Center Method is not recommended for non-redundant bearing connections.

The applied load, P, acts in a plane at an eccentricity e. The overall nominal strength of the weld group is the sum of the differential weld elements.

The differential nominal strength of an element within the weld group is based on the product of three functions. These functions are strength, angular orientation, and deformation compatibility. This differential strength is:

$$dR_n = \{f(F_{exx})[g(\theta)][h(\Delta)]\} dA \quad \text{(Eq. 6.7.5.8)}$$

where:
$f(F_{exx})$ = functional capacity based on electrode strength
$g(\theta)$ = functional variation of load and orientation
$h(\Delta)$ = functional variation of load and deformation

The total capacity of the weld group:

$$R_n = \int dR_n$$
$$R_n = \int_{A_{weld}} [f(F_{exx})(g(\theta))(h(\Delta))] dA \quad \text{(Eq. 6.7.5.9)}$$

The resultant force of each element acts through a radial line r to the instantaneous center of the weld group. The instantaneous center is defined as the point at which both force and moment equilibrium occur.

$$\sum F_x = 0$$
$$\sum F_y = 0$$
$$\sum M_{IC} = 0$$

Discrete Solution:

Consider the free-body diagram of Figure 6.7.5.4 in which the group weld is divided into a number of discrete elements.

The strength of each discrete element is:

$$R_{n_i} = \left(0.60 F_{exx} \left(1.0 + 0.5 \sin^{\frac{3}{2}} \theta_i \right) \left\{ \frac{\Delta(\theta_i)}{\Delta_{max}(\theta_i)} \left[1.9 - 0.9 \frac{\Delta(\theta_i)}{\Delta_{max}(\theta_i)} \right] \right\}^{0.3} \right) A_{e_i}$$

(Eq. 6.7.5.10)

where:
$\Delta(\theta_i)$ = discrete form of deformation linearly proportional to critical deformation
F_{exx} = weld electrode strength
θ = angle of resistance force to longitudinal axis of weld element (degrees) [11]
A_{ei} = area of each element

For a weld group, the maximum deformation limit based on an ultimate load acting at an angle is:

$$\Delta_{max}(\theta) = 0.209(\theta + 2)^{-0.32} a \quad \text{(Eq. 6.7.5.11)}$$

where:
$\Delta_{max}(\theta)$ = maximum deformation for an element as a function of θ

The element's deformation is linearly proportional to the critical deformation. The deformation of each element is:

$$\Delta(\theta) = \Delta_u \frac{r}{r_{critical}} \quad \text{(Eq. 6.7.5.12)}$$

where:
r = radial distance from instantaneous center
$r_{critical}$ = radial distance of critical weld element based on smallest ratio of r to ultimate deformation
$= \left[\frac{r}{\Delta_u(\theta)} \right]$

The ultimate deformation of a weld element with respect to the loading angle is:

$$\Delta_u = 1.087(\theta + 6)^{-0.65}(a) \leq 0.17(a)$$

(Eq. 6.7.5.13)

Two primary equations of static equilibrium are:

$$\sum F_y = 0 \qquad \sum_{i=1}^{N_{el}} Rn_{i_y} - P_n = 0$$

$$\sum M_{IC} = 0 \qquad \sum_{i=1}^{N_{el}} (R_{n_i} r_i) - P_n(e + r_o) = 0$$

(Eq. 6.7.5.14)

The solution is iterated to solve for both the location of the instantaneous center, r_o, and the maximum ultimate load, P_n. It is important to note that the radial distance is also a function of r_o. The strength reduction factor for this analysis is $\phi = 0.75$ and it is based on shear capacity through the effective throat.

Example 6.7.4.1
Strength Analysis of Weld Group

Given:
The plate connection is shown at the right.
Electrode strength: $F_{exx} = 70$ ksi
Plate material: ASTM A-36

Problem:
Determine the design strength of the weld group using (1) Elastic Vector Method and (2) Instantaneous Center Method. Compare the results.

Solution:
(1) Elastic Vector Method:

Center of gravity of the weld:

$$\bar{x} = \frac{b^2}{2b+d} = \frac{(2)^2}{2(2)+5} = 0.444 \text{ in.}$$

$\phi V_n \leq V_u$, solve for ϕV_n

$$\bar{y} = \frac{d}{2} = \frac{5}{2} = 2.5 \text{ in.}$$

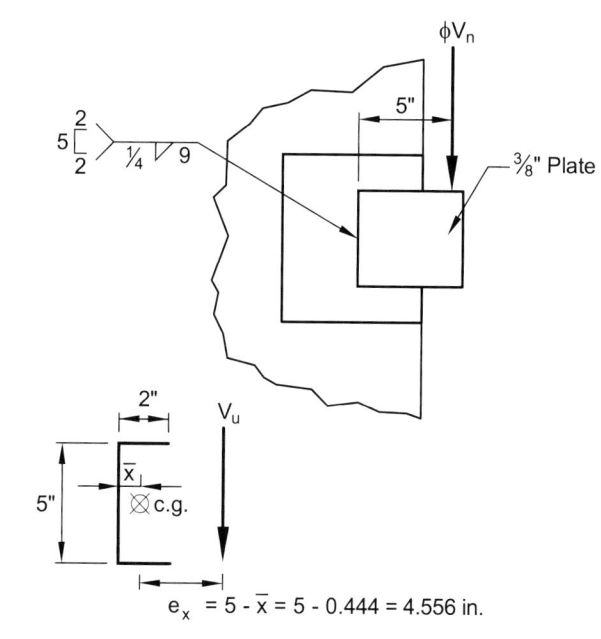

$e_x = 5 - \bar{x} = 5 - 0.444 = 4.556$ in.

Allowable stress: $\phi f_n = \phi(0.6 F_y) t_w = 0.75(0.6)(70)\left(\frac{0.25}{\sqrt{2}}\right) = 5.57$ kips/in.

Applied stress: $f = \frac{Mc}{I_p} + \frac{P}{A_w}$

$A_w = \ell_w = 2 + 2 + 5 = 9$ in.

$$I_p = \frac{8b^3 + 6bd^2 + d^3}{12} - \frac{b^4}{2b+d} = \frac{8(2)^3 + 6(2)(5)^2 + 5^3}{12} - \frac{2^4}{2(2)+5} = 39.0 \text{ in.}^4$$

Based on a unit weld thickness:

$$f_y = \frac{\phi V_n e_x (b - \bar{x})}{I_p} + \frac{\phi V_n}{\ell_w} = \frac{\phi V_n (4.556)(2 - 0.444)}{39.0} + \frac{\phi V_n}{9} = 0.293(\phi V_n) \text{ kips/in.}$$

Example 6.7.4.1 (Cont.)
Strength Analysis of Weld Group

$$f_x = \frac{\phi V_n e_x \left(\frac{d}{2}\right)}{I_p} = \frac{\phi V_n (4.556)(2.5)}{39.0} = 0.292(\phi V_n) \text{ kips/in.}$$

Resultant:
$$f = \sqrt{f_x^2 + f_y^2} = \sqrt{(0.293)^2 + (0.292)^2} = 0.414(\phi V_n) \text{ kips/in.}$$

Strength: $0.414(\phi V_n) = 5.57$

$$\phi V_n = \frac{5.57}{0.414} = 13.4 \text{ kips}$$

(2) Instantaneous Center Method:
Use ½ in. weld element lengths. Origin for element geometry is at the Instantaneous Center. Applied Eccentricities: $e_x = 4.556$ in. (same as Elastic Vector method). Initial estimate of $R_o = 1.5$ in.

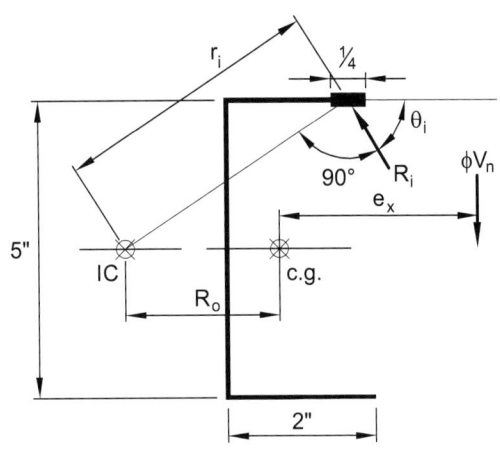

Weld Element Geometry

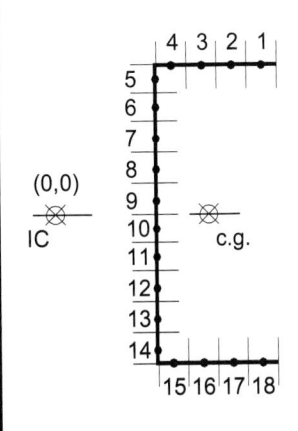

Initial estimate for $R_o = 1.5$ c.g. = 0.444

Element	Length	x (in.)	y (in.)	r (in.)	θ (degrees)
1	0.5	2.806	2.5	3.7581	48.3006
2	0.5	2.306	2.5	3.4011	42.6884
3	0.5	1.806	2.5	3.0841	35.8443
4	0.5	1.306	2.5	2.8206	27.5826
5	0.5	1.056	2.25	2.4855	64.8578
6	0.5	1.056	1.75	2.0439	58.8920
7	0.5	1.056	1.25	1.6363	49.8089
8	0.5	1.056	0.75	1.2952	35.3834
9	0.5	1.056	0.25	1.0852	13.3191
10	0.5	1.056	-0.25	1.0852	-13.3191
11	0.5	1.056	-0.75	1.2952	-35.3834
12	0.5	1.056	-1.25	1.6363	-49.8089
13	0.5	1.056	-1.75	2.0439	-58.8920
14	0.5	1.056	-2.25	2.4855	-64.8578
15	0.5	1.306	-2.5	2.8206	-27.5826
16	0.5	1.806	-2.5	3.0841	-35.8443
17	0.5	2.306	-2.5	3.4011	-42.6884
18	0.5	2.806	-2.5	3.7581	-48.3006

Example 6.7.4.1 (Cont.)
Strength Analysis of Weld Group

Force in each weld elements:

$$R_{n_i} = \left(0.60F_{exx}\left(1.0 + 0.5\sin^{\frac{3}{2}}\theta_i\right)\left\{\frac{\Delta_i}{\Delta_{max}}\left[1.9 - 0.9\frac{\Delta_i}{\Delta_{max}}\right]\right\}^{0.3}\right)A_{e_i}$$

Ultimate Deformation:
$\Delta_u = 1.087a(\theta+6)^{-0.65}$
$\Delta_u \leq 0.17a$

Limiting Displacement:
$\Delta_{max} = 0.209a(\theta+2)^{-0.32}$

Deformation Geometry

Weld Size, a = 0.25 r_{crit} = 3.7581

Resultant Element Forces

e_x = 4.556

Element	Δ_{max}	Δ_u	Δ_u Limit	Δ_u/R	Δ_i	Δ_i/Δ_m	R_i (kips)	R_x (kips)	R_y (kips)	M_i (kips)
1	0.0149	0.0203	0.0425	0.0054	0.0203	1.3582	9.5782	6.3716	7.1515	35.9961
2	0.0155	0.0217	0.0425	0.0064	0.0197	1.2705	9.3851	6.8985	6.3632	13.6045
3	0.0163	0.0240	0.0425	0.0078	0.0197	1.2054	9.0403	7.3282	5.2939	12.3364
4	0.0177	0.0277	0.0425	0.0098	0.0208	1.1755	8.5680	7.5942	3.9672	11.2823
5	0.0136	0.0170	0.0425	0.0069	0.0113	0.8276	10.4797	9.4868	4.4525	9.9419
6	0.0140	0.0180	0.0425	0.0088	0.0098	0.6994	10.0048	8.5661	5.1690	8.1757
7	0.0148	0.0199	0.0425	0.0122	0.0087	0.5865	9.2783	7.0877	5.9877	6.5454
8	0.0164	0.0242	0.0425	0.0187	0.0083	0.5079	8.2537	4.7793	6.7292	5.1809
9	0.0218	0.0397	0.0425	0.0365	0.0115	0.5248	7.1851	1.6553	6.9918	4.3408
10	0.0218	0.0397	0.0425	0.0365	0.0115	0.5248	7.1851	−1.6553	6.9918	4.3408
11	0.0164	0.0242	0.0425	0.0187	0.0083	0.5079	8.2537	−4.7793	6.7292	5.1809
12	0.0148	0.0199	0.0425	0.0122	0.0087	0.5865	9.2783	−7.0877	5.9877	6.5454
13	0.0140	0.0180	0.0425	0.0088	0.0098	0.6994	10.0048	−8.5661	5.1690	8.1757
14	0.0136	0.0170	0.0425	0.0069	0.0113	0.8276	10.4797	−9.4868	4.4525	9.9419
15	0.0177	0.0277	0.0425	0.0098	0.0208	1.1755	8.5680	−7.5942	3.9672	11.2823
16	0.0163	0.0240	0.0425	0.0078	0.0197	1.2054	9.0403	−7.3282	5.2939	12.3364
17	0.0155	0.0217	0.0425	0.0064	0.0197	1.2705	9.3851	−6.8985	6.3632	13.6045
18	0.0149	0.0203	0.0425	0.0054	0.0203	1.3582	9.5782	−6.3716	7.1515	15.0326
								0.0000	104.21	193.84

R (based on Moment) | 32.01

Factored Load = | 24.01

The sum of the vertical component R_y does not equal R (based on moment)

Where $R = \dfrac{\sum M_i}{e_x + (R_o - \bar{x})}$

After trial and error R_o = 0.4975, and the resultant factored load is, ϕV_n = 22.7 kips

(3) *Tabulated Solution (Alternate Technique):*

From AISC LRFD, [6] Tables 8-5 to 8-12:

$\phi V_n = C(C_1)(D)(\ell)$

Example 6.7.4.1 (Cont.)
Strength Analysis of Weld Group

where:
- D = number of 16ths of weld size: a = ¼ in.; D = 4
- C = tabulated value, includes ϕ
- C_1 = based on weld electrode strength for F_{exx} = 70 ksi, C_1 = 1.0
- ℓ = 5 in.

Independent Values:

$k(\ell)$ = 2 in. $\quad\quad$ a $= \dfrac{4.556}{5} = 0.9112$

k $= \dfrac{2}{5} = 0.4$ $\quad\quad$ $a(\ell) = 4.556$

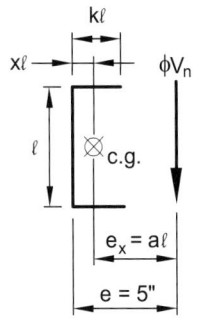

From AISC Tables (round out for conservative values):

a = 1.0, k = 0.4, Thus, C = 1.08
$\phi V_n = C(C_1)(D)(\ell)$
$\phi V_n = 1.08(1.0)(4)(5) = 21.6$ kips

Comparison of Results

Method of Analysis	Design Strength
Elastic Vector Method	13.4 kips
Instantaneous Center Numerical Solution	22.7 kips
Instantaneous Center Tabulated Solution	21.6 kips

For this example, the Instantaneous Center method provides about a 70% increase in design capacity. As long as the components are not sensitive to small rotational displacements at the design load, the higher capacity may be used.

6.8 Concrete Brackets or Corbels

There are two methods of analysis for a concrete corbel: Cantilever Beam and Strut-and-Tie. The Cantilever Beam Method is described in Chapter 11 of ACI 318-02 and the Strut-and-Tie method follows Appendix A of ACI 318-02.

6.8.1 Cantilever Beam Design Method

ACI 318-02 prescribes the design method for corbels, based on Refs. 15 and 16. The equations in this section follow those recommendations, and are subject to the following limitations (see Figures 6.8.1.1 and 6.8.1.2):

1. a/d ≤ 1
2. $N_u \le V_u$
3. Use $\phi = 0.75$ for all calculations.
4. Anchorage at the front face must be provided by welding or other positive means.
5. Concentrated loads on continuous corbels may be distributed similar to a beam ledge, Section 4.5.

The area of primary tension reinforcement, A_s, is the greater of $A_f + A_n$ or $\dfrac{2}{3}A_{vf} + A_n$ as calculated in Section 4.3.6:

$$A_f = \dfrac{V_u a + N_u(h-d)}{\phi f_y d} \quad \text{(Eq. 6.8.1.1)}$$

$$A_n = \dfrac{N_u}{\phi f_y} \quad \text{(Eq. 6.8.1.2)}$$

For convenience, the equations can be rewritten so that A_s is the greater of Eq. 6.8.1.3 and 6.8.1.4:

$$A_s = \frac{1}{\phi f_y}\left[V_u\left(\frac{a}{d}\right) + N_u\left(\frac{h}{d}\right)\right] \quad \text{(Eq. 6.8.1.3)}$$

$$A_s = \frac{1}{\phi f_y}\left[\frac{2V_u}{3\mu_e} + N_u\right] \quad \text{(Eq. 6.8.1.4)}$$

where (see Figure 6.8.1.1):
- a = applied load eccentricity
- d = depth of tension steel
- h = height of corbel
- μ_e = effective shear-friction coefficient as defined in Chapter 4 (satisfies Section 11.7.3 of ACI 318)
- ϕ = strength reduction factor = 0.75
- f_y = yield strength of tension steel
- V_u and N_u = applied factored loads

The minimum required tension steel is:

$$A_{s,min} = 0.04\frac{f'_c}{f_y}bd \quad \text{(Eq. 6.8.1.5)}$$

where:
- b = width of corbel (in.)

Where the corbel depth is larger than required by design, a reduced depth, d, may be used provided it is used in all the design equations:

$$A_h \geq 0.5(A_s - A_n) \quad \text{(Eq. 6.8.1.6)}$$

A_h should be distributed within the upper ⅔d.

The shear strength of a corbel is limited by the maximum values given in Table 4.3.6.1.

Figure 6.16.10 tabulates values based on the Cantilever Beam Method, suitable for preliminary design.

Figure 6.8.1.1 Design of concrete corbels

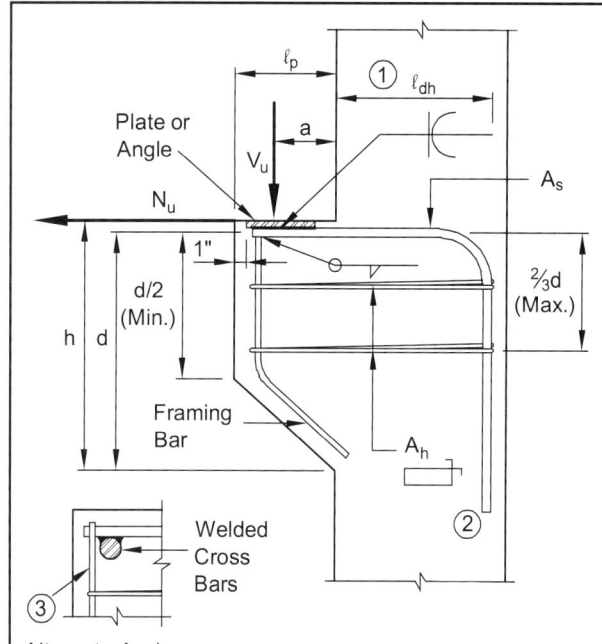

Alternate Anchorage

1. A_s bar should be extended to the far face of the column. Provide fully developed A_s bars by selecting the number of bars so that the bar size remains small enough to ensure that the length, ℓ_{dh}, is provided (Design Aid 11.2.9).
2. Vertical length of a standard 90° hook is 12d (Design Aid 11.2.9). The horizontal length of A bar, ℓ_{dh}, must be provided in order to use a standard 90° hook. It
3. may be assumed that the A_s bar is developed at the outside face of the welded anchor bar when the A_h bars are outside that point. Size of welded cross bar is same as A_s bar.

Figure 6.8.1.2 Wall corbels

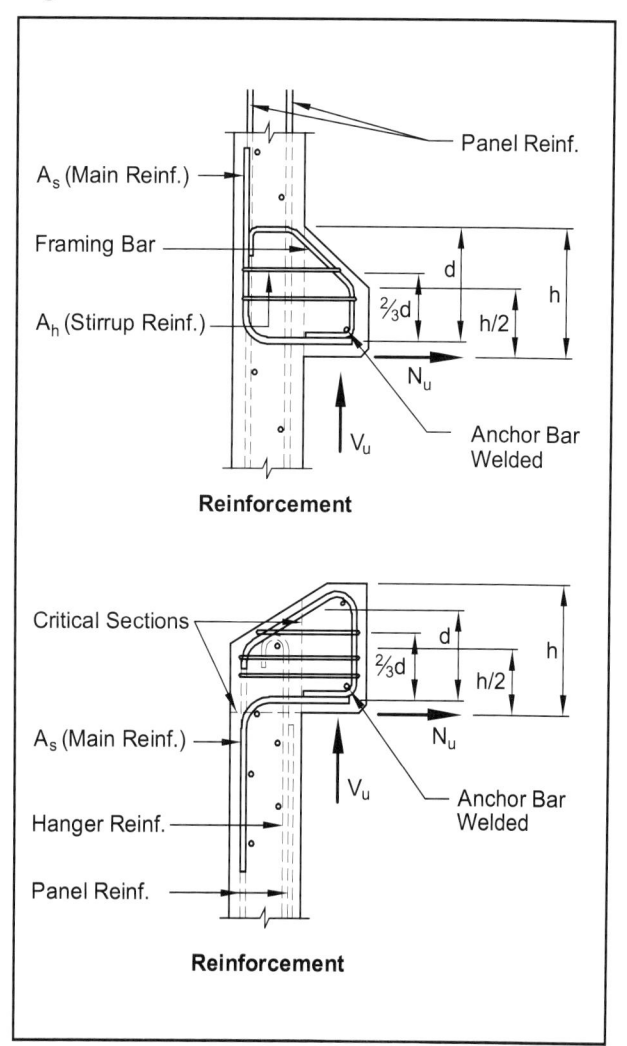

Example 6.8.1.1
Reinforced Concrete Corbel (Cantilever Beam Design Method)

Given:
A concrete corbel similar to that shown.
V_u = 80 kips (includes all loads and overload factors)
N_u = 15 kips (includes all loads and overload factors)
f_y = Grade 60 (weldable)
f'_c = 5000 psi (normal weight concrete)
Bearing pad – 12 in. x 6 in.
b = 14 in.
ℓ_p = 8 in.

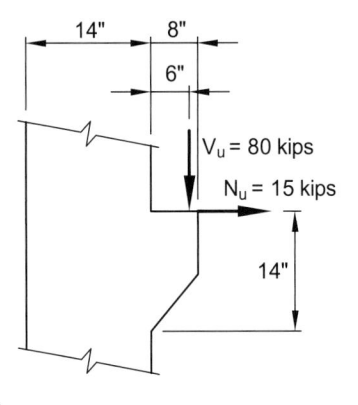

Problem:
Find the corbel depth and reinforcement by the Cantilever Beam Method.

Solution:
Try h = 14 in.
d = 13 in.
a = ¾ ℓ_p = 6 in.

From Table 4.3.6.1:

$$\max V_u = 1000\lambda^2 A_{cr} = \frac{1000(1)^2(14)(14)}{1000} = 196 \text{ kips} > 80.0 \text{ kips} \quad \text{OK}$$

By Eq. 6.8.1.3: $\quad A_s = \frac{1}{\phi f_y}\left[V_u\left(\frac{a}{d}\right) + N_u\left(\frac{h}{d}\right)\right] = \frac{1}{0.75(60)}\left[80\left(\frac{6}{13}\right) + 15\left(\frac{14}{13}\right)\right] = 1.18 \text{ in.}^2$

By Eq. 4.3.6.2: $\quad \mu_e = \frac{1000\lambda b h \mu}{V_u} = \frac{1000(1)(14)(14)(1.4)}{80,000} = 3.43 > 3.4, \text{ Use } 3.4$

By Eq. 6.8.1.4: $\quad A_s = \frac{1}{\phi f_y}\left[\frac{2V_u}{3\mu_e} + N_u\right] = \frac{1}{0.75(60)}\left[\frac{2(80)}{3(3.4)} + 15\right] = 0.68 \text{ in.}^2 < 1.18 \text{ in.}^2$

By Eq. 6.8.1.5: $\quad A_{s,\min} = 0.04 bd\left(\frac{f'_c}{f_y}\right) = 0.04(14)(13)\left(\frac{5}{60}\right) = 0.61 \text{ in.}^2 < 1.18 \text{ in.}^2$

Provide 2 – #7 bars = 1.20 in.²
The A_s reinforcement could also be estimated from Figure 6.16.10:
For:
b = 14 in.
ℓ_p = 8 in.
h = 14 in.

The corbel would have a design strength of about 57 kips with 2 – #6 bars and 78 kips with 2 – #7 bars. Use 2 – #7 bars for V_u = 80 kips.

By Eq. 6.8.1.6: $\quad A_h = 0.5(A_s - A_n) = 0.5\left[1.18 - \frac{15}{0.75(60)}\right] = 0.42 \text{ in.}^2$

Provide 2 – #3 closed ties = 0.44 in.², distributed within the upper two-thirds of the corbel.

Figure 6.8.2.1 Strut and tie geometry

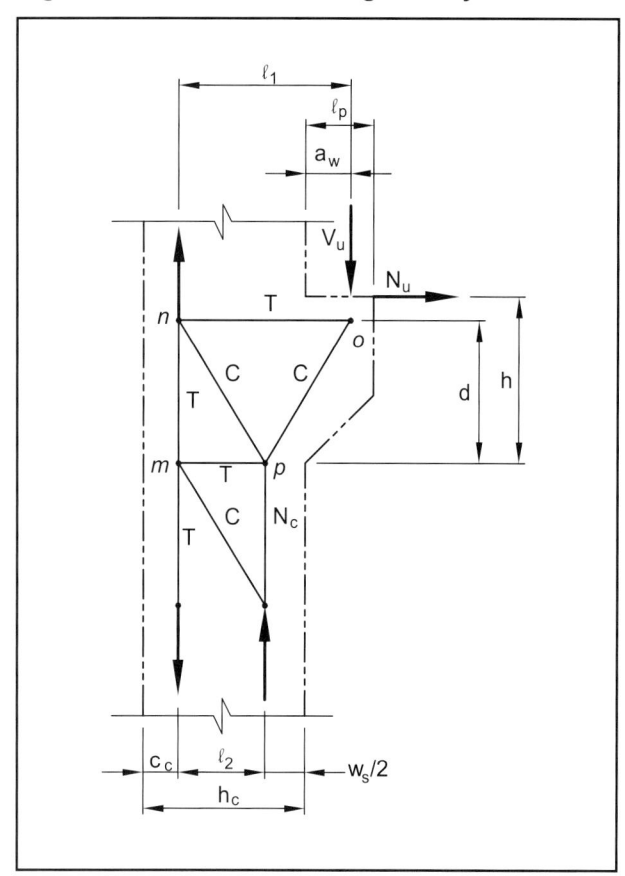

6.8.2 Strut-and-Tie Design Method

ACI 318-02, Appendix A, permits strut-and-tie, or truss modeling of "D-regions," defined as "the portion of a member within a distance equal to the member height h or depth d from a force discontinuity or a geometric discontinuity." One such member is a reinforced concrete corbel.

There are eight basic steps for the strut-and-tie method of analysis:

1. Determination of bearing plate dimension and protection for the corner against spalling.
2. Determination of truss geometry.
3. Determination of forces in the members of the truss.
4. Design of tension ties.
5. Design of nodal zone.
6. Check for compressive struts.
7. Determine area of surface reinforcement.
8. Consider detailing to ensure design technique.

Section 9.3.2.6 of ACI 318-02 requires that the strength reduction factor, ϕ, be 0.75 for "Strut and tie models, and struts, ties, nodal zones and bearing areas in such models."

6.8.2.1 Bearing Area

Area required for bearing plate:

$$A_{pl} = \frac{V_u}{\phi(0.85)f'_c} \qquad \text{(Eq. 6.8.2.1)}$$

Bearing pads can be placed away from the edge to avoid spalling.
See Section 4.6 for concrete bearing.

6.8.2.2 Truss Geometry

The truss geometry has been determined by locating the nodes. According to the flow of stresses, the corbel has a four-node (m, n, o, p) truss geometry as shown in Figure 6.8.2.1.

Node 'm' is located at the intersection of the tension reinforcement of the column and the tie at the bottom fiber of the corbel.

Node 'n' is located at the intersection of the tension bar of the column and the upper tension tie of the corbel.

Node 'o' is located at the intersection of the resultant of the applied reactions (V_u and N_u) and the upper tension tie of the corbel.

Node 'p' is located at the intersection of the longitudinal compression strut of the column and bottom tension tie. The width of the compression strut w_s, which is the governing factor to locate 'p', is determined by moment equilibrium about 'm'.

The maximum compression stress at a nodal zone is given in ACI 318-02, Section A 5.2, as:

$$f_{cu} = 0.85\beta_n f'_c \qquad \text{(Eq. 6.8.2.2)}$$

where:
β_n = 1.0 in nodal zones bounded by structural or bearing areas
= 0.8 in nodal zones anchoring one tie (e.g., 'p' in Figure 6.8.2.1)
= 0.6 in nodal zones anchoring two or more ties (e.g., 'm' in Figure 6.8.2.1)

6.8.3 Comparison of Corbel Design Methods

The following general observations can be made regarding the two different design methods (see Figure 6.8.3.1):

1. The primary reinforcement required for the strut-and-tie design method is typically greater than the cantilever beam method; however, additional ties are not required since the

moment of the applied load in the strut-and-tie method is effectively taken about the center of compressive force in the column; while in the cantilever beam method, it is taken about the face of the column. It should be noted that the cantilever beam method has been verified by tests. [15]

2. By using the lower value of β_s in Step 6, no transverse surface reinforcement is therefore required.
3. The strut-and-tie method requires a tension tie at the bottom of the corbel which is not a requirement of the cantilever beam method. Note that this bar must be anchored.

Example 6.8.2.1
Reinforced Concrete Corbel (Strut-and-Tie Design Method)

Given:
Same as Example 6.8.1.1.

Problem:
Find corbel depth and reinforcement by the strut-and-tie design method.

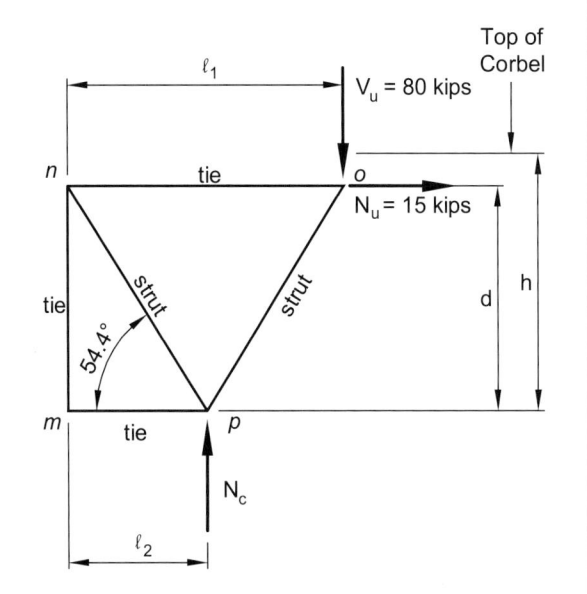

Solution:
Note: ACI 318-02, Section 9.3.2.6, requires a capacity reduction factor, ϕ, of 0.75 for analysis of the model and design of struts, ties and nodal zones.

Step 1:

Check the bearing area:
Required plate area:

$$= \frac{V_u}{\phi(0.85)f'_c} = \frac{80}{0.75(0.85)(5)} = 25.1 \text{ in.}^2$$

Use 12 in. by 6 in. plate, area = 72 in.^{2} > 25.1 in.^{2}

Step 2:

Refer to Figure 6.8.2.1:
To determine compressive force, N_c, at node 'p':

$$\sum M_m = 0$$

$$V_u(\ell_1) + N_u(d) - N_c(\ell_2) = 0 \quad \text{(Eq. 1)}$$

$$\tan \theta_R = \frac{N_u}{V_u} = \frac{15}{80} = 0.19$$

$$\ell_1 = (h-d)\tan\theta_R + a_w + (h_c - c_c) = 1(0.19) + 6 + (14 - 2.25) = 17.94 \text{ in.}$$

$$\ell_2 = h_c - c_c - \frac{w_s}{2} = 14 - 2.25 - \frac{w_s}{2} = 11.75 - \frac{w_s}{2}$$

Example 6.8.2.1 (Cont.)
Reinforced Concrete Corbel (Strut-and-Tie Design Method)

Substituting ℓ_1 and ℓ_2 into Eq. 1: $80(17.94) + 15(13) - N_c\left(11.75 - \dfrac{w_s}{2}\right) = 0$

$1630 - N_c\left(11.75 - \dfrac{w_s}{2}\right) = 0$ (Eq. 2)

Maximum compressive stress at the nodal zone p (anchors one tie, $\beta_n = 0.8$):
  f_{cu} = $0.85\beta_n f'_c = 0.85(0.8)(5) = 3.4$ ksi
  A_n = area of the nodal zone = $bw_s = 14w_s$

Maximum allowable force on strut:
  $N_c = \phi f_{cu} A_n = 0.75(3.4)(14w_s)$

  $w_s = \dfrac{N_c}{0.75(3.4)(14)} = 0.028 N_c$

Substituting in Eq. 2 and rearranging:
  $0.014 N_c^2 - 11.75 N_c + 1630 = 0$

  N_c = 175 kips

  w_s = $0.028(175) = 4.90$ in.

  $\ell_2 = 11.75 - \dfrac{4.90}{2} = 9.30$ in.

With w_s determined, the node 'p' has been located, and the entire truss geometry has been fixed.

Step 3:

Solving the truss '*mnop*' by statics, the member forces are as follows:
  Strut *op* = 96.0 kips compression
  Tie *no* = 68.2 kips tension
  Strut *np* = 116.8 kips compression
  Tie *mp* = 14.9 kips tension
  Tie *mn* = 95.0 kips tension

Step 4:

Reinforcement requirements
  For top tension tie '*no*'

  Required $A_s = \dfrac{F_{nt}}{\phi f_y} = \dfrac{68.2}{0.75(60)} = 1.52$ in.^{2}

  Provide 2 – #8 = 1.58 in.^{2} at the top

Example 6.8.2.1 (Cont.)
Reinforced Concrete Corbel (Strut-and-Tie Design Method)

For bottom tension tie '*mp*':

$$A_s = \frac{14.9}{0.75(60)} = 0.33 \text{ in.}^2$$

Step 5:

The width "w_s" of the nodal zone '*p*' has been chosen in Step 2 to satisfy the stress limit on this zone. The stress at nodal zone '*o*' must be checked against the compressive force in strut '*op*' and the applied reaction, V_u. From the compressive stress flow in struts of the corbel, Figure 6.8.2.1, it is obvious that the nodal zone '*p*' is under the maximum compressive stress due to force N_c. As it is within the acceptable limit, all nodal zones are acceptable.

Step 6:

Strut '*np*' is the most critical strut at node '*p*'. The nominal compressive strength of a strut without compressive reinforcement is:

$$F_{ns} = f_{cu} A_c$$

where A_c = width of corbel × width of strut

$$\text{Width of strut } np = \frac{w_s}{\sin 54.4°} = \frac{4.90}{\sin 54.4°} = 6.03 \text{ in.}$$

From ACI 318-02, Section A.3.2:
$f_{cu} = 0.85\beta_s f'_c$ for a "bottle shaped strut" (see A.3.2.2).

Note: The intermediate ties provided by the shear-friction method do not meet the requirements of A.3.3.2 for surface reinforcement, therefore:

$\beta_s = 0.60\lambda$
$f_{cu} = 0.85(0.60)(1.0)(5) = 2.55 \text{ ksi}$
$\phi F_{ns} = (0.75)(2.55)(14)(6.03) = 161.5 \text{ kips} > 116.8 \text{ kips}$

Step 7:
Since the lowest value of β_s was used, surface reinforcement is not required based on ACI 318 Appendix D.

Figure 6.8.3.1 Comparison of Corbel Design Methods

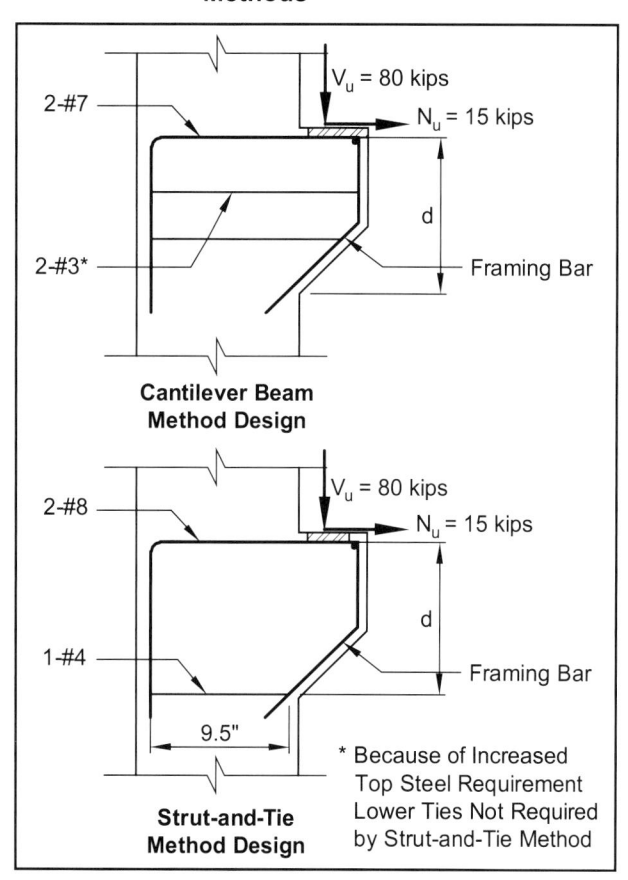

Cantilever Beam Method Design
- 2-#7
- 2-#3*
- V_u = 80 kips
- N_u = 15 kips
- d
- Framing Bar

Strut-and-Tie Method Design
- 2-#8
- 1-#4
- 9.5"
- V_u = 80 kips
- N_u = 15 kips
- d
- Framing Bar

* Because of Increased Top Steel Requirement Lower Ties Not Required by Strut-and-Tie Method

6.8.4 Development of Corbel Reinforcement

The development length of each reinforcing bar is based on ACI 318. Tension development lengths are measured from the assumed failure plane. Typically, concrete corbels are a part of components (columns or walls) that do not have room to develop the bar by embedment, so hooked bars are used. As shown in Figures 6.8.1.1 and 6.8.1.2, the main tension steel within the concrete corbel is usually developed at the outer edge of the corbel by welding to a cross bar or plate.

6.9 Structural Steel Corbels

Structural steel shapes, such as wide flange beams, double channels, tubes or vertical plates, often serve as haunches or brackets as illustrated in Figure 6.9.1. The concrete-based capacity of these members can be calculated by statics, using the assumptions shown in Figures 6.9.2 and 6.9.3. [17]

The nominal strength of the section is:

$$V_c = \frac{0.85 f'_c b \ell_e}{1 + 3.6 e / \ell_e} \quad \text{(Eq. 6.9.1)}$$

where:
  V_c = nominal strength of section controlled by concrete, lb

Figure 6.9.1 Structural steel corbels

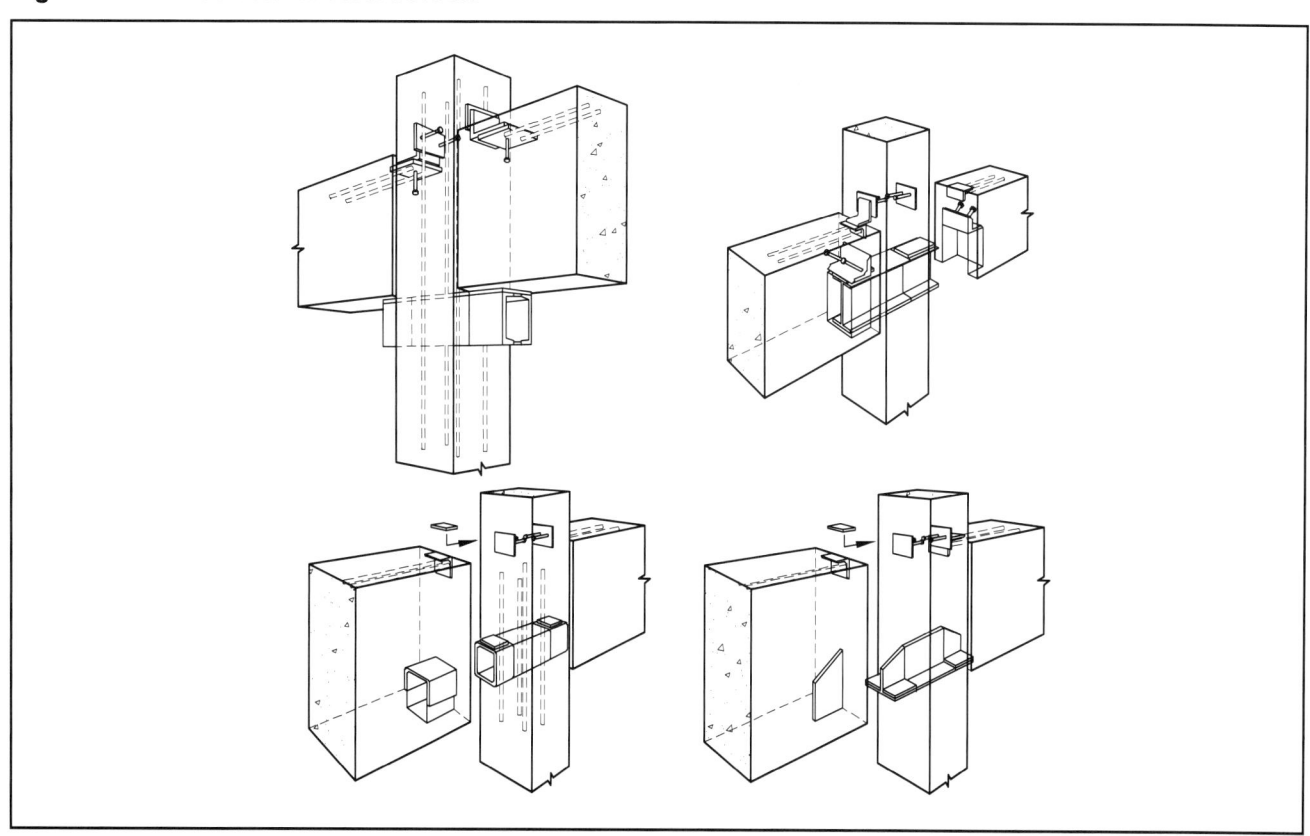

Figure 6.9.2 Stress-strain relationships

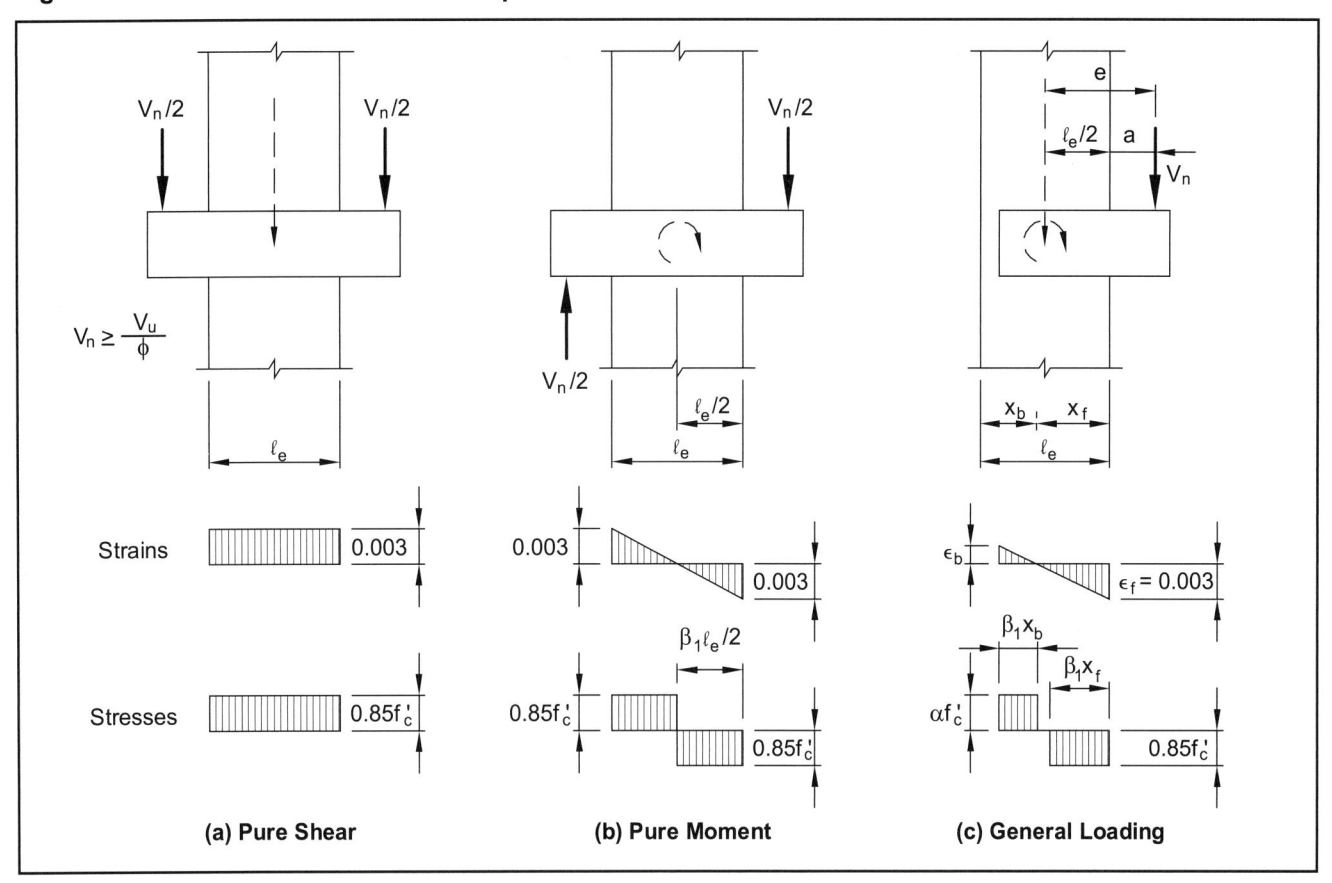

(a) Pure Shear

(b) Pure Moment

(c) General Loading

Figure 6.9.3 Assumptions and notations – steel haunch design

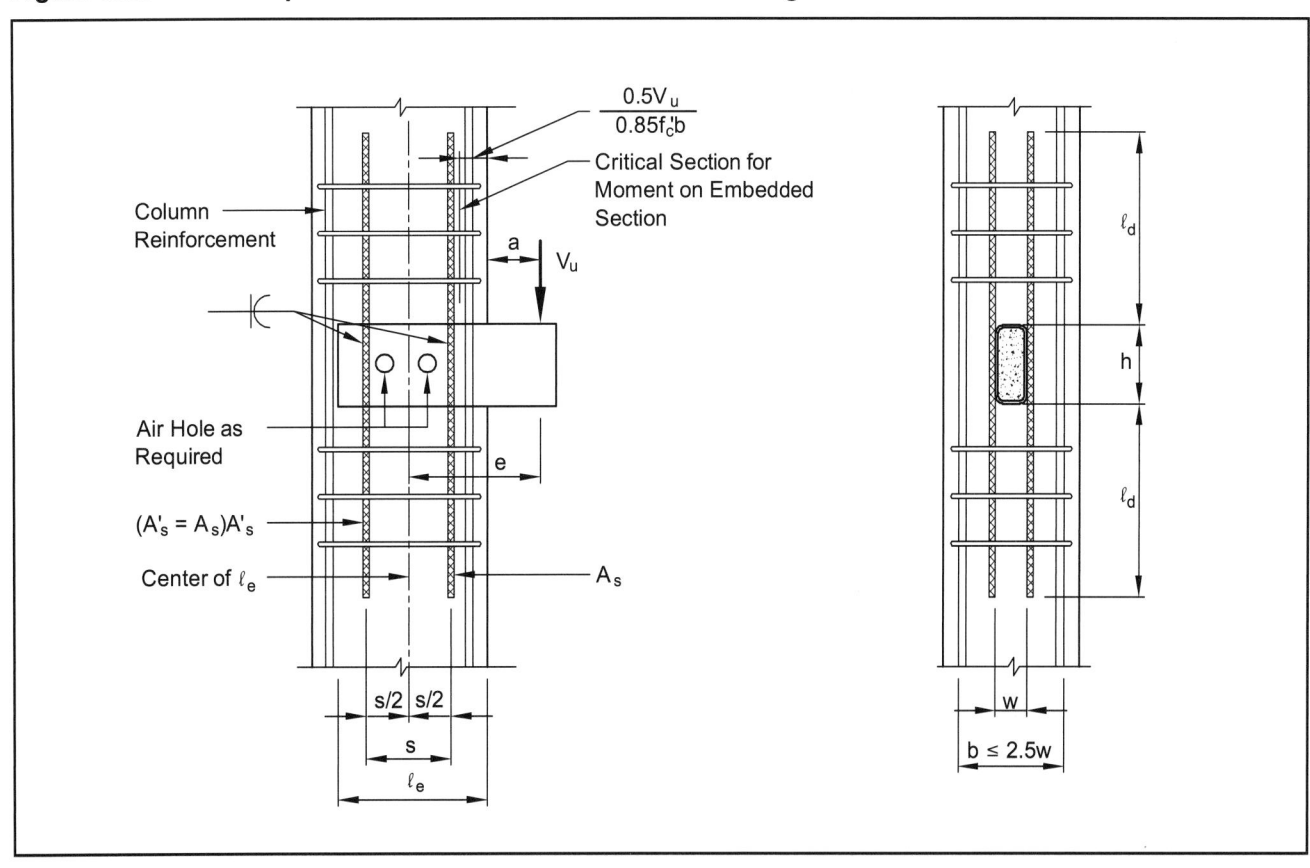

$e = a + \dfrac{\ell_e}{2}$, in.

a = shear span, in.
ℓ_e = embedment length, in.
b = effective width of compression block, in.

For the additional contribution of reinforcement welded to the embedded shape, and properly developed in the concrete, and with $A'_s = A_s$:

$$V_r = \dfrac{2A_s f_y}{1 + \dfrac{6e/\ell_e}{(4.8s/\ell_e)-1}} \quad \text{(Eq. 6.9.2)}$$

where:
f_y = yield strength of reinforcement

Then $V_n = (V_c + V_r)$; and $V_u \le \phi V_n$, where $\phi = 0.75$. Other notation for Eqs. 6.9.1 and 6.9.2 are shown in Figure 6.9.3.

The following assumptions and limitations are recommended:

1. In a column with closely spaced ties (spacing ≈ 3 in.) above and below the haunch, the effective width, b, can be assumed as the width of the confined region, i.e., outside to outside of ties, or 2.5 times the width of the steel section, whichever is less.
2. Thin-walled members, such as the tube shown in Figure 6.9.3, may require filling with concrete to prevent local buckling.
3. When the supplemental reinforcement, A_s and A'_s, is anchored both above and below the members, as in Figure 6.9.3, it can be counted twice, assuming adequate weld for the total force.
4. The critical section for bending of the steel member is located a distance $V_u/(0.85 f'_c b)$ inward from the face of the column.

If the steel section projects from both sides, as in Figure 6.9.2, the eccentricity factor, e/ℓ_e, in Eq. 6.9.1 should be calculated from the total unbalanced live load. Conservatively, e/ℓ_e may be taken equal to 0.5.

The design strength of the steel section can be determined by:

Flexural design strength:

$$\phi V_n = \dfrac{\phi Z_s F_y}{a + 0.5 V_u/(0.85 f'_c b)} \quad \text{(Eq. 6.9.3)}$$

Shear design strength:

$$\phi V_n = \phi(0.6 F_y h t) \quad \text{(Eq. 6.9.4)}$$

where:
Z_s = plastic section modulus of steel section (see Design Aid 11.5.2, Chapter 11)
F_y = yield strength of the steel
h, t = depth and thickness of steel web
ϕ = 0.90

The horizontal forces, N_u, are resisted by bond on the perimeter of the embedded section. If the bond stress resulting from factored loads exceeds 250 psi, then headed studs or reinforcing bars should be welded to the embedded steel section to ensure sufficient load transfer.

Example 6.9.1
Structural Steel Corbel

Given:
The structural steel corbel shown at right.

f'_c = 5000 psi

f_y (reinforcement) = 60,000 psi
(weldable)

F_y (structural steel) = 46,000 psi

Problem:
Find the design strength.

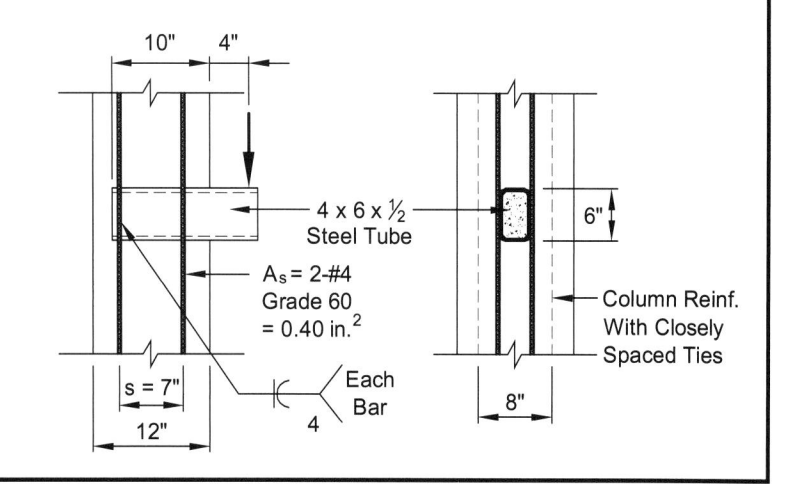

4 x 6 x ½ Steel Tube

A_s = 2-#4 Grade 60 = 0.40 in.²
Each Bar

Column Reinf. With Closely Spaced Ties

s = 7", 12", 10", 4", 6", 8"

Example 6.9.1 (Cont.)
Structural Steel Corbel

Solution:

Effective width, b = confined width (8 in.) or

    b = 2.5w = 2.5(4) = 10 in.

Use b = 8 in.

    e = 4 + 10/2 = 9 in.

$$V_c = \frac{0.85 f'_c b \ell_e}{1 + 3.6 e / \ell_e} = \frac{0.85(5)(8)(10)}{1 + 3.6(9)/(10)} = 80.2 \text{ kips}$$

Since the A_s bars are anchored above and below, they can be counted twice.

    A_s = 2 – #4 = 2(2)(0.2) = 0.80 in.²

$$V_r = \frac{2 A_s f_y}{1 + \dfrac{6 e / \ell_e}{(4.8 s / \ell_e) - 1}} = \frac{2(0.80)(60)}{1 + \dfrac{6(9)/(10)}{[4.8(7)/(10)] - 1}} = 29.2 \text{ kips}$$

    ϕV_n = 0.75(80.2 + 29.2) = 82.0 kips

Alternatively, using Figures 6.16.11 and 6.16.12:

    For b = 8 in.; a = 4 in.; ℓ_e = 10 in.
    Read ϕV_c = 60 kips

For A_s = 2 – #4, anchored above and below: V_r = 4(7) = 28 kips
    ϕV_n = 60 + 0.75(28) = 81 kips

Steel section flexure capacity:

From AISC LRFD Manual: [6]

    Z_s = 15.4

Assume V_u = 85 kips.

$$\phi V_n = \frac{\phi Z_s F_y}{a + 0.5 V_u / (0.85 f'_c b)} = \frac{0.9(15.4)(46)}{4 + 1.25} = 121.4 \text{ kips} > 82.0 \text{ kips}$$

Since the bar must be anchored for forces above and below, twice the minimum length of weld from Figure 6.16.3 is required: 2(2) = 4 in. each bar.

Figure 6.10.1 Hanger Connections

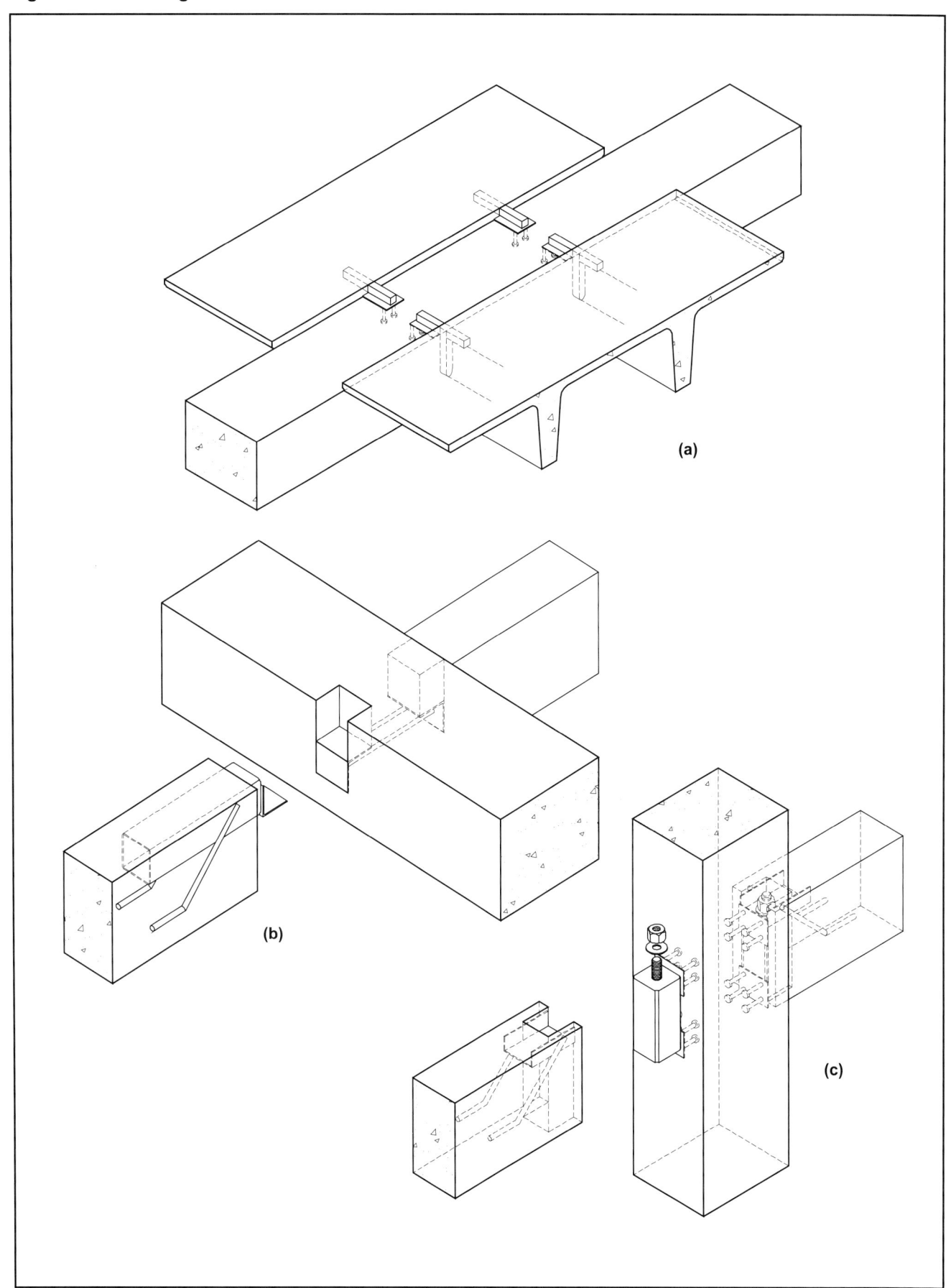

Figure 6.10.1.1 Cazaly hanger

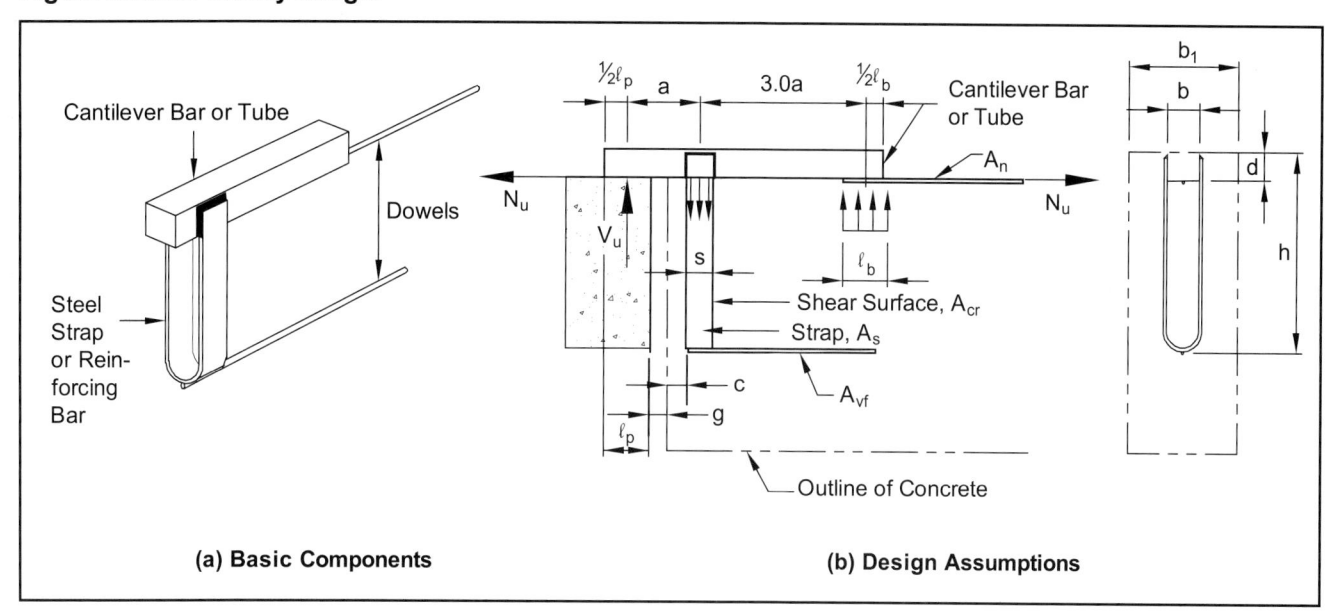

6.10 Hanger Connections

Hangers are similar to dapped ends, except that the extended or bearing end is steel instead of concrete. They are used when it is desired to keep the structural depth very shallow. Examples are shown in Figure 6.10.1. These connections typically have short bearings and may be particularly sensitive to tolerances and volume change movements. The need for accuracy in dimensioning and installation must be emphasized.

6.10.1 Cazaly Hanger

The Cazaly hanger [18] has three basic components [see Figure 6.10.1.1(a)]. Design assumptions are as follows [see Figure 6.10.1.1(b)]:

1. The cantilevered bar is usually proportioned so that the interior reaction from the concrete is $0.33V_u$. The hanger strap should then be proportioned to yield under a tension of $1.33V_u$:

$$A_s = \frac{1.33V_u}{\phi F_y}$$
(Eq. 6.10.1.1)

where:
F_y = yield strength of strap material
ϕ = 0.90

2. V_u may be assumed to be applied $\ell_p/2$ from the face of the seat. The remaining part of the moment arm is the width of the joint, g, and the cover, c, from the end of the member to the edge of the strap. Since moment is sensitive to this dimension, it is important that this dimension be kept as small as feasible and the value used in analysis is not exceeded in the field. Most hangers in practice have exterior cantilever lengths, $(\ell_p + g + c)$, of 3 to 4 in.

3. The moment in the cantilevered bar is then given by:

$$M_u = V_u a = V_u (0.5\ell_p + g + c + 0.5s)$$
(Eq. 6.10.1.2)

where:
ℓ_p = bearing length of exterior cantilever
$a = 0.5\ell_p + g + c + 0.5s$

Other notation is shown in Figure 6.10.1.1(b).

The bar should be proportioned to carry this moment in combination with shear and tensile forces. Alternatively, if the bar is proportioned to take this moment at the yield stress, but using elastic section properties (i.e., $M_u = \phi F_y \frac{bd^2}{6}$), the shear and tensile forces can usually be neglected.

4. The conservative and simplifying assumption that strap weld forces are concentrated at the strap centerline is implicit in the 0.5s factor in Eq. 6.10.1.2.

5. The bearing pressure creating the interior reaction may be calculated as in Section 4.6.1. Conservatively, if the width of the member in which the hanger is cast equals b_1; then:

$$f_{bu} = 0.85\phi f'_c \sqrt{b_1/b} \le 1.1 f'_c \quad \text{(Eq. 6.10.1.3)}$$

where:
$\quad \phi = 0.65$

The bearing length, ℓ_b, is then given by:

$$\ell_b = \frac{V_u/3}{b f_{bu}} \quad \text{(Eq. 6.10.1.4)}$$

6. To maintain the conditions of equilibrium assumed, the interior cantilever must have a length:

$$3.0a = (1.5\ell_p + 3.0g + 3.0c + 1.5s)$$

7. The minimum total length of bar is then:

$$0.5\ell_p + a + 3.0a + 0.5\ell_b$$
$$= (2.5\ell_p + 4.0g + 4.0c + 2.0s + 0.5\ell_b) \text{ in.}$$
$$\text{(Eq. 6.10.1.5)}$$

8. Longitudinal dowels, A_n, are welded to the cantilevered bar to transmit the axial force, N_u:

$$A_n = \frac{N_u}{\phi f_y} \quad \text{(Eq. 6.10.1.6)}$$

where:
$\quad f_y$ = yield strength of dowel
$\quad \phi = 0.90$

9. The lower dowel area, A_{vf}, can be proportioned using effective shear-friction described in Section 4.3.6:

$$A_{vf} = \frac{1.33 V_u}{\phi f_y \mu_e} \quad \text{(Eq. 6.10.1.7)}$$

where:
$\quad \phi = 0.75$
$\quad f_y$ = yield strength of lower dowels

$$\mu_e = \frac{1000 \lambda b_1 h \mu}{V_u} \le \text{values in Table 4.3.6.1}$$
$$\text{(Eq. 6.10.1.8)}$$

The nominal shear strength is limited by the values in Table 4.3.6.1.

Example 6.10.1
Design of a Cazaly Hanger

Given:
Hanger is similar to that shown in Figures 6.10.1(a) and 6.10.1.1 (not exposed to earth or weather).

$\quad f'_c$ = 5000 psi (both member and support)
$\quad f_y$ (reinforcing bars) = 60 ksi
$\quad f_y$ (structural steel straps) = 36 ksi
$\quad f_y$ (tubes) = 46 ksi
$\quad V_u$ = 36 kips
$\quad\quad$ (includes all load and overload factors)
$\quad N_u$ = 5 kips
$\quad\quad$ (includes all load and overload factors)
$\quad b_1$ = 6 in.
$\quad c$ = ⅝ in. (minimum cover)
$\quad g$ = 1 in.
$\quad \ell_p$ = 4 in.

Problem:
Size the hanger components.

Solution:
By Eq. 6.10.1.1:

$$A_s(\text{strap}) = \frac{1.33 V_u}{\phi F_y} = \frac{1.33(36)}{0.9(36)} = 1.48 \text{ in.}^2$$

Use ⅜ x 2 in. strap; A_s = 0.375(2)(2) = 1.50 in.^{2}

Use ⁵⁄₁₆ in. fillet weld

Figure 6.16.2 E70 electrode:

$$\ell_w = \frac{1.33(36)}{2(6.96)} = 3.44 \text{ in.}$$

Weld 2 in. across top, 1 in. down sides = 4.0 in.
$\quad a$ = 0.5 (strip width) + $g + c + 0.5\ell_p$
$\quad\quad$ = 0.5 (2) + 1 + ⅝ + 0.5 (4) = 4.625 in.

By Eq. 6.10.1.2:

$$M_u = V_u a = 36(4.625) = 166.5 \text{ kip-in.}$$

$$Z_{req'd} = \frac{M_u}{\phi F_y} = \frac{166.5}{0.9(46)} = 4.02 \text{ in.}^3$$

**Example 6.10.1 (Cont.)
Design of a Cazaly Hanger**

Try structural tube
HSS 4 x 4 x ¼
$Z = 4.97$ in.^{3} OK

Min. interior cantilever = $3a = 3(4.625)$
= 13.875 in.

By Eqs. 6.10.1.3 and 6.10.1.4:

$$f_{bu} = 0.85\phi f'_c \sqrt{\frac{b_1}{b}} = 0.85(0.65)(5)\sqrt{\frac{6}{4}}$$

$$= 3.38 \text{ ksi}$$

$$\ell_b = \frac{V_u/3}{f_{bu}(b)} = \frac{36/3}{3.38(4)} = 0.89 \text{ in.}$$

Min. total length (Eq. 6.10.1.5):

$$= 0.5(2) + 4.625 + 3(4.625) + 0.5(0.89)$$

$$= 19.95 \text{ in.}$$

Use HSS 4 x 4 x ¼ x 20 in. long

By Eq. 6.10.1.6:
$$A_n = \frac{N_u}{\phi f_y} = \frac{5}{0.9(60)} = 0.09 \text{ in.}^2$$

Use 1 – #3 dowel

Try $h = 16$ in.; by Eqs. 4.3.6.2 and 6.10.1.7:

$$\mu_e = \frac{1000\lambda b_1 h \mu}{V_u}$$

$$= \frac{1000(1.0)(6)(16)(1.4)(1.0)}{24,000}$$

$$= 5.6 > 3.4$$

Use $\mu_e = 3.4$

$$A_{vf} = \frac{1.33 V_u}{\phi f_y \mu_e} = \frac{1.33(36)}{0.75(60)(3.4)} = 0.31 \text{ in.}^2$$

Use 1 – #5 dowel. $A_{vf} = 0.31$ in.^{2} OK

Also check welding requirements.

6.10.2 Loov Hanger

The hanger [19] illustrated in Figure 6.10.2.1 is designed using the following equations:

$$A_{sh} = \frac{V_u}{\phi f_y \cos\alpha} \quad \text{(Eq. 6.10.2.1)}$$

where:
$\phi = 0.75$
f_y = yield strength of A_{sh}

$$A_n = \frac{N_u}{\phi f_y}\left(1 + \frac{h-d}{d-a/2}\right) \quad \text{(Eq. 6.10.2.2)}$$

where:
$\phi = 0.90$
f_y = yield strength of A_n

The steel bar or tube is proportioned so that the bearing strength of the concrete is not exceeded, and to provide sufficient weld length to develop the diagonal bars. Bearing strength is discussed in Section 4.6.1.

$$f_{bu} = 0.85\phi f'_c = 0.55 f'_c \quad \text{(Eq. 6.10.2.3)}$$

where:
$\phi = 0.65$

The connection should be detailed so that the reaction, the center of compression and the center of the diagonal bars meet at a common point, as shown in Figure 6.10.2.1. The compressive force, C_u, is assumed to act at a distance $a/2$ from the top of the bearing plate. Thus:

$$a = \frac{C_u}{b f_{bu}} \quad \text{(Eq. 6.10.2.4)}$$

where:

$$C_u = V_u \tan\alpha + \frac{N_u(h-d)}{d-a/2} \quad \text{(Eq. 6.10.2.5)}$$

For most designs, the horizontal reinforcement, A_n, is placed very close to the bottom of the steel bar. Thus, the term $(h - d)$ can be assumed as equal to zero, simplifying Eqs. 6.10.2.2 and 6.10.2.5.

Tests have indicated a weakness in shear in the vicinity of the hangers, so it is recommended that stirrups in the beam end be designed to carry the total shear.

6.11 Bearing Pads

Bearing pads are used to distribute concentrated loads and reactions over the bearing area and to allow limited horizontal and rotational movements to provide stress relief. Their use has proven beneficial and often may be necessary for satisfactory performance of precast concrete structures.

Several materials are commonly used for bearing pads:

1. AASHTO – grade chloroprene pads are made with 100% chloroprene (neoprene) as the only elastomer and conform to the requirements of the AASHTO Standard Specifications for Highway Bridges (1996), Section 18. Inert fillers are used with the chloroprene and the resulting pad is black in color and of a smooth uniform texture. While allowable compressive stresses are somewhat lower than other pad types, these pads allow the greatest freedom in movement at the bearing. Note: chloroprene pads which do not meet the AASHTO Specifications are not recommended for use in precast concrete structures.

2. Pads reinforced with randomly oriented fibers have been used successfully for many years. These pads are usually black, and the short reinforcing fibers are clearly visible. Vertical load capacity is increased by the reinforcement, but the capability of rotations and horizontal movement is somewhat less than chloroprene pads. Some random oriented fiber pads possess different properties in different directions in the plane of the pad. Therefore, unless proper planning and care is used in their installation, it may be prudent to specify those pads that have been tested to exhibit similar properties in different directions. No national standard specifications for this material exist. Manufacturers have developed appropriate design and performance documentation.

3. Cotton duck fabric reinforced pads are generally used where a higher compressive strength is desired. These pads are often yellow-orange in color and are reinforced with closely spaced, horizontal layers of fabric, bonded in the

Figure 6.10.2.1 Loov hanger

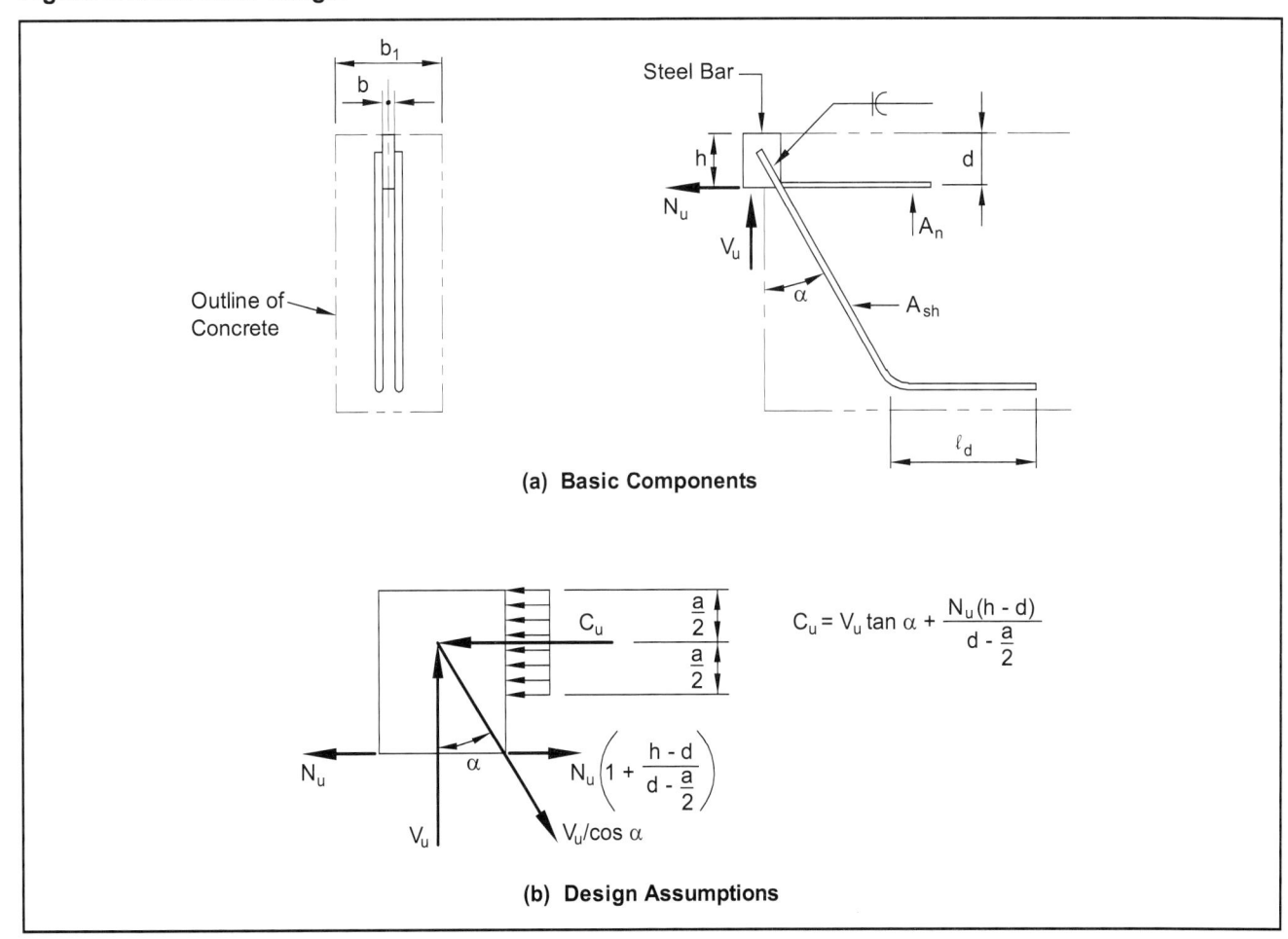

(a) Basic Components

$$C_u = V_u \tan \alpha + \frac{N_u(h - d)}{d - \frac{a}{2}}$$

(b) Design Assumptions

Example 6.10.2
Design of a Loov Hanger

Given:
    Hanger is similar to that shown in Figure 6.10.2.1.
    f'_c = 5000 psi (both member and support)
    f_y (reinforcing bars) = 60 ksi
    f_y (structural steel) = 36 ksi
    V_u = 24 kips (includes all load and overload factors)
    N_u = 4 kips (includes all load and overload factors)
    b_1 = 6 in.
    α = 30 deg

Problem:
    Size the hanger connections.

Solution:

$$A_{sh} = \frac{V_u}{\phi f_y \cos\alpha} = \frac{24}{0.75(60)\cos 30°} = 0.62 \text{ in.}^2$$

Use 2 – #5 bars = 0.62 in.² say OK
    Minimum weld length, #5 bar, E70 electrode (Figure 6.16.3) = 2½ in.
    Detail A_n so it is near the bottom of the steel bar.

$$h - d \approx 0$$

$$A_n = \frac{N_u}{\phi f_y} = \frac{4}{0.9(60)} = 0.07 \text{ in.}^2$$

Provide end bearing plate as shown:
    Use 1 – #3 dowel

By Eq. 6.10.2.3:
    $f_{bu} = 0.85\phi f'_c = 0.85(0.65)(5) = 2.76$ ksi

By Eq. 6.10.2.5:
    $C_u = V_u \tan\alpha = 24\tan 30° = 13.9$ kips

Assume b = 1 in.

$$a = \frac{C_u}{b f_{bu}} = \frac{13.9}{1(2.76)} = 5.04 \text{ in.}$$

a/2 = 2.52 in.

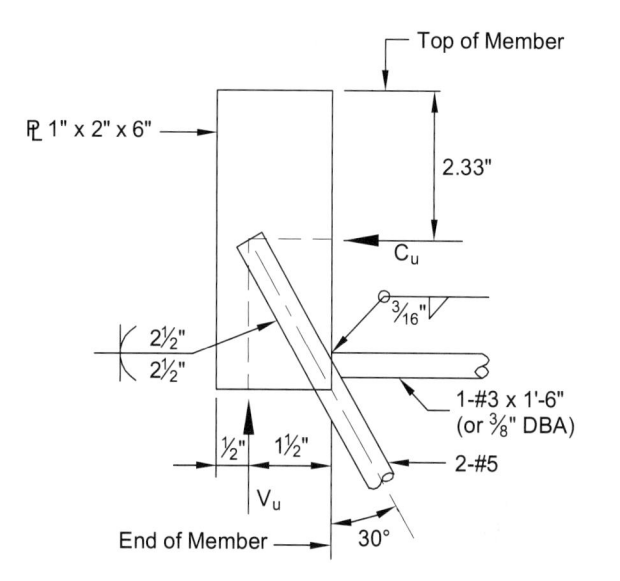

elastomer. The horizontal reinforcement layers are easily observed at the edge of the pad. Section 18.10.2 of the AASHTO Standard Specifications for Bridges and Military Specification MIL-C-882D discuss this material.

4. Chloroprene pads laminated with alternate layers of bonded steel or fiberglass are often used in bridges, but seldom in building construction. The above mentioned AASHTO Specifications cover these pads.
5. A multimonomer plastic bearing strip is manufactured expressly for bearing purposes. It is a commonly used material for the bearing support of hollow-core slabs, and is highly suitable for this application. It is also often used for bearing of architectural precast concrete cladding panels.
6. Tempered hardboard strips are also used with hollow-core slabs. These pads should be used with caution under moist conditions. In addition to progressive deterioration of the pad, staining of the precast concrete unit may occur.
7. TFE (trade name Teflon) coated materials are often used in bearing areas when large horizontal movements are anticipated, for example at "slip" joints or expansion joints. The TFE is normally reinforced by bonding to an appropriate backing material, such as steel.

Figure 6.11.1 Typical TFE bearing pad detail

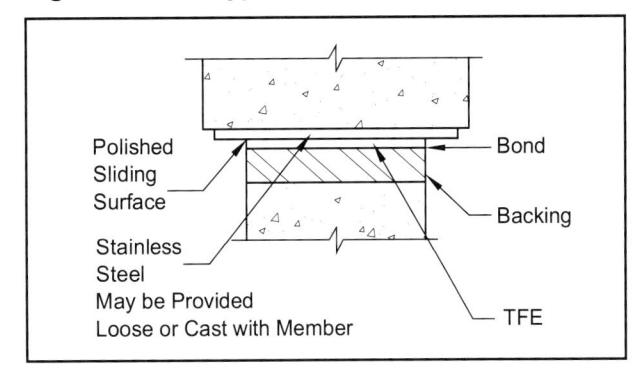

Figure 6.11.2 TFE friction coefficients

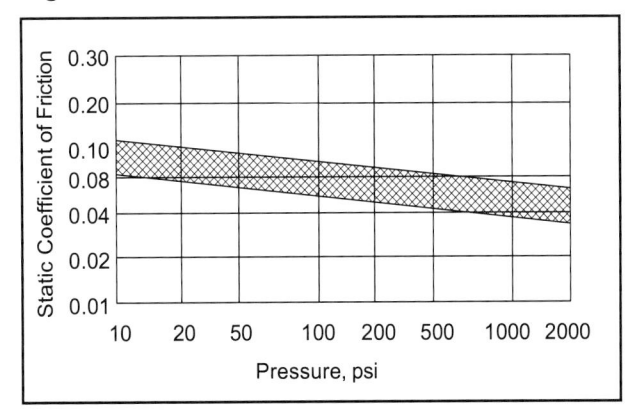

Figure 6.11.1 shows a typical bearing detail using TFE, and Figure 6.11.2 shows the range of friction coefficients that may be used for design. Typical allowable stress is about 1000 psi for virgin TFE and up to 2000 psi for filled material with reinforcing agents such as glass fibers.

6.11.1 Design Recommendations

Bearing pads provide stress relief due to a combination of slippage and pad deformation. In elastomeric bearing pads (1 through 4 above), research [20] has shown that slippage is the more significant factor. This research has also shown that the ratio of shear to compressive stress on the pad reduces significantly under slow cyclic movements, such as those produced by temperature variations. The following recommendations, along with Figures 6.11.1.1 and 6.11.1.2, can be used to select bearing pads:

1. Use unfactored service loads for design.
2. At the suggested maximum uniform compressive stress, instantaneous vertical strains of 10 to 20% can be expected. This number may double if the bearing surfaces are not parallel. In addition, the time-dependent creep will typically increase the instantaneous strains by 25 to 100%, depending on the magnitude of sustained dead load.
3. For stability of the pad, the length and width of unreinforced pads should be at least five times the thickness.
4. A minimum pad thickness of ⅜ in. is recommended for all precast members except solid slabs and hollow-core slabs.
5. Figure 6.11.1.2 may be used to estimate the shear resistance of chloroprene, random fiber reinforced and cotton duck pads.
6. The portion of pad outside of the covered bearing surface as well as the portion which is not under load because of rotation of member should be ignored in calculating shape factors, pad stresses, stability and movements. Pads should be centered under the bearing area.
7. Shape factors, S, for unreinforced pads should be greater than 2 when used under tee stems, and greater than 3 under beams.
8. The sustained dead load compressive stress on unreinforced chloroprene pads should be limited to the range of 300 to 500 psi.
9. The volume change strains shown in Section 3.4 may be significantly reduced when calculating horizontal movement, Δ, because of compensating creep and slip in the bearing pad.

Figure 6.11.1.1 Single layer bearing pads free to slip

Shape Factor $= S = \dfrac{wb}{2(w+b)t}$

D = Durometer (Shore A Hardness)

Δ = Design Horizontal Movement at End of Member

Pad Material	Allowable[a] Compressive Stress (psi)	Shore A Hardness D	Recommended Minimum Thickness[b]	Recommended Maximum Rotation[b]
Unreinforced Chloroprene or Rubber	4DS ≤ 800	50 through 70	1.4Δ	$\dfrac{0.3t}{b \text{ or } w}$
Random Fiber Reinforced Elastometric	1000 + 100S ≤ 1500	80 ± 10	1.4Δ	$\dfrac{0.3t}{b \text{ or } w}$
Cotton Duck Fabric Reinforced	≤ 2500 (uniform) ≤ 4000 (nonuniform)	90 ± 10	2.0Δ	$\dfrac{0.12t}{b \text{ or } w}$

a. Allowable compressive stresses may be increased based on test data supplied by the bearing pad manufacturer.
b. The values in the table are based on sliding criteria. If sliding is not critical or testing indicates more advantageous conditions, thinner pads may be used. The minimum thickness and maximum rotation values for the cotton duck pad account for the effects of creep.

Figure 6.11.1.2 Shear resistance of bearing pads

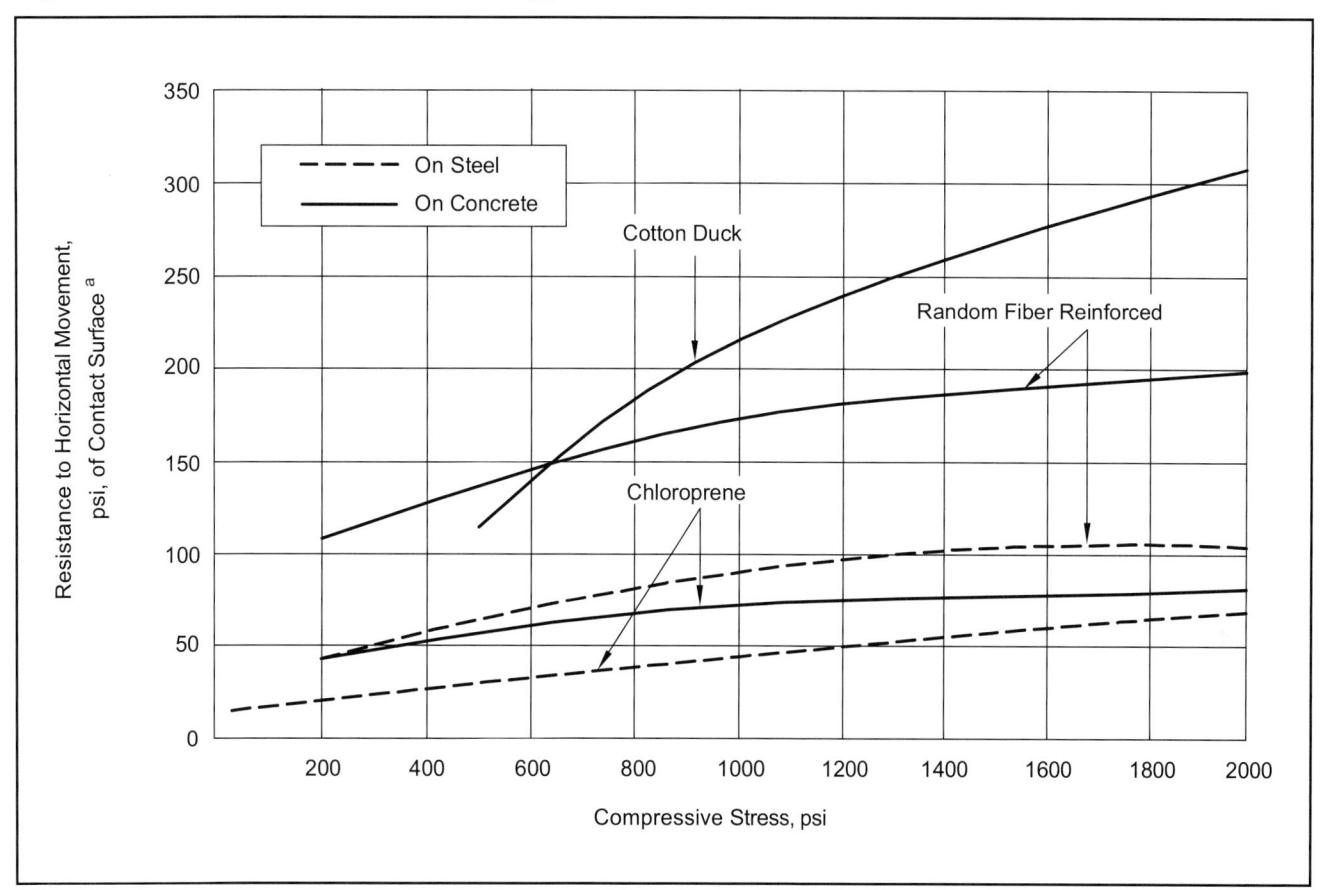

a. Average values based on tests at 70% shear and slippage strain.

10. Certain fiber reinforced bearing pads are reinforced in one direction only; for these types of pads, orientation in the field may be critical.

6.12 Column Bases

Column bases must be designed for both erection loads and loads which occur in service, the former often being more critical. Two commonly used base plate details are shown in Figure 6.12.1 although many other details are also used.

6.12.1 Base Plates

If in the analysis for erection loads or temporary construction loads, *before* grout is placed under the plate, all the anchor bolts are in compression, the base plate thickness required to satisfy bending about line Z-Z is determined from:

$$t = \sqrt{\frac{C_u(4)x}{\phi(B)F_y}} \quad \text{(Eq. 6.12.1.1)}$$

where:
- ϕ = 0.90
- x, B = dimension as shown in Figure 6.12.1
- F_y = yield strength of base plate
- C_u = total factored forces on anchor bolts

If the analysis indicates the anchor bolts on one or both sides of the column are in tension, the base plate thickness is determined by:

$$t = \sqrt{\frac{T_u(4)x}{\phi(B)F_y}} \quad \text{(Eq. 6.12.1.2)}$$

where:
- ϕ = 0.90
- x, B = dimension as shown in Figure 6.12.1
- T_u = total factored forces on anchor bolts

6.12.2 Anchor Bolts

The anchor bolt diameter is determined by the tension or compression on the stress area of the threaded portion of the bolt. Anchor bolts may be ASTM A 307 bolts or, more frequently, threaded rods of ASTM A 36 steel. ASTM F1554 may also be used. The requirements for structural integrity (Chapter 3) must also be satisfied.

In most cases, both base plate and anchor bolt stresses can be significantly reduced by using properly placed shims during erection.

When the bolts are near a free edge, as in a pier or wall, the buckling of the bolts before grouting may be a consideration. Confinement reinforcement, as shown in Figure 6.12.1, should be provided as specified in ACI 318-02, Section 7.10.5.6.

6.12.2.1 Embedment Strength of Anchor Bolts in Tension

ACI 318-2002, Appendix D, procedures for the strength of anchorages are applicable for anchor bolts in tension.

The procedure is identical to the procedure for headed studs, and the equations presented in Section 6.5.3 are valid for the design of headed anchor bolts with the following addition:

For headed anchor bolts [Figure 6.12.2.1(a) and (b)]: when 11 in. < h_{ef} < 25 in., Eq. 6.12.2.1 may be used as an alternative to Eq. 6.5.4.2:

$$C_{bs} = 2.22 \frac{\sqrt{f'_c}}{\sqrt[3]{h_{ef}}} \quad \text{(Eq. 6.12.2.1)}$$

For hooked anchor bolts [Figure 6.12.2.1(c)], the pullout strength may not exceed:

$$N_p = 1.26 f'_c e_h d_o C_{crp} \quad \text{(Eq. 6.12.2.2)}$$

where:
- e_h = hook projection $\geq 3d_o$
- d_o = bolt diameter
- C_{crp} = cracking factor (Section 6.5.4.2)

Figure 6.12.2.1 Typical anchor bolts

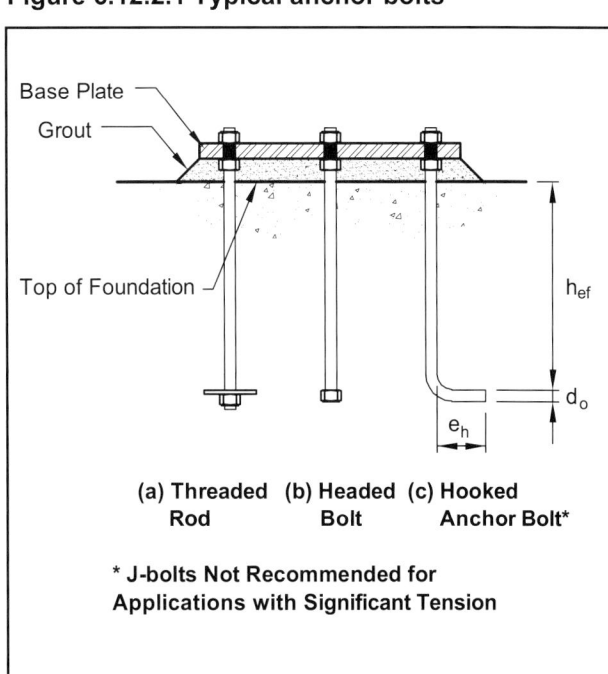

(a) Threaded Rod (b) Headed Bolt (c) Hooked Anchor Bolt*

*J-bolts Not Recommended for Applications with Significant Tension

Figure 6.12.1 Typical column base plate detail

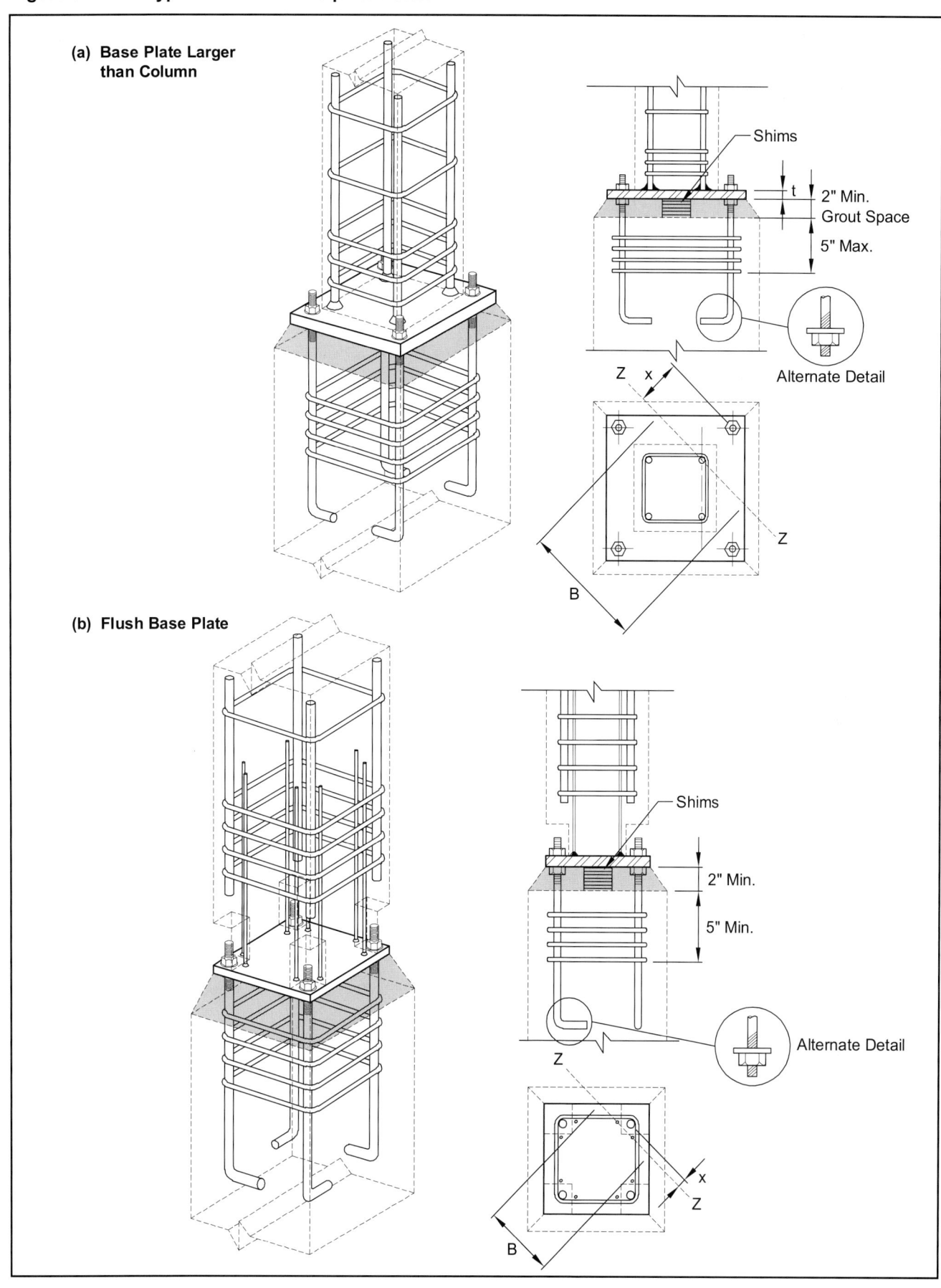

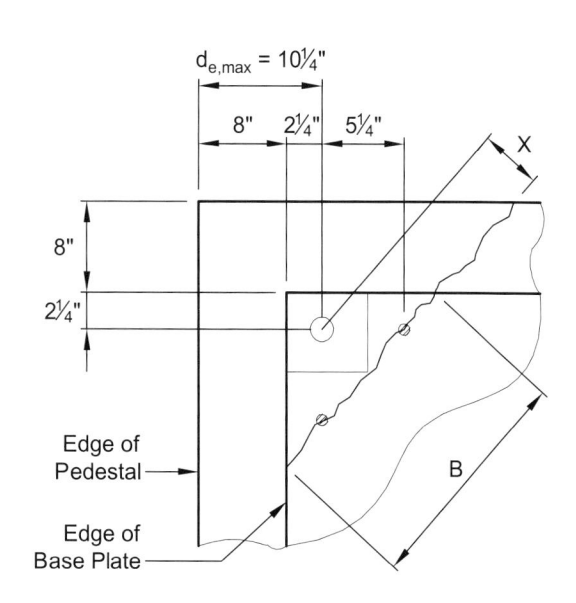

Example 6.12.2.1
Column Connection – Base Plate and Anchor Bolt Design

Given:
   20 in. square column anchored with headed threaded rod anchor bolts into a 36 in. by 36 in. pedestal. The column is designed as pinned at the base, thus, there are no tension requirements other than structural integrity (see Section 3.3).

Factored axial load = 500 kips, including overload factors.

Concrete: Pedestal f'_c = 4000 psi
           Column f'_c = 5000 psi

Steel: Base plate and anchor bolts: F_y = 36 ksi
        Deformed bar anchor (DBA): f_y = 60 ksi

Problem:
   Determine the required base plate thickness and DBAs to anchor the plate. Check anchor bolts in the pedestal.

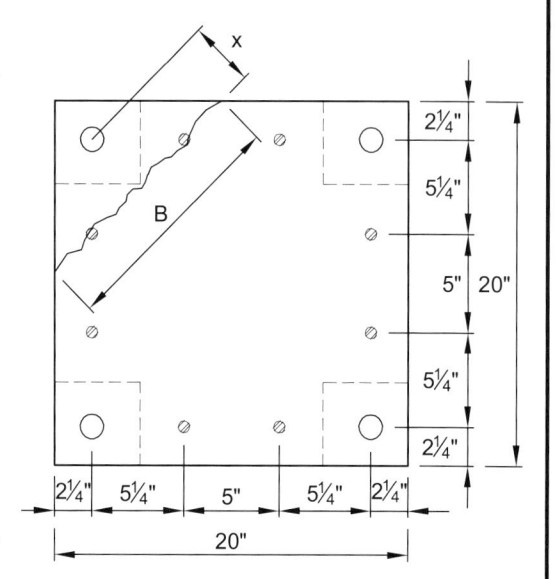

Solution:
Structural integrity requirement (Section 3.3.2) (ϕ = 1.0):
   T_u = 200(A_g) = 0.2(20)² = 80 kips, or 20 kips per bolt

Required base plate thickness, Eq. 6.12.1.2:

$$x = \frac{5.25}{\sqrt{2}} = 3.71 \text{ in.}$$

$$B = 2(2.25\sqrt{2} + 5.25/\sqrt{2}) = 13.79 \text{ in.}$$

$$t = \sqrt{\frac{T_u(4)x}{\phi BF_y}} = \sqrt{\frac{20(4)(3.71)}{1.0(13.79)(36)}} = 0.77 \text{ in.}$$

To minimize handling damage, use 1 in. thick plate.

Deformed bar anchors (DBA):

$$A_s = \frac{T_u}{\phi F_y} = \frac{80}{1.0(60)} = 1.33 \text{ in.}^2$$

Use 8 – ½ in. diameter DBA

8(0.20) = 1.6 in.² > 1.33 in.²   OK

Example 6.12.2.1 (Cont.)
Column Connection – Base Plate and Anchor Bolt Design

Anchor Bolt Requirements:

Steel strength from Figure 6.16.6:

Use 1 in. diameter bolts = 4(25.6) = 102.4 kips > 80 kips   OK

If a hooked anchor bolt is chosen (Figure 6.12.2.1), the hook projection can be determined by rearranging Eq. 6.12.2.2:

$$e_h = \frac{N_p}{1.26 f'_c d_o C_{crp}} = \frac{20}{1.26(4)(1)(1.0)} = 3.97 \text{ in. Use 4 in. projection.}$$

Assume headed bolt with 11 in. minimum embedment. The spacing between anchor bolts is 15.5 in. which is less than $3h_{ef}$ (see notes to Figure 6.5.4.2) so use $h_{ef} = \frac{d_{e,max}}{1.5} = \frac{10.25}{1.5} = 6.83$ in. for Eqs. 6.5.4.1 through 6.5.4.5.

Concrete breakout strength, Eq. 6.5.4.1: $N_{cb} = C_{bs} A_N C_{crb} \psi_{ed,N}$

From Eq. 6.5.4.2: $C_{bs} = 3.33\lambda \sqrt{\frac{f'_c}{h_{ef}}} = 3.33(1.0)\sqrt{\frac{4000}{6.83}} = 80.6$ psi

or From Eq. 6.12.2.1: $C_{bs} = 2.22\lambda \left(\frac{\sqrt{f'_c}}{\sqrt[3]{h_{ef}}} \right) = 2.22(1.0) \left(\frac{\sqrt{4000}}{\sqrt[3]{6.83}} \right) = 74.0$ psi

Use $C_{bs} = 80.6$ psi

From Figure 6.5.4.1: $d_{e1} = d_{e2} = d_{e3} = d_{e4} < 1.5 h_{ef}$

$A_N = 36(36) = 1296$ in.^{2}

Edge distance factor (Eq. 6.5.4.3):

$$\psi_{ed,N} = 0.7 + 0.3 \left(\frac{d_{e,min}}{1.5 h_{ef}} \right) = 0.7 + 0.3 \left(\frac{10.25}{1.5(6.83)} \right) = 1.00$$

Since this is a corner of a pedestal, corner cracking may occur. Therefore, $C_{crb} = 0.8$.

$$\phi C_{bs} A_N C_{crb} \psi_{ed,N} = (1.0) \frac{80.6}{1000} (1296)(0.8)(1.0) = 83.6 \text{ kips}$$

(The design is based on the minimum structural integrity tension requirement, $\phi = 1.0$.)

Figure 6.13.1 Moment connections

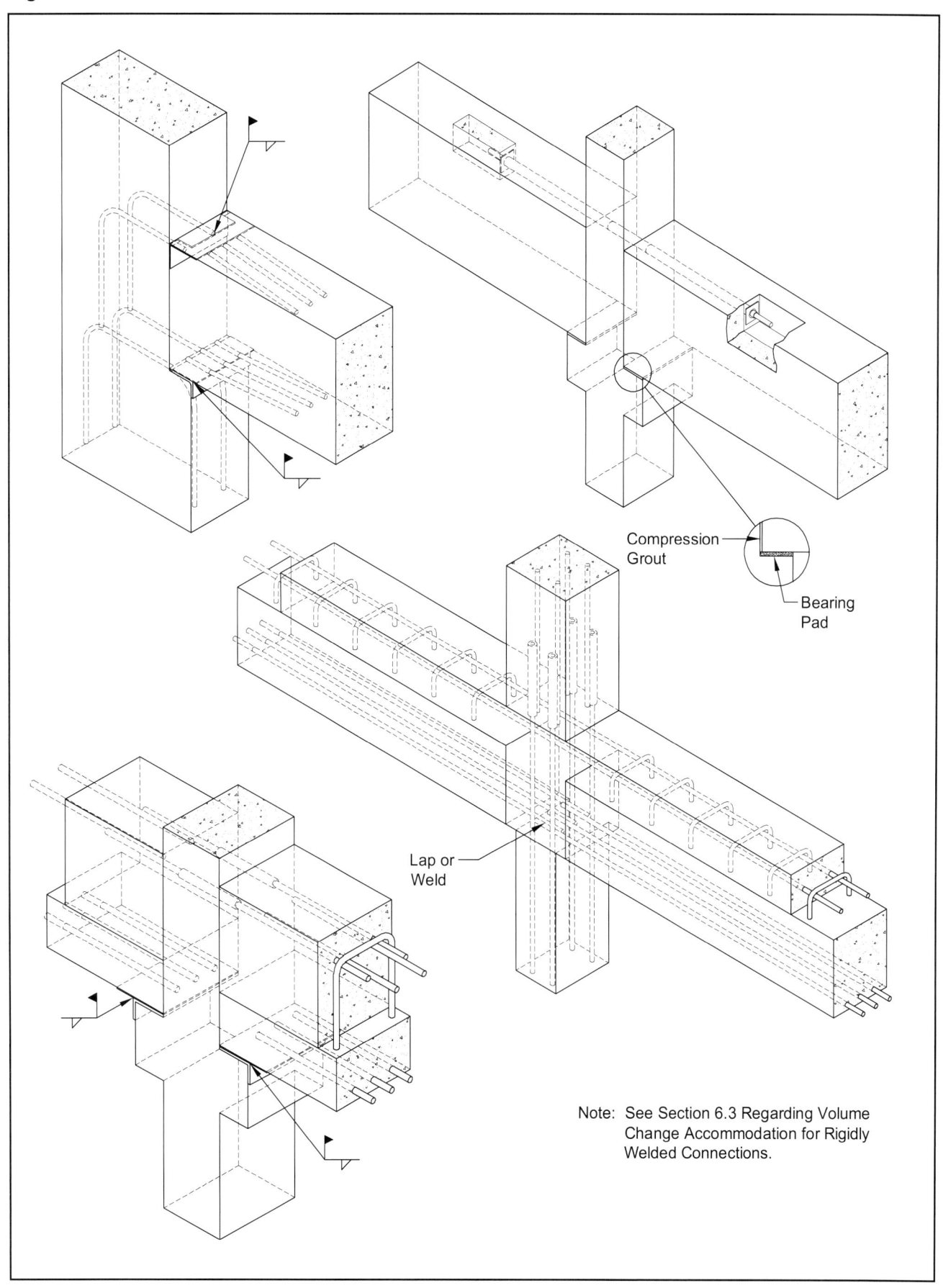

Compression Grout

Bearing Pad

Lap or Weld

Note: See Section 6.3 Regarding Volume Change Accommodation for Rigidly Welded Connections.

6.13 Moment Connections

When lateral stability of precast, prestressed concrete buildings is achieved by frame action or by a combination of shear wall and frame action, the connections developing frame action must be designed for appropriate moment and shear transfer capabilities.

The tension force for the moment resistance within a connection can be resisted by various types of cast-in embedments, such as headed studs and deformed bar anchors. These inserts must be properly anchored to preclude failure of the concrete and thus ensure a ductile mode of failure. Post-tensioning can also be used to develop moment resistance at joints between interconnected members. Where a high degree of moment resistance and ductility are required, composite construction is frequently used to achieve connections that are similar to monolithic concrete joints in their behavior.

Achieving rigid connections can be costly. In most cases, it may not be desirable to build-in a high degree of fixity, since the restraint of volume changes could result in large forces in the connections and the members. It is, therefore, preferable that the design of moment-resisting connections be based on the concept that the desired moment resistance is achieved with some deformation/rotation at the connection. The deformation should be controlled to provide for the desired ductility.

A few examples of different types of moment-resisting connections are shown in Figure 6.13.1. Additional examples can be found in Chapter 3 and Ref. 2.

Moment-curvature analysis of precast and prestressed concrete members is readily done based on established analytical methods. PCI funded research, [21] as well as other research in progress, is leading to improved knowledge of moment-resisting connections and improved analytical procedures.

Example 6.14.6 illustrates the design of a moment connection.

6.14 Typical Connection Designs for Lateral Load Resisting Systems

Chapter 3 of this Handbook describes the analysis of precast concrete structures for various loadings, including lateral loads. The resistance to these lateral loads depends largely on the connections used. Examples of designs for typical such connections follow.

Example 6.14.1
Cladding Connections – Reference Examples 3.2.3.1 and 3.2.4.2

Given:
Cladding tension-compression tieback connection of Examples 3.2.3.1 and 3.2.4.2.

Loads – from "Summary of Factored Loads to Connections (lb)" Example 3.2.4.2:

Seismic: 3368 lb (in or out)

Wind: 4184 lb (out)

Concrete: $f'_c = 5000$ psi

Steel: $F_y = 36$ ksi; $F_u = 58$ ksi

Weld: E70

Problem:
Size the components for the design loads listed.

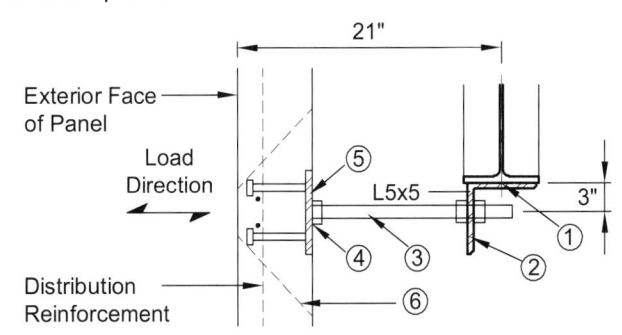

Seismic Connection Factors:

Component	Component	Design Load*
①	Shear of Weld	1.50 S
②	Flexure of Angle	0.48 S
③	Buckling of Rod	1.50 S
④	Shear of Weld	1.50 S
⑤	Flexure of Plate	0.48 S
⑥	Concrete Pull Out	1.50 S

* Or Wind if Larger

Example 6.14.1 (Cont.)
Cladding Connections – Reference Examples 3.2.3.1 and 3.2.4.2

Solution:
Apply connection factors (seismic only) shown at right:

$$\frac{3368}{1000}(1.5) = 5.05 \text{ kips}$$

$$\frac{3368}{1000}(0.48) = 1.62 \text{ kips}$$

Factored wind load: $\frac{4184}{1000} = 4.18 \text{ kips}$

Component	Mode of Failure	Seismic Multiplier	Seismic Design Load	Wind Design Load	Critical Design Load
1	Shear of Weld	1.50	5.05 kips	4.18 kips	5.05 kips
2	Flexure of Angle	0.48	1.62 kips	4.18 kips	4.18 kips
3	Buckling of Rod	1.50	5.05 kips	4.18 kips	5.05 kips
4	Shear of Weld	1.50	5.05 kips	4.18 kips	5.05 kips
5	Flexure of Plate	0.48	1.62 kips	4.18 kips	4.18 kips
6	Flexure of Pullout	1.50	5.05 kips	4.18 kips	5.05 kips

Component 2:
Try angle 5 x 5 x 7/16 x 6 in.   Design shear load = 5.05 kips

Shear Yielding (Eq. 6.6.2.1):
$\phi V_n = \phi (0.6 F_y) A_w = 0.9(0.6)(36)(7/16)(6) = 51.0$ kips > 5.05 kips

Flexure: Moment arm = 3 kip-in. of angle
$M_u = 4.18(3 - 0.94) = 8.61$ kip-in.
$Z = \frac{bt^2}{4} = \frac{6(0.44)^2}{4} = 0.29$ in.^{3}
$\phi M_n = \phi Z F_y = 0.9(0.29)(36) = 9.40$ kip-in. > 8.61 kip-in.

Component 3 – Connection rod (see Ref. 6)
$C_u = T_u = 5.05$ kips
Try ¾ in. A36 All-thread rod. $A_s = 0.44$ in.^{2}

Unsupported length, $\ell_u = 21 - 5 - 2.5 = 13.5$ in.; $r = \frac{D}{4} = \frac{0.75}{4} = 0.19$ in.

$$\lambda_c = \frac{K\ell_u}{r\pi}\sqrt{\frac{F_y}{E}} = \frac{1.0(13.5)}{0.19\pi}\sqrt{\frac{36}{29000}} = 0.80 < 1.5; \; \lambda_c^2 = 0.64$$

$$\phi C_n = \phi(0.658)^{\lambda_c^2} F_y A_s = 0.85(0.658)^{0.64}(36)(0.44) = 10.5 \text{ kips} > 5.05 \text{ kips}$$

$$\phi T_n = \phi F_y A = 0.9(36)\left(\frac{\pi D^2}{4}\right) = 0.9(36)\left(\frac{\pi(0.75)^2}{4}\right) = 14.3 \text{ kips} > 5.05 \text{ kips}$$

Component 4 – Weld design – Nut weld
$T_u = 5.05$ kips

Example 6.14.1 (Cont.)
Cladding Connections – Reference Examples 3.2.3.1 and 3.2.4.2

Assumed length of weld is circumference of all thread = πD. Try 3/16 in. throat size.

$$\phi T_s = \phi(0.6F_{exx})A = \phi(0.6F_{exx})\frac{1}{\sqrt{2}}a\ell_w = (0.75)(0.6)(70)\left(\frac{1}{\sqrt{2}}\right)\left(\frac{3}{16}\right)\left[(\pi)\frac{3}{4}\right] = 9.84 \text{ kips} > 5.05 \text{ kips}$$

Component 1, Weld Design – Connection Angle
$T_u = 5.05$ kips at eccentricity of 3 in.
Using the Elastic Vector Method:
Assume two line welds, one at each end of angle welded to structural steel shape.
Weld section properties based on a unit weld size:

$$S_x = \frac{d^2}{3} = \frac{(5)^2}{3} = 8.33 \text{ in.}^2$$

Maximum Tension Stress at end points:

$$f_y = \frac{T_u e_r}{S_x} = \frac{5.05(3)}{8.33} = 1.82 \text{ kips/in.}$$

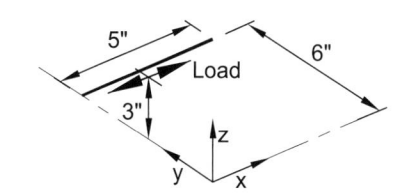

Maximum Shear Stress:

$$f_z = \frac{T_w}{\ell_w} = \frac{5.05}{5+5} = 0.51 \text{ kip/in.}$$

Resultant Stress:

$$f_{result} = \sqrt{f_x^2 + f_y^2 + f_z^2} = \sqrt{0^2 + 1.82^2 + 0.51^2} = 1.89 \text{ kips/in.}$$

Reduced Design Strength ¼ in. Fillet Weld:

$$\phi f_n = \phi(0.6F_{exx})\frac{1}{\sqrt{2}}(0.25) = 0.75(0.6)(70)\frac{1}{\sqrt{2}}0.25 = 5.56 \text{ kips/in.} > 1.89 \text{ kips/in.}$$

Component 5 – Embedded Plate Flexure:

Try PL 3/8 x 6 x 6 in. w/(4) ½ in. diameter x 3 in. studs at 4 in. on center
$T_u = 4.18$ kips

$$M_u = \frac{PL}{4} = \frac{(4.18)(4)}{4} = 4.18 \text{ kip-in.}$$

$$\phi M_n = \phi(F_y)Z_p = \phi(F_y)\left(\frac{bd^2}{4}\right) = 0.9(36)\left(\frac{6(0.375)^2}{4}\right) = 6.83 \text{ kip-in.} > 4.18 \text{ kip-in.}$$

Component 6 – Headed Stud Capacity
$N_u = 5.05$ kips

Effective Length $h_{ef} = L + t_{plate} - t_{head} - 1/8 = 3 + 0.375 - 5/16 - 1/8 = 2.94$ in.

Assume not near free edge: $C_{min} \geq 1.5 h_{ef}$

Breakout Strength:

$$\phi N_{cb} = C_{bs} A_N C_{crb} \psi_{ed,N}$$

Example 6.14.1 (Cont.)
Cladding Connections – Reference Examples 3.2.3.1 and 3.2.4.2

$$C_{bs} = 3.33\sqrt{\frac{f'_c}{h_{ef}}} = 3.33\sqrt{\frac{5000}{2.94}} = 137.4$$

$$A_N = (s + 3h_{ef})(s + 3h_{ef}) = 164.2 \text{ in.}^2 = [4 + 3(2.94)][4 + 3(2.94)] = 164.2 \text{ in.}^2$$

$$C_{crb} = 1.0 \, C_{crb} = 1.0$$

$$\psi_{ed,N} = 1.0$$

Panel reinforcement goes through failure surface: $\phi = 0.75$

$$\phi N_{cb} = 0.75(137.4)(164.2)(1.0)(1.0) = 16{,}921 \text{ lb} = 16.9 \text{ kips} > 5.05 \text{ kips}$$

Connection Detail:

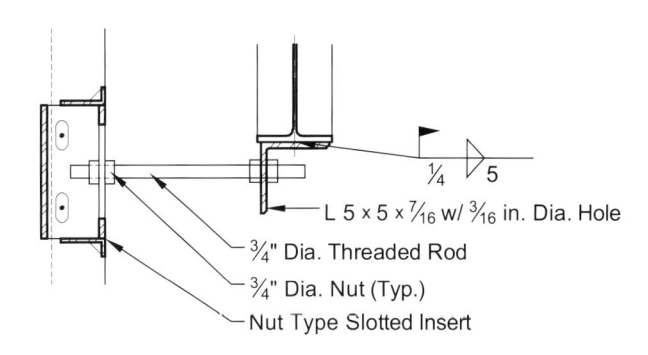

- L 5 × 5 × 7/16 w/ 3/16 in. Dia. Hole
- 3/4" Dia. Threaded Rod
- 3/4" Dia. Nut (Typ.)
- Nut Type Slotted Insert

Example 6.14.2 [26]
Plate-and-Bar Diaphragm Shear Connection – Reference Figure 3.8.1.2

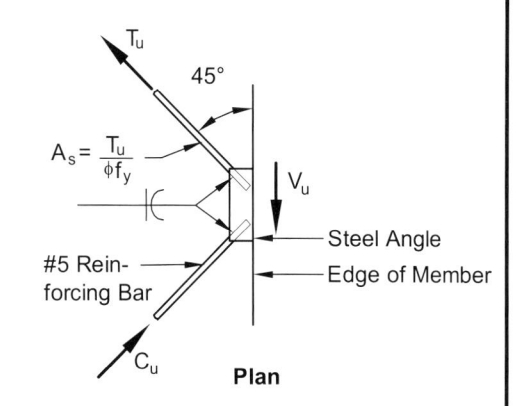

Given:
    Double tee to double tee diaphragm shear connection shown in Figure 3.8.1.2(a).
    Concrete f'_c: 5000 psi
    Reinforcing bar f_y: 60 ksi
    Plate: $F_y = 36$ ksi; $F_u = 58$ ksi
    Welds: E70

Reinforcing bars fully developed and welds as shown in Figure 6.16.3.

Problem:
    Determine the shear strength of the connection.

Solution:
Strength based on tension and compression in the bars:

$$\phi T_n = \phi A_s f_y = 0.9(0.31)(60) = 16.7 \text{ kips}$$
$$\phi C_n = \phi A_s f_y = 0.65(0.31)(60) = 12.1 \text{ kips}$$
$$\phi V_n = \phi T_n \cos(45°) + \phi C_n \cos(45°)$$
$$= 16.7(0.707) + 12.1(0.707) = 20.4 \text{ kips}$$

Example 6.14.2 [26] (Cont.)
Plate-and-Bar Diaphragm Shear Connection – Reference Figure 3.8.1.2

Strength based on weld shear through the effective throat with applied eccentricity based on erection plate size and analysis based on erection Instantaneous Center Method from AISC-LRFD Manual. [6] (see also Example 6.7.4.1).

$$x\ell = \frac{(k\ell)^2}{2(k\ell)+\ell} = \frac{(1)^2}{2+4} = 0.167 \text{ in.}$$

$$k\ell = 1 \quad k = \frac{1}{4} = 0.25 \quad \text{Round to 0.2 for conservative results.}$$

$$a\ell = \frac{4}{2} - x\ell; \quad a = \frac{2-0.167}{4} = 0.458$$

Round to 0.5 for conservative results.

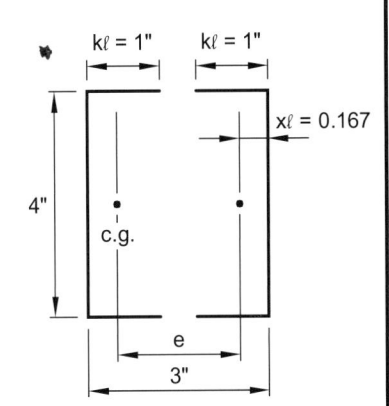

From Ref 6:
- C = 1.35
- C_1 = 1.0
- D = 4
- ℓ = 4
- $V_u = C\, C_1\, D\, \ell = 1.35(1.0)(4)(4) = 21.6$ kips

Plate Strength:

Shear Rupture:
$$\phi V_n = \phi(0.6F_u)A_w$$
$$\phi V_n = 0.75(0.6)(58)(0.25)(4) = 26.1 \text{ kips}$$

Shear Yielding:
$$\phi V_n = \phi(0.6F_y)A_w$$
$$\phi V_n = 0.9(0.6)(36)(0.25)(4) = 19.4 \text{ kips}$$

Moment Strength:

$$\phi V_n = \frac{\phi(F_y)Z_p}{e} = \frac{\phi(F_y)\left(\frac{bd^2}{4}\right)}{e}$$

$$\phi V_n = \frac{0.9(36)\left(\frac{0.25(4^2)}{4}\right)}{3-2(0.167)} = 12.1 \text{ kips}$$

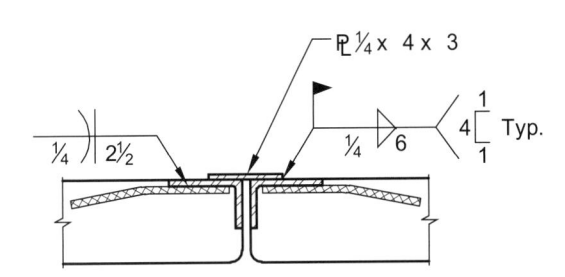

Failure Mode	Design Strength
Reinforcing Bar	20.4 kips
Weld Failure	21.6 kips
Plate Failure	12.1 kips

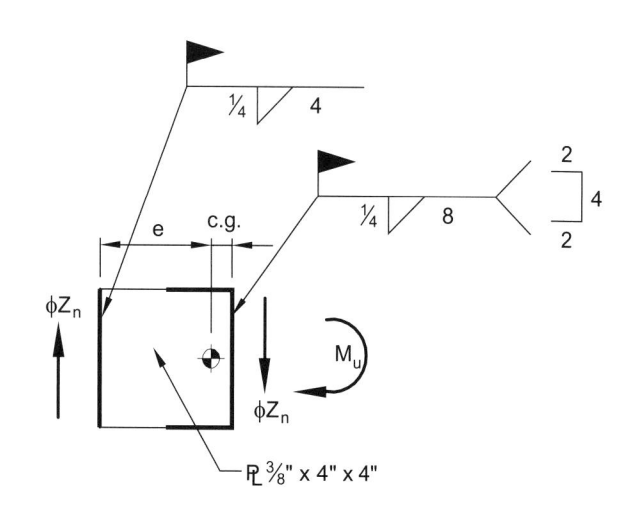

Example 6.14.3
Diaphragm-to-Wall Shear Connection – Reference Sections 3.5 and 3.8

Given:

The typical diaphragm connection to shear wall shown.
- Wall thickness: 8 in.
- Studs: $f_{ut} = 65$ ksi
- Reinforcing Bar Grade 60: $f_y = 60$ ksi
- Plates ASTM A-36: $F_y = 36$ ksi
- Weld Material: E70
- Concrete: $f'_c = 5000$ psi

Problem:

Determine the shear capacity in the Z direction.

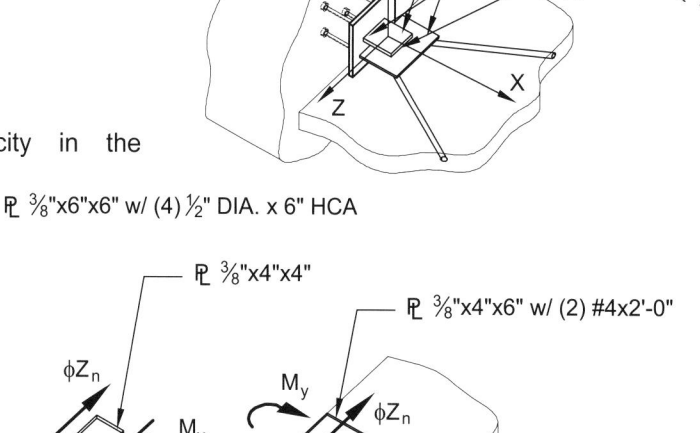

Solution:

The flange plate assembly should be detailed so that the lines of the diagonal bars intersect at a point on the face of the wall. If this is done properly, there will theoretically be no moment transferred to the wall. Because of residual moments due to tolerances, the wall and flange plate assemblies should have some moment resisting capacity.

Because of the manner in which the erection plate is welded to the two assemblies, there is a moment in the erection plate and the weld group. The following assumptions are used:
- The erection plate is fixed at one end.
- The eccentricity of the connection is the distance between the centroids of the weld groups.

Erection Plate Moment Strength:

$$\text{c.g.} = \frac{2b\left(\frac{b}{2}\right)}{d+2b} = \frac{2(2)\left(\frac{2}{2}\right)}{4+2(2)} = 0.5 \text{ in.}$$

$$e = 4 - \text{c.g.} = 4 - 0.5 = 3.5 \text{ in.}$$

$$M_u \le \phi M_p$$

$$M_u = \phi Z_n(e)$$

$$\phi M_p = \phi(f_y)(Z_p) = \phi(f_y)\left(\frac{bd^2}{4}\right)$$

$$= 0.9(36)\left(\frac{0.375(4)^2}{4}\right) = 48.6 \text{ kip-in.}$$

Example 6.14.3 (Cont.)
Diaphragm-to-Wall Shear Connection – Reference Sections 3.5 and 3.8

$\phi Z_n(e) = \phi M_p = 48.6$ kip-in.

$\phi Z_n = \dfrac{48.6}{e} = \dfrac{48.6}{3.5 \text{ in.}} = 13.9$ kips

Erection Plate Shear Strength:

$\phi Z_n = \phi(0.6 f_y)A_{plate} = 0.9(0.6)(36)\left(\dfrac{3}{8}\right)(4) = 29.2$ kips

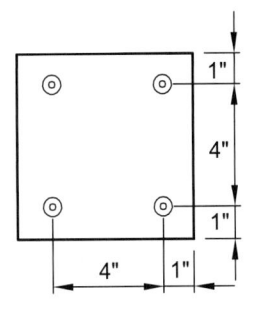

PL $\frac{3}{8}$" x 6" x 6" w/ (4) $\frac{1}{2}$" DIA. x 6" HCA

Weld Strengths
  Wall Embed to Erection Plate, ¼ in. fillet weld, 4 in. long.
  From Figure 6.16.2: $\phi Z_n = 4(5.57) = 22.3$ kips

Double Tee Embed to Erection Plate
  This C-shaped weld can be designed by either of the methods shown in Section 6.7.5. The AISC-LRFD Manual also contains tabulated values and design aids for determining the shear strength of such welds. Using one of those methods yields:

From the AISC-LRFD tables:

$\ell = 4$ in., $k = \dfrac{2}{4} = 0.5$, $a = \dfrac{3.5}{4} = 0.875$

Shows C = 1.40
  $\phi Z_n = 1.40(1.0)(4)(4) = 22.4$ kips

Wall plate strength:
Steel shear from Eq. 6.5.2.1:
  $\phi Z_n = \phi V_s = \phi n A_{se} f_{ut} = 0.65(4)(0.20)(65) = 33.8$ kips

Concrete shear: The connection is far enough from any edge to be considered "in the field."

From Eq. 6.5.2.1:
  $\phi V_s = \phi(n)(A_{se} f_{ut}) = 0.75(4)(0.2)(65) = 39.0$ kips

Double tee deck plate (see previous example):
  $\phi V_n = 12.1$ kips

Summary:

Failure Mode	Design Strength
Weld on Erection Plate	22.3 kips
Moment in Erection Plate	13.9 kips
Shear in Erection Plate	29.2 kips
C-Shaped Weld on Erection Plate	22.4 kips
Wall Plate Studs Steel Shear	33.8 kips
Wall Plate Studs Concrete Shear	28.0 kips
Double Tee Deck Plate	12.1 kips

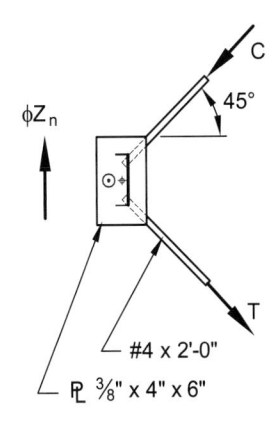

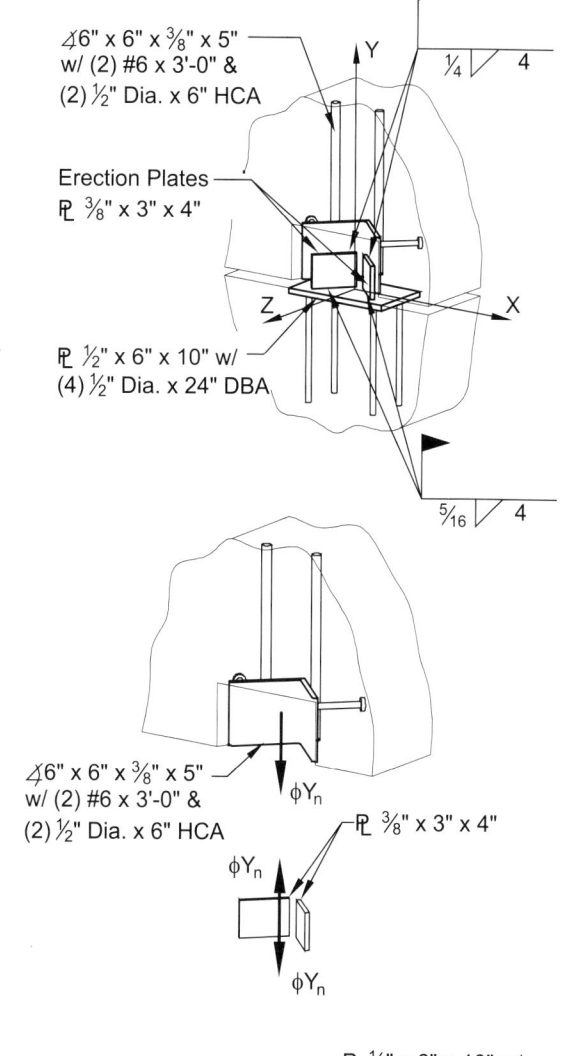

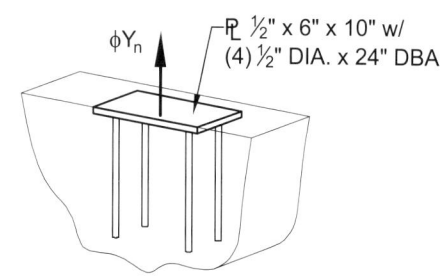

Example 6.14.4
Wall-to-Wall Tension Connection – Reference Section 3.5

Given:
The typical wall-to-wall tension connection shown.
    Wall thickness: 8 in.
    Deformed Bar Anchors (DBA): $f_y = 60$ ksi
    Reinforcing Bar Grade 60: $f_y = 60$ ksi
    Plates ASTM A-36: $F_y = 36$ ksi
    Weld Material: E70
    Concrete: $f'_c = 5000$ psi

Problem:
    Determine the tension design strength in the Y direction.

Solution:
Erection plate tension strength:
    $\phi Y_n = \phi A_{pl} F_y = 0.9(3/8)(4)(2 \text{ plates})(36) = 97.2$ kips

Weld strength:
Erection plate to lower plate:
From Figure 6.16.2:
    Strength of 5/16 in. fillet weld = 6.96 kips/in.
    $\phi Y_n = 6.96(4)(2) = 55.7$ kips

Erection plate to angle:
From Figure 6.16.2:
    $\phi Y_n = 6.96(4)(2) = 55.7$ kips

Example 6.14.4 (Cont.)
Wall-to-Wall Tension Connection – Reference Section 3.5

Lower plate capacity – controlled by 4 – ½ in. DBA
    Assume all four DBA yield to resist applied force
    $\phi Y_n = \phi A_b f_y = 0.9(0.20)(60)(4) = 43.2$ kips

Development length of DBA is the same as reinforcing bars.
From Design Aid 11.2.9:

Required $\ell_d = 17$ in. < 24 in. OK

Pocket angle capacity:
Neglect headed studs, strength is controlled by #6 reinforcing bars:
    $\phi Y_n = \phi A_b f_y = 0.9(2)(0.44)(60) = 47.5$ kips

From Design Aid 11.2.9:
Required length of #6 bar = 25 in.
Development length provided = 32 in. OK

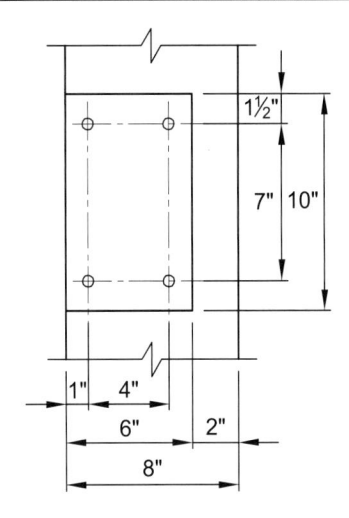

Summary:

Failure Mode	Design Strength
Weld on Erection Plate	55.7 kips
Tension in Erection Plate	97.2 kips
L-Shaped Weld Erection Material	55.7 kips
Deformed Bar Anchor Tension	43.2 kips
Angle Rebar	47.5 kips

Note: It is good practice to force the failure mode into the tensile plate. This can be done by putting a hole in the erection plate or using other methods of reducing the area.

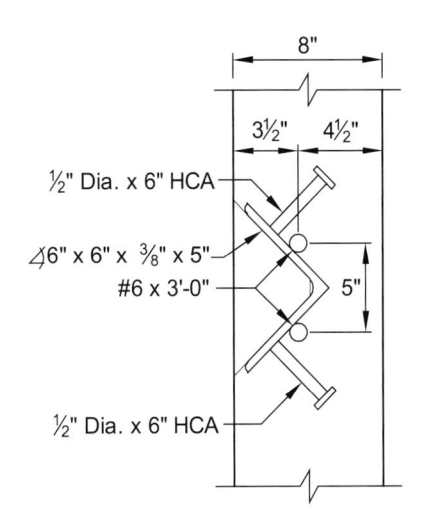

Example 6.14.5
Wall-to-Wall Shear Connection – Reference Section 3.5

Given:
    The typical wall-to-wall shear connection shown on next page.
    Wall thickness: 8 in.
    Deformed Bar Anchors (DBA): $f_y = 60$ ksi   (see Section 6.4.5)
    Headed Studs (HCA): $f_{ut} = 65$ ksi
    Reinforcing Bar Grade 60: $f_y = 60$ ksi, $f_{ut} = 90$ ksi
    Plates ASTM A-36: $F_y = 36$ ksi
    Weld Material: E70
    Concrete: $f'_c = 5000$ psi
    Connection plate is recessed 2 in.

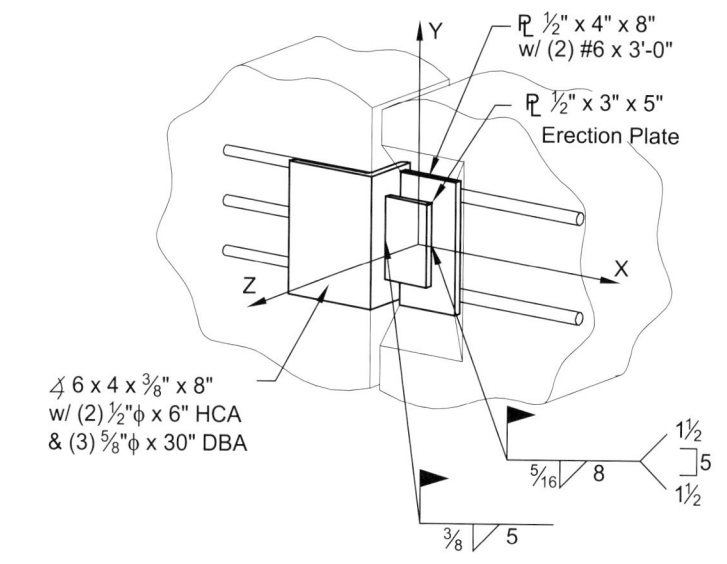

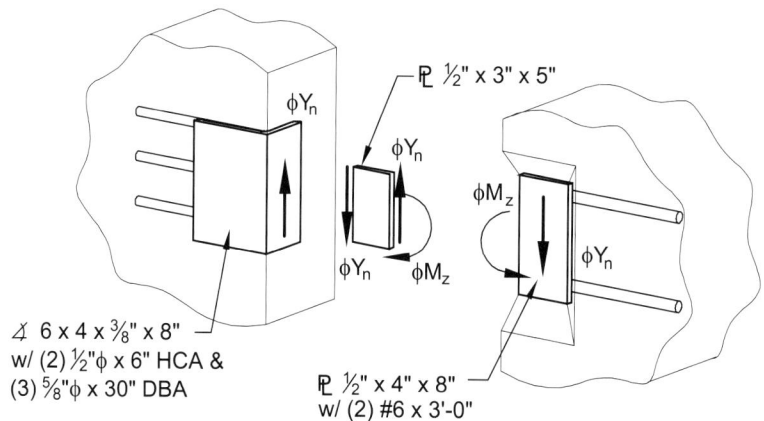

Example 6.14.5 (Cont.)
Wall-to-Wall Shear Connection – Reference Section 3.5

Problem:
  Determine the tension design strength in Y direction.

Solution:

Erection plate shear strength:

$$\phi Y_n = \phi A_{pl}(0.6F_y)$$
$$= 0.9(\tfrac{1}{2})(5)(0.6)(36)$$
$$= 48.6 \text{ kips}$$

Weld strength:

Erection plate to angle (left side):
From Figure 6.16.2:

Strength of ⅜ in. fillet weld
  = 8.35 kips/in.
ϕY_n = 8.35 (5) = 41.8 kips

Erection plate to plate with #6 bars (right side):
  This C-shaped weld can be designed by either of the methods shown in Section 6.7.4.
  The AISC-LRFD manual also contains tabulated values and design aids for determining the shear strength of such welds. Using one of these methods yields:

ϕY_n = 35.3 kips

Erection plate moment capacity:

$$\text{c.g.} = \frac{2b\left(\dfrac{b}{2}\right)}{d+2b} = \frac{2(1.5)\left(\dfrac{1.5}{2}\right)}{5+2(1.5)}$$

  = 0.28 in.

ℓ_e = 3 − C.G. = 3 − 0.28 = 2.72 in.

Example 6.14.5 (Cont.)
Wall-to-Wall Shear Connection – Reference Section 3.5

$$\phi M_z = \phi f_y Z_p = \phi f_y \left(\frac{bd^2}{4}\right) = 0.9(36)\left(\frac{0.5(5)^2}{4}\right) = 101.25 \text{ kip-in.}$$

$$\phi Y_n = \frac{\phi M_z}{\ell_e} = \frac{101.25}{2.72} = 37.2 \text{ kips}$$

Angle capacity:
The angle strength will be controlled by the shear strength of the studs and the deformed bar anchor. Since the connection is not located near a free edge, the steel strength will govern.

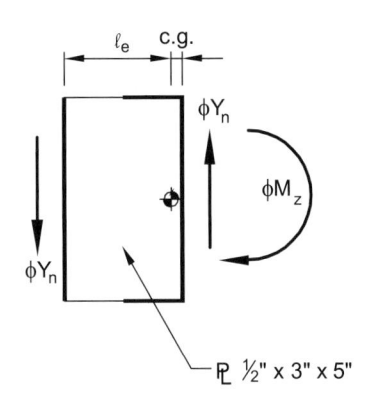

From Eq. 6.5.2.1:
$$\phi V_s = \phi n A_s f_y = 0.75(3)(0.31)(60) = 41.8 \text{ kips}$$

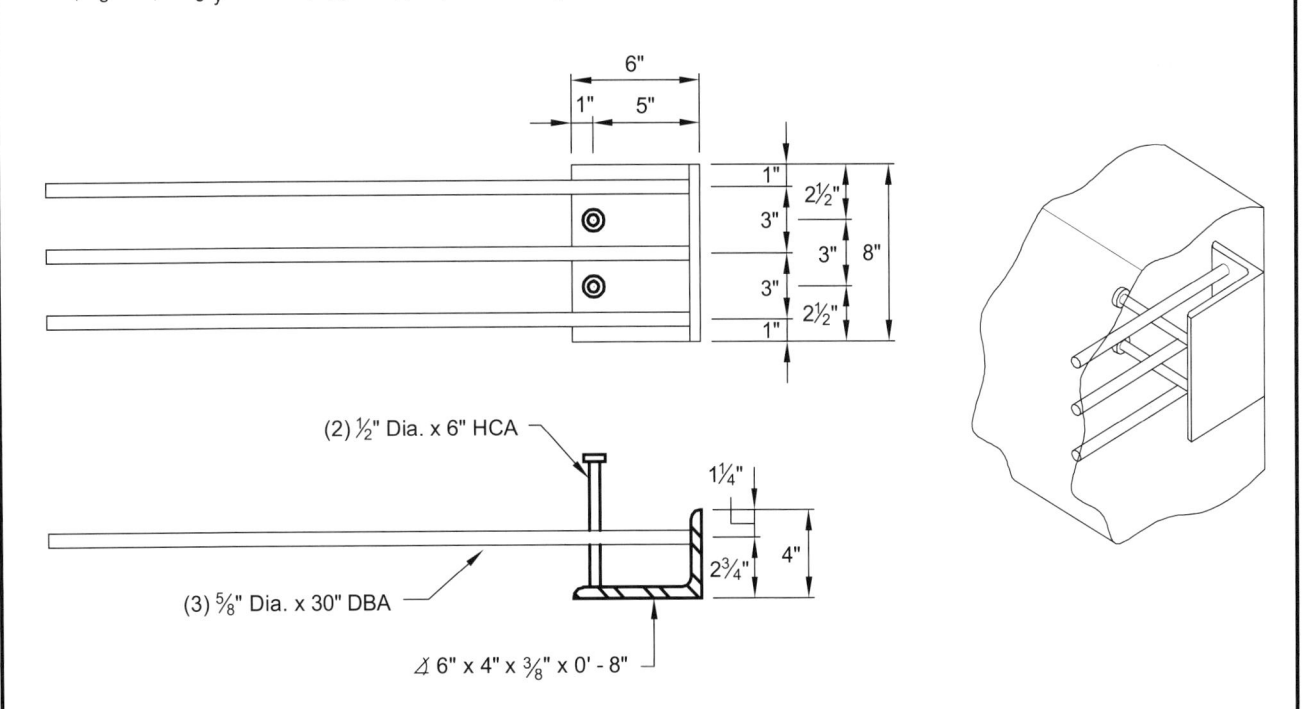

Embed Plate with #6 reinforcing bar (right side)

Steel Shear Strength:

Assume the bar is developed in the wall. Strength based on combined loading of steel.
Assume shear rupture of reinforcing bar is $0.6 f_{ut}$.
Moment arm, d, is equal to distance between reinforcing bars.

$$\frac{1}{\phi}\left(\left(\frac{V_u}{V_n}\right)^2 + \left(\frac{M_u}{M_p}\right)^2\right) \leq 1.0$$

Example 6.14.5 (Cont.)
Wall-to-Wall Shear Connection – Reference Section 3.5

$$\frac{1}{\phi}\left[\left(\frac{\phi Y_n}{0.6F_y A_{rb}}\right)^2 + \left(\frac{\phi Y_n e}{f_y A_{rb} d}\right)^2\right] \leq 1.0$$

$$\frac{1}{0.9}\left[\left(\frac{\phi Y_n}{0.6(60)(2)(0.44)}\right)^2 + \left(\frac{\phi Y_n (5)}{60(0.44)(6)}\right)^2\right] \leq 1.0$$

$$\phi Y_n^2 \left(\left(\frac{1}{31.68}\right)^2 + \left(\frac{5}{158.4}\right)^2\right) \leq 0.9$$

$\phi Y_n^2 (0.0020) \leq 0.9$
$\phi Y_n = 21.2$ kips

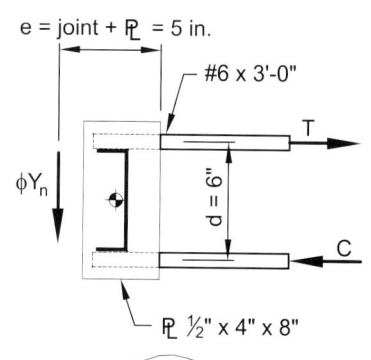

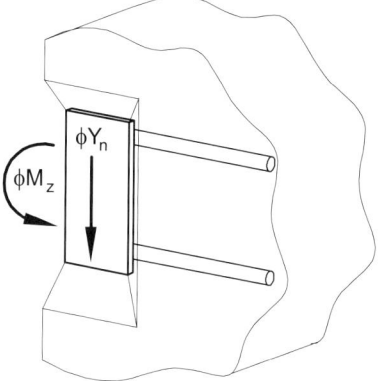

$e = \text{joint} + P_L = 5$ in.
#6 × 3'-0"
$d = 6"$
$P_L \; \frac{1}{2}" \times 4" \times 8"$

Summary:

Failure Mode	Design Strength
Weld on Erection Plate	41.8 kips
Moment of Erection Plate	37.2 kips
Shear of Erection Plate	48.6 kips
C-Shaped Weld on Erection Plate	35.3 kips
DBA	41.8 kips
Wall Plate Reinforcing Bars	21.2 kips

Example 6.14.6
Ordinary Moment Frame Connection – Reference Section 3.6

Given:
The spandrel-to-column ordinary moment frame connection is shown below. Assume the spandrel and column reinforcement is adequate to resist gravity loads. The connection is not required to resist out of plane forces caused by spandrel beam rotation.

Reinforcing Bar Grade 60: $f_y = 60$ ksi
Plates ASTM A-36 $F_y = 36$ ksi
Weld Material E70
Concrete: $f'_c = 5000$ psi

Example 6.14.6 (Cont.)
Ordinary Moment Frame Connection – Reference Section 3.6

Problem:
Determine the tension capacity in the X direction.

Solution:
Plate tension capacity:
$\phi X_n = \phi f_y A_{pl} = 0.9(36)(0.5)(7) = 113.4$ kips

Weld Capacity from Figure 6.16.2:
$\phi X_n = 8.36(16) = 133.8$ kips

Reinforcement strength (4 – #7):
$\phi A_s f_y = 0.9(4)(0.60)(60) = 129.6$ kips

From Figure 6.16.3, provide 3¼ in. of weld each side of bar to ½ in. thick plate.

From Design Aid 11.2.9, a minimum of 25 in. of development length is required.

Limiting strength is plate tension = 113.4 kips

Note: To ensure yielding in the plate, the length between welds should be 1.2 times the length of the plate. [26]

Note: These connections are susceptible to restraint forces caused by creep, shrinkage, and temperature strain in the connecting beam. Creep strains are greater in prestressed concrete beams than in non-prestressed members. Design and detailing should take these forces into consideration.

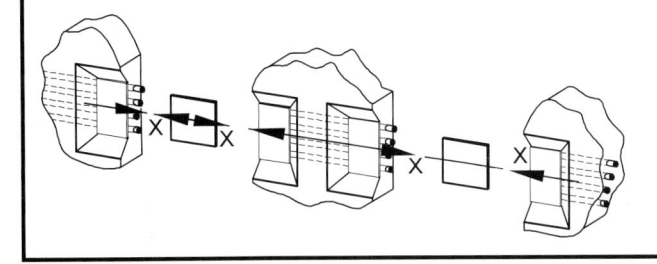

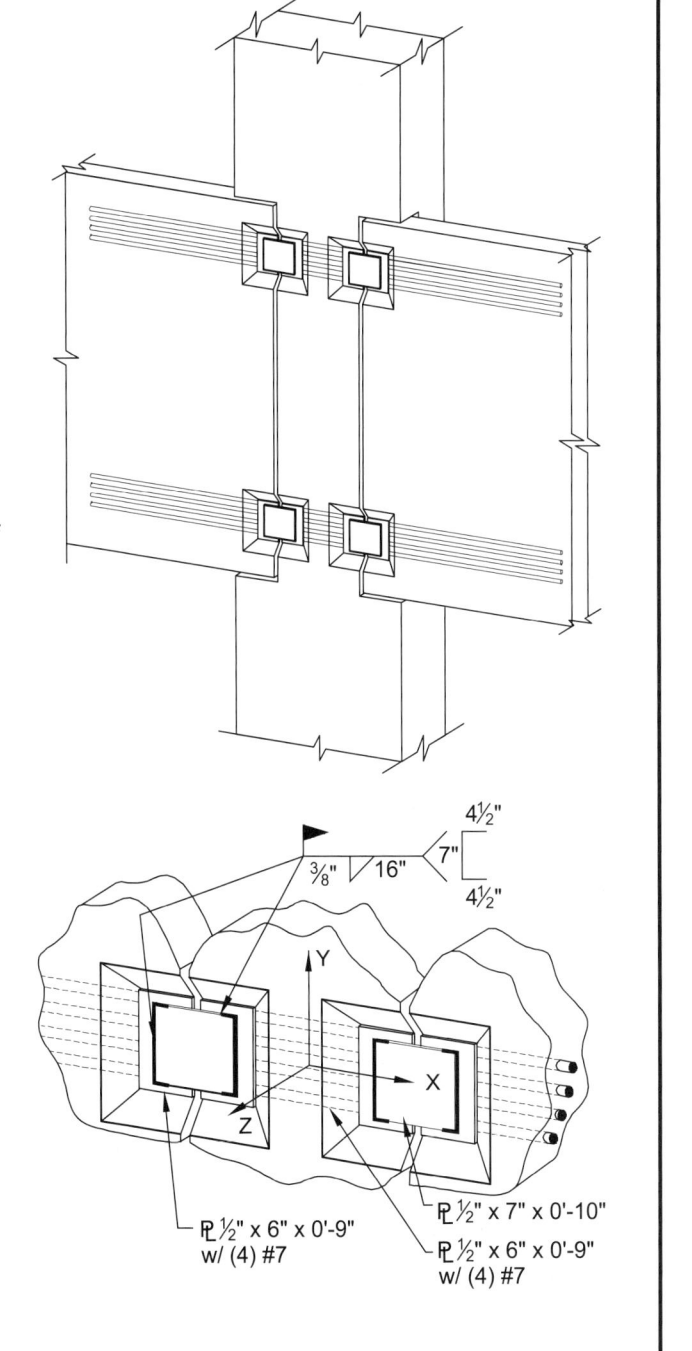

6.15　　References

1. ACI Committee 318, "Building Code Requirements for Structural Concrete (ACI 318-02) and Commentary (ACI 318R-02)," American Concrete Institute, Farmington Hills, MI, 2002.

2. *Design and Typical Details of Connections for Precast and Prestressed Concrete*, Second Edition, MNL-123-88, Precast/Prestressed Concrete Institute, Chicago, IL, 1988.

3. *Reinforcement, Anchorages, Lap Splices and Connections*, Concrete Reinforcing Steel Institute, Schaumburg, IL, 1990.

4. ACI Committee 439, "Mechanical Connections of Reinforcing Bars," *Concrete International*, V. 5, No. 1, January 1983.

5. *Manual of Steel Construction — Allowable Stress Design*, Ninth Edition, American Institute of Steel Construction, Chicago, IL, 2001.

6. *Manual of Steel Construction — Load and Resistance Factor Design*, Third Edition, American Institute of Steel Construction, Chicago, IL, 2001.

7. Anderson, Neal S., and Meinheit, Donald F., "Design Criteria for Headed Stud Groups in Shear: Part 1 — Steel Capacity and Back Edge Effects," PCI JOURNAL, V. 45, No. 5, September-October 2000.

8. *Structural Welding Code — Steel*, AWS D1.1-02, American Welding Society, Miami, FL, 2002.

9. *Structural Welding Code — Reinforcing Steel*, AWS D1.4-02, American Welding Society, Miami, FL, 2002.

10. *The Procedure Handbook of Arc Welding*, Fourteenth Edition, The Lincoln Electric Company, Cleveland, OH, 1999.

11. Salmon, Charles G., and Johnson, John E., *Steel Structures: Design and Behavior*, Fourth Edition, Harper and Collins, New York, NY, 1996.

12. *Welding Handbook — Materials and Applications*, Eighth Edition, American Welding Society, Miami, FL, 1998.

13. Chambers, H. A., "Principles and Practices of Stud Welding," PCI JOURNAL, V. 46, No. 5, September-October 2001.

14. Boresi, Schmidt, Sidebottom, *Advanced Mechanics of Materials*, Fifth Edition, John Wiley and Sons, New York, NY, 1993.

15. Kriz, L. B., and Raths, C. H., "Connections in Precast Concrete Structures — Strength of Corbels," PCI JOURNAL, V. 10, No. 1, February 1965.

16. Mattock, A. H., "Design Proposals for Reinforced Concrete Corbels," PCI JOURNAL, V. 21, No. 3, May-June 1976.

17. Marcakis, K., and Mitchell, D., "Precast Concrete Connections with Embedded Steel Members," PCI JOURNAL, V. 25, No. 4, July-August 1980.

18. Cazaly, L., and Huggins, M. W., *Canadian Prestressed Concrete Institute Handbook*, Third Edition, Canadian Prestressed Concrete Institute, Ottawa, Ontario, Canada, 1996.

19. Loov, Robert, "A Precast Beam Connection Designed for Shear and Axial Load," PCI JOURNAL, V. 13, No. 3, June 1968.

20. Iverson, J. K., and Pfeifer, D. W., "Criteria for Design of Bearing Pads," Technical Report, TR-4-85, Precast/Prestressed Concrete Institute, PCI JOURNAL, V. 30, No. 5, September-October 1985.

21. Stanton, J. F., Anderson, R. G., Dolan, D. W., and McCleary, D. E., "Moment Resistant Connections and Simple Connections," Research Project No 1/4, See also PCI JOURNAL, V. 32, No. 2, March-April 1987.

22. Bickel, T. S., and Shaikh, A. F., "Shear Strength of Adhesive Anchors," PCI JOURNAL, V. 47, No. 5, September-October 2002.

23. Cook, R. A., Kunz, J., Fuchs, W., and Konz, R. C., "Behavior and Design of Single Adhesive Anchors Under Tensile Load in Uncracked Concrete," *ACI Structural Journal*, V. 95, No. 1, January-February 1998.

24. *Design Guidelines for the Selection and Use of Stainless Steel, Designer's Handbook*. Specialty Steel Industry of North America (SSINA), Washington, DC, 1995.

25. Concrete Technology Associates, "Ductile Pullout Connections," *CTA Technical Bulletins, Vol. II*, Precast/Prestressed Concrete Institute, Chicago, IL, 2000. See also PCI JOURNAL, V. 44, No. 5, September-October 1999.

26. Blodgett, Omer W., *Design of Welded Structures,* Eighth Printing, The James F. Lincoln Arc Welding Foundation, Cleveland, OH, 1976.

6.16 DESIGN AIDS

Figure 6.16.1 Allowable and design stress for fillet and partial penetration welds[a]

Electrode	Allowable[b] working stress (ksi)	Design[c] strength (ksi)
E60	18	27.0
E70	21	31.5
E80[d]	24	36.0
E90[d]	27	40.5
E100[d]	30	45.0

a. For partial penetration welds loaded in shear parallel to the axis of the weld.
b. Based on AISC Allowable Stress Design Manual of Steel Construction. [5]
c. Based on AISC Load and Resistance Factor Design Manual of Steel Construction. [6] Includes $\phi = 0.75$.
d. Check yield strength of base metal for compatibility with selected electrode.

Figure 6.16.2 Strength of fillet welds for building construction

Fillet[a] weld size	E60 Electrode Allowable stress design (kips/in.)	E60 Electrode Design[b] strength (kips/in.)	E70 Electrode Allowable stress design (kips/in.)	E70 Electrode Design[b] strength (kips/in.)
1/8	1.59	2.39	1.86	2.78
3/16	2.39	3.58	2.78	4.18
1/4	3.18	4.77	3.71	5.57
5/16	3.98	5.96	4.64	6.96
3/8	4.77	7.16	5.57	8.35
7/16	5.57	8.35	6.50	9.74
1/2	6.36	9.54	7.42	11.14
9/16	7.16	10.74	8.35	12.53
5/8	7.95	11.93	9.28	13.92

a. Assumes 45-degree fillet.
b. Based on AISC Load and Resistance Factor Design Manual of Steel Construction. [6] Includes $\phi = 0.75$.

Figure 6.16.3 Minimum length of weld to develop full strength of bar. Weld parallel to bar length[a,b]

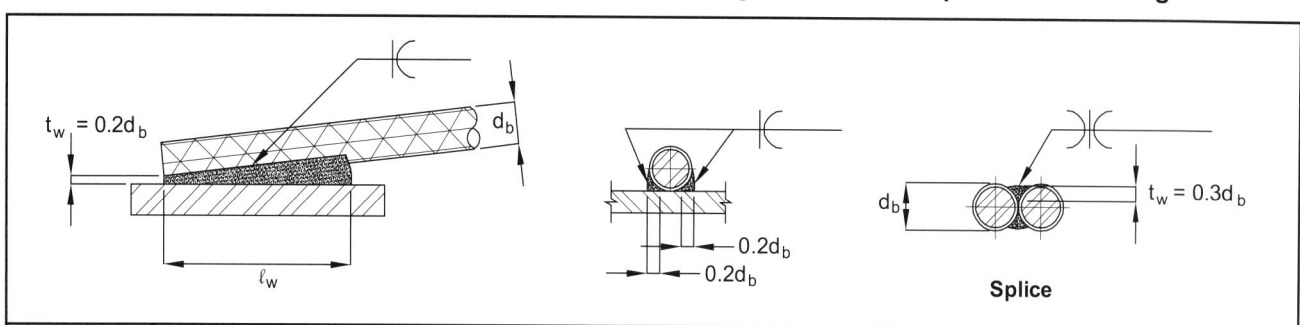

Electrode	Bar size	Plate thickness, in. 1/4	5/16	3/8	7/16	1/2	Min. splice length, in.
E70	3	1½	1½	1½	1½	1½	1
	4	2	2	2	2	2	1½
	5	2½	2½	2½	2½	2½	1¾
	6	3	3	3	3	3	2
	7	3¾	3¼	3¼	3¼	3¼	2¼
	8	5	4	3¾	3¾	3¾	2½
	9	6¼	5	4¼	4¼	4¼	3
	10	8	6¼	5¼	4¾	4¾	3¼
	11	9¾	7¾	6½	5½	5¾	3½
E80	3	1¼	1¼	1¼	1¼	1¼	1
	4	1¾	1¾	1¾	1¾	1¾	1¼
	5	2¼	2¼	2¼	2¼	2¼	1½
	6	2¾	2½	2½	2½	2½	1¾
	7	3¾	3	3	3	3	2
	8	5	4	3½	3½	3½	2¼
	9	6¼	5	4¼	3¾	3¾	2½
	10	8	6¼	5¼	4¼	4¼	3
	11	9¾	7¾	6½	5½	4¾	3¼
E90	3	1¼	1¼	1¼	1¼	1¼	1
	4	1½	1½	1½	1½	1½	1
	5	2	2	2	2	2	1¼
	6	2¾	2¼	2¼	2¼	2¼	1½
	7	3¾	3	2¾	2¾	2¾	1¾
	8	5	4	3¼	3	3	2
	9	6¼	5	4¼	3¾	3½	2¼
	10	8	6¼	5¼	4½	4	2½
	11	9¾	7¾	6½	5½	5	2¾

a. Lengths above heavy line are governed by weld strength. Lengths below heavy line are governed by plate shear.
Basis: bar f_y = 60 ksi; plate F_y = 36 ksi; shear on plate limited to 0.9(0.6)(36) = 19.44 ksi.
b. Weld length listed is the required "effective" length of weld. Engineer should consider whether weld at start and stop is fully effective.

Figure 6.16.4 Size of fillet weld required to develop full strength of bar

BAR PERPENDICULAR TO PLATE, WELDED ONE SIDE

$$\ell_w = \pi \left(d_b + \frac{a}{2} \right)$$

Plate F_y = 36 ksi

Plate area = $\pi(d_b + 2a)t_{pl}$

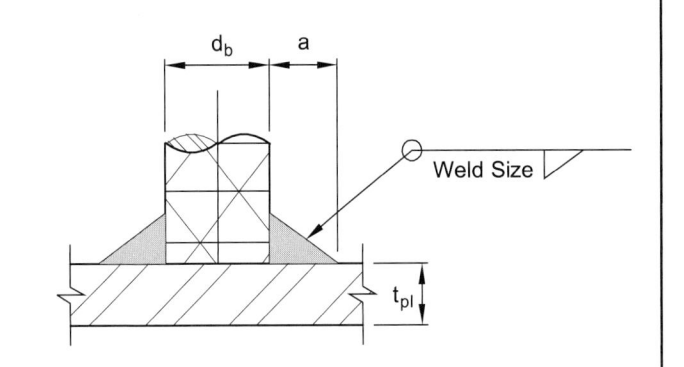

	Grade 40 Bar					
	E70 Electrode		E80 Electrode[c]		E90 Electrode[c]	
Bar Size	Weld size[a] (in.)	Min. plate thickness[b] (in.)	Weld size[a] (in.)	Min. plate thickness[b] (in.)	Weld size[a] (in.)	Min. plate thickness[b] $t_{p\ell}$ (in.)
3	3/16	1/4	3/16	1/4	3/16	1/4
4	1/4	1/4	3/16	1/4	3/16	1/4
5	1/4	1/4	1/4	1/4	1/4	1/4
6	5/16	1/4	1/4	1/4	1/4	1/4
7	3/8	5/16	5/16	5/16	5/16	5/16
8	7/16	5/16	3/8	5/16	5/16	3/8
9	7/16	3/8	7/16	3/8	3/8	3/8
10	1/2	3/8	7/16	7/16	7/16	7/16
11	9/16	7/16	1/2	7/16	7/16	1/2
	Grade 60 Bar					
3	1/4	1/4	3/16	1/4	3/16	1/4
4	5/16	1/4	1/4	1/4	1/4	1/4
5	3/8	1/4	5/16	1/4	5/16	5/16
6	7/16	5/16	3/8	5/16	3/8	5/16
7	1/2	3/8	7/16	3/8	3/8	3/8
8	9/16	3/8	1/2	7/16	7/16	7/16
9	5/8	7/16	9/16	1/2	1/2	1/2
10	11/16	1/2	5/8	1/2	9/16	9/16
11	3/4	9/16	11/16	9/16	5/8	5/8

a. A minimum of 3/16 in. weld size is suggested.
b. Theoretical thickness for shear stress on base metal = 0.9(0.6)(36) ksi. A more practical thickness might be taken as 1/2d_b as used with headed studs. A minimum of 1/4 in. plate thickness is suggested.
c. Check yield strength of base metal for compatibility with selected electrode.

Figure 6.16.5 Size of fillet weld required to develop full strength of bar

---

BAR PERPENDICULAR TO PLATE, WELDED BOTH SIDES

$$\ell_w = \pi\left[d_b + \frac{a}{2}\right]^2$$

Plate F_y = 36 ksi

Plate area = $\pi(d_p + 2_a)t_{pl}$

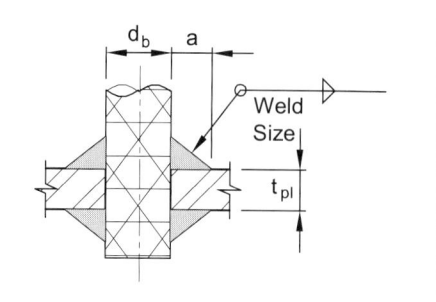

Bar Size	E70 Electrode Weld size[a] (in.)	E70 Electrode Min. plate thickness[b] (in.)	E80 Electrode[c] Weld size[a] (in.)	E80 Electrode[c] Min. plate thickness[b] (in.)	E90 Electrode[c] Weld size[a] (in.)	E90 Electrode[c] Min. plate thickness[b] (in.)
	Grade 40 Bar
3	3/16	1/4	3/16	1/4	3/16	1/4
4	3/16	1/4	3/16	1/4	3/16	1/4
5	3/16	1/4	3/16	1/4	3/16	1/4
6	3/16	5/16	3/16	5/16	3/16	5/16
7	3/16	3/8	3/16	3/8	3/16	3/8
8	1/4	3/8	3/16	7/16	3/16	7/16
9	1/4	7/16	1/4	7/16	3/16	7/16
10	5/16	1/2	1/4	1/2	1/4	1/2
11	5/16	9/16	5/16	9/16	1/4	9/16
	Grade 60 Bar
3	3/16	1/4	3/16	1/4	3/16	1/4
4	3/16	1/4	3/16	5/16	3/16	5/16
5	3/16	5/16	3/16	3/8	3/16	3/8
6	1/4	3/8	1/4	7/16	3/16	7/16
7	5/16	7/16	1/4	1/2	1/4	1/2
8	5/16	1/2	5/16	9/16	1/4	9/16
9	3/8	9/16	5/16	5/8	5/16	5/8
10	3/8	5/8	3/8	11/16	5/16	11/16
11	7/16	11/16	3/8	3/4	3/8	3/4

a. A minimum of 3/16 in. weld size is suggested.
b. Theoretical thickness for shear stress on base metal = 0.9(0.6)(36) ksi. A more practical thickness might be taken as ½d_b as used with headed studs. A minimum of ¼ in. plate thickness is suggested.
c. Check yield strength of base metal for compatibility with selected electrode.

Figure 6.16.6 Strength of bolts and threaded fasteners[a]

Bolt Diameter, in.	Nominal Area A, in.²	A36 $F_u = 58$ ksi Tension Design	A36 Tension Service	A36 Shear Design	A36 Shear Service	A307 $F_u = 60$ ksi Tension Design	A307 Tension Service	A307 Shear Design	A307 Shear Service
½	0.196	6.4	3.8	3.4	1.9	6.6	3.9	3.5	2.0
⅝	0.307	10.0	5.9	5.3	3.0	10.4	6.1	5.5	3.1
¾	0.442	14.4	8.5	7.7	4.4	14.9	8.8	8.0	4.4
⅞	0.601	19.6	11.5	10.5	5.9	20.3	12.0	10.8	6.0
1	0.785	25.6	15.0	13.7	7.7	26.5	15.7	14.1	7.9
1¼	1.227	40.0	23.5	21.3	12.1	41.4	24.5	22.1	12.3
1½	1.767	57.6	33.8	30.7	17.4	59.6	35.3	31.8	17.7
2	3.142	102.4	60.1	54.7	31.0	106.0	62.8	56.6	31.4

Bolt Diameter, in.	Nominal Area A, in.²	ASTM A193 Gr. B5 $F_u = 100$ ksi Tension Design	Tension Service	Shear Design	Shear Service	ASTM A193 Gr. B7 $F_u = 125$ ksi Tension Design	Tension Service	Shear Design	Shear Service
½	0.196	11.0	6.5	5.9	3.3	13.8	8.1	7.4	4.2
⅝	0.307	17.3	10.1	9.2	5.2	21.6	12.7	11.5	6.5
¾	0.442	24.9	14.6	13.3	7.5	31.1	18.2	16.6	9.4
⅞	0.601	33.8	19.8	18.0	10.2	42.3	24.8	22.5	12.8
1	0.785	44.2	25.9	23.6	13.3	55.2	32.4	29.4	16.7
1¼	1.227	69.0	40.5	36.8	20.9	86.3	50.6	46.0	26.1
1½	1.767	99.4	58.3	53.0	30.0	124.2	72.9	66.3	37.5
2	3.142	176.7	103.7	94.3	53.4	220.9	129.6	117.8	66.8

a. AISC Allowable Stress Design [5] or AISC Load and Resistance Factor Design Third Edition. [6] See these manuals for shear-tension interaction.

Figure 6.16.7 High strength coil bolt and coil threaded rod selection chart

Coil Rod Dia. (in.)	Safe Working Load[a] Tension (lb)	Shear (lb)	Minimum Root Area (in.²)	Tensile Strength (psi)	Yield Strength (psi)	Minimum Coil Penetration (in.)
½[b]	9,000	6,000	0.1385	130,000	110,000	2
¾[b]	18,000	12,000	0.3079	117,000	100,000	2¼
1[b]	38,000	25,300	0.5410	140,000	120,000	2½
1¼[b]	56,000	37,500	0.9161	123,000	105,000	2½
1½	68,000	45,300	1.3892	98,000	85,000	3

a. Factor of safety is approximately 2 to 1.
b. Strength requirements similar to ASTM A 325.

Figure 6.16.8 Strength of connection angles

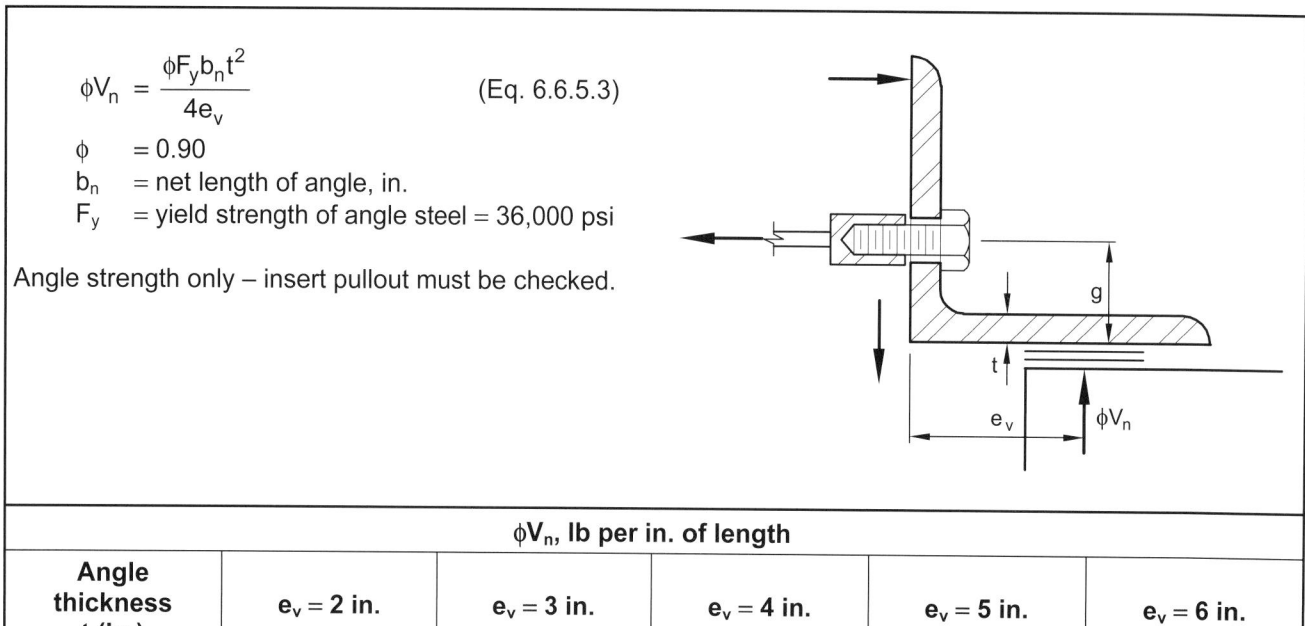

$$\phi V_n = \frac{\phi F_y b_n t^2}{4e_v} \quad \text{(Eq. 6.6.5.3)}$$

$\phi = 0.90$
b_n = net length of angle, in.
F_y = yield strength of angle steel = 36,000 psi

Angle strength only – insert pullout must be checked.

		ϕV_n, lb per in. of length			
Angle thickness t (in.)	e_v = 2 in.	e_v = 3 in.	e_v = 4 in.	e_v = 5 in.	e_v = 6 in.
3/8	570	380	285	228	190
1/2	1013	675	506	405	338
3/4	2278	1519	1139	911	759
1	4050	2700	2025	1620	1350

Figure 6.16.9 Strength of connection angles

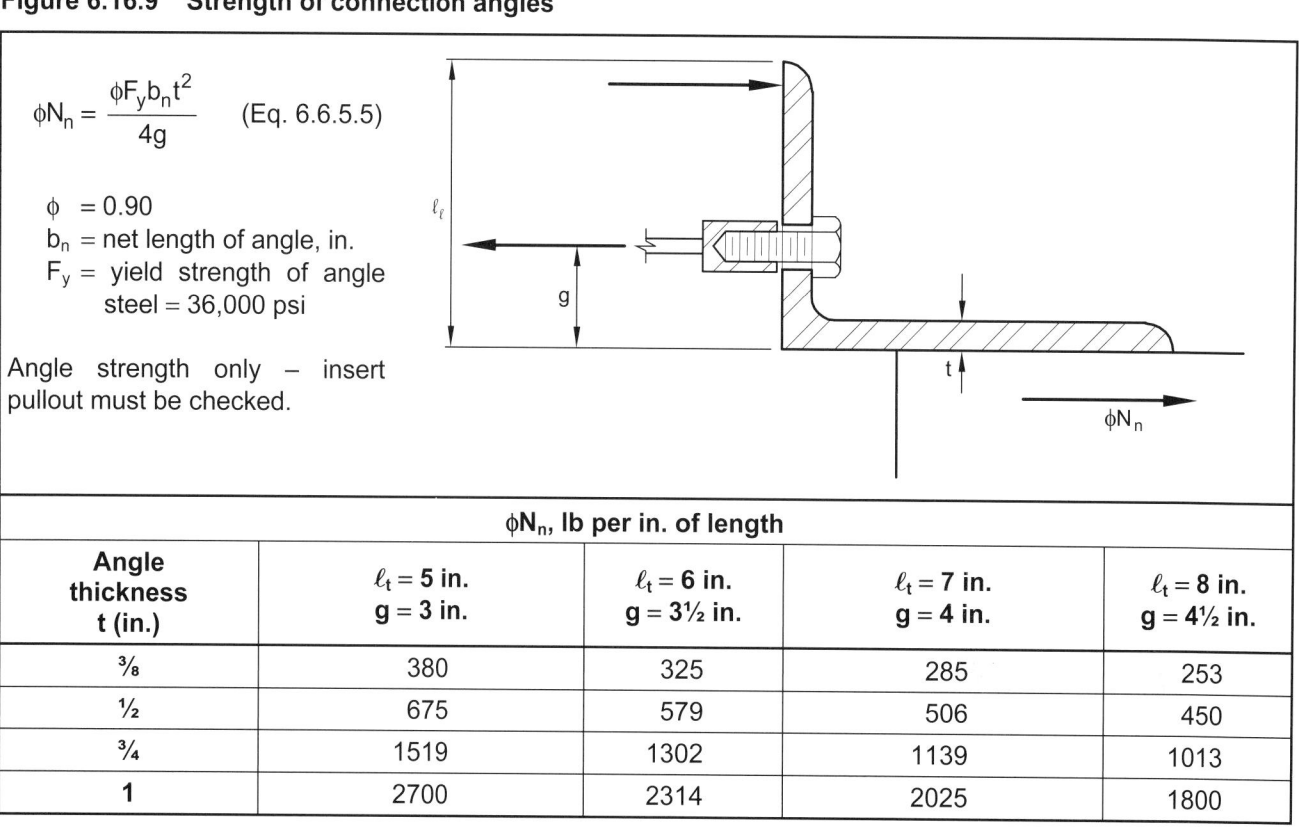

$$\phi N_n = \frac{\phi F_y b_n t^2}{4g} \quad \text{(Eq. 6.6.5.5)}$$

$\phi = 0.90$
b_n = net length of angle, in.
F_y = yield strength of angle steel = 36,000 psi

Angle strength only – insert pullout must be checked.

		ϕN_n, lb per in. of length		
Angle thickness t (in.)	ℓ_t = 5 in. g = 3 in.	ℓ_t = 6 in. g = 3½ in.	ℓ_t = 7 in. g = 4 in.	ℓ_t = 8 in. g = 4½ in.
3/8	380	325	285	253
1/2	675	579	506	450
3/4	1519	1302	1139	1013
1	2700	2314	2025	1800

Figure 6.16.10 Design strength of concrete brackets, corbels or haunches

Design strength by Section 6.8.1 – Cantilever Beam Method for following criteria:

- $\phi V_n = \leq \phi 1000 \lambda^2 bd$
- $f'_c = 5000$ psi
- $f_y = 60,000$ psi
- $N_u = 0.2 V_u$
- b = width of bearing, in.
- d = h − 1.25 (h in.)
- $\phi = 0.75$
- a = ¾ ℓ_p
- $\lambda = 1.0$ (normal weight concrete)

ℓ_p = Projection

$V_u \leq \phi V_n$

Notes:
1. This table is suitable for preliminary design.
2. The strength listed in the table immediately above a blank entry is the limiting value for that depth. The blank entries in a given column will have this same limiting values.
3. Zero entries indicate that reinforcement is less than $A_{s,\,min}$

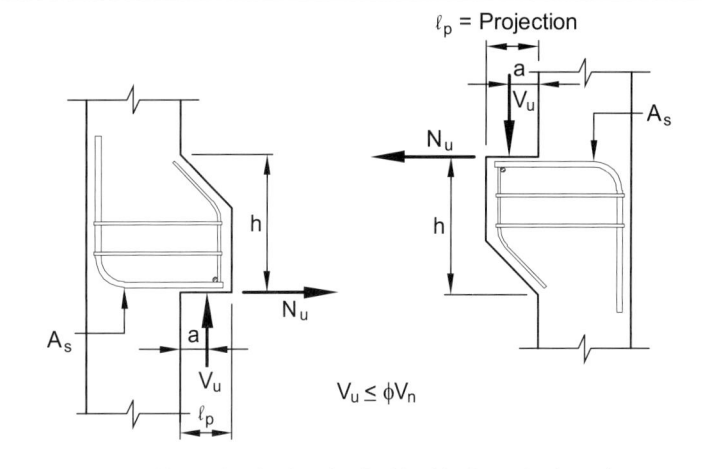

Values of ϕV_n (kips)

b	ℓ_p (in.)		6							8							10						
	h / A_s	6	8	10	12	14	16	18	20	6	8	10	12	14	16	18	20	6	8	10	12	14	16	18	20
6	2 – #4	15	20	24	28	31	34	37	40	12	16	20	23	26	29	31	34	10	13	17	20	22	25	27	29
	2 – #5	23	31	38	43	49	53	58	62	18	25	31	36	40	45	49	52	15	21	26	30	35	38	42	45
	2 – #6	33	44	53	62	68	71	75	78	26	35	43	51	57	63	69	74	22	29	36	43	49	55	60	65
	2 – #7	45	60	71	76	82	86	90	94	36	48	59	69	78	86	90	94	29	40	50	59	67	74	81	88
8	2 – #4	15	20	24	28	31	34			12	16	20	23	26	29			10	13	17	20	22	25		
	2 – #5	23	31	38	43	49	53	58	62	18	25	31	36	40	45	49	52	15	21	26	30	35	38	42	45
	2 – #6	33	44	53	62	69	76	82	87	26	35	43	51	57	63	69	74	22	29	36	43	49	55	60	65
	2 – #7	45	60	73	84	92	97	101	105	36	48	59	69	78	87	94	101	29	40	50	59	67	74	81	88
	2 – #8	59	79	94	101	108	114	119	125	47	63	78	91	103	114	119	125	39	53	65	77	88	98	107	116
	2 – #9	75	97	107	116	124	131	137	144	59	80	98	115	124	131	137	144	49	67	83	98	111	124	136	144
10	2 – #4	15	20	24	28					12	16	20	23					10	13	17	20				
	2 – #5	23	31	38	43	49	53	58		18	25	31	36	40	45	49		15	21	26	30	35	38	42	
	2 – #6	33	44	53	62	69	76	82	87	26	35	43	51	57	63	69	74	22	29	36	43	49	55	60	65
	2 – #7	45	60	73	84	94	103	110	114	36	48	59	69	78	87	94	101	29	40	50	59	67	74	81	88
	2 – #8	59	79	96	111	118	125	130	136	47	63	78	91	103	114	124	133	39	53	65	77	88	98	107	116
	2 – #9	75	100	118	127	136	144	150	157	59	80	98	115	130	144	150	157	49	67	83	98	111	124	136	147
12	2 – #4	15	20	24						12	16	20						10	13	17					
	2 – #5	23	31	38	43	49	53			18	25	31	36	40	45			15	21	26	30	35	38		
	2 – #6	33	44	53	62	69	76	82	87	26	35	43	51	57	63	69	74	22	29	36	43	49	55	60	65
	2 – #7	45	60	73	84	94	103	112	119	36	48	59	69	78	87	94	101	29	40	50	59	67	74	81	88
	2 – #8	59	79	96	111	124	134	140	145	47	63	78	91	103	114	124	133	39	53	65	77	88	98	107	116
	2 – #9	75	100	121	137	146	154	162	168	59	80	98	115	130	144	157	168	49	67	83	98	111	124	136	147
	3 – #4	23	30	36	42	47	52			18	24	30	35	39	43			15	20	25	29	33	37		
	3 – #5	35	46	56	65	73	80	87	92	28	37	46	54	61	67	73	78	23	31	39	45	52	58	63	68
	3 – #6	50	66	80	93	104	114	123	130	39	53	65	76	86	95	104	111	32	44	55	65	74	82	90	97
	3 – #7	68	90	109	126	137	145	152	158	53	72	89	104	117	130	141	152	44	60	75	88	100	112	122	132
	3 – #8	89	118	141	152	162	171	179	187	70	95	117	136	155	171	179	187	58	79	98	116	132	147	161	174
	3 – #9	113	146	161	174	186	196	206	215	89	120	148	173	186	196	206	215	74	100	124	147	167	186	204	215

Figure 6.16.10 (Cont.) Design strength of concrete brackets, corbels or haunches

| | | \multicolumn{21}{c}{Values of ϕV_n (kips)} |
| ℓ_p (in.) | | \multicolumn{8}{c}{6} | \multicolumn{7}{c}{8} | \multicolumn{7}{c}{10} |
b	h / A_s	6	8	10	12	14	16	18	20	6	8	10	12	14	16	18	20	6	8	10	12	14	16	18	20
14	2 – #4	15	20							12	16							10	13						
	2 – #5	23	31	38	43	49				18	25	31	36	40				15	21	26	30	35			
	2 – #6	33	44	53	62	69	76	82	87	26	35	43	51	57	63	69	74	22	29	36	43	49	55	60	65
	2 – #7	45	60	73	84	94	103	112	119	36	48	59	69	78	87	94	101	29	40	50	59	67	74	81	88
	2 – #8	59	79	96	111	124	136	147	154	47	63	78	91	103	114	124	133	39	53	65	77	88	98	107	116
	2 – #9	75	100	121	140	156	164	172	179	59	80	98	115	130	144	157	169	49	67	83	98	111	124	136	147
	3 – #4	23	30	36	42	47				18	24	30	35	39				15	20	25	29	33			
	3 – #5	35	46	56	65	73	80	87	92	28	37	46	54	61	67	73	78	23	31	39	45	52	58	63	68
	3 – #6	50	66	80	93	104	114	123	131	39	53	65	76	86	95	104	111	32	44	55	65	74	82	90	97
	3 – #7	68	90	109	126	141	154	161	167	53	72	89	104	117	130	141	152	44	60	75	88	100	112	122	132
	3 – #8	89	118	144	162	172	182	190	198	70	95	117	136	155	171	186	198	58	79	98	116	132	147	161	174
	3 – #9	113	149	172	186	198	209	220	229	89	120	148	173	196	209	220	229	74	100	124	147	167	186	204	220
16	2 – #6	33	44	53	62	69	76			26	35	43	51	57	63			22	29	36	43	49	55		
	2 – #7	45	60	73	84	94	103	112	119	36	48	59	69	78	87	94	101	29	40	50	59	67	74	81	88
	2 – #8	59	79	96	111	124	136	147	157	47	63	78	91	103	114	124	133	39	53	65	77	88	98	107	116
	2 – #9	75	100	121	140	157	172	181	188	59	80	98	115	130	144	157	169	49	67	83	98	111	124	136	147
	3 – #6	50	66	80	93	104	114	123	131	39	53	65	76	86	95	104	111	32	44	55	65	74	82	90	97
	3 – #7	68	90	109	126	141	155	168	176	53	72	89	104	117	130	141	152	44	60	75	88	100	112	122	132
	3 – #8	89	118	144	166	182	192	201	209	70	95	117	136	155	171	186	200	58	79	98	116	132	147	161	174
	3 – #9	113	149	182	196	209	221	232	241	89	120	148	173	196	216	232	241	74	100	124	147	167	186	204	220
	4 – #6	66	88	107	123	138	152	164	173	52	70	87	101	115	127	138	149	43	59	73	86	98	109	120	129
	4 – #7	90	120	145	168	183	193	202	210	71	96	118	138	156	173	188	203	59	80	99	117	134	149	163	176
	4 – #8	119	157	188	202	216	228	239	249	94	126	156	182	206	228	239	249	78	105	131	154	176	196	215	232
	4 – #9	150	195	215	232	248	262	275	287	119	160	197	230	248	262	275	287	98	134	166	195	223	248	272	287
18	2 – #6	33	44	53	62	69				26	35	43	51	57				22	29	36	43	49			
	2 – #7	45	60	73	84	94	103	112	119	36	48	59	69	78	87	94	101	29	40	50	59	67	74	81	88
	2 – #8	59	79	96	111	124	136	147	157	47	63	78	91	103	114	124	133	39	53	65	77	88	98	107	116
	2 – #9	75	100	121	140	157	172	186	196	59	80	98	115	130	144	157	169	49	67	83	98	111	124	136	147
	3 – #6	50	66	80	93	104	114	123	131	39	53	65	76	86	95	104	111	32	44	55	65	74	82	90	97
	3 – #7	68	90	109	126	141	155	168	179	53	72	89	104	117	130	141	152	44	60	75	88	100	112	122	132
	3 – #8	89	118	144	166	186	201	210	218	70	95	117	136	155	171	186	200	58	79	98	116	132	147	161	174
	3 – #9	113	149	182	206	220	232	243	253	89	120	148	173	196	216	236	253	74	100	124	147	167	186	204	220
	4 – #6	66	88	107	123	138	152	164	175	52	70	87	101	115	127	138	149	43	59	73	86	98	109	120	129
	4 – #7	90	120	145	168	189	202	211	220	71	96	118	138	156	173	188	203	59	80	99	117	134	149	163	176
	4 – #8	119	157	191	213	226	239	250	261	94	126	156	182	206	228	248	261	78	105	131	154	176	196	215	232
	4 – #9	150	199	226	244	260	275	288	301	119	160	197	230	260	275	288	301	98	134	166	195	223	248	272	293
20	2 – #6	33	44	53	62	69				26	35	43	51	57				22	29	36	43	49			
	2 – #7	45	60	73	84	94	103	112		36	48	59	69	78	87	94		29	40	50	59	67	74	81	
	2 – #8	59	79	96	111	124	136	147	157	47	63	78	91	103	114	124	133	39	53	65	77	88	98	107	116
	2 – #9	75	100	121	140	157	172	186	199	59	80	98	115	130	144	157	169	49	67	83	98	111	124	136	147
	3 – #6	50	66	80	93	104	114	123	131	39	53	65	76	86	95	104	111	32	44	55	65	74	82	90	97
	3 – #7	68	90	109	126	141	155	168	179	53	72	89	104	117	130	141	152	44	60	75	88	100	112	122	132
	3 – #8	89	118	144	166	186	204	218	227	70	95	117	136	155	171	186	200	58	79	98	116	132	147	161	174
	3 – #9	113	149	182	210	229	241	253	263	89	120	148	173	196	216	236	253	74	100	124	147	167	186	204	220
	4 – #6	66	88	107	123	138	152	164	175	52	70	87	101	115	127	138	149	43	59	73	86	98	109	120	129
	4 – #7	90	120	145	168	189	207	220	229	71	96	118	138	156	173	188	203	59	80	99	117	134	149	163	176
	4 – #8	119	157	191	222	236	249	261	272	94	126	156	182	206	228	248	267	78	105	131	154	176	196	215	232
	4 – #9	150	199	236	255	272	287	301	314	119	160	197	230	261	287	301	314	98	134	166	195	223	248	272	293
24	2 – #6	33	44	53	62					26	35	43	51					22	29	36	43				
	2 – #7	45	60	73	84	94	103			36	48	59	69	78	87			29	40	50	59	67	74		
	2 – #8	59	79	96	111	124	136	147	157	47	63	78	91	103	114	124	133	39	53	65	77	88	98	107	116
	2 – #9	75	100	121	140	157	172	186	199	59	80	98	115	130	144	157	169	49	67	83	98	111	124	136	147
	3 – #6	50	66	80	93	104	114			39	53	65	76	86	95			32	44	55	65	74	82		
	3 – #7	68	90	109	126	141	155	168	179	53	72	89	104	117	130	141	152	44	60	75	88	100	112	122	132
	3 – #8	89	118	144	166	186	204	221	235	70	95	117	136	155	171	186	200	58	79	98	116	132	147	161	174
	3 – #9	113	149	182	210	236	259	271	282	89	120	148	173	196	216	236	253	74	100	124	147	167	186	204	220
	4 – #6	66	88	107	123	138	152	164	175	52	70	87	101	115	127	138	149	43	59	73	86	98	109	120	129
	4 – #7	90	120	145	168	189	207	223	238	71	96	118	138	156	173	188	203	59	80	99	117	134	149	163	176
	4 – #8	119	157	191	222	248	267	280	291	94	126	156	182	206	228	248	267	78	105	131	154	176	196	215	232
	4 – #9	150	199	242	275	293	309	323	337	119	160	197	230	261	289	314	337	98	134	166	195	223	248	272	293

Figure 6.16.11 Design of structural steel haunches — concrete

Values are for design strength of concrete by Eq. 6.9.1 for following criteria:
$f'_c = 5000$ psi, for other strengths multiply values by
$f'_c / 5000$

Adequacy of structural steel section should be checked.

Additional design strength, ϕV_r, can be obtained with reinforcing bars – see Figure 6.16.12.

$$V_u \leq \phi(V_c + V_r)$$
$$\phi = 0.75$$

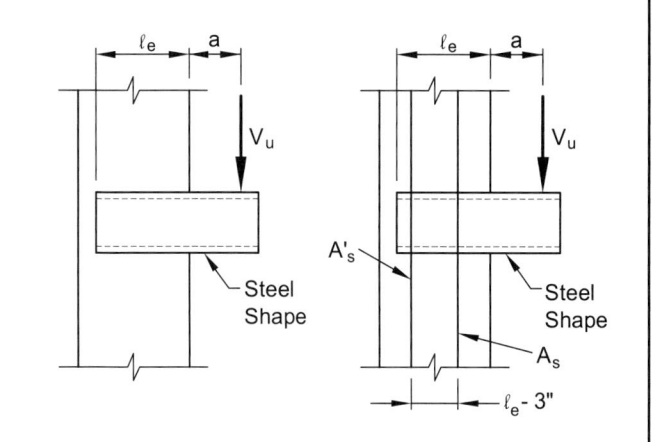

Values of ϕV_c (kips)

Shear span a (in.)	Embedment ℓ_e (in.)	6	7	8	9	10	11	12	13	14	15	16	17
2	6	29	33	38	43	48	53	57	62	67	72	77	81
2	8	41	48	55	62	69	76	83	90	96	103	110	117
2	10	54	63	72	81	91	100	109	118	127	136	145	154
2	12	68	79	90	101	113	124	135	146	158	169	180	191
2	14	81	94	108	121	135	148	162	175	189	202	215	229
2	16	94	110	126	141	157	173	188	204	220	235	251	267
2	18	108	126	143	161	179	197	215	233	251	269	287	305
2	20	121	141	161	182	202	222	242	262	282	303	323	343
2	22	135	157	179	202	224	247	269	292	314	336	359	381
4	6	22	26	29	33	37	40	44	48	51	55	59	63
4	8	33	39	44	50	55	61	67	72	78	83	89	94
4	10	45	53	60	68	75	83	90	98	105	113	120	128
4	12	57	67	77	86	96	105	115	124	134	143	153	163
4	14	70	82	93	105	117	128	140	152	163	175	186	198
4	16	83	96	110	124	138	152	165	179	193	207	221	234
4	18	96	112	128	143	159	175	191	207	223	239	255	271
4	20	109	127	145	163	181	199	217	235	254	272	290	308
4	22	122	142	162	183	203	223	244	264	284	304	325	345
6	6	18	21	24	27	30	33	36	39	42	45	48	51
6	8	28	32	37	42	46	51	56	60	65	70	74	79
6	10	39	45	51	58	64	71	77	84	90	96	103	109
6	12	50	58	67	75	83	91	100	108	116	125	133	141
6	14	62	72	82	92	103	113	123	134	144	154	164	175
6	16	74	86	98	111	123	135	147	160	172	184	197	209
6	18	86	100	115	129	143	158	172	186	201	215	230	244
6	20	99	115	131	148	164	181	197	214	230	246	263	279
6	22	111	130	148	167	185	204	223	241	260	278	297	315
8	6	15	18	20	23	25	28	30	33	35	38	40	43
8	8	24	28	32	36	40	44	48	52	56	60	64	68
8	10	34	39	45	51	56	62	67	73	79	84	90	95
8	12	44	51	59	66	74	81	88	96	103	110	118	125
8	14	55	64	74	83	92	101	110	119	129	138	147	156
8	16	67	78	89	100	111	122	133	144	155	166	177	188
8	18	78	91	104	117	130	143	156	170	183	196	209	222
8	20	90	105	120	135	150	165	180	195	210	226	241	256
8	22	102	119	137	154	171	188	205	222	239	256	273	290

Figure 6.16.12 Design of structural steel haunches — reinforcement

Values are for additional design strength of concrete obtained from reinforcement by Eq. 6.9.2 for following criteria:
  A_s = two bars welded to steel shape
  $A'_s = A_s$
Reinforcement anchored in only one direction.

When reinforcement A_s and A'_s is anchored both above and below steel shape, it can be counted twice (values may be doubled).

For design strength of concrete, ϕV_c — see Figure 6.16.11.
  $V_u \leq \phi V_n$

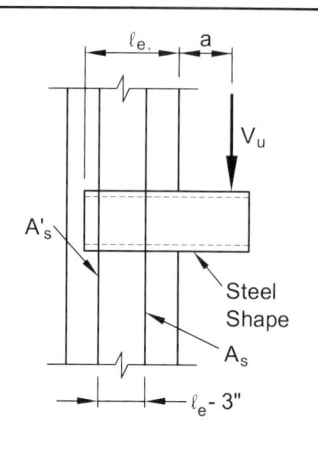

Shear span a (in.)	Embedment ℓ_e (in.)	\#4	\#5	\#6	\#7	\#8	\#9	\#4	\#5	\#6	\#7	\#8	\#9
		\multicolumn{6}{c}{Reinforcing bar size f_y = 40 ksi}	\multicolumn{6}{c}{Reinforcing bar size f_y = 60 ksi}
2	6	5	8	12	16	21	26	8	12	17	24	31	39
	8	7	11	16	22	29	37	11	17	24	33	44	55
	10	9	13	19	26	34	43	13	20	28	39	51	65
	12	9	15	21	28	37	47	14	22	31	43	56	71
	14	10	16	22	30	40	50	15	23	33	45	59	75
	16	10	16	23	31	41	52	16	24	35	47	62	78
	18	11	17	24	32	43	54	16	25	36	49	64	81
	20	11	17	24	33	44	55	17	26	37	50	66	83
	22	11	17	25	34	45	56	17	26	37	51	67	85
4	6	4	6	9	12	16	20	6	9	13	18	24	30
	8	6	9	13	18	24	30	9	14	20	27	36	45
	10	7	11	16	22	29	36	11	17	24	33	43	55
	12	8	13	18	25	32	41	12	19	27	37	49	62
	14	9	14	20	27	35	44	13	21	29	40	53	67
	16	9	15	21	28	37	47	14	22	31	42	56	71
	18	10	15	22	29	39	49	15	23	32	44	58	74
	20	10	16	22	30	40	51	15	24	34	46	60	76
	22	10	16	23	31	41	52	16	24	34	47	62	78
6	6	3	5	7	10	13	16	5	8	11	15	19	24
	8	5	8	11	15	20	25	8	12	17	23	30	38
	10	6	10	14	19	25	32	9	15	21	28	37	47
	12	7	11	16	22	29	36	11	17	24	33	43	54
	14	8	12	18	24	31	40	12	19	26	36	47	60
	16	9	13	19	26	34	43	13	20	28	38	51	64
	18	9	14	20	27	36	45	14	21	30	41	53	68
	20	9	15	21	28	37	47	14	22	31	42	56	70
	22	10	15	21	29	38	49	15	23	32	44	57	73
8	6	3	4	6	8	11	14	4	6	9	12	16	20
	8	4	7	10	13	17	22	7	10	14	20	26	33
	10	6	9	12	17	22	28	8	13	18	25	33	42
	12	7	10	14	20	26	33	10	15	21	29	39	49
	14	7	11	16	22	29	36	11	17	24	33	43	54
	16	8	12	17	23	31	39	12	18	26	35	46	59
	18	8	13	18	25	33	42	12	19	27	37	49	62
	20	9	14	19	26	34	44	13	20	29	39	52	65
	22	9	14	20	27	36	45	14	21	30	41	54	68

Values of ϕV_r (kips)

Figure 6.16.13 Column base plate thickness requirements

Thickness required for concrete bearing

f_{bu} (psi)	x_0 = 3 in.	x_0 = 4 in.	x_0 = 5 in.
500	⅝	¾	⅞
1000	¾	1	1¼
1500	1	1¼	1⅝
2000	1⅛	1½	1¾
2500	1¼	1⅝	2
3000	1⅜	1¾	2¼
3500	1½	1⅞	2⅜
4000	1½	2	2½

Plate F_y = 36 ksi

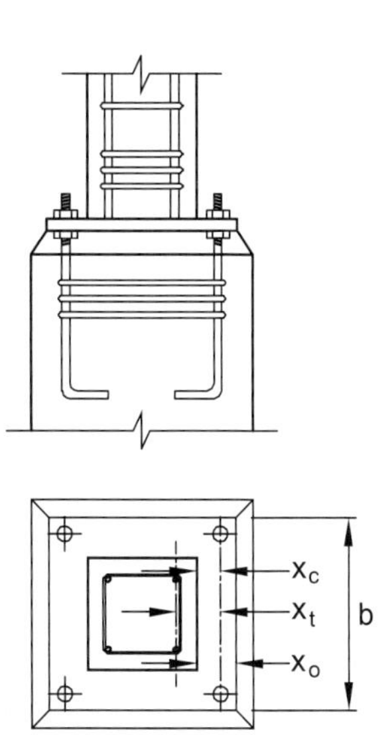

External Anchor Bolts

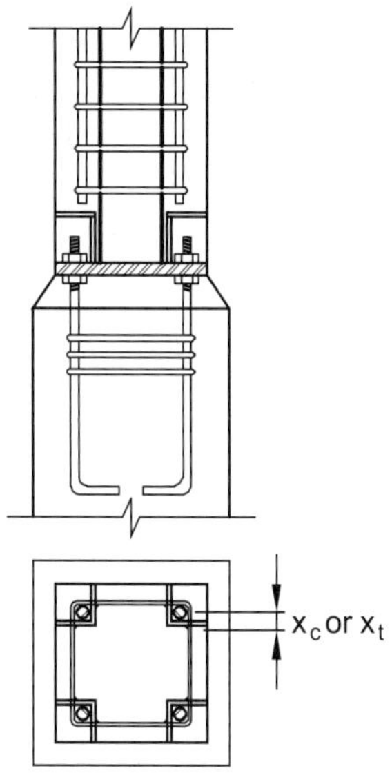

Internal Anchor Bolts

Thickness required for bolt loading

Tension on external anchor bolts

b (in.)	No. and diameter of A-36 or A-307 anchor bolts per side							
	2–¾ in. x_t = 3.75 in.	2–¾ in. x_t = 4.25 in.	2–1 in. x_t = 3.75 in.	2–1 in. x_t = 4.25 in.	2–1¼ in. x_t = 3.75 in.	2–1¼ in. x_t = 4.25 in.	2–1½ in. x_t = 3.75 in.	2–1½ in. x_t = 4.25 in.
12	1⅛	1⅛	1½	1½	1⅞	2	2⅛	2¼
14	1	1⅛	1⅜	1½	1¾	1¾	2	2⅛
16	1	1	1¼	1⅜	1⅜	1⅝	1⅞	2
18	⅞	1	1¼	1¼	1½	1⅝	1¾	1⅞
20	⅞	1	1⅛	1¼	1⅜	1½	1¾	1¾
22	⅞	⅞	1⅛	1⅛	1⅜	1½	1⅝	1¾
24	¾	⅞	1	1⅛	1¼	1⅜	1½	1⅝
26	¾	⅞	1	1⅛	1¼	1⅜	1½	1⅝
28	¾	¾	1	1	1⅛	1¼	1½	1½

Compression on anchor bolts or tension on internal anchor bolts

b (in.)	No. and diameter of A-36 or A-307 anchor bolts per side							
	2–¾ in. x_c = 1.5 in.	2–¾ in. x_c = 2.0 in.	2–1 in. x_c = 1.5 in.	2–1 in. x_c = 2.0 in.	2–1¼ in. x_c = 1.5 in.	2–1¼ in. x_c = 2.0 in.	2–1½ in. x_c = 1.5 in.	2–1½ in. x_c = 2.0 in.
12	¾	⅞	1	1⅛	1¼	1⅜	1⅜	1⅝
14	¾	¾	⅞	1	1⅛	1¼	1¼	1½
16	¾	¾	⅞	1	1	1⅛	1¼	1⅜
18	¾	¾	¾	⅞	1	1⅛	1⅛	1⅜
20	¾	¾	¾	⅞	⅞	1	1⅛	1¼
22	¾	¾	¾	⅞	⅞	1	1	1¼
24	¾	¾	¾	¾	⅞	1	1	1⅛
26	¾	¾	¾	¾	⅞	⅞	1	1⅛
28	¾	¾	¾	¾	¾	⅞	⅞	1⅛

Figure 6.16.14 Strength of welded headed studs based on steel strength

Stud diameter, in.	¼	⅜	½	⅝	¾	⅞
Tension strength, kips	2.25	5.05	9.57	14.96	21.54	29.31
Shear strength, kips	1.95	4.38	8.30	12.96	18.67	25.41

Figure 6.16.15 Concrete tension strength of welded headed stud assemblies

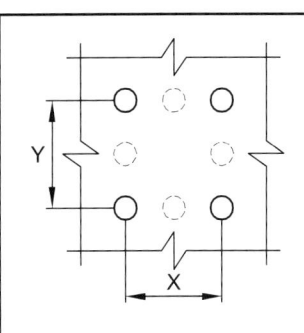

CASE 1: Not near a free edge
Applies to rectangular stud patterns with outside dimensions X and Y. See Figure 6.5.5.1. If stud spacing is $\geq 3h_{ef}$, single stud capacity multiplied by number of studs may govern design.

$$\phi N_{cb} = \phi C_{bs}(X + 3h_{ef})(Y + 3h_{ef})C_{crb}$$

Assumptions:
- ϕ = 0.7 (no confinement reinforcement – multiply Table A values by 0.75/0.7 when confinement reinforcement is used.)
- C_{crb} = 1.0 (uncracked concrete – multiply Table A values by 0.8 when concrete may be cracked.)
- $C_{bs} = 3.33\sqrt{(f'_c/h_{ef})}$ (normal weight concrete – multiply Table A values by λ when lightweight concrete is used.)
- f'_c = 5000 psi (multiply Table A values by $\sqrt{f'_c/5000}$ for other concrete strengths)
- $h_{ef} = L_s + t_p - t_{hs}$
- t_p (plate thickness) = ⅜ in. t_{hs} (stud head thickness) = ⅜ in.

Design Tensile Strength, ϕN_{cb}

h_{ef}	Y \ X	2	4	6	8	10	12	14	16	18	20	22	24	26	28	30
3	0	9.4	11.1	12.8	14.6	16.3	18.0	19.7	21.4	23.1	24.8	26.6	28.3	30.0	31.7	33.4
	2	11.5	13.6	15.7	17.8	19.9	22.0	24.1	26.2	28.3	30.4	32.5	34.5	36.6	38.7	40.8
	4	13.6	16.1	18.6	21.0	23.5	26.0	28.5	30.9	33.4	35.9	38.4	40.8	43.3	45.8	48.2
	6	15.7	18.6	21.4	24.3	27.1	30.0	32.8	35.7	38.5	41.4	44.3	47.1	50.0	52.8	55.7
	8	17.8	21.0	24.3	27.5	30.7	34.0	37.2	40.4	43.7	46.9	50.2	53.4	56.6	59.9	63.1
	10	19.9	23.5	27.1	30.7	34.4	38.0	41.6	45.2	48.8	52.4	56.1	59.7	63.3	66.9	70.5
	12	22.0	26.0	30.0	34.0	38.0	42.0	46.0	50.0	54.0	58.0	62.0	65.9	69.9	73.9	77.9
	14	24.1	28.5	32.8	37.2	41.6	46.0	50.3	54.7	59.1	63.5	67.9	72.2	76.6	81.0	85.4
4	0	13.8	15.8	17.8	19.8	21.8	23.7	25.7	27.7	29.7	31.6	33.6	35.6	37.6	39.6	41.5
	2	16.2	18.5	20.8	23.1	25.4	27.7	30.0	32.3	34.6	36.9	39.2	41.5	43.8	46.2	48.5
	4	18.5	21.1	23.7	26.4	29.0	31.6	34.3	36.9	39.6	42.2	44.8	47.5	50.1	52.7	55.4
	6	20.8	23.7	26.7	29.7	32.6	35.6	38.6	41.5	44.5	47.5	50.4	53.4	56.4	59.3	62.3
	8	23.1	26.4	29.7	33.0	36.3	39.6	42.9	46.2	49.4	52.7	56.0	59.3	62.6	65.9	69.2
	10	25.4	29.0	32.6	36.3	39.9	43.5	47.1	50.8	54.4	58.0	61.6	65.3	68.9	72.5	76.1
	12	27.7	31.6	35.6	39.6	43.5	47.5	51.4	55.4	59.3	63.3	67.2	71.2	75.2	79.1	83.1
	14	30.0	34.3	38.6	42.9	47.1	51.4	55.7	60.0	64.3	68.6	72.9	77.1	81.4	85.7	90.0
6	0	24.2	26.6	29.1	31.5	33.9	36.3	38.8	41.2	43.6	46.0	48.4	50.9	53.3	55.7	58.1
	2	26.9	29.6	32.3	35.0	37.7	40.4	43.1	45.8	48.4	51.1	53.8	56.5	59.2	61.9	64.6
	4	29.6	32.6	35.5	38.5	41.5	44.4	47.4	50.3	53.3	56.3	59.2	62.2	65.1	68.1	71.1
	6	32.3	35.5	38.8	42.0	45.2	48.4	51.7	54.9	58.1	61.4	64.6	67.8	71.1	74.3	77.5
	8	35.0	38.5	42.0	45.5	49.0	52.5	56.0	59.5	63.0	66.5	70.0	73.5	77.0	80.5	84.0
	10	37.7	41.5	45.2	49.0	52.8	56.5	60.3	64.1	67.8	71.6	75.4	79.1	82.9	86.7	90.4
	12	40.4	44.4	48.4	52.5	56.5	60.6	64.6	68.6	72.7	76.7	80.7	84.8	88.8	92.9	96.9
	14	43.1	47.4	51.7	56.0	60.3	64.6	68.9	73.2	77.5	81.8	86.1	90.4	94.7	99.1	103.4
	16	45.8	50.3	54.9	59.5	64.1	68.6	73.2	77.8	82.4	86.9	91.5	96.1	100.7	105.2	109.8
8	0	36.4	39.2	42.0	44.8	47.6	50.3	53.1	55.9	58.7	61.5	64.3	67.1	69.9	72.7	75.5
	2	39.4	42.4	45.5	48.5	51.5	54.5	57.6	60.6	63.6	66.7	69.7	72.7	75.8	78.8	81.8
	4	42.4	45.7	49.0	52.2	55.5	58.7	62.0	65.3	68.5	71.8	75.1	78.3	81.6	84.8	88.1
	6	45.5	49.0	52.4	55.9	59.4	62.9	66.4	69.9	73.4	76.9	80.4	83.9	87.4	90.9	94.4
	8	48.5	52.2	55.9	59.7	63.4	67.1	70.9	74.6	78.3	82.1	85.8	89.5	93.2	97.0	100.7
	10	51.5	55.5	59.4	63.4	67.4	71.3	75.3	79.3	83.2	87.2	91.1	95.1	99.1	103.0	107.0
	12	54.5	58.7	62.9	67.1	71.3	75.5	79.7	83.9	88.1	92.3	96.5	100.7	104.9	109.1	113.3
	14	57.6	62.0	66.4	70.9	75.3	79.7	84.1	88.6	93.0	97.4	101.9	106.3	110.7	115.2	119.6
	16	60.6	65.3	69.9	74.6	79.3	83.9	88.6	93.2	97.9	102.6	107.2	111.9	116.6	121.2	125.9

Figure 6.16.15 Concrete tension strength of welded headed stud assemblies (Cont.)

CASE 2: Near one free edge

Applies to rectangular stud patterns with outside dimensions X and Y. See Figure 6.5.5.1. If stud spacing is $\geq 3h_{ef}$, single stud capacity multiplied by number of studs may govern design.

$\phi N_{cb} = \phi N'_{cb}(\psi_{ed,N})$ [$\phi N'_{cb}$ from Table A ; $\psi_{ed,N}$ from Table B]

Assumptions for Table A:
- ϕ = 0.7 (no confinement reinforcement – multiply Table A values by 0.75/0.7 when confinement reinforcement is used)
- C_{crb} = 1.0 (uncracked concrete – multiply Table A values by 0.8 when concrete may be cracked)
- $C_{bs} = 3.33\sqrt{(f'_c/h_{ef})}$ (normal weight concrete – multiply Table A values by λ when lightweight concrete is used)
- f'_c = 5000 psi (multiply Table A values by $\sqrt{f'_c/5000}$ for other concrete strengths)
- $h_{ef} = L_s + t_p - t_{hs}$
- t_p (plate thickness) = 3/8 in. t_{hs} (stud head thickness) = 3/8 in.

Table A: $\phi N'_{cb} = \phi C_{bs}(d_{e1} + X + 1.5h_{ef})(Y + 3h_{ef})C_{crb}$

h_{ef}	Y \ b	2	4	6	8	10	12	14	16	18	20	22	24	26	28	30
3	0	5.6	7.3	9.0	10.7	12.4	14.1	15.8	17.6	19.3	21.0	22.7	24.4	26.1	27.8	29.5
	2	6.8	8.9	11.0	13.1	15.2	17.3	19.4	21.5	23.6	25.6	27.7	29.8	31.9	34.0	36.1
	4	8.0	10.5	13.0	15.5	17.9	20.4	22.9	25.4	27.8	30.3	32.8	35.3	37.7	40.2	42.7
	6	9.3	12.1	15.0	17.8	20.7	23.6	26.4	29.3	32.1	35.0	37.8	40.7	43.5	46.4	49.2
	8	10.5	13.8	17.0	20.2	23.5	26.7	29.9	33.2	36.4	39.6	42.9	46.1	49.3	52.6	55.8
	10	11.8	15.4	19.0	22.6	26.2	29.8	33.4	37.1	40.7	44.3	47.9	51.5	55.1	58.8	62.4
	12	13.0	17.0	21.0	25.0	29.0	33.0	37.0	41.0	45.0	49.0	53.0	57.0	61.0	64.9	68.9
	14	14.2	18.6	23.0	27.4	31.7	36.1	40.5	44.9	49.2	53.6	58.0	62.4	66.8	71.1	75.5
4	0	7.9	9.9	11.9	13.8	15.8	17.8	19.8	21.8	23.7	25.7	27.7	29.7	31.6	33.6	35.6
	2	9.2	11.5	13.8	16.2	18.5	20.8	23.1	25.4	27.7	30.0	32.3	34.6	36.9	39.2	41.5
	4	10.5	13.2	15.8	18.5	21.1	23.7	26.4	29.0	31.6	34.3	36.9	39.6	42.2	44.8	47.5
	6	11.9	14.8	17.8	20.8	23.7	26.7	29.7	32.6	35.6	38.6	41.5	44.5	47.5	50.4	53.4
	8	13.2	16.5	19.8	23.1	26.4	29.7	33.0	36.3	39.6	42.9	46.2	49.4	52.7	56.0	59.3
	10	14.5	18.1	21.8	25.4	29.0	32.6	36.3	39.9	43.5	47.1	50.8	54.4	58.0	61.6	65.3
	12	15.8	19.8	23.7	27.7	31.6	35.6	39.6	43.5	47.5	51.4	55.4	59.3	63.3	67.2	71.2
	14	17.1	21.4	25.7	30.0	34.3	38.6	42.9	47.1	51.4	55.7	60.0	64.3	68.6	72.9	77.1
6	0	13.3	15.7	18.2	20.6	23.0	25.4	27.9	30.3	32.7	35.1	37.5	40.0	42.4	44.8	47.2
	2	14.8	17.5	20.2	22.9	25.6	28.3	31.0	33.6	36.3	39.0	41.7	44.4	47.1	49.8	52.5
	4	16.3	19.2	22.2	25.2	28.1	31.1	34.0	37.0	40.0	42.9	45.9	48.9	51.8	54.8	57.7
	6	17.8	21.0	24.2	27.5	30.7	33.9	37.1	40.4	43.6	46.8	50.1	53.3	56.5	59.8	63.0
	8	19.2	22.7	26.2	29.7	33.2	36.7	40.2	43.7	47.2	50.7	54.2	57.7	61.2	64.7	68.2
	10	20.7	24.5	28.3	32.0	35.8	39.6	43.3	47.1	50.9	54.6	58.4	62.2	65.9	69.7	73.5
	12	22.2	26.2	30.3	34.3	38.4	42.4	46.4	50.5	54.5	58.5	62.6	66.6	70.7	74.7	78.7
	14	23.7	28.0	32.3	36.6	40.9	45.2	49.5	53.8	58.1	62.4	66.8	71.1	75.4	79.7	84.0
	16	25.2	29.7	34.3	38.9	43.5	48.0	52.6	57.2	61.8	66.3	70.9	75.5	80.1	84.7	89.2
8	0	19.6	22.4	25.2	28.0	30.8	33.6	36.4	39.2	42.0	44.8	47.6	50.3	53.1	55.9	58.7
	2	21.2	24.2	27.3	30.3	33.3	36.4	39.4	42.4	45.5	48.5	51.5	54.5	57.6	60.6	63.6
	4	22.8	26.1	29.4	32.6	35.9	39.2	42.4	45.7	49.0	52.2	55.5	58.7	62.0	65.3	68.5
	6	24.5	28.0	31.5	35.0	38.5	42.0	45.5	49.0	52.4	55.9	59.4	62.9	66.4	69.9	73.4
	8	26.1	29.8	33.6	37.3	41.0	44.8	48.5	52.2	55.9	59.7	63.4	67.1	70.9	74.6	78.3
	10	27.7	31.7	35.7	39.6	43.6	47.6	51.5	55.5	59.4	63.4	67.4	71.3	75.3	79.3	83.2
	12	29.4	33.6	37.8	42.0	46.2	50.3	54.5	58.7	62.9	67.1	71.3	75.5	79.7	83.9	88.1
	14	31.0	35.4	39.9	44.3	48.7	53.1	57.6	62.0	66.4	70.9	75.3	79.7	84.1	88.6	93.0
	16	32.6	37.3	42.0	46.6	51.3	55.9	60.6	65.3	69.9	74.6	79.3	83.9	88.6	93.2	97.9

Table B: Modification for Edge Distance ($\psi_{ed,N}$)

L_s \ d_{min}	<1.25	1.25	1.5	2	2.5	3	3.5	4	4.5	5	5.5	6	7	8	10	12
3	a	0.700	0.700	0.700	0.700	0.700	0.700	0.700	0.700	b	b	b	b	b	b	b
4	a	a	a	0.700	0.700	0.700	0.700	0.700	0.700	0.700	0.700	0.700	b	b	b	b
6	a	a	a	a	0.700	0.700	0.700	0.700	0.700	0.700	0.700	0.700	0.700	0.700	b	b
8	a	a	a	a	a	a	0.700	0.700	0.700	0.700	0.700	0.700	0.700	0.700	0.700	0.700

a – $d_{min} < 0.4h_{ef}$ thus, side-face blowout governs (see Section D.5.4 of Appendix D, ACI 318-02).
b – $d_{min} > 1.5h_{ef}$ thus, stud group is no longer considered near that free edge (another case may govern).

Figure 6.16.15 Concrete tension strength of welded headed stud assemblies (Cont.)

CASE 3: Free edges on two opposite sides

Applies to rectangular stud patterns with outside dimensions X and Y. See Figure 6.5.5.1. If stud spacing is $\geq 3h_{ef}$, single stud capacity multiplied by number of studs may govern design.

$\phi N_{cb} = \phi N'_{cb}(\psi_{ed,N})$ [$\phi N'_{cb}$ from Table A ; $\psi_{ed,N}$ from Table B]

Assumptions for Table A:

ϕ = 0.7 (no confinement reinforcement – multiply Table A values by 0.75/0.7 when confinement reinforcement is used)

C_{crb} = 1.0 (uncracked concrete – multiply Table A values by 0.8 when concrete may be cracked)

C_{bs} = $3.33\sqrt{(f'_c/h_{ef})}$ (normal weight concrete – multiply Table A values by λ when lightweight concrete is used)

f'_c = 5000 psi (multiply Table A values by $\sqrt{f'_c/5000}$ for other concrete strengths)

h_{ef} = $L_s + t_p - t_{hs}$

t_p (plate thickness) = ⅜ in. t_{hs} (stud head thickness) = ⅜ in.

Table A: $\phi N'_{cb} = \phi C_{bs}(d_{e1} + X + d_{e2})(Y + 3h_{ef})C_{crb}$

h_{ef}	b → Y ↓	2	4	6	8	10	12	14	16	18	20	22	24	26	28	30
3	0	1.7	3.4	5.1	6.9	8.6	10.3	12.0	13.7	15.4	17.1	18.8	20.6	22.3	24.0	25.7
	2	2.1	4.2	6.3	8.4	10.5	12.6	14.7	16.7	18.8	20.9	23.0	25.1	27.2	29.3	31.4
	4	2.5	4.9	7.4	9.9	12.4	14.8	17.3	19.8	22.3	24.7	27.2	29.7	32.2	34.6	37.1
	6	2.9	5.7	8.6	11.4	14.3	17.1	20.0	22.8	25.7	28.5	31.4	34.3	37.1	40.0	42.8
	8	3.2	6.5	9.7	12.9	16.2	19.4	22.6	25.9	29.1	32.4	35.6	38.8	42.1	45.3	48.5
	10	3.6	7.2	10.8	14.5	18.1	21.7	25.3	28.9	32.5	36.2	39.8	43.4	47.0	50.6	54.2
	12	4.0	8.0	12.0	16.0	20.0	24.0	28.0	32.0	36.0	40.0	44.0	48.0	52.0	56.0	60.0
	14	4.4	8.8	13.1	17.5	21.9	26.3	30.6	35.0	39.4	43.8	48.2	52.5	56.9	61.3	65.7
4	0	2.0	4.0	5.9	7.9	9.9	11.9	13.8	15.8	17.8	19.8	21.8	23.7	25.7	27.7	29.7
	2	2.3	4.6	6.9	9.2	11.5	13.8	16.2	18.5	20.8	23.1	25.4	27.7	30.0	32.3	34.6
	4	2.6	5.3	7.9	10.5	13.2	15.8	18.5	21.1	23.7	26.4	29.0	31.6	34.3	36.9	39.6
	6	3.0	5.9	8.9	11.9	14.8	17.8	20.8	23.7	26.7	29.7	32.6	35.6	38.6	41.5	44.5
	8	3.3	6.6	9.9	13.2	16.5	19.8	23.1	26.4	29.7	33.0	36.3	39.6	42.9	46.2	49.4
	10	3.6	7.3	10.9	14.5	18.1	21.8	25.4	29.0	32.6	36.3	39.9	43.5	47.1	50.8	54.4
	12	4.0	7.9	11.9	15.8	19.8	23.7	27.7	31.6	35.6	39.6	43.5	47.5	51.4	55.4	59.3
	14	4.3	8.6	12.9	17.1	21.4	25.7	30.0	34.3	38.6	42.9	47.1	51.4	55.7	60.0	64.3
6	0	2.4	4.8	7.3	9.7	12.1	14.5	17.0	19.4	21.8	24.2	26.6	29.1	31.5	33.9	36.3
	2	2.7	5.4	8.1	10.8	13.5	16.1	18.8	21.5	24.2	26.9	29.6	32.3	35.0	37.7	40.4
	4	3.0	5.9	8.9	11.8	14.8	17.8	20.7	23.7	26.6	29.6	32.6	35.5	38.5	41.5	44.4
	6	3.2	6.5	9.7	12.9	16.1	19.4	22.6	25.8	29.1	32.3	35.5	38.8	42.0	45.2	48.4
	8	3.5	7.0	10.5	14.0	17.5	21.0	24.5	28.0	31.5	35.0	38.5	42.0	45.5	49.0	52.5
	10	3.8	7.5	11.3	15.1	18.8	22.6	26.4	30.1	33.9	37.7	41.5	45.2	49.0	52.8	56.5
	12	4.0	8.1	12.1	16.1	20.2	24.2	28.3	32.3	36.3	40.4	44.4	48.4	52.5	56.5	60.6
	14	4.3	8.6	12.9	17.2	21.5	25.8	30.1	34.5	38.8	43.1	47.4	51.7	56.0	60.3	64.6
	16	4.6	9.2	13.7	18.3	22.9	27.5	32.0	36.6	41.2	45.8	50.3	54.9	59.5	64.1	68.6
8	0	2.8	5.6	8.4	11.2	14.0	16.8	19.6	22.4	25.2	28.0	30.8	33.6	36.4	39.2	42.0
	2	3.0	6.1	9.1	12.1	15.2	18.2	21.2	24.2	27.3	30.3	33.3	36.4	39.4	42.4	45.5
	4	3.3	6.5	9.8	13.1	16.3	19.6	22.8	26.1	29.4	32.6	35.9	39.2	42.4	45.7	49.0
	6	3.5	7.0	10.5	14.0	17.5	21.0	24.5	28.0	31.5	35.0	38.5	42.0	45.5	49.0	52.4
	8	3.7	7.5	11.2	14.9	18.6	22.4	26.1	29.8	33.6	37.3	41.0	44.8	48.5	52.2	55.9
	10	4.0	7.9	11.9	15.9	19.8	23.8	27.7	31.7	35.7	39.6	43.6	47.6	51.5	55.5	59.4
	12	4.2	8.4	12.6	16.8	21.0	25.2	29.4	33.6	37.8	42.0	46.2	50.3	54.5	58.7	62.9
	14	4.4	8.9	13.3	17.7	22.1	26.6	31.0	35.4	39.9	44.3	48.7	53.1	57.6	62.0	66.4
	16	4.7	9.3	14.0	18.6	23.3	28.0	32.6	37.3	42.0	46.6	51.3	55.9	60.6	65.3	69.9

Table B: Modification for Edge Distance ($\psi_{ed,N}$)

L_s \ d_{min}	<1.25	1.25	1.5	2	2.5	3	3.5	4	4.5	5	5.5	6	7	8	10	12
3	a	0.700	0.700	0.700	0.700	0.700	0.700	0.700	0.700	b	b	b	b	b	b	b
4	a	a	a	0.700	0.700	0.700	0.700	0.700	0.700	0.700	0.700	0.700	b	b	b	b
6	a	a	a	a	0.700	0.700	0.700	0.700	0.700	0.700	0.700	0.700	0.700	0.700	b	b
8	a	a	a	a	a	a	0.700	0.700	0.700	0.700	0.700	0.700	0.700	0.700	0.700	0.700

a – d_{min} < $0.4h_{ef}$ thus, side-face blowout governs (see Section D.5.4 of Appendix D, ACI 318-02).
b – d_{min} > $1.5h_{ef}$ thus, stud group is no longer considered near that free edge (another case may govern).

Figure 6.16.15 Concrete tension strength of welded headed stud assemblies (Cont.)

CASE 4: Free edges on two adjacent sides

Applies to rectangular stud patterns with outside dimensions X and Y. See Figure 6.5.5.1. If stud spacing is $\geq 3h_{ef}$, single stud capacity multiplied by number of studs may govern design.

$\phi N_{cb} = \phi N'_{cb}(\psi_{ed,N})$ [$\phi N'_{cb}$ from Table A ; $\psi_{ed,N}$ from Table B]

Assumptions for Table A:

- ϕ = 0.7 (no confinement reinforcement – multiply Table A values by 0.75/0.7 when confinement reinforcement is used)
- C_{crb} = 1.0 (uncracked concrete – multiply Table A values by 0.8 when concrete may be cracked)
- $C_{bs} = 3.33\sqrt{(f'_c/h_{ef})}$ (normal weight concrete – multiply Table A values by λ when lightweight concrete is used)
- f'_c = 5000 psi (multiply Table A values by $\sqrt{f'_c/5000}$ for other concrete strengths)
- $h_{ef} = L_s + t_p - t_{hs}$
- t_p (plate thickness) = 3/8 in. t_{hs} (stud head thickness) = 3/8 in.

Table A: $\phi N'_{cb} = \phi C_{bs}(d_{e1} + X + 1.5h_{ef})(d_{e3} + Y + 1.5h_{ef})C_{crb}$

h_{ef}	b → a ↓	2	4	6	8	10	12	14	16	18	20	22	24	26	28	30
3	0	2.8	3.6	4.5	5.4	6.2	7.1	7.9	8.8	9.6	10.5	11.3	12.2	13.1	13.9	14.8
	2	4.0	5.3	6.5	7.7	9.0	10.2	11.4	12.7	13.9	15.2	16.4	17.6	18.9	20.1	21.3
	4	5.3	6.9	8.5	10.1	11.7	13.3	15.0	16.6	18.2	19.8	21.4	23.1	24.7	26.3	27.9
	6	6.5	8.5	10.5	12.5	14.5	16.5	18.5	20.5	22.5	24.5	26.5	28.5	30.5	32.5	34.5
	8	7.7	10.1	12.5	14.9	17.2	19.6	22.0	24.4	26.8	29.1	31.5	33.9	36.3	38.7	41.0
	10	9.0	11.7	14.5	17.2	20.0	22.8	25.5	28.3	31.0	33.8	36.6	39.3	42.1	44.8	47.6
	12	10.2	13.3	16.5	19.6	22.8	25.9	29.0	32.2	35.3	38.5	41.6	44.8	47.9	51.0	54.2
	14	11.4	15.0	18.5	22.0	25.5	29.0	32.6	36.1	39.6	43.1	46.7	50.2	53.7	57.2	60.7
4	0	4.0	4.9	5.9	6.9	7.9	8.9	9.9	10.9	11.9	12.9	13.8	14.8	15.8	16.8	17.8
	2	5.3	6.6	7.9	9.2	10.5	11.9	13.2	14.5	15.8	17.1	18.5	19.8	21.1	22.4	23.7
	4	6.6	8.2	9.9	11.5	13.2	14.8	16.5	18.1	19.8	21.4	23.1	24.7	26.4	28.0	29.7
	6	7.9	9.9	11.9	13.8	15.8	17.8	19.8	21.8	23.7	25.7	27.7	29.7	31.6	33.6	35.6
	8	9.2	11.5	13.8	16.2	18.5	20.8	23.1	25.4	27.7	30.0	32.3	34.6	36.9	39.2	41.5
	10	10.5	13.2	15.8	18.5	21.1	23.7	26.4	29.0	31.6	34.3	36.9	39.6	42.2	44.8	47.5
	12	11.9	14.8	17.8	20.8	23.7	26.7	29.7	32.6	35.6	38.6	41.5	44.5	47.5	50.4	53.4
	14	13.2	16.5	19.8	23.1	26.4	29.7	33.0	36.3	39.6	42.9	46.2	49.4	52.7	56.0	59.3
6	0	6.7	7.9	9.1	10.3	11.5	12.7	13.9	15.1	16.4	17.6	18.8	20.0	21.2	22.4	23.6
	2	8.1	9.6	11.1	12.6	14.1	15.5	17.0	18.5	20.0	21.5	22.9	24.4	25.9	27.4	28.9
	4	9.6	11.4	13.1	14.9	16.6	18.4	20.1	21.9	23.6	25.4	27.1	28.9	30.6	32.4	34.1
	6	11.1	13.1	15.1	17.2	19.2	21.2	23.2	25.2	27.3	29.3	31.3	33.3	35.3	37.3	39.4
	8	12.6	14.9	17.2	19.4	21.7	24.0	26.3	28.6	30.9	33.2	35.5	37.7	40.0	42.3	44.6
	10	14.1	16.6	19.2	21.7	24.3	26.8	29.4	32.0	34.5	37.1	39.6	42.2	44.7	47.3	49.9
	12	15.5	18.4	21.2	24.0	26.8	29.7	32.5	35.3	38.2	41.0	43.8	46.6	49.5	52.3	55.1
	14	17.0	20.1	23.2	26.3	29.4	32.5	35.6	38.7	41.8	44.9	48.0	51.1	54.2	57.3	60.4
	16	18.5	21.9	25.2	28.6	32.0	35.3	38.7	42.1	45.4	48.8	52.1	55.5	58.9	62.2	65.6
8	0	9.8	11.2	12.6	14.0	15.4	16.8	18.2	19.6	21.0	22.4	23.8	25.2	26.6	28.0	29.4
	2	11.4	13.1	14.7	16.3	17.9	19.6	21.2	22.8	24.5	26.1	27.7	29.4	31.0	32.6	34.3
	4	13.1	14.9	16.8	18.6	20.5	22.4	24.2	26.1	28.0	29.8	31.7	33.6	35.4	37.3	39.2
	6	14.7	16.8	18.9	21.0	23.1	25.2	27.3	29.4	31.5	33.6	35.7	37.8	39.9	42.0	44.1
	8	16.3	18.6	21.0	23.3	25.6	28.0	30.3	32.6	35.0	37.3	39.6	42.0	44.3	46.6	49.0
	10	17.9	20.5	23.1	25.6	28.2	30.8	33.3	35.9	38.5	41.0	43.6	46.2	48.7	51.3	53.8
	12	19.6	22.4	25.2	28.0	30.8	33.6	36.4	39.2	42.0	44.8	47.6	50.3	53.1	55.9	58.7
	14	21.2	24.2	27.3	30.3	33.3	36.4	39.4	42.4	45.5	48.5	51.5	54.5	57.6	60.6	63.6
	16	22.8	26.1	29.4	32.6	35.9	39.2	42.4	45.7	49.0	52.2	55.5	58.7	62.0	65.3	68.5

Table B: Modification for Edge Distance ($\psi_{ed,N}$)

L_s \ d_{min}	<1.25	1.25	1.5	2	2.5	3	3.5	4	4.5	5	5.5	6	7	8	10	12
3	a	0.700	0.700	0.700	0.700	0.700	0.700	0.700	0.700	b	b	b	b	b	b	b
4	a	a	a	0.700	0.700	0.700	0.700	0.700	0.700	0.700	0.700	0.700	b	b	b	b
6	a	a	a	a	a	0.700	0.700	0.700	0.700	0.700	0.700	0.700	0.700	0.700	b	b
8	a	a	a	a	a	a	a	0.700	0.700	0.700	0.700	0.700	0.700	0.700	0.700	0.700

a – $d_{min} < 0.4h_{ef}$ thus, side-face blowout governs (see Section D.5.4 of Appendix D, ACI 318-02).
b – $d_{min} > 1.5h_{ef}$ thus, stud group is no longer considered near that free edge (another case may govern).

Figure 6.16.15 Concrete tension strength of welded headed stud assemblies (Cont.)

CASE 5: Free edges on three sides

Applies to rectangular stud patterns with outside dimensions X and Y. See Figure 6.5.5.1. If stud spacing is ≥ 3h_{ef}, single stud capacity multiplied by number of studs may govern design.

$\phi N_{cb} = \phi N'_{cb}(\psi_{ed,N})$ [$\phi N'_{cb}$ from Table A ; $\psi_{ed,N}$ from Table B]

Assumptions for Table A:
- ϕ = 0.7 (no confinement reinforcement – multiply Table A values by 0.75/0.7 when confinement reinforcement is used)
- C_{crb} = 1.0 (uncracked concrete – multiply Table A values by 0.8 when concrete may be cracked)
- C_{bs} = $3.33\sqrt{(f'_c / h_{ef})}$ (normal weight concrete – multiply Table A values by λ when lightweight concrete is used)
- f'_c = 5000 psi (multiply Table A values by $\sqrt{f'_c / 5000}$ for other concrete strengths)
- h_{ef} = $L_s + t_p - t_{hs}$
- t_p (plate thickness) = ⅜ in. t_{hs} (stud head thickness) = ⅜ in.

Table A: $\phi N'_{cb} = \phi C_{bs}(d_{e1} + X + d_{e2})(d_{e3} + Y + 1.5h_{ef})C_{crb}$

h_{ef}	b / a	2	4	6	8	10	12	14	16	18	20	22	24	26	28	30
3	0	0.9	1.7	2.6	3.4	4.3	5.1	6.0	6.9	7.7	8.6	9.4	10.3	11.1	12.0	12.8
	2	1.2	2.5	3.7	4.9	6.2	7.4	8.7	9.9	11.1	12.4	13.6	14.8	16.1	17.3	18.6
	4	1.6	3.2	4.9	6.5	8.1	9.7	11.3	12.9	14.6	16.2	17.8	19.4	21.0	22.6	24.3
	6	2.0	4.0	6.0	8.0	10.0	12.0	14.0	16.0	18.0	20.0	22.0	24.0	26.0	28.0	30.0
	8	2.4	4.8	7.1	9.5	11.9	14.3	16.7	19.0	21.4	23.8	26.2	28.5	30.9	33.3	35.7
	10	2.8	5.5	8.3	11.0	13.8	16.6	19.3	22.1	24.8	27.6	30.4	33.1	35.9	38.6	41.4
	12	3.1	6.3	9.4	12.6	15.7	18.8	22.0	25.1	28.3	31.4	34.5	37.7	40.8	44.0	47.1
	14	3.5	7.0	10.6	14.1	17.6	21.1	24.6	28.2	31.7	35.2	38.7	42.3	45.8	49.3	52.8
4	0	1.0	2.0	3.0	4.0	4.9	5.9	6.9	7.9	8.9	9.9	10.9	11.9	12.9	13.8	14.8
	2	1.3	2.6	4.0	5.3	6.6	7.9	9.2	10.5	11.9	13.2	14.5	15.8	17.1	18.5	19.8
	4	1.6	3.3	4.9	6.6	8.2	9.9	11.5	13.2	14.8	16.5	18.1	19.8	21.4	23.1	24.7
	6	2.0	4.0	5.9	7.9	9.9	11.9	13.8	15.8	17.8	19.8	21.8	23.7	25.7	27.7	29.7
	8	2.3	4.6	6.9	9.2	11.5	13.8	16.2	18.5	20.8	23.1	25.4	27.7	30.0	32.3	34.6
	10	2.6	5.3	7.9	10.5	13.2	15.8	18.5	21.1	23.7	26.4	29.0	31.6	34.3	36.9	39.6
	12	3.0	5.9	8.9	11.9	14.8	17.8	20.8	23.7	26.7	29.7	32.6	35.6	38.6	41.5	44.5
	14	3.3	6.6	9.9	13.2	16.5	19.8	23.1	26.4	29.7	33.0	36.3	39.6	42.9	46.2	49.4
6	0	1.2	2.4	3.6	4.8	6.1	7.3	8.5	9.7	10.9	12.1	13.3	14.5	15.7	17.0	18.2
	2	1.5	3.0	4.4	5.9	7.4	8.9	10.4	11.8	13.3	14.8	16.3	17.8	19.2	20.7	22.2
	4	1.7	3.5	5.2	7.0	8.7	10.5	12.2	14.0	15.7	17.5	19.2	21.0	22.7	24.5	26.2
	6	2.0	4.0	6.1	8.1	10.1	12.1	14.1	16.1	18.2	20.2	22.2	24.2	26.2	28.3	30.3
	8	2.3	4.6	6.9	9.2	11.4	13.7	16.0	18.3	20.6	22.9	25.2	27.5	29.7	32.0	34.3
	10	2.6	5.1	7.7	10.2	12.8	15.3	17.9	20.5	23.0	25.6	28.1	30.7	33.2	35.8	38.4
	12	2.8	5.7	8.5	11.3	14.1	17.0	19.8	22.6	25.4	28.3	31.1	33.9	36.7	39.6	42.4
	14	3.1	6.2	9.3	12.4	15.5	18.6	21.7	24.8	27.9	31.0	34.0	37.1	40.2	43.3	46.4
	16	3.4	6.7	10.1	13.5	16.8	20.2	23.6	26.9	30.3	33.6	37.0	40.4	43.7	47.1	50.5
8	0	1.4	2.8	4.2	5.6	7.0	8.4	9.8	11.2	12.6	14.0	15.4	16.8	18.2	19.6	21.0
	2	1.6	3.3	4.9	6.5	8.2	9.8	11.4	13.1	14.7	16.3	17.9	19.6	21.2	22.8	24.5
	4	1.9	3.7	5.6	7.5	9.3	11.2	13.1	14.9	16.8	18.6	20.5	22.4	24.2	26.1	28.0
	6	2.1	4.2	6.3	8.4	10.5	12.6	14.7	16.8	18.9	21.0	23.1	25.2	27.3	29.4	31.5
	8	2.3	4.7	7.0	9.3	11.7	14.0	16.3	18.6	21.0	23.3	25.6	28.0	30.3	32.6	35.0
	10	2.6	5.1	7.7	10.3	12.8	15.4	17.9	20.5	23.1	25.6	28.2	30.8	33.3	35.9	38.5
	12	2.8	5.6	8.4	11.2	14.0	16.8	19.6	22.4	25.2	28.0	30.8	33.6	36.4	39.2	42.0
	14	3.0	6.1	9.1	12.1	15.2	18.2	21.2	24.2	27.3	30.3	33.3	36.4	39.4	42.4	45.5
	16	3.3	6.5	9.8	13.1	16.3	19.6	22.8	26.1	29.4	32.6	35.9	39.2	42.4	45.7	49.0

Table B: Modification for Edge Distance ($\psi_{ed,N}$)

L_s \ d_{min}	<1.25	1.25	1.5	2	2.5	3	3.5	4	4.5	5	5.5	6	7	8	10	12
3	a	0.700	0.700	0.700	0.700	0.700	0.700	0.700	0.700	b	b	b	b	b	b	b
4	a	a	a	0.700	0.700	0.700	0.700	0.700	0.700	0.700	0.700	0.700	b	b	b	b
6	a	a	a	a	a	0.700	0.700	0.700	0.700	0.700	0.700	0.700	0.700	0.700	b	b
8	a	a	a	a	a	a	a	0.700	0.700	0.700	0.700	0.700	0.700	0.700	0.700	0.700

a – d_{min} < 0.4h_{ef} thus, side-face blowout governs (see Section D.5.4 of Appendix D, ACI 318-02).
b – d_{min} > 1.5h_{ef} thus, stud group is no longer considered near that free edge (another case may govern).

Figure 6.16.15 Concrete tension strength of welded headed stud assemblies

CASE 6: Free edges on four sides

Applies to rectangular stud patterns with outside dimensions X and Y. See Figure 6.5.5.1. If stud spacing is $\geq 3h_{ef}$, single stud capacity multiplied by number of studs may govern design.

$\phi N_{cb} = \phi N'_{cb}(\psi_{ed,N})$ [$\phi N'_{cb}$ from Table A ; $\psi_{ed,N}$ from Table B]

Assumptions for Table A:

- ϕ = 0.7 (no confinement reinforcement – multiply Table A values by 0.75/0.7 when confinement reinforcement is used)
- C_{crb} = 1.0 (uncracked concrete – multiply Table A values by 0.8 when concrete may be cracked)
- $C_{bs} = 3.33\sqrt{(f'_c/h_{ef})}$ (normal weight concrete – multiply Table A values by λ when lightweight concrete is used)
- f'_c = 5000 psi (multiply Table A values by $\sqrt{f'_c/5000}$ for other concrete strengths)
- $h_{ef} = L_s + t_p - t_{hs}$
- t_p (plate thickness) = 3/8 in. t_{hs} (stud head thickness) = 3/8 in.

$a = d_{e3} + Y + d_{e4}$
$b = d_{e1} + X + d_{e2}$

Table A: $\phi N'_{cb} = \phi C_{bs}(d_{e1} + X + d_{e2})(d_{e3} + Y + d_{e4})C_{crb}$

h_{ef}	a \ b	2	4	6	8	10	12	14	16	18	20	22	24	26	28	30
3	0	0.0	0.0	0.0	0.0	0.0	0.0	0.0	0.0	0.0	0.0	0.0	0.0	0.0	0.0	0.0
	2	0.4	0.8	1.1	1.5	1.9	2.3	2.7	3.0	3.4	3.8	4.2	4.6	4.9	5.3	5.7
	4	0.8	1.5	2.3	3.0	3.8	4.6	5.3	6.1	6.9	7.6	8.4	9.1	9.9	10.7	11.4
	6	1.1	2.3	3.4	4.6	5.7	6.9	8.0	9.1	10.3	11.4	12.6	13.7	14.8	16.0	17.1
	8	1.5	3.0	4.6	6.1	7.6	9.1	10.7	12.2	13.7	15.2	16.7	18.3	19.8	21.3	22.8
	10	1.9	3.8	5.7	7.6	9.5	11.4	13.3	15.2	17.1	19.0	20.9	22.8	24.7	26.6	28.5
	12	2.3	4.6	6.9	9.1	11.4	13.7	16.0	18.3	20.6	22.8	25.1	27.4	29.7	32.0	34.3
	14	2.7	5.3	8.0	10.7	13.3	16.0	18.7	21.3	24.0	26.6	29.3	32.0	34.6	37.3	40.0
4	0	0.0	0.0	0.0	0.0	0.0	0.0	0.0	0.0	0.0	0.0	0.0	0.0	0.0	0.0	0.0
	2	0.3	0.7	1.0	1.3	1.6	2.0	2.3	2.6	3.0	3.3	3.6	4.0	4.3	4.6	4.9
	4	0.7	1.3	2.0	2.6	3.3	4.0	4.6	5.3	5.9	6.6	7.3	7.9	8.6	9.2	9.9
	6	1.0	2.0	3.0	4.0	4.9	5.9	6.9	7.9	8.9	9.9	10.9	11.9	12.9	13.8	14.8
	8	1.3	2.6	4.0	5.3	6.6	7.9	9.2	10.5	11.9	13.2	14.5	15.8	17.1	18.5	19.8
	10	1.6	3.3	4.9	6.6	8.2	9.9	11.5	13.2	14.8	16.5	18.1	19.8	21.4	23.1	24.7
	12	2.0	4.0	5.9	7.9	9.9	11.9	13.8	15.8	17.8	19.8	21.8	23.7	25.7	27.7	29.7
	14	2.3	4.6	6.9	9.2	11.5	13.8	16.2	18.5	20.8	23.1	25.4	27.7	30.0	32.3	34.6
6	0	0.0	0.0	0.0	0.0	0.0	0.0	0.0	0.0	0.0	0.0	0.0	0.0	0.0	0.0	0.0
	2	0.3	0.5	0.8	1.1	1.3	1.6	1.9	2.2	2.4	2.7	3.0	3.2	3.5	3.8	4.0
	4	0.5	1.1	1.6	2.2	2.7	3.2	3.8	4.3	4.8	5.4	5.9	6.5	7.0	7.5	8.1
	6	0.8	1.6	2.4	3.2	4.0	4.8	5.7	6.5	7.3	8.1	8.9	9.7	10.5	11.3	12.1
	8	1.1	2.2	3.2	4.3	5.4	6.5	7.5	8.6	9.7	10.8	11.8	12.9	14.0	15.1	16.1
	10	1.3	2.7	4.0	5.4	6.7	8.1	9.4	10.8	12.1	13.5	14.8	16.1	17.5	18.8	20.2
	12	1.6	3.2	4.8	6.5	8.1	9.7	11.3	12.9	14.5	16.1	17.8	19.4	21.0	22.6	24.2
	14	1.9	3.8	5.7	7.5	9.4	11.3	13.2	15.1	17.0	18.8	20.7	22.6	24.5	26.4	28.3
	16	2.2	4.3	6.5	8.6	10.8	12.9	15.1	17.2	19.4	21.5	23.7	25.8	28.0	30.1	32.3
8	0	0.0	0.0	0.0	0.0	0.0	0.0	0.0	0.0	0.0	0.0	0.0	0.0	0.0	0.0	0.0
	2	0.2	0.5	0.7	0.9	1.2	1.4	1.6	1.9	2.1	2.3	2.6	2.8	3.0	3.3	3.5
	4	0.5	0.9	1.4	1.9	2.3	2.8	3.3	3.7	4.2	4.7	5.1	5.6	6.1	6.5	7.0
	6	0.7	1.4	2.1	2.8	3.5	4.2	4.9	5.6	6.3	7.0	7.7	8.4	9.1	9.8	10.5
	8	0.9	1.9	2.8	3.7	4.7	5.6	6.5	7.5	8.4	9.3	10.3	11.2	12.1	13.1	14.0
	10	1.2	2.3	3.5	4.7	5.8	7.0	8.2	9.3	10.5	11.7	12.8	14.0	15.2	16.3	17.5
	12	1.4	2.8	4.2	5.6	7.0	8.4	9.8	11.2	12.6	14.0	15.4	16.8	18.2	19.6	21.0
	14	1.6	3.3	4.9	6.5	8.2	9.8	11.4	13.1	14.7	16.3	17.9	19.6	21.2	22.8	24.5
	16	1.9	3.7	5.6	7.5	9.3	11.2	13.1	14.9	16.8	18.6	20.5	22.4	24.2	26.1	28.0

Table B: Modification for Edge Distance ($\psi_{ed,N}$)

L_s \ d_{min}	<1.25	1.25	1.5	2	2.5	3	3.5	4	4.5	5	5.5	6	7	8	10	12
3	a	0.700	0.700	0.700	0.700	0.700	0.700	0.700	0.700	b	b	b	b	b	b	b
4	a	a	a	0.700	0.700	0.700	0.700	0.700	0.700	0.700	0.700	0.700	b	b	b	b
6	a	a	a	a	0.700	0.700	0.700	0.700	0.700	0.700	0.700	0.700	0.700	0.700	b	b
8	a	a	a	a	a	a	0.700	0.700	0.700	0.700	0.700	0.700	0.700	0.700	0.700	0.700

a – d_{min} < 0.4h_{ef} thus, side-face blowout governs (see Section D.5.4 of Appendix D, ACI 318-02).
b – d_{min} > 1.5h_{ef} thus, stud group is no longer considered near that free edge (another case may govern).

CHAPTER 7
SELECTED TOPICS FOR ARCHITECTURAL PRECAST CONCRETE

7.1 Introduction 7–2

7.2 Types and Functions of Architectural Precast Concrete Panels 7–2
    7.2.1 Non-Loadbearing Panels 7–2
        7.2.1.1 Spandrels 7–3
        7.2.1.2 Column Covers and Mullions 7–4
        7.2.1.3 Window Panels 7–4
        7.2.1.4 Wall Panels 7–5
    7.2.2 Loadbearing Panels 7–5
        7.2.2.1 Spandrels 7–5
        7.2.2.2 Window Panels 7–6
        7.2.2.3 Wall Panels 7–6

7.3 Structural Design Considerations 7–7
    7.3.1 Design Loads 7–7
        7.3.1.1 Volume Changes 7–7
    7.3.2 Support Deflections and Building Movements 7–7
    7.3.3 Handling, Shipping and Erection 7–9
    7.3.4 Tolerances 7–9
    7.3.5 Crack Control 7–10
    7.3.6 Reinforcement 7–10
        7.3.6.1 Welded Wire Reinforcement 7–10
        7.3.6.2 Reinforcing Bars 7–11
        7.3.6.3 Prestressing 7–11
    7.3.7 Stacking Non-Loadbearing Panels 7–11

7.4 Connections 7–12
    7.4.1 General 7–12
    7.4.2 Design Considerations 7–13

7.5 Veneered Panels 7–13
    7.5.1 General 7–13
    7.5.2 Clay Products 7–14
        7.5.2.1 Clay Product Properties 7–14
        7.5.2.2 Clay Product Selection 7–14
        7.5.2.3 Design Considerations 7–15
        7.5.2.4 Other Considerations 7–17
    7.5.3 Natural Stone 7–17
        7.5.3.1 Properties 7–17
        7.5.3.2 Sizes 7–18
        7.5.3.3 Anchorage of Stone Facing 7–18
        7.5.3.4 Veneer Jointing 7–21

7.6 Precast Concrete Used as Forms 7–21

7.7 Structural Design of Architectural Precast Concrete 7–21

7.8 References 7–22

SELECTED TOPICS FOR ARCHITECTURAL PRECAST CONCRETE

7.1 Introduction

Architectural Precast Concrete refers to any precast concrete unit of special (or occasionally standard) shape that through application of finish, color and texture contribute to the architectural form and finished effect of the structure. Units may be part of the structural building frame carrying gravity, wind, seismic, or other forces; non-structural cladding, or simply small, decorative accent units. They may be conventionally reinforced or prestressed.

The successful design of architectural precast concrete requires a thorough knowledge of the product. The designer must be aware of the types and functions of a wide variety of products used in architectural precast concrete and understand production and erection procedures. In most cases, the products and connections will be designed by a precast specialty engineer, employed or retained by the precast manufacturer. The building frame is usually the responsibility of the engineer of record.

The primary purpose of this chapter is to provide a brief discussion of the major aspects of architectural precast concrete structural design for the benefit of the architect and the engineer of record. It is apparent that close cooperation between the precast engineer, the contractor, the architect and the engineer of record is essential for a successful building cladding system. Communication should be kept open at all times during the design and construction of the building.

7.2 Types and Functions of Architectural Precast Concrete Panels

7.2.1 Non-Loadbearing Panels

Non-loadbearing (cladding) panels are those precast concrete units that transfer negligible load from other units of the structure. They are designed to resist wind, seismic forces generated from the self-weight, and forces required to transfer the weight of the panel to the support. The forces generated during manufacturing and erection will often govern the design of the panel. Wind may control in areas of high design loads such as hurricane regions, building corner zones, or panels adjacent to large openings. Earthquake forces usually control in areas of high seismicity.

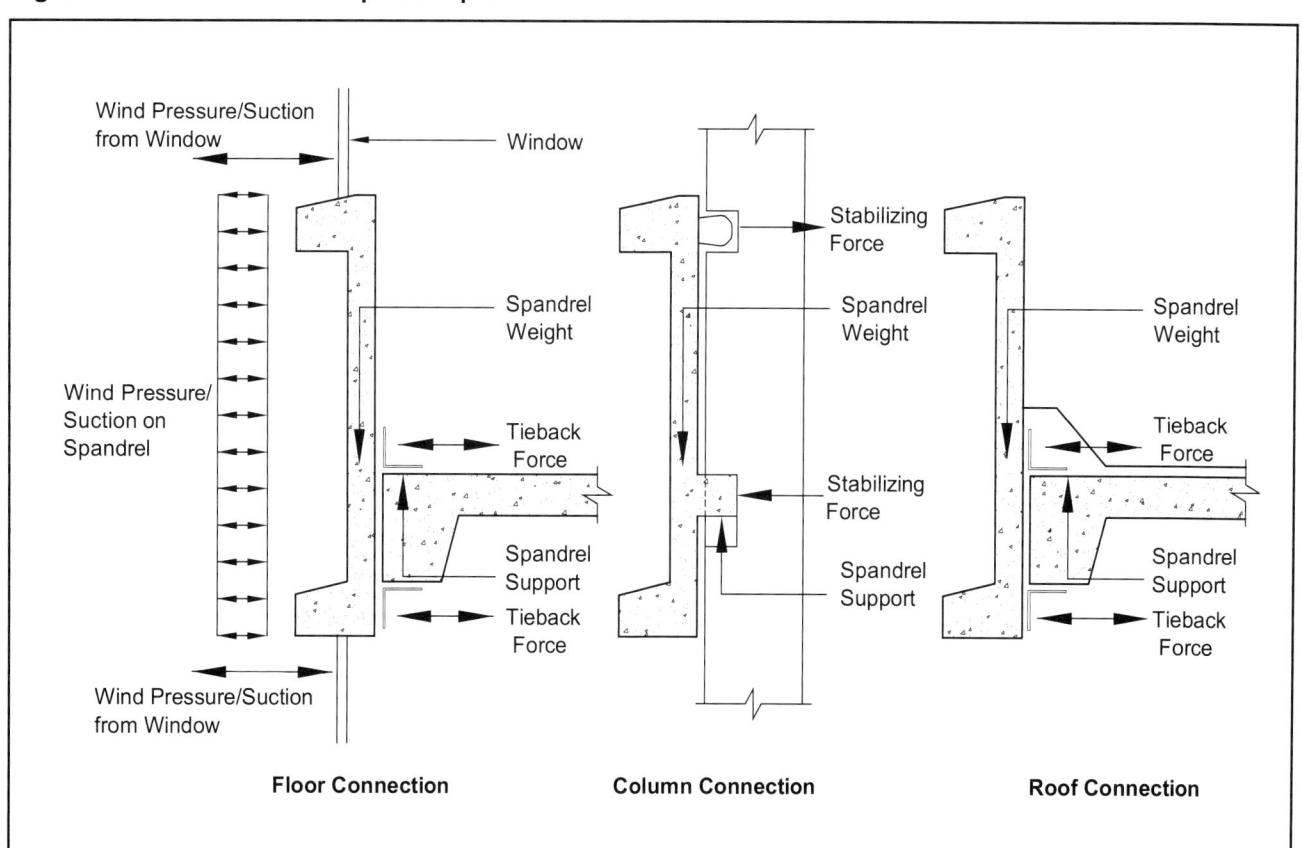

Figure 7.2.1 Forces on a spandrel panel

7.2.1.1 Spandrels

Non-loadbearing spandrels are precast concrete elements that are less than a story in height, made up either as a series of individual units or as one unit extending between columns. Support for the spandrel weight may be provided by either the floor or the columns, and stability against eccentric loading is usually achieved by connections to the floor or to the column (see Figure 7.2.1).

When spandrels are part of a window wall, consideration should be given to the effect of deflections and rotations of the spandrel on the windows.

Since the panels are generally designed to remain uncracked under design loads, deformation calculations should be based on the gross concrete section. For spandrels that extend in one piece between columns, it is preferable that the gravity supports be located near the ends of the panel. This will minimize interaction and load transfer between the floor and the spandrel.

While structural steel frames tend to have good dimensional control, rotation and deflection of the steel support beams is more severe than in concrete framing, and should be considered by the engineer of record. The flexural and torsional flexibility of the steel framing must be considered when determining the placement of the gravity and tieback connections. The engineer of record should clearly specify connection points when flexibility is of concern. Otherwise, the precast engineer may assume that the structure has sufficient rigidity to allow spandrels to be erected without subsequent re-alignment and to allow deformations to be limited to those specified for the spandrels.

Gravity supports for spandrels are most efficiently located at columns. Column sections are generally very stiff vertically resulting in minimal deformations and can be readily designed to resist the eccentricity of panel weight. When spandrels must be supported on edge beams, it is generally not practical to extend the spandrel gravity connections to the beam centerline. The large eccentricity created results in high flexural stresses in the spandrel and large connections that are difficult to conceal. It is preferable to cantilever a slab or steel bracket from the side of the beam to allow gravity support close to the back of the spandrel. This may cause torsion in the support beam that must be recognized from both a strength and stiffness perspective. Similarly, when tieback connections are made to the underside of a steel beam, the frame designer should make provision for torsional loading by specifying heavier members, braces or an intersecting beam. Tiebacks to columns may require that stiffeners be added to resist local bending. The engineer of record needs to understand the effects of localized tieback connections so such stiffeners can be detailed and supplied with the structural steel.

Figure 7.2.2 Auxiliary strut support

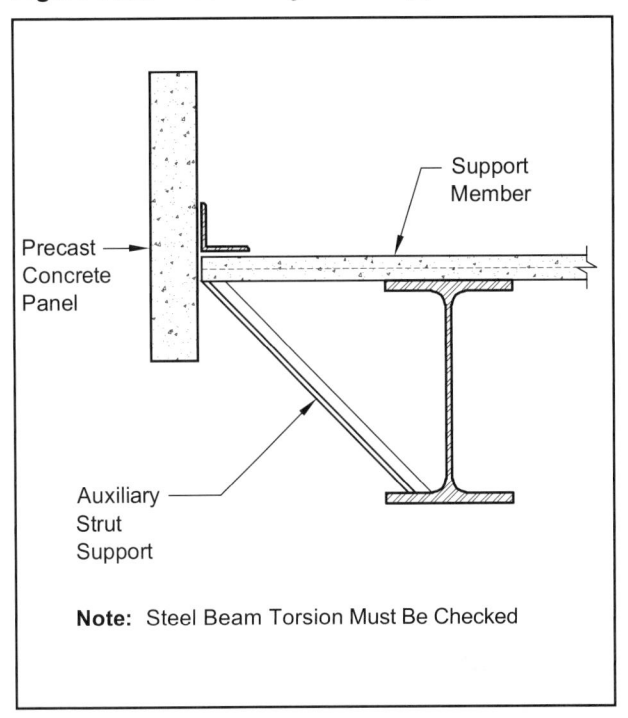

Note: Steel Beam Torsion Must Be Checked

Connections may be detailed to allow for final adjustment after initial erection. However, normal erection procedures assume a panel can be set and aligned without returning for later adjustment. Often the best way to deal with this condition is to use a support scheme that does not rely on cantilever action, such as is shown in Figure 7.2.2. The auxiliary strut supports in Figure 7.2.2 are normally supplied and erected by the structural steel or miscellaneous steel subcontractor prior to precast erection.

Since the contract bid documents (drawings and specifications) are the first line of communication between the designers of record and the precast specialty engineer, it is imperative that the documents clearly indicate the connection concept to the precast engineer. For example, if spandrel loads are to be resisted by the columns rather than the floor beams, it must be indicated on the contract documents. This intent can be shown schematically and further described in applicable notes. Also, the division of responsibilities for providing and installing items such as miscellaneous steel used to stabilize the structural members that support precast elements must be clearly indicated in the contract bid documents. The stability of a structure for lateral loads is the responsibility of the engineer of record.

7.2.1.2 Column Covers and Mullions

The use of precast concrete panels as covers over steel or concrete columns and beams, and as mullions, is a common method of achieving architectural expression, special shapes, or fire rating. [1]

Column covers are usually supported by the structural column or the floor, and are themselves designed to transfer no vertical load other than their own weight. The vertical load of each length of column cover section is usually supported at one elevation, and tied back at the top and bottom for lateral load transfer and stability. In order to minimize erection costs and horizontal joints, it is desirable to make the cover or mullion as long as possible, subject to limitations imposed by weight and handling.

Mullions are vertical elements serving to separate glass areas. They generally resist only wind loads applied from the adjacent glass and must be stiff enough to maintain deflections within the limitations imposed by the window manufacturer. Since mullions are often thin, they are sometimes prestressed to prevent cracking.

Column covers and mullions are usually major focal points in a structure, and aesthetic success requires that careful thought be given to all facets of design and erection. The following are some items that should be considered:

1. Since column covers and mullions are often isolated elements forming a long vertical line, any variation from a vertical plane is readily observable. This variation is usually the result of the tolerances allowed in the structural frame. To some degree, these variations can be handled by precast connections that provide adjustability. The designer should plan adequate clearance between panel and structure. For steel columns, the designer should consider clearances around splice plates and projecting bolts.
2. Gravity support should be provided by two bearing points (or one for narrow precast members such as vertical mullions) at only one elevation and connections provided at additional locations for lateral loads and stability. When access is available, consider providing an intermediate connection for lateral support and restraint of bowing.
3. Column covers and mullions which project from the facade will be subjected to shearing wind loads. The connection design must account for these forces.
4. Members which are exposed to the environment will be subjected to temperature and humidity change. Horizontal joints between abutting precast column covers and mullions should be wide enough and connections ductile enough to permit length changes and rotation from temperature gradients. The behavior of thin flexible members can be improved by prestressing.
5. Due to vertical loads and the effects of creep and shrinkage in cast-in-place concrete columns, structural columns will tend to shorten. The width of the horizontal joint between abutting precast covers should be sufficient to accommodate this shortening.
6. The designer must envision the erection process. Column cover and mullion connections are often difficult to reach and, once made, difficult to adjust. This difficulty of access is compounded when column covers must be stacked around all four sides of a column. Sometimes this condition can be solved by welding the lower piece to the column and anchoring the upper piece to the lower piece with dowels or by a mechanical device that does not require access.
7. Use of insulation on the interior face of the column cover reduces heat loss at these locations. Such insulation will also minimize temperature differentials between exterior columns and the interior of the structure.

7.2.1.3 Window Panels

Window panels are usually one bay long and one floor high with large single or multiple openings for windows (see Figure 7.2.7). Non-loadbearing panels that contain openings such as window panels may develop stress concentrations at these openings, resulting from unintended loading or restrained bowing. Hairline cracks radiating from the corners can result (see Figure 7.2.3). While these stress concentrations may be partially resisted by reinforcement, the designer should consider methods of eliminating imposed restraints. Areas of abrupt change in cross section should be reinforced and should be rounded or chamfered whenever possible.

If adequate clearance is not provided between the precast panels and the support structure, or if the connections do not allow for unrestrained movement, loads from adjacent floors can be imposed on non-loadbearing panels (see Figure 7.2.4). These loads can cause excessive stresses at the "beam" portion of an opening.

Unless a method of preventing load transfer, such as slotted angle connections, can be developed and permanently maintained, the "beam" should be designed for some loads from the floor. The magnitude of such loads requires engineering judgment.

Figure 7.2.3   Corner cracking due to restrained bowing

Figure 7.2.5   Shearing wind on ribbed panels

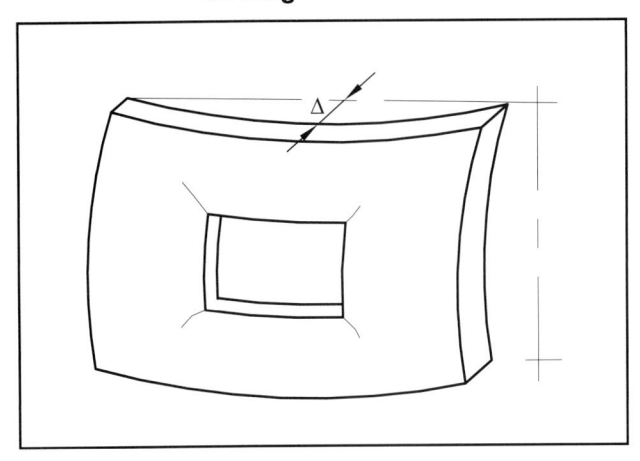

Figure 7.2.4   Unanticipated loading on a non-loadbearing panel

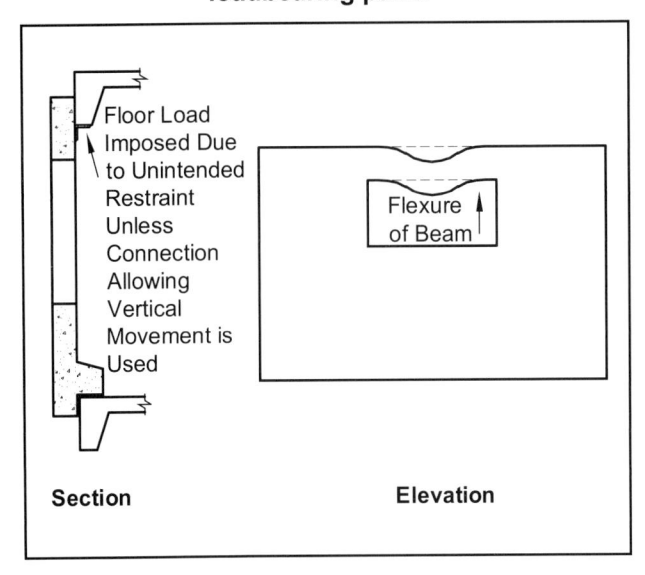

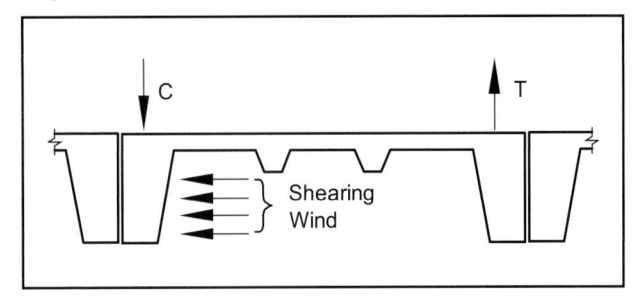

Figure 7.2.6   Loadbearing spandrel

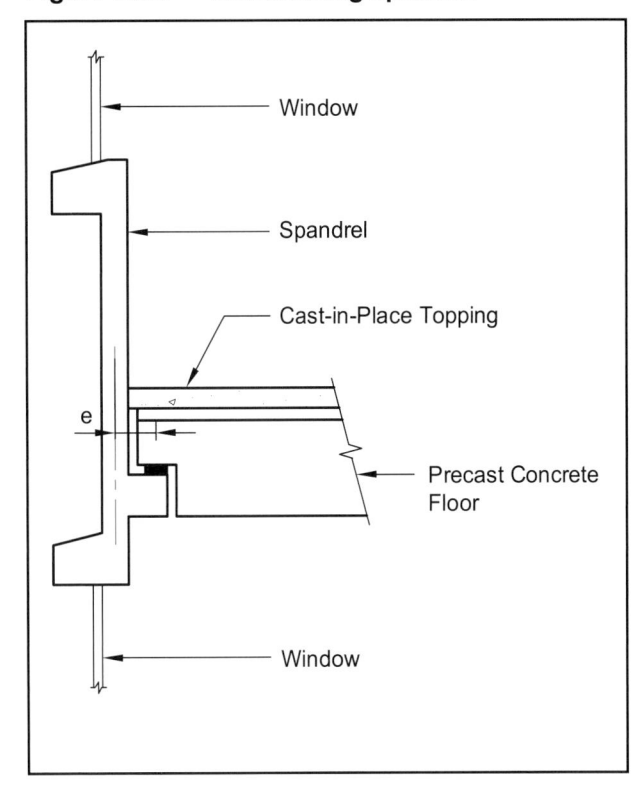

7.2.1.4   Wall Panels

Non-loadbearing wall panels may be strictly cladding or may be used as shear walls as part of the lateral force resisting system. In-place stresses are seldom a problem for cladding panels. The panel cross section is generally chosen for architectural or aesthetic reasons. A panel that is too thin may bow or deflect excessively, thereby creating caulking problems at the building corners or fit-up and leakage problems at attached windows.

Panels with deep protruding ribs may require analysis for shearing winds, as indicated in Figure 7.2.5, and the connections may have to be designed for the resulting twist. Whether cladding or shear walls, the design of connections is critical and care should be taken to account for all forces generated by lateral loads, eccentricity of weight, thermal bowing, and volumetric changes. All non-loadbearing panels should be designed to accommodate movement freely and, whenever possible, with no redundant supports, except where necessary to restrain bowing.

7.2.2   Loadbearing Panels

Loadbearing panels may support either roof or floor loads and may be either horizontal or vertical. Horizontal panels are designed as beams with proper consideration given to cracking, ultimate moment, shear, and torsion. Design of vertical panels is similar to that of columns. Because of the large height to thickness ratios and the magnitude and eccentricity of the loads, the in-plane stresses may be critical.

7.2.2.1   Spandrels

Loadbearing spandrels are panels that span horizontally between columns and support floor or roof loads. Except for the magnitude and location of these additional loads, the design is the same as for non-loadbearing spandrels.

Loadbearing spandrels support gravity loads that are often applied eccentrically with respect to the support. A typical arrangement of spandrel and supported floor is shown in Figure 7.2.6.

Potential rotation due to eccentricity is usually resisted by a horizontal couple developed in the floor system or by a coupled connection to suppporting columns. In order to prevent rotation, the details must provide for a compressive force transfer at the top of the floor and a tensile force transfer at the bearing of the precast floor element. The load path of these floor forces must be followed through the structure and considered in the design of other members in the building. Because of the stresses that may develop due to volumetric changes, tensile connections at the bottom surface of both ends of a floor member should not be used. Even when torsion is resisted by a couple in the floor elements in the completed structure, twisting on the spandrel prior to completion of connections must be considered.

If torsion cannot be accommodated by floor connections, the spandrel panel should be designed for induced stresses. See Chapter 4 for torsion design.

While adding unnecessary tiebacks is not advisable due to the resulting build-up of restraint forces, many precasters will choose to provide additional connections along the length of a spandrel to prevent cracking under anticipated gravity plus lateral loads.

7.2.2.2 Window Panels

Most of the items in the previous section must also be considered in the analysis of loadbearing window panels. Panels may be designed to span horizontally between columns or vertically. When spanning horizontally, they are designed as beams, or if they have frequent regularly spaced window openings, as shown in Figure 7.2.7(a), as Vierendeel trusses. When so designed there must be a space or horizontal joint between panels to ensure that they will not transfer loads to panels below.

If a large portion of the panel is a window opening, as in Figure 7.2.7(c), it may be necessary to analyze it as a rigid frame.

7.2.2.3 Wall Panels

When the panels are placed vertically, they are usually designed similar to columns (Section 4.9),

Figure 7.2.7   Horizontal and vertical rib panels

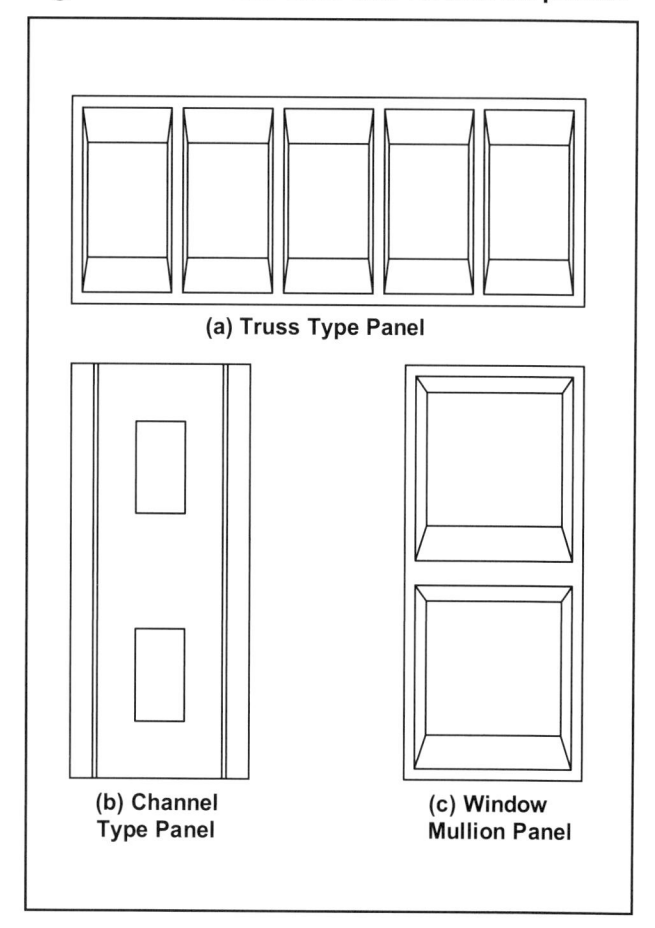

(a) Truss Type Panel

(b) Channel Type Panel

(c) Window Mullion Panel

and slenderness, as described in Section 4.9.3, should be considered. Because of the height-to-thickness ratios commonly used, a second-order analysis is often incorporated for design. Since the deflections associated with a cracked panel are generally unacceptable, the panels are usually designed to remain uncracked (see Section 4.2.2.1). This may require prestressing of the panels.

Figure 7.2.7 shows architectural wall panels, generally used with relatively short vertical spans, although they may sometimes span continuously over two or more floors. They are usually custom-made for each project and reinforced with mild steel. Standard flat, hollow-core and stemmed members are also used as wall panels, and are frequently prestressed. When stemmed floor or roof members are used, the width of the loadbearing walls should be selected in modules of the floor unit. That is, for 10-ft double tees, the walls should be 10, 20, or 30 ft wide. Local precast concrete producers should be contacted for their recommended modules.

Dimensions of architectural panels are usually selected based on a desired appearance. When these panels are also used to carry loads, or act as shear walls, it is important to have some engineering input in the preliminary stages of the project.

7.3 Structural Design Considerations

7.3.1 Design Loads

The forces that must be considered in the design of architectural precast concrete components include:
1. Those caused by the precast member itself, e.g., self-weight and earthquake forces.
2. Loads during handling and erection (prior to installation on the building).
3. Externally applied loads, such as wind, earthquake, bumper forces, snow, roof and floor live loads, soil or fluid pressure, or construction loads. These loads should be shown on the contract drawings.
4. Loads resulting from volume change restraint or support system movement. These loads are generally concentrated at the connections.
5. Loads from adjacent members.

The designer should provide simple load paths through the connections and ductility within the connections. This will reduce the sensitivity of the connection and the necessity to precisely calculate loads and forces from, for example, volume changes and frame distortions. The number of load transfer points should be kept to a practical minimum. It is recommended that no more than two connections per panel be used to transfer gravity loads.

7.3.1.1 Volume Changes

Volume changes of precast concrete are caused by variations in temperature, shrinkage, and creep. If precast members are free to move or deform, volume changes are of little consequence. For example, unrestrained thermal bowing of a panel induces no stress in the panel. However, if the bowing is restrained by attachment to the foundation or frame, significant stresses and cracking may develop (see Section 4.8.5). Since the time between casting and final connection of precast members is usually over 30 days, most creep and shrinkage will have taken place.

Temperature

The coefficient of thermal expansion of concrete varies with the aggregate. Ranges for normal weight concrete are 5 to 7×10^{-6} in./in./°F when made with siliceous aggregate and 3.5 to 5×10^{-6} in./in./°F when made with calcareous aggregate.

Since the thermal coefficient for steel is approximately 6×10^{-6} in./in./°F, the addition of steel reinforcement does not significantly affect the concrete coefficient.

Different temperatures on the interior and exterior of the building may cause the panel to bow. This bowing can be resisted by center connections, but this causes stresses in the panel (see Section 4.8.5).

Shrinkage

Precast concrete members are subject to air-drying as soon as they are removed from the forms. During exposure to the atmosphere, the concrete slowly loses some of its original water, causing shrinkage to occur. About 40% of the drying shrinkage occurs by an age of 30 days and about 60% occurs by an age of 90 days.

During the first year, total unit length change due to drying shrinkage of normal weight concrete typically ranges from about 400 to 650×10^{-6} in./in./°F when exposed to air at 50% relative humidity.

Differential shrinkage or expansion between panels with face mixes and or material (brick, tile, etc.) prone to shrinkage or expansion must be carefully evaluated to avoid excessive bowing.

Creep

When concrete is subjected to a sustained load, the deformation may be divided into two parts: (1) an elastic deformation which occurs immediately, and (2) a creep deformation which begins immediately and continues over time.

For design, it is convenient to refer to specific creep, which is defined as the creep strain per unit of sustained stress. The specific creep of architectural precast concrete panels made with normal weight aggregates per unit stress (psi) can range from 0.5 to 1.0×10^{-6} in./in./psi. About 40% of the creep occurs within 30 days of load application, and about 60% occurs within 90 days.

7.3.2 Support Deflections and Building Movements

All elements should be designed to accommodate movement and, whenever possible, with no redundant supports, except where necessary to restrain bowing. Relatively simple analyses provide the forces required for connection design. The calculations required for movement accommodation are more complex. The designer can use the simplified methods discussed in Chapter 4, or computer analysis programs.

When redundant supports are necessary or when movement is to be resisted, the load-deformation characteristics of the element, connections, and support system should be taken into account.

Figure 7.3.1   Deformation of panels on flexible beam

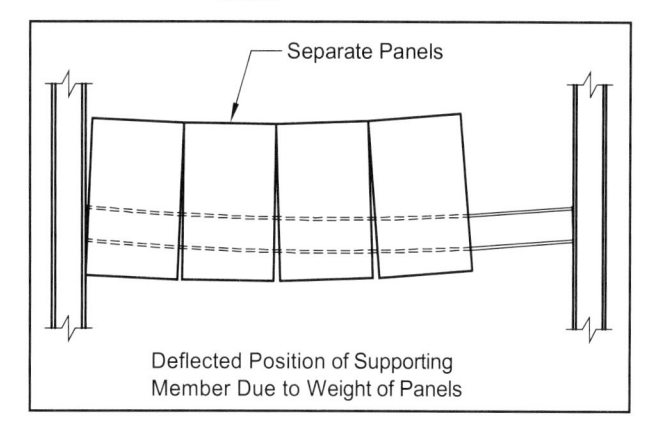

Figure 7.3.2   Effect of cantilever supports

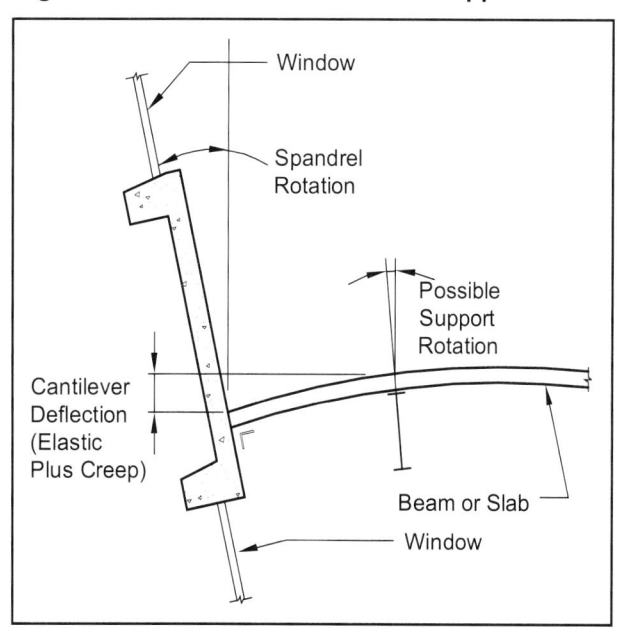

Figure 7.3.3   Optimum handling sequence of precast concrete units

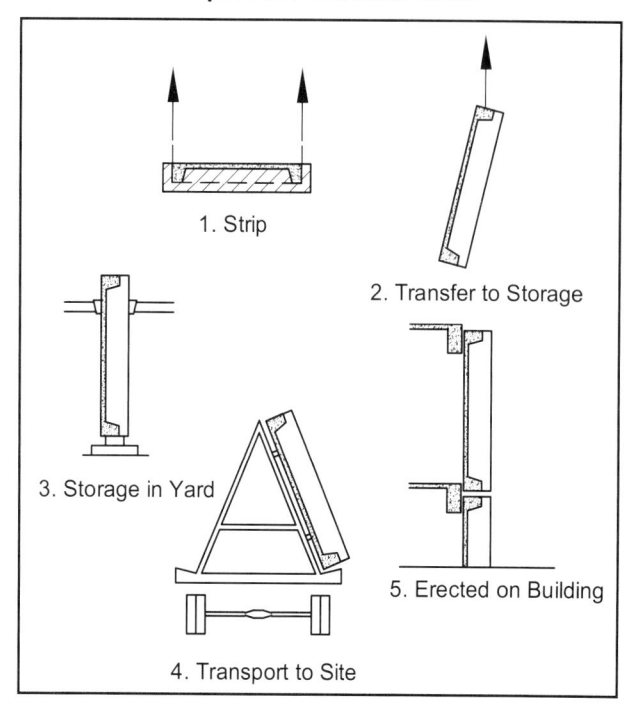

Figure 7.3.4   Hoisting and turning multistory units

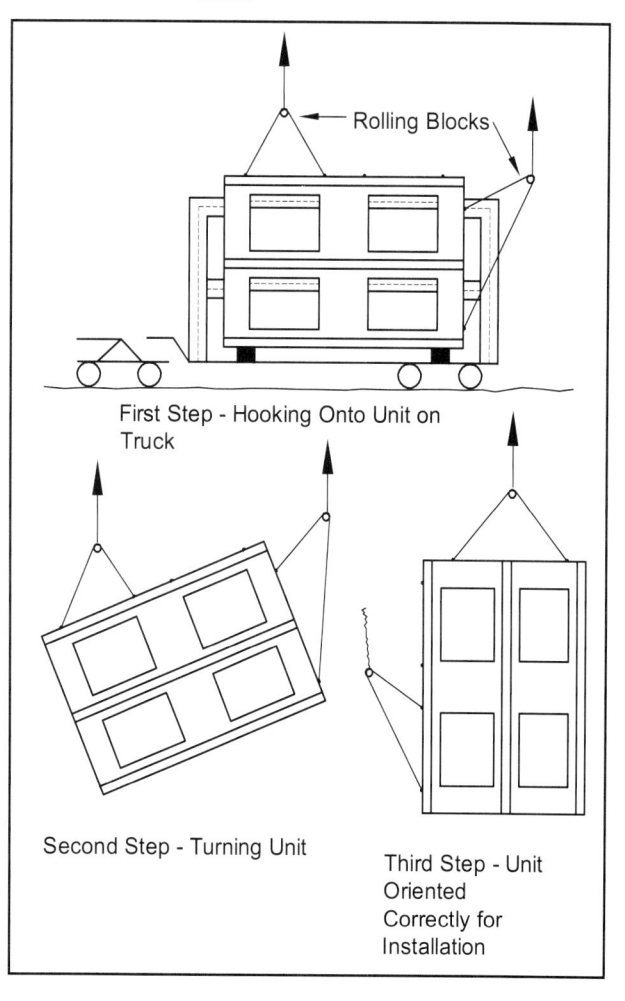

The weight of a series of wall panels supported on a flexible beam will cause deflection (or rotation) of the beam (see Figure 7.3.1). To prevent imposing loads on the panel, the connections must be designed and installed to permit these deformations to freely occur. In addition, the engineer of record should design sufficient stiffness in the edge beam so that specified tolerances can be attained without having to re-set or re-align panels. Because of the difficulty in detailing these connections and the difficulty of erecting panels on a flexible support, it is generally preferable to provide spandrels that span from column to column. Consideration should also be given to spandrels that are supported at the ends of long cantilevers. The engineer of record must determine the effect of deflection and rotation of the support, including the effects of creep, arrange the details of all attachments and provide adequate stiffness to accommodate this condition (see Figure 7.3.2).

A particularly critical condition can occur at the corners of a building, especially when there is a cantilever on both faces adjacent to the corner.

When panels supported on cantilevers are adjacent to panels supported in a different manner, joint tapers and jogs in alignment may occur.

Architectural precast concrete panels can be used as shear walls to provide all or a portion of the lateral stability of a structure, regardless of the type of structure. If so, the connections and the panels themselves must be designed to do so. The structural engineer of record should provide the required design forces to the precast specialty engineer. If the structure includes architectural precast concrete panels that could act as shear walls, but are not intended to do so, the connections must be designed so as not to attract unintended forces into the panels.

Lateral drift must be accommodated in the design of architectural precast concrete panels. Joint sizes and connection details must be developed to allow differential horizontal movement between floors without causing distress in the panels and their connections. Also, tall buildings can have significant axial shortening. The shortening must also be accommodated in the design of the joints and connections.

7.3.3 Handling, Shipping and Erection

Design consideration must be given to installation conditions for the project with respect to handling, transportation and hoisting. This must be done before sizes, shapes and other features are finalized, as erection equipment will frequently influence panel size.

Excessive re-handling and turning of units between stripping and final installation adds to the cost and increases the danger of accidental damage. Consequently, handling should be reduced to a minimum. The precast concrete manufacturer is generally responsible for designing the panels for handling stresses and for the design of the handling inserts.

The optimum solution for economical handling is the ability to strip a unit from the form and tilt it into a vertical position similar to the position of the unit in its final location on the building. Figure 7.3.3 illustrates this point. This procedure has the added advantage that during prolonged storage, the panel will weather much as it would following final installation.

The preceding solution is obviously not possible when more important structural or connection considerations dictate units several stories high. Such units may be stored and shipped on their long side and handled at the jobsite by turning in the air as shown in Figure 7.3.4. In most cases, this will be a more economical solution than single-story units.

One set of lifting devices is required for handling and erection, and another may be needed for stripping. Lifting devices should be designed for actual loads plus an impact factor which depends on configurations and finishes. A safety factor of 4 should be used for all handling inserts. Their location in the panel should be chosen to result in crack-free handling. See Chapter 5 for additional information on handling and erection.

7.3.4 Tolerances

Tolerances set the limits of size and shape within which the actual precast concrete units must lie. The reason for specifying tolerances is to establish construction criteria that will ensure that the parts will fit together without having to be modified. A more long-term reason is to be certain the structure will perform as intended. The architect establishes the tolerances required to make the building concept work and must temper the desire for close tolerances with knowledge of what can be attained in the field at a reasonable cost and what tolerances are practically achievable in the supporting frame. By specifying reasonable tolerances, the architect has strengthened and simplified his standards for acceptance. The architect must recognize that unrealistically tight tolerances are costly, particularly for custom produced elements. The cost of manufacturing to close tolerances decreases with increased repetition.

The designer of architectural precast concrete must recognize that erection and manufacturing tolerances apply to this product as they do to other building materials. When tolerances are understood and allowances made for them in the design stage, the task of determining and specifying them becomes fairly simple.

Three groups of tolerances should be established as part of the precast concrete design: product tolerances, erection tolerances, and interfacing tolerances. Precast concrete tolerances in manufacture and erection do not usually cause site difficulty. The interfacing of precast concrete with other building materials is the usual reason for site difficulty.

Tolerances should be established for the following reasons:

Structural. To ensure that structural design properly accounts for factors sensitive to variations in dimensional control. Examples include eccentric loading conditions, bearing areas, hardware and hardware anchorage locations, and locations of reinforcing or prestressing steel.

Feasibility. To ensure acceptable performance of joints and interfacing materials in the finished structure, such as glazing between panels.

Visual. To ensure that the variations will result in an aesthetically pleasing structure.

Economic. To ensure ease and speed of production and erection with a known degree of accuracy in the dimensions of precast concrete products.

Legal. To avoid encroaching on property lines and to establish tolerance standards against which the work can be compared in the event of a dispute.

Contractual. To establish a known acceptability range and to establish responsibility for developing, achieving and maintaining mutually agreed tolerances.

It should be understood by those involved in the design and construction process that tolerances shown in this Handbook must be considered as guidelines for an acceptability range and not limits for rejection. If specified tolerances are met, the member should be accepted. If these tolerances are exceeded, the member may still be acceptable if it meets any of the following criteria:

1. Exceeding the tolerances does not affect the structural integrity or architectural performance of the member.
2. The member can be brought within tolerance by structurally and architecturally satisfactory means.
3. The total erected assembly can be modified to meet all structural and architectural requirements.

The enforcement of tolerances should be based on the judgment of the designer. This design professional is able to decide whether a deviation from the allowable tolerances affects safety, appearance, or other trades. In building construction, very little out of tolerance work, whether it is concrete, masonry, cast-in-place concrete, steel, or precast concrete, has been rejected and removed solely because it was "out of tolerance." Additional information on tolerances is given in Chapter 8 and the references cited in that chapter.

7.3.5 Crack Control

In designing architectural precast concrete panels, it is desirable that there not be any discernible cracking. In some cases, cracking may be permitted, but the crack width must be limited. When a reinforced concrete element is subjected to flexural tension, the amount and location of reinforcing steel has a negligible effect on member performance until a crack has developed. As stresses increase, hairline cracks may develop and extend a distance into the element. If cracks are narrow, the structural adequacy of the element will remain unimpaired.

Even when concrete stresses during service are less than the allowable flexural tension, distributed reinforcement is needed to control cracking. Such cracking may occur unintentionally during fabrication, handling or erection. Reinforcement also provides ductility in the event of an unexpected overload. In members in which the stresses are expected to be greater than the allowable flexural tension, conventional or prestressed reinforcement is required for satisfactory service load performance, adequate safety and satisfying aesthetic requirements.

Sufficient reinforcement must be used in each unit to control the distribution of any shrinkage cracking. Where units have complex shapes, and particularly where they have unbalanced volumes, unsymmetrical reinforcement, large protrusions or changes of section, the risk of shrinkage cracking is increased. Distortion (bowing or warping) of the panel can also occur from these causes.

7.3.6 Reinforcement

Architectural precast concrete panels can be reinforced with all of the common reinforcing methods used in the precast concrete industry. Most cladding panels are conventionally reinforced with either welded wire reinforcement or individual bars. Loadbearing panels and insulated panels are often prestressed; however, conventional reinforcement may also be used. Many producers also have the ability to post-tension architectural precast concrete panels. Structural design is governed by the provisions of ACI 318.

7.3.6.1 Welded Wire Reinforcement

Welded wire reinforcement (WWR) is the most common type of reinforcement used in architectural precast concrete. One or more layers of WWR is often used to satisfy the minimum reinforcement requirements and is then supplemented with reinforcing bars as required to meet structural requirements. WWR used in architectural concrete is required to comply with requirements of the ACI Building Code (ACI 318). WWR is available in a wide range of sizes and spacings, making it possible to furnish almost exactly the cross-sectional steel area required. Two sizes, a heavy and a light WWR, will usually suffice for most architectural precast concrete. WWR 4x4–W4xW4 to 6x6–W2.5xW2.5 are common sizes of welded wire reinforcement used by precast manufacturers. Other sizes may be standard in various geographical areas.

Welded wire reinforcement for architectural precast concrete is supplied in flat sheets. Reinforcement from rolls, if used in thin precast concrete sections, must be flattened to the required tolerances. Stock sizes of flat sheets vary, but 8 and 10-ft widths are very common. Many precast concrete plants have the capability of accurately bending wire reinforcement to desired shapes, increasing its usefulness in large members. Because the wire reinforcement is closely and uniformly spaced, it is well suited to control cracking. Furthermore, the welded intersections ensure that the reinforcement will be effective close to the edge of the member, resisting cracking that may be caused by handling, and making it easier to repair damaged edges.

7.3.6.2 Reinforcing Bars

Deformed reinforcing bars are also used extensively in architectural precast concrete. Deformed reinforcing bars are hot-rolled from steels with varying carbon content. Deformed bars conforming to ASTM A615, *Deformed and Plan Billet Steel Bars for Concrete Reinforcement*, are generally available in #3 through #11, Grade 60. Selection of grades of reinforcing steel is determined by the structural design of the precast concrete units. For bars that are to be welded, ASTM A706, *Low-Alloy Steel Deformed Bars for Concrete Reinforcement* specifies a bar with controlled chemistry that is weldable without preheating.

As a general rule, bar sizes should be kept reasonably small (#3 through #6) even where this will increase the number of bars. Closely spaced bars of small diameter will control crack width and improve the distribution of temperature stresses better than fewer, larger bars at a wider spacing. The use of small diameter, closely spaced bars becomes more important in thin concrete sections. Since the sum of the widths of cracks in concrete is essentially the same for a given set of conditions, the more cracks there are, the smaller and less visible they will be. The recommended maximum spacing of reinforcement in panels exposed to the environment is 6 in. for welded wire reinforcement and three times the panel thickness (18 in. maximum) for reinforcing bars. Large reinforcing bars generally are not used. They require anchorage lengths and hook sizes that may be impractical.

Good bond between the reinforcing bar and the concrete is essential if the bar is to perform its functions of resisting tension and keeping cracks small. Therefore, the reinforcing bar must be free of materials injurious to bond, including loose rust. Mill scale that withstands hard wire brushing or a coating of tight rust is not detrimental to bond. Epoxy coated reinforcement does not control cracks as well as uncoated reinforcement. If uncracked design is used, then there is no reason for epoxy coating. Thus, epoxy coated reinforcement should not be specified for architectural precast concrete.

7.3.6.3 Prestressing

Prestressing may be used to minimize cracking of members by applying a precompression in the concrete that counteracts the tensile stresses generated by the self-weight and applied loads. Prestressing may be either pretensioning or post-tensioning. In either case, the prestressing force should generally be concentric with the effective cross section. It is recommended that prestressing in a panel, after all losses, be limited to the range of 150 to 600 psi.

Concrete panels may be prestressed for the following reasons:

1. Structural requirements for in-service loads.
2. Units are supported near the top and it is desired to maintain the concrete in compression.
3. Units are slender and prestressing is the method chosen to facilitate handling without undue tensile stresses.
4. For general crack control.

In order for prestressing to reduce or control bowing or warping, it is essential that tendons be located accurately (concentric with the effective cross section) and securely maintained in that location. Because panels tend to bow outwards due to thermal and shrinkage effects, and because outward bowing is generally more objectionable than inward, some precasters choose to force a slight inward initial bow by adjusting the location of the prestressing force.

7.3.7 Stacking Non-Loadbearing Panels

Architectural precast concrete cladding panels are usually independently supported. That is, each panel has its own set of gravity and lateral connections to secure it to the building's structural frame. Gravity load is supported by the building columns and/or beams and, thus, is transferred to the foundations. In most buildings, it is usually preferable to support panels in the above described manner.

However, there are some building types where it is beneficial to take advantage of concrete's

Figure 7.3.7 Stacked panels

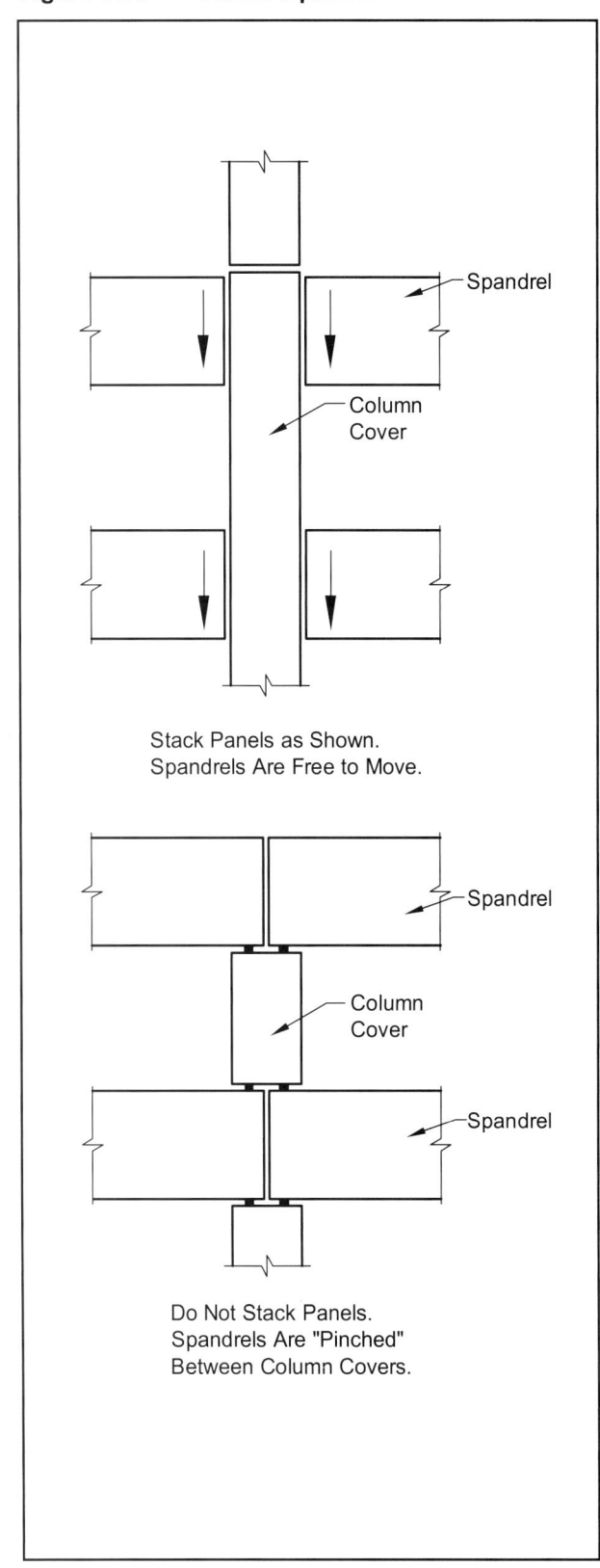

Stack Panels as Shown. Spandrels Are Free to Move.

Do Not Stack Panels. Spandrels Are "Pinched" Between Column Covers.

inherent strength and make the architectural concrete cladding self supporting. Only lateral tie back connections are made to the building's frame. One such building type is the suburban office building which is usually a two to six-story steel braced frame building. The exterior cladding may consist of horizontal spandrels and vertical column covers that are very repetitive in size, shape and layout. The vertical column covers may be able to be designed to support the weight of the spandrels, glass, and any other gravity loads being carried by the spandrels.

There are several advantages to stacking precast concrete cladding panels. The building frame can be lighter, thus saving the customer expense, since it does not have to carry the weight of the precast panels. Bearing brackets welded to the structure that are required to individually support the panels are very expensive. These are eliminated when panels are stacked. The eccentricity in the bearing connections is greatly minimized which makes the panels much easier to erect and reduces the load on the lateral connections. Overall, the design, fabrication, and erection of the cladding panels are made easier and thus more economical.

There are some precautions to take when using the stacking method. In the vertical direction, all of the thermal volume change movements accumulate at the top of the building. One must make sure that precast joint sizes and building details allow this movement without causing detrimental effects. The horizontal spandrel panels should hang on the side of the column covers and not bear on top of them (see Figure 7.3.7). This allows the spandrels to move freely and accommodate axial volume change forces. Since it is not desirable to grout joints between architectural panels, all of the gravity loads should be transferred through shims. Shim sizes must be such that bearing pressures are kept to an acceptable level. The joint at the bottom of the column covers should, however, be grouted if bearing pressures are high. Cladding panels should not be stacked in high seismic risk areas where large in-plane building movements can occur. Braced frames in low seismic risk areas typically do not have appreciable lateral drift, making stacking panels acceptable.

7.4 Connections

7.4.1 General

It is the intent of this section to briefly focus on connections for non-structural, architectural cladding panels. Although the design of connections for architectural precast concrete panels follows the principles presented in Chapter 6 of this Handbook, most of the connection types presented in that chapter apply to structural, loadbearing members. A more complete discussion of the design of connections for architectural precast concrete panels can be found in Chapter 4 of Ref. 1.

7.4.2 Design Considerations

The purpose of cladding panel connections is to transfer loads that the panel must resist to the supporting structure. Generally, connections are divided into two types, those that transfer gravity loads such as panel weights and other materials that the panel may be supporting, and those that transfer lateral loads such as those caused by eccentric bearing, wind, or seismic loads. Although it is possible, it may not be desirable to design connections that transfer both types of loads at the same time. The simpler the connection, the less likely it is that it will be subjected to unintended forces.

Connections for architectural precast concrete cladding panels are designed using strength design methods as presented in Chapter 6 of this Handbook. Forces on a connection are calculated based on the various types of loads that the connection must resist. The resulting forces are then multiplied by the appropriate load factors and combined as prescribed by ACI 318-02. The resulting "worst case" combination is compared to the design strength of the connection as calculated by the methods in Chapter 6. Connections are designed using the above described method because it is compatible with the design method for the panels themselves. In addition, the strength design method results in more economical connection details. The use of the additional load factor of 1.0 to 1.33 as discussed in Chapter 6 should be left up to the precast specialty engineer, who is the most experienced in connection design.

Connections must meet certain design and performance considerations. These considerations are shown in Chapter 6.

Connections for architectural precast concrete panels should be designed to be unaffected by weather conditions. In steel framed buildings, connections should be designed and located such that they do not induce torsion into the structural steel members of the building. If torsion cannot be avoided, the steel members of the building must be adequately braced against rotation.

The following guidelines may be used for locating and designing connections:

1. Provide only two bearing connections per panel. It is virtually impossible to erect a panel such that the load is distributed to more than two bearing connections in the same manner that was assumed in the design. Inaccurate prediction of load can result in a connection that is under-designed.
2. Never assume that a panel imparts a uniform load on to a perimeter building beam.
3. Design connections for worst-case loads. Designing connections for the actual load on every panel results in far too many different connection details which can lead to unintended errors.
4. Design connections using a limited number of different steel shapes and plate thicknesses. Minimize the number of bolt or threaded rod sizes throughout the project.
5. Make sure that the interior finish of the structure covers all connections.
6. Do not assume that inserts used for connections can also be used to hoist the panel. Use different inserts for each purpose.

7.5 Veneered Panels

7.5.1 General

Precast concrete panels faced with brick, tile, terra cotta or natural stone combine the rich beauty of traditional materials with the strength, versatility, and economy of precast concrete. [21] Some applications are shown in Figure 7.5.1.

Structural design of veneered precast concrete units is the same as for other precast concrete wall panels, except that consideration must be given to the veneer material and its attachment to the concrete. The physical properties of the facing material must be compared with the properties of the concrete back-up when facing material is bonded to the concrete back-up. These properties include tensile, compressive and shear strength, modulus of elasticity, coefficient of thermal expansion, variable shrinkage and volume change.

Because of the difference in material properties, veneered panels are somewhat more susceptible to bowing than homogeneous concrete panels. The subsequent sections of this chapter provide for the consideration of different properties. Bowing can be reduced using one or more of the following techniques:

1. Use of bond breakers and flexible anchors with thin stone veneers.
2. Minimum thickness of 5 to 6 in. back-up concrete, depending on panel length.
3. Use of prestressing in long, flat panels.
4. Concrete ribs formed on the back of the panel.
5. Intermediate tie-back connections to the supporting structure.
6. Avoid veneer material that has large expansion or shrinkage characteristics.

Cracking in the veneer may occur if the bonding or anchoring details force the veneer pieces to follow the panel curvature. This is particularly critical where the face materials are large pieces (cut stone) and the differential movement between concrete and veneer is significant. A good mix design with a low water content, quality control, and prolonged curing will help reduce shrinkage.

The subsequent sections of this chapter provide further guidance for veneered panels.

7.5.2 Clay Products

Clay products which are bonded directly to concrete include brick, tile, and architectural terra cotta (ceramic veneer). [18] The clay product facing may cover the entire exposed panel surface or only part of the face, serving as an accent band.

7.5.2.1 Clay Product Properties

Physical properties of brick vary considerably depending on the source and grade of brick. Table 7.5.1 shows the range of physical properties of clay products. Since clay products are subject to local variation, the designer should seek property values from suppliers that are being considered.

As the temperature or length of burning period is increased, clays burn to darker colors, and compressive strength and modulus of elasticity are increased. In general, the modulus of elasticity of brick increases with compressive strength up to approximately 5000 psi, after which, there is little change. The thermal expansions of individual clay units are not the same as the thermal expansion of clay product-faced precast concrete panels due to mortar joints.

7.5.2.2 Clay Product Selection

Clay product manufacturers or distributors should be consulted early in the design stage to determine available colors, textures, shapes, sizes, and size deviations as well as manufacturing capability for special shapes, sizes and tolerances. [7] In addition to standard facing brick shapes and sizes (conforming to ASTM C 216, Type FBX), thin brick veneer units ⅜ to ¾ in. thick are available in various sizes, colors and textures. Thin brick units should conform to Type TBX of ASTM C1088. Thin bricks minimize the effects of differential shrinkage and expansion and, therefore, have less tendency to cause panel bowing.

When a preformed grid is used to position bricks for a precast concrete panel, a brick tolerance of ±1/16 in. is necessary. Tighter tolerances may be obtained by saw cutting each brick, which increases costs.

Figure 7.5.1   Applications of veneer faced precast concrete

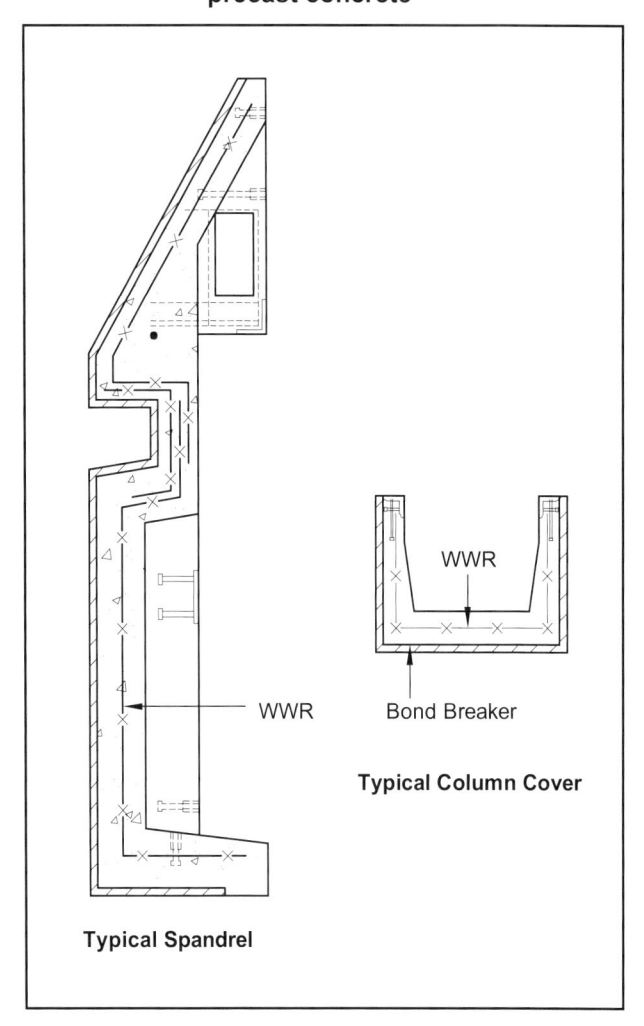

Table 7.5.1   Range of physical properties of clay products[c]

Type of unit	Compressive strength, psi	Modulus of elasticity, psi (×10⁶)	Tensile strength, psi[a]	Coefficient of thermal expansion, in./in./°F
Brick	3,000 – 15,000	1.4 – 5.0	See Note a	4.0×10⁻⁶
Ceramic Tile Quarry	10,000 – 30,000	7.0	See Note a	2.2 – 4.1×10⁻⁶
Glazed Wall Tile[b]	8,000 – 22,000	1.4 – 5.0	See Note a	4.0 – 4.7×10⁻⁶
Terra Cotta	8,000 – 11,000	2.8 – 6.1	See Note a	4.0×10⁻⁶

a. Usually approximated at 10% of the compressive strength.
b. See Ref. 5.
c. This is Table 1 in Ref. 18.

Thin brick should be ½ in. thick minimum to ensure proper location and secure fit in the template during casting operations.

Glazed and unglazed ceramic tile units should conform to American National Standards Institute (ANSI) A137.1, which includes ASTM test procedures and provides a standardized system for evaluating a tile's key characteristics. Tiles are typically ⅜ to ½ in. thick with a 1½% tolerance on the length and width measurements. Within one shipment, the maximum tolerance is ±1/16 in. When several sizes or sources of tile are used to produce a pattern on a panel, the tiles must be manufactured on a modular sizing system in order to have grout joints of the same width.

Architectural terra cotta is a custom product and, within limitations, is produced in sizes for specific jobs. Two thicknesses and sizes are usually manufactured: 1¼ in. thick units, including dovetails spaced 5 in. on centers, size may be 20 x 30 in.; 2¼ in. thick units including dovetails spaced at 7 in. on centers, size may be 32 x 48 in. Other sizes used are 4 or 6 ft x 2 ft. Tolerances on length and width are a maximum of ±1/16 in., with warpage tolerance on the exposed face (variation from a plane surface) of not more than 0.005 in. per in. of length.

7.5.2.3 Design Considerations

Clay products are bonded to backup concrete. The back side of clay product units should preferably have keyback or dovetail configurations or be grooved or ribbed to develop adequate bond.

Latex additives in the concrete or latex bonding materials provide high bond and high strength, but have limitations. They are water sensitive, losing as much as 50% of their strength when wet (although they regain that strength when dry). The lower strength of the concrete is usually sufficient to sustain low shear stress like the weight of the clay product, but when differential movements cause additional stress, problems can occur.

Generally, clay products cast integrally with the concrete have bond strengths exceeding that obtained when laying units in the conventional manner in the field (clay product to mortar). It is necessary, in either case, to use care to avoid entrapped air or excess water-caused voids which could reduce the area of contact between the units and the concrete, thereby reducing bond.

Bond between the facing and the concrete varies depending on the absorption of the clay product. Low absorption will result in poor bond, as will high absorption due to the rapid loss of the mixing water preventing proper hydration of the cement and the development of good bond strength. Bricks with a water absorption by boiling (ASTM C 216) of about 6% to 9% provide good bonding potential. Bricks with an initial rate of absorption (suction) less than 30 g per min. per 30 in.^{2}, when tested in accordance with ASTM C 67, are not required to be wetted. With a higher rate of absorption they should be wetted prior to placement of the concrete, to reduce the amount of mix water absorbed, and thereby improve bond. Terra cotta units should be soaked in water for at least one hour and be damp at the time of concrete placement to reduce suction.

Clay bricks, when removed from the kiln after firing, will begin to permanently increase in size as a result of absorption of atmospheric moisture. Such expansion can cause serious bowing of brick-faced panels. The design coefficient of moisture expansion of clay bricks as recommended by the Brick Institute of America is 500×10^{-6} in./in. but is specified as 300×10^{-6} by Ref. 6. A value of 500×10^{-6} is typical for ceramic quarry tile. The environmental factors affecting moisture expansion of clay brick are:

1. Time since manufacturing. Expansion increases linearly with the logarithm of time. It is estimated that, as a percentage of total potential moisture expansion, 25% occurs within two weeks and 60% will have occurred approximately one year after the bricks have been fired.
2. Time of placement. Expansion subsequent to placement in a panel depends on the portion of total potential for expansion which has already occurred.
3. Temperature. The rate of expansion increases with increased temperature when moisture is present.
4. Humidity. The rate of expansion increases with an increase in relative humidity (RH). Bricks exposed to a relative humidity of 70% have a moisture expansion two to four times as large as those exposed to a RH of 50% over a four-month interval. The 70% RH bricks also exhibit almost all of their expansion within the first twelve months of exposure, while the 50% RH bricks generally exhibit a gradual continuous moisture expansion.
5. The expansion characteristics of the selected brick must be determined to avoid excessive bowing.

Seasonal expansion and contraction of clay bricks will occur due to changes in the ambient air temperature. It is not uncommon for the exterior surface to reach temperatures of 150°F with dark-colored brick, 130°F for medium color, or 120°F for light colored brick on a hot summer day with an ambient air temperature well below 100°F. Surface

ambient air temperature well below 100°F. Surface temperatures as low as −30°F can be reached on a cold night.

The expansion of clay products can be absorbed by dimensional changes of the clay product and grout (mortar) or concrete due to:

1. Drying shrinkage of the mortar or concrete.
2. Elastic deformation of the mortar or concrete under stress.
3. Creep of the mortar or concrete under stress.
4. Elastic deformation of the clay product under stress.

In general, strains imposed slowly and evenly will not cause problems. Consider the first six months to a year after panel production (see Figure 7.5.2). Tile expansion is small (rate of strain application is slow) but mortar shrinkage is nearly complete. The mortar or concrete creeps under load to relieve the tensile stress generated in the tile by the mortar or concrete shrinkage since the tile is relatively rigid. After this time period, the tile has years to accommodate the additional moisture expansion.

Bond failures occur when strain rates exceed creep relief rates. This can occur when:

1. Total shrinkage is higher than normal because overly rich concrete or mortar was used.
2. Sudden rise in temperature or drop in humidity causes shrinkage to proceed faster than creep relieves the stresses that are generated.
3. Bond between clay product and concrete was never adequately achieved.
4. Sudden temperature drop imposes a sudden differential strain because the clay product and mortar (or concrete) have different thermal coefficients of expansion.

The difference in creep characteristics between concrete and clay products, along with the differences in their respective modulus of elasticity, do not pose a problem to the production of small (less than 30 ft) panels when good quality clay products are used. Some producers have used larger panels after first conducting static load tests simulating differential creep.

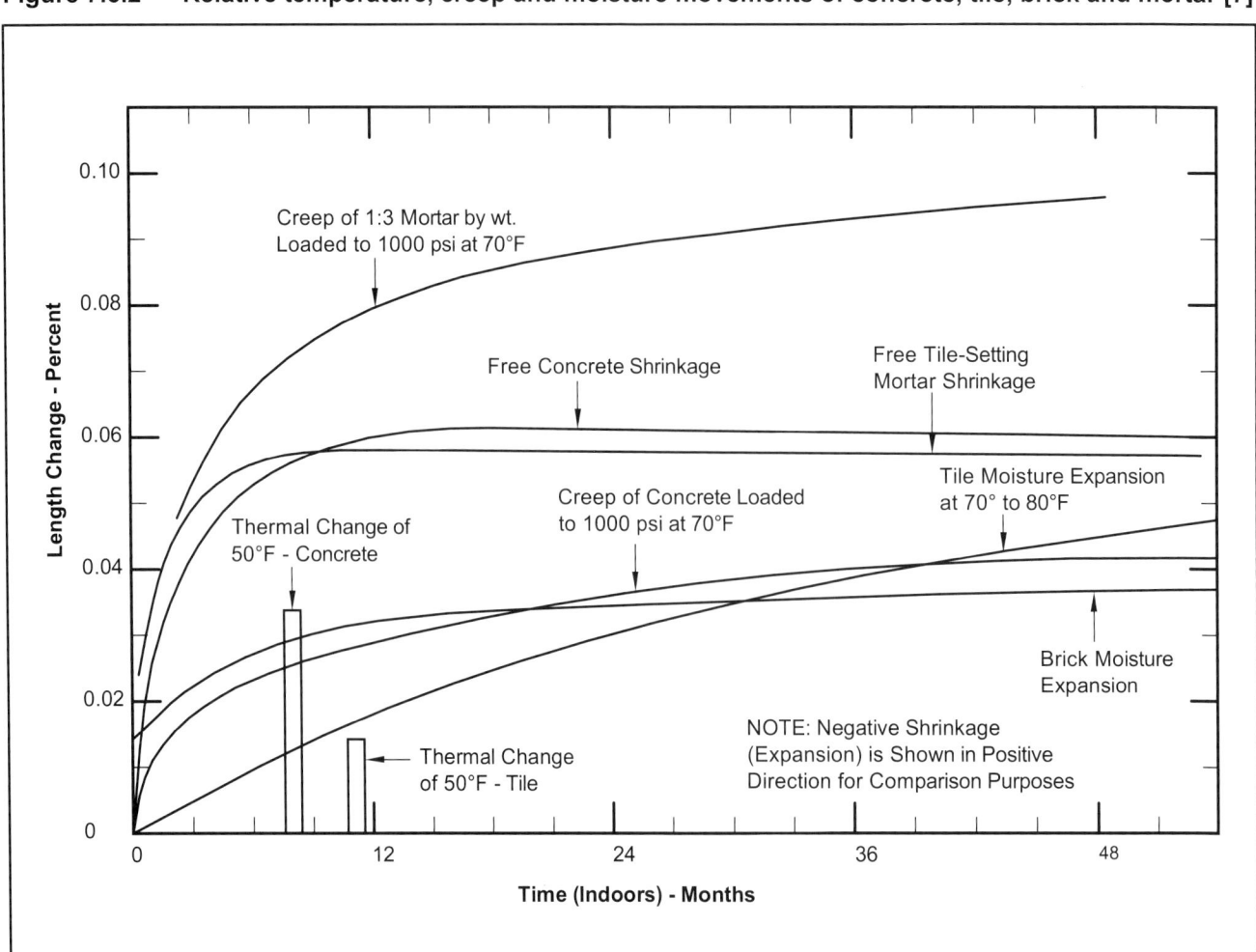

Figure 7.5.2   Relative temperature, creep and moisture movements of concrete, tile, brick and mortar [7]

Clay product faced precast panels may be designed as concrete members, neglecting for design purposes, the structural action of the face veneer. The thickness of the precast panel is reduced by the thickness of the veneer and design assumptions usually exclude consideration of differential shrinkage or differential thermal expansion. However, if the panel is to be prestressed, the effect of composite behavior and the resulting prestress eccentricity should be considered in design.

7.5.2.4 Other Considerations

Bowing of panels in storage can be minimized by providing blocking so that panels remain plane in storage. Storing panels so that flexure is resisted about the strong axis will minimize adverse stresses and bowing. Where feasible, panels should be oriented in the yard so that the sun does not overheat one side. Storing with length in north-south orientation should improve the condition.

Good detailing practice will improve the probability that the finished product satisfies the owner. Use of "picture frame" panels where clay products need not relate directly to those in adjacent panels will avoid alignment difficulties. All dimensions should accommodate brick coursing and use of half-bricks will minimize waste.

7.5.3 Natural Stone

Natural stone facings are used in various sizes, shapes and colors to provide an infinite number of pattern and color possibilities. [8, 9, 10, 11, 12, 21]

7.5.3.1 Properties

The strength of natural stone depends on several factors: the size, rift and cleavage of crystals, the degree of cohesion, the interlocking geometry of crystals, and the nature of cementing materials present. The properties of the stone will vary with the locality from which it is quarried. Sedimentary and metamorphic rocks such as limestone and marble will exhibit different strengths when measured parallel and perpendicular to their original bedding planes. Igneous rocks such as granite may exhibit relatively uniform strength characteristics on the various planes. In addition, the surface finish, freezing and thawing, and large temperature fluctuations will affect the strength and in turn influence the anchorage system.

Information on the durability of the specified stone should be obtained from the supplier or from observations of existing installations of that particular stone. This information should include such factors as tendency to warp, reaction to weathering forces, resistance to chemical pollutants, resistance to chemical reaction from adjacent materials, and reduction in strength from the effects of weathering.

Testing for mechanical properties should be done on stone with the same finish and thickness as will be used on the structure. An adequate number of test samples, usually 20, should be selected and statistical methods should be used to evaluate the properties and obtain design values. Also, modulus of rupture tests (ASTM C 99) are required to demonstrate compliance of the stone to minimum properties specified by ASTM for the particular stone type [limestone (ASTM C 568); marble (ASTM C 503); granite (ASTM C 615)].

The process used to obtain a thermal or flame finish on granite veneers reduces the effective thickness by about 1/8 in. and the flexural strength by 25 to 30%. [10,12] Bushhammered and other similar surface finishes also reduce the effective thickness.

Thermal finishing of granite surfaces causes micro-fracturing, particularly of quartz and feldspars. These micro-cracks permit absorption of water to a depth of about 1/4 in. in the distress surface region of the stone which can result in degradation by cyclic freezing and a further reduction in bending strength.

Most natural stone loses strength as a result of exposure to thermal cycling, (i.e., heating to 170°F and cooling to −10°F), and wet/dry cycling. The modulus of rupture of building stone can be affected by freezing and thawing of the stone. Flexural tests (ASTM C 880) should be conducted on the selected stone, at the thickness and surface finish to be used, in both the new condition and the condition after 100 cycles of laboratory freeze-thaw testing to determine the reduction in strength, if any. Suggested freeze-thaw test procedures include (1) dry cycling between 170°F and −10°F, and (2) freezing in water −10°F and thawing in water at room temperature. Also, stones with high absorption should also be tested in a saturated condition as their flexural, shear and tensile properties may be significantly lower when wet. An approximate indication of good durability is a saturated modulus of rupture of at least 70% of the dry modulus.

For most types of stone, temperature induced movements are theoretically reversible. However, certain stones, particularly uniform-textured, fine-grained, relatively pure marble, when subjected to a large number of thermal cycles, develop an irreversible expansion in the material amounting to as much as 20% of the total original thermal expansion. This residual growth is caused by slipping of individual calcite crystals with respect to each other. [13,14] Such growth, if not considered in the stone size, design of the anchors, or the stone veneer joints may result in curling or bowing of thin marble. For relatively thick marble veneers, the expansion effects are restrained or accommodated by the unaffected portion of the veneer. Tests should be performed to establish the minimum thickness required to obtain satisfactory serviceability.

Volume changes due to moisture changes in most stones are relatively small and not a critical item in design, except that bowing of the stone can occur. Moisture permeability of stone veneers is generally not a problem (see Table 7.5.2). However, as stone veneers become thinner, water may penetrate in greater amounts and at faster rates than normally expected, and damp appearing areas of moisture on the exterior surface of thin stone veneers may occur. These damp areas result when the rate of evaporation of water from the stone surface is slower than the rate at which the water moves to the surface.

7.5.3.2 Sizes

Stone veneers used for precast facing are usually thinner than those used for conventionally set stone with the maximum size generally determined by strength of stone. Table 7.5.3 summarizes typical dimensions. Veneers thinner than those listed can result in anchors being reflected on the exposed surface, excessive breakage or permeability problems.

The length and width of veneer materials should be sized to a tolerance of ±1/16 in. This tolerance becomes important when trying to line up the false joints on one panel with the false joints on an adjacent panel, particularly when there are a large number of pieces of stone on a panel. Tolerance allowance for out-of-square is ±1/16 in. difference in length of the two diagonal measurements. Flatness tolerances for finished surfaces depend on the type of finish. For example, the granite industry tolerances vary from 3/64 in. for a polished surface to 3/16 in. for a flame (thermal) finish when measured with a 4 ft straightedge. [15] Thickness variations are less important since concrete will provide a uniform back face, except at corner butt joints. In such cases, the finished edges should be within ±1/16 in. of specified thickness. However, large thickness variations may lead to the stone being encased with concrete and thus being unable to move.

7.5.3.3 Anchorage of Stone Facing

It is recommended that a bond breaker be used between stone veneer and concrete backup in order to minimize bowing, cracking and staining of the veneer.

Table 7.5.2 Permeability of commercial building stones, [13] cu in./ft^2/hr for ½ in. thickness

Stone type	Pressure, psi		
	1.2	50	100
Granite	0.06 – 0.08	0.11	0.28
Limestone	0.36 – 2.24	4.2 – 44.8	0.9 – 109
Marble	0.06 – 0.35	1.3 – 16.8	0.9 – 28.0
Sandstone	4.2 – 174.0	51.2	221
Slate	0.006 – 0.008	0.08 – 0.11	0.11

Table 7.5.3 Dimensional parameters of various stone materials

Stone type	Min. recommended thickness (in.)	Length range (ft)	Width range (ft)	Maximum area (sq ft)
Marble	1.25	3 – 5	2 – 5	20
Granite	1.25	3 – 7	1 – 5	30
Limestone	1.75	4 – 5	2 – 4	15

Even with concrete shrinkage kept to the lowest possible level, there may still be some interaction with the facing material either through bond or the mechanical anchors of the facing units. This interaction will be minimized if a bond-breaker is used between the facing material and the concrete. Connections of natural stone to the concrete should be made with mechanical anchors which can accommodate some relative in-plane movement, a necessity if bond-breakers are used. One exception is the limestone industry which uses rigid, rather than flexible connectors. See Ref. 1 for recommendations on bond breakers and anchorages.

The following methods have been used to break the bond between the veneer and concrete: (1) a liquid bond breaker, of sufficient thickness to provide a low shear modulus, applied to the veneer back surface prior to placing the concrete; (2) a 6 mil polyethylene sheet; (3) a 1/8 in. polyethylene foam pad; and (4) a one component polyurethane coating. The use of a compressible bond breaker is preferred in order to have movement capability with uneven stone surfaces, either on individual pieces or between stone pieces on a panel. However, during shipment, consideration must be given to preventing cracking of the stone due to compressibility of the pad.

Mechanical anchors should be used to secure the veneer. The details of anchorage will depend on a number of items, including:

1. Shrinkage of concrete during curing. Stresses imposed during handling and erection.
2. Thermal response caused by different coefficients of expansion and by thermal gradients (see Chapter 3).
3. Moisture expansion of veneer.
4. Service loads.

For those veneers rigidly attached or bonded to the backup, the differential shrinkage of the concrete and veneer will cause outward bowing in a simple span panel. The flat surfaces of some veneers, such as cut stone, reveal bowing more prominently than other finishes. Therefore, bowing may be a critical consideration even though the rigidity of the cut stone helps to resist bowing.

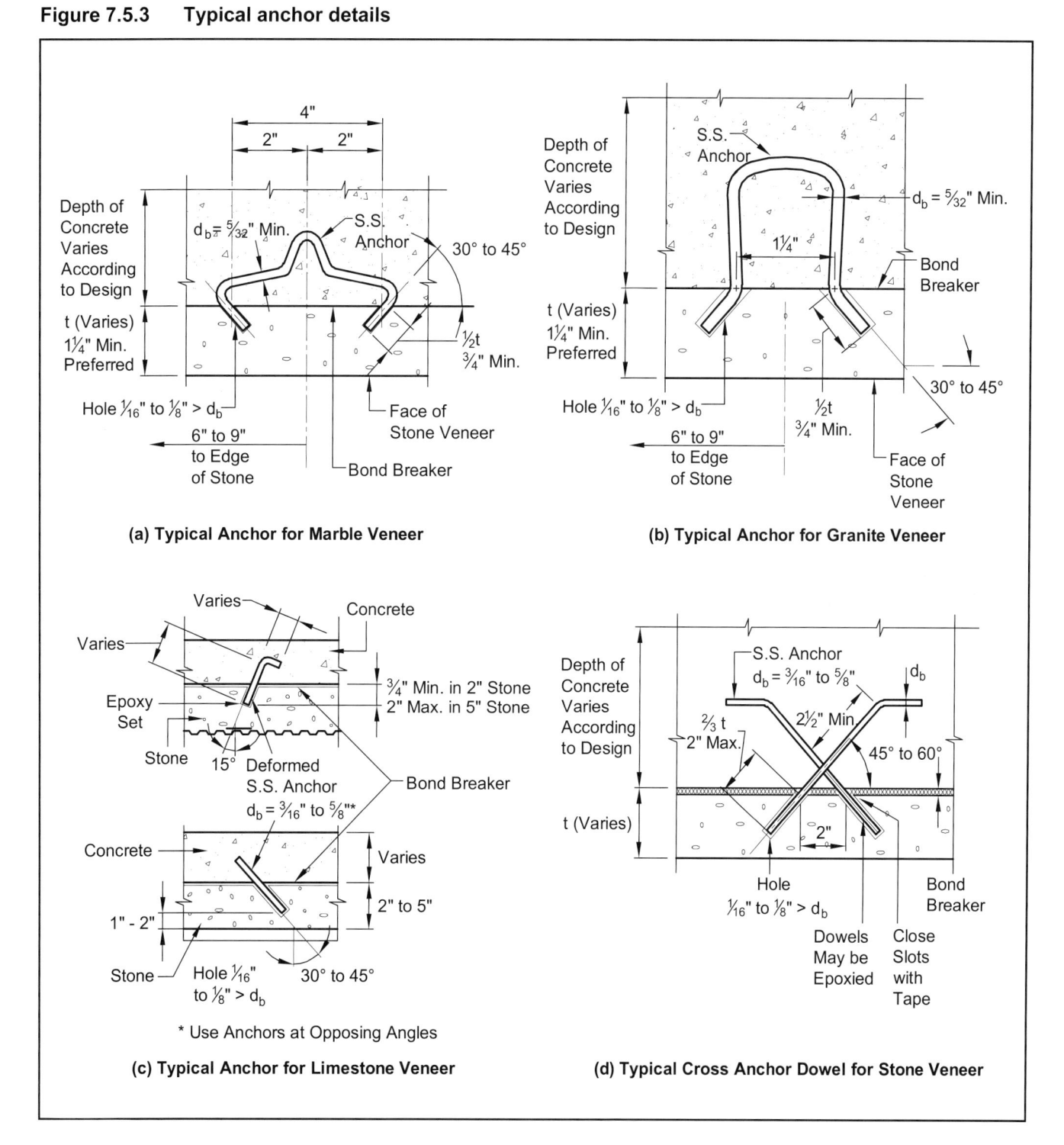

Figure 7.5.3   Typical anchor details

(a) Typical Anchor for Marble Veneer

(b) Typical Anchor for Granite Veneer

(c) Typical Anchor for Limestone Veneer

(d) Typical Cross Anchor Dowel for Stone Veneer

Stone veneer is supplied with holes predrilled in the back surface for the attachment of mechanical anchors. Preformed anchors fabricated from stainless steel, Type 304, are usually used. The number and location of anchors should be pre-determined by shear and tension tests conducted on the anchors embedded in a stone/precast concrete test sample and the anticipated applied loads (both normal and transverse) to the panel. Anchor size and spacing in veneers of questionable strengths or with natural planes of weakness may require special analysis.

Four anchors are usually used per stone piece with a minimum of two recommended. The number of anchors has varied from one per 1½ ft² of stone to one per 6 ft² with one per 2 to 3 ft² being the most common. [1] Anchors should be 6 to 12 in. from an edge with not over 30 in. between anchors. The shear capacity of the spring clip (hairpin) anchors perpendicular to the anchor legs is greater than when they are parallel and depends on the strength of the stone. A typical marble veneer anchor detail with a toe-in spring clip (hairpin) is shown in Figure 7.5.3(a) and a typical granite veneer anchor detail is shown in Figure 7.5.3(b). The toe-out anchor in granite may have as much as 50% more tensile capacity than a toe-in anchor depending on the stone strength.

Depth of anchor holes should be approximately one-half the thickness of the veneer (minimum depth of ¾ in.), and are often drilled at an angle of 30 to 40 degrees to the plane of the stone. Holes which are approximately 50% oversize have been used to allow for differential movement between the stone and the concrete. However, in most cases, holes 1/16 to 1/8 in. larger than the anchor are common as excessive looseness in a hole reduces holding power. Anchor holes should be within ±3/16 in. of the specified hole spacing, particularly for the spring clip anchors.

For other stone veneers, stainless steel dowels, smooth or threaded, are installed to a depth of two-thirds the stone thickness with a maximum depth of 2 in. at angles of 45 to 60 degrees to the plane of the stone. Dowel size varies from 3/16 to 5/8 in. for most stones, except that it varies from ¼ to 5/8 in. for soft limestone and sandstone and depends on thickness and strength of stone. The dowel hole is usually 1/16 to 1/8 in. larger in diameter than the anchor [see Figures 7.5.3(c) and 7.5.3(d)].

Limestone has traditionally been bonded and anchored to the concrete, because it has the lowest coefficient of expansion. Limestone has also traditionally been used in thicknesses of 3 to 5 in. but it is now being used as thin as 1¾ in. When limestone is 2 in. or thinner, it is prudent to use a bond-breaker, along with mechanical anchors. Dowels and spring clip anchors have been used to anchor limestone. Typical dowel details for limestone veneers are shown in Figures 7.5.3(c) and 7.5.3(d). The dowels in Figure 7.5.3(c) should be inserted at angles alternately up and down to secure stone facing to back-up concrete.

Some flexibility should be introduced with all anchors of stone veneer to precast concrete panels, e.g., by keeping the diameter of the anchors to a minimum, to allow for the inevitable relative movements which occur with temperature variations and concrete shrinkage. Unaccommodated relative movements can result in excessive stresses and eventual failure at an anchor location. Depending on the size of the project, consideration may be given to accelerated cyclic temperature tests to determine the effects on the anchors.

Some designers use epoxy to fill the spring clip anchor or dowel holes in order to eliminate intrusion of water into the holes and the possible dark, damp appearance of moisture on the exposed stone surface. The epoxy increases the shear capacity and rigidity of the anchor. The rigidity may be partially overcome by using compressible rubber or elastomeric grommets or sleeves on the anchor at the back surface of the stone. Differential thermal expansion of the stone and epoxy may cause cracking of the stone veneer; this may be overcome by keeping the oversizing of the hole to a minimum, thereby reducing epoxy volume. It may be preferable to fill the anchor hole with low-modulus, polyurethane sealant, which has been proven to be non-staining to light colored stones, or a low modulus polyurethane sealant.

The overall effect of either epoxy or sealant materials on the behavior of the entire veneer should be evaluated prior to their use. At best, the long-term service of epoxy is questionable, therefore, any increase in long term pull-out strength should not be considered in the calculation of long-term anchor capacity.

Design of anchorage and size of the stone should be based on specific test values for the actual stone to be installed. Anchor test procedures have not been standardized. Test samples for anchor tests should be a typical panel section of about 1 ft² and approximate, as closely as possible, actual panel anchoring conditions. A bond-breaker should be placed between the stone and concrete during sample manufacture to eliminate any bond between veneer and concrete surface. Each test sample should contain one anchor connecting stone to concrete back-up and a minimum of five tests are needed to determine tensile (pull-out) and shear strength of each anchor. Depending on the size of the project, it may be desirable to perform shear and tensile tests of the anchors at intervals during the fabrication period.

Safety factors are recommended by the stone trade associations and the suppliers of different kinds of building stones. Because of the expected variation in the physical properties of natural stones and the effects of weathering, recommended safety factors are larger than those used for manufactured building materials, such as steel and concrete. The minimum recommended safety factor, based on the average of the test results, is 4 for anchorage components. [14] If the range of test values exceeds the average by more than ±20%, then the safety factor should be applied to the lower bound value. See Appendix to ASTM C1242 for a discussion on safety factors.

7.5.3.4 Veneer Jointing

Joints between veneer pieces on a precast element should be a minimum of ¼ in. The veneer pieces may be spaced with a non-staining, compressible spacing material, or a chemically neutral, resilient, non-removable gasket which will not adversely affect the sealant to be applied later. The gaskets are of a size and configuration that will provide a recess to receive the sealant and also prevent any backing concrete from entering the joint between the veneer units. Shore A hardness of the gasket should be less than 20.

When stone veneer is used as an accent or feature strip on precast concrete panels, a ½ in. space is left between the edge of the stone and the precast concrete to allow for differential movements of the materials. This space is then caulked as if it were a conventional joint.

Caulking should be of a type that will not stain the veneer material. In some projects, caulking may be installed more economically and satisfactorily at the same time as the caulking between precast elements.

7.6 Precast Concrete Used as Forms

Occasionally, architectural precast concrete is used as the formwork for cast-in-place structural concrete. [3]

In most cases, the architectural precast concrete is considered solely as a form, serving only decorative purposes after the cast-in-place concrete has achieved design strength. This is accomplished by providing open or compressible joints between abutting precast panels, and cast-in-place concrete. Using this arrangement, the architectural precast concrete unit is non-composite with the cast-in-place concrete, and the reinforcing steel extending from the precast members into the cast-in-place concrete need only be of sufficient strength to support the formwork unit.

In other cases, it may be desirable to detail the structure so that the precast members and cast-in-place concrete act compositely, thus combining the strength of both. It is then necessary to provide shear transfer as for other composite assemblies. For a more detailed discussion including design examples, refer to the Fifth Edition of this Design Handbook. [20]

7.7 Structural Design of Architectural Precast Concrete

Architectural Precast Concrete wall panels and other components are designed using the principles outlined in Chapters 3, 4 and 5 of this Handbook. Examples in those chapters illustrate these principles.

7.8 References

1. *Architectural Precast Concrete,* Second Edition, MNL-122-89, Precast/Prestressed Concrete Institute, Chicago, IL, 1989.

2. Sheppard, D. A., and Phillips, W. R., *Plant-Cast Precast and Prestressed Concrete — A Design Guide*, Third Edition, McGraw-Hill Publishing Co., New York, NY, 1989.

3. ACI Committee 347, "Precast Concrete Units Used as Forms for Cast-in-Place Concrete," *Journal of the American Concrete Institute,* V. 66, No. 10, October 1969.

4. *Formwork for Concrete*, Sixth Edition, Special Publication SP-4, American Concrete Institute, Farmington Hills, MI, 1995.

5. Fitzgerald, J. V., and Kastenbein, E. L., "Tests for Engineering Properties of Ceramic Tile," ASTM Bulletin No. 231, American Society for Testing and Materials, Philadelphia, PA, July 1958.

6. *Building Code Requirements for Masonry Structures*, ACI 530-02/ASCE 5-TMS 402, and *Specifications for Masonry Structures*, ACI 530.1-02/ASCE 6-TMS 602, American Concrete Institute, Farmington Hills, MI/American Society of Civil Engineers, Reston, VA, 2002.

7. Bernett, Frank E., "Effects of Moisture Expansion in Installed Quarry Tile," *Ceramic Bulletin*, V. 5, No. 12, The American Ceramic Society, Westerville, OH, 1976.

8. *Marble-Faced Precast Panels*, National Association of Marble Producers, 1966.

9. Chin, I. R., Stecich, J. P., and Erlin, B., "Design of Thin Stone Veneers on Buildings," *Building Stone Magazine*, May-June 1986.

10. Merritt, Frederick S., and Rickets, Jonathan T. (Editors), *Building Construction Handbook*, Fifth Edition, McGraw-Hill Book Co., New York, NY, 1994.

11. *Dimensional Stone: Design Manual VI*, Marble Institute of America, Columbus, OH, 2003.

12. *Specifications for Architectural Granite*, 2004 Edition, National Building Granite Quarries Association, Washington, DC, 2004.

13. *Indiana Limestone Handbook*, 21st Edition, Indiana Limestone Institute of America, Bedford, IN, 2002.

14. Kulka, Felix, Lin, T. Y., and Yang, Y. C., "Prestressed Concrete Building Construction Using Precast Wall Panels," PCI JOURNAL, V. 20, No. 1, January-February 1975.

15. Freedman, Sidney, "Clay Product-Faced Precast Concrete Panels," PCI JOURNAL, V. 39, No. 1, January-February 1994.

16. *Architectural Precast Concrete Cladding — Its Contribution to Lateral Resistance of Buildings*, Proceedings, SP-CP, Precast/Prestressed Concrete Institute, Chicago, IL, 1990.

17. *PCI Design Handbook, Precast and Prestressed Concrete*, Fifth Edition, MNL-120-99, Precast/Prestressed Concrete Institute, Chicago, IL, 1999.

18. Freedman, Sidney, "Stone Veneer-Faced Precast Concrete Panels," PCI JOURNAL, V. 45, No. 4, July-August 2000.

CHAPTER 8
TOLERANCES FOR PRECAST AND PRESTRESSED CONCRETE

8.1 General 8–2
    8.1.1 Definitions 8–2
    8.1.2 Purpose 8–3
    8.1.3 Responsibility 8–4
    8.1.4 Tolerance Acceptability Range 8–4
    8.1.5 Relationships Between Different Tolerances 8–4

8.2 Product Tolerances 8–4
    8.2.1 General 8–4
    8.2.2 Overall Dimensions 8–5
    8.2.3 Sweep or Horizontal Alignment 8–5
    8.2.4 Position of Strands 8–6
    8.2.5 Camber and Differential Camber 8–6
    8.2.6 Weld Plates 8–7
    8.2.7 Haunches of Columns and Wall Panels 8–7
    8.2.8 Warping and Bowing 8–7
    8.2.9 Smoothness 8–7
    8.2.10 Architectural Panels vs. Structural Walls 8–7

8.3 Erection Tolerances 8–9
    8.3.1 General 8–9
    8.3.2 Recommended Erection Tolerances 8–9
    8.3.3 Mixed Building Systems 8–9
    8.3.4 Connections and Bearing 8–9

8.4 Clearances 8–26
    8.4.1 General 8–26
    8.4.2 Joint Clearance 8–26
    8.4.3 Procedure for Determining Clearance 8–26
    8.4.4 Clearance Examples 8–26

8.5 Interfacing Tolerances 8–32
    8.5.1 General 8–32
    8.5.2 Interface Design Approach 8–32
    8.5.3 Characteristics of the Interface 8–33

8.6 References 8–34

TOLERANCES FOR PRECAST AND PRESTRESSED CONCRETE

8.1 General

The intent of this Chapter is to briefly present the subject of tolerances and to provide the designer with some of the most basic tolerances that should be considered during the layout and design of structures. This chapter is a cursory presentation of information that has been previously published by PCI and reviewed by the precast concrete industry. Because of this fact, other PCI documents (see Refs. 1 to 6), which discuss more fully the subject of tolerances, need to be specified in contract documents and referred to when resolving questions about tolerances.

8.1.1 Definitions

Bowing. An overall out-of-planeness condition which differs from warping in that while two edges of the panel may fall in the same plane, the portion of the plane between the edges is out of plane (see Warping).

Camber. (1) The deflection that occurs in prestressed concrete members due to the net bending resulting from stresses associated with the effects of the prestress force (not including dimensional inaccuracies); and (2) a built-in curvature to improve appearance.

Connection. Device for the attachment of precast concrete members to each other, to the building or to the structure. Connection design must often account for the cumulative effects of all allowed tolerance variations.

Contract Documents. General conditions, project specifications and design drawings issued on behalf of the owner by the design professionals of record (architect/engineer) and from which the project shop drawings and production drawings are developed.

Clearance. Interface space (distance) between two elements. Clearance is normally specified to allow for the effects of product and erection tolerances and for anticipated movement such as deflection, volume change movement, etc.

Clear Distance. The least distance between the surface of the reinforcement and the referenced surface. The referenced surface may be the form, adjacent reinforcement, embedments, concrete surface, or other surfaces.

Concealed Surface. Surface not visible during normal use of the member.

Dimensions. The following are several different categories of dimensions relevant to precast concrete fabrication.

Actual Dimension. The measured dimension of the precast member after casting. The actual or as-built dimension may differ from the working dimension due to construction and material induced variation.

Basic Dimension. The dimensions shown on the contract drawings or called for in the specifications. The basic dimension applies to size, location, and relative location. It may also be called the "nominal" dimension.

Working Dimension. The planned dimension of the precast member obtained from its basic dimension, the necessary joint or clearance dimensions, and other adjustments.

Discrepancy. Indicates the difference between planned dimension and actual dimension. The existence of a discrepancy frequently reveals the need for closer monitoring. Less precise measurement techniques tend to obscure problems that more precise techniques may reveal.

Draft. The taper given to features of a mold or form to allow the precast piece to be removed from the mold or form without damage. Draft can result in different feature dimensions between the front and back of a piece.

Flatness. The degree to which a surface approximates a plane (see Smoothness). This tolerance is most important in wall and slab members.

Pre-topped Systems. A construction approach, such as may be used for the floor system in parking structures, in which the flange for the floor member, often a double tee, is constructed to its final thickness in the plant, resulting in no cast-in-place topping being required in the field.

Quality. The appearance, strength, durability, and dimensional conformance which is appropriate for the specific product, its particular application and its expected performance requirements. Quality also refers to the totality of features and characteristics of a product and on its ability to satisfy stated needs.

Quality Assurance (QA). All those planned or systematic actions necessary to ensure that the final product or service will satisfy given requirements for quality and performance of intended function. Typically, the quality assurance effort will focus on the requirements of the overall project, thus identifying the tolerance quality control requirements for member fabrication.

Quality Control (QC). Those planned actions, which provide a means to measure and control the characteristics of members and materials to predetermined quantitative criteria.

Relative Alignment. The distance between two or more elements in any plane, or the distance between adjacent elements, or the distance between an element and a defined point or plane.

Setup. The process of preparing molds or forms for casting, including installation of materials (reinforcement and hardware) prior to the actual placing of concrete. The setup process is second only to the mold or form construction in its importance in the achievement of specified member tolerances.

Shrinkage. The volume change in precast concrete members caused by drying that normally occurs during the curing and initial life of concrete members. The expected shrinkage must be subtracted from the form setup dimensions to determine the as-cast dimensions of a member.

Shop Drawings. (1) Collective term used for erection drawings, production drawings and hardware details; and (2) Diagrams of precast concrete members and their connecting hardware, developed from information in the contract documents. Shop drawings show information needed for both field assembly (erection) and manufacture (production) of the precast concrete members.

Erection Drawings. Those drawings which show the relationship of the precast members and their connections in the erected structure and which provide such information as is necessary to properly erect and connect the various members.

Production Drawings. A set of instructions in the form of diagrams and text which contain all the information necessary for the manufacturer to produce the precast member. These documents are usually produced by or under the direction of the precast plant engineering department or by a party hired by the producer to do this.

Smoothness. The absence of local irregularity or roughness. It does not refer to the overall shape of the member.

Specially Finished Structural Precast Concrete. A product fabricated using forms and techniques common to the production of structural members and having specified surface finishes that require uniformity and detailing more demanding than the typical requirements for structural members.

Sweep. A global variation in member horizontal alignment. This can sometimes be caused by horizontally eccentric prestress in narrow members.

Tolerance. Specified permissible variation from specified requirements such as dimensions, location and alignment such as:
- The permitted variation from a basic dimension or quantity, as in the length, width, and depth of a member.
- The range of variation permitted in maintaining a basic dimension, as in an alignment tolerance.
- A permitted variation from location or aligment.

Product Tolerances. Those allowable variations in dimensions relating to individual precast concrete members.

Erection Tolerances. Those allowable variations in dimensions of member placement in the completed structure required for acceptable matching of precast members after they are erected.

Interfacing Tolerances. Those allowable variations in dimensions associated with other materials or systems in contact with or in close proximity to the precast concrete.

Warping. Twisting of a member, resulting in overall out-of-plane curvature of surfaces characterized by non-parallel edges. Warping is most often a concern in panel members, although it can occur in other types of members.

8.1.2 Purpose

Tolerances are normally established by economical and practical production, erection and interfacing considerations. They are based on Refs. 1 to 4. Once established, they should be shown in the contract documents, and used in design and detailing of components and connections. Architectural and structural concepts should be developed with the practical limitations of dimensional control in mind, as the tolerances will affect the dimensions of the completed structure.

Tolerances are required for the following reasons:

Structural. To ensure that structural design accounts for factors sensitive to variations in dimension. Examples include eccentric loadings, bearing areas, and locations of reinforcement and embedded items.

Feasibility. To ensure acceptable performance of joints and interfacing materials in the finished structure.

Visual. To ensure that the variations will be controllable and result in an acceptable looking structure.

Economic. To ensure ease and speed of production and erection.

Legal. To avoid encroaching on property lines and to establish a standard against which the work can be compared.

Contractual. To establish an acceptability range and also to establish responsibility for developing and maintaining specified tolerances.

8.1.3 Responsibility

While the responsibility for specifying and maintaining tolerances of the various elements may vary among projects, it is important that these responsibilities be clearly assigned. The conceptual design phase of a precast project is the place to begin consideration of dimensional control. The established tolerances or required performance should fall within generally accepted industry limits and should not be made more restrictive than necessary.

Once the tolerances have been specified, and connections that consider those tolerances have been designed, the production and erection of the elements must be organized to assure tolerance compliance.

A quality control program that emphasizes dimensional control is necessary. Likewise, an erection quality assurance program which includes a clear definition of responsibilities will aid in assuring that the products are assembled in accordance with the specified erection tolerances.

Responsibility should include dimension verification and adjustment, if necessary, of both precast components and any interfacing structural elements.

8.1.4 Tolerance Acceptability Range

Tolerances must be used as guidelines for acceptability and not limits for rejection. If specified tolerances are met, the member should be accepted. If not, the member may be accepted if it meets any of the following criteria:

1. Exceeding the tolerance does not affect the structural integrity or architectural performance of the member.
2. The member can be brought within tolerance by structurally and architecturally satisfactory means.
3. The total erected assembly can be modified to meet all structural and architectural requirements.

8.1.5 Relationships Between Different Tolerances

A precast member is erected so that its primary control surface is in conformance with the established erection and interfacing tolerances. The secondary control surfaces are generally not directly positioned during erection, but are controlled by the product tolerances. Thus, if the primary control surfaces are within erection and interfacing tolerances, and the secondary surfaces are within product tolerances, the member is erected within tolerance. The result is that the tolerance limit for the secondary surface may be the sum of the product and erection tolerances.

Since tolerances for some features of a precast member may be additive, it must be clear to the erector which are the primary control surfaces. If both primary and secondary control surfaces must be controlled, provisions for adjustment should be included. The accumulated tolerance limits may have to be accommodated in the interface clearance. Surface and feature control requirements should be clearly outlined in the plans and specifications.

On occasion, the structure may not perform properly if the tolerances are allowed to accumulate. Which tolerance takes precedence is a question of economics. The costs associated with each of the three tolerances must be evaluated, recognizing unusual situations. This may include difficult erection requirements, connections which are tolerance sensitive, or production requirements which are set by the available equipment. Any special tolerance requirements should be clearly noted in the contract documents.

It is important for the designer of record to be aware of and take into consideration the tolerances of other building materials and systems used in the project. [6, 7, 8]

8.2 Product Tolerances

8.2.1 General

Product tolerances are listed in the reports of the PCI Committee on Tolerances. [1, 2, 3.] These reports contain more complete discussion of tolerances and should be referred to for more specific details, such as location tolerances for inserts, voids, haunches and corbels and for such things as warping tolerances and local smoothness requirements. Tolerances are also presented more completely in the *Manual for Quality Control for Plants and Production of Structural Precast Concrete Products,* [2] the *Manual for Quality Control for Plants and Production of Architectural Precast Concrete Products,* [3] and the *Erectors' Manual – Standards and Guidelines for the Erection of Precast Concrete Products.* [4] The products included are listed in Table 8.2.1. Discussion of the more critical tolerances are given in the following sections. The values shown have become the consensus standards of the precast concrete industry and these values are occasionally different from tolerances published by other organizations. [7] More restrictive tolerances may significantly increase costs, so they should not be specified unless absolutely necessary.

8.2.2 Overall Dimensions

Typical tolerances for most precast/prestressed concrete products are given in Table 8.2.1. Architectural precast panels have plan dimension tolerances that vary with panel size from ±⅛ in. for a dimension under 10 ft to ±¼ in. for a dimension of 20 to 40 ft. Table 8.2.2 lists erection tolerances for interface design of precast and cast-in-place concrete members. Clearances are listed in Table 8.2.3.

Top and bottom slabs (flanges) of box beams and hollow-core slabs are dependent on the position of cores. Flange thickness tolerances are not given for hollow-core slabs. Instead, measured flange areas cannot be less than 85% of the nominal calculated area.

The *PCI Tolerance Manual* [1] emphasizes that the recommended tolerances are only guidelines. Different values may be applicable in some cases, and each project should be considered individually.

8.2.3 Sweep or Horizontal Alignment

Horizontal misalignment, or sweep, usually occurs as a result of form and member width tolerances. It can also result from prestressing with lateral eccentricity, which should be considered in the design. Joints should be dimensioned to accommodate such variations.

Sweep tolerances generally vary with length of unit, for example, ±⅛ in. per 10 ft. The upper limit of sweep varies from ±⅜ in. for wall panels and hollow-core slabs to ±¾ in. for joists usually used in composite construction.

Table 8.2.1 Typical tolerances for precast prestressed concrete products

Product Tolerances	Products
Length —	
±¼ in.	18
±⅜ in.	16,17
±½ in.	6,7,8,9,13,15
±¾ in.	3,5
±1 in.	1,2,4,11,12,14
Width —	
±¼ in.	1,2,3,5,6,7,8,9,12,15,16,18
+⅜ in.	14
+⅜ in., –¼ in.	4
±⅜ in.	11,13
±½ in.	17
Depth —	
+¼ in., –⅛ in.	10,18
±¼ in.	1,2,3,5,6,7,8,9,12,13,14,15
+½ in., –¼ in.	4
±⅜ in.	11
Flange thickness —	
+¼ in., –⅛ in.	1,2,8,10,12,15
±¼ in.	3,4
Web thickness —	
±⅛ in.	1,8,10,12,15
±¼ in.	2,3
+⅜ in., –¼ in.	4
±⅜ in.	5
Position of tendons —	
±¼ in	1,2,3,4,5,6,8,9,11,12,14,15,18
±⅛ in.	10
Camber, variation from design —	
±¼ in. per 10 ft, ±¾ in. max.	1,2,12,15
±⅛ in. per 10 ft, ±1 in. max.	4
±¾ in. max.	3
±½ in. max.	5,15
Camber, differential —	
¼ in. per 10 ft, ¾ in. max.	1,2,5
±¼ in. per 10 ft, ±½ in. max.	15
Bearing plates, position —	
±½ in.	1,2,3,12,15
±⅝ in.	4
Bearing plates, topping and flushness —	
±⅛ in.	1,2,3,4,12,13,15

Key:

1 = double tee
2 = single tee
3 = building beam (rect. and ledger)
4 = I-beam
5 = box beam
6 = column
7 = hollow-core slab
8 = ribbed wall panel
9 = insulated wall panel
10 = architectural wall panel
11 = pile
12 = joist
13 = step unit
14 = sheet piling
15 = single riser bleacher slabs
16 = prison cell module – single
17 = prison cell module – double
18 = prestressed concrete panels for storage tanks

Table 8.2.2 Erection tolerances for interface design of precast and cast-in-place concrete members

Item	Recommended Tolerances
Variation in plan location (any column or beam, any location)	±½ in. for column, ±1 in. for beam
Variation in plan parallel to specified building lines	+¹⁄₄₀ in. per ft for any beam less than 20 ft or adjacent columns less than 20 ft apart
	½ in. maximum for adjacent columns 20 ft or more apart
Difference in relative position of adjacent columns from specified relative position (at any check level)	+½ in.
Variation from plumb	+¼ in. any 10 ft of height
	1 in. maximum for the entire height
Variation in elevation of bearing surfaces from specified elevation (any column or beam, any location)	+½ in.
Variation of top of spandrel from specified elevation (any location)	+½ in.
Variation in elevation of bearing surfaces from lines parallel to specified grade lines	+¹⁄₄₀ in. per ft for any beam less than 20 ft or adjacent columns less than 20 ft apart
	½ in. maximum any beam 20 ft or more in length or adjacent columns 20 ft or more apart
Variation from specified bearing length on support	±¾ in.
Variation from specified bearing width on support	±½ in.
Jog in alignment of matching edges	½ in., maximum

Table 8.2.3 Recommended clearances

Item	Recommended Minimum Clearance
Precast to precast	½ in. (1 in. preferred)
Precast to cast-in-place	1 in. (2 in. preferred)
Precast to steel	1 in. (2 in. preferred)
Precast column covers	1½ in. (3 in. preferred for tall buildings)

8.2.4 Position of Strands

It is a common practice to use ⅝ in. diameter holes in end dividers (bulkheads or headers) for ⅜ to 0.6 in. diameter strands, since it is costly to switch end dividers for different strand diameters. Thus, better accuracy is achieved when using larger diameter strands.

Generally, individual strands must be positioned within ±¼ in. of design position and bundled strands within ±½ in. Hollow-core slabs have greater individual strand tolerances as long as the center of gravity of the strand group is within ±¼ in. and a minimum cover of ¾ in. is maintained.

8.2.5 Camber and Differential Camber

Design camber is generally based on camber at release of prestress; thus, camber measurements on products should be made as soon after stripping as possible. Differential camber refers to the final in-place condition of adjacent products.

It is important that cambers are measured at the same time of day, preferably in the early hours before the sun has begun to warm the members. Cambers for all units used in the same assembly should be checked at the same age.

If a significant variation in camber from calculated values is observed, the cause should be determined and the effect of the variation on the performance of the member evaluated. If differential cambers exceed recommended tolerances, additional effort is often required to erect the members in a manner which is satisfactory for the intended use.

The final installed differential between two adjacent cambered members erected in the field may be the combined result of member differential cambers, variations in support elevations, and any adjustments made to the members during erection.

For most flexural members, maximum camber variation from design camber is ±¾ in., and maximum differential camber between adjacent units of the same design is ¾ in. This may be increased for joists used in composite construction. Recommendations for camber and differential camber of hollow-core slabs are not listed, because production variations between hollow-core systems result in different tolerances for each type.

8.2.6 Weld Plates

In general, it is easier to hold plates to closer tolerances at the bottom of the member (as cast) (or against the side form) than with plates cast on top of the member. Bottom and side plates can be fastened to the form and hence are less susceptible to movement caused by vibration. This applies to position of weld plates as well as tipping and flushness.

The tolerance on weld plates is less restrictive than for bearing plates. The position tolerance is ±1 in. for all products. Tipping and flushness tolerance is ±¼ in.

8.2.7 Haunches of Columns and Wall Panels

The importance of corbel or haunch location tolerances depends on the connection at the base of the member. Since base connections usually allow some flexibility, it is more important to control dimensions from haunch to haunch in multilevel columns or walls than from haunch to end of member.

The haunch to haunch tolerance is ±⅛ to ±¼ in. Bearing surface squareness tolerance is ±⅛ in. per 18 in. with a maximum of ±¼ in., except for architectural precast concrete panels, with a tolerance of ±⅛ in., and columns, with a maximum of ±⅛ in. in short direction and ±⅜ in. in long direction.

8.2.8 Warping and Bowing

Warping and bowing tolerances affect panel edge matchup during erection, and the appearance of the erected members. They are especially critical with architectural panels.

Warping is a variation from plane in which the corners of the panel do not fall within the same plane. Warping tolerances are given in terms of corner variations, as shown in Figure 8.2.1. The allowable variation from the nearest adjacent corner is 1/16 in. per ft.

Bowing differs from warping in that two opposite edges of a panel may fall in the same plane, but the portion between is out of plane (Figure 8.2.2). Bowing tolerance is L/360, where L is the length of bow. Maximum tolerance on differential bowing between panels of the same design is ½ in.

The effects of differential temperature and moisture absorption between the inside and outside of a panel and the prestress eccentricity should be considered in design of the panel and its connections. Pre-erection storage conditions may also affect warping and bowing (see Sections 4.9.7 and 5.2.10).

Thin panels are more likely to bow, and the tolerances should be more liberal. Table 8.2.4 gives thicknesses, related to panel dimensions, below which the warping and bowing tolerances given above may not apply. (Note: Table 8.2.4 is not intended to limit panel thickness.) For example, a panel that is 16 x 8 ft, and less than 6 in. thick, may require greater warping and bowing tolerance than indicated above.

Similarly, panels made from concrete with over ¾ in. aggregate, panels using two significantly different concrete mixes, and veneered and insulated panels may require special consideration. In all cases, the local precaster should be consulted regarding overall economic and construction feasibility.

Table 8.2.4 Minimum panel thickness (in.) to maintain bowing and warping within suggested normal tolerances[a,b]

Panel dimensions, ft	8	10	12	16	20	24	28	32
4	3	4	4	5	5	6	6	7
6	3	4	4	5	5	6	6	7
8	4	5	5	6	6	7	7	8
10	5	5	6	6	7	7	8	8

a. Do not use this table for panel thickness selection.
b. For ribbed panels, the overall thickness of ribs may be used for comparison with this table if the ribs are continuous from one end of the panel to the other.

8.2.9 Smoothness

Local smoothness describes the condition where small areas of the surface may be out of plane, as shown in Figure 8.2.3. The tolerance for this type of variation is ¼ in. per 10 ft for all products. The tolerance is usually checked with a 10 ft straight edge or the equivalent, as explained in Figure 8.2.3.

8.2.10 Architectural Panels vs. Structural Walls

When discussing tolerances, "architectural panel" refers to a class of tolerances specified, and not necessarily to the use of the member in the final structure. Architectural panels usually require more restrictive tolerances than structural members for aesthetic reasons.

Double tees, hollow-core slabs and solid slabs are often used for exterior facades, but are not classed as "architectural panels." Since the above listed products are manufactured the same whether they are architectural or structural products, the manufacturing accuracy for the product when used architecturally should not be expected to meet

Figure 8.2.1　Definition of panel warping

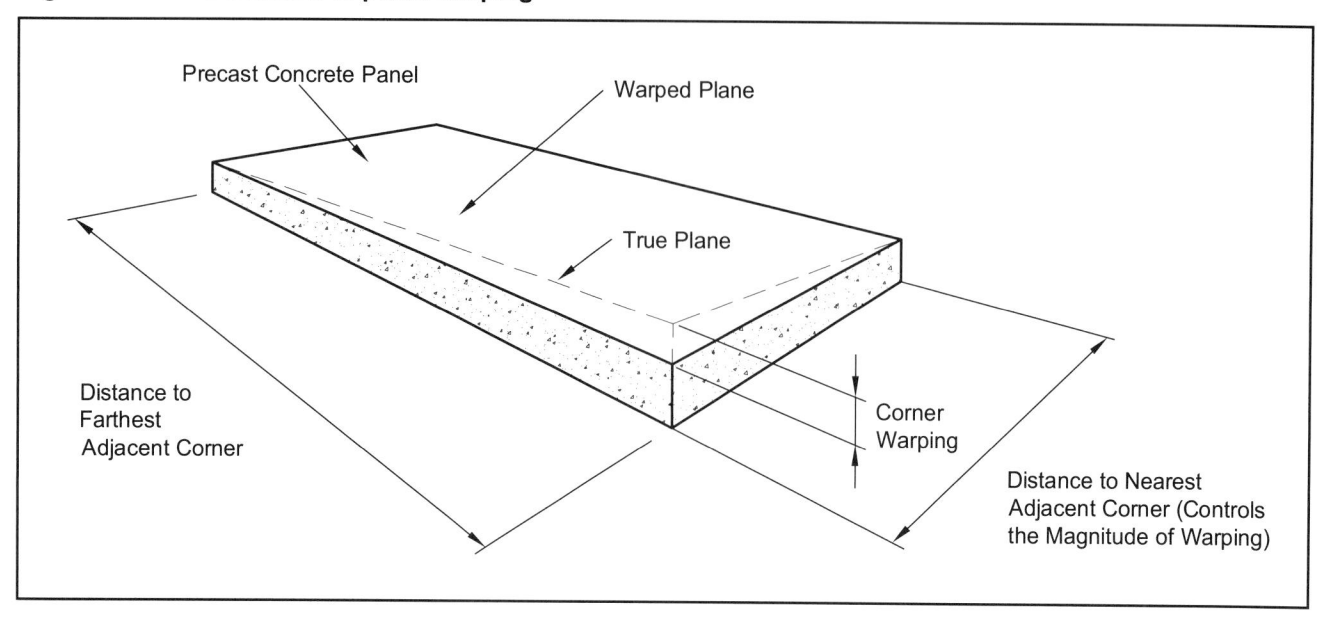

Figure 8.2.2　Definition of panel bowing

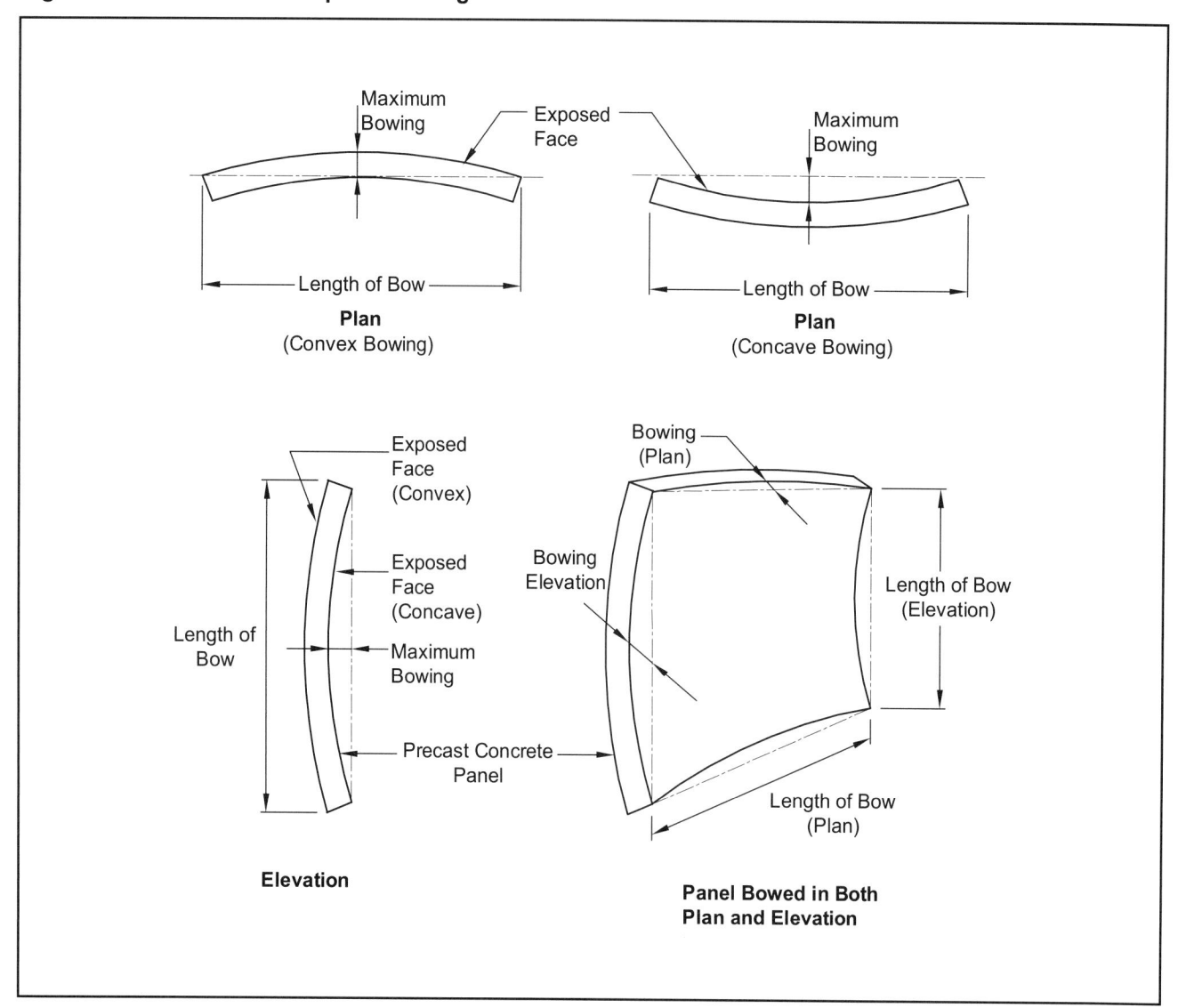

Figure 8.2.3   Local smoothness variation

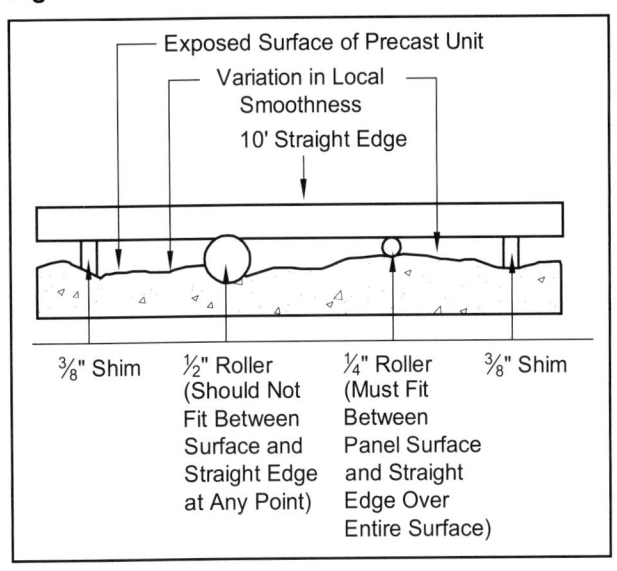

```
         Exposed Surface of Precast Unit
            Variation in Local
              Smoothness
             10' Straight Edge
```

3/8" Shim | 1/2" Roller (Should Not Fit Between Surface and Straight Edge at Any Point) | 1/4" Roller (Must Fit Between Panel Surface and Straight Edge Over Entire Surface) | 3/8" Shim

"architectural panel" tolerances. If more restrictive tolerances are required, they must be clearly indicated in the contract documents, and subsequently increased costs anticipated.

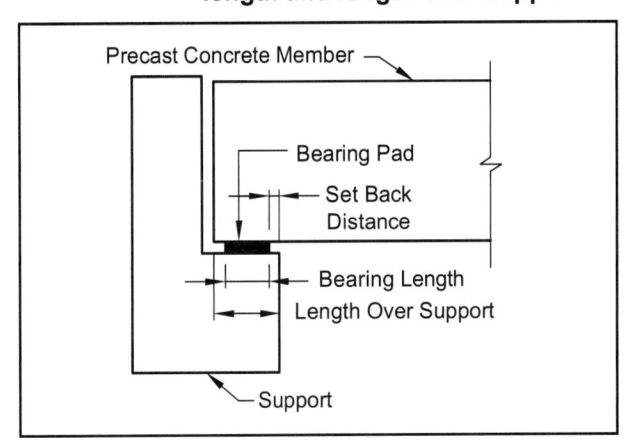

8.3   Erection Tolerances

8.3.1   General

Erection tolerance values are those to which the primary control surfaces of the member should be set. The final location of other features and surfaces will be the result of the combination of the erection tolerances and the product tolerances given in Section 8.2.

Because erection is equipment and site dependent, there may be good reason to vary some of the recommended tolerances to account for unique project conditions. In general, the more restrictive the erection tolerances, the higher the cost of erection will be. Combining liberal product tolerances with restrictive erection tolerances may place an unrealistic burden on the erector. Thus, the designer should review proposed tolerances with manufacturers and erectors prior to deciding on the final project tolerances.

To minimize erection problems, the dimensions of the in-place structure should be checked prior to starting precast erection. After erection, and before other trades interface with the precast concrete members, it should be verified that the precast elements are erected within tolerances. [7]

8.3.2   Recommended Erection Tolerances

Figures 8.3.2 to 8.3.9 show erection tolerances for the following four mixed building systems:

- Precast element to precast element.
- Precast element to cast-in-place concrete.
- Precast element to masonry.
- Precast element to structural steel construction.

See Ref. 4 for recommended erection tolerances for additional precast members.

These tolerances should be considered guidelines for the development of project specific tolerances for erection.

8.3.3   Mixed Building Systems

Mixed building systems combine precast and prestressed concrete with other materials, usually cast-in-place concrete, masonry or steel. Each industry has its own recommended erection tolerances which apply when its products are used exclusively. The compatibility of those tolerances with the precast tolerances should be checked and adjusted when necessary.

Example 8.4.2 shows one problem that can occur when erection tolerances are chosen for each system without considering the project as a whole.

8.3.4   Connections and Bearing

The details of connections must be considered when specifying erection tolerances. Space must be provided to make the connection under the most adverse combination of tolerances.

Bearing length is measured in the direction of the span, and bearing width is measured perpendicular to the span. Bearing length is often not the same as the length of the end of a member over the support, as shown in Figure 8.3.1. When they differ, it should be noted on erection drawings.

The Engineer may wish to specify a minimum bearing for various precast products. For further information, see Ref. 7.

Figure 8.3.1   Relationship between bearing length and length over support

```
   Precast Concrete Member
                  Bearing Pad
                  Set Back Distance
                  Bearing Length
                  Length Over Support
                  Support
```

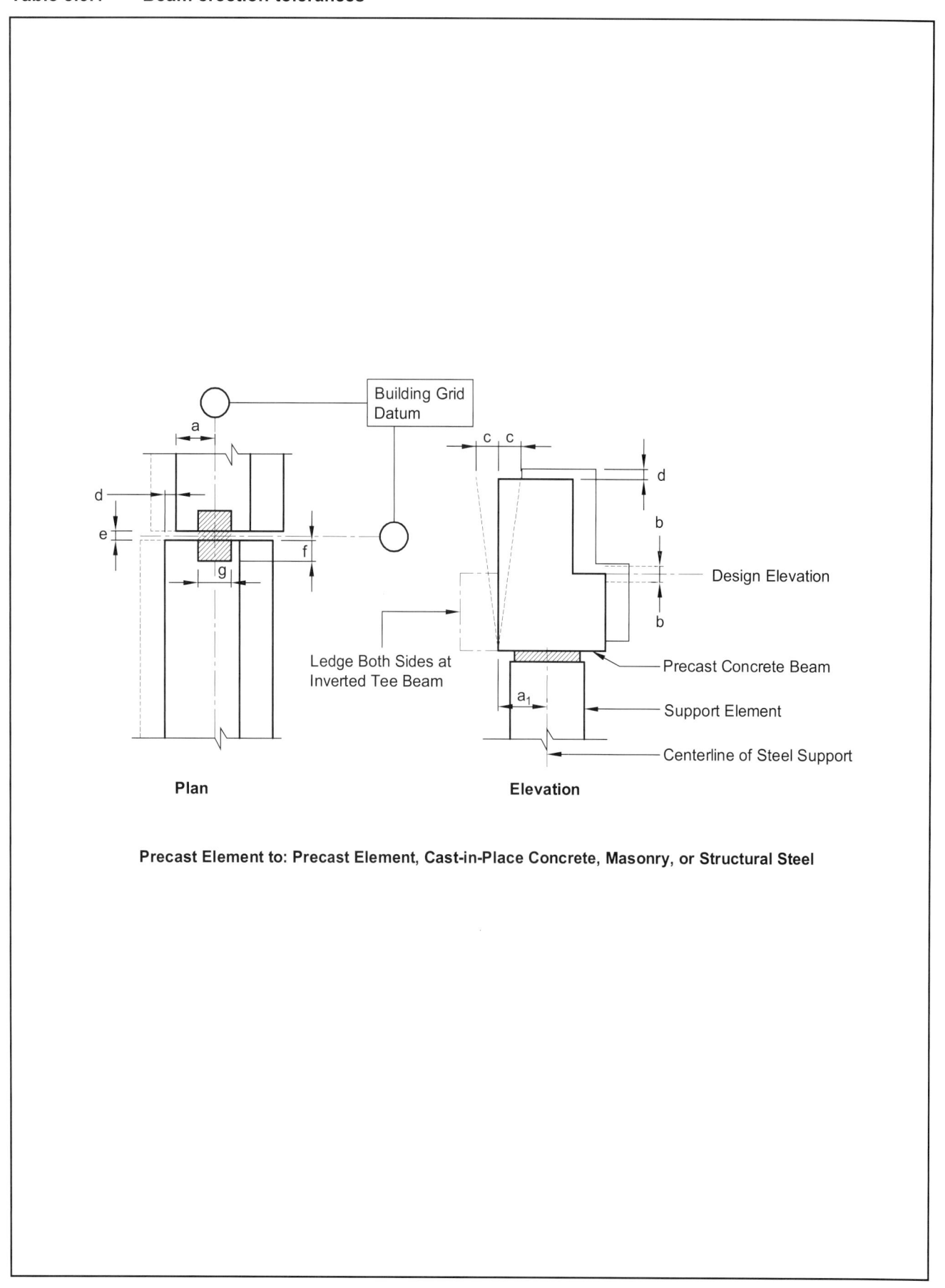

Table 8.3.1   Beam erection tolerances

Plan

Elevation

Precast Element to: Precast Element, Cast-in-Place Concrete, Masonry, or Structural Steel

Table 8.3.1 Beam erection tolerances (Cont.)

The primary control surfaces are usually as shown, although this needs to be confirmed on a job-by-job basis.

a = Plan location from building grid datum ....±1 in.

a_1 = Plan location from centerline of steel* ....±1 in.

b = Bearing elevation[†] from nominal elevation at support:
   Maximum low ....½ in.
   Maximum high ....¼ in.

c = Maximum plumb variation over height of element:
   Per 12 in. height ....⅛ in.
   Maximum at rectangular or L-beam ....½ in.
   Maximum at inverted tee beam ....¾ in.

d = Maximum jog in alignment of matching edges:
   Architectural exposed edges ....¼ in.
   Visually non-critical edges ....½ in.

e = Joint width:
   Architectural exposed joints ....±¼ in.
   Hidden joints ....±¾ in.
   Exposed structural joint not visually critical ....±½ in.

f = Bearing length[‡] (span direction) ....±¾ in.

g = Bearing width[‡] ....±½ in.

Note: When bearing pads are used at unarmored edges, they should be set back a minimum of ½ in. from the face of the support or at least the chamfered dimension at chamfered edges.

---

[*] For precast elements on a steel frame, this tolerance takes precedence over tolerance on dimension "a".

[†] Or member top elevation where member is part of a frame without bearing ledges.

[‡] This is a setting tolerance and should not be confused with structural performance requirements set by the architect/engineer. The nominal bearing dimensions and the allowable variations in the bearing length and width should be specified by the engineer and shown on the erection drawings.

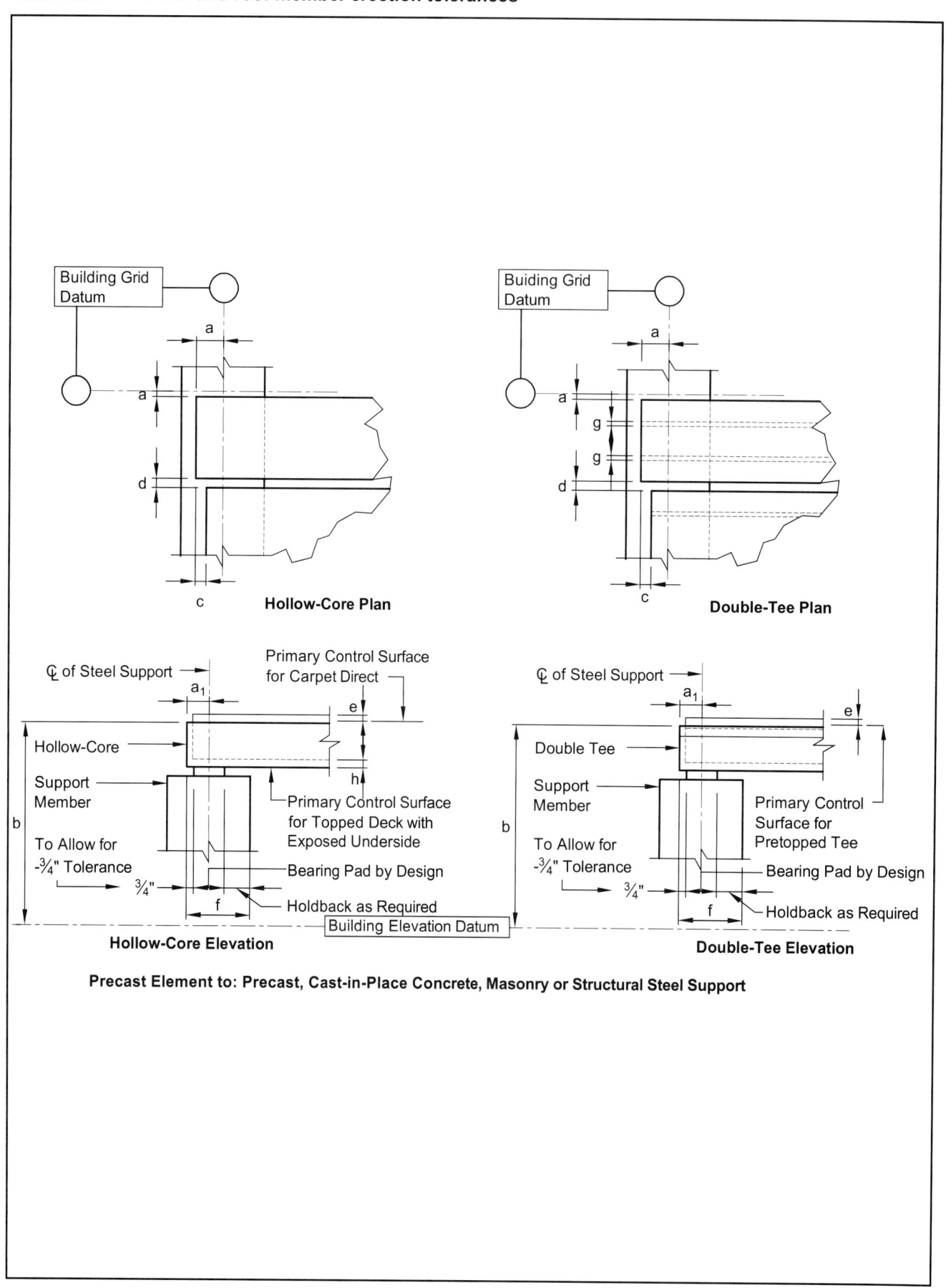

Table 8.3.2   Floor and roof member erection tolerances

Table 8.3.2 Floor and roof member erection tolerances (Cont.)

The primary control surfaces are usually as shown. A majority of the time, there is no designated vertical primary control surface, and in some scenarios there are no primary control surfaces at all. This needs to be determined on a job-by-job basis.

a = Plan location from building grid datum .... ±1 in.

a_1 = Plan location from centerline of steel support* .... ±1 in.

b = Top elevation from building elevation datum at member ends:
Covered with topping .... ±¾ in.
Pretopped tee/carpet direct hollow-core slab .... ±¼ in.
Untopped roof .... ±¾ in.

c = Maximum jog in alignment of matching edges
(both topped and untopped construction) .... 1 in.

d = Joint width:
0 to 40 ft member .... ±½ in.
41 to 60 ft member .... ±¾ in.
61 ft plus member .... ±1 in.

e = Differential top elevation as erected (for units of same design and length):
Field topped .... ¾ in.
Pretopped tees at driving lanes/carpet direct hollow-core slabs .... ¼ in.
Untopped roof[†] .... ¾ in.

f = Bearing length[‡] (span direction) .... ±¾ in.

g = Bearing width[‡] (n/a for hollow-core slabs) .... ±½ in.

h = Differential bottom elevation of
exposed hollow-core slabs[§] .... ¼ in.

Note: When bearing pads are used at unarmored edges they should be set back a minimum of ½ in. from the face of the support or at least the chamfered dimension at chamfered edges.

---

[*] For precast concrete erected on a steel frame building, this tolerance takes precedence over tolerance on dimension "a".

[†] It may be necessary to feather the edges to ±¼ in. to properly apply some roof membranes.

[‡] This is a setting tolerance and should not be confused with structural performance requirements set by the architect/engineer. The nominal bearing dimensions and the allowable variations in the bearing length and width should be specified by the engineer and shown on the erection drawings.

[§] Untopped installations will require a larger tolerance.

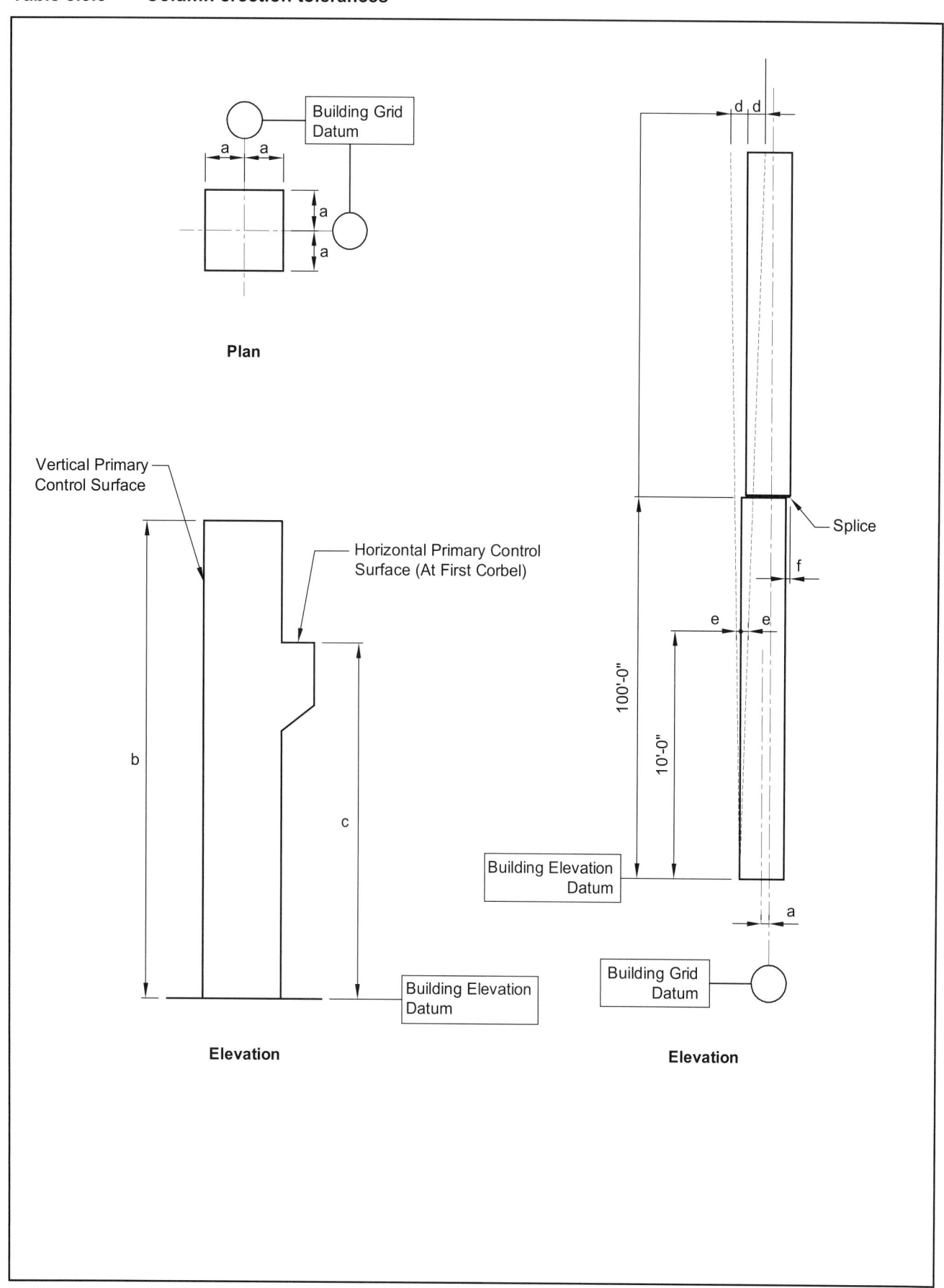

Table 8.3.3   Column erection tolerances

Plan

Elevation

Elevation

Table 8.3.3   Column erection tolerances (Cont.)

The primary control surfaces are usually as shown, although this needs to be confirmed on a job-by-job basis.

a = Plan location from building grid datum:
  Structural applications ....±½ in.
  Architectural applications ....±⅜ in.

b = Top elevation from nominal top elevation:
  Maximum low ....−½ in.
  Maximum high ....+¼ in.

c = Bearing haunch elevation from nominal elevation:
  Maximum low ....−½ in.
  Maximum high ....+¼ in.

d = Maximum plumb variation over height of element (element
  in structure of maximum height of 100 ft)..±1 in.

e = Plumb in any 10 ft of element height..±¼ in.

f = Maximum jog in alignment of matching edges:
  Architectural exposed edges..±¼ in.
  Visually non-critical edges..±½ in.

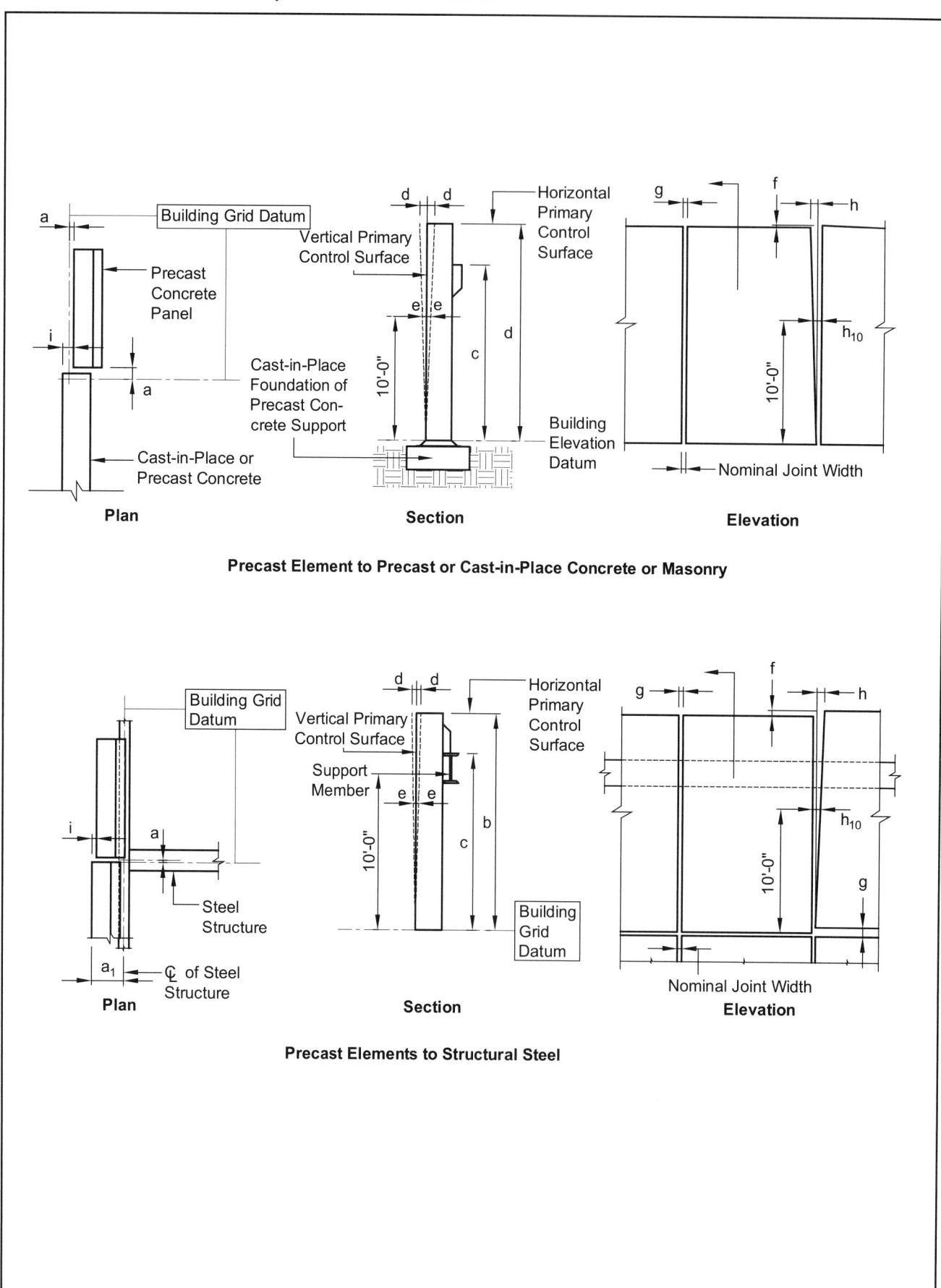

Table 8.3.4   Structural wall panel erection tolerances

Precast Element to Precast or Cast-in-Place Concrete or Masonry

Precast Elements to Structural Steel

Table 8.3.4   Structural wall panel erection tolerances (Cont.)

The primary control surfaces are usually as shown, although this needs to be confirmed on a job-by-job basis.

a = Plan location from building grid datum* .... ±½ in.

a_1 = Plan location from centerline of steel support .... ±½ in.

b = Top elevation from nominal top elevation:
  Exposed individual panel .... ±½ in.
  Non-exposed individual panel .... ±¾ in.
  Exposed relative to adjacent panel .... ±½ in.
  Non-exposed relative to adjacent panel .... ±¾ in.

c = Support elevation from nominal elevation:
  Maximum low .... ½ in.
  Maximum high .... ¼ in.

d = Maximum plumb variation over height of
  structure or over 100 ft which ever is less* .... ±1 in.

e = Plumb in any 10 ft of element height .... ±¼ in.

f = Maximum jog in alignment of matching edges .... ±½ in.

g = Joint width (governs over joint taper) .... ±⅜ in.

h = Joint taper over height of panel .... ±½ in.

h_{10} = Joint taper over 10 ft height .... ±⅜ in.

i = Maximum jog in alignment of matching faces:
  Exposed to view .... ±⅜ in.
  Not exposed to view .... ±¾ in.

j = Differential bowing or camber as erected
  between adjacent members of the same design[†] .... ±½ in.

---

[*] For precast buildings in excess of 100 ft tall, Tolerances "a" and "d" can increase at the rate of ⅛ in. per story to a maximum of 2 in.

[†] Refer to Section 8.2.8 for description of bowing tolerance.

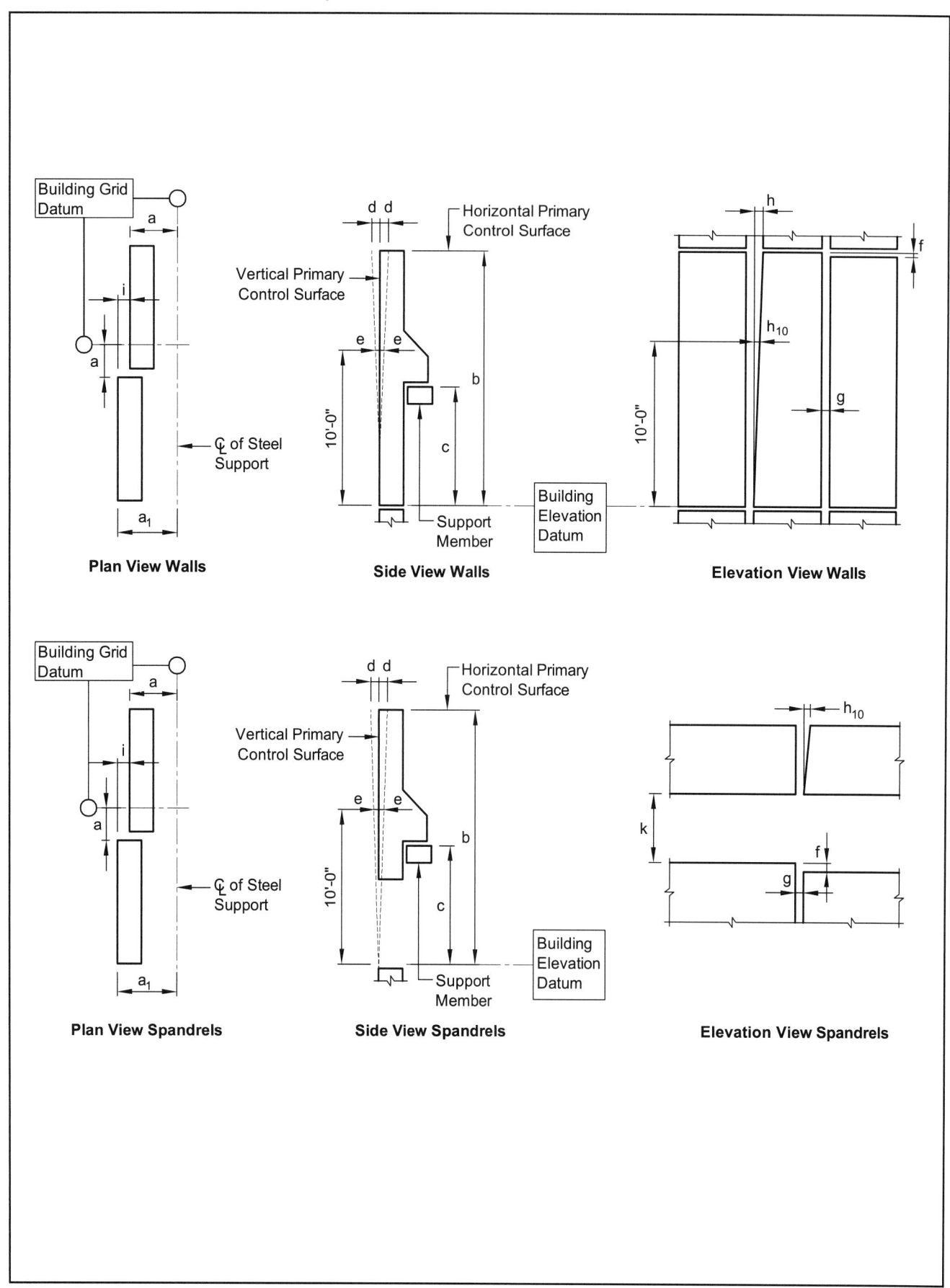

Table 8.3.5   Architectural walls/spandrel erection tolerances

Plan View Walls

Side View Walls

Elevation View Walls

Plan View Spandrels

Side View Spandrels

Elevation View Spandrels

Table 8.3.5   Architectural walls/spandrel erection tolerances (Cont.)

The primary control surfaces are usually as shown, although this needs to be confirmed on a job-by-job basis.

a = Plan location from building grid datum* .... ±½ in.

a_1 = Plan location from centerline of steel support† .... ±½ in.

b = Top elevation from nominal top elevation:
  Exposed individual panel .... ±¼ in.
  Non-exposed individual panel .... ±½ in.

c = Support elevation from nominal elevation:
  Maximum low .... ½ in.
  Maximum high .... ¼ in.

d = Maximum plumb variation over height of structure
  or 100 ft whichever is less* .... ±1 in.

e = Plumb in any 10 ft of element height .... ±¼ in.

f = Maximum jog in alignment of matching edges:
  Exposed relative to adjacent panel .... ±¼ in.
  Non-exposed relative to adjacent panel .... ±½ in.

g = Joint width (governs over joint taper) .... ±¼ in.

h = Joint taper maximum .... ±⅜ in.

h_{10} = Joint taper over 10 ft length .... ±¼ in.

i = Maximum jog in alignment of matching faces .... ±¼ in.

j = Differential bowing or camber as erected between
  adjacent members of the same design .... ±¼ in.

k = Opening height between spandrels .... ±¼ in.

---

* For precast buildings in excess of 100 ft tall, Tolerances "a" and "d" can increase at the rate of ⅛ in. per story to a maximum of 2 in.

† For precast elements erected on a steel frame, this tolerance takes precedence over tolerance on Dimension "a".

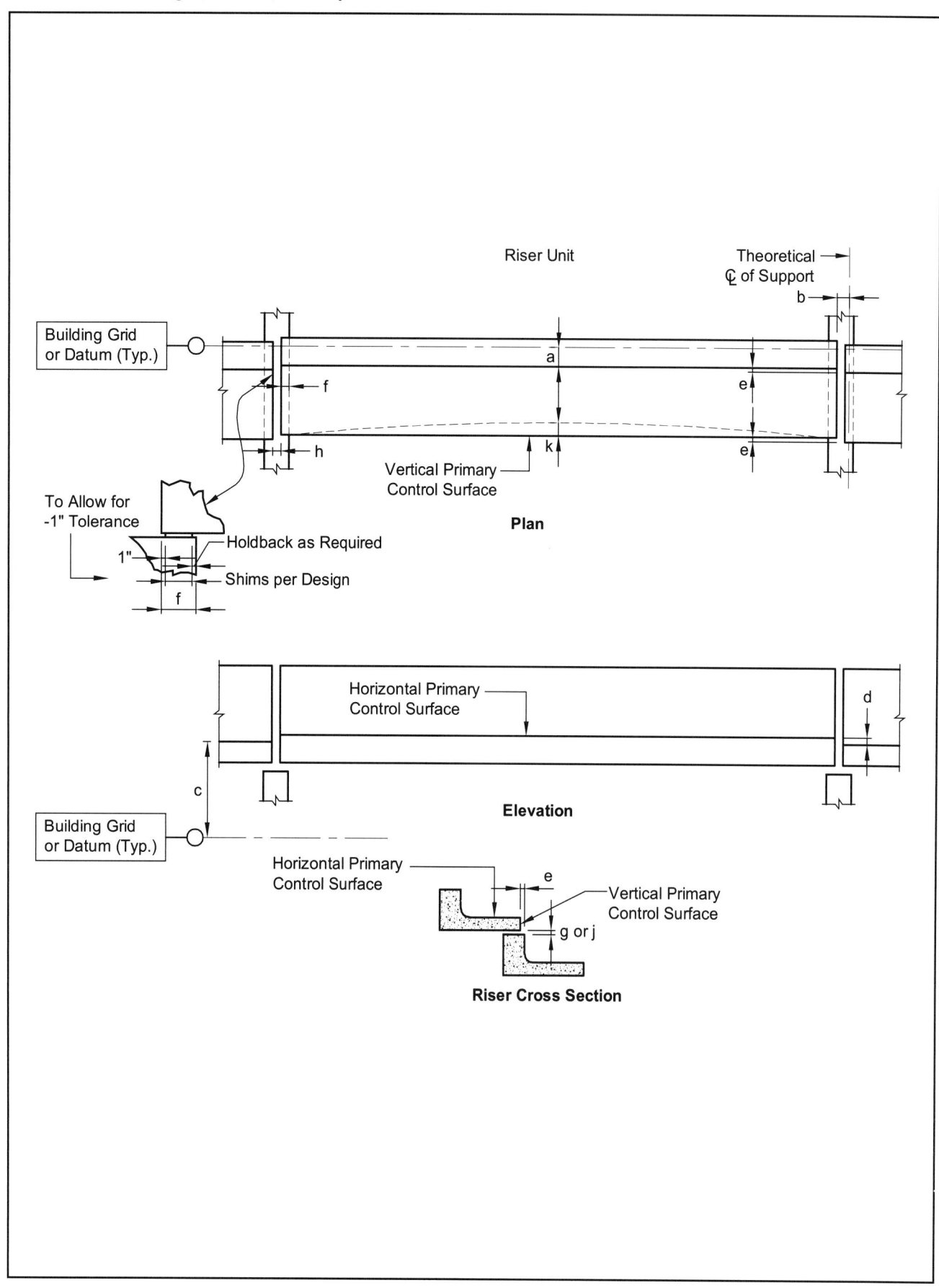

Table 8.3.6   Single, double, and triple stadium riser erection tolerances

Plan

- Building Grid or Datum (Typ.)
- Riser Unit
- Theoretical ℄ of Support
- a, b, e, f, h, k
- To Allow for -1" Tolerance
- 1"
- Holdback as Required
- Shims per Design
- f
- Vertical Primary Control Surface

Elevation

- Horizontal Primary Control Surface
- c, d
- Building Grid or Datum (Typ.)

Riser Cross Section

- Horizontal Primary Control Surface
- Vertical Primary Control Surface
- e
- g or j

8–20   PCI Design Handbook/Sixth Edition

Table 8.3.6   Single, double and triple riser erection tolerances (Cont.)

The primary control surfaces are usually as shown, although this is something that needs to be confirmed with the contractor on a job-by-job basis. Local building codes may require more restrictive riser height tolerances which could also affect product tolerance.

a = Plan location from building grid line datum .... ±1 in.

b = Plan location from theoretical centerline of
    support structure .... ±1 in.

c = Top elevation from building elevation datum at
    member's end. (This datum may be adjusted to
    accommodate existing field conditions.) .... ±½ in.

d = Maximum jog in alignment of matching edges
    at the horizontal primary control surface .... ¼ in.

e = Maximum jog in alignment of matching edges
    at the vertical primary control surface .... ½ in.

f = Bearing in span direction .... −1 in.

g = Joint width (horizontal) at end of piece. (Joint
    width needs to be ¼ in. minimum.) .... ±½ in.

h = Joint width (Joint width needs to be ¼ in. minimum in either case.)
    90-deg angle .... ±½ in.
    Joint width at skewed ends .... ±⅝ in.

j = Differential camber (at midspan as erected)
    between adjacent members of the same design .... ±³⁄₁₆ in. per 10 ft of member length.

k = Differential sweep (at midspan as erected)
    between adjacent members of the same design .... ±³⁄₁₆ in. per 10 ft of member length.

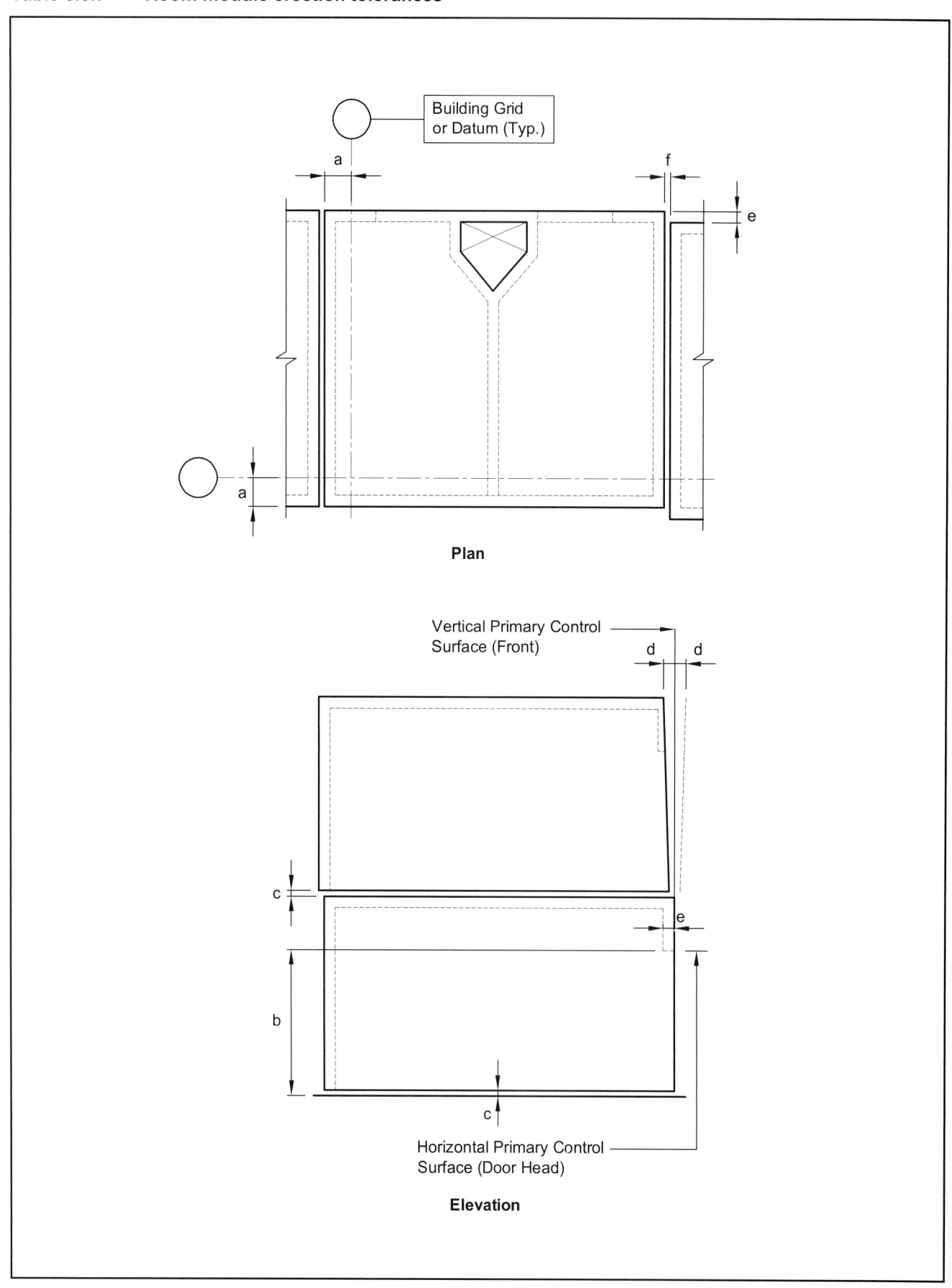

Table 8.3.7   **Room module erection tolerances**

Plan

Elevation

Table 8.3.7   Room module erection tolerances (Cont.)

The tolerances listed below are used at the primary control surfaces only, and only those tolerances that are applicable to that surface. Normally, the primary control surfaces are the front face of the cell unit as the vertical primary control surface, and either the head of the door (as shown at left), top of cell, or the bottom of balcony as the horizontal primary control surface. Note: on jobs where pre-topped balconies are cast as part of the cell unit, the horizontal primary control surface may be the top surface of the balcony.

a = Plan location from building grid line datum ....±½ in.

b = Vertical control (at primary control surface)
   from a horizontal datum ....±⅜ in.

c = Actual grout joint.. ½ in. minimum

d = Plumb at element height .... ¼ in.

e = Maximum jog in alignment of matching edges... ¼ in.

f = Vertical joint width ....±⅜ in.

g = Joint taper...Not applicable

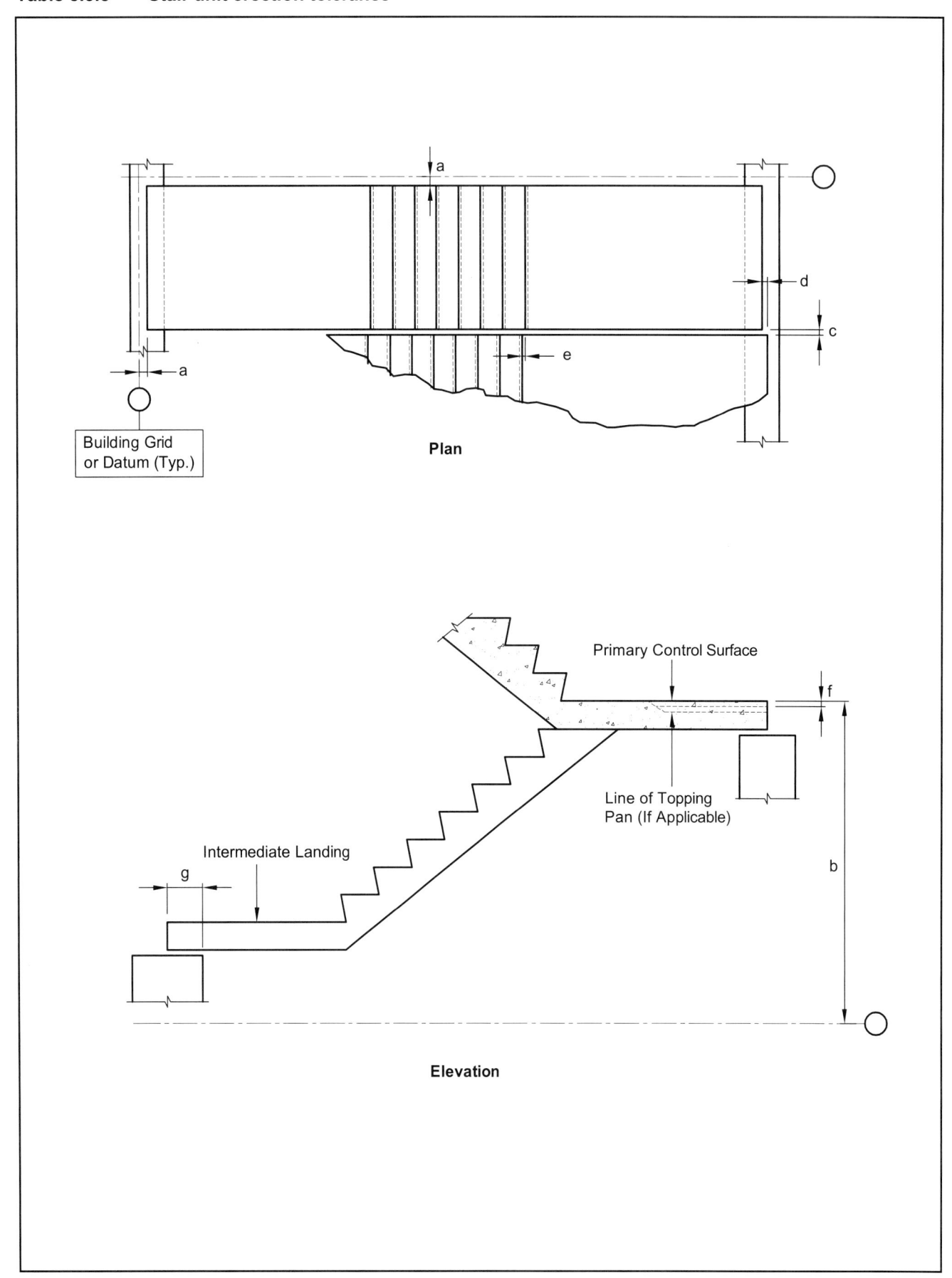

Table 8.3.8   Stair unit erection tolerance

Plan

Elevation

- Primary Control Surface
- Line of Topping Pan (If Applicable)
- Intermediate Landing
- Building Grid or Datum (Typ.)

Table 8.3.8   Stair unit erection tolerance (Cont.)

The primary control surface for stair units is the top of landing at floor levels. Tolerances listed below are the same whether landings are monolithic or separate pieces. Local building codes may require more restrictive riser height tolerance which could also affect the product tolerance.

a = Plan location from building grid line datum ....±½ in.

b = Differential elevation as erected* ....±⅜ in.

c = Joint width ....±¾ in.

d = Maximum jog in alignment of matching edges.. 1 in.

e = Maximum jog in alignment of stair tread nosings
    (This tolerance overrides "d" if needed.)... ½ in.

f = Maximum jog in alignment of matching edges
    at the primary control surface* .... ⅜ in.

g = Bearing (in span direction) ....±¾ in.

---

* At stair units that have pre-topped precast landings, the maximum jog between stair units as well as from stair unit to finish floor cannot exceed ¼ in. However, units which have landings that are topped have more leeway. This needs to be discussed and agreed upon with the general contractor.

8.4 Clearances

8.4.1 General

Clearance is space between adjacent members and provides a buffer area where erection and production tolerance variations can be absorbed. The following items should be addressed when determining the appropriate clearance to provide in the design:

- Product tolerance
- Type of member
- Size of member
- Location of member
- Member movement
- Function of member
- Erection tolerance
- Fireproofing of steel
- Thickness of plates, bolt heads, and other projecting elements

Of these factors, product tolerances and member movement are the most significant. As shown in the examples, it may not always be practical to account for all possible factors in the clearance provided. Table 8.4.1 provides recommended minimum clearances for various mixed building systems.

Table 8.4.1 Recommended clearances

Item	Recommended Minimum Clearance
Precast to precast	½ in. (1 in. preferred)
Precast to cast-in-place	1 in. (2 in. preferred)
Precast to steel	1 in. (2 in. preferred)
Precast column covers	1½ in. (3 in. preferred for tall buildings)

8.4.2 Joint Clearance

Joints between architectural panels must accommodate variations in the panel dimensions and the erection tolerances for the panels. They must also provide a good visual line and sufficient width to allow for a proper sealant joint. Generally, the larger the panels, the wider the joints should be. For most situations, architectural panel joints should be designed to be not less than ¾ in. wide. Tolerances in overall building width and length are normally accommodated in panel joints.

8.4.3 Procedure for Determining Clearance

The following is a systematic approach for making a trial selection of a clearance value and then testing that selection to ensure that it will allow practical erection to occur:

Step 1:
Determine the maximum size of the members involved (basic or normal dimension plus additive tolerances). This should include not only the precast and prestressed members, but also other materials.

Step 2:
Add to the maximum member size the minimum space required for member movement resulting from deflection and thermal variations.

Step 3:
Check if the resulting clearance allows the member to be erected within the erection and interfacing tolerances, such as plumbness, face alignment, etc. Adjust the clearance as required to meet all the needs.

Step 4:
Check if the member can physically be erected with this clearance. Consider the size and location of members in the structure and how connections will be made. Adjust the clearance as required.

Step 5:
Review the clearance to see whether increasing its dimensions will allow easier, more economical erection without adversely affecting aesthetics. Adjust the clearance as required.

Step 6:
Review structural considerations such as types of connections involved, sizes required, bearing area requirements, and other structural issues.

Step 7:
Check design to ensure adequacy in the event that minimum member size should occur. Adjust clearance as required for minimum bearing and other structural considerations.

Step 8:
Select final clearance which will satisfy all of the conditions considered.

8.4.4 Clearance Examples

The following examples are given to show the thought process, and may not be the only correct solutions for the situations described.

Figure 8.4.1 Clearance example

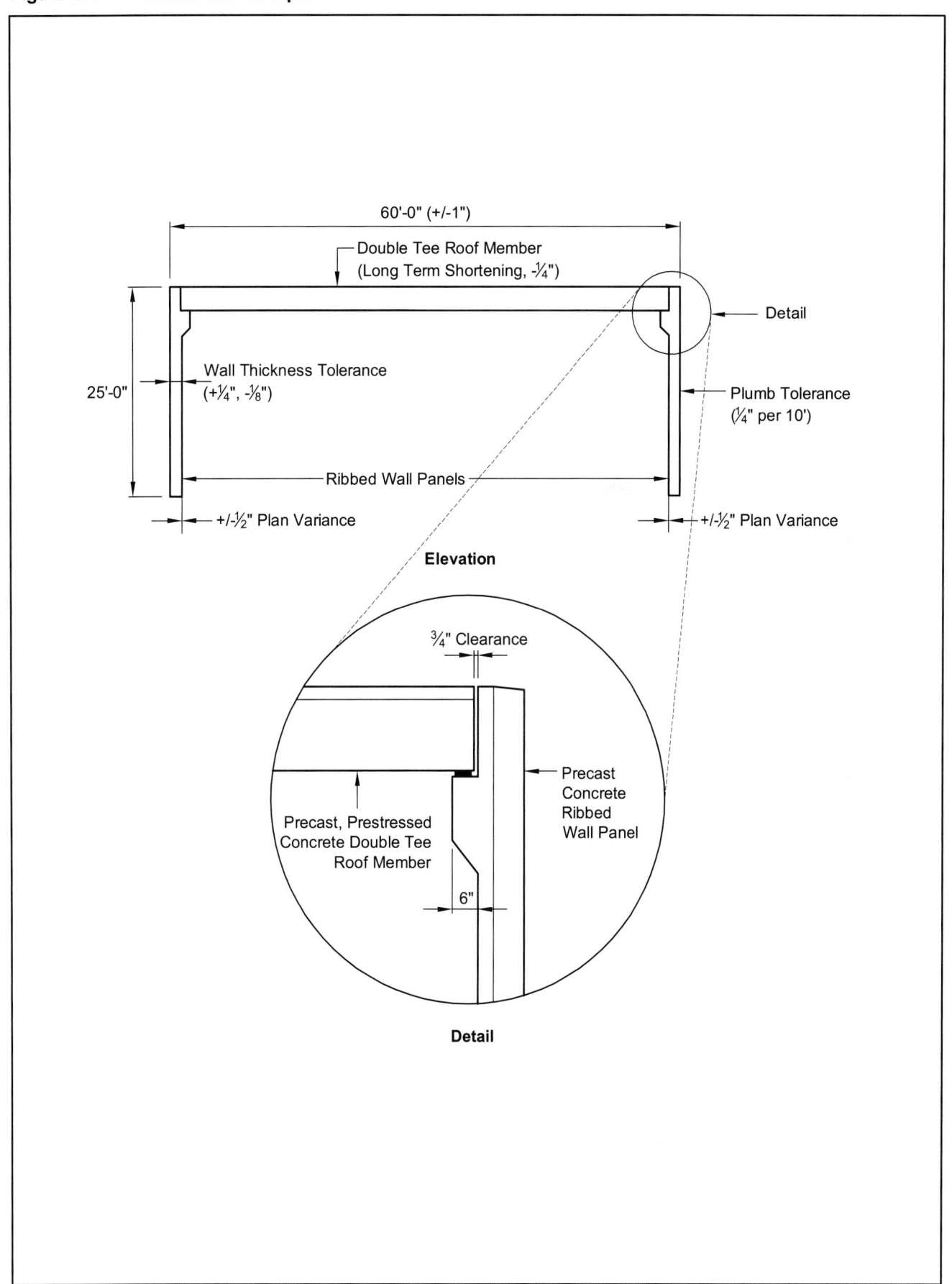

Example 8.4.1
Clearance Determination – Single-Story Industrial Building

Given:

A double tee roof member 60 ft long, ±1 in. length tolerance, bearing on ribbed wall members 25 ft high, maximum plan variance ±½ in., variation from plumb ¼ in. per 10 ft, haunch depth 6 in. beyond face of panel, long term roof movement ¼ in. Refer to Figure 8.4.1.

Problem:
Find the minimum acceptable clearance.

Solution:

Step 1: Determine Maximum Member Sizes.

(refer to product tolerances)
Maximum tee length   +1 in.
Wall thickness   +¼ in.
Initial clearance chosen   ¾ in. per end

Step 2: Member Movement.

Long term shrinkage and creep will increase the clearance so this movement can be neglected in the initial clearance determination, although it must be considered structurally.

Required clearance adjustment
as a result of member movement   0
Clearance chosen (from Step 1)   ¾ in.

Step 3: Other Erection Tolerances.

If the wall panel is set inward toward the building interior ½ in. and erected plumb, this would suggest the clearance should be increased by ½ in. However, if the panel is erected out of plumb outward ½ in., no clearance adjustment is needed.

Clearance adjustment required
for erection tolerances   0
Clearance chosen (from Step 1)   ¾ in.

Step 4: Erection Considerations.

If all members are fabricated perfectly, then the joint clearance is ¾ in. at either end (1½ in. total). This is ample space for erection. If all members are at maximum size variance, maximum inward plan variance, and maximum inward variance from plumb, then the total clearance is zero. This is undesirable as it would require some rework during erection. A judgment should be made as to the likelihood of maximum product tolerances all occurring in one location. If the likelihood is low, the ¾ in. clearance needs no adjustment, but, if the likelihood is high, the engineer might increase the clearance to 1 in. In this instance, the likelihood has been judged low; therefore, no adjustment has been made.

Clearance chosen (from Step 1)   ¾ in.

Example 8.4.1 (Cont.)
Clearance Determination – Single-Story Industrial Building

Step 5: Economy.

In single-story construction, increasing the clearance beyond ¾ in. is not likely to speed up erection as long as product tolerances remain within allowables. No adjustment is required for economic considerations.

Step 6: Review Structural Considerations.

Allowing a setback from the edge of the corbel, assumed in this instance to have been set by the engineer at ½ in., plus the clearance, the bearing is 4 in. and there should be space to allow member movement. The engineer judges this to be acceptable from a structural and architectural point of view and no adjustment is required for structural considerations.

Step 7: Check for Minimum Member Sizes.

(refer to Table 8.2.1)
Tee length	−1 in. (½ in. each end)
Wall thickness	−⅛ in.
Bearing haunch	No change
Clearance chosen	¾ in.
Minimum bearing with setback	+4 in.
(OK in this instance)	

Wall plumbness would also be considered in an actual application.

Step 8: Final Solution.

Minimum clearance used   ¾ in. per end
(Satisfies all conditions considered.)

(Note: For simplicity in this example, end rotation, flange skew, and global skew tolerances have not been considered. In an actual situation, these issues should also be considered.)

Figure 8.4.2 Clearance example

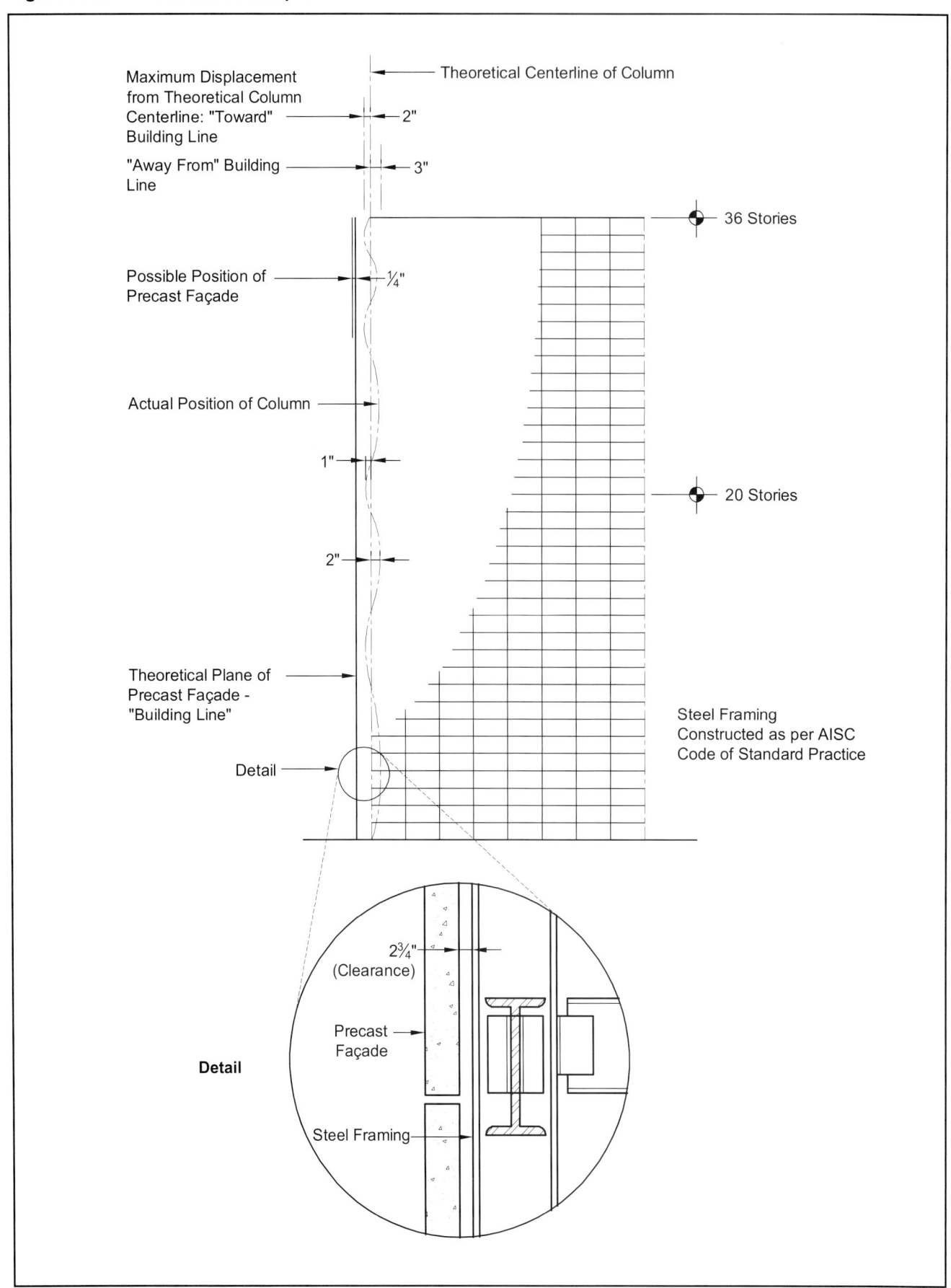

Example 8.4.2
Clearance Determination – High Rise Frame Structure

Given:
    A 36-story steel frame structure, precast concrete cladding, steel tolerances per AISC, member movement negligible. In this example, precast tolerance for variation in plan is ±¼ in. Refer to Figure 8.4.2.

Problem:
    Determine whether or not the panels can be erected plumb and determine the minimum acceptable clearance at the 36th story.

Solution:
Step 1: Product Tolerances.

    (refer to Table 8.2.1)
Precast cladding thickness	+¼ in., –⅛ in.
Steel width	+¼ in., –³⁄₁₆ in.
Steel sweep (varies)	±¼ in. assumption
Initial clearance chosen	¾ in.

Step 2: Member Movement.

    For simplification, assume this can be neglected in this example.

Step 3: Other Erection Tolerances.

Steel variance in plan, maximum	2 in.
Initial clearance	¾ in.
Clearance chosen	2¾ in.

Step 4: Erection Considerations.

    No adjustment required for erection considerations.

Step 5: Economy.

    Clearance chosen   2¾ in.
    Increasing clearance will not increase economy.
    No adjustment for economic considerations.

Step 6: Structural Considerations.

    Clearance chosen   2¾ in.
    Expensive connection but possible. No adjustment.

Step 7: Check Minimum Member Sizes at 36th Story.

(refer to product tolerances)	
Initial clearance	2¾ in.
Precast thickness	⅛ in.
Steel width	³⁄₁₆ in.
Steel sweep	¼ in.
Steel variance in plan minimum	<u>3 in.</u>
Clearance calculated	6⁵⁄₁₆ in.

Example 8.4.2 (Cont.)
Clearance Determination – High Rise Frame Structure

Step 8: Final Solution.

A clearance of over 6 in. would require an extremely expensive connection for the precast panel, and would produce high torsional stresses in the steel supporting beams. The 6 in. clearance is not practical, although the 2¾ in. minimum initial clearance is still needed. Either the precast panels should be allowed to follow the steel frame or the tolerances for the exterior columns need to be made more stringent, such as the AISC requirements for elevator columns. The most economical choice will likely be for the panels to follow the steel frame.

Minimum clearance used: 2¾ in.
Allow panels to follow the steel frame.

8.5 Interfacing Tolerances

8.5.1 General

In interfacing with other materials, the tolerances may be very system-dependent. For example, different brands of windows may have different tolerance requirements. If substitutions are made after the initial design is complete, the interface details must be reviewed for the new system. Following is a partial checklist for consideration in determining interfacing requirements:

- Structural requirements
- Volume change
- Exposure and corrosion protection
- Waterproofing
- Drainage requirements
- Architectural requirements
- Dimensional considerations
- Vibration considerations
- Fire-rating requirements
- Acoustical considerations
- Economics
- Manufacturing/Erection considerations

8.5.2 Interface Design Approach

The following approach is one method of organizing the task of designing the interface between two systems:

Step 1:
Review the interface between the two systems, show shape and location, and determine contractual responsibilities. For example, the precast panel is furnished by the precaster, the window is furnished and installed by the window manufacturer, and the sealant between the window and the precast concrete is furnished and installed by the general contractor.

Step 2:
Review the functional requirements of each interfacing system. For example, the building drain line must have a flow line slope which allows adequate drainage. This will place limits on where the line must penetrate precast units. Note whether this creates problems such as conflict with prestressing strands.

Step 3:
Review the tolerances of each interfacing system. For example, determine from manufacturer's specifications the external tolerances on the door jamb. Determine from the precast concrete product tolerances the tolerance on a large panel door blockout. For the door installation, determine the floor surface tolerance in the area of the door swing path.

Step 4:
Review the operational clearances required. For example, determine the magnitude of operational clearances which are needed to align the door to function properly. Then choose dimensions which include necessary clearances.

Step 5:
Review compatibility of the interface tolerances. Starting with the least precise system, check the tolerance requirements and compare against the minimum and maximum dimensions of the interfacing system. If interferences result, alter the nominal

dimension of the appropriate system. For example, it is usually more economical to make a larger opening than to specify a non-standard window size.

Step 6:

Review assembly and installation procedures for the interfacing systems to ensure compatibility. Show the preferred adjustments to accommodate the tolerances of the systems. Consider such things as minimum bearing areas, minimum and maximum joint gaps, and other dimensions which will vary as a result of interface tolerances. Consider economic trade-offs such as in-plant work versus field work, and minor fit-up rework versus more restrictive tolerances.

Step 7:

Review the final project specifications as they relate to interfacing. Be aware of subsystem substitutions which might be made during the final bidding and procurement.

8.5.3 Characteristics of the Interface

The following list of questions will help to define the nature of the interface:

1. What specifically is to be interfaced?

2. How does the interface function?

3. Is there provision for adjustment upon installation?

4. How much adjustment can occur without rework?

5. What are the consequences of an interface tolerance mismatch?
   — Rework requirements (labor and material)
   — Rejection limits

6. What are the high material cost elements of the interface?

7. What are the high labor cost elements of the interface?

8. What are the normal tolerances associated with the systems to be interfaced?

9. Are the system interface tolerances simple planar tolerances or are they more complex and three-dimensional?

10. Do all of the different products of this type have the same interface tolerance requirements?

11. Does the designer of the precast system have control overall the aspects of the interfaces involved? If not, what actions need to be taken to accommodate this fact?

Listed below are common characteristics to be considered for most systems of the type listed:

1. Windows and Doors
   — No gravity load transfer through window element.
   — Compatible with air and moisture sealant system.
   — Open/close characteristics (swing or slide).
   — Compatibility with door locking mechnisms.

2. Mechanical Equipment
   — Duct clearances for complex prefabricated ductwork.
   — Large diameter prefabricated pipe clearance requirements.
   — Deflections from forces associated with large diameter piping and valves.
   — Expansion/contraction allowances for hot/cold piping.
   — Vibration isolation/transfer considerations.
   — Acoustical shielding considerations.
   — Hazardous gas/fluid containment requirements.

3. Electrical Equipment
   — Multiple mating conduit runs.
   — Prefabricated cable trays.
   — Embedded conduits and outlet boxes.
   — Corrosion related to DC power.
   — Special insert placement requirements for isolation.
   — Location requirements for embedded grounding cables.
   — Shielding clearance requirements for special "clean" electrical lines.

4. Elevators and Escalators
   — Elevator guide location requirements.
   — Electrical conduit location requirements.
   — Elevator door mechanism clearances.
   — Special insert placement requirements.
   — Door opening size.

5. Architectural Cladding
   — Joint tolerance for sealant system.
   — Flashing and reglet fit-up. (Lining up reglets from panel to panel is very difficult and often costly. Surface-mounted flashing should be considered.)
   — Expansion and contraction provisions for dissimilar materials.

- Effects of rotation, deflection, and differential thermal gradients.

6. Structural Steel and Miscellaneous Steel
   - Details to prevent rust staining of concrete.
   - Details to minimize potential for corrosion at field connections between steel and precast concrete.
   - Coordination of structural steel expansion/contraction provisions with those of the precast system.
   - Special provisions for weld plates or other attachment features for steel structures.
   - Consideration of thermal insulation and fire proofing requirements.

7. Masonry
   - Coordination of masonry expansion/contraction provisions with those of the precast system.
   - Detailing to ensure desired contact bearing between masonry and precast units.
   - Detailing to ensure desired transfer of load between masonry shear wall and precast frame.
   - Requirements for dovetail anchors-field installation preferred.

8. Roofing
   - Roof camber, both upon erection and long-term, as it relates to roof drain placement.
   - Fit-up of prefabricated flashing.
   - Dimensional effects of added material during reroofing.
   - Coordination of structural control joint locations with roofing system expansion/contraction provisions.
   - Location of embedded HVAC unit supports.
   - Deflections due to live load and added equipment dead loads.

9. Waterproofing
   - Location and dimensions of flashing reglets grooves.
   - Coordination of waterproofing system requirements with structural system expansion/provisions.
   - Special details around special penetrations.

10. Interior Finishes
    - Floors, Walls, and Ceilings.
    - Joints between precast members for direct-carpet overlay.
    - Visual appearance of joints for exposed ceilings.
    - Fit-up details to ensure good appearance of interior corners.
    - Appearance of cast-in-place to precast concrete interfaces.

11. Interior Walls and Partitions
    - Clearance for prefabricated cabinetry.
    - Interfacing of mating embedded conduit runs.

8.6 References

1. PCI, *Tolerance Manual for Precast and Prestressed Concrete Construction*, MNL-135-00, Precast/Prestressed Concrete Institute, Chicago, IL, 2000

2. PCI, *Manual for Quality Control for Plants and Production of Structural Precast Concrete Products*, Fourth Edition, MNL-116-99, Precast/Prestressed Concrete Institute, Chicago, IL, 1999.

3. PCI, *Manual for Quality Control for Plants and Production of Architectural Precast Concrete Products*, Third Edition, MNL-117-96, Precast/Prestressed Concrete Institute, Chicago, IL, 1996.

4. PCI, *Erectors Manual — Standards and Guidlines, for the Erection of Precast Concrete Products*, MNL-127-99, Precast/Prestressed Concrete Institute, Chicago, IL, 1999.

5. "Proposed Design Requirements for Precast Concrete," PCI JOURNAL, V. 31, No. 6, November-December 1986.

6. AISC, *Code of Standard Practice for Steel Buildings and Bridges*, American Institute of Steel Construction, Chicago, IL, March 2000.

7. ACI, *Standard Specifications for Tolerances for Concrete Construction and Materials*, ACI 117-90, American Concrete Institute, Farmington Hills, MI, 1990.

8. NCMA, *Specification for Masonry Structures*, NCMA-TEK 1-2A, National Concrete Masonry Association, Herndon, VA, 1995.

CHAPTER 9
THERMAL, ACOUSTICAL, FIRE AND OTHER CONSIDERATIONS

9.1	Thermal Properties of Precast Concrete	9–4
	9.1.1 Notation	9–4
	9.1.2 Definitions	9–4
	9.1.3 General	9–5
	9.1.4 Thermal Properties of Materials, Surfaces, and Air Spaces	9–6
	9.1.5 Computation of Thermal Transmittance Values	9–6
	9.1.6 Thermal Storage Effects	9–13
	9.1.7 Condensation Control	9–15
	9.1.7.1 Air Barriers	9–17
	9.1.7.2 Vapor Retarders	9–18
	9.1.7.3 Prevention of Condensation on Wall Surfaces	9–20
	9.1.7.4 Condensation Prevention Within Wall Construction	9–20
	9.1.8 Thermal Bridges	9–20
	9.1.9 References	9–21
9.2	Acoustical Properties of Precast Concrete	9–24
	9.2.1 Definitions	9–24
	9.2.2 General	9–24
	9.2.3 Dealing with Sound Levels	9–24
	9.2.4 Sound Transmission Loss	9–24
	9.2.5 Impact Noise Reduction	9–24
	9.2.6 Acoustical Test Results	9–25
	9.2.7 Establishment of Noise Insulation Objectives	9–25
	9.2.8 Composite Wall Considerations	9–25
	9.2.9 Leaks and Flanking	9–25
	9.2.10 References	9–27
9.3	Fire Resistance	9–28
	9.3.1 Notation	9–28
	9.3.2 Definitions	9–28
	9.3.3 Introduction	9–28
	9.3.4 Standard Fire Tests	9–29
	9.3.4.1 Fire Endurance, End Point Criteria, and Fire Rating	9–30
	9.3.4.2 Restrained Fire Ratings	9–32
	9.3.5 Fire Tests of Prestressed Concrete Assemblies	9–32
	9.3.5.1 Fire Tests of Flexural Elements	9–32
	9.3.5.2 Fire Tests of Walls and Columns	9–32
	9.3.6 Designing for Heat Transmission	9–33
	9.3.6.1 Single Course Slabs or Wall Panels	9–33
	9.3.6.2 Floors, Roofs or Walls Faced with Gypsum Wallboard	9–34
	9.3.6.3 Ribbed Panels	9–34
	9.3.6.4 Multi-Course Assemblies	9–34
	9.3.6.5 Sandwich Panels	9–37
	9.3.6.6 Treatment of Joints Between Wall Panels	9–38
	9.3.7 Fire Endurance by Rational Design	9–39
	9.3.7.1 Simply Supported Members	9–39
	9.3.7.2 Continuous Members	9–44
	9.3.7.3 Members Restrained Against Thermal Expansion	9–45
	9.3.7.4 Shear Resistance	9–51
	9.3.8 Protection of Connections	9–51

	9.3.9 Precast Concrete Column Covers	9–51
	9.3.10 Code and Economic Considerations	9–53
	9.3.11 References	9–54
9.4	Sandwich Panels	9–55
	9.4.1 General	9–55
	9.4.2 Structural Design	9–55
	9.4.3 Connections	9–56
	9.4.3.1 Panel Connections	9–56
	9.4.3.2 Wythe Connectors	9–58
	9.4.4 Insulation	9–60
	9.4.5 Thermal Bowing	9–60
	9.4.6 Typical Details	9–60
	9.4.7 References	9–60
9.5	Quality Assurance and Control	9–61
	9.5.1 Introduction	9–61
	9.5.2 Plant Certification Program	9–61
	9.5.3 Plant Quality Personnel Certification	9–62
	9.5.4 Field Qualification Program	9–62
	9.5.5 References	9–63
9.6	Concrete Coatings and Joint Sealants	9–64
	9.6.1 Coatings for Horizontal Deck Surfaces	9–64
	9.6.2 Clear Surface Sealers for Architectural Precast Panels	9–64
	9.6.3 Joint Sealants	9–64
	9.6.4 References	9–65
9.7	Vibration in Concrete Structures	9–66
	9.7.1 Notation	9–66
	9.7.2 Human Response to Building Vibrations	9–67
	9.7.3 Types of Vibration Analysis	9–67
	9.7.3.1 Walking	9–67
	9.7.3.2 Rhythmic Activities	9–67
	9.7.3.3 Mechanical Equipment	9–67
	9.7.3.4 Analysis Methods	9–67
	9.7.3.5 Using Consistent Units	9–67
	9.7.4 Natural Frequency of Vibration	9–67
	9.7.4.1 Computing the Natural Frequency	9–67
	9.7.4.2 Computing Deflection	9–68
	9.7.4.3 Effect of Supporting Girders	9–68
	9.7.4.4 Minimum Natural Frequency	9–68
	9.7.4.5 Graphs of Natural Frequency	9–68
	9.7.5 Damping	9–69
	9.7.5.1 Types of Damping	9–69
	9.7.5.2 Estimation of Damping	9–69
	9.7.6 Vibrations Caused by Walking	9–69
	9.7.6.1 Minimum Natural Frequency	9–69
	9.7.6.2 Effective Weight	9–69
	9.7.6.3 Recommended Values	9–69
	9.7.7 Design for Rhythmic Excitation	9–70
	9.7.7.1 Harmonics	9–70
	9.7.7.2 Recommended Minimum Natural Frequency	9–70
	9.7.7.3 Higher Harmonics	9–70
	9.7.7.4 Adjacent Activities	9–71
	9.7.8 Stadium Seating	9–71
	9.7.9 Vibration Isolation for Mechanical Equipment	9–73

		9.7.10 References	9–74
9.8	Cracking, Repair and Maintenance		9–75
	9.8.1	Cracking	9–75
	9.8.2	Repair	9–75
		9.8.2.1 Patching	9–75
		9.8.2.2 Crack Repair	9–76
		9.8.2.3 Connection Repair	9–76
		9.8.2.4 Member Strengthening	9–77
	9.8.3	Maintenance	9–78
	9.8.4	References	9–78
9.9	Precast Segmental Construction		9–79
	9.9.1	General	9–79
	9.9.2	Joints	9–79
		9.9.2.1 Open Joints	9–79
		9.9.2.2 Closed Joints	9–79
	9.9.3	Design	9–80
	9.9.4	Post-Tensioning	9–80
		9.9.4.1 Tendon and Duct Placement	9–80
		9.9.4.2 Couplers	9–80
		9.9.4.3 Bearing Areas	9–80
	9.9.5	References	9–81
9.10	Coordination with Mechanical, Electrical and Other Sub-Systems		9–82
	9.10.1	Introduction	9–82
	9.10.2	Lighting and Power Distribution	9–82
	9.10.3	Electrified Floors	9–82
	9.10.4	Ductwork	9–82
	9.10.5	Other Sub-Systems	9–84
	9.10.6	Systems Building	9–85
	9.10.7	Prison Cell Box Module	9–86

9.1 THERMAL PROPERTIES OF PRECAST CONCRETE

9.1.1 Notation

a = thermal conductance of an air space

Btu = British thermal unit

C = thermal conductance for specified thickness, Btu/(hr)(ft^2)(°F)

d = clear cover to ends of metal ties in sandwich panels

D = heating degree day (65°F base)

E_z = width of affected zone around solid concrete

f = film or surface conductance

H_c = heat capacity

k = thermal conductivity, (Btu-in.)/(hr)(ft^2)(°F)

k_{con} = concrete conductivity

k_{in} = insulation conductivity

ℓ = thickness of material, in.

m = width or diameter of metal ties in sandwich panels

M = permeance, perms

R = thermal resistance, (hr)(ft^2)(°F)/Btu

R_a = thermal resistance of air space

R_{fi}, R_{fo} = thermal resistances of inside and outside surfaces, respectively

$R_{materials}$ = summation of thermal resistance of opaque material layers

R_p = thermal resistance of perfectly insulated region

R_s = thermal resistance of solid concrete region

R.H. = relative humidity, %

t_{cb} = thickness of concrete back wythe, in.

t_{cf} = thickness of concrete face wythe, in.

t_i = indoor temperature, °F

t_{in} = thickness of insulation, in.

t_o = outdoor temperature, °F

t_s = dew-point temperature, room air, at design maximum relative humidity, °F

U = heat transmittance value Btu/(hr)(ft^2)(°F)

U_{ow}, U_{or} = U of outside and inside walls, respectively

W = width of Zone A in sandwich panels, Section 9.1.8

α = a parameter to account for insulation conductivity, K_{in}. See Eqs. 9.1.8.1 and 9.1.8.2

β = a parameter to account for concrete conductivity, k_{con}. See Eqs. 9.1.8.1 and 9.1.8.3

μ = permeability, permeance of a unit thickness, given material, perm-in.

9.1.2 Definitions

Affected zone. Area around a concrete thermal bridge in an insulated panel, larger than the bridge, that affects the R-value of the panel.

British thermal unit (Btu). Approximate amount of heat to raise one pound of water from 59°F to 60°F.

Degree day (D). A unit, based on temperature difference and time, used in estimating fuel consumption and specifying nominal heating load of a building in winter. For any one day, when the mean temperature is less than 65°F, there are as many degree days as there are degrees F difference in temperature between the mean temperature for the day and 65°F.

Dew-point temperature (t_s). The temp-erature at which condensation of water vapor begins for a given humidity and pressure as the vapor temperature is reduced. The temperature corresponding to saturation (100% R.H.) for a given absolute humidity at constant pressure.

Film or surface conductance (f). The time rate of heat exchange by radiation, conduction, and convection of a unit area of a surface with its surroundings. Its value is usually expressed in (Btu per hr)(sq ft of surface area)(°F temperature difference). Subscripts "i" and "o" are usually used to denote inside and outside surface conductances, respectively.

Heat capacity (H_c). The amount of heat required to raise the temperature of a given mass one degree. Numerically, the mass multiplied by the specific heat.

Heat transmittance (U). Overall coefficient of heat transmission or thermal transmittance (air-to-air); the time rate of heat flow usually expressed in (Btu per hr)(sq ft of surface area)(°F temperature difference between air on the inside and air on the outside of a wall, floor, roof, or ceiling). The term is applied to the usual combinations of materials and also single materials such as window glass, and includes the surface conductance on both sides. This term is frequently called the U-value.

Perm. A unit of permeance. A perm is 1 grain per (sq ft of area)(hr)(in. of mercury vapor pressure difference).

Permeability, water vapor (μ). The property of a substance which permits the passage of water vapor. It is equal to the permeance of 1 in. of a substance. Permeability is measured in perm-in. The permeability of a material varies with barometric pressure, temperature and relative humidity conditions.

Permeance (M). The water vapor permeance of any sheet or assembly is the ratio of the water vapor flow per unit area per hour to the vapor pressure difference between the two surfaces. Permeance is measured in perms. Two commonly used test methods are the Wet Cup and Dry Cup Tests. Specimens are sealed over the tops of cups containing either water or desiccant, placed in a controlled atmosphere, usually at 50% R.H., and weight changes measured.

Relative humidity (R.H.). The ratio of the water vapor present in air to the water vapor present in saturated air at the same temperature and pressure.

Thermal conductance (C). The time rate of heat flow expressed in (Btu per hr)(sq ft of area)(°F average temperature difference between two surfaces). The term is applied to specific materials as used, either homogeneous or heterogeneous for the thickness of construction stated, not per in. of thickness.

Thermal conductance of an air space (a). The time rate of heat flow through a unit area of an air space per unit temperature difference between the boundary surfaces. Its value is usually expressed in (Btu per hr)(sq ft of area)(°F).

Thermal conductivity (k). The time rate of heat flow by conduction only through a unit thickness of a homogeneous material under steady-state conditions per unit area per unit temperature gradient in the direction perpendicular to the isothermal surface. Its unit is (Btu-in. per hr)(sq ft of area)(°F).

Thermal mass. Characteristic of materials with mass heat capacity and surface area capable of affecting building heating and cooling loads by storing and releasing heat as the interior and/or exterior temperature and radiant conditions fluctuate.

Thermal resistance (R). The reciprocal of a heat transmission coefficient, as expressed by U, C, f, or a. Its unit is (°F)(hr)(sq ft of area) per Btu. For example, a wall with a U-value of 0.25 would have a resistance value of R = 1/U = 1/0.25 = 4.0.

9.1.3 General

Thermal codes and standards specify the heat transmission requirements for buildings in many different ways. Prescriptive standards specify U or R-values for each building component, whereas with performance standards, two buildings are equivalent if they use the same amount of energy, regardless of the U or R-values of the components. This allows the designer to choose conservation strategies that provide the required performance at the least cost.

In ASHRAE Standard 90.1, [2] these two approaches are used to determine the maximum allowable U for a wall or roof assembly:

1. Prescriptive Criteria (ASHRAE Standard 90.1, Section 8.5) — Table of Alternate Component Packages provides a limited number of complying combinations of building variables for a set of climate variable ranges. For most climate locations and building assemblies, the Prescriptive Criteria may be slightly more stringent than the System Performance Criteria.
2. System Performance Criteria (ASHRAE Standard 90.1, Section 8.6) — Mathematical models and computer programs based on these models provide a system approach (criteria and compliance values) to comply with the envelope requirements of the Standard. It provides more flexibility than the prescriptive approach, but it requires more analysis.

Precast and prestressed concrete construction, with its high thermal inertia and thermal storage properties, has an advantage over lightweight materials. Procedures to account for the benefits of heavier materials are presented in ASHRAE Standard 90.1. [2]

The trend is toward more insulation with little regard given to its total impact or energy used. Mass effects, glass area, air infiltration, ventilation, building orientation, exterior color, shading or reflections from adjacent structures, surrounding surfaces or vegetation, building aspect ratio, number of stories, wind direction and speed, all have an effect on insulation requirements.

This section is condensed from a more complete treatment given in Ref. 4. Except where noted, the information and design criteria are taken or derived from the ASHRAE Handbook, [1] and from the ASHRAE Standard 90.1. [2] All design criteria are not given in this section, and may change as the ASHRAE Standard and Handbook are revised. Local codes and latest references must be used for specified values and procedures.

9.1.4 Thermal Properties of Materials, Surfaces, and Air Spaces

The thermal properties of materials and air spaces are based on steady state tests, which measure the heat that passes from the warm side to the cool side of the test specimen. The tests determine the conductivity, k, or, for non-homogeneous sections, compound sections and air spaces, the conductance, C, for the total thickness. The values of k and C do not include surface conductances, f_i and f_o.

The overall thermal resistance of wall, floor, and roof sections is the sum of the resistances, R (reciprocal of k, C, f_i, and f_o). The R-values of construction materials are not influenced by the direction of heat flow, but the R-values of surfaces and air spaces differ depending on whether they are vertical, sloping, or horizontal. Also, the R-values of surfaces are affected by the velocity of air at the surfaces and by their reflective properties.

Tables 9.1.4.1 and 9.1.4.2 give the thermal resistances of surfaces and 3½ in. air space. Table 9.1.4.3 gives the thermal properties of most commonly used building materials. Only U-values are given for glass because the surface resistances and air space between panes account for nearly all of the U-value. Table 9.1.4.4 gives the thermal properties of various weight concretes and some standard precast, prestressed concrete products in the "normally dry" condition. Normally dry is the condition of concrete containing an equilibrium amount of free water after extended exposure to warm air at 35 to 50% relative humidity.

Thermal conductances and resistances of other building materials are usually reported for oven dry conditions. Normally dry concrete in combination with insulation generally provides about the same R-value as equally insulated oven dry concrete, but because of the moisture content, has the ability to store a greater amount of heat than oven dry concrete. However, higher moisture content in concrete causes higher thermal conductance.

9.1.5 Computation of Thermal Transmittance Values

The heat transmittance (U-values) of a building wall, floor or roof is computed by adding together the R-values of the materials in the section, the surfaces (R_{fi} and R_{fo}) and air spaces (R_a) within the section. The reciprocal of the sum of the Rs is the U-value:

$$U = \frac{1}{R_{fi} + R_{materials} + R_a + R_{fo}} \quad \text{(Eq. 9.1.5.1)}$$

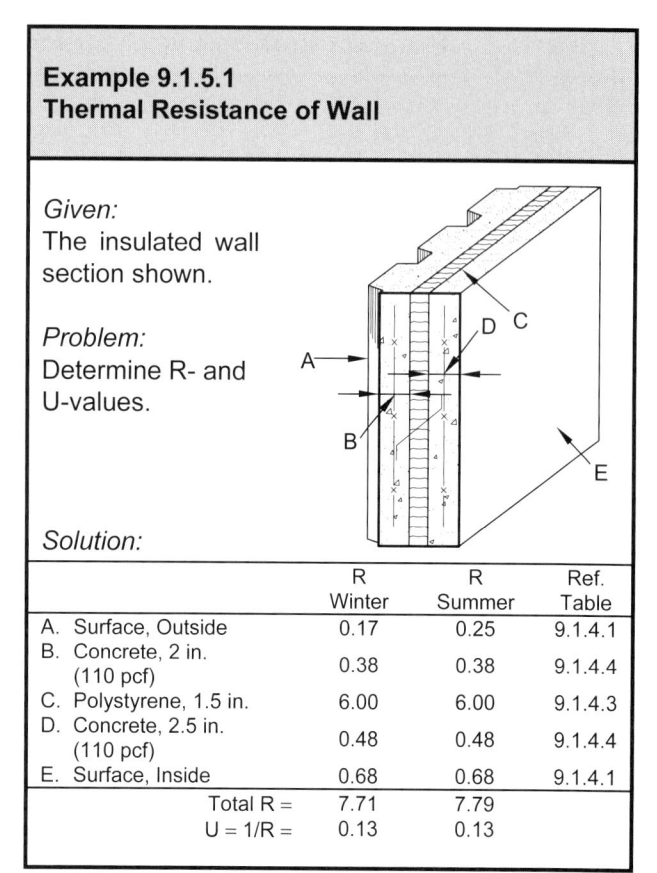

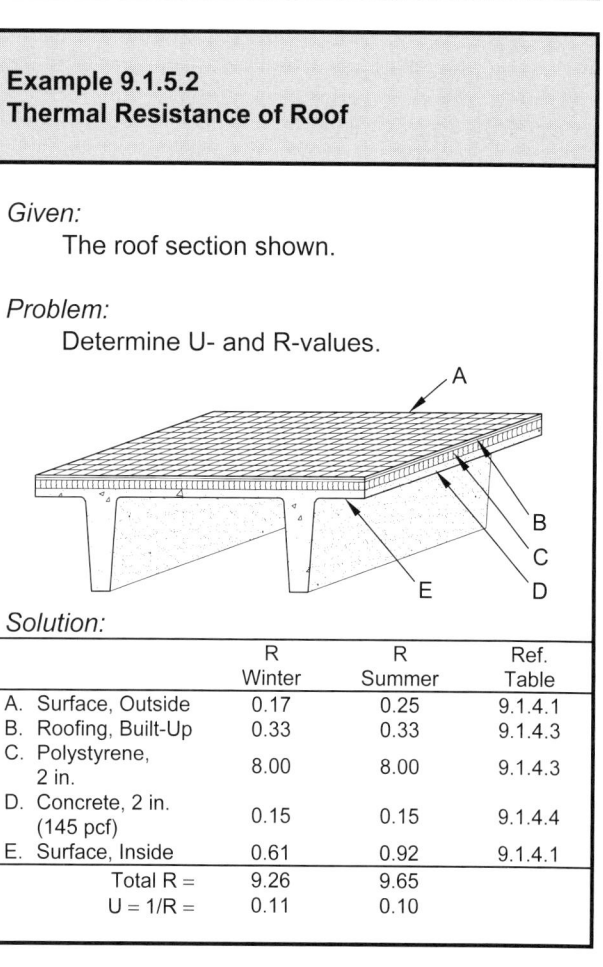

Example 9.1.5.1
Thermal Resistance of Wall

Given:
The insulated wall section shown.

Problem:
Determine R- and U-values.

Solution:

	R Winter	R Summer	Ref. Table
A. Surface, Outside	0.17	0.25	9.1.4.1
B. Concrete, 2 in. (110 pcf)	0.38	0.38	9.1.4.4
C. Polystyrene, 1.5 in.	6.00	6.00	9.1.4.3
D. Concrete, 2.5 in. (110 pcf)	0.48	0.48	9.1.4.4
E. Surface, Inside	0.68	0.68	9.1.4.1
Total R =	7.71	7.79	
U = 1/R =	0.13	0.13	

Example 9.1.5.2
Thermal Resistance of Roof

Given:
The roof section shown.

Problem:
Determine U- and R-values.

Solution:

	R Winter	R Summer	Ref. Table
A. Surface, Outside	0.17	0.25	9.1.4.1
B. Roofing, Built-Up	0.33	0.33	9.1.4.3
C. Polystyrene, 2 in.	8.00	8.00	9.1.4.3
D. Concrete, 2 in. (145 pcf)	0.15	0.15	9.1.4.4
E. Surface, Inside	0.61	0.92	9.1.4.1
Total R =	9.26	9.65	
U = 1/R =	0.11	0.10	

Table 9.1.4.1 Thermal resistances, R_f, of surfaces

Position of surface	Direction of heat flow	Still air, R_{fi} Non-reflective surface	Still air, R_{fi} Reflective surface Aluminum painted paper	Still air, R_{fi} Reflective surface Bright aluminum foil	Moving air, R_{fo} Non-reflective surface 15 mph winter design	Moving air, R_{fo} Non-reflective surface 7½ mph summer design
Vertical	Horizontal	0.68	1.35	1.70	0.17	0.25
Horizontal	Up (Winter)	0.61	1.10	1.32	0.17	0.25
Horizontal	Down (Summer)	0.92	2.70	4.55	0.17	0.25

Table 9.1.4.2 Thermal resistances, R_a, of air spaces[a]

Position of air space	Direction of heat flow	Air space Mean Temp. °F	Air space Temp. Diff. °F	Non-reflective surfaces	Reflective surface One side[b]	Reflective surface One side[c]	Reflective surface Both sides[c]
Vertical	Horizontal (walls)	Winter 50 50	10 30	1.01 0.91	2.32 1.89	3.40 2.55	3.69 2.67
Vertical	Horizontal (walls)	Summer 90	10	0.85	2.15	3.40	3.69
Horizontal	Up (roofs)	Winter 50 50	10 30	0.93 0.84	1.95 1.58	2.66 2.01	2.80 2.09
Horizontal	Down (floors)	50	30	1.22	3.86	8.17	9.60
Horizontal	Down (roofs)	Summer 90	10	1.00	3.41	8.19	10.07

a. For 3½ in. air space thickness. The values with the exception of those for reflective surfaces, heat flow down, will differ about 10% for air space thickness of ¾ in. Refer to the ASHRAE Handbook for values of other thicknesses, reflective surfaces, heat flow down.
b. Aluminum painted paper.
c. Bright aluminum foil.

Table 9.1.4.3 Thermal properties of various building materials[a,f]

Material	Unit weight, pcf	Resistance, R Per inch of thickness, 1/k	Resistance, R For thickness shown, 1/C	Transmittance, U	Specific heat, Btu/(lb)(°F)
Insulation, rigid					
Cellular glass	8.5	2.86			0.18
Glass fiber, organic bonded	4 – 9	4.00			0.23
Mineral fiber, resin binder	15	3.45			0.17
Mineral fiberboard, wet felted, roof insulation	16 – 17	2.94			—
Cement fiber slabs (shredded wood with magnesia oxysulfide binder)	22	1.75			0.31
Expanded polystyrene extruded smooth skin surface	1.8 – 3.5	5.00			0.29
Expanded polystyrene molded bead	1.0	3.85 – 4.17			—
Cellular polyurethane	1.5	6.25 – 5.56			0.38
Miscellaneous					
Acoustical tile (mineral fiberboard, wet felted)	18	2.86			0.19
Carpet, fiberous pad			2.08		0.34
Carpet, rubber pad			1.23		0.33
Floor tile, asphalt, rubber, vinyl			0.05		0.30
Gypsum board	50	0.88[b]			0.26
Particle board	50	1.06			0.31
Plaster					
cement, sand aggregate	116	0.20			0.20
gypsum, L.W. aggregate	45	0.63[b]			—
gypsum, sand aggregate	105	0.18			0.20
Roofing, ⅜ in. built-up	70		0.33		0.35
Wood, hard	38 – 47	0.94 – 0.80			0.39
Wood, soft	24 – 41	1.00 – 0.89			0.33
Plywood	34	1.25			0.29
Glass doors and windows[c]					
Single, winter				1.11	
Single, summer				1.04	
Double, winter[d]				0.50	
Double, summer[d]				0.61	
Doors, metal[e]					
Insulated				0.40	

a. See Table 9.1.4.4 for all concretes, including insulating concrete for roof fill.
b. Average value.
c. Does not include correction for sash resistance. Refer to the ASHRAE Handbook for sash correction.
d. ¼ in. air space; coating on either glass surface facing air space.
e. Urethane foam core without thermal break.
f. See manufacturer's data for specific values.

Table 9.1.4.4 Thermal properties of concrete[a]

Description	Concrete weight, pcf	Thickness, in.	Resistance, R Per inch of thickness, 1/k	Resistance, R For thickness shown, 1/C	Specific heat,[c] Btu/(lb)(°F)
Concretes Including normal weight, lightweight and lightweight insulation concretes	140 120 100 80 60 40 30 20		0.10 – 0.05 0.18 – 0.09 0.27 – 0.17 0.40 – 0.29 0.63 – 0.56 1.08 – 0.90 1.33 – 1.10 1.59 – 1.20		0.19
Normal weight tees[b] and solid slabs	145	2 3 4 5 6 8		0.15 0.23 0.30 0.38 0.45 0.60	0.19
Normal weight hollow-core slabs	145	6 8 10 12		1.07 1.34 1.73 1.91	0.19
Structural lightweight tees[b] and solid slabs	110	2 3 4 5 6 8		0.38 0.57 0.76 0.95 1.14 1.52	0.19
Structural lightweight hollow-core slabs	110	8 12		2.00 2.59	0.19

a. Based on normally dry concrete (see Chapter 4 of Ref. 3).
b. Thickness for tees is thickness of slab portion including topping, if used. The effect of the stems generally is not significant, therefore, their thickness and surface area may be disregarded.
c. The specific heat shown is the mean value from test data complied in Ref. 4.

Table 9.1.5.1 Wall U-values: prestressed tees, hollow-core slabs, solid and sandwich panels; winter and summer conditions[a]

Concrete weight, pcf	Type of wall panel	Thickness, t, and resistance, R, of concrete		Winter $R_{fo} = 0.17$, $R_{fi} = 0.68$					Summer $R_{fo} = 0.25$, $R_{fi} = 0.68$				
				Insulation resistance, R									
		t	R	None	4	6	8	10	None	4	6	8	10
145	Solid walls, tees,[b] and sandwich panels	2	0.15	1.00	0.20	0.14	0.11	0.09	0.93	0.20	0.14	0.11	0.09
		3	0.23	0.93	0.20	0.14	0.11	0.09	0.86	0.19	0.14	0.11	0.09
		4	0.30	0.87	0.19	0.14	0.11	0.09	0.81	0.19	0.14	0.11	0.09
		5	0.38	0.81	0.19	0.14	0.11	0.09	0.76	0.19	0.14	0.11	0.09
		6	0.45	0.77	0.19	0.14	0.11	0.09	0.72	0.19	0.14	0.11	0.09
		8	0.60	0.69	0.18	0.13	0.11	0.09	0.65	0.18	0.13	0.10	0.09
145	Hollow core slabs[c]	6(o)	1.07	0.52	0.17	0.13	0.10	0.08	0.50	0.17	0.13	0.10	0.08
		(f)	1.86	0.37	0.15	0.11	0.09	0.08	0.36	0.15	0.11	0.09	0.08
		8(o)	1.34	0.46	0.16	0.12	0.10	0.08	0.44	0.16	0.12	0.10	0.08
		(f)	3.14	0.25	0.13	0.10	0.08	0.07	0.25	0.12	0.10	0.08	0.07
		10(o)	1.73	0.39	0.15	0.12	0.09	0.08	0.38	0.15	0.12	0.09	0.08
		(f)	4.05	0.20	0.11	0.09	0.08	0.07	0.20	0.11	0.09	0.08	0.07
		12(o)	1.91	0.36	0.15	0.11	0.09	0.08	0.35	0.15	0.11	0.09	0.08
		(f)	5.01	0.17	0.10	0.08	0.07	0.06	0.17	0.10	0.08	0.07	0.06
110	Solid walls, tees,[b] and sandwich panels	2	0.38	0.81	0.19	0.14	0.11	0.09	0.76	0.19	0.14	0.11	0.09
		3	0.57	0.70	0.18	0.13	0.11	0.09	0.67	0.18	0.13	0.11	0.09
		4	0.76	0.62	0.18	0.13	0.10	0.09	0.59	0.18	0.13	0.10	0.09
		5	0.95	0.56	0.17	0.13	0.10	0.09	0.53	0.17	0.13	0.10	0.08
		6	1.14	0.50	0.17	0.13	0.10	0.08	0.48	0.16	0.12	0.10	0.08
		8	1.52	0.42	0.16	0.12	0.10	0.08	0.41	0.16	0.12	0.10	0.08
110	Hollow-core slabs[c]	8(o)	2.00	0.35	0.15	0.11	0.09	0.08	0.34	0.14	0.11	0.09	0.08
		(f)	4.41	0.19	0.11	0.09	0.08	0.07	0.19	0.11	0.09	0.07	0.07
		12(o)	2.59	0.29	0.13	0.11	0.09	0.07	0.28	0.13	0.11	0.09	0.07
		(f)	6.85	0.13	0.09	0.07	0.06	0.06	0.13	0.08	0.07	0.06	0.06

a. When insulations having other R-values are used, U-values can be interpolated with adequate accuracy, or U can be calculated as shown in Section 9.1.5. When a finish, air space or any other material layer is added, the new U-value is:

$$\frac{1}{\frac{1}{U \text{ from table}} + R \text{ of added finish, air space or material}}$$

b. Thickness for tees is thickness of slab portion. For sandwich panels, t is the sum of the thicknesses of the wythes.

c. For hollow panels (o) and (f) after thickness designates cores open or cores filled with insulation.

Table 9.1.5.2 Roof U-values: concrete units with built-up roofing, winter conditions, heat flow upward[a]

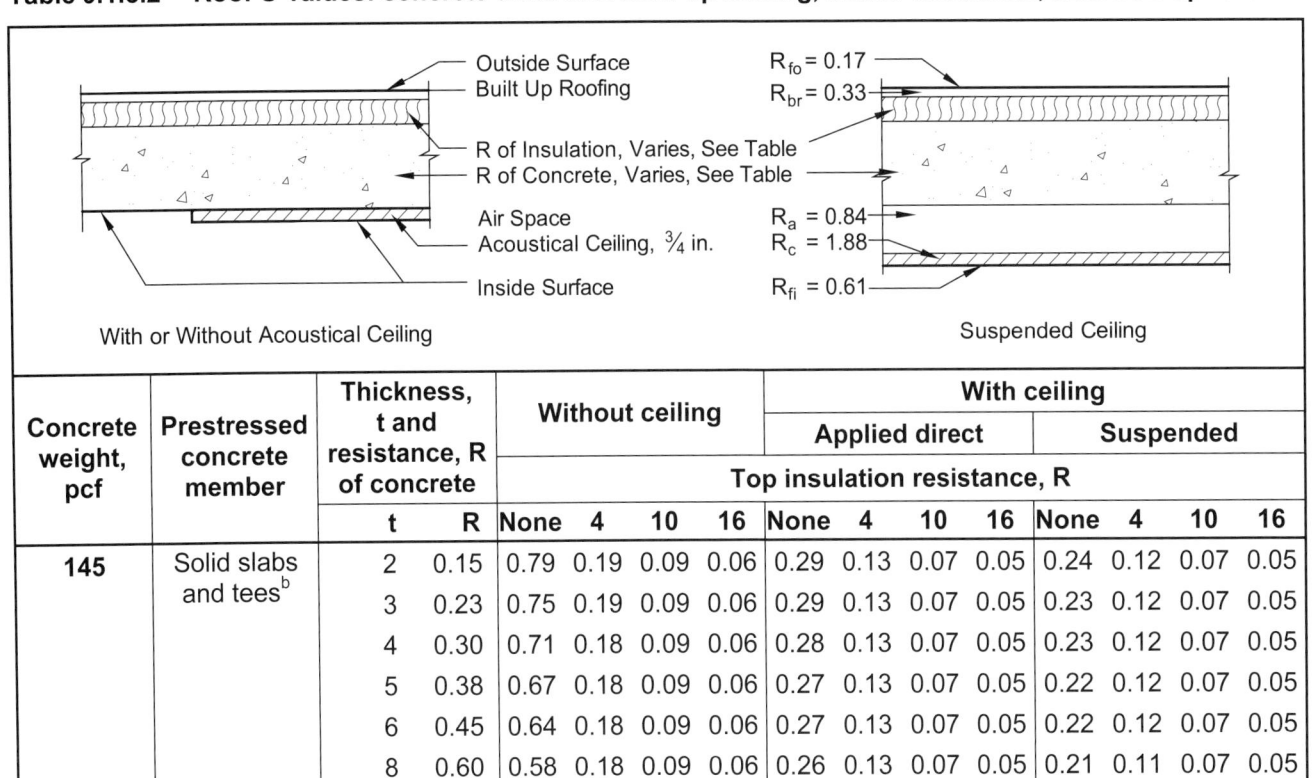

Diagram labels:
- Outside Surface
- Built Up Roofing
- R of Insulation, Varies, See Table
- R of Concrete, Varies, See Table
- Air Space
- Acoustical Ceiling, ¾ in.
- Inside Surface
- $R_{fo} = 0.17$
- $R_{br} = 0.33$
- $R_a = 0.84$
- $R_c = 1.88$
- $R_{fi} = 0.61$

With or Without Acoustical Ceiling | Suspended Ceiling

Concrete weight, pcf	Prestressed concrete member	Thickness, t and resistance, R of concrete		Without ceiling				With ceiling — Applied direct				With ceiling — Suspended			
		t	R	None	4	10	16	None	4	10	16	None	4	10	16
145	Solid slabs and tees[b]	2	0.15	0.79	0.19	0.09	0.06	0.29	0.13	0.07	0.05	0.24	0.12	0.07	0.05
		3	0.23	0.75	0.19	0.09	0.06	0.29	0.13	0.07	0.05	0.23	0.12	0.07	0.05
		4	0.30	0.71	0.18	0.09	0.06	0.28	0.13	0.07	0.05	0.23	0.12	0.07	0.05
		5	0.38	0.67	0.18	0.09	0.06	0.27	0.13	0.07	0.05	0.22	0.12	0.07	0.05
		6	0.45	0.64	0.18	0.09	0.06	0.27	0.13	0.07	0.05	0.22	0.12	0.07	0.05
		8	0.60	0.58	0.18	0.09	0.06	0.26	0.13	0.07	0.05	0.21	0.11	0.07	0.05
145	Hollow-core slabs[c]	6(o)	1.07	0.46	0.16	0.08	0.06	0.23	0.12	0.07	0.05	0.19	0.11	0.07	0.05
		(f)	1.86	0.34	0.14	0.08	0.05	0.20	0.11	0.07	0.05	0.17	0.10	0.06	0.05
		8(o)	1.34	0.41	0.16	0.08	0.05	0.22	0.12	0.07	0.05	0.18	0.11	0.06	0.05
		(f)	3.14	0.24	0.12	0.07	0.05	0.16	0.10	0.06	0.04	0.14	0.09	0.06	0.04
		10(o)	1.73	0.35	0.15	0.08	0.05	0.20	0.11	0.07	0.05	0.17	0.10	0.06	0.05
		(f)	4.05	0.19	0.11	0.07	0.05	0.14	0.09	0.06	0.04	0.12	0.08	0.06	0.04
		12(o)	1.91	0.33	0.14	0.08	0.05	0.19	0.11	0.07	0.05	0.17	0.10	0.06	0.05
		(f)	5.01	0.16	0.10	0.06	0.05	0.12	0.08	0.05	0.04	0.11	0.08	0.05	0.04
110	Solid slabs and tees[b]	2	0.38	0.67	0.18	0.09	0.06	0.27	0.13	0.07	0.05	0.22	0.12	0.07	0.05
		3	0.57	0.60	0.18	0.09	0.06	0.26	0.13	0.07	0.05	0.21	0.12	0.07	0.05
		4	0.76	0.53	0.17	0.08	0.06	0.25	0.12	0.07	0.05	0.21	0.11	0.07	0.05
		5	0.95	0.49	0.17	0.08	0.06	0.24	0.12	0.07	0.05	0.20	0.11	0.07	0.05
		6	1.14	0.44	0.16	0.08	0.05	0.23	0.12	0.07	0.05	0.19	0.11	0.07	0.05
		8	1.52	0.38	0.15	0.08	0.05	0.21	0.11	0.07	0.05	0.18	0.10	0.06	0.05
110	Hollow-core slabs[c]	8(o)	2.00	0.32	0.14	0.08	0.05	0.19	0.11	0.07	0.05	0.16	0.10	0.06	0.05
		(f)	4.41	0.18	0.11	0.06	0.05	0.13	0.09	0.06	0.04	0.12	0.08	0.05	0.04
		12(o)	2.59	0.27	0.13	0.07	0.05	0.17	0.10	0.06	0.05	0.15	0.09	0.06	0.04
		(f)	6.85	0.13	0.08	0.06	0.04	0.10	0.07	0.05	0.04	0.09	0.07	0.05	0.04

a. When insulations having other R-values are used, U-values can be interpolated with adequate accuracy, or U can be calculated as shown in Section 9.1.5. When a finish, air space or any other material layer is added, the new U-value is:

$$\frac{1}{\frac{1}{U \text{ from table}} + R \text{ of added finish, air space or material}}$$

b. Thickness for tees is thickness of slab portion.
c. For hollow panels (o) and (f) after thickness designates cores open or cores filled with insulation.

Table 9.1.5.3 Roof U-values: concrete units with built-up roofing, summer conditions, heat flow downward[a]

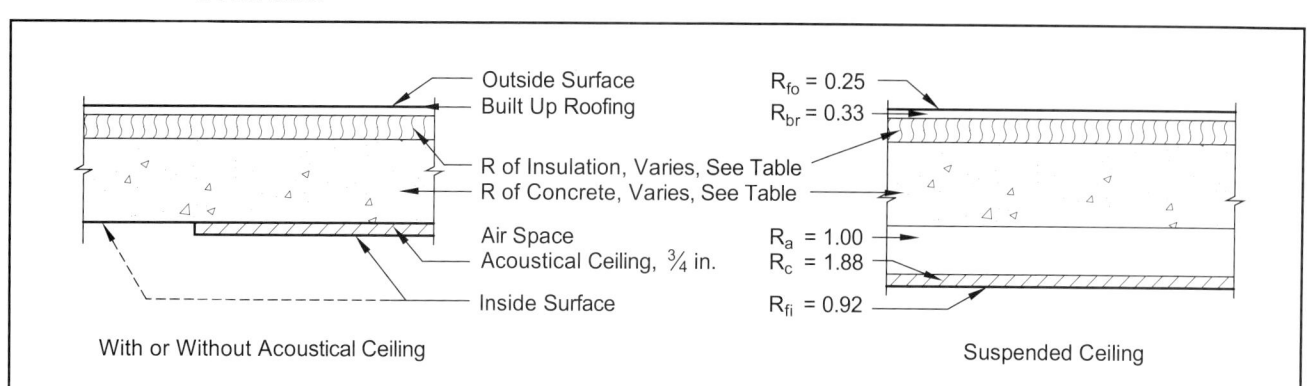

Diagram legend:
- Outside Surface, $R_{fo} = 0.25$
- Built Up Roofing, $R_{br} = 0.33$
- R of Insulation, Varies, See Table
- R of Concrete, Varies, See Table
- Air Space, $R_a = 1.00$
- Acoustical Ceiling, ¾ in., $R_c = 1.88$
- Inside Surface, $R_{fi} = 0.92$
- With or Without Acoustical Ceiling | Suspended Ceiling

Concrete weight, pcf	Prestressed concrete member	Thickness, t	Resistance, R of concrete	Without ceiling None	4	10	16	With ceiling Applied direct None	4	10	16	Suspended None	4	10	16
145	Solid slabs and tees[b]	2	0.15	0.61	0.18	0.09	0.06	0.26	0.13	0.07	0.05	0.21	0.11	0.07	0.05
		3	0.23	0.58	0.17	0.09	0.06	0.26	0.13	0.07	0.05	0.20	0.11	0.07	0.05
		4	0.30	0.56	0.17	0.08	0.06	0.25	0.13	0.07	0.05	0.20	0.11	0.07	0.05
		5	0.38	0.53	0.17	0.08	0.06	0.25	0.12	0.07	0.05	0.20	0.11	0.07	0.05
		6	0.45	0.51	0.17	0.08	0.06	0.24	0.12	0.07	0.05	0.20	0.11	0.07	0.05
		8	0.60	0.48	0.16	0.08	0.06	0.24	0.12	0.07	0.05	0.19	0.11	0.07	0.05
145	Hollow-core slabs[c]	6(o)	1.07	0.39	0.15	0.08	0.05	0.21	0.11	0.07	0.05	0.17	0.10	0.06	0.05
		(f)	1.86	0.30	0.14	0.07	0.05	0.18	0.11	0.06	0.05	0.15	0.10	0.06	0.04
		8(o)	1.34	0.35	0.15	0.08	0.05	0.20	0.11	0.07	0.05	0.17	0.10	0.06	0.05
		(f)	3.14	0.22	0.12	0.07	0.05	0.15	0.09	0.06	0.04	0.13	0.08	0.06	0.04
		10(o)	1.73	0.31	0.14	0.08	0.05	0.19	0.11	0.07	0.05	0.16	0.10	0.06	0.04
		(f)	4.05	0.18	0.10	0.06	0.05	0.13	0.09	0.06	0.04	0.11	0.08	0.06	0.04
		12(o)	1.91	0.29	0.13	0.07	0.05	0.18	0.10	0.06	0.05	0.15	0.09	0.06	0.04
		(f)	5.01	0.15	0.10	0.06	0.04	0.12	0.08	0.05	0.04	0.10	0.07	0.05	0.04
110	Solid slabs and tees[b]	2	0.38	0.53	0.17	0.08	0.06	0.25	0.12	0.07	0.05	0.20	0.11	0.07	0.05
		3	0.57	0.48	0.16	0.08	0.06	0.24	0.12	0.07	0.05	0.19	0.11	0.07	0.05
		4	0.76	0.44	0.16	0.08	0.05	0.23	0.12	0.07	0.05	0.18	0.11	0.06	0.05
		5	0.95	0.41	0.16	0.08	0.05	0.22	0.12	0.07	0.05	0.18	0.10	0.06	0.05
		6	1.14	0.38	0.15	0.08	0.05	0.21	0.11	0.07	0.05	0.17	0.10	0.06	0.05
		8	1.52	0.33	0.14	0.08	0.05	0.19	0.11	0.07	0.05	0.16	0.10	0.06	0.05
110	Hollow-core slabs[c]	8(o)	2.00	0.29	0.13	0.07	0.05	0.18	0.10	0.06	0.05	0.15	0.09	0.06	0.04
		(f)	4.41	0.17	0.10	0.06	0.05	0.12	0.08	0.06	0.04	0.11	0.08	0.05	0.04
		12(o)	2.59	0.24	0.12	0.07	0.05	0.16	0.10	0.06	0.04	0.14	0.09	0.06	0.04

a. When insulations having other R-values are used, U-values can be interpolated with adequate accuracy, or U can be calculated as shown in Section 9.1.5. When a finish, air space or any other material layer is added, the new U-value is:

$$\frac{1}{\frac{1}{U \text{ from table}} + R \text{ of added finish, air space or material}}$$

b. Thickness for tees is thickness of slab portion.
c. For hollow panels (o) and (f) after thickness designates cores open or cores filled with insulation.

where $R_{materials}$ is the sum of all opaque materials in the wall. A number of typical wall and roof U-values are given in Tables 9.1.5.1, 9.1.5.2, and 9.1.5.3.

Examples 9.1.5.1 and 9.1.5.2 show the use of Tables 9.1.4.1 through 9.1.4.2 in calculating R- and U-values for wall and roof assemblies.

9.1.6 Thermal Storage Effects

In years past, the U-factor was considered the most significant indication of heat gain, principally because laboratory tests have shown that thermal transmission is directly proportional to the U-factor during steady-state heat flow. However, the steady-state condition is rarely realized in actual practice.

External conditions (temperature, position of the sun, presence of shadows, etc.) vary throughout a 24-hour day, and heat gain is not instantaneous through most solid materials, resulting in the phenomenon of time lag (thermal inertia).

As temperatures rise on one side of a wall, heat begins to flow towards the cooler side. Before heat transfer can be achieved, the wall must undergo a temperature increase. The amount of thermal energy necessary to achieve this increase is directly proportional to the specific heat and density of the wall.

Due to its density, concrete has the capacity to absorb and store large quantities of heat. This thermal mass allows concrete to react very slowly to changes in outside temperature. This characteristic of thermal mass reduces peak heating and cooling loads and delays the time at which these peak loads occur by several hours. This delay improves the performance of heating and cooling equipment, since the peak cooling loads are delayed until the evening hours, when the outside temperature has dropped.

Energy use differences between light and heavy materials are illustrated in the hour-by-hour computer analyses shown in Figure 9.1.6.1.

Figure 9.1.6.1(A) compares the heat flow through three walls having the same U-value, but made of different materials. The concrete wall consists of a layer of insulation sandwiched between inner and outer wythes of 2 in. concrete and weighing 48.3 psf. The metal wall, weighing 3.3 psf, has insulation sandwiched between an exterior metal panel and ½ in. drywall. The wood frame wall weighs 7.0 psf and has wood siding on the outside, insulation between 2 x 4 studs, and ½ in. drywall on the inside. The walls are exposed to simulated outside temperatures that represent a typical spring day in a moderate climate. The massive concrete wall has lower peak loads by about 13% for heating and 30% for cooling than the less massive walls.

Normal weight concrete walls of various thicknesses that are exposed to the same simulated outside temperatures, are compared in Figure 9.1.6.1(B). The walls have a layer of insulation sandwiched between concrete on the outside and ½ in. drywall on the inside; U-values are the same. The figure shows that the more massive the wall the lower the peak loads and the more the peaks are delayed.

Figure 9.1.6.1(C) compares concrete sandwich panels having an outer wythe of 2 in., various thicknesses of insulation, and various thicknesses of inner wythes. All walls have U-values of 0.091 and are exposed to the same simulated outside temperatures. The figure shows that by increasing the thickness of the inner concrete wythes, peak loads are reduced and delayed.

A metal roof is compared to a concrete roof in Figure 9.1.6.1(D). Both roof systems have built-up roofing on rigid board insulation on the outside and acoustical tile on the inside. The concrete roof weighs 48.3 psf and the metal roof 1.5 psf. The roofs have identical U-values of 0.10 and are exposed to the same simulated outside temperatures. The figure shows that the concrete roof has lower peak loads by 68% for heating and by 94% for cooling, and peaks are delayed by about 1.8 hours for heating and about 4 hours for cooling.

Another factor affecting the behavior of thermal mass is the availability of so-called "free heat." This includes heat generated inside the building by lights, equipment, appliances, and people. It also includes heat from the sun entering through windows. Generally, during the heating season, benefits of thermal mass increase with the availability of "free heat" as shown in Tables 9.1.6.1 and 9.1.6.2. Thus, office buildings which have high internal heat gains from lights, people, and large glass areas represent an ideal application for thermal mass designs. This is especially true if the glass has been located to take maximum advantage of the sun. Building codes and standards now provide for the benefits of thermal mass. In increasing numbers, they are beginning to acknowledge the effect of the greater heat storage capacity in buildings having high thermal mass.

The rates of many utilities are structured so that lower peaks and delayed peaks can result in significant cost savings.

Other studies have shown that concrete buildings have lower average heating and cooling loads than lightweight buildings for a given insulation level. Thus, life-cycle costs will be lower, or less insulation can be used for equivalent performance.

Figure 9.1.6.1 Heating and cooling load comparisons

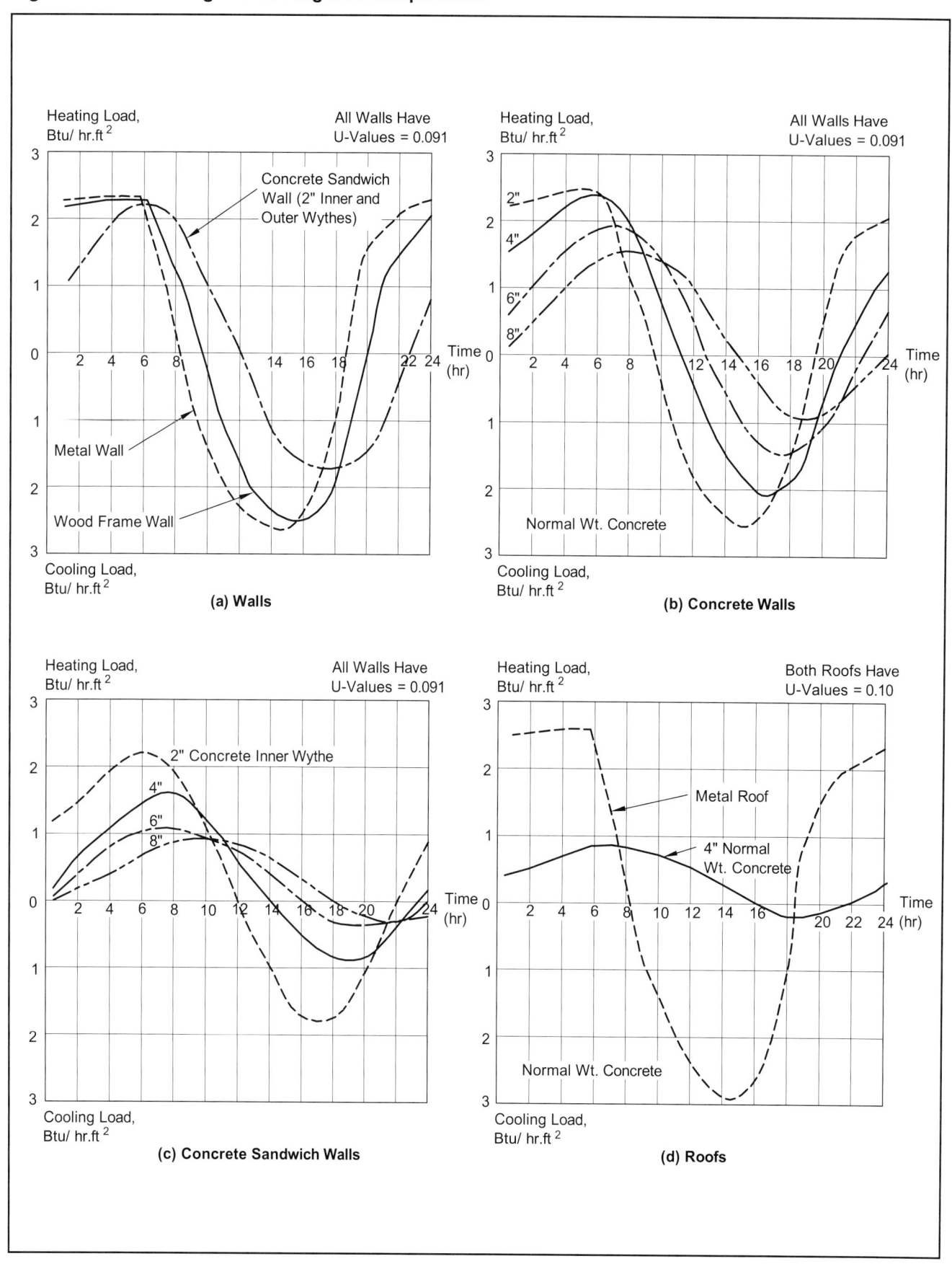

Table 9.1.6.1 Design considerations for building with high available free heat[a]

Climate Classification		Relative Importance of Design Considerations[b]				Daylighting	Reduce Infiltration
		Thermal Mass	Increase Insulation	External Fins[c]	Surface Color (Light / Dark)		
Winter							
Long Heating Season (6000 degree days or more)	With sun[d] and wind[e]	1	2	2	/ 2	1	3
	With sun without wind	1	2		/ 2	1	3
	Without sun and wind		2		1 /		3
	Without sun with wind	1	2	2	1 /		3
Moderate Heating Season (3000 to 6000 degree days)	With sun and wind	2	2	1	/ 1	2	2
	With sun without wind	2	2		/ 1	2	2
	Without sun and wind	1	2				2
	Without sun with wind	1	2	1			2
Short Heating Season (3000 degree days or less)	With sun and wind	3	1			2	1
	With sun without wind	3	1			2	1
	Without sun and wind	2	1				1
	Without sun with wind	2	1				1
Summer							
Long Cooling Season (1500 hr @ 80°F)	Dry or humid	3		3	3 /	3	3
Moderate Cooling Season (600 to 1500 hr @ 80°F)	Dry or humid	3		2	2 /	2	3
Short Cooling Season (Less than 600 hr @ 80°F)	Dry or humid	2		1	1 /	1	2

a. Includes office buildings, factories, and commercial buildings.
b. Higher numbers indicate greater importance.
c. Provide shading and protection from direct wind.
d. With sun: Sunshine during at least 60% of daylight time.
e. With wind: Average wind velocity over 9 mph.

9.1.7 Condensation Control

Moisture which condenses on the interior of a building is unsightly and can cause damage to the building or its contents. Even more undesirable is the condensation of moisture within a building wall or ceiling assembly where it is not readily noticed until damage has occurred. All air in buildings contains water vapor, with warm air carrying more moisture than cold air. In many buildings moisture is added to the air by industrial processes, cooking, laundering, or humidifiers. If the inside surface temperature of a wall, floor or ceiling is too cold, the air contacting this surface will be cooled below its dew-point temperature and leave its excess water on that surface. Condensation occurs on the surface with the lowest temperature.

Condensation on interior room surfaces can be controlled both by suitable construction and by precautions such as: (1) reducing the interior dew point temperature; (2) raising the temperatures of interior surfaces that are below the dew point, generally by use of insulation; and (3) using vapor retarders.

The interior air dew point temperature can be lowered by removing moisture from the air, either through ventilation or dehumidification. Adequate surface temperatures can be maintained during the winter by incorporating sufficient thermal insulation, using double glazing, circulating warm air over the

Table 9.1.6.2 Design considerations for building with low available free heat[a]

Climate Classification		Relative Importance of Design Considerations[b]					
		Thermal Mass	Increase Insulation	External Fins[c]	Surface Color Light	Surface Color Dark	Reduce Infiltration
Winter							
Long Heating Season (6000 degree days or more)	With sun[d] and wind[e]		3	2		3	3
	With sun without wind		3			3	3
	Without sun and wind		3			2	3
	Without sun with wind		3	2		2	3
Moderate Heating Season (3000 to 6000 degree days)	With sun and wind	1	2	1		2	3
	With sun without wind	1	2			2	3
	Without sun and wind		2			1	3
	Without sun with wind	1	2	1		1	3
Short Heating Season (3000 degree days or less)	With sun and wind	2	1			1	2
	With sun without wind	2	1			1	2
	Without sun and wind	1	1				2
	Without sun with wind	1	1				2
Summer							
Long Cooling Season (1500 hr @ 80°F)	Dry[f] or humid[g]	3		2	2		3
Moderate Cooling Season (600 to 1500 hr @ 80°F)	Dry	2		1	1		2
	Humid	2		1	1		3
Short Cooling Season (Less than 600 hr @ 80°F)	Dry or humid	1					1

a. Includes low-rise residential buildings and some warehouses.
b. Higher numbers indicate greater importance.
c. Provide shading and protection from direct wind.
d. With sun: Sunshine during at least 60% of daylight time.
e. With wind: Average wind velocity over 9 mph.
f. Dry: Daily average relative humidity less than 60% during summer.
g. Humid: Daily average relative humidity greater than 60% during summer.

surfaces or directly heating the surfaces, and by paying proper attention during the design phase to the prevention of thermal bridging (see Section 9.1.8).

Infiltration and exfiltration are air leakage into and out of a building through cracks or joints between infill components and structural elements, interstices around windows and doors, through floors and walls and openings for building services. They are often a major source of energy loss in buildings.

Moisture can move into or across a wall assembly by means of vapor diffusion and air movement. If air, especially exfiltrating, hot, humid air, can leak into the enclosure, then this will be the major source of moisture. Air migration occurs from air pressure differentials independent of moisture pressure differentials.

Atmospheric air pressure differences between the inside and outside of a building envelope exist because of the action of wind, the density difference between outside cold heavy air and inside warm light air creating a "stack effect," and the operation of equipment such as fans. The pressure differences will tend to equalize, and the air will flow through holes or cracks in the building envelope carrying with it the water vapor it contains.

A thorough analysis of air leakage is very complex, involving many parameters, including wall construction, building height and orientation.

Many condensation-related problems in building enclosures are caused by exfiltration and subsequent condensation within the enclosure assembly. Condensation due to air movement is usually much greater than that due to vapor diffusion for most buildings. However, when air leakage is controlled or avoided, the contribution from vapor diffusion can still be significant. In a well designed wall, attention must, therefore, be paid to the control of air flow and vapor diffusion.

An air barrier and vapor retarder are both needed, and in many instances a single material can be used to provide both of these as well as other functions. The principal function of the air barrier is to stop outside air from entering the building through the walls, windows or roof, and inside air from exfiltrating through the building envelope to the outside. This applies whether the air is humid or dry, since air leakage can result in problems other than the deposition of moisture in cavities. Exfiltrating air carries away heating and cooling energy, while incoming air may bring in pollution as well as reduce the effectiveness of a rain screen wall system.

9.1.7.1 Air Barriers

Materials and the method of assembly chosen to build an air barrier must meet several requirements if they are to perform the air leakage control function successfully.

1. There must be continuity throughout the building envelope. The low air permeability materials of the wall must be continuous with the low air barrier materials of the roof (e.g., the roofing membrane) and must be connected to the air barrier material of the window frame, etc.
2. Each membrane or assembly of materials intended to support a differential air pressure load must be designed and constructed to carry that load, inward or outward, or it must receive the necessary support from other elements of the wall. If the air barrier system is made of flexible materials, then it must be supported on both sides by materials capable of resisting the peak air pressure loads; or it must be made of self-supporting materials, such as board products adequately fastened to the structure. If an air pressure difference can not move air, it will act to displace the materials that prevent the air from flowing
3. The air barrier system must be virtually air-impermeable. A value for maximum allowable air permeability has not yet been determined. However, materials such as polyethylene, gypsum board, precast concrete panels, metal sheeting or glass qualify as low air-permeable materials when joints are properly sealed, whereas concrete block, acoustic insulation, open cell polystyrene insulation or fiberboard would not qualify. The metal and glass curtain wall industry in the United States has adopted a value of 0.06 CFM/ft^2 at 1.57 lb/ft^2 as the maximum allowable air leakage rate for these types of wall construction.
4. The air barrier assembly must be durable in the same sense that the building is durable, and be made of materials that are known to have a long service life or be positioned so that it may be serviced from time to time.

In climates where the heating season dominates, it is strongly recommended that the visible interior surface of a building envelope be installed and treated as the primary air barrier and vapor retarder. Where floors and cross walls are of solid concrete, it is necessary to seal only the joints, as floors and walls themselves do not constitute air paths. Where hollow partitions, such as steel stud or hollow masonry units are used, the interior finish of the envelope should be first made continuous. Where this is impractical, polyethylene film should be installed across these junctions and later sealed to the interior finish material. An interior finish of gypsum wallboard or plaster painted with two coats of enamel paint will provide a satisfactory air barrier/vapor retarder in many instances if the floor/wall and ceiling/wall joints are tightly fitted and sealed with caulking.

A recent development is an air barrier and vapor retarder system consisting of panel joints sealed from the inside with a foam backer rod and sealant, plus a thermal fusible membrane (TFM) seal around the panel, covering the gap between the structure and the panel. Surfaces should be clean, as dry as possible, smooth, and free of foreign matter which may impede adhesion. The bond between the concrete and membrane may be improved by priming the concrete before fusing the membrane to it.

While it is preferable that the air barrier system be placed on the warm side of an insulated assembly, where thermal stresses will be at a minimum, it is not an essential requirement. (This does not necessarily mean on the inside surface of the wall.) The position of the air barrier in a wall is more a matter of suitable construction practice and the type of materials to be used. However, if this barrier is positioned on the outside of the insulation, consideration must be given to its water vapor permeability in case it should also act as a barrier to vapor which is on its way out from inside the wall assembly. This situation may be prevented by choosing an air barrier material that is ten to twenty times more permeable to water vapor diffusion than the vapor barrier material.

In the case of construction assemblies which do not lend themselves to the sealing of interior surfaces, or where it is desirable to limit condensation to very small amounts, such as sandwich wall panel construction (which may have no air leaks through the panels themselves) or any air space which can ensure venting and drying out in summer, the use of a separate vapor retarder must be considered. In such cases, the insulation material itself, if it is a rigid closed-cell type, can be installed on a complete bed of adhesive applied to the interior of the inner wythe of the wall with joints fully sealed with adhesive to provide a barrier to both air and vapor movement.

While the discussion above has been concerned with the flat areas of walls, the joints between them may well present the most important design and construction problems. There are many kinds of joints and of them the following are considered the most critical: the roof/wall connection, the wall/foundation connection, soffit connections, corner details, and connections between different types of exterior wall systems, such as brick and precast concrete, or curtain wall and precast concrete.

9.1.7.2 Vapor Retarders

Water vapor diffusion, another way in which indoor water can move through a building envelope to condense in the colder zones, occurs when water vapor molecules diffuse through solid interior materials at a rate dependent on the permeability of the materials, the vapor pressure and temperature differentials. Generally, the colder the outside temperature the greater the pressure of the water vapor in the warm inside air to reach the cooler, drier outside air.

The principal function of a vapor retarder made of low permeability materials is to stop or, more accurately, to retard the passage of moisture as it diffuses through the assembly of materials in a wall. Vapor diffusion control is simple to achieve and is primarily a function of the water vapor diffusion resistance of the chosen materials and their position within the building envelope assembly. The vapor retarder should be clearly identified by the designer and also be clearly identifiable by the general contractor.

In temperate climates, vapor retarders should be applied on or near the warm side (inner surface) of assemblies. Vapor retarders may be structural, or in the form of thin sheets, or as coatings. Vapor retarders may also be positioned part way into the insulation but, to avoid condensation, they should be no further than the point at which the dew-point temperature is reached.

In climates with high humidities and high temperatures, especially where air-conditioning is virtually continuous, the ingress of moisture may be minimized by a vapor retarder system in the building envelope near the outer surface. For air-conditioned buildings in hot and humid climates without extended cold periods, it may be more economical to use only adequate air infiltration retarding systems rather than vapor retarders since the interior temperature is very rarely below the dew point of the outside air.

Where warm and cold sides may reverse, with resulting reversal of vapor flow, careful analysis of the condition is recommended rather than to ignore the problem and omit any vapor retarders. The designer should refer to the ASHRAE Handbook Fundamentals [1] or ASTM C 755, Selection of Vapor Barriers for Thermal Insulation. [5] In general, a vapor retarder should not be placed at both the inside and outside of wall assemblies.

High thermal conductance paths reaching inward from or near the colder surfaces may cause condensation within the construction. High conductance paths may occur at junctions of floors and walls, walls and ceilings, and walls and roofs, around wall or roof openings, at perimeters of slabs on the ground, and at connections.

Fittings installed in outer walls, such as electrical boxes without holes and conduits, should be completely sealed against moisture and air passage, and they should be installed on the warm side of unbroken vapor retarders or air barriers that are completely sealed.

Example 9.1.7.1
Condensation Prevention

Given:
A room in which the temperature and relative humidity are to be maintained at 70°F and 40%, respectively. During heating season, $t_o = -10°F$.

Problem:
Determine the heat transmission resistance, R_t required to prevent condensation.

Solution:
From Table 9.1.7.1 the dew-point temperature t_s is 45°F, and from Table 9.1.4.1, $R_{fi} = 0.68$.
From Eq. 9.1.7.1:

$$R_t = R_{fi}\left(\frac{t_i - t_o}{t_i - t_s}\right) = \frac{0.68[70 - (-10)]}{[70 - 45]} = 2.18$$

$$U = 1/R_t = 0.46$$

Table 9.1.7.1 Dew-point temperatures,[a] t_s, °F

Dry bulb or room temperature, °F	Relative humidity (%)									
	10	20	30	40	50	60	70	80	90	100
40	−7	6	14	19	24	28	31	34	37	40
45	−3	9	18	23	28	32	36	39	42	45
50	−1	13	21	27	32	37	41	44	47	50
55	5	17	26	32	37	41	45	49	52	55
60	7	21	30	36	42	46	50	54	57	60
65	11	24	33	40	46	51	55	59	62	65
70	14	27	38	45	51	56	60	63	67	70
75	17	32	42	49	55	60	64	69	72	75
80	21	36	46	54	60	65	69	73	77	80
85	23	40	50	58	64	70	74	78	82	85
90	27	44	55	63	69	74	79	83	85	90

a. Temperatures are based on barometric pressure of 29.92 in. Hg.

Figure 9.1.7.1 Relative humidity at which visible condensation occurs on inside surface. Inside temperature, 70°F

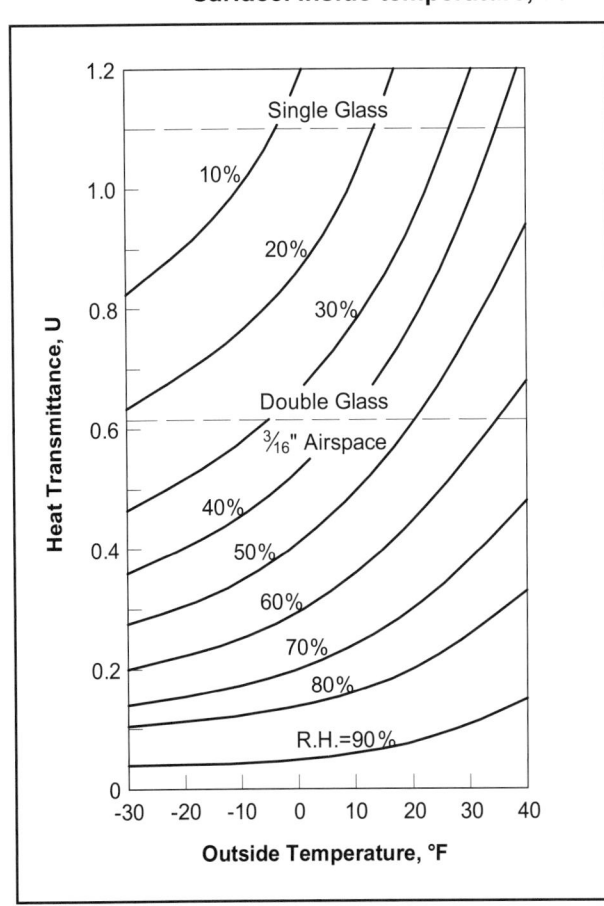

Table 9.1.7.2 Typical permeance (M) and permeability (μ) values[a]

Material	M perms	μ perm-in.
Concrete (1:2:4 mix)[b]	—	3.2
Wood (sugar pine)	—	0.4–5.4
Expanded polystyrene (extruded)	—	1.2
Expanded polystyrene (bead)	—	2.0–5.8
Paint–two coats		
Asphalt paint on plywood	0.4	
Enamels on smooth plaster	0.5–1.5	
Various primers plus one coat flat oil paint on plaster	1.6–3.0	
Plaster on gypsum lath (with studs)	20.00	
Gypsum wallboard, 0.375 in.	50.00	
Polyethylene, 2 mil	0.16	
Polyethylene, 10 mil	0.03	
Aluminum foil, 0.35 mil	0.05	
Aluminum foil, 1 mil	0.00	
Built-up roofing (hot mopped)	0.00	
Duplex sheet, asphalt laminated aluminum foil one side	0.002	

a. ASHRAE Handbook. [1]
b. Permeability for concrete varies depending on the concrete's water-cementitious materials ratio and other factors.

9.1.7.3 Prevention of Condensation on Wall Surfaces

The U-value of a wall must be such that the surface temperature will not fall below the dew-point temperature of the room air in order to prevent condensation on the interior surface of a wall. See Example 9.1.7.1.

Figure 9.1.7.1 gives U-values for any combination of outside temperatures and inside relative humidities above which condensation will occur on the interior surfaces. For example, if a building were located in an area with an outdoor design temperature of 0°F and it was desired to maintain a relative humidity within the building of 25%, the wall must be designed so that all components have a U-value less than 0.78; otherwise there will be a problem with condensation. In many designs the desire to conserve energy will dictate the use of lower U-values than those required to avoid the condensation problem.

The degree of wall heat transmission resistance that must be provided to avoid condensation may be determined from the following relationship:

$$R_t = R_{fi}\left(\frac{t_i - t_o}{t_i - t_s}\right) \qquad \text{(Eq. 9.1.7.1)}$$

Dew-point temperatures to the nearest °F for various values of t_i and relative humidity are shown in Table 9.1.7.1.

9.1.7.4 Condensation Prevention Within Wall Construction

Water vapor in air behaves as a gas and will diffuse through building materials at rates which depend on vapor permeability of materials and vapor pressure differentials. The colder the outside temperature the greater the pressure of the water vapor in the warm inside air to reach the cooler, drier outside air. Also, leakage of moisture laden air into an assembly through small cracks may be a greater problem than vapor diffusion. The passage of water vapor through material is in itself generally not harmful. Water vapor passage becomes harmful when the vapor flow path encounters a temperature below the dew-point. Condensation then results within the material.

Building materials have water vapor permeances from very low to very high (see Table 9.1.7.2). When properly used, low permeance materials keep moisture from entering a wall or roof assembly. Materials with higher permeance allow construction moisture and moisture which enters inadvertently or by design to escape.

When a material such as plaster or gypsum board has a permeance which is too high for the intended use, one or two coats of paint is frequently sufficient to lower the permeance to an acceptable level, or a vapor barrier can be used directly behind such products. Polyethylene sheet, aluminum foil and roofing materials are commonly used. Proprietary vapor barriers, usually combinations of foil and polyethylene or asphalt, are frequently used in freezer and cold storage construction.

Concrete is a relatively good vapor retarder, provided it remains crack free. Permeance is a function of the water-cementitious materials ratio of the concrete. A low ratio, such as that used in most precast concrete members, results in concrete with low permeance.

Where climatic conditions require insulation, a vapor retarder is generally necessary in order to prevent condensation. A closed cell insulation, if properly applied, will serve as its own vapor retarder. For other insulation materials, a vapor retarder should be applied to the warm side of the insulation.

For a more complete treatment of the subject of condensation within wall or roof assemblies, see Refs. 1 and 4.

9.1.8 Thermal Bridges

Metal ties through walls or solid concrete paths through sandwich panels as described in Section 9.4 may cause localized cold spots. The most significant effect of these cold spots is condensation which may cause annoying or damaging wet streaks. Other types of ties are available.

The effect of metal tie thermal bridges on the heat transmittance can be calculated with reasonable accuracy by the zone method described in Chapter 22 of Ref. 1. With the zone method, the panel is divided into Zone A, which contains the thermal bridge, and Zone B, where thermal bridges do not occur, as shown in Figure 9.1.8.1. The width of Zone A is calculated as W = m + 2d, where m is the width or diameter of the metal or other conductive bridge material, and d is the distance from the panel surface to the metal.

After the width (W) and area of Zone A are calculated, the heat transmissions of the zonal sections are determined and converted to area resistances, which are then added to obtain the total resistance (R_t) of that portion of the panel. The resistance of Zone A is combined with that of Zone B to obtain the overall resistance and the gross transmission value U_o, where U_o is the overall weighted average heat transmission coefficient of the panel.

The net effect of metal ties is to increase the U-value by 10 or 15%, depending on type, size and spacing. For example, a wall as shown in Figure 9.1.8.1 would have a U-value of 0.13 if the effect of

Figure 9.1.8.1 Metal tie thermal bridges

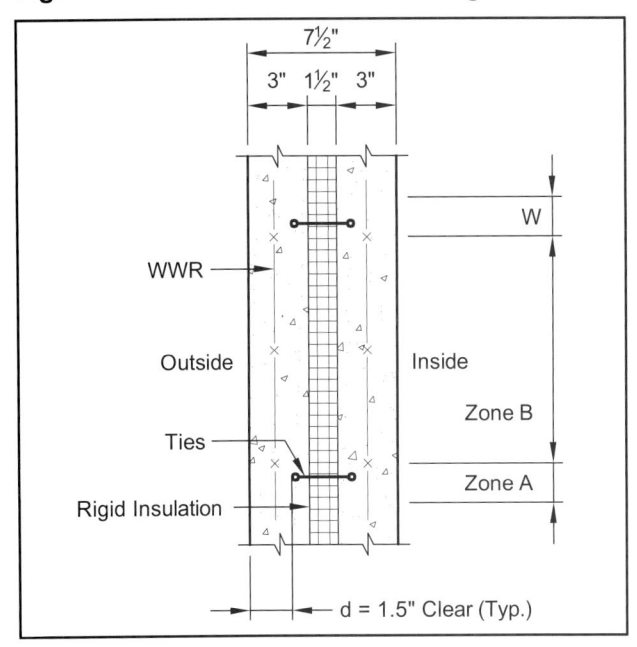

the ties is neglected. If the effect of ¼ in. diameter ties at 16 in. on center is included, U = 0.16; at 24 in. spacing, U = 0.15. Ongoing research indicates these numbers are conservative.

The effect of solid concrete path thermal bridges can be calculated by the characteristic section method described in Ref. 6. In this method, the panel is divided into two regions. The first is treated as a perfectly insulated panel without any thermal bridge. The second is treated as a solid concrete panel without insulation. The total thermal resistance of the panel is calculated as the resistances of these two regions added in parallel.

The portion of the panel that is treated as a solid concrete panel without any insulation is larger than the actual solid concrete region that exists in the panel. There is an affected zone around each solid concrete region that is added to the actual area of the solid concrete region to obtain the size of the concrete region used in the calculation. The size of the affected zone E_z is computed as:

$$E_z = 1.4 - 0.1t_{in}\alpha + [0.4t_{cf} + 0.1(t_{cb} - t_{cf})]\beta$$
(Eq. 9.1.8.1)

In this equation, t_{in}, t_{cf} and t_{cb} are the thicknesses of the insulation layer, concrete face wythe, and concrete back wythe, respectively. This is an empirical equation with all dimensions expressed in inches. The parameters α and β account for the insulation and concrete conductivity values (k_{in} and k_{con}) that are used to construct the panel. Their values are computed as:

$$\alpha = 1 + 2.25\left(\frac{k_{in} - 0.26}{0.26}\right)$$
(Eq. 9.1.8.2)

and

$$\beta = 1 + 1.458\left(\frac{k_{con} - 12.05}{12.05}\right)$$
(Eq. 9.1.8.3)

In these equations, k_{in} and k_{con} have units of Btu(in./hr)(ft²)(°F).

To calculate an R-value, a panel is divided into two regions: a solid concrete region and a perfectly insulated region, as explained previously. Eq. 9.1.8.1 is used to calculate E_z, and the area of each region is then calculated. The thermal resistance of the solid concrete region (R_s) is then added in parallel with the thermal resistance of the perfectly insulated region (R_p) to obtain the thermal resistance of the panel R:

$$\frac{1}{R} = \frac{A'_s}{R_s} + \frac{A'_p}{R_p}$$
(Eq. 9.1.8.4)

where A'_s and A'_p represent the areas of the solid concrete region (A_s) and perfect panel region (A_p) divided by the total panel area A_t (i.e., $A'_s = A_s/A_t$, $A'_p = A_p/A_t$). The procedure is illustrated in Example 9.1.8.1.

9.1.9 References

1. ASHRAE Handbook 2003, *Fundamentals*, American Society of Heating, Refrigerating, and Air-Conditioning Engineers, New York, NY, 1997.

2. ASHRAE Standard 90.1-2001, *Energy Efficient Design of New Buildings Except Low-Rise Residential Buildings*, American Society of Heating, Refrigerating, and Air-Conditioning Engineers, New York, NY, 1989.

3. "Simplified Thermal Design of Building Envelopes," *Bulletin EB089*, Portland Cement Association, Skokie, IL, 1981.

4. Balik, J. S., and Barney, G. B., "Thermal Design of Precast Concrete Buildings," PCI JOURNAL, V. 29, No. 6, November-December 1984.

5. *Recommended Practice for Selection of Vapor Barriers for Thermal Insulation*, ASTM C 755, American Society for Testing and Materials, Philadelphia, PA.

6. Lee, Y. J., and Pessiki, S., "Development of the Characteristic Section Method to Estimate Thermal R-Values for Precast Concrete Sandwich Wall Panels," ATLSS Report No. 03-06, Center for Advanced Technology for Large Structural Systems, Lehigh University, April 2003, 73 pp.

Example 9.1.8.1
Determination of R-Value for Sandwich Panel

Problem:
Determine the R-value for the sandwich panel shown below for conductivities of 10.0 Btu·(in./hr)(ft²)°F and 0.15 Btu·(in./hr)(ft²)(°F) for the concrete and insulation, respectively. Face and back wythe thicknesses are 3 in., and the insulation layer thickness is 2 in.

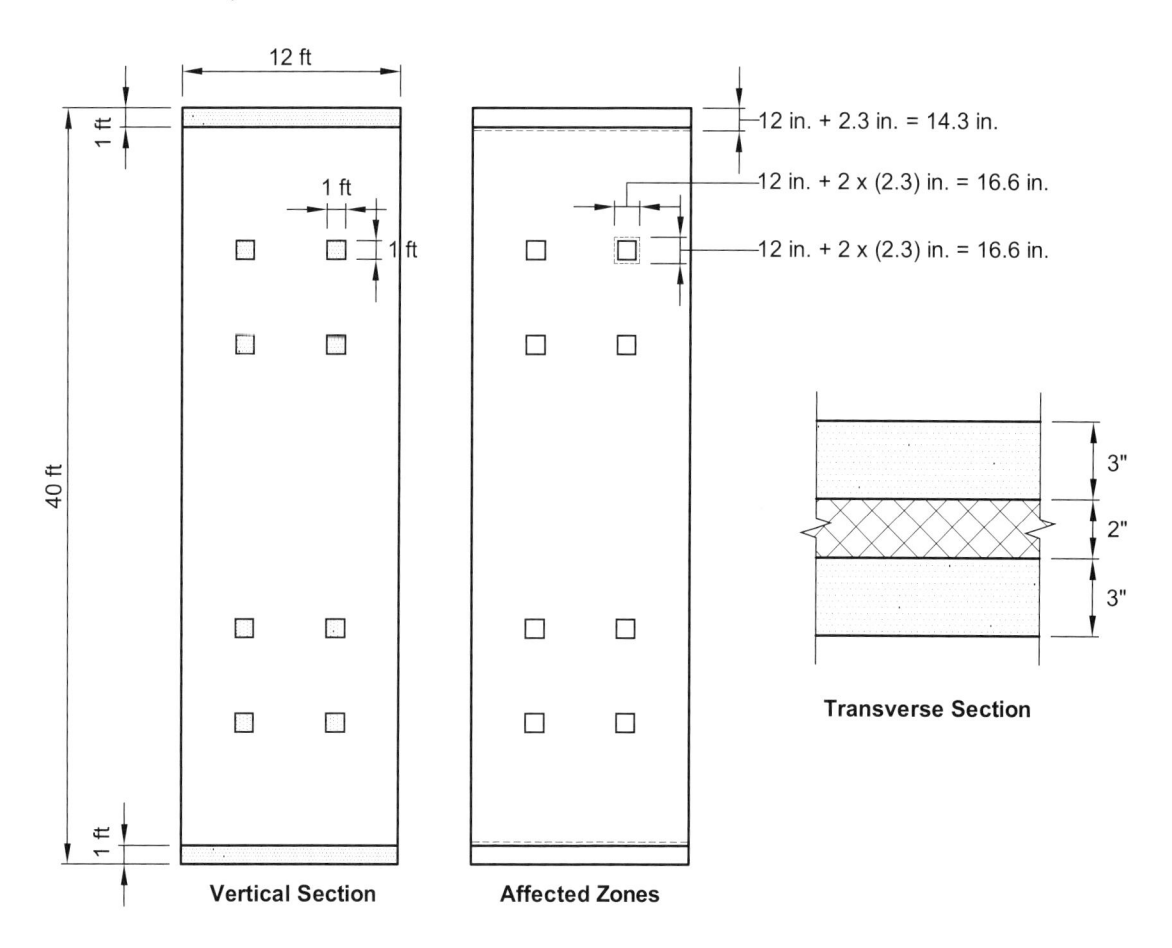

Vertical Section

Affected Zones

12 in. + 2.3 in. = 14.3 in.
12 in. + 2 x (2.3) in. = 16.6 in.
12 in. + 2 x (2.3) in. = 16.6 in.

Transverse Section

Solution:
Calculate the parameters α and β:

$$\alpha = 1 + 2.25 \left(\frac{k_{in} - 0.26}{0.26} \right) = 1 + 2.25 \left(\frac{0.15 - 0.26}{0.26} \right) = 0.05$$

$$\beta = 1 + 1.458 \left(\frac{k_{con} - 12.05}{12.05} \right) = 1 + 1.458 \left(\frac{10.00 - 12.05}{12.05} \right) = 0.75$$

From the panel thicknesses, the affected zone dimension E_z is computed as:

$$E_z = 1.4 - 0.1(t_{in})(\alpha) + \left[0.4 t_c + 0.1(t_{cb} - t_{cf}) \right] \beta$$
$$= 1.4 - 0.1(2)(0.05) + 0.4(3)(0.75)$$
$$= 2.3 \text{ in.}$$

Add E_z to the actual solid concrete areas to obtain the areas of the panel to treat as solid concrete (shown as dashed lines above).

Example 9.1.8.1 (Cont.)
Determination of R-Value for Sandwich Panel

Calculate the areas of the panel (A_t), solid concrete region (A_s), and perfectly insulated region (A_p):

A_t = panel area = (40)(12) = 480 ft² = 69,120 in.²
A_s = concrete area = 2(14.3)(144) + 8(16.6)(16.6) = 6,323 in.²
A_p = insulated area = 69120 − 6323 = 62,797 in.²

The resistance of that portion of the panel that is treated as perfectly insulated is calculated from the resistances of the concrete, insulation, and surfaces in series.

Insulated path

		k	Thickness	U = k/t	R = 1/U Winter	R = 1/U Summer
A	Outside surface	—	—	—	0.17	0.25
B	Concrete	10.00	3	3.33	0.30	0.30
C	Insulation	0.15	2	0.08	12.50	12.50
D	Concrete	10.00	3	3.33	0.30	0.30
E	Inside surface	—	—	—	0.68	0.68
					13.95	14.03

The resistance of that portion of the panel that is treated as solid concrete is calculated from the resistances of the concrete and surfaces in series.

Concrete path

		k	Thickness	U = k/t	R = 1/U Winter	R = 1/U Summer
A	Outside surface	—	—	—	0.17	0.25
B	Concrete	10.00	8	1.25	0.80	0.80
C	Insulation	—	—	—	0.68	0.68
					1.65	1.73

Calculate the fractional areas of the panel that are treated as solid concrete and as insulated:

$A_s / A_t = 6323 / 69120 = 0.091$
$A_p / A_t = 62797 / 69120 = 0.909$

Compute the R-value of the panel treating the solid concrete and perfectly insulated regions in parallel.

Winter:
$$\frac{1}{R} = \frac{0.909}{13.95} + \frac{0.091}{1.65}$$
$R = 8.31 \text{ hr} \cdot \text{ft}^2(°F)/\text{Btu}$

Summer:
$$\frac{1}{R} = \frac{0.909}{14.03} + \frac{0.091}{1.73}$$
$R = 8.52 \text{ hr} \cdot \text{ft}^2(°F)/\text{Btu}$

9.2 ACOUSTICAL PROPERTIES OF PRECAST CONCRETE

9.2.1 Definitions

Hertz (Hz). A measure of sound wave frequency, i.e., the number of complete vibration cycles per second.
  STC. Sound Transmission Class
  IIC. Impact Insulation Class

9.2.2 General

The basic purpose of architectural acoustics is to provide a satisfactory environment in which desired sounds are clearly heard by the intended listeners and unwanted sounds (noise) are isolated or absorbed.

Under most conditions, the architect/engineer can determine the acoustical needs of the space and then design the building to satisfy those needs. Good acoustical design utilizes absorptive and reflective surfaces, sound barriers and vibration isolators. Some surfaces must reflect sound so that the loudness will be adequate in all areas where listeners are located. Other surfaces absorb sound to avoid echoes, sound distortion and long reverberation times. Sound is isolated from rooms where it is not wanted by selected wall and floor-ceiling constructions. Vibration generated by mechanical equipment must be isolated from the structural frame of the building.

Information is provided on the commonly used acoustical properties of some of the more common precast concrete products used in building construction. This information can be incorporated into the acoustical design of a building and/or can show compliance with local ordinances or other minimum acoustical requirements (see Section 9.2.7).

For buildings or occupancies that require more sophisticated acoustical analysis, such as churches, concert halls, auditoriums, recording studios, and the like, it may be desirable to use the services of a competent acoustical design consultant or specialist.

9.2.3 Dealing with Sound Levels

The problems of sound insulation are usually considerably more complicated than those of sound absorption. The former involves reductions of sound level, which are of greater orders of magnitude than can be achieved by absorption. These large reductions of sound level from space to space can be achieved only by continuous, impervious barriers. If the problem also involves structure-borne sound, it may be necessary to introduce resilient layers or discontinuities into the barrier.

Figure 9.2.4.1 Sound transmission class as a function of weight of floor or wall

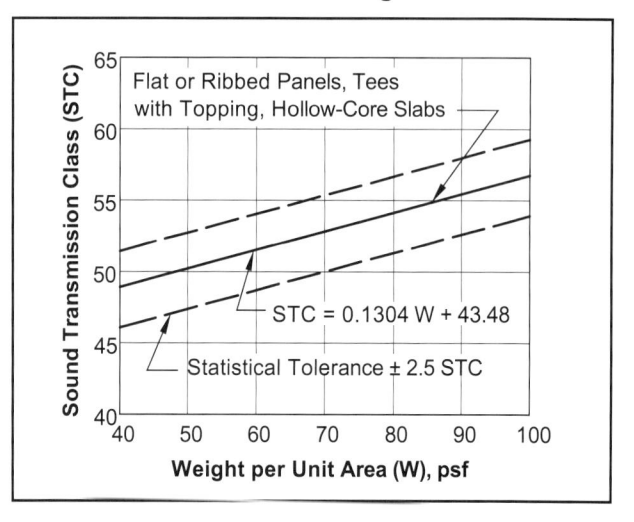

Sound absorbing materials and sound insulating materials are used for different purposes. There is not much sound absorption from an 8 in. concrete wall; similarly, high sound insulation is not available from a porous lightweight material that may be applied to room surfaces. It is important to recognize that the basic mechanisms of sound absorption and sound insulation are quite different.

9.2.4 Sound Transmission Loss

The ability of a barrier to reduce the intensity of airborne sound is commonly designated by its Sounds Transmission Class (STC). The influence of the concrete weight on STC is shown in Figure 9.2.4.1. Precast concrete walls, floors and roofs usually do not need additional treatments in order to provide adequate sound insulation. If desired, greater sound insulation can be obtained by using a resiliently attached layer or layers of gypsum board or other building material. The increased transmission loss occurs because the energy flow path is now increased to include a dissipative air column and additional mass.

9.2.5 Impact Noise Reduction

Footsteps, dragged chairs, dropped objects, slammed doors, and plumbing generate impact noise. Even when airborne sounds are adequately controlled there can be severe impact noise problems.

The test method used to evaluate systems for impact sound insulation is described in ASTM E 492, Laboratory Measurement of Impact Sound Transmission Using the Tapping Machine. The impact

Table 9.2.5.1 Impact insulation class for concrete slabs

Thickness, in.	Unit weight of concrete, pcf	IIC
5	79	23
	114	24
	144	24
10	79	28
	114	30
	144	31

Table 9.2.7.1 HUD recommendations for STC and IIC

Location	STC	IIC
Between living units	45	45
Between living units and public space	50	50

Table 9.2.7.2 Sound insulation criteria for Ref. 3

	Grade I Suburban	Grade II Residential Urban and Suburban	Grade III Urban
Walls	STC 55	STC 52	STC 48
Floor-ceiling assemblies	STC 55 IIC 55	STC 52 IIC 55	STC 48 IIC 48

sound pressure levels are measured in the 16 contiguous one-third octave bands with center frequencies in the range from 100 to 3150 Hz. For performance specification purposes, the single number Impact Insulation Class (IIC) is used (see Table 9.2.5.1).

In general, thickness or unit weight of concrete does not greatly affect the transmission of impact sounds as shown in the above table.

Structural concrete floors in combination with resilient materials effectively control impact sound. One simple solution consists of good carpeting on resilient padding. So-called resilient flooring materials, such as linoleum, rubber, asphalt, vinyl, etc., or parquet or strip wood floors are not entirely satisfactory when applied directly on concrete.

Impact sound also may be controlled by providing a discontinuity in the structure such as would be obtained by adding a resilient-mounted plaster or dry-wall suspended ceiling or a floating floor consisting of a second layer of concrete cast over resilient pads, insulation boards or mastic. The thickness of floating slabs is usually controlled by structural requirements; however, a thickness providing as little as 8 psf would be acoustically adequate in most instances.

9.2.6 Acoustical Test Results

Table 9.2.6.1 presents the ratings for various precast concrete walls and floor-ceiling assemblies.

9.2.7 Establishment of Noise Insulation Objectives

Often, acoustical control is specified as to the minimum insulation values of the dividing partition system. Local building codes, lending institutions and the Department of Housing and Urban Development (HUD) list both airborne STC and impact IIC values for different living environments. For example, the HUD recommendations [2] are given in Table 9.2.7.1.

Other community ordinances are more specific, listing the sound insulation criteria with relation to particular ambient environments (Table 9.2.7.2). [3]

Once the objectives are established, the designer then should refer to available data, e.g., Figure 9.2.4.1 or Table 9.2.6.1, and select the system which best meets these requirements. In this respect, concrete systems have superior properties and can with minimal effort comply with these criteria.

9.2.8 Composite Wall Considerations

An acoustically composite wall is made up of elements of varying acoustical properties. Doors and windows are often the weak link in an otherwise effective sound barrier. Minimal effects on sound transmission loss will be achieved in most cases by a proper selection of glass (plate vs. insulating). [4] Mounting of the glass in its frame should be done with care to eliminate noise leaks and to reduce the glass plate vibrations.

Sound transmission loss of a door depends upon its material and construction, and the sealing between the door and the frame.

Acoustical design must consider the acoustical properties of doors, windows or other elements, the ratio of openings in the wall system, and the distances between openings. Installation procedures and materials must be specified to ensure desired acoustical seals.

9.2.9 Leaks and Flanking

The performance of a building section with an otherwise adequate STC can be seriously reduced by a relatively small hole or any other path which allows sound to bypass the acoustical barrier. All noise which reaches a space by paths other than through the primary barrier is called flanking. Common flanking paths are gaps between floor perimeters and curtain walls, openings around doors or windows, at electrical outlets, telephone and television connections, and pipe and duct penetrations. Sealants, safing and/or closure plates are useful in reducing flanking.

Suspended ceilings in rooms where walls do not extend from the ceiling to the roof or floor above allow sound to travel to adjacent rooms.

Use of full height walls will alleviate this source of leakage. In general, the probability of flanking paths in a concrete structure is much lower than in a structure of steel or wood.

Table 9.2.6.1 Airborne sound transmission loss (STC) and impact insulation class (IIC) ratings from tests of precast concrete assemblies

Assembly No.	Description	STC	IIC
	Wall Systems		
1	4 in. flat panel, 54 psf	49	–
2	6 in. flat panel, 75 psf	55	–
3	Assembly 2 with "Z" furring channels, 1 in. insulation and ½ in. gypsum board, 75.5 psf	62	–
4	Assembly 2 with wood furring, 1½ in. insulation and ½ in. gypsum board, 73 psf	63	–
5	Assembly 2 with ½ in. space, 1⅝ in. metal stud row, 1½ in. insulation and ½ in. gypsum board	63[a]	–
6	8 in. flat panel, 95 psf	58	–
7	14 in. prestressed tees with 4 in. flange, 75 psf	54	–
	Floor-Ceiling Systems		
8	8 in. hollow-core prestressed units, 57 psf	50	28
9	Assembly 8 with carpet and pad, 58 psf	50	73
10	8 in. hollow-core prestressed units with ½ in. wood block flooring adhered directly, 58 psf	51	47
11	Assembly 10 except ½ in. wood block flooring adhered to ½ in. sound-deadening board underlayment adhered to concrete, 60 psf	52	55
12	Assembly 11 with acoustical ceiling, 62 psf	59	61
13	Assembly 8 with quarry tile, 1¼ in. reinforced mortar bed with 0.4 in. nylon and carbon black spinerette matting, 76 psf	60	54
14	Assembly 13 with suspended 5/8 in. gypsum board ceiling with 3½ in. insulation, 78.8 psf	61	62
15	14 in. prestressed tees with 2 in. concrete topping, 75 psf	54	24
16	Assembly 15 with carpet and pad, 76 psf	54	72
17	Assembly 15 with resiliently suspended acoustical ceiling with 1½ in. mineral fiber blanket above, 77 psf	59	51
18	Assembly 17 with carpet and pad, 78 psf	59	82
19	4 in. flat slabs, 54 psf	49	25
20	5 in. flat slabs, 60 psf	52[a]	24
21	5 in. flat slab concrete with carpet and pad, 61 psf	52[a]	68
22	6 in. flat slabs, 75 psf	55	34
23	8 in. flat slabs, 95 psf	58	34[a]
24	10 in. flat slabs, 120 psf	59[a]	31
25	10 in. flat slab concrete with carpet and pad, 121 psf	59[a]	74

a. Estimated values.

9.2.10 References

1. *2000 ASHRAE Handbook — HVAC Systems and Equipment*, American Society of Heating, Refrigerating and Air-Conditioning Engineers, Atlanta, GA, 2000.

2. Berendt, R. D., Winzer G. E., and Burroughs, C. B., *A Guide to Airborne, Impact and Structure-Borne Noise Control in Multi-family Dwellings*, prepared for Federal Housing Administration, U.S. Government Printing Office, Washington, DC, 1975.

3. Sabine, H. J., Lacher, M. B., Flynn, D. R., and Quindry, T.L., *Acoustical and Thermal Performance of Exterior Residential Walls, Doors and Windows*, National Bureau of Standards, U.S. Government Printing Office, Washington, DC, 1975.

4. IITRI, *Compendium of Materials for Noise Control*, U.S. Department of Heath, Education and Welfare, U.S. Government Printing Office, Washington, DC, 1980.

5. Harris, C. M., *Handbook of Acoustical Measurements and Noise Control*, Acoustical Society of America, Melville, NY, 1997.

6. Litvin, A., and Belliston, H. W., "Sound Transmission Loss Through Concrete and Concrete Masonry Walls," *Journal of the American Concrete Institute*, V. 75, No. 12, December 1978.

7. "Acoustical Properties of Precast Concrete," PCI JOURNAL, V. 23, No. 2, March-April 1978.

9.3 FIRE RESISTANCE

9.3.1 Notation

Note: Subscript θ indicates the property as affected by elevated temperatures.

- a = depth of equivalent rectangular compression stress block
- A_{ps} = area of uncoated prestressing steel
- A_s = area of non-prestressed reinforcement
- A_s^- = area of reinforcement in negative moment region
- b = width of member
- d = distance from centroid of prestressing steel to extreme compression fiber
- f_c' = compressive strength of concrete
- f_{ps} = stress in uncoated prestressing steel at nominal strength
- f_{pu} = ultimate tensile strength of uncoated prestressing steel
- h = total depth of member
- ℓ = span length
- M_n = nominal moment strength
- $M_{n\theta}^+$ = positive nominal moment strength at elevated temperatures
- $M_{n\theta}^-$ = negative nominal moment strength at elevated temperatures
- R = fire endurance of element or composite assembly
- R_1, R_2, R_n = fire endurance of individual courses
- s = spacing of ribs in ribbed panels
- t = minimum thickness of ribbed panels
- t_e = equivalent thickness of ribbed panels
- u = distance from prestressing steel to fire exposed surface
- w = uniform total load
- w_d = uniform dead load
- w_ℓ = uniform live load
- x, x_0, x_1, x_2 = horizontal distances as shown in Figures 9.3.7.11, 9.3.7.12, and 9.3.7.12
- y_s = distance from centroid of prestressing steel to bottom fiber
- ϕ = strength reduction factor
- θ_s = temperature of steel

9.3.2 Definitions

Carbonate aggregate concrete. Concrete made with aggregates consisting mainly of calcium or magnesium carbonate, e.g., limestone or dolomite.

Fire endurance. A measure of the elapsed time during which a material or assembly continues to exhibit fire resistance under specified conditions of test and performance. As applied to elements of buildings it shall be measured by the methods and to the criteria defined in ASTM E 119 (defined in ASTM E 176).

Fire resistance. The property of a material or assembly to withstand fire or to give protection from it. As applied to elements of buildings, it is characterized by the ability to confine a fire or to continue to perform a given structural function, or both (defined in ASTM E 176).

Fire resistance rating (sometimes called **fire rating, fire resistance classification**, or **hourly rating**). A legal term defined in building codes, usually based on fire endurances. Fire resistance ratings are assigned by building codes or building officials for various types of construction and occupancies and are usually given in half-hour increments.

Lightweight aggregate concrete. Concrete made with lightweight, coarse and fine aggregate (expanded clay, shale, slag, or slate, or sintered fly ash) and having a 28-day air-dry unit weight of 95 to 105 pcf.

Sand-lightweight concrete. Concrete made with lightweight, coarse aggregate (expanded clay, shale, slag, or slate, or sintered fly ash) and normal weight fine aggregate and having a 28-day air-dry unit weight of 105 to 120 pcf.

Siliceous aggregate concrete. Concrete made with normal weight aggregates consisting mainly of silica or compounds other than calcium or magnesium carbonate.

9.3.3 Introduction

Precast and prestressed concrete members can be designed to meet any degree of fire resistance that may be required by building codes, insurance companies, and other authorities. Fire resistance ratings of building assemblies can be determined from ASTM E 119 standard fire tests, code-approved empirical data, or by calculation procedures detailed in Section 9.3.7. The calculation method is based on engineering principles, taking into account the time-temperature condition of the standard fire test. This method, hereafter referred to as the Rational Design Method of determining fire resistance, was formulated as a result of extensive research sponsored in part by the Precast/Prestressed Concrete Institute (PCI) and conducted by the Portland Cement Association (PCA) and other laboratories. [1]

Examples using the Rational Design Method are provided in Section 9.3.7 along with a brief explanation of the method's underlying principles. For additional examples, design charts, and an in-depth explanation of the method, refer to the PCI Manual, MNL-124-89, *Design for Fire Resistance of Precast Prestressed Concrete* [1] as well as the CRSI Manual, *Reinforced Concrete Fire Resistance.* [2] These references have for years been recognized by the model codes as acceptable resource documents for determining fire resistance ratings of concrete by other than prescriptive means.

High strength concrete (compressive strengths up to 10,000 psi) will perform under fire conditions as described herein provided minimum cover and other dimensional requirements are adhered to. [3]

Underwriters Laboratories (UL) provides certification of fire resistance ratings of some building assemblies for precast concrete manufacturers that subscribe to the service. These certifications are based on standard fire tests. UL certification is not required by PCI or any model building code.

9.3.4 Standard Fire Tests

The fire resistance of building components is measured in standard fire tests defined by ASTM E 119, Standard Test Methods for Fire Tests of Building Construction and Materials. [11] During these tests, the building assembly, such as a portion of floors, walls, roofs or columns, is subjected to increasing temperatures that vary with time as shown in Figure 9.3.4.1. This time-temperature relationship is used as a standard to represent the combustion of about 10 lb of wood (with a heat potential of 8,000 Btu per lb) per ft^2 of exposed area per hour of test. Actually, the fuel consumption to maintain the standard time-temperature relationship during a fire test depends on the design of the furnace and on the test specimen. When fire tested, assemblies with exposed concrete members, such as double tees and hollow-core slabs, require considerably more fuel than other assemblies due to their favorable heat absorption capacity. This fact is not recognized when evaluating fire resistance by current standard test methods.

In addition to defining a standard time-temperature relationship, standard fire tests involve regulations concerning the size of the assemblies,

Figure 9.3.4.1 Standard time-temperature curve

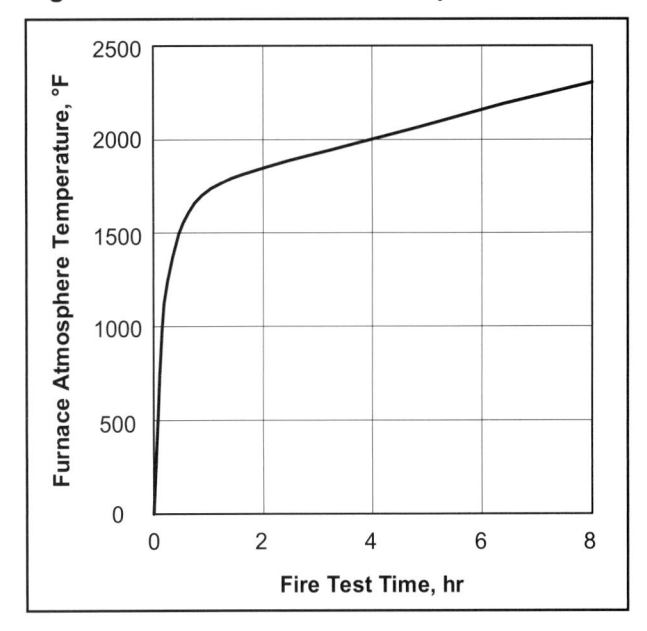

the amount of externally applied load, the region of the assembly to be exposed to fire, and the end point criteria on which fire resistance (duration) is based.

ASTM E 119 [11] specifies the minimum sizes of specimens to be exposed in fire tests, although much valuable data have been developed from tests on specimens smaller than the ASTM minimum sizes. For floors and roofs, at least 180 ft^2 must be exposed to fire from beneath, and neither dimension can be less than 12 ft. For tests of walls, either loadbearing or non-loadbearing, the minimum specified area is 100 ft^2 with neither dimension less than 9 ft. The minimum length for columns is specified to be 9 ft, while for beams it is 12 ft.

During fire tests of floors, roofs, beams, loadbearing walls, and columns, loads are applied which, along with the dead weight of the specimen, closely approximate the maximum loads required by the model codes. Tests that have been conducted on prestressed members were generally done with full design loads, whereas other materials and members have been tested utilizing less than the maximum loads. This, in effect, limits those test results to the specific loading conditions.

Floor and roof specimens are exposed to fire from beneath, beams from the bottom and sides, walls from one side, and columns from all sides.

ASTM E 119 distinguishes between "restrained" and "unrestrained" assemblies and defines them as follows:

"Floor and roof assemblies and individual beams in buildings shall be considered restrained when the surrounding or supporting structure is capable of resisting substantial thermal expansion throughout the range of anticipated elevated temperatures. Constructions not complying with this definition are assumed to be free to rotate and expand and shall therefore be considered as unrestrained."

While the focus of this definition is mainly on the axial resistance of the supporting or surrounding structure to thermal expansion, the intent of "restraint" actually can be expanded for concrete members, as opposed to other materials, to include rotational restraint and continuity as well (see Section 9.3.7.2 on continuous members).

During early standard ASTM-E119 fire tests performed on precast concrete floors, the gaps between the ends of the members and the furnace frame were grouted, which prevented the members from expanding or rotating as the furnace temperature increased. Some observers concluded that this degree of restraint could not happen in real buildings. However, field experience and observation of real fires have shown that precast concrete buildings have consistently performed as well as would be expected if the precast members were assumed to be restrained.

It should also be noted that E119 testing requires the specimens to be representative of actual building construction. However, because of the test furnace size, the results derived from members tested in laboratory furnaces cannot accurately reflect their behavior in a real building. In a 1999 manual produced by The International Conference of Building Officials (ICBO) entitled "Fire Resistive Workbook," the following is stated:

> "Although UBC Standard 7-1 [ASTM-E119] is frequently described as a large scale test, it clearly is not a full-scale test. Most floor slabs and roof decks are continuous over supports. Beams, girders and trusses are framed into columns and other structural members in a variety of ways. As a result, testing laboratories are faced with the difficult problem of providing both end support or restraint for test assemblies representative of actual field conditions."

ASTM-E119 includes a guide for classifying types of construction as restrained or unrestrained and is reproduced in Table 9.3.4.1. The guide indicates that, typically, cast-in-place and precast concrete construction is considered restrained. Section 9.3.4.2 provides further information on restraint.

9.3.4.1 Fire Endurance, End Point Criteria, and Fire Rating

The *Fire Resistance* of an assembly is measured by its fire endurance, defined as the period of time elapsed before a prescribed condition of failure or end point is reached during a standard fire test. A *Fire Rating* or *Classification* is a legal term for a fire endurance required by a building code authority.

End point criteria defined by ASTM E119 include:

1. Loadbearing specimens must sustain the applied loading. Collapse is an obvious end point (structural end point).
2. Holes, cracks, or fissures through which flames or gases hot enough to ignite cotton waste must not form (flame passage end point).
3. The temperature increase of the unexposed surface of floors, roofs, or walls must not exceed an average of 250°F or a maximum of 325°F at any one point (heat transmission end point).

After a dual rating system of fire rating classification of restrained and unrestrained was introduced around 1970 by ASTM Committee E-5, the decision was made to apply structural end-point criteria to concrete that had been tested in a restrained test procedure so as to avoid the expense of retesting a large number of assemblies.

The original ratings determined by the restrained test procedure were kept intact and unrestrained ratings were added based on temperature criteria. Thus, unrestrained assembly classifications can be derived from tests of restrained floor, roof, or beam specimens provided that the average temperature of the tension steel at any section does not exceed 800°F for cold-drawn prestressing steel or 1100°F for reinforcing bars, the structural end point assumed for an unrestrained condition. These temperatures may be exceeded when rational design is used (See Section 9.3.7.)

Additional end point criteria for restrained specimens are:

1. Beams more than 4 ft on centers: the above steel temperatures must not be exceeded for classifications of 1 hr or less; for classifications longer than 1 hr, the above temperatures must not be exceeded for the first half of the classification period or 1 hr, whichever is longer.
2. Beams 4 ft or less on centers or slabs are not subjected to steel temperature limitations.

Walls and partitions must meet the same structural, flame passage, and heat transmission end points described above. In addition, they must withstand a hose stream test (simulating, in a specified manner, a fire fighter's hose stream).

Table 9.3.4.1 Considerations of restraint for common construction[a]

I. Wall bearing:

   Single span and simply supported end spans of multiple bays:[b]
   1. Open-web steel joists or steel beams supporting concrete slab, precast units, or metal decking .... Unrestrained
   2. Concrete slabs, precast units, or metal decking .... Unrestrained

   Interior spans or multiple bays:
   1. Open-web steel joists, steel beams or metal decking supporting continuous concrete slab .... Restrained
   2. Open-web steel joists or steel beams supporting precast units or metal decking .... Unrestrained
   3. Cast-in-place concrete slab systems .... Restrained
   4. Precast concrete where the potential thermal expansion is resisted by adjacent construction [c] .... Restrained

II. Steel framing:

   1. Steel beams welded, riveted, or bolted to the framing members .... Restrained
   2. All types of cast-in-place floor and roof systems (such as beam-and-slabs, flat slabs, pan joists, and waffle slabs) where the floor or roof systems is secured to the framing members .... Restrained
   3. All types of prefabricated floor or roof systems where the structural members are secured to the framing members and the potential thermal expansion of the floor or roof system is resisted by the framing system or the adjoining floor or roof construction[c] .... Restrained

III. Concrete framing:

   1. Beams securely fastened to the framing members .... Restrained
   2. All types of cast-in-place floor or roof systems (such as beam-and-slabs, flat slabs, pan joists, and waffle slabs) where the floor system is cast with the framing members .... Restrained
   3. Interior and exterior spans of precast systems with cast-in-place joints resulting in restraint equivalent to that which would exist in Condition III-1 .... Restrained
   4. All types of prefabricated floor or roof systems where the structural members are secured to such systems and the potential thermal expansion of the floor or roof systems is resisted by the framing system or the adjoining floor or roof construction[c] .... Restrained

IV. Wood construction:

   All types .... Unrestrained

a. Source: ASTM E 119, "Standard Test Methods for Fire Tests of Building Construction and Materials," Table X3.1.
b. Floor and roof systems can be considered restrained when they are tied into walls with or without tie beams, the walls being designed and detailed to resist thermal thrust from the floor or roof system.
c. For example, resistance to potential thermal expansion is considered to be achieved when:
   1. Continuous structural concrete topping is used;
   2. The space between the ends of precast units or between the ends of units and the vertical face of supports is filled with concrete or mortar; or
   3. The space between the ends of precast units and the vertical faces of supports, or between the ends of solid or hollow-core slab units does not exceed 0.25% of the length for normal weight concrete members or 0.1% of the length for structural lightweight concrete members.

9.3.4.2 Restrained Fire Ratings

The dual rating system mentioned previously has caused much confusion over the years to designers and building officials alike. In response to this, ASTM Committee E-5 developed a table that greatly simplified the decision making process. This table is reproduced as Table 9.3.4.1. The footnotes at the bottom of Table 9.3.4.1 list the requirements that must be satisfied for a structure to be considered restrained. ICBO's "Fire Resistive Workbook" paraphrases the table as follows:

"Concrete framing may be classified as restrained. Examples of restrained conditions for concrete framing are (1) beams securely fastened to the structural framing members, (2) all types of cast-in-place floor or roof systems where the system is cast with the structural framing members, and (3) all types of pre-fabricated floor or roof systems where the structural members are secured to such systems and the potential thermal expansion of the floor or roof system is resisted by the framing system or the adjoining floor or roof construction."

Restraint is often defined as the ability of the adjacent structure to resist thermal expansion of structural members. In continuous members restraint is provided by the adjacent bays so continuous members have an obvious advantage over simply supported ones. Almost all precast/prestressed buildings are designed with the floor and roof systems acting as diaphragms transferring lateral loads to the resisting elements. This is accomplished by tying the deck elements together and to the framing elements with welded connections or with reinforcement when cast-in-place topping is used. Note that this is also done to meet the ACI 318 requirements for structural integrity. The total structural system will, in most buildings, provide the necessary resistance to thermal expansion during a fire. The third edition of MNL-124 "Design for Fire Resistance of Precast Prestressed Concrete" [13] (expected in 2005) will discuss this subject in more detail. It is also the subject of ongoing PCI sponsored research.

9.3.5 Fire Tests of Prestressed Concrete Assemblies

The first fire test of a prestressed concrete assembly in America was conducted in 1953 and, since then, more than 150 prestressed concrete assemblies have been subjected to standard fire tests in America. Although many of the tests were conducted for the purpose of deriving specific fire ratings, most of the tests were performed in conjunction with broad research studies whose objectives have been to understand the behavior of prestressed concrete subjected to fire. The knowledge gained from these tests has resulted in the development of (1) lists of fire resistive prestressed concrete building components, and (2) procedures for determining the fire endurance of prestressed concrete members by calculation.

9.3.5.1 Fire Tests of Flexural Elements

Reports of a number of tests sponsored by the Precast/Prestressed Concrete Institute (PCI) have been issued by Underwriters Laboratories, Inc (UL). Most of the reports have been reprinted by PCI, and the results of the tests are the basis for UL's listings and specifications for non-proprietary products such as double tee and single tee floors and roofs, wet cast hollow-core and solid slabs, and prestressed concrete beams.

The Portland Cement Association (PCA) conducted many fire tests of prestressed concrete assemblies. PCA's unique furnaces have made it possible to study in depth the effects of support conditions. Four series of tests dealt with simply supported slabs and beams; two series dealt with continuous slabs and beams; and one major series dealt with the effects of restrained thermal expansion on the behavior during fire of prestressed concrete floors and roofs. PCA has also conducted a number of miscellaneous fire tests of prestressed and reinforced concrete assemblies. Test results that have been published as Research and Development Bulletins are available from PCA. [4,5]

9.3.5.2 Fire Tests of Walls and Columns

A test was conducted by Underwriters Laboratories, Inc., on a double tee wall assembly for research purposes in which fire was applied to the flat surface of the 1½ in. thick flange. A gravity load of about 10 kips/ft was applied at the top of the wall. The wall withstood a 2-hour fire and a subsequent hose stream test followed by a load test with the design load doubled. No distress was observed. Because the flange was only 1½ in. thick, the heat transmission requirement was exceeded for most of the test. By providing adequate flange thickness or insulation, the heat transmission requirement would have been met in addition to the structural requirement.

Fire tests of reinforced concrete columns have been conducted by PCA and the National Research Council of Canada. While no tests have been conducted for prestressed concrete columns, results from these tests are considered to be equally applicable to prestressed concrete columns with adjustment made for the difference in thermal properties between mild reinforcing steel and prestressing strand as may be appropriate.

9.3.6 Designing for Heat Transmission

As noted in Section 9.3.4.2, ASTM E 119 imposes heat transmission criteria for floor, roof, and wall assemblies. Thus, floors, roofs, or walls requiring a fire-resistance rating must satisfy the heat transmission requirements as well as the various structural criteria. The heat transmission fire endurance of a concrete assembly is essentially the same whether the assembly is tested as a floor (oriented horizontally) or as a wall (tested vertically). Because of this, and unless otherwise noted, the information which follows is applicable to floors, roofs, or walls.

9.3.6.1 Single Course Slabs or Wall Panels

For concrete slabs or wall panels, the temperature rise of the unexposed surface depends mainly on the thickness and aggregate type of the concrete. Other less important factors include unit weight, moisture condition, air content, and maximum aggregate size. Within the usual ranges, water-cementitious materials ratio, strength, and age have only insignificant effects.

Figure 9.3.6.1 Fire endurance (heat transmission) of concrete slabs or panels

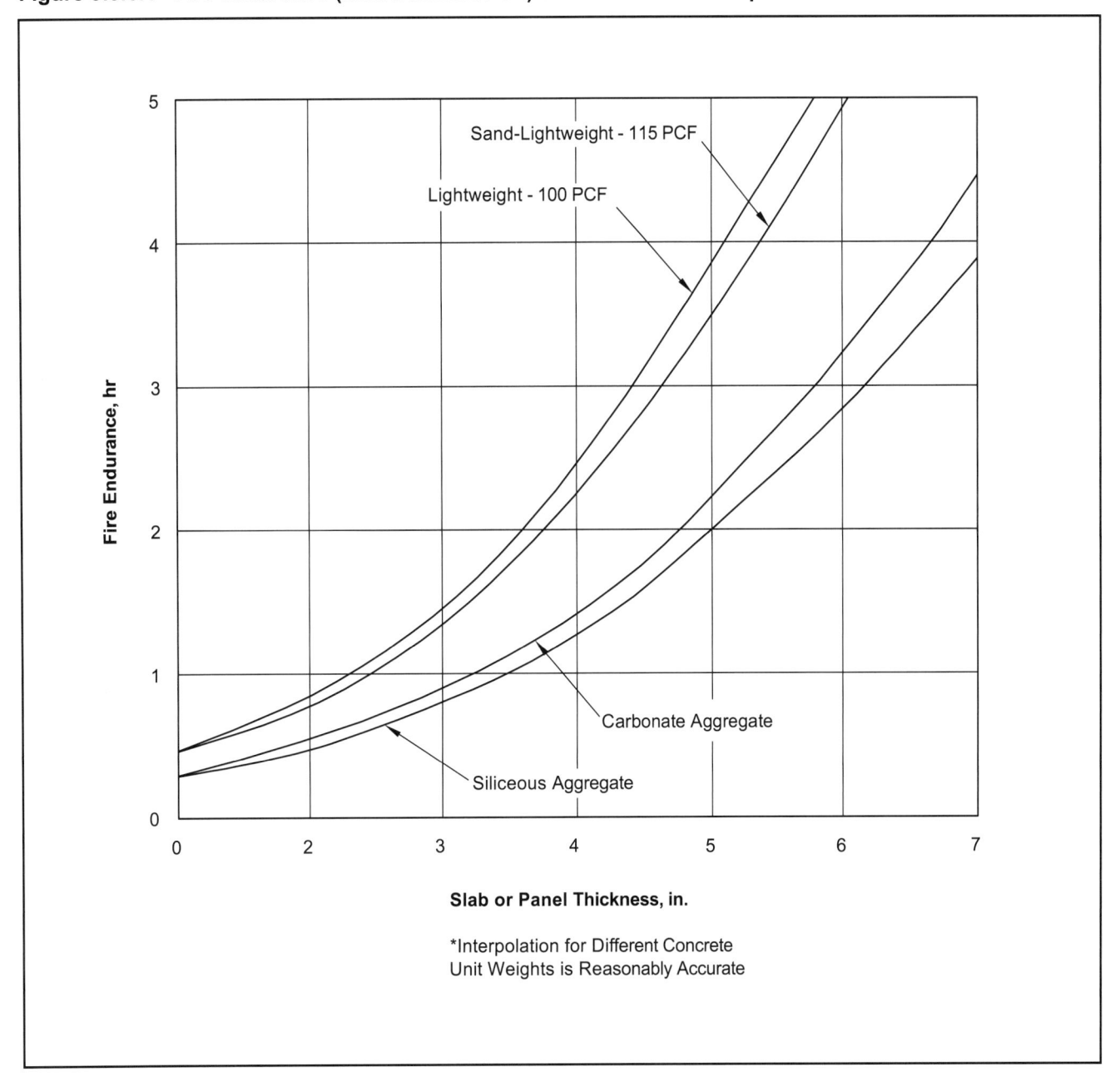

*Interpolation for Different Concrete Unit Weights is Reasonably Accurate

Table 9.3.6.1 Thickness of concrete slabs or wall panels faced with ⅝ in. Type X gypsum wallboard to provide fire endurances of 2 and 3 hours

	Thickness (in.) of concrete panel for fire endurance of			
	With no air space		With 6 in. air space	
Aggregate	2 hr	3 hr	2 hr	3 hr
Sand-lightweight	2.5	3.6	2.0	2.5
Carbonate	2.8	4.0	2.0	2.7
Siliceous	2.9	4.2	2.0	2.8

Figure 9.3.6.2 Cross sections of ribbed wall panels

(Neglect Hatched Area in Calculation of Equivalent Thickness)
(a)   (b)

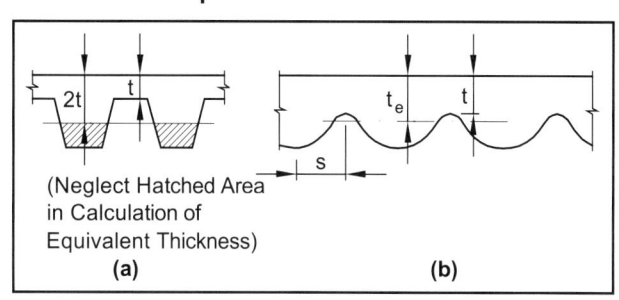

Figure 9.3.6.1 shows the fire endurance (heat transmission) of concrete slabs as influenced by aggregate types and thickness. For a hollow-core slab, this thickness may be obtained by dividing the net cross-sectional area by its width. The curves represent air-entrained concrete made with air-dry aggregates having a nominal maximum size of ¾ in. and fire tested when the concrete was at the standard moisture condition (75% R.H. at mid-depth). On the graph, concrete aggregates are designated as lightweight, sand-lightweight, carbonate, or siliceous. Lightweight aggregates include expanded clay, shale, and slate which produce concretes having unit weights of about 95 to 105 pcf without sand replacement. Lightweight concretes, in which sand is used as part or all of the fine aggregate and weigh no more than about 120 pcf, are designated as sand-lightweight. Carbonate aggregates include limestone and dolomite, i.e., those consisting mainly of calcium and/or magnesium carbonate. Siliceous aggregates include quartzite, granite, basalt, and most hard rocks other than limestone and dolomite.

9.3.6.2 Floors, Roofs or Walls Faced with Gypsum Wallboard

Table 9.3.6.1 shows the fire endurance of concrete slabs with ⅝ in. gypsum wallboard (Type X) for two cases: (1) a 6 in. air space between the wallboard and slab, and (2) no space between the wallboard and slab. Materials and techniques of attaching the wallboard should be similar to those used in the UL test on which the data are based.

9.3.6.3 Ribbed Panels

Heat transmission through a ribbed or corrugated panel is influenced by the thinnest portion of the panel and by the panel's "equivalent thickness," t_e. Here, equivalent thickness is defined as the net cross-sectional area of the panel divided by the width of the cross section. In calculating the net cross-sectional area of the panel, portions of ribs that project beyond twice the minimum thickness should be neglected, as shown in Figure 9.3.6.2(A).

The heat transmission fire endurance can be governed by either the thinnest section, or by the average thickness, or by a combination of the two. The following rule-of-thumb expressions appear to give a reasonable guide for selecting R:

If $t \leq \dfrac{s}{4}$, fire endurance R is governed by t and is equal to R_t

If $t \geq \dfrac{s}{2}$, fire endurance R is governed by t_e and is equal to R_{te}

If $\dfrac{s}{2} > t > \dfrac{s}{4}$:

$$R = R_t + \left(\dfrac{4t}{s} - 1\right)(R_{te} - R_t) \quad \text{(Eq. 9.3.6.1)}$$

where:
t = minimum thickness
t_e = equivalent thickness of panel
s = rib spacing

and where R is the fire endurance of a concrete panel and subscripts t and t_e relate the corresponding R-values to concrete slab thicknesses t and t_e, respectively.

These expressions apply to ribbed and corrugated panels, but for panels with widely spaced grooves or rustications they give excessively low results. Consequently, engineering judgment must be used when applying the above expressions.

9.3.6.4 Multi-Course Assemblies

Floors and roofs often consist of concrete base slabs with overlays or undercoatings of other types of concrete or insulating materials. In addition, roofs generally have built-up roofing.

If the fire endurance of the individual courses is known, the fire endurance of the composite assembly can be estimated from the formula:

$$R^{0.59} = R_1^{0.59} + R_2^{0.59} + ... + R_n^{0.59} \quad \text{(Eq. 9.3.6.2)}$$

Example 9.3.6.1
Fire Endurance of a Ribbed Panel

Given:
The section of a wall panel shown. The panel is made of sand-lightweight concrete (115 pcf).

Problem:
Estimate the fire endurance.

Solution:
$H = 5.5$ in. $< 2t$, therefore, full section can be considered:
$t_e = [5.5(12) - 1.5(4 + 15)]/12 = 4.8$ in.
$t = 4$ in., $\frac{s}{2} = 6$ in., $\frac{s}{4} = 3$ in.

Therefore, $s/2 > 4 > s/4$

From Eq. 9.3.6.1:
$$R = R_t + \left(\frac{4t}{s} - 1\right)(R_{te} - R_t) \quad \text{(Eq. 9.3.6.1)}$$

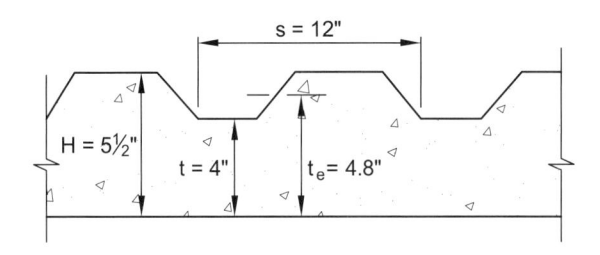

From Figure 9.3.6.1:
R_t = fire endurance of 4 in. sand-lightweight panel = 135 min
R_{te} = fire endurance of 4.8 in. sand-lightweight panel = 193 min

$$R = 135 + \left[\left(\frac{4(4)}{12} - 1\right)(193 - 135)\right] = 154 \text{ min}$$

where:
R = fire endurance of composite assembly in minutes
R_1, R_2, R_n = fire endurances of individual courses in minutes

Figure 9.3.6.3 gives R-values raised to the 0.59 power for insulating materials (in the table) and for concrete of various types and thicknesses (in the graph). Table 9.3.6.2 gives R-values which can be used in this equation for certain insulating materials. For heat transmission, three-ply built-up roofing contributes 10 min. to the fire endurance. Either set of data can be used in solving Eq. 9.3.6.2 as shown in Example 9.3.6.2.

Eq. 9.3.6.2 has certain shortcomings in that it does not account for the location of the individual courses relative to the fired surface. Also, it is not possible to directly obtain the fire endurances of many insulating materials. Nevertheless, in a series of tests, the formula estimated the fire endurances within about 10% for most assemblies.

A report on two-course floors and roofs [6] gives results of many fire tests. The report also shows graphically the fire endurances of assemblies consisting of various thicknesses of two materials. Tables 9.3.6.3 through 9.3.6.5, which are based on test results, can be used to estimate the required thicknesses of two-course materials for various fire endurances.

Table 9.3.6.2   Values of R of various insulating materials for use in Eq. 9.3.6.2

Roof insulation material	Thickness (in.)	R (min)
Cellular plastic	>1	5
Glass fiber board	¾	11
Glass fiber board	1½	35
Foam glass	2	55
Mineral board	1	19
Mineral board	2	62
Mineral board	3	123

Note: Some of these materials may be used only where combustible construction is permitted.

Figure 9.3.6.3   Design aid for use in solving Eq. 9.3.6.2

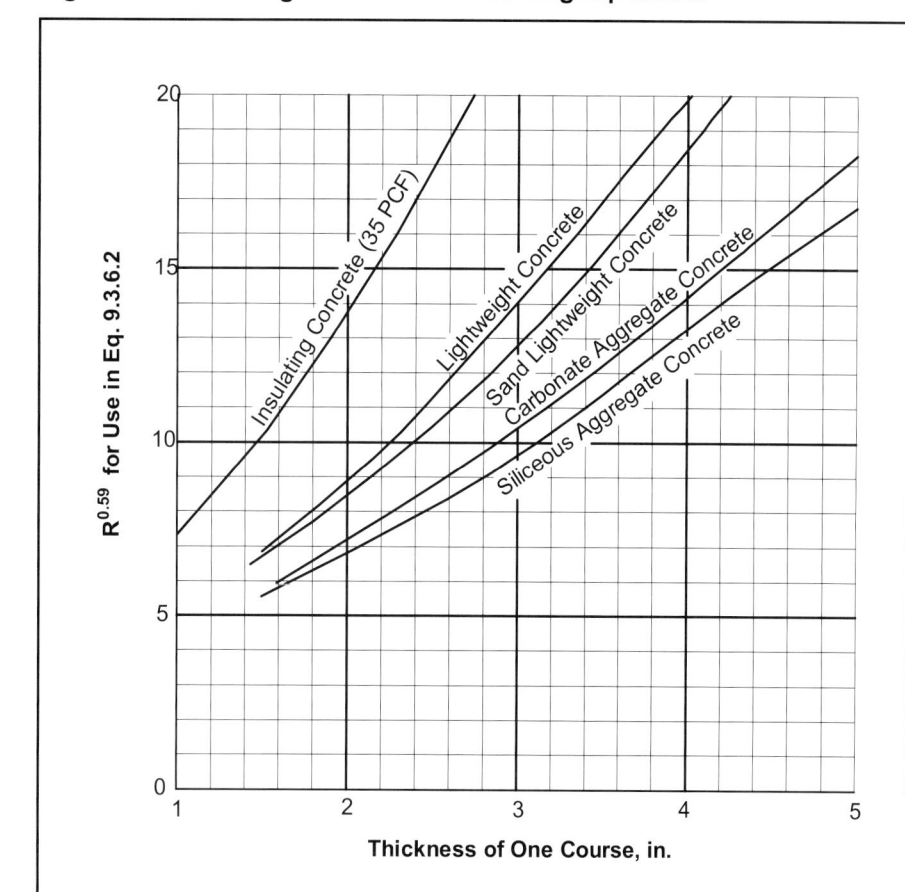

R, Minutes	$R^{0.59}$
60	11.20
120	16.85
180	21.41
240	25.37

Material	$R^{0.59}$
Cellular Plastic (1 in. or Thicker)	2.57
¾ in. Glass Fiber Board	4.03
1½ in. Glass Fiber Board	8.57
Continuous Air Space	3.33
Two Continuous Air Spaces	6.67
2 in. Foam Glass	10.61

Example 9.3.6.2
Fire Endurance of an Assembly

Problem:
Determine the fire endurance of a slab consisting of a 2 in. base slab of siliceous aggregate concrete with a 2½ in. topping of sand-lightweight concrete (115 pcf).

Solution:
From Figure 9.3.6.1, the fire endurances of a 2 in. thick slab of siliceous aggregate concrete and 2½ in. of sand-lightweight aggregate concrete are 25 min and 54 min, respectively.

$R^{0.59} = (25)^{0.59} + (54)^{0.59}$
$R^{0.59} = 6.68 + 10.52 = 17.20$
$R = (17.20)^{1/0.59} = 124$ min = 2 hr 4 min

Alternatively, from Figure 9.3.6.3:
$R^{0.59} = 6.7 + 10.4 = 17.1$

By linear interpolation from table in Figure 9.3.6.3:
$R = 120 + \dfrac{0.25}{4.56}(60) = 123.3$
$R = 123$ min = 2 hr 3 min

Table 9.3.6.3 Thickness of two-course roof assemblies consisting of concrete[a] slabs with insulating concrete overlays[b]

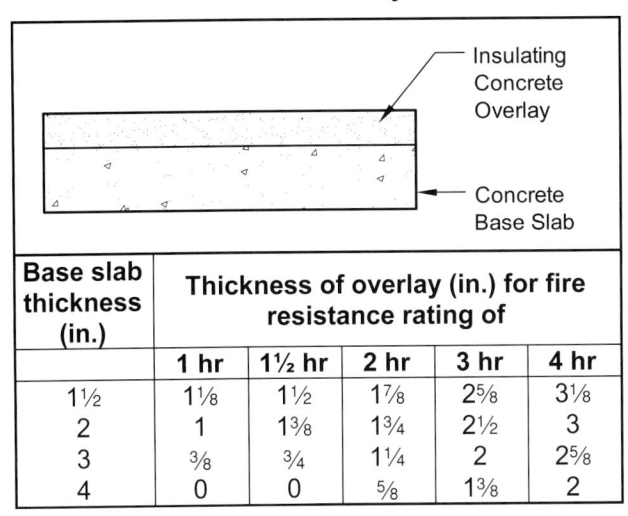

Base slab thickness (in.)	Thickness of overlay (in.) for fire resistance rating of				
	1 hr	1½ hr	2 hr	3 hr	4 hr
1½	1⅛	1½	1⅞	2⅝	3⅛
2	1	1⅜	1¾	2½	3
3	⅜	¾	1¼	2	2⅝
4	0	0	⅝	1⅜	2

a. Values shown are for siliceous aggregate concrete, and are conservative for other concretes.
b. Insulating concrete having a dry density less than 35 pcf.

9.3.6.5 Sandwich Panels

Some wall panels are made by sandwiching an insulating material between two face slabs of concrete (see Section 9.4).

The IBC requires that where noncombustible construction is specified, combustible elements in walls are limited to thermal and sound insulation having a flame spread index of 25 or less. When the insulation is sandwiched between two layers of noncombustible material such as concrete, the maximum flame spread index allowed is 100, except that it shall not exceed 75 for foam plastic insulation. Data on flame spread classifications are available from insulation manufacturers.

When insulation is not installed in this manner, it is required to have a flame spread of not more than 25. Data on flame spread classification are available from insulation manufacturers.

A fire test was conducted of one such panel that consisted of a 2 in. base slab of carbonate aggregate concrete, a 1 in. layer of cellular polystyrene insulation, and a 2 in. face slab of carbonate aggregate concrete. The resulting fire endurance was 2 hr 00 min. From Eq. 9.3.6.2, the contribution of the 1 in. layer of polystyrene was calculated to be 5 min.

It is likely that the comparable R value for a 1 in. layer of cellular polyurethane would be somewhat greater than that for a 1 in. layer of cellular polystyrene, but test values are not available. Until more definitive data are obtained, it is suggested that 5 min. be used as the value for R for any layer of cellular plastic 1 in. or greater.

It should be noted that the cellular plastics melt and are consumed at about 400 to 600°F. Thus, additional thickness or changes in composition probably have only a minor effect on the fire endurance of sandwich panels. The danger of toxic fumes caused by burning cellular plastics is practically eliminated when the plastics are completely encased within concrete sandwich panels. [7]

Table 9.3.6.6 lists fire endurances of sandwich panels with either cellular plastic, glass fiber board, or insulating concrete used as the insulating material. The fire resistance values were obtained from Eq. 9.3.6.2.

Table 9.3.6.4 Thickness of spray-applied insulation on fire-exposed surface of concrete[a] slabs or panels to resist transfer of heat through the assemblies

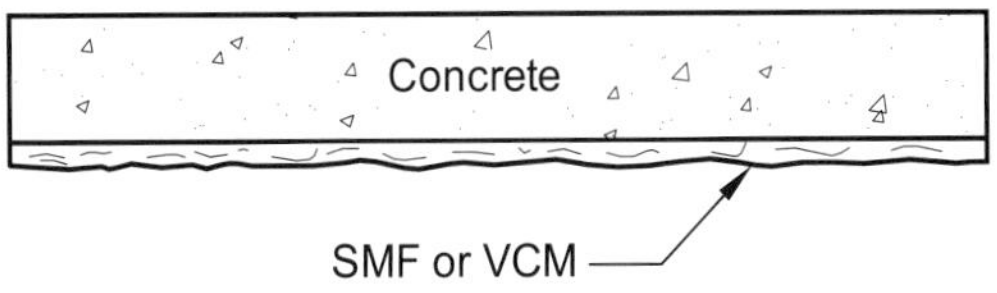

Slab equivalent thickness (in.)	Type of insulation[b]	Thickness (in.) for fire-resistance rating of				
		1 hr	1½ hr	2 hr	3 hr	4 hr
1½	SMF	½	¾	1⅛	N.A.	N.A.
2	SMF	⅜	⅝	⅞	1⅜	N.A.
2½	SMF	¼	½	⅝	1⅛	1⅝
3	SMF	⅛	¼	½	⅞	1⅜
4	SMF	0	0	¼	⅝	1
1½	VCM	½	¾	1⅛	1¾	N.A.
2	VCM	⅜	⅝	⅞	1⅜	1¾
2½	VCM	¼	½	¾	1¼	1⅝
3	VCM	⅛	⅜	⅝	⅞	1⅜
4	VCM	0	0	¼	⅝	⅞

a. Values shown are for siliceous aggregate concrete, and are conservative for other concretes.
b. SMF = Sprayed mineral fiber consists of refined mineral fibers with inorganic binders and water added during the spraying operation. The density of the oven-dry material should be at least 13 pcf.
VCM = Vermiculite cementitious material consists of expanded vermiculite with inorganic binders and water. The density of the oven-dry material should be at least 14 pcf.
N.A. = Not applicable.

Table 9.3.6.5 Thickness of roof assemblies consisting of concrete[a] slabs with insulation and built-up roofing

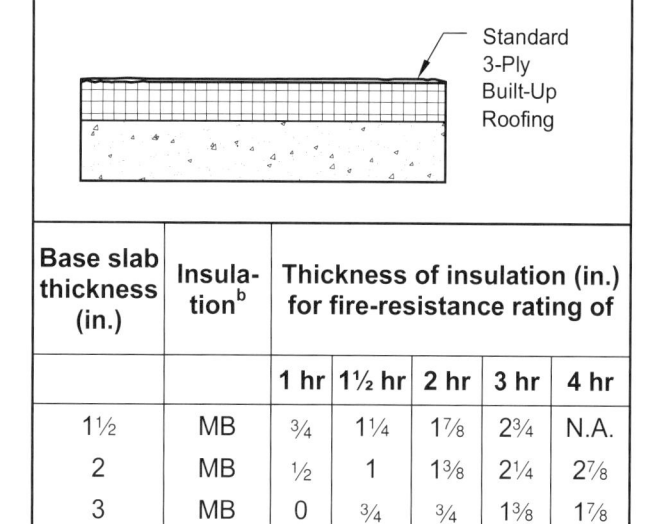

Base slab thickness (in.)	Insulation[b]	Thickness of insulation (in.) for fire-resistance rating of				
		1 hr	1½ hr	2 hr	3 hr	4 hr
1½	MB	¾	1¼	1⅞	2¾	N.A.
2	MB	½	1	1⅜	2¼	2⅞
3	MB	0	¾	¾	1⅜	1⅞
4	MB	0	0	¼	¾	1¼
1½	GFB	⅝	1⅜	2	N.A.	N.A.
2	GFB	¼	⅞	1½	2⅞	N.A.
3	GFB	0	⅜	¾	1½	2⅛
4	GFB	0	0	¼	¾	1¼

a. Values shown are for siliceous aggregate concrete, and are conservative for other concretes.
b. MB = Mineral board insulation composed of spherical cellular beads of expanded aggregate and fibers formed into rigid flat rectangular units with an integral waterproofing treatment.
GFB = Glass fiber board fibrous glass roof insulation consisting of inorganic glass fibers formed into rigid boards using a binder. The board has a top surface faced with glass fiber reinforced with asphalt and kraft.
N.A. = Not applicable.

9.3.6.6 Treatment of Joints Between Wall Panels

Joints between wall panels should be detailed so that passage of flame or hot gases is prevented, and transmission of heat does not exceed the limits specified in ASTM E 119. Concrete wall panels expand when heated, so the joints tend to close during fire exposure. Non-combustible materials that are flexible, such as ceramic fiber blankets, provide thermal, flame, and smoke barriers, and, when used in conjunction with caulking materials can provide the necessary weather-tightness while permitting normal volume change movements. Joints that do not move can be filled with mortar. For a more detailed discussion and additional information, refer to PCI MNL-124. [1]

The IBC addresses joints in exterior walls in various sections of the code:

Table 9.3.6.6 Fire endurance of precast concrete sandwich walls (calculated, based on Eq. 9.3.6.2)

Outside and inside wythes	Insulation	Fire endurance hr:min
1½ in. Sil	1 in. CP	1:23
1½ in. Carb	1 in. CP	1:23
1½ in. SLW	1 in. CP	1:45
2 in. Sil	1 in. CP	1:50
2 in. Carb	1 in. CP	2:00
2 in. SLW	1 in. CP	2:32
3 in. Sil	1 in. CP	3:07
1½ in. Sil	¾ in. GFB	1:39
2 in. Sil	¾ in. GFB	2:07
2 in. SLW	¾ in. GFB	2:52
1½ in. Sil	1½ in. GFB	2:35
2 in. Sil	1½ in. GFB	3:08
2 in. SLW	1½ in. GFB	4:00
1½ in. Sil	1 in. IC	2:12
1½ in. SLW	1 in. IC	2:39
2 in. Carb	1 in. IC	2:56
2 in. SLW	1 in. IC	3:33
1½ in. Sil	1½ in. IC	2:54
1½ in. SLW	1½ in. IC	3:24
2 in. Sil	2 in. IC	4:25
1½ in. SLW	2 in. IC	4:19

Carb = carbonate aggregate concrete
Sil = siliceous aggregate concrete
SLW = sand-lightweight concrete (115 pcf maximum)
CP = cellular plastic (polystyrene or polyurethane)
IC = lightweight insulating concrete (35 pcf maximum)
GFB = glass fiber board

a. In Section 713, joints are specified to have the same fire-resistance rating as the wall. Walls that are permitted to have unprotected openings are noted as an exception to this requirement.
b. Table 704.8 is used to determine if unprotected openings are allowed as well as designating the percentage of unprotected openings allowed.
c. Section 704.13 also addresses joints and has the same exception as Section 713.
d. Section 721.2.1.3 addresses joints in precast walls. This section requires that unprotected joints be included as openings in the calculation of opening percentages for comparison to the allowed opening percentage of Table 704.8.

Where no openings are permitted, the fire resistance required for the wall should be provided at the joints.

Table 9.3.6.7 is based on results of fire tests of panels with butt joints. [8] The tabulated values apply to one-stage butt joints and are conservative for two-stage and ship-lap joints. [1]

Joints between adjacent precast floor or roof elements may be ignored in calculating the slab thickness provided that a concrete topping at least 1½ in. thick is used. Where no concrete topping is used, joints should be grouted to a depth of at least one-third the slab thickness at the joint, or the joints made fire-resistive in a manner acceptable to the authority having jurisdiction. No joint treatment is required in pre-topped double tee parking structures.

9.3.7 Fire Endurance by Rational Design

It was noted above that many fire tests and related research studies have been directed toward an understanding of the structural behavior of prestressed concrete subjected to fire. The information gained from that work has led to the development of calculation procedures that can be used in lieu of fire tests. The purpose of this section is to introduce these calculation procedures.

Because the method of support is the most important factor affecting structural behavior of flexural elements during a fire, the discussion that follows deals with three conditions of support: simply supported members, continuous slabs and beams, and members which are restrained from thermal expansion. For additional examples and more detailed information, refer to PCI MNL-124. [1]

The fire endurance of concrete walls, as determined by fire tests is normally governed by the ASTM criteria for temperature rise of the unexposed surface rather than by structural behavior during fire tests. This is probably due to the low level of stresses, even in concrete bearing walls, and the fact that reinforcement generally does not perform a primary structural function. In most cases, the amount of cover protection required by code exceeds that required for fire protection so there is, in effect, reserve structural fire endurance within the concrete wall.

9.3.7.1 Simply Supported Members

To understand the effects of fire, let us assume that a simply supported prestressed concrete slab is exposed to fire from below, that the ends of the slab are free to rotate, and that expansion can occur without restriction. Also, assume that the

Figure 9.3.7.1 Moment diagrams for simply supported beam or slab.

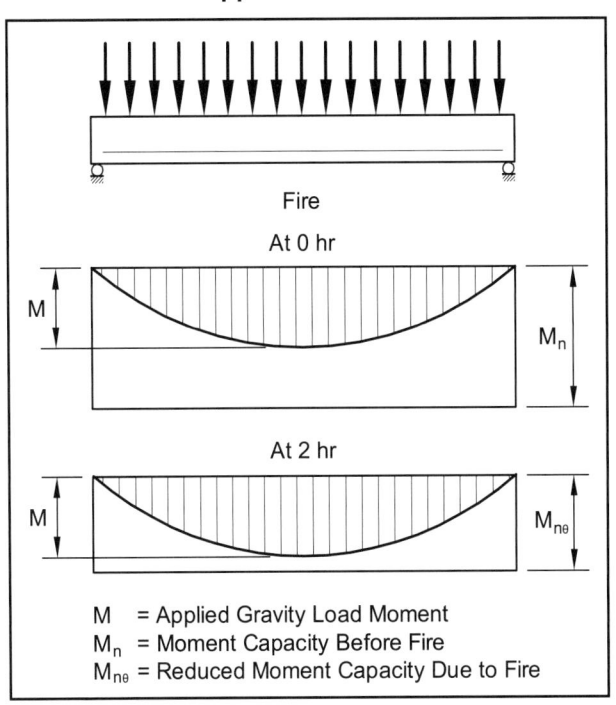

M = Applied Gravity Load Moment
M_n = Moment Capacity Before Fire
$M_{n\theta}$ = Reduced Moment Capacity Due to Fire

Table 9.3.6.7 Protection of joints between wall panels utilizing ceramic fiber felt

Panel equivalent thickness[b] (in.)	Thickness of ceramic fiber felt (in.) required for fire resistance ratings and joint widths[a] shown							
	Joint width = ⅜ in.				Joint width = 1 in.			
	1 hr	2 hr	3 hr	4 hr	1 hr	2 hr	3 hr	4 hr
4	¼	N.A.	N.A.	N.A.	¾	N.A.	N.A.	N.A.
5	0	¾	N.A.	N.A.	½	2⅛	N.A.	N.A.
6	0	0	1⅛	N.A.	¼	1¼	3½	N.A.
7	0	0	0	1	¼	⅞	2	3¾

N.A. = Not applicable.
a. Interpolation may be used for joint width between ⅜ in. and 1 in. The tabulated values apply to one-stage butt joints and are conservative for two-stage and ship-lap joints.
b. Panel equivalent thicknesses are for carbonate concrete. For siliceous aggregate concrete change "4, 5, 6, and 7" to "4.3, 5.3, 6.5, and 7.5." For sand-lightweight concrete change "4, 5, 6, and 7" to "3.3, 4.1, 4.9, and 5.7."

Figure 9.3.7.2 Strength-temperature relationships for various steels

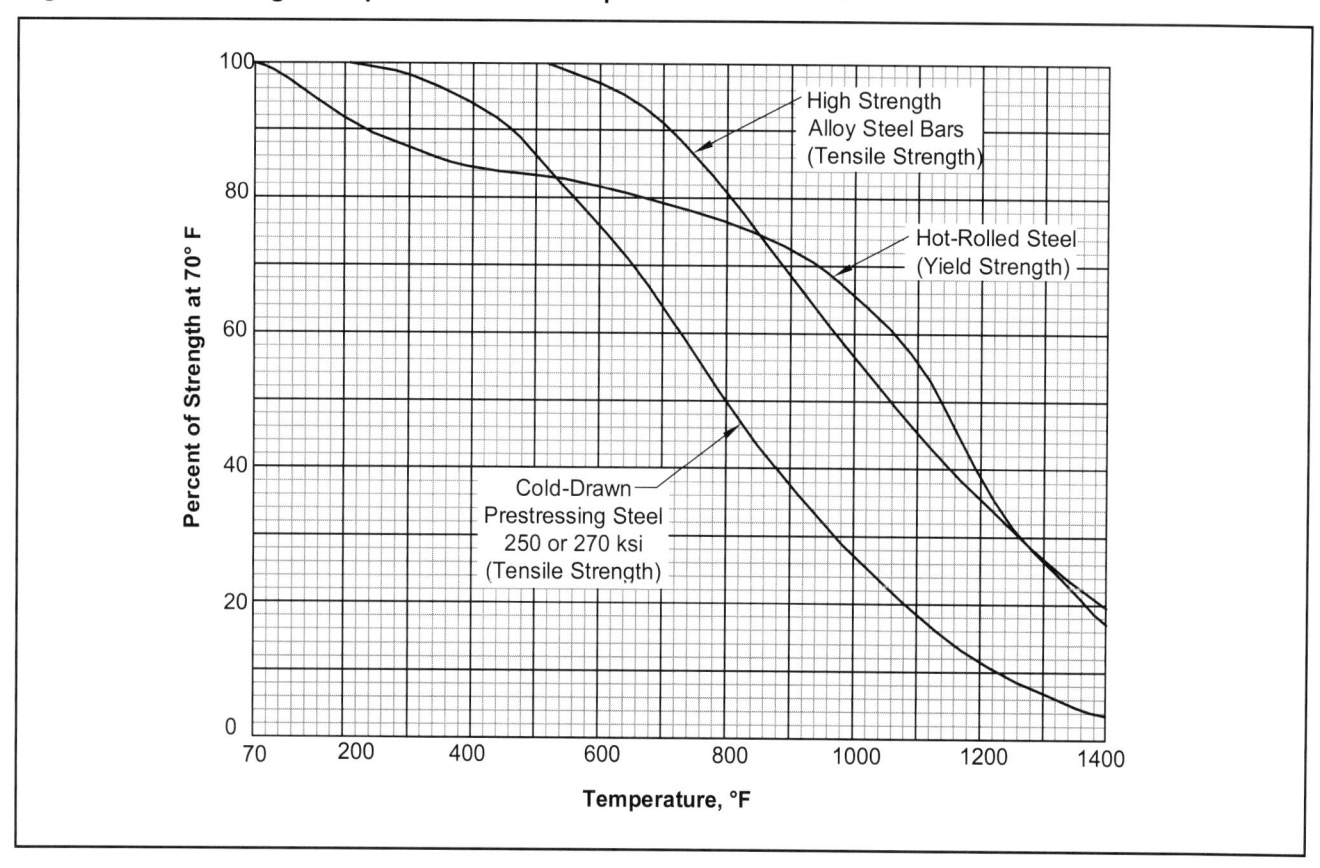

Figure 9.3.7.3 Compressive strength of concrete at high temperatures

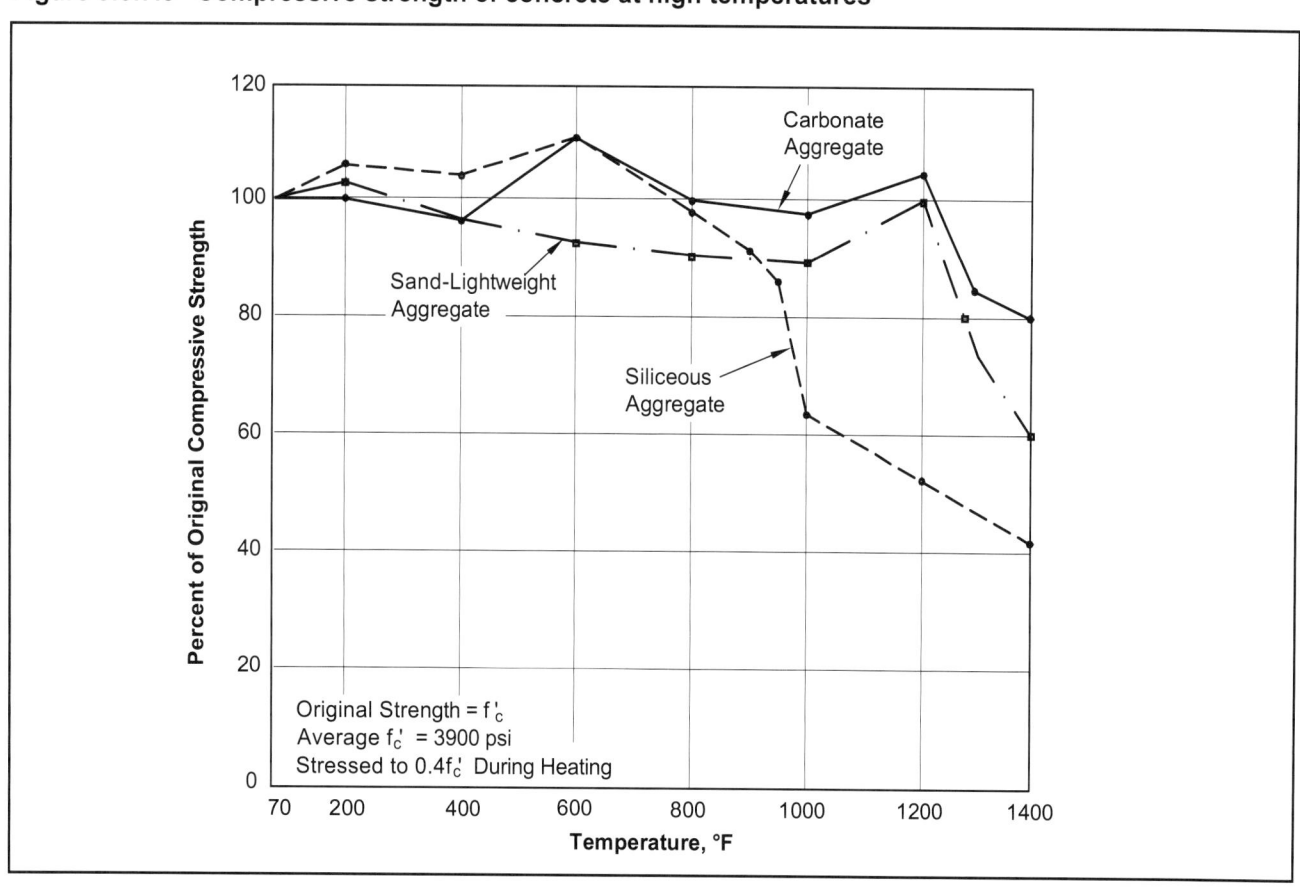

Figure 9.3.7.4 Temperature within concrete slabs or panels during fire tests — normal weight concrete

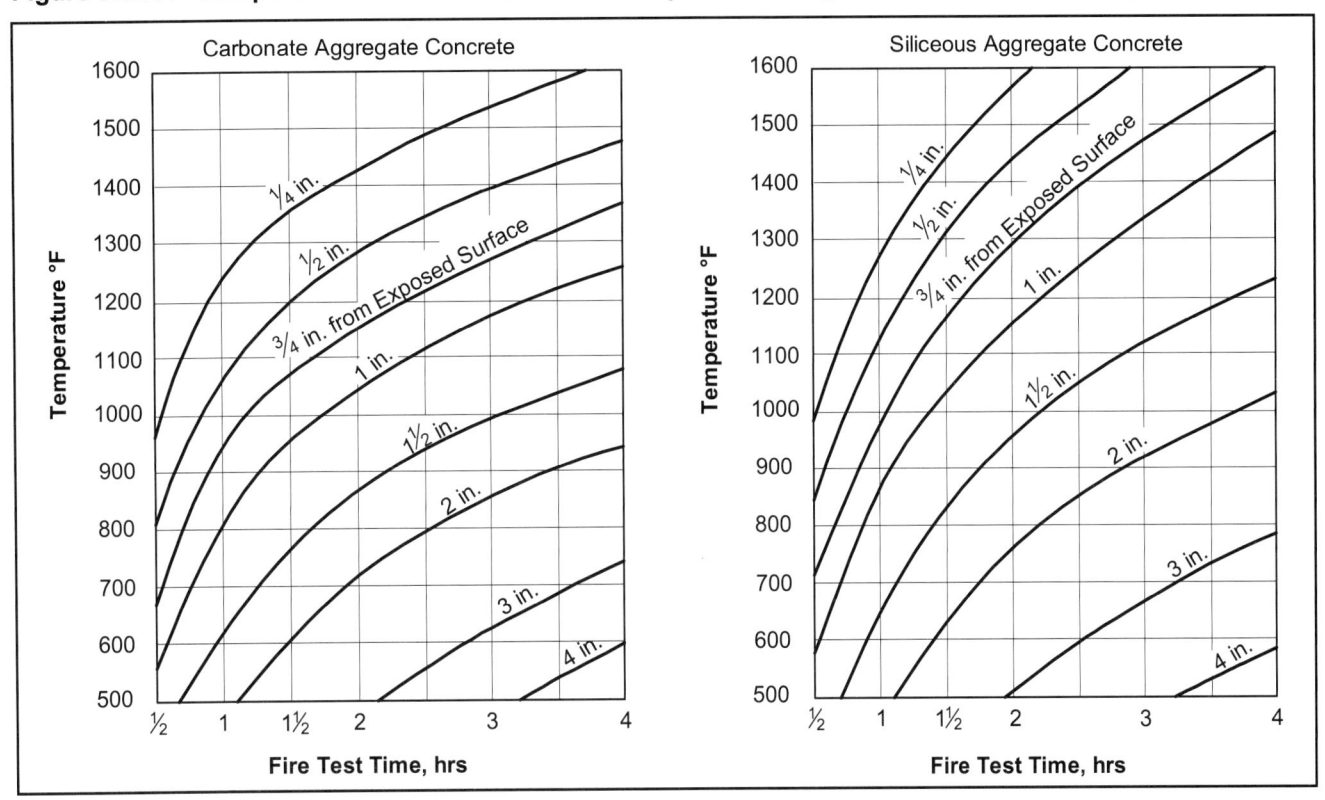

Figure 9.3.7.5 Temperatures within concrete slabs or panels during fire tests — sand-lightweight concrete

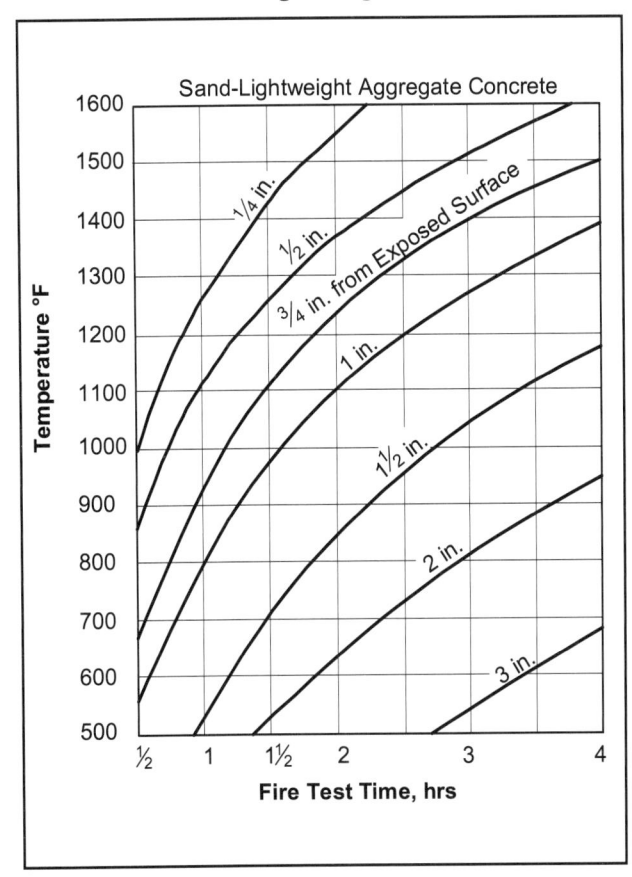

reinforcement consists of straight uncoated strands located near the bottom of the slab. With the underside of the slab exposed to fire, the bottom will expand more than the top causing the slab to deflect downward; also, the strength of the steel and concrete near the bottom will decrease as the temperature rises. When the strength of the steel diminishes to less than that required to support the slab, flexural collapse will occur. In essence, the applied moment remains practically constant during the fire exposure, but the resisting moment capacity is reduced as the steel strand weakens.

Figure 9.3.7.1 illustrates the behavior of a simply supported slab exposed to fire from beneath, as described above. Because strands are parallel to the axis of the slab, the design moment strength is constant throughout the length:

$$M_n = A_{ps}f_{ps}\left(d - \frac{a}{2}\right) \quad \text{(Eq. 9.3.7.1)}$$

f_{ps} can be determined from Figure 4.12.3 or Eq. 18-3 of ACI 318-02. [9]

If the slab is uniformly loaded, the moment diagram will be parabolic with a maximum value at midspan of:

$$M = \frac{w\ell^2}{8} \quad \text{(Eq. 9.3.7.2)}$$

where:
w = dead plus live load per unit of length, k/in.
ℓ = span length, in.

Figure 9.3.7.6 Temperatures on vertical centerline of stemmed units at 1 hr of exposure

Stemmed Units 1 hr

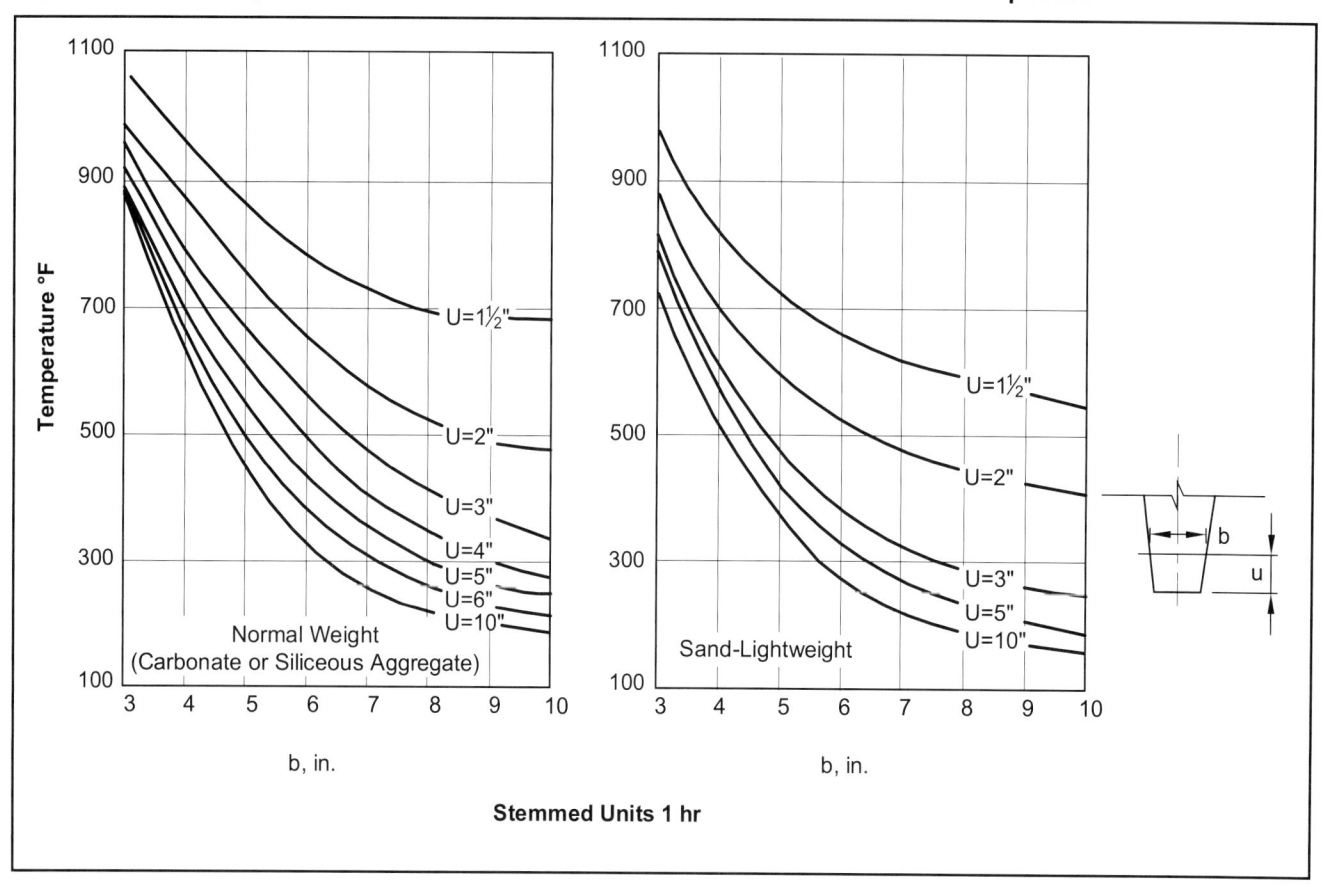

Figure 9.3.7.7 Temperatures on vertical centerline of stemmed units at 2 hr of exposure

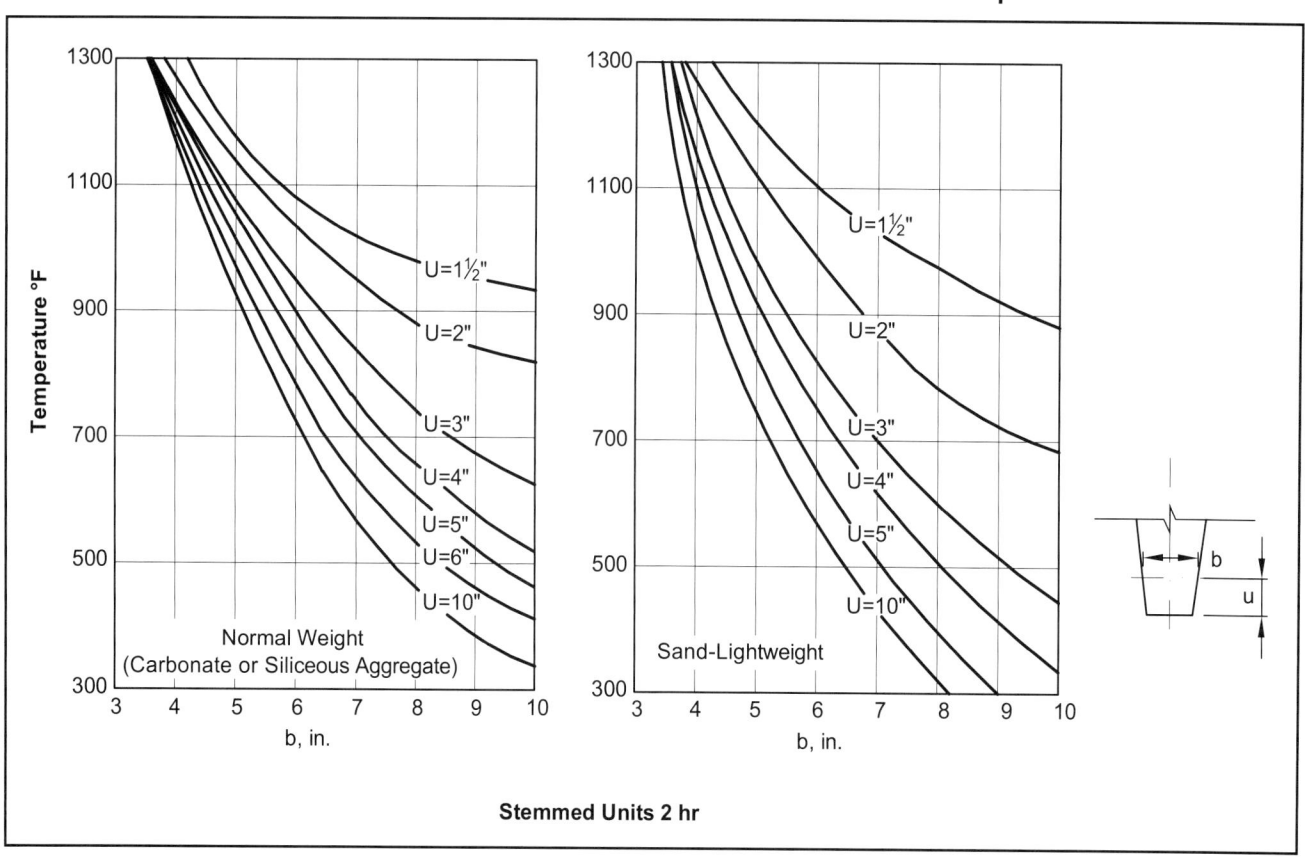

Stemmed Units 2 hr

Figure 9.3.7.8 Temperatures on vertical centerline of stemmed units at 3 hr of exposure

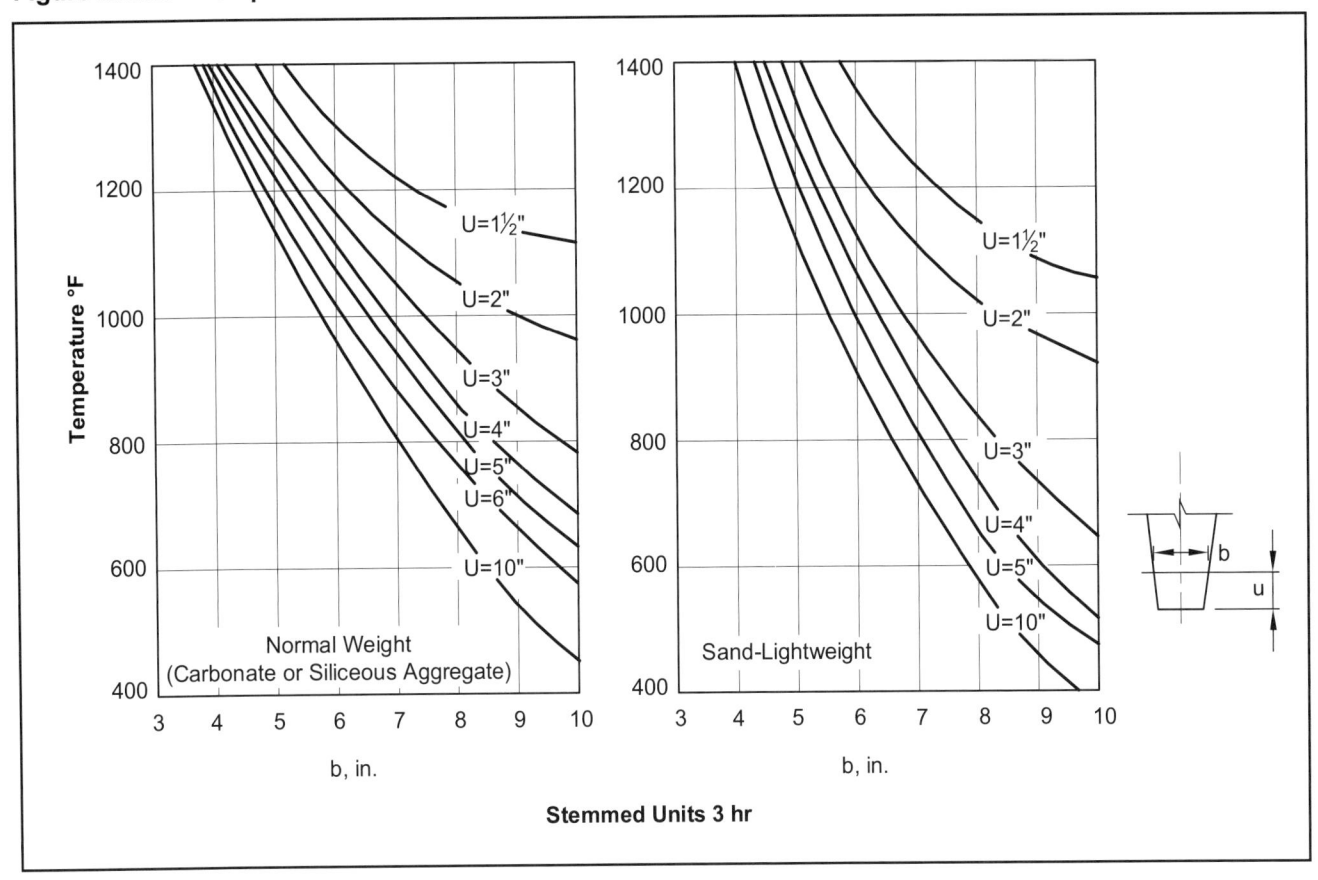

Figure 9.3.7.9 Temperatures on vertical centerline of stemmed units at 4 hr of exposure

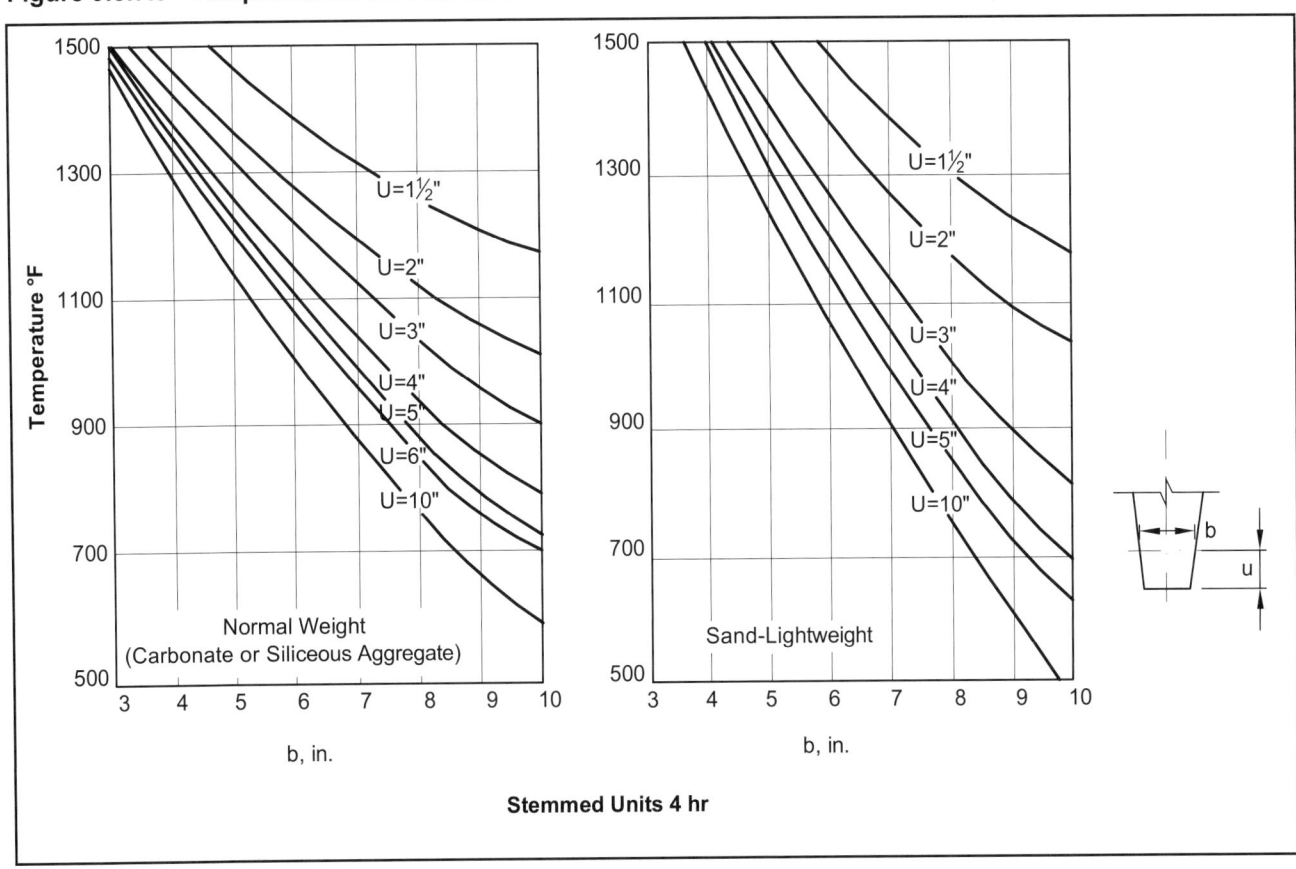

PCI Design Handbook/Sixth Edition

As the material strengths decrease with elevated temperatures, the remaining nominal strength becomes:

$$M_{n\theta} = A_{ps} f_{ps\theta} \left(d - \frac{a_\theta}{2} \right) \quad \text{(Eq. 9.3.7.3)}$$

in which θ signifies the effects of high temperatures. Note that A_{ps} and d are not affected, but f_{ps} is reduced. Similarly, a is reduced, but the concrete strength at the top of the slab, f'_c, is generally not reduced significantly because of its lower temperature.

Flexural failure can be assumed to occur when $M_{n\theta}$ is reduced to M. ACI load factors and strength reduction factors, ϕ, are not applied because a safety factor is included in the required ratings. [1] From this expression, it can be seen that the fire endurance depends on the applied loading and on the strength-temperature characteristics of the steel.

In turn, the duration of the fire before the "critical" steel temperature is reached depends on the protection afforded to the reinforcement.

To solve problems involving the above equations, it is necessary to utilize data on the strength-temperature relationships for steel and concrete, and information on temperature distributions within concrete members during fire exposures. Figure 9.3.7.2 shows strengths of certain steels at elevated temperatures, and Figure 9.3.7.3 shows similar data for concrete.

Data on temperature distribution in concrete slabs during fire tests are shown in Figures 9.3.7.4 and 9.3.7.5. These figures can also be used for beams wider than about 10 in. An "effective u," $\bar{u}$, is used, which is the average of the distances between the centers of the individual strands or bars and the nearest fire-exposed surface. The values for corner strands or bars are reduced one-half to account for the exposure from two sides. The procedure does not apply to bundled bars or strands. Data on temperature distribution for stemmed members during fire tests are shown in Figures 9.3.7.6 through 9.3.7.9.

9.3.7.2 Continuous Members

Continuous members undergo changes in stresses when subjected to fire. These stresses result from temperature gradients within the structural members, or changes in strength of the materials at high temperatures, or both.

Figure 9.3.7.10 shows a two-span continuous beam whose underside is exposed to fire. The bottom of the beam becomes hotter than the top and tends to expand more than the top. This differential temperature effect causes rotation that increases the reaction at the interior support. This action results in a redistribution of moments, i.e., the negative moment at the interior support increases while the positive moments decrease.

During a fire, the negative moment reinforcement (Figure 9.3.7.10) remains cooler than the positive moment reinforcement because it is better protected from the fire. In addition, the redistribution that occurs is sufficient to cause yielding of the negative moment reinforcement. Thus, a relatively large increase in negative moment can be accommodated throughout the test. The resulting decrease in positive moment means that the positive moment reinforcement can be heated to a higher temperature before failure will occur. Therefore, the fire endurance of a continuous concrete beam is generally significantly longer than that of a simply supported beam having the same cover and the same applied loads.

It is possible to design the reinforcement in a continuous beam or slab for a particular fire endurance period. From Figure 9.3.7.10, the beam can be expected to collapse when the positive moment capacity, $M_{n\theta}^+$, is reduced to the value of the maximum redistributed positive moment at a distance x_1 from the outer support.

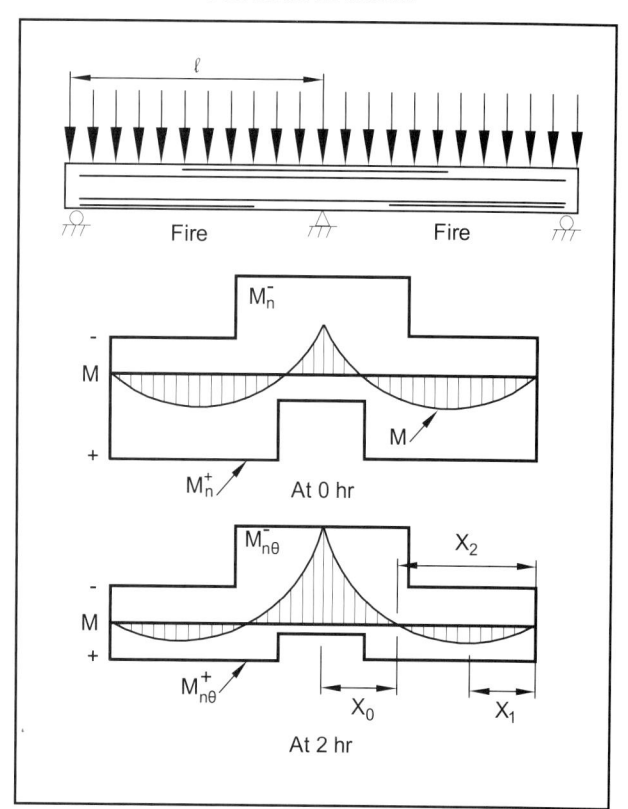

Figure 9.3.7.11 shows a uniformly loaded beam or slab continuous (or fixed) at one support and simply supported at the other. Also shown is the redistributed applied moment diagram at failure.

Figure 9.3.7.10 Moment diagram for a two-span continuous beam

Figure 9.3.7.11 Uniformly loaded member continuous (or fixed) at one support

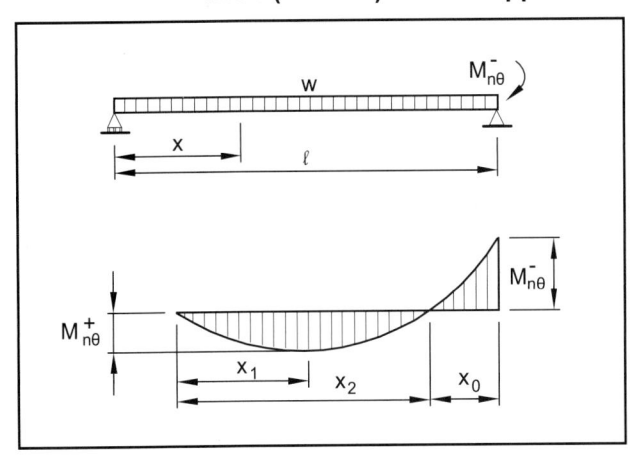

Figure 9.3.7.12 Uniformly loaded member continuous at supports

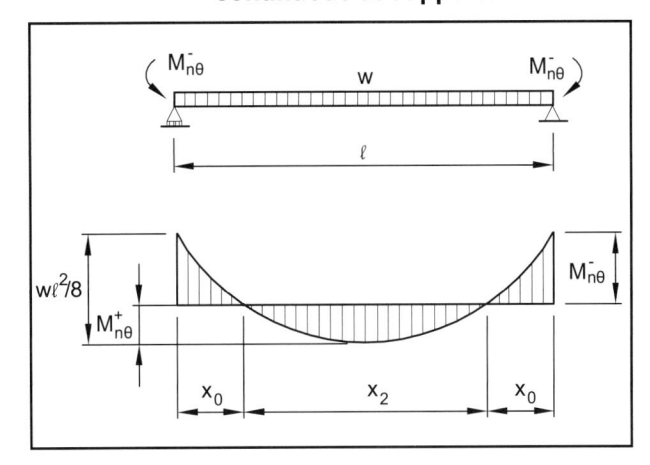

It can be shown that at the point of positive moment, x_1,

$$x_1 = \frac{\ell}{2} - \frac{M_{n\theta}^-}{w\ell} \quad \text{(Eq. 9.3.7.4)}$$

at $x = x_2$, $M_x = 0$ and $x_2 = 2x_1$

$$x_o = \frac{2M_{n\theta}^-}{w\ell} \quad \text{(Eq. 9.3.7.5)}$$

$$M_{n\theta}^- = \frac{w\ell^2}{2} \pm w\ell^2 \sqrt{\frac{2M_{n\theta}^+}{w\ell^2}} \quad \text{(Eq. 9.3.7.6)}$$

In most cases, redistribution of moments occurs early during the course of a fire and the negative moment reinforcement can be expected to yield before the negative moment capacity has been reduced by the effects of fire. In such cases, the length of x_o is increased, i.e., the inflection point moves toward the simple support. If the inflection point moves beyond the point where the bar stress cannot be developed in the negative moment reinforcement, sudden failure may result.

Figure 9.3.7.12 shows a symmetrical beam or slab in which the end moments are equal:

$$M_{n\theta}^- = \frac{w\ell^2}{8} - M_{n\theta}^+ \quad \text{(Eq. 9.3.7.7)}$$

$$\frac{wx_2^2}{8} = M_{n\theta}^+ \quad \text{(Eq. 9.3.7.8)}$$

$$x_2 = \sqrt{\frac{8M_{n\theta}^+}{w}} \quad \text{(Eq. 9.3.7.9)}$$

$$x_0 = \tfrac{1}{2}(\ell - x_2) = \frac{\ell}{2} - \frac{1}{2}\sqrt{\frac{8M_{n\theta}^+}{w}} \quad \text{(Eq. 9.3.7.10)}$$

To determine the maximum value of x_0, the value of w should be the minimum service load anticipated, and $(w\ell^2/8 - M_n^-)$ should be substituted for $M_{n\theta}^+$ in Eq. 9.3.7.10.

For any given fire endurance period, the value of $M_{n\theta}^+$ can be calculated by the procedures given above. Then the value of M_n^- can be calculated by the use of Eqs. 9.3.7.6 or 9.3.7.7 and the necessary lengths of the negative moment reinforcement can be determined from Eqs. 9.3.7.5 or 9.3.7.10. Use of these equations is illustrated in Example 9.3.7.1.

The amount of moment redistribution that can occur is dependent on the amount of negative moment reinforcement. Tests have clearly demonstrated that in most cases the negative moment reinforcement will yield, so the negative moment capacity is reached early during a fire test, regardless of the applied loading. The designer must exercise care to ensure that a secondary type of failure will not occur. To avoid a compression failure in the negative moment region, the amount of negative moment reinforcement should be small enough so that $\omega_\theta = (A_s f_{y\theta})/(b_\theta d_\theta f'_{c\theta})$ is less than 0.30, before and after reductions in f_y, b, d and f'_c are taken into account. Furthermore, the negative moment bars or welded wire reinforcement must be long enough to accommodate the complete redistributed moment and change in the inflection points. It should be noted that the worst condition occurs when the applied loading is smallest, such as the dead load plus partial or no live load. It is recommended that at least 20% of the maximum negative moment reinforcement extend throughout the span.

9.3.7.3 Members Restrained Against Thermal Expansion

If a fire occurs beneath a portion of a large concrete floor (or roof), the heated portion will tend to expand and push against the surrounding portion and creates restraint. In turn, the unheated portion exerts compressive forces on the heated portion and creates restraint.

Table 9.3.7.1(1) Cover thickness for reinforced concrete floor or roof slabs (in.)

Concrete Aggregate Type	Fire-Resistance Rating (hours)									
	Restrained					Unrestrained				
	1	1½	2	3	4	1	1½	2	3	4
Siliceous	¾	¾	¾	¾	¾	¾	¾	1	1¼	1⅝
Carbonate	¾	¾	¾	¾	¾	¾	¾	¾	1¼	1¼
Sand-lightweight or lightweight	¾	¾	¾	¾	¾	¾	¾	¾	1¼	1¼

Table 9.3.7.1(2) Cover thickness for prestressed concrete floor or roof slabs (in.)

Concrete Aggregate Type	Fire-Resistance Rating (hours)									
	Restrained					Unrestrained				
	1	1½	2	3	4	1	1½	2	3	4
Siliceous	¾	¾	¾	¾	¾	1⅛	1½	1¾	2⅜	2¾
Carbonate	¾	¾	¾	¾	¾	1	1⅜	1⅝	2⅛	2¼
Sand-lightweight or lightweight	¾	¾	¾	¾	¾	1	1⅜	1½	2	2¼

Table 9.3.7.1(3) Minimum cover for main reinforcing bars of reinforced concrete beams[c] (applicable to all types of structural concrete)

Restrained or Unrestrained[a]	Beam Width[b] (in.)	Fire-Resistance Rating (hours)				
		1	1½	2	3	4
Restrained	5	¾	¾	¾	1[a]	1¼[a]
	7	¾	¾	¾	¾	¾
	≥ 10	¾	¾	¾	¾	¾
Unrestrained	5	¾	1	1¼	—	—
	7	¾	¾	¾	1¾	3
	≥ 10	¾	¾	¾	1	1¾

a. Tabulated values for restrained assemblies apply to beams spaced more than 4 ft on centers. For restrained beams spaced 4 ft or less on centers, minimum cover of ¾ in. is adequate for ratings of 4 hours or less.
b. For beam widths between the tabulated values, the minimum cover thickness can be determined by direct interpolation.
c. The cover for an individual reinforcing bar is the minimum thickness of concrete between the surface of the bar and the fire-exposed surface of the beam. For beams in which several bars are used, the cover for corner bars used in the calculation shall be reduced to one-half of the actual value. The cover for an individual bar must be not less than one-half of the value given in Table 9.3.7.1(3) nor less than ¾ in.

Table 9.3.7.1(4) Minimum cover for prestressed concrete beams 8 in. or greater in width

Restrained or Unrestrained[a]	Concrete Aggregate Type	Beam Width[b] (in.)	Fire-Resistance Rating (hours)				
			1	1½	2	3	4
Restrained	Carbonate or Siliceous	8	1½	1½	1½	1¾[a]	2½[a]
	Carbonate or Siliceous	≥ 12	1½	1½	1½	1½	1⅞[a]
	Sand lightweight	8	1½	1½	1½	1½	2[a]
	Sand lightweight	≥ 12	1½	1½	1½	1½	1⅝[a]
Unrestrained	Carbonate or siliceous	8	1½	1¾	2½	5[c]	—
	Carbonate or siliceous	≥ 12	1½	1½	1⅞	2½	3
	Sand lightweight	8	1½	1½	2	3¼	—
	Sand lightweight	≥ 12	1½	1½	1⅝	2	2½

a. Tabulated values for restrained assemblies apply to beams spaced more than 4 ft on centers. For restrained beams spaced 4 ft or less on centers, minimum cover of ¾ in. is adequate for ratings of 4 hours or less.
b. For beam widths between 8 in. and 12 in., minimum cover thickness can be determined by direct interpolation.
c. Not practical for 8 in. wide beam but shown for purposes of interpolation.

Table 9.3.7.1(5) Minimum cover for prestressed concrete beams of all widths

Restrained or Unrestrained[a]	Concrete Aggregate Type	Beam Area[b] A (sq in.)	Fire-Resistance Rating (hours) 1	1½	2	3	4
Restrained	All	40 ≤ A ≤ 150	1½	1½	2	2½	—
	Carbonate or Siliceous	150 < A ≤ 300	1½	1½	1½	1¾	2½
		300 < A	1½	1½	1½	1½	2
	Sand lightweight	150 < A	1½	1½	1½	1½	2
Unrestrained	All	40 ≤ A ≤ 150	2	2½	—	—	—
	Carbonate or Siliceous	150 < A ≤ 300	1½	1¾	2½	—	—
		300 < A	1½	1½	2	3[c]	4[c]
	Sand lightweight	150 < A	1½	1½	2	3[c]	4[c]

a. Tabulated values for restrained assemblies apply to beams spaced more than 4 ft on centers. For restrained beams spaced 4 ft or less on centers, minimum cover of ¾ in. is adequate for ratings of 4 hours or less.
b. The cross-sectional area of a stem is permitted to include a portion of the area in the flange, provided the width of flange used in the calculation does not exceed three times the average width of the stem.
c. U-shaped or hooped stirrups spaced not to exceed the depth of the member and having a minimum cover of 1 in. shall be provided.

There are two commonly used methods for determining the fire endurance of restrained precast concrete flexural members. The simplest method is use of the reinforcement cover tables (restrained case) that are provided in most of the model building codes. Table 9.3.7.1(1–5) show the cover requirements of IBC 2003. [12] Heat transmission must also be checked as previously described. (The restraint condition does not affect heat transmission.)

The other method for determining the fire endurance of a restrained precast concrete flexural member is to determine the endurance of the member as if it is unrestrained (Section 9.3.7.1). That value is then doubled to obtain the restrained fire endurance. This method of determining fire endurance is not valid for assemblies with an unrestrained fire rating of less than one hour. Conversely stated, for a given required fire rating, the restrained member would be checked for one-half of the required rating (but not less than one hour) as an unrestrained member (Example 9.3.7.3). Full value of the heat transmission requirement must also be checked.

The rationale for this method is the ASTM E119 criteria for deriving unrestrained member ratings from restrained fire tests (see Section 9.3.4.1). This method is further supported by Tables 9.3.7.1(1–5) since, for all cases, the cover required for any particular restrained rating is equal to or less than the cover required for the unrestrained rating at half the time period.

Example 9.3.7.1
Fire Endurance for Hollow-Core Slab with Topping

Problem:
Design a floor using hollow-core slabs and topping for a 22 ft span for a 4-hr fire endurance. Service loads = 175 psf dead (including structure) and 150 psf live. Use 4 ft wide, 10 in. deep slabs with 2 in. topping, carbonate aggregate concrete. Continuity can be achieved at both ends. Use f'_c (precast) = 5000 psi, f_{pu} = 250 ksi, and f'_c (topping) = 3000 psi, eight – ½ in., 270 ksi strands at u = 1.75 in. Provide negative moment reinforcement needed for fire resistance.

Solution:
A_{ps} = 8(0.153) = 1.224 in.^{2}
u = 1.75 in.
d = 12 − 1.75 = 10.25 in.

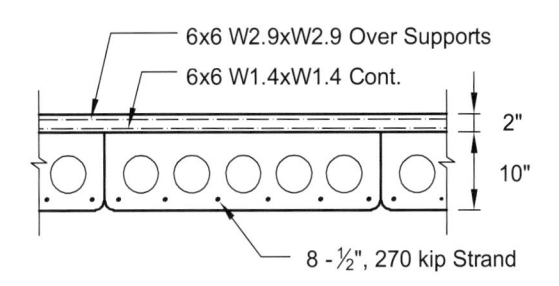

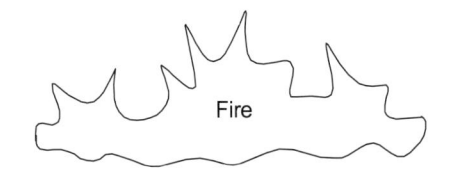

Example 9.3.7.1 (Cont.)
Fire Endurance for Hollow-Core Slab with Topping

From Figure 9.3.7.4: $\theta_s = 1010°F$
From Figure 9.3.7.2: $f_{pu\theta} = 0.24 f_{pu} = 60$ ksi
Use Figure 4.12.2. Note that values for K'_u in Figure 4.12.2 include $\phi = 0.9$. Since in the design for fire, $\phi = 1.0$, the value of K'_u must be divided by 0.9.

$$\omega_{pu\theta} = \frac{1.224(60)}{48(10.25)(3)} = 0.050$$

$$K'_u = \frac{131}{0.9} = 146$$

$$M^+_{n\theta} = \frac{K'_u b d^2}{12000} = \frac{146(48)(10.25)^2}{12000} = 61.4 \text{ kip-ft/unit}$$
$$= 15.3 \text{ kip-ft/ft}$$

For simply supported members:

$$M = \frac{w\ell^2}{8} = \frac{(175+150)(22)^2}{8(1000)} = 19.7 \text{ kip-ft/ft}$$

$$\text{Req'd } M^-_{n\theta} = \frac{w\ell^2}{8} - M_{n\theta} = 19.7 - 15.3 = 4.4 \text{ kip-ft/ft}$$

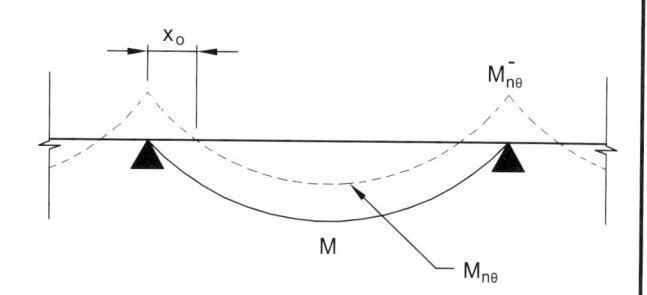

For topping reinforcement, assume $d - \frac{a_\theta}{2} = 10.25$ in., and $f_y = 60$ ksi

$$A^-_s = \frac{M^-_{n\theta}}{f_y(d - \frac{a_\theta}{2})} = \frac{4.4(12)}{60(10.25)} = 0.086 \text{ in.}^2/\text{ft}$$

Use 20% of required A_s throughout span:
Try 6x6–W1.4xW1.4 cont. plus 6x6–W2.9xW2.9 over supports.
A^-_s = 0.29 in.²/ft continuous
A^-_s = 0.029 + 0.058 = 0.087 in.²/ft > 0.086 in.²/ft over supports

Neglect concrete above 1400°F in negative moment region, i.e., from Figure 9.3.7.4, neglect bottom ⅝ in. Also, concrete within compressive zone will be about 1350 to 1400°F; from Figure 9.3.7.3, use $f'_{c\theta} = 0.81 f'_c = 4.05$ ksi.

Check $M^-_{n\theta}$. Because of thickness of the slab, the temperature of the negative steel will not rise above 200°F. If greater than 200°F, steel strength should be reduced according to Figure 9.3.7.2.

$$a_\theta = \frac{A_s f_y}{0.85 f'_{c\theta} b} = \frac{0.087(60)}{0.85(4.05)(12)} = 0.126 \text{ in.}$$

$$M^-_{n\theta} = A_s f_y \left(d - \frac{a_\theta}{2}\right) = \frac{0.087(60)(10.37 - 0.063)}{12} = 4.48 \text{ kips/ft}$$

With dead load + ½ live load, w = 0.25 ksf, M = 15.12 kip-ft/ft, and M^-_n = 4.76 kip-ft/ft (calculated for room temperature):

$$M^+_{min} = 15.12 - 4.76 = 10.36 \text{ kip-ft/ft}$$

From Eq. 9.3.7.10: $\max x_o = \frac{22}{2} - \frac{1}{2}\sqrt{\frac{8(10.36)}{0.25}} = 1.90$ ft

Use 6x6–W1.4xW1.4 continuous throughout plus 6x6–W2.9xW2.9 for a distance of 3 ft from the support. Welded wire reinforcement must extend into walls which must be designed for the moment induced at the top.

Example 9.3.7.2
Fire Endurance by Code Tables

Determine if the 10DT24+3 (stem as shown) meets a 2-hour restrained fire rating.
Strands are ½ in. diameter.
Both double tee and topping are of siliceous aggregate.

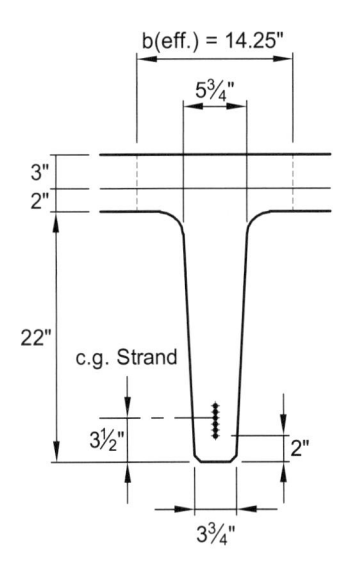

Reinforcement Cover:
Refer to Table 9.3.7.1(5)

See Footnote b for flange width allowed to be included in beam area.

Avg stem width = (3.75 + 5.75)/2 = 4.75 in.

Flange width to be considered = 3(4.75) = 14.25 in.

$A = 22(4.75) + 14.25(5) = 175.75$ in.^{2}

Therefore 150 < A < = 300

From the table, required strand cover = 1½ in. (sides and bottom)

Min. bottom cover provided = 2 − 0.5/2 = 1.75 in.

Min. side cover provided = [3.75 + (3.5/22)(5.75 − 3.75) − 0.5]/2 = 1.78 in.

Both exceed 1.50 in.   OK

Heat Transfer:
    Per Figure 9.3.6.1, slab thickness required = 5.0 in.
    Slab thickness provided = 2 + 3 = 5 in.   OK

Example 9.3.7.3
Fire Endurance by Rational Design

Given:
The 12RB24 shown.
Span = 30 ft
Dead load (including beam weight) = 1.20 klf
Live load = 1.30 klf

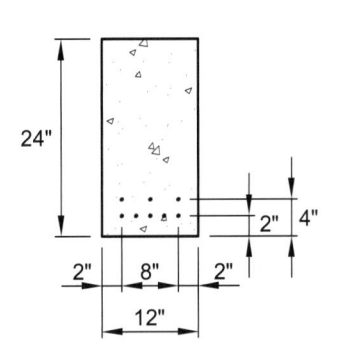

Part of a precast concrete frame well-connected to meet lateral load and structural integrity requirements; therefore, consider the member restrained for fire endurance determination.

Siliceous aggregate concrete, f'_c = 5000 psi
Stress-relieved ½ in. dia. strand, f_{pu} = 270 ksi
Eight strands as shown provide a satisfactory design without fire considerations.

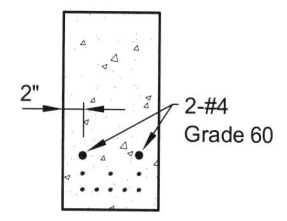

Example 9.3.7.3 (Cont.)
Fire Endurance by Rational Design

Problem:
Determine the necessary reinforcement for a 4-hr fire endurance rating.

Solution:
Since the member is considered restrained, use rational design procedures for unrestrained member at one-half the rating, i.e., 2 hours.

Try eight strands as shown:
$A_{ps} = 8(0.153) = 1.224$ in.^{2}
$y_s = \dfrac{5(2) + 3(4)}{8} = 2.75$ in.
$d = 24 - 2.75 = 21.25$ in.
$\bar{u} = \dfrac{5(2) + 1(4) + 2(2)(0.5)}{8} = 2.00$ in.

From Figure 9.3.7.4, siliceous aggregate: at 2 hr, strand temperature = 750°F

From Figure 9.3.7.2:
$f_{pu\theta} = 0.54(270) = 145.8$ ksi

From Figure 4.2.1.2 with $\gamma_p = 0.4$, $\beta_1 = 0.8$, $\rho = \dfrac{A_{ps}}{bd}$, $f_{pu} = f_{pu\theta}$, $d = 21.25$ in.

$f_{ps\theta} = f_{pu\theta}\left[1 - \dfrac{\gamma_p A_{ps} f_{pu\theta}}{\beta_1 bd f'_c}\right] = 145.8\left[1 - \dfrac{0.4(1.224)(145.8)}{0.8(12)(21.25)(5)}\right] = 135.6$ ksi

$a_\theta = \dfrac{A_{ps} f_{ps\theta}}{0.85 f'_c b} = \dfrac{1.224(135.6)}{0.85(5)(12)} = 3.25$ in.

$M_{n\theta} = A_{ps} f_{ps\theta}\left(d - \dfrac{a_\theta}{2}\right) = 1.224(135.6)\left(21.25 - \dfrac{3.25}{2}\right) = 3257$ kip-in. $= 271$ kip-ft

Service load moment:
$w = 1.20 + 1.30 = 2.50$ kips/ft
$M = \dfrac{w\ell^2}{8} = \dfrac{2.50(30)^2}{8} = 281$ kip-ft > 271 kip-ft

Therefore, the beam will not satisfy criteria for a 4-hr fire endurance.
Try adding 2 – #4 bars, $f_y = 60$ ksi, at 6 in. from bottom as shown:
$A_s = 0.40$ in.^{2}; $f_y = 60$ ksi
d (bars) $= 24 - 6 = 18$ in.
u (bars) $= 2$ in.

From Figure 9.3.7.4:
Bar temperature at 2 hr = 750°F

From Figure 9.3.7.2:
$f_{y\theta} = 0.77$, $f_y = 46.2$ ksi

Example 9.3.7.3 (Cont.)
Fire Endurance by Rational Design

Re-evaluate $f_{ps\theta}$ considering the effect of bars:

$$f_{ps\theta} = 145.8 \left\{ 1 - \left(\frac{0.4}{0.8}\right) \left[\frac{(1.224)(145.8)}{(12)(21.25)(5)} + \frac{\left(\frac{18}{21.25}\right)(0.40)(46.2)}{12(18)(5)} \right] \right\} = 134.5 \text{ ksi}$$

$$a_\theta = \frac{1.224(134.5) + 0.40(46.2)}{0.85(5)(12)} = 3.59 \text{ in.}$$

$$M_{n\theta} \text{ (strand)} = 1.224(134.5)\left(21.25 - \frac{3.59}{2}\right) = 3203 \text{ kip-in.} = 267 \text{ kip-ft}$$

$$M_{n\theta} \text{ (bars)} = 0.40(46.2)\left(18 - \frac{3.59}{2}\right) = 299 \text{ kip-in.} = 25 \text{ kip-ft}$$

$$M_{n\theta} \text{ (total)} = 267 + 25 = 292 \text{ kip-ft} > 281 \text{ kip-ft} \quad \text{OK}$$

9.3.7.4 Shear Resistance

Many fire tests have been conducted on simply supported reinforced or prestressed concrete elements as well as on elements in which restraint to thermal expansion occurred. Shear failures did not occur in any of those tests.

It should be noted that when beams which are continuous over one support (e.g., as shown in Figure 9.3.7.10) are exposed to fire, both the moment and the shear at the interior support increase. Such a redistribution of moment and shear results in a severe stress condition. However, of the several fire tests of reinforced concrete beams in which that condition was simulated, shear failure occurred only in one beam. [5] In that test, the shear reinforcement was inadequate, even for service load conditions without fire, as judged by the shear requirements of ACI 318-02. Thus, it appears from available test data that members which are designed for shear strength in accordance with ACI 318-02 will perform satisfactorily in fire situations, i.e., failure will not occur prematurely due to a shear failure.

9.3.8 Protection of Connections

Many types of connections in precast concrete construction are not vulnerable to the effects of fire, and consequently, require no special treatment. For example, connections such as the bearing between precast concrete panels and concrete footings or beams which support them, do not generally require special fire protection.

If the panels rest on elastomeric pads or other combustible materials, protection of the pads is not generally needed because deterioration of the pads will not cause collapse.

Connections that can be weakened by fire and thereby jeopardize the structure's load carrying capacity should be protected to the same degree as that required for the supported member. For example, an exposed steel bracket supporting a panel or spandrel beam will be weakened by fire and might fail causing the panel or beam to collapse. Such a bracket should be protected.

The amount of protection depends on (a) the stress-strength ratio in the steel at the time of the fire and (b) the intensity and duration of the fire. The thickness of protection materials required is greater as the stress level and fire severity increase.

Figure 9.3.8.1 shows the thickness of various commonly used fire protection materials required for fire endurances up to 4 hr. The values shown are based on a critical steel temperature of 1000°F, i.e., a stress-strength ratio (f_s/f_y) of about 65%. Values in Figure 9.3.8.1(B) are applicable to concrete or dry-pack mortar encasement of structural steel shapes used as brackets or lintels.

9.3.9 Precast Concrete Column Covers

Steel columns are often clad with precast concrete panels or covers for architectural reasons. Such covers also provide fire protection for the columns.

Figure 9.3.9.1 shows the relationship between the thickness of concrete column covers and fire endurance for various steel column sections. The fire endurances shown are based on an empirical relationship developed by Lie and Harmathy. [10]

The above authors also found that the air space between the steel core and the column covers has only a minor effect on the fire endurance. An air space will probably increase the fire endurance but only by an insignificant amount.

Most precast concrete column covers are 3 in. or more in thickness, but some are as thin as 2½ in. From Figure 9.3.9.1, it can be seen that they can qualify the column for fire endurances of at least 2½ hr, and usually more than 3 hr. For steel column sections other than those shown, including shapes other than wide flange beams, interpolation between the curves on the basis of weight per foot will generally give reasonable results.

For example, the fire endurance afforded by a 3 in. thick column cover of normal weight concrete for a 8 x 8 x ½ in. steel tube column, section weight 47.35 lb/ft, can be determined by using the data for a W10x49. By Figure 9.3.9.1, the fire endurance is about 3 hr 20 min.

Figure 9.3.8.1 Thickness of protection materials applied to connections of structural steel shapes

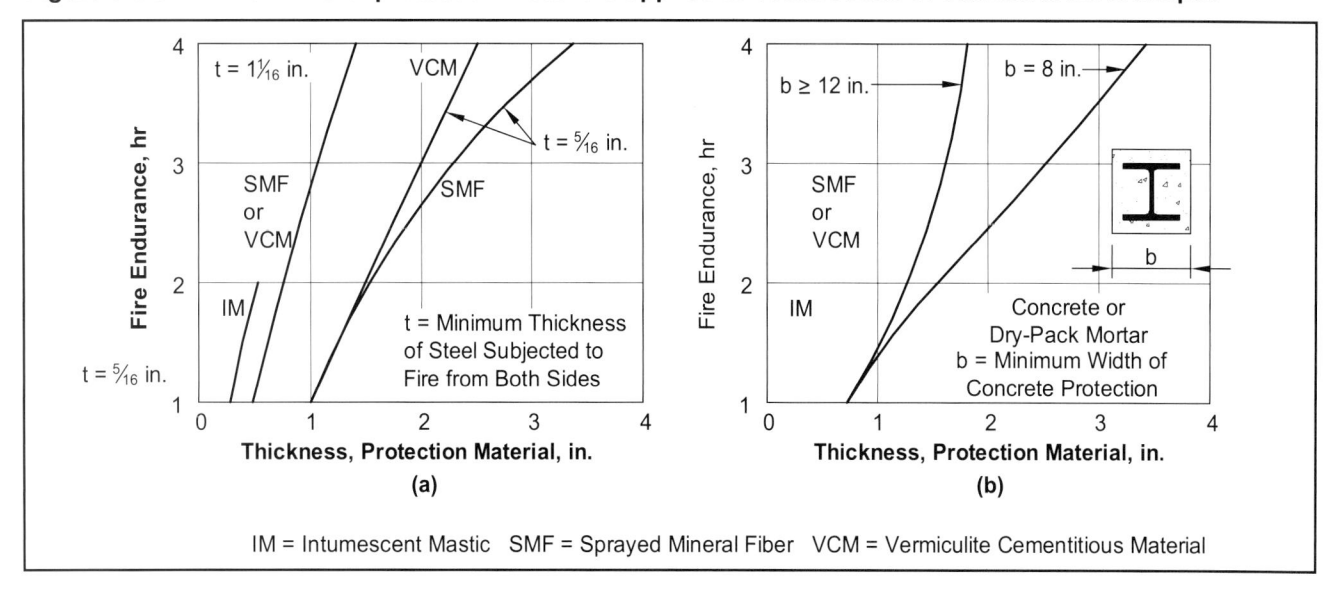

IM = Intumescent Mastic   SMF = Sprayed Mineral Fiber   VCM = Vermiculite Cementitious Material

Figure 9.3.9.1 Fire endurance of steel columns afforded protection by concrete column covers

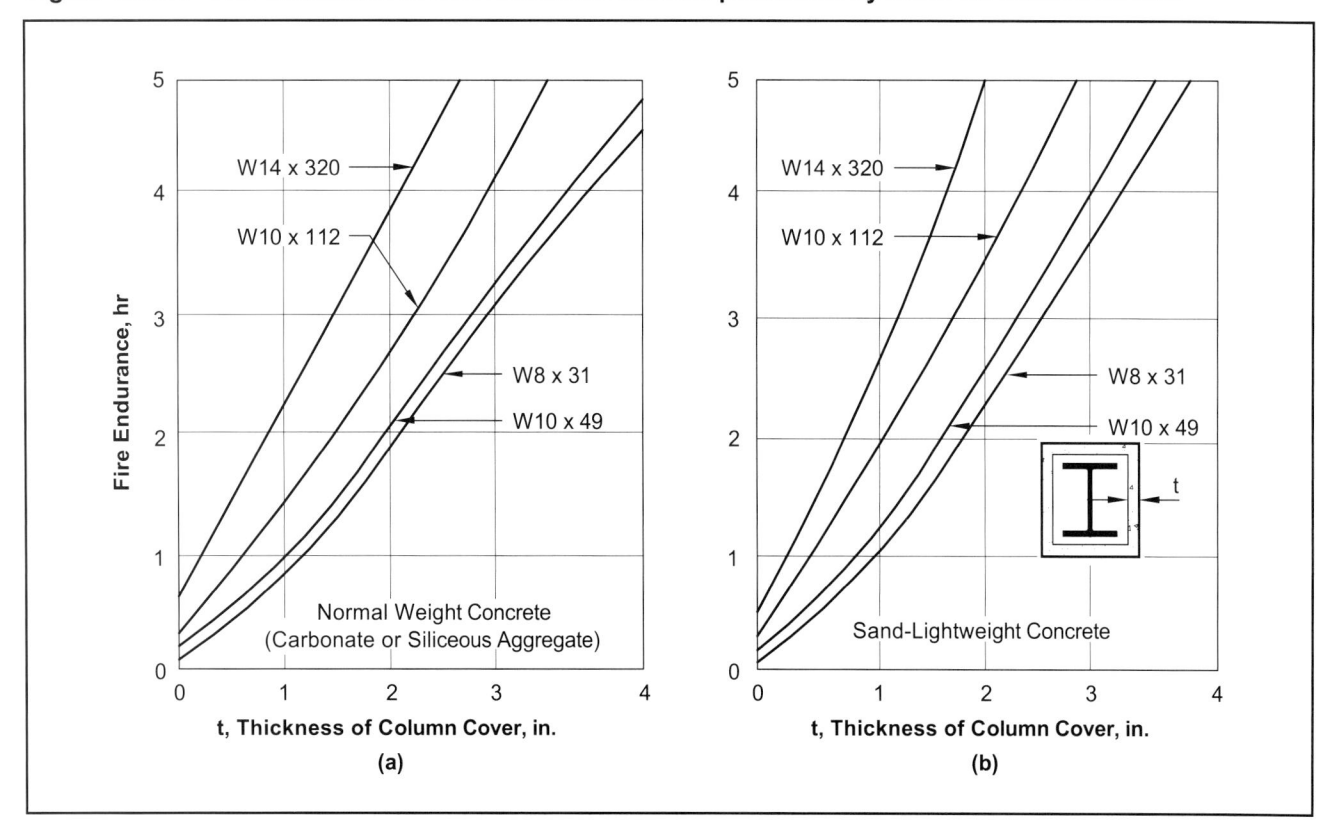

Precast concrete column covers (Figure 9.3.9.2) are made in various shapes such as (a) four flat panels with butt or mitered joints that fit together to enclose the steel column, (b) four L-shaped units, (c) two L-shaped units, (d) two U-shaped units, and (e) and (f) U-shaped units and flat closure panels. Type (a) would probably be most vulnerable to bowing during fire exposure while Type (f) would probably be the least vulnerable. There are, of course, many combinations to accommodate isolated columns, corner columns, and column walls.

To be fully effective, the column covers must remain in place without severe distortion. Many types of connections are used to hold the column covers in place. Some connections consist of bolted or welded clip angles attached to the tops and bottoms of the covers. Others consist of steel plates embedded in the covers that are welded to angles, plates, or other shapes which are, in turn, welded or bolted to the steel column. In any case, the connections are used primarily to position the column covers and as such are not highly stressed. As a result, temperature limits need not be applied to the steel in most column cover connections.

If restrained, either partially or fully, concrete panels tend to deflect or bow when exposed to fire. For example, for a steel column that is clad with four flat panels attached top and bottom, the column covers will tend to bulge at midheight thus tending to open gaps along the sides. The gap size decreases as the panel thickness increases.

With L, C, or U-shaped panels, the gap size is further reduced. The gap size can be further minimized by connections at midheight. In some cases, ship-lap joints can be used to minimize the effects of joint openings.

Joints should be sealed in such a way to prevent passage of flame to the steel column. A non-combustible material such as sand-cement mortar or ceramic fiber blanket can be used to seal the joint.

Precast concrete column covers should be installed in such a manner that, if they are exposed to fire, they will not be restrained vertically. As the covers are heated they tend to expand. Connections should accommodate such expansion without subjecting the cover to additional loads.

Fire resistive compressible materials, such as mineral fiber safing, can be used to seal the tops or bases of the column covers, thus permitting the column covers to expand.

9.3.10 Code and Economic Considerations

An important aspect of dealing with fire resistance is to understand what the benefits are to the owner of a building in the proper selection of materials incorporated in his structure. These benefits fall into two areas: codes and economics.

Building codes are laws that must be satisfied regardless of any other considerations and the manner in which acceptance of code requirements is achieved is explained in the preceding pages. The designer or representing the owner, has no option in the code regulations, but does have a number of options in the materials and assemblies that meet these regulations, such as those included in this Handbook.

Economic benefits associated with increased fire resistance should be considered by the designer/owner team at the time decisions are made on the structural system. Proper consideration of fire resistive construction through a life-cycle cost analysis will provide the owner economic benefits over other types of construction in many areas, e.g., lower insurance costs, larger allowable gross area under certain types of construction, fewer stairwells and exits, increased value for loan purposes, longer mortgage terms, and better resale value.

To ensure an owner of the best return on his investment, a life-cycle cost analysis using fire resistive construction should be prepared.

Beyond the theoretical considerations is the history of excellent performance of prestressed concrete in actual fires. Structural integrity has been maintained, fires are contained in the area of origin, and, in many instances, repairs consist of "cosmetic" treatment only, leading to early re-occupancy of the structure.

Figure 9.3.9.2 Types of precast concrete column covers

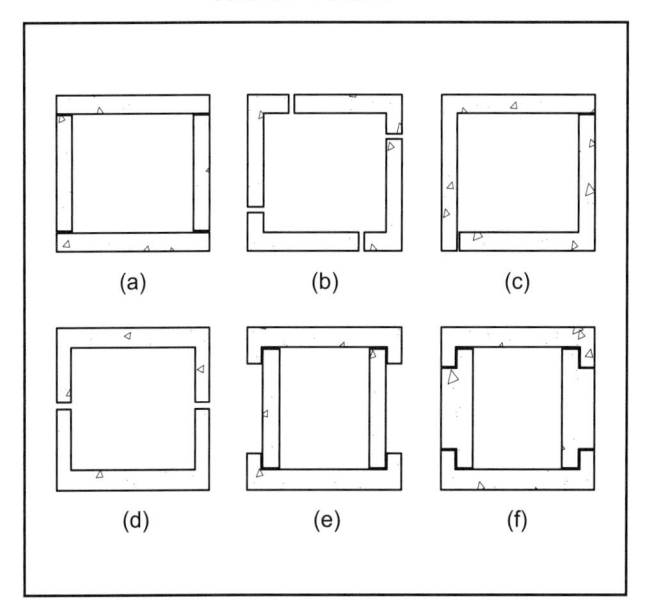

9.3.11 References

1. *Design for Fire Resistance of Precast Prestressed Concrete*, Second Edition, MNL-124-89, Precast/Prestressed Concrete Institute, Chicago, IL, 1989.

2. *Reinforced Concrete Fire Resistance*, Concrete Reinforcing Steel Institute, Schaumburg, IL, 1980.

3. Shirley, Scott T., Burg, Ronald G., and Fiorato, Anthony E., "Fire Endurance of High Strength Concrete Slabs," *ACI Materials Journal*, V. 85, No. 2, March-April 1988.

4. Abrams, M. S., Gustaferro, A. H., and Salse, E. A. B., "Fire Tests of Concrete Joist Floors and Roofs," *RD Bulletin 006B*, Portland Cement Association, Skokie, IL, 1971.

5. "Fire Endurance of Continuous Reinforced Concrete Beams," *RD Bulletin 072B*, Portland Cement Association, Skokie, IL.

6. Abrams, M. S., and Gustaferro, A. H., "Fire Endurance of Two-Course Floors and Roofs," *Journal of the American Concrete Institute*, V. 66, No. 2, February 1969.

7. Lie, T. T., "Contribution of Insulation in Cavity Walls to Propagation of Fire," *Fire Study No. 29*, Division of Building Research, National Research Council of Canada, Ottawa, Ontario, Canada.

8. Gustaferro, A. H., and Abrams, M. S., "Fire Tests of Joints Between Precast Concrete Wall Panels: Effect of Various Joint Treatments," PCI JOURNAL, V. 20, No. 5, September-October 1975.

9. ACI Committee 318, "Building Code Requirements for Reinforced Concrete, ACI 318-02, and Commentary, ACI 318R-02," American Concrete Institute, Farmington Hills, MI, 2002.

10. Lie, T. T., and Harmathy, T. Z., "Fire Endurance of Concrete-Protected Steel Columns," *Journal of the American Concrete Institute*, V. 71, No. 1, January 1974.

11. *Standard Test Methods for Fire Tests of Building Construction and Materials*, ASTM E119-00, ASTM International, West Conshohocken, PA, 2000.

12. *International Building Code* (IBC 2003) International Conference of Building Officials, Falls Church, VA, 2003.

13. *Design for Fire Resistance of Precast Prestressed Concrete*, Third Edition, MNL-124-05, Precast/Prestressed Concrete Institute, Chicago, IL (scheduled for 2005).

9.4 SANDWICH PANELS

9.4.1 General

Insulated wall panels, commonly known as sandwich panels, are typically composed of two concrete wythes separated by a layer of insulation. One of the concrete wythes may be a standard shape, as shown in Figure 9.4.1.1, or any architectural concrete section produced for a single project. In place, sandwich panels provide the dual function of transferring load and insulating the structure, and can provide the interior and exterior finished wall surfaces. They may be used only for cladding, or they may act as beams, bearing walls or shear walls.

9.4.2 Structural Design

The structural design of sandwich panels is the same as the design of other wall panels once the section properties of the panel have been determined. Three different assumptions may be used for the properties, depending on the construction:

1. **Non-composite for full life cycle of panel.** The two concrete layers act independently (Figure 9.4.2.1(a)). The wythes are connected by ties which are flexible enough that they offer insignificant resistance to shrinkage and temperature movement between the wythes in the plane of the panel. Positive steps may be taken to ensure that one wythe does not bond to the insulation. Typical techniques are by placing a sheet of polyethylene or reinforced paper over the insulation before the final concrete wythe is placed, by applying a retarder or form release agent to one side of the insulation, or by placing insulation in two layers (Figure 9.4.4.1). One wythe is usually assumed to be "structural" and all loads are carried by that wythe, both during handling and in service, although, when designing for wind loads and slenderness effects, the panel may in some cases be designed as in (4) below.
2. **Composite for full life cycle of panel.** The two wythes act as a fully composite unit for the full life of the structure [Figure 9.4.2.1(b)]. In order to provide for composite behavior, positive measures must be taken to effect shear transfer between the wythes in the direction of panel span. This may be accomplished by rigid ties, longitudinal welded wire trusses, or regions of solid concrete which join both wythes. [4] Consult with the precast producer for tie performance capabilities.
3. **Partially-composite for full life cycle of panel.** The two wythes do not act independently nor as a fully composite unit, but rather somewhere in between. In a partially-composite panel, the ties provide significant shear transfer between the wythes in the direction of the span. Consult with the precast producer for tie performance capabilities.
4. **Composite during handling, but non-composite during service life.** Composite action results largely because of early bond between insulation and adjacent wythes, in combination with flexible ties. The bond is considered unreliable for the long term; thus, the panel is considered as non-composite for service life loads. Wind loads are distributed to each wythe in proportion to wythe stiffness. For effects of slenderness, the sum of the individual moments of inertia may be used. See Example 9.4.1.

For some machine-produced sandwich panels, tests and experience have demonstrated that a degree of composite action can be anticipated for the full life cycle. Manufacturer's recommendations for these panels should be followed.

A composite panel will have a stiffness significantly greater than a non-composite panel or partially composite panel of the same thickness. Thus, a composite panel can span greater heights.

Figure 9.4.1.1 Typical precast sandwich wall panels

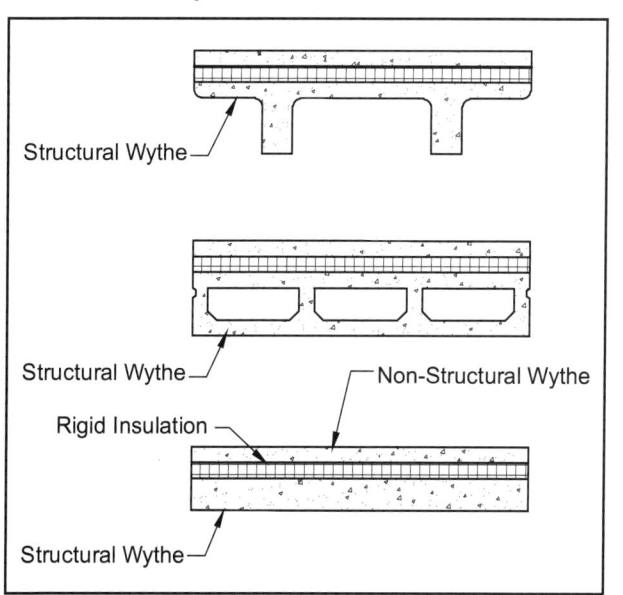

Figure 9.4.2.1 Non-composite and composite panels

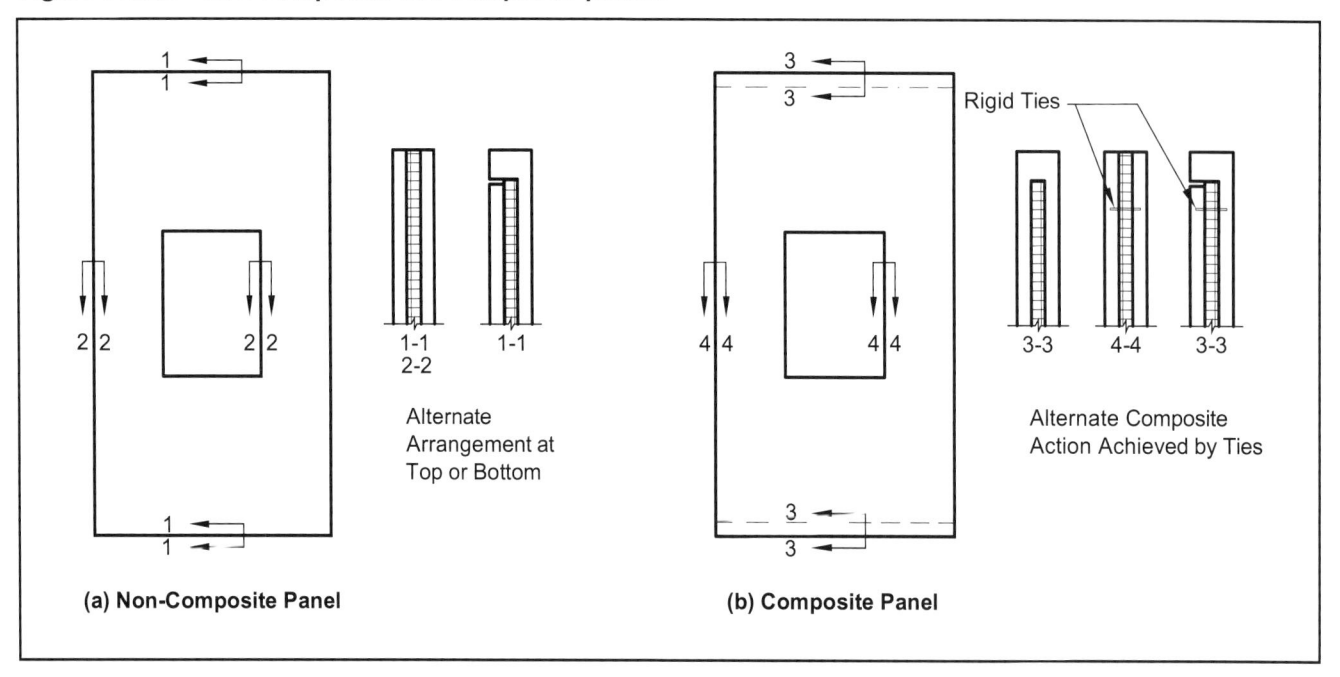

Thermal bowing will tend to be greater in composite panels than in non-composite panels due to the inherent restraint of differential movements between wythes.

Panel size is limited only by stresses due to handling and service loads, and slenderness in loadbearing panels, modified by the experience of the producer. Wythes should be no thinner than three times the maximum aggregate size. Panels with ¾ in. aggregate and 2 in. thick wythes have been successfully used.

Lifting points should be chosen to limit stresses below the modulus of rupture during stripping and handling so that cracks do not occur (see Chapter 5). Prestressing in the long direction is particularly effective as a method of providing for virtually crack-free panels; cracking due to restrained shrinkage in the short direction of the panel usually need not be considered, provided the dimension in that direction does not exceed about 15 ft. Composite panels should be concentrically prestressed. In non-composite or partially composite panels, prestressing should be concentric within each wythe which is stressed. The designer may consider prestressing the non-structural exterior wythe to the same or higher prestress level as the structural wythe, in order to counteract panel bowing.

In some cases, panel openings are so extensive that an insufficient section modulus is left to keep the applied stresses below the cracking limit. In these cases, the designer can only satisfy the strength criteria, and some cracking is to be expected. This is normally accomplished by the use of longitudinal steel reinforcement.

The maximum strand diameter that may be used is related to wythe thickness. Satisfactory results have been achieved using ⅜ in. diameter strand in 2 in. thick wythes. A wythe thickness of 3 in. is sufficient when using a ½ in. diameter strand.

Transverse reinforcement should be provided at the ends of the panel to limit splitting cracks over the strands during detensioning. With low thickness-to-strand diameter ratios, special attention during production should be given to the as-placed strand location in order to maintain required cover and to minimize the risk of splitting cracks.

Ref. 3 includes design examples for composite and non-composite sandwich panels used as cladding, bearing walls and shear walls. A design example for a semi-composite panel (design assumption 4, above) is also included.

9.4.3 Connections

9.4.3.1 Panel Connections

Analysis and design of panel connections to resist normal and transverse wind and seismic shears as well as gravity and alignment loads are described in Chapters 3 and 6. Consideration must also be given to anticipated bowing, particularly of composite panels (see Section 4.8.5).

The effects of differential bowing between adjacent panels of similar span may be minimized by carefully aligning the panel during erection, and providing a sealant on both the inside and outside of panel joints. Where adjacent panels are of significantly different stiffness, connections should be

Example 9.4.1
Section Properties of Sandwich Panels

Given:
   The sandwich panel shown.

Problem:
   Calculate the moment of inertia and section modulus if the section is (a) non-composite and (b) composite. Also find the load distribution for the non-composite case. The lateral load applied to the non-composite panel will be resisted by each wythe in proportion to its stiffness.

Solution:
(a) The non-composite properties (per ft width) are as follows:

   Interior (structural) wythe:
   $I_{int} = bd^3/12 = 12(4)^3/12 = 64$ in.^{4}/ft width
   $S_{int} = I/c = 64/2 \qquad = 32$ in.^{3}/ft width

   Exterior (non-structural) wythe:
   $I_{ext} = 12(2.5)^3/12 \qquad = 15.6$ in.^{4}/ft
   $S_{ext} = 15.6/1.25 \qquad = 12.5$ in.^{3}/ft

   Distribution of lateral load:
   $I_{int} + I_{ext} = 64 + 15.6 = 79.6$ in.^{4}/ft

   Lateral load resisted by:
   Interior: 64(100)/79.6 $\quad = 80\%$
   Exterior: 15.6(100)/79.6 $\; = 20\%$

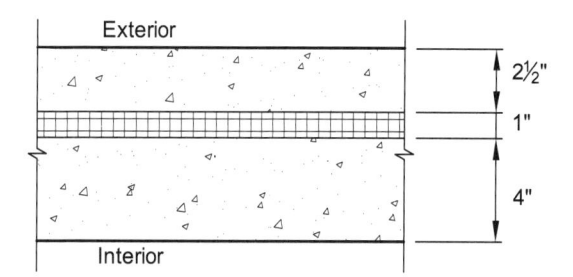

(b) The composite properties are as follows:

	A	y	Ay	y – y_b	Ay²	I
Interior	48	2.00	96.0	1.63	127.5	64.0
Exterior	30	6.25	187.5	2.62	205.9	15.6
	78		283.5		333.4	79.6

$y_b = 283.5/78 \qquad = 3.63$ in.
$I_c = 333.4 + 79.6 = 413.0$ in.^{4}/ft width

provided across the adjacent joint, to maintain alignment. Such connections should be located in the center third of the span. The connection should provide for anticipated movement to prevent build-up of volumetric restraint forces.

Panel embedments may be limited by the wythe thickness, and hence may not be fully effective. Confinement reinforcement should be placed around any embedment. When bowing does occur, it is generally outward. Section 4.8.5 discusses the effects on connections. Recommendations stated should be followed in order to prevent the opening of joints and the resultant possibility of joint sealant failure. Special attention should also be paid to panels that are hung from adjacent panels. If a hung panel is supported by full height panels that are supported on a foundation, the assembly of panels can move as a unit, without distress to the panel or joint sealant. However, if the hung panel is also rigidly attached to the structure, e.g., at a mezzanine level, this restraint may result in differential bowing of the hung panel with respect to the adjacent long panels. A detail to permit the hung panel to move with the adjacent panels should be provided. Attention should be paid to the position of the connection near the bottom of a panel supported on a foundation, in order to minimize the possibility of spalling due to thermal bowing (see Figure 9.4.3.1).

9.4.3.2 Wythe Connectors

When one wythe is non-structural, its weight must be transferred to the structural wythe. This may be accomplished by using shear connectors or solid concrete ribs at either the top or bottom of the panel. This permits the non-structural wythe to contract and expand with the least restraint.

Figure 9.4.3.1 Restraint at foundation

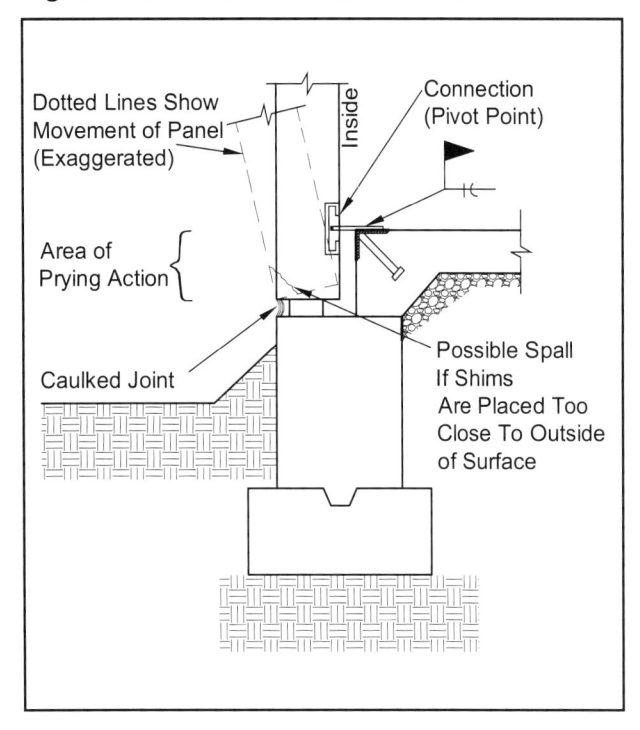

Figure 9.4.3.2 Typical shear connectors

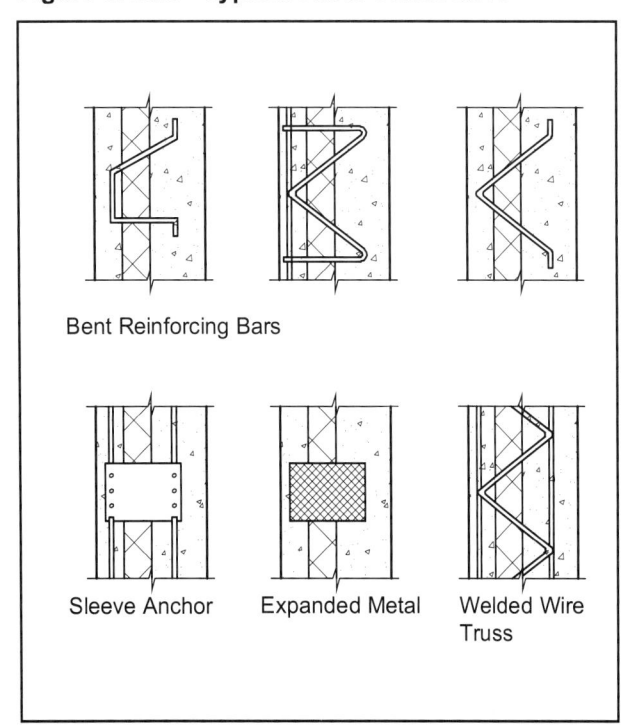

Figure 9.4.3.3 Anchorage for stemmed panels

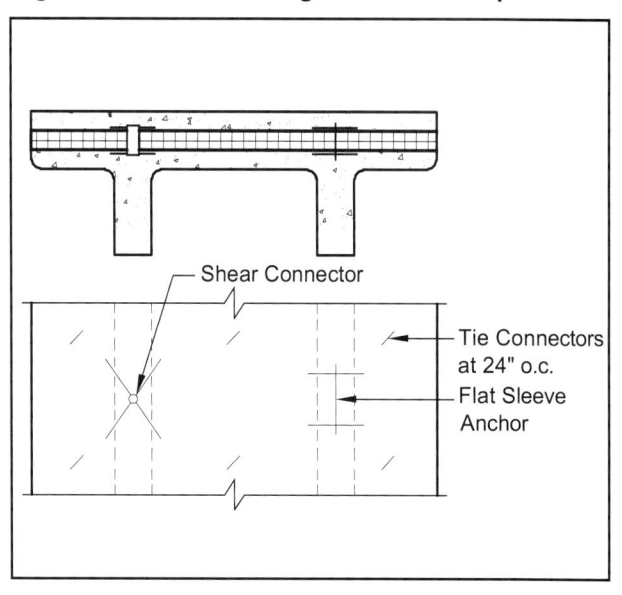

Figure 9.4.3.4 Functional behavior of connectors

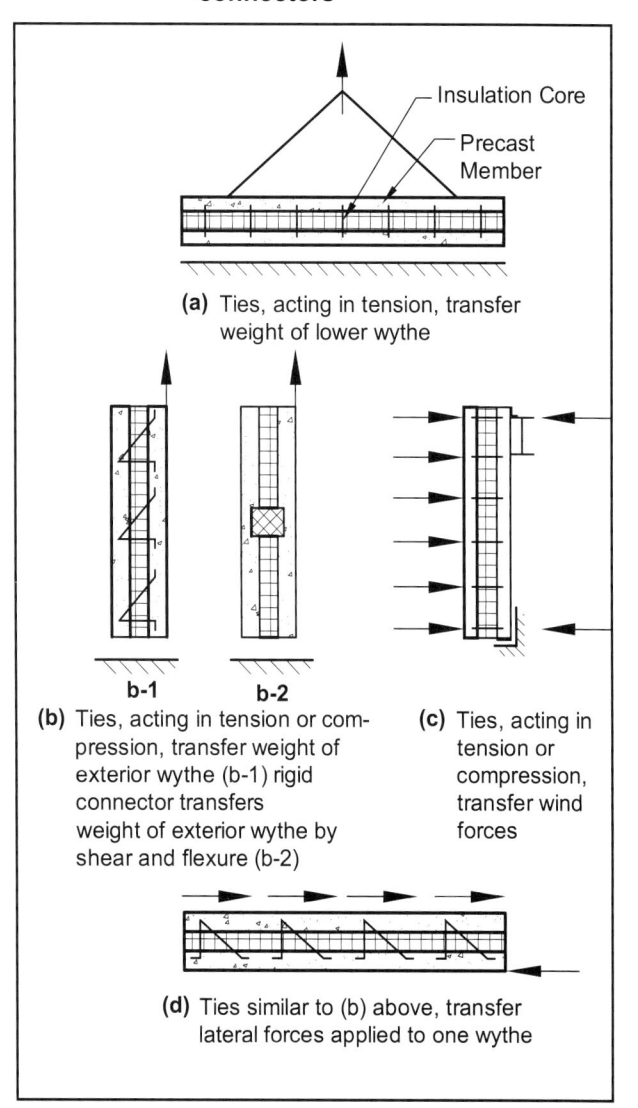

(a) Ties, acting in tension, transfer weight of lower wythe

(b) Ties, acting in tension or compression, transfer weight of exterior wythe (b-1) rigid connector transfers weight of exterior wythe by shear and flexure (b-2)

(c) Ties, acting in tension or compression, transfer wind forces

(d) Ties similar to (b) above, transfer lateral forces applied to one wythe

Figure 9.4.3.5 Tension/compression ties

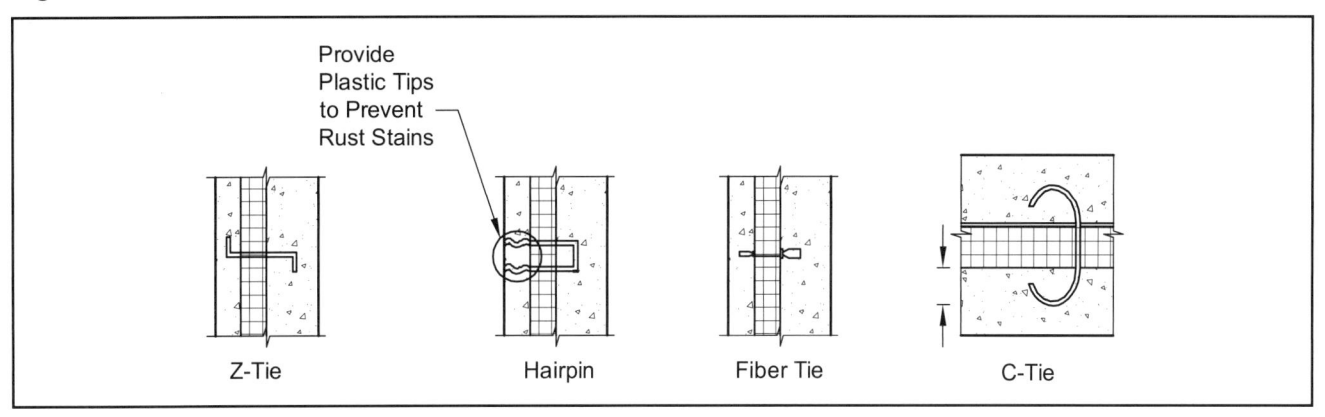

Z-Tie — Hairpin (Provide Plastic Tips to Prevent Rust Stains) — Fiber Tie — C-Tie

Table 9.4.1 Properties of insulation

	Polystyrene					Polyisocyanurate		Phenolic	Cellular Glass
	Expanded			Extruded			Unfaced	Faced		
Density (pcf)	0.7–0.9	1.1–1.4	1.8	1.3–1.6	1.8–2.2	3.0	2.0–6.0	2.0–6.0	2.0–3.0	6.7–9.2
Water absorption (% volume)	< 4.0	< 3.0	< 2.0	< 0.3			< 3.0	1.0–2.0	< 3.0	< 0.5
Compressive strength (psi)	5–10	13–15	25	15–25	40–60	100	16–50	16	10–16	65
Tensile strength (psi)	18–25			25	50	105	45–140	500	60	50
Linear coefficient of expansion (in./in./°F) × 10^{-6}	25–40			25–40			30–60		10–20	1.6–4.6
Shear strength (psi)	20–35			—	35	50	20–100		12	50
Flexural strength (psi)	10–25	30–40	50	40–50	60–75	100	50–210	40–50	25	60
Thermal conductivity (Btu-in./hr/ft^2/°F) at 75°F	0.32–0.28	0.26–0.25	0.23	0.20			0.18	0.10–0.15	0.16–0.23	0.35
Max. use temp.	165°F			165°F			250°F		300°F	900°F

Figure 9.4.4.1 Preferred installation of insulation sheets

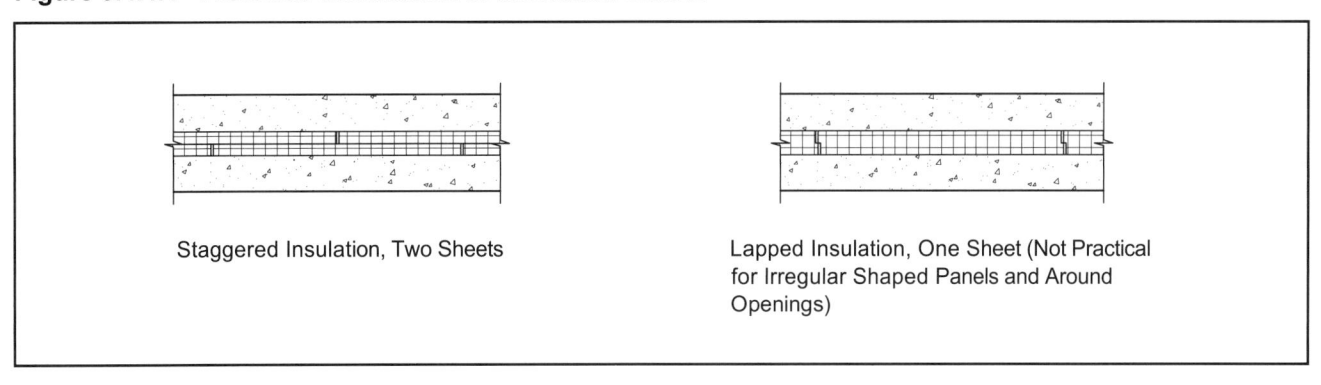

Staggered Insulation, Two Sheets

Lapped Insulation, One Sheet (Not Practical for Irregular Shaped Panels and Around Openings)

Shear connectors may include bent reinforcing bars, sleeve anchors, expanded metal, fiber connectors, or welded wire trusses, as illustrated in Figure 9.4.3.2.

For stemmed panels, the shear connector is placed in the stem to ensure proper embedment depth. In non-composite panels, it is preferable to have only one anchoring center. In a panel with two stems, the shear connector can be positioned in either one of the stems, and a flat anchor with the same vertical shear capacity is used in the other stem (Figure 9.4.3.3). Since the flat anchor has little or no horizontal shear capacity, restraint of the exterior wythe is minimized. In a multi-ribbed panel, the shear connector is placed as near the center of rigidity as is possible, and flat anchors are used in the other stems.

To complete the connection, tension/compression ties passing through the insulation are spaced at regular intervals to prevent the wythes from separating. Functions of wythe connectors are shown in Figure 9.4.3.4. Typical tie details are shown in Figure 9.4.3.5. Wire tie connectors are usually 12 to 14 gauge, and preferably of stainless steel. Galvanized metal or plastic ties may also be acceptable. Ties of welded wire reinforcement and reinforcing bars are sometimes used.

Tension/compression ties should be flexible enough so as not to significantly resist temperature and shrinkage parallel to the panel surface, yet strong enough to resist a lifetime of stress reversals caused by temperature strains. Follow the manufacturer's recommendations when available, otherwise the spacing of ties should be approximately 2 ft on centers.

9.4.4 Insulation

Physical properties of insulation materials are listed in Table 9.4.1. Thermal properties are discussed in Section 9.1. The insulation should have low absorption or a water-repellent coating to minimize absorption of water from fresh concrete.

The thickness of insulation is determined as described in Section 9.1. A minimum of 1 in. is recommended. While there is no upper limit on the thickness, the deflection characteristics of the wythe connectors should be considered.

Openings in the insulation around connectors should be packed with insulation to avoid forming thermal bridges between wythes.

Using the maximum standard size of insulation sheets, consistent with the panel shape, is recommended. This will minimize joints and the resulting thermal links. Lapped abutting ends of single layer insulation, or staggered joints with double layer insulation, will effectively remove thermal links at joints (Figure 9.4.4.1). Insulation may expand when subject to curing temperatures greater than 140°F; proper precautions should be taken, such as leaving room for the insulation to expand where solid concrete ribs and insulation are in contact.

9.4.5 Thermal Bowing

Thermal bowing of exterior composite panels must be considered. Calculation of bow and the forces required to restrain it are discussed in Section 4.8.5. Differential movement between panels is seldom a problem, except at corners or where abrupt changes in the building occur. Crazing or cracking is sometimes caused by bowing, but will be minimized if the recommendations of Section 9.4.2 are followed; such cracking is seldom of structural significance.

9.4.6 Typical Details

As with all precast concrete construction, satisfactory performance depends on proper detailing. The typical wall panel connections and details shown in Chapter 6 are equally applicable to sandwich panels. Common details and connections for sandwich wall panels can be found in Ref. 3.

9.4.7 References

1. Einea, Amin, Salmon, David C., Fogarasi, Gyula J., Culp, Todd D., and Tadros, Maher K., "State-of-the-Art of Precast Concrete Sandwich Panels," PCI JOURNAL, V. 36, No. 6, November-December 1991.

2. Bush, Thomas D., and Stine, Gregory L., "Flexural Behavior of Composite Precast Concrete Sandwich Panels with Continuous Truss Connectors," PCI JOURNAL, V. 39, No. 2, March-April 1994.

3. PCI Committee on Precast Sandwich Wall Panels, "State-of-the-Art of Precast/Prestressed Sandwich Wall Panels," PCI JOURNAL, V. 42, No. 2, March-April 1997 and V. 42, No. 3, May-June 1997.

4. Pessiki, Stephen, and Mlynarczyk, Alexandar, "Experimental Evaluation of the Composite Behavior of Precast Concrete Sandwich Wall Panels," PCI JOURNAL, V. 48, No. 2, March-April 2003.

9.5 QUALITY ASSURANCE AND CONTROL

9.5.1 Introduction

The successful use and application of precast and prestressed concrete demands a high level of attention to quality in design, production and installation.

Each manufacturer (producer) must have in place a documented quality assurance program with the goals to assure structural integrity, ease of construction and the desired appearance in the final structure. Further, the owner or architect/engineer must be confident that materials, methods, products and the producer's quality control meet the requirements of the project. This assurance is available by requiring in the project specification that:

(a) The precast concrete manufacturing facility be certified by the Precast/Prestressed Concrete Institute's (PCI) Plant Certification Program.
(b) The precast manufacturer have personnel certified in the appropriate levels of the PCI Plant Quality Personnel Certification Program.
(c) The precast concrete erector be Qualified by the PCI Field Qualification Program or the precast concrete manufacturer have a qualified person to oversee the work of the erector.

In those instances when site-precasting is done, the same degree of quality should be expected by the owner or architect/engineer and rigorous inspections and/or quality assurance must be provided. These expectations can only be achieved if they are clearly written and included in the original project specifications.

9.5.2 Plant Certification Program

Producers registered under the PCI Plant Certification Program have demonstrated that their processes for production and quality assurance meet or exceed industry-wide standards. These plants maintain a comprehensive, documented and approved quality system that is present in every aspect of their business. Each certified plant conducts a formal quality control program with a trained, permanent quality control staff. The staff must meet the current requirements of the PCI Quality Control Personnel Certification Program. Conformance to the nationally accepted requirements is determined by a minimum of two quality audits per year. All audits are unannounced and most audits take a two-day period. The audits are conducted by specially trained personnel employed by a national structural consulting engineering firm under contract to and accredited by PCI.

The Plant Certification Program identifies the general types of products in which the producer has demonstrated acceptable experience. This is done through the program's Product Group & Category provisions. There are four groups of products: A – Architectural; B – Bridges; C – Commercial (Structural) and G – Glass Fiber Reinforced Concrete.

Product Group A has two categories: A1 for major, primary architectural panels and products; and AT for miscellaneous architectural trim elements.

Groups B and C each have four separate categories that identify the experience and equipment available at the certified plant. A description of groups and categories B1 through B4 and C1 through C4 may be found in PCI's Certified Quality for People, Products and Performance [1] as follows:

B1 Precast Bridge Products (no prestressed reinforcement)
B2 Prestressed Miscellaneous Bridge Products
B3 Prestressed Straight-Strand Bridge Beams
B4 Prestressed Deflected-Strand Bridge Beams
C1 Precast Concrete Products (no prestressed reinforcement)
C2 Prestressed Hollow-Core and Repetitively Produced Products
C3 Prestressed Straight-Strand Structural Members
C4 Prestressed Deflected-Strand Structural Members

Many producers have experience with applying architectural finishes to typical structural products. This certified architectural capability is identified with an "A" following the Group and Category designation (e.g., C3A for a spandrel panel for a parking structure made with straight strands and having a special architectural pattern and finish on the face). Therefore, Group and Category designations B1 through B4 and C1 through C4 may be modified with an "A" suffix. Structural products with simple architectural finishes, such as warehouse panels, parking structure spandrel panels, etc., should not be required to meet the specification of PCI MNL 117. Specifications of MNL 116 meet the requirement of CA certification and will assure an acceptable simple architectural finish without requiring unduly stringent and more costly tolerances.

Auditing criteria and grading are based on the industry's standards for quality and quality control. These standards are presented in PCI's three quality control manuals, one for structural precast and prestressed concrete products, [2] one for architectural precast products, [3] and one for glass fiber reinforced concrete products. [4] Audits cover all phases of production including shop drawings, materials, production methods, product handling and storage, appearance, testing, record keeping, quality control, personnel training, and safety practices. Failure to maintain a production plant at or above required standards results in loss of certification. A current listing of plants holding certification is published a minimum of four times each year and may be obtained by calling PCI or by viewing www.pci.org.

Care should be exercised in evaluating the effectiveness of other certification programs. Criteria should include:

1. Unannounced quality audits are fundamental to the program.
2. The auditor should be recognized as experienced in the field of precast and prestressed concrete.
3. The auditor should have particular experience with the products and production methods involved.
4. The auditor should be independent and not be hired by the precast concrete manufacturer.
5. The program should be based on the industry approved and nationally recognized quality standards set forth in Refs. 2, 3 and 4.
6. The auditor should view the entire fabrication cycle.
7. The program should be executed by a single auditing agency that ensures uniformity for all size companies throughout the United States.
8. The program and the auditing agency should be recognized by major public and private agencies and organizations.

9.5.3 Plant Quality Personnel Certification

Conducting an effective quality control program requires knowledgeable and motivated testing and inspection personnel. Each must understand quality basics, the necessity for quality control, how products are manufactured, and precisely how to conduct tests and inspections. PCI has been training quality control personnel since 1974. There are three levels of Plant Quality Personnel Certification.

- **Plant Quality Personnel Certification, Level I** requires a basic level of understanding of the many quality control issues normally encountered in a precast plant. It also requires current certification by the American Concrete Institute (ACI) Concrete Field Testing Technician Program, Grade I. A candidate must have at least six months of industry experience.

- **Plant Quality Personnel Certification, Level II** requires Level I as a prerequisite. Other requirements for Level II include a greater level of knowledge of the topics for Level I, as well as at least one year of industry experience. Certification at Levels I and II is accomplished by passing a written examination. A manual for Levels I and II, TM-101, is available from PCI for training and self-study. [5]

- **Plant Quality Personnel Certification, Level III** provides significant instruction in concrete materials and technology. Certification at this level requires attendance at a four-day course and Level II certification as a prerequisite. The candidate must have two years of industry experience or equivalent. There is a training manual, TM-103, available from PCI that covers all course material. [6]

9.5.4 Field Qualification Program

The Field Qualification Program extends PCI quality assurance from fabrication (Plant Certification) to field installation. It has been found that participation in the program is highly beneficial in improving the quality of the installation and the safety and efficiency of the installation crews.

The PCI Field Qualification Program evaluates producer-erectors and independent erectors of precast concrete products against nationally adopted standards. Specifying a PCI Qualified Erector or a precast concrete manufacturer with Certified Field Auditors ensures the project specifier and owner that the installation meets the requirements of the Field Qualification Program.

An erector may be qualified in up to three (3) categories (see Ref. 8 for more detailed descriptions):

S1 Simple Structural Systems
(horizontal decking members, bridge beams, single-lift walls)
S2 Complex Structural Systems
(category S1 plus all other structural products)
A Architectural Systems
(non-loadbearing cladding and GFRC products)

The auditing criteria for the Field Qualification Program are based on the industry's quality, procedural and safety standards as presented in Refs. 7 and 8. Audits are conducted by Certified Field Auditors accredited by PCI. Precast concrete manufacturers who do not have their own erection crews may still have Certified Field Auditors. Audits are conducted semiannually on all of the erector's primary crews and cover all phases of the erection process including pre-construction planning, practices and procedures, equipment, safety, erection tolerances and quality control. Failure to maintain field work at or above required standards results in mandatory loss of PCI Qualified Erector status. A current list of PCI Qualified erectors is published a minimum of four times each year or may be obtained by calling PCI or by viewing www.pci.org.

9.5.5 References

1. *Certified Quality for People, Products and Performance*, PC-4-01, Precast/Prestressed Concrete Institute, Chicago, IL, 2001.

2. *Manual for Quality Control for Plants and Production of Structural Precast Concrete Products*, Fourth Edition, MNL-116-99, Precast/Prestressed Concrete Institute, Chicago, IL, 1999.

3. *Manual for Quality Control for Plants and Production of Architectural Precast Concrete Products*, Third Edition, MNL-117-96, Precast/Prestressed Concrete Institute, Chicago, IL, 1996.

4. *Manual for Quality Control for Plants and Production of Glass Fiber Reinforced Concrete Products*, MNL-130-91, Precast/Prestressed Concrete Institute, Chicago, IL, 2003.

5. *Quality Control Technician/Inspector Level I & II Training Manual*, TM-101, Precast/Prestressed Concrete Institute, Chicago, IL, 1987.

6. *Quality Control Personnel Certification Level III Training Manual*, TM-103, Precast/Prestressed Concrete Institute, Chicago, IL, 1996.

7. *Erection Safety for Precast and Prestressed Concrete*, First Edition, MNL-132-95, Precast/Prestressed Concrete Institute, Chicago, IL, 1995.

8. *Erectors' Manual — Standards and Guidelines for the Erection of Precast Concrete Products*, Second Edition, MNL-127-99, Precast/Prestressed Concrete Institute, Chicago, IL, 1999.

9.6 CONCRETE COATINGS AND JOINT SEALANTS

9.6.1 Coatings for Horizontal Deck Surfaces

Concrete surface sealers reduce moisture and salt (chloride) penetration, which is particularly important in parking structures. While these sealers can enhance the durability of any concrete surface, they do not substitute for basic durable concrete design, nor do they provide protection against penetration of moisture and chlorides through cracks longer than about 0.01 in. Research has shown that the performance of concrete sealers will vary, depending on the product and application. Products should be evaluated against the criteria established in the NCHRP 244 study. [3] Sealers may be classified into two groups: penetrants and surface sealers. [1], [2]

Penetrants. These are generally silanes or siloxanes. They penetrate the surface, reacting with cementitious materials and making the concrete hydrophobic. They do not have crack-bridging capabilities. These materials do not appreciably affect the appearance or characteristics of the surface to which they are applied. While generally more expensive than other types of sealers they are typically longer lasting and less subject to wear under traffic or deterioration from exposure to sun.

Surface Sealers. These are generally polymer resins such as urethanes, epoxies, acrylics, or other proprietary blends. They provide protection by penetrating the surface slightly, and/or by providing a tough film over the surface; these materials also do not bridge cracks. Surface sealers are generally less expensive than penetrants, and performance characteristics of many of these products compare favorably with the penetrating sealers, as demonstrated by the NCHRP 244 criteria. These sealers are more likely to be subject to wear under traffic. They may be more slippery than the bare concrete surface and thus require a grit surface.

9.6.2 Clear Surface Sealers for Architectural Precast Panels

Clear surface coatings or sealers [4] are sometimes used on precast concrete wall panels to improve weathering qualities or to reduce attack of the concrete surface by airborne pollutants. Because of the quality of concrete normally achieved in plant-cast precast concrete, even with very thin sections, sealers are not required for waterproofing. Because the results are uncertain, use of sealers in locations having little or no air pollution is not recommended.

A careful evaluation should be made before deciding on the type of sealer. This includes consultation with the local precasters. In the absence of nearly identical experience, it is desirable to test sealers on reasonably sized samples of varying age to verify performance over a suitable period of exposure or usage, based on prior experience under similar exposure conditions.

Sealers are usually applied to wall panels after joints are caulked to avoid problems with adhesion of joint sealants.

Any coating used should be guaranteed by the supplier or applicator not to stain, soil, or discolor the precast concrete finish. Also, some clear coatings may cause joint sealants to stain concrete. Consult manufacturers of both sealants and coatings or pretest before applying the coating.

Sealers should be applied in accordance with manufacturer's recommendations. Generally, good airless spray equipment is used for uniformity and to prevent surface rundown. Two coats are usually required to provide a uniform coating, because the first coat is absorbed into the concrete. The second coat does not penetrate as much and provides a more uniform surface color. Care must be taken to keep sealers off glass and other adjacent surfaces.

9.6.3 Joint Sealants

Successful performance of wall, roof, and parking deck systems depend on the proper sealant selection and the attention given to the details. Generally, a joint sealant is any material used to prevent the passage of moisture, air, heat, or dirt into or through a joint.

Joint sealants include viscous liquids, mastics or pastes, and tapes, gaskets and foams. Viscous liquid sealants are typically poured in horizontal joint applications. Mastics are applied with a gun and are compounded to prevent sagging or flowing when used in vertical joints. Tapes are most often used around glass. Elastomeric gaskets are used in joints that experience considerable movement. Foams are used as air seals or backup for more durable surface seals.

Most of the raw materials are manufactured by a few major chemical companies and then compounded with other ingredients by numerous sealant manufacturers. The quality of the finished products varies widely so proven performance should be viewed as a key indicator of future success. Refs. 5 through 11 will aid in the design and selection of joint sealants.

Parking structures in particular require special attention to joint detailing and material selection. A high quality traffic-bearing polyurethane or silicone sealant is necessary to prevent intrusion of waterborne salts into the joints and, therefore, prevent the subsequent deterioration of embedded metals. For conditions of severe UV exposure, silicone may be worth consideration from the standpoint of longevity, however two-part polyurethane sealants manufactured specifically for vehicular traffic are typically recommended for parking structure applications.

For precast systems that are topped in the field, joints should be hand-tooled, not saw-cut at a later time, so as to reflect the joints between all of the precast members under the topping. The typical tooled joint is ¾ in. deep and ½ in. wide and is filled with sealant to provide water tightness.

For systems that require no topping, the joints must be prepared with an edging tool in the plant during final finish to provide a rounded edge and then ground or sand-blasted in the field to remove any potentially sharp or rough edges. Some precasters form an indented edge along the continuous edge of the flange to facilitate the installation of the sealant. After a backer rod is installed, the joint is then primed to enhance the bond between the sealant and the concrete and to provide added moisture protection at the bond line. The sealant should be installed in a concave configuration, slightly recessed below the deck surface, to protect it from wheel impact.

9.6.4 References

1. "A Guide to the Use of Waterproofing, Dampproofing, Protective, and Decorative Barrier Systems for Concrete," ACI 515.1R-85, *ACI Manual of Concrete Practice*, Part 5, American Concrete Institute, Farmington Hills, MI.

2. Pfeifer, D. W., and Perenchio, W. F., "Coatings, Penetrants and Specialty Concrete Overlays for Concrete Surfaces," National Association of Corrosion Engineers Seminar, Chicago, IL, September 1982.

3. "Concrete Sealers for Protection of Bridge Structures," NCHRP Report 244, Transportation Research Board, Washington, DC.

4. Litvin, Albert, "Clear Coatings for Exposed Architectural Concrete," *Development Department Bulletin D137*, Portland Cement Association, Skokie, IL, 1968.

5. *Guide Specifications, Section 07900 (Sealants)*, Sealant, Waterproofing and Restoration Institute, Kansas City, MO, 1982.

6. *Sealants: The Professionals' Guide*, Sealant, Waterproofing and Restoration Institute, Kansas City, MO, 1995.

7. *Architectural Precast Concrete*, Second Edition, MNL-122-89, Precast/Prestressed Concrete Institute, Chicago, IL, 1989.

8. "Guide to Sealing Joints in Concrete Structures," ACI 504R-90, *ACI Manual of Concrete Practice*, Part 5, American Concrete Institute, Farmington Hills, MI, 1990 (Reapproved 1997).

9. Cook, J. P., *Construction Sealants and Adhesives*, Wiley-Interscience, New York, NY.

10. Kubal, Michael T., *Waterproofing the Building Envelope*, McGraw-Hill Companies, Inc., New York, NY, 1992.

11. *Standard Guide for Use of Joint Sealants*, ASTM C 1193-90, American Society for Testing and Materials, West Conshohocken, PA, 2000.

9.7 Vibration in Concrete Structures

9.7.1 Notation

a_o = acceleration limit

a_p = peak acceleration

B = width of floor affected by a point load (see 9.7.6.2)

E = dynamic modulus of elasticity = static modulus per ACI 318 x 1.2 (see Ref. 2)

f = forcing frequency, Hertz (see Table 9.7.3)

f_d = driving frequency of equipment, Hertz

f_n = natural frequency of fundamental mode of vibration, Hertz (cycles per sec)

f_{step} = step frequency

g = acceleration due to gravity, 386 in./sec^2 (9,800 mm/sec^2)

i = number of harmonic (see Table 9.7.3)

I = gross moment of inertia

k = a dimensionless constant (1.3 for dancing, 1.7 for lively concert or sports event, 2.0 for aerobics)

K = a constant, given in Table 9.7.1

ℓ = span length

P_o = constant force representing walking force

w = uniform load, per unit length

w_p = effective distributed weight of participants per unit area

w_t = effective total distributed weight per unit area (weight of participants plus weight of floor system)

W = weight of floor panel affected by a point load (see Section 9.7.6.2)

α_i = dynamic coefficient (see Table 9.7.7.2)

β = modal damping ratio (fraction of critical damping)

Δ_f = static deflection of floor system caused by weight of equipment, including inertial block, at location of equipment

Δ_g = instantaneous deflection of a supporting girder

Δ_i = static deflection of isolator

Δ_j = instantaneous simple-span deflection of a floor panel due to dead load plus actual (not code) live load

9.7.2 Human Response to Building Vibrations

This section is a condensation of the material contained in Ref. 1, which is based on information in Refs. 2 to 6.

Limits are stated as a minimum natural frequency of a structural system. These, in turn, depend on the permissible peak accelerations (as a fraction of gravitational acceleration), on the mass engaged during an activity, the degree of continuity of the floor system, the environment in which the vibration occurs, the effectiveness of interaction between connected structural components, and the degree of damping. Much vibration theory derives from experience with steel and wood floors. In general, floor vibrations are much less likely to be a problem with stiffer, more massive, concrete floors.

Some building types common in precast construction are not dealt here, because of a lack of source information. Choice of limits for usage not listed may be selected, with judgment, from other types listed here.

It must be emphasized that the calculations presented are very approximate. The actual natural frequency of a floor can be estimated to a reasonable degree of accuracy, but the calculation of the required frequency is based on damping and on human response, both of which are subject to much variation. When in doubt about the acceptability of a proposed floor system, the best way to decide is to compare it to existing similar systems that are known to be acceptable or unacceptable, using the same method of analysis.

9.7.3 Types of Vibration Analysis

Three types of vibration analysis are described. These analyses differ because the inputs causing the vibration differ.

9.7.3.1 Walking

As a walking person's foot touches the floor, a vibration of the floor system is caused. This vibration may be annoying to other persons sitting or lying in the same area, such as an office, a church, or a residence. Although more than one person may be walking in the same area at the same time, their footsteps are normally not synchronized. Therefore, the analysis is based on the effect of the impact of the steps of individual walking persons.

9.7.3.2 Rhythmic Activities

In some cases, several or many people may engage in a coordinated activity that is at least partially synchronized. Spectators at sporting events, rock concerts, and other entertainment events often move in unison in response to music, a cheer, or other stimuli. The people engaged in the rhythmic activity have a higher level of tolerance for the induced vibrations, while those nearby will have a lower level of tolerance.

9.7.3.3 Mechanical Equipment

Mechanical equipment may produce a constant impulse at a fixed frequency, causing the structure to vibrate.

9.7.3.4 Analysis Methods

Each of the three input types described above requires a somewhat different solution. But, all require knowledge of an important response parameter of the floor system, its natural frequency of vibration.

9.7.3.5 Using Consistent Units

All the equations in this section are dimensionally correct. Provided one is careful to be sure that the units used cancel out to produce the desired units for the answer, a correct result will be obtained using either customary or SI units.

9.7.4 Natural Frequency of Vibration

The natural frequency of a floor system is important in determining how human occupants will perceive vibrations. It has been found that certain frequencies seem to set up resonance with internal organs of the human body, making these frequencies more annoying to people.

The human body is most sensitive to frequencies in the range of 4 to 8 Hertz (cycles per sec). This range of natural frequencies is commonly found for typical floor systems.

9.7.4.1 Computing the Natural Frequency

The natural frequency of a vibrating beam is determined by the ratio of its mass (or weight) to its stiffness. The deflection of a simple span beam is also dependent on its weight and stiffness. A simple relationship exists between deflection and natural frequency of a uniformly loaded simple span beam on rigid supports: [2,3]

$$f_n = 0.18\sqrt{\frac{g}{\Delta_j}} \qquad \text{(Eq. 9.7.4.1)}$$

Figure 9.7.4.1 Natural frequency of selected floor units

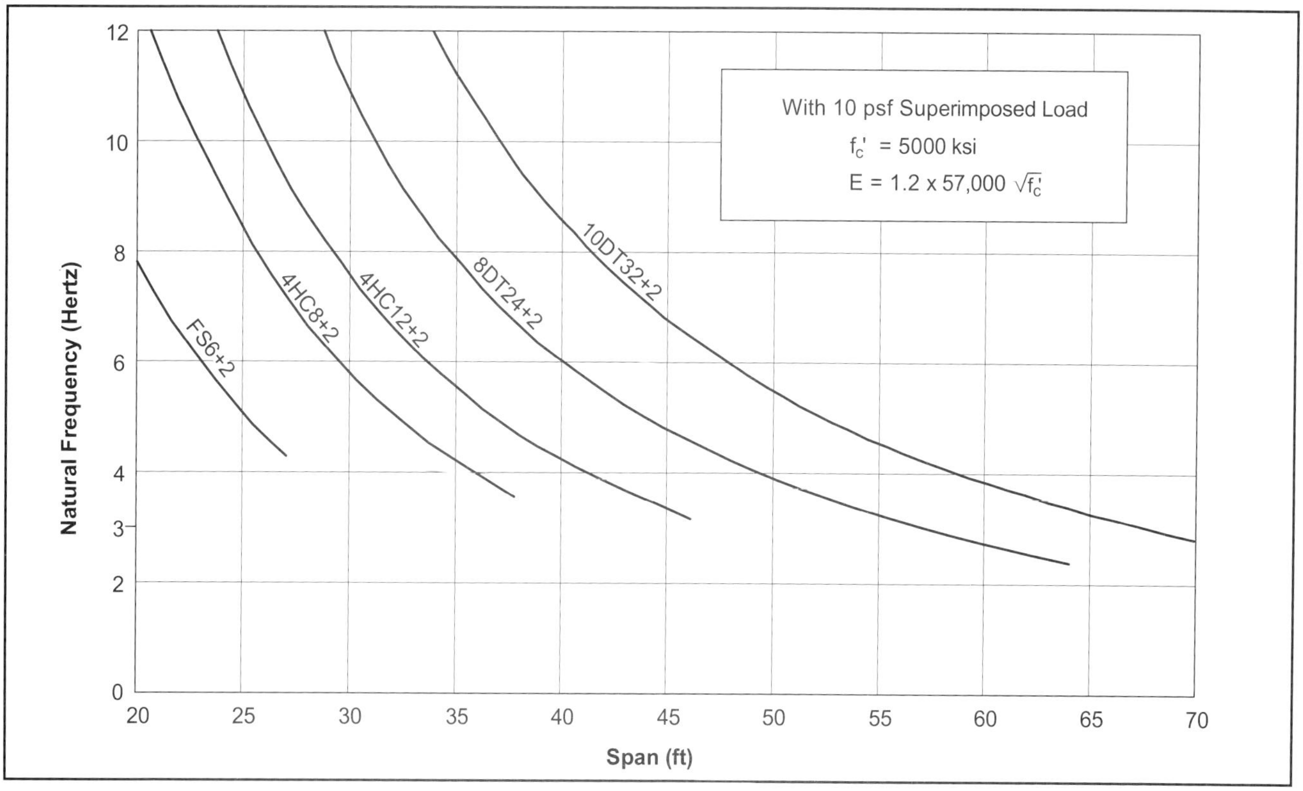

[Graph: Natural Frequency (Hertz) vs Span (ft), showing curves for FS6+2, 4HC8+2, 4HC12+2, 8DT24+2, and 10DT32+2. With 10 psf Superimposed Load, $f_c' = 5000$ ksi, $E = 1.2 \times 57{,}000 \sqrt{f_c'}$]

9.7.4.2 Computing Deflection

The deflection, Δ_j, for a uniformly loaded simple span floor unit is:

$$\Delta_j = \frac{5w\ell^4}{384EI} \quad \text{(Eq. 9.7.4.2)}$$

Many vibration problems are more critical when the mass (or weight) is low. When computing Δ_j, use a minimum realistic live load when computing w, not the maximum live load.

For continuous spans of equal length, the natural frequency is the same as for simple spans. During vibration, one span deflects down while the adjacent spans deflect upward. An inflection point exists at the supports, and the deflection and natural frequency are the same as for a simple span.

For unequal continuous spans, and for partial continuity with supports, the natural frequency may be increased by a small amount. Refs. 2 and 3 suggest how this increase may be computed.

9.7.4.3 Effect of Supporting Girders

The deflection of beams or girders supporting the floor system also affect the natural frequency of the floor system. The simple-span deflection, Δ_g, of the floor girder may be calculated in the same manner as Δ_j. The natural frequency of the floor system may then be estimated by the following formula: [2,3]

$$f_n = 0.18 \sqrt{\frac{g}{\Delta_j + \Delta_g}} \quad \text{(Eq. 9.7.4.3)}$$

For concrete floor systems supported on walls, Δ_g may be assumed to be zero. For concrete floor systems supported by concrete girders, Δ_g is normally small, and is often neglected, unless the girders are unusually long or flexible. For concrete floor units supported on steel beams, the beam deflection can have a significant effect, and should usually be included in computing f_n.

9.7.4.4 Minimum Natural Frequency

Floors with natural frequencies lower than 3 Hertz are not recommended, because people may more readily synchronize their actions at lower frequencies. [3]

9.7.4.5 Graphs of Natural Frequency

Eqs. 9.7.4.1 and 9.7.4.2 may be combined to produce the following Eq. 9.7.4.4, for a floor unit on stiff supports:

$$f_n = \left(\frac{1.58}{\ell^2}\right)\sqrt{\frac{EIg}{w}} \quad \text{(Eq. 9.7.4.4)}$$

Figure 9.7.4.1 shows the relation between span and expected natural frequency for various topped floor units given in Chapter 2.

9.7.5 Damping

Damping usually is expressed is a fraction or percent of critical damping. Real building structures have damping from 1 percent to a few percent of critical.

9.7.5.1 Types of Damping

Damping is not a well understood phenomenon. In the literature, differing methods are used for calculation. This section and its references are based on modal damping. Do not mix values of damping from other sources with damping values in the equations of this section, as they may be based on a different calculation method.

9.7.5.2 Estimation of Damping

Damping of a floor system is highly dependent on the non-structural items (partitions, ceilings, furniture, etc.) present. The modal damping ratio of a bare structure can be very low, on the order of 0.01. Non-structural elements may increase this, up to 0.05.

The results of a vibration analysis are highly influenced by the choice of the assumed damping, which can vary widely. Yet, this choice is based more on judgment than science.

9.7.6 Vibrations Caused by Walking

Vibrations caused by walking are seldom a problem in concrete floor systems because of their mass and stiffness. When using concrete floor systems of ordinary proportions, it is usually not necessary to check for vibrations caused by walking.

When designing concrete floor systems of long-span or slender proportions, this section may be used to evaluate their serviceability with respect to vibrations.

9.7.6.1 Minimum Natural Frequency

An empirical formula, based on resonant effects of walking, has been developed to determine the minimum natural frequency of a floor system needed to prevent disturbing vibrations caused by walking: [4]

$$f_n \geq 2.86 \left[\ln\left(\frac{K}{\beta W}\right) \right] \quad \text{(Eq. 9.7.6.1)}$$

The constant 2.86 has the units 1/sec.

9.7.6.2 Effective Weight

The effect of an impact such as a footfall is strongly influenced by the mass (or weight) of the structure affected by the impact. This weight, W, is normally taken as the unfactored dead load (per square foot) of the floor units plus some (not full code) live load, multiplied by the span and by a width B. For solid or hollow-core slabs, which are stiff in torsion, it is recommended to take B equal to the span. [2] For double tees, it is recommended to take B varying from 0.8ℓ for 18-in. double tees with 3-in. topping to 0.6ℓ for 32-in. double tees with 3-in. topping. [5] For continuous spans, W may be increased 50 percent. [2,3] At an unstiffened edge of a floor, the width B used for estimating floor system weight should be halved. [2]

9.7.6.3 Recommended Values

The recommended values of K and β for use in Eq. 9.7.6.1 are given in Table 9.7.6.1 below.

Table 9.7.6.1 Values of K and β for use in Eq. 9.7.6.1 (based on Table 3 of Ref. 4)

Occupancies Affected by the Vibrators	K Kips	kN	β
Offices, Residences, Churches	13	58	0.02[a] 0.03[b] 0.05[c]
Shopping Malls	4.5	20	0.02
Outdoor Footbridges	1.8	8	0.01

a. For floors with few non-structural components and furnishings, open work area, and churches.
b. For floors with non-structural components and furnishings, cubicles.
c. For floors with full-height partitions.

Example 9.7.6.1
Vibrations Caused by Walking

Given:
   10DT32+2 (see Chapter 2)
   Open office area : 60-ft span

Problem:
   Check for vibration caused by walking.

Solution:
Use Eq. 9.7.6.1 to find minimum required f_n:

$$f_n \geq 2.86\left[\ln\left(\frac{K}{\beta W}\right)\right] \quad \text{(Eq. 9.7.6.1)}$$

Estimate:
   β = 0.02   (Table 9.7.6.1)
   K = 13 kips   (Table 9.7.6.1)

Estimate effective weight W:
   w = 89 psf + 10 psf assumed superimposed

Estimate effective width B = 0.6ℓ = 36 ft:

   W = w (B)(ℓ) = 0.099 kips/ft² (36)(60)
      = 214 kips

Note that in this case, it was not necessary to convert ft to in., because the ft units cancel out.

Minimum requirement from Eq. 9.7.6.1:

$$f_n = 2.86\left[\ln\left(\frac{13}{0.02(214)}\right)\right] = 3.18 \text{ Hertz}$$

From Figure 9.7.4.1 for a 10DT32+2 with a 60 ft span, the expected fundamental frequency is 3.8 Hertz, which is greater than the required minimum value of 3.18 Hertz.

9.7.7 Design for Rhythmic Excitation

Rhythmic excitation occurs when a group of people move in unison to a cadence. A resonance can occur when the input frequency is at or near the fundamental frequency of vibration, and the fundamental frequency of the floor must be sufficiently higher than the input frequency to prevent resonance.

Table 9.7.7.1 Recommended acceleration limits for rhythmic activities (based on Table 2-3, Ref. 2)

Occupancies Affected by the Vibration	Acceleration Limit, Fraction of Gravity, a_o/g
Office or residential	0.004 – 0.007
Dining	0.015 – 0.025
Weightlifting	0.015 – 0.025
Rhythmic activity only	0.04 – 0.07

9.7.7.1 Harmonics

A harmonic of a frequency is any higher frequency that is equal to the first or fundamental frequency multiplied by an integer. If the fundamental frequency of a floor system is equal to a harmonic of the exciting frequency, resonance may occur.

9.7.7.2 Recommended Minimum Natural Frequency

The following design criterion for minimum natural frequency for a floor subjected to rhythmic excitation is based on dynamic response of the floor system to dynamic loading: [2, 3]

$$f_n \geq f\sqrt{1+\left(\frac{k}{a_o/g}\right)\left(\frac{\alpha_i w_p}{w_t}\right)} \quad \text{(Eq. 9.7.7.1)}$$

See Table 9.7.7.1 for limiting values of a_o/g and Table 9.7.7.2 for α_i and f.

The computation of the natural frequency of the floor system f_n is done as discussed in Section 9.7.4. If rhythmic activities take place on the upper floors of a tall building, it is sometimes necessary to take the elastic shortening of the columns into effect, in a manner similar to Eq. 9.7.4.3 for girder flexibility. This is discussed in Ref 7.

Recommended values for all of the parameters on the right side of Eq. 9.7.7.1 are given in Tables 9.7.7.1 and 9.7.7.2, except for w_t, which includes the actual distributed dead weight of the floor system. Note that Eq. 9.7.7.1 uses the distributed weight w_t, not the total weight of a panel W that was used in Eq. 9.7.6.1.

9.7.7.3 Higher Harmonics

Eq. 9.7.7.1 will always require a higher natural frequency f_n than the forcing frequency f. Thus, a crucial decision is the determination of whether the forcing frequencies for higher harmonics need be

Table 9.7.7.2 Estimated loading during rhythmic events (based on Table 5.2, Ref. 3)

Activity	Forcing Frequency f, Hertz	Weight of Participants,* w_p psf	Weight of Participants,* w_p kPa	Dynamic Coefficient α_i	Dynamic Load $\alpha_i w_p$ psf	Dynamic Load $\alpha_i w_p$ kPa
Dancing: First Harmonic	1.5 – 3	12	0.6	0.5	6	0.3
Lively concert or sports event: First Harmonic Second Harmonic	1.5 – 3 3 – 5	30 30	1.5 1.5	0.25 0.05	8 1.5	0.4 0.07
Jumping exercises: First Harmonic Second Harmonic Third Harmonic	2 – 2.75 4 – 5.5 6 – 8.25	4 4 4	0.2 0.2 0.2	1.5 0.6 0.1	6 2.4 0.4	0.3 0.1 0.02

* Based on maximum density of participants on the occupied area of the floor for commonly encountered conditions. For special events, the density of participants can be greater.

considered. Eq. 9.7.7.2 gives the peak acceleration a_p/g for a condition of resonance: [3]

$$\frac{a_p}{g} = \left(\frac{1.3}{2\beta}\right)\left(\frac{\alpha_i w_p}{w_t}\right) \quad \text{(Eq. 9.7.7.2)}$$

In applying Eq. 9.7.7.2, Ref. 3 recommends a value for the damping ratio β as follows. "Because participants contribute to the damping, a value of approximately 0.06 may be used, which is higher than … for walking vibration."

If the damping ratio β or the total distributed weight, w_t, is high enough, the dynamic load $\alpha_i w_p$ from Table 9.7.7.2 for higher harmonics may result in a peak acceleration a_p/g within the acceleration limits a_o/g given in Table 9.7.7.1. If this is so, that harmonic need not be considered.

Most topped or pre-topped concrete floors weigh 75 psf or more. For these floors, the weight, w_t, is such that the resonant acceleration at the third harmonic frequency will usually be within limits. Usually, only the first and second harmonics need be considered for topped concrete floors.

9.7.7.4 Adjacent Activities

A space with a quiet activity may be located next to a space with rhythmic activity. In such cases, it is desirable to have a rigid wall between the two spaces, supporting the floor system in each space. If this is not practical, the acceleration limits for the quiet activity should be used in combination with the rhythmic loading for the rhythmic activity. This combination can often be critical for concrete floor systems, requiring a stiffer floor than needed for supporting gravity loads.

9.7.8 Stadium Seating

Precast, prestressed concrete seating slabs, such as that shown in Example 9.7.8.1, are often used in stadiums and arenas. They are usually manufactured in units that are two or three seats wide. Connections are usually provided between the upper and lower units, to prevent differential deflection of the adjacent units.

These seating slabs are subjected to rhythmic excitation, as a crowd responds in unison to a cheer or song. The response of the seating slabs is different from that of an ordinary flat floor. The seating slabs have a three-dimensional nature, and vibrate and deflect about their weakest principal axis, as shown in the example. Furthermore, the bays in stadiums are often of non-uniform width. This causes each seating slab to have a different span and thus, a different natural frequency. This helps to prevent resonance.

Fifty-six different seating slabs made by PCI members have been examined. All are known to have satisfactory performance in service. This examination produced the following recommendations:

1. The slabs should be sufficiently interconnected, with a minimum of three connections per bay, to prevent differential deflection between adjacent units. If people sit on one slab, with their feet resting on another slab below, they are much more sensitive to differential deflections of the two units.

Example 9.7.8.1
Stadium Seat

Given:
Stadium seat section shown below
f'_c = 5000 psi (Normal Weight Concrete)
E = 4031 ksi × 1.2 = 4837 ksi
w = 474 lb/ft
I_{min} = 12,422 in.^{4} on inclined weak axis

Problem:
Find maximum span governed by vibration.

Solution:
Use Eq. 9.7.7.1 to find minimum natural frequency:

$$f_n \geq f\sqrt{1+\left(\frac{k}{a_o/g}\right)\left(\frac{\alpha_i w_p}{w_t}\right)}$$

where k = 1.7 for sports events.
Refer to Tables 9.7.7.1 and 9.7.7.2.

For first harmonic use the following assumptions:

Forcing frequency f = 2.5 Hertz   (Table 9.7.7.2)
Acceleration limit a_o/g = 0.06   (Table 9.7.7.1)
Weight of participants w_p = 30 psf   (Table 9.7.7.2)
Dynamic load $\alpha_i w_p$ = 8 psf   (Table 9.7.7.2)

Dynamic load component in weak direction = $8\left(\frac{64}{12}\right)(\cos 31.8°)$ = 36.3 lb/ft

Total weight (a measure of mass) = 474 lb/ft + $30 \times \frac{64}{12}$ ft = 634 lb/ft

Note that the mass is not reduced by cos 31.8 degrees, because mass is the same in all directions.
For bays of uniform width (Figure 9.7.8.1)

$$f_n \geq 2.5\sqrt{1+\left(\frac{1.7}{0.06}\right)\left(\frac{36.3}{634}\right)} = 4.0 \text{ Hertz}$$

Find maximum span from Eq. 9.7.4.4:

$$f_n \geq \frac{1.58}{\ell^2_{max}}\sqrt{\frac{EIg}{w}} = \frac{1.58}{\ell^2_{max}}\sqrt{\frac{4837(12,422)(386)}{0.634/12}} = \frac{1,047,000}{\ell^2_{max}}$$

$$\ell^2_{max} = \frac{1,047,000}{f_n} = \frac{1,047,000}{4.0} = 261,750 \text{ in.}^2$$

$$\ell_{max} = \sqrt{261,750} = 512 \text{ in.} = 43 \text{ ft}$$

For bays of non-uniform width (Figure 9.7.8.1), assume f_n = 0.75(4) = 3 Hertz

$$\ell_{max} = \sqrt{\frac{1,047,000}{3.0}} = 591 \text{ in.} = 49 \text{ ft}$$

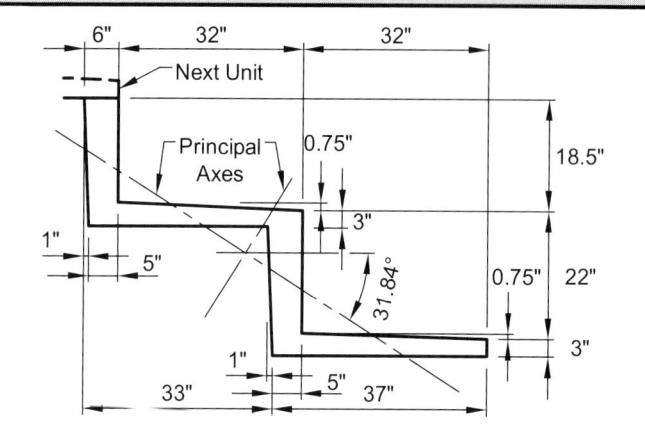

Figure 9.7.8.1 Uniform and non-uniform width bays in a stadium

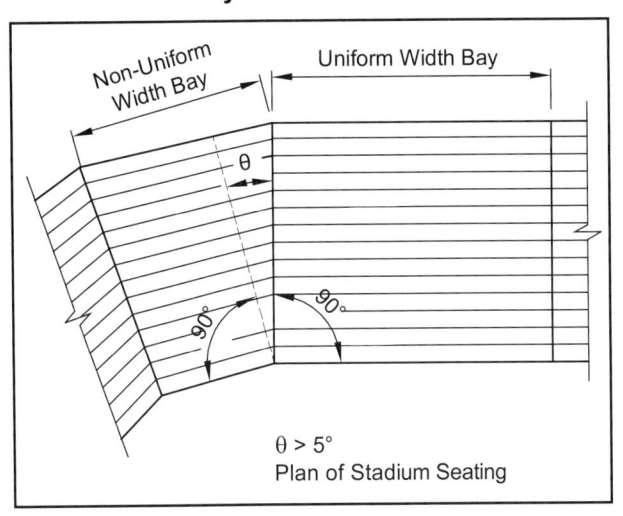

Plan of Stadium Seating
θ > 5°

2. For bays of uniform width with an angle in plan, θ, less than 5 degrees (see Figure 9.7.8.1), the minimum natural frequency requirement for the first harmonic should be satisfied. For bays of non-uniform width with an angle in plan, θ, of 5 degrees or more, the minimum frequency requirement may be reduced 25 percent.

3. For interconnected units, the minimum natural frequency requirement for the second harmonic need not be applied. Apparently, vibration in a diagonal plane caused by the second harmonic of the input motion is not occurring in these seating slabs.

9.7.9 Vibration Isolation for Mechanical Equipment

Vibration produced by equipment with unbalanced operating or starting forces can usually be isolated from the structure by mounting on a heavy concrete slab placed on resilient supports. This type of slab, called an inertia block, provides a low center of gravity to compensate for thrusts, such as those generated by large fans.

For equipment with less unbalanced weight, a "housekeeping" slab is sometimes used below the resilient mounts to provide a rigid support for the mounts and to keep them above the floor so they are easier to clean and inspect. This slab may also be mounted on pads of precompressed glass fiber or neoprene.

The natural frequency of the total load on resilient mounts must be well below the frequency generated by the equipment. The required weight of an inertia block depends on the total weight of the machine and the unbalanced force. For a long-stroke compressor, five to seven times the compressor weight might be needed. For high pressure fans, one to five times the fan weight is usually sufficient.

A floor supporting resiliently mounted equipment must be stiffer than the isolation system. If the static deflection of the floor approaches the static deflection of the mounts, the floor becomes a part of the vibrating system, and little vibration isolation is achieved. In general, the floor deflection should be limited to about 15 percent of the deflection of the mounts.

Simplified theory shows that for 90 percent vibration isolation, a single resilient supported mass (isolator) should have a natural frequency of about one-third the driving frequency of the equipment. The natural frequency of this mass can be calculated by: [8]

$$f_n = 0.16\sqrt{\frac{g}{\Delta_i}} \quad \text{(Eq. 9.7.9.1)}$$

From the above, the required static deflection of an isolator can be determined as follows:

$$f_n = \frac{f_d}{3} = 0.16\sqrt{\frac{g}{\Delta_i}}$$

or $\quad \Delta_i = 0.23\left(\dfrac{g}{f_d^2}\right) \quad$ (Eq. 9.7.9.2)

and $\quad \Delta_f \leq 0.15\Delta_i \quad$ (Eq. 9.7.9.3)

Example 9.7.9
Vibration Isolation

Given:
A piece of mechanical equipment has a driving frequency of 800 cycles per min.

Problem:
Determine the approximate minimum deflection of the isolator and the maximum deflection of the floor system that should be allowed.

Solution:

Use Eq. 9.7.9.2:
$f_d = 800/60 = 13.33$ Hertz

$\Delta_i = 0.23\dfrac{g}{(f_d)^2} = 0.23\dfrac{386 \text{ in./sec}^2}{(13.33 \text{ Hertz})^2}$

$\Delta_i = 0.50$ in.

From Eq. 9.7.9.3:
$\Delta_f \leq 0.15(0.5) = 0.075$ in.

9.7.10 References

1. Mast, R. F., "Vibration of Precast Prestressed Concrete Floors," PCI JOURNAL, V. 46, No. 6, November-December 2001, pp. 76-86.

2. ATC, *ATC Design Guide 1, Minimizing Floor Vibration*, Applied Technology Council, Redwood City, CA, 1999, 49 pp.

3. AISC/CISC, *Steel Design Guide Series 11. Floor Vibrations Due to Human Activity*, American Institute of Steel Construction, Chicago, IL, 1997.

4. Allen, D. E., and Murray, T. M., "Design Criterion for Vibrations Due to Walking," *Engineering Journal*, Fourth Quarter, 1993, American Institute of Steel Construction, pp. 117-129.

5. Chen, Y., and Aswad, A., "Vibration Characteristics of Double Tee Building Floors," PCI JOURNAL, V. 39, No. 1, January-February 1994, pp. 84-95.

6. Murray, T. M., "Floor Vibrations: Tips for Designers of Office Buildings," *Structure*, Fall 2000, National Council of Structural Engineers, Washington, DC, pp. 26-30.

7. Allen, D. E., "Building Vibration From Human Activities," *Concrete International*, V. 12, No. 6, June 1990, pp. 66-73.

8. Harris, C. M., and Crede, C. E., *Shock and Vibration Handbook*, Second Edition, McGraw-Hill Book Co., New York, NY, 1976.

9.8 CRACKING, REPAIR AND MAINTENANCE

9.8.1 Cracking

Minor cracking may occur in precast concrete without being detrimental, and it is impractical to impose specifications that prohibit all cracking. [5, 6] However, in addition to being unsightly, cracks are potential locations of moisture intrusion and possible later concrete deterioration, and should be avoided if possible. Prestressing and proper handling procedures are two of the best methods of keeping cracks to a minimum. To evaluate the acceptability of a crack, the cause and service conditions of the precast unit should be determined. (Note: In other sections, the term "crack-free" design is used. This refers to a design in which concrete tension is kept within limits, and should not be construed as being a guarantee that there will be no cracking.)

Crazing, i.e., fine, random (commonly called "hairline") cracks, may occur in the cement film on the surface of concrete. The primary cause is the shrinkage of the surface with respect to the mass of the unit. Crazing has no structural significance and does not significantly affect durability, but may be visually accentuated if dirt settles in these minute cracks. Crazing should not be cause for rejection.

Tension cracks are sometimes caused by temporary loads during production, transportation, or erection of these products (see Chapter 5). These cracks may extend through to the reinforcement. If the crack width is narrow, the structural adequacy of the casting will remain unimpaired, as long as corrosion of the reinforcement is prevented. The acceptability of cracks wider than recommended maximums should be governed by the function of the unit. Most cracks can be effectively repaired and sealed.

Long-term volume changes can also cause cracking after the member is in place in the building, if the connections provide enough restraint to the member (see Section 3.4). Internal causes, such as corrosion of reinforcement or cement-aggregate reactivity, can also lead to long-term cracking and should be considered when materials are selected.

Refs. 1 to 3 provide general information on cracking in concrete. Refs. 4 to 6 provide information on cracking in certain specific precast and/or prestressed concrete members.

9.8.2 Repair

Precast concrete members may occasionally require repairing on the job site. A number of different time tested repair methods are used, as discussed in the following. With proper design and installation, repairs can be made to satisfy the original requirements of the member. To ensure a permanent repair, analysis to determine the cause of the defect or distress should be made. If the defect has structural significance, it may be possible to transfer loads to other members or connections. Considerations relating to the type of repair include evaluation of the following:

- Environmental conditions during installation and service.
- Fire protection requirements.
- Cosmetic requirements.
- Access to the repair and impact on adjacent construction.
- Requirements and limitations for supplementary supports.
- Load testing, if required to evaluate the repair.
- Requirements for design approval and inspection.

9.8.2.1 Patching

Patches may be either structural or cosmetic. Structural patches may be required to satisfy some load transfer requirements, while cosmetic patches are related to appearance only. Large spalls require repair by structural patching. Honeycomb may require no patching or only cosmetic patching. Procedures for the preparation of the materials and application of the patch are important. Surface preparation is about the same for all types of material chosen for the patch. All loose concrete must be removed from the patch area. If possible, the edges of the repair should be chiseled with a slight undercut to allow keying action for the repair material. Feathered edges should be avoided. The surface must be clean, sound and rough to accept the patch, with any laitance removed from any exposed reinforcement.

Occasionally, mechanical anchorage may be advisable in addition to a bonding agent to anchor the patch. Ideally, the concrete substrate should be saturated surface dry (unless using an epoxy bonding agent or patching mortar, in which case the presence of moisture may not be allowed). The bonding agent, if used, should be applied thoroughly over the entire area to be patched, following manufacturer's instructions. Following is information on the various commonly used patching materials:

Cementitious Mortars. They are composed of cement, sand (usually about two to three parts sand per part cement), and water. Various admixtures may be incorporated. Water quantity can vary from minimal, creating a dry pack mortar, to a moderate amount for a more workable mixture. The advantage of dry pack mortars is that they develop high strengths with very little shrinkage. Wetter mortars have better bond characteristics and are easier to place. Proper curing is essential.

Prepackaged Cementitious Mortars. Prepackaged grouts are a mixture of cement, water, aggregate (usually sand), and often one or more admixtures. They are mixed with water according to the manufacturer's instructions to create the patching mortar. The more commonly used admixtures reduce shrinkage, increase strength and/or strength gain and enhance durability. Patching materials that contain calcium chloride should be avoided. The manufacturer's recommendations should always be followed.

Concrete. As a patching material, this is especially desirable in larger repairs. Concrete should be properly proportioned mixes of cement, aggregate and water to produce a workable mix with minimum water-cement ratio to minimize shrinkage. Ideally, the concrete repair material should be the same as the concrete in the member. Proper curing is essential.

Epoxy Mortars. This is a mixture of epoxy resin and aggregate (usually sand). The mortar is prepared by thoroughly mixing as much aggregate as possible into the mixed epoxy, consistent with workability needs. The sand must be "oven dry." Maximizing sand content minimizes heat buildup from the chemical reaction and the associated tendency towards shrinkage and cracking. Epoxy mortars are strong and cure rapidly. Dimensional compatibility due to differing modulus of elasticity with the base concrete should be considered. [7]

Bonding Agents. They are used to increase the bond between a hardened concrete and fresh mortar. Bonding agents may be an unsanded cement paste, a commercially available latex-based material, or an epoxy. Epoxy bonding agents must have a long pot life to prevent hardening prior to curing of the patching mortar. It is important to follow the manufacturer's recommendations for the use of commercially prepared bonding agents to ensure their effectiveness. Bonding agents are not required for all patches. They are most advisable when using a stiff mortar or when the patch surface is large and the thickness is relatively thin.

Cosmetic patching should match the surrounding concrete in form, color and texture. The materials used in the repair should weather in a manner similar to the base concrete. Substrate preparation and bonding of the patch to the substrate are essentially the same as for structural patching. The patching mixture consists of the basic ingredients of cement, fine aggregate, and water.

Coarse aggregate, color additives and various admixtures may also be required. The proportions are usually best determined by an experienced finisher at the plant or the site. It is recommended that records be kept on patching mixtures and techniques to establish proper patching procedures on future projects.

9.8.2.2 Crack Repair

Cracks do not always require repair, or may need only cosmetic treatment. Refs. 5 and 6 discuss cracking in precast members, and should be reviewed before a repair progam is initiated. If repair of a crack is necessary, low viscosity epoxy is usually the best material. There are many available epoxy materials that can fill and repair cracks. As with patching, surface preparation is important. The crack should be clean and clear of loose material, and oil, grease, and dust. If moisture is present, a moisture resistant epoxy should be selected. It is important to follow the manufacturer's directions in mixing and applying epoxies to concrete. Epoxy may be introduced by either pressure injection or by gravity flow.

Gravity flow involves forming a small ridge of sand or other material along each side of the crack, or a V-groove may be chipped out along the top of the crack. This serves to direct the epoxy into the crack. If the crack extends to the underside of the member, it is recommended to patch the bottom of the crack with plaster of Paris or other suitable material. Then low viscosity epoxy is poured into the crack until no more is accepted. Cracks should be wider than 0.005 in. and less than about 4 in. deep to allow epoxy entry.

Injection involves forcing epoxy (normally low viscosity) into the crack under pressure. Injection applications generally start with setting injection ports into the crack, spaced 6 to 10 in., but no further apart than the crack depth. Then the surface of the crack is sealed between the ports, usually with a high viscosity epoxy. Injection progresses from one end of the crack to the other, pumping material into each port until epoxy begins exiting the next port. After sealing that port, the pumping operation is shifted to the next one. After the epoxy is cured, the injection ports are removed and the sealing material ground flush to the surface. [14]

9.8.2.3 Connection Repair

Connections may require repair or modification for such reasons as placement errors, tolerance issues, damage, or omission of an embedment. Prior to choosing a repair, the magnitude of the design load should be verified. Eccentricities, load paths, and effect of edge distances on connection strength should be considered. Connection repair methods include, but are not limited to, the following methods:

- Up-sizing welds on connection materials to increase connection strength when embedded plates are misaligned.
- Increasing the size of connection material.
- Adding reinforcement in a topping slab to replace a deficient connection.
- Adding through bolts or post-installed anchors with appropriate connection materials.
- Installing a new connection in the vicinity of the repair.
- Adding stiffeners to a connection.

When choosing a repair method, the requirements of fire rating should be considered. When selecting proprietary expansion anchors and epoxy anchors, some have load limitations that require special inspection. Note that epoxy anchors should be used with caution when adjacent to welding, which could heat the epoxy. Epoxy anchors should not be used in connections where fire rating is a consideration.

9.8.2.4 Member Strengthening

Capacity of existing precast/prestressed members can be increased by different methods. The following discusses a few of the common methods:

Drilling and Installing Reinforcement. Reinforcing bars can be installed by drilling holes and anchoring the bars in the holes by use of epoxy and grout. This method is of limited usefulness because of the relatively short lengths of holes that can be drilled in concrete. This is most commonly used to install missing dowels or shear ties.

Incorporating Reinforcement in Hollow-Core Voids. Hollow-core slabs can be strengthened by grouting or concreting reinforcing bars into the voids. If shear strengthening alone is required, then filling the voids with grout or concrete may be sufficient to increase the effective width. The procedure consists of cutting into the desired void(s), cleaning out and verifying that the surfaces are sound enough to bond new concrete, placing reinforcement for the length required, and filling the void around the reinforcement with concrete of strength similar to that of the slab.

External Post-tensioning. By adding post-tensioning externally to existing members, the flexural and shear strengths can be increased. This usually consists of installing anchorages on the sides or bottom of each end of the member, installing deflection devices (if required) at points along the member, installing the post-tensioning strand attached to these points, and stressing. Corrosion protection, and fire resistance protection must be considered. Refer to ACI 318, [8] Section 18.22, for applicable regulations for external post-tensioning.

Fiber Reinforced Polymers (FRP). FRP systems offer a simple solution to restore or supplement the strength of concrete members. These polymers are a composite system comprised of fibers embedded in an epoxy resin matrix. Fibers of carbon, glass, or aramid are formed into a unidirectional or bi-directional fabric. The fabric is then saturated in an epoxy resin during manufacture or during application. The resin transfers load from the structural member to and between the fibers. FRP can be bonded to a concrete member in strips or sheets to increase flexural and/or shear strength. By wrapping a column, the axial strength, shear strength, and ductility can be upgraded. It is recommended that design proceed in close consultation with the manufacturer's technical staff. Inspection should be provided for all aspects of application. [9] Additional fire protection may be required.

Structural Plate Bonding and Steel Jacketing. Structural plate bonding and jacketing are methods for attaching steel plates or other steel elements to concrete members, to increase strength. Steel plates are attached to the concrete by adhesives and mechanical fasteners. The process requires removing damaged concrete from the surface, repairing the substrate, roughening the concrete surface, installing drilled mechanical anchors (used for positioning the steel and for load transfer), and epoxying the steel to the concrete surface. Cosmetic treatment and corrosion resistant and fire resistant coatings may be applied as necessary. Design is as for reinforced concrete. The interface should be analyzed for horizontal shear using the shear capacity of the epoxy and of the mechanical anchors. [10]

Shotcrete. Shotcrete is mortar or concrete conveyed through a hose and pneumatically projected at high velocity onto a surface. This process can be used to add concrete to existing members for various purposes including: to increase the size of a concrete member, to add reinforcement cover for increasing fire or corrosion resistance, or to encase added reinforcement. Shotcrete can be an advantageous solution for either tight access or high volume repairs.

Before shotcreting, all loose and damaged concrete is removed. A bonding agent is usually not required if the surface is roughened by sandblasting or needle scaling. Micro-cracking of the substrate, which is detrimental to bond, can be caused by over aggressively removing loose concrete with bush hammers or jackhammers. The substrate should be at saturated surface dry condition just prior to shotcrete application.

Shotcrete can be applied by either the dry or wet method. In the dry method, water is added to the mixed dry ingredients at the nozzle. In the wet method, water is added to the mixer prior to entering the hose. Advantages of the dry method include thinner placement layers, suitable for intermittent application and better for remote locations (longer pumping distances). Advantages of the wet method include ability to use air entrainment and some other admixtures, a higher placement rate, less rebound, and requires less cleanup.

Shotcrete should be installed by certified nozzlemen and should be monitored by a qualified inspector. Prior to shotcreting, test panels mimicking the actual application should be prepared at the site by the same personnel, using the same mix and equipment to be used in the actual installation.

9.8.3 Maintenance

Concrete surfaces normally require little maintenance except for those that are subject to harsh environments, such as parking structures in northern climates. These should receive regular inspections and washings to help in identifying and preventing long-term maintenance problems. Surfaces that have had sealers applied to them should have a program for re-applying the sealer at specified intervals. It is important that building drains be kept open to keep water from ponding and potentially freezing on the concrete. Routine inspection and maintenance can do much to extend the life of precast concrete buildings. Refs. 11 and 12 provide guidance for maintenance of parking structures and Ref. 13 provides maintenance information for architectural precast concrete.

9.8.4 References

1. "Control of Cracking in Concrete Structures, (ACI 224R-90)," *ACI Manual of Concrete Practice*, Part 3, American Concrete Institute, Farmington Hills, MI, 1990.

2. "Causes, Evaluation, and Repair of Cracks in Concrete Structures (ACI 224.1R-93)," *ACI Manual of Concrete Practice*, Part 3, American Concrete Institute, Farmington Hills, MI, 1993.

3. "Guide to Durable Concrete (ACI 201.2R-92)," *ACI Manual of Concrete Practice*, Part 1, American Concrete Institute, Farmington Hills, MI.

4. *Manual for Quality Control for Plants and Production of Architectural Precast Concrete Products*, Third Edition, MNL-117-96, Precast/Prestressed Concrete Institute, Chicago, IL, 1996.

5. PCI Committee on Quality Control Performance Criteria, "Fabrication and Shipment Cracks in Prestressed Hollow-Core Slabs and Double Tees," PCI JOURNAL, V. 28, No. 1, January-February 1983.

6. PCI Committee on Quality Control Performance Criteria, "Fabrication and Shipment Cracks in Precast or Prestressed Beams and Columns," PCI JOURNAL, V. 30, No. 3, May-June 1985.

7. "Use of Epoxy Compounds with Concrete (ACI 503R-93)," *ACI Manual of Concrete Practice*, Part 5, American Concrete Institute, Farmington Hills, MI, 1993 (Reapproved 1998).

8. ACI Committee 318, "Building Code Requirements for Structural Concrete and Commentary (ACI 318-02/318R-02)," American Concrete Institute, Farmington Hills, MI, 2002.

9. "State-of-the-Art Report on Fiber Reinforced Plastic (FRP) Reinforcement for Concrete Structures (ACI 440R-96)," *ACI Manual of Concrete Practice*, Part 5, American Concrete Institute, Farmington Hills, MI, 1996.

10. *Repair and Strengthening of Concrete Members with Adhesive Bonded Plates*, SP-165, American Concrete Institute, Farmington Hills, MI.

11. *Precast, Prestressed Concrete Parking Structures: Recommended Practice For Design and Construction,* MNL-129-98, Precast/Prestressed Concrete Institute, Chicago, IL, 1998.

12. *Parking Garage Maintenance Manual*, National Parking Association, Washington, DC, 1996.

13. *Architectural Precast Concrete*, Second Edition, MNL-122-89, Precast/Prestressed Concrete Institute, Chicago, IL, 1989.

14. Trout, John F., *Epoxy Injection in Construction*, The Aberdeen Group, Addison, IL, 1997.

9.9 PRECAST SEGMENTAL CONSTRUCTION

9.9.1 General

Segmental construction is a method of construction in which primary load carrying members are composed of individual segments post-tensioned together. This allows precast concrete to be used for long horizontal or vertical spans constructed of smaller units within size limitations imposed by manufacturing, transportation, and handling equipment. With proper planning and element selection, a large re-use of forms is possible, with the resulting economy.

The method is best known for a number of landmark bridges [1,6,10], but has also been used in the construction of airport control towers, storage tanks for various solids and liquids, including natural gas, long-span exterior Vierendeel trusses on high-rise office buildings, an Olympic stadium and other non-bridge structures. [2–9]

Segmental construction requires that the designer give special consideration to:

1. Size and weight of the precast elements.
2. Configuration and behavior of the joints between elements.
3. Construction sequence, and the loads and deflections imposed at various stages.
4. The effect of normal tolerances and deviations upon the joints.

9.9.2 Joints

Joints may be either "open," to permit completion by a field placed grout, or "closed," where the joint is either dry or bonded by a thin layer of adhesive (Figure 9.9.2.1).

9.9.2.1 Open Joints

The individual segments are separated by an amount sufficient to place (usually by pressure) a grout mix, but not more than about 2 in. Prior to placing the segments, the joint surface is thoroughly cleaned and wire brushed or sand-blasted.

The perimeter of the joint is sealed with a gasket which is compressed by use of "come-alongs" or by a small amount of prestress. Gaskets are also provided around the post-tensioning elements to prevent leakage into the ducts, blocking passage of the tendons. Vents are provided at the top to permit escape of entrapped air during grouting. Prior to filling the joint, the surfaces should be thoroughly wetted, or coated with a bonding agent. Grout strength should be at least equal to that of the precast segments, but not less than 4000 psi.

After grouting, vents are closed and pressure increased to ensure full grout intrusion. After a few days, the vent is reopened, and filled with grout as required.

9.9.2.2 Closed Joints

If a closed joint is used, the segment is usually "match-cast," i.e., each segment is cast against its previously cast neighbor. A bond breaker is applied to the joint during casting. Thus, the connecting surfaces fit each other accurately, so that little or no filling material is needed at the joint. The sharpness of line of the assembled construction depends mainly on the accuracy of the manufacture of the segments.

Match-cast segments are usually joined by coating the abutting surfaces with a thin layer of epoxy adhesive, and then using the post-tensioning to draw the elements together and hold them in position. In some structures, dry joints have been used successfully, in which case the compression provided by the post-tensioning is relied upon for weatherproofing, as well as for transfer of forces.

Figure 9.9.2.1 Types of joint

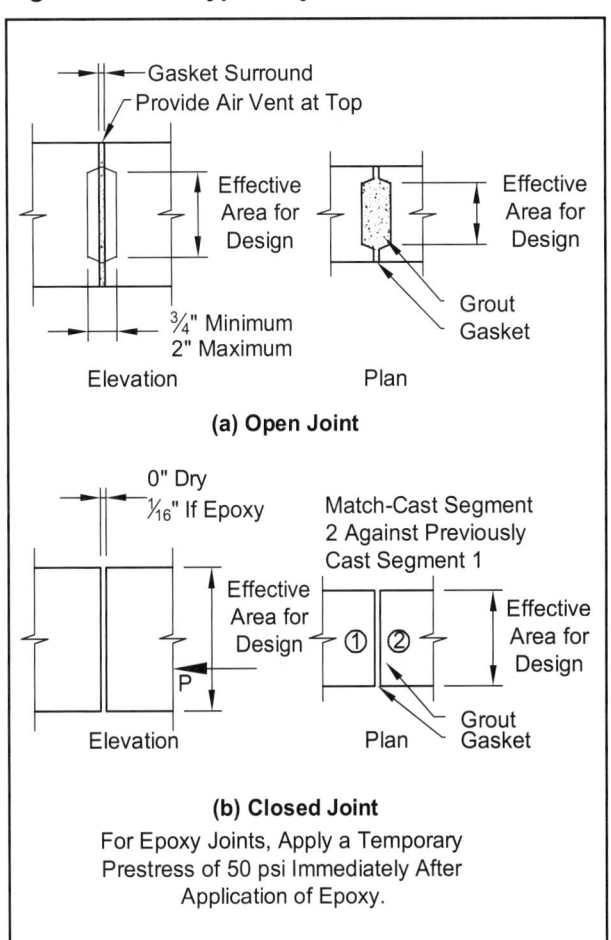

(a) Open Joint

(b) Closed Joint
For Epoxy Joints, Apply a Temporary Prestress of 50 psi Immediately After Application of Epoxy.

Surface preparation of closed joints is extremely important. They should be sound and clean, free from all traces of form release agents, curing compounds, laitance, oil, dirt and loose concrete.

A small piece of foreign material in a joint, or imperfect alignment will frequently cause the concrete to spall around the edges of the contact area of a closed joint. Thus, care in joint preparation and segment alignment cannot be over-emphasized.

9.9.3 Design

Analysis of precast segmental structures usually assumes monolithic behavior of the members under service loads, except that if dry joints are used, tension is not permitted between segments. Some designers also prefer to not allow tension in grouted joints, unless tests indicate otherwise.

The behavior during construction is of particular importance in segmental structures. Stresses which may be caused by settlement and shortening of scaffolding, temperature changes, elastic shortening from post-tensioning and other construction related stresses and movements may need to be considered in the design.

Shear stresses across joints are resisted by epoxy adhesives or grout, sometimes in combination with shear keys. In dry joints, shear stresses are resisted by friction (and shear keys if present), with the post-tensioning providing the normal force. In the absence of test data, the coefficient of friction may be assumed to be 0.8.

Individual segments are designed for handling and erection stresses (Chapter 5). They may be pretensioned or reinforced with mild steel, depending on size and manufacturing procedure.

9.9.4 Post-Tensioning

Information on various post-tensioning systems and their applications is given in the *PTI Post-Tensioning Manual.* [10] Nearly any type of bonded system can be used.

9.9.4.1 Tendon and Duct Placement

In addition to the design parameters for in-service conditions, the tendon layout must consider the sequence of construction and the changes in load conditions during the various construction stages.

Ducts for the tendons are placed in the segments prior to concrete placement. In some cases, tendons are installed in the ducts before the segment is erected, or even before concrete is placed, and subsequently stressed at each joint. In others, the tendons are placed after the segments are erected, either full length of the member or in shorter lengths, which may be coupled at the joints. Special attention must be given to the alignment of the ducts, especially at the joints. They must be large enough in diameter to adequately place the tendon, allowing some tolerance in alignment, and receive the grout placed subsequent to stressing. Special attention must be given to the corrosion protection of the post-tensioning steel, if it is to be unbonded at any stage of construction. Also, drains should be installed in the ducts at low points of the tendon profile.

Post-tensioning tendons crossing joints should be approximately perpendicular to the joint surface in the smaller dimension of the segment, e.g., flange or web thickness, but may be inclined to the direction of the larger dimension (e.g., depth). This is to minimize any unbalanced shearing force which could lead to dislocation of edges at the joints.

9.9.4.2 Couplers

Couplers are designed to develop the full strength of the tendons they connect. Adjacent to the coupler, the tendons should be straight, or have very minor curvature for a minimum length of 12 times the diameter of the coupler. Adequate provisions must be made to ensure that the couplers can move during prestressing.

9.9.4.3 Bearing Areas

Prestress force is transferred to the joined segments through bearing plates or other anchorage devices. This causes high load concentrations at the bearing areas, usually requiring vertical and horizontal reinforcement in several locations:

- Under end surfaces, not more than ¾ in. deep, to control possible surface cracking around anchorage.
- Internally, to prevent splitting between individual anchors. Size and location of this area and the magnitude of splitting (bursting) force depends on the type of anchorage and the force in the post-tensioning tendon.
- Internally, to prevent splitting between groups of anchors.
- Adjacent to joint surfaces, to decrease the possibility of damage to segments during post-tensioning or handling.

Grouting of the ducts is provided for corrosion protection and to develop bond between the prestressing steel and the surrounding concrete. Grouting procedures should follow the "Recommended Practice for Grouting of Post-Tensioned Prestressed Concrete," contained in Ref. 10.

9.9.5 References

1. Joint PCI-PTI Committee on Segmental Construction "Recommended Practice for Precast Post-Tensioned Segmental Construction," PCI JOURNAL, V. 27, No. 1, January-February 1982.

2. Anderson, Arthur R., "World's Largest Prestressed LPG Floating Vessel," PCI JOURNAL, V. 22, No. 1, January-February 1977.

3. Perry, William E., "Precast Prestressed Clinker Storage Silo Saves Time and Money," PCI JOURNAL, V. 21, No. 1, January-February 1976.

4. Martynowicz, A., and McMillan, C. B., "Large Precast Prestressed Vierendeel Trusses Highlight Multistory Building," PCI JOURNAL, V. 20, No. 6, November-December 1975.

5. Arafat, M. Z., "Giant Precast Prestressed LNG Storage Tanks at Staten Island," PCI JOURNAL, V. 20, No. 3, May-June 1975.

6. Muller, Jean, "Ten Years of Experience in Precast Segmental Construction," PCI JOURNAL, V. 20, No. 1, January-February 1975.

7. Lamberson, E. A., "Post-Tensioned Structural Systems–Dallas–Ft. Worth Airport," PCI JOURNAL, V. 18, No. 6, November-December 1973.

8. Zielinski, Z. A., "Prestressed Slabs and Shells Made of Prefabricated Components," PCI JOURNAL, V. 9, No. 4, August 1964.

9. Zielinski, Z. A., "Prefabricated Building Made of Triangular Prestressed Components," *Journal of the American Concrete Institute*, V. 61, No. 4, April 1964.

10. *PTI Post-Tensioning Manual*, Fifth Edition, Post-Tensioning Institute, Phoenix, AZ, 1990.

9.10 COORDINATION WITH MECHANICAL, ELECTRICAL AND OTHER SUB-SYSTEMS

9.10.1 Introduction

Prestressed concrete is used in a wide variety of buildings, and its integration with lighting, mechanical, plumbing, and other services is of importance to the designer. Because of increased environmental demands, the ratio of costs for mechanical and electrical installations to total building cost has increased substantially in recent years. This section is intended to provide the designer with the necessary perspective to economically satisfy mechanical and electrical requirements, and to describe some standard methods of providing for the installation of other sub-systems.

9.10.2 Lighting and Power Distribution

For many applications, the designer can take advantage of the fire resistance, reflective qualities and appearance of prestressed concrete by leaving the columns, beams, and ceiling structure exposed. To achieve uniform lighting free from distracting shadows, the lighting system should parallel the stems of tee members.

By using a reflective paint and properly spaced high-output florescent lamps installed in a continuous strip, the designer can achieve a high level of illumination at a minimum cost. In special areas, lighting troffers can be enclosed with diffuser panels fastened to the bottom of the tee stems providing a flush ceiling (see Figure 9.10.2.1). By using reflective paints, these precast concrete lighting channels can be made as efficient as conventional fluorescent fixtures.

9.10.3 Electrified Floors

The increasing use of office machines, computer networks, telephones, and other communication systems stresses the need for adequate and flexible means of supplying electricity and communication service. Since a cast-in-place topping is usually placed on prestressed floor members, conduit runs and floor outlets can be readily buried within this topping. Burying conduits in toppings of parking structures is not recommended because of the possibility of conduit corrosion. Most systems can easily be included in a 2 to 4 in. thick slab. When the system is placed in a structural composite slab, the effect of ducts and conduits must be carefully examined and their location coordinated with reinforcing steel. Voids in hollow-core slabs can also be used as electrical raceways.

Areas requiring large quantities of wiring are usually addressed with the use of computer floors. These are raised floors of metal gridwork installed above the structural floor that allow wiring to be placed between the finished floor and the structural floor in any horizontal direction.

Because of the high load-carrying capacity of prestressed concrete members, it is possible to locate high-voltage substations, with heavy transformers, near the areas of consumption with little or no additional expense. For extra safety, distribution feeds can also be run within those channels created by stemmed members. Such measures also aid the economy of the structure by reducing the overall story height and minimizing maintenance expenses.

9.10.4 Ductwork

The designer may also utilize the space within stemmed members or the holes inside hollow-core slabs for distribution ducts for heating, air-conditioning, or exhaust systems. In stemmed members, three sides of the duct are provided by the bottom of the flange and the sides of the stems. The bottom of the duct is completed by attaching a metal panel to the tee stems in the same manner as the lighting diffusers (see Figure 9.10.2.1).

Figure 9.10.2.1 Metal panels attached at the bottoms of the stems create ducts, and diffuser panels provide a flush ceiling

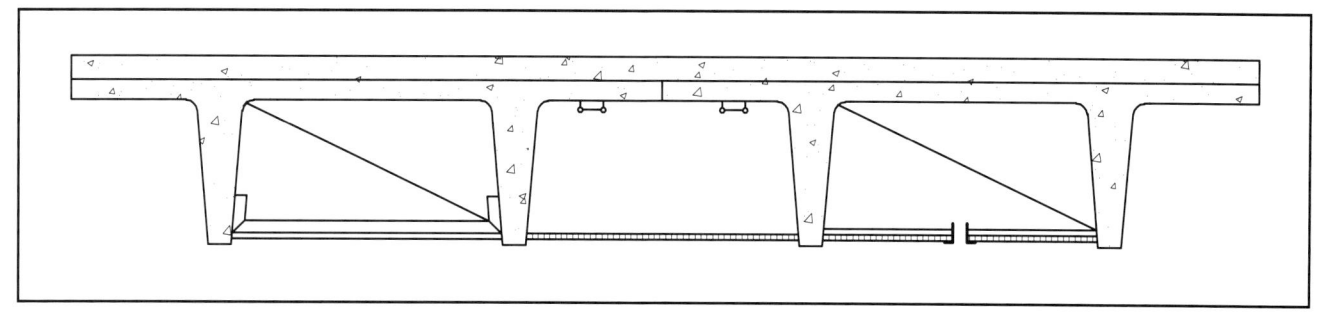

Connections can be made by several means, among them powder-activated fasteners, cast-in inserts or reglets. Powder-activated fasteners and other field installed devices should be located so as not to damage reinforcement. Check with the precast producer as necessary. Field installed devices generally offer the best economy and ensure placement in the exact location where the connecting devices will be required. Inserts should only be cast-in when they can be located in the design stage of the job, well in advance of casting the precast members.

If high velocity air movement is utilized, the enclosed space becomes a long plenum chamber with uniform pressure throughout its length. Diffusers are installed in the ceiling to distribute the air. Branches, when required, can be standard ducts installed along the column lines.

When ceilings are required, proper selection of precast components can result in shallow ceiling spaces to accommodate required ducts, piping and lighting fixtures.

Branch ducts of moderate size can also be accommodated by providing block-outs in the stems of tees or beams. To achieve best economy and performance in prestressed concrete members, particularly stemmed members, such block-outs should be repeated in size and location to handle all conditions demanded by mechanical, electrical or plumbing runs. While this may lead to slightly larger openings in some cases, the end result will probably be more economical. It should also be noted that sufficient tolerance should be allowed in sizing the openings to provide the necessary field assembly considerations (see Chapter 8).

Prestressed concrete box girders have been used to serve a triple function as air conditioning distribution ducts, conduit for utility lines and structural supporting members for the roof deck units. Conditioned air can be distributed within the void area of the girders and then introduced into the building work areas through holes cast into the sides and bottoms of the box girders. The system is balanced by pluging selected holes.

Vertical supply and return air trunks can be carried in the exterior walls, with only small ducts needed to branch out into the ceiling space. In some cases, the exterior wall cavities are replaced with three or four-sided precast boxes stacked to provide vertical runs for the mechanical and electrical systems. These stacked boxes can also be used as columns or lateral bracing elements for the structure.

In some cases, it may be required to provide openings through floor and roof units. Large openings are usually made by block-outs in the forms during the manufacture; smaller ones (up to about 8 in.) are usually field drilled or cored. When field drilling, care must be taken not to cut prestressing strand. Openings in flanges of stemmed members should be limited to the "flat" portion of the flange, that is, beyond 1 in. of the edge of the stem on double tees. Angle headers are often used for framing large openings in hollow-core floor or roof systems (see Figure 9.10.4.1).

Figure 9.10.4.1 Large openings in floors and roofs are made during manufacture of the units; small openings are field drilled. Some common types of openings are shown here.

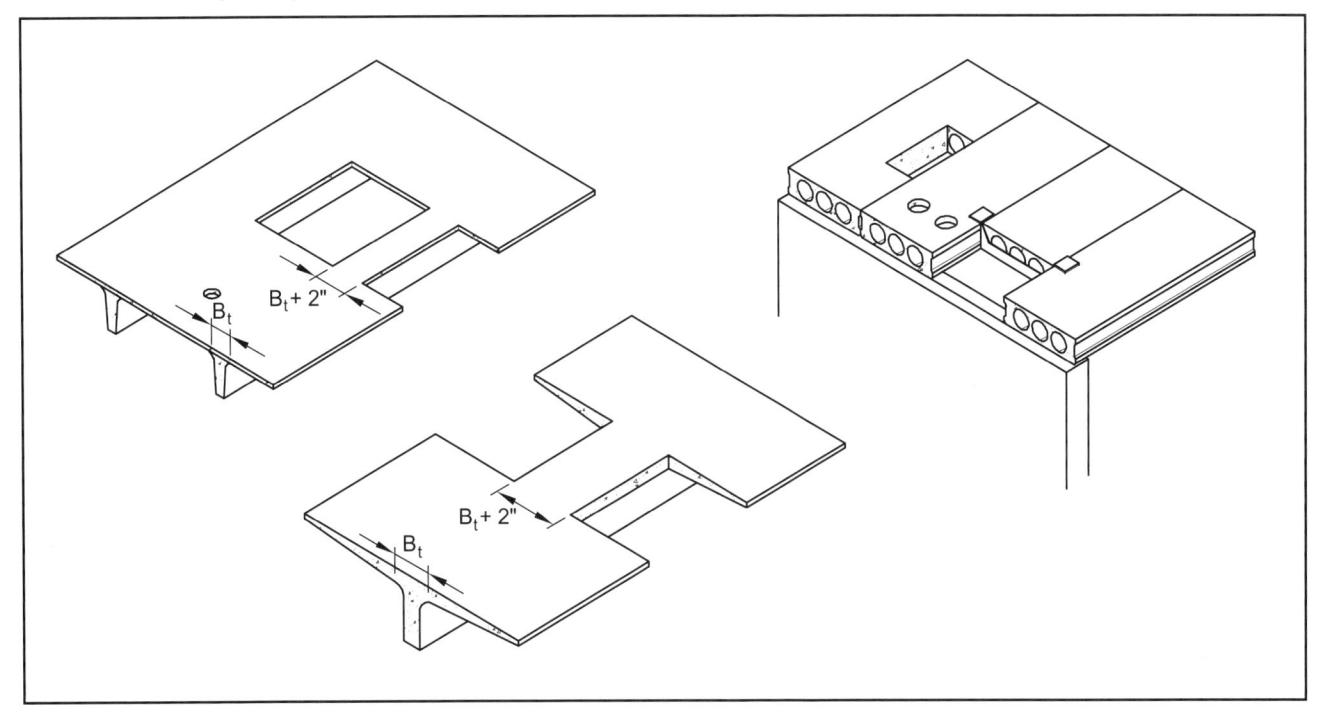

9.10.5 Other Sub-Systems

Suspended ceilings, crane rails, and other sub-systems can be easily accommodated with standard manufactured hardware items and embedded plates as shown in Figure 9.10.5.1.

Architectural precast wall panels can be adapted to combine with pre-assembled window or door units. Door or window frames, properly braced to prevent bowing during concrete placement, can be cast in the panels and then the glazing or doors can be installed prior to or after delivery to the job site. If the glazing or doors are properly protected, they can also be cast into the panel at the plant. When casting in aluminum window frames, particular attention should be made to properly coat the aluminum so that it will not react with the concrete. It should be noted that repetition is one of the real keys to economy in a precast concrete wall assembly. For example, windows and doors should be located in identical places for all panels wherever possible.

Insulated wall panels can be produced by embedding an insulating material such as expanded polyurethane between layers of concrete. These "sandwich panels" are described in detail in Section 9.4. Sandwich panels are normally cast on flat beds or tilting tables. The inside surface of the concrete panel can be given a factory troweled finish followed by minor touch-up work. The interior face is completed by painting, or by wall papering to achieve a finished wall. The formed surface of the panel can also be treated in a similar manner when used as the interior face.

Figure 9.10.5.1 Methods of attaching suspended ceilings, crane rails, and other sub-systems

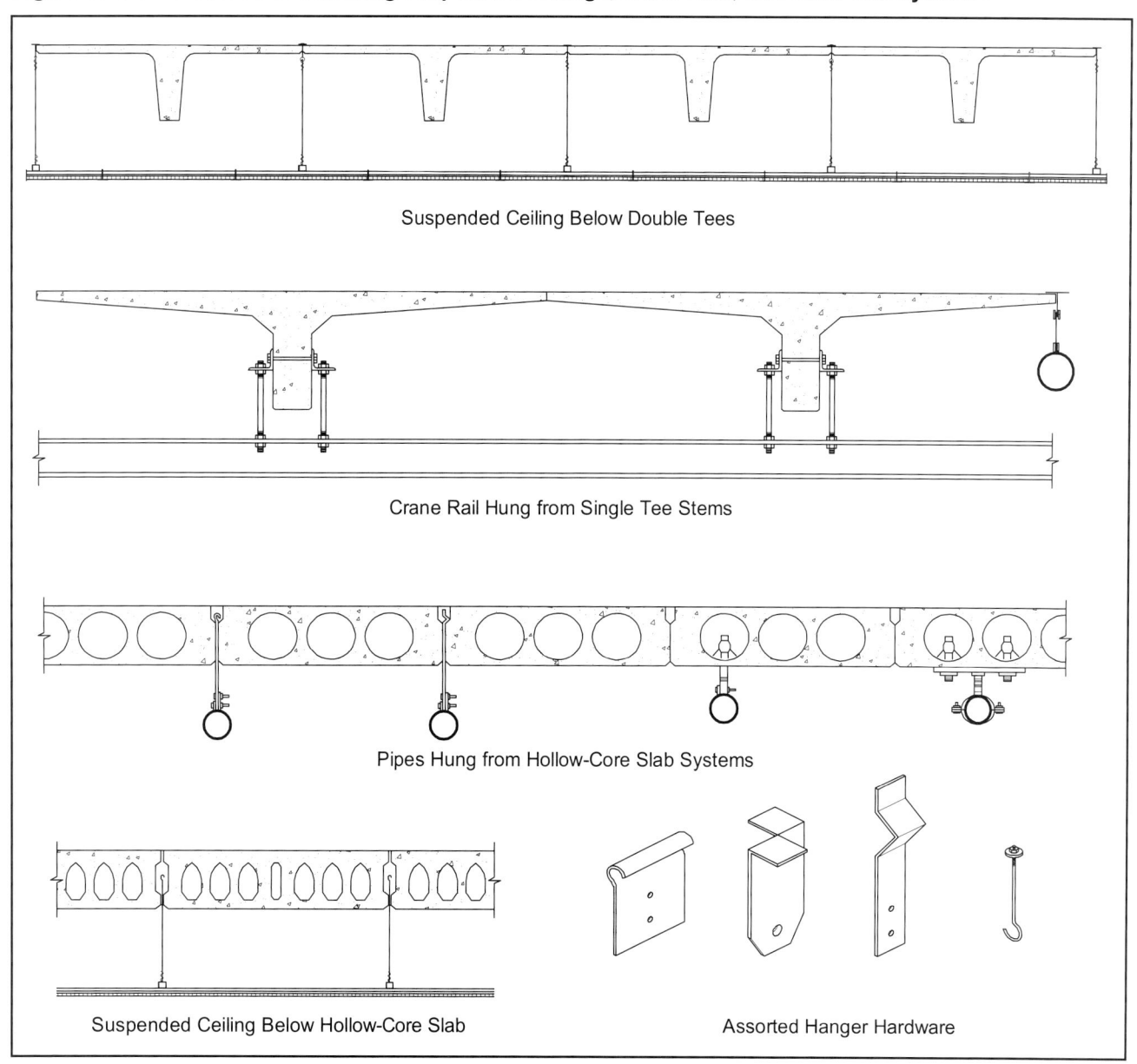

Suspended Ceiling Below Double Tees

Crane Rail Hung from Single Tee Stems

Pipes Hung from Hollow-Core Slab Systems

Suspended Ceiling Below Hollow-Core Slab

Assorted Hanger Hardware

9.10.6 Systems Building

As more and more complete systems buildings are built with precast and prestressed concrete, and as interest in this method of construction increases, we can expect that more of the building sub-systems will be prefabricated and pre-coordinated with the structure.

This leads to the conclusion that those parts of the structure that require the most labor skills should logically be prefabricated prior to installation in the field. The prefabricated components can be pre-assemblies of basic plumbing systems or electrical/mechanical systems plus lighting.

For housing systems, electrical conduits and boxes can be cast in the precast wall panels. This process requires coordination with the electrical contractor; and savings on job-site labor and time are possible. The metal or plastic conduit is usually pre-bent to the desired shape and delivered to the casting bed already connected to the electrical boxes. It is essential that all joints and connections be thoroughly sealed and the boxes enclosed prior to casting in order to prevent the system from becoming clogged. The wires are usually pulled through at the job site. Television antennas and telephone conduits have also been cast-in using the same procedure.

To reduce on-site labor, prefabricated bathroom units or combination bathroom/kitchen modules have been developed (see Figure 9.10.6.1). Such units include bathroom fixtures, kitchen cabinets and sinks, as well as wall, ceiling, floor surfaces, and other fixtures.

Plumbing units are often connected and assembled prior to delivery to the job sites. These bathroom/kitchen modules can be molded plastic units or fabricated from drywall components. To eliminate a double floor, the module can be plant built on the structural member or the walls of the unit can be designed strong enough for all fixtures to be wall hung.

Figure 9.10.6.1 Kitchen/bathroom modules can be pre-assembled on precast slabs ready for installation in systems buildings

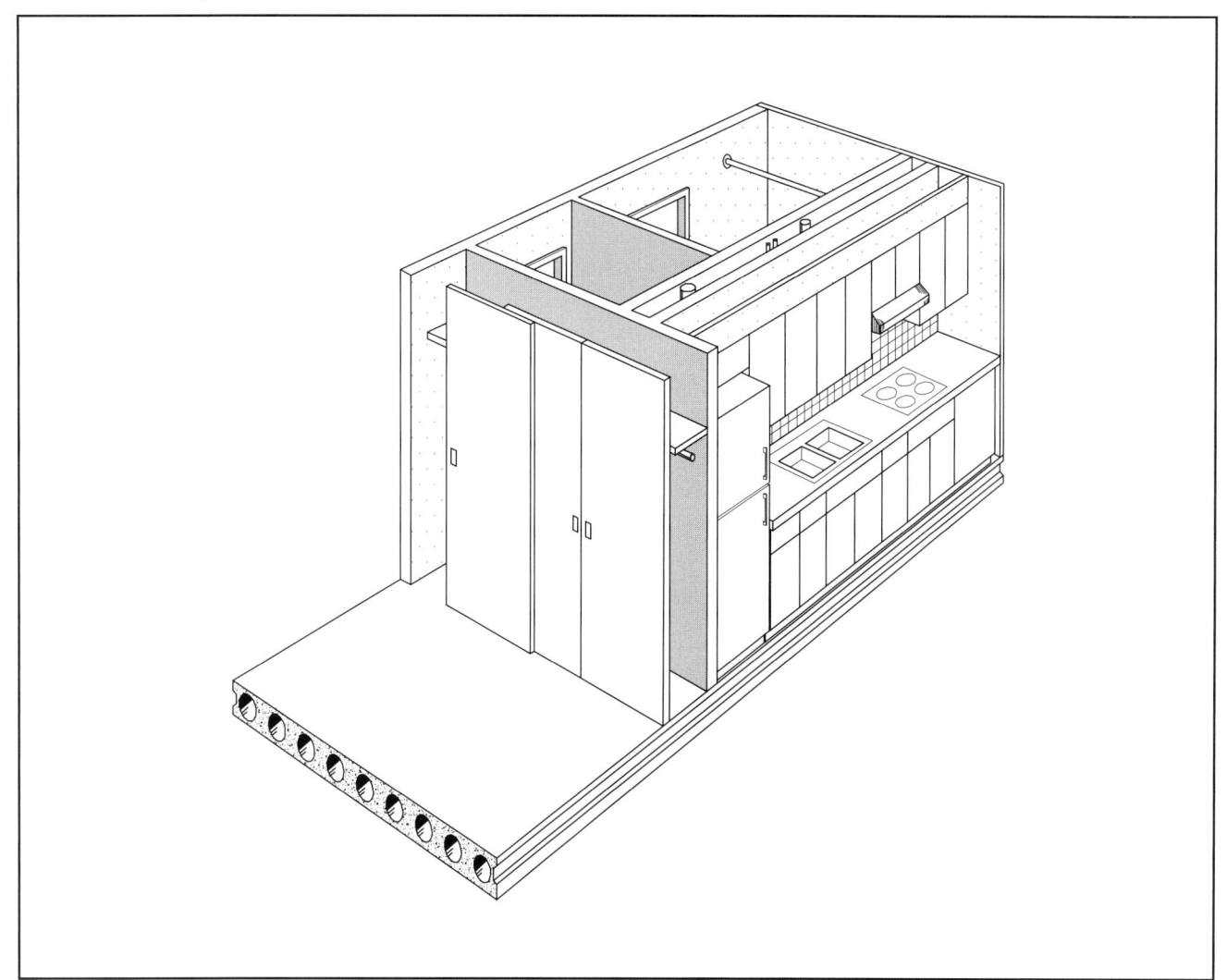

Figure 9.10.6.2 Prefabricated wet-wall plumbing systems incorporate pre-assembled piping

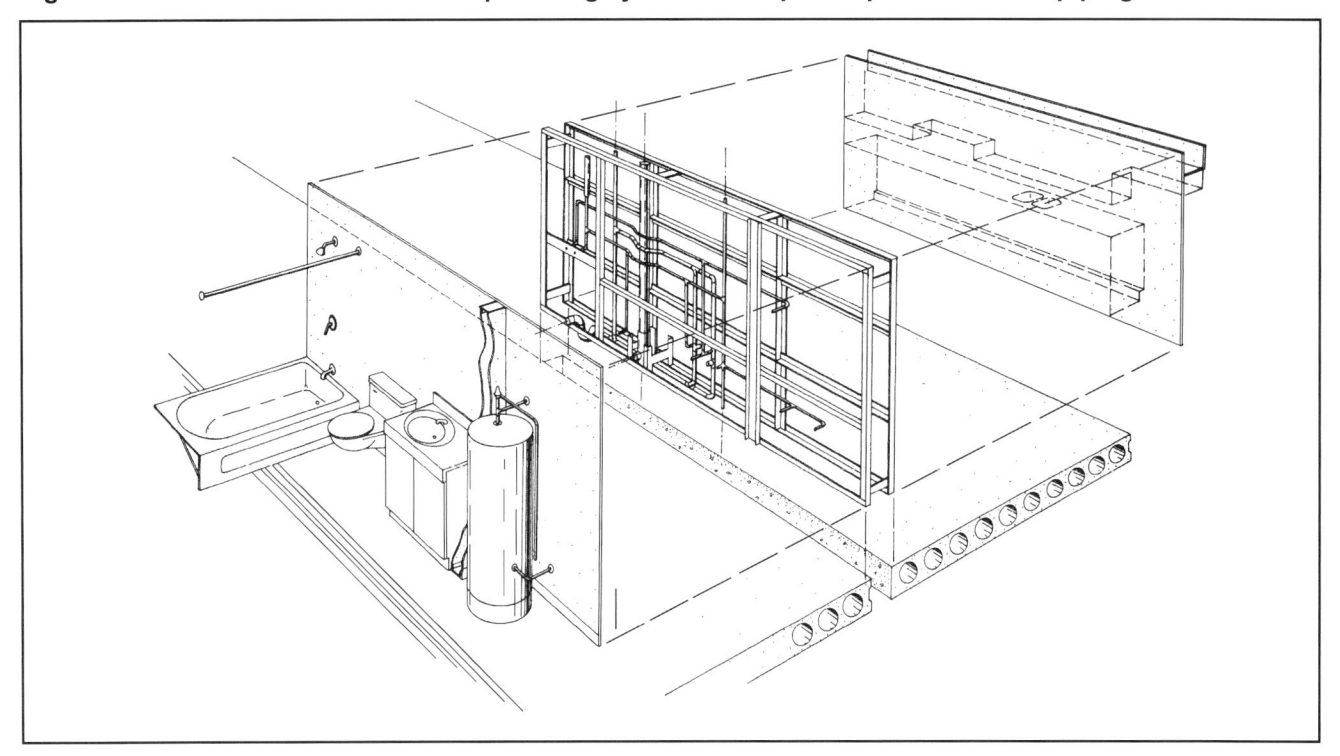

In the latter case, the units are placed directly on a precast floor and, in multistory construction, are located in a stack fashion with one bathroom directly over the one below. A block-out for a chase is provided in the precast floor and connections are made from each unit to the next to provide a vertical plumbing stack. Prefabricated wet-wall plumbing systems, as shown in Figure 9.10.6.2, incorporate pre-assembled piping systems using snap-on or no-hub connections made up of a variety of materials. These units only require a block-out in the prestressed flooring units and are also arranged in a stack fashion. Best economy results when bathrooms are backed up to each other, since a common vertical run can service two bathrooms.

Some core modules not only feature bath and kitchen components, but also HVAC components which are all packaged in one unit. These modules can also be easily accommodated in prestressed structural systems by placing them directly on the prestressed members with shimming and grouting as required.

9.10.7 Prison Cell Box Module

Prisons represent a building type that makes maximum use of the systems building and sub-

Figure 9.10.7.1 Precast concrete modular two-cell box unit for prison construction

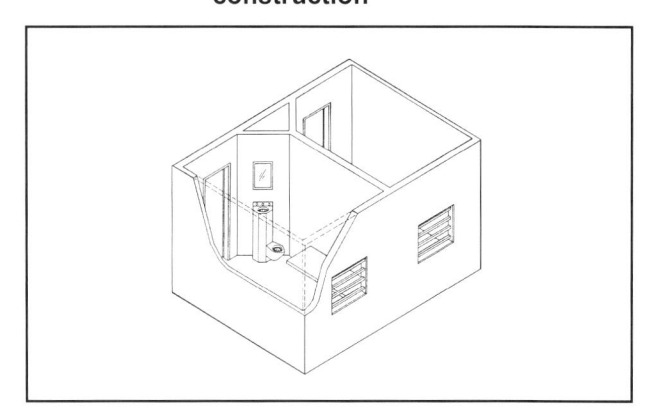

systems coordination concepts. Modular precast concrete boxes, typically multi-cell units, generally consist of cell walls, chase walls and a floor and/or ceiling (see Figure 9.10.7.1). All components of each cell — windows, beds, mirrors, desks, air vents, light fixtures, sinks, water closets and associated chase plumbing and wiring — are completely installed in each module at the manufacturer's plant. Erection at the site is rapid and field work is greatly reduced leading to a safe, functional, and economical prison facility in the shortest time of construction.

CHAPTER 10
SPECIFICATIONS AND STANDARD PRACTICES

10.1 Guide Specification for Structural Precast Concrete
and Structural Precast Concrete with Commercial Architectural Finish 10–2

10.2 Guide Specification for Architectural Precast Concrete 10–2

10.3 Standard Practice Recommendations for Precast Concrete 10–3

10.4 Recommendations on Responsibility for Design and
Construction of Precast Concrete Structures 10–14

10.5 PCI Standard Design Practice 10–19

10.1 GUIDE SPECIFICATION FOR STRUCTURAL PRECAST CONCRETE AND STRUCTURAL PRECAST CONCRETE WITH COMMERCIAL ARCHITECTURAL FINISH

This Guide Specification is intended to be used as a basis for the development of an office master specification or in the preparation of specifications for a particular project. In either case, this Guide Specification must be edited to fit the conditions of use.

Particular attention should be given to the deletion of inapplicable provisions. Necessary items related to a particular project should be included. Also, appropriate requirements should be added where blank spaces have been provided.

The actual specification section is on the CD mounted at the end paper of this book.

10.2 GUIDE SPECIFICATION FOR ARCHITECTURAL PRECAST CONCRETE

This Guide Specification is intended to be used as a basis for the development of an office master specification or in the preparation of specifications for a particular project. In either case, this Guide Specification must be edited to fit the conditions of use.

Particular attention should be given to the deletion of inapplicable provisions. Necessary items related to a particular project should be included. Also, appropriate requirements should be added where blank spaces have been provided.

The actual specification section is on the CD mounted at the end paper of this book.

10.3 STANDARD PRACTICE RECOMMENDATIONS FOR PRECAST CONCRETE

The precast/prestressed concrete industry has grown rapidly and certain practices relating to the design, manufacture and erection of precast concrete have become standard in many areas of North America. This "Code of Standard Practice" is a compilation of these practices presented in the form of recommendations for the guidance of those involved with the use of structural and architectural precast concrete.

The goal of this Code is to build a better undestanding by suggesting standards which more clearly define procedures and responsibilities, thus resulting in fewer problems for everyone involved in the planning and execution of projects.

As the precast/prestressed concrete industry continues to evolve, additional practices will become standard in the industry and current standards will require modification. The Precast/Prestressed Concrete Institute will continue updating this Code.

1. DEFINITIONS OF PRECAST CONCRETE

1.1 Structural Precast Concrete

Structural precast concrete usually includes beams, tees, joists, purlins, girders, lintels, columns, spandrels, poles, posts, piers, piles, slab or deck members, and wall panels. In order to avoid misunderstandings, it is important that the contract documents for each project list all the elements that are considered to be structural precast concrete. Some structural members may be left exposed in the structure for desired aesthetic appearance. High quality, attractive architectural treatments may be provided on the surfaces of these structural elements, and these should be specially listed in the contract documents. Quality assurance for structural precast and structural precast with an architectural finish is defined in PCI MNL-116.

1.2 Architectural Precast Concrete

Architectural precast concrete is characterized by a higher standard of uniformity of appearance with respect to surface details, color and texture. Typical architectural precast elements fall into two groups: major primary elements including wall panels, window wall panels, and column covers; other elements are decorative pieces and trim units including copings, mullions, sills and appurtenances such as benches and bollards. In order to avoid misunderstandings, it is important that the contract documents for each project list all the elements that are considered to be architectural precast concrete. Quality assurance for architectural precast concrete is defined in PCI MNL-117.

1.3 Prestressed Concrete

Both structural and architectural precast concrete may be prestressed or non-prestressed. All structural precast concrete products referred to herein which are prestressed are specifically referred to as prestressed concrete.

2. SAMPLES, MOCKUPS, AND QUALIFICATION OF MANUFACTURERS

2.1 Samples and Mockups

Samples, mockups, etc., are rarely required for structural prestressed concrete. If samples are required, they should be described in the contract documents. Samples should be manufactured in accordance with Section 3.2, of *Architectural Precast Concrete*, Second Edition, MNL-122.*

2.2 Qualification of Manufacturer

Manufacture, transportation, erection and testing should be provided by a company, firm, corporation, or similar organization specializing in providing precast products and services normally associated with structural or architectural precast concrete construction.

The manufacturer should be certified in the PCI Plant Certification Program. Because plant certification identifies the types of products for which the manufacturer has demonstrated capability and experience, project specifications should require the appropriate group and category defined in Section 9.5, "Quality Assurance and Control." In addition, for special and unique projects, the manufacturer may be required to list similar and comparable work successfully completed.

Standards of performance are given in the latest editions of the PCI manuals for quality control, MNL-116, MNL-117 and MNL-130.*

3. CONTRACT DOCUMENTS AND DESIGN RESPONSIBILITY

3.1 Contract Documents

Prior to beginning engineering and drafting, the

---

* Available from the Precast/Prestressed Concrete Institute.

manufacturer should have the following contract documents at his disposal:

1. Architectural drawings.
2. Structural drawings.
3. Specifications (complete with addenda).

Other pertinent drawings may also be desirable, such as approved shop drawings from other trades, roofing requirements and alternates.

3.2 Design Responsibilities

It is the responsibility of the owner* to keep the manufacturer supplied with up-to-date contract documents and written information. The manufacturer should not be held responsible for problems arising from the use of outdated or obsolete contract documents. If revised documents are furnished, it may also be necessary to modify the contract.

The contract documents should clearly define the following:

1. Structural design criteria including floor live loads, roof live and snow loads, earthquake design data, and any loading that may be special to the project.
2. Items designed and/or furnished by the manufacturer.
3. Size, location and function of all openings, blockouts, and cast-in items.
4. Design requirements including connections and reinforcement.†
5. Allowable tolerances. Normal field tolerances should be recommended by the manufacturer.
6. Dimensions and material requirements.
7. General and supplemental general conditions.
8. Any other special requirements and conditions.
9. Interface requirements with other trades.
10. Governing codes.
11. Lateral load resisting system as it relates to the precast concrete components.

Other design responsibility relationships are described in Section 10.4 of this Handbook.

4. SHOP DRAWINGS

Shop drawings consist of erection and production drawings. Different areas of the country may use different terminology.

Normal practices for the preparation of drawings for precast concrete are described in the PCI *Drafting Handbook — Precast and Prestressed Concrete, Second Edition*, MNL-119-90.‡

4.1 Erection Drawings

The information provided in the contract documents is used by the manufacturer to prepare erection drawings for approval and field use. The erection drawings should contain:

1. Plans and/or elevations locating and dimensioning all members furnished by the manufacturer.
2. Sections and details showing connections, finishes, openings, blockouts and cast-in items and their relationship to the structure.
3. Description of all loose and cast-in hardware including designation of who furnishes it.
4. Drawings showing location of anchors installed in the field.
5. Erection sequences and bracing when required to satisfy erection stability if the lateral resisting system is comprised of precast components.

4.2 Production Drawings

The contract documents and erection drawings are used to prepare production drawings for manufacturing showing all dimensions together with locations and quantities for all cast-in materials (reinforcement, inserts, etc.) and completely defining all finish requirements.

4.3 Discrepancies

When discrepancies or omissions are discovered on the contract documents, the manufacturer has the responsibility to check with the design and construction team to resolve the problem. If this is not possible, the following procedures are normally followed (unless the contract specifically states otherwise):

1. Anything handwritten and initialed by authorized parties governs over anything printed.
2. Contract terms govern over specifications and drawings.
3. Specifications govern over drawings.
4. Structural drawings govern over architectural drawings.

---

* The owner of the proposed structure or his designated representatives, who may be the architect, engineer, general contractor, public authority or others contracting with the precast manufacturer.

† When the manufacturer accepts design responsibility, the area or amount of responsibility must be clearly defined in the contract documents. See Section 10.4 of this Handbook.

‡ Available from PCI.

Written or graphic verification should be requested for any unclear condition.

4.4 Approvals

Completed erection drawings and supporting calculations when required should be submitted to the owner or his representation for approval. The exact sequence is dictated by construction schedules and erection sequences, and is determined when the contract is awarded.

Preferably, production drawings should not be started prior to receipt of "approved" or "approved as noted" erection drawings. Production drawings are normally not submitted for approval.*

Corrections should be noted on the reproducible erection drawings and copies made for distribution.

The following approval interpretations are normal practice:[†]

1. **Approved** — The approvers[‡] have completely checked and verified the drawings and calculations for conformance with contract documents and all expected loading conditions. Such approval should not relieve the manufacturer from responsibility for this design when that responsibility is so placed by the contract. The manufacturer may then proceed with production drawings and production without resubmitting. Erection drawings may then be released for field use and plant use.
2. **Approved as Noted** — Same as above except that noted changes should be made and corrected erection drawings issued. Production drawings and production may be started after noted changes have been made.
3. **Not Approved** — Drawings must be corrected and resubmitted. Production drawings can not be completed correctly until "approved" or "approved as noted" erection drawings are returned.

5. MATERIALS

The relevant AWS and ASTM Standards that apply to materials for a project should be listed in the contract documents together with any special requirements that are not included in the ASTM Standards.

---

* When production drawings are the only drawings showing reinforcement, representative drawings or design summaries should be submitted for approval.
† When production drawings are submitted, the same applies.
‡ The contract should state who has approval authority.

Note: Additional information regarding material specifications can be found in Section 10.1, "*Guide Specification for Structural Precast Concrete and Structural Precast Concrete with Commercial Architectural Finish,*" and Section 10.2, "*Guide Specification for Architectural Precast Concrete.*" (Disc included with this Handbook)

6. TESTS AND INSPECTIONS

6.1 Tests of Materials

Manufacturers are required to keep test records in accordance with the PCI manuals for quality control (MNL-116, MNL-117 and MNL-130). The contract documents may require the precast concrete manufacturer to make these records available for inspection by the owner's representative upon his request.

When the manufacturer is required to submit copies of test records to the owner and/or required to perform or have performed tests not required by the PCI manuals for quality control, these special testing requirements should be clearly described in the contract documents along with the responsibility for payment.

6.2 Inspections

On certain projects the owner may require inspection of precast concrete products in the manufacturer's yard by persons other than the manufacturer's own quality control personnel. Such inspections are normally made at the owner's expense. The contract documents should describe how, when and by whom the inspections are to be made, the responsibility of the inspection agency, and who is to pay for them. Alternatively, the owner may accept plant certification, such as provided in the PCI Plant Certification Program, in lieu of outside inspection.

6.3 Fire Rated Products

The Contract Documents must specify the required fire endurance ratings for the products used in the project. Where no ratings are specified, no specific fire rating will be provided. Section 9.3 of Chapter 9 describes procedures for determining fire resistance and ratings.

7. FINISHES

Finishes of precast concrete products, both structural and architectural, are probably the cause of more misunderstandings between the various members of the building team than any other question concerning product quality.

It is, therefore, extremely important that the contract documents describe clearly and completely the required finishes for all surfaces of all members, and that the erection drawings also include this information. When finish is not specified, the standard grade finish described in *"Guide Specification for Structural Precast Concrete and Structural Precast Concrete with Commercial Architectural Finish"* should normally be furnished. Architect's approval of a reasonably sized sample is recommended.

For descriptions of the usual finishes for structural precast concrete, see Section 10.1, *"Guide Specification for Structural Precast Concrete and Structural Precast Concrete with Commercial Architectural Finish,"* and for architectural precast concrete, see Section 10.2, *"Guide Specification for Architectural Precast Concrete."* Where special or critical requirements exist or where large expanses of exposed precast concrete will occur on a project, samples are essential and, if required, should be so stated and described in the contract documents.

8. DELIVERY OF MATERIALS

8.1 Manner of Delivery

The manufacturer should deliver the precast concrete to the erector* on a schedule to facilitate the planned of erection schedule for the building. Special requirements of the owner for the delivery of materials or the mode of transport, should be stated in the contract documents.

8.2 Marking and Shipping of Materials

The precast concrete members should be separately marked in accordance with approved drawings in such a manner as to distinguish varying pieces and to facilitate erection (including erection sequence) of the structure. The owner should give the manufacturer sufficient time to fabricate and ship any special plates, bolts, anchorage devices, etc., contractually agreed to be furnished by the manufacturer.

8.3 Precautions During Delivery

Special precautions beyond that required in MNL-116, MNL-117 and MNL-130 should not be expected unless stated in the bid invitation or specifications. The manufacturer is not responsible for the product, including loose material, after delivery to the site unless required by the contract documents.

8.4 Access to Jobsite

Free and easy access to the delivery site should be provided to the manufacturer, including backfilling and compacting, drainage and snow removal, so that delivery trucks and cranes can operate under their own power. The responsibility for access should be clearly defined in the contract.

8.5 Unloading Time Allowance

Delivery of product should include a reasonable unloading time allowance. Any delay beyond a reasonable time is normally paid for by the party which is responsible for the delay.

9. ERECTION

9.1 Special Erection Requirements

When the owner requires a particular erection method or sequence of erection for his own needs, this information should be stated in the contract documents. In the absence of such stated restrictions, the erector will proceed using the most efficient and economically safe method and sequence available to him, consistent with the contract documents, as directed by the manufacturer.

9.2 Tolerances

Some variation is to be expected in the overall dimensions of any building or other structure. It is common practice for the manufacturer and erector to work within the tolerances shown in PCI's *"Tolerance Manual for Precast and Prestressed Concrete Construction"*, MNL-135.

The owner, by whatever agencies he may elect, immediately upon completion of any portion of the erection, should determine if the work is plumb, level, aligned and properly fastened. Discrepancies should immediately be brought to the attention of the erector so that proper corrective action can be taken.

The work is complete once the precast product has been properly plumbed, leveled and aligned within the established tolerances. Acceptance for this work should be secured from the authorized representative of the owner (see Section 11.2).

9.3 Foundations, Piers, Abutments and Other Bearing Surfaces

The invitation to bid should state the anticipated time when all foundations, piers, abutments and

---

* The erector may be either the manufacturer or a subcontractor engaged by the manufacturer, or the general contractor.

other bearing surfaces will be ready and accessible to the erector.

Before authorizing the commencement of precast concrete erection, the owner/general contractor should ensure the precast concrete erector is provided with the following written notifications:

1. The concrete in the footings, piers and walls and/or mortar in the masonry piers has attained, on the basis of appropriate ASTM standard test methods of field-cured sample, either 75% of the specified minimum compressive strength or sufficient strength to support the loads imposed during precast concrete erection.
2. Any repairs, replacements and modifications to the anchor bolts were conducted as directed by the Engineer of Record.
3. The structural elements to receive the precast concrete have been accepted by the designated approval authority.

9.4 Building Lines and Bench Marks

The precast erector should be furnished all building lines and bench marks at each floor level of the structure, as appropriate.

9.5 Anchor Bolts and Bearing Devices

9.5.1 The precast manufacturer normally furnishes, but does not install, anchor bolts, plates, bearing pads, etc. that are to be installed in cast-in-place concrete or masonry for connection with precast members. However, if this is to be the responsibility of the precast manufacturer, it should be so defined in the specifications. It is important that such items be installed true to line and grade, and that installation be completed in time to avoid delays or interference with the precast concrete erection.

9.5.2 Anchor bolts and foundation bolts are set by the general contractor in accordance with an approved drawing. They must not vary from the dimensions shown on the erection drawings by more than the following:

1. ⅛ in. center to center of any two bolts within an anchor bolt group, where an anchor bolt group is defined as the set of anchor bolts which receive a single fabricated shipping piece.
2. ¼ in. center to center of adjacent anchor bolt groups.
3. Maximum accumulation of ¼ in. per hundred ft along the established column line of multiple anchor bolt groups, but not to exceed a total of 1 in., where the established column line is the actual field line most representative of the centers of the as-built anchor bolt groups along a line of columns.
4. ¼ in. from the center of any anchor bolt group to the established column line through that group.
5. The tolerances of Items 2, 3 and 4 apply to offset dimensions shown on the plans, measured parallel and perpendicular to the grid lines.
6. Special embeds or hardware, such as grout sleeves for reinforcing bars or proprietary corbel hangers, may require tighter tolerances as specified by the specialty embed manufacturer.
7. A template designed and placed to achieve the required tolerance should always be used for setting anchor bolts.
8. Where ties at anchor bolts are required by code, the contractor installing the bolts is responsible for placing the ties.

9.5.3 Erectors should check both line and grade in sufficient time before erection is scheduled to permit any necessary corrections. Proposed corrections by the General Contractor should be submitted to the owner for approval. Responsibility for corrections is that of the general contractor.

9.6 Grout at Bearing Areas

At critical bearing locations, as shown on the drawings, grout shall be placed as soon as possible, not later than the end of the day the precast piece is set, unless otherwise directed by the engineer.

9.7 Utilities

Water and electricity should be furnished by the owner for erection and grouting operations.

9.8 Working Space and Crane Access

The owner should furnish adequate, properly drained, graded, and convenient working space for the erector and access for his equipment necessary to assemble the structure. The owner should provide adequate storage space for the precast concrete products to enable the erector to operate at the speed required to meet the established schedule. Unusual hazards such as high voltage lines, buried utilities, or areas of restricted access should be stated in the invitation to bid. Hazards should be marked on site.

9.9 Materials of Other Trades

Other building materials or work of other trades should not be built up above the bearing of the precast concrete until after erection of the precast concrete members.

9.10 Correction of Errors

Corrections of minor misfits are considered a part of erection even if the precast concrete is not erected by the manufacturer. Any error in manufacturing which prevents proper connection or fitting should be immediately reported to the manufacturer and the owner so that corrective action can be taken. The manufacturer should approve any alterations or corrections to the product.

9.11 Field Assembly

The size of precast concrete pieces may be limited by transportation requirements for weight and clearance dimensions. Unless agreed upon between the manufacturer and owner, the manufacturer should provide for such field connections that will meet required loads and forces without altering the function or appearance of the structure.

The manufacturer furnishes those items embedded in the precast members, and generally furnishes all loose materials for temporary and permanent connection of precast members. Temporary guys, braces, falsework, shims, and cribbing are the property of the erector and are removed only by the erector or with the erector's approval upon completion of the erection of the structure, unless otherwise agreed.

9.12 Blockouts, Cuts and Alterations

Neither the manufacturer nor the erector is responsible for the blockouts, cuts or alterations by or for other trades unless so specified in the contract documents. Whenever such additional work is required, all information regarding size, location and number of alterations is furnished by the owner prior to preparation of the precast production and erection drawings.

The general contractor is responsible for warning other trades against cutting of precast concrete members without prior approval of the Engineer of Record.

9.13 Temporary Floors and Access

The manufacturer or erector is not required to furnish temporary flooring for access unless so specified in the contract documents.

9.14 Painting, Caulking and Closure Panels

Painting, caulking and placing of closure panels between stems of flanged concrete members are services not ordinarily supplied by the manufacturer or erector. If any of these services are required of the manufacturer, it should be stated in the contract documents.

9.15 Patching

A certain amount of patching of product is to be expected. Patching should meet the finish requirements of the project and color should be reasonably matched. Responsibility for accomplishing this work should be resolved between the manufacturer and erector.

9.16 Safety

Safety procedures for the erection of the precast concrete members is the responsibility of the erector and must be in accordance with all local, state or Federal rules and regulations which have jurisdiction in the area where the work is to be performed, but not less than required in ANSI Standard A10.9, *Safety Requirements for Concrete Construction and Masonry Work*.* See PCI's *Erection Safety Manual MNL-132.*

9.17 Security Measures

Security protection at the jobsite should be the responsibility of the general contractor.

10. INTERFACE WITH OTHER TRADES

Coordination of the requirements for other trades to be included in the precast concrete members should be the responsibility of the owner unless clearly defined otherwise in the contract documents.

The PCI manuals for quality control (MNL-116, MNL-117 and MNL-130) and the *PCI Tolerance Manual MNL-135* specify manufacturing tolerances for precast concrete members. Interfaces with other materials and trades must take these tolerances into account. Unusual requirements or allowances for interfacing should be stated in the contract documents.

---

* American National Standards Institute, New York, NY.

11. WARRANTY AND ACCEPTANCE

11.1 Warranties

Warranties of product and workmanship have become a widely accepted practice in this industry, as in most others. Warranties given by the precast concrete manufacturer and erector should indicate that their product and work meet the specifications for the project.

In no case should the warranty of the manufacturer and erector be in excess of the warranty required by the specifications. Warranties should in all instances include a time limit and it is recommended that this should not exceed one year.

In order to protect the interests of all parties concerned, warranties should also state that any deviations in the designed use of the product, modifications of the product by the owner and/or contractor or changes in other products used in conjunction with the precast concrete will cause said warranty to become null and void.

Warranty may be included as a part of the conditions of the contract agreement, or it may be presented in letter form as requested by the owner. A sample warranty follows:

> Manufacturer warrants that all materials furnished have been manufactured in accordance with the specifications for this project. Manufacturer further warrants that if erection of said material is to be performed by those subject to his control and direction, work will be completed in accordance with the same specifications.
>
> In no event shall manufacturer be held responsible for any damages, liability or costs of any kind or nature occasioned by or arising out of the actions or omissions of others, or for work, including design, done by others; or for material manufactured, supplied or installed by others; or for inadequate construction of foundations, bearing walls, or other units to which materials furnished by the precast manufacturer are attached or affixed.
>
> This warranty ceases to be in effect beyond the date of _____. Should any defect develop during the contract warranty period, which can be directly attributed to defect in quality of product or workmanship, precast manufacturer shall, upon written notice, correct defects or replace products without expense to owner and/or contractor.

COMPANY NAME

_____
Signature   Title

11.2 Acceptance

Manufacturer should request approval and acceptance for all materials furnished and all work completed by him periodically and in a timely manner as deemed necessary in order to adequately protect the interests of everyone involved in the project. The size and nature of the project will dictate the proper intervals for securing approval and acceptance. Periodic approval in writing should be requested as soon as it appears that such action will minimize possible problems which would seriously affect the progress of the project. A sample acceptance form follows:

FIELD INSPECTION REPORT

Project # _____

On this _____ day of _____, 20 _____

_____ of
Company Field Superintendent

_____ and
Precast Manufacturer

_____ of
General Contractor Superintendent

_____
General Contractor
have inspected _____
           portion of building being inspected

All of the work performed by the above indicated company in the above described portion of the project has been performed to the satisfaction of the above named General Contractor's Superintendent with the exception of the following:

_____
_____

The General Contractor's Superintendent hereby releases the Precast Manufacturer of its responsibility to perform any other work in the above described portion of the project except as detailed herein.

The Precast Manufacturer in turn hereby releases the above described portion of the project to the General Contractor.

_____
Precaster's Superintendent

_____
General Contractor Superintendent

Final inspection and acceptance of erected precast and prestressed concrete should be made by the owner's representative within a reasonable time after the work is completed.

12. CONTRACT ADMINISTRATION

12.1 General Statement

Contract agreements may vary widely from area to area, but the objective should be the same in all instances. The contract agreement should be written to protect the interests of all parties concerned and, at the same time, be specific enough in content to avoid misunderstandings once the project begins.

Information relative to invoicing, payment, bonding and other data pertinent to a project or material sale should be specifically provided for in the major provisions of the contract documents or in the special terms and conditions applicable to all contractual agreements between manufacturer and owner.

The intent of this section is to recommend those matters which ought to be considered, but not necessarily the form in which they should be expressed. The final contract document should be the result of careful consideration of all pertinent factors as well as of the normal practices in the area.

12.2 Retentions

Although retentions have been used for many years as a means of ensuring a satisfactory job performance, it is apparent that they directly contribute to the cost of construction, frequently lead to disputes, and often result in job delays. In view of the unfavorable consequences of retentions and possible abuse, it is recommended that the following procedure be followed:

1. Wherever possible, retentions should be eliminated and bonding should be used as the single, best source of protection. This should apply to prime contractors and subcontractors equally.
2. Where there are no bonding requirements, the retention percentage should be as low as possible. It is recommended that this be not more than 5 percent of the work invoiced.
3. The percentage level of any retention should be the same for subcontractors as for prime contractors on a job.
4. Release of retained funds and final payment, as well as computing the point of reduction of the retention, should be done on a line item basis, that is, each contractor or subcontractor's work considered as a separate item and the retention reduced by 50 percent upon substantial completion and the balance released within 30 days after final completion of that work.
5. Retained funds should be held in an escrow account with interest accruing to the benefit of the party to whom the funds are due.
6. When materials are furnished FOB plant or jobsite, it is recommended that there be no retentions.
7. Retention must be paid in full within 90 days after the last precast item has been delivered.

12.3 Contract Agreement

1. Contract agreement should fully describe the project involved, including job location, project name, name of owner/developer, architect or other design professionals and all reference numbers identifying job relation information such as plans, specifications, addenda, bid number, etc.
2. Contract agreement should fully describe the materials to be furnished and/or all work to be completed by the seller. This can be done in a scope of work document attached to and referenced in the contract agreement.
3. All inclusions and exclusions should be stated to avoid the possibility of any misunderstanding.
4. Price quoted should be stated to eliminate any possibility of misunderstanding.
5. Reference should be made to the terms and conditions governing the proposed contract agreement. The terms and conditions may best be stated on the reverse side of the contract form. Special terms or conditions should be stated in sufficient detail to avoid the possibility of misunderstanding.
6. The terms of payment should be specifically detailed so there is no doubt as to intent. Special care should be exercised where the terms of payment will differ from those normally in effect or where they deviate from the general terms and conditions appearing on the reverse side of the contract form.
7. A statement of policy should be made with reference to the inclusion or exclusion of taxes in the stated price.
8. The proposal form stating the full conditions under which the project will be performed may contain an acceptance clause to be signed by the purchaser. At such time as said acceptance clause is signed, the proposal form then becomes the contract agreement.
9. Seller should clearly state the limits of time within which an accepted proposal will be recognized as a binding contract.
10. A statement indicating the classification of labor to perform the work in the field is advisable to eliminate later dispute over jurisdiction of work performed.

12.4 Terms and Conditions

The terms and conditions stated on the proposal contract agreement should include, but are not necessarily limited to, the following:

1. **Lien Laws** — Where the lien laws of a state specifically require advance notice of intent, it is advisable to include the required statement in the general terms and conditions.

2. **Specifications** — Seller should make a specific declaration of material and/or work specifications, but normally this should not be in excess of the specifications required by the contract agreement.

3. **Contract Control** — A statement should be made indicating that the agreement when duly signed by both parties supersedes and invalidates any verbal agreement and can only be modified in writing with the approval of those signing the original agreement.

4. **Terms of Payment** — Terms of payment should be specifically stated either on the face of the contract or in the general terms and conditions. Mode and frequency of invoicing should be so stated, indicating time within which payment is expected.

5. **Late Payment Charges** — The contract may provide for legal interest charges for late payments not made in accordance with contract terms, and if this is desired, it should be stated in the general terms and conditions. A statement indicating seller is entitled to reasonable attorney's fees and related costs should collection proceedings be necessary may also be included.

6. **Overtime Work** — Prices quoted in the proposal should be based on an 8-hour day and a 5-day week, excluding holidays, under prevailing labor regulations. Provisions should be included in the contract agreement to provide for recovery of overtime costs plus a reasonable markup when the seller is requested to provide such service.

7. **Financial Responsibility** — General terms and conditions may indicate the right of the seller to suspend or terminate material delivery and/or work on a project if there is a reasonable doubt of the ability of the purchaser to fulfill his financial responsibility.

8. **Payment for Inventory**

    a. It has become common practice to include in the contract terms and conditions provisions for the invoicing and payment of all materials stored at the plant. When deliveries or placement of said materials are delayed for more than a stipulated time beyond the originally scheduled date because of purchaser's inability either to accept delivery of materials or to provide proper job access, terms for costs of extended storage should be stated in the contract.

    b. Under certain conditions, it may be necessary to purchase special materials or to produce components well in advance of job requirements to ensure timely deliveries. When job requirements are of such a nature, it is advisable to include provisions for payment of such raw and finished inventories stored in seller's plant or on jobsites on a current basis.

9. **Payment for Suspended or Discontinued Projects** — The terms and conditions should provide that in the event of a discontinued or suspended project, seller shall be entitled to payment for all material purchased and/or manufactured including costs, overhead and profit, and not previously billed, as well as reasonable engineering and other costs incurred, including overhead caused by plant being idle from late notification of suspension of work.

10. **Job Extras** — Requests for job extras should be confirmed in writing. Invoicing should be presented immediately following completion of the extra work with payment subject to the terms and conditions of the contract agreement, or as otherwise stated in the change order.

11. **Claims for Shortages, Damages or Delays** — Seller should, upon immediate notification in writing on the face of the delivery ticket of rejected material or shortage, acknowledge and furnish replacement material, when appropriate, at no cost to purchaser. It is normal practice that the seller should not be responsible for any loss, damage, detention or delay caused by fire, accident, labor dispute, civil or military authority, insurrection, riot, flood or by occurrences beyond his control.

12. **Back Charges** — Back charges should not be binding on the seller, unless the condition is a part of the contract and promptly reported in writing, and opportunity is given seller to inspect and correct the problem.

13. **Permits, Fees and Licenses** — Costs of permits, fees, licenses and other similar expenses are normally assumed by the purchaser.

14. **Bonds** — Cost of bonds is normally assumed by the purchaser.

15. **Taxes** — Federal, state, county or municipal occupation or similar taxes which may be imposed are normally paid by the purchaser and, in the case of sales taxes, through the seller (varies by locale). For tax-exempt projects, purchaser issues to the seller a tax-exempt certificate when the purchase agreement is finalized.

16. **Insurance** — Seller shall carry Workmen's Compensation, Public Liability, Property Damage and Auto Insurance and certificates of insurance will be furnished to purchaser upon request. Additional coverage required over and above that provided by the seller is normally paid by the purchaser. A statement specifically excluding both additional insured status and a waiver of subrogation should be included.

17. **Services** — Heat, water, light, electricity, toilet, telephone, watchmen and general services of a similar nature are normally the responsibility of the purchaser unless specifically stated otherwise in the contract agreement.

18. **Safety Equipment** — The purchaser is normally responsible for necessary barricades, guard rails and warning lights for the protection of vehicular and pedestrian traffic and responsible for furnishing, installing and maintaining all safety appliances and devices required on the project under U.S. Department of Labor, *Safety and Health Standards for Construction Industry* (OSHA 2207), as well as all other safety regulations imposed by other agencies having jurisdiction over the project.

19. **Warranty** — Seller should provide specific information relative to warranties given, including limitations, exclusions and methods of settlement. Warranties should not be in excess of warranty required by the specific project.

20. **Title** — Contract should provide for proper identification of title to material furnished. It is normal practice for title and risk of loss or damage to the product furnished to pass to the purchaser at the point of delivery, except in cases of FOB plant, in which event title to and risk of loss or damage to the product normally should pass to purchaser at plant pickup.

21. **Shop Drawing Approval** — Seller should prepare and submit to purchaser for approval all shop drawings* necessary to describe the work to be completed. Shop drawing approval should constitute final agreement to quantity and general description of material to be supplied. No work should be done upon material to be furnished by seller until approved shop drawings are in the seller's possession.

22. **Delivery** — Delivery times or schedules set forth in contract agreements should be computed from the date of delivery to the seller of approved shop drawings. Where materials are specified to be delivered FOB to jobsite, the purchaser should provide labor, cranes or other equipment to remove the materials from the trucks and should pay seller for truck expense for time at the jobsite in excess of a specified time for each truck. On shipments to be delivered by trucks, delivery should be made as near to the construction site as the truck can travel under its own power. In the event delivery is required beyond the curb line, the purchaser should assume full liability for damages to sidewalks, driveways or other properties and should secure in advance all necessary permits or licenses to effect such deliveries.

23. **Builder's Risk Insurance** — Purchaser should provide Builder's Risk Insurance without cost to seller, protecting seller's work, materials and equipment at the site from loss or damage caused by fire or the standard perils of extended coverage, including vandalism and malicious acts.

24. **Erection** — Purchaser should ensure that the proposed project will be accessible to all necessary equipment including man lifts, cranes and trucks, and that the operation of this equipment will not be impeded by construction materials, water, presence of wires, pipes, poles, fences or framings. Purchaser should further indemnify and save harmless the seller and his

---

* See Section 4 of this *Code of Standard Practice* for definition of shop drawings.

respective representatives, including subcontractors, vendors, assigns and successors from any and all liability, fine, penalty or other charge, cost or expense and defend any action or claim brought against seller for any failures by purchaser to provide suitable access for work to be performed. Seller also reserves the right to discontinue the work for failure of purchaser to provide suitable access and the purchaser should be responsible for all expenses and costs incurred.

25. **Exclusions of Work to be Performed** — Unless otherwise stated in the contract, all shoring, forming, framing, cutting holes, openings for mechanical trades and other modifications of seller's products should not be performed by the seller nor are they included in the contract price. Seller should not be held responsible for modifications made by others to his product unless said modifications are previously approved by him.

26. **Sequence of Erection** — Sequence of erection, when required to satisfy stability of the structure, should be as agreed upon between seller and purchaser and expressly stated in the contract agreement. Purchaser should have ready all foundations, bearing walls or other units to which seller's material is to be affixed, connected or placed, prior to start of erection. Purchaser should be responsible for the accuracy of all job dimensions, bench marks, and true and level bearing surfaces. Claims or expenses arising from the purchaser's neglect to fulfill this responsibility should be assumed by the purchaser.

27. **Arbitration** — In view of the difficulties and misunderstandings which may occur due to misinterpretation of contractual documents, it is recommended that the seller stipulate that all claims, disputes and other matters in question, arising out of or related to the contract, be decided by mediation or arbitration in accordance with the Construction Industry Rules of the American Arbitration Association then in effect, or some other rules acceptable to both parties. The location for such alternate dispute resolution should be stipulated.

28. **Contract Form** — Contract documents should stipulate policy governing acceptance of proposal on other than the seller's form. In the event purchaser does not accept the seller's proposal and/or contract agreement, but requires the execution of a contract on his own form, it is advisable that the seller stipulate in writing on the contract agreement that the contract will be fulfilled according to his proposal originally submitted. All identifying information such as proposal number, dates, etc. should be included so there can be no question of what the document referred to. Standard subcontract forms are available from AIA, CASE, AGC, Association of Subcontractors and other organizations.

29. **Indemnification** — Any indemnification or hold harmless obligation of the seller should extend only to claims relating to bodily injury and property damage and then only to that part or proportion of any claim, damage, loss or defect that results from the negligence or intentional act of the seller or someone for whom it is responsible. Seller shall have no duty to defend.

10.4 RECOMMENDATIONS ON RESPONSIBILITY FOR DESIGN AND CONSTRUCTION OF PRECAST CONCRETE STRUCTURES

1. INTRODUCTION

Design and construction of structures is a complex process. Defining the scope of work and the responsibilities of the parties involved in this process, by contract, is necessary to achieve a safe, high quality structure.

Besides the Owner, the parties involved in the design and construction of precast concrete structures or other structures containing precast concrete members may include the Engineer, Architect, Construction Manager, General Contractor, Manufacturer, Precast Engineer, Erector and Inspector. These and other terms related to precast concrete construction are defined in Section 2.

This Section is taken primarily from Ref. 1. Excerpts from Ref. 2 have been added where appropriate. Other organizations have published documents on the same subject.[3,4,5]

2. TERMINOLOGY

Terms used in engineering practice and the construction industry may have meanings that differ somewhat from ordinary dictionary definitions. The following definitions are commonly understood in precast concrete construction.

Approval (Shop Drawings or Submittals). Action with respect to shop drawings, samples and other data which the General Contractor is required to submit, but only for conformance with the design requirements and compliance with the information given in the contract documents. Such action does not extend to means, methods, techniques, sequences or procedures of construction, or to safety precautions and programs incident thereto, unless specifically required in the contract documents.

Authority. The power, conferred or implied by contract, to exercise effective direction and control over an activity for which a party has responsibility.

Connection. A structural assembly or component that transfers forces from one precast member to another, or from one precast member to another type of structural member.

Contract Documents. The design drawings and specifications, as well as general and supplementary conditions and addenda, that define the construction and the terms and conditions for performing the work. These documents are incorporated by reference into the contract.

Contractor. A person or firm that enters into an agreement to construct all or part of a project.

Construction Manager. A person or firm engaged by the Owner to manage and administer the construction.

Designated Engineer. In this document, same as "Precast Engineer" below. Used in Ref. 6 to indicate any engineer designated by the Engineer of Record to perform specific tasks. Required to be a Licensed Professional Engineer.

Design (as a transitive verb). The process of applying the principles of structural mechanics and materials science, and interpreting code regulations to determine the geometry, composition, and arrangement of members and their connections in order to establish the composition and configuration of a structure.

Design (as a noun). The product of the design process, as usually expressed by design drawings and specifications.

Design Drawings. Graphic diagrams with dimensions and accompanying notes that describe the structure.

Detail (as a transitive verb). The process of utilizing the principles of geometry and the art of graphics to develop the dimensions of structural components.

Detail (as a noun). The product of the detailing process, shown on either design or shop drawings, such as the graphic depiction of a connection.

Engineer/Architect. A person or firm engaged by the Owner or the Owner's representative to design the structure and/or to provide services during the construction process. In some cases the Engineer may be a Subcontractor to the Architect, or vice versa. The Engineer of Record is usually an individual employed by the Engineer or Architect.

Engineer of Record (EOR). The registered professional engineer (or architect) who is responsible for developing the design drawings and specifications in such a manner as to meet the applicable requirements of governing state laws and of local building authorities. The EOR is commonly identified by the professional engineer's seal on the design drawings and specifications. See also "Structural Engineer of Record."

Erector. Usually the Subcontractor who erects the precast concrete components at the site. The General Contractor may also be the Erector.

General Contractor. A person or firm engaged by the Owner to construct all or part of the project. The General Contractor supervises the work of its Subcontractors and coordinates the work with other Contractors.

Inspector. The person or firm retained by the Owner or the Owner's representative to observe and report on compliance of the construction with the contract documents.

Manufacturer (Producer, Precaster, Fabricator). The firm that manufactures the precast concrete components.

Member. In this document, a precast concrete piece or component.

Owner. The public body or authority, corporation, association, firm or person for whom the structure is designed and constructed.

Precast Concrete. Concrete cast elsewhere than in its final position. Includes prestressed and non-prestressed components used in structural or non-structural applications.

Precast Engineer. The person or firm who designs precast members for specified loads and who may also direct the preparation of the shop drawings. The Precast Engineer may be employed by the Manufacturer or be an independent person or firm to whom the Manufacturer subcontracts the work.

Responsibility. Accountability for providing the services and/or for performing the work required by contract.

Shop Drawings. Graphic diagrams of precast members and their connecting hardware, developed from information in the contract documents. The contract documents show information needed for both field assembly (erection) and manufacture (production) of the precast concrete. Shop drawings for precast concrete may be separated into erection and production drawings. Erection drawings typically describe the location and assembly details of each precast member at the construction site. Connection hardware is detailed on erection drawings and may be shown on production drawings. Production drawings contain all information necessary for the manufacturer to cast the member.

Specialty Structural Engineer (SSE). In this document, same as "Precast Engineer" above. Used by Refs. 2 and 5 to indicate any engineer designated by the Engineer of Record to perform specific tasks.

Structural Engineer of Record (SER). Same as "Engineer of Record (EOR)" above. Used by Refs. 2 and 5.

Specifications. Written requirements for materials and workmanship that complement the design drawings.

Subcontractor. A person or firm contracting to perform all or part of another person or firm's contract.

3. DESIGN PRACTICES

Practices vary throughout North America with respect to design of structures using precast concrete members. However, it is of fundamental importance that every aspect of the design be in accordance with the requirements of all state laws governing the practice of engineering and/or architecture, and also meet all requirements of local regulatory authorities. This generally requires that a registered professional engineer or architect accept responsibility as the Engineer of Record (EOR) for ensuring that these requirements are met. The EOR seals the contract documents. These documents constitute the structural design and are customarily submitted to regulatory authorities for a building permit. In addition, the EOR ordinarily approves shop drawings. The EOR may also have other ongoing responsibilities during construction to satisfy local authorities.

"The division of responsibility between the SER [EOR] and SSE's [Precast Engineer] on a project is controversial. While legally and contractually the SER is usually ultimately responsible for a project, much of the value of using a 'specialist' is lost if the specialist is not willing to stand behind his/her work. It is not at all unusual for a structural engineer or consulting structural engineering firm to serve as SER on some projects and SSE on others. In some cases, with the knowledge of the Owner and Client, the SER may perform specialty engineering tasks on the same project." [2]

Structures are usually designed by an engineering firm that is retained by or on behalf of the Owner. A person within the firm is selected to prepare and/or to supervise the preparation of the contract documents. This person normally is registered to practice engineering in the state where the structure will be built and becomes the EOR when the design drawings and specifications are approved by the local regulatory authority. An individual in private practice may also be retained by an Owner to prepare the design and become the EOR. The design drawings and the specifications become a part of the contract documents used by contractors to construct the structure. A critical function of the contract documents is to clearly define responsibility among involved design professionals. The contract drawings, at a minimum, should include all of the items listed in Section 10.3, Subsection 3.2.

Design-Build

Owners may allow, or even encourage, consideration of alternative construction schemes on a project. In such cases, the design drawings are frequently not as definitive as in fully developed designs. When such alternatives use precast concrete as the primary structural system, the Precast Engineer, or another qualified consulting engineer may be designated as the EOR. This additional responsibility should be recognized by the Owner, and compensation adjusted accordingly and appropriate responsibilities defined in the contract.

If a fully developed design is included in the contract documents, a contractor proposing an alternate for some part of the structure is expected to consider the effect of the alternate on all other parts of the structure, and to provide all necessary design changes. It is common to require that an alternate design be prepared under the direction of an engineer registered to practice in the state where the structure will be built. However, the EOR still has the responsibility to review and accept the alternate and to submit the alternate to the regulatory authorities.

Owners may also directly seek proposals from General Contractors who are willing to prepare the design (design-build). The General Contractor may use an employed registered professional engineer who is the EOR, or subcontract the design to a firm or individual who becomes the EOR. In some instances, a Manufacturer may perform the role of the General Contractor. Since the Owner may already have a contract with the selected General Contractor, any additional contracts will be between the General Contractor and selected designers. Under this arrangement, the Owner often retains an engineering consultant to review the proposals and the design related work of the selected General Contractor.

Normally the Manufacturer is a Subcontractor to the General Contractor. Most Manufacturers are willing to accept responsibility for component design of the members that they produce, provided that sufficient information is contained in the contract documents. This design work is commonly done by a Precast Engineer. The Manufacturer may also accept responsibility for design of the connections when the forces acting on the connections are defined by the EOR. At the time of bidding, if there is insufficient information in the contract documents to fully cover all the reinforcement requirements, the Manufacturer customarily assumes that industry minimum standards are acceptable.

Local regulatory authorities may approve design documents for starting construction without final design of the precast members. In some cases, the design can be performed and submitted at a later time, often in conjunction with preparation of shop drawings. The EOR may require that a registered engineer seal the documents that depict the component design of the precast members. This does not relieve the EOR of the responsibility or reviewing the designs to ensure that the designated loading requirements have been properly applied and interactive forces with other construction fully coordinated. However, some states accept designs made by these delegated engineers, [6] relieving the EOR of some responsibility.

4. RECOMMENDATIONS

The Precast/Prestressed Concrete Institute is keenly aware of the competitiveness in the marketplace for building systems. It believes precast concrete products provide a high quality structure. Along with quality, it is essential that a structure be both safe and serviceable, i.e., have structural integrity and perform as intended. Because the construction process involves many parties, it is essential that work assignments and responsibilities be clearly defined in the contractual arrangements. The Institute offers the following comments and recommendations towards achieving quality and integrity in precast concrete structures.

4.1 To the Owner

At the outset, the Owner must decide whether to enter into a contract with an Engineer or Architect to design and prepare contract documents for the structure which subsequently can be used to obtain bids by a General Contractor, or with a General Contractor (or Manufacturer willing to assume that responsibility) to both design and build the structure. These two arrangements may be summarized as follows:

1. Owner retains an Engineer or Architect who will be the EOR:

    a. To design and prepare contract documents sufficient for construction without further design by the General Contractor or Subcontractors.

    b. To prepare contract documents sufficient for construction with further design by the General Contractor or Subcontractor. (Design-Build)

2. Owner contracts with a General Contractor for design and construction.

Under either arrangement, structural integrity will be best insured if the EOR assumes responsibility for the lateral stability of the structure, including relevant connection details, even if the entire structure has been designed by the Precast Engineer. Under Arrangement 1b, the contract documents must establish the loadings and identify the criteria for design to be used by the Contractor. Design work by any Contractor should be submitted to and approved by the EOR.

The EOR may require the design and shop drawings to be sealed by a registered professional engineer as a demonstration of qualifications. Under Arrangement 2, where the EOR will be engaged or employed by the Contractor, it may be desirable for the Owner to retain another Engineer or Architect for consultation on design criteria and verification that the design intent is achieved.

The Owner should consider the experience and qualifications of both the Engineer or Architect and the General Contractor with precast construction similar to that for the intended project. If the Owner plans to have an active role in the project, it is essential this role and lines of communication with the other parties be clearly expressed in written documents pertaining to the project. The Owner must allow sufficient time in the various phases of the construction process to achieve the necessary review, coordination, and implementation of the project requirements.

Changes initiated after the start of construction invariably add cost to the structure. A thorough review by the Owner after the design is completed, before construction is started, is essential.

Producer members of the Precast/Prestressed Concrete Institute are required to adhere to a formal and rigorous Plant Certification Program. This program is recognized in the master specifications of the American Institute of Architects. It is also recommended and recognized in standard specifications by numerous federal, state and local government agencies. PCI is recognized by the Council of American Building Officials (CABO) as a Quality Assurance Inspection Agency. CABO includes the three national building codes. The Owner or Engineer/Architect should require that PCI Plant Certification be written into the project specifications.

4.2 To the EOR (Engineer/Architect)

As discussed in Section 3, the role of the EOR (Engineer/Architect) varies significantly depending on whether this party is retained by the Owner or by other parties. If retained by the Owner, the EOR's lines of communication among the parties involved in the project should be established by written documents.

Prebid and preconstruction conferences should be held, during which lines of communication and responsibilities are reviewed and the design requirements are discussed. "The primary failure in projects involving SSE's is the lack of coordination and delineation of responsibility…When interfacing with the SSE, the SER should always be the one who delineates responsibility for the various structural requirements." [2]

If retained or employed by the General Contractor or Manufacturer, the EOR must remain cognizant of the professional responsibility to satisfy state laws and local regulatory authorities. Procedures should be established to allow the EOR to discharge these responsibilities without restriction by other parties involved in the construction.

Review and approval of shop drawings by the EOR is essential to ensure the design intent is achieved. Timeliness of review may be crucial to the Manufacturer. The EOR must be responsive to the shop drawing submittal and review schedules for the various parties. These parties should accept that the approval of the EOR does not relieve them of their responsibilities.

The EOR should make clear whether the contract documents, specifications or drawings prevail in the event of conflicts. This subject is addressed in Section 10.3, Subsection 4.3.

Where the design involves a non-self-supporting precast concrete frame, the contract documents should indicate which party is responsible for the design of the construction bracing and how long it needs to remain in place.

Where erection procedures require design and calculations for erection stability, the contract documents should specify that a registered engineer perform these services. The EOR's review of this work should be limited to its effect on the integrity of the completed structure.

Interfaces between precast components and other construction materials require special attention. The EOR is responsible for considering these interface conditions during the design of the structure including temporary loading conditions during erection.

4.3 To the General Contractor

When the construction of a building that was independently designed for the Owner is undertaken, the General Contractor's responsibility is to build the structure in accordance with the contract documents.

When design responsibility is accepted by the General Contractor, the professional nature of this function must be recognized and allowed to be achieved in a manner that does not compromise integrity or impair quality.

The responsibilities of the Manufacturer must be clearly defined. It must be understood that all design is submitted through the EOR for approval or acceptance. The Manufacturer's responsibility is usually limited to product design and preparation of shop drawings. However, it may include the design of the entire structure if agreed to by the Manufacturer.

When the Manufacturer is responsible only for product design, all loads which are applied to the precast members, including forces developed by restraint, must be provided by the EOR. The Manufacturer should be given responsibility and authority for properly implementing the design drawings, properly furnishing materials and workmanship, maintaining the specified fabrication and erection tolerances, and for fit and erectibility of the structure.

The Manufacturer must be given all of the drawings and specifications that convey the full requirements for the precast members. Drawings are commonly divided into architectural, structural, electrical, mechanical, fire protection and other groupings depending on the size and scope of the project. Other pertinent drawings may also be desirable, such as approved shop drawings from other trades, roofing requirements, alternates, etc.

Timely review and approval of shop drawings and other pertinent information submitted by the Manufacturer is essential.

4.4 To the Manufacturer

After award of the contract, the Manufacturer and Precast Engineer should meet with the EOR to review design requirements. The Manufacturer should prepare shop drawings in accordance with the design information supplied in the contract documents and subsequent instructions from the EOR.

In cases where the Manufacturer requests permission to revise connections to facilitate manufacture and/or erection, information supporting the revision should be submitted to the EOR for review and approval. The Manufacturer and/or the Precast Engineer should have a direct channel of communication with the EOR as the project goes forward and should keep the General Contractor informed. The Manufacturer should request clarification in writing from the EOR on special connections or unusual structural conditions not clearly defined by the design drawings or specifications.

4.5 References

1. PCI Board Ad-hoc Committee for Responsibility for Design of Precast Concrete Structures, "Recommendations on Responsibility for Design and Construction of Precast Concrete Structures," PCI JOURNAL, V. 33, No. 4, July-August 1988.

2. CASE Task Group on Specialty Engineering, *National Practice Guidelines for Specialty Structural Engineers*, Coalition of American Structural Engineers, Washington, DC, 1996.

3. ACI Committee on Responsibility in Concrete Construction, "Guidelines for Authorities and Responsibilities in Concrete Design and Construction," *Concrete International*, V. 17, No. 9, September 1995.

4. "Quality in the Constructed Project — Chapter 16: Construction Contract Submittals," Manuals and Reports on Engineering Practice No. 73, American Society of Civil Engineers, New York, NY, 2001.

5. *National Practice Guidelines for the Structural Engineer of Record,* Coalition of American Structural Engineers, Washington, DC, 1997.

6. Florida Administrative Code, Chapter 61G15, State of Florida.

10.5 PCI STANDARD DESIGN PRACTICE

Precast, prestressed concrete design is based on the provisions of the ACI Building Code. In most cases, these provisions are followed literally. Occasionally, though, there is disagreement as to the interpretation of some sections of the ACI Code. Also, in some situations, research may support other design and construction practices. In such cases, strict compliance with the ACI provisions can cause design, production and performance problems that may unnecessarily increase the cost of a structure or may actually result in an inferior product.

In most cases, the practices reported herein are supported by many years of good performance and/or research. Members of the PCI Building Code Committee, along with other experienced precast concrete design engineers, have identified these code provisions as detailed herein. The list of provisions represents a starting point for discussion, and complete agreement with the positions taken is not expected. Nevertheless, a listing of the design practices followed by a majority of precast concrete design engineers is anticipated to be helpful in producing safe, economical precast, prestressed concrete structures by minimizing conflict among the members of the design and construction team.

This list of provisions is based on ACI 318-02, and the numbers refer to sections in that document. References to the PCI Design Handbook are to the Sixth Edition, unless otherwise noted. Excerpts from ACI 318-02 are reprinted here with permission of the American Concrete Institute, Farmington Hills, Michigan. This information was first published in the PCI JOURNAL, January-February 2003.

ACI CODE	PCI PRACTICE
CHAPTER 1 – GENERAL REQUIREMENTS	
1.2.1(e) – Size and location of all structural elements, reinforcement, and anchors.	**1.2.1(e)** – "reinforcement" in this case does not refer to prestressing steel. In precast concrete members, reinforcement may be shown only on the piece drawings. (Ref. Handbook Section 10.3.3.2)
1.2.1(g) – Magnitude and location of prestressing forces.	**1.2.1(g)** – For pretensioned concrete products, the prestressing design and detailing may be left to an engineer employed or retained by the manufacturer. (Ref. Handbook Sections 10.3 and 10.4)
1.2.2 – Calculations pertinent to design shall be filed with the drawings when required by the building official. Analyses and designs using computer programs shall be permitted provided design assumptions, user input, and computer-generated output are submitted. Model analysis shall be permitted to supplement calculations.	**1.2.2** – Product calculations and frequently other items such as connections are usually done by the manufacturer's engineer. They are then submitted to the Engineer or Architect of Record, who is responsible for filing these documents with the building official. (Ref. Handbook Sections 10.3 and 10.4)
1.3.1 – Concrete construction shall be inspected as required by the legally adopted general building code. In the absence of such inspection requirements, concrete construction shall be inspected throughout the various work stages by or under the supervision of a registered design professional or by a qualified inspector.	**1.3.1** – Precast concrete products are inspected by internal quality control inspectors under the guidance of the "Manual for Quality Control for Plants and Production of Structural Precast Concrete Products" (PCI MNL-116) or "Manual for Quality Control for Plants and Production of Architectural Precast Concrete Products" (PCI MNL-117). PCI member producers are required to follow these procedures and are periodically monitored by independent Quality Certification inspectors. PCI Certified Plants are "Approved

ACI CODE	PCI PRACTICE
	Fabricators," as defined in the model codes, and thus work done in the plant and approved by the building official is exempt from "special inspection" requirements.

CHAPTER 2 – DEFINITIONS

Moment frame – Frame in which members and joints resist forces through flexure, shear, and axial force. Moment frames shall be categorized as follows:

   Intermediate moment frame – A cast-in-place frame complying with the requirements of 21.2.2.3 and 21.12 in addition to the requirements for ordinary moment frames.

   Ordinary moment frame – A cast-in-place or precast concrete frame complying with the requirements of Chapters 1 through 18.

   Special moment frame – A cast-in-place frame complying with the requirements of 21.2 through 21.5, or a precast frame complying with the requirements of 21.2 through 21.6. In addition, the requirements for ordinary moment frames shall be satisfied.

Design of precast concrete moment frames is discussed in the PCI Design Handbook, Chapter 3.

Registered design professional – An individual who is registered or licensed to practice the respective design profession as defined by the statutory requirements of the professional registration laws of the state or jurisdiction in which the project is to be constructed.

Precast concrete shop drawings and product design calculations are normally done under the supervision of a licensed professional engineer, depending on project requirements.

Structural walls – Walls proportioned to resist combinations of shears, moments, and axial forces induced by earthquake motions. A shear wall is a structural wall. Structural walls shall be categorized as follows:

   Intermediate precast structural wall – A wall complying with all applicable requirements of Chapters 1 through 18 in addition to 21.13.

   Special precast structural wall – A precast wall complying with the requirements of 21.8. In addition, the requirements of ordinary reinforced concrete structural walls and the requirements of 21.2 shall be satisfied.

   Special reinforced concrete structural wall – A cast-in-place wall complying with the requirements of 21.2 and 21.7 in addition to the requirements for ordinary reinforced concrete structural walls.

Design of precast concrete shear wall buildings is discussed in the PCI Design Handbook, Chapter 3.

ACI CODE	PCI PRACTICE

Tendon – In pretensioned applications, the tendon is the prestressing steel. In post-tensioned applications, the tendon is a complete assembly consisting of anchorages, prestressing steel, and sheathing with coating for unbonded applications or ducts with grout for bonded applications.

Precast, prestressed concrete products are nearly always pretensioned with seven-wire strand. Thus, the terms "tendon," "prestressing steel" and "strand" are used interchangeably.

CHAPTER 3 – MATERIALS

3.5.2 – Welding of reinforcing bars shall conform to "Structural Welding Code — Reinforcing Steel," ANSI/AWS D1.4 of the American Welding Society. Type and location of welded splices and other required welding of reinforcing bars shall be indicated on the design drawings or in the project specifications. ASTM reinforcing bar specifications, except for ASTM A 706, shall be supplemented to require a report of material properties necessary to conform to the requirements in ANSI/AWS D1.4.

3.5.2 – A significant amount of connection field welding is common in precast concrete construction. The American Welding Society (AWS) and the American Institute of Steel Construction (AISC) recommendations are generally followed, with some modifications as shown in the PCI Design Handbook and the PCI manual "Design and Typical Details of Connections for Precast and Prestressed Concrete." Other connection devices such as welded headed studs and deformed bar anchors are also shown in these publications. Special precaution is necessary when welded stainless steel reinforcing bars or plates are used. (Ref. Handbook Section 6.5.1)

3.5.5 – Prestressing steel

3.5.5.1 – Steel for prestressing shall conform to one of the following specifications:

(a) Wire conforming to "Specification for Uncoated Stress-Relieved Steel Wire for Prestressed Concrete" (ASTM A 421);

(b) Low-relaxation wire conforming to "Specification for Uncoated Stress-Relieved Steel Wire for Prestressed Concrete" including Supplement "Low-Relaxation Wire" (ASTM A 421);

(c) Strand conforming to "Specification for Steel Strand, Uncoated Seven-Wire for Prestressed Concrete" (ASTM A 416);

(d) Bar conforming to "Specification for Uncoated High Strength Steel Bars for Prestressing Concrete" (ASTM A 722).

3.5.5.2 – Wire, strands, and bars not specifically listed in ASTM A 421, A 416, or A 722 are allowed provided they conform to minimum requirements of these specifications and do not have properties that make them less satisfactory than those listed in ASTM A 421, A 416, or A 722.

3.5.5 – Nearly all strand used in precast, prestressed concrete products is seven-wire strand conforming with ASTM A 416, manufactured with low-relaxation wire conforming with the supplement to ASTM A 421.

ACI CODE	PCI PRACTICE
3.8.8 – "Acceptance Criteria for Moment Frames Based on Structural Testing (ACI T1.1-01)," is declared to be part of this code as if fully set forth herein.	*3.8.8 – ACI T1.1-01 is primarily intended to address precast concrete moment frames.*
CHAPTER 4 – DURABILITY REQUIREMENTS	
4.2 – Freezing and thawing exposures	
4.2.1 – Normal weight and lightweight concrete exposed to freezing and thawing or deicing chemicals shall be air-entrained with air content indicated in Table 4.2.1. Tolerance on air content as delivered shall be ±1.5 percent. For specified compressive strength f'_c greater than 5000 psi, reduction of air content indicated in Table 4.2.1 by 1.0 percent shall be permitted.	*4.2.1 – Some studies have shown that the very low water-cementitious materials ratios used in most precast concrete products require less air entrainment than cast-in-place concrete. (Ref. "Some Physical Properties of High Strength Concrete," Portland Cement Association, 1978; "Frost and Scaling Resistance of High Strength Concrete," PCA, 2001)*
4.2.2 – Concrete that will be subject to the exposures given in Table 4.2.2 shall conform to the corresponding maximum water-cementitious materials ratios and minimum specified concrete compressive strength requirements of that table. In addition, concrete that will be exposed to deicing chemicals shall conform to the limitations of 4.2.3.	*4.2.2 – (Note: See ACI 318-02 for the table referenced in this section.) The exposures discussed in this section affect the cover requirements given in Chapter 7 of the Code. While the high quality concrete produced in precasting plants is generally resistant to severe exposure, the use of deicing chemicals directly on all concrete surfaces is strongly discouraged.*
4.4.1 – For corrosion protection of reinforcement in concrete, maximum water soluble chloride ion concentrations in hardened concrete at ages from 28 to 42 days contributed from the ingredients including water, aggregates, cementitious materials, and admixtures shall not exceed the limits of Table 4.4.1. When testing is performed to determine water-soluble chloride ion content, test procedures shall conform to ASTM C1218.	*4.4.1 – Calcium chloride or other admixtures containing chlorides are rarely used in precast concrete, and never in prestressed concrete, as required in Section 3.6.3. The requirements of this section regarding prestressed concrete are assumed to be met when all materials used in the concrete meet the appropriate ASTM specifications. See report by Donald W. Pfeifer, J. R. Landgren, and William Perenchio, "Concrete, Chlorides, Cover and Corrosion," PCI JOURNAL, V. 31, No. 4, July-August 1986, pp. 42–53. (Ref. Handbook Section 1.3.4)*
CHAPTER 5 – CONCRETE QUALITY, MIXING, AND PLACING	
5.2.3 – Concrete proportions shall be established in accordance with 5.3 or, alternatively, 5.4, and shall meet applicable requirements of Chapter 4.	*5.2.3 – Most producers of precast concrete products use standard mixes which have been designed and substantiated in accordance with this section.*

ACI CODE	PCI PRACTICE
5.11.3.2 – Accelerated curing shall provide a compressive strength of the concrete at the load stage considered at least equal to required design strength at that load stage.	*5.11.3.2* – *The Commentary states "...the elastic modulus, E_c, of steam-cured specimens may vary from that of specimens moist-cured at normal temperatures." It is, however, most common for the ACI equation to be used to calculate E_c even when accelerated curing is used. Some producers may recommend other values based on testing. (Ref. Handbook Section 1.3.1.4) Also note that curing by direct exposure to steam is seldom used in precasting plants.*

CHAPTER 7 – DETAILS OF REINFORCEMENT

ACI CODE	PCI PRACTICE
7.5.2 – Unless otherwise specified by the registered design professional, reinforcement, including tendons, and post-tensioning ducts shall be placed within the tolerances in 7.5.2.1 and 7.5.2.2.	*7.5.2* – *Precast concrete products will normally conform to PCI tolerance standards specified in PCI MNL-135-00, and Chapter 8 of the PCI Design Handbook. Closer tolerances should not be specified except for special situations. (Ref. Handbook Section 8.2.4)*
7.7.5 – Corrosive environments In corrosive environments or other severe exposure conditions, amount of concrete protection shall be suitably increased, and denseness and nonporosity of protecting concrete shall be considered, or other protection shall be provided.	
7.7.5.1 – For prestressed concrete members exposed to corrosive environments or other severe exposure conditions, and which are classified as Class T or C in 18.3.3, minimum cover to the prestressed reinforcement shall be increased 50 percent. This requirement shall be permitted to be waived if the precompressed tensile zone is not in tension under sustained loads.	*7.7.5.1* – *Nearly all precast, prestressed concrete members will be in compression in the precompressed tensile zone under sustained loads. Because of the compression and the high concrete quality achieved in precasting plants, members which meet this section have met the requirement of Section 7.7.5.*
7.10.3 – It shall be permitted to waive the lateral reinforcement requirements of 7.10, 10.16, and 18.11 where tests and structural analysis show adequate strength and feasibility of construction.	*7.10.3* – *Section 7.10.3 waives minimum lateral ties with "tests and calculations..." Section 18.11.2.3 specifically excludes prestressed walls with a minimum average prestress of 225 psi (1.55 MPa) from lateral reinforcement requirements. (Ref. Handbook Example 4.7.1)*

ACI CODE	PCI PRACTICE
7.10.4 – Spirals Spiral reinforcement for compression members shall conform to 10.9.3 and to the following: **7.10.4.1** – Spirals shall consist of evenly spaced continuous bar or wire of such size and so assembled to permit handling and placing without distortion from designed dimensions. **7.10.4.2** – For cast-in-place construction, size of spirals shall not be less than ⅜ in. diameter.	*7.10.4 and 7.10.5 – Precast, prestressed concrete columns frequently use continuously wound rectangular wire for lateral reinforcement. Section 7.10.4.2 specifically applies to only cast-in-place construction and Section 7.10.5.1 refers to "non-prestressed bars" so the minimum size requirements for ties clearly do not apply. The usual practice is to design such columns as tied columns under Section 18.11.2.2, with the wire sized and spaced to provide an area equal to the minimum requirement for ties. There are several research reports to support reduced tie requirements for prestressed concrete columns. For further information on this topic, see report by PCI Prestressed Concrete Columns Committee, "Recommended Practice for the Design of Prestressed Concrete Columns and Walls," PCI JOURNAL, V. 33, No. 4, July-August 1988, pp. 56-95.*
7.10.5 – Ties Tie reinforcement for compression members shall conform to the following: **7.10.5.1** – All non-prestressed bars shall be enclosed by lateral ties, at least No. 3 in size for longitudinal bars No. 10 or smaller, and at least No. 4 in size for No. 11, No. 14, No. 18, and bundled longitudinal bars. Deformed wire or welded wire reinforcement of equivalent area shall be permitted.	
7.12 – Shrinkage and temperature reinforcement	*7.12 – This section does not apply to flanges of precast, prestressed stemmed members. The flexural reinforcement in the flange of a stemmed member is transverse to the stems, and is usually welded wire reinforcement. Practice varies, but the Wire Reinforcement Institute requires that the longitudinal wires have an area at least 0.4 times that of the transverse wires. Section 7.12.3 is intended for post-tensioned slabs.*
CHAPTER 8 – ANALYSIS AND DESIGN – GENERAL CONSIDERATIONS **8.1.3** – Anchors within the scope of Appendix D, Anchoring to Concrete, installed in concrete to transfer loads between connected elements shall be designed using Appendix D. **8.3.2** – Except for prestressed concrete, approximate methods of frame analysis shall be permitted for buildings of usual types of construction, spans, and story heights.	*8.1.3 – Appendix D has specific provisions allowing modifications based on research. PCI has sponsored research on connections using welded headed studs that meet those provisions.* *8.3.2 – The intent of this section is to prohibit Section 8.3.3 to be used for post-tensioned concrete framing. Approximate (e.g., "portal") methods are sometimes used to design precast "litewalls" in parking structures.*

ACI CODE	PCI PRACTICE

ACI CODE

8.10.2 – Width of slab effective as a T-beam flange shall not exceed one-quarter of the span length of the beam, and the effective overhanging flange width on each side of the web shall not exceed:

(a) Eight times the slab thickness;

(b) One-half the clear distance to the next web.

CHAPTER 9 – STRENGTH AND SERVICEABILITY REQUIREMENTS

9.2 – Required strength

9.2.1 – Required strength U shall be at least equal to the effects of factored loads in Eq. (9-1) through (9-7). The effect of one or more loads not acting simultaneously shall be investigated.

$$U = 1.4(D + F) \tag{9-1}$$

$$U = 1.2(D + F + T) + 1.6(L + H) + 0.5(L_r \text{ or } S \text{ or } R) \tag{9-2}$$

$$U = 1.2D + 1.6(L_r \text{ or } S \text{ or } R) + (1.0L \text{ or } 0.8W) \tag{9-3}$$

$$U = 1.2D + 1.6W + 1.0L + 0.5(L_r \text{ or } S \text{ or } R) \tag{9-4}$$

$$U = 1.2D + 1.0E + 1.0L + 0.2S \tag{9-5}$$

$$U = 0.9D + 1.6W + 1.6H \tag{9-6}$$

$$U = 0.9D + 1.0E + 1.6H \tag{9-7}$$

except as follows:

(a) The load factor on L in Eqs. (9-3) to (9-5) shall be permitted to be reduced to 0.5 except for garages, areas occupied as places of public assembly, and all areas where the live load L is greater than 100 lb/ft^2.

9.2.3 – Estimations of differential settlement, creep, shrinkage, expansion of shrinkage-compensating concrete, or temperature change shall be based on a realistic assessment of such effects occurring in service.

R9.2.3 – The designer should consider the effects of differential settlement, creep, shrinkage, temperature, and shrinkage-compensating concrete. The term realistic assessment is used to indicate that the most probable values rather than the upper bound values of the variables should be used.

PCI PRACTICE

8.10.2 – Although Section 18.1.3 excludes this section, eight times the slab thickness is often used as a guide for determining the topping width to be used in designing composite beams. Thin flange members are commonly designed including the entire flange width in the compression block. (Ref. Handbook Examples 4.2.1.6 and 4.3.5.1)

9.2.1 – It should be emphasized that volume changes, settlement, and other movements, T, are not to be considered simultaneously with wind or earthquake forces. Structural effects of T need only be considered when the structural element is restrained and can produce internal forces as a result of T.

The load factor modification of 9.2.1(a) must be distinguished from live load reductions allowed in local codes. Where allowed, the reduced live loads establish a value for L to be used in the load combinations of 9.2.1 and 9.2.1(a). While the code provision does not allow a reduced load factor for garages, live load reductions are allowed, for example, under the International Building Code (IBC) and the Uniform Building Code (UBC) for parking garages.

9.2.3 – Chapter 3 of the PCI Design Handbook provides guidelines for estimating creep, shrinkage and temperature changes in precast concrete structures.

ACI CODE

9.3.2.7 – Flexure sections without axial load in pretensioned members where strand embedment is less than the development length
as provided in 12.9.1.1.. 0.75

9.5 – Control of deflections

9.5.4 – Prestressed concrete construction

9.5.4.1 – For flexural members designed in accordance with provisions of Chapter 18, immediate deflection shall be computed by usual methods or formulas for elastic deflections, and the moment of inertia of the gross concrete section shall be permitted to be used for Class U flexural members, as defined in 18.3.3.

9.5.4.2 – For Class C and Class T flexural members, as defined in 18.3.3, deflection calculations shall be based on a cracked transformed section analysis. It shall be permitted to base computations on a bilinear moment-deflection relationship, or an effective moment of inertia as defined by Eq. (9-8).

9.5.4.3 – Additional long-term deflection of prestressed concrete members shall be computed taking into account stresses in concrete and steel under sustained load and including effects of creep and shrinkage of concrete and relaxation of steel.

9.5.4.4 – Deflection computed in accordance with 9.5.4.1 or 9.5.4.2, and 9.5.4.3 shall not exceed limits stipulated in Table 9.5(b).

CHAPTER 10 – FLEXURE AND AXIAL LOADS

10.4.1 – Spacings of lateral supports for a beam shall not exceed 50 times the least width b of compression flange or face.

10.6.4 – The spacing s of reinforcement closest to a surface in tension shall not exceed that given by

$$s = \frac{540}{f_s} - 2.5c_c \qquad (10\text{-}4)$$

but not greater than $12(36/f_s)$.

PCI PRACTICE

9.3.2.7 – Section 4.2.3 of the PCI Design Handbook shows examples of designing for partially developed strands.

9.5.4 – Deflections are always calculated for precast, prestressed concrete members. Calculations will usually include both instantaneous and long-term camber and dead and live load deflection. The Engineer or Architect of Record will determine if this meets requirements, e.g., Table 9.5(b). Satisfactory performance may depend on many non-structural considerations. (Ref. Handbook Section 4.8)

10.4.1 – The spans of non-loadbearing spandrels on parking structures have frequently exceeded 50 times the width of the top of the member, and no problems have been observed. This is undoubtedly because they typically carry only their own weight, which is concentric. (see ACI 318 Commentary to this section) Where lateral (bumper) loads are applied to the spandrel, lateral supports at mid-height of the spandrel into the deck are typical.

10.6.4 – Note that Section 10.6 is specifically excluded for prestressed concrete (Section 18.1.3), except as specified in 18.4.4.4. (Ref. Handbook Section 4.2.2.1)

ACI CODE

10.9.3 – Ratio of spiral reinforcement ρ_s shall be not less than the value given by

$$\rho_s = 0.45 \left(\frac{A_g}{A_c} - 1 \right) \frac{f'_c}{f_y} \qquad (10\text{-}5)$$

where f_y is the specified yield strength of spiral reinforcement but not more than 60,000 psi.

10.10 – Slenderness effects in compression members

10.10.1 – Except as allowed in 10.10.2, the design of compression members, restraining beams, and other supporting members shall be based on the factored forces and moments from a second-order analysis considering material nonlinearity and cracking, as well as the effects of member curvature and lateral drift, duration of the loads, shrinkage and creep, and interaction with the supporting foundation. The dimensions of each member cross section used in the analysis shall be within 10 percent of the dimensions of the members shown on the design drawings or the analysis shall be repeated. The analysis procedure shall have been shown to result in prediction of strength in substantial agreement with the results of comprehensive tests of columns in statically indeterminate reinforced concrete structures.

10.10.2 – As an alternate to the procedure prescribed in 10.10.1, it shall be permitted to base the design of compression members, restraining beams, and other supporting members on axial forces and moments from the analyses described in 10.11.

CHAPTER 11 – SHEAR AND TORSION

11.0 – Notation
b_w = web width, or diameter of circular section, in.

11.1.3.2 – For prestressed members, sections located less than a distance h/2 from face of support shall be permitted to be designed for the same shear V_u as that computed at a distance h/2.

11.5.5 – Minimum shear reinforcement

11.5.5.1 – A minimum area of shear reinforcement shall be provided in all reinforced concrete flexural members (prestressed and non-prestressed) where factored shear force V_u exceeds

PCI PRACTICE

10.9.3 – See discussion of Sections 7.10.4 and 18.11.2.2.

10.10 – The PCI Design Handbook, Chapter 4, addresses the application of this section to precast and prestressed concrete columns. (Ref. Handbook Section 4.9)

10.10.2 – The "Moment Magnifier" method is not recommended for prestressed concrete compression members.

11.0 – The quantity b_w is the sum of the average stem width of all stems in tapered stem members such as double-tees. This is critical in Eq. 11-14. In hollow-core slabs, b_w is the minimum web width.

11.1.3.2 – In beams with loads applied near the bottom, such as L-beams or inverted tees, h is taken as the depth of the ledge for shear calculations, but not necessarily for torsion. (Ref. Handbook Section 4.3)

11.5.5 – If V_u is less than ϕV_c, shear reinforcement may be omitted in prestressed double tees, with a nominal minimum provided for 5 to 10 ft (1.5 to 3 m) from the ends. This is based on research by Alex Aswad and George Burnley, "Omission of Web Reinforcement in

ACI CODE

one-half the shear strength provided by concrete ϕV_c, except:

    (a) Slabs and footings;

    (b) Concrete joist construction defined by 8.11;

    (c) Beams with total depth not greater than 10 in., 2.5 times thickness of flange, or 0.5 the width of web, whichever is greatest.

11.6 – Design for torsion

11.7 – Shear-friction

11.7.3 – A crack shall be assumed to occur along the shear plane considered. The required area of shear-friction reinforcement A_{vf} across the shear plane shall be designed using either 11.7.4 or any other shear transfer design methods that result in prediction of strength in substantial agreement with results of comprehensive tests.

11.7.7 – Net tension across shear plane shall be resisted by additional reinforcement. Permanent net compression across shear plane shall be permitted to be taken as additive to the force in the shear-friction reinforcement, $A_{vf}f_y$, when calculating required A_{vf}.

11.9.1 – Brackets and corbels with a shear span-to-depth ratio a/d less than 2 shall be permitted to be designed using Appendix A. Design shall be permitted using 11.9.3 and 11.9.4 for brackets and corbels with:

    (a) a/d not greater than 1, and

    (b) subject to horizontal tensile force for N_{uc} not larger than V_u.

PCI PRACTICE

Prestressed Double Tees," PCI JOURNAL, V. 34, No. 2, March-April 1989, pp. 48-65. The approach is permitted by Section 11.5.5.2. (Ref. Handbook Section 4.3 and 4.3.4)

Prestressed hollow-core and flat slab units fall under (a) slabs and footings, and require no shear reinforcement, provided $V_u \leq \phi V_c$.

11.6 – Torsion design has typically been done using the Zia-McGee method (PCI Design Handbook, Second Edition) or Zia-Hsu method (Fourth Edition), with excellent results. See also Zia-Hsu article, "Design for Torsion and Shear in Prestressed Concrete Flexural Members," in May-June 2004 PCI JOURNAL. The "thin-walled tube" model which has been in ACI 318 since 1995 typically requires significantly greater reinforcement than the previous methods in, for example, spandrel beams in parking structures. Based on performance of beams designed by the previous methods, this additional reinforcement is unnecessary and uneconomical, and most precast engineers are using the methods indicated above. (Ref. Handbook Section 4.4)

11.7.3 – The "effective shear-friction" method described in the PCI Design Handbook is most often used. Use is permitted under Section 11.7.3. (Ref. Handbook Section 4.3.6)

11.7.7 – At shear wall bases, for example, the sustained dead load on the wall (including the weight of the wall) is added to the force developed in the bars across the shear plane. The minimum positive anchorage requirements of Chapter 16 still apply.

11.9.1 – Section 6.8 of the PCI Design Handbook, Sixth Edition, describes a method of corbel design that has been used successfully. It is consistent with the strut-and-tie method of Appendix A permitted by this section.

ACI CODE	PCI PRACTICE
The requirements of 11.9.2, 11.9.3.2.1, 11.9.3.2.2, 11.9.5, 11.9.6, and 11.9.7 shall apply to design of brackets and corbels. Distance d shall be measured at the face of the support.	
11.9.3.2 – Design of shear-friction reinforcement A_{vf} to resist shear V_u shall be in accordance with 11.7.	
11.9.3.2.1 – For normal weight concrete, shear strength V_n shall not be taken greater than $0.2f'_c b_w d$ nor $800 b_w d$, in lb.	**11.9.3.2.1** – The PCI Design Handbook allows V_n up to $1000 b_w d$. This is consistent with the "effective shear-friction" approach when concrete strengths of 5000 psi (34 MPa) and greater are used. (Ref. Handbook Table 4.3.6.1)
11.9.3.2.2 – For all-lightweight or sand-lightweight concrete, shear strength V_n shall not be taken greater than $(0.2 - 0.07a/d) f'_c b_w d$ nor $(800 - 280a/d) b_w d$, in lb.	**11.9.3.2.2** – The equations given here are more conservative in relation to normal weight concrete than the use of the λ factor in the effective shear-friction coefficient would produce.
11.9.3.4 – Reinforcement A_n to resist tensile force N_{uc} shall be determined from $N_{uc} \le \phi A_n f_y$. Tensile force N_{uc} shall not be taken less than $0.2V_u$ unless special provisions are made to avoid tensile forces. Tensile force N_{uc} shall be regarded as a live load even when tension results from creep, shrinkage, or temperature change.	**11.9.3.4** – Bearing pads are used to "avoid tensile forces." The PCI Design Handbook suggests that a value of N_{uc} which will cause the pad to slip is the maximum that can occur, or, alternatively, a value of $0.2V_{dead}$ is used as a guide. (Ref. Handbook Chapter 6)
11.9.6 – At front face of bracket or corbel, primary tension reinforcement, A_s, shall be anchored by one of the following: (a) By a structural weld to a transverse bar of at least equal size; weld to be designed to develop specified yield strength f_y of A_s bars; (b) By bending primary tension bars A_s back to form a horizontal loop; or (c) By some other means of positive anchorage.	**11.9.6** – Frequently, front face anchorage is by welding to an angle or a plate with vertical anchors. This is permitted by Section 11.9.6(c). (Ref. Handbook Section 6.13)
11.9.7 – Bearing area of load on bracket or corbel shall not project beyond straight portion of primary tension bars A_s, nor project beyond interior face of transverse anchor bar (if one is provided).	**11.9.7** – If primary tension bars are anchored by welding (Section 11.9.6), the bearing area can be considered to extend to the exterior face of the anchoring bar or plate. This section is not typically applied to beam ledges, where ledge reinforcement is typically anchored by bending bars near the front face. Research sponsored by PCI Specially Funded Research and Development Project No. 5, "Design of Spandrel Beams" addressed this issue and found that placement of bars is critical. (Ref. Handbook Section 6.13)
11.10.8 – When factored shear force V_u is less than $\phi V_c/2$, reinforcement shall be provided in accordance	**11.10.8** – For precast walls, the reference should be to Section 16.4.2 rather than to Chapter 14.

ACI CODE	PCI PRACTICE

with 11.10.9 or in accordance with Chapter 14. When V_u exceeds $\phi V_c/2$, wall reinforcement for resisting shear shall be provided in accordance with 11.10.9.

11.10.9 – Design of shear reinforcement for walls

11.10.9.1 – Where factored shear force V_u exceeds shear strength ϕV_c, horizontal shear reinforcement shall be provided to satisfy Eqs. (11-1) and (11-2), where shear strength V_s shall be computed by

$$V_s = \frac{A_v f_y d}{s_2} \quad (11\text{-}31)$$

where A_v is area of horizontal shear reinforcement within a distance s_2 and distance d is in accordance with 11.10.4. Vertical shear reinforcement shall be provided in accordance with 11.10.9.4.

11.10.9 – Sections 11.10.9.2 through 11.10.9.4 apply only when the in-plane shear, $V_u > \phi V_c$, as described in Section 11.10.9.1. Otherwise, minimum reinforcement required by Section 16.4.2 applies (0.001 times the gross cross-sectional area).

CHAPTER 12 – DEVELOPMENT AND SPLICES OF REINFORCEMENT

12.5 – Development of standard hooks in tension

12.5.1 – Development length ℓ_{dh}, in inches, for deformed bars in tension terminating in a standard hook (see 7.1) shall be determined from 12.5.2 and the applicable modification factors of 12.5.3, but ℓ_{dh} shall not be less than $8d_b$, nor less than 6 in.

12.5.1 – Bars in beam ledges are assumed to be developed with a hook, even when the straight portion is less than 6 in. (152 mm), measured to the stem face. See the research project listed as Ref. 11-46 in ACI 318-02.

12.11.1 – At least one-third the positive moment reinforcement in simple members and one-fourth the positive moment reinforcement in continuous members shall extend along the same face of member into the support. In beams, such reinforcement shall extend into the support at least 6 in.

12.11.1 – Does not apply to precast construction. Excluded by Section 16.6.2.3.

12.13.2.4 – For each end of a single leg stirrup of welded plain or deformed wire fabric, two longitudinal wires at a minimum spacing of 2 in. and with the inner wire at least the greater of d/4 or 2 in. from d/2. Outer longitudinal wire at tension face shall not be farther from the face than the portion of primary flexural reinforcement closest to the face.

12.13.2.4 – Figure R12.13.2.4 shows how WWR is used as shear reinforcement in double tee stems. For further information, see the Joint PCI/WRI Ad Hoc Committee on Welded Wire Fabric for Shear Reinforcement report, "Welded Wire Fabric for Shear Reinforcement," PCI JOURNAL, V. 25, No. 4, July-August 1980, pp. 32-36.

CHAPTER 14 – WALLS

14.3 – Minimum reinforcement

14.3 – Minimum reinforcement for precast walls is specified in Section 16.4.2.

14.6.1 – Thickness of nonbearing walls shall not be less than 4 in., nor less than 1/30 the least distance between members that provide lateral support.

14.6.1 – Minimum thickness is not applicable to prestressed walls. (see Section 18.1.3)

ACI CODE	PCI PRACTICE

CHAPTER 15 – FOOTINGS

15.8.3.1 – Connection between precast columns or pedestals and supporting members shall meet the requirements of 16.5.1.3(a).

15.8.3.1 – Note reference to Chapter 16.

CHAPTER 16 – PRECAST CONCRETE

16.2.4 – In addition to the requirements for drawings and specifications in 1.2, (a) and (b) shall be included in either the contract documents or shop drawings:

(a) Details of reinforcement, inserts and lifting devices required to resist temporary loads from handling, storage, transportation, and erection;

(b) Required concrete strength at stated ages or stages of construction.

16.2.4 – Connection design is typically a part of the precast contract and connection forces are typically developed by the precast engineer, or sometimes listed on the Contract Drawings. (Ref. Handbook Sections 10.3 and 10.4)

16.5.1.3 – Vertical tension tie requirements of 7.13.3 shall apply to all vertical structural members, except cladding, and shall be achieved by providing connections at horizontal joints in accordance with (a) through (c):

(a) Precast columns shall have a nominal strength in tension not less than $200A_g$, in lb. For columns with a larger cross section than required by consideration of loading, a reduced effective area A_g, based on cross section required but not less than one-half the total area, shall be permitted;

(b) Precast wall panels shall have a minimum of two ties per panel, with a nominal tensile strength not less than 10,000 lb per tie;

(c) When design forces result in no tension at the base, the ties required by 16.5.1.3(b) shall be permitted to be anchored into an appropriately reinforced concrete floor slab on grade.

16.5.1.3 – This section applies to structures composed of many elements which must be tied together. Structures which use modules, or "boxes" will require different details to ensure integrity.

16.5.1.3(b) – Some panels may be too narrow to accommodate two connections. In such cases, as judged by the engineer, one connection may be adequate.

16.5.1.4 – Connection details that rely solely on friction caused by gravity loads shall not be used.

16.5.1.4 – This section should not be interpreted to prohibit connections designed using shear-friction principles.

16.6.2.2 – Unless shown by test or analysis that performance will not be impaired, (a) and (b) shall be met:

(a) Each member and its supporting system shall have design dimensions selected so that, after consideration of tolerances, the distance from the edge of the support to the end of the precast member in the direction

16.6.2.2 – When shorter bearing lengths occur in the field, analysis is usually the basis for acceptability. When designing bearing lengths, the effects of member shortening and movement at expansion joints should be considered.

ACI CODE

of the span is at least 1/180 of the clear span ℓ, but not less than:

For solid or hollow-core slabs .... 2 in.
For beams or stemmed members .... 3 in.

(b) Bearing pads at unarmored edges shall be set back a minimum of ½ in. from the face of the support, or at least the chamfer dimension at chamfered edges.

16.6.2.3 – The requirements of 12.11.1 shall not apply to the positive bending moment reinforcement for statically determinate precast members, but at least one-third of such reinforcement shall extend to the center of the bearing length, taking into account permitted tolerances in 7.5.2.2 and 16.2.3.

CHAPTER 17 – COMPOSITE CONCRETE FLEXURAL MEMBERS

17.5.2.1 – When contact surfaces are clean, free of laitance, and intentionally roughened, shear strength V_{nh} shall not be taken greater than $80b_v d$, in lb.

17.5.2.3 – When ties are provided in accordance with 17.6, and contact surfaces are clean, free of laitance, and intentionally roughened to a full amplitude of approximately ¼ in., shear strength V_{nh} shall be taken equal to $(260 + 0.6\rho_v f_y)\lambda b_v d$, in lb, but not greater than $500 b_v d$, in lb. Values for λ in 11.7.4.3 shall apply.

17.6.3 – All ties shall be fully anchored into interconnected elements in accordance with 12.13.

PCI PRACTICE

17.5.2.1 – The $80b_v d$ horizontal shear strength level can be obtained by many finishes that appear smooth when compared with the roughness required in 17.5.2.2. Examples include floated, light broomed, or machine extruded surfaces. (Ref. Handbook Section 4.3.5) Because the strength of the interface in this case is developed by cementitious bond, proper preparation of the surface is of utmost importance.

17.5.2.3 – The surface should not be so rough as to allow bridging of the cast-in-place coarse aggregate and the formation of voids at the interface. The most important element is the "clean, free of laitance."

17.6.3 – Ties for horizontal shear in precast concrete members are typically U-shaped reinforcing bars which are embedded (mucked in) after the member has been cast and the top surface has been intentionally roughened and finished. The anchorage of the tie in the precast member is achieved by embedding the bar for the required development length without hooks. (see also R16.7.1)

Anchorage of hooked or bent ties in CIP topping is considered adequate if a minimum distance of 2¼, 2¾ and 3¼ in. (55, 70 and 80 mm) is provided between the shear transfer interface and the outside ends of standard hooks or U-bends of No. 3, No. 4, and No. 5 ties,

ACI CODE	PCI PRACTICE
	respectively, based on research cited in PCI Handbook Section 4.3.5.

ACI CODE

CHAPTER 18 – PRESTRESSED CONCRETE

18.4.1 – Stresses in concrete immediately after prestress transfer (before time-dependent prestress losses) shall not exceed the following:

(a) Extreme fiber stress in compression . $0.60f'_{ci}$

(b) Extreme fiber stress in tension except as permitted in (c) .... $3\sqrt{f'_{ci}}$

(c) Extreme fiber stress in tension at ends of simply supported members .... $6\sqrt{f'_{ci}}$

Where computed tensile stresses exceed these values, bonded additional reinforcement (non-prestressed or prestressed) shall be provided in the tensile zone to resist the total tensile force in concrete computed with the assumption of an uncracked section.

R18.4.1(b) and (c) – ...Where tensile stresses exceed the permissible values, the total force in the tensile stress zone may be calculated and reinforcement proportioned on the basis of this force at a stress of $0.6f_y$ but not more than 30,000 psi...

18.4.2 – For Class U and Class T prestressed flexural members, stresses in concrete at service loads (based on uncracked section properties, and after allowance for all prestress losses) shall not exceed the following:

(a) Extreme fiber stress in compression due to prestress plus sustained load .... $0.45 f'_c$

(b) Extreme fiber stress in compression due to prestress plus total load .... $0.60 f'_c$

PCI PRACTICE

18.4.1 – Recent research (see "Strength Design of Pretensioned Flexural Concrete Members at Prestress Transfer" by Noppakunwijai, Tadros, Ma, and Mast, PCI JOURNAL, January-February 2001, pp. 34-52) has shown that the compression limitations at transfer are more conservative than necessary, and have an effect on economy and safety. It has been common practice to allow compression up to $0.70f'_{ci}$. Other sections of the code define cracking stress as $7.5\sqrt{f'_{ci}}$, so the $6\sqrt{f'_{ci}}$ is not consistent. There also does not seem to be a logical reason for limiting the transfer tension at midspan to less than at the ends, since service load compression in the top is higher at midspan. Thus, at all sections, tension limits of $7.5\sqrt{f'_{ci}}$ are more consistent with Code philosophy. It is recommended that nominal reinforcement (at least two No. 4 or nominally tensioned strands) be provided in tops of beams even when tension stress is less than $7.5\sqrt{f'_{ci}}$.

R18.4.1(b) and (c) – Where beam tops are in tension at transfer of prestress forces, but are in compression under service load, and are not exposed to weather, the large amounts of steel indicated by this commentary item is excessive. Experience has shown that nominal top reinforcing bars or prestressing strand will adequately control temporary top cracking. Use of f_y (up to 60,000 psi) for the steel stress has been shown to be adequate. Bars must be detailed (e.g., hooks, C-bars, U-bars) to ensure development in the top tensile region.

18.4.2 – Table R18.3.3 clearly defines the differences between Class U (uncracked), Class T (transition) and Class C (cracked) prestressed concrete members. Most members are designed as Class U or T, and thus the design procedures are essentially unchanged from previous editions of the Code. For special cases, a member may be designed as Class C, but more attention must be paid to performance. Chapter 4 of the PCI Design Handbook, Sixth Edition, provides examples of detailed transformed cracked section analysis.

ACI CODE	PCI PRACTICE
18.4.4 – For Class C prestressed flexural members not subject to fatigue or to aggressive exposure, the spacing of bonded reinforcement nearest the extreme tension face shall not exceed that given by 10.6.4. For structures subject to fatigue or exposed to corrosive environments, special investigations and precautions are required. **18.4.4.1** – The spacing requirements shall be met by nonprestressed reinforcement and bonded tendons. The spacing of bonded tendons shall not exceed ⅔ of the maximum spacing permitted for nonprestressed reinforcement. Where both reinforcement and bonded tendons are used to meet the spacing requirement, the spacing between a bar and a tendon shall not exceed ⅚ of that permitted by 10.6.4. (see also Section 18.4.4.3) **18.4.4.3** – The magnitude of Δf_{ps} shall not exceed 36 ksi. When Δf_{ps} is less than or equal to 20 ksi, the spacing requirements of 18.4.4.1 and 18.4.4.2 shall not apply.	*18.4.4 – This section refers to maximum spacing requirements by applying Eq. (10-4) to prestressed concrete:* $$s = 540/f_s - 2.5c_c$$ *where s is the maximum spacing and c_c is the clear cover to the reinforcement nearest the tension face. Note that these requirements are for Class C members only (that is, those in which the concrete tension under service loads exceeds $12\sqrt{f'_c}$). In checking this requirement, first check Section 18.4.4.3. If the spacing requirements can be met by substituting 36 ksi (250 MPa) for f_s in Eq. (10-4), no further check is necessary. If not, check if Δf_{ps} is less than 20 ksi (140 MPa) (which will often be the case). If so, no further check is necessary.*
18.6 – Loss of prestress **18.6.1** – To determine effective prestress f_{se}, allowance for the following sources of loss of prestress shall be considered: (a) Prestressing steel seating at transfer; (b) Elastic shortening of concrete; (c) Creep of concrete; (d) Shrinkage of concrete; (e) Relaxation of prestressing steel stress; (f) Friction loss due to intended or unintended curvature in post-tensioning tendons.	*18.6 – Most structural engineers who specialize in the design of prestressed concrete follow the recommendations of ACI-ASCE Committee 423 task force given in Ref. 18.6. (Ref. Handbook Section 4.7)*
18.7.2 – As an alternative to a more accurate determination of f_{ps} based on strain compatibility, the following approximate values of f_{ps} shall be permitted to be used if f_{se} is not less than $0.5f_{pu}$. (a) For members with bonded tendons: $$f_{ps} = f_{pu}\left\{1 - \frac{\gamma_p}{\beta_1}\left[\rho_p\frac{f_{pu}}{f'_c} + \frac{d}{d_p}(\omega - \omega')\right]\right\} \quad (18\text{-}3)$$	*18.7.2 – Many engineers and most computer programs use strain compatibility analysis for determining f_{ps}. Others use Eq. (18-3). With low-relaxation strand, the results are not substantially different. (Ref. Handbook Section 4.2.1)*

ACI CODE

If any compression reinforcement is taken into account when calculating f_{ps} by Eq. (18-3), the term

$$\left[\rho_p \frac{f_{pu}}{f'_c} + \frac{d}{d_p}(\omega - \omega')\right]$$

shall be taken not less than 0.17 and d' shall be no greater than $0.15 d_p$.

18.8.2 – Total amount of prestressed and non-prestressed reinforcement shall be adequate to develop a factored load at least 1.2 times the cracking load computed on the basis of the modulus of rupture f_r specified in 9.5.2.3. This provision shall be permitted to be waived for:

(a) Two-way, unbonded post-tensioned slabs; and

(b) Flexural members with shear and flexural strength at least twice that required by 9.2.

18.11.2 – Limits for reinforcement of prestressed compression members

18.11.2.1 – Members with average prestress f_{pc} less than 225 psi shall have minimum reinforcement in accordance with 7.10, 10.9.1 and 10.9.2 for columns, or 14.3 for walls.

18.11.2.2 – Except for walls, members with average prestress f_{pc} equal to or greater than 225 psi shall have all tendons enclosed by spirals or lateral ties in accordance with (a) through (d):

(a) Spirals shall conform to 7.10.4;

(b) Lateral ties shall be at least No. 3 in size or welded wire reinforcement of equivalent area, and shall be spaced vertically not to exceed 48 tie bar or wire diameters, or the least dimension of the compression member;

(c) Ties shall be located vertically not more than half a tie spacing above top of footing or slab in any story, and not more than half a tie spacing below the lowest horizontal reinforcement in members supported above;

PCI PRACTICE

18.8.2 – For simple span members, this provision is generally assumed to apply only at critical flexural sections. (Ref. Handbook Section 4.2.1). This provision is intended to ensure adequate ductility in flexural members.

18.11.2.1 – Columns which are larger than required for architectural purposes will use the level of prestress for the size of column needed. For example, if a 16 x 16 in. (406 x 406 mm) column will carry the load, but a 24 x 24 in. (610 x 610 mm) column is used, the total prestress force necessary is 225(16 x 16) = 57,600 lb (26127 kg). This practice is supported by Sections 10.8.4 and 16.5.1.3(a).

18.11.2.2 – The PCI Prestressed Concrete Columns Committee report, "Recommended Practice for the Design of Prestressed Concrete Columns and Walls," recommends that column capacity be reduced to 85 percent of calculated if ties do not meet all of the requirements. Most producers use some ties, but may modify the size and spacing based on research. Note that walls are excluded from the lateral tie requirements. Column ties are required in seismic regions. (Ref. Handbook Example 4.9.1.2)

ACI CODE	PCI PRACTICE

ACI CODE

(d) Where beams or brackets frame into all sides of a column, ties shall be terminated not more than 3 in. below lowest reinforcement in such beams or brackets.

CHAPTER 21 – SPECIAL PROVISIONS FOR SEISMIC DESIGN

PCI PRACTICE

Chapter 3 of the PCI Design Handbook, the PCI Seismic Design Manual (in progress), and other publications and research reports are available to assist the designer in the design of precast concrete structures in seismic areas.

ACI CODE

1 APPENDIX A – STRUT-AND-TIE MODELS

PCI PRACTICE

Strut-and-tie modeling may have many applications in precast concrete construction, including corbels and dapped ends of beams. The design procedures for these elements given in the PCI Design Handbook have certain limits of applicability, and strut-and-tie methods may be used for cases which fall outside these limits.

ACI CODE

2 APPENDIX D – ANCHORING TO CONCRETE

PCI PRACTICE

The PCI Design Handbook and the PCI Connections Manual have for many years given design recommendations for connections which use welded headed studs and other anchorage devices. Connections designed by these recommendations have performed satisfactorily.

D4.2 – The nominal strength for any anchor or group of anchors shall be based on design models that result in predictions of strength in substantial agreement with results of comprehensive tests. The materials used in the tests shall be compatible with the materials used in the structure. The nominal strength shall be based on the 5 percent fractile of the basic individual anchor strength. For nominal strengths related to concrete strength, modifications for size effects, the number of anchors, the effects of close spacing of anchors, proximity to edges, depth of the concrete member, eccentric loadings of anchor groups, and presence or absence of cracking shall be taken into account. Limits on edge distances and anchor spacing in the design models shall be consistent with the tests that verified the model.

D4.2 *– PCI has sponsored tests of stud assemblies which meet these requirements and result in design criteria which can be used in lieu of the requirements of this Appendix.*

CHAPTER 11
GENERAL DESIGN INFORMATION

11.1 Design Information 11–2
    11.1.1 Dead Weights of Floors, Ceilings, Roofs, and Walls 11–2
    11.1.2 Recommended Minimum Uniformly Distributed and
           Concentrated Live Loads 11–3
    11.1.3 Beam Design Equations and Diagrams 11–5
    11.1.4 Camber (Deflection) and Rotation Coefficients for
           Prestress Force and Loads 11–23
    11.1.5 Moments in Beams with Fixed Ends 11–25
    11.1.6 Torsion Diagrams, Reactions, and Rotations 11–26
    11.1.7 Moving Load Placement for Maximum Moment and Shear 11–27
    11.1.8 Moments, Shears, and Deflections in Beams with Overhangs 11–28

11.2 Material Properties 11–29
    11.2.1 Table of Concrete Stresses 11–29
    11.2.2 Concrete Modulus of Elasticity as Affected by
           Unit Weight and Strength 11–29
    11.2.3 Properties and Design Strengths of Prestressing Strand and Wire 11–30
    11.2.4 Properties and Design Strengths of Prestressing Bars 11–31
    11.2.5 Typical Stress-Strain Curve, 7-Wire Low-Relaxation Prestressing Strand ... 11–32
    11.2.6 Transfer and Development Lengths for 7-Wire Uncoated Strand 11–33
    11.2.7 Reinforcing Bar Data 11–34
    11.2.8 Location of Reinforcement Confined by Stirrups or Ties 11–35
    11.2.9 Required Development Lengths for Reinforcing Bars 11–36
    11.2.10 Common Styles of Structural Welded Wire Reinforcement 11–38
    11.2.11 Wire Used in Structural Welded Wire Reinforcement 11–39
    11.2.12 Bar Area Equivalents in a One Foot Wide Section 11–40
    11.2.13 ACI Required Minimum Reinforcement Areas Per Foot Width of Section 11–41

11.3 Standard Bolts, Nuts and Washers 11–42
    11.3.1 Dimensions of Nuts and Bolts 11–42
    11.3.2 Dimensions of Standard Washers 11–44

11.4 Welding Information 11–45
    11.4.1 Weld Symbols Commonly Used in Precast Construction 11–45
    11.4.2 Typical Welded Joints in Precast Construction 11–46
    11.4.3 Properties of Weld Groups Treated as Lines 11–47

11.5 Section Properties 11–48
    11.5.1 Properties of Geometric Sections 11–48
    11.5.2 Plastic Section Moduli and Shape Factors 11–53

11.6 Metric Conversion 11–54
    11.6.1 Metric Calculations and Example 11–54
    11.6.2 Conversion from U.S. Customary Units to International System 11–55
    11.6.3 Preferred SI Units and U.S. Customary Equivalents 11–57

11.1 DESIGN INFORMATION

Design Aid 11.1.1 Dead Weights of Floor, Ceilings, Roofs and Walls[a]

Component	Load (psf)	Component	Load (psf)	Component	Load (psf)
Ceilings		**Floor Fill**		**Masonry walls**[b]	
Acoustical fiber board	1	Cinder concrete, per inch	9	Clay brick wythes:	
Gypsum board (per 1/8 in. thickness)	0.55	Lightweight concrete, per inch	8	4 in.	39
Mechanical duct allowance	4	Sand, per inch	8	8 in.	79
Plaster on tile or concrete	5	Stone concrete, per inch	12	12 in.	115
Plaster on wood lath	8			16 in.	155
Suspended steel channel system	2	**Floors and floor finishes**			
Suspended mtl lath and cem plaster	15	Asphalt block (2 in.) 1/2 in. mortar	30	**Hollow concrete masonry unit wythe:**	
Suspended mtl lath and gyp plaster	10	Cement finish (1 in.), on stone concrete fill	32	**Wythe thickness (in.)**	
Wood furring suspension system	2.5	Ceramic or quarry tile (3/4 in.) on 1/2 in. bed	16		
		Ceramic or quarry tile (3/4 in.) on 1 in. bed	23		4 / 6 / 8 / 10 / 12
Coverings, roof and wall		Concrete fill finish (per inch thickness)	12		
		Hardwood flooring, 7/8 in.	4	Density of unit (105 pcf)	
Asbestos – cement shingles	4	Linoleum or asphalt tile, 1/4 in.	1		
Asphalt shingles	2	Marble and mortar on stone-concrete fill	33	No grout	22 / 24 / 31 / 37 / 43
Cement tiles	16			48 in. o.c.	29 / 38 / 47 / 55
		Slate (per inch thickness)	15	40 in. o.c.	30 / 40 / 49 / 57
Clay tile (for mortar add 10 lb.)		Solid flat tile on 1 in. mortar base	23	(grout spacing)	
Book tile, 2 in.	12	Subflooring, 3/4 in.	3	32 in. o.c.	32 / 42 / 52 / 61
Book tile, 3 in.	20	Terazzo (1 1/2 in.) directly on slab	19	24 in. o.c.	34 / 46 / 57 / 67
Ludowici	10	Terazzo (1 in.) on stone-concrete fill	32	16 in. o.c.	40 / 53 / 66 / 79
Roman	12	Terazzo (1 in.) 2 in. stone concrete	32	Full grout	55 / 75 / 95 / 115
Spanish	19	Wood block (3 in.) on mastic, no fill	10	Density of unit (125 pcf):	
Composition:		Wood block (3 in.) on 1/2 in. mortar base	16		
Three-ply ready roofing	1			No grout	26 / 28 / 36 / 44 / 50
Four-ply felt and gravel	5.5			48 in. o.c.	33 / 44 / 54 / 62
Five-ply felt and gravel	6	**Floors, wood joist (no plaster)**		40 in. o.c.	34 / 45 / 56 / 65
		Double wood floor		(grout spacing)	
Copper on tin	1			32 in. o.c.	36 / 47 / 58 / 68
Corrugated asbestos-cement roofing	4		12 in. / 16 in. / 24 in.	24 in. o.c.	39 / 51 / 63 / 75
Deck, metal 20 gage	2.5	Joist Size (in.)	spacing (psf)	16 in. o.c.	44 / 59 / 73 / 87
Deck, metal 18 gage	3			Full grout	59 / 81 / 102 / 123
Decking, 2 in. wood (Douglas fir)	5	2 x 6	6 / 5 / 5	Density of unit (135 pcf)	
Decking, 3 in. wood (Douglas fir)	8	2 x 8	6 / 6 / 5	No grout	29 / 30 / 39 / 47 / 54
Fiberboard, 1/2 in.	0.75	2 x 10	7 / 6 / 6	48 in. o.c.	36 / 47 / 57 / 66
Gypsum sheating, 1/2 in.	2	2 x 12	8 / 7 / 6	40 in. o.c.	37 / 48 / 59 / 69
				(grout spacing)	
Insulation, roof boards (per in. thickness)				32 in. o.c.	38 / 50 / 62 / 72
Cellular glass	0.7	**Frame partitions**		24 in. o.c.	41 / 54 / 67 / 78
Fibrous glass	1.1	Movable steel partitions	4	16 in. o.c.	46 / 61 / 76 / 90
Fiberboard	1.5	Wood or steel studs, 1/2 gyp board each side	8	Full grout	62 / 83 / 105 / 127
Perlite	0.8	Wood studs, 2 x 4, unplastered	4		
Polystyrene foam	0.2	Wood studs, 2 x 4, plastered one side	12	**Solid concrete masonry unit wythe:**	
Urethane foam with skin	0.5	Wood studs, 2 x 4, plastered two sides	20	**Wythe thickness (in.)**	
Plywood (per 1/8 in. thickness)	0.4				
Rigid insulation, 1/2 in.	0.75	**Frame walls**			4 / 6 / 8 / 10 / 12
Skylight, metal frame, 3/8 in. wire glass	8	Exterior stud walls:		Density of unit (105 pcf)	32 / 51 / 69 / 87 / 105
Slate, 3/16 in.	7	2 x 4 @ 16 in., 5/8 in. gyp., insulated, 3/8 in. siding	11		
Slate, 1/4 in.	10	2 x 6 @ 16 in., 5/8 in. gyp., insulated, 3/8 in. siding	12	Density of unit (125 pcf)	38 / 60 / 81 / 102 / 124
Waterproofing membranes		Exterior stud walls with brick veneer	48		
Bituminous, gravel-coated	5.5	Windows, glass, frame, and sash	8	Density of unit (135 pcf)	41 / 64 / 87 / 110 / 133
Bituminous, smooth surface	1.5				
Liquid applied	1				
Single-ply, sheet	0.7				
Wood sheathing (per inch thickness)	3				
Wood shingles	3				

a. Source: "Minimum Design Loads for Buildings and Other Structures," ASCE 7-02, American Society of Civil Engineers, Reston, VA.
b. Weights of masonry include mortar but not plaster. For plaster, add 5 lb/ft^2 for each face plasterEdition, Values given represent averages. In some cases, there is a considerable range of weight for the same construction.

DESIGN INFORMATION

Design Aid 11.1.2   Recommended Minimum Uniformly Distributed and Concentrated Live Loads[a]

Occupancy or use	Uniform load (psf)	Concentrated load (lb)
Apartments (see residential)		
Access floor systems		
Computer use	100	2,000
Office use	50	2,000
Armories and drill rooms	150	
Assembly areas and theaters		
Fixed seats (fastened to floor)	60	
Lobbies	100	
Movable seats	100	
Platforms (assembly)	100	
Stage floors	150	
Balconies (exterior)	100	
On one- and two-family residences only, and not exceeding 100 ft^2	60	
Bowling alleys, poolrooms and similar recreational areas	75	
Corridors		
First floor	100	
Other floors, same as occupancy served except as indicated		
Dance halls and ballrooms	100	
Decks (patio and roof)		
Same as area served, or for the type of occupancy accommodated		
Dining rooms and restaurants	100	
Dwellings (see residential)		
Elevator machine room grating (on area of 4 in.^{2})		300
Finish light floor plate construction (on area of 1 in.^{2})		200
Fire escapes	100	
On single-family dwellings only	40	
Garages (passenger cars only)	50	Note b
Truck and buses	Note c	
Grandstands (see stadium and arena bleachers)		
Gymnasiums, main floors and balconies (see note e)	100	
Handrails, guardrails and grab bars		Note i
Hospitals		
Corridors above first floor	80	1,000
Operating room, laboratories	60	1,000
Private rooms	40	1,000
Wards	40	1,000
Hotels (see residential)		
Libraries		
Corridors above first floor	80	1,000
Reading rooms	60	1,000
Stack rooms (see note d)	150	1,000
Manufacturing		
Heavy	250	2,000
Light	125	3,000
Marquees and Canopies	75	
Office Buildings		
Corridors above first floor	80	
File and computer rooms shall be designed for heavier loads based on anticipated occupancy		2,000
Lobbies and first floor corridors	100	2,000
Offices	50	2,000

See following page for all notes.

DESIGN INFORMATION

Design Aid 11.1.2 Recommended Minimum Uniformly Distributed and Concentrated Live Loads[a] (Cont.)

Occupancy or use	Uniform load (psf)	Concentrated load (lb)
Penal institutions		
Cell Blocks	40	
Corridors	100	
Residential		
Dwellings (one- and two-family)		
Habitable attics and sleeping areas	30	
Uninhabitable attics with storage	20	
Uninhabitable attics without storage	10	
All other areas except balconies	40	
Hotels and multifamily houses		
Private rooms and corridors serving them	40	
Public rooms and corridors serving them	100	
Reviewing stands, grandstands and bleachers (see Note e)	100	
Roofs		Note j
Schools		
Classrooms	40	1,000
Corridors above first floor	80	1,000
First floor corridors	100	1,000
Scuttles, skylight ribs, and accessible ceilings		200
Sidewalks, vehicular driveways, and yards, subject to trucking (see Note f, g)	250	8,000
Stadiums and arenas		
Bleachers (see Note e)	100	
Fixed seats, fastened to floor (see Note e)	60	
Stairs and exitways	100	Note h
Storage areas above ceilings	20	
Heavy	250	
Light	125	
Storage warehouses (shall be designed for heavier loads if required for anticipated storage)		
Stores		
Retail		
First floor	100	1,000
Upper floors	75	1,000
Wholesale, all floors	125	1,000
Vehicle barriers		Note i
Walkways and elevated platforms (other than exitways)	60	
Yards and terraces, pedestrians	100	

a. Source: "Minimum Design Loads for Buildings and Other Structures," ASCE 7-02, American Society of Civil Engineers, Reston, VA.
b. Floors in garages or portions of buildings used for the storage of motor vehicles shall be designed for the uniformly distributed live loads of Design Aid 11.1.2 or the following concentrated load: (1) for passenger cars accommodating not more than nine passengers 3,000 lb. acting on an area of 20 in.^{2}; and (2) mechanical parking structures without slab or deck, passenger car only, 1,500 lb/wheel.
c. Garages accommodating trucks and buses shall be designed in accordance with an approved method which contains provisions for truck and bus loadings.
d. The weight of books and shelving shall be computed using an assumed density of 65 pcf and converted to a uniformly distributed load; this load shall be used if it exceeds 150 pcf.
e. In addition to the vertical live loads, horizontal swaying forces parallel and normal to the length of seats shall be included in the design according to the requirements of ANSI/NFPA 102.
f. Other uniform loads in accordance with an approved method which contains provisions for truck loadings shall also be considered where appropriate.
g. The concentrated wheel load shall be applied on an area of 20 in.^{2}.
h. Minimum concentrated load on stair treads on area of 4 in.^{2} is 300 lb.
i. See ASCE 7-02, Section 4.4.
j. See ASCE 7-02, Sections 4.3 and 4.9.

DESIGN INFORMATION

Design Aid 11.1.3 Beam Design Equations and Diagrams

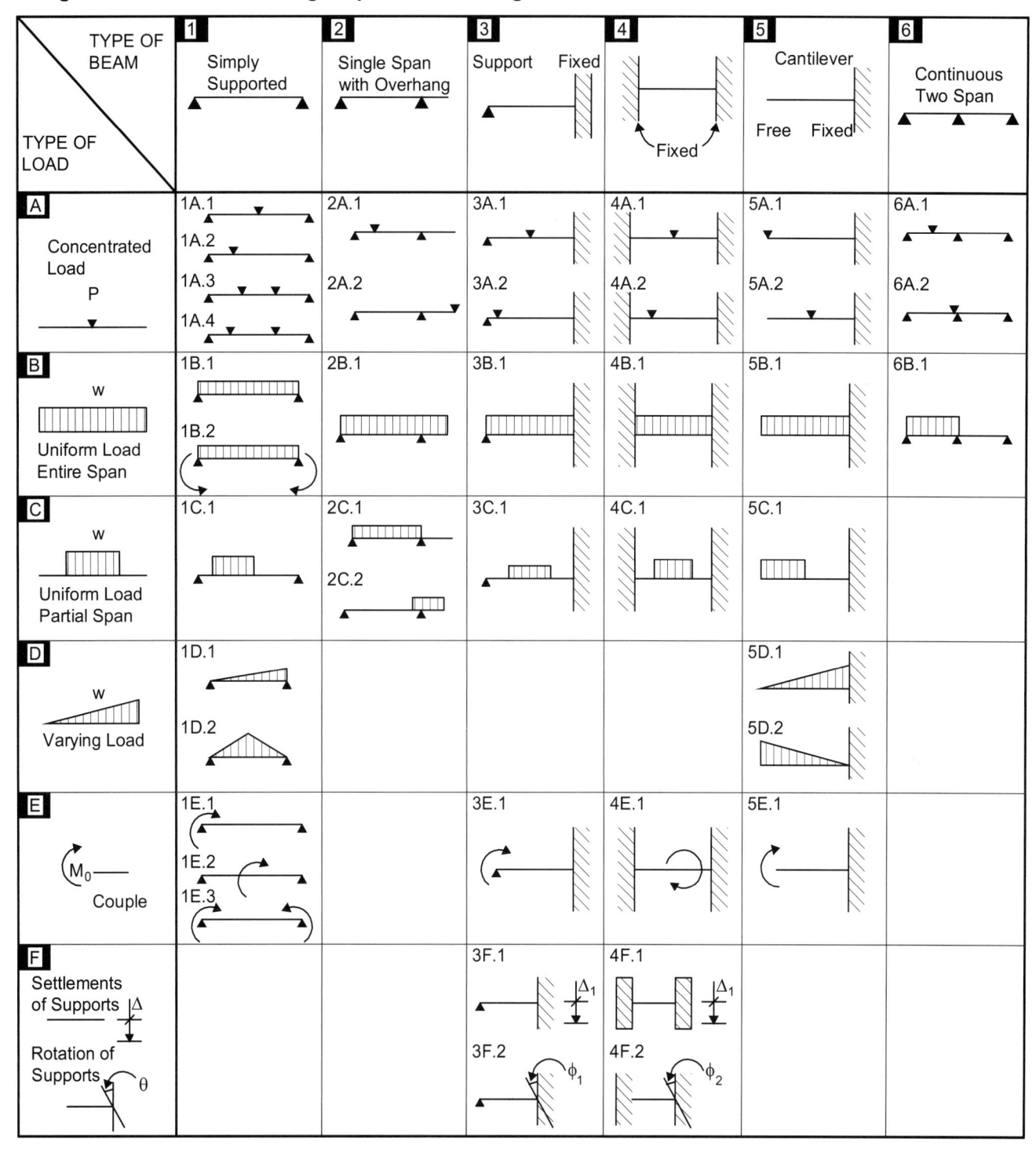

TYPE OF LOAD \ TYPE OF BEAM	1 Simply Supported	2 Single Span with Overhang	3 Support—Fixed	4 Fixed—Fixed	5 Cantilever Free—Fixed	6 Continuous Two Span
A Concentrated Load P	1A.1, 1A.2, 1A.3, 1A.4	2A.1, 2A.2	3A.1, 3A.2	4A.1, 4A.2	5A.1, 5A.2	6A.1, 6A.2
B Uniform Load Entire Span w	1B.1, 1B.2	2B.1	3B.1	4B.1	5B.1	6B.1
C Uniform Load Partial Span w	1C.1	2C.1, 2C.2	3C.1	4C.1	5C.1	
D Varying Load w	1D.1, 1D.2				5D.1, 5D.2	
E Couple M_0	1E.1, 1E.2, 1E.3		3E.1	4E.1	5E.1	
F Settlements of Supports Δ / Rotation of Supports θ			3F.1, 3F.2	4F.1, 4F.2		

This visual index assists in quickly locating the desired beam equations. The top row shows the support type, and the left column shows the load. For example, to find the equations for a beam fixed on both ends with a uniform load, go down from Column 4 and right on Row B. Locate 4B on upper right corner (page 11–17).

PCI Handbook/Sixth Edition   11–5

DESIGN INFORMATION

Design Aid 11.1.3 Beam Design Equations and Diagrams (Cont.) 1A

1A.1 SIMPLE BEAM – CONCENTRATED LOAD AT CENTER

$R = V = \dfrac{P}{2}$

M_{max} (at point of load) $= \dfrac{P\ell}{4}$

M_x (when $x < \dfrac{\ell}{2}$) $= \dfrac{Px}{2}$

Δ_{max} (at point of load) $= \dfrac{P\ell^3}{48EI}$

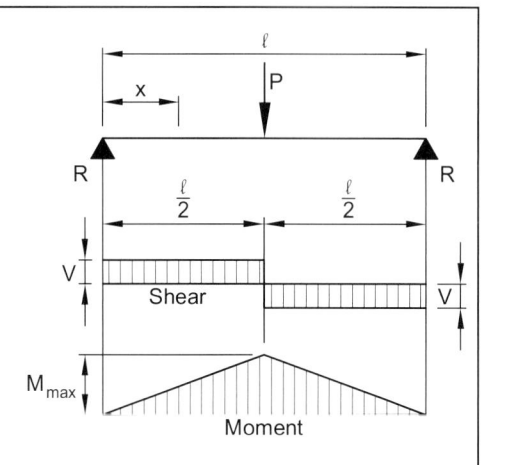

Δ_x (when $x < \dfrac{\ell}{2}$) $= \dfrac{Px}{48EI}(3\ell^2 - 4x^2)$

1A.2 SIMPLE BEAM – CONCENTRATED LOAD AT ANY POINT

$R_1 = V_1$ (max when $a < b$) $= \dfrac{Pb}{\ell}$

$R_2 = V_2$ (max when $a > b$) $= \dfrac{Pa}{\ell}$

M_{max} (at point of load) $= \dfrac{Pab}{\ell}$

M_x (when $x < a$) $= \dfrac{Pbx}{\ell}$

$\Delta_{max} \left(\text{at } x = \sqrt{\dfrac{a(a+2b)}{3}} \text{ when } a > b \right) = \dfrac{Pab(a+2b)\sqrt{3a(a+2b)}}{27EI\ell}$

Δ_a (at point of load) $= \dfrac{Pa^2 b^2}{3EI\ell}$

Δ_x (when $x < a$) $= \dfrac{Pbx}{6EI\ell}(\ell^2 - b^2 - x^2)$

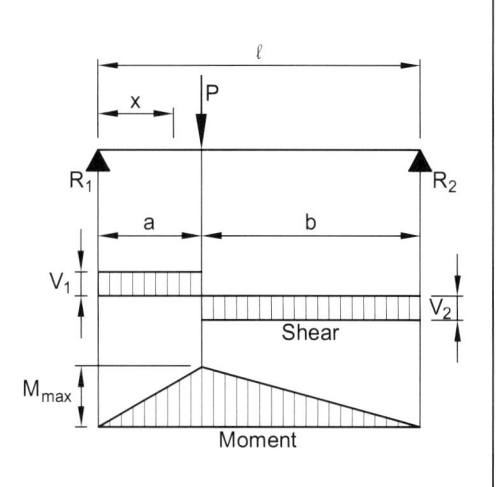

1A.3 SIMPLE BEAM – TWO EQUAL CONCENTRATED LOADS SYMMETRICALLY PLACED

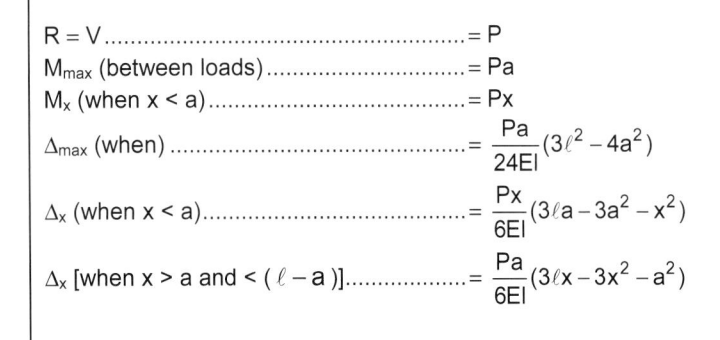

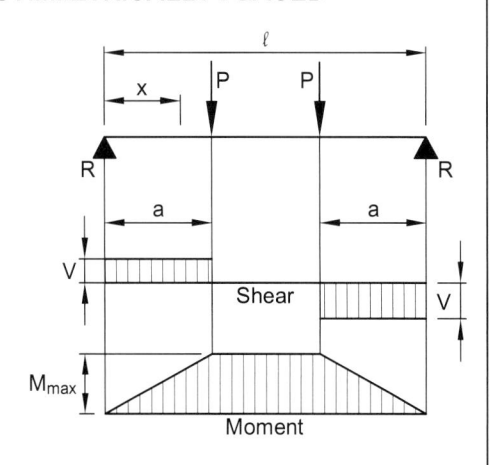

$R = V = P$

M_{max} (between loads) $= Pa$

M_x (when $x < a$) $= Px$

Δ_{max} (when) $= \dfrac{Pa}{24EI}(3\ell^2 - 4a^2)$

Δ_x (when $x < a$) $= \dfrac{Px}{6EI}(3\ell a - 3a^2 - x^2)$

Δ_x [when $x > a$ and $< (\ell - a)$] $= \dfrac{Pa}{6EI}(3\ell x - 3x^2 - a^2)$

DESIGN INFORMATION

Design Aid 11.1.3   Beam Design Equations and Diagrams (Cont.)   1A,1B

1A.4   SIMPLE BEAM – TWO UNEQUAL CONCENTRATED LOADS UNSYMMETRICALLY PLACED

$R_1 = V_1 = \dfrac{P_1(\ell - a) + P_2 b}{\ell}$

$R_2 = V_2 = \dfrac{P_1 a + P_2(\ell - b)}{\ell}$

V_x [when $x > a$ and $< (\ell - b)$] $= R_1 - P_1$

M_1 (max when $R_1 < P_1$) $= R_1 a$

M_2 (max when $R_2 < P_2$) $= R_2 b$

M_x (when $x < a$) $= R_1 x$

M_x [when $x > a$ and $< (\ell - b)$] $= R_1 x - P_1(x - a)$

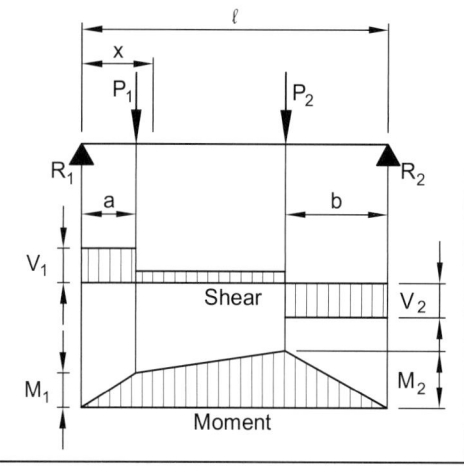

1B.1   SIMPLE BEAM – UNIFORMLY DISTRIBUTED LOAD

$R = V = \dfrac{w\ell}{2}$

$V_x = w\left(\dfrac{\ell}{2} - x\right)$

M_{max} (at center) $= \dfrac{w\ell^2}{8}$

$M_x = \dfrac{wx}{2}(\ell - x)$

Δ_{max} (at center) $= \dfrac{5w\ell^4}{384EI}$

$\Delta_x = \dfrac{wx}{24EI}(\ell^3 - 2\ell x^2 + x^3)$

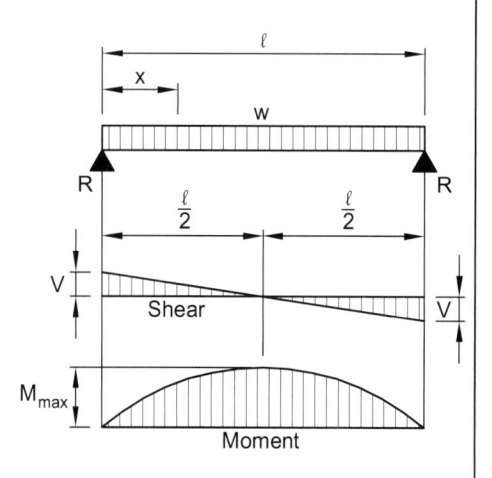

1B.2   SIMPLE BEAM – UNIFORMLY DISTRIBUTED LOAD AND VARIABLE END MOMENTS

$R_1 = V_1 = \dfrac{w\ell}{2} + \dfrac{M_1 - M_2}{\ell}$

$R_2 = V_2 = \dfrac{w\ell}{2} - \dfrac{M_1 - M_2}{\ell}$

$V_x = w\left(\dfrac{\ell}{2} - x\right) + \dfrac{M_1 - M_2}{\ell}$

$M_3 \left(\text{at } x = \dfrac{\ell}{2} + \dfrac{M_1 - M_2}{w\ell}\right) = \dfrac{w\ell^2}{8} - \dfrac{M_1 + M_2}{2} + \dfrac{(M_1 - M_2)^2}{2w\ell^2}$

$M_x = \dfrac{wx}{2}(\ell - x) + \left(\dfrac{M_1 - M_2}{\ell}\right)x - M_1$

b (to locate inflection points) $= \sqrt{\dfrac{\ell^2}{4} - \left(\dfrac{M_1 + M_2}{w}\right) + \left(\dfrac{M_1 - M_2}{w\ell}\right)^2}$

$\Delta_x = \dfrac{wx}{24EI}\left[x^3 - \left(2\ell + \dfrac{4M_1}{w\ell} - \dfrac{4M_2}{w\ell}\right)x^2 + \dfrac{12M_1}{w}x + \ell^3 - \dfrac{8M_1\ell}{w} - \dfrac{4M_2\ell}{w}\right]$

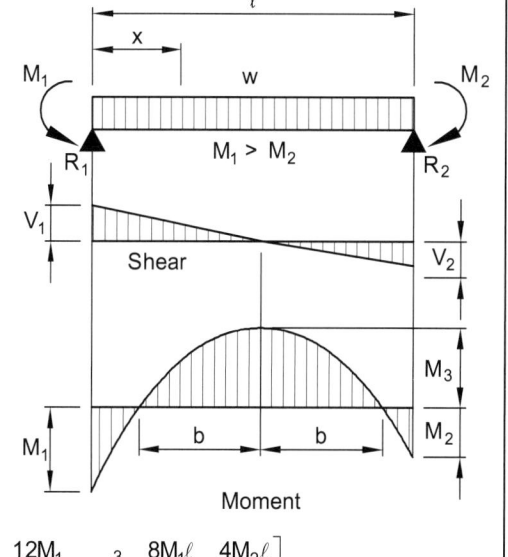

DESIGN INFORMATION

Design Aid 11.1.3 Beam Design Equations and Diagrams (Cont.) 1C,1D

1C.1 SIMPLE BEAM – UNIFORM LOAD PARTIALLY DISTRIBUTED

$R_1 = V_1$ (max when $a < c$) $= \dfrac{wb}{2\ell}(2c + b)$

$R_2 = V_2$ (max when $a > c$) $= \dfrac{wb}{2\ell}(2a + b)$

V_x [when $x > a$ and $< (a + b)$] $= R_1 - w(x - a)$

$M_{max}\left(\text{at } x = a + \dfrac{R_1}{w}\right) = R_1\left(a + \dfrac{R_1}{2w}\right)$

M_x (when $x < a$) $= R_1 x$

M_x [when $x > a$ and $< (a + b)$] $= R_1 x - \dfrac{w}{2}(x - a)^2$

M_x [when $x > (a + b)$] $= R_2(\ell - x)$

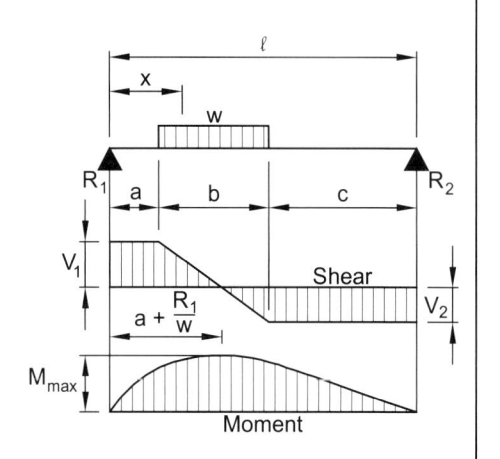

1D.1 SIMPLE BEAM – LOAD INCREASING UNIFORMLY TO ONE END (W IS TOTAL LOAD)

$W = \dfrac{w\ell}{2}$

$R_1 = V_1 = \dfrac{W}{3}$

$R_2 = V_2$ (max) $= \dfrac{2W}{3}$

$V_x = \dfrac{W}{3} - \dfrac{Wx^2}{\ell^2}$

$M_{max}\left(\text{at } x = \dfrac{\ell}{\sqrt{3}} = 0.5774\ell\right) = \dfrac{2W\ell}{9\sqrt{3}} = 0.1283W\ell$

$M_x = \dfrac{Wx}{3\ell^2}(\ell^2 - x^2)$

$\Delta_{max}\left(\text{at } x = \ell\sqrt{1 - \sqrt{\dfrac{8}{15}}} = 0.5193\ell\right) = 0.01304\dfrac{W\ell^3}{EI}$

$\Delta_x = \dfrac{Wx}{180EI\ell^2}(3x^4 - 10\ell^2 x^2 + 7\ell^4)$

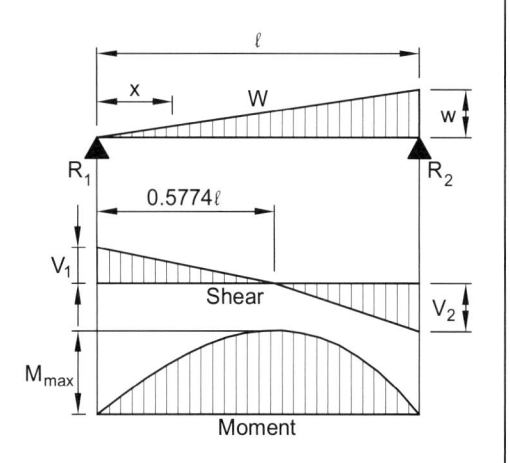

1D.2 SIMPLE BEAM – LOAD INCREASING UNIFORMLY TO CENTER (W IS TOTAL LOAD)

$W = \dfrac{w\ell}{2}$

$R = V = \dfrac{W}{2}$

$V_x\left(\text{when } x < \dfrac{\ell}{2}\right) = \dfrac{W}{2\ell^2}(\ell^2 - 4x^2)$

M_{max} (at center) $= \dfrac{W\ell}{6}$

$M_x\left(\text{when } x < \dfrac{\ell}{2}\right) = Wx\left(\dfrac{1}{2} - \dfrac{2x^2}{3\ell^2}\right)$

Δ_{max} (at center) $= \dfrac{W\ell^3}{60EI}$

$\Delta_x\left(\text{when } x < \dfrac{\ell}{2}\right) = \dfrac{Wx}{480EI\ell}(5\ell^2 - 4x^2)^2$

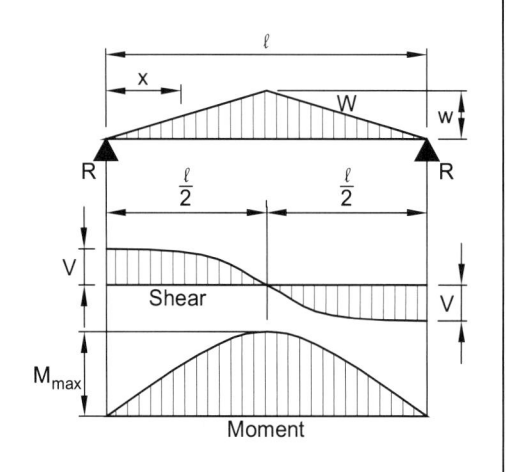

DESIGN INFORMATION

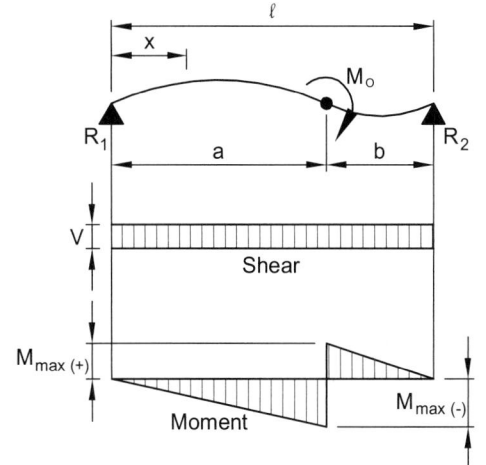

Design Aid 11.1.3 Beam Design Equations and Diagrams (Cont.) 1E

1E.1 BEAM SIMPLY SUPPORTED AT BOTH ENDS – MOMENT APPLIED AT ONE END

$-R_1 = R_2 = V = \dfrac{M_o}{\ell}$

M_{max} (at R_1) $= M_o$

$M_x = M_o - R_1 x = M_o\left(1 - \dfrac{x}{\ell}\right)$

Δ_{max} (when $x = 0.422\ell$) $= 0.0642 \dfrac{M_o \ell^2}{EI}$

$\Delta_x = \dfrac{M_o}{6EI}\left(3x^2 - \dfrac{x^3}{\ell} - 2\ell x\right)$

θ_1 (at R_1) $= \dfrac{M_o \ell}{3EI}$

θ_2 (at R_2) $= \dfrac{M_o \ell}{6EI}$

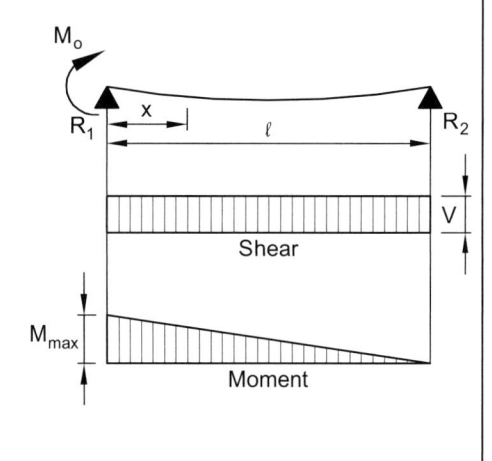

1E.2 BEAM SIMPLY SUPPORTED AT BOTH ENDS – MOMENT APPLIED AT ANY POINT

$R_1 = V$ (when $a > b$) $= \dfrac{M_o}{\ell}$

R_2 (when $a > b$) $= \dfrac{M_o}{\ell}$

$M_{max(-)}$ (at $x = a$) $= \dfrac{M_o a}{\ell}$

$M_{max(+)}$ (at $x = a$) $= M_o\left(1 - \dfrac{a}{\ell}\right)$

M_x (when $x < a$) $= \dfrac{M_o x}{\ell}$

M_x (when $x > a$) $= M_o\left(1 - \dfrac{x}{\ell}\right)$

Δ_x (when $x < a$) $= \dfrac{M_o x}{6EI\ell}\left(\ell^2 - 3b^2 - x^2\right)$

Δ_x (when $x > a$) $= \dfrac{M_o(\ell - x)}{6EI\ell}\left(3a^2 - 2\ell x + x^2\right)$

$\Delta_{max}\left(\text{at } x = \sqrt{\dfrac{\ell^2 - 3b^2}{3}} \text{ if } a > 0.4226\ell\right) = \dfrac{M_o}{3EI\ell}\left(\dfrac{\ell^2 - 3b^2}{3}\right)^{3/2}$

$\Delta_{max}\left(\text{at } x = \ell - \sqrt{\dfrac{\ell^2 - 3a^2}{3}} \text{ if } a > 0.5774\ell\right) = \dfrac{M_o}{3EI\ell}\left(\dfrac{\ell^2 - 3a^2}{3}\right)^{3/2}$

$M_{C\!L}$ (at center) $= -\dfrac{M_o}{2}$

$\Delta_{C\!L}$ (at center) $= \dfrac{M_o}{16EI}(\ell^2 - 4b^2)$

Δ_{max} (when $a = b = \dfrac{\ell}{2}$, at $x = \dfrac{\sqrt{3}}{6}\ell = 0.28867\ell$) $= \dfrac{M_o \ell^2}{124.71 EI}$

$\theta_{C\!L}$ (at center) $= \dfrac{M_o \ell}{12EI}$

DESIGN INFORMATION

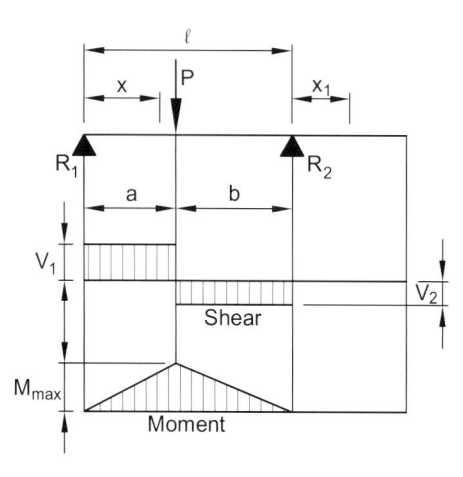

Design Aid 11.1.3   Beam Design Equations and Diagrams (Cont.)   1E,2A

1E.3   BEAM SIMPLY SUPPORTED AT BOTH ENDS – MOMENTS APPLIED AT EACH END

$R_1 = -R_2 = V \dots = \dfrac{M_2 - M_1}{\ell}$

$M_x \dots = (M_2 - M_1)\dfrac{x}{\ell} + M_1$

$\Delta_x \dots = \dfrac{x(\ell - x)}{6EI\ell}[M_1(2\ell - x) + M_2(\ell + x)]$

at $x_1 = \dfrac{6M_1\ell \pm \sqrt{36M_1^2\ell^2 - 12(M_1 - M_2)\ell^2(2M_1 + M_2)}}{6(M_1 - M_2)}$,

Δ = max and $\theta = 0$

θ_1 (at end) $\dots = -\dfrac{\ell}{6EI}(2M_1 + M_2)$

θ_2 (at end) $\dots = \dfrac{\ell}{6EI}(M_1 + 2M_2)$

If M_1 and M_2 are of opposite signs, the above formulas hold; just use actual sign of moment.

M_x (at point of contraflexure),

$\left(\text{where } x = \dfrac{M_1\ell}{M_2 - M_1}\right) \dots = 0$

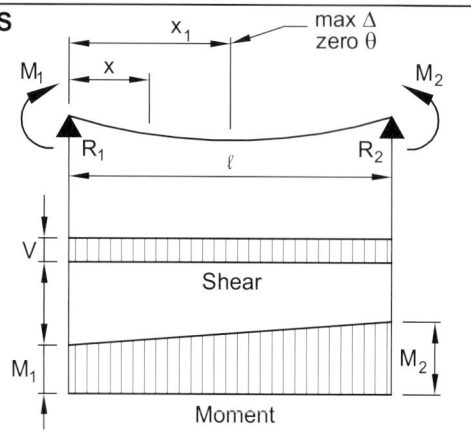

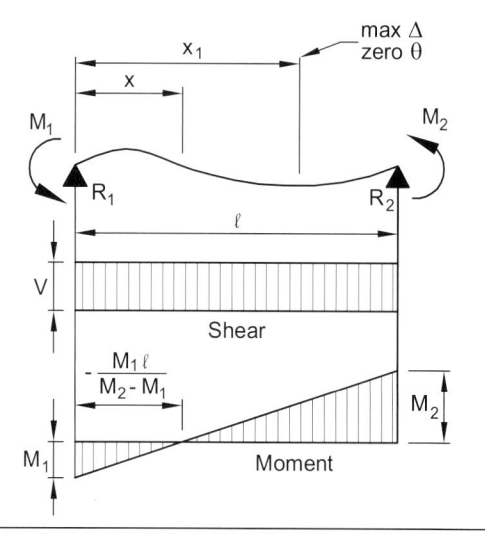

2A.1   BEAM OVERHANGING ONE SUPPORT – CONCENTRATED LOAD AT ANY POINT BETWEEN SUPPORTS

$R_1 = V_1$ (max when $a < b$) $\dots = \dfrac{Pb}{\ell}$

$R_2 = V_2$ (max when $a > b$) $\dots = \dfrac{Pa}{\ell}$

M_{max} (at point of load) $\dots = \dfrac{Pab}{\ell}$

M_x (when $x < a$) $\dots = \dfrac{Pbx}{\ell}$

$\Delta_{max} \left(\text{AT } x = \sqrt{\dfrac{a(a + 2b)}{3}} \text{ when } a > b \right) \dots = \dfrac{Pab(a + 2b)\sqrt{3a(a + 2b)}}{27EI\ell}$

Δ_a (at point of load) $\dots = \dfrac{Pa^2b^2}{3EI\ell}$

Δ_x (when $x < a$) $\dots = \dfrac{Pbx}{6EI\ell}(\ell^2 - b^2 - x^2)$

Δ_x (when $x > a$) $\dots = \dfrac{Pa(\ell - x)}{6EI\ell}(2\ell x - x^2 - a^2)$

$\Delta_{x_1} \dots = \dfrac{Pabx_1}{6EI\ell}(\ell + a)$

DESIGN INFORMATION

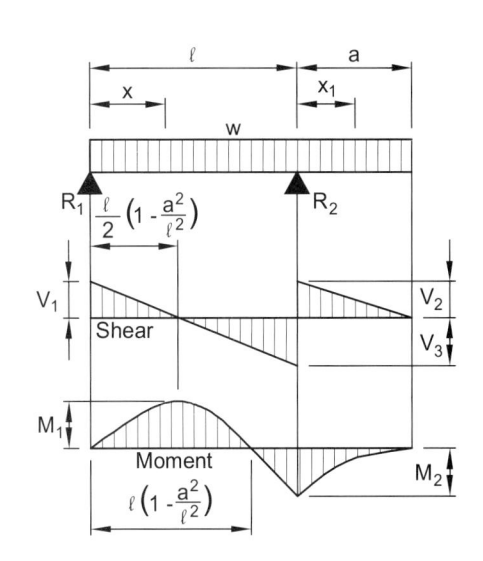

Design Aid 11.1.3   Beam Design Equations and Diagrams (Cont.)   2A,2B

2A.2 BEAM OVERHANGING ONE SUPPORT – CONCENTRATED LOAD AT END OF OVERHANG

$R_1 = V_1 = \dfrac{Pa}{\ell}$

$R_2 = V_1 + V_2 = \dfrac{P}{\ell}(\ell + a)$

$V_2 = P$

M_{max} (at R_2) $= Pa$

M_x (between supports) $= \dfrac{Pax}{\ell}$

M_{x_1} (for overhang) $= P(a - x_1)$

$\Delta_{max}\left(\text{between supports } x = \dfrac{\ell}{\sqrt{3}}\right) = \dfrac{Pa\ell^2}{9\sqrt{3}EI} = 0.06415 \dfrac{Pa\ell^2}{EI}$

Δ_{max} (for overhang at $x_1 = a$) $= \dfrac{Pa^2}{3EI}(\ell + a)$

Δ_x (between supports) $= \dfrac{Pax}{6EI\ell}(\ell^2 - x^2)$

Δ_{x_1} (for overhang) $= \dfrac{Px_1}{6EI}(2a\ell + 3ax_1 - x_1^2)$

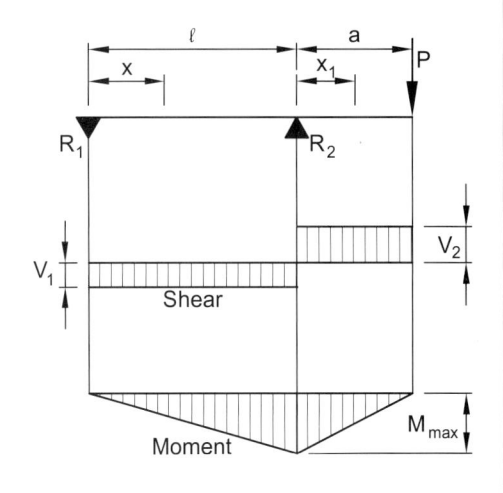

2B.1 BEAM OVERHANGING ONE SUPPORT – UNIFORMLY DISTRIBUTED LOAD

$R_1 = V_1 = \dfrac{w}{2\ell}(\ell^2 - a^2)$

$R_2 = V_2 + V_3 = \dfrac{w}{2\ell}(\ell + a)^2$

$V_2 = wa$

$V_3 = \dfrac{w}{2\ell}(\ell^2 + a^2)$

V_x (between supports) $= R_1 - wx$

V_{x_1} (for overhang) $= w(a - x_1)$

$M_1\left(\text{at } x = \dfrac{\ell}{2}\left[1 - \dfrac{a^2}{\ell^2}\right]\right) = \dfrac{w}{8\ell^2}(\ell + a)^2(\ell - a)^2$

M_2 (at R_2) $= \dfrac{wa^2}{2}$

M_x (between supports) $= \dfrac{wx}{2\ell}(\ell^2 - a^2 - x\ell)$

M_{x_1} (for overhang) $= \dfrac{w}{2}(a - x_1)^2$

Δ_x (between supports) $= \dfrac{wx}{24EI\ell}(\ell^4 - 2\ell^2 x^2 + \ell x^3 - 2a^2\ell^2 + 2a^2 x^2)$

Δ_{x_1} (for overhang) $= \dfrac{wx_1}{24EI}(4a^2\ell - \ell^3 + 6a^2 x_1 - 4ax_1^2 + x_1^3)$

DESIGN INFORMATION

Design Aid 11.1.3 Beam Design Equations and Diagrams (Cont.) 2C

2C.1 BEAM OVERHANGING ONE SUPPORT – UNIFORMLY DISTRIBUTED LOAD BETWEEN SUPPORTS

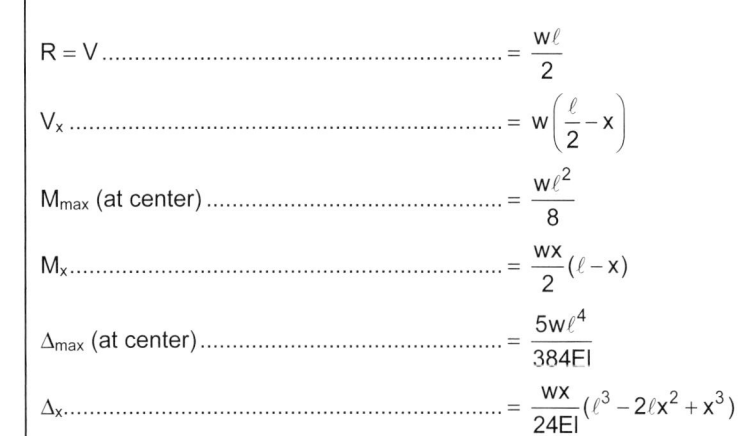

$R = V \dotso = \dfrac{w\ell}{2}$

$V_x \dotso = w\left(\dfrac{\ell}{2} - x\right)$

M_{max} (at center) $\dotso = \dfrac{w\ell^2}{8}$

$M_x \dotso = \dfrac{wx}{2}(\ell - x)$

Δ_{max} (at center) $\dotso = \dfrac{5w\ell^4}{384EI}$

$\Delta_x \dotso = \dfrac{wx}{24EI}(\ell^3 - 2\ell x^2 + x^3)$

$\Delta_{x_1} \dotso = \dfrac{w\ell^3 x_1}{24EI}$

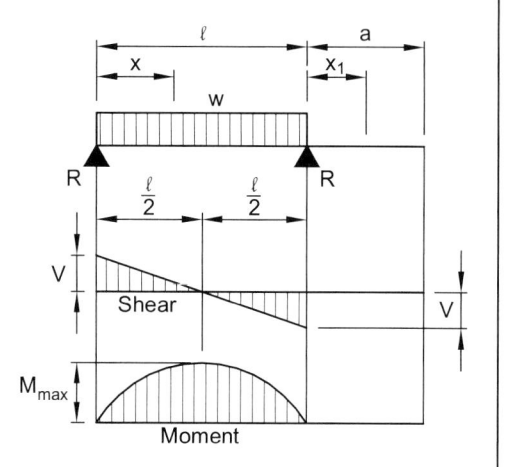

2C.2 BEAM OVERHANGING ONE SUPPORT – UNIFORMLY DISTRIBUTED LOAD ON OVERHANG

$R_1 = V_1 \dotso = \dfrac{wa^2}{2\ell}$

$R_2 = V_1 + V_2 \dotso = \dfrac{wa}{2\ell}(2\ell + a)$

$V_2 \dotso = wa$

V_{x_1} (for overhang) $\dotso = w(a - x_1)$

M_{max} (at R_2) $\dotso = \dfrac{wa^2}{2}$

M_x (between supports) $\dotso = \dfrac{wa^2 x}{2\ell}$

M_{x_1} (for overhang) $\dotso = \dfrac{w}{2}(a - x_1)^2$

Δ_{max} (between supports at $x = \dfrac{\ell}{\sqrt{3}}$) $\dotso = \dfrac{wa^2 \ell^2}{18\sqrt{3}EI} = 0.03208 \dfrac{wa^2 \ell^2}{EI}$

Δ_{max} (for overhang at $x_1 = a$) $\dotso = \dfrac{wa^3}{24EI}(4\ell + 3a)$

Δ_x (between supports) $\dotso = \dfrac{wa^2 x}{12EI\ell}(\ell^2 - x^2)$

Δ_{x_1} (for overhang) $\dotso = \dfrac{wx_1}{24EI}(4a^2\ell + 6a^2 x_1 - 4ax_1^2 + x_1^3)$

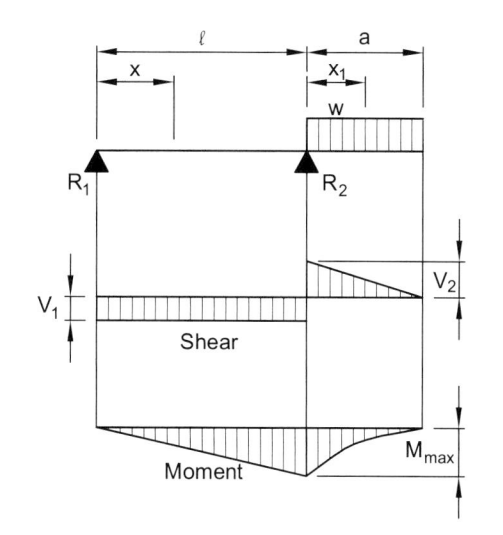

DESIGN INFORMATION

Design Aid 11.1.3   Beam Design Equations and Diagrams (Cont.)   3A

3A.1   BEAM FIXED AT ONE END, SIMPLY SUPPORTED AT THE OTHER END – CONCENTRATED LOAD AT CENTER

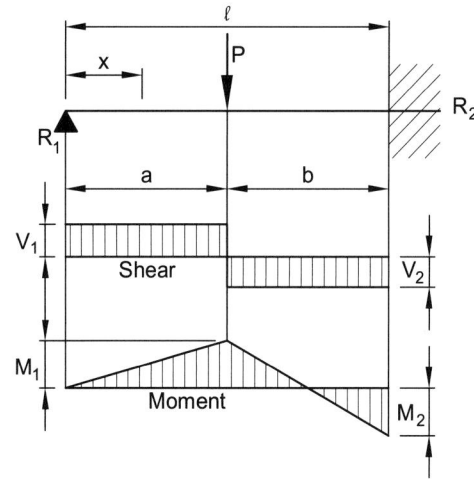

$R_1 = V_1 = \dfrac{5P}{16}$

$R_2 = V_2 \text{ (max)} = \dfrac{11P}{16}$

M_{max} (at fixed end) $= \dfrac{3P\ell}{16}$

M_1 (at point of load) $= \dfrac{5P\ell}{32}$

M_x (when $x < \dfrac{\ell}{2}$) $= \dfrac{5Px}{16}$

M_x (when $x > \dfrac{\ell}{2}$) $= P\left(\dfrac{\ell}{2} - \dfrac{11x}{16}\right)$

Δ_{max} (at $x = \ell\sqrt{\dfrac{1}{5}} = 0.4472\ell$) $= \dfrac{P\ell^3}{48EI\sqrt{5}} = 0.009317\dfrac{P\ell^3}{EI}$

Δ_x (at point of load) $= \dfrac{7P\ell^3}{768EI}$

Δ_x (when $x < \dfrac{\ell}{2}$) $= \dfrac{Px}{96EI}\left(3\ell^2 - 5x^2\right)$

Δ_x (when $x > \dfrac{\ell}{2}$) $= \dfrac{P}{96EI}(x-\ell)^2(11x-2\ell)$

3A.2   BEAM FIXED AT ONE END, SIMPLY SUPPORTED AT THE OTHER END – CONCENTRATED LOAD AT ANY POINT

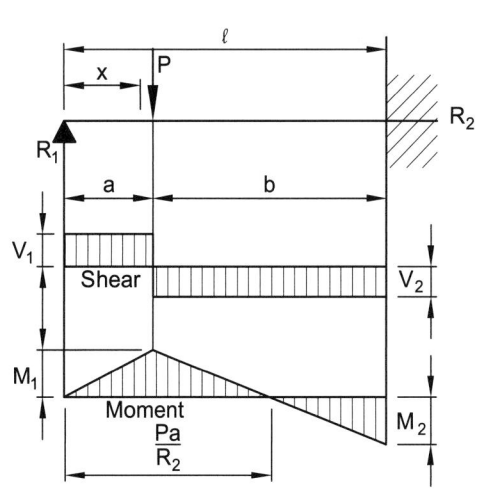

$R_1 = V_1 = \dfrac{Pb^2}{2\ell^3}(a+2\ell)$

$R_2 = V_2 = \dfrac{Pa}{2\ell^3}(3\ell^2 - a^2)$

M_1 (at point of load) $= R_1 a$

M_2 (at fixed end) $= \dfrac{Pab}{2\ell^2}(a+\ell)$

M_x (when $x < a$) $= R_1 x$

M_x (when $x > a$) $= R_1 x - P(x-a)$

Δ_{max} (when $a < 0.414\ell$, AT $x = \ell\dfrac{\ell^2 + a^2}{3\ell^2 - a^2}$) $= \dfrac{Pa\left(\ell^2 - a^2\right)^3}{3EI\left(3\ell^2 - a^2\right)^2}$

Δ_{max} (when $a > 0.414\ell$, AT $x = \ell\sqrt{\dfrac{a}{2\ell+a}}$) $= \dfrac{Pab^2}{6EI}\sqrt{\dfrac{a}{2\ell+a}}$

Δ_a (at point of load) $= \dfrac{Pa^2b^3}{12EI\ell^3}(3\ell + a)$

Δ_x (when $x < a$) $= \dfrac{Pb^2 x}{12EI\ell^3}\left(3a\ell^2 - 2\ell x^2 - ax^2\right)$

Δ_x (when $x > a$) $= \dfrac{Pa}{12EI\ell^2}(\ell-x)^2(3\ell^2 x - a^2 x - 2a^2\ell)$

DESIGN INFORMATION

Design Aid 11.1.3   Beam Design Equations and Diagrams (Cont.)   3B,3C

3B.1   BEAM FIXED AT ONE END, SIMPLY SUPPORTED AT THE OTHER END – UNIFORMLY DISTRIBUTED LOAD

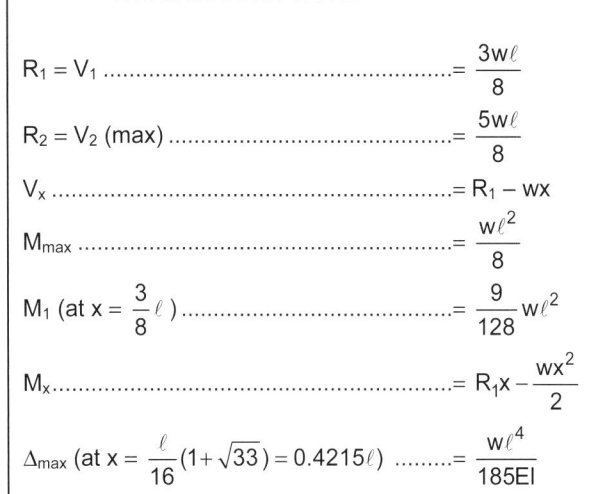

$R_1 = V_1 = \dfrac{3w\ell}{8}$

$R_2 = V_2 \text{ (max)} = \dfrac{5w\ell}{8}$

$V_x = R_1 - wx$

$M_{max} = \dfrac{w\ell^2}{8}$

$M_1 \left(\text{at } x = \dfrac{3}{8}\ell\right) = \dfrac{9}{128}w\ell^2$

$M_x = R_1 x - \dfrac{wx^2}{2}$

$\Delta_{max}\left(\text{at } x = \dfrac{\ell}{16}(1+\sqrt{33}) = 0.4215\ell\right) = \dfrac{w\ell^4}{185EI}$

$\Delta_x = \dfrac{wx}{48EI}(\ell^3 - 3\ell x^2 + 2x^3)$

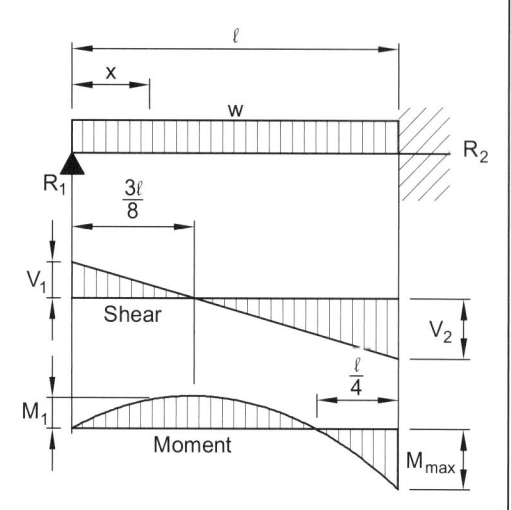

3C.1   BEAM FIXED AT ONE END, SIMPLY SUPPORTED AT THE OTHER END – UNIFORM LOAD PARTIALLY DISTRIBUTED OVER SPAN

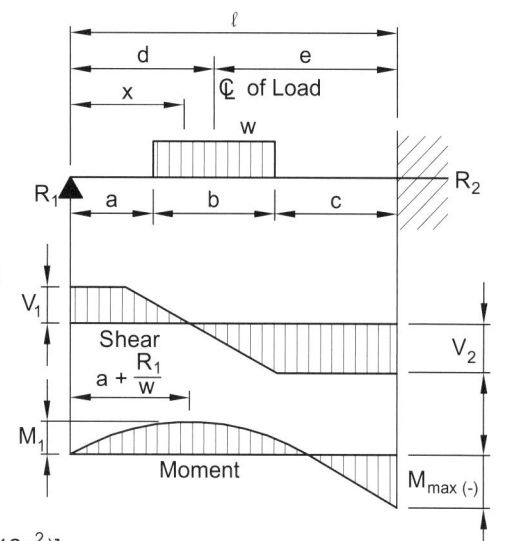

$R_1 = V_1 = \dfrac{wb}{8\ell^3}(12e^2\ell - 4e^3 + b^2 d)$

$R_2 = V_2 = wb - R_1$

$M_{max(-)} = \dfrac{wb}{8\ell^2}(12e^2\ell - 4e^3 + b^2 d - 8e\ell^2)$

$M_1 = R_1\left(a + \dfrac{R_1}{2w}\right)$

$M_x \text{ (when } x < a) = R_1 x$

$M_x \text{ [when } x > a \text{ and } x < (a+b)] = R_1 x - \dfrac{w}{2}(x-a)^2$

$M_x \text{ (when } x > (a+b) \text{ and } x < \ell) = R_1 x - wb(x-d)$

$\Delta_x \text{ (when } x < a) = \dfrac{x}{24EI}[4R_1(x^2 - 3\ell^2) + wb(b^2 + 12e^2)]$

$\Delta_x \text{ [when } x > a \text{ and } x < (a+b)] = \dfrac{1}{24EI}[4R_1 x(x^2 - 3\ell^2) + wbx(b^2 + 12e^2) - w(x-a)^4]$

$\Delta_x \text{ (when } x > (a+b) \text{ and } x < c) = \dfrac{1}{6EI}[3M_{MAX}(\ell-x)^2 + R_2(\ell-x)^3]$

DESIGN INFORMATION

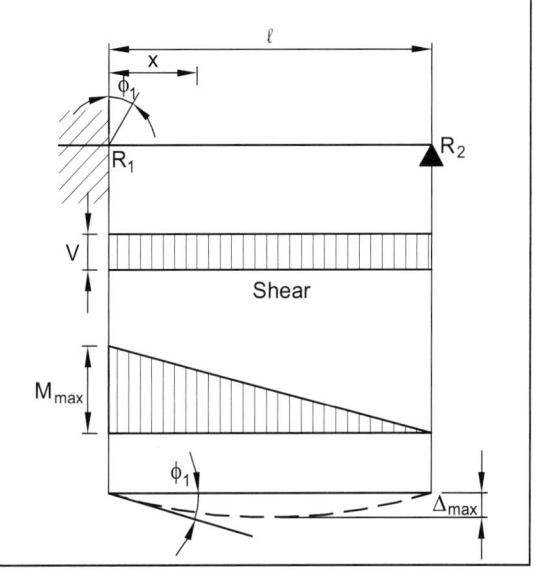

Design Aid 11.1.3　Beam Design Equations and Diagrams (Cont.)　　　　3E,3F

3E.1　BEAM FIXED AT ONE END, SIMPLY SUPPORTED AT THE OTHER END – MOMENT APPLIED AT THE FLEXIBLE END

$R_1 = -R_2 = V = \dfrac{3M_o}{2\ell}$

$M_1 = M_o$

$M_2 = \tfrac{1}{2} M_o$

$M_x = \dfrac{M_o}{2\ell}(2\ell - 3x)$

Δ_{max} (at $x = \dfrac{\ell}{3}$) $= \dfrac{M_o \ell^2}{27 EI}$

$\Delta_x = \dfrac{M_o x}{4 EI \ell}(\ell - x)^2$

θ (at supported end) $= \dfrac{M_o \ell}{4 EI}$

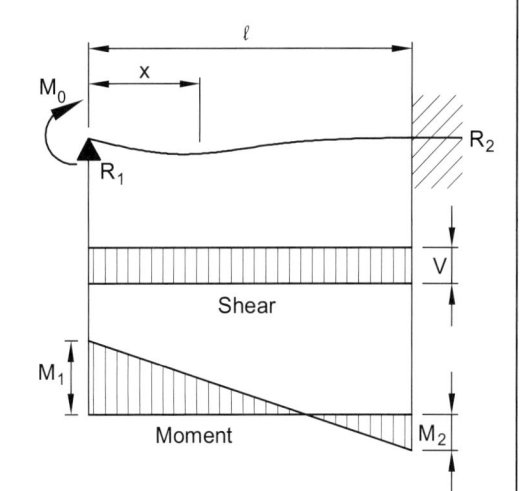

3F.1　BEAM FIXED AT ONE END – DIFFERENTIAL SETTLEMENT OF SUPPORTS

$V = R_1 = R_2 = \dfrac{3EI}{\ell^3}(\Delta_2 - \Delta_1)$

$M_{max} = \dfrac{3EI}{\ell^2}(\Delta_2 - \Delta_1)$

$M_x = M_{max}\left(1 - \dfrac{x}{\ell}\right)$

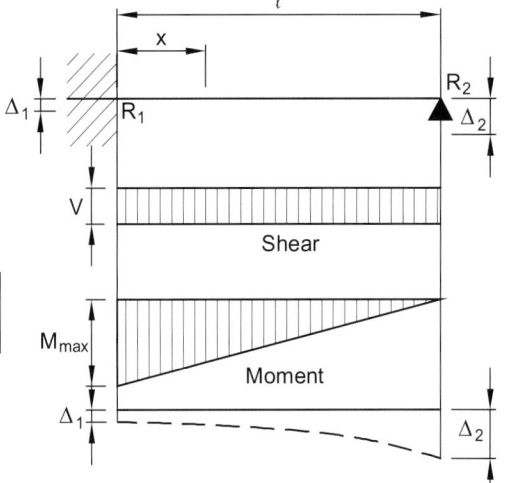

$\Delta_x = \Delta_1 + \dfrac{\Delta_2 - \Delta_1}{2}\left[3\left(\dfrac{x}{\ell}\right)^2 - \left(\dfrac{x}{\ell}\right)^3\right]$

3F.2　BEAM FIXED AT ONE END – ROTATION OF SUPPORT

$V = -R_1 = R_2 = \dfrac{3EI}{\ell^2}\phi_1$

$M_{max} = \dfrac{3EI}{\ell}\phi_1$

$M_x = M_{max}\left(1 - \dfrac{x}{\ell}\right)$

$\Delta_{max} = \phi_1\left[\dfrac{\ell}{5.196}\right]$

$\Delta_x = \phi_1\left[-x + \dfrac{3x^2}{2\ell} - \dfrac{x^3}{2\ell^2}\right]$

DESIGN INFORMATION

Design Aid 11.1.3 Beam Design Equations and Diagrams (Cont.) 4A

4A.1 BEAM FIXED AT BOTH ENDS – CONCENTRATED LOAD AT CENTER

total equivalent uniform load $= P$

$R = V = \dfrac{P}{2}$

M_{max} (at center and ends) $= \dfrac{P\ell}{8}$

$M_x \left(\text{when } x < \dfrac{\ell}{2} \right) = \dfrac{P}{8}(4x - \ell)$

Δ_{max} (at center) $= \dfrac{P\ell^3}{192EI}$

$\Delta_x \left(\text{when } x < \dfrac{\ell}{2} \right) = \dfrac{Px^2}{48EI}(3\ell - 4x)$

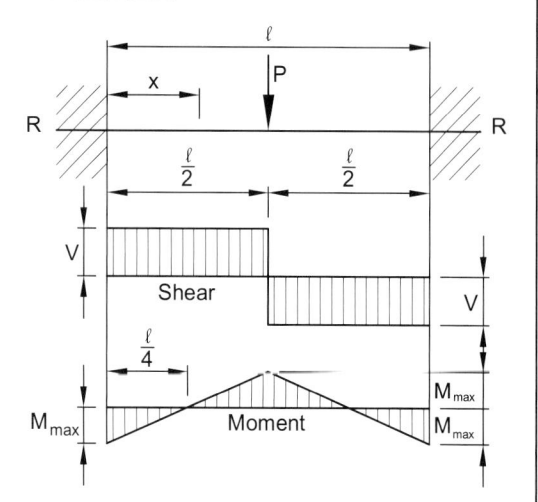

4A.2 BEAM FIXED AT BOTH ENDS – CONCENTRATED LOAD AT ANY POINT

$R_1 = V_1$ (max when $a < b$) $= \dfrac{Pb^2}{\ell^3}(3a + b)$

$R_2 = V_2$ (max when $a > b$) $= \dfrac{Pa^2}{\ell^3}(a + 3b)$

M_1 (max when $a < b$) $= \dfrac{Pab^2}{\ell^2}$

M_2 (max when $a > b$) $= \dfrac{Pa^2b}{\ell^2}$

M_a (at point of load) $= \dfrac{2Pa^2b^2}{\ell^3}$

M_x (when $x < a$) $= R_1 x - \dfrac{Pab^2}{\ell^2}$

Δ_{max} (when $a > b$, at $x = \dfrac{2a\ell}{3a+b}$) $= \dfrac{2Pa^3b^2}{3EI(3a+b)^2}$

Δ_a (at point of load) $= \dfrac{Pa^3b^3}{3EI\ell^3}$

Δ_x (when $x < a$) $= \dfrac{Pb^2x^2}{6EI\ell^3}(3a\ell - 3ax - bx)$

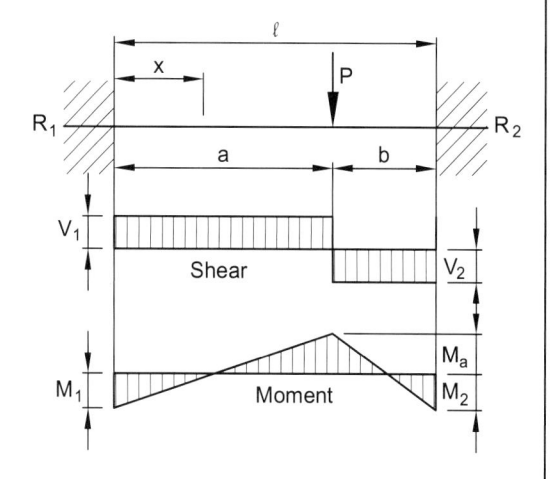

DESIGN INFORMATION

Design Aid 11.1.3 Beam Design Equations and Diagrams (Cont.) 4B, 4C

4B.1 BEAM FIXED AT BOTH ENDS – UNIFORMLY DISTRIBUTED LOADS

$R = V = \dfrac{w\ell}{2}$

$V_x = w\left(\dfrac{\ell}{2} - x\right)$

M_{max} (at ends) $= \dfrac{w\ell^2}{12}$

M_1 (at center) $= \dfrac{w\ell^2}{24}$

$M_x = \dfrac{w}{12}(6\ell x - \ell^2 - 6x^2)$

Δ_{max} (at center) $= \dfrac{w\ell^4}{384EI}$

$\Delta_x = \dfrac{wx^2}{24EI}(\ell - x)^2$

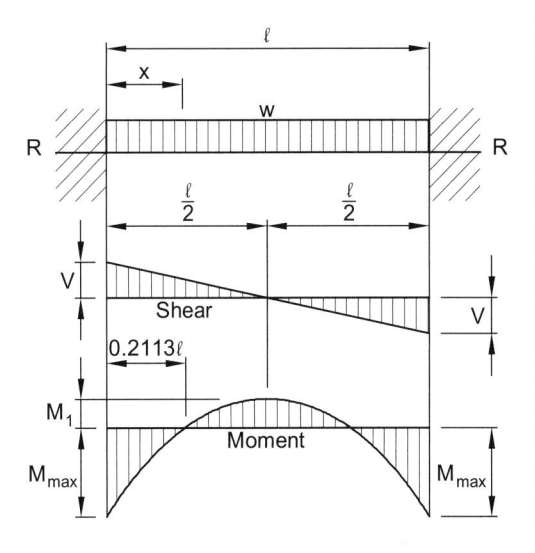

4C.1 BEAM FIXED AT BOTH ENDS – UNIFORM LOAD PARTIALLY DISTRIBUTED OVER SPAN

$R_1 = V_1 = \dfrac{wb}{4\ell^3}[4e^2(\ell + 2d) - b^2(c - a)]$

$R_2 = V_2 = wb - R_1$

$M_1 = \dfrac{wb}{24\ell^2}\{b^2[\ell + 3(c - a)] - 24e^2 d\}$

$M_2 = R_1\ell - wbe + M_1$

$M_{max(+)}\left(\text{at } x = a + \dfrac{R_1}{w}\right) = M_1 + R_1\left(a + \dfrac{R_1}{2w}\right)$

M_x (when $x < a$) $= M_1 + R_1 x$

M_x [when $x > a$ and $x < (a + b)$] $= M_1 + R_1 x - \dfrac{w}{2}(x - a)^2$

Δ_x (when $x < a$) $= \dfrac{1}{6EI}(3M_1 x^2 + R_1 x^3)$

Δ_x [when $x > a$ and $x < (a + b)$] $= \dfrac{1}{24EI}[12M_1 x^2 + 4R_1 x^3 - w(x - a)^4]$

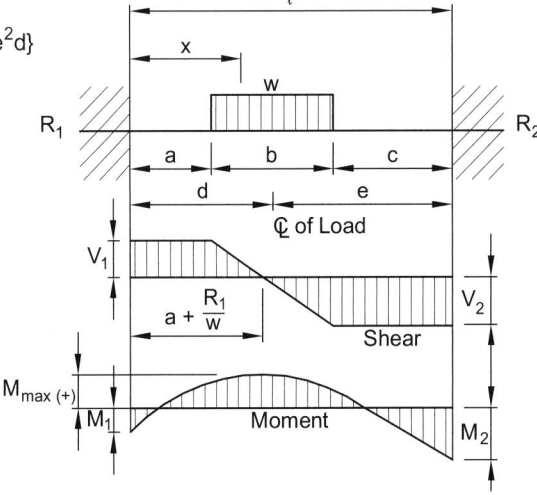

DESIGN INFORMATION

Design Aid 11.1.3 Beam Design Equations and Diagrams (Cont.) 4E,4F

4E.1 BEAM FIXED AT BOTH ENDS – MOMENT APPLIED AT ANY POINT

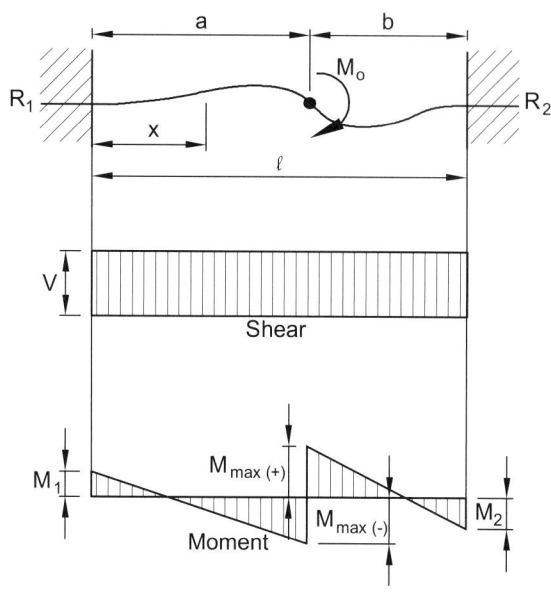

$R_1 = V = -\dfrac{6M_o ab}{\ell^3}$

$R_2 = \dfrac{6M_o ab}{\ell^3}$

$M_1 = -\dfrac{M_o b}{\ell^2}(\ell - 3a)$

$M_2 = -\dfrac{M_o a}{\ell^2}(2\ell - 3a)$

$M_x \text{ (when } x < a) = -\dfrac{M_o}{\ell^2}\left[\dfrac{6abx}{\ell} + b(\ell - 3a)\right]$

$M_x \text{ (when } x > a) = \dfrac{M_o a}{\ell^2}\left(6b - \dfrac{6bx}{\ell} - 2\ell + 3a\right)$

$M_{max(-)} \text{ (at } x = a \text{ on left side)} = M_{max(+)} - M_o$

$M_{max(+)} \text{ (at } x = a \text{ on right side)} = M_o\left[-\dfrac{6a^2 b}{\ell^3} - \dfrac{b}{\ell^2}(\ell - 3a) + 1\right]$

$\Delta_x \text{ (when } x < a) = -\dfrac{M_o b x^2}{2EI\ell^2}\left(\ell - 3a + \dfrac{2ax}{\ell}\right)$

$\Delta_x \text{ (when } x > a) = \dfrac{M_o a(\ell - x)^2}{2EI\ell^2}\left(3a - 2\ell + 2b - \dfrac{2bx}{\ell}\right)$

$M_\mathrm{C\!\!\!\!L} \text{ (at center)} = -\dfrac{M_o}{\ell^2}[3ab + b(\ell - 3a)]$

$\Delta_\mathrm{C\!\!\!\!L} \text{ (at center)} = -\dfrac{M_o b}{8EI}(\ell - 2a)$

$\Delta_{max} \text{ (when } a = 0.2324l\,\ell\text{)} = -\dfrac{0.01615 M_o \ell^2}{EI}$

4F.1 BEAM FIXED AT BOTH ENDS – DIFFERENTIAL SETTLEMENT OF SUPPORTS

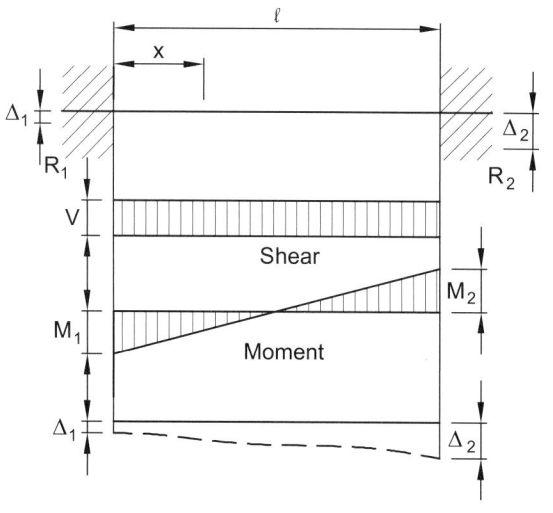

$V = -R_1 = R_2 = \dfrac{12EI}{\ell^3}(\Delta_2 - \Delta_1)$

$M_1 = -M_2 = \dfrac{6EI}{\ell^2}(\Delta_2 - \Delta_1)$

$M_x = \dfrac{6EI}{\ell^2}(\Delta_2 - \Delta_1)\left(1 - \dfrac{2x}{\ell}\right)$

$\Delta_x = \Delta_1 + (\Delta_2 - \Delta_1)\left[3\left(\dfrac{x}{\ell}\right)^2 - 2\left(\dfrac{x}{\ell}\right)^3\right]$

DESIGN INFORMATION

Design Aid 11.1.3   Beam Design Equations and Diagrams (Cont.) 4F,5A

4F.2   BEAM FIXED AT BOTH ENDS – ROTATION OF SUPPORT

$V = -R_1 = R_2 = \dfrac{6EI}{\ell^2}\phi_2$

$M_1 = \dfrac{2EI}{\ell}\phi_2$

$M_2 = \dfrac{4EI}{\ell}\phi_2$

$M_x = \dfrac{2EI}{\ell}\phi_2\left(1 - \dfrac{3x}{\ell}\right)$

Δ_{max} (at $x = \dfrac{2}{3}\ell$) $= -\dfrac{4}{27}\ell\phi_2$

$\Delta_x = -\ell\phi_2\left[\left(\dfrac{x}{\ell}\right)^2 - \left(\dfrac{x}{\ell}\right)^3\right]$

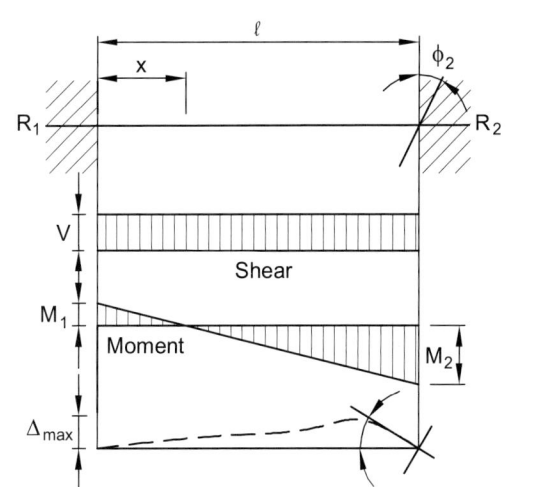

5A.1   CANTILEVER BEAM – CONCENTRATED LOAD AT FREE END

$R = V = P$

M_{max} (at fixed end) $= P\ell$

$M_x = Px$

Δ_{max} (at free end) $= \dfrac{P\ell^3}{3EI}$

$\Delta_x = \dfrac{P}{6EI}(2\ell^3 - 3\ell^2 x + x^3)$

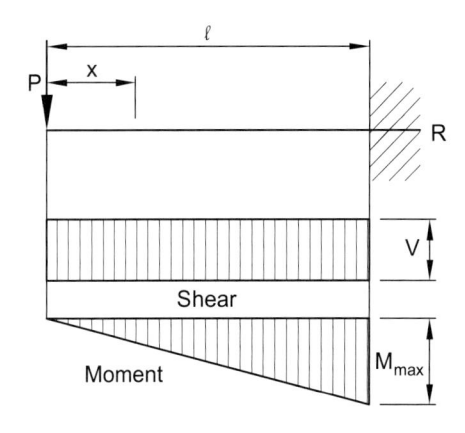

5A.2   CANTILEVER BEAM – CONCENTRATED LOAD AT ANY POINT

$R = V = P$

M_{max} (at fixed end) $= Pb$

M_x (when $x > a$) $= P(x - a)$

Δ_{max} (at free end) $= \dfrac{Pb^2}{6EI}(3\ell - b)$

Δ_a (at point of load) $= \dfrac{Pb^3}{3EI}$

Δ_x (when $x < a$) $= \dfrac{Pb^2}{6EI}(3\ell - 3x - b)$

Δ_x (when $x > a$) $= \dfrac{P(\ell - x)^2}{6EI}(3b - \ell + x)$

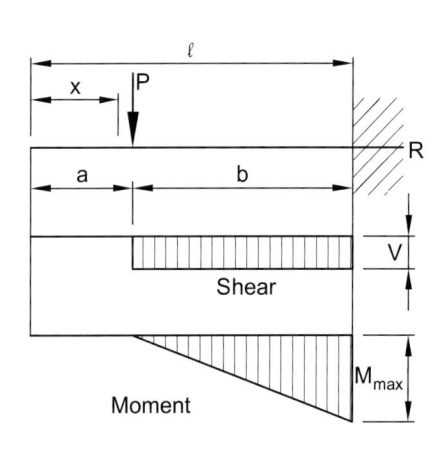

DESIGN INFORMATION

Design Aid 11.1.3 Beam Design Equations and Diagrams (Cont.) 5B,5C,5D

5B.1 CANTILEVER BEAM – UNIFORMLY DISTRIBUTED LOAD

$R = V = w\ell$

$V_x = wx$

M_{max} (at fixed end) $= \dfrac{w\ell^2}{2}$

$M_x = \dfrac{wx^2}{2}$

Δ_{max} (at free end) $= \dfrac{w\ell^4}{8EI}$

$\Delta_x = \dfrac{w}{24EI}(x^4 - 4\ell^3 x + 3\ell^4)$

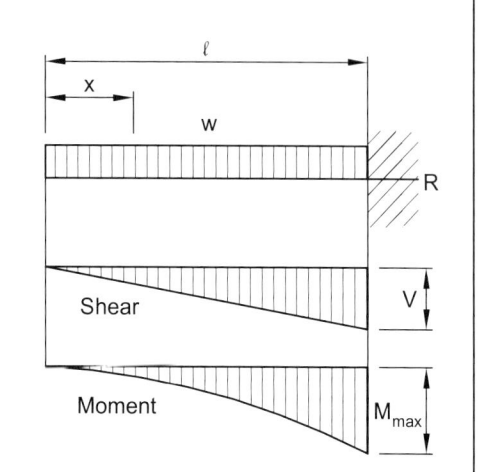

5C.1 CANTILEVER BEAM – UNIFORM LOAD PARTIALLY DISTRIBUTED AT FREE END

$R = V = wb$

M_{max} (at support) $= wbe$

M_x (when $x < b$) $= \dfrac{wx^2}{2}$

M_x (when $x > b$) $= \dfrac{wb}{2}(b - 2x)$

Δ_{max} (at free end) $= \dfrac{wb}{48EI}(8e^3 - 24e^2\ell - b^3)$

Δ_x (when $x < b$) $= \dfrac{w}{48EI}[8be^3 - 24be^2(\ell - x) + 2b^3 x - b^4 - 2x^4]$

Δ_x (when $x > b$) $= \dfrac{wb}{48EI}[8e^3 - 24e^2(\ell - x) - (2x - b)^3]$

θ (at free end) $= \dfrac{wb}{24EI}(b^2 + 12e^2)$

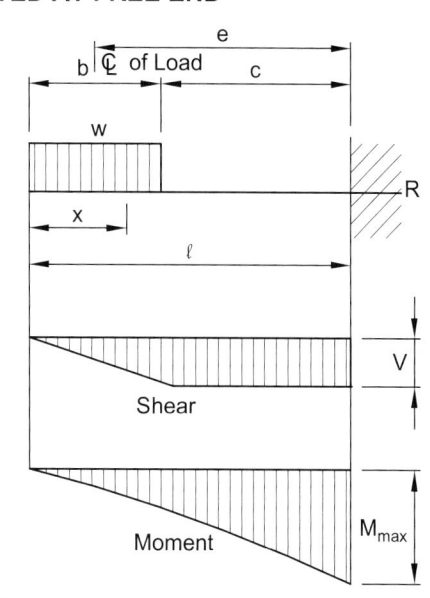

5D.1 CANTILEVER BEAM – LOAD INCREASING UNIFORMLY TO FIXED END (W IS TOTAL LOAD)

$W = \dfrac{w\ell}{2}$

$R = V = W$

$V_x = W \dfrac{x^2}{\ell^2}$

M_{max} (at fixed end) $= \dfrac{W\ell}{3}$

$M_x = \dfrac{Wx^3}{3\ell^2}$

Δ_{max} (at free end) $= \dfrac{W\ell^3}{15EI}$

$\Delta_x = \dfrac{W}{60EI\ell^2}(x^5 - 5\ell^4 x + 4\ell^5)$

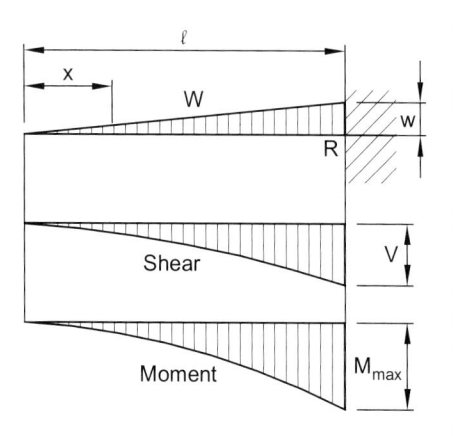

DESIGN INFORMATION

Design Aid 11.1.3 Beam Design Equations and Diagrams (Cont.) 5D, 5E

5D.2 CANTILEVER BEAM – VARYING LOAD INCREASING UNIFORMLY FROM SUPPORT TO FREE END (W IS TOTAL LOAD)

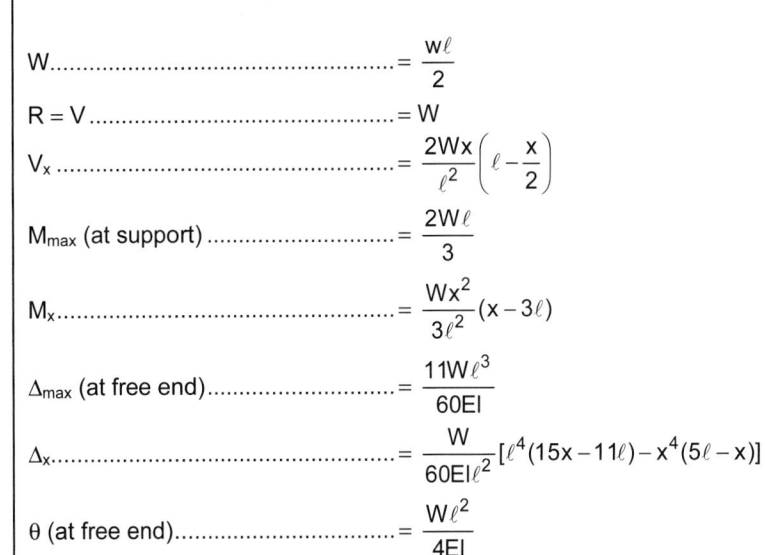

$W = \dfrac{w\ell}{2}$

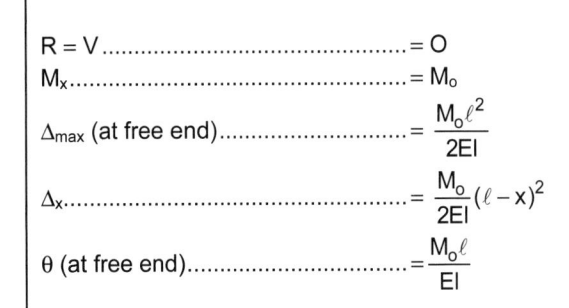

$R = V = W$

$V_x = \dfrac{2Wx}{\ell^2}\left(\ell - \dfrac{x}{2}\right)$

M_{max} (at support) $= \dfrac{2W\ell}{3}$

$M_x = \dfrac{Wx^2}{3\ell^2}(x - 3\ell)$

Δ_{max} (at free end) $= \dfrac{11W\ell^3}{60EI}$

$\Delta_x = \dfrac{W}{60EI\ell^2}[\ell^4(15x - 11\ell) - x^4(5\ell - x)]$

θ (at free end) $= \dfrac{W\ell^2}{4EI}$

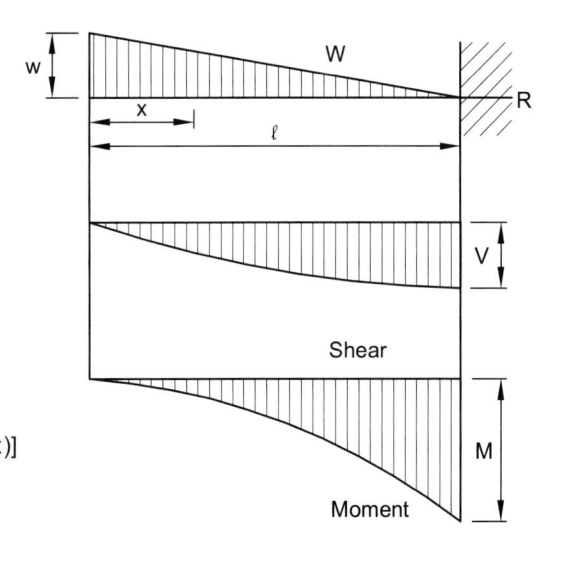

5E.1 CANTILEVER BEAM – MOMENT APPLIED AT FREE END

$R = V = 0$

$M_x = M_o$

Δ_{max} (at free end) $= \dfrac{M_o \ell^2}{2EI}$

$\Delta_x = \dfrac{M_o}{2EI}(\ell - x)^2$

θ (at free end) $= \dfrac{M_o \ell}{EI}$

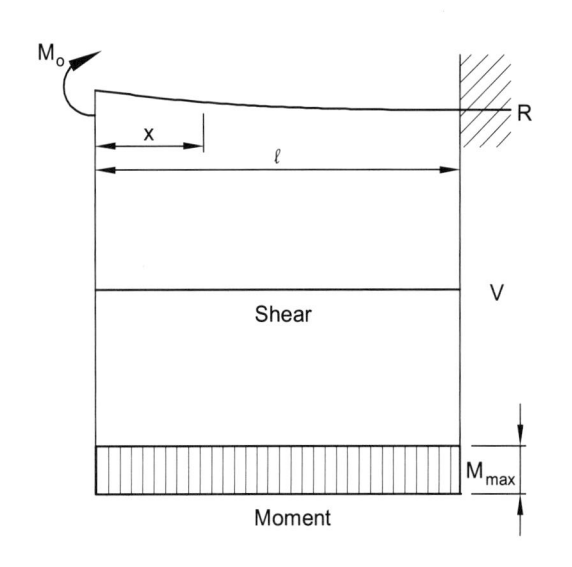

DESIGN INFORMATION

Design Aid 11.1.3   Beam Design Equations and Diagrams (Cont.)   6A,6B

6A.1   TWO SPANS, CONTINUOUS BEAM – CONCENTRATED LOAD AT CENTER OF ONE SPAN ONLY

$R_1 = V_1 = \dfrac{13}{32}P$

$R_2 = V_2 + V_3 = \dfrac{11}{16}P$

$R_3 = V_3 = -\dfrac{3}{32}P$

$V_2 = \dfrac{19}{32}P$

M_{max} (at point of load) $= \dfrac{13}{64}P\ell$

M_2 (at R_2) $= \dfrac{3}{32}P\ell$

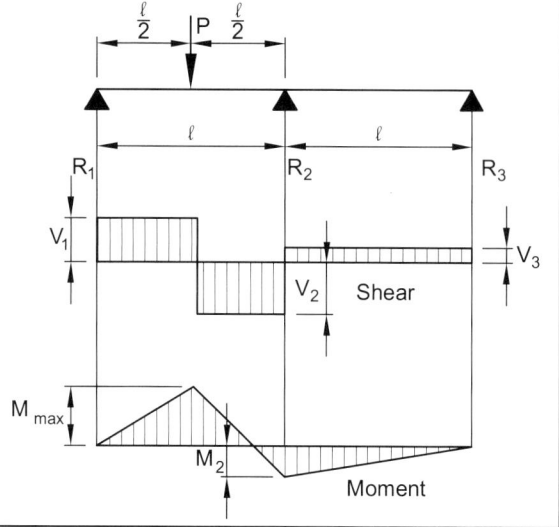

6A.2   TWO SPANS, CONTINUOUS BEAM – CONCENTRATED LOAD AT ANY POINT OF ONE SPAN ONLY

$R_1 = V_1 = \dfrac{Pb}{4\ell^3}[4\ell^2 - a(\ell+a)]$

$R_2 = V_2 + V_3 = \dfrac{Pa}{2\ell^3}[2\ell^2 + b(\ell+a)]$

$R_3 = V_3 = -\dfrac{Pab}{4\ell^3}(\ell+a)$

$V_2 = -\left(\dfrac{Pa}{4\ell^3}\right)[4\ell^2 - b(\ell+a)]$

M_{max} (at point of load) $= \dfrac{Pab}{4\ell^3}[4\ell^2 - a(\ell+a)]$

M_2 (at R_2) $= \dfrac{Pab}{4\ell^2}(\ell+a)$

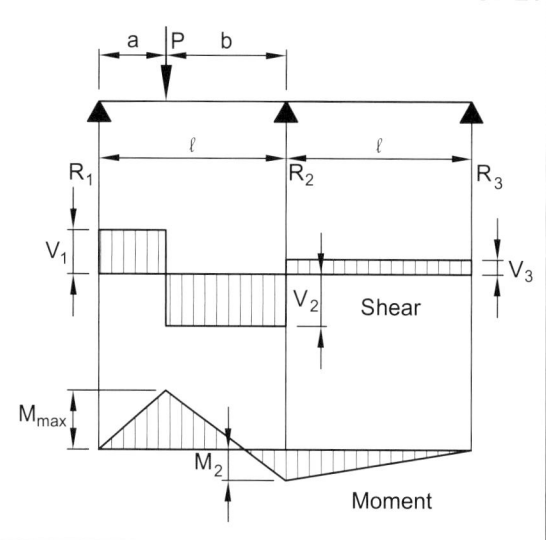

6B.1   TWO SPAN, CONTINUOUS BEAM – UNIFORM LOAD OVER ONE SPAN ONLY

$R_1 = V_1 = \dfrac{7}{16}w\ell$

$R_2 = V_2 + V_3 = \dfrac{5}{8}w\ell$

$R_3 = V_3 = -\dfrac{1}{16}w\ell$

$V_2 = \dfrac{9}{16}w\ell$

$M_{max}\left(\text{at } x = \dfrac{7}{16}\ell\right) = \dfrac{49}{512}w\ell^2$

M_1 (at R_2) $= \dfrac{w\ell^2}{16}$

M_x (when $x < \ell$) $= \dfrac{wx}{16}(7\ell - 8x)$

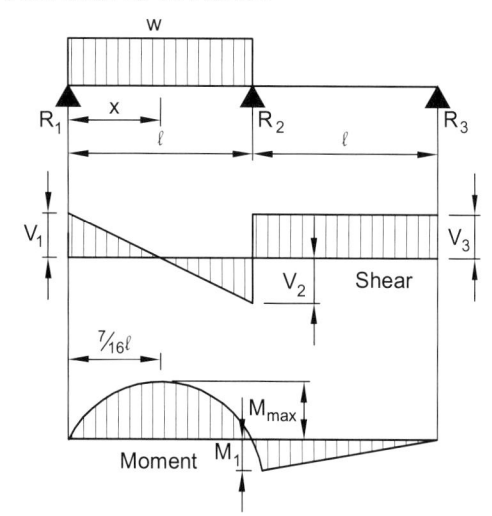

DESIGN INFORMATION

Design Aid 11.1.4 Camber (Deflection) and Rotation Coefficients for Prestress Force and Loads

PRESTRESS PATTERN	EQUIVALENT MOMENT OR LOAD	EQUIVALENT LOADING	CAMBER	END ROTATION (left)	END ROTATION (right)
(1)	$M = Pe$		$+\dfrac{M\ell^2}{16EI}$	$+\dfrac{M\ell}{3EI}$	$-\dfrac{M\ell}{6EI}$
(2)	$M = Pe$		$+\dfrac{M\ell^2}{16EI}$	$+\dfrac{M\ell}{6EI}$	$-\dfrac{M\ell}{3EI}$
(3)	$M = Pe$		$+\dfrac{M\ell^2}{8EI}$	$\dfrac{M\ell}{2EI}$	$-\dfrac{M\ell}{2EI}$
(4)	$N = \dfrac{4Pe'}{\ell}$		$+\dfrac{N\ell^3}{48EI}$	$\dfrac{N\ell^2}{16EI}$	$-\dfrac{N\ell^2}{16EI}$
(5)	$N = \dfrac{Pe'}{b\ell}$		$\dfrac{b(3-4b^2)N\ell^3}{24EI}$	$\dfrac{b(1-b)N\ell^2}{2EI}$	$-\dfrac{b(1-b)N\ell^2}{2EI}$
(6)	$w = \dfrac{8Pe'}{\ell^2}$		$\dfrac{5w\ell^4}{384EI}$	$\dfrac{w\ell^3}{24EI}$	$-\dfrac{w\ell^3}{24EI}$

DESIGN INFORMATION

Design Aid 11.1.4 Camber (Deflection) and Rotation Coefficients for Prestress Force and Loads (Cont.)

PRESTRESS PATTERN	EQUIVALENT MOMENT OR LOAD	EQUIVALENT LOADING	CAMBER	END ROTATION	
(7)	$w = \dfrac{8Pe'}{\ell^2}$		$\dfrac{5w\ell^4}{768EI}$	$\dfrac{9w\ell^3}{384EI}$	$-\dfrac{7w\ell^3}{384EI}$
(8)	$w = \dfrac{8Pe'}{\ell^2}$		$\dfrac{5w\ell^4}{768EI}$	$\dfrac{7w\ell^3}{384EI}$	$-\dfrac{9w\ell^3}{384EI}$
(9)	$w = \dfrac{4Pe'}{(0.5-b)\ell^2}$ $w_1 = \dfrac{w}{b}(0.5-b)$		$\left[\dfrac{5}{8} - \dfrac{b}{2}(3-2b^2)\right]$ $\dfrac{w\ell^4}{48EI}$	$\dfrac{(1-b)(1-2b)w\ell^3}{24EI}$	$-\dfrac{(1-b)(1-2b)w\ell^3}{24EI}$
(10)	$w = \dfrac{4Pe'}{(0.5-b)\ell^2}$ $w_1 = \dfrac{w}{b}(0.5-b)$		$\left[\dfrac{5}{16} - \dfrac{b}{4}(3-2b^2)\right]$ $\dfrac{w\ell^4}{48EI}$	$\left[\dfrac{9}{8} - b(2-b)^2\right]$ $\dfrac{w\ell^3}{48EI}$	$\left[-\dfrac{7}{8} + b(2-b)^2\right]$ $\dfrac{w\ell^3}{48EI}$
(11)	$w = \dfrac{4Pe'}{(0.5-b)\ell^2}$ $w_1 = \dfrac{w}{b}(0.5-b)$		$\left[\dfrac{5}{16} - \dfrac{b}{4}(3-2b^2)\right]$ $\dfrac{w\ell^4}{48EI}$	$\left[\dfrac{7}{8} - b(2-b)^2\right]$ $\dfrac{w\ell^3}{48EI}$	$\left[-\dfrac{9}{8} + b(2-b)^2\right]$ $\dfrac{w\ell^3}{48EI}$

Determination of camber along length of member based on camber at midspan:

Camber at midspan = y_c $y_x = y_c - y_c \dfrac{\left(\dfrac{\ell}{2} - x\right)^2}{\left(\dfrac{\ell}{2}\right)^2}$

DESIGN INFORMATION

Design Aid 11.1.5 Moments in Beams with Fixed Ends

LOADING	MOMENT AT A	MOMENT AT CENTER	MOMENT AT B
(1) P at center, $\ell/2$ each side	$\dfrac{-P\ell}{8}$	$\dfrac{P\ell}{8}$	$\dfrac{-P\ell}{8}$
(2) P at $a\ell$ from A	$-P\ell a(1-a)^2$		$-P\ell a^2(1-a)$
(3) Two P loads at third points	$\dfrac{-2P\ell}{9}$	$\dfrac{P\ell}{9}$	$\dfrac{-2P\ell}{9}$
(4) Three P loads at quarter points	$\dfrac{-5P\ell}{16}$	$\dfrac{3P\ell}{16}$	$\dfrac{-5P\ell}{16}$
(5) Uniform load W over full span	$\dfrac{-W\ell}{12}$	$\dfrac{W\ell}{24}$	$\dfrac{-W\ell}{12}$
(6) Uniform load W centered, $a\ell$ from each end	$\dfrac{-W\ell(1+2a-2a^2)}{12}$	$\dfrac{W\ell(1+2a-2a^2)}{24}$	$\dfrac{-W\ell(1+2a-2a^2)}{12}$
(7) Two uniform loads W/2 at each end, length $a\ell$	$\dfrac{-W\ell(3a-2a^2)}{12}$	$\dfrac{W\ell a^2}{6}$	$\dfrac{-W\ell(3a-2a^2)}{12}$
(8) Uniform load W from A over length $a\ell$	$\dfrac{-W\ell a(6-8a+3a^2)}{12}$		$\dfrac{-W\ell a^2(4-3a)}{12}$
(9) Triangular load W peak at center	$\dfrac{-5W\ell}{48}$	$\dfrac{3W\ell}{48}$	$\dfrac{-5W\ell}{48}$
(10) Triangular load W peak at A	$\dfrac{-W\ell}{10}$		$\dfrac{-W\ell}{15}$

W = Total load on the beam.

DESIGN INFORMATION

Design Aid 11.1.6 Torsion Diagrams, Reactions, and Rotations

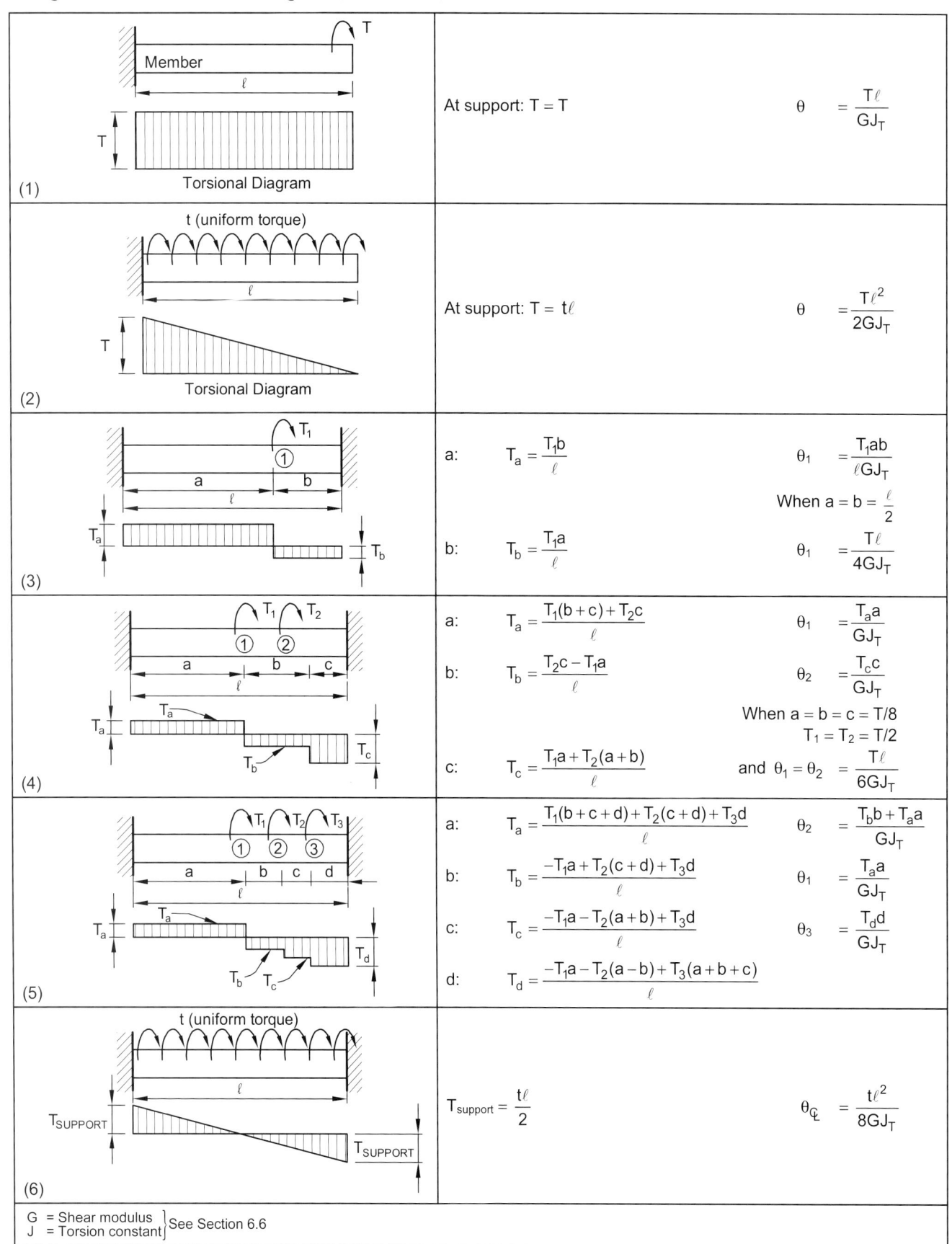

Diagram	Reactions	Rotations
(1) Member with torque T at free end, length ℓ; Torsional Diagram constant = T	At support: $T = T$	$\theta = \dfrac{T\ell}{GJ_T}$
(2) Uniform torque t over length ℓ; Torsional Diagram triangular, peak T at support	At support: $T = t\ell$	$\theta = \dfrac{T\ell^2}{2GJ_T}$
(3) Fixed-fixed with T_1 at distance a from left, b from right	a: $T_a = \dfrac{T_1 b}{\ell}$ b: $T_b = \dfrac{T_1 a}{\ell}$	$\theta_1 = \dfrac{T_1 ab}{\ell GJ_T}$ When $a = b = \dfrac{\ell}{2}$ $\theta_1 = \dfrac{T\ell}{4GJ_T}$
(4) Fixed-fixed with T_1, T_2 at points 1, 2; segments a, b, c	a: $T_a = \dfrac{T_1(b+c) + T_2 c}{\ell}$ b: $T_b = \dfrac{T_2 c - T_1 a}{\ell}$ c: $T_c = \dfrac{T_1 a + T_2(a+b)}{\ell}$	$\theta_1 = \dfrac{T_a a}{GJ_T}$ $\theta_2 = \dfrac{T_c c}{GJ_T}$ When $a = b = c = T/8$ $T_1 = T_2 = T/2$ and $\theta_1 = \theta_2 = \dfrac{T\ell}{6GJ_T}$
(5) Fixed-fixed with T_1, T_2, T_3 at points 1, 2, 3; segments a, b, c, d	a: $T_a = \dfrac{T_1(b+c+d) + T_2(c+d) + T_3 d}{\ell}$ b: $T_b = \dfrac{-T_1 a + T_2(c+d) + T_3 d}{\ell}$ c: $T_c = \dfrac{-T_1 a - T_2(a+b) + T_3 d}{\ell}$ d: $T_d = \dfrac{-T_1 a - T_2(a-b) + T_3(a+b+c)}{\ell}$	$\theta_2 = \dfrac{T_b b + T_a a}{GJ_T}$ $\theta_1 = \dfrac{T_a a}{GJ_T}$ $\theta_3 = \dfrac{T_d d}{GJ_T}$
(6) Fixed-fixed with uniform torque t over length ℓ	$T_{support} = \dfrac{t\ell}{2}$	$\theta_{\mathcal{C}} = \dfrac{t\ell^2}{8GJ_T}$

G = Shear modulus
J = Torsion constant } See Section 6.6

DESIGN INFORMATION

Design Aid 11.1.7 Moving Load Placement for Maximum Moment and Shear[a]

---

(1) SIMPLE BEAM – ONE CONCENTRATED MOVING LOAD

$R_{1\,max} = V_{1\,max}$ (AT $x = 0$) .... $= P$

M_{max} (at point of load, when $x = \dfrac{\ell}{2}$) .... $= \dfrac{P\ell}{4}$

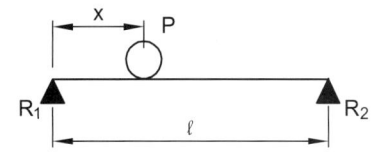

---

(2) SIMPLE BEAM – TWO EQUAL CONCENTRATED MOVING LOADS

$R_{1\,max} = V_{1\,max}$ (AT $x = 0$) .... $= P\left(2 - \dfrac{a}{\ell}\right)$

when $a < (2 - \sqrt{2})\ell$ .... $= 0.586\ell$

M_{max} = under load 1, at $x = \dfrac{1}{2}\left(\ell - \dfrac{a}{2}\right)$ .... $= \dfrac{P}{2\ell}\left(\ell - \dfrac{a}{2}\right)^2$

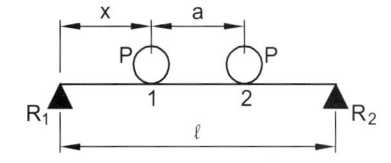

when $a > (2 - \sqrt{2})\ell$ .... $= 0.586\ell$

M_{max} = with one load at center of span .... $= \dfrac{P\ell}{4}$

---

(3) SIMPLE BEAM – TWO UNEQUAL CONCENTRATED MOVING LOADS

$R_{1\,max} = V_{1\,max}$ (at $x = 0$) .... $= P_1 + P_2 \dfrac{\ell - a}{\ell}$

M_{max} under P_1, at $x = \dfrac{1}{2}\left(\ell - \dfrac{P_2 a}{P_1 + P_2}\right)$ .... $= (P_1 + P_2)\dfrac{x^2}{\ell}$

M_{max} may occur with larger load at center of span and other load off span .... $= \dfrac{P_1 \ell}{4}$

---

a. Source "Manual of Steel Construction, Allowable Stress Design," Ninth Edition, 1989 American Institute of Steel Construction, Chicago, IL.

DESIGN INFORMATION

Design Aid 11.1.8 Moments, Shears, and Deflections in Beams with Overhangs

LOADING AND SUPPORT	REACTIONS AND VERTICAL SHEAR	BENDING MOMENT M AND MAXIMUM BENDING MOMENT	DEFLECTION y, MAXIMUM DEFLECTION, AND END SLOPE θ
EQUAL OVERHANGS, UNIFORM LOAD $W = w\ell$	$R_B = R_C = \dfrac{W}{2}$ (A to B) $V = -\dfrac{W(c-x)}{\ell}$ (B to C) $V = W\left(\dfrac{1}{2} - \dfrac{x+c}{\ell}\right)$ (C to D) $V = \dfrac{W(c+d-x)}{\ell}$	(A to B) $M = -\dfrac{W}{2\ell}(c-x)^2$ (B to C) $M = -\dfrac{W}{2\ell}[c^2 - x(d-x)]$ $M = -\dfrac{Wc^2}{2\ell}$ at B and C $M = -\dfrac{W}{2\ell}\left(c^2 - \dfrac{d^2}{4}\right)$ at $x = \dfrac{d}{2}$ if $d > 2c$, $M = 0$ at $x = \dfrac{d}{2} \pm \sqrt{\dfrac{d^2}{4} - c^2}$ if $c = 0.207\,\ell$, $M = -\dfrac{W\ell}{46.62}$ at $x = 0 = d$ and $M = \dfrac{W\ell}{46.62}$ ar $x = \dfrac{d}{2}$ x is considered positive on both sides of the origin.	(A to B) $y = -\dfrac{Wx}{24EI\ell}[6c^2(d+x) - x^2(4c - x) - d^3]$ (B to C) $y = -\dfrac{Wx(d-x)}{24EI\ell}[x(d-x) + d^2 - 6c^2]$ $y = -\dfrac{Wc}{24EI\ell}[3c^2(c+2d) - d^3]$ at A and D $y = -\dfrac{Wd^2}{384EI\ell}(5d^2 - 24c^2)$ AT $x = \dfrac{d}{2}$ if $2c < d < 2.449c$, the maximum deflection between supports is: $y = \dfrac{W}{96EI\ell}(6c^2 - d^2)^2$ AT $x = \dfrac{d}{2} \pm \sqrt{3\left(\dfrac{d^2}{4} - c^2\right)}$ $\theta = \dfrac{W}{24EI\ell}(6c^2d + 4c^3 - d^3)$ AT A $\theta = -\dfrac{W}{24EI\ell}(6c^2d + 4c^3 - d^3)$ AT D
UNEQUAL OVERHANGS, UNIFORM LOAD $W = w\ell$	$R_B = \dfrac{W}{2d}(c+d-e)$ $R_C = \dfrac{W}{2d}(d+e-c)$ (A to B) $V = -\dfrac{W}{\ell}(c-x)$ (B to C) $V = R_B - \dfrac{W}{\ell}(c+x)$ (C to D) $V = -\dfrac{W}{\ell}(d+e-x)$	(A to B) $M = -\dfrac{W}{2\ell}(c-x)^2$ (B to C) $M = -\dfrac{W}{2\ell}(c-x)^2 + R_B x$ (C to D) $M = -\dfrac{W}{2\ell}(e+d-x)^2$ $M = -\dfrac{Wc^2}{2\ell}$ at B $M = -\dfrac{We^2}{2\ell}$ at C M_{max} between supports $= \dfrac{W}{2\ell}(c^2 - x_1^2)$ at $x = x_1$ $= \dfrac{c^2 + d^2 - e^2}{2d}$ if $x_1 > c$, $M = 0$ AT $x = x_1 \pm \sqrt{x_1^2 - c^2}$ x is considered positive on both sides of the origin.	(A to B) $y = -\dfrac{Wx}{24EI\ell}$ $[2d(e^2 + 2c^2) + 6c^2x - x^2(4c - x) - d^3]$ (B to C) $y = -\dfrac{Wx(d-x)}{24EI\ell}$ $\left\{x(d-x) + d^2 - 2(c^2 + e^2) - \dfrac{2}{d}[e^2x + c^2(d-x)]\right\}$ (C to D) $y = -\dfrac{W(x-d)}{24EI\ell}[2d(c^2 + 2e^2) + 6e^2(x-d)$ $-(x-d)^2(4e + d - x) - d^3]$ $y = -\dfrac{Wc}{24EI\ell}[2d(e^2 + 2c^2) + 3c^3 - d^3]$ AT A $y = -\dfrac{We}{24EI\ell}[2d(c^2 + 2e^2) + 3e^3 - d^3]$ AT D This case is too complicated to obtain a general expression for critical deflections between the supports. $\theta = \dfrac{W}{24EI\ell}(4c^3 + 4c^2d - d^3 + 2de^2)$ AT A $\theta = -\dfrac{W}{24EI\ell}(2c^2d + 4de^2 - d^3 + 4e^3)$ AT D

11.2 MATERIAL PROPERTIES
CONCRETE

Design Aid 11.2.1 Table of Concrete Stresses (psi)

f'_c	$0.45f'_c$	$0.6f'_c$	$\sqrt{f'_c}$	$2\sqrt{f'_c}$	$3.5\sqrt{f'_c}$	$4\sqrt{f'_c}$	$5\sqrt{f'_c}$	$6\sqrt{f'_c}$	$7.5\sqrt{f'_c}$	$12\sqrt{f'_c}$
2500	1125	1500	50	100	175	200	250	300	375	600
3000	1350	1800	55	110	192	219	274	329	411	657
3500	1575	2100	59	118	207	237	296	355	444	710
4000	1800	2400	63	126	221	253	316	379	474	759
5000	2250	3000	71	141	247	283	354	424	530	849
6000	2700	3600	77	155	271	310	387	465	581	930
7000	3150	4200	84	167	293	335	418	502	627	1004
8000	3600	4800	89	179	313	358	447	537	671	1073
9000	4050	5400	95	190	332	379	474	569	712	1138
10000	4500	6000	100	200	350	400	500	600	750	1200

Design Aid 11.2.2 Concrete Modulus of Elasticity as Affected by Unit Weight (pcf) and Strength (psi)

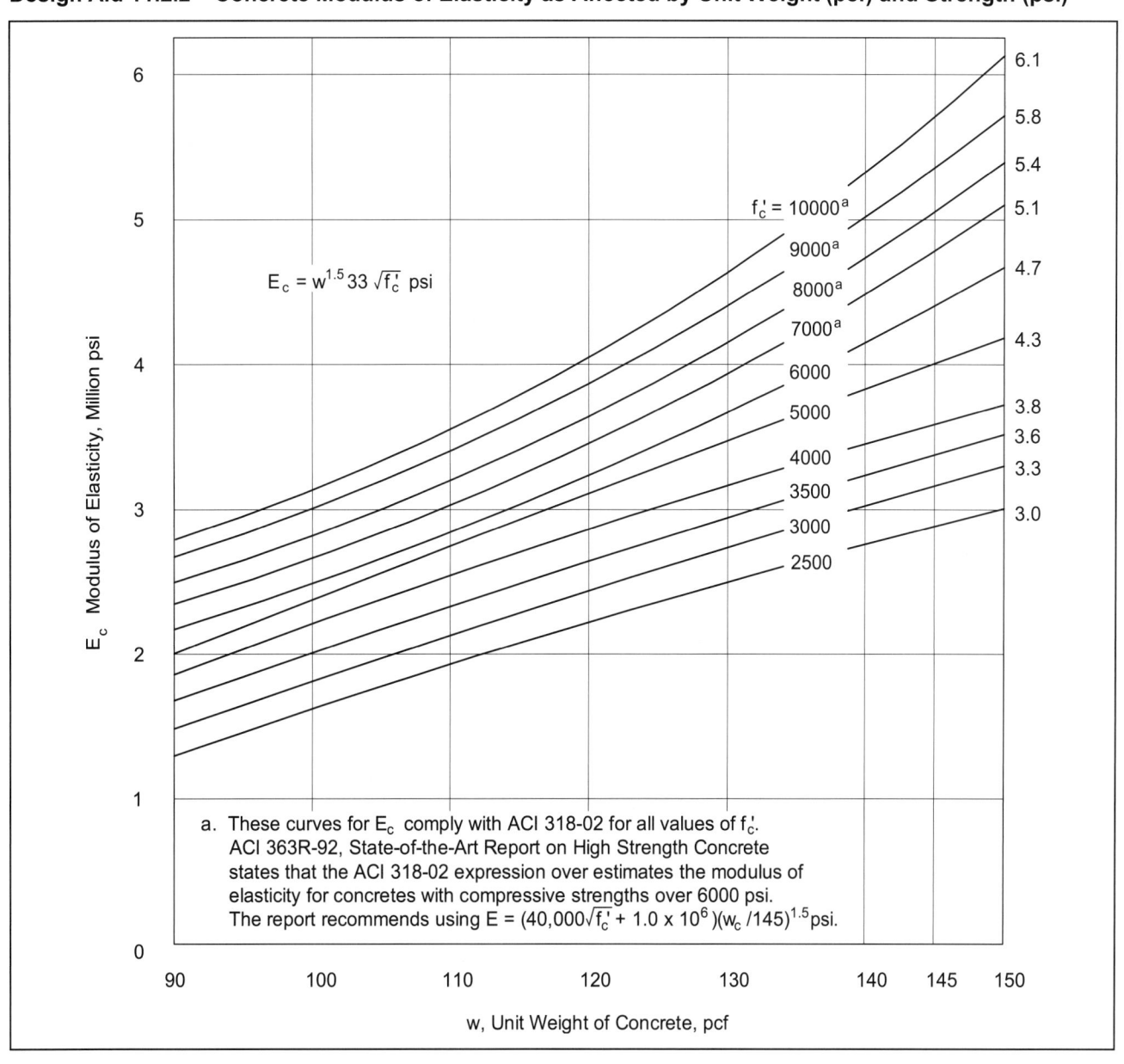

$E_c = w^{1.5} 33 \sqrt{f'_c}$ psi

a. These curves for E_c comply with ACI 318-02 for all values of f'_c. ACI 363R-92, State-of-the-Art Report on High Strength Concrete states that the ACI 318-02 expression over estimates the modulus of elasticity for concretes with compressive strengths over 6000 psi. The report recommends using $E = (40{,}000\sqrt{f'_c} + 1.0 \times 10^6)(w_c/145)^{1.5}$ psi.

MATERIAL PROPERTIES
PRESTRESSING STEEL

Design Aid 11.2.3 Properties and Design Strengths of Prestressing Strand and Wire

Seven-Wire Strand, f_{pu} = 270 ksi						
Nominal Diameter, in.	3/8	7/16	1/2	1/2 Special[a]	9/16	3/5
Area, sq in.	0.085	0.115	0.153	0.167	0.192	0.217
Weight, plf	0.29	0.40	0.52	0.53	0.65	0.74
$0.7f_{pu}A_{ps}$, kips	16.1	21.7	28.9	31.6	36.3	41.0
$0.75f_{pu}A_{ps}$, kips	17.2	23.3	31.0	33.8	38.9	43.0
$0.8f_{pu}A_{ps}$, kips	18.4	24.8	33.0	36.1	41.5	46.9
$f_{pu}A_{ps}$, kips	23.0	31.1	41.3	45.1	51.8	58.6

Seven-Wire Strand, f_{pu} = 250 ksi						
Nominal Diameter, in.	1/4	5/16	3/8	7/16	1/2	3/5
Area, sq in.	0.036	0.058	0.080	0.108	0.144	0.216
Weight, plf	0.12	0.20	0.27	0.37	0.49	0.74
$0.7f_{pu}A_{ps}$, kips	6.3	10.2	14.0	18.9	25.2	37.8
$0.8f_{pu}A_{ps}$, kips	7.2	11.6	16.0	21.6	28.8	43.2
$f_{pu}A_{ps}$, kips	9.0	14.5	20.0	27.0	36.0	54.0

Three- and Four-Wire Strand, f_{pu} = 250 ksi				
Nominal Diameter, in.	1/4	5/16	3/8	7/16
No. of wires	3	3	3	4
Area, sq in.	0.036	0.058	0.075	0.106
Weight, plf	0.12	0.20	0.26	0.36
$0.7f_{pu}A_{ps}$, kips	6.3	10.2	13.1	18.6
$0.8f_{pu}A_{ps}$, kips	7.2	11.6	15.0	21.2
$f_{pu}A_{ps}$, kips	9.0	14.5	18.8	26.5

Prestressing Wire									
Diameter	0.105	0.120	0.135	0.148	0.162	0.177	0.192	0.196	0.250	0.276
Area, sq in.	0.0087	0.0114	0.0143	0.0173	0.0206	0.0246	0.0289	0.0302	0.0491	0.0598
Weight, plf	0.030	0.039	0.049	0.059	0.070	0.083	0.098	0.10	0.17	0.20
Ult. strength, f_{pu} ksi	279	273	268	263	259	255	250	250	240	235
$0.7f_{pu}A_{ps}$, kips	1.70	2.18	2.68	3.18	3.73	4.39	5.06	5.29	8.25	9.84
$0.8f_{pu}A_{ps}$, kips	1.94	2.49	3.07	3.64	4.27	5.02	5.78	6.04	9.43	11.24
$f_{pu}A_{ps}$, kips	2.43	3.11	3.83	4.55	5.34	6.27	7.23	7.55	11.78	14.05

a. The 1/2 in. special strand has a larger actual diameter than the 1/2 in. regular strand. The table values take this difference into account.

MATERIAL PROPERTIES
PRESTRESSING STEEL

Design Aid 11.2.4   Properties and Design Strengths of Prestressing Bars

Plain Prestressing Bars, f_{pu} = 145 ksi[a]						
Nominal Diameter, in.	3/8	7/8	1	1 1/8	1 1/4	1 3/8
Area, sq in.	0.442	0.601	0.785	0.994	1.227	1.485
Weight, plf	1.50	2.04	2.67	3.38	4.17	5.05
$0.7f_{pu}A_{ps}$, kips	44.9	61.0	79.7	100.9	124.5	150.7
$0.8f_{pu}A_{ps}$, kips	51.3	69.7	91.0	115.3	142.3	172.2
$f_{pu}A_{ps}$, kips	64.1	87.1	113.8	144.1	177.9	215.3
Plain Prestressing Bars, f_{pu} = 160 ksi[a]						
Nominal Diameter, in.	3/8	7/8	1	1 1/8	1 1/4	1 3/8
Area, sq in.	0.442	0.601	0.785	0.994	1.227	1.485
Weight, plf	1.50	2.04	2.67	3.38	4.17	5.05
$0.7f_{pu}A_{ps}$, kips	49.5	67.3	87.9	111.3	137.4	166.3
$0.8f_{pu}A_{ps}$, kips	56.6	77.0	100.5	127.2	157.0	190.1
$f_{pu}A_{ps}$, kips	70.7	96.2	125.6	159.0	196.3	237.6
Deformed Prestressing Bars						
Nominal Diameter, in.	5/8	1	1	1 1/4	1 1/4	1 3/8
Area, sq in.	0.28	0.85	0.85	1.25	1.25	1.58
Weight, plf	0.98	3.01	3.01	4.39	4.39	5.56
Ult. strength, f_{pu}, ksi	157	150	160[a]	150	160[a]	150
$0.7f_{pu}A_{ps}$, kips	30.5	89.3	95.2	131.3	140.0	165.9
$0.8f_{pu}A_{ps}$, kips	34.8	102.0	108.8	150.0	160.0	189.6
$f_{pu}A_{ps}$, kips	43.5	127.5	136.0	187.5	200.0	237.0

Stress-strain characteristics (all prestressing bars):

For design purposes, the following assumptions are satisfactory:

E_s = 29,000 ksi

f_y = 0.95f_{pu}

a. Verify availability before specifying.

MATERIAL PROPERTIES
PRESTRESSING STEEL

Design Aid 11.2.5 Typical Stress-Strain Curve, 7-Wire Low-Relaxation Prestressing Strand

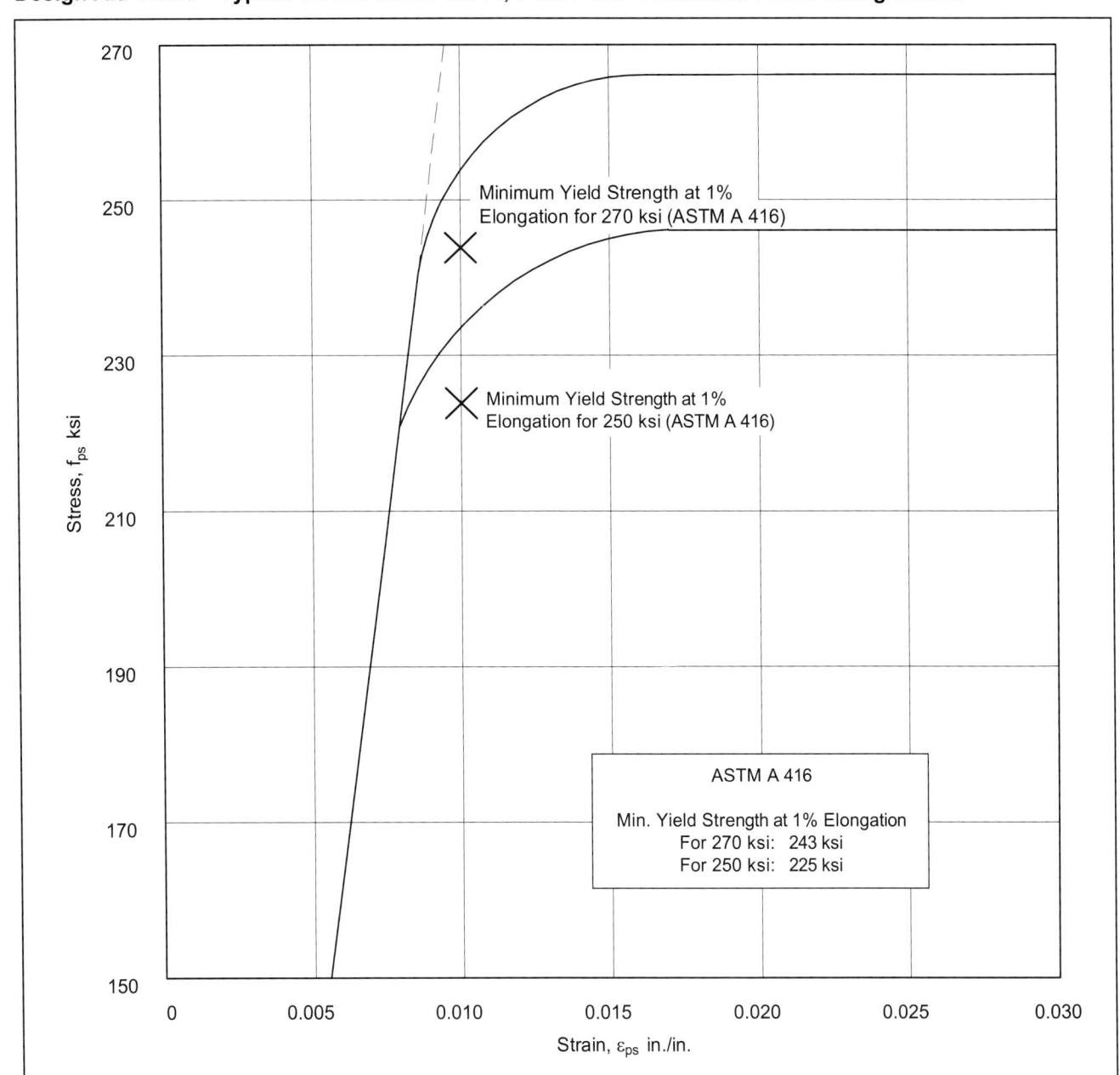

[Graph: Stress f_{ps} (ksi) vs. Strain ε_{ps} (in./in.), showing curves for 270 ksi and 250 ksi strand. Markers indicate Minimum Yield Strength at 1% Elongation for 270 ksi (ASTM A 416) and for 250 ksi (ASTM A 416).]

ASTM A 416

Min. Yield Strength at 1% Elongation
For 270 ksi: 243 ksi
For 250 ksi: 225 ksi

Note: approximate strain at rupture is 0.05 to 0.07 in./in.

These curves can be approximated by the following equations:

250 ksi strand:

$\varepsilon_{ps} \leq 0.0076$: $f_{ps} = 28{,}500\varepsilon_{ps}$ (ksi)

$\varepsilon_{ps} > 0.0076$: $f_{ps} = 250 - \dfrac{0.04}{\varepsilon_{ps} - 0.0064}$ (ksi)

270 ksi strand:

$\varepsilon_{ps} \leq 0.0086$: $f_{ps} = 28{,}500\varepsilon_{ps}$ (ksi)

$\varepsilon_{ps} > 0.0086$: $f_{ps} = 270 - \dfrac{0.04}{\varepsilon_{ps} - 0.007}$ (ksi)

MATERIAL PROPERTIES
PRESTRESSING STEEL

Design Aid 11.2.6 Transfer and Development Lengths for 7-Wire Uncoated Strand

The ACI 318-02 (Section 12.9.1) equation for required development length may be rewritten as:

$$\ell_d = (f_{se}/3)d_b + (f_{ps} - f_{se})d_b$$

where:

ℓ_d = required development length, in.
f_{se} = effective prestress, ksi
f_{ps} = stress in prestressing steel at nominal strength, ksi
d_b = nominal diameter of strand, in.

The first term in the equation is the transfer length and the second term is the additional length required for the stress increase ($f_{ps} - f_{se}$) corresponding to the nominal strength.

Transfer and development length[a] in inches

Nominal Diameter, in.	f_{se} = 150 ksi Transfer Length	Development Length f_{ps}, ksi 240	250	260	270	f_{se} = 160 ksi Transfer Length	Development Length f_{ps}, ksi 240	250	260	270	f_{se} = 170 ksi Transfer Length	Development Length f_{ps}, ksi 240	250	260	270
3/8	18.8	52.5	56.3	60.0	63.8	20.0	50.0	53.8	57.5	61.3	21.3	47.5	51.3	55.0	58.8
7/16	21.9	61.3	65.6	70.0	74.4	23.3	58.3	62.7	67.0	71.4	24.8	55.4	59.8	64.2	68.5
1/2	25.0	70.0	75.0	80.0	85.0	26.7	66.7	71.7	76.7	81.7	28.3	66.3	68.3	73.3	78.3
1/2 S[b]	26.1	73.1	78.4	83.6	88.8	27.9	69.7	74.9	80.1	85.4	29.6	66.1	71.3	76.6	81.8
9/16	28.1	78.8	84.4	90.0	95.6	30.0	75.0	80.6	86.3	91.9	31.9	71.3	76.9	82.5	88.1
3/5	30.0	84.0	90.0	96.0	102.0	32.0	80.0	86.0	92.0	98.0	34.0	76.0	82.0	88.0	94.0

a. The development length values given in the table must be doubled where bonding of the strand does not extend to the member end and the member is designed such that tension in the precompressed tensile zone is produced under service loads (see ACI 318-02, Section 12.9.3).

b. The ½ in. special (½ S) strand has a larger nominal diameter than the ½ in. regular (½) strand. The table values for transfer and development length reflect this difference in diameters.

MATERIAL PROPERTIES
REINFORCING BARS

Design Aid 11.2.7   Reinforcing Bar Data

ASTM STANDARD REINFORCING BARS							
BAR SIZE[a] DESIGNATION		NOMINAL DIMENSIONS					
^	^	DIAMETER		AREA		WEIGHT OR MASS	
U.S. CUSTOMARY	SI	in.	mm	in.²	mm²	lb/ft	kg/m
#3	#10	0.375	9.5	0.11	71	0.376	0.560
#4	#13	0.500	12.7	0.20	129	0.668	0.994
#5	#16	0.625	15.9	0.31	199	1.043	1.552
#6	#19	0.750	19.1	0.44	284	1.502	2.235
#7	#22	0.875	22.2	0.60	387	2.044	3.042
#8	#25	1.000	25.4	0.79	510	2.670	3.973
#9	#29	1.128	28.7	1.00	645	3.400	5.060
#10	#32	1.270	32.3	1.27	819	4.303	6.404
#11	#36	1.410	35.8	1.56	1006	5.313	7.907
#14	#43	1.693	43.0	2.25	1452	7.650	11.380
#18	#57	2.257	57.3	4.00	2581	13.600	20.240

a. Many mills will mark and supply bars only with metric (SI) designation, which is a soft conversion. Soft conversion means that the metric (SI) bars have exactly the same dimensions and properties as the equivalent U.S. customary designation.

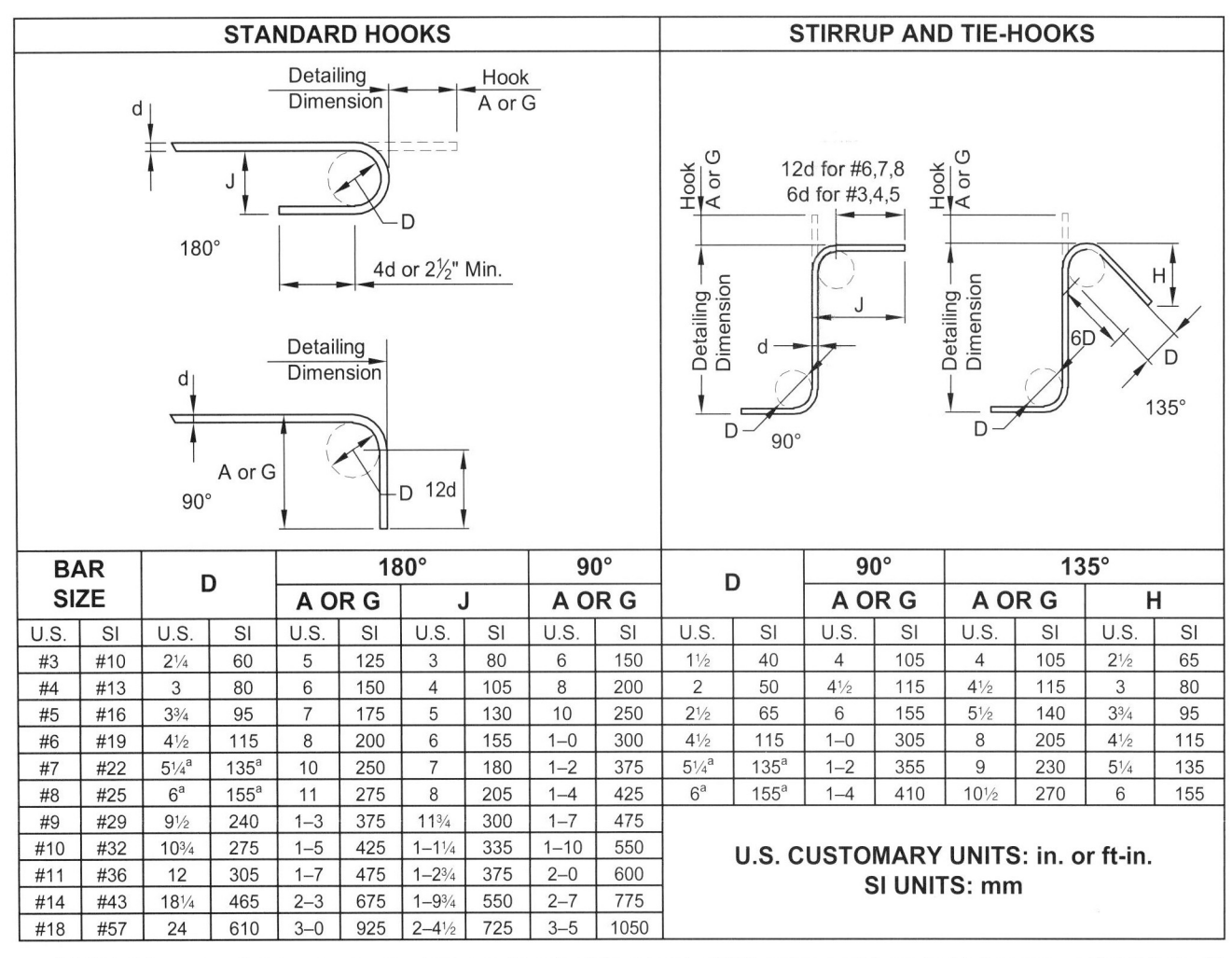

STANDARD HOOKS

STIRRUP AND TIE-HOOKS

BAR SIZE		D		180°				90°			D		90°		135°			
^	^	^	^	A OR G		J		A OR G		^	^	A OR G		A OR G		H	
U.S.	SI	U.S.	SI	U.S.	SI	U.S.	SI	U.S.	SI	U.S.	SI	U.S.	SI	U.S.	SI	U.S.	SI
#3	#10	2¼	60	5	125	3	80	6	150	1½	40	4	105	4	105	2½	65
#4	#13	3	80	6	150	4	105	8	200	2	50	4½	115	4½	115	3	80
#5	#16	3¾	95	7	175	5	130	10	250	2½	65	6	155	5½	140	3¾	95
#6	#19	4½	115	8	200	6	155	1–0	300	4½	115	1–0	305	8	205	4½	115
#7	#22	5¼[a]	135[a]	10	250	7	180	1–2	375	5¼[a]	135[a]	1–2	355	9	230	5¼	135
#8	#25	6[a]	155[a]	11	275	8	205	1–4	425	6[a]	155[a]	1–4	410	10½	270	6	155
#9	#29	9½	240	1–3	375	11¾	300	1–7	475								
#10	#32	10¾	275	1–5	425	1–1¼	335	1–10	550								
#11	#36	12	305	1–7	475	1–2¾	375	2–0	600								
#14	#43	18¼	465	2–3	675	1–9¾	550	2–7	775								
#18	#57	24	610	3–0	925	2–4½	725	3–5	1050								

U.S. CUSTOMARY UNITS: in. or ft-in.
SI UNITS: mm

a. ASTM A767 requires that bars bent cold <u>prior to hot dip galvanizing</u> must be fabricated to a minimum bend diameter equal to 7 in. for #7 bar and 8 in. for #8 bar.

MATERIAL PROPERTIES
REINFORCING BARS

Design Aid 11.2.8   Location of Reinforcement Confined by Stirrups or Ties

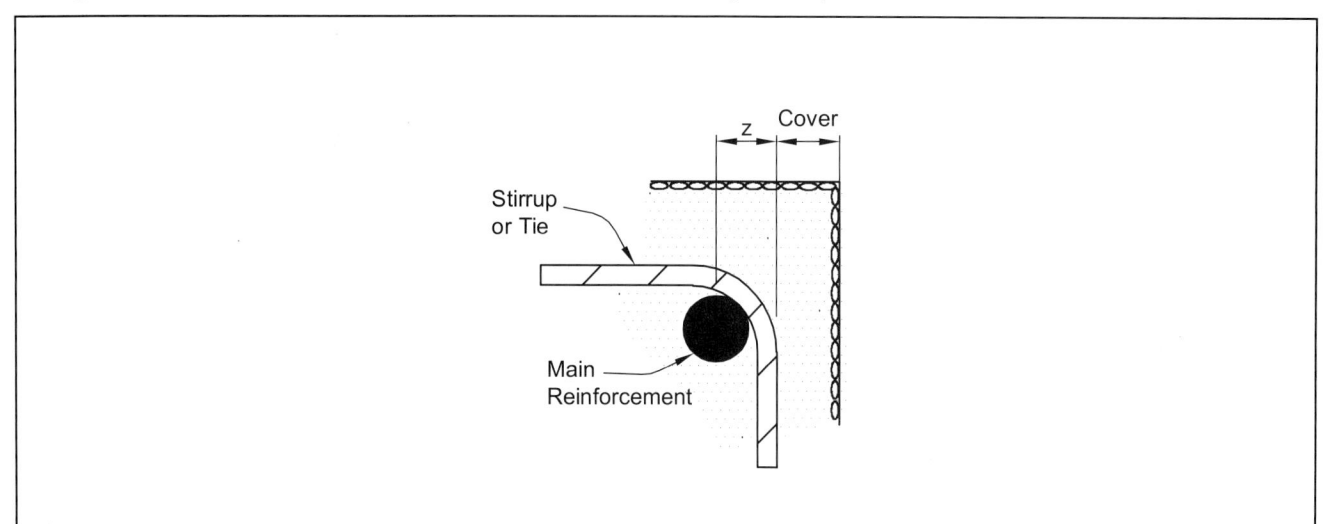

	Z Dimension (in.)		
	Stirrup or Tie Size		
Main Reinforcement Size	#3	#4	#5
#4	7/8	1 1/16	1 1/4
#5	7/8	1 1/8	1 5/16
#6	15/16	1 3/16	1 3/8
#7	1	1 3/16	1 7/16
#8	1 1/16	1 1/4	1 7/16
#9	1 1/16	1 5/16	1 1/2
#10	1 1/8	1 5/16	1 1/2
#11	1 3/16	1 3/8	1 9/16

To determine location of main reinforcement, add specified cover to the "z" dimension from above table.

MATERIAL PROPERTIES
REINFORCING BARS

Design Aid 11.2.9 Required Development Lengths[a] for Reinforcing Bars (Grade 60)

Tension Development Length:

$\ell_d = 2400 \dfrac{d_b}{\sqrt{f'_c}}$; min. 12 in. (#6 and smaller)

$\ell_d = 3000 \dfrac{d_b}{\sqrt{f'_c}}$; min. 12 in. (#7 and larger)

(Note: for Grade 40 bars, replace 2400 and 3000 with 1600 and 2000, respectively.)
Multiply ℓ_d values by:
(a) 1.3 for lightweight concrete
(b) 1.3 for "top bars"
(c) 1.5 for epoxy coated bars with cover < $3d_b$ or clear spacing < $6d_b$, otherwise multiply by 1.2
(Note: Product of factors (b) and (c) need not exceed 1.7)
(d) 1.5 for bars with less than minimum stirrups or ties, clear spacing less than $2d_b$ or clear cover less than d_b.
(e) A_s (required)/A_s (provided) for excess reinforcement unless development of f_y is specially required. This multiplier is not to be applied to lap splices per ACI 318-02, Section R12.15.1.

Compression Development Length:

$\ell_d = \dfrac{1200 d_b}{\sqrt{f'_c}}$; min. $18d_b$ and 8 in.

(Note: For Grade 40 bars, replace 1200 with 800 and 18 with 12)

Multiply ℓ_d values by:
(a) A_s(required)/A_s (provided) for excess reinforcement
(b) 0.75 for adequate spiral or tie enclosure (see ACI 318-02, Section 12.3.3b)

Compression Splice Lap Length:
Lap length = $30d_b$; min. 12 in.

The values of $\sqrt{f'_c}$ used in these equations shall not exceed 100 psi (see Section 12.1.2, ACI 318-02)

Bar Size	$f'_c = 3000$ psi Tension ℓ_d	$1.3\ell_d$	$1.5\ell_d$	Compression ℓ_d	$f'_c = 4000$ psi Tension ℓ_d	$1.3\ell_d$	$1.5\ell_d$	Compression ℓ_d	$f'_c = 5000$ psi Tension ℓ_d	$1.3\ell_d$	$1.5\ell_d$	Compression ℓ_d	$f'_c = 6000$ psi Tension ℓ_d	$1.3\ell_d$	$1.5\ell_d$	Compression ℓ_d	Min. Comp. Splice
3	16	21	25	8	14	18	21	8	13	17	19	8	12	15	17	8	12
4	22	28	33	11	19	25	28	9	17	22	25	9	15	20	23	9	15
5	27	36	41	14	24	31	36	12	21	28	32	11	19	25	29	11	19
6	33	43	49	16	28	37	43	14	25	33	38	14	23	30	35	14	23
7	48	62	72	19	42	54	62	17	37	48	56	16	34	44	51	16	26
8	55	71	82	22	47	62	71	19	42	55	64	18	39	50	58	18	30
9	62	80	93	25	54	70	80	21	48	62	72	20	44	57	66	20	34
10	70	90	104	28	60	78	90	24	54	70	81	23	49	64	74	23	38
11	77	100	116	31	67	87	100	27	60	78	90	25	55	71	82	25	42

Bar Size	$f'_c = 7000$ psi Tension ℓ_d	$1.3\ell_d$	$1.5\ell_d$	Compression ℓ_d	$f'_c = 8000$ psi Tension ℓ_d	$1.3\ell_d$	$1.5\ell_d$	Compression ℓ_d	$f'_c = 9000$ psi Tension ℓ_d	$1.3\ell_d$	$1.5\ell_d$	Compression ℓ_d	$f'_c = 10,000$ psi Tension ℓ_d	$1.3\ell_d$	$1.5\ell_d$	Compression ℓ_d	Min. Comp. Splice
3	12	14	16	8	12	13	15	8	12	12	14	8	9	12	14	8	12
4	14	19	22	9	13	17	20	9	13	16	19	8	12	16	18	8	15
5	18	23	27	11	17	22	25	11	16	21	24	8	15	20	23	8	19
6	22	28	32	14	20	26	30	14	19	25	28	9	18	23	27	9	23
7	31	41	47	16	29	38	44	16	28	36	42	11	26	34	39	11	26
8	36	47	54	18	34	44	50	18	32	41	47	13	30	39	45	12	30
9	40	53	61	20	38	49	57	20	36	46	54	14	34	44	51	14	34
10	46	59	68	23	43	55	64	23	40	52	60	16	38	49	56	15	38
11	51	66	76	25	47	61	71	25	45	58	67	17	41	54	62	17	42

a. For limitations and items related to hooked bars, stirrups or ties in excess of minimum, and spacing of non-contact lap splices, etc., see ACI 318-02, Chapter 12.

MATERIAL PROPERTIES
REINFORCING BARS

Design Aid 11.2.9 Required Development Lengths for Reinforcing Bars (Grade 60) (Cont.)

Bar Size	Minimum tension embedment lengths, ℓ_{dh}, for standard hooks, in. General use (non-seismic) [see ACI 318-02, Section 12.5.2 and 12.5.3(a)]							
	Normal weight concrete, f'_c (psi)							
	3,000	4,000	5,000	6,000	7,000	8,000	9,000	10,000
3	6	6	6	6	6	6	6	6
4	8	7	6	6	6	6	6	6
5	10	9	8	7	7	6	6	6
6	12	10	9	8	8	7	7	6
7	14	12	11	10	9	9	8	7
8	16	14	12	11	10	10	9	8
9	18	15	14	13	12	11	10	9
10	20	17	15	14	13	12	11	11
11	22	19	17	16	14	14	13	12

Notes: 1. Side Cover ≥ 2½ in.
2. End Cover (90° hooks) ≥ 2 in.

Bar Size	Minimum tension embedment lengths, ℓ_{dh}, for standard hooks, in. Special confinement (non-seimic) [see ACI 318-02, Section 12.5.3(b)]							
	Normal weight concrete, f'_c (psi)							
	3,000	4,000	5,000	6,000	7,000	8,000	9,000	10,000
3	6	6	6	6	6	6	6	6
4	6	6	6	6	6	6	6	6
5	8	7	6	6	6	6	6	6
6	10	8	7	7	6	6	6	6
7	11	10	9	8	7	7	6	6
8	13	11	10	9	8	8	7	6
9	14	12	11	10	9	9	8	7
10	16	14	12	11	11	10	9	9
11	18	15	14	13	12	12	10	10

Notes: 1. Side Cover ≥ 2½ in.
2. End Cover (90° hooks) ≥ 2 in.

BARS WITH STANDARD HOOKS

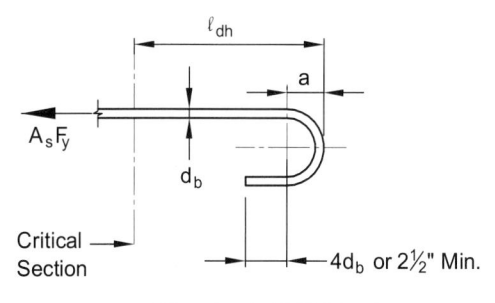

Standard 180° Hook

Standard 90° Hook

Dimension a = $4d_b$ for #3 through #8, = $5d_b$ for #9, #10 and #11
Modification factors:
  Grade 40 bars = 0.67
  Lightweight concrete = 1.3
  Epoxy coated reinforcement = 1.2

MATERIAL PROPERTIES
STRUCTURAL WELDED WIRE REINFORCEMENT (WWR)

Design Aid 11.2.10 Common Styles of Structural Welded Wire Reinforcement[a]

Style designation		Steel area in.² per ft		Approximate Weight lb per 100 ft²
Former designation (By steel wire gage)	Current Designation (By W-number)	Longit.	Trans.	
6x6–10x10	6x6–W1.4xW1.4	0.029	0.029	21
4x12–8x10[c]	4x12–W2.1xW1.4	0.062	0.014	26
6x6–8x8	6x6–W2.1xW2.1	0.041	0.041	30
4x4–10x10	4x4–W1.4xW1.4	0.043	0.043	31
4x12–7x10[c]	4x12–W2.5xW1.4	0.074	0.014	31
6x6–6x6[b]	6x6–W2.9xW2.9	0.058	0.058	42
4x4–8x8	4x4–W2.1xW2.1	0.062	0.062	44
6x6–4x4[b]	6x6–W4.0xW4.0	0.080	0.080	58
4x4–6x6	4x4–W2.9xW2.9	0.087	0.087	62
6x6–2x2[b]	6x6–W5.5xW5.5[d]	0.110	0.110	80
4x4–4x4[b]	4x4–W4.0xW4.0	0.120	0.120	85
4x4–3x3[b]	4x4–W4.7xW4.7	0.141	0.141	102
4x4–2x2[b]	4x4–W5.5xW5.5[d]	0.165	0.165	119

Industry Method of Designating Style

Example: 6 X 12-W16 X W8

- Longitudinal Wire Spacing
- Transverse Wire Spacing
- Longitudinal Wire Size
- Transverse Wire Size

(Substitute "D" for "W" when Specifying Deformed Wire)

Specify:
ASTM A 185 for Plain WWR
ASTM A 497 for Deformed WWR

End overhangs may differ. The sum of the two end overhangs, however, should equal the transverse wire spacing.

Side overhangs may be varied as required and do not need to be equal. Overhang lengths limited only by overall sheet width.

NOTE 1: Wires may also be deformed (see Design Aid 11.2.12), except where plain wire is required by building codes (usually less than W4). Wire diameters can be specified in 0.001 in. increments. Check with local manufacturers on minimum quantities for customized orders.

NOTE 2: Since most double tees meet the requirement of ACI 318-02 Sect. 7.12.3, shrinkage and temperature reinforcement in the longitudinal direction typically is not required. In such cases, only a nominal amount is used to facilitate shipping and handling of the reinforcement.

NOTE 3: The wire reinforcement institute requires that the smaller wire must have an area at least equal to 0.4 times the larger wire area.

a. Source: Manual of Standard Practice – Structural Welded Wire Reinforcement, Wire Reinforcement Institute, 1992, Findlay, Ohio.
b. Commonly available in 8 ft x 12 ft or 8 ft x 15 ft sheets.
c. These items may be carried in sheets by various manufacturers in certain parts of the U.S. and Canada.
d. Exact W-number size for 2 gage is 5.4.

MATERIAL PROPERTIES
STRUCTURAL WELDED WIRE REINFORCEMENT (WWR)

Design Aid 11.2.11 Wires Used in Structural Welded Wire Reinforcement[a]

Wire Size Number		Nominal Diameter in.	Nominal Weight plf	Area – in.² per ft of width — Center to Center Spacing, in.						
Plain[b]	Deformed[c]			2	3	4	6	8	10	12
W45	D45	0.757	1.530	2.700	1.800	1.350	0.900	0.675	0.540	0.450
W31	D31	0.628	1.054	1.860	1.240	0.930	0.620	0.465	0.372	0.310
W30	D30	0.618	1.020	1.800	1.200	0.900	0.600	0.450	0.360	0.300
W28	D28	0.597	0.952	1.680	1.120	0.840	0.560	0.420	0.336	0.280
W26	D26	0.575	0.934	1.560	1.040	0.780	0.520	0.390	0.312	0.260
W24	D24	0.553	0.816	1.440	0.960	0.720	0.480	0.360	0.288	0.240
W22	D22	0.529	0.748	1.320	0.880	0.660	0.440	0.330	0.264	0.220
W20	D20	0.504	0.680	1.200	0.800	0.600	0.400	0.300	0.240	0.200
W18	D18	0.478	0.612	1.080	0.720	0.540	0.360	0.270	0.216	0.180
W16	D16	0.451	0.544	0.960	0.640	0.480	0.320	0.240	0.192	0.160
W14	D14	0.422	0.476	0.840	0.560	0.420	0.280	0.210	0.168	0.140
W12	D12	0.390	0.408	0.720	0.480	0.360	0.240	0.180	0.144	0.120
W11	D11	0.374	0.374	0.660	0.440	0.330	0.220	0.165	0.132	0.110
W10.5	D10.5	0.366	0.357	0.630	0.420	0.315	0.210	0.157	0.126	0.105
W10	D10	0.356	0.340	0.600	0.400	0.300	0.200	0.150	0.120	0.100
W9.5	D9.5	0.348	0.323	0.570	0.380	0.285	0.190	0.143	0.114	0.095
W9	D9	0.338	0.306	0.540	0.360	0.270	0.180	0.135	0.108	0.090
W8.5	D8.5	0.329	0.289	0.510	0.340	0.255	0.170	0.128	0.102	0.085
W8	D8	0.319	0.272	0.480	0.320	0.240	0.160	0.120	0.096	0.080
W7.5	D7.5	0.309	0.255	0.450	0.300	0.225	0.150	0.113	0.090	0.075
W7	D7	0.298	0.238	0.420	0.280	0.210	0.140	0.105	0.084	0.070
W6.5	D6.5	0.288	0.221	0.390	0.260	0.195	0.130	0.098	0.078	0.065
W6	D6	0.276	0.204	0.360	0.240	0.180	0.120	0.090	0.072	0.060
W5.5	D5.5	0.264	0.187	0.330	0.220	0.165	0.110	0.083	0.066	0.055
W5	D5	0.252	0.170	0.300	0.200	0.150	0.100	0.075	0.060	0.050
W4.5	D4.5	0.240	0.153	0.270	0.180	0.135	0.090	0.068	0.054	0.045
W4	D4	0.225	0.136	0.240	0.160	0.120	0.080	0.060	0.048	0.040
W3.5		0.211	0.119	0.210	0.140	0.105	0.070	0.053	0.042	0.035
W3		0.195	0.102	0.180	0.120	0.090	0.060	0.045	0.036	0.030
W2.9		0.192	0.098	0.174	0.116	0.087	0.058	0.044	0.035	0.029
W2.5		0.178	0.085	0.150	0.100	0.075	0.050	0.038	0.030	0.025
W2.1		0.162	0.070	0.126	0.084	0.063	0.042	0.032	0.025	0.021
W2		0.159	0.068	0.120	0.080	0.060	0.040	0.030	0.024	0.020
W1.5		0.138	0.051	0.090	0.060	0.045	0.030	0.022	0.018	0.015
W1.4		0.135	0.049	0.084	0.056	0.042	0.028	0.021	0.017	0.014

a. Source: Manual of Standard Practice—Structural Welded Wire Reinforcement, Wire Reinforcement Institute, 1992, Findlay, Ohio.
b. ASTM A 82, Available f_y = 65,000 psi to 80,000 psi in 2500 psi increments.
c. ASTM A 496, Available f_y = 70,000 psi to 80,000 psi in 2500 psi increments.

MATERIAL PROPERTIES
BAR AREA EQUIVALENTS

Design Aid 11.2.12 Bar Area Equivalents in a One Foot Wide Section

Bar Spacing c/c (in.)	#3 (0.375)	#4 (0.500)	#5 (0.625)	#6 (0.750)	#7 (0.875)	#8 (1.000)	#9 (1.128)	#10 (1.270)	#11 (1.410)	Area Range
2	0.66	1.20	1.86	2.64	3.60	4.74	Exceeds min. bar clear spacing of d_b			
2½	0.53	0.96	1.49	2.11	2.88	3.79	4.80			
3	0.44	0.80	1.24	1.76	2.40	3.16	4.00	5.08	6.24	
3½	0.38	0.69	1.06	1.51	2.06	2.71	3.43	4.35	5.35	
4	0.33	0.60	0.93	1.32	1.80	2.37	3.00	3.81	4.68	≥ 3.0 sq in.
4½	0.29	0.53	0.83	1.17	1.60	2.11	2.67	3.39	4.16	
5	0.26	0.48	0.74	1.06	1.44	1.90	2.40	3.05	3.74	
5½	0.24	0.44	0.68	0.96	1.31	1.72	2.18	2.77	3.40	
6	0.22	0.40	0.62	0.88	1.20	1.58	2.00	2.54	3.12	
6½	0.20	0.37	0.57	0.81	1.11	1.46	1.85	2.34	2.88	
7	0.19	0.34	0.53	0.75	1.03	1.35	1.71	2.18	2.67	
7½	0.18	0.32	0.50	0.70	0.96	1.26	1.60	2.03	2.50	2.0 to 3.0 sq in.
8	0.17	0.30	0.47	0.66	0.90	1.19	1.50	1.91	2.34	
8½	0.16	0.28	0.44	0.62	0.85	1.12	1.41	1.79	2.20	
9	0.15	0.27	0.41	0.59	0.80	1.05	1.33	1.69	2.08	
9½	0.14	0.25	0.39	0.56	0.76	1.00	1.26	1.60	1.97	
10	0.13	0.24	0.37	0.53	0.72	0.95	1.20	1.52	1.87	
10½	0.13	0.23	0.35	0.50	0.69	0.90	1.14	1.45	1.78	1.5 to 2.0 sq in.
11	0.12	0.22	0.34	0.48	0.65	0.86	1.09	1.39	1.70	
11½	0.11	0.21	0.32	0.46	0.63	0.82	1.04	1.33	1.63	
12	0.11	0.20	0.31	0.44	0.60	0.79	1.00	1.27	1.56	
13	0.10	0.18	0.29	0.41	0.55	0.73	0.92	1.17	1.44	
14	0.09	0.17	0.27	0.38	0.51	0.68	0.86	1.09	1.34	
15	0.09	0.16	0.25	0.35	0.48	0.63	0.80	1.02	1.25	1.0 to 1.5 sq in.
16	0.08	0.15	0.23	0.33	0.45	0.59	0.75	0.95	1.17	
17	0.08	0.14	0.22	0.31	0.42	0.56	0.71	0.90	1.10	
18	0.07	0.13	0.21	0.29	0.40	0.53	0.67	0.85	1.04	
19	0.07	0.13	0.20	0.28	0.38	0.50	0.63	0.80	0.99	
20	0.07	0.12	0.19	0.26	0.36	0.47	0.60	0.76	0.94	
21	0.06	0.11	0.18	0.25	0.34	0.45	0.57	0.73	0.89	0.75 to 1.0 sq in.
22	0.06	0.11	0.17	0.24	0.33	0.43	0.55	0.69	0.85	
23	0.06	0.10	0.16	0.23	0.31	0.41	0.52	0.66	0.81	
24	0.06	0.10	0.16	0.22	0.30	0.40	0.50	0.64	0.78	
Area Range	≤ 0.25 sq in.				0.25 to 0.50 sq in.			0.50 to 0.75 sq in.		

NOTE: Check minimum requirements for temperature and shrinkage steel.

How to use this design aid. Given a design (or minimum temperature/shrinkage) reinforcement are required per foot, enter the design aid along right column or bottom row. Select one of the bar area ranges from that given sq in./ft, and follow the range band upward and/or to the left. Select the combination of bar size and spacing satisfying the design and spacing requirements for the section.

Example. A design that requires reinforcement at 0.62 in.^{2}/ft, with bar size restricted to No. 7 or smaller. Enter the design aid in the 0.50 to 0.75 sq in. range located along the bottom row. Follow the shaded band up to the top of the table. Select one of the following combinations: one layer of No.4 at about 3.5 in. o.c., No. 5 at 6 in. o.c., No. 6 at 8.5 in. o.c., or No. 7 at 11.5 in. o.c. Similar spacing(s) could be determined for two reinforcement layers, if desired.

MATERIALS PROPERTIES

Design Aid 11.2.13 ACI Required Minimum Reinforcement Areas Per Foot Width of Section

Element/Category		A_s/A_{gross} Ratio[a]	\multicolumn{8}{c}{Member Thickness or Vertical Height – h (in.)}
			2	3	4	5	6	7	8	9
Structural Slabs										
	$f_y < 60$ ksi	0.0020	0.048	0.072	0.096	0.120	0.144	0.168	0.192	0.216
	$f_y = 60$ ksi	0.0018	0.043	0.065	0.086	0.108	0.130	0.151	0.173	0.194
	$f_y = 70$ ksi	0.0015	0.036	0.054	0.072	0.090	0.108	0.126	0.144	0.162
	Minimum[b]	0.0014	0.034	0.050	0.067	0.084	0.101	0.118	0.134	0.151
Walls **One layer permissible in walls**										
	Vertical[c]	0.0012	0.029	0.043	0.058	0.072	0.086	0.101	0.115	0.130
Cast-in-place	Vertical[d]	0.0015	0.036	0.054	0.072	0.090	0.108	0.126	0.144	0.162
	Horizontal[c]	0.0020	0.048	0.072	0.096	0.120	0.144	0.168	0.192	0.216
	Horizontal[d]	0.0025	0.060	0.090	0.120	0.150	0.180	0.210	0.240	0.270
Precast[(5)]	Vert & Horiz	0.0010	0.024	0.036	0.048	0.060	0.072	0.084	0.096	0.108

Element/Category		A_s/A_{gross} Ratio[a]	\multicolumn{10}{c}{Member thickness or Vertical Height – h (in.)}
			10	11	12	15	18	21	24	30	36	42	48
Structural Slabs													
	$f_y < 60$ ksi	0.0020	0.240	0.264	0.288	0.360	0.432	0.504	0.546	0.720	0.864	1.008	1.152
	$f_y = 60$ ksi	0.0018	0.216	0.238	0.259	0.324	0.389	0.454	0.518	0.648	0.778	0.907	1.037
	$f_y = 70$ ksi	0.0015	0.180	0.198	0.216	0.270	0.324	0.378	0.432	0.540	0.648	0.756	0.864
	Minimum[b]	0.0014	0.168	0.185	0.202	0.252	0.302	0.353	0.403	0.504	0.605	0.706	0.806
Walls **Two layers required in walls**													
	Vertical[c]	0.0012	0.144	0.158	0.173	0.216	0.259	0.302	0.346	0.432	0.518	0.605	0.691
Cast-in-place	Vertical[d]	0.0015	0.180	0.198	0.216	0.270	0.324	0.378	0.432	0.540	0.648	0.756	0.864
	Horizontal[c]	0.0020	0.240	0.264	0.288	0.360	0.432	0.504	0.576	0.720	0.864	1.008	1.152
	Horizontal[d]	0.0025	0.300	0.330	0.360	0.450	0.540	0.630	0.720	0.900	1.080	1.260	1.440
Precast[(e)]	Vert. & Horiz.	0.0010	0.120	0.132	0.144	0.180	0.216	0.252	0.288	0.360	0.432	0.504	0.576

Notes:
a. Ratio is reinforcement area to gross concrete area.
b. Minimum controls for $f_y > 77$ ksi.
c. For deformed bars not larger than #5, with a specified yield strength not less than 60,000 psi, or for welded wire reinforcement (plain or deformed) not larger than W31 or D31.
d. For other deformed bars.
e. Refer to ACI 318-02, Section 16.4.2.

11.3 STANDARD BOLTS, NUTS AND WASHERS

Design Aid 11.3.1 Dimensions of Nuts and Bolts

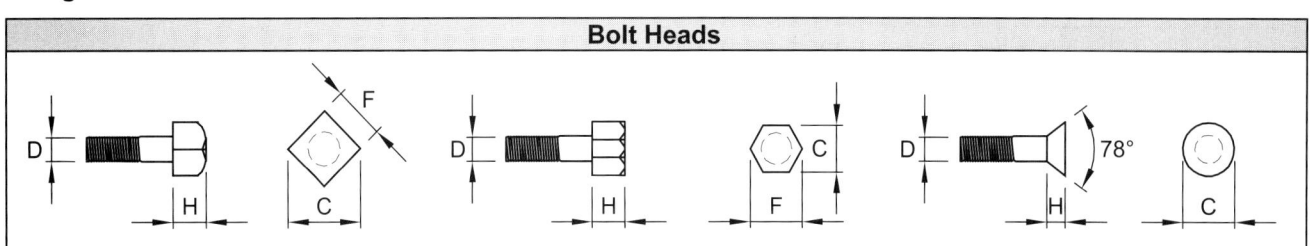

Bolt Heads

Bolt head dimensions, rounded to nearest 1/16 inch, are in accordance with ANSI B18.2.1 – 1972 (Square and Hex) and ANSI 18.5 – 1971 (Countersunk)

Standard Dimensions for Bolt Heads

Dia. of Bolt D	Square Width F	Square Width C	Square Height H	Hex Width F	Hex Width C	Hex Height H	Heavy Hex Width F	Heavy Hex Width C	Heavy Hex Height H	Countersunk Diam. D	Countersunk Height H
in.	in.	in.	in.	in.	in.	in.	in.	in.	in.	in.	in.
1/4	3/8	1/2	3/16	7/16	1/2	3/16	...	...	...	1/2	1/8
3/8	9/16	13/16	1/4	9/16	5/8	1/4	...	...	...	11/16	3/16
1/2	3/4	1 1/16	5/16	3/4	7/8	3/8	7/8	1	3/8	7/8	1/4
5/8	15/16	1 5/16	7/16	15/16	1 1/16	7/16	1 1/16	1 1/4	7/16	1 1/8	5/16
3/4	1 1/8	1 9/16	1/2	1 1/8	1 5/16	1/2	1 1/4	1 7/16	1/2	1 3/8	3/8
7/8	1 5/16	1 7/8	5/8	1 5/16	1 1/2	9/16	1 7/16	1 11/16	9/16	1 9/16	7/16
1	1 1/2	2 1/8	11/16	1 1/2	1 3/4	11/16	1 5/8	1 7/8	11/16	1 13/16	1/2
1 1/8	1 11/16	2 3/8	3/4	1 11/16	1 15/16	3/4	1 13/16	2 1/16	3/4	2 1/16	9/16
1 1/4	1 7/8	2 5/8	7/8	1 7/8	2 3/16	7/8	2	2 5/16	7/8	2 1/4	5/8
1 3/8	2 1/16	2 15/16	15/16	2 1/16	2 3/8	15/16	2 3/16	2 1/2	15/16	2 1/2	11/16
1 1/2	2 1/4	3 3/16	1	2 1/4	2 5/8	1	2 3/8	2 3/4	1	2 11/16	3/4
1 3/4	...	...	...	2 5/8	3	1 3/16	2 3/4	3 3/16	1 3/16	...	...
2	...	...	...	3	3 7/16	1 3/8	3 1/8	3 5/8	1 3/8	...	...
2 1/4	...	...	...	3 3/8	3 7/8	1 1/2	3 1/2	4 1/16	1 1/2	...	...
2 1/2	...	...	...	3 3/4	4 5/16	1 11/16	3 7/8	4 1/2	1 11/16	...	...
2 3/4	...	...	...	4 1/8	4 3/4	1 13/16	4 1/4	4 15/16	1 13/16	...	...
3	...	...	...	4 1/2	5 3/16	2	4 5/8	5 5/16	2	...	...
3 1/4	...	...	...	4 7/8	5 5/8	2 3/16	...	...	...	...	...
3 1/2	...	...	...	5 1/4	6 1/16	2 5/16	...	...	...	...	...
3 3/4	...	...	...	5 5/8	6 1/2	2 1/2	...	...	...	...	...
4	...	...	...	6	6 15/16	2 11/16	...	...	...	...	...

Source: AISC Manual of Steel Construction, Allowable Stress Design, Ninth Edition, 1989.

STANDARD BOLTS, NUTS AND WASHERS

Design Aid 11.3.1   Dimensions of Nuts and Bolts (Cont.)

Nuts

Nut dimensions, rounded to nearest 1/16 inch, are in accordance with ANSI B18.2.2 – 1972

Dimensions for Nuts

Nut Size	Square Width F	Square Width C	Square Height H	Hex Width F	Hex Width C	Hex Height H	Heavy Square Width F	Heavy Square Width C	Heavy Square Height H	Heavy Hex Width F	Heavy Hex Width C	Heavy Hex Height H
in.	in.	in.	in.	in.	in.	in.	in.	in.	in.	in.	in.	in.
1/4	7/16	5/8	1/4	7/16	1/2	1/4	1/2	11/16	1/4	1/2	9/16	1/4
3/8	5/8	7/8	5/16	9/16	5/8	5/16	11/16	1	3/8	11/16	13/16	3/8
1/2	13/16	1 1/8	7/16	3/4	7/8	7/16	7/8	1 1/4	1/2	7/8	1	1/2
5/8	1	1 7/16	9/16	15/16	1 1/16	9/16	1 1/16	1 1/2	5/8	1 1/16	1 1/4	5/8
3/4	1 1/8	1 9/16	11/16	1 1/8	1 5/16	5/8	1 1/4	1 3/4	3/4	1 1/4	1 7/16	3/4
7/8	1 5/16	1 7/8	3/4	1 5/16	1 1/2	3/4	1 7/16	2 1/16	7/8	1 7/16	1 11/16	7/8
1	1 1/2	2 1/8	7/8	1 1/2	1 3/4	7/8	1 5/8	2 5/16	1	1 5/8	1 7/8	1
1 1/8	1 11/16	2 3/8	1	1 11/16	1 15/16	1	1 13/16	2 9/16	1 1/8	1 13/16	2 1/16	1 1/8
1 1/4	1 7/8	2 5/8	1 1/8	1 7/8	2 3/16	1 1/16	2	2 13/16	1 1/4	2	2 5/16	1 1/4
1 3/8	2 1/16	2 15/16	1 1/4	2 1/16	2 3/8	1 3/16	2 3/16	3 1/8	1 3/8	2 3/16	2 1/2	1 3/8
1 1/2	2 1/4	3 3/16	1 5/16	2 1/4	2 5/8	1 5/16	2 3/8	3 3/8	1 1/2	2 3/8	2 3/4	1 1/2
1 3/4	...	...	...	...	...	...	...	...	...	2 3/4	3 3/16	1 3/4
2	...	...	...	...	...	...	...	...	...	3 1/8	3 5/8	2
2 1/4	...	...	...	...	...	...	...	...	...	3 1/2	4 1/16	2 3/16
2 1/2	...	...	...	...	...	...	...	...	...	3 7/8	4 1/2	2 7/16
2 3/4	...	...	...	...	...	...	...	...	...	4 1/4	4 15/16	2 11/16
3	...	...	...	...	...	...	...	...	...	4 5/8	5 5/16	2 15/16
3 1/4	...	...	...	...	...	...	...	...	...	5	5 3/4	3 3/16
3 1/2	...	...	...	...	...	...	...	...	...	5 3/8	6 3/16	3 7/16
3 3/4	...	...	...	...	...	...	...	...	...	5 3/4	6 5/8	3 11/16
4	...	...	...	...	...	...	...	...	...	6 1/8	7 1/16	3 15/16

Source: AISC Manual of Steel Construction, Allowable Stress Design, Ninth Edition, 1989.

STANDARD BOLTS, NUTS AND WASHERS

Design Aid 11.3.2   Dimensions of Standard Washers

Size of Bolt	Size of Hole	Outside Diameter	Thickness Wire Gage No.	Weight per 1,000 in Pounds	Size of Bolt	Size of Hole	Outside Diameter	Thickness Wire Gage No.
\multicolumn{5}{c|}{**Plain Washers, U.S Standard (Dimensions in Inches)**}	\multicolumn{4}{c}{**Plain Washers, S.A.E. Standard (Dimensions in Inches)**}

Size of Bolt	Size of Hole	Outside Diameter	Thickness Wire Gage No.	Weight per 1,000 in Pounds	Size of Bolt	Size of Hole	Outside Diameter	Thickness Wire Gage No.
½	9/16	1 3/8	12 (3/32)	38.4	½	17/32	1 1/16	3/32
9/16	5/8	1 ½	12 (3/32)	44.4	9/16	19/32	1 3/16	3/32
5/8	11/16	1 ¾	10 (1/8)	77.0	5/8	21/32	1 5/16	3/32
¾	13/16	2	10 (1/8)	111.0	¾	13/16	1 ½	1/8
7/8	15/16	2 ¼	9 (5/32)	153.0	7/8	15/16	1 ¾	1/8
1	1 1/16	2 ½	9 (5/32)	176.0				
1 1/8	1 ¼	2 ¾	9 (5/32)		\multicolumn{4}{c}{**Narrow – Gage Washers (Dimensions in Inches)**}
1 ¼	1 3/8	3	9 (5/32)					
					½	9/16	1 ¼	12 (3/32)
					9/16	5/8	1 3/8	12 (3/32)
					5/8	11/16	1 ½	10 (1/8)
					¾	13/16	1 ¾	10 (1/8)
					7/8	15/16	2	9 (5/32)
					1	1 1/16	2 ¼	9 (5/32)
					1 1/8	1 ¼	2 ½	9 (5/32)
					1 ¼	1 3/8	2 ¾	9 (5/32)
					1 3/8	1 ½	3	8 (11/64)

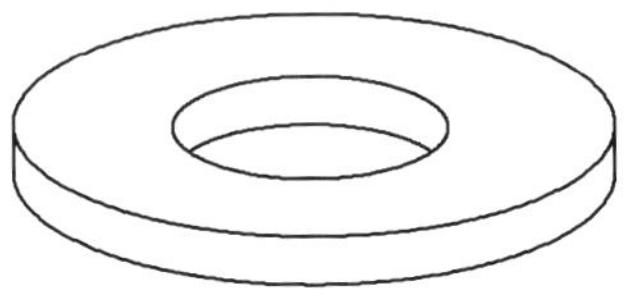

Plain Washers

Source: AISC Manual of Steel Construction, Allowable Stress Design, Ninth Edition, 1989.

11.4 WELDING INFORMATION

Design Aid 11.4.1 Weld Symbols Commonly Used in Precast Construction

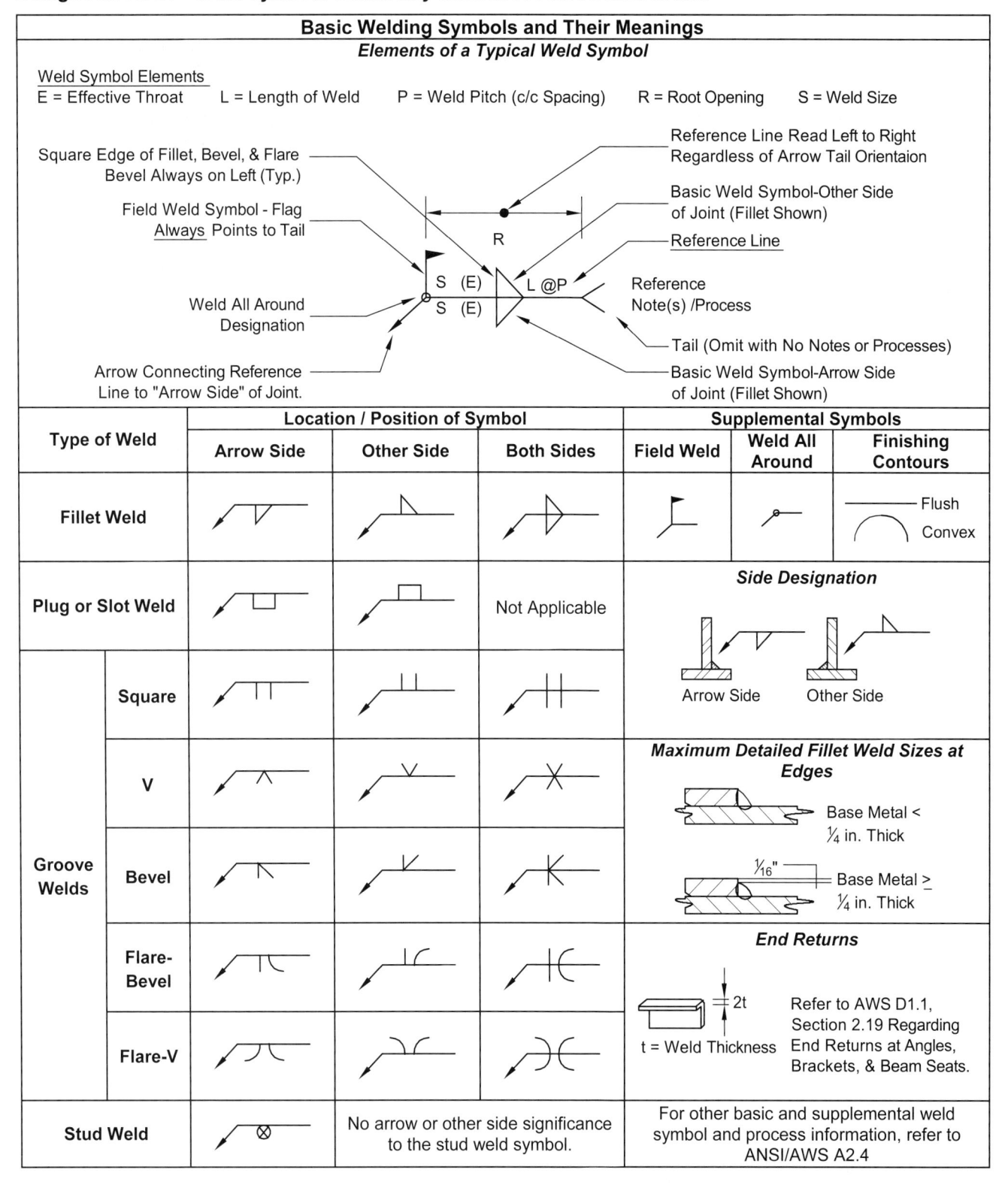

Basic Welding Symbols and Their Meanings

Elements of a Typical Weld Symbol

Weld Symbol Elements
E = Effective Throat   L = Length of Weld   P = Weld Pitch (c/c Spacing)   R = Root Opening   S = Weld Size

- Square Edge of Fillet, Bevel, & Flare Bevel Always on Left (Typ.)
- Field Weld Symbol - Flag <u>Always</u> Points to Tail
- Weld All Around Designation
- Arrow Connecting Reference Line to "Arrow Side" of Joint.
- Reference Line Read Left to Right Regardless of Arrow Tail Orientaion
- Basic Weld Symbol-Other Side of Joint (Fillet Shown)
- Reference Line
- Reference Note(s) /Process
- Tail (Omit with No Notes or Processes)
- Basic Weld Symbol-Arrow Side of Joint (Fillet Shown)

Type of Weld		Location / Position of Symbol			Supplemental Symbols		
		Arrow Side	Other Side	Both Sides	Field Weld	Weld All Around	Finishing Contours
Fillet Weld		╱╲	╱╲	╱╳	⌐	○	─ Flush / ⌒ Convex
Plug or Slot Weld				Not Applicable	*Side Designation* — Arrow Side / Other Side		
Groove Welds	Square						
	V				*Maximum Detailed Fillet Weld Sizes at Edges* — Base Metal < ¼ in. Thick; 1/16" Base Metal ≥ ¼ in. Thick		
	Bevel						
	Flare-Bevel				*End Returns* — 2t, t = Weld Thickness. Refer to AWS D1.1, Section 2.19 Regarding End Returns at Angles, Brackets, & Beam Seats.		
	Flare-V						
Stud Weld		⊗	No arrow or other side significance to the stud weld symbol.		For other basic and supplemental weld symbol and process information, refer to ANSI/AWS A2.4		

References
a. AWS, *Standard Symbols for Welding, Brazing and Nondestruction Examination*, ANSI/AWS A2.4-86, American Welding Society, Miami, Florida, 1986.
b. AWS, *Structural Welding Code – Steel*, ANSI/AWS D1.1:2000, 17th Edition, American Welding Society, Miami, Florida, 2000.
c. AISC, *Load & Resistance Factor Design,* – Manual of Steel Construction, Third Edition, American Institute of Steel Construction, Chicago, Illinois, 2001.

WELDING INFORMATION

Design Aid 11.4.2 Typical Welded Joints in Precast Construction

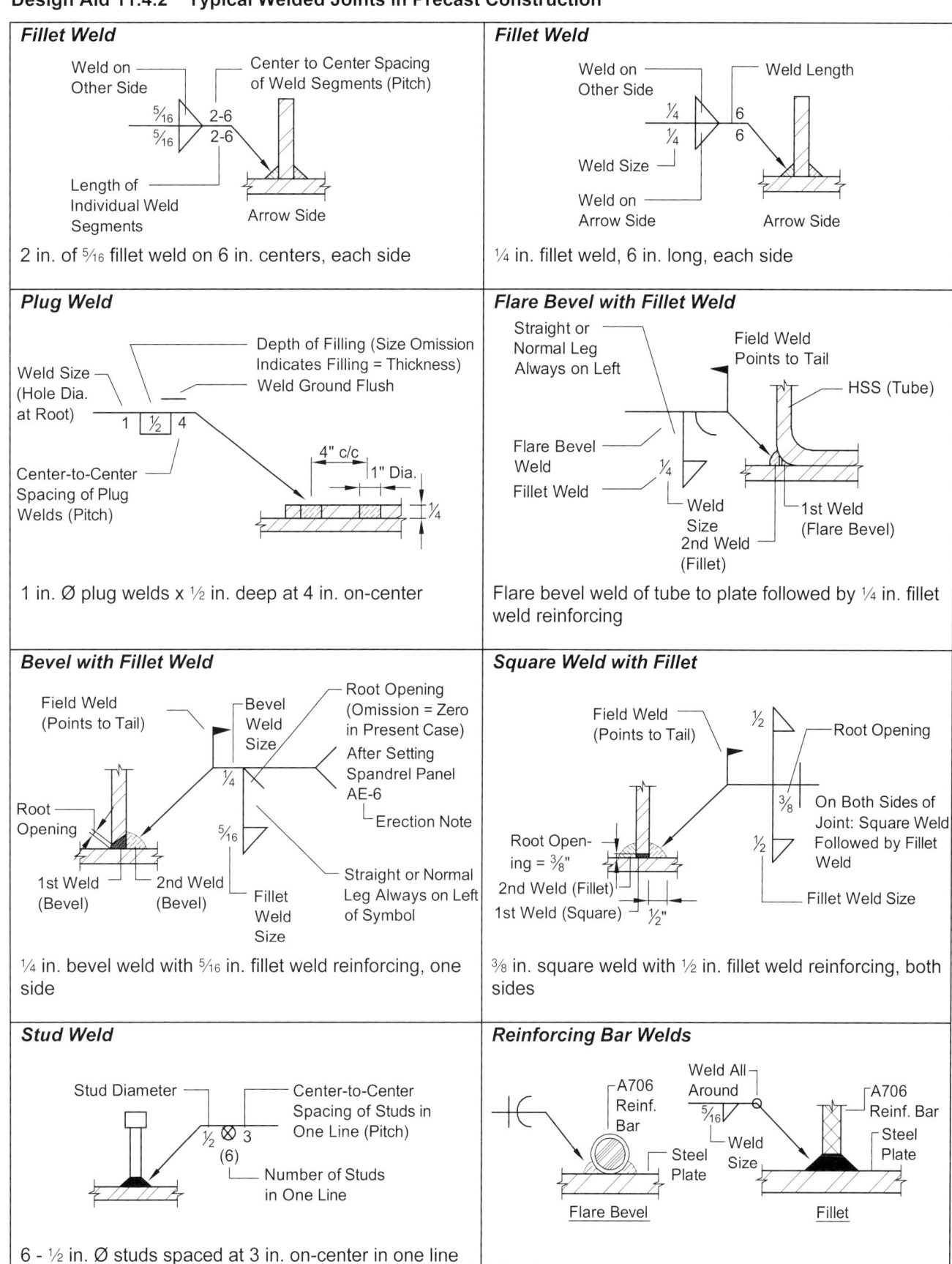

Fillet Weld

2 in. of 5/16 fillet weld on 6 in. centers, each side

Fillet Weld

1/4 in. fillet weld, 6 in. long, each side

Plug Weld

1 in. Ø plug welds x 1/2 in. deep at 4 in. on-center

Flare Bevel with Fillet Weld

Flare bevel weld of tube to plate followed by 1/4 in. fillet weld reinforcing

Bevel with Fillet Weld

1/4 in. bevel weld with 5/16 in. fillet weld reinforcing, one side

Square Weld with Fillet

3/8 in. square weld with 1/2 in. fillet weld reinforcing, both sides

Stud Weld

6 - 1/2 in. Ø studs spaced at 3 in. on-center in one line

Reinforcing Bar Welds

WELDING INFORMATION

Design Aid 11.4.3 Properties of Weld Groups Treated as Lines ($t_w = 1$)

Section b = width; d = depth	Distance to centroid	Section modulus $I_x / \bar{y}$	Polar moment of inertia I_P, about center of gravity
(1)		$S = \dfrac{d^2}{6}$	$I_p = \dfrac{d^3}{12}$
(2)		$S = \dfrac{d^2}{3}$	$I_p = \dfrac{d(3b^2 + d^2)}{6}$
(3)		$S = bd$	$I_p = \dfrac{b(3d^2 + b^2)}{6}$
(4)	$\bar{y} = \dfrac{d^2}{2(b+d)}$ $\bar{x} = \dfrac{b^2}{2(b+d)}$	$S_{top} = \dfrac{4bd + d^2}{6}$ $S_{bott} = \dfrac{d^2(4b+d)}{6(2b+d)}$	$I_p = \dfrac{(b+d)^4 - 6b^2d^2}{12(b+d)}$
(5)	$\bar{x} = \dfrac{b^2}{2b+d}$	$S = bd + \dfrac{d^2}{6}$	$I_p = \dfrac{8b^3 + 6bd^2 + d^3}{12} - \dfrac{b^4}{2b+d}$
(6)	$\bar{y} = \dfrac{d^2}{b+2d}$	$S_{top} = \dfrac{2bd + d^2}{3}$ $S_{bott} = \dfrac{d^2(2b+d)}{3(b+d)}$	$I_p = \dfrac{b^3 + 6b^2d + 8d^3}{12} - \dfrac{d^4}{2d+b}$
(7)		$S = bd + \dfrac{d^2}{3}$	$I_p = \dfrac{(b+d)^3}{6}$
(8)	$\bar{y} = \dfrac{d^2}{b+2d}$	$S_{top} = \dfrac{2bd + d^2}{3}$ $S_{bott} = \dfrac{d^2(2b+d)}{3(b+d)}$	$I_p = \dfrac{b^3 + 8d^3}{12} - \dfrac{d^4}{b+2d}$
(9)		$S = bd + \dfrac{d^2}{3}$	$I_p = \dfrac{b^3 + 3bd^2 + d^3}{6}$
(10)		$S = \pi r^2$	$I_p = 2\pi r^3$

11.5 SECTION PROPERTIES

Design Aid 11.5.1 Properties of Geometric Sections [a]

SQUARE — Axis of Moments Through Center

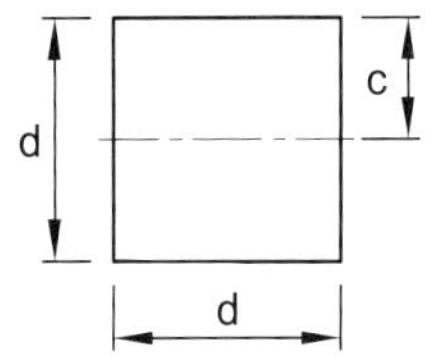

$A = d^2$
$c = \dfrac{d}{2}$
$I = \dfrac{d^4}{12}$
$S = \dfrac{d^3}{6}$
$r = \dfrac{d}{\sqrt{12}} = 0.288675d$

RECTANGLE — Axis of Moments on Diagonal

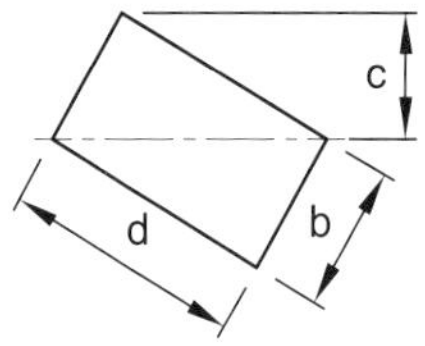

$A = bd$
$c = \dfrac{bd}{\sqrt{b^2 + d^2}}$
$I = \dfrac{b^3 d^3}{6(b^2 + d^2)}$
$S = \dfrac{b^2 d^2}{6\sqrt{b^2 + d^2}}$
$r = \dfrac{bd}{\sqrt{6(b^2 + d^2)}}$

SQUARE — Axis of Moments on Base

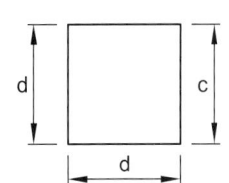

$A = d^2$
$c = d$
$I = \dfrac{d^4}{3}$
$S = \dfrac{d^3}{3}$
$r = \dfrac{d}{\sqrt{3}} = 0.577350d$

RECTANGLE — Axis of Moments Any Line Through Center of Gravity

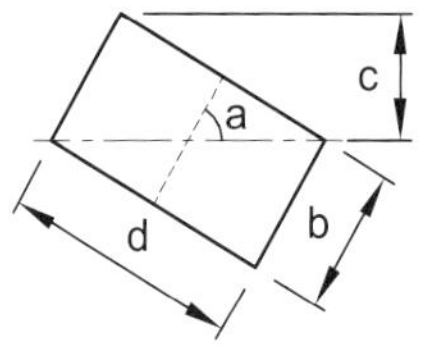

$A = bd$
$c = \dfrac{b\sin a + d\cos a}{2}$
$I = \dfrac{bd(b^2 \sin^2 a + d^2 \cos^2 a)}{12}$
$S = \dfrac{bd(b^2 \sin^2 a + d^2 \cos^2 a)}{6(b\sin a + d\cos a)}$
$r = \sqrt{\dfrac{b^2 \sin^2 a + d^2 \cos^2 a}{12}}$

SQUARE — Axis of Moments on Diagonal

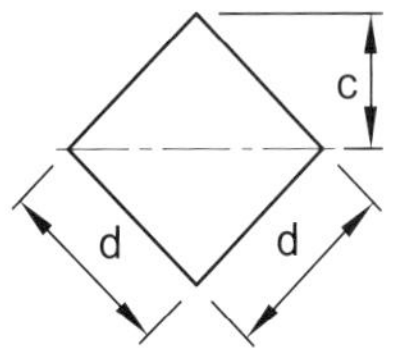

$A = d^2$
$c = \dfrac{d}{\sqrt{2}} = 0.707107d$
$I = \dfrac{d^4}{12}$
$S = \dfrac{d^3}{6\sqrt{2}} = 0.117851 d^3$
$r = \dfrac{d}{\sqrt{12}} = 0.288675d$

HOLLOW RECTANGLE — Axis of Moments Through Center

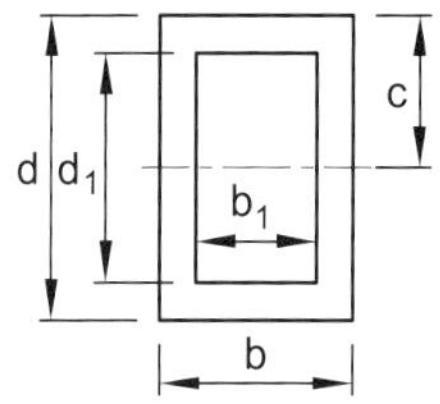

$A = bd - b_1 d_1$
$c = \dfrac{d}{2}$
$I = \dfrac{bd^3 - b_1 d_1^3}{12}$
$S = \dfrac{bd^3 - b_1 d_1^3}{6d}$
$r = \sqrt{\dfrac{bd^3 - b_1 d_1^3}{12A}}$

RECTANGLE — Axis of Moments Through Center

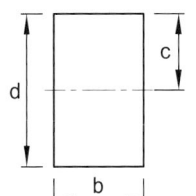

$A = bd$
$c = \dfrac{d}{2}$
$I = \dfrac{bd^3}{12}$
$S = \dfrac{bd^2}{6}$
$r = \dfrac{d}{\sqrt{12}} = 0.288675d$

EQUAL RECTANGLES — Axis of Moments Through Center of Gravity

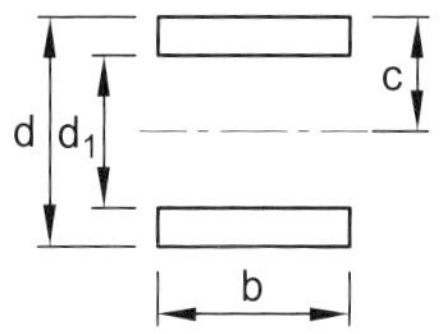

$A = b(d - d_1)$
$c = \dfrac{d}{2}$
$I = \dfrac{b(d^3 - d_1^3)}{12}$
$S = \dfrac{b(d^3 - d_1^3)}{6d}$
$r = \sqrt{\dfrac{d^3 - d_1^3}{12(d - d_1)}}$

---

a. Source: "Manual of Steel Construction, Allowable Stress Design," Ninth Edition, American Institute of Steel Construction, Chicago, IL, 1989.

SECTION PROPERTIES

Design Aid 11.5.1   Properties of Geometric Sections (Cont.)

RECTANGLE
Axis of Moments on Base

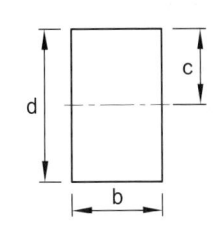

$A = bd$
$c = \dfrac{d}{2}$
$I = \dfrac{bd^3}{3}$
$S = \dfrac{bd^2}{3}$
$r = \dfrac{d}{\sqrt{3}} = 0.577350d$

UNEQUAL RECTANGLES
Axis of Moments Through Center of Gravity

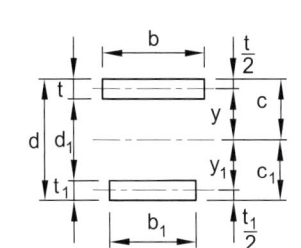

$A = bt + b_1 t_1$
$c = \dfrac{\tfrac{1}{2}bt^2 + b_1 t_1 (d - \tfrac{1}{2}t_1)}{A}$
$I = \dfrac{bt^3}{12} + bty^2 + \dfrac{b_1 t_1^3}{12} + b_1 t_1 y_1^2$
$S = \dfrac{I}{c} \quad S_1 = \dfrac{I}{c_1}$
$r = \sqrt{\dfrac{I}{A}}$

TRIANGLE
Axis of Moments Through Center of Gravity

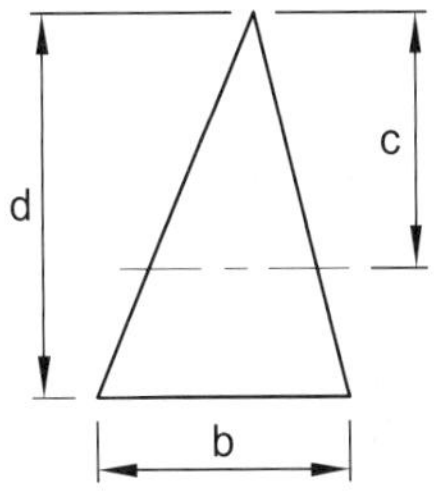

$A = \dfrac{bd}{2}$
$c = \dfrac{2d}{3}$
$I = \dfrac{bd^3}{36}$
$S = \dfrac{bd^2}{24}$
$r = \dfrac{d}{\sqrt{18}}$

HALF CIRCLE
Axis Through Moments of Center of Gravity

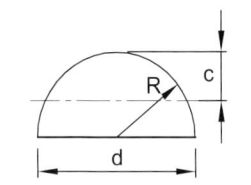

$A = \dfrac{\pi R^2}{2}$
$c = R\left(1 - \dfrac{4}{3\pi}\right)$
$I = R^4 \left(\dfrac{\pi}{8} - \dfrac{8}{9\pi}\right)$
$S = \left[\dfrac{R^3}{24}\right]\left[\dfrac{(9\pi^2 - 64)}{(3\pi - 4)}\right]$
$r = R\dfrac{\sqrt{9\pi^2 - 64}}{6\pi}$

TRIANGLE
Axis of Moments on Base

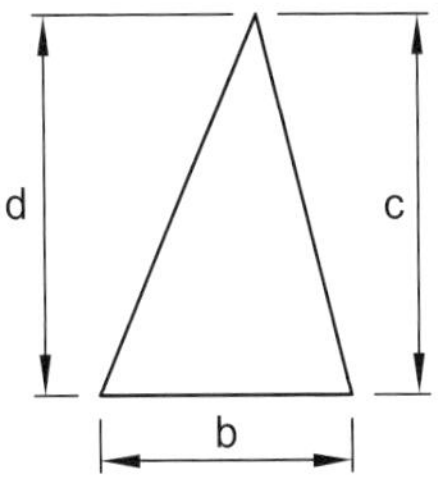

$A = \dfrac{bd}{2}$
$c = d$
$I = \dfrac{bd^3}{12}$
$S = \dfrac{bd^2}{12}$
$r = \dfrac{d}{\sqrt{6}}$

PARTIAL CIRCLE
Axis of Moments Through Circle Center

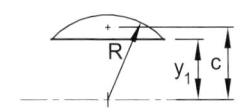

NOTE: Angles in Radians

$I = \dfrac{\pi R^4}{8} + \dfrac{y_1}{2}\sqrt{(R^2 - y_1^2)^3}$
$\quad - \dfrac{R^2}{4}\left(y_1\sqrt{R^2 - y_1^2} + R^2 \sin^{-1}\dfrac{y_1}{R}\right)$

$A = \dfrac{\pi R^2}{2} - y_1\sqrt{R^2 - y_1^2}$
$\quad - R^2 \sin^{-1}\left(\dfrac{y_1}{R}\right)$

$c = \dfrac{2(R^2 - y_1^2)^{3/2}}{3} / A$

TRAPEZOID
Axis of Moments Through Center of Gravity

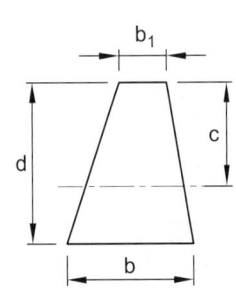

$A = \dfrac{d(b + b_1)}{2}$
$c = \dfrac{d(2b + b_1)}{3(b + b_1)}$
$I = \dfrac{d^3(b^2 + 4bb_1 + b_1^2)}{36(b + b_1)}$
$S = \dfrac{d^2(b^2 + 4bb_1 + b_1^2)}{12(2b + b_1)}$
$r = \dfrac{d}{6(b + b_1)} \times \sqrt{2(b^2 + 4bb_1 + b_1^2)}$

PARABOLA

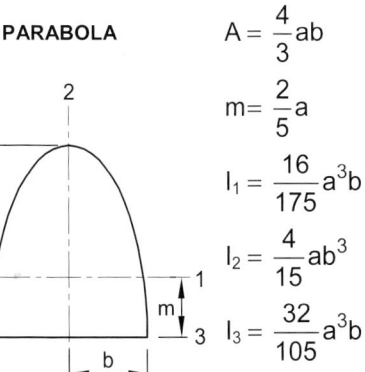

$A = \dfrac{4}{3}ab$
$m = \dfrac{2}{5}a$
$I_1 = \dfrac{16}{175}a^3 b$
$I_2 = \dfrac{4}{15}ab^3$
$I_3 = \dfrac{32}{105}a^3 b$

PCI Handbook/Sixth Edition   11–49

SECTION PROPERTIES

Design Aid 11.5.1   Properties of Geometric Sections (Cont.)

CIRCLE — Axis of Moments Through Center

$$A = \frac{\pi d^2}{4} = \pi R^2$$

$$c = \frac{d}{2} = R$$

$$I = \frac{\pi d^4}{64} = \frac{\pi R^4}{4}$$

$$S = \frac{\pi d^3}{32} = \frac{\pi R^3}{4}$$

$$r = \frac{d}{4} = \frac{R}{2}$$

HALF PARABOLA

$$A = \frac{2}{3}ab$$

$$m = \frac{2}{5}a$$

$$n = \frac{3}{8}b$$

$$I_1 = \frac{8}{175}a^3 b$$

$$I_2 = \frac{19}{480}ab^3$$

$$I_3 = \frac{16}{105}a^3 b$$

$$I_4 = \frac{2}{15}ab^3$$

HOLLOW CIRCLE — Axis of Moments Through Center

$$A = \frac{\pi(d^2 - d_1^2)}{4}$$

$$c = \frac{d}{2}$$

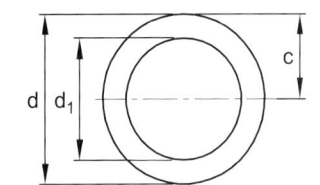

$$I = \frac{\pi(d^4 - d_1^4)}{64}$$

$$S = \frac{\pi(d^4 - d_1^4)}{32d}$$

$$r = \frac{\sqrt{d^2 - d_1^2}}{4}$$

COMPLEMENT OF HALF PARABOLA

$$A = \frac{1}{3}ab$$

$$m = \frac{7}{10}a$$

$$n = \frac{3}{4}b$$

$$I_1 = \frac{37}{2100}a^3 b$$

$$I_2 = \frac{1}{80}ab^3$$

PARABOLIC FILLET IN RIGHT ANGLE

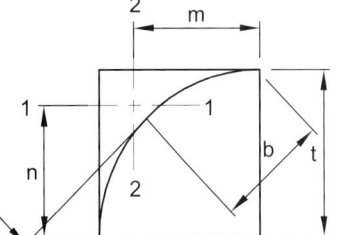

$$a = \frac{t}{2\sqrt{2}}$$

$$b = \frac{t}{\sqrt{2}}$$

$$A = \frac{1}{6}t^2$$

$$m = n = \frac{4}{5}t$$

$$I_1 = I_2 = \frac{11}{2100}t^4$$

*ELLIPTIC COMPLEMENT

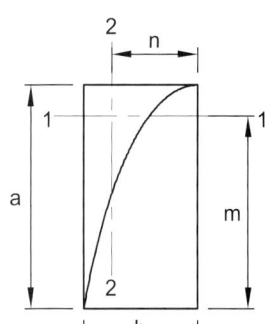

$$A = ab\left(1 - \frac{\pi}{4}\right)$$

$$m = \frac{a}{6\left(1 - \frac{\pi}{4}\right)}$$

$$n = \frac{b}{6\left(1 - \frac{\pi}{4}\right)}$$

$$I_1 = a^3 b\left(\frac{1}{3} - \frac{\pi}{16} - \frac{1}{36\left(1 - \frac{\pi}{4}\right)}\right)$$

$$I_2 = ab^3\left(\frac{1}{3} - \frac{\pi}{16} - \frac{1}{36\left(1 - \frac{\pi}{4}\right)}\right)$$

* See note on next page

SECTION PROPERTIES

Design Aid 11.5.1   Properties of Geometric Sections (Cont.)

*HALF ELLIPSE

$$A = \frac{1}{2}\pi ab$$

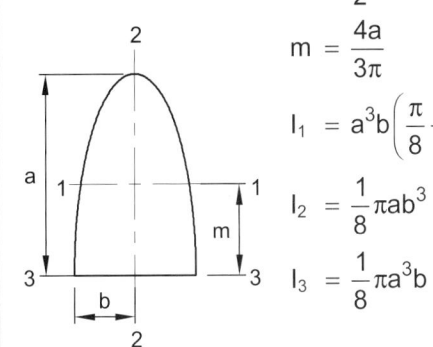

$$m = \frac{4a}{3\pi}$$

$$I_1 = a^3 b\left(\frac{\pi}{8} - \frac{8}{9\pi}\right)$$

$$I_2 = \frac{1}{8}\pi ab^3$$

$$I_3 = \frac{1}{8}\pi a^3 b$$

REGULAR POLYGON

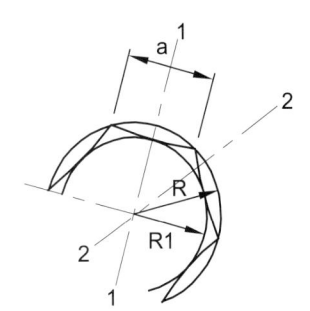

n = number of sides

$$\phi = \frac{180°}{n}$$

$$a = 2\sqrt{R^2 - R_1^2}$$

$$R = \frac{a}{2\sin\phi}$$

$$R_1 = \frac{a}{2\tan\phi}$$

$$A = \frac{1}{4}na^2 \cot\phi$$

$$= \frac{1}{2}nR^2 \sin 2\phi = nR_1^2 \tan\phi$$

$$I_1 = I_2 = \frac{A(6R^2 - a^2)}{24}$$

$$= \frac{A(12R_1^2 + a^2)}{48}$$

$$r_1 = r_2 = \sqrt{\frac{6R^2 - a^2}{24}}$$

$$= \sqrt{\frac{12R_1^2 + a^2}{48}}$$

*QUARTER ELLIPSE

$$A = \frac{1}{4}\pi ab$$

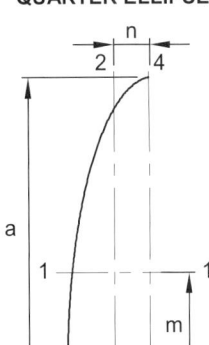

$$m = \frac{4a}{3\pi}$$

$$n = \frac{4b}{3\pi}$$

$$I_1 = a^3 b\left(\frac{\pi}{16} - \frac{4}{9\pi}\right)$$

$$I_2 = ab^3 \left(\frac{\pi}{16} - \frac{4}{9\pi}\right)$$

$$I_3 = \frac{1}{16}\pi a^3 b$$

$$I_4 = \frac{1}{16}\pi ab^3$$

BEAMS AND CHANNELS

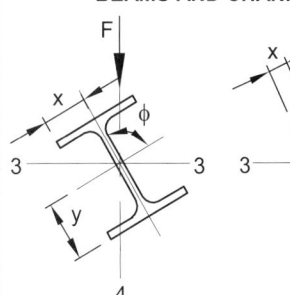

$$I_3 = I_x \sin^2\phi + I_y \cos^2\phi$$

$$I_4 = I_x \cos^2\phi + I_y \sin^2\phi$$

$$f_b = M\left(\frac{y}{I_x}\sin\phi + \frac{x}{I_y}\cos\phi\right)$$

Where M is bending moment due to force F.

* To obtain properties of half circles, quarter circle and circular complement, substitute $a = b = R$.

SECTION PROPERTIES

Design Aid 11.5.1   Properties of Geometric Sections (Cont.)

ANGLE
AXIS OF MOMENTS THROUGH CENTER OF GRAVITY

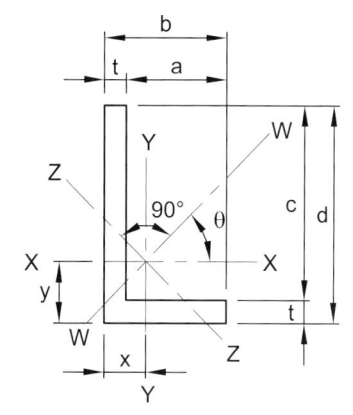

Z-Z is Axis of Minimum I

$$\tan 2\theta = \frac{2k}{I_y - I_x}$$

$A = t(b-c)$ $x = \dfrac{b^2 + ct}{2(b+c)}$ $y = \dfrac{d^2 + at}{2(b+c)}$

K = product of inertia about X-X & Y-Y

$$= \pm \frac{abcdt}{4(b+c)}$$

$I_x = \dfrac{1}{3}[t(d-y)^3 + by^3 - a(y-t)^3]$

$I_y = \dfrac{1}{3}[t(b-x)^3 + dx^3 - c(x-t)^3]$

$I_z = I_x \sin^2\theta + I_y \cos^2\theta + K \sin 2\theta$

$I_w = I_x \cos^2\theta + I_y \sin^2\theta + K \sin 2\theta$

K is negative when heel of angle, with respect to center of gravity, is in first or third quadrant, positive when in second or fourth quadrant.

SECTION PROPERTIES

Design Aid 11.5.2   Plastic Section Moduli and Shape Factors

SECTION	PLASTIC MODULUS, Z_3, in.^{3}	SHAPE FACTOR
Rectangle (b × h)	$\dfrac{bh^2}{4}$	1.5
I-section	x-x axis: $bt(h-t)+\dfrac{w}{4}(h-2t)^2$	1.12 (approx.)
	y-y axis: $\dfrac{b^2 t}{2}+\dfrac{(h-2t)w^2}{4}$	1.55 (approx.)
Channel (t average)	$bt(h-t)+\dfrac{w(h-2t)^2}{4}$	1.12 (approx.)
Solid circle	$\dfrac{h^3}{6}$	1.70
Hollow circle	$\dfrac{h^3}{6}\left[1-\left(1-\dfrac{2t}{h}\right)^3\right]$ th^2 for $t \ll h$	$\dfrac{16}{3\pi}\left[\dfrac{1-\left(1-\dfrac{2t}{h}\right)^3}{1-\left(1-\dfrac{2t}{h}\right)^4}\right]$ 1.27 for $t \ll h$
Hollow rectangle	$\dfrac{bh^2}{4}\left[1-\left(1-\dfrac{2w}{b}\right)\left(1-\dfrac{2t}{h}\right)^2\right]$	1.12 (approx.) for thin walls
Diamond	$\dfrac{bh^2}{12}$	2

For TS shapes, refer to AISC-LRFD Manual, Third Edition for Z_s values.

11.6 METRIC CONVERSION

Design Aid 11.6.1 Metric Calculations and Example

"Hard" SI (metric) Calculations for Precast Concrete

Design Aid 11.4.1 shows direct conversion from inch-lb. (U.S. customary) to SI units. Calculations made in SI units are usually rounded to even numbers. These are known as "hard" metric calculations. The metric version of the Code is 318M-02. Some of the common SI units used are:

Concrete Strength:
  20 MPa is approximately equivalent to 3000 psi
  35 MPa is approximately equivalent to 5000 psi

Concrete weight (or mass):
  Normal weight concrete may be assumed to weigh 2400 kg/m^3 (149.8 pcf). Lightweight concrete may be assumed to weigh 1900 kg/m^3 (118.6 pcf).

Reinforcement:
  Most U. S. reinforcing bar manufacturers mark bars with metric designations, although the actual sizes have not changed (see Design Aid 11.2.7).
  Grade 420 reinforcing bar (f_y = 420 MPa) is equivalent to Grade 60 reinforcing bar.

Prestressing Strand:
  Strand diameters and areas are rounded to the nearest mm, e.g., 13 mm is equivalent to ½ in. diameter, area = 99 mm^2.

Relationships in SI
  Structural engineering calculations in SI units involve forces which include gravitational effects, rather than just weights, or mass. Thus one kilogram (kg) of mass converts to 9.8 newtons (N) of force. For example, a 50 mm thick concrete topping 1 meter wide weighs (50/1000)m × 1m × 2400 kg/m^3 = 120 kg/m. It exerts 120 × 9.8 = 1176 N/m or 1.18 kN/ m of force.
  Pressure or stress is expressed in pascals (P). 1P = 1N/m^2. It is more common to work in megapascals (MPa). 1 MPa =1N/mm^2. Bending moments are expressed in newton-meters (N-m) or kilonewton-meters (kN-m).
  Design Aid 11.6.2 lists quantities that are frequently encountered in the design of precast concrete as well as in general structural engineering practice. Most U.S. Government agencies that require SI unit dimensioning of contract documents will also require consistent use of the listed SI units given in the table. The equivalent U.S. customary units listed are those traditionally used by the design professions in the U.S.
  Conversion of frequently encountered concrete stress coefficients used in ACI 318-02 and the PCI Handbook are tabulated in Design Aid 11.6.4.

Example   Use of Eq. 18-3 from 318M-02 (Similar to Example 4.2.2.3)

Given:
Double tee similar to PCI standard 8DT24+2

Concrete:
Precast f'_c = 35 MPa   Topping f'_c = 20 MPa

Reinforcement:
12-13 mm dia. 1860 MPa low relaxation strands (6 ea. stem). Area per strand = 99 mm^2
A_{ps} = 12(99) = 1188 mm^2 A_s = 2 – #19 = 568 mm^2
f_y = 420 MPa

Problem:
Find the design flexural strength of the composite section using Eq. 18-3 from ACI 318M-02.

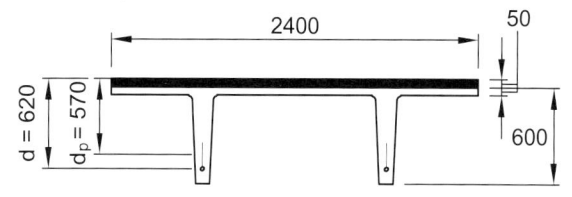

Dimensions in Millimeters (mm)

$$f_{ps} = f_{pu}\left(1 - \frac{\gamma_p}{\beta_1}\left[\rho_p \frac{f_{pu}}{f'_p} + \frac{d}{d_p}(\omega - \omega')\right]\right)$$

$\omega' = 0$ in this example

$$\rho_p = \frac{A_{ps}}{bd_p} + \frac{1188}{2400(570)} = 0.00087$$

$$\omega = \frac{A_s}{bd}\left(\frac{f_y}{f'_c}\right) = \frac{568}{2400(620)}\left(\frac{420}{20}\right) = 0.0080$$

$$f_{ps} = 1860\left(1 - \frac{0.28}{0.85}\left[0.00087\frac{1860}{20} + \frac{620}{570}(0.0080)\right]\right)$$

$= 1805$ MPa

$$a = \frac{A_{ps}f_{ps} + A_s f_y}{0.85 f'_c b} = \frac{1188(1805) - 568(420)}{0.85(20)2400}$$

$a = 47$ mm

$$M_n = A_{ps}f_{ps}\left(d_p - \frac{a}{2}\right) + A_s f_y\left(d - \frac{a}{2}\right) =$$

$$1188(1805)\left(\frac{570 - 24}{1000}\right) + 568(420)\left(\frac{620 - 24}{1000}\right)$$

$\{x(d-x) + d^2 - 2(c^2 + e^2) - \frac{2}{d}[e^2 x + c^2(d-x)]\}$

$-(x-d)^2(4e + d - x) - d^3] = 1{,}312{,}991$ N-m = 1313 kN-m

$\phi M_n = 0.9(1313) = 1182$ kN-m

METRIC CONVERSION

Design Aid 11.6.2 Conversion from U.S. Customary Units to International System (SI)

To convert from	to	multiply by
Length		
inch (in.)	millimeter (mm)	25.4
inch (in.)	meter (m)	0.0254
foot (ft)	meter (m)	0.3048
yard (yd)	meter (m)	0.9144
mile (mi)	kilometer (km)	1.6093
Area		
square foot (ft^2)	square meter (m^2)	0.09290
square inch (in.^{2})	square millimeter (mm^2)	645.2
square inch (in.^{2})	square meter (m^2)	0.0006452
square yard (yd^2)	square meter (m^2)	0.8361
acre (A)	hectare (ha) = 10,000 m^2	0.4047
square mile	square kilometer	2.590
Volume		
cubic inch (in.^{3})	cubic meter (m^3)	0.00001639
cubic foot (ft^3)	cubic meter (m^3)	0.02832
cubic yard (yd^3)	cubic meter (m^3)	0.7646
gallon (gal) Can. liquid[a]	liter	4.546
gallon (gal) Can. liquid[a]	cubic meter (m^3)	0.004546
gallon (gal) U.S. liquid[a]	liter	3.785
gallon (gal) U.S. liquid[a]	cubic meter (m^3)	0.003785
Force		
kip	kilogram (kgf)	453.6
kip	newton (N)	4448.0
pound (lb)	kilogram (kgf)	0.4536
pound (lb)	newton (N)	4.448
Pressure or Stress		
kips/square inch (ksi)	megapascal (MPa)[b]	6.895
pound/square foot (psf)	kilopascal (kPa)[b]	0.04788
pound/square inch (psi)	kilopascal (kPa)[b]	6.895
pound/square inch (psi)	megapascal (MPa)[b]	0.006895
pound/square foot (psf)	kilogram/square meter (kgf/m^2)	4.882
Mass		
pound (avdp)	kilogram (kg)	0.4536
ton (short, 2000 lb)	kilogram (kg)	907.2
ton (short, 2000 lb)	tonne (t)	0.9072
grain	kilogram (kg)	0.00006480
tonne (t)	kilogram (kg)	1000

a. One U.S. gallon equals 0.8321 Canadian gallon.
b. A pascal equals one newton/square meter; 1 Pa = 1 N/m^2.

Note: To convert from SI units to U.S. customary units (except for temperature), divide by the factors given in this table.

METRIC CONVERSION

Design Aid 11.6.2   Conversion from U.S. Customary Units to International System (SI) (Cont.)

To convert from	to	multiply by
Mass (weight) per Length		
kip/linear foot (klf)	kilogram/meter (kg/m)	1488
pound/linear foot (plf)	kilogram/meter (kg/m)	1.488
pound/linear foot (plf)	newton/meter (N/m)	14.593
Mass per volume (density)		
pound/cubic foot (pcf)	kilogram/cubic meter (kg/m^3)	16.02
pound/cubic yard (pcy)	kilogram/cubic meter (kg/m^3)	0.5933
Bending Moment or Torque		
pound-inch (lb-in.)	newton-meter	0.1130
pound-foot (lb-ft)	newton-meter	1.356
kip-foot (kip-ft)	newton-meter	1356
Temperature		
degree Fahrenheit (°F)	degree Celsius (°C)	$t_C = (t_F - 32)/1.8$
degree Fahrenheit (°F)	degree Kelvin (K)	$t_K = (t_F + 459.7)/1.8$
Energy		
British thermal unit (Btu)	joule (j)	1056
kilowatt-hour (kwh)	joule (j)	3,600,000
Power		
horsepower (hp) (550 ft lb/sec.)	watt (W)	745.7
Velocity		
mile/hour (mph)	kilometer/hour	1.609
mile/hour (mph)	meter/second (m/s)	0.4470
Other		
Section modulus (in.^{3})	mm^3	16,387
Moment of inertia (in.^{4})	mm^4	416,231
Coefficient of heat transfer (Btu/ft^2/h/°F)	W/m^2/°C	5.678
Modulus of elasticity (psi)	MPa	0.006895
Thermal conductivity (Btu/in./ft^2/h/°F)	Wm/m^2/°C	0.1442
Thermal expansion in./in./°F	mm/m^2/°C	1.800
Area/length (in.^{2}/ft)	mm^2/m	2116.80

METRIC CONVERSION

Design Aid 11.6.3   Preferred SI Units and U.S. Customary Equivalents

Quantity	SI	U.S. customary
Area, cross section	mm^2	in.^{2}
Area, plan dimension	mm^2	ft^2
Bending moment	kN-m	kip-ft
Coefficient of thermal expansion	mm/(mm-°C)	in./in./°F
Deflection	mm	in.
Density, linear	kg/m	lb/ft, kip/ft
Density, area	kg/m^2	lb/ft^2, kip/ft^2
Density, mass	kg/m^3	lb/ft^3, kip/ft^3
Force	kN	lb, kip
Force, per unit length	kN/m	lb/ft, kip/ft
Force, per unit area	kN/m^2	lb/ft^2, kip/ft^2
Length, cross section	mm	in.
Length, plan dimension	mm	ft
Mass	kg	lb, kip
Modulus of elasticity	MPa	psi, ksi
Moment of inertia	10^6 mm^4	in.^{4}
Section modulus	10^6 mm^3	in.^{3}
Stress	MPa	psi, ksi
Temperature	°C	°F
Torque	kN-m	lb-ft, kip-ft

Design Aid 11.6.4   Concrete stress coefficents

U.S. customary coefficient	SI coefficient	U.S. customary coefficient	SI coefficient
0.5	0.04	3.3	0.27
0.6	0.05	3.5	0.29
0.667	0.06	4.0	0.33
1.0	0.08	4.4	0.37
1.1	0.09	5.0	0.42
1.2	0.10	5.5	0.46
1.25	0.10	6.0	0.50
1.5	0.12	6.3	0.52
1.6	0.13	6.5	0.54
1.7	0.14	7.0	0.58
1.9	0.16	7.5	0.62
2.0	0.17	8.0	0.66
2.4	0.20	10.0	0.83
3.0	0.25	12.0	1.00
Examples: U.S. customary: psi	SI equivalent: 0.08 MPa	U.S. customary: 10 psi	SI equivalent: 0.83 MPa

INDEX

ACI 318 .... 3–9, 3–13, 4–8
Acoustical properties 9–24
Adhesive anchors 6–10
Affected zone, thermal 9–4, 9–20
Air barriers 9–17
Air entrainment 1–24
Airborne sound 9–24
Aircraft cable 5–11
Allowable stresses
    (see service load stresses) 4–22
Anchor bolts 3–55, 6–66, 6–95, 10–9
Anchorage seating loss 4–84
Anchors in conduit 6–8
Anchors, stone facing 7–18
Angle stiffeners 6–34
Angles, unstiffened 6–32, 6–90
Appendix D, ACI 318–02 6–12
Applications 1–5
Approvals 10–7
Architectural component, seismic 3–17, 3–113
Architectural precast concrete 1–12, 1–29, 7–1
Arenas 1–14
ASCE 7 .... 3–8, 3–10
Asymmetrical sections 4–45

Base plates 3–55, 6–66, 6–95
Base shear 3–13, 3–41
Beam design equations
    and diagrams .... 11–5 to 11–22
Beam erection tolerances 8–10
Beams with overhangs 11–28
Bearing 4–77
Bearing pads 6–62
Bearing wall construction 2–4
Bearing walls 3–24
Bending, out-of-plane 4–67
Bilinear behavior 4–89
Bolts .... 6–9, 6–89, 11–42 to 11–44
Bond, strand 4–41
Bowing 4–95, 4–130, 8–7
Box elements 3–6
Box modules 1–11
Bracing equipment 5–28
Brackets
    (see corbels) 6–6, 6–47, 6–54, 6–91, 6–93
Brick veneers 7–14
Bridges 1–16
British thermal units (Btu) 9–4
Buckling 4–107
Building movements 7–7
Built-up steel sections 6–28

Cable guys 5–28
Camber 4–86, 4–93, 4–128
Camber coefficients 11–23, 11–24

Cantilevered Columns 3–6
Cantilevers 4–117
Carbon equivalent 6–38
Cazaly hanger 6–59
Ceramic fiber felt 9–39
Characteristic section method 9–21
Chemical resistance 1–24
Chloride ions 1–24
Chloroprene bearing pads 6–62
Chord forces, diaphragm 3–84
Cladding connections, example 6–71
Cladding, precast 1–12, 1–31, 7–1
Clay product veneers 7–14
Clearance, example 8–28, 8–31
Clearances 8–26
Code requirements 3–8, 3–34, 3–88
Coefficient of thermal expansion 1–20
Coil rods 6–10, 6–89
Column bases 6–66
Column bases, moment resistance 3–54
Column covers 7–4
Column covers, fire resistance 9–52
Column erection tolerances 8–14
Column splices 3–23
Columns 4–97, 4–105
Columns,
    interaction curves 2–48, 2–49, 2–50, 2–51
Columns, precast 2–50, 2–51
Columns, prestressed 2–48, 2–49
Combined shear and tension, HCA 6–26
Combined shear and tension, steel 6–32
Common products 1–3
Composite members 4–25, 4–51
Compression members 4–97, 4–106
Compressive strength, concrete 1–19
Compressive strength, effect of temperature 9–40
Computer models 3–40, 3–59
Concentrated live loads 11–3, 11–27
Concentrated loads 4–18, 4–47, 4–114
Concrete Capacity Design (CCD) 6–11
Concrete coatings 9–64
Concrete corbels 6–6, 6–47, 6–91
Concrete properties 11–29
Concrete sealers 9–64
Concrete shear, HCA 6–17
Concrete tension, HCA .... 6–13, 6–96 to 6–101
Concrete, material properties 1–19
Concrete, modulus of elasticity 1–20
Concrete, strength 2–9
Concrete, tensile strength 1–20
Concrete, unit weights 2–9
Condensation 9–15, 9–20
Confinement reinforcement, HCA 6–12, 6–18
Connection design criteria 6–7
Connection hardware 6–8

Connection repair	9–76
Connections	6–1
Connections, architectural panels	7–12
Connections, fire protection	9–51
Connections, spandrel	7–3
Construction loads	5–26
Continuity	3–23, 4–117
Contract administration	10–12
Contract documents	10–5
Corbels, concrete	6–6, 6–47, 6–91
Corbels, steel	6–54, 6–93
Corrosion inhibitors	1–26
Corrosion protection	1–25
Coupled shear walls	3–39
Crack control	4–22, 5–7, 5–9
Crack control, architectural precast concrete	7–10
Crack repair	9–76
Cracked section	4–22, 4–24
Cracked section analysis	4–25, 4–31, 4–89, 4–129
Cracking	9–75
Cracking moment	4–10, 4–25, 4–89
Cracking, HCA connections	6–12, 6–19
Cracking, safety factor	5–9
Cracking, welded connections	6–41
Crane access	10–8
Creep	3–6, 3–25
Creep loss	4–84
Critical section	4–11, 4–46, 4–57, 4–85
Cylinders, test	1–19
Damping	3–13
Damping of vibrations	9–69
Dapped-end bearing	4–79
Dead loads	3–9, 11–2
Debonding, strand	4–41, 4–122
Deck openings	4–115
Deck panels	1–16
Definitions, precast concrete	10–5
Deflection coefficients	11–23, 11–24
Deflections	3–16, 3–40, 3–112, 3–123, 4–86, 4–89, 4–94
Deflections, wall support	7–7
Deformations, shear wall	3–35, 3–123
Deformed bar anchors	6–9
Degree day	9–4
Deicers	1–24
Delivery of materials	10–8
Design criteria, connections	6–7
Design practices	10–21
Design responsibility	10–5, 10–6, 10–16
Design-build	10–18
Detailing, diaphragm	3–98
Development length	11–33, 11–36, 11–37
Development, corbel reinforcement	6–54
Development, strand	4–41, 4–122
Dew point	9–4, 9–19
Diagrams, beam design	11–5 to 11–22
Diaphragm connections, examples	6–74, 6–76
Diaphragms	3–22, 3–84, 3–98
Differential camber	8–6
Distribution of lateral loads	3–37
Double tee wall panels, interaction curves	2–52
Double tee, flange strength	4–16
Double tees, load tables	2–12 to 2–30
Double tees, pretopped, load tables	2–24 to 2–30
Dowels, anchorage in conduit	6–8
Drift amplification factor	3–16, 3–112
Drift, seismic	3–16, 3–41, 3–112, 3–113
Drifting, snow	3–9, 3–106, 3–107
Dry connections	3–6
Drypack	1–21
Ductility	3–23
Ductility, connections	6–7
Ductwork	9–82
Durability	1–23
Durability, connections	6–7
Dy-Core	2–35
Dynaspan	2–35
Earthquake (see seismic)	3–13
Eccentrically loaded columns	4–106
Economic considerations, fire	9–53
Effective moment of inertia	4–89
Effective shear-friction	4–56
Effective width	4–107
Elastic deflections	4–89
Elastic design, diaphragms	3–90
Elastic shortening	4–84
Elastic vector method, weld analysis	6–42
Elastomeric bearing pads	6–62
Elastomeric pads, fire	9–51
Electrified floors	9–82
Emulation	3–6
Elematic	2–38
End point criteria, fire	9–30
End region – columns	4–101
End stresses at transfer	4–45
Engineer of record	10–16, 10–19
Epoxy coated reinforcement	1–26
Epoxy coated strand	1–26
Epoxy grout	1–22
Equivalent volume change	3–29, 3–120
Erection	5–22, 10–8
Erection analysis	5–28
Erection bracing	5–26
Erection drawings	10–6
Erection loads	5–26
Erection tolerances	8–9
Erection, architectural panels	7–9
Expansion anchors	6–10
Expansion joints	3–28, 3–118
Exposure category	3–11

Factors of safety	5–9, 5–28
Features of precast/ prestressed concrete	1–2
Fiber reinforced polymers	9–77
Field qualification	9–62
Fillet welds	6–38
Finishes	10–7
Finite element analysis	3–39, 3–100
Fire endurance	9–30
Fire rating	9–30
Fire resistance	9–28
Fire resistance, connections	6–7
Five-percent fractile	6–13
Fixed-end moments	11–25
Fixity of column bases	3–54
Flanged elements	4–8
Flanking	9–25
Flexible conduit, dowels	6–8
Flexible diaphragms	3–87
Flexicore	2–36
Flexure design	4–8, 4–120, 4–121, 4–122,
Floor openings	4–115
Floors, erection tolerances	8–12
Flow chart, strength design	4–9
Force-resisting systems	3–6
Form draft	5–3
Form suction	5–9
Forms	1–29, 5–3
Forms, architectural precast concrete	7–21
Frame analysis and design	3–61, 3–69, 3–77
Frame classifications, seismic	3–60, 3–112
Freeze-thaw	1–24
Full penetration welds	6–38
Galvanizing	1–25, 6–36
General contractor	10–17, 10–20
Geometric sections properties	11–48 to 11–52
Glass Fiber Reinforced Concrete (GFRC)	1–14
Gravity loads	3–9
Group welds	6–41
Grout	1–21
Gusset plates	6–34
Gypsum wall board	9–34
Handling devices	5–10
Handling equipment	5–26
Handling of panels, diagrams	5–5, 5–6
Handling, architectural panels	7–9
Hanger connections	6–59
Hanger reinforcement	4–67, 4–79
Hardware, connection	6–8
Hauling (see transportation)	5–20
Haunches, concrete	6–6, 6–47, 6–91
Haunches, steel	6–54, 6–93
Headed concrete anchors (HCA)	6–8, 6–11, 6–96
Heat capacity	9–4
Heat flow	9–4
Heat transmission	9–4, 9–33

Heat transmittance	9–4, 9–6, 9–20
High-performance concrete	1–19
Holes through decks	4–115
Holes through webs	4–115
Hollow-core load distribution	4–115
Hollow-core, openings	4–115
Hollow-core, load tables	2–31 to 2–34
Hollow-core, properties	2–36 to 2–38
Hollow-core wall panels, interaction curves	2–53
Hooks, standard	11–37
Horizontal alignment	8–5
Horizontal shear	4–51
Human response to vibrations	9–67
Hybrid structures	3–25
IBC 2003	3–8
Impact factors	5–9
Impact Insulation Class (IIC)	9–25
Impact noise reduction	9–24
Importance factors	3–103
Industrial building, example	3–41
Industrial buildings	1–6
Initial camber	4–88, 4–128
Inserts	6–10
Inspections	10–7
Instantaneous Center Method, weld analysis	6–43
Insulated panels	1–6
Insulated wall panels	9–55
Insulation	9–55, 9–60
Integrity, structural	3–23
Interaction curves	2–48 to 2–55
Interaction, frames and shear walls	3–84
Interaction, tension and shear, HCA	6–26
Interfacing tolerances	8–32
Intermediate moment frames	3–60
Intermediate shear walls	3–33
International Building Code	3–8
Inverted tee beams, load tables	2–45 to 2–47
Joint clearance	8–26
Joint sealants	9–65
Jointed precast construction	3–8
Joints, expansion	3–28, 3–118
Joints, segmental construction	9–79
Joints, thermal treatment	9–38
Justice facilities	1–10
Laminated bearing pads	6–64
Large-panel bearing wall systems	3–24
Lateral force distribution	3–16, 3–37
Lateral force resisting system	2–4, 3–23
Lateral stability	5–19
Ledger beam	4–65
Ledges, beam	4–65
Lifting configurations	5–3
Lifting loops	5–11
Light wall, parking structure	1–10, 3–47

Lighting	9–82
Lightweight concrete	1–20, 4–51
Live loads	3–9, 11–3
Load combination	3–22
Load distribution, decks	4–115
Load factors	2–8, 2–9, 3–22, 6–6
Load tables, explanation	2–7
Load tables, limiting criteria	2–8
Load transfer devices	6–8
Loadbearing panels	7–5
Loadbearing wall panels	4–107
Loadbearing walls	7–5
Loads	7–7
Loads, connections	6–6
Longitudinal reinforcement, torsion	4–58
Long-line forms	1–29
Long-term camber	4–93
Loov hanger	6–61
Loss of prestress	2–10, 4–84
Low-relaxation strand	1–22
L-shaped beams, load tables	2–43, 2–44
Maintenance	9–78
Mechanical systems, coordination	9–82
Member strengthening	9–77
Metric conversion	11–54 to 11–57
M-factors	4–69
Minimum clearances	8–26
Minimum live loads	11–3
Minimum reinforcement	4–11, 4–51, 4–58, 4–126, 11–41
Minimum shear reinforcement	4–51, 4–126
Mixed building systems, tolerances	8–9
Modulus of elasticity, concrete	1–20, 11–29
Modulus of rupture	4–22, 5–9
Modulus of rupture, concrete	1–20
Moment connections	6–71
Moment diagrams, fire	9–39, 9–44, 9–45
Moment frame connections, example	6–82
Moment of Inertia	4–25, 4–89, 4–129
Moment-resisting frame examples	3–61, 3–68, 3–77
Moment-resisting frames	2–4, 3–6, 3–28, 3–54, 3–60
Moments in panels, handling	5–5, 5–6
Mortar	1–21
Moving loads	11–27
Multipliers, static load	5–10
Natural frequency	9–67
NEHRP	3–9
Neoprene bearing pads	6–62
NFPA	3–9
Noise insulation objectives	9–24
Non-composite, sandwich panels	9–55
Non-loadbearing panels	7–2
Non-prestressed panels	4–22

Non-shrink grout	1–22
Nuts	11–42 to 11–44
Office buildings	1–6
One-story building example	3–41
Openings in panels	7–4
Openings, Deck	4–115
Openings, shear walls	3–40
Openings, Web	4–115
Ordinary moment frames	3–60, 3–61
Ordinary shear walls	3–33
Out-of-plane bending	4–67
Over Load Factors (OLF)	6–7
Overhangs, beam equations	11–28
Overjacking	4–84
Overstrength factor	3–23, 3–112
Overturning	3–36, 3–40
Pads, bearing	6–62
Panels, loadbearing	7–5
Panels, non-loadbearing	7–4
Parking structure example	3–47
Parking structures	1–9
Partial prestress	4–8, 4–19
Partially-composite, sandwich panels	9–55
Patching	9–75, 10–10
Payment	10–13
PCI standard design practice	10–21
P-delta analysis	4–107
P-delta effects	3–17, 3–41
Penetrating sealers	1–23, 9–64
Performance based design	3–97
Period	3–14, 3–15
Perm	9–5
Permeability	9–5, 9–17, 9–18
Permeance	9–5
Permits	10–14
Personnel certification	9–62
Pick-up points	5–5, 5–6
Piles	1–3, 2–10, 4–111
Piles, section properties	2–56
Pipe braces	5–28
Plain concrete bearing	4–77
Plant certification	1–3, 9–62
Plastic bearing pads	6–64
Plastic section modulus	11–53
Pocketed beams	4–73
Poisson's ratio	1–20
Polar moment of inertia, welds	11–47
Poles	1–17
Post-installed anchors	6–10
Post-tensioned connections	6–71
Post-tensioning	1–22, 1–30
Post-tensioning, segmental	9–80
Precast cladding	1–12, 7–1
Precast columns, interaction curves	2–50, 2–51

Precast wall panels,
    interaction curves .... 2–52 to 2–55
Prefabricated modules 9–85
Preliminary design 2–3
Premixed grout 1–22
PRESSS 3–8
Pressure zones, wind 3–11
Prestress loss 4–84
Prestressed columns,
    interaction curves 2–48, 2–49
Prestressed wall panels 5–10
Prestressed wall panels,
    interaction curves .... 2–52 to 2–54
Prestressing steel poperties .... 11–30 to 11–33
Prestressing strands .... 2–10, 11–30 to 11–33
Prestressing tendons 1–22
Prestressing, architectural panels 7–11
Pretensioning 1–30
Pretopped diaphragms 3–99
Prison cell modules 1–12, 9–86
Product handling 5–3
Production drawings 10–6
Production limitations 1–30
Production process 1–29
Protection of reinforcement 1–25
Pullout strength, HCA connections 6–17

Quality control 9–61

Railroad ties 1–17
Raker beams 1–14
Rational design, fire endurance 9–39
Ratios, span-to-depth 2–4
Rayleigh's formula 3–14
Rectangular beams, load tables 2–42
Redundancy 3–22, 3–23
Reentrant corner, daps 4–81
Reinforced concrete bearing 4–77
Reinforcement 1–22
Reinforcement limitations 4–11, 4–51,
    4–58, 4–126
Reinforcement spacing 4–22
Reinforcement, architectural panels 7–10
Reinforcement, confinement, HCA 6–12, 6–18
Reinforcement, epoxy coated 1–26
Reinforcing steel
    properties .... 11–34 to 11–37, 11–40
Reinforcing bar anchorage 6–8
Reinforcing bar couplers 6–9
Reinforcing bar splicing systems 6–9
Reinforcing bar welding 6–38, 6–86, 6–87, 6–88
Reinforcing bars 1–23
Relative humidity 9–5, 9–20
Relaxation loss 4–84
Release strength 2–9
Release stresses 4–24, 4–45
Repair 9–75

Residential buildings 1–5
Response accelerations 3–14
Response spectrum 3–14
Restrained fire ratings 9–32
Restrained members, fire 9–45
Restraint forces 3–29
Retaining walls 1–17
Ribbed panels, fire resistance 9–34
Rigging configurations 5–7
Rigid diaphragms 3–87
Rolling blocks 5–8
Roof member erection tolerances 8–12
Room module erection tolerances 8–22
Rotation coefficients 11–23, 11–24
Roth (Hollow-core) 2–38
R-values, insulating materials 9–35
Rhythmic activities, vibrations 9–70

Safe superimposed loads 2–8
Samples 10–5
Sand-cement mixtures 1–21
Sand-lightweight concrete 1–20, 4–51
Sandwich panels 9–55
Sandwich panels, fire resistance 9–37
Sandwich panels, thermal 9–6, 9–20
Sealers 1–23
Section modulus, weld 11–47
Section properties .... 11–48 to 11–53
Segmental construction 9–79
Seismic base shear 3–14
Seismic connections, examples 6–71, 6–78, 6–79
Seismic design categories 3–13, 3–15
Seismic drift 3–16, 3–41, 3–112, 3–113
Seismic forces 3–13
Seismic response coefficient 3–14, 3–112
Self-consolidating concrete 1–19
Self-stressing forms 1–30
Service load class, prestressed members 4–24
Service load stresses 4–24
Shape factor, bearing pads 6–65
Shape factors 11–53
Shear 4–46, 4–123, 4–124, 4–125
Shear reinforcement 4–51, 4–126, 4–127
Shear resistance, fire 9–51
Shear strength, HCA connections 6–12, 6–17
Shear wall connections,
    examples 6–76, 6–78, 6–79
Shear wall examples 3–41, 3–47
Shear wall systems 3–33, 3–40
Shear wall-frame interaction 3–84
Shear walls 3–6, 3–33
Shear walls, coupled 3–39
Shear, horizontal 4–51
Shear, steel 6–28
Shear-friction 4–55
Sheet piles 4–111
Sheet piles, section properties 2–57

Shielding strand (see debonding)	4–41, 4–122
Shop drawings	10–6
Shrinkage	1–22, 3–25
Shrinkage loss	4–84
Single-story building	3–41
Site classification	3–14, 3–111
Slenderness	4–107
Smoothness	8–7
Snow loads	3–9, 3–104, 3–106, 3–107
Soil loads, lateral	3–22
Soil properties	3–111
Solid flat slabs, load tables	2–39 to 2–41
Sound absorption	9–24
Sound insulation	9–24
Sound Transmission Class (STC)	9–24
Sound transmission loss	9–24
Sound wall	1–17
Span Deck	2–37
Spancrete	2–36
Spandrel connections	7–3
Spandrels	7–3, 7–5
Span-to-depth ratios	2–4
Special moment frames	3–61, 3–69, 3–77
Special shear walls	3–34
Specifications	10–2, 10–3, 10–4
Spreader beam	5–8
Stability	3–17, 3–56
Stacking panels	7–11
Stadium riser erection tolerances	8–20
Stadium risers	4–45
Stadium seating, vibrations	9–71
Stadiums	1–14
Stainless steel	1–28, 6–37
Stair erection tolerances	8–24
Standard design practice	10–21
Standard fire tests	9–29
Standard practice	10–5, 10–21
Standard time-temperature curve	9–29
Static load multipliers	5–10
Steel corbels	6–54, 6–93, 6–94
Steel shapes	6–8, 6–11, 6–28
Stemmed units, fire resistance	9–42
Stiffeners, angle	6–34
Stiffness	3–36, 3–54
Stone veneers	7–17
Storage	5–19
Strain compatibility	4–9, 4–11, 4–19, 4–122
Strand debonding	4–41
Strand development	4–41
Strand, epoxy coated	1–26
Strands	2–10, 11–30 to 11–33
Strength design	4–8
Strength reduction factor	4–10, 2–8, 2–9, 3–23, 6–6, 6–13
Strength-temperature relationship, steel	9–40
Stress block	4–8
Stress limits	4–24, 5–9
Stress, concrete	11–29
Stress-relieved strand	1–22
Stripping	5–3
Strong connections	3–8
Strong-back	5–3, 5–4
Structural integrity	3–23
Structural steel, connections	6–28
Structural wall panel tolerances	8–16
Structural walls (see shear walls)	3–8
Strut-and-tie design	6–50
Strut-and-tie, diaphragms	3–100
Studs	6–8, 6–11, 6–96
Subgrade reaction	3–55
Sulphate attack	1–28
Supplemental reinforcement	6–12, 6–18
Surface conductance	9–4
Surface sealers	9–64
Surveying	5–26
Suspended ceilings	9–84
Sweep	8–5
Swivel plate	5–11
Systems building	9–85
Tanks	1–17
Teflon (TFE) bearing pads	6–64
Temperature change	3–6, 3–25, 3–114
Temporary loads	5–26
Tensile strength, concrete	1–19
Tests	10–7
Thermal bowing	4–95, 4–130
Thermal bowing, sandwich panels	9–60
Thermal bridges	9–20
Thermal conductance	9–5, 9–18
Thermal conductivity	9–5
Thermal expansion, concrete	1–20
Thermal loads, comparisons	9–14
Thermal mass	9–5, 9–13
Thermal properties	9–6
Thermal resistance	9–5, 9–6, 9–21
Thermal storage	9–13
Thermal transmittance	9–6
Threaded connectors	6–9, 6–89
Threaded inserts	5–11
Tie-back connections	6–71
Tilt tables	5–7
Tilted beam, equilibrium	5–19
Tolerance, architectural panels	7–9
Tolerances	8–2, 10–8
Tolerances, acceptability range	8–4
Tolerances, camber variation	8–6
Tolerances, connections	8–9
Tolerances, erection	8–9
Tolerances, interfacing	8–32
Tolerances, product	8–4
Tolerances, smoothness	8–7
Topped diaphragm	3–99
Torsion	4–57

Torsion equations and diagrams	11–26
Torsion, lateral force	3–8, 3–16
Torsion, steel	6–30
Transfer length	4–41, 11–33
Transfer of prestress	4–25, 4–41
Transfer stresses	4–25, 4–45
Transformed cracked section	4–25, 4–31, 4–89, 4–129
Transportation	5–20
Transporting architectural panels	7–9
Transverse reinforcement, torsion	4–58
Tripping, diagrams	5–23
Ultra Span	2–37
Uncracked sections	4–22
Underdeveloped strand	4–41, 4–122
Unstiffened angles	6–32, 6–90
Untopped diaphragm	3–99
Utility poles	1–17
U-values	9–6
U-values, roofs	9–11, 9–12
U-values, walls	9–10
Vapor retarder	9–18
Veneer jointing	7–21
Veneered panels	7–13
Vertical force distribution	3–16
Vibration analysis	9–67
Vibration isolation	9–73
Volume change accommodation	6–7
Volume changes	1–20, 7–7, 3–6, 3–25, 3–28, 3–115, 3–116, 3–117, 3–120, 3–121

Walking, vibrations	9–67, 9–69
Wall connections, examples	6–76, 6–78, 6–79
Wall panel erection tolerances	8–16, 8–18
Wall panels	7–5, 7–6
Wall panels, effective width	4–107
Wall panels, post-tensioned	5–9, 5–10
Wall panels, prestressed	5–10
Wall support deflections	7–7
Walls, fire endurance	9–38
Walls, structural (see shear walls)	3–8
Warpage	5–19
Warping	8–3, 8–7
Warranties	10–11
Washers	11–44
Water-repellant coating, insulation	9–60
Web openings	4–115
Weight, concrete	11–29
Weights	11–2, 11–3
Weld design	6–38
Weld group properties	11–47
Weld plate tolerances	8–9
Weld symbols	11–45
Welded Wire Reinforcement	1–23, 11–38, 11–39
Welding	6–36, 6–85, 11–45, 11–46
Welding, reinforcing bar	6–38, 6–86, 6–87, 6–88
Wet connections	3–6
Wind design	3–10
Wind loads	3–10, 3–108, 3–110
Window panels	7–4, 7–6
Wire rope	5–28
Wythe connectors	9–58
X-bracing	3–6